O9-BUD-474

Medieval cosmology was a fusion of pagan Greek ideas and biblical descriptions of the world, especially the creation account in Genesis. Because cosmology was based on discussions of the relevant works of Aristotle, primary responsibility for its study fell to scholastic theologians and natural philosophers in the universities of western Europe from the thirteenth to the seventeenth century. The present work describes the extraordinary range of themes, ideas, and arguments that constituted scholastic cosmology for approximately five hundred years, from around 1200 to 1700. Primary emphasis is placed on the world as a whole, what might lie beyond it, and the celestial region, which extended from the Moon to the outermost convex surface of the cosmos.

During the late Middle Ages (ca. 1200–1500), Aristotelian cosmology met little opposition or challenge. By the time rival interpretations appeared in the sixteenth century – for example, Platonism, atomism, Stoicism, Neoplatonism, Hermeticism, and especially Copernicanism – Aristotelian cosmology was firmly entrenched. By the seventeenth century, however, Copernican heliocentric cosmology and the geoheliocentric variant of it proposed by Tycho Brahe offered significant alternatives and thereby challenged medieval Aristotelian cosmology as never before. How scholastic natural philosophers of the sixteenth and seventeenth centuries responded to the new interpretations is an important aspect of this study.

Planets, stars, and orbs

PLANETS, STARS, AND ORBS

The Medieval Cosmos, 1200–1687

EDWARD GRANT
Indiana University

This book has been supported by a grant from the National Endowment for the Humanities, an independent federal agency.

Published by the Press Syndicate of the University of Cambridge
The Pitt Building, Trumpington Street, Cambridge CB2 1RP
40 West 20th Street, New York, NY 10011-4211, USA
10 Stamford Road, Oakleigh, Melbourne 3166, Australia

First published 1994

Printed in the United States of America

Library of Congress Cataloging-in-Publication Data

Grant, Edward. 1926 –

Planets, stars, and orbs : the medieval cosmos, 1200–1687 / Edward Grant.

p. cm.

Includes bibliographical references and index.

ISBN 0-521-43344-4

1. Cosmology – History. 2. Astronomy – History. I. Title.

QB981.G664 1993

523.1'09–dc20 93-25899
CIP

A catalog record for this book is available from the British Library

ISBN 0-521-43344-4 hardback

In remembrance

ALBERTO COFFA (1935–1984)

Professor of History and Philosophy of Science
Indiana University–Bloomington

WILLIAM COLEMAN (1934–1988)

Dickson–Bascom Professor of Humanities and
Professor of History of Science and Medicine
University of Wisconsin–Madison

VICTOR E. THOREN (1935–1991)

Professor of History and Philosophy of Science
Indiana University–Bloomington

Contents

Illustrations

Preface

The links between medieval and modern cosmology

Few areas of science seem more exciting and spectacular than cosmology, which has become the speculative science par excellence. We have come to expect extraordinary pronouncements about the universe from cosmologists, whether they be astrophysicists or physicists. According to a relatively recent popular discussion (*Newsweek*, June 13, 1988, 60), "modern cosmology" is "a child of this century." Indeed, it was only "in the 1920s that scientists realized that our Milky Way is not the only galaxy." By the 1960s, two major opposition theories had emerged: the steady state theory, which maintains that the universe is "infinite and is forever the same" (Overbye, 1991, 2), a theory that required no search for origins, since the world is assumed to be of infinite duration, without beginning or end; and the big bang theory, which "holds that the universe began as an infinitely dense, infinitely hot point called a singularity. Then, 10 billion to 20 billion years ago, the singularity exploded. This was not an explosion *into* space, as popularly thought, but a smooth, slow explosion *of* space itself" (*Newsweek*, June 13, 1988, 60). As the dominant theory, the effects of the big bang hypothesis are still unfolding today. A momentous implication of this theory is that the universe is expanding, as a consequence of the original explosion, and creating ever-new spaces and universes.

The big bang theory is controversial, and there is much about it that seems almost beyond comprehension. Would the expanding universe one day cease to expand and collapse back into itself and vanish? What of the mind-boggling black holes, those "regions of extremely dense matter in which gravity is so strong that nothing, not even light can escape" (*Newsweek*, June 13, 1988, 57)? Indeed, what is a black hole? Is it a burned-out star that has collapsed, as astrophysicists have conjectured? And what was the state of cosmic affairs before the big bang? Was it a condition in which "there was no gravity, no matter, no space. Not even time," but "simply nothingness," from which "the Universe spontaneously appeared"? The theory suggests that "eventually the Universe may collapse and simply disappear into the nothingness from which it sprang" (Gino del Guercio, UPI Science Writer, Oct. 21, 1985, in *Bloomington Herald Times*). Here was "creation *ex nihilo*," without appeal to a creator. Modern cosmology has other fantastic events to contemplate, but what has already been mentioned is surely sufficient to establish its mind-boggling nature. Modern cosmologists have conditioned us to expect spectacular and incredible claims about the physical universe. They have reduced the unthinkable and incomprehensible to the routine and expected.

Those who are comfortable with the fantastic ideas that make up the fabric of modern cosmology should suffer little or no distress confronting the claims of medieval cosmology. Indeed, modern readers will be surprised, perhaps even shocked, to discover that medieval cosmologists – or better "natural philosophers" – were, in their own ways, as imaginative, spectacular, and seemingly "far-out" as their modern counterparts. Some medieval speculations and proposed solutions – though in a radically different context, to be sure – bear a striking resemblance to the speculative assumptions and problems of modern cosmology. We might appropriately inquire whether medieval cosmologists were as imaginative in their own way as are modern cosmologists in theirs. With the help of God (as we shall see), they were.

The basic antithesis between the steady state universe and that of the big bang has an immediate counterpart in medieval cosmology in the momentous thirteenth-century struggle between supporters of the creation account of Genesis and those who in some sense sought to uphold Aristotle's denial of a creation by insisting on the possibility that the world might have had no beginning and would have no end. On theological grounds, the world was assumed to have been created from nothing rather than from a preexistent matter. Although the Aristotelian cosmos was a nonexpanding finite sphere, some located it in an infinite, nondimensional, void space which they identified with God's infinite immensity and omnipresence. If other worlds existed in those spaces beyond our finite world, would they be like ours, with the same structure and laws? Others remained faithful to Aristotle and argued that beyond the last convex surface of our world, nothing at all existed: neither time, nor matter (and, therefore, no worlds), nor void, nor place. Could matter that was subject to change exist in a celestial region composed of an ether that was allegedly incorruptible? And, like their modern counterparts, medieval cosmologists were fond of imagining thought experiments and proposing paradoxical problems. Could God move the spherical cosmos with a rectilinear motion if that implied that a vacuum, which was thought naturally impossible, would be left behind? If God annihilated all the matter between the earth and the orb of the Moon, how would bodies placed in that void behave? Is it possible that the world had no beginning and was yet created? Was the world created all at once, in an instant, or was it created over time, say in six days?

During the approximately five centuries covered by this study, Aristotelian scholars speculated about cosmological events and possibilities that went far beyond anything Aristotle had described. In this sense, they extended the frontiers of cosmological speculation. They did what cosmologists have always done, from the time of the ancient Greeks to the present:

they challenged the prevailing dogma and imagined things as they might otherwise be. Copernicus, Tycho Brahe, and Galileo continued this tradition. But Tycho and Galileo intruded a new element into the history of cosmology: the use of instruments on which to base their arguments and speculations.

Since the introduction of these instruments, technological developments have served to advance the cause of cosmology, the data for which is now derived almost exclusively from the use of instruments that grow more powerful and subtle year by year. Without these instruments to produce an ever-expanding body of sophisticated data, modern cosmology would be simply unimaginable. By contrast, medieval Aristotelian cosmology was constructed without the use of instruments (except in its last phase, when Brahe's astronomical instruments and Galileo's telescope played a role). The driving forces in medieval cosmology were primarily metaphysics and, to a lesser extent, astronomy and theology. In general, medieval cosmology was a product of natural philosophers whose Aristotelian training and instincts prompted them to construct a world based on a priori assumptions and principles. Within the fundamental constraints of those principles and assumptions, they constructed a cosmos as they thought it had to be. Because it was God's handiwork, they assumed a beautiful, diverse, and harmonious world, although they disagreed on whether it was simple or complex.

Despite vast differences caused by the centuries that separate them, medieval and modern cosmologists seem agreed on their ultimate goal: (1) to describe and make sense of the origin of the world and its structure and operation, and (2) to do so by responding to the fundamental questions of the day with imaginative answers that are as consistent as possible with whatever passes for relevant data.

On reading this book

Because no study has yet provided us with a detailed description of the nature and sources of medieval cosmology, I have sought to do so in this work. It was therefore essential to include more than a broad description and analysis of the arguments. As a consequence, the first three chapters and the "Catalog of Questions" in Appendix I and its explication and analysis in Appendix II are devoted to background details that provide a useful and, I believe, necessary guide to the methodology of this study and to the context of medieval cosmology. Readers who are impatient with such details and eager to confront the arguments and opinions right away should begin with Chapter 4 and read on through Chapter 20. Many of the translations are my own. Where a translation is by another, the form of the reference should make this apparent.

Acknowledgments

In the approximately fifteen years it took to complete this study, I have accumulated a number of debts and obligations. None is greater than that owed to the National Science Foundation (Division of Social Sciences, Program in History and Philosophy of Science), which awarded me three separate grants between 1980 and 1992 to pursue research on the medieval cosmos. By offering me an ideal environment for research and contemplation during the academic year 1983–1984, the Institute for Advanced Study, in Princeton, enabled me to make the kind of progress that would not otherwise have been possible. That progress was facilitated by the intellectual and social companionship of Marshall Clagett, Professor of the School of Historical Studies of the Institute and my teacher and friend since 1951. I am delighted to express my gratitude to Indiana University for its annual research support over a period of eight or nine years and for providing an exceptional work environment.

Without the availability of two special projects, my research efforts would have been considerably more complicated and difficult. The first is the *Manuscripta Microfilms of Rare and Out-of- Print Books* compiled by the editors of the journal *Manuscripta* at the Pius XII Memorial Library of Saint Louis University. The extensive collection of microfilms of early printed editions by medieval and early modern scholastic natural philosophers which they assembled provided an enormous bibliographical base for the five centuries covered by my study. To all who were instrumental in the production of that marvelous series I offer my gratitude and happily acknowledge my huge debt. To Professor Charles H. Lohr (Raimundus-Lullus-Institut der Universität Freiburg), I am equally indebted for his splendid alphabetized compilation of Latin commentaries on Aristotle by medieval and Renaissance authors (see my Bibliography). Through his Herculean labors, Professor Lohr provided indispensible bibliographical and biographical information about virtually all scholastic authors from the thirteenth century to approximately 1650.

For the illustrations included in this volume, I offer grateful thanks to the following libraries and institutions: the Bibliothèque Nationale (Figures 7 and 16); the Hofmuseum, Vienna (Figure 2); the Nassauische Landesbibliothek, Wiesbaden, Federal Republic of Germany (Figure 3); the Staatliche Museen, Berlin (Figure 4); Österreichische Nationalbibliothek, Vienna (Figure 5); the Lilly Library of Indiana University (Figures 6, 9, and 11); the

British Library (Figures 10 and 14); and Martinus Nijhoff Publishers (Figure 1).

Of those who contributed directly to the final version of this volume, I owe my greatest debt to Mr. Danny Burton, my student and research assistant, who read the entire book, proposing valuable suggestions and critical queries. For help on a few astronomical matters, I thank my student, Mr. James Voelkel. David C. Lindberg (Evjue-Bascom Professor of the History of Science, University of Wisconsin-Madison) and John E. Murdoch (Professor of the History of Science, Harvard University), friends and fellow historians of medieval science, generously offered assistance on specific problems.

Over the years, colleagues in the Department of History and Philosophy of Science have played a role in shaping my views and opinions and have aided me in various ways. I have benefited greatly from innumerable conversations with Noretta Koertge, whose inexhaustible patience, keen insight, and extraordinary intelligence were always at my disposal. As a specialist on the theory of relativity, John Winnie always responded to questions on cosmology, but more significantly he kept my computers and printers operating and thus enabled me to convert this study into hard copy. Over a period of years, my late departmental colleagues, Alberto Coffa and Victor E. Thoren, to whom this book is dedicated, responded to numerous questions on cosmology and related matters and did so with unfailing good humor and insight.

Finally, I must thank Syd (Sydelle), my wife for over forty years, for her intelligent and sensible responses to many questions associated with this volume. Her contributions were genuinely a labor of love.

Abbreviations

The following abbreviations of Latin terms (and one English term) are frequently used in the footnotes. With the exception of *proportio* they appear in printed editions from the late fifteenth to the late seventeenth centuries and occasionally in modern editions of medieval works. Virtually all of them represent a division or subdivision of a treatise and are therefore difficult to translate. The scholastic penchant for subdivisions is nowhere better illustrated than in the *Summa theologica* of Alexander of Hales, who used seven terms for textual subdivision in the following descending order: *inquisitio, tractatus, quaestio, titulus, membrum, caput*, and *articulus*. Within this scheme, *membrum* seems to signify a division of a question, which is how I render it here.

art.	*articulus* (article)
comment.	*commentarius, commentarium* (commentary)
conclus.	*conclusio* (conclusion)
contro.	*controversia* (topic or theme)
differ.	*differentia* (question)
disp.	*disputatio* (disputation)
dist.	*distinctio* (distinction)
dub.	*dubium* (doubt)
dubit.	*dubitatio* (doubt)
fasc.	*fasciculus* (fascicule)
inquis.	*inquisitio* (inquiry or investigation)
lec.	*lectio* (lecture)
memb.	*membrum* (division)
part.	*particula* (part)
pt.	*parts, partis* (part)
post.	*posterior* (last)
prop.	*proportio* (proportion)
punc.	*punctum* (chapter)
qu.	*questio* (question)
sec.	*sectio* (section)
sig.	signature
tit.	*titulus* (title)
tract.	*tractatio* (tractate)
ult.	*ultima* (last)

INTRODUCTION

Scope, sources, and social context

I

Pierre Duhem, medieval cosmology and the scope of the present study

I. Duhem and *Le Système du monde*

No study of medieval cosmology could proceed without taking cognizance of Pierre Duhem's monumental ten-volume study on that subject, which bore the title *Le Système du monde: Histoire des doctrines cosmologiques de Platon à Copernic* (1913–1959).[1] Duhem's contributions to the history of medieval science and cosmology were nothing less than extraordinary. He was not only the discoverer of much of what is today deemed significant in the history of medieval natural philosophy, but he was also a brilliant interpreter of texts and ideas (Fig. 1). In order to define the scope of the present study, it will be useful to describe and characterize *Le Système du monde.*[2]

In his ten volumes, subdivided into six parts, Duhem began with Greek cosmology and subsequently treated the Church Fathers, Islamic authors, and a large number of scholastic authors who wrote in Latin, ranging from the early Middle Ages to around 1500 and occasionally extending his survey to approximately 1550. Duhem included a formidable range of topics, often presenting them for the first time. Among the many themes he discussed in detail were the dimensions of the world; the possibility of a plurality of worlds; eccentrics and epicycles; the Cabbala; the fate of Aristotelianism in the Latin West; and Parisian physics in the fourteenth century (including

1. For a brief summary of the history of the completion and publication of *Le Système du monde*, see Stanley Jaki's foreword in Duhem, 1985, xi–xviii.
2. My discussion is based upon the ten volumes as they appeared in print. When Duhem died in 1916, he was correcting proofs of the fifth volume (Duhem, 1985, xii), and he left manuscript versions of what would become the last five volumes. Thus it is unclear whether Duhem's chapter and section headings were embedded in his manuscript versions or added by subsequent editors. The truth of either of these alternatives is irrelevant to our purposes, since readers of Duhem's volumes must judge them and their subtitles and divisions on the basis of what they find in the printed version. Hence my interpretation is based solely on the published edition, which, for convenience, we shall assume represents Duhem's own titles and subtitles. Duhem formulated some of the subtitles for the various subdivisions of *Le Système du monde* in a series of lectures he delivered between 1909 and 1916 (see Jaki, 1984, 195). Indeed, the title of the lectures for 1911–1912, "Les doctrines cosmologiques de Platon à Copernic," was almost identical with the subtitle that Duhem later adopted for the whole of *Le Système du monde*.

Figure 1. Pierre Duhem in his study (ca. 1905).

the Aristotelian doctrines of place and void, projectile motion, the fall of bodies, the tides, and the various possible motions of the earth).

Despite its great significance, there are serious problems with Duhem's *Le Système du monde*. A major drawback derives from Duhem's failure to provide readers with any criteria for determining what counts as cosmology. Consequently he was free to choose any topics whatever, a freedom that resulted in a strange potpourri. Topics that are more appropriate for physics than cosmology form a significant portion of the whole, as is obvious in the part he called "Parisian Physics in the Fourteenth Century," which extends over volumes 7 through 9.[3] Topics in *Le Système du monde* often seem largely unrelated to one another, and their organization leaves much

3. That Duhem mixed together cosmology and physics in a work that bears in its subtitle the phrase "cosmological doctrines from Plato to Copernicus" may have resulted from the fact that "after the *Origines de la statique* he wanted to write a similar book on dynamics. It was that book, projected as 'Origines de la dynamique,' that finally became *Le Système du monde*" (Jaki, 1984, 195, n. 88).

to be desired.[4] Moreover, many specific topics are unsynthesized, because Duhem chose to proceed largely by describing in chronological order the opinions of one author after another, producing considerable repetition. Although Duhem usually retained chronological treatment, he did not always include all relevant authors in one chapter or section. Occasionally he spread the treatment of a given topic over several volumes, filling the intervening space with quite different topics.[5] Rarely does Duhem offer an overview of concepts and ideas on a given topic but usually ends a section abruptly with the opinions of the last author.

Because he offered no rationale or justification for the topics he included, Duhem was free to interpret cosmology in the broadest sense as the study of the world and all that occurs and exists within it. In this unrestricted approach, Duhem chose to include topics that were of interest to him regardless of their overall interrelationships. Without any apparent guiding principles, he could, with equal justification, have selected other topics. He might, for example, have included sections on Aristotle's biological works and the commentaries thereon. These would have been no less appropriate than the latitude of forms and the accelerated fall of bodies. Biology, after all, also forms part of *Le Système du monde*.

To avoid the problems and pitfalls inherent in Duhem's approach, it seems wise to impose some order on the concept of medieval cosmology. A rationale for the inclusion and exclusion of topics and themes is desirable, as is the establishment of reasonable and appropriate bounds. These are formidable tasks, where decisions may sometimes seem arbitrary.

II. The scope of the present study

1. The three parts

Because medieval cosmology was essentially derived from the natural philosophy of Aristotle, it had one fundamental feature that bears heavily on the scope of this volume. In his conception of the world, Aristotle sharply distinguished between the celestial region, which embraced everything from

4. For example, Duhem divided part 2 (vols. 2–4), on Latin astronomy in the Middle Ages, into secular, Dominican, Franciscan, Parisian, and Italian astronomy, as if all these were somehow different astronomies. He placed his discussion on the plurality of worlds at the end of part 5, following immediately after a lengthy consideration of the earth's properties and behavior. Although Duhem was probably influenced in this decision by the fact that one aspect of the plurality problem involved the question as to whether the earth of one world would move to the center of another world, it seems inappropriate to link the plurality of worlds with questions about the earth.
5. The most extreme form of this practice may be seen in Duhem's rigid separation of fourteenth- and fifteenth-century authors who treated the same problem. To discover medieval views on celestial movers, the reader has to peruse distinct sections in volumes 5, 6, 8, and 10. The same pattern was employed for the themes of celestial matter, the creation of the world, the eternity of the world, and others. By proceeding in this manner, Duhem could hardly avoid a fragmented, unintegrated, and repetitious account.

the Moon to the outermost limits of the world, and the sublunar, or terrestrial, region, which included all matter and activity below the Moon – that is, between the Moon and the center of the world, which coincided with the center of the earth.

Although Aristotle treated aspects of the celestial region in his *Physics* and *Metaphysics*, his most extensive and systematic treatment of it was in the first two books of *De caelo*. His consideration of the sublunar region was spread over a number of treatises. In the *De generatione et corruptione* (*On Generation and Corruption*), he described and analyzed the formation of the four elements and their combinations; in the third and fourth books of *De caelo*, he focused on "the four elements considered in respect of their heaviness and lightness, i.e. their tendencies to locomotion" (Ross, 1949, 98); and, finally, in the *Meteorology*, Aristotle studied phenomena of the upper regions of the atmosphere just below the Moon (thunder, lightning, rain, snow, etc.; also comets and the Milky Way, which he believed were sublunar, or meteorological, occurrences rather than celestial), as well as the properties of metals and compounds.

The primary focus in this study is on the celestial region, as it was characterized by Albert of Saxony (see the beginning of Ch. 20). Although I shall have numerous occasions to mention the behavior and nature of the sublunar elements, I shall not devote specific chapters or sections to them. In a sense, I have followed Pierre d'Ailly's definition of "universe" (*universum*), which includes the four elements and the compound (or "mixed") bodies formed from them only insofar as they are thought to be governed by the celestial motions (as described in Ch. 18). In keeping with Albert of Saxony's approach, the discussion of the earth in Chapter 20 emphasizes its cosmological role as the center of the world, not its behavior as an element and source of compounds. By emphasizing the celestial region over the terrestrial, both of which are treated in Aristotle's *De caelo*, this study remains faithful to medieval practice. As will become evident, medieval commentators on the four books of *De caelo* devoted far more space to the celestial region, covered in the first two books, than to the four elements of the sublunar realm, discussed in the last two. In the second half of the fifteenth century, Johannes de Magistris omitted questions on the third book of *De caelo* entirely (skipping from the second to the fourth book), with the explanation that this book was not commonly considered by Parisian scholars because it was of little utility.[6]

With Aristotle's distinction between celestial and terrestrial regions in mind, I can now briefly describe the three divisions of this study. The first,

6. "Nota quod a Parisiensibus communiter non ponitur liber tertius quia parve est utilitatis. Ergo finito secundo *Libro celi et mundi* incipit quartus." Johannes de Magistris [*De celo*, bk. 2], 1490, 27 (unfoliated; pages numbered from beginning of *De celo*). Throughout these footnotes, bracketed citation forms (used only in the first citation within a chapter) indicate the genre type of the work: *De caelo* (or *De coelo*, or *De celo*, depending on the language); *Sphere*; *Physics*; *Sentences*, etc. For the original title of each work, consult the Bibliography.

which has been labeled an "Introduction," includes not only Chapters 1 to 3, but also Appendixes I and II. It provides the relevant background for understanding medieval cosmology, including a detailed description and analysis of its sources; the social context in which it was developed; and the questions and problems that medieval and Renaissance scholars discussed. The second and third divisions – actually Parts I and II of the book – form the actual intellectual content of the study. Of the two actual parts into which cosmology proper has been divided, Part I concerns the kinds of problems that pertain to the world as a whole and what may lie beyond it, rather than to either of its two major subdivisions, the celestial and terrestrial regions. Part II ranges over the celestial region but also includes the earth, "insofar as it is the center of the heavens and of the whole world."[7]

2. *Terminology*

The *Oxford English Dictionary* defines the word "cosmos" as "the world or universe as an ordered and harmonious system,"[8] a definition that shall serve as our guide in this study. Although the words "cosmos" and "cosmology" are derived from the Greek terms *kosmos* and *kosmologikos*, Latin cognates for these terms were rarely if ever used during the Middle Ages.[9] In 1603, Nicolaus Taurellus (1547–1606) used the Greek term for *cosmologia* in a work titled Κοσμολογια, *hoc est, De mundo libri II* (Amberg, 1603).[10] According to the *Oxford English Dictionary*, the first use of "cosmos" in English occurred in 1650 and of "cosmology" in 1656. The medieval Latin words that best describe what we understand by "cosmos" are *mundus, caelum*, and *universum*. The first, *mundus*, could have as many as four mean-

7. Based on Albert of Saxony's questions on *De celo* (described below in the present volume at the beginning of Chapter 20). How does this twofold division of Albert's book compare with the fourfold division of the nineteen themes embracing 400 questions in the "Catalog of Questions" in Appendix I? The four parts of the catalog (titled, respectively, "The World as a Whole"; "The Celestial Region"; "Questions Relevant to the Celestial and Terrestrial Regions"; and "The Terrestrial [or Sublunar] Region" represent a slightly more refined version of the twofold division of Albert's study. The titles of the two parts of his study and of the first two parts of my catalog are identical. The title of the third part of my catalog is represented by Chapter 19 in Part II of this study, while Chapter 20 of this study corresponds to certain questions in the fourth part of my catalog (qus. 383, 387, 389, 396) that Albert and I have interpreted as relevant to the celestial region because they focus on the earth's cosmic relations – that is, its centrality, immobility, sphericity (or shape), and habitability (in the title of Chapter 20, however, I have replaced habitability, which I treat briefly, with size, which Albert ignored). The remaining questions in the fourth part of the catalog are concerned with the terrestrial, or sublunar, region, which, because of space limitations, has been largely ignored in this study.
8. Virtually the same definition appears in the *Encyclopaedia Britannica* (see the article "Cosmology" (1968), 6:582).
9. In the large body of cosmological literature that I examined for this study, they do not appear. Indeed, the term *cosmologia* does not occur as a Latin word in *Harper's Latin Dictionary*. The word *cosmographia*, however, does appear, with the appropriate meaning of "a description of the universe." But it was not used by the numerous authors who are discussed in this volume.
10. See Lohr, 1988, 450.

ings but usually embraced at least the heavens and earth and all that lay between.[11] The second, *caelum*, usually translated as "heaven" or "heavens," was more limited in scope. In its narrowest signification, it could represent a single planetary sphere (and is appropriately translated as "heaven"), but it could also signify the entire celestial region, which included the totality of celestial orbs, everything from the sphere of the Moon to the sphere of the fixed stars and whatever orbs might lie beyond. (In this context, *caelum* is better translated as "heavens," but in historical usage the terms "heaven" and "heavens" have usually been interchangeable, as they often are in this study.) And, as we shall see, *caelum* was even sometimes used for subdivisions of the world that excluded only the earth.[12] The third term, *universum*, was probably synonymous with the broadest meanings of *mundus* but, in the definitions provided by Pierre d'Ailly, takes on special significance in this study. In his *14 Questions on the Sphere of Sacrobosco*, d'Ailly gives three definitions of *universum*. The first takes as its domain the totality of celestial bodies;[13] the second equates *universum* with the prime mover, a definition that was not to be taken seriously;[14] the third definition at first glance embraces everything as part of the universe, thus including "the aggregate of celestial bodies, the intelligences that are applied to them, all the mixed bodies and the four elements contained under the Moon."[15] But then d'Ailly adds that this totality is not to be taken absolutely, but only relatively. By this he meant that *universum* represents all the celestial bodies that are moved with circular motions and also includes the elements with the mixed bodies formed from them, but only insofar as these elements and mixed bodies are ruled or governed by the motions of the celestial bodies.[16] D'Ailly's qualified third definition of *universum* comes closest to a description of what I shall attempt to encompass in this volume. And it should be emphasized that the three terms – *mundus*, *caelum*, and *universum* – could be, and were, used interchangeably.[17] Two of the three terms are reflected in the title of what was in the Middle Ages the most fundamental

11. For the four meanings, see Chapter 7, Section I and note 4.
12. Here *caelum* signifies more than one planetary sphere, and even the whole celestial region, and is therefore better translated as "heavens." There is little consistency in modern translations of *caelum*, because the precise intent of ancient and medieval authors is often unclear. Although the terms "heaven" and "heavens" are usually interchangeable, and both frequently occur here, I have preferred "heavens."
13. "Primo modo pro aggregato ex corporibus coelestibus." D'Ailly, *14 Questions*, qu. 1, 1531, 147v.
14. "Secundo modo pro primo motore." Ibid.
15. "Et tertio modo capitur universum pro aggregato ex corporibus coelestibus et intelligentiis eis applicatis et omnibus mixtis et quattuor elementis sub orbe lunae contentis." Ibid.
16. "Et isto modo capiendo ille terminus universum non significat tale aggregatum absolute sed respective per istum modum quod significat corpora coelestia inquantum sunt circulariter mobilia et elementa cum mixtis in eis contentis inquantum a corporibus coelestibus reguntur per suos motus." Ibid.
17. In Chapter 2, Section III.2, of the present book, see Buridan's discussion of the term *mundus*, which, in one version of his lengthy definition, he equates with *universum*. For more on definitions of "world," see Chapter 7, Section I, and Chapter 7, note 4.

cosmological treatise, Aristotle's *De caelo* (*On the Heavens*), or as it was often called, *De caelo et mundo* (*On the Heavens and the World*). What was written about part or all of *mundus, caelum*, and *universum* would, taken together, represent medieval cosmology, that is, represent a description of "the world or universe as an ordered system."

As the primary cosmological treatise available during the Middle Ages, Aristotle's *De caelo* serves as the most suitable focal point for the study of medieval cosmology. The collection of commentaries and questions on *De caelo*, along with commentaries and questions on certain themes in a few other works, constitutes the basis for this investigation. The nature of these treatises and their organization are considered in Chapter 2.

III. Temporal limits

The temporal limits for medieval cosmology as I have interpreted them differ considerably from those of Duhem, who began with the ancient Greeks and extended his account into the sixteenth century, ranging, on occasion, to the middle of that century. Because my objective is to describe and analyze medieval Latin Aristotelian cosmology over the whole of its viable existence, my study begins at approximately 1200, when Aristotle's works on natural philosophy effectively made their entrance into western Europe, and terminates in 1687, or toward the end of the seventeenth century. Although scholastic natural philosophers made significant alterations in Aristotle's cosmology, the latter's cosmological views provided the dominant world view for some five centuries. Indeed, judging by the large number of printed editions of Aristotle's physical treatises and the commentaries thereon made in the two centuries after the introduction of printing, we may infer that the study of Aristotle became even more extensive and intensive between, say, 1450 and 1650, or 1687, the terminal date of this study, than it had ever been in the thirteenth and fourteenth centuries, the climactic centuries of the Middle Ages.[18]

Within the seventeenth century, 1687, the year of Isaac Newton's epoch-making treatise *The Mathematical Principles of Natural Philosophy (Philosophiae Naturalis Principia Mathematica)*,[19] seemed an ideal terminal date for Aristotelian cosmology. When, in 1632, Galileo published his *Dialogue on the Two Chief World Systems*, his monumental assault on Aristotelian cosmology, the majority of natural philosophers, if not the majority of astronomers, were probably defenders of some form of geocentric cosmology, a

18. Schmitt, 1983, 14, asserts that "at rough estimate, three to four thousand editions of *Aristotelica* were published between the invention of printing and the year 1600." The same may be said about Avicenna, whose *Canon of Medicine* was studied more thoroughly and extensively in the age of printing, say in the period between 1500 and about 1625, than it was in the thirteenth and fourteenth centuries (Siraisi, 1987, 6, 356–357).
19. For the remainder of the paragraph, I draw upon Grant, 1985a, 417.

cosmology that had been shaped in the thirteenth and fourteenth centuries and associated with the names of Aristotle and Ptolemy. Although Aristotelian cosmology undoubtedly suffered some erosion in the fifty-five years that intervened between Galileo's *Dialogue* and Newton's *Principia*, that same period saw a number of Aristotelian natural philosophers attempt to incorporate some of the new cosmological ideas into the traditional framework. Seventeenth-century scholastic authors were usually knowledgeable about the basic ideas of Copernicus, Tycho Brahe, and Galileo. Indeed, the impact of their ideas was not wholly confined to their scholastic colleagues. Occasionally a nonscholastic reveals a respectful interest in their opinions, as was the case with Otto von Guericke.[20] Aristotelian cosmology was thus still alive, and even viable.

The appearance of Newton's *Principia* in 1687 changed all this. Copernicus's heliocentric system, with its Keplerian modifications, was at last provided a mechanical and mathematical foundation that made continued support for Aristotle's geocentric cosmology untenable. After 1687, medieval cosmology became irrelevant, because it no longer represented even a minimally plausible alternative to Newtonian cosmology. Unlamented, it simply faded away. Because the final century of Aristotelian cosmology has thus far been virtually ignored, it will form an important aspect of this study, which takes for its objective the whole of late medieval scholastic cosmology. Thus, whereas Duhem provided extensive coverage for the period prior to 1200, the point at which I begin, I have thought it advantageous and important to carry my deliberations to the time when Aristotelian cosmology was effectively dead, thus ranging approximately 150 years beyond Duhem.

20. In his famous *Experimenta nova* of 1672, von Guericke mentioned numerous scholastic authors who play a role in this study. For example, see page 46, col. 1 (Roderigo de Arriaga); 47, col. 2 (Conimbricenses); 48, cols. 1–2 (Thomas Aquinas, Saint Bonaventure, Durandus de Sancto Porciano, and Thomas Compton-Carleton); 205, 217, 219, 228 (Giovanni Baptista Riccioli); 205 (Christopher Scheiner); and 229 (Raphael Aversa).

2

The sources of cosmology in the late Middle Ages

I. The twofold significance of the term "sources"

Within the context of medieval cosmology, the term "sources" has a twofold signification. The first concerns the inherited literature that profoundly shaped the framework and content of late medieval cosmology. Literature in this category embraces primarily Greco-Arabic treatises that were translated into Latin and, to a lesser extent, the commentary literature of the Church Fathers on the six days of creation (known as "hexaemeral" literature), along with a small cluster of Latin works that came from late antiquity and the early Middle Ages. What late medieval natural philosophers did with this received body of literature forms the second signification of the term "source." Using their inheritance as a foundation and frequently building upon their own ideas and interpretations, scholastic authors produced a large body of cosmological literature between the thirteenth and the late seventeenth century, the types and forms of which are examined in this chapter. Only through a detailed study of these texts can we come to understand and appreciate the medieval world view, both by determining what remained reasonably constant, what ideas were altered, and what the major opinions were on a wide range of problems.

II. The cosmological inheritance

1. Latin cosmological literature from the early Middle Ages

During the early Middle Ages, prior to the great age of translation in the twelfth and thirteenth centuries, science was truly at a low ebb. Without the seminal works of Greek science and their subsequent elaboration in the Islamic world, which occurred in the ninth and tenth centuries, early medieval cosmology and cosmological ideas were largely based on Latin encyclopedic treatises associated with the names of the Roman authors Pliny the Elder (ca. 23–79), *Natural History* in thirty-seven books, and Seneca (ca. 4 B.C.–A.D. 65), *Natural Questions*, and with the following Latin encyclopedic authors: Solinus (fl. 3rd or 4th c.), *Collection of Remarkable Facts*; Chalcidius (fl. 4th or 5th c.), translated, and commented on approximately two-thirds of Plato's *Timaeus*; Macrobius (fl. early 5th c.), *Commentary on the Dream of Scipio*; Martianus Capella (ca. 365–440), *The Marriage of Philology and Mercury*; Cassiodorus (ca. 480–ca. 575), *Introduction to Divine and Human Readings*; Isidore of Seville (ca. 560–636), *The Etymologies* and *On the Nature of Things*; and Venerable Bede (672–735), *On the Nature of Things, On the Division of Time, On the Reckoning of Time.*[1]

The cosmology embedded in these treatises was meager, superficial, and often unreliable, largely because many of their authors lacked comprehension of the material they included, material that was itself derived from handbooks that were copies of earlier encyclopedic collections. More reliable and cohesive was the approximately two-thirds of Plato's *Timaeus* that had been translated and commented upon by Chalcidius. But the *Timaeus* lacked a sufficiently detailed natural philosophy with adequate physical and metaphysical principles. However deficient these early treatises may appear to us in retrospect, their authors were generally regarded as venerable authorities, and a number of these old works continued to play a role in the later Middle Ages.

Despite the unreliability and meagerness of the traditional scientific and cosmological literature, twelfth-century scholars, such as Adelard of Bath, Bernard Silvester, Thierry of Chartres, William of Conches, and Clarenbaldus of Arras began to exhibit an interest in nature for its own sake. They interpreted natural phenomena and theories about natural phenomena, including cosmological ideas, with a surprisingly critical spirit. Even biblical texts relevant to cosmology and nature generally were subjected to the same penetrating scrutiny. Allegorical interpretations were often preferred over literal explications that had satisfied the faithful during the preceding centuries.[2] Whether, if given sufficient time, this bold intellectual venture would have generated a new natural philosophy will never be known. The influx of Greco-Arabic science into western Europe had already begun and would soon overwhelm the incipient rational science that had been evolving within the context of the old learning.

2. Greco-Arabic sources translated in the twelfth and thirteenth centuries

The great wave of translations that flowed into the Latin West during the twelfth and thirteenth centuries marks a significant transition in the intellectual life of western Europe and in the history of science. Although a number of significant medical, and perhaps a few minor arithmetical and

1. For a description of the contributions of most of these authors, see Stahl, 1962. Samples from the writings of Macrobius, Chalcidius, Martianus Capella, and Isidore of Seville appear in Grant, 1974.
2. On these twelfth-century scholars, see Chenu, 1968, 4–18 ("The Discovery of Nature"), 33; Stock, 1972, 271–273. For the strongest claims in behalf of the critical objectivity of twelfth-century scholars, see Stiefel, 1985.

astronomical, treatises had been translated during the eleventh century, scholars in the early Middle Ages did not have available the great mass of scientific and philosophical learning that was available in Greek and Arabic manuscripts in the Greek- and Arabic-speaking parts of the Mediterranean and Middle East. By the twelfth century, however, knowledge of the intellectual treasures available in the Greek and Arabic languages, coupled with an intense yearning to improve the level of learning in the Latin language, induced an international brigade of translators to seek out the new Arabic learning in Spain and Sicily; similar motives led others to search after Greek manuscripts for translation into Latin.[3]

Within a period of a little over one hundred years, from approximately 1125 to 1230, the translators achieved monumental results. If we extend our terminus to 1280, and include William of Moerbeke, the sum total of new works made available increases considerably, both quantitatively and qualitatively. The efforts of the two greatest translators, Gerard of Cremona, who translated from Arabic to Latin, and William of Moerbeke, who translated from Greek to Latin, would alone have transformed the course of Western science and natural philosophy.[4] Between them they rendered into Latin some, most, or all of the important works of Aristotle, Galen, Euclid, Archimedes, Ptolemy, Geminus of Rhodes, Hypsicles, Autolycus, Theodosius of Bithynia, as well as works of Greek commentators and authors in late antiquity, such as Menelaus, Proclus, Alexander of Aphrodisias, Eutocius, Hero of Alexandria, Themistius, John Philoponus, and Simplicius. To this list we must add Gerard's translations of numerous Arab scientific authors, such as al-Khwarizmi, Al Farabi, Alkindi, Alhazen (ibn al-Haytham), Anaritius (al-Nairizi), Thabit ibn Qurra, Alfraganus (al-Farghani), Messahala, the Banu Musa (the sons of Moses), Rhazes, Avicenna, and Isaac Israeli. Numerous other translations were made by Adelard of Bath, Alfred of Sareshel, Michael Scot, Plato of Tivoli, and John of Seville. But it was not only a matter of rendering works from Greek and Arabic into Latin. Occasionally, the line of translation went from Arabic to Hebrew to Latin, or simply from Hebrew to Latin.[5]

With this vast body of scientific knowledge, we may appropriately say that science took root in western Europe and began its long period of evolution culminating in the Scientific Revolution of the seventeenth century. The translations of Greco-Arabic science, within which Aristotle's natural books formed the core,[6] laid the foundation for the continuous development of science to the present and also provided a curriculum that made possible the development of the university as we recognize it today. Without the well-articulated body of theoretical science made possible by

3. For an excellent account of this translating activity, see Lindberg, 1978a.
4. See Grant, 1974, 35–38 (for the works translated by Gerard of Cremona) and 39–41 (for those by William of Moerbeke).
5. See Lindberg, 1978a, 67–69.
6. For the works embraced by the expression "natural books," see Section III.1 of this chapter.

the translations, the great scientists of the sixteenth and seventeenth centuries would have had little to reflect upon, little that could have focused their attention on significant physical problems. Many of the burning issues that were resolved in the Scientific Revolution of the seventeenth century entered western Europe with the translations or were developed by university-trained medieval natural philosophers in the process of commenting upon that impressive body of knowledge or in independent treatises. The overthrow of one world system by another need not imply a lack of continuity. Medieval science, based on the translations of the twelfth and thirteenth centuries, furnished the physicists and natural philosophers of the seventeenth century with issues, theories, and principles much of which had to be rejected in order that significant advances be made. Galileo did not invent the problems of motion that he tried to resolve; he inherited them from the Middle Ages. To emphasize the importance of the Greco-Arabic scientific corpus that shaped medieval science for almost five centuries, we need only ask ourselves whether the Scientific Revolution would have been possible if the level of medieval science had remained where it was at the beginning of the twelfth century prior to the influx of the new translations. One can scarcely imagine it.

a. Aristotle

With such a large number of works available in Latin for the first time, we must ascertain which among them were especially relevant for cosmology. In light of what has already been said, the answer is obvious: the physical works of Aristotle, among which the *De caelo* was by far the most important. In this work of four books, the first two, which are of primary importance for medieval cosmology, concern the celestial region, while the last two describe the behavior of the four elements in the terrestrial, or sublunar, region. In the first book, Aristotle treats the celestial ether, or fifth element, and its properties, including its natural, circular motion. He argues for the finitude of the world and for the impossibility of other worlds, as well as for the impossibility of the existence of place, void, or time beyond our world. He concludes the first book with a demonstration that the world could have had neither beginning nor end; that is, it is ungenerated, incorruptible and indestructible. In the second book, Aristotle considers whether the celestial body, or sphere, could be said to have a top and bottom, as well as a right and a left side; why there are more revolving bodies than one; and why the planets and stars have different circular motions in the heavens. He also demonstrates why the whole heavens must be spherical and why the motions of its parts must be circular and uniform. Also included are sections about the composition and effects of the stars and planets and the cause of their motions, each carried by the sphere in which it is embedded. Aristotle also refutes the Pythagorean claim that the motions of the stars and planets generate harmonious music. Other prominent themes are

Figure 2. Aristotle. From a bust in the Hofmuseum, Vienna.

concerned with whether the planets move more slowly in proportion to their proximity to the outermost celestial sphere; whether the stars are spherical; whether the order of the planets correlates with the complexity of the planetary motions. Aristotle concludes the second book with a discussion of the earth's shape, centrality, size, and whether it rests or moves.

In the third and fourth books, where Aristotle turns to the terrestrial elements that exist below the orb of the Moon, he considers the number of elements, their heaviness and lightness (devoting the brief fourth book entirely to this problem), their motions, the definition of an element, and whether they are eternal or generated.[7]

The *De caelo* is the only cosmological treatise that Aristotle wrote, although, as we shall see, parts of other Aristotelian treatises contained sections that were substantively cosmological or at least relevant to cosmology. The enormous importance of Aristotle's *De caelo* in the history of cosmology will become more evident later in this chapter when we examine the comp-

7. A good brief account of the contents of the *De caelo* appears in Ross, 1949, 95–99. The four elements were also the subject matter of *On Generation and Corruption*, where Aristotle treats the elements as the constituent units of compound bodies and explains how such compounds come into being and pass away.

mentaries and questions that it elicited. It was a powerful influence in cosmological thought well into the seventeenth century, even as its authority began to crumble under the weight of criticism from some of the greatest figures of the Scientific Revolution: Copernicus, Brahe, Galileo, and Kepler. But even these great critics continued to respect the cosmology embedded in the *De caelo*. As late as 1620, Kepler characterized the fourth book of his *Epitome of Copernican Astronomy* (*Epitome astronomiae Copernicanae*) as "designed to serve as a supplement to Aristotle's *On the Heavens*."[8]

b. *Other Greco-Arabic treatises*

Except for commentaries or questions on the cosmological books of *De caelo*,[9] there were no specifically cosmological treatises that entered the Latin Middle Ages by translation from Greek and Arabic sources. Although Averroës' *De substantia orbis* may at first glance seem an exception, it treats only a single problem – a comparison of matter and form in the celestial substance with that in terrestrial substances – and thus does not qualify as a general cosmological treatise.[10] Despite the absence of such works, some treatises translated from the Greco-Arabic tradition contained ideas relevant to cosmology; for example, *The Guide of the Perplexed*, wherein Moses Maimonides presents difficulties arising in Aristotelian cosmology from the assumption of epicycles and eccentrics.[11]

Works on astrology and astronomy might also include relevant arguments or descriptions. The first chapter of Ptolemy's *Almagest* contained arguments about the immobility of the earth, and his *Hypotheses of the Planets* described ways in which eccentric and epicyclic orbs might be arranged in the heavens.[12] Some Arabic astronomers had similarly included occasional cos-

8. Cited from Van Helden, 1985, 82, who drew it from Glen Wallis's translation of Kepler's *Epitome*, books 4 and 5, in Kepler, 1952, 845.
9. The most important of all such commentaries was by Averroës (see Averroës [*De caelo*] 1562–1574, vol. 5). Another important, though not widely used, commentary was that of Simplicius, a Greek Aristotelian commentator, whose treatise was translated from Greek into Latin by William of Moerbeke in 1271 and printed in 1540 (see Simplicius [*De celo*], 1540). A brief commentary by Avicenna was also available, though it was inadequately translated (see Avicenna, 1508).
10. See Averroës, 1562–1574, vol. 9, where it is titled *Sermo de substantia orbis*. Just where the text ends is a puzzle. According to appearances, it extends over fols. 3r–14v. But it seems to end on 5v, which terminates what is called "chapter 1," where we read "Set hic finitus est *Sermo*" (this ending coincides with that of Arthur Hyman's translation into English from a Hebrew version; see Averroës [Hyman and Walsh], 1973, 307–314). But the work continues until we read, at the end of chapter 5 on 11r, that the next two chapters, 6 and 7, had been separated from the rest of the work but were here added to it by the translator, Abraham de Balmes. The treatise then continues on to 14v.
11. See Maimonides, *Guide*, pt. 2, ch. 24, 1963, 2:322–327.
12. For the Greek text of the first book and a German translation of both books, see Ptolemy *Hypotheses of the Planets*, 1907; for the translation of a part missing in the previous edition and translation, see Ptolemy [Goldstein], 1967.

mological information.[13] But most technical astronomers paid little attention to cosmological arguments. Detailed discussions about the structure and operation of the world were generally left to natural philosophers. This division of duties was based upon a general medieval belief that although astronomers could devise hypotheses about the celestial motions that were not necessarily true but saved the astronomical phenomena, physicists, or natural philosophers, had the duty to seek the true principles that governed celestial motions and operations.[14] With the important exception of descriptions and ideas about eccentric and epicyclic orbs,[15] concepts that did not even exist until sometime after the death of Aristotle, the primary source for medieval cosmological ideas and arguments was, and remained, Aristotle's *De caelo*.

3. The advent of printing and Greco-Latin additions to the medieval heritage in the fifteenth and sixteenth centuries

During the fifteenth and sixteenth centuries, two significant occurrences affected the course of events in the history of science: a second wave of translation, now almost exclusively from Greek to Latin, of works largely unknown in the Middle Ages, and the invention of printing around 1460. The new translations and the printing press mark the definite beginnings of a challenge to traditional science and natural philosophy. These events would affect the course of scholastic cosmology by facilitating the emergence of the new cosmology that would eventually replace the Aristotelian scholastic world view.

Before Greek scientific treatises could be translated, however, manu-

13. For example, al-Bitrūjī (see al-Bitrūjī [Carmody], 1952, 80–83, 92–94; and al-Bitrūjī [Goldstein], 1971, 63–66, 74, 76–79). See Section III.7, in this chapter, for more on the distinction between astronomers and physicists.
14. Perhaps the most extensive statement of this distinction between the objectives of astronomers and physicists, or natural philosophers, was by Simplicius, in the latter's *Commentary on Aristotle's Physics* (the passage appears in Cohen and Drabkin, 1948, 90–91). Aristotle laid the basis for this important distinction when, in *Physics* 2.2.193b.21–194b.15, he describes the difference between the mathematician and the physicist, or natural philosopher. The latter treats of the essential attributes of bodies, whereas the former ignores those attributes and considers only the limits of bodies in terms of solids, planes, lines, and points. These mathematical limits of real bodies are studied in abstraction as immobile entities. Similar sentiments were voiced in the Middle Ages by Averroës (see [*De caelo*, bk. 2, comment. 35], 1562–1574, 5:118v, col. 2); Bacon ([*De celestibus*, pt. 5, ch. 17], *Opera*, fasc. 4, 1913, 443); and Thomas Aquinas (for references, see Litt, 1963, 359). For a detailed discussion in which all of these individuals are mentioned, see Duhem, 1969, chs. 1–3, 5–45.
15. With the exception of Ptolemy's *Hypotheses of the Planets*, I have mentioned no specific Greco-Arabic treatises as sources for knowledge about eccentric and epicyclic orbs in the Latin West because, as we shall see, none of them seems a suitable candidate. There is little doubt, however, that the ultimate source for such knowledge was Ptolemy's *Hypotheses of the Planets*, composed in the second century but unavailable in Latin translation during the Middle Ages. Although the main concepts of the *Hypotheses* reached the Latin West, probably from Arabic sources, the manner in which this was achieved remains largely unknown.

scripts of them had to be obtained. This was a significant preoccupation of numerous fifteenth-century humanists, who were aided by the fact that Greek manuscripts were brought to Italy by Byzantine Greeks, many of whom fled the Turkish siege and then the capture of Constantinople in 1453. Most, if not all, of these works were translated into Latin and subsequently printed during the fifteenth or sixteenth century. Among the Greek scientific, and science-related, treatises that were unknown in the Middle Ages but became available during the fifteenth and sixteenth centuries in Greek editions, and usually in Latin translations as well, were the following: the works of Plato, translated by Marsilius Ficino, which included the complete *Timaeus*;[16] the Corpus Hermeticum, a series of fourteen Greek works on magic attributed falsely to Hermes Trismegistus and also translated by Ficino;[17] a handbook on astronomy (*Contemplation of the Highest Orbs*) in two parts by Cleomedes (ca. 1st c.); the first book of Alexander of Aphrodisias's (fl. 2nd–3rd c.) *On the soul* (*De anima*); Aristarchus of Samos's (ca. 320–230 B.C.) *On the Sizes and Distances of the Sun and Moon*; the *Lives of the Eminent Philosophers* of Sextus Empiricus; the *Cosmographia* of Claudius Ptolemy; the *Geographia* of Strabo; *On the History of Plants* and *On the Causes of Plants* by Theophrastus, one of Aristotle's students; the *Paraphrase of Aristotle* by Themistius; Hero of Alexandria's *Pneumatica*; commentaries on various works of Aristotle by Greek commentators of late antiquity, among which were the commentaries on the *Physics* by John Philoponus and Simplicius, commentaries on the *De anima* by Alexander of Aphrodisias and Simplicius, a commentary on the *Metaphysics* by Alexander of Aphrodisias, and a commentary on the *Meteorology* by John Philoponus. Of importance were the Greek treatises *De placitis philosophorum* (falsely ascribed to Plutarch) and Plutarch's *De facie quae in orbe lunae apparet* (*Concerning the Face Which Appears in the Orb of the Moon*), where the earth's axial rotation and orbital motions are mentioned.[18]

The advent of printing not only brought works to the fore that had been previously unavailable, but it made possible the widespread dissemination of works that were known to scholars in the Middle Ages. In the case of great authorities such as Aristotle and Galen, new translations were made and numerous editions published. But many printed editions were made of medieval translations, especially of the great Arabic authorities, Avicenna and Averroës. Works by medieval and early modern scholastic authors were enormously popular, as is evidenced by the many editions used in this study. Indeed, the great impact of Albert of Saxony's ideas on sixteenth- and seventeenth-century scholastics is due solely to the publication of his books

16. During the Middle Ages, approximately two-thirds of the *Timaeus* was known, in the fourth-century translation of Chalcidius.
17. The Hermetic treatises, only a few of which were known during the Middle Ages, assumed a large and significant role in shaping the world view of the late fifteenth and sixteenth centuries. For their content and impact, see Yates, 1969.
18. On the earth's axial rotation, see Chapter 20, Section V.

of questions on the *Physics, De caelo,* and *On Generation and Corruption.* Great medieval figures of the caliber of Thomas Aquinas, Bonaventure, Duns Scotus, Albertus Magnus, William of Ockham, and others retained their influence because some or most of their works were printed. By printing numerous medieval texts from manuscript sources, fifteenth- and sixteenth-century publishers solidified and perpetuated the medieval world view. Printing from fixed type largely guaranteed that an edition of, say Ptolemy's *Almagest,* would be uniform: readers in London confronted the same text and diagrams as readers in Rome. Such constancy and uniformity could not be attained by manuscript copies, where virtually every one was a unique and idiosyncratic version of a given text.[19]

III. The medieval Latin literature of scholastic cosmology, 1200–1687

With Aristotle's treatises available in Latin translation and with a large body of additional relevant Greco-Arabic literature at their disposal, medieval natural philosophers began the difficult process of comprehending and reacting to this new world view. Beginning with the thirteenth century, they learned about it in the universities, which emerged on the scene right along with the new learning – indeed, in no small measure because of it. The lectures on the works of Aristotle and other authors given at institutions like the universities of Paris and Oxford presented various aspects of cosmology. The masters discussed a great variety and number of cosmological problems. Their students, in turn, did much the same thing. This process, as we shall see, continued on into the late seventeenth century. In this way, cosmological problems came to be discussed regularly in certain kinds of treatises, which are already identifiable by the end of the thirteenth century. Only by a careful study of these scientific literary genres can we hope to comprehend medieval ideas about the structure and operation of the world. Before doing so, however, we must first introduce the medieval university into our story, in order to understand and appreciate the institutional, pedagogical, and literary matrices within which medieval cosmological ideas were developed and embedded.[20]

1. The medieval university, Aristotle, and the use of the terms "natural philosophy," "scholastic," and "Aristotelian"

With the introduction of Latin translations of Aristotle's scientific and philosophical works in the late twelfth century, along with translations of numerous other Greco-Arabic works, the basis for an extensive curriculum

19. For the impact of printing on science and natural philosophy, see Eisenstein, 1979, vol. 2 (pt. 3, "The Book of Nature Transformed").
20. Here I am partially dependent on Grant [Kittelson and Transue], 1984a, 78–79.

was at hand. By 1200, two of the greatest universities of Christendom, those of Oxford and Paris, were already in existence. Their curricula would eventually be based on the new science and natural philosophy. At the heart of the arts curriculum at Oxford and, by 1255, Paris lay Aristotle's logic and "natural books" (*libri naturales*). Comprising the natural books were his *De caelo* (*On the Heavens*); *Physics*; *De generatione et corruptione* (*On Generation and Corruption*); *De anima* (*On the Soul*); *Meteorology*; *Parva naturalia* (*The Small Natural Books*), as well as the biological works, namely the *Generation of Animals*; the *Parts of Animals*; and the *History of Animals*. Numerous other university texts were written in the Middle Ages in Latin, although they were usually based on the Greco-Arabic inheritance. In astronomy, for example, two prime texts were the *De sphera* (*On the Sphere*) by John of Sacrobosco (or Holywood), the most popular astronomical text of the Middle Ages, and the *Theorica planetarum* (*The Theory of the Planets*).[21]

During the thirteenth to fifteenth centuries, students matriculating for a bachelor of arts degree were required to take four years of specified course work. Those who completed the requirements in four years were usually in the fifteen- to nineteen-year-old age bracket. As undergraduates, they spent most of their time on grammar and Aristotelian logic, the latter deemed especially important for disputations, which formed an essential component of a medieval university education. At Paris, undergraduates were required to take only a bare minimum of natural philosophy, whereas the universities of Oxford and Toulouse required more courses from the "quadrivium," which embraced arithmetic, geometry, astronomy, music, and natural philosophy. With only a bachelor's degree, students were candidates for positions as teachers of logic or grammar or as royal or papal bureaucrats. But without the master of arts degree, they were not eligible for regular university teaching posts, nor were they eligible for matriculation to the higher faculties of theology, law, and medicine.

For students who wished to teach regularly at the university level, it was necessary to obtain a teaching license (*licentia docendi*), which was a formal prerequisite for becoming a master of arts and usually required an additional two years of course work. During this period, the student had to attend lectures on natural philosophy, which meant studying Aristotle's natural books, and study the quadrivial sciences, especially astronomy, geometry, and arithmetic. The six years of matriculation toward a master of arts degree saw an increasingly greater emphasis on science and natural philosophy. Indeed, an arts education was designed to convey in-depth knowledge about the structure and operation of the physical cosmos. Overall, the full six-

21. For the Latin text and an English translation of Sacrobosco's *De sphera*, see Sacrobosco, *Sphere*, 1949, 76–117 (Latin) and 118–142 (English). An English translation of the *Theorica planetarum* by Olaf Pedersen appears in Grant, 1974, 451–465. Although medieval technical astronomy was based on the famous *Almagest* of Ptolemy, which appears on some curriculum lists, it is unlikely that anything more than the descriptive sections of the first book could have served as a text.

year program may be appropriately characterized as a logic–physics–cosmology curriculum. Between the thirteenth and the seventeenth century, that curriculum constituted the regular fare for legions of students who passed through the universities of Europe.

As a direct consequence of the university curriculum, with its overwhelming emphasis on Aristotle's logic and natural philosophy, and by virtue of the nature of the educational methods employed at medieval institutions of higher learning, certain terms have come into common usage to identify the subject matter of study *and* the ones who did the studying, namely the terms "natural philosophy" and its correlative "natural philosopher"; "scholastic"; and "Aristotelian," with its correlative "Aristotelianism." These terms appear frequently in this study – indeed, "scholastic" or "scholastics" may occur a few hundred times – and it seems appropriate to convey briefly some idea of their meaning. Unfortunately, this is easier said than done.

The term "natural philosophy" (*philosophia naturalis*), or even "natural science" (*scientia naturalis*), which was an occasional synonym, was in use during the Middle Ages and on up into the seventeenth century. The subject matter of natural philosophy was generally identified with Aristotle's natural books, namely *On the Heavens* (*De caelo*), *Physics, Meteorology, On Generation and Corruption, On the Soul* (*De anima*), the *Parva naturalia* (*The Small Natural Books*), and probably his biological works as well. As one of three major subdivisions of speculative philosophy, natural philosophy was concerned with mobile bodies and their changes (or lack of change).[22] Although they were distinct from mathematics, sciences such as optics and astronomy, which used mathematics but were also concerned with mobile bodies, might also fall within the domain of natural philosophy.[23] The term was still popular in the seventeenth century, when Isaac Newton used it in 1687 in the title of his epochal study on mathematical physics, *The Mathematical Principles of Natural Philosophy*. Cosmology is thus an integral part of natural philosophy and is best represented by Aristotle's *De caelo*, although relevant problems turn up in other of Aristotle's treatises.

When modern scholars use the term "scholastic," they usually employ it with reference to the medieval "schoolmen." Although the term "scholastic" probably gained currency in the seventeenth century and was then sometimes used pejoratively, generally it simply referred to a commentator on the logical and natural books of Aristotle who was trained at a European university and often also taught at one for some period of time during which he probably wrote his commentaries. Such individuals could be either

22. The other two subdivisions are moral philosophy and metaphysics. See Weisheipl, 1964, 173–176.

23. Robert Kilwardby, 1976, 15–29, includes virtually all of these under natural philosophy but chooses to subsume optics (perspective) and astronomy under mathematics. For Domingo Gundisalvo's division of natural philosophy, or natural science, see Grant, 1974, 62–65.

arts masters – that is, secular masters, who were untrained in theology – or theologians, trained in both natural philosophy and theology. During the Renaissance, the emergence of the humanistic movement, which emphasized grammar, rhetoric, poetry, and history, also influenced early modern scholasticism, largely by broadening its approach. But those influences failed to alter the basic medieval tradition. "Even when the links with humanism were rather strong, as they were for [Jacopo] Zabarella [1533–1589] or the Coimbra commentators, the ties with the medieval and traditional philosophical concerns as dictated by university curriculum are very evident. Thus Zabarella still debated questions in the terms framed by Thomas Aquinas, Albert the Great, Buridan, and John of Jandun."[24] In this study, therefore, the term "scholastic" refers to someone – customarily a master of arts or theologian – who was educated at a European university and probably had also taught at one for some period of time, and, most importantly, had commented on the natural books of Aristotle. (He might also have written on logic, but that is irrelevant here.)

The most difficult term to define is "Aristotelian" and its correlative "Aristotelianism." In an article that I wrote a few years ago (Grant, 1987d), I sought to determine different ways to interpret these terms. The fundamental issue concerned the capaciousness of Aristotelianism, the degree to which it could be stretched and extended and yet continue to have real meaning. How far could an Aristotelian stray from Aristotle's basic tenets and still be called an Aristotelian? This was especially a problem in the sixteenth and seventeenth centuries, when the Copernican system and the new astronomical discoveries by Tycho Brahe and Galileo were having a profound impact on traditional cosmology. How, then, can we identify an Aristotelian?[25]

At the very least, an Aristotelian ought to be someone whose education involved a reasonable degree of familiarity with the works of Aristotle and earlier Aristotelians and who had himself probably written one or more *questiones* (treatise in the form of questions) or commentaries on the natural books of Aristotle, for the purpose of presenting opinions about the operation of the physical world. If our hypothetical Aristotelian failed to write commentaries or supercommmentaries (that is, commentaries on commentaries) on the relevant works of Aristotle, his Aristotelianism might nonetheless be evidenced in other kinds of treatises.

But we must assume that an Aristotelian in natural philosophy is so identified because of a more than cursory familiarity with Aristotle's natural books, and we ought to assume further that Aristotelians form a continuous community of scholars bound by a common interest in the works of Aristotle. Thus an alien being in another galaxy who might have independently written verbatim all the Aristotelian treatises that Thomas Aquinas or Nicole

24. Schmitt, 1983, 17.
25. For the remainder of this section, I follow Grant, 1987d, 349–350, 352.

Oresme had written would not, for that reason alone, be considered an Aristotelian unless he had also been trained in the tradition and was part of a community of scholars who had studied some or all of the works of Aristotle and even one or more subsequent commentaries on those works.

But what about natural philosophers who proposed ideas radically at variance with the thought of Aristotle and his followers, for example, Andrea Cesalpino and Thomas White, who adopted elements of the Copernican theory (see Ch. 20)? If they were either self-proclaimed Aristotelians or obvious commentators on Aristotle who were not commenting on him for the sole purpose of refuting him, they ought to be admitted to the company of Aristotelians. Broad inclusion fits the historical development of medieval Aristotelianism, which from its entry into Europe was an extraordinarily encompassing natural philosophy. Over the centuries during which it was a force to be reckoned with, much that was deemed fundamental to Aristotelianism was challenged, though rarely abandoned. Aristotelianism often included conflicting earlier and later opinions simultaneously. The regular coexistence of traditional and innovative opinions made it inevitably elastic and absorbent.

2. *The basic form of cosmological literature: the* questiones

Two basic methods were employed to teach natural philosophy, and therefore also cosmology, and each method had its counterpart in the scholarly literature. The less widely used method was the commentary. Since lectures were at first largely sequential, section-by-section expositions of an authoritative text (for example, this or that work of Aristotle), the analogous written commentary simply followed the same procedure. A section of text was cited and then explained with occasional elaborations by the commentator; upon completion, the next section or unit of text was expounded; and so on. It was not unusual in these commentaries to provide numerous references to other texts in support of an argument.[26]

But the most widely and regularly used format for medieval cosmology, and for natural philosophy in general, was the *questio*, or question format. The concept of a medieval "scholastic method" is more intimately associated with the *questio* form of literature and analysis than with anything else.[27] The practice of dealing with questions may have arisen from the commentary. When a master had finished reading and explaining a number of passages or sections (*textus*) of an Aristotelian work, it became customary to pose a question based on those passages and to present the pros and cons of the answers to it, followed by a solution.[28] These questions frequently formed the basis of the master's formal questions, or *questiones*, on that particular Aristotelian work. In time, however, the questions previously

26. See Chapter 3, Section II, for a further brief discussion of cosmological commentaries.
27. In what follows on the *questio*, I draw on Grant, 1984a, 79–82.
28. See Weisheipl, 1964, 154.

posed toward the end of a lecture came to displace the commentary on the text itself. Eventually the teaching masters focused on specific questions or problems that followed the order of the required text and developed from it.[29] The written forms of this pedagogical technique that have survived – questions treatises, or *questiones* – are often associated with the names of well-known masters, who presumably presented as lectures some version of the surviving text.

During the Middle Ages and even into the seventeenth century, scholastic natural philosophers centered their teaching and analysis of authoritative texts around the *questio*. Natural philosophy, which constituted the core of the curriculum, was thus taught by the analysis of a series of questions posed by a master and eventually determined – that is, resolved – by him. Collections of these *questiones* on the different works of Aristotle, and other works as well, were written down by the masters and published through university auspices.

Although occasional variations in the arrangement of the constituent elements of the *questio* occurred, scholastics tended to present their arguments in a rather standard format that remained remarkably constant over the centuries. First came the enunciation of the problem or question, usually beginning with a phrase such as "Let us inquire whether" or simply "Whether" (*utrum*): for example, "Whether the earth is spherical," or "Whether the earth moves," or "Whether it is possible that several worlds exist." This was followed by one or more – sometimes five or six – solutions supporting either the negative or affirmative position. If arguments for the affirmative side appeared first, the reader could confidently assume that the author would ultimately adopt the negative position; conversely, if the negative side appeared first, it could be assumed that the author would subsequently adopt and defend the affirmative side. The initial opinions, which would ultimately be rejected, were called the "principal arguments" (*rationes principales*).

Immediately following the description of the principal arguments, the author might further clarify and qualify his understanding of the question or explain particular terms in it. For example, Jean Buridan, asking "Whether it is possible that several worlds exist," explains not only his interpretation of the term "world" but also what he means by "several [i.e., a plurality of] worlds":

> "world" [*mundus*] can be taken in many ways. In one way as the totality [*universitate*] of all beings; thus the world is called "universe" [*universum*]. "World" is taken in another way for generable and corruptible things and in another way for perpetual things; and so it is that we distinguish world into this inferior world and into a superior world. And yet "world" is taken in many other ways that are not relevant

29. This was true for lectures in both the arts and theology. For the former, see Weisheipl, 1964, 154 and for the latter, McLaughlin, 1977, 208.

to our present discussion. But "world" is taken in another way that is pertinent, [namely] as the totality [*congregato*] of heavy and light [bodies] which appear to us *and* [also] the celestial spheres that contain those heavy and light [bodies]. And it is about such a world that the question – Whether it is possible that several worlds exist – inquires. And with regard to this, it must be noted that a plurality of such worlds can be imagined in two ways: in one way existing simultaneously, as if outside this world one other such world existed now; in another way they exist successively, namely one after the other.[30]

With the necessary qualifications completed, the author was ready to present his own opinions, usually by way of one or more detailed conclusions or propositions. In order to anticipate objections, the master might choose to raise doubts about his own conclusions and subsequently resolve them. To conclude the question, he would respond sequentially to each of the "principal arguments" enunciated at the outset.

If Jean Buridan's *Questions on De caelo* represents a typical medieval *questiones* of the thirteenth and fourteenth centuries, as it probably does, an analysis of it may reveal important features of this genre of scholastic literature. Two types of citations are distinguishable in the questions literature. The first is a reference to other parts of the same Aristotelian work on which the author was commenting or to other works of Aristotle. These are commonplace and may even be construed as integral parts of a given question, since they are cited in behalf of both the affirmative and negative sides of the question. Buridan included many such citations in his *Questions on De caelo*.[31]

The second kind of citation is to other of the author's own questions in the same treatise or on other treatises of Aristotle, or his citation of medieval authors. These references are significant because they are a measure of the extent to which medieval authors were concerned about linking their own questions to a broader context.

In his *Questions on De caelo*, Buridan included a total of 59 questions (see Appendix II, Sec. IV. 3). Of these, only 3 contained references to other questions in the same treatise,[32] and 4 contained references to Buridan's own *Questions on the Physics* but to no other work.[33] As for other scholastic authors, Buridan found occasion to cite them in 4 of his 59 questions, naming Aegidius Romanus, Thomas Aquinas, Petrus de Alvernia, and Albertus Magnus.[34] Although medieval scholastics from the thirteenth to fif-

30. Buridan [*De caelo*, bk. 1, qu. 19: "Consequenter quaeritur: Utrum possibile est esse plures mundos"], 1942, 88. On the term *mundus* and its synonyms, see this volume, Chapter 1, Section II.2, and Chapter 7, Section I.
31. For example, in book 1, question 7, Buridan cites Aristotle's *Physics, On Generation and Corruption*, and *Meteorology* for a total of seven times. Four of these citations appear in the negative side and three in the main body of the question.
32. Bk. 1, qu. 1, and bk. 4, qus. 4 and 6.
33. Bk. 1, qus. 2, 4, and 8, and bk. 4, qu. 3.
34. Bk. 1, qu. 11, and bk. 2, qus. 9, 16, and 18.

teenth centuries found relatively few occasions to cite their scholastic predecessors and contemporaries, their successors in the late sixteenth and the seventeenth century did so frequently.

If Buridan's *Questions on De caelo* is a typical *questiones* treatise of the medieval period, what tentative conclusions may we draw? There was apparently a tendency to treat each question independently from other questions. Although there were numerous linkages between a large number of questions within a given treatise, and also to questions in other treatises, these connections were usually ignored. Cross-references were the exception rather than the custom. Such a practice might have tended to hinder, and perhaps even to prevent, the formulation of larger syntheses beyond the individual question.

Over the centuries, scholastics departed from Aristotle on many issues.[35] What is curious, however, is that these departures were rarely acknowledged.[36] The prevailing tendency was to reconcile one's conclusions with Aristotle's interpretation, which was itself often in dispute. Departures were frequently made by way of an interpretation that seems to us quite different from anything Aristotle could have had in mind, and yet, such was the authority of Aristotle that scholastic authors usually chose to emphasize their essential agreement with him, even when it seems apparent to us that they had actually departed from him. Thus we are left to ponder whether they were inwardly aware of their departure but opted to remain silent out of respect for Aristotle's authority, or whether they assumed that their interpretation was Aristotle's genuine opinion. Thus Aristotle not only furnished the sequence of questions, by virtue of the order in which he chose to consider the subjects in his works, but his powerful authority made it difficult to proclaim new departures.

By its very nature, the *questio* form encouraged differences of opinion. It was a good vehicle for dispute and argumentation. Medieval scholastics were trained to dispute and therefore often disagreed among themselves. In almost all questions, there were at least two, and often more, previously distinguished positions. The *questio* format may thus have served to reduce the tendency to follow slavishly the authority of Aristotle. Had the *questiones* been of a form in which a single conclusion was provided that merely required a rationale and a defense, medieval and Renaissance scholasticism would have been less interesting and productive than it was.

The scholars who wrote the *questiones* were masters of arts, or else masters or doctors of theology. They were trained to evaluate arguments critically and by a process of elimination arrive at the most plausible response under the circumstances. Their skill and ingenuity were displayed by the introduction of subtle distinctions that, upon further development, might yield

35. For some of these departures, see Grant, 1989.
36. Buridan made an emphatic departure when he declared (*De caelo*, bk. 1, qu. 7, 1942, 35): "And so finally, it seems to me that the authority of Aristotle ought to be corrected, [namely] that a mixed body is moved according to the nature of the dominant element."

new opinions on this or that question. For a medieval master to do cosmology within the larger context of natural philosophy, and within the university system as a whole, was to present an oral or written analysis of some of the many questions that collectively formed the basis of the medieval world view. In each question, the master would arrive at a conclusion that usually had traditional support but occasionally represented a new departure.

We must now investigate the kinds of treatises in which cosmological questions were normally included, and the nature and range of those questions.

3. Questions and commentaries on Aristotle's De caelo (On the Heavens)

If Aristotle's *De caelo* was the most fundamental Greco-Arabic source for late medieval conceptions of the cosmos, so also are the Latin commentaries and *questiones* on that treatise our most significant body of scholastic cosmological literature from the late Middle Ages to the seventeenth century. To obtain a good indication of the temporal range and number of *De caelo* commentaries and *questiones*, we are indeed fortunate to possess Charles Lohr's catalog of manuscripts and printed editions of commentaries and *questiones* on the entire range of Aristotle's works from the Middle Ages to around 1650.[37] Because of their importance, an analysis of his data on *De caelo* will prove helpful.

From approximately 1200 to around 1650, the span of Lohr's catalog, the number of authors who wrote extant expositions, commentaries or questions on Aristotle's *De caelo*, or who are cited as having done so, is approximately 270. Since a few authors wrote more than one *De caelo* treatise, the total number of works that are mentioned, and that are perhaps extant, is slightly higher, say around 275 to 280. In what follows, however, I assume for convenience one treatise for every author and also assume, again for convenience, that the number of authors in Lohr's catalog who are recorded as having written on *De caelo* is precisely 270. At this stage, the count of authors and treatises can only be approximate, because the titles of some works fail to indicate whether all or part of a manuscript or printed book was devoted to *De caelo*. Fortunately, in a number of such instances, the full title page or a table of contents reveals whether a section on *De caelo* is included. For example, Balthasar Tellez (1596–1675) titled his two-volume treatise *Summa universae philosophiae, cum quaestionibus theologicis, quae hodie inter philosophos agitantur* (*A Summa of the Whole of Philosophy with Questions That Philosophers Argue about Today*). Without additional information about the book's contents, we could not determine whether Tellez included a section on *De caelo*. Fortunately, Tellez chose to provide

37. See Lohr, 1967–1974; and Lohr, 1988.

further information on his title page, where he explains that the first volume is divided into two parts, the first of which includes a discussion of Aristotle's *Logica, Physica, De caelo*, and *De meteoris.*[38]

Where the contents are not printed on the title page, and therefore not reported by Lohr, only a personal inspection of the manuscript or printed book can determine with certainty whether some part of the treatise was devoted to *De caelo*. Thus Nicolaus Diel, who lived in the first half of the seventeenth century but of whom little is otherwise known, wrote a treatise titled *Disputatio in universam philosophiam naturalem* (*Disputation on the Whole of Natural Philosophy*), which apparently exists in a single manuscript in Salzburg (Lohr, 1988, 125). Without examination of that manuscript, we cannot know whether Diel included a section on *De caelo*. Investigation of numerous treatises with similar titles in the sixteenth and seventeenth centuries, however, reveals that many did indeed contain sections on *De caelo*. Thus it is not unreasonable to incorporate such works into our overall list of *De caelo* treatises (approximately 7 or 8 in this category have been included). We may reasonably infer that the approximately 270 *De caelo* treatises derived from Lohr is a minimum figure for the total number of *De caelo* authors and treatises that are still extant or of which there is some record.

Although a number of medieval natural philosophers wrote commentaries and questions on more than one Aristotelian treatise, it was customary to disseminate each commentary or set of questions (*questiones*) as a single work.[39] With the advent of printing, it became common to publish a number of Aristotelian commentaries by a single author in the same volume. During the sixteenth and seventeenth centuries, especially the latter, commentaries and questions on Aristotle's physical works were often, and quite logically, grouped together. General or umbrella titles of the kind mentioned in the preceding paragraph were found convenient and appropriate.[40] Usually the

38. See Lohr, 1988, 451. Lengthy title pages were rather common in the seventeenth century.
39. For example, Nicole Oresme's questions on *De caelo, De anima, Physics, Meteorology*, and *On Generation and Corruption* were all copied and disseminated as single, independent treatises. An exception to this practice in the Middle Ages were the summas on natural philosophy. Here an author would comment on four or five Aristotelian physical treatises, treating each one in briefer compass than if it had been dealt with separately. Examples of medieval summas are the *Summa naturalium* of Albertus de Orlamunde, (fl. 2nd half 13th c.), the *Summa logicae et philosophiae naturalis* of John of Dumbleton (ca. 1310–1349), and the *Summa philosophie naturalis*, or *Summa naturalium*, of Paul of Venice (ca. 1370–1429). Summas on natural philosophy were relatively rare during the Middle Ages but became commonplace (under other titles) during the sixteenth and seventeenth centuries.
40. For example, in 1481, under the general title of *Quaestiones super tota philosophia naturali cum explanatione textus secundum mentem Doctoris subtilis Scoti*, the Scotistic commentator Johannes de Magistris (fl. 2nd half 15th c.) included questions on Aristotle's *Physics, De caelo, De generatione et corruptione, De meteoris, De anima*, and *Parva naturalia*. In 1586, Conrad Gesner's (1516–1565) commentaries on *Physics, De caelo, Meteora, De anima*, were printed under the general title *Physicarum meditationum, annotationum et scholiorum libri V*; and in 1615, Pedro Hurtado de Mendoza (1578–1651) published a treatise under the title *Universa philosophia*, which has no indication of content on the title page but does include a section on *De caelo*. Many other examples could be given (for two additional works

order of the Aristotelian treatises in such individual volumes followed some plan. But from this, we may not properly infer that the discussions were therefore interrelated and that an integrated world view emerged. To the contrary, the commentary on each Aristotelian work was customarily independent of the other commentaries in the same volume.[41]

A chronological analysis of Lohr's catalog entries proves significant.[42] Taking all 270 identifiable and probable *De caelo* authors who wrote treatises between 1200 and 1650 that are either extant or of which there is some record, and adding to this 8 treatises composed between 1650 and 1695, we arrive at a grand total of 278. If we arrange these works by the centuries in which they were composed, we obtain the following distribution:[43]

13th century	18
14th century	21
15th century	32
16th century	108
17th century	99
Total	278

Of the 278 authors, the works of approximately 40 are presently represented neither by extant manuscripts nor printed editions. Knowledge that these treatises once existed is derived largely from manuscript catalogs and from citations of such works by contemporary or later authors.[44] Although this leaves some 248 extant *De caelo* treatises, we may plausibly conjecture that if 40 treatises were mentioned between 1200 and 1650 (or 1687, the terminal date of this volume), for which no extant texts have yet been found, many others – perhaps numbering in the hundreds – were also written but left no record or trace.

In the group of 248 extant authors and works, 21 are represented by both

with general titles each of which contains a section on *De caelo*, see Appendix II, note 6). Indeed, of the estimated 270 treatises on *De caelo* composed between 1200 and 1687, approximately 50 are embedded in works with general titles that fail to indicate whether *De caelo* is included.

41. For further discussion, see Grant, 1978b, 99–100.
42. The count of treatises and authors, as well as the division into categories, is my own, not Lohr's.
43. The totals for each century are only approximate. Because precise birth and death dates for many authors are lacking and because so many treatises lack specific dates, it is often difficult if not impossible to determine whether a treatise was composed in the late thirteenth or the early fourteenth century, the late fourteenth or the early fifteenth, and so on.
44. Among the 40 are Boethius of Dacia, Siger of Brabant, and Albertus de Orlamunde, in the thirteenth century; Richardus de Lavenham and Johannes de Sancti Fide, in the fourteenth century; Andreas de Bilius, Hieronymus de Janua, and Johannes Snerveding de Hamburg, in the fifteenth century; Ludovicus de Molina, Franciscus Suarez, Adam Higgins, and Franciscus Verinus, in the sixteenth century; and Melchior Coronado, Johannes Dannemeyr, and Johannes a Sancto Thomas (Johannes Poinsot), in the seventeenth century.

manuscripts and printed editions; 139 by manuscripts only; and 76 by printed editions with no recorded manuscripts. By adding 21 and 76 we obtain a total of 97 extant printed books, virtually all of which appeared between the late fifteenth and the end of the seventeenth century.[45]

A remarkable feature of these *De caelo* treatises is their temporal distribution. During the Middle Ages, that is, from the thirteenth through the fifteenth century, a total of 71 (18 [13th c.] + 21 [14th c.] + 32 [15th c.]) *De caelo* works were composed. But during the Renaissance – that is, the sixteenth and seventeenth centuries – we count 207 (108 [16th c.] + 99 [17th c.]), a roughly three-to-one ratio of Renaissance to medieval texts. Because printing in the Renaissance proved a boon to the dissemination and preservation of texts, it is implausible to assume that three times as many commentaries and questions on *De caelo* were composed in the Renaissance as in the Middle Ages. But the multiplication of universities in the Renaissance and the establishment of numerous Jesuit colleges and other centers of study do make it reasonable to suppose that at least as many questions and commentaries on *De caelo* were composed in the Renaissance as in the Middle Ages, and perhaps even more. Thus even as rival cosmologies and world views were emerging to challenge Aristotelian-based cosmology in the sixteenth and seventeenth centuries, the latter remained vigorous and dominant through most of this period.

With one or two exceptions, the present study is based exclusively on printed editions. The enormous range of source materials made any other decision impractical. But this does not undermine the validity of the conclusions and results. The probability that the majority of manuscripts of *De caelo* treatises differ substantively and qualitatively from the published versions is small. Although we cannot discount the possibility that an obscure author's unpublished version of a *De caelo* treatise may have startling conclusions and arguments that would alter our conception of medieval cosmology, we can afford to ignore that extreme contingency because of its overwhelming implausibility. The printed books provide us with a full sense of the nature and content of medieval cosmology as encompassed within the *De caelo* tradition. Indeed, even some 97 books is more than can readily be digested.

The distribution of these books, however, is quite uneven: 4 from authors of the thirteenth century; 5 from the fourteenth century; 13 from the fifteenth century; 40 from the sixteenth century; and 35 from the seventeenth century. Although there is a paucity of printed *De caelo* treatises for the thirteenth and fourteenth centuries (from Lohr's catalog, we can name the following: Albertus Magnus, Thomas Aquinas, and Thomas de Bungeye in the thirteenth century, to which I add Roger Bacon;[46] John of Jandun, Jean Buridan,

45. Exceptions are, for example, the edition of Nicole Oresme's French commentary on *De caelo* published in 1968 and the first book of Thomas de Bungeye's commentary, edited in a Ph.D. dissertation in 1969.

46. Although, strictly speaking, Bacon left no work on *De caelo*, his *Liber secundus communium*

Nicole Oresme, Albert of Saxony, and Jacobus de Blanchis in the fourteenth century), the 9 examples from those two centuries when joined to the 13 from the fifteenth century furnish a total of 22 to cover the late Middle Ages. When cosmological ideas from other types of treatises (described later in this chapter) are also taken into account,[47] the sources provide a reasonably good overall picture of medieval cosmology. Let us now briefly consider the other kinds of works that supplement the basic core of cosmological material in the *De caelo* tradition.

4. *Questions and commentaries on cosmological themes in Aristotle's other physical treatises*

Although *De caelo* was Aristotle's primary contribution to cosmology, we have already noted that some of his other works – the *Physics*, the *Metaphysics*, and the *Meteorology* – included topics relevant to the subject. Thus in his *Physics*, Aristotle discussed the concepts of vacuum and place in the fourth book; continuity and contiguity in the fifth; and the prime mover in the eighth; and, in the twelfth book of the *Metaphysics*, he discussed the number of celestial spheres and the substances that move them. The *Meteorology* was essentially "a study of the combinations and mutual influences of the four elements" (Ross, 1949, 109), but it has some bearing on cosmology, because Aristotle considered its subject matter to be natural phenomena that "take place in the region nearest to the motion of the stars," such as "the milky way, and comets, and the movements of meteors."[48]

As with *De caelo*, scholastic natural philosophers commented upon these texts or discussed questions that emerged from them and thus expanded the range of medieval cosmological literature (the nature of these questions is described below). Judging from Charles Lohr's catalog of medieval and Renaissance Aristotelian commentaries, the number of commentaries and questions on the *Physics* and *Metaphysics* between 1200 and 1650 was at least twice that on *De caelo*, whereas the number on the *Meteorology* approximates that for *De caelo*. With their emphasis on general principles and concepts, the *Metaphysics* and *Physics* were seemingly deemed more fundamental than *De caelo*.

5. *Questions on the* Sentences *of Peter Lombard*

Sometime around 1150, Peter Lombard (d. ca. 1160) wrote a treatise titled *Sentences* (*Sententiae*, or opinions). Divided into four books devoted, re-

naturalium, titled *De celestibus*, includes a number of typical problems from *De caelo*. See Bacon, *De celestibus, Opera*, fasc. 4, 1913.

47. For example, encyclopedias by Vincent of Beauvais and Bartholomew the Englishman (Bartholomaeus Anglicus); commentaries on the *Sentences* of Peter Lombard; commentaries on Sacrobosco's *Sphere*; and commentaries on Aristotle's *Physics, Metaphysics*, and *Meteorology*. Separate treatises like William of Auvergne's *De universo* are also relevant and helpful.

48. Aristotle, *Meteorology* 1.1.338b.21–23, in Aristotle [Webster], 1984, 1:555.

spectively, to God, the Creation, the Incarnation, and the Sacraments, the *Sentences* served for some four centuries as the standard text on which all theological students were required to lecture and comment.[49] Because it treated of the creation account in Genesis, the second book was highly relevant for medieval cosmology.[50] In this sense, the second book functioned as a medieval version of earlier hexaemeral treatises associated with such eminent Church Fathers as Basil, Ambrose, and Augustine. Medieval commentaries on the second book of the *Sentences* often included *questiones* on themes of great moment to theologians: the nature of light, the problem of the supracelestial waters, and the order and motion of the celestial spheres and planets. Although the amount of space given over to cosmological questions in the second book of the *Sentences* varied from author to author, commentaries on that work form an important source of knowledge on a somewhat limited range of topics. But it was largely those topics that introduced into medieval cosmology a Christian component. Insofar as there was a "Christian cosmology" in the Middle Ages, it developed from problems that evolved in commentaries on the second book of the *Sentences* of Peter Lombard.

At least one major theme, or "distinction," as they were called, in the first book of the *Sentences* also played a significant role in cosmology. In book 1, distinction 37, theologians confronted the question as to the way God exists in things. Among the problems that frequently arose in this context was the manner in which God could be omnipresent. Since omnipresence signified God's presence beyond the world as well as within it, the manner in which God could exist beyond the world and in what he existed, posed perplexing problems. Within this framework, a few fourteenth-century scholastics suggested a linkage between God's omnipresence and an infinite space. By the seventeenth century, many scholastic authors pushed beyond their medieval predecessors and considered at some length the manner in which God could be omnipresent and how that might be conceived spatially.[51]

Because theologians were required to lecture and comment upon Lombard's *Sentences*, hundreds of lengthy tomes of this genre have been preserved in manuscript form, of which a small percentage has been published. With the advent of printing, medieval commentaries on the *Sentences* began to issue from the presses.[52] Among works of this genre printed between

49. For details about the development and structure of commentaries on the *Sentences*, see Glorieux, 1941, cols. 1860–1884.

50. For the first two books, see Peter Lombard, *Sentences*, 1971.

51. A detailed discussion of the relations between God and space as it evolved within the commentary tradition on book 1, distinction 37, of Peter Lombard's *Sentences* is given in Grant, 1981a, pt. 2.

52. In his *Reportorium* of commentaries on the *Sentences*, Stegmüller, 1947, sought to list all the commentaries on the *Sentences* composed between the late twelfth and the sixteenth century. He also organized an *Index chronologicus* in the second volume in which he cites editions from the late fifteenth to the twentieth century. What follows is based on Stegmüller.

the late fifteenth century and the twentieth century are those of Alexander of Hales, Albertus Magnus, Thomas Aquinas, Bonaventure, Duns Scotus, Aegidius Romanus (Giles of Rome), Petrus de Tarantasia, and Richard of Middleton, all of whom wrote in the thirteenth century; Hervaeus Natalis, Adam Wodeham, Robert Holkot, William of Ockham, Marsilius of Inghen, Petrus Aureoli, Franciscus de Mayronis, Thomas de Argentina, Johannes Hus, Petrus de Palude, Durandus de Sancto Porciano, Gregory of Rimini, Johannes de Bassolis, and Petrus de Aquila, who composed their treatises in the fourteenth century; William of Vorillon, Nicolaus de Orbellis, Stefanus Brulefer, Johannes Capreolus, Gabriel Biel, and Dionysius Carthusianus, who wrote in the fifteenth century; and John Major and Paulus Cortesius,[53] who completed their commentaries in the sixteenth century.[54] Although not all of these authors included cosmological discussions in their second book, enough did – including the seven from whose *Sentences* commentaries questions have been drawn for my "Catalog of Questions" – to provide us with a considerable range of relevant source material.

6. Other relevant literature

Mention must be made of a few other relevant works, or types of works, that were concerned with specific cosmological subjects. None is of overriding significance, but each forms a legitimate part of the source literature.

Perhaps the most important of these is the *Treatise on the Sphere* (*Tractatus de sphera*) by John of Sacrobosco (d. ca. 1244–1256), who, in the first of four chapters, included brief descriptions of the heavens and the earth.[55] Because it served as a university undergraduate text for some four centuries, the *Sphere* was widely known and evoked a small number of commentaries, of which those by Michael Scot and Robertus Anglicus in the thirteenth century,[56] Pierre d'Ailly in the fourteenth,[57] and Christopher Clavius[58] in the late sixteenth century are noteworthy. Subsequent commentators used

53. Paulus Cortesius (1465–1510), wrote his commentary on the *Sentences* before 1504, and therefore perhaps in the fifteenth century (see Stegmüller, 1957, 1:298).
54. Few if any *Sentences* commentaries were written in the seventeenth century. Stegmüller's list seems to extend only to the sixteenth century. By that time, most theologians were writing commentaries or questions on the *Summa theologiae* of Thomas Aquinas. Many of the Jesuits whose cosmological opinions are cited in this study did so, including Suarez and Toletus in the sixteenth century and Arriaga, Compton-Carleton, de Rhodes, and Oviedo in the seventeenth century.
55. See the Latin text and English translation in Sacrobosco, *Sphere*, 1949.
56. For the Latin texts of both and an English translation of Robertus's treatise, see Michael Scot, *Sphere*, 1949 and Robertus Anglicus, *Sphere*, 1949.
57. D'Ailly, *14 Questions*, 1531, 146v–165r.
58. In Clavius [*Sphere*], 1593, and Clavius [*Sphere*], *Opera*, 1611, 3:1–307. According to Randles, 1980, 48, 95, there were eight editions in the sixteenth century and nine more in the seventeenth, the last in 1618. In Sommervogel, 1890–1911, vol. 2, cols. 1212–1213, I count 14 editions, plus 1 version in Clavius's *Opera*, 2 abbreviated editions, and a translation into Chinese by P. Ricci, for a total of 18. In general, I shall cite from both the fourth edition of Lyon, 1593, and the third volume of the *Opera*.

Sacrobosco's brief remarks as points of departure for more detailed discussions. Indeed, Sacrobosco's treatise may have served as the model for other authors of the caliber of Robert Grosseteste and Nicole Oresme, both of whom composed works with the same title.[59]

Although the works of Averroës (ibn Rushd) (1126–1198) were usually employed as aids for the study of Aristotle's thought, one of Averroës' brief works became the focus of a number of commentaries. In his *De substantia orbis* (*On the Substance of the [Celestial] Orb*), Averroës investigated whether the form and matter of the total celestial body – that is, the heavens as a whole – was like the form and matter in terrestrial bodies.[60] He also speaks briefly of celestial motions. John of Jandun, for example, composed an extensive commentary of some thirty folios on the *De substantia orbis* (see John of Jandun, 1552). Among a number of topics, he considered the composition and motion of the heavens, whether they are animated, and whether they are corruptible. With the exception of a work by John of Jandun and a few others, most of the questions and commentaries on *De substantia orbis* were written in the fifteenth through seventeenth centuries. Indeed, there appears to have been as much interest in it during the seventeenth century as in the Middle Ages, if not more.

The composition of the celestial region formed the subject matter of another kind of treatise, which usually bore the title *De materia celi*. Here the major concern was whether the incorruptible heavens consisted of the same kind of matter as existed in the corruptible terrestrial region. Although few authors wrote distinct treatises titled *De materia celi*,[61] a number of scholastics wrote one or more questions on the possible existence of matter in the heavens.[62]

Although few in number, medieval encyclopedias also provide information and insight into medieval views of the cosmos. Three are note-

59. See Baur's edition of Grosseteste's *De sphaera* in Grosseteste, 1912, 10–32; for Oresme's treatise, see Oresme [*De spera*], 1966a. Although Oresme's treatise includes relevant cosmological material, Grosseteste's does not.

60. A Latin translation of the *De substantia orbis* was made by Abraham de Balmes (2nd half 15th c.) from a Hebrew text and was published in Averroës, 1562–1574, 9:3r–5v. According to information on the verso of the title page of volume 9, de Balmes added two chapters, which appear on folios 5v–14v (actually, five additional chapters were added, chs. 2–7). For an English translation from a Hebrew version by Arthur Hyman, see Averroës [Hyman and Walsh], 1973, 307–314. The medieval Latin translations may have also been made from Hebrew, rather than Arabic, versions. Without citing a specific volume or page, Thorndike and Kibre, 1963, col. 681, report that J. Valentinelli (*Bibliotheca manuscripta ad S. Marci Venetiarum*, 6 vols. [Venice, 1868–1876]) attributes a Hebrew translation to Abraham de Balneis around 1178. According to the dates, the two Abrahams lived a few centuries apart. But if the dates are incorrect, the similarity in names raises the possibility that Abraham de Balmes and Abraham de Balneis are one and the same.

61. Four who did were Hervaeus Natalis and Aegidius Romanus in the late thirteenth century and Pietro Pomponazzi and Simon Portius (or Porta) in the sixteenth century. For references to the first two, see Grant, 1983, 162–170; for editions and manuscripts for Pomponazzi and Simon Portius, see Lohr, 1988, 355 and 366, respectively.

62. For example, Buridan, *De caelo*, bk. 1, qu. 11, 1942, 49–54: "Whether the sky [or heaven] has matter."

worthy. The first, by Bartholomew the Englishman (Bartholomaeus Anglicus) (fl. 1220–1250), published in 1601, was titled *On the Properties of Things* (*De rerum proprietatibus*). Among his eighteen books, Bartholomew devoted the eighth to "The Properties of the World and Celestial Bodies."[63] Here, within a variety of topics, he includes the meaning of the word "world" (*mundus*) and also presents the number and order of the planetary heavens. The second, and perhaps most important and influential, of all medieval encyclopedias was *The Mirror of Natural Things*, or *The Mirror of Nature* (*Speculum naturale*), by Vincent of Beauvais (ca. 1190–ca. 1264).[64] Following a general scheme based on the six days of creation, Vincent divided the *Speculum naturale* into thirty-two books. Topics of cosmological interest abound: light (bk. 2), the heavens and their motions (bk. 3), and much on the stars and planets (bk. 15). Finally, in 1503, Gregor Reisch (ca. 1470–1525) published his famous *Margarita philosophica*, or *Philosophical Pearl*, and followed it with a number of expanded editions, the last appearing in 1517 (see Reisch, 1517). Of the twelve books into which the *Margarita* is divided, the first seven are devoted to the seven liberal arts (grammar, logic, and rhetoric, which comprise the trivium; and arithmetic, music, geometry, and astronomy, which make up the quadrivium). The seventh book treats astronomy, the seventh liberal art, and is divided into two tractates, one on astronomy and one on astrology. Although the astronomical part is mostly astronomical, Reisch includes subsections that are clearly cosmological, concerning, for example, the number of heavens, the nature and influences of the heavens, including the empyrean sphere, and the spot on the Moon.

The astrological part is virtually devoid of relevance or utility for understanding medieval cosmology. Indeed, this holds true for the genre of astrological treatises. At most we learn about certain properties of the planets: for example, that Saturn is cold and dry and Jupiter moist and hot. But these properties are rarely discussed or explained (see Reisch, 1517, 295–297). Little effort was made to show how they were related to their terrestrial counterparts. Since astrology depended on the assumption that celestial bodies influenced terrestrial bodies and events, astrologers necessarily emphasized this important aspect of cosmic interrelations.[65] But it is unusual to find them providing explanations and arguments to justify this indispensable tenet of their discipline. By contrast, natural philosophers frequently found occasion to consider that important theme, especially in commentaries on Aristotle's *De caelo* (see Ch. 19). The divergence in treatment derives essentially from the fundamental differences between astrol-

63. "De proprietatibus mundi et corporibus coelestibus liber octavuus." Bartholomew the Englishman, 1601, 367–433.
64. See Vincent of Beauvais, *Speculum naturale*, 1624. The *Speculum morale* was not written by Vincent but was added sometime between 1310 and 1325 (see Wallace, 1976, 35).
65. For Reisch's discussion, see *Margarita philosophica*, bk. 7, tract. 2, chs. 8 and 9, 301–303. Reisch's brief section is more informative than the accounts that most astrologers would have provided.

ogers and natural philosophers. Cosmology and the interrelationships between the celestial and terrestrial regions were primarily the domain of natural philosophers. Astrologers assumed the influence of the celestial region on the terrestrial and passed over such matters quickly in order to proceed directly to prognostication, which was, after all, the real basis of astrology. Natural philosophers were not astrologers, and the two groups appear to have had little mutual influence. In this study, astrological treatises play little role.[66]

7. *Technical astronomical literature as a source for scholastic cosmology*

But what of astronomical treatises? As indicated earlier, technical astronomers only occasionally found reason to discuss cosmological ideas and problems. The explanation is perhaps attributable to a long-standing tradition that assigned different objectives to astronomers and natural philosophers, or "physicists" as the latter were sometimes called. Simplicius (6th c.), a Greek commentator on the works of Aristotle, described the different objectives in his commentary on Aristotle's *Physics*, which was translated from Greek to Latin in the sixteenth century:[67]

It is the business of physical inquiry to consider the substance of the heaven and the stars, their force and quality, their coming into being and their destruction, nay, it is in a position even to prove the facts about their size, shape, and arrangement; astronomy, on the other hand, does not attempt to speak of anything of this kind, but proves the arrangement of the heavenly bodies by considerations based on the

66. Astrology was, of course, very popular. But this does not of itself signify that it was important to natural philosophy. Astrologers and natural philosophers may have shared the Aristotelian conviction that celestial bodies were the ultimate causes of terrestrial effects, but natural philosophers largely excluded the prognosticative aspects of astrology from their deliberations. Except for the attribution of certain qualities to certain planets, astrological details and concepts are virtually ignored in questions on Aristotle's natural books, especially on *De caelo*. The properties, positions, and relationships of the celestial bodies used for astrological prognostication were of little significance for the scholastic tradition in natural philosophy. The always-dubious status of astrology from the theological standpoint may also have worked against its inclusion in traditional natural philosophy, which was so closely linked to the universities. Generally, natural philosophers appear to have found no good cosmological reasons for introducing traditional astrological arguments into their descriptions and interpretations of the celestial region.

    Physicians were usually trained in both medicine and natural philosophy and some – perhaps many – engaged in medical astrology. As physicians, however, they did not find occasion to write questions and commentaries on Aristotle's natural books. On certain aspects of medical astrology, see Lynn White, Jr., "Medical Astrologers and Late Medieval Technology," in Lynn White, Jr., *Medieval Religion and Technology, Collected Essays* (Berkeley/Los Angeles: University of California Press, 1978), 297–315.

67. Although Simplicius's commentary on the *Physics* was unknown in the Middle Ages, he repeated similar, if less explicit and less obvious, sentiments in his commentary on *De caelo*, which William of Moerbeke translated into Latin in 1271. For a discussion of these passages as cited from the Greek text, see Duhem, 1969, 23. The distinction was drawn ultimately from Geminus of Rhodes (fl. ca. 70 B.C.), who, in turn, was summarizing the *Meteorologica* of Posidonius (ca. 135 B.C.–ca. A.D. 51).

> view that the heaven is a real kosmos, and further, it tells us of the shapes and sizes and distances of the earth, Sun, and Moon, and of eclipses and conjunctions of the stars, as well as of the quality and extent of their movements. . . . The things, then, of which alone astronomy claims to give an account it is able to establish by means of arithmetic and geometry. . . . For it is no part of the business of an astronomer to know what is by nature suited to a position of rest, and what sort of bodies are apt to move, but he introduces hypotheses under which some bodies remain fixed, while others move, and then considers to which hypotheses the phenomena actually observed in the heaven will correspond. But he must go to the physicist for his first principles, namely that the movements of the stars are simple, uniform and ordered, and by means of these principles he will then prove that the rhythmic motion of all alike is in circles, some being turned in parallel circles, others in oblique circles.[68]

Astronomers were thus primarily concerned with the prediction and determination of planetary and stellar positions, for which they invoked a variety of mechanisms, some real and some imaginary. Geometry and arithmetic were the major instruments. By contrast, natural philosophers, or more specifically cosmologists, were little interested in the positions of the planets but sought rather to describe the nature of the heavens and the causes of its various motions. They were expected to explain the nature of the celestial substance, that is, to determine whether it is incorruptible and indivisible; whether it is equally perfect throughout its extent or differentially so; whether its properties are similar to matter in the terrestrial region; what causes it to move, and so on.

This important distinction was largely preserved during the Middle Ages and Renaissance, although an important qualification must be mentioned. Natural philosophers, or physicists, were rarely competent in technical astronomy. Consequently, there was little likelihood that they would, or could, inject technical astronomy into their cosmological questions. In this, they differed little from Galileo, who discussed a variety of cosmological issues but ignored technical astronomy, presumably because he was largely ignorant of it. Like scholastic natural philosophers, Galileo was a cosmologist, not an astronomer. But what of technical astronomers? Did they choose to integrate cosmological material within the body of their astronomical treatises? For the most part, they did not, preferring, apparently, to adhere to the traditional division. Thus in his famous *Almagest*, the most fundamental astronomical treatise between antiquity and the sixteenth century, Ptolemy introduced only a modicum of cosmology, providing reasons for the earth's immobility in the first book and briefly discussing the order of the planets in the ninth book.

When he chose to consider the way in which the world might be constructed from concentric, eccentric, and epicyclic orbs that reflected real celestial motions, Ptolemy composed a separate treatise in two books, titled

68. The translation is by Heath, 1913, 275–276.

Hypotheses of the Planets, which had a significant subsequent influence on medieval cosmology.[69] In this treatise Ptolemy not only discusses the order of the planets and their distances from the earth, deciding that the least distance of a particular planet from the earth is equal to the greatest distance of the planet immediately below it: for example, the greatest distance of the Moon is equal to the least distance of Mercury; the greatest distance of Mercury is equal to the least distance of Venus; and so on.[70] Because of this configuration of planetary surfaces, Ptolemy concluded that no vacuum or anything else can lie between the surfaces of any two successive planets.

Despite its apparent unavailability in Latin translation during the Middle Ages, the *Hypotheses of the Planets* nevertheless exerted an indirect influence on the overall medieval conception of the orbs and their relationships. Its major contribution to cosmology was the "three-orb" compromise system and the general denial that vacua or matter could intervene between successive planetary surfaces. Apart from these important contributions, however, there is little else in the *Almagest* or *Hypotheses of the Planets* that is cosmological, as that term is understood in this study.

Most medieval astronomical treatises were overwhelmingly technical, but the extent to which they included cosmological material varied considerably. The old *Theorica planetarum*, probably composed in the latter half of the thirteenth century, includes nothing relevant to cosmology, omitting even discussion of the orbs.[71] Apart from a discussion of the order and distances of the planets, there is little that is cosmological in Campanus of Novara's similarly named *Theorica planetarum*.[72] The same judgment applies to Georg Peurbach's *Theoricae novae planetarum* of 1460–1461, which is virtually devoid of cosmological content, although Peurbach does discuss the orbs as if they were real physical bodies.[73] Bernard of Verdun chose to include more cosmology than most astronomers. In his *Tractatus super totam astrologiam*, he devoted the first eleven brief chapters to cosmological themes, approximately 4 percent of the entire treatise.[74] If astronomers chose to confine themselves to technical matters, scholastic natural philosophers, in their role as cosmologists, rarely discussed planetary positions or technical astronomy but concerned themselves with the properties of the heavens, the causes of celestial motions, and the relations between celestial orbs.

Technical astronomy and cosmology did not remain isolated from each

69. Of the original Greek, only part of the first book is extant, while the first and second books have survived in Arabic. For the partial Greek text of the first book and a German translation from the two books of the Arabic version, see Ptolemy, *Hypotheses*, 1907, 2:69–145. A part of the first book omitted from Heiberg's Greek text has been translated from Arabic into English and published along with the entire Arabic text in Ptolemy [Goldstein], 1967.
70. For these passages, see Ptolemy [Goldstein], 1967, 7, col. 1.
71. For the translation, see Grant, 1974, 452–465.
72. See Campanus of Novara, *Theorica planetarum*, 1971.
73. See Peurbach, *Theoricae*, 1987.
74. See Bernard of Verdun, 1961, 19–25. My estimate is based on a printed text of 126 pages of which approximately 5 are devoted to cosmology.

other indefinitely. Astronomers eventually merged the two approaches. In his *Apologia pro Tychone contra Ursum (A Defense of Tycho against Ursus)*, Kepler attributes the first fusion of the two disciplines to Copernicus and Tycho Brahe, although Jardine (1984, 247–248) has cogently argued that it was Kepler himself who first achieved the fusion in the very same *Apologia*. Copernicus, however, briefly treated cosmological problems in the first book of *On the Revolutions*. He confined himself to unavoidable problems involving the heliocentric system, namely the order of the planets and the motions of the earth. With his demonstration of the celestial nature of new stars and comets, Tycho Brahe also inevitably introduced cosmological considerations into his astronomy when he rejected solid, hard orbs and argued that the heavens were fluid. In his attempt to provide a causal explanation for the planetary motions, Kepler was perhaps the first astronomer to see the need to merge cosmology and astronomy and to treat them as of equal importance. Among scholastic astronomers, Christopher Clavius, in his *Commentary on the Sphere of Sacrobosco*, which went through many editions between 1570 and 1618, and Giovanni Baptista Riccioli, in his *Almagestum novum* of 1651, merged technical astronomy and cosmology.

Finally, questions on cosmology also appear in individual works that were neither encyclopedias nor commentaries nor questions on any particular treatise. In this category, we may include *quodlibeta*, or quodlibetal treatises, which contained occasional relevant questions, as is evident from the quodlibetal questions of Godfrey of Fontaines, included in Chapter 3.[75] Astronomical treatises, such as Riccioli's *Almagestum novum* sometimes contained valuable questions, although straightforward technical astronomical treatises usually did not.[76] As its title indicates, the *De universo* of William of Auvergne in the thirteenth century was in considerable measure cosmological. In the domain of theology, there was also the summa, which often included questions on creation, as was the case in the treatises by Alexander of Hales and Thomas Aquinas.

IV. Idiosyncratic cosmologies

Because this is a study of cosmology as it was formulated by university-trained scholastic natural philosophers, whose interpretations and opinions were largely based on a few physical works of Aristotle and to a much lesser extent on biblical sources, especially Genesis, ideas and even works that do not fit into this pattern will either be ignored or treated cursorily.

75. *Quodlibeta* were held at fixed times of the year (Lent and Advent) and consisted of questions posed to masters of theology and masters of arts from members of the audience. The questions could be on any topic whatever. See Glorieux, 1925–1935. For a brief description, see Leff, 1968, 171–173.

76. Riccioli's seventeenth-century treatise contains numerous and often quite lengthy questions on cosmology and refers to an extensive range of scholastic authors.

Numerous "cosmologies," both written and unwritten, were probably conceived during the Middle Ages. Although their concepts were largely unrecorded, peasants and uneducated people generally must have held some views, however primitive and crude, about the structure and operation of the world. Occasionally we gain a glimpse into the minds that fashioned these divergent, though unpublished – indeed unwritten – cosmogonies and cosmologies, as when Carlo Ginzburg discovered the strange conceptions of an Italian miller, Domenico Scandella (1532–1599/1600?), called Menocchio. What was learned about Menocchio derives from the records of two trials for heresy in Italy during the sixteenth century. Menocchio, who was apparently a pantheist ("all things in the world are God"),[77] believed that God and the angels and all living things are spontaneously generated by nature from a primordial chaos, "*just as worms are produced from a cheese*" (Ginzburg, 1982, 57). The chaos was originally composed of the four elements (earth, air, water, and fire) from which a solid mass formed, "just as cheese is made out of milk." Within the mass of cheese, worms appeared, and "the most holy majesty decreed that these should be God and the angels," as Ginzburg describes it (ibid., 6). For such heretical opinions, and many others, Menocchio was judged guilty by the Inquisition and executed around 1599. We may confidently assume that such extraordinary views exercised no influence on scholastic cosmogonical or cosmological thought. Among the literate, authors and poets sometimes reveal considerable knowledge about cosmological theories and ideas, as, for example, Dante and Chaucer do.[78] Chaucer not only included much cosmology, astronomy, and astrology in his poetry, but was himself the author of two astronomical works. But the cosmology these great poets included was usually tied to their literary and poetic needs and resembles but little the detailed, prosaic mode of presentation of academic natural philosophers.

Even authors who wrote on science and natural philosophy occasionally presented idiosyncratic world views that lay outside the mainstream of medieval cosmological thought, as, for example, did Hildegarde of Bingen (1098–1179) and Robert Grosseteste (ca. 1168–1253). Hildegarde wrote as the translations were under way but before they had had significant impact. Although she was undoubtedly aware of some traditional cosmological literature and was perhaps acquainted with some of the newly translated works, her views about the structure of the world were nonetheless exceptional. Her most striking presentation appears in an early treatise titled *Scivias*, composed around 1141–1150. A diagram in one of the manuscript versions of the *Scivias* reveals how radically different was her image of the physical world (Fig. 3).[79]

77. Ginzburg, 1982, 71.
78. For astronomy and cosmology in Dante's works, see Orr, 1969; in Chaucer's, see North, 1988.
79. The diagram appears in the manuscript version of the *Scivias* in the Nassauische Landes-

Figure 3. The egg-shaped cosmos of Hildegarde of Bingen. (Nassauische Landesbibliothek, Wiesbaden, Federal Republic of Germany, MS. B, fol. 14r.)

The most striking feature of Hildegarde's world is its oval, or egg, shape. Also extraordinary is the pure ether in which are located the fixed stars, the inner planets, Mercury and Venus, the Moon, and at the center of the oval a mixture of the four elements, rather than the earth alone (although Hildegarde assumed a spherical earth; Singer, 1917, 22). Beyond the fixed stars and the inferior planets is a shell of "dark fire" (*ignis niger*) or a "dark skin" (*umbrosa pellis*). And beyond this lies a mass of bright fire (*lucidus*

bibliothek in Wiesbaden, 14r. Singer (1917, 7, 9) also reproduced Hildegarde's figure in a "slightly simplified" version.

ignis) within which are located, in ascending order from the earth, the Sun, Mars, Jupiter, and Saturn. Thus the Sun and the three outer planets are not only separated from the inner planets and the fixed stars, but they lie beyond the fixed stars.

We need not pursue further Hildegarde's extraordinary conception of the cosmos. Her oval world with its strange distribution of planets, stars, and elements, which differed so radically from the Aristotelian–Ptolemaic system that took root in the thirteenth century, struck no responsive chord in the scholastic tradition that followed.[80]

Robert Grosseteste is of particular interest because, in his *De luce* (*On Light*), he seems to fuse the creation account in Genesis with ideas from Aristotle and thus seems to fall into the very framework of our study. But, as we shall now see, Grosseteste chose to focus narrowly on cosmogony and to ignore almost all of the cosmological issues that were subsequently deemed important in questions on *De caelo*.

Grosseteste's creation account is unique and extraordinary.[81] The universe is formed from light, no doubt the light created on the first day. Not only is light nobler than all corporeal things and similar to forms separated from matter, such as intelligences, but it is the first form and corporeity itself. By virtue of its corporeity, "a point of light will produce instantaneously a sphere of light of any size whatsoever, unless some opaque object stands in the way."[82] As the first form, light is created in matter. In "the beginning of time," light multiplied itself from a single point an infinite number of times and in so doing created a material sphere of finite size.[83] The sphere terminates at the point where the expanded matter or light is so rare that it is incapable of further rarefaction. Grosseteste calls this outermost, most highly rarefied region the firmament, or "first body." Thus was formed the bare structure of the world.

"When the first body, which is the firmament, has in this way been completely actualized," Grosseteste continues ([Riedl], 13), "it diffuses its light [*lumen*] from every part of itself to the center of the universe." But

80. In later life, Hildegarde abandoned her oval world for concentric spheres (see Singer, 1917, 18–19). But her world was still quite different from the versions proposed in subsequent centuries.
81. For a translation of the *De luce*, see Grosseteste, *On Light* [Riedl], 1942; and for a summary of it, McEvoy, 1982, 151–167.
82. Grosseteste, ibid., 10.
83. Grosseteste held that an "extension of matter could not be brought about through a finite multiplication of light, because the multiplication of a simple being a finite number of times does not produce a quantity, as Aristotle shows in the *De caelo et mundo*. However, the multiplication of a simple being an infinite number of times must produce a finite quantity" (Grosseteste, ibid., 11). In a note, Riedl explains that in *De caelo* "Aristotle is at pains to show that a quantity cannot be produced by combining things which are without quantity . . . (3.1.299a.25–30). Grosseteste, however, interprets Aristotle to mean only that a *finite* multiplication of the simple will *not* produce a quantity, thereby leaving the way open for Grosseteste's own notion that an *infinite* multiplication of the simple *will* produce a quantity." On Grosseteste's conception of infinites and their relationships, see McEvoy, 1982, 152–153.

the first immediate effect of the light [*lumen*] from the first body, or firmament, is to rarefy, or perfect, as much as possible the more dense matter directly below it and thus to form a second sphere that is necessarily more dense than the first. The light from the second sphere then rarefies to the extent possible the next, or third, sphere below it. The result is that the third sphere is more dense than the second sphere. As the light, or *lumen*, descends toward the center of the world, each successive celestial sphere is actualized, or perfected, into a denser mass than its immediately superior sphere, a process that continues until the ninth, or lunar, sphere (ibid., 14).[84] Because the matter of each of these spheres was perfected to the highest degree possible for its level of matter, these nine celestial spheres are not further changeable. In sum, they are incorruptible.

During this process, the power of the light gradually diminished until the lunar sphere was unable to transmit sufficient light to rarefy, or perfect, the matter below the Moon. As a result, the matter in this region is capable of further rarefaction and condensation and is therefore capable of change. This is the region of the four elements, which is organized into the four descending spheres of fire, air, water, and earth.[85] The thirteen sensible spheres of the world form a continuous sequence without void spaces. Moreover, all of these bodies existed *virtually* in the first body, which set off the chain reaction that produced them all.[86]

Because the nine heavenly spheres are fully actualized and therefore incapable of condensation and rarefaction, they have no capacity to move away from, or toward, the center of the world – that is, they have no propensity to move rectilinearly upward or downward. By contrast, the four elements, which are subject to condensation and rarefaction, are naturally capable of moving up or down – that is, toward or away from the center of the world.

By emphasizing light, presumably the light created on the first day, Grosseteste sought to provide a creation account that was compatible with Genesis. And by determining a way to distinguish radically between the celestial region of the planets and stars and the sublunar region of the four elements, he sought to reconcile himself with Aristotle's account in *De caelo*.[87] Indeed, he also agrees with Aristotle on the finitude of the cosmos and rejection of vacua.[88]

84. Traditionally, only eight physical spheres are assumed for the celestial bodies, namely, the firmament of the fixed stars and the respective spheres of the seven visible planets. Since Grosseteste fails to name any spheres, the identity of the ninth sphere he mentions is left unclear.
85. "In this way," says Grosseteste (*On Light* [Riedl], 15), "the thirteen spheres of this sensible world were brought into being. Nine of them, the heavenly spheres, are not subject to change, increase, generation or corruption because they are completely actualized. The other four spheres have the opposite mode of being, that is, they are subject to change, increase, generation and corruption, because they are not completely actualized."
86. Ibid., 17; also McEvoy, 1982, 183.
87. For the sources of Grosseteste's *De luce*, see McEvoy, 1982, 158–162.
88. Ibid., 163.

But major differences separate the two, largely because Grosseteste chose to make light the unifying entity for the celestial and terrestrial regions.[89] Light is thus the "ultimate unity of matter."[90] By contrast, Aristotle distinguished the celestial ether, or fifth element, radically from the four terrestrial, or sublunar elements.

Moreover, by his "emanationlike" Neoplatonic generation of the celestial spheres (McEvoy, 1982, 183), Grosseteste made the successively descending spheres more and more dense. Whatever differences Aristotle may have assigned to the hierarchy of spheres, he did not differentiate between them with respect to physical properties such as density or rarity. As physical entities, they were all identical.

We may agree that Grosseteste wrote an ingenious and important cosmogony,[91] but should his extremely brief *De luce* qualify as a treatise on cosmology? I think not.[92] Apart from how the celestial orbs were formed and that they move with circular motion, we learn virtually nothing about the operation of the celestial region. Indeed, Grosseteste makes no mention of any planet and therefore does not even hint at the relationships between planets and their orbs. Without such discussions, it seems inappropriate to categorize Grosseteste's treatise as essentially cosmological. Almost any medieval questions on *De caelo* is more characteristically cosmological than Grosseteste's *De luce*.

Although idiosyncratic cosmologies are worthy of independent investigation, they play only an incidental role here. Our focus is on the mainstream of medieval cosmology, which was basically Aristotelian and found overwhelmingly in the treatises described in this chapter. It was this scholastic Aristotelian cosmology that was gradually replaced in the seventeenth century and which was the main adversary of the new cosmology that emerged in the sixteenth and seventeenth centuries.

With the completion of this general survey of the cosmological literature of the late Middle Ages, the next logical move is to identify and discuss the specific works, authors, and questions on which this study is primarily based. This aspect of the Introduction is complex because it involves much detailed information: a list of 400 of the most important cosmological questions discussed by scholastic cosmologists. To avoid overburdening the text with these questions and the relevant material about them, I have placed this data in two appendixes at the end of this volume. Appendix I contains a "Catalog of Questions," with the 400 questions and the discussants of each question; Appendix II includes a chronological list of the 52 authors,

89. Ibid., 182.
90. Ibid. On p. 183, McEvoy declares that "Light, the *species et perfectio* of all bodies, accounts for both the unity and the diversity found at the level of material being."
91. Because he did not believe the world had a beginning, Aristotle could not have composed a cosmogony.
92. In this assessment, I disagree with McEvoy, 1982, 151, who characterizes the *De luce* as "one of the few scientific cosmologies . . . written between the *Timaeus* and early modern times."

a description of the selection and distribution of the authors, an analysis of the works from which the questions were drawn, and a detailed examination of the questions themselves, including their organization, relationships, and what they reveal by way of statistical analysis and tables. The interested reader should consult the appendixes.

Before leaving this Introduction, one task remains. We must briefly inquire about the interaction of medieval cosmology with the society within which it was developed.

3

The social and institutional matrix of scholastic cosmology

Thus far, in the first two chapters (with Appendixes I and II), this study has been largely confined to describing texts and their arguments. But is there some larger social context for our 52 authors and their 400 cosmological questions? Did social forces help shape the questions posed?

I. Social factors

To cope with such problems, it seems appropriate to begin with a fundamental question: in order to understand the evolving content and history of medieval cosmology between 1200 and approximately 1687, what must we know? The answer seems obvious. We must first determine what constituted cosmology in the Middle Ages and who were its practitioners. The 400 questions in my "Catalog of Questions" (App. I) and the explanation of my reasons for choosing them (App. II) represent a detailed response about the substantive content of cosmology. Its practitioners, as described in this volume, were scholastic natural philosophers, almost all of whom had one important attribute in common: they had been educated at universities or colleges that based their curricula on Aristotelian natural philosophy, of which cosmology was a major part. Over the centuries, scholastic natural philosophers came from all social classes: peasants, burghers, nobles, and merchants. And yet, in reading any questions on *De caelo*, or any other kind of questions treatise, we find virtually no indications of the social class of the author.

There seems a simple explanation for this. Scholastic natural philosophers were professionals who, despite diverse backgrounds, were molded by their training into a homogeneous intellectual group. When they responded to questions and problems in Aristotelian natural philosophy, they were expected to avoid the introduction of personal data and feelings. Rarely do they cite the names of contemporaries or introduce autobiographical elements into the elucidation of a question.[1] In this regard they resemble

1. Personal disclosures were more apt to intrude into questions concerning magic or astrolo-

modern professionals. From a research article in a physics journal, we can determine nothing about the social background of the physicists who were its authors.

But if we can determine nothing about the social background of scholastic natural philosophers from the content of their individual questions, it is still possible that social factors may have influenced the kinds of questions they posed, as well as their content. First, however, I must characterize my understanding of "social factors." For our purposes, I shall distinguish three possible kinds of external social influences on medieval cosmology: (1) the more usual kinds of social influences represented by the state, the economy, and society in general; (2) the universities; and (3) the Catholic Church as represented by its theologians and theological faculties.

What, however, shall serve as a measure or indicator of social influence? Obviously, the questions themselves. Only an examination of the titles and content of the questions over the period of our study can show whether scholastic natural philosophers reveal, explicitly or implicitly, social influences, that is, whether their responses to questions were substantively influenced by politics, the economy, or generally by social conditions of any kind. Of the three types just mentioned, the more general and usual type will be considered first; the second and third kinds will be treated together, namely the universities, where medieval cosmology was fashioned and perpetuated, and theology, which had a special relationship to the universities and the cosmology taught therein.

II. General societal influences

A brief examination of question 97 ("On the number of spheres, whether there are eight or nine, or more or less"), in the "Catalog of Questions" may reveal something about the possible role of general societal influences. The catalog listing shows that between the thirteenth and the seventeenth century, 22 of our sample of 52 scholastic authors included this question. Most of them assumed the existence of between eight and eleven spheres, although a few extended the number to twelve, and at least two invoked fourteen orbs.[2] Despite slight variations in the number of spheres, or heavens, the responses are linked in a powerful tradition, as is made evident when later authors such as Galileo and Riccioli cite numerous medieval predecessors. Among seventeenth-century scholastic authors who believed

gy. For examples from works by Nicole Oresme, see Caroti [Curry], 1987, 77, and, especially, 89, where Oresme admits that in his youth he not only diligently studied astrology but even tried to become an astrologer.

2. In a lengthy discussion, Riccioli (*Almagestum novum*, pars post., bk. 9, sec. 3, ch. 1, 1651, 271–276) identifies authors who assumed anywhere from one to twelve orbs or heavens and two who even assumed fourteen (one was Fracastoro). Riccioli's list of authors includes the ancient Greeks, the Church Fathers, and authors from the early Middle Ages up to his own time.

in the existence of real, physical celestial orbs (by then some had joined Tycho Brahe in rejecting them), arguments on the number of spheres differed little in kind from that of their medieval predecessors. If interpretations of this most popular of questions could remain so remarkably similar and stable over a period of four hundred years, during which enormous political, religious, social, and economic changes had occurred, it would appear that such external events played no detectable role in cosmology. An examination of the responses to the question fully supports this conclusion. Nothing in them pertains to external social forces or suggests in any way that such external events influenced the responses.

For almost all of the hundreds of questions that scholastics posed in their cosmological discussions, it is difficult to imagine how societal conditions could have exerted any influence, and equally difficult to conjure up the forms that such influences might have taken. How, for example, could societal actions have affected such typical questions as the following, from my "Catalog of Questions" (Appendix I):

1. Whether the universe could have existed from eternity.
12. Whether the sky [or heaven] is generable and corruptible, augmentable and diminishable, and alterable.
46. How many and which heavens were created by God on the first day?
55. What are imaginary spaces?
62. Whether there are, or could be, more worlds.
70. Whether the world is a finite or infinite magnitude.
74. Whether there are five distinct simple bodies in the world that differ in species, namely the four elements and the heavens or fifth essence.
79. Whether the heaven is composed of matter and form.
195. Whether the heavens or planets [*sidera*] are moved by intelligences or intrinsically by a proper form or nature.
290. Whether the orbs of the planets have a place *per se* or *per accidens*.
308. Whether it is possible that a vacuum can exist naturally.
378. Whether or not elements have heaviness or lightness in their [natural] places.

Although most questions seem as immune to societal changes and activities as those just cited, one question, number 388, on the terraqueous sphere ("Whether water and earth make one globe"), was apparently affected by new empirical data that was obtained from Portuguese explorations of the Southern Hemisphere in the early sixteenth century.[3] The discovery that land and sea were interspersed in that hemisphere, which was previously thought to be covered by water, helped to gain acceptance for the existence of a terraqueous sphere, that is, a single globe of earth and water that lay at the center of the universe and which replaced Aristotle's concept of two

3. For a discussion, see Grant, 1984b, 22–32, and Chapter 20, Section IV.3.

separate spheres, one for earth and another for the water that surrounded it. All four seventeenth-century discussants of question 388 adopted the terraqueous sphere, as had Copernicus in the preceding century. Whether the new knowledge acquired by maritime explorers of the sixteenth century, which produced a change of opinion on a point of Aristotelian cosmology, qualifies as a "societal influence" is difficult to say. But it was surely a definitive change that was generated, if only indirectly, from outside the scholastic tradition.[4]

One other question deserves brief mention. A glance at question 135, "Whether the heavens are fluid or solid," shows that all 11 authors who discussed it lived and published in the seventeenth century. Although a few Church Fathers and an occasional author in the Middle Ages indicated an opinion as to the solidity or fluidity of the heavenly orbs, no scholastic natural philosophers devoted a specific question to this subject in their *De caelo* treatises or any other scholastic work until the seventeenth century, or the late sixteenth century at the earliest.[5] Why and how did this new question arise? Was it because of external societal influences? Not at all. It arose because of Tycho Brahe's observations about the comet of 1577, whose path he placed in the celestial region beyond the Moon. Tycho inferred that in order for comets to move within the celestial region, the heavens must be fluid, not rigid and hard. As a direct consequence of Tycho's claim, scholastic natural philosophers thought it necessary to consider this important problem. Thus did the eleven authors represented in question 135 in the catalog take up a problem their medieval predecessors had virtually ignored. Societal concerns, however, played no role. The fluidity or hardness of the heavens derived from considerations internal to the history of astronomy, cosmology and natural philosophy.

Societal influences, we may plausibly conclude, did not shape the cosmological opinions and judgments of scholastic natural philosophers. What did shape them was the bookish tradition they inherited and to which they responded. It was a tradition formed from two disparate components. The first was the natural philosophy of Aristotle and, to a much lesser extent, the Greco-Arabic scientific tradition that entered western Europe along with it. The second derived from the Christian tradition, which had sought, soon after its establishment, to integrate the creation account in Genesis into the

4. Patronage was also a possible source of external societal influence, although there are few instances of it. One that is known involves Oresme and King Charles V of France (1364–1380). The latter commissioned Oresme to translate four of Aristotle's treatises from Latin to French, one of which was the *De caelo*, which Oresme rendered into French under the title *Le Livre du ciel et du monde* and to which he added one of the most fascinating, profound, and original commentaries on the work made during the Middle Ages and Renaissance (see Oresme, *De prop. prop.*, 1966b, 9). Without the patronage of Charles V, Oresme might not have produced his brilliant French translation and commentary on Aristotle's *De caelo*. An external factor may therefore have played a major role in making possible a significant cosmological work. Thus far, however, the content of Oresme's treatise reveals no influences deriving from the patronage of Charles V and the French state.

5. I have discussed this problem in Grant, 1987a, 155–156.

overall complex of Greek pagan cosmology. Although this process was well under way in the early Middle Ages, when the natural books of Aristotle were unknown, it was completed in a thorough and sophisticated manner during the course of the thirteenth century. Indeed, it formed part of the overall assimilation of Aristotelian natural philosophy into Christian culture. The locus of this important process was the medieval university, especially the University of Paris, in the thirteenth to sixteenth centuries. It was continued into the sixteenth and seventeenth centuries in various Catholic colleges and schools, especially in those of the Jesuit order.[6]

If there are any significant "external" factors in medieval scholastic cosmology, we must seek them in the universities, where cosmology was studied and produced. And because the Church, represented by its university-based theologians, elucidated and developed the theological aspects of medieval cosmology, we must also investigate its role.

III. The influence of university and church

1. The 1260s and 1270s prior to the Condemnation of 1277

Because the universities formed the general institutional matrix of medieval scholasticism, they also served as the particular locus for the study and development of the whole range of cosmology. Natural philosophy was the province of the faculty of arts, which devised the curriculum, the schedule for courses and lectures, and possessed the authority, when it chose to exercise it, to formulate rules as to what could and could not be included in university lectures. Thus in 1272, the faculty of arts of the University of Paris voted to "decree and ordain that no master or bachelor of our faculty should presume to determine or even to dispute any purely theological question." But if perchance a question should be considered that touched both philosophy and faith, the question had to be resolved in favor of the faith. The statute further declared that "if any master or bachelor of our faculty reads or disputes any difficult passages or any questions which seem to undermine the faith, he shall refute the arguments or text as far as they are against the faith or concede that they are absolutely false and entirely erroneous, and he shall not presume to dispute or lecture further upon this sort of difficulties, either in the text or in authorities, but shall pass over them entirely as erroneous."[7]

After 1272 and during the course of the fourteenth century, arts masters were required to swear an oath to uphold the statute of 1272. Because certain

6. Protestants also produced a scholastic tradition that often dealt with similar cosmological problems. Among Protestant scholastics were Philip Melanchthon (1497–1560), Rudolphus Goclenius (1547–1628), Bartholomew Keckermann (1571–1609) and Johannes Magirus (d. 1596). See Schmitt, 1983, Appendix B ("Biographical Guide"); for Goclenius, see Lohr, 1988, 169–170.
7. From Thorndike, 1944, 64–65; reprinted in Grant, 1974, 44–45.

Figure 4. A classroom in fourteenth-century Germany. (Staatliche Museen, Berlin.)

questions touched faith and philosophy – for example, questions about the eternity of the world and whether God could create vacua or actual infinites – such questions could pose problems for the more venturesome arts masters, of whom Jean Buridan was one of the most prominent. In a question as to whether God could create a vacuum supernaturally (see "Catalog of Questions," qu. 321), Buridan, citing the statute of 1272, complains about the criticisms of some theologians, who apparently accused him of intruding forbidden theological matters into his questions.[8] Despite the oath he was required to uphold, Buridan insists on the need to introduce theology to deal properly with certain questions. Elsewhere Buridan defers to the theologians concerning the possibility that a celestial impetus, rather than an intelligence, moves each heavenly orb[9] and on the question about the ex-

8. For the relevant passages, see Grant, 1974, 50–51; for a further discussion, see Grant, 1978c, 107, 118, nn. 12, 13.

9. In his *Questions on the Physics*, bk. 8, qu. 12 (see Buridan [*Physics*], 1509, 120r, col. 2–121r, col. 2), as translated by Clagett, 1959, 532–538 and reprinted in Grant, 1974, 277. Because the section on celestial impetus forms but a small part of Buridan's twelfth question, the

istence of things beyond the sky (qu. 54). Except for theological problems and questions of faith, university authorities did not, to my knowledge, otherwise impinge on the freedom of arts masters to treat questions freely within the *questio* framework or within a commentary. Indeed, an innate conservatism of the corporation of arts masters was as likely a cause for the restriction of inquiry of its members as were external forces.[10] Despite some obstacles, however, the arts masters had a considerable, if not remarkable, degree of intellectual freedom.[11]

One of these "obstacles" was the Church and its theologians. Most of the medieval authors included in our study taught and wrote either in the arts faculty or in the theological faculty of the University of Paris, which could rightly proclaim itself the greatest school of theology in Christendom, numbering among its teachers the likes of Thomas Aquinas, Saint Bonaventure, and Duns Scotus. As we have seen, the major problem at the University of Paris during the thirteenth century was the assimilation and incorporation of the natural works of Aristotle and the commentaries thereon of his greatest commentator, Averroës. The turmoil that arose from this process would pit arts masters against theologians, theologian against theologian, and arts master against arts master.[12]

Aristotelian principles and assumptions about the world were not easily subordinated to Christian theology. Aristotle's philosophy and metaphysics were found to be useful not only for the physical world but also for the analysis of problems in theology and Scripture. Some theologians and Church authorities found this state of affairs disturbing and feared that Aristotle's concepts and principles might eventually dominate theology and become the arbiter and validator of theological truths and dogmas. During the first forty to fifty years of the thirteenth century, Christian authorities at Paris sought to ban and then edit Aristotle's texts to eliminate objectionable concepts. By the 1250s, these efforts had failed, and Aristotle's natural philosophy became firmly entrenched at the University of Paris, where it formed the basis of the arts curriculum.[13]

A second phase of the struggle developed in the 1260s and 1270s, when an intense effort was made to prevent the by-then accepted pagan Aristotelian natural philosophy from subverting traditional theological doctrines

latter, which asks "Whether a projectile after leaving the hand of the projector is moved by the air, or by what it is moved" (Clagett, 1959, 532) has been omitted from the "Catalog of Questions." The 1509 edition of Buridan's *Questions on the Physics* was based on the longer (and presumably later) version of his *Physics* (see Maier, 1958, 127, n. 87).

10. Even if the "conservatism of the corporation of arts masters" is interpreted as an external influence, it would be difficult to relate that conservatism to an identifiable influence on any given question.
11. For a brilliant summary of the intellectual milieu at the University of Paris, see chapter 6 ("Freedom, Truth and Fellowship") in McLaughlin, 1977, 305–317, especially page 312, on the conservatism of the corporation.
12. Here I follow Grant, 1985b, 76–78.
13. For documents depicting the rejection and then gradual acceptance of Aristotle's natural books, see Grant, 1974, 42–44.

and interpretations. When warnings about the perils of secular philosophy by conservative theologians, many of whom were neo-Augustinian Franciscans (for instance, Saint Bonaventure), went for nought, these same traditional theologians carried their appeal to the bishop of Paris, Etienne Tempier, who obliged them in 1270 with a condemnation of thirteen propositions. When this proved unavailing, he responded once again in 1277 with a massive condemnation of 219 propositions, any one of which could thereafter be held or taught only under penalty of excommunication.

2. *The Condemnation of 1277*

A number of the condemned propositions are of direct concern for cosmology.[14] Depending on how one counts, anywhere from fifteen to twenty-seven, or even more, articles specifically or implicitly condemned the hypothesis of the eternity of the world, as, for example, article 9, which asserted that "there was no first man, nor will there be a last; on the contrary, there always was and always will be generation of man from man"; or article 98, which declared that "the world is eternal because that which has a nature by [means of] which it could exist through the whole future [surely] has a nature by [means of] which it could have existed through the whole past"; or article 107, which held that the elements are eternal but that "they have been made [or created] anew in the relationship which they now have."[15]

Other condemned articles relevant to cosmology concerned "God's absolute power" (*potentia Dei absoluta*) to do as he pleased, short of doing anything that involved a logical contradiction.[16] That is, some articles were damned because natural philosophers were thought to have restricted God's infinite power as they zealously sought to interpret the world in accordance with Aristotelian principles. For example, under Aristotle's influence natural philosophers had regularly denied the possibility of other worlds, and therefore, by implication, had denied that God could create other worlds. Article 34, which denounced the opinion "That the first cause [God] could not make several worlds," made it mandatory to concede that God could indeed create other worlds, as many as he pleased. Although no one was required to believe that God had created a plurality of worlds – until John Major in

14. The Latin text of the 219 articles appears in Denifle and Chatelain, 1889–1897, 1:543–555. For another reprinting of the Latin texts of the condemned articles and an attempt to determine their sources, see Hissette, 1977. A complete English translation, based on a reorganization of the questions, appears in Fortin and O'Neill, 1963, 337–354 (reprinted in Hyman and Walsh, 1973, 540–549). The questions relevant to natural philosophy and science appear in Grant, 1974, 45–50.

15. The translations are from Grant, 1974, 48, 49. In addition to articles 9, 98, and 107, other condemned articles that concern the eternity of the physical world or some of the immaterial substances associated with it are articles 4–6, 12, 31, 48, 52, 72, 80, 87, 89–90, 91, 93–95, 99, 101, 109, 125, 200, 202–203, 205.

16. Here again I follow my article, Grant, 1985b, 81–84.

the sixteenth century, no one in the Middle Ages did so believe[17] – the effect of article 34 was to encourage speculation about the conditions and circumstances that would obtain if God had created other worlds.

By declaring that "God could not move the heavens [that is, the world] with rectilinear motion; and the reason is that a vacuum would remain," article 49 effectively denied to God the power to move the outermost heaven, and therefore the world itself, because a vacuum would be left when the world departed from its present position. After the condemnation of article 49 in 1277, scholastics routinely conceded that God could indeed move the world rectilinearly.

Other condemned articles were also pertinent to cosmology.[18] The Condemnation of 1277 may not have substantively altered cosmological opinions, but it did cause changes in attitude and approach, especially in discussions that involved God's absolute power.[19] The latter became a convenient vehicle for the introduction of subtle and imaginative questions, which generated novel replies. Although these novel, speculative responses did not lead to the overthrow of the Aristotelian world view, they did challenge some of its fundamental principles, as we shall see. They made many aware that things might be quite otherwise than were dreamt of in Aristotle's philosophy.

In another sense, however, the Condemnation of 1277 was a restrictive document that affected both arts masters and theologians. We may rightly conclude, therefore, that an action by Parisian theologians ostensibly to protect the faith had an impact on the development of medieval cosmology. It not only affected responses to certain problems but also altered attitudes toward basic Aristotelian principles. May we justifiably construe the Condemnation of 1277 as an external social influence? And does the statute of 1272, executed by the arts faculty of the University of Paris and which excluded theological considerations from cosmological questions treated by arts masters, also qualify as an external social influence?

On the face of it, we may reasonably assume so, because there is no doubt that Christianity, through the institution of the Church, influenced Aristotelian cosmology and the interpretation of it by scholastic natural philosophers, whether arts masters or theologians. But the theologians who compiled the list of condemned articles that affected cosmology could hardly be construed as "external" factors. Almost all of them, if not all, were trained in Aristotelian natural philosophy, which they absorbed when they pursued the master of arts degree, a necessary prerequisite for entry into the school of theology. They were an integral part of the university system, although they constituted a different, and sometimes hostile, faculty from

17. For Major's opinion, see Chapter 8, Section II.2.
18. In the list presented by Denifle and Chatelain (1889–1897, 1:544–555), see, for example, articles 6, 14, 30, 38, 59, 61, 66, 74, 88, 92, 95, 100, 102, 106, 110, 137, 143, 156, 161–162, 186, 201, and 212.
19. I depend here on Grant, 1979, 241–242.

that of the arts masters. Moreover insofar as the theologians were themselves trained natural philosophers, they formed part of the same intellectual tradition as did the arts masters, who were professional natural philosophers. From this standpoint, it is difficult to view the Church as a genuinely external social force that helped shape late medieval cosmology.

By the time of the Church's condemnation of the theory that the earth moves, in 1616 and its condemnation of Galileo's views in 1633, all this may have changed, aided and abetted perhaps by the Protestant Reformation and the attitudes generated by the Council of Trent. Moreover, the new cosmology was not centered in the universities and colleges of the seventeenth century but was largely taught and propagated outside their jurisdiction. Within that jurisdiction, however, Aristotelian natural philosophy continued its overwhelming dominance, although, as will be evident, it was hardly immune to the force of certain ideas and concepts of the new cosmology.

From the initial entry of Aristotelian natural philosophy into the Latin West in the late twelfth century, university and church were integral to its development. As parts of the same intellectual tradition, university arts masters and university theologians approached cosmological problems in much the same manner, with one significant difference: theologians were free to introduce theological matters into their discussions, while arts masters, who were untrained in theology, were not.[20] Although prior to the twelfth century the Christian religion may be viewed as an external force that significantly affected the character of pagan Greek cosmology, this situation had drastically altered by the thirteenth century, when, despite the conflict between arts masters and theologians, and between the requirements of natural philosophy and the dogmas of theology, there was a much higher level of sophistication about the relations between Christianity and natural philosophy, which by then had become a potent intellectual force. Indeed, the efforts to reconcile theological dogma and Aristotelian natural philosophy in the thirteenth century was as much a struggle between conservative theologians such as Saint Bonaventure, who feared the inroads of Aristotelian thought, and theologians such as Thomas Aquinas, who wished to harmonize theology and natural philosophy. It is no coincidence that many of the articles condemned in 1277 by the conservative bishop and theologian Etienne Tempier were directed against Thomas and his followers.[21]

20. Theological masters who had written questions on *De caelo* before earning their theological degrees would have included little or no theological material in their work. Indeed Thomas Aquinas, who wrote his commentary on *De caelo* at the end of his life, found little occasion to introduce theological matters into his comments. Oresme marks a striking contrast. When, as a master of arts, he composed his *Questiones super De celo*, he ignored theological problems. But when, as a theologian late in life, he added a French commentary to his French translation of *De caelo*, he inserted numerous biblical references and theological discussions.

21. Only some fifty years later, on February 14, 1325, did the bishop of Paris annul the penalty of excommunication for those articles that affected the opinions of Thomas Aquinas. See Grant, 1974, 47.

IV. The intellectual tradition as social context

The best candidate for a "social context" for scholastic natural philosophy between the thirteenth and seventeenth centuries is the company of Aristotelian natural philosophers (embracing both secular and theological masters) whose loci were the higher-educational institutions of Europe. What they shared by virtue of a common higher education was a bookish and learned tradition that flourished in the universities and colleges and was centered on the works of Aristotle set within a larger matrix of Greco-Arabic learning. For more than four centuries, Aristotelian natural philosophers were a distinctive group, who, despite all individual differences, formed a continuous community of scholars bound together by a common focus on the works of Aristotle. The continuity of Aristotelianism was guaranteed by the community of scholars who transmitted it generation after generation.

The continuity of Aristotelianism over nearly five centuries conveys a remarkable message. Although medieval cosmology and natural philosophy evolved modestly, they remained, for the most part, fairly constant between the thirteenth and the seventeenth century. During this same period, however, the economy and political structure of Europe changed radically, and religion underwent a revolutionary reformation. What is remarkable here is not that social conditions changed but that Aristotelian natural philosophy remained so relatively stable and traditional while so much was changing around it. Why did this occur? Perhaps the answer lies in the fact that Aristotelianism had no significant rivals until the sixteenth century, when new works reached Europe for translation and a new science began to develop. And even then the Aristotelian tradition remained a formidable body of knowledge as its guardians sought to preserve and defend it by making some adjustments toward the new science while simultaneously upholding the bulk of Aristotelian natural philosophy.

Who were these scholastic scholars? We know relatively little about them as individuals. As teachers and students, they came from all over Europe and from all social classes. As natural philosophers, their primary mode of scholarly expression was by means of questions or commentaries on the natural books of Aristotle. Each succeeding generation of scholars added to the legacy. As some works disappeared, or remained little known, others were widely disseminated. With the advent of printing, ever more works became available. Over the centuries, scholastic natural philosophers reacted to a variety of opinions embedded in the available works and occasionally departed from them. The enterprise was self-contained and exhibited both a remarkable degree of constancy and a surprising range of diverse opinion.

The independence from external influences of the scholastic tradition in natural philosophy seems to confer upon it an aspect of timelessness. As support for this characterization, one need only point to the manner in which scholastics tended to cite authorities, as if they were all contemporaries and of equal importance. A seventeenth-century scholastic, for example, might cite Aristotle, Thomas Aquinas, Albert of Saxony, the Coimbra Jesuits, and one or more contemporaries, who might even be labeled "moderns," as if all were of equal reliability and weight. He might side with this or that earlier authority on one issue and disagree with him on another. As a class, authorities were respected and venerated, and the historical period to which they belonged was seemingly irrelevant. A thirteenth-century authority was equal in status to any scholastic commentator whose reputation was made in the sixteenth or seventeenth century. In short, authorities were utilized as if what they had written was timeless and perennially current.

The seemingly timeless feature of scholastic natural philosophy and cosmology is further underscored when one realizes that a seventeenth-century author (for example, Bartholomew Amicus) had little difficulty understanding Albert of Saxony's fourteenth-century questions on *De caelo*. More significantly, it is likely that Albert of Saxony would have understood most, if not all, of the cosmological arguments engendered by his seventeenth-century successors.

Medieval and Renaissance scholastic cosmology was thus seemingly immune from external social conditions. Virtually all the changes that occurred during its history are explicable from within its own intellectual tradition, with no need to go beyond that tradition. To invoke social conditions outside the university environment within which Aristotelian natural philosophy was produced and preserved would be to seek explanations in events and conditions which bear no obvious connection to the cosmological questions that were regularly discussed. As we distance ourselves farther and farther from the intellectual tradition itself, causal influences grow more remote and become less and less relevant to that tradition.

Indeed, intellectual traditions are themselves often problematic, and even enigmatic. Nothing could illustrate this better than certain differences between the universities of Oxford and Cambridge during the Middle Ages. In his fine book *The Medieval English Universities* (1988), Alan C. Cobban corrects a general misapprehension about medieval Oxford and Cambridge whereby Cambridge is often viewed as little more than a lesser clone of Oxford, perhaps because "the decisive factor in the rise of the *studium generale* at Cambridge was the exodus of masters and scholars from Oxford in 1209."[22] But, as Cobban shows, Cambridge was anything but a minor facsimile of Oxford. Despite Oxford's greater international reputation, Cambridge was recognized as equal to it in status (Cobban, 1988, 60) and indeed to all other *studia generalia* (Paris, Bologna, Montpellier, Padua, and Orleans). "It is no longer tenable," Cobban concludes (ibid., 110), "to view Cambridge as merely a lesser version of Oxford, even in the first half century

22. Cobban, 1988, 53. Here I follow my review of Cobban's book (see Grant, 1990, 276–277).

of its existence." Thus Cambridge University was every bit as developed as Oxford and was so recognized during the Middle Ages.

But if this is true, and there seems no reason to doubt it, how can we explain a curious anomaly that developed between these two venerable institutions? Some of the greatest minds of the Middle Ages are associated with Oxford University: Robert Grosseteste, Duns Scotus, William of Ockham, and the great Merton College Calculators, among whom we need only mention Thomas Bradwardine, Richard Swineshead, and John Dumbleton. During that same period, not a single logician, natural philosopher, or theologian of any note is associated with Cambridge. How could such a gross disparity have occurred? How can we explain this? Not in terms of wealth, since neither was significantly wealthier than the other. Nor can we attribute this extraordinary disparity to the fact that Oxford University attracted approximately two to three times as many students as did Cambridge; nor does it plausibly follow from the less attractive physical location of Cambridge.

If societal conditions manifestly shape and even determine intellectual achievements, we ought to be able to explain this striking discrepancy between these two world-renowned universities. In truth, it is highly unlikely that any explanation rooted in societal or environmental conditions will convincingly account for the intellectual discrepancy between the universities of medieval Cambridge and Oxford. By chance alone we might expect that in the course of three centuries at least one great and famous scholar would have been associated with Cambridge University.

Even if one assumes that all intellectual concepts and ideas are ultimately explicable by social and environmental conditions because we are, at any point in our lives, the sum total of all our previous experiences and activities, it does not follow that we can produce evidence for the influence of any particular social activity, or collection of such activities. Indeed, for many cosmological questions, one would scarcely know which social factors to invoke. What, for example, would count as a social factor in explaining responses to questions such as "What are imaginary spaces?" or "Whether the heaven is composed of matter and form"? Intellectual matters, it seems, are not readily reducible to, or explicable by, social conditions.[23]

23. Bechler, 1987, argues that social conditions and physical causes are utterly incapable of explaining ideas, concepts, and theories. In a section titled "It is Impossible to Explain Theories by their Physical Causes" and using Isaac Newton's *Principia* as his example, Bechler declares (ibid., 88–89):

> It is not my business, vocation or training as a professional person to psychoanalyze Newton's mind as a prelude to an analysis of his *Principia*. However let us suppose it was, and I was trained as a multidimensional physicist, that is, chemist, psychoanalyst, and sociologist. Let us suppose that the *whole* (I am in my generous mood) of the *Principia* could be shown to be causally connected to Newton's unconsciousness (tough childhood, libidinal hangups, the whole bag). What would this imply for my practice in the history of ideas?
>
> It seems clear to me it would mean nothing at all. The main reason is that "ideas," "theories," "concepts," and "conceptual and logical connections" are not and cannot be

V. The impact of medieval cosmology on society

One other aspect of the relationship between medieval cosmology and society remains to be mentioned, namely the impact, if any, of medieval cosmology on society as a whole. Here the evidence is overwhelming, as manifested by the enormous outpouring of printed editions in the late fifteenth and sixteenth centuries which reflected medieval Aristotelian cosmology and served as the window on the world for society as a whole. Society's concept of the origin, structure, and operation of the world was drawn almost exclusively from the Aristotelian–Ptolemaic astronomical and cosmological tradition. As a dramatic example, I draw attention to the information about the size of the earth that Christopher Columbus gleaned from Pierre d'Ailly's *Ymago mundi* (see Ch. 20, Sec. I). All who learned to read and write absorbed at least the skeletal frame of scholastic cosmology, which was itself virtually synonymous and coextensive with medieval cosmology as a whole. Indeed, it is this virtual coextensivity between medieval scholastic cosmology and medieval cosmology as a whole that makes it unnecessary to qualify this study by the insertion of the term "scholastic" somewhere in the title. Although the utilization of scholastic cosmology within society is a legitimate theme, it falls outside the scope of this study. It will be considered here only where it seems relevant, as in the example about Columbus just cited.

For all of the reasons presented in this chapter, my interpretations and analyses of medieval cosmology are drawn from the intellectual tradition that gave birth to it and within which it developed between the thirteenth and the seventeenth century.

either causes or effects of physical things, such as states of human biology, psychology, or cosmic physics. The reason for this is that whereas Newton's biological and psychological states are events with exact temporal and spatial locations, his theories, concepts, and structures are not. They are completely different entities, belonging to different ontological strata of the world. Thus, one cannot be connected to the other via any known causality: theories cannot be caused by brain chemistry, nor can they give rise to psychological complexes. Denying this implies that theories, concepts, and structures have spatial and temporal locations, and as such must have states of motion, kinetic energy, and some momenta. If this absurdity is to be avoided, the project of the physical (biological, psychological, economic, sociological, etc.) explanation of theories and concepts must be rejected as well, since present-day physics is able to explain physical events only.

PART I

The cosmos as a whole and what, if anything, lies beyond

4

Is the world eternal, without beginning or end?

With the introduction of the natural books of Aristotle in the thirteenth century, medieval scholastics confronted a serious dilemma. From Scripture they learned that the world was created and brought into being by supernatural means and by that same method would eventually suffer destruction and pass away.[1] From Aristotle's *De caelo* (*On the Heavens*) came a powerful contrary message: the world – that is, the whole cosmos – could not have had a beginning and could never come to an end.[2] Aristotle's strong and reasoned conviction that the world was without beginning or end was in direct conflict with one of the most powerful messages of Christianity: our world was divinely created, is unique, and is destined to endure for only a finite time. Although the Church and its theologians would guarantee the triumph of the doctrine of the world's creation and temporal existence, such an outcome may not have seemed as clear in the thirteenth century at the University of Paris as it does today with hindsight.

I. Did the world begin by creation, or has it existed without a beginning through an eternal past?

The ancient Greeks found the idea of a material beginning for the world untenable. But it was not the beginning of the cosmos as such to which they objected – indeed, numerous Greeks thought the cosmos in its present arrangement had a beginning. What they, especially Aristotle, unanimously rejected was the idea that matter itself could have had a beginning.[3] Aristotle, whose arguments on the eternity of the world would be of momentous significance during the Middle Ages, insisted (*De caelo* 1.10.279b.17–20

1. For the creation, see Genesis 1.1–2.1, John 1.2–3 and 17.5; for a description of the end of the world (the heavens and the earth), see Revelation 21.1, Isaiah 65.17 and 66.22, and perhaps Romans 8.19–21.
2. On the basis of the arguments he gave at the end of the first book of *De caelo* (bk. 1, chs. 10–12), Aristotle concludes, at the very beginning of the second book (*De caelo* [Guthrie], 2.1.283b.26–30), "that the world as a whole was not generated and cannot be destroyed, as some allege, but is unique and eternal, having no beginning or end of its whole life."
3. For a discussion of Greek and Christian attitudes, see Sorabji, 1983, 193–283 (chs. 13–17). Aristotle's opinions are described on pages 193–197, 210–213, 226–228, 233, and 276–283.

[Stocks], 1984) that to claim that the world "was generated and yet is eternal is to assert the impossible; for we cannot reasonably attribute to anything any characteristics but those which observation detects in many or all instances. But in this case the facts point the other way: generated things are seen always to be destroyed." Our single world could not have been generated from any prior state of material existence. For if so, then the preceding material state which produced it would have terminated and therefore could not have been eternal; therefore it must have been generated, and we must seek for its generator. In this manner we would be led to an infinite regress of generations and never reach a first beginning. But if the prior state of material existence was of infinite duration, then it could not have been altered to produce our quite different world (*De caelo* 1.10.20–31). Creation from nothing would have made even less sense to Aristotle, who fails to mention, much less discuss, it.

In another important argument, Aristotle insists that anything that is generable and destructible – that is, is subject to being and not-being – is so because it possesses a pair of contrary qualities, such as hot and cold, wet and dry, and so forth. These contrary qualities will cause it to change and become something else. Within our cosmos, only the always-changing region below the sphere of the Moon contains contrary qualities. Such qualities are, however, absent from the celestial region, which is therefore incorruptible.[4] Indeed, the world as a whole lacks contrary qualities that could cause it to pass out of existence (see *De caelo* 1.12.283b.16–20). By coalescing these arguments, Aristotle concluded that the world could not have had a beginning, because it could not have been derived from a prior state of material existence; and he also concluded that it would not come to an end, because the heaven, and indeed the world, do not have contrary qualities that could destroy them. The world must therefore be eternal.[5]

The advent of Christianity eventually changed this. Matter and the world composed of it were assumed to have been created by God out of nothing – *ex nihilo* – and therefore to have had a beginning.[6] For some time Christians sought to repudiate arguments that the universe could not have had a beginning. Eventually John Philoponus, in his *De aeternitate mundi contra Proclum* of 529, abandoned this defensive posture when he attempted to demonstrate that the cosmos must have had a beginning.[7] His arguments, which were repeated and refined in later works, were attacks against Aristotle's defense of a beginningless world and had a powerful effect on both Christian and Islamic authors.[8]

4. For a detailed discussion of the role of contrary qualities in change, see Chapter 10, Section II.1.
5. As will be seen later in this chapter, Aristotle's concept of potential and actual infinites would also influence discussions about the eternity of the world.
6. For a brief discussion of the development of the concept of creation from nothing, see Chapter 5, Section II ("Was Creation from Nothing?").
7. See Sorabji, 1983, 198.
8. For detailed discussions of the problem of eternity and creation, see the references to Sorabji, 1983, in note 3 of this chapter, and Wolfson, 1976, chapter 5 ("Creation of the World").

During the centuries of the early Middle Ages, the issue of the eternity of the world was of little moment, largely because Aristotle's natural books were absent from western Europe. Not until the thirteenth century, when Aristotle's natural philosophy became the basis of the curriculum in the arts faculties of European universities, especially at the University of Paris, did it emerge as a significant problem.

The problem of the eternity of the world was perhaps the most controversial issue at the University of Paris during the thirteenth century.[9] It was to the relations between science and religion in the Middle Ages what the Copernican theory was to the sixteenth and seventeenth centuries and the Darwinian theory to the nineteenth and twentieth centuries. The extent of its significance may be gauged by the Condemnation of 1277, where, among the 219 condemned articles, it was denounced, explicitly or by implication, in at least 27.[10] That large number is undoubtedly a reflection of the numerous guises under which the opinion of the eternity of the world could masquerade. It will be instructive to identify a few of the different kinds of claims for eternity that so provoked Parisian ecclesiastical authorities.

The most sweeping denunciation of the thesis of the eternity of the world is found in the first part of article 87 (Grant, 1974, 48), which specifically condemned the eternity of all species, as well as the eternity of time, motion, matter, agent, and what is effected by an agent.

A common argument sought to establish an equality between past and future: if the one was eternal, so was the other. Thus article 4 condemned the opinion "That nothing is eternal with respect to its end [or goal] which is not eternal with respect to its beginning," and article 98 condemned the claim "That the world is eternal because that which has a nature by [means of] which it could exist through the whole future [surely] has a nature by [means of] which it could have existed through the whole past." Thus, according to Siger of Brabant, Aristotle believed that the intellective, or rational, soul will exist through an eternal future; therefore it must also have existed through an eternal past, because "everything that is eternal into the future is [also] eternal in the past, and conversely."[11]

A number of condemned articles (5, 31, 72, and 80) assumed that certain created immaterial and/or separated substances – such as the human intellect

9. I am here following Grant, 1985b, 77–81.
10. For a list of articles as they are numbered in Denifle and Chatelain, 1889–1897, 1:544–555, see Chapter 3, note 15. Under the heading "Sur l'éternité du monde," Hissette, 1977, 147–160, lists only 10 condemned articles, namely articles 4, 6, 87, 89, 98–99, 101, 200, 203, and 205. He includes articles 5, 9, 12, 31, 48, 52, 72, 80, 90–91, 93–95, 107, 109, 125, and 202 under other headings. For a translation of articles 9, 98, and 107, see Chapter 3, Section III.2. For various lists and translations of the 219 condemned articles, see Chapter 3, note 14. As further evidence of the significance of the question of the eternity of the world, see the "Catalog of Questions," qus. 1–34 (qus. 35–53 are also relevant) in Appendix I.
11. "Omne aeternum in futuro est aeternum in praeterito, et e converso." Quoted from Siger's *De anima intellectiva* (ch. 5) by Hissette, 1977, 150. Siger asserted the same claim in his commentary on Aristotle's *Metaphysics*. However, Siger declares that although Aristotle's opinion is contrary to the truth, he [Siger] has no intention of concealing it. Boethius of Dacia held a similar viewpoint (Hissette, 1977, 150–151).

or celestial intelligences – were eternal because of the immutability of the being that created them and because as actualized substances they could not have been in potentiality to their actualized state. For example, article 31 condemns the opinion "That the human intellect is eternal because it always remains constant and because it has no matter by which it could be previously in potentiality and then in actuality," and article 72 condemns the argument "That because separated substances do not have matter by which they are in potentiality before they are in actuality and because they are always in the same state, they are, therefore eternal."[12] These articles were probably directed against Siger of Brabant, although Siger readily acknowledged that such opinions are contrary to the Christian faith and said that his only purpose in citing them was to present the authentic thought of Aristotle (Hissette, 1977, 80).

Using God's acknowledged immutability, article 48 denied that he could create our world, because "God cannot be the cause of a new act [or thing], nor can he produce something anew." The rationale underlying this claim is that in creating or producing a new effect, say the creation of the world, God would have to take some creative action that would cause him to change, however slightly. Indeed, the second part of article 87 makes this emphatic when it declares that "because the world is [derived] from the infinite power of God, it is impossible that there be novelty in an effect without novelty in the cause."[13]

Finally, we must note article 205, which condemned the opinion "That time is infinite with respect to each extreme [that is, past and future]. For although it is impossible that an infinite have been traversed which was actually traversed [in some finite time], it is however not impossible that an infinite could be traversed which was not traversed [in some finite time]." That is, it follows logically that an infinite could be traversed in an infinite time. Here the authorities may have had in mind Pseudo–Siger of Brabant, who, in his *Questions on the Physics*, cites this argument but offers a solution by observing that one could properly assume the traversal of an infinite number of revolutions, provided that the completion is not assumed to have been achieved in any particular time. In effect, Pseudo-Siger rightly implies that it is logically possible to traverse an infinite series of revolutions in an infinite time, an argument which Aristotle used against two of Zeno of Elea's paradoxes against motion.[14]

12. Article 5 denounces the view "That all separate things are coeternal with the First Principle [God]," and article 80 condemns the judgment "That everything which lacks matter is eternal because what was not made by a transmutation of matter was previously not in potentiality; therefore it is eternal." See also Hissette, 1977, 78–82, 208.
13. For articles 48 and 87, see Grant, 1974, 48.
14. See Pseudo–Siger of Brabant [*Physics*, bk. 8, qu. 6], 1941, 198, 201; for Aristotle, *Physics* 8.8.263a.11–15. Hissette, 1977, 55, also discusses article 205. Peter of Auvergne has been suggested as the possible author of the *Questions on the Physics* attributed to Siger of Brabant by Philippe Delhaye and cited under "Pseudo-Siger" in this study. For more on this, see Grant, 1981a, 267, n. 4 and Lohr, 1972, 345 (Lohr's "Note").

II. Bonaventure's defense of temporal creation and rejection of creation from eternity

The inclusion in the Condemnation of 1277 of numerous articles on the eternity of the world was probably due in no small measure to Saint Bonaventure, who was convinced that the temporal creation of the world was demonstrable. Like John Philoponus before him, Bonaventure attacked what he saw as inherent absurdities in the concept of the infinite. Article 101 may serve as the focus of Bonaventure's reaction. It condemns the opinion "That an infinite [number] of celestial revolutions has preceded, which it was not impossible for the first cause [that is, God] to comprehend, but [which is impossible of comprehension] by a created intellect."[15] Among Bonaventure's numerous arguments against eternity and in favor of a creation beginning in time, two may be seen as directly relevant to article 101: the impossibility of traversing an infinite, and the impossibility of adding to an infinite.[16]

For Bonaventure, reason and self-evident notions were sufficient to destroy these two unavoidable aspects of the infinite. Against the first,[17] he argues that if the world had no beginning, then an infinite number of celestial revolutions must have occurred to the present. But an infinite number of revolutions cannot be traversed, therefore the present revolution could not have been reached. Bonaventure then tries to anticipate objections. If one argues that a numerically infinite number of revolutions has not been traversed, because there was no first revolution in our beginningless universe, then either a particular revolution could have infinitely preceded today's revolution, or none could. If none, then all past revolutions are distant from the present one by a finite number of revolutions, however large the number, and the world must have had a beginning. But if a particular revolution is infinitely distant, Bonaventure asks "whether the revolution immediately following it is infinitely distant. If not, then neither is the former (infinitely) distant since there is a finite distance between the two of them. But if it (i.e., the one immediately following) is infinitely distant, then I ask in a similar way about the third, the fourth, and so on to infinity. Therefore, one is no more distant than another from this present one, one is not before another, and so they are all simultaneous," an absurd consequence.[18]

In the second argument, Bonaventure insists that if an actual infinite

15. Grant, 1974, 49 (I have changed "have" to "has" and "which are" to "which is"). Hissette, 1977, 157, found no identifiable source for this article. But there is no doubt that Bonaventure had something like it in mind in his attacks on the eternity of the world.
16. The ultimate source of Bonaventure's arguments is John Philoponus in the sixth century. The latter's ideas were known to the Arabs and passed into the Latin West via translations from Arabic to Latin (see Sorabji, 1987, 167).
17. From Bonaventure [*Sentences*, bk. 2, dist. 1, pt. 1, art. 1, qu. 2], *Opera*, 2:19–25, as translated in Bonaventure, 1964, 105–113.
18. Bonaventure, 1964, 108. Because "he was not aware of all the subtleties of infinite sets that were revealed last century by Bolzano, Cantor and others," Whitrow, 1978, 40, holds that Bonaventure was "definitely mistaken" on this argument.

number of revolutions has occurred to the present, then all additional revolutions will have to be added to it. But adding to an infinite cannot make it larger, because "nothing is more than an infinite." To further illustrate the absurdity of the infinite, Bonaventure compares the revolutions of Sun and Moon in a world that has an infinite past. The latter will have made twelve times as many revolutions as the former. Therefore the Moon's infinite will be larger than the Sun's, which is impossible.[19] Because of such apparent absurdities, Bonaventure and many Augustinians rejected the possibility of an eternal world with its infinite past time and its infinite distances to be traversed.[20]

There were interesting responses to Bonaventure's arguments. One response, against his first, concerning the traversal of an infinite distance, was that in a beginningless universe there could be no first day from which to begin the measurement. Thus one could assume that an infinite sequence of days has intervened to the present and further assume that more days can be added indefinitely. It can work backward as well. An infinite sequence of days can be imagined to extend backward in time, since there is no first day to halt the sequence. The justification for this approach lies in the concept of a potential infinite, which will be described shortly.

A solution to the second of Bonaventure's puzzles was presented in 1344 by Gregory of Rimini in his lectures on the *Sentences* in Paris, when he ingeniously showed that one infinite could indeed be larger than another or serve as a whole with respect to another infinite multitude. In sum, he had the concept of an infinite set and infinite subset, as is the case with the Moon's infinity of revolutions as compared to that of the Sun.[21] But Gregory's solution was little understood or appreciated.[22] Other solutions were also offered. Averroës argued that although the Sun makes thirty annual revolutions to every one for Saturn, it does not follow that if the world were eternal the infinite distance traversed by the Sun would be thirty times the infinite distance traversed by Saturn. Indeed, there is no ratio between

19. Whitrow, ibid., note 1, explains that although Bonaventure rejected this conclusion, his "argument anticipates the modern idea that an infinite set, unlike a finite set, can be put in one-one correlation with a sub-set of itself." In truth, Bonaventure ought not to be credited with anticipation of this important concept. Long before him, the Stoics had clearly expressed the idea when they declared that "Man does not consist of more parts than his finger, nor the cosmos of more parts than man" (Sambursky, 1959, 97). During the Middle Ages, this concept was expressed by Roger Bacon (not Bonaventure) and others (for references, see Oresme, *De prop. prop.*, 1966b, 42, n. 54). As will be seen shortly, in the fourteenth century Gregory of Rimini expressed the same idea in a specific discussion on the infinite.

20. Whitrow, 1978, 41–43, offers a "disproof of the possibility of an infinite sequence of discrete past events" and concludes (43) that "just as every future sequence of discrete events . . . will always be finite, . . . so in every past sequence the total number of events, however large, can never be infinite." This was, of course, what Bonaventure and like-minded colleagues had argued, though in ways very different from Whitrow.

21. For Gregory's resolutions and other aspects of the problem of infinites, see Murdoch, 1969, 222–224.

22. Ibid., 224.

them at all, since they are not wholes but only potential infinites, that is, "they have neither beginning nor end."[23] Such indefinite entities are not comparable, although their finite parts are.

Some fourteenth-century scholastics, such as Nicole Oresme and Albert of Saxony, also denied that infinites could be compared with respect to "equal," "greater than," or "less than."[24] Jean Buridan too denied the comparability of infinites. In what could be taken as a reply to Bonaventure, Buridan mentions the argument of those who reject the possibility of an infinite time because the absurd consequence would follow that one infinite could be twice or many times greater than another. He cites the revolutions of the Sun and Mars over an infinite past time. Since the Sun makes two revolutions for every one made by Mars, over an infinite past time the Sun should have made twice as many revolutions as Mars. Consequently one infinite would be twice as great as another, which is absurd.[25] Buridan rebuts this claim by observing that "elsewhere it has been said that there are no more parts in the whole world than in a millet seed. And so I say that the traversal of an infinity of days does not exceed the traversal of an infinity of years; but however great a finite time we take, days will be proportional to years and revolutions of the moon to revolutions of the Sun, and so with other comparisons."[26] Bonaventure's absurdity vanishes.

Those who disagreed with Bonaventure's arguments and wished to allow for at least the possibility of an infinite past time or of an infinitely enduring motion resorted to Aristotle's distinction between an "actual infinite" and a "potential infinite." Aristotle rejected the existence of actual infinites. There could be no actually infinite number or magnitude. But the infinite does exist potentially. "The infinite," Aristotle declares, "has this mode of existence: one thing is always being taken after another, and each thing that is taken is always finite, but always different."[27] As Richard Sorabji has explained it (1987, 168), Aristotle criticized his predecessors because "they thought of infinity as something which is so all-embracing that it has nothing outside it. But the very opposite is the case: infinity is what always has something outside it." Bonaventure's attacks on the infinite seem to associate him with the first way, wheras most medieval scholastics would follow Aristotle and the second way.

During the Middle Ages, Aristotle's actual infinite was described as a

23. Averroës offers this argument in his *Tahafut al-Tahafut*, "First Discussion," pars. 16–19 (see Averroës, 1954, 1:8–10). The argument is formulated against al-Ghazali and is summarized in Wolfson, 1976, 430–431.
24. Murdoch, 1969, 222; see also Murdoch, 1982, 570, where he says much the same thing about Gregory, Oresme, and Albert of Saxony. For Oresme's discussion, see Oresme, *Le Livre du ciel*, bk. 1, ch. 33, 1968, 236, 237.
25. Buridan [*De caelo*, bk. 1, qu. 10], 1942, 44–45.
26. Ibid., 47–48. John of Jandun ([*De coelo*, bk. 1, qu. 29], 1552, fols. 19v, col. 1 and 19v, col. 2–20r, col. 1) came to a similar conclusion in comparing the revolutions of the Sun and the Moon. This argument is similar to that of Averroës, cited in the immediately preceding paragraph.
27. Aristotle, *Physics* 3.5.206a.27–29 [Hardie and Gaye], 1984. See also Sorabji, 1987, 168.

quantity "so great that it could not be greater" (*tantum quod non maius*), whereas the potential infinite was described as a quantity "not so great that it could not be greater" (*non tantum quin maius*).[28] The former was called a "categorematic infinite," the latter a "syncategorematic infinite."[29] Jean Buridan shows how all these elements could be brought together to allow for a potentially eternal motion and time. He first explains that he is taking the terms "eternal" (*eternus*) and "infinite" (*infinitus*) syncategorematically, "because according to Aristotle it could be said that there is no motion so great that it could not be greater, nor any time but that it could not be greater; and with respect to faith [as well] there is no time but that it could not be greater, and similarly for motion. Therefore both time and motion can exist into infinity and perpetuity. Therefore a finite motion can be infinite because it cannot be so great a finite [motion] but that it could not become greater."[30] An infinite is therefore traversible because one can always add to it.

III. On the philosophical and theological reconciliation of creation and eternity

Thus Buridan and many other scholastics of the late Middle Ages allowed for the possibility that time, motion, and the world might be eternal in the syncategorematic sense just described. But even as they allowed for the possibility of an eternal world, many if not most scholastics, such as Thomas Aquinas and Godfrey of Fontaines in the late thirteenth century, were convinced that neither the creation of the world nor the eternity of the world was a demonstrable proposition.[31] Indeed, each was equally probable. "That the world had a beginning," Thomas insisted, "is an object of faith . . . not of demonstration or science."[32] Although time, motion, and other entities might possibly be eternal, how could such eternal existents be reconciled with the Christian dogma of a divine creation?

28. From Murdoch, 1982, 567.
29. Ibid.
30. "Hec ergo secunda conclusio patet capiendo hec nomina eternus et infinitus sincathegoreumatice quia secundum Aristotelem diceretur quod non esset tantus motus quin erit maior nec tantum tempus quin erit maius; et secundum fidei veritatem non potest esse tantum tempus quin possit esse maius et sic de motu. Igitur in infinitum et perpetuum possunt esse et tempus et motus; igitur infinitus potest esse motus finitus quia non potest esse tantus finitus quin possit esse maior finitus." Buridan [*Physics*, bk. 8, qu. 3 ("Utrum sit aliquis motus eternus")], 1509, 111v, col. 1. In the text, "quin" is spelled "quim."
31. Thomas argues that only probable or sophistical arguments can be given for either side (see Thomas Aquinas [*Sentences*, bk. 2, dist. 1, qu. 1, art. 5], 1929–1947, 2:33). Godfrey says much the same thing (*Quodlibet II*, qu. 3: "Utrum mundus sive aliqua creatura potuit esse vel existere ab aeterno," 1904, 80). Also see Wippel, 1981, 167–168.
32. Thomas Aquinas, *Summa theologiae*, pt. 1, qu. 46, art. 2, as translated in Thomas Aquinas, 1964, 66. Thomas also rejected alleged demonstrations for the eternity, or beginninglessness, of the world. Boethius of Dacia and Siger of Brabant held similar opinions (Grant, 1985b, 78–80).

At least three ways were devised to allow for both a creation of the world and its eternity. The first assumed the eternity of the matter from which the world was constituted but allowed that, in its present configuration, the world was created. This was clearly contrary to the Christian faith and was condemned in articles 107 and 202.[33]

A second assumed that time began when God created the first motion and that eternity is equivalent to the whole duration of time. Thus, according to the Franciscan theologian Alexander of Hales, "the world may be said to be eternal in the sense that its motion is commensurate with the total duration of time, which has a beginning but not an end, since its existence is only from the will of the Creator."[34] Or, as Roger Bacon would express it, if one understands " 'eternity' as the whole extension of time from the beginning of the motion of the sky, which could be made perpetual by the divine will, one can posit that the world is eternal because there was not a time in which there was not motion, as he [Aristotle] himself argues in book 8 of the *Physics*."[35] But the dilemma for Christians concerning Aristotle's arguments about the eternity of the world would not vanish by such verbal conjuring. For the real issue of the world's eternity would center on the possibility of a beginningless world, not on a world in which time began with the creation and was assumed to exist through an eternal future.

The third way, historically the most important, derived from the conviction that no demonstrative proof for a beginning of the world had been, or could be, formulated. This did not, however, signify that the world was uncreated. In his treatise *On the Eternity of the World* (*De aeternitate mundi*), Thomas Aquinas argues that God could have willed the existence of creatures – and therefore the world – without a temporal beginning. That is, a thing may have been created by God and yet have had eternal existence.[36] Thomas insists that "the statement that something was made by God and nevertheless was never without existence . . . does not involve any logical contradiction."[37] Because of his absolute power, God, as an efficient cause of the world, "need not precede His effect in duration, if that is what He Himself should wish."[38] God can achieve this because he produces effects instantaneously and so could have created an eternally existent world.[39] The

33. Article 107: "That the elements are eternal. However, they have been made [or created] anew in the relationship which they now have" (Grant, 1974, 49). Article 202: "That the elements have been made in a previous generation from chaos; but they are eternal" (ibid., 50).
34. See Dales, 1984, 172. Dales traces this interpretation from William of Conches in the twelfth century to the 1260s and 1270s.
35. Ibid., 175. Dales's translation is from Bacon's *De viciis contractis in studio theologie*.
36. Thomas Aquinas, 1964, 20.
37. Ibid., 23.
38. Ibid., 20. Godfrey of Fontaines assumed much the same thing. See *Quodlibet II*, qu. 3 ("Utrum mundus sive aliqua creatura potuit esse vel existere ab aeterno"), 1904, 72–73.
39. Wippel, 1981, 21–37, argues that in the *De aeternitate* Thomas thought a universe without a beginning was possible, although in earlier works he had assumed such a universe neither proven nor disproven. See also Sorabji, 1983, 197, n. 32. Weisheipl, 1974, 288, conjectures that in the early spring of 1270 Thomas wrote the *De aeternitate* against John

mutable world, however, is totally dependent on an immutable God, thus guaranteeing that the former cannot be coequal with the latter. With one possible exception, no article condemned in 1277 seems to have been designed specifically to denounce this point of view.[40] And yet its status may have been doubtful in light of the clear intent of the bishop of Paris and his colleagues to condemn the thesis of the eternity of the world in all its manifestations.

Despite the condemnation of the eternity of the world in 1277, the idea that it was not a logical contradiction to hold that the world could have existed from eternity and also have been created was surprisingly popular during the Middle Ages and Renaissance. Thomas's clearly stated concept that an eternal though created universe was possible because a cause need not necessarily precede its effect was accepted by numerous scholastic authors, including Marsilius of Inghen, and later by Galileo,[41] the Coimbra Jesuits (Conimbricenses),[42] Franciscus de Oviedo,[43] and others. Not only did scholastic authors discuss whether the world could have existed from eternity, but they proposed numerous other questions that suggested the possibility of an eternal world. For example, they asked whether the generation of human beings and other creatures could have occurred through an eternal past; or whether motion, time, or other successive things could have existed from eternity.[44] On all of these questions, scholastic opinion was divided. The literature is large, and the numerous arguments cannot

Pecham, who had opposed the concept of a world that was coeternal with its immutable creator.

40. The possible exception is article 99, the first part of which condemns the view "That the world, though it was made from nothing, was not, however, made anew" (Grant, 1974, 49). This allows for a creation that is eternal.

41. Galileo [*De caelo*, qu. 4 (F)], 1977, 55–56. Galileo seems to have been somewhat ambiguous in his treatment of the problem. God, who "has been omnipotent from eternity," could indeed have willed the existence of the universe from eternity, although, under these circumstances, the world would not have possessed a perfection in the same sense that God's eternity is a perfection. This follows because the world's temporal perfection "would . . . have always been dependent on the divine duration." But then Galileo seems to depart from Thomas and argue that with respect to the creatures themselves, whether they are "corruptible or incorruptible, permanent or successive, it is impossible that the universe could have existed from eternity." Why? Because God, who created the world from nothing, must have preceded it in duration. "Since there is eternity in God, however, it implies [a contradiction to make him coeternal with creatures]." Thus Galileo seems here to abandon the idea that God need not temporally precede any effect he creates.

42. Conimbricenses [*Physics*, bk. 8, ch. 2, qu. 6, art. 2], 1602, cols. 446–449. On the Conimbricenses and their commentaries, see Appendix II, note 2.

43. Oviedo [*Physics*, bk. 8, contro. 19, punc. 1 ("An creatura permanens potuerit esse ab aeterno")], 1640, 445, col. 1–447, col. 1. Among those who believed that God could create a creature, including a world, from eternity, Oviedo cites (447, col. 1) Thomas Aquinas, Francisco Suarez, the Coimbra Jesuits, Hurtado de Mendoza, and Roderigo de Arriaga.

44. For the range and types of questions, see Part I, questions 1–20, of my "Catalog of Questions" in Appendix I. Franciscus de Oviedo is an example of a scholastic author who took up these basic questions. Thus in his *Physics*, book 8, *punctum* 1 (each *punctum* is really a question), 1640, he asks "Whether a permanent creature could exist from eternity" (445, col. 1–447, col. 1); in *punctum* 2, he inquires "Whether there could be a series of generations from eternity" (447, col. 1–448, col. 1); and in *punctum* 3, he considers

be described here. What emerges, however, is the fact that there was a considerable body of scholastic opinion that upheld the possibility that not only was the world created but that it might also be eternal. In medieval parlance, they held with Thomas Aquinas that the possibility of an eternal world was a *problema neutrum*, whereby logical argument and evidence favored neither a temporal beginning of the world nor the absence of a beginning.

But at least one scholastic would go further. During the fourteenth century, Marsilius of Inghen went beyond Thomas and argued that an eternal world was probable.[45] Basing his arguments on theories of motion and time, Marsilius defends as probable Aristotle's view that the world and motion are eternal. He does this in a series of eight arguments in which it is assumed that the prime mover, God, is absolutely immutable. Let us summarize a few of these arguments that were so typical among those that were used to justify eternity.

The first assumes that either God could create the world from eternity or not. If he could, then he either wished to create it from eternity or not. If he wished to create it from eternity, and since we have assumed that he was capable of so doing, it follows that he created it from eternity and therefore did not create it anew (that is, again). But if he did not wish to create the world from eternity, but afterward wished to do so and did so, it follows that God changed from not wishing to create the world to wishing to create it. But this is contrary to the standard assumption that God is immutable. Should it be said, however, that he could not create the world from eternity but afterward could, then it follows that God suffered a mutation.[46] On this argument, it seems more probable that God created the world from eternity.

In the fifth argument, Marsilius says that if the world and motion were created, that is, began anew (*de novo*), then it is true to say that the world and motion exist "now" (*nunc*) but did not exist before (*ante*). But the term "before" reflects a difference in time. Therefore, before this "now," time existed and consequently so did motion (presumably because, in Aristotelian terms, time is the measure of motion). Therefore the first motion associated with the creation of the world was not the first motion.

The seventh argument appeals to the priority of something with respect to its nature. The world could exist from eternity because of the following: if something, by its very nature, is prior to another thing, it does not follow that it must therefore exist without the existence of the other thing to which it is prior in nature. But having assumed that God is eternal and that the world has been created, it is necessary that God be prior without the ex-

"Whether motion, time, or another successive being could exist from eternity" (448, col. 1–451, col. 1).

45. Marsilius of Inghen [*Physics*, bk. 8, qu. 1], 1518, 79r, col. 2–80r, col. 2. On the doubt concerning Marsilius's authorship, see this volume, Appendix II, note 1.

46. Marsilius of Inghen, ibid., 79v, col. 1. For a similar argument, see Conimbricenses, *Physics*, bk. 8, ch. 2, qu. 4, art. 2, 1602, col. 436.

istence of the world. Therefore God is prior to the world with respect to nature and therefore must be prior in time. But this does not follow, because the Sun is prior to its light with respect to nature, but it does not follow from this that the Sun exists without its light.

Although Marsilius considers these and similar arguments to be inconclusive, he judges them probable on the basis of natural principles; indeed, more probable than arguments in favor of a beginning of the world. Moreover, they are not contrary to the faith.[47] Marsilius's interpretation is significant because it represents the ultimate reconciliation of Aristotle and the Christian faith. It argued for the probability of a beginningless, eternal world, thus supporting Aristotle against those who denied eternality, and it simultaneously upheld a created world in support of the faith. No Christian could go further.

In the fifteenth century, Johannes Versor made an important distinction between philosophers and theologians. Both groups, he explains, agree that God produces the heavens by a simple flow from himself rather than by some series of successive motions.[48] They disagree, however, as to a beginning of the world. The philosophers hold that God created the heavens – and therefore the world – coeternally with himself, whereas the theologians say that "the heavens are produced in some determinate beginning of time."[49] Since, as we have seen, Thomas Aquinas and other theologians accepted the possibility of the coeternity of the world with God, Versor misrepresents (perhaps inadvertently) the differences between philosophers and theologians when, without qualification, he attributes to the philosophers the view that the world is coeternal with God. Most indicated that it was only possible, or at best probable.

But Versor's concern about the reconciliation between the philosophical and theological positions was real, and many authors made an effort to bring the two viewpoints into harmony. Indeed, this is precisely what Thomas had attempted. But how was the reconciliation effected by scholastics who chose a different tactic than his, namely those who believed in the possibility of the world's eternity as an uncreated and ungenerated entity, as Aristotle had described it, and who also accepted on faith the supernatural creation of the world in time? Within this group of scholastic authors, one of the most interesting attempts was formulated by Boethius of Dacia, an arts master and contemporary of Thomas's. In his *Questions on the Physics*, Boethius inquires "whether prime matter was made anew [that is, cre-

47. "Notandum quod iste rationes sunt probabiles et evidentes satis suppositis principiis naturalibus nec alique sunt ita probabiles ad oppositum. Non tamen sunt demonstrative, nec similiter concludunt aliquid contra fides." Marsilius, ibid., bk. 8, qu. 1, 79v, col. 2.

48. "Dicendum quod tam philosophi quam theologi ponunt celum a deo productum non per motum successivum, sed per simplicem effluxum a primo principio a quo processit totum esse rerum." Versor [*De celo*, bk. 1, qu. 6], 1493, 4v, col. 2.

49. "Philosophi dicunt deum produxisse celum sibi coeternum; theologi vero dicunt celum esse productum in aliquo determinato principio temporis." Ibid., 5r, col. 1.

ated]."[50] He decides, in a manner similar to that of Aquinas, that prime matter is coeternal with God but that the latter had nonetheless created it. The eternity of prime matter was for Boethius of Dacia a conclusion that followed logically from the application of reason (*per rationem*).

In his *De aeternitate*, Boethius accepted a creation, although, like Thomas, he believed that human reason was incapable of demonstrating the creation of the world or its eternity. Despite its indemonstrability, however, the creation of the world was an article of faith. Indeed, if it were demonstrable it would no longer be a matter of faith but of science. "There are many things in the faith," Boethius insists, "which cannot be demonstrated by reason, as [for example] that a dead person comes to life again numerically the same as he is now, and that a generable thing returns without generation. And who does not believe these things is a heretic; [and] whoever seeks to know these things by reason is a fool."[51]

Boethius of Dacia's approach differed from that of Thomas. Boethius assumed that faith demanded an interpretation of the creation of the world as a temporal act, not a creation in which the world was coeternal with the creator. Despite this commitment, Boethius also believed that reason would lead one to assume an eternal world, an assumption that was forbidden by faith. Between the two options, Boethius felt compelled to accept the supernatural creation of the world in time, or with time,[52] as the only true response.

Siger of Brabant argued in a similar fashion. The world and its species cannot have been created, because no species of being could be actualized from a previous state of potentiality. Consequently, all species must have existed previously.[53] Therefore, the world could not have been created. To protect himself against charges of heresy, Siger justified his conclusions by statements such as this: "we say these things as the opinion of the Philosopher [Aristotle], although not asserting them as true."[54] Where the pronouncements of faith conflicted with Aristotle's conclusions, the former were assumed true. Under such circumstances, Aristotle's conclusions, which were the product of reason based on sense experience, had to be judged inadequate, even though they were the best that could be achieved by the unaided reason.

This attitude was typical of many scholastics, especially arts masters, of the fourteenth century, who also ignored Thomas's solution of a possible eternal, created world. When Church doctrine conflicted directly with the

50. Boethius of Dacia [*Physics*, bk. 1, qu. 30], 1974, 186. The question extends over pages 186–188. On Boethius, I follow Grant, 1985b, 78–79.
51. Boethius of Dacia, *De aeternitate mundi*, 1976, 355.
52. Strictly speaking, most scholastics would have held that the world was created simultaneously with time, since time depends on the motion of the heavens. Therefore time began with the creation of the heavens. See Michael Scot [*Sphere*, lec. 1], 1949, 255.
53. This is the theme of Siger's *De aeternitate mundi*. See Siger of Brabant, 1964, 12, 91–93.
54. Ibid., 93.

conclusions of Aristotelian natural philosophy – as it did in the question about the eternity of the world – they yielded to theology and faith. But there was always an underlying dilemma. Natural reason seemed to indicate an eternal world, one without a beginning, because there were seemingly no good arguments that could account for a generated world. But faith decreed otherwise. John of Jandun expressed the dilemma quite clearly when he declared ([*De coelo*, bk. 1, qu. 14], 1552, fol. 10v, col. 1):

> It must be understood that although according to nature, the heaven is not generable and corruptible . . . nevertheless it is, in another way, produced by a creation and by a creation without motion. And this is faith and truth. But the philosophers, because they prove production [of things] from sensible things, ignore this mode. But one cannot be convinced of the production of something from nothing from sensible things because it is beyond nature. And they [the philosophers] do not accept these things.

Similarly, Jean Buridan argued ([*De caelo*, bk. 1, qu. 10], 1942, 46) that according to faith, the heaven, and therefore the world, were created supernaturally and could be annihilated in the same manner. But logical analysis of the celestial region indicates that the heaven could not have been generated by natural means nor corrupted by such means. In his *Questions on De celo*, Nicole Oresme expresses a similar opinion when he declares (bk. 1, qu. 10, 1965, 149 [Latin], 150 [English]) that "naturally speaking" (*loquendo naturaliter*) the heaven could not have been generated, because "it is not apparent by what generating agent it would be created nor from what matter nor also by what alteration nor in what manner this would take place." In natural terms, the world should be eternal. Nevertheless, it was created *ex nihilo* by the will of God and cannot have had an eternal motion.

In the fifteenth century, Johannes Versor concluded that the world was ungenerated and incorruptible. It is ungenerated because "by generation Aristotle understood physical generation which presupposes physical matter subject to privation and alteration" ([*De celo*, bk. 1, qu. 14], 1493, 12v, col. 1). From this standpoint, Versor argues, the notion of an ungenerated world is not repugnant to the faith, since the faith assumes a different conception of generation than did Aristotle. Where Aristotle assumed generation from preexisting matter, the faith decrees the generation, or creation, of things and the world itself from nothing.

All attempts by scholastics to reconcile a temporal creation from nothing with Aristotle's strong arguments for the impossibility of a generated world were in vain. Nevertheless, scholastic arguments about the eternity of the world were quite remarkable in a number of respects. First, there is the general conviction of many scholastics who followed Thomas Aquinas that God might have created the world from all eternity. Moreover, many believed that reason and experience alone would lead to the conclusion that the world is eternal, without beginning or end, a conclusion that was reached in the various ways just described. Some even found the eternity of the world a more probable option than its creation. It was in this connection that Pierre Ceffons declared that "Although it is erroneous to say that God is not three and one and that the world had no beginning, it is nevertheless not erroneous to assert that, faith aside, it is more *probable* that God is not three and one or that the world never began than to assert their opposites. For nothing prevents some false propositions from being more probable than some true ones."[55] The straightforward dictates of faith, however, triumphed over reason and probabilistic arguments. Most Christian scholastics accepted the supernatural creation of the world from nothing, even though as Aristotelians they viewed such an action as unintelligible and implausible.

IV. Is the world incorruptible, or will it cease to be?

Intimately linked with the possibility of an ungenerated world was Aristotle's other claim that the world as a whole is incorruptible. As we have seen, Aristotle argued that if the world had no beginning, it could have no end and was therefore indestructible. If scholastics were compelled to accept the beginning of the world by supernatural creation, were they also committed to the rejection of a world that could endure through an infinite future? Opinion was divided on the perpetuation of the world and the form it would take.

Most were agreed that at the very least God, by virtue of his absolute power, could, if he wished, preserve the world through all eternity. They differed as to whether our created world could by its natural powers – the powers God conferred upon it – endure through an eternal future. Roger Bacon, for example ([*Physics*, bk. 8], *Opera*, fasc. 13, 1935, 383), insisted that the world could not endure infinitely by its own powers, but only by the will of God. At the end of the sixteenth century, the Coimbra Jesuits argued in a similar vein. The world could remain forever in its present state by the ordinary powers of conservation which God currently employs.[56] Bacon and the Coimbra Jesuits probably understood by this a world on which God conferred, at its creation, the capacity for infinite duration rather than a world which God had constantly to preserve. Galileo seems to have expressed this viewpoint when he declared ([*De caelo*, qu. 4 (J)], 1977, 97) that "it is more probable that the heavens are incorruptible by nature," citing Aristotle's argument that circular motion has no contrary and that what has no contrary is incorruptible. As further evidence, Galileo observed (ibid., 99) that the speed of the heavens is so great "that it could destroy

55. Translated by Weinberg, 1948, 116–117; also see Sylla, 1991, 232–233.
56. "Posset mundus sibi relictus sub eo statu, quem modo habet, solo communi et ordinario concursu, quo nunc a Deo conservatur, in perpetuum durare." Conimbricenses [*De coelo*, bk. 1, ch. 12, qu. 1, art. 2], 1598, 155.

even the most solid bodies, and since it has endured for such a long time, preserving the stars always at the same distance, opposition and magnitude, . . . it is most certain that the heavens are incorruptible." Thus did Galileo move, within the span of two pages, from the probability of celestial incorruptibility to the certainty of it. Others conceded powers to parts of the world that enabled them to endure eternally. Thomas Bricot argued ([*De celo*, bk. 2], 1486, 13v, cols. 1–2) that something made anew by generation or creation which does not have a contrary can be perpetuated, that is, be eternal. As created entities that lack potentially destructive contrary qualities he names the heavens (*celum*), the world (*mundus*), and light (*lumen*). By contrast, things that have contrary qualities would, in the course of nature, be destroyed. Although individual things that possess contrary qualities would perish and new things would be generated, Bricot holds that the world as a whole could endure forever, since it lacks contraries.[57]

Bartholomew Amicus held much the same opinion but distinguished between things that are generated in the manner described by Aristotle and those that are created anew by God. The former, such as the elements and the compound bodies they form, are destructible, but the latter, among which Amicus includes prime matter, the heavens, and angels, are not.[58] Although the celestial region is incorruptible, whereas the individual bodies of the terrestrial, or elementary, region are corruptible, "The whole world," as Amicus explains, "is not corruptible, because if one considers the nature of things the world cannot be deficient with respect to its species, although individuals of the species vary in their corruptibility."[59] That is, all mundane species of things remain incorruptible, even though the individuals of each species come into being and pass away.

Bricot and Amicus reveal an important trend by arguing that just because Aristotle had shown that everything that begins by generation is corruptible, it does not follow that everything that begins *de novo* by creation is corruptible.

This important challenge to Aristotle had already been effectively made in the fourteenth century by Nicole Oresme, who challenged both aspects of Aristotle's claims about eternity, namely, that whatever had a beginning must have an end; and that what has no end cannot have had a beginning.[60] To refute Aristotle, Oresme assumed the eternity of the world. In an important passage, worthy of citation, Oresme declared:

Aristotle tries to prove that everything, whether substance or accident or any tendency whatsoever which had a beginning, will have an end and will cease of necessity and cannot possibly last forever; and that it is likewise impossible that anything

57. On the Aristotelian and medieval doctrine of contraries, see Chapter 10, Section II.1.
58. Amicus [*De caelo*, tract. 3, qu. 4, dubit. 2, art. 4, conclus. 3–4], 1626, 125, cols. 1–2.
59. Ibid., 125, col. 2.
60. Here I follow Grant, 1978c, 112–113. For Aristotle's discussion, see *De caelo*, 1.10 and 1.12.

which will ultimately perish can always have been there without a beginning. Since this is not true and is, in the first part, against the faith, I want to demonstrate the opposite according to natural philosophy and mathematics. In this way it will become clear that Aristotle's arguments are not conclusive. In the first place I posit with Aristotle, although it is false, that the world and the motions of the heavens are eternal by necessity, without beginning or end.[61]

To achieve his objective, Oresme resorted to his doctrine of the probable incommensurability of celestial and terrestrial motions, a doctrine based on his demonstration that any two unknown ratios are probably incommensurable.[62] Since ratios could represent magnitudes, Oresme extended his demonstration to any two continuous magnitudes of the same kind, such as time or velocity, and assumed that they were also probably incommensurable and would form an irrational ratio.[63] As applied to celestial velocities or motions, probable incommensurability signified that celestial events are inherently unique and nonrepetitive. That is, no two or more celestial bodies could ever be in an identical configuration or relationship more than once. To counter Aristotle, Oresme had only to imagine one or more unique celestial events, each of which terminated one cosmic condition that had existed from all past eternity and which immediately thereafter began a new cosmic condition that would last through all future eternity.

The Sun's motion provided more than one such illustration. Because the Sun's motion as a whole was assumed by astronomers to be the consequence of at least three constituent motions, any one of which is probably incommensurable to the others, it followed that the Sun's center, and therefore the apex of the earth's shadow, which depends on the Sun's position, could never occupy the same celestial point twice. Thus the apex of the earth's shadow will continually and necessarily occupy one point after another in which it had never before been and to which it would never again return. In every point it occupied, then, some portion of the sunlight in the sky, which by assumption had existed there from all eternity and was therefore without a beginning, would be darkened and come to an end. And at the moment when the apex of the shadow shifted to the next point, sunlight would once again shine on the previously darkened place and exist thereafter through all eternity, from which it follows that something which had a

61. Oresme, *Le Livre du ciel*, bk. 1, ch. 29, 1968, 195–197. In the *De proportionibus proportionum*, Oresme declares that when he assumes an eternity of future motion, he is "speaking naturally" (*naturaliter loquendo*), that is, speaking solely in terms of natural philosophy and ignoring the faith and theology. See Oresme, *De prop. prop.*, 1966b, 305–307.

62. For more on Oresme's discussion of celestial incommensurability, see Chapter 18, Section I.3.a.

63. Oresme's most fundamental discussions occur in three treatises: *De proportionibus proportionum*; *Ad pauca respicientes*; and *Tractatus de commensurabilitate vel incommensurabilitate motuum celi* (for the first two, see Oresme, *De prop. prop.*, 1966b; for the third, see Oresme, *De commensurabilitate*, 1971. A summary account appears in the last-named work, 56–76, nn. 89–114. For a description of the demonstration of probable incommensurability, see n. 113, 73–76.

beginning need not necessarily terminate.[64] Oresme offered other examples as well.[65] Indeed, on the universally held assumption that the planets and stars cause physical events in the terrestrial region, Oresme argued the possibility that a unique celestial aspect, say a conjunction or opposition, could cause the existence of a new species never known before, which thereafter might exist through an infinite future.[66]

Oresme's unusual attack on Aristotle's concept of eternity had relatively little influence. Marsilius of Inghen, however, may have known of it, because he agreed with Oresme that if the celestial motions were incommensurable, a new species could be generated that had never existed before and which might thereafter exist through all eternity.[67] However, whereas Oresme was convinced that the celestial motions were probably incommensurable and could therefore produce new effects, Marsilius left no opinion, except to say that experience could not decide the issue.

Scholastics were thus prepared to break with Aristotle and hold that a created world – that is, something that had a beginning – could indeed exist through an eternal future. But if many thought that the created world could endure through an infinite future, there was a considerable body of opinion that believed it would not exist forever in its present form. Already in the thirteenth century, Michael Scot was willing to concede that by natural powers the celestial motions would never cease, but in terms of the ultimate end of the world, which is to attain a state of blessedness, all mundane motions will cease. For as it is written in Isaiah 65, "behold, I create new heavens and a new earth."[68] Thus the old world would cease, and a new

64. Oresme, *Le Livre du ciel*, bk. 1, ch. 29, 1968, 199.
65. Thus he applied similar reasoning to lunar eclipses, showing that the earth's shadow could never return twice to the same point on the lunar surface. See Oresme [*De celo*, bk. 1, qu. 24], 1965, 421–424, and Oresme, *De commensurabilitate*, 1971, 63, n. 97. On the assumption of incommensurability, Oresme further conceived of a perpetual circular motion which had a beginning but no end (*Le Livre du ciel*, bk. 1, ch. 29, 1968, 203). He also imagined the fall of a heavy body through a successively more resistant medium where the motion had a beginning but no end (ibid., 205).
66. See Oresme, *Le Livre du ciel*, bk. 1, ch. 34, 1968, 243, and Oresme, *De commensurabilitate*, 1971, 57, and n. 90.
67. Marsilius of Inghen, *Physics*, bk. 8, qu. 3 ("Whether a new action could arise from an eternal and immutable mover"), 1518, 81r, col. 2. For the Latin text and translations, see Oresme, *De commensurabilitate*, 1971, 127–130. As a younger contemporary at the University of Paris, Marsilius – if he was the author (see Appendix II, note 1) – probably knew Oresme's work, although he does not mention it.
68. Michael Scot, *Sphere*, lec. 1, 1949, 255. The line from Isaiah 65 reads in the Vulgate: "Ecce enim ego creo [Michael has *commovebo*] caelos novos [Michael has *novos celos*] et terram novam." Others who cited Isaiah 65 in the same context are Conimbricenses, *De coelo*, bk. 1, ch. 12, qu. 1, art. 2, 1598, 156; Aversa [*De caelo*, qu. 31, sec. 10], 1627, 39, col. 2; and Amicus, *De caelo*, tract. 3, qu. 4, dubit. 5, 1626, 129, col. 1. Conimbricenses and Amicus cite Isaiah 66. Aversa has Isaiah 65. Also cited in the same context was Apocrypha 21, which, according to the Conimbricenses (*De coelo*, bk. 1, ch. 12, qu. 1, art. 2, 1598, 156), declares, "I saw the new heaven and new earth" (Vidi coelum novum et terram novam), signifying the end of the present world and the creation of a new one. Others who cited the same text are Aversa, ibid.; Amicus, ibid; and Cornaeus [*De coelo*, tract. 4, disp. 1, qu. 2, dub. 4], 1657, 488. I am ignorant of the location of "Apocrypha 21" among the fifteen different works of the Apocrypha.

one would be created to receive the blessed and the elect. Although God could preserve the present world through all eternity,[69] Michael and others – including the Coimbra Jesuits, Raphael Aversa, Bartholomew Amicus, and Melchior Cornaeus – found powerful scriptural support for assuming he would not. With some minor differences, the nature of those changes, which would be made on the Day of Judgment, were fairly well agreed on.

On that day, the Coimbra Jesuits believed that God would cause the intelligences to cease moving the celestial spheres.[70] Therefore time and motion would be no more. Animals, plants, and mixed bodies, which now sustain human life, would cease to exist, although the elements, left unmentioned, would almost certainly continue to exist, but in a more perfect state than before. And as God declared in Isaiah 30, "the moon shall shine with a brightness like the sun's, and the sun with seven times his wonted brightness."[71] Man would reach his greatest perfection and be like the angels.[72]

Raphael Aversa assumed most of these opinions but tended to see a greater continuity between the old and "new" worlds ([*De caelo*, qu. 31, sec. 10], 1627, 38–40). He agreed with the Conimbricenses that the celestial motions would cease and that a new brilliance would shine forth from the heavenly bodies. The heavens and the elements would not be destroyed but only perfected. The cessation of celestial motions would prevent the actions and passions of the elements and negate the action of contraries.[73] Consequently no generations or corruptions could occur. Aversa departed from the Conimbricenses (and Amicus) by allowing that some mixed bodies would continue to exist forever without suffering any change. Because he saw the structure of the universe remaining the same on Judgment Day, despite significant changes, Aversa interpreted the scriptural passages just cited (Isaiah 65 and Apocrypha 21), which declared that a new heaven and a new earth would be created, as reflective of accidental, rather than substantial, changes.

Would the conservation powers that God had applied to the old world prove adequate to the new, or altered, world? Amicus ([*De caelo*, tract. 3, qu. 4, dubit. 5], 1626, 129, col. 2) thought not. Without a different kind of divine assistance, to sustain and preserve the world altered on Judgment Day, chaos would intrude, because in a motionless celestial region the Sun's rays would always heat only one part of the world, while the other side would always be cold. God would have to take this into account and establish laws and rules that would enable such a world to exist forever.

69. The Coimbra Jesuits, for example, held that God could, if he wished, preserve the present state of the world forever. For the Latin text, see note 56 of this chapter.
70. Conimbricenses, *De coelo*, bk. 1, ch. 12, qu. 1, 1598, 156–157.
71. Translation from the *New English Bible*, 1976.
72. Amicus shared all these opinions (*De caelo*, tract. 3, qu. 4, dubit. 5, 1626, 129, cols. 1–2).
73. On the action of contraries, see Chapter 10, Section II.1.

Amicus appears to have interpreted the degree of difference between the worlds before and after Judgment Day as sufficiently great to warrant new divine regulations for the eternal perpetuation of the altered world. Does this imply that he further believed that the prejudgment and postjudgment worlds were substantially different? If so, he would have differed radically from his contemporary, Aversa, who, as we saw, viewed the differences between those worlds as only accidental. Amicus and Aversa seem to represent the two basic interpretations toward which scholastic opinion gravitated.

V. Medieval ambivalence

The debate over the eternity of the world concerned both ends of the temporal spectrum. At the front end, so to speak, the question posed was whether the world was a temporal creation or was without a beginning. At the back end, the question focused on whether a supernaturally created world could endure forever and, if so, whether God intended to let it do so. Conventional views of scholastics prepare us for a simple, direct response to these questions: the world was created in time and would come to an end on the Day of Judgment. The actual responses to these difficult questions, however, subvert our ordinary expectations. We are surprised to find a considerable sentiment for a world that might be logically conceived as coeternal with God or for one that would have been without a beginning, "naturally speaking," if God had not created the world supernaturally. It seems no exaggeration to suggest that medieval natural philosophers sought to have both faith and Aristotle simultaneously. The outcome of this ambivalence generated the kinds of arguments that we have examined in this chapter.

5

The creation of the world

On one aspect of the creation account, Genesis is explicit: the world was created in six days, and God rested on the seventh. Despite the apparent straightforwardness of the account of a six-day creation, numerous commentators on Genesis, beginning with the commentaries of Philo Judaeus (in Greek) and Saint Augustine (in Latin), imposed radically different interpretations on the seemingly plain text. The widespread medieval belief in syncategorematic infinites and in the possibility of an eternal and created world, described in Chapter 4, may already have alerted readers to expect something other than a literal exposition.

I. Was creation simultaneous, in six days, or both?

The various interpretations of the creation may be reduced to three basic types, all deriving from the first five centuries of Christianity. The first was a literal interpretation in which God was assumed to have created the world in six successive, natural days of twenty-four hours. The second assumed that the world and all the things in it were created simultaneously and instantaneously. The third opinion combined the first two by assuming the creation of an elementary, unformed matter at the beginning which was subsequently formed into our world over a period of six days.[1] We must now investigate the manner in which scholastic natural philosophers and theologians interpreted the six days of creation.

The departure from the literal account was given Church sanction in the first of seventy canons that issued from the Fourth Lateran Council, presided over by Pope Innocent III in 1215. In the relevant canon, the faithful were told that thereafter they were to believe that God was "the creator of all visible and invisible things, spiritual and corporeal, who, by His omnipotent power created each creature, spiritual and corporeal, namely angelic and mundane, at the beginning of time simultaneously [*simul*]) from nothing; and then [*deinde*] made man from spirit and body."[2]

1. My tripartite division follows that of William A. Wallace, tr., in Thomas Aquinas, *Summa theologiae*, 1967, vol. 10, app. 7: "Hexaemeron: Patristic Accounts," pp. 203–204.
2. "Pater generans . . . creator omnium invisibilium et visibilium, spiritualium et corporalium, qui sua omnipotenti virtute simul ab initio temporis utramque de nihilo condidit creaturam, spiritualem et corporalem, angelicam videlicet et mundanam, ac deinde humanum quasi communem ex spiritu et corpore constitutam." Hefele, 1913, 1324.

Although the council of 1215 seems at first glance to have proclaimed a simultaneous creation for everything, both corporeal and spiritual, it then spoke of the advent of man as a subsequent creation, in which "spirit and body" were combined. The proclamation of a simultaneous creation at the council indicates not only the probable influence of the book of Ecclesiasticus but also that of Saint Augustine. In the former we are told: "He that lives forever created all things together [*simul*]" – that is, at the same time, or instantaneously.[3] Was there not a conflict between Ecclesiasticus and Genesis, which describes a creation spread over six days, not one made simultaneously? Augustine took note of the seeming dilemma when he declared that "In this narrative of creation Holy Scripture has said of the Creator that He completed His works in six days; and elsewhere, without contradicting this, it has been written of the same Creator that He created all things together."[4] What exegesis did Augustine employ to avoid what seems a straightforward contradiction between a simultaneous creation and one extending over six days? Augustine explained that God did indeed create all things simultaneously but chose to narrate the creation day by day, because "those who cannot understand the meaning of the text, *He created all things together*, cannot arrive at the meaning of Scripture unless the narrative proceeds slowly step by step."[5]

But it was not merely for our convenience that the scriptural narrative specifies six days. Augustine insisted that God not only created all things simultaneously but also in six days – that is, "the creation of things took place all at once" but there was also a "before" and "after."[6] For despite the simultaneity, God followed the order described in Genesis. To illustrate how this might be envisioned, Augustine invokes the rising Sun. Although we see the rising Sun in a virtually instantaneous moment, the ray that goes from our eyes to the Sun passes over all the intervening spaces – that is, passes over things in a certain order, nearer things first and then more remote things until it reaches the Sun. And so it was with the creation of the world. All things were created in the order described in Genesis, but in an instant, so quickly that "before" and "after" were indistinguishable.[7]

3. Ecclus. 18.1: "Qui vivit in aeternum creavit omnia simul." From the Latin of the Vulgate Bible.
4. Augustine, *Genesis*, bk. 4, ch. 33, 1982, 1:142. In a note to this passage, the editor (Taylor) explains (1:254, n. 69) that "the word *simul* ('at one time,' 'all together') in the Latin version seems to be a mistranslation of the Greek κοινη ('commonly,' 'without exception'). A more accurate translation, therefore, would probably be: "He who lives forever created the whole universe."
5. Augustine, ibid.
6. Ibid., ch. 34, 1:143–145. The title of chapter 34 is "All things were made both simultaneously and in six days."
7. God effected a simultaneous creation by first creating seeds, or *rationes seminales*, of everything all at once and then allowing each of them to develop later at different times. For example, in *The Literal Meaning of Genesis* (bk. 7, ch. 28), 1982, 2:30–31, Augustine declares: "I have been let to hold that God first created all things simultaneously at the beginning of the ages, creating some in their own substances and others in pre-existing causes. Hence the all-powerful God has made not only what is existing at the present, but also what is

Of the three basic interpretations of the creation that I have mentioned, Augustine's seems to approximate most closely to the third. All things were created simultaneously in the seeds, or *rationes seminales*, and these came into being in the order described in Genesis.

As is obvious, the Fourth Lateran Council did not rigorously follow Ecclesiasticus or Saint Augustine. In the council's statement, the creation of man is said to have followed the creation of everything else. Without mentioning six days, this statement did at least allow for successive creations: the first, of everything except man; the second, of man. Indeed, it could even be taken as supporting creation in six days, as it was by Pedro Hurtado de Mendoza.[8] Perhaps because of its vagueness, the council's brief statement about creation did not become doctrine, and various opinions were tolerated.[9]

Augustine's conception of a simultaneous creation of all things was probably the most widely held opinion on creation during the Middle Ages. Theologians of the stature of Peter Lombard, Alexander of Hales, Saint Bonaventure, and Thomas Aquinas supported it. Although Peter Lombard seems to have upheld simultaneous creation, he does report that although Augustine held that belief, others (such as Gregory, Jerome, and Bede) assumed that God first created a crude matter composed of a mixture of the four elements from which the different kinds of corporeal things were formed according to their proper species.[10]

to be." Frederick Copleston explains (1957, 77) that "In this way, God created in the beginning all the vegetation of the earth before it was actually growing on the earth, and even man himself. He [Augustine] would thus solve the apparent contradiction between Ecclesiasticus and Genesis by making a distinction. If you are speaking of actual formal completion, then Ecclesiasticus is not referring to this, whereas Genesis is: if you are including germinal or seminal creation, then this is what Ecclesiasticus refers to." Augustine describes the *rationes seminales* in his commentary on Genesis, books 5 and 6. From the time of Augustine, Ecclesiasticus 18.1 was frequently cited in contrast to Genesis. For example, see Thomas Aquinas, *Summa theologiae*, pt. 1, qu. 74, art. 2, 1967, 10:154 (Latin).

8. Hurtado de Mendoza [*De coelo*, disp. 3, sec. 1], 1615, 376, col. 2; 377, col. 2; 378, col. 1. Hurtado argues that the relevant text from the Fourth Lateran Council supports the opinion that "the world was produced successively in six natural days" (377, col. 2) and was not simultaneous and instantaneous. As evidence of a successive creation, he cites the use of the adverbs *simul* and *deinde* in the council's text. These terms, he argues, "always signify the priority and posteriority of time." Thus Hurtado took the separate creations as equivalent to a successive, six-day creation and thus completely ignored the significance of *simul*.
9. The description of creation issued by the Fourth Lateran Council was frequently quoted in the seventeenth century in a number of contexts. Thus Mastrius and Bellutus [*De coelo*, disp. 1, qu. 5, art. 1], 1727, 3:484, col. 2, par. 40, declare that the temporal creation of the world "is had from ch. 1 of Genesis and is defined in the Lateran Council and in the chapter *Firmiter*"; Franciscus Bonae Spei [comment. 3, *De coelo*, disp. 3, dub. 3], 1652, 6, col. 2, cites the Fourth Lateran Council in support of the claim that contrary to the opinions of Aristotle and Averroës, the world was created in time (on Bonae Spei, see this volume, Appendix II, n. 3). Finally, Illuminatus Oddus [*De coelo*, disp. 1, dub. 5], 1672, 9, col. 1, invokes the text to uphold the judgment that the world had its existence from God and not from itself (on Oddus, see Appendix II, note 5).
10. Peter Lombard, *Sentences*, bk. 2, dist. 2, ch. 5, 1971, 340, for what appears to be his

Figure 5. God, with compass in hand, designing the universe. (Österreichische Nationalbibliothek, Vienna, Latin MSS, MS. 2554, fol. 1r. See also Murdoch, 1984, 330.)

In apparent agreement with Saint Augustine, Alexander of Hales believed that the world was created simultaneously. Alexander distinguishes between "the making" (*factio*) and "the creation" (*creatio*) of something. The former

acceptance of simultaneous creation and bk. 2, dist. 12, ch. 2, 385, for the different opinions.

concerns distinct forms that produce all the essential features of things. The addition of forms to matter involves a temporal process – a before and after. By contrast, a creation concerns unformed matter and does not involve before and after. Therefore it lies outside of time and is simultaneous. Alexander believed that God first created the unformed matter from nothing[11] and then created the material and visible heavens and the earth from that unformed, or prime, matter.[12] He also asks whether "all created things were created in one indivisible 'now' [*nunc*] or in more, or whether each thing was created in a particular now."[13] Alexander concludes that "all things were created in a single 'now,' " because God's power is more readily manifest by creating many things in one "now" than many things in more "nows" or only one thing in each "now."[14] Alexander of Hales has obviously adopted a version of Augustine's concept of a simultaneous creation.

Although Thomas Aquinas adopted a similar position on simultaneous creation, he sought to reconcile Augustine's interpretation more explicitly with the literal six days of creation in Genesis. Thomas held that God created all things simultaneously with respect to their unformed substance but did not create them simultaneously with respect to their differentiation and ornamentation, which occurred over the six days.[15] God could, of course, have differentiated and ornamented all things simultaneously but chose instead to follow an ordered pattern. "And thus," Thomas explains, "there was good reason that different days be made to serve for the different states of the world."[16]

Over the centuries, variations on the theme of simultaneous creation appeared. In the seventeenth century, Bartholomew Amicus observed ([*De caelo*, tract. 2, qu. 4, dubit. 3, art. 2], 1626, 84, col. 2) that those who believe in a simultaneous creation are divided on how it occurred. Some think that all things, both simple (i.e., elemental) and mixed, were created simultaneously in an instant; others hold that all the simple bodies were produced in an instant and all the mixed bodies were produced in time. Still others believe that all things were created successively, but all in one day, while others deny a simultaneous creation, whether in an instant or in a day, and insist that the world was created in real time over six days.

11. For these ideas, see Alexander of Hales, *Summa theologica*, inquis. 3, tract. 1, qu. 1, ch. 3 ("Utrum omne res corporales sint simul creatae"), 1928, 313, col. 2, and tract. 2, qu. 1, ch. 4 ("Utrum omnia naturalia sint facta simul in genere vel in specie"), 321, col. 2; also tract. 1, ch. 2 ("Utrum eadem sit informis materia caeli et terrae"), 311, col. 2, for the statement about creation from nothing.
12. Ibid., 311, col. 2–312, col. 1.
13. Ibid., inquis. 3, tract. 1, qu. 1, ch. 3 ("Utrum omnes res corporales sint simul creatae"), 314, col. 1.
14. "Quod concedimus quod omnia creata sunt in unico 'nunc' et illud 'nunc' non est aeternitatis nec 'nunc' temporis, sed 'nunc' aevi. Et hoc dicimus, tum quia magis manifestatur Dei potentia creando in uno 'nunc' multa." Ibid., col. 2.
15. Thomas Aquinas, *Summa theologiae*, pt. 1, qu. 74, art. 2, 1967, 10:161.
16. Ibid. Indeed, Thomas seems here to follow a Greek tradition that derived from Philo Judaeus and Saint Basil and passed into Latin with Gregory the Great.

In the opinion he eventually adopted, Amicus argued (ibid., art. 3, 85, cols. 1–2) that the things produced on the first day – the heavens and the elements with all their places and properties – were created in an instant. All other things were created on subsequent days. Moreover, the six days of creation are real, natural days, not spiritual entities (ibid., 87, cols. 1–2). This is evident from Exodus 20.11, where it is said that "in six days the Lord made heaven and earth, the sea, and all that is in them."[17] But why were some things made simultaneously and others not? Amicus confesses ignorance and invokes the free will of God. Basic things like the elements were made in the first instant, but mixed bodies composed of those elements were made later (ibid., 85, col. 2).

In one or another of its various guises, medieval and Renaissance scholastics found Augustine's idea of a simultaneous creation congenial. Although it found its strongest support during the thirteenth century, when Augustine's theology and philosophy shaped the thought of conservative Franciscans, it continued to play a significant role. Like Thomas Aquinas and Bartholomew Amicus, however, most sought to reconcile it more closely with the six days of creation. Thus Amicus confined simultaneous creation to the things created on the first day. Indeed, between the thirteenth and seventeenth centuries many would have agreed with him that, whatever the order of created things, the creation was made over six natural days. In the 1230s, William of Auvergne argued against simultaneous creation when he declared that "the blessed creator does not say that things are done otherwise than they are done. Indeed he says that everything is done in its time, order, and place."[18] Bartholomaeus Mastrius and his coauthor Bonaventura Bellutus declared that "with respect to permanent beings, the world could have existed from eternity, although not with respect to successive things. But in fact it was produced by God in time at the vernal equinox and in six days."[19] Theologians who commented on the six days of creation, whether responding to the second book of Peter Lombard's *Sentences* or another work, had little choice but to treat the days sequentially, as if each was an ordinary day.[20]

17. Translation from the *New English Bible* (1976).

18. "Creator benedictus non aliter dixit res fieri quam factae sint. Immo dixit unumquodque fieri sub tempore, ordine, et loco." William of Auvergne, *De universo*, pt. 1 of pt. 1, ch. 22, 1:617, col. 2. On page 617, column 1, William declares that "not all things could be made simultaneously" (non omnia simul effici potuerunt).

19. Mastrius and Bellutus, *De coelo*, disp. 1, qu. 5, art. 1, 1727, 3:484, col. 2, par. 39. In summarizing Walter Burley's definitions of *res permanens* and *res successiva*, Wilson, 1956, 32–33, describes the former "as a thing all the parts of which can exist at one time" and the latter, by contrast, as something "such as a motion or time, the different parts of which must exist at different times."

20. Those who followed Saint Augustine would not have thought of the six days of creation as ordinary days determined by the Sun's course. Augustine observed that God did not create the heavenly bodies until the fourth day. Therefore the first three days could hardly have been "ordinary." Because he thought it would be inappropriate that the fourth through seventh days should be different from the first three, Augustine concluded that all seven days were identical and in no way ordinary (see Augustine, *Genesis*, bk. 4, ch.

II. Was creation from nothing?

That creation was "from nothing" (*ex* or sometimes *de nihilo*) had become part of Christian belief perhaps as early as the second century, when "both the notion and the formula of the creation *ex nihilo*" are found in Theophilus of Antioch's (fl. 181) apologetic treatise *To Autolycus*, "for the first time in words which preclude all hesitation on the meaning of the doctrine" (Gilson, 1955, 20). As Gilson explains, "The God of Theophilus is not a Greek 'maker' of the world; he is its creator" (ibid.). Indeed, some defined the creative act as the "production of something from nothing."[21]

Despite the early shift to the concept of a creation from nothing, no explicit statement in the Jewish, Christian, or Muslim Scripture declares that creation was out of nothing – that is, *ex nihilo*.[22] According to Harry Wolfson, the phrase *ex nihilo* derives ultimately from the Second Book of Maccabees, 7.28,

> where God is said to have made heaven and earth and all that is therein οὐκ ἐξ ὄντων, "not from things existent", on the basis of which Church Fathers, to mention only the earliest one, the Pastor of Hermas, and the latest one, John of Damascus, in their formulation of the doctrine of creation, describe creation as being "from the nonexistent" (ἐκ τοῦ μὴ ὄντοζ). From the context, however, it can be shown that by "non-existent" they mean "nothing".[23]

Although certain biblical passages suggested a world created from a preexisting chaos,[24] the *ex nihilo* doctrine triumphed. Its victory is perhaps at-

26, 1982, 1:134). Were the seven days each created in turn? Augustine denied this, because he could not account for the creation of the seventh day, on which God rested. If God did not create on the seventh day, how was the seventh day created? Augustine concluded that God created only one day, the first, and then multiplied it to produce each of the remaining six days (ibid., chs. 20–21, 1:127–129).

21. In the seventeenth century, Sigismundus Serbellonus ([*De caelo*, disp. 1, qu. 1], 1663, 2:1, col. 2), in an article titled "What is creation" (Quid sit creatio), said that "creation is commonly called *the production of a thing from nothing*" (Communiter dicitur creatio *productio rei ex nihilo*). (On Serbellonus, see Appendix II, n. 4.) Much the same definition is offered by the Conimbricenses ("Creatio est alicuius e nihilo productio") in Conimbricenses [*Physics*, bk. 8, ch. 2, qu. 1, art. 1], 1602, col. 417, where they cite as sources John Damascene's *De fide orthodoxa*, ch. 8, and Augustine's *De civitate Dei*, bk. 12, ch. 25. Neither, however, presents the definition in this succinct version. I have not found it in the Middle Ages.
22. Wolfson, 1976, 355.
23. Ibid. The translation of 2 Maccabees 7.28 in the *New English Bible* (1976) reads: "I beg you, child, look at the sky and the earth; see all that is in them and realize that God made them out of nothing, and that man comes into being in the same way." In a note, the translators say that "this is the first biblical mention of creation from nothingness." Wolfson's point appears to be that the assertion that God made them "not from things existent" does not, strictly speaking, imply that the world was made from nothing, although it is easy to see how the Church Fathers could have adopted that interpretation. Sorabji, 1983, 194, believes that the opening lines of Genesis "strongly" suggest a beginning of the material universe.
24. Job, 28 and 38; Wisdom of Solomon 11.17 (Sorabji, 1983, 194).

tributable to a powerful inner dynamic that made creation *ex nihilo* almost irresistible. A deity who could create a world from nothing would have, prima facie, appeared more powerful than one who could only create it from preexistent matter. We may plausibly assume that Church Fathers from Hermas to John of Damascus found some form of this argument appealing. Thus was creation *ex nihilo* widely adopted long before it was made explicit Christian doctrine in the first canon of the Fourth Lateran Council in 1215. Despite frequent repetition of the doctrine of creation *ex nihilo* by medieval and early modern scholastic natural philosophers, considerable ambivalence toward it is detectable by the strong arguments in favor of a beginningless world, which avoided creation *ex nihilo*. Nicole Oresme's interpretation was probably typical (*De celo*, bk. 1, qu. 10, 1965, 149 [Latin], 150 [English]): naturally speaking, the world should be eternal, because no natural agent could be invoked to account for a beginning of it. Nevertheless, it was created by the will of God from nothing.

III. Scriptural exegesis: Augustine and Thomas Aquinas

Before we consider the days of creation, it will prove helpful to describe an important attitude toward medieval biblical interpretation and exegesis that was proposed by Saint Augustine and repeated by Thomas Aquinas and which was probably characteristic of scholastic attitudes over the entire period covered by this study.

As a biblical exegete who analyzed the creation account in Genesis, Thomas is a valuable source because he often chose to summarize conflicting traditions and interpretations on particular issues. Not infrequently, he declined to choose between them. This approach may have been shaped by at least two attitudes toward Scripture, the importance of which for any description of the creation account is obvious. Thomas explains that

> There are some things that are by their very nature the substance of faith, as to say of God that he is three and one, and other similar things, about which it is forbidden for anyone to think otherwise. . . . There are other things that relate to the faith only incidentally . . . and, with respect to these, Christian authors have different opinions, interpreting the Sacred Scripture in various ways. Thus with respect to the origin of the world, there is one point that is of the substance of faith, viz. to know that it began by creation, on which all the authors in question are in agreement. *But the manner and the order according to which creation took place concerns the faith only incidentally, in so far as it has been recorded in Scripture*, and of these things the aforementioned authors, safeguarding the truth by their various interpretations, have reported different things.[25]

25. Thomas Aquinas [*Sentences*, bk. 2, dist. 12, qu. 1, art. 2], 1929–1947, 2:305–306. Translated by William Wallace, in Thomas Aquinas, *Summa theologiae*, 1967, 10:222. The italics are mine. Thomas here echoes Augustine, who declared: "in matters that are obscure and

With these general guidelines, Thomas became more specific in his *Summa theologiae*, where, in discussing the question as to whether the firmament was made on the second day, he observes:

> Augustine teaches that two points should be kept in mind when resolving such questions. First, the truth of Scripture must be held inviolable. Secondly, when there are different ways of explaining a Scriptural text, no particular explanation should be held so rigidly that, if convincing arguments show it to be false, anyone dare to insist that it is still the definitive sense of the text. Otherwise unbelievers will scorn Sacred Scripture, and the way to faith will be closed to them.[26]

Thomas's approach, drawn from Saint Augustine, may be taken as typical of medieval scholastic interpretations of creation and explains the proliferation of opinions concerning the different days of the creation account in Genesis. Although Augustine believed that the literal truth of the Scriptures should be accepted unless there were good and overriding reasons for abandoning the literal text, we saw earlier that he himself chose to ignore the clear intent of Genesis and to interpret the six days of creation as a simultaneous and instantaneous act of divine creation. Despite Augustine's failure to provide help in determining when the literal text should be abandoned in favor of an allegorical, metaphorical, or scientific interpretation, no serious problems arose until the Church and Galileo collided in the seventeenth century. In that conflict, the latent ambiguity in Augustine's position became apparent, as "both Galileo and his opponents called upon Augustine when the question arose whether the Copernican doctrine was invalidated by the frequent Scriptural mentions of the Sun's motion" (McMullin, 1970, 336).

IV. On the first four days of creation

Of the six days of creation, only the first four are relevant for cosmology, and of these the first, second, and fourth are the most important. What specific entities were created on each of these days? On the first day, heaven (*caelum*), earth (*terra*), and light (*lux*); on the second, the firmament (*firmamentum*) that divides the waters above from those below and which God

far beyond our vision, even in such as we may find treated in Holy Scripture, different interpretations are sometimes possible without prejudice to the faith we have received. In such a case, we should not rush in headlong and so firmly take our stand on one side that, if further progress in the search of truth justly undermines this position, we too fall with it." *Genesis*, bk. 1, ch. 18, par. 37, 1982, 1:41.

26. *Summa theologiae*, pt. 1, qu. 68, art. 1, 1967, 10:71–73. Based upon Augustine, *Genesis*, 1982, vol. 1 (bk. 1, chs. 18–19, 21). In book 2, chapter 5, where Augustine discusses the water above the firmament, he declares that "whatever the nature of that water and whatever the manner of its being there, we must not doubt that it does exist in that place. The authority of Scripture in this matter is greater than all human ingenuity" (Augustine, ibid., 1:52).

called "heaven" (*caelum*);[27] on the third day God turned his attention to the earth, where he gathered the seas together in one place, exposing the dry land on which he then placed plants and trees capable of reproducing themselves; and finally, on the fourth day, he made the physical light of the heavens by creating all the celestial bodies, assigning the Sun to provide the light of day and the Moon to provide the light of night.

Within these brief passages commentators were obliged to resolve some basic dilemmas, obscurities, and seeming inconsistencies. How, for example, does the heaven (*caelum*), or firmament, created on the second day, differ from the heaven (*caelum*) created on the first day? How does the light created on the first day compare to the light created on the fourth day? How could plants come forth on the third day if the Sun, whose warmth and light are required, was not created until the fourth day? What are the waters above and below the firmament? Do they differ?

And then there were problems that arose as a consequence of the need to reconcile the Christian creation account with contemporary physics and cosmology, which, in the period we are discussing, was overwhelmingly Aristotelian. How do prime matter and the four elements relate to creation? How do mixed, or compound, bodies formed from those elements fit into the creation account? What aspect of creation embraces Aristotle's celestial ether? And so on.

To obtain a reasonable sense of the range of opinions and the manner in which scholastic theologian–natural philosophers coped with these and similar questions, we shall focus on the responses of Thomas Aquinas in his commentary on the *Sentences* and, especially, in his *Summa theologiae*.[28] Thomas is an appropriate choice because, in the absence of clear-cut interpretations of most relevant scriptural passages, he frequently cited two or more alternative opinions, often declining to choose between them.

Thomas and other commentators on Genesis divided creation into three aspects or distinctions. On the basis of Genesis 2.1 ("So the heavens and the earth were finished and all the furniture of them"),[29] Thomas identifies the three distinctions as follows: "the work of creation" (*opus creationis*), in which heaven, water, and earth were made, though in an incomplete state; "the work of differentiation" (*opus distinctionis*), which involved the completion and ordering of heaven, earth, and water from the beginnings just described; and "the work of adornment" (*opus ornatus*), in which the elements were separated and things that move in heaven and on earth were produced.[30]

27. The Latin terms are from the Vulgate Bible.
28. Thomas composed his commentary on the four books of the *Sentences* in Paris, between 1252 and 1256 (Weisheipl, 1974, 358–359) and completed the first part of the *Summa theologiae*, which contains the section on creation, between 1266 and 1268 at Viterbo (ibid., 360–362).
29. Thomas quotes Genesis 2.1 in *Summa theologiae*, pt. 1, qu. 70, art. 1, 1967, 10:109.
30. Thomas describes the three aspects of creation most extensively in *Summa theologiae*, ibid.,

The work of creation occurred on the first day, when heaven, water, and earth were made. As evidence of the incompleteness of these three basic cosmic entities, Thomas observes that all three lacked a vital form: the heaven lacked light, for "darkness was on the face of the deep" (tenebrae erant super faciem abyssi); water was formless, because it is referred to as "the deep" (*abyssus*); and "earth was empty" or "invisible" (terra erat inanis et vacua) and also "void" or "uncomposed" (vacua vel incomposita), signifying that it was not only covered by waters and totally hidden but also lacked shrubs and plants.[31]

To complete – that is, distinguish or differentiate – heaven, water, and earth was the work of the first three days. Heaven was substantially completed on the first day, presumably by the creation of light, which produced night and day; water on the second day, when the firmament separated the waters above from those below; and earth on the third day, when dry land was exposed, following the gathering of the waters.

The next three days, the fourth through sixth of creation, were taken up with adornment, that is, with the placing of various inanimate and animate things that would fill the universe with moving things. Thus on the fourth day, the planets were placed in the firmament and thereby "adorned" it; on the fifth day, birds and fishes were made to move about and thus adorned the intermediate region comprised of air and water, which are treated as one; and on the sixth day animals were placed on earth to move about and adorn it.[32]

Let us examine further the first, or incomplete, phase of creation. In its incomplete state, earth was subject to different interpretations in which Aristotelian concepts were involved. Thomas explains that Augustine interpreted earth and water as primary matter, whereas other Church Fathers assumed that one or more (or even all) of the elements were already there. Indeed, although only earth and water are mentioned, Thomas seems to believe that all four elements were intended. That Moses omitted the elements air and fire is explicable by the fact that Moses spoke to the unlettered to whom "it was not so plain that these are bodies as it was that earth and water are."[33]

Thomas believed that those who considered matter as originally formless[34] really meant that matter was originally created with various

10:109–111, and to a lesser extent in qu. 66, art. 1, 10:29–33, and qu. 69, art. 1, 10:95–97.

31. For heaven and earth, see ibid., qu. 66, art. 1, 10:28 (Latin); for water, see qu. 69, art. 1, 10:95.
32. Despite these detectable distinctions in the six days of creation, Thomas, as we have seen, adopted Augustine's concept of *rationes seminales* and argued that "all bodies were created immediately by God," which Moses is alleged to have described when he said, "In the beginning God created heaven and earth." Ibid., qu. 65, art. 3, 10:17. Aquinas ([*Sentences*, bk. 2, dist. 13, qu. 1, art. 1], 1929–1947, 2:327) accepts the *rationes seminales*. The Vulgate text reads, "In principio creavit Deus caelum et terram."
33. Thomas Aquinas, *Summa theologiae*, pt. 1, qu. 66, art. 1, 1967, 10:31–33.
34. Ibid., 10:27.

substantial forms that would later be differentiated by accidental forms (of which light was the first).[35] Because God "produces being in actuality out of nothing,"[36] he did not create by bringing something from potentiality to actuality, as normally happens in nature. Thus from the general state of formlessness in the initial stages of creation – that is, from the things originally created with matter and substantial forms – things would be differentiated and made more specific.

The first of these accidental forms to alter the nature of the universe was light, which Thomas defines as "an active quality deriving from the substantial form of the Sun, or of any other body that is self-illuminating should any exist."[37] As the quality of a primary body, the Sun, light was essential in shaping the universe and also was a common feature of lower and higher bodies. "It was right, therefore, that the orderliness of divine wisdom be manifested, with light, among the works of diversification, being produced first in that it is the form of a primordial body, and is something more general."[38] Moreover, as Basil observed, light makes all things visible. It also seems that light had to be produced on the first day because "there can be no day without light."[39]

If light is a quality of the Sun and the Sun was not created until the fourth day, what kind of light was created on the first day? Augustine's theory of simultaneous creation, to which Thomas subscribed, provides the explanation. Since all things were in existence simultaneously, the substance of the Sun already existed, but in the course of the first day it had only a general power to illuminate. It was this general power that was utilized on the first day in order to separate day and night, which are produced by the daily motion, the most common motion of the entire heaven. Thus did light differentiate the heaven on the first day. Only with the creation of the luminaries – and all the planets – on the fourth day would the Sun assume its full powers, dividing the day and the night, as well as the seasons, days, and years.[40]

V. What is the heaven created on the first day?

In the seventeenth century, Bartholomew Amicus mentioned two interpretations that were applied to the heaven created on the first day.[41] Both involved the empyrean heaven, which was conceived as an immobile orb

35. Ibid., qu. 67, art. 4, 10:67.
36. "Sed Deus producit ens actu ex nihilo." Ibid., qu. 66, art. 1, 10:31 (Latin on p. 30).
37. Ibid., qu. 67, art. 3, 10:61.
38. Ibid., qu. 67, art. 4, 10:65–67.
39. Ibid., 10:67.
40. Ibid., 10:67–69; also Genesis 1.14.
41. Augustine, and those who followed him, held that the words "In the beginning God created heaven and earth" signified things that were created before the "beginning of days." See Augustine, *Genesis*, bk. 1, ch. 9, 1982, 1:27.

surrounding the cosmos, wherein dwelled God and all the elect. One interpretation held that the empyrean heaven alone was created on the first day, while the other assumed that not only was the empyrean heaven created on the first day, but all the heavens, from the outermost movable sphere to the sphere of the Moon.[42] The latter interpretation, according to Amicus, was the most common scholastic opinion, drawn largely from Saint Basil, who understood by the creation of "earth" (*terra*) on the first day the creation of the four elements and by the creation of "heaven" (*caelum*) everything above the four elements, that is from the Moon outward.[43] Although the empyrean heaven is not explicitly mentioned in Genesis, most Christian authors not only agreed that it was the heaven created on the first day but also assumed its immobility. Because of the latter property and the importance of the empyrean heaven, I shall devote a separate chapter (Ch. 15) to it, emphasizing its unique cosmic role as an all-encompassing, immobile orb.

VI. On the firmament of the second day

Commentaries on the second day of creation eventually generated two major orbs that were essentially theological in character, namely the crystalline orb and the firmament.[44] Let us examine the latter first and use Thomas Aquinas as our primary guide so that we may come to know some of the interpretations that were generated.

42. Amicus [*De caelo*, tract. 2, qu. 5 ("An caelum empyreum fuerit initio creatum"), dubit. 3, 1626, 94, col. 2–95, col. 1 and dubit. 4 ("An coeli aetherei fuerint creati primo die"), 95, col. 1–96, col. 1. Among those who assumed that only the empyrean sphere was signified by the creation of the heaven on the first day, Amicus mentions (ibid., 95, col. 1) among early commentators, Bede, Origen, Anselm, and Bonaventure; in the sixteenth century, Pererius; and in the seventeenth century, Molina. Those who held the other opinion, which Amicus supports, are Basil and Thomas. Thomas, however, declares that Strabo, Bede, and Basil are the only authorities – presumably ancient authorities – who considered the heaven of the opening statement in Genesis to be the empyrean heaven. *Summa theologiae*, pt. 1, qu. 66, art. 3, 1967, 10:41. Thus Basil shows up as supporting both opinions. In truth, they are compatible. Although Amicus does not mention Augustine in this context, the latter held that "by the expression 'heaven' we must understand a spiritual created work already formed and perfected, which is, as it were, the heaven of this heaven which is the loftiest in the material world." Augustine, *Genesis*, bk. 1, ch. 9, 1982, 1:27. Actually, Augustine identified the heaven created in the beginning with angels, hence its spirituality (see Taylor's note, ibid., 1:227, n. 33). On this, Thomas, mistakenly it seems, interprets Augustine to mean by heaven "a spiritual nature yet unformed." *Summa theologiae*, pt. 1, qu. 67, art. 4, 1967, 10:65. If Augustine identified the heaven with angels, why should Thomas consider the spiritual nature of angels as unformed?
43. Amicus, *De caelo*, tract. 2, qu. 5, dubit. 4, 1626, 95, col. 1.
44. Based on Genesis 1.6–8 (King James Version): "[6.] And God said, Let there be a firmament in the midst of the waters, and let it divide the waters from the waters. [7.] And God made the firmament and divided the waters which were under the firmament from the waters which were above the firmament; and it was so. [8.] And God called the firmament Heaven. And the evening and the morning were the second day."

Thomas explains that the *firmamentum* created on the second day can be understood in two ways: either as the sphere in which the fixed stars are located,[45] or "as that part of the atmosphere where clouds undergo condensation."[46] Within the first interpretation, in a manner reminiscent of Grosseteste's commentary on the creation, Thomas distinguishes three opinions as to the composition of the firmament. Some held that it was composed of the four elements, as did Empedocles; others that it was made of a single element (for example, Plato, who thought it was composed of fire); still others insisted that the firmament was composed of a fifth element different from the other four, as did Aristotle.[47]

Thomas judged all of these opinions compatible with the substantial formation of the firmament on the second day. None was sufficiently compelling, however, and he rested content merely to describe them.

1. The firmament as air

As for the second interpretation, that the firmament is "that part of the atmosphere where clouds undergo condensation," the term *firmamentum* is relevant "because of the density of the air in that part, since the dense and solid is said to be a firm body, to distinguish it from a mathematical body, as Basil observes."[48] Although he did not accept it as the final explanation, Thomas believed that identifying the firmament with the air was compatible with faith and compatible with the evidence. Minute drops of water could rise above the air and form clouds: here, then, were the waters above the firmament. When these fine vapors condensed to the point where the air could no longer support them in the form of drops, the latter would fall as rain and join the seas and rivers below: here were the waters below the firmament. Moreover, passages from Scripture speaking of "flying creatures of the heaven" (*volucres coeli*)[49] seemed to give further credibility to air as the firmament.[50]

Although Durandus de Sancto Porciano may have accepted the air as the

45. Thomas Aquinas, *Summa theologiae*, pt. 1, qu. 68, art. 1, 1967, 10:73–75. Although Thomas says only "uno modo, de firmamento in quo sunt sidera" (ibid., 72), it is clear that he means fixed stars when he refers to the second interpretation of *firmamentum* and says: "Potest autem et alio modo intelligi, ut per firmamentum quod legitur secunda die factum, non intelligitur firmamentum illud in quo fixae sunt stellae, sed illa pars aeris in qua condensantur nubes." Ibid., 74.
46. Ibid., 75.
47. Robert Grosseteste had earlier described these same three opinions (*Hexaëmeron*, part. 3, ch. 6, 1, 1982, 106, lines 7–8). Whereas Thomas merely reported them, Grosseteste viewed them as a sign that "The philosophers write mutually contrary statements about these things" (Scribunt enim super hiis philosophi sibi invicem contraria). For my translation of the relevant passage, see Grant, 1987a, 165.
48. Thomas Aquinas, *Summa theologiae*, pt. 1, qu. 68, art. 1, 1967, 10:75.
49. See Richard of Middleton [*Sentences*, bk. 2, dist. 14, art. 1, qu. 1], 1591, 2:167, col. 1. For more on "flying creatures of the heaven," see note 68 of this chapter.
50. Thomas Aquinas, *Summa theologiae*, pt. 1, qu. 68, art. 1, 1967, 10:75.

firmament that divided the waters,[51] most scholastics rejected the air (or any part of it) as the firmament. Richard of Middleton thought that the theory conflicted with Scripture, which says that the firmament was made on the second day and the celestial luminaries were placed in it on the fourth day. Air, however, was made on the third day, and no stars were placed in it. That air alone could be the firmament was dubious also because, as Hurtado de Mendoza explained, Genesis speaks of the waters above the firmament. If the latter was air, how could waters – Hurtado specified "elemental" waters – lie above the air?[52] Moreover, when God made the firmament on the second day, there were no clouds and fog to divide the air from the sea. Indeed, God did not make the seas until the third day, when he commanded all the waters to gather in one place.[53] In a similar vein, Bartholomew Amicus insisted that "the lowest region of the air, which divides the clouds from the waters and rivers, is not understood by the name of *firmamentum*, because there was no rain during the six days of Scripture; therefore there were no clouds in the air, which are the matter of rain."[54] Echoing Richard of Middleton, Amicus argued that the lowest region of the air is not a candidate for the firmament, because God expressly placed the luminaries in the firmament – not in the air – on the fourth day.[55]

2. *The firmament as a single heaven embracing all the planets and the fixed stars*

It was natural to inquire about how the heaven (*caelum*) created on the first day was related to the heaven (*caelum*) created on the second day and specifically called the firmament. If one adopted the interpretation of John Chrysostom, it was not essential to distinguish between the two. By declaring that "In the beginning God created heaven and earth," Moses first proclaimed what God did and then elaborated its implications. As Thomas Aquinas explained, "this would be like saying, 'this builder made that house', and later to add, 'he first made the foundation, and afterward put up the walls, and thirdly placed the roof on top'. And thus we need not

51. In his *Sentences*, Durandus argued that if the firmament were composed of the fifth celestial element, no elementary waters could exist above the heavens, because such waters would suffer generation and corruption by rarefaction caused by the heat from the light of the celestial ether. But if the air, or more particularly that part of the air where the clouds are condensed, is taken as the firmament, elementary waters could indeed lie above the firmament, because the power of the Sun and stars can draw watery vapors up above the clouds. These clouds possess a certain thickness, which also makes it appropriate to call them a firmament. Thus the air can be considered as a proper and true firmament. Whether Durandus believed this is left ambiguous. Durandus [*Sentences*, bk. 2, dist. 14], 1571, 156r, col. 2.
52. Hurtado de Mendoza [*De coelo*, disp. 3, sec. 3, par. 43], 1615, 381, col. 1. In this argument, Hurtado takes for granted that in the natural scheme of things air lies above water, as Aristotle assumed.
53. Ibid., 380, col. 2.
54. Amicus, *De caelo*, tract. 2, qu. 5, dubit. 4, 1626, 95, col. 2.
55. Ibid., 96, col. 1.

maintain a difference between the heaven of which it is said 'In the beginning God created heaven and earth,' and the firmament of which it is said that it was made on the second day."[56]

But Thomas observes that differences between the heavens created on the first two days could be distinguished and proceeds to cite a number of opinions and their proponents.[57] For Augustine, the heaven created on the first day is of an unformed spiritual nature, whereas the heaven of the second day is the corporeal heaven. Venerable Bede and Walafrid Strabo interpreted the heaven of the first day as the empyrean heaven, while they construed the heaven of the second day as the sidereal heaven. By contrast, Damascene identified the heaven of the first day as a starless, transparent orb, which the philosophers called the "ninth sphere" (*nona sphaera*) and which was the first, or outermost, body in motion.[58] For Damascene and the philosophers, the heaven of the second day is the sidereal heaven. As was his custom, Thomas selected none of these opinions as his own but was content merely to report them.

Since God called the firmament heaven, what sense of heaven was assigned to the firmament? What celestial body (or bodies) did it embrace? Some were prepared to argue that there was only one heaven, just as there was only one earth.[59] According to Thomas, John Chrysostom considered the entire body above the earth and water as one heaven,[60] an opinion that found favor with some theologians because, as we shall see, it seemed to accord best with the scriptural text, though it clashed with Aristotle. Because it included air and fire and all the visible celestial bodies, Chrysostom's heaven linked, in Aristotelian terms, the mutable upper elements, air and fire, to the incorruptible celestial bodies. It was thus a combination that was unacceptable to Aristotelian natural philosophers but found favor with some theologians.

Thomas, one theologian who did not follow Chrysostom, kept the ter-

56. Thomas Aquinas, *Summa theologiae*, pt. 1, qu. 68, art. 1, 1967, 10:75. Thomas seems to have reversed the divine method of creation described by Chrysostom, who says that God "executes his creation in a way contrary to human procedures, first stretching out the heavens and then laying out the earth beneath, first the roof and then the foundation." *Homilies on Genesis* [Hill], homily 2, 1986, 35.
57. Thomas Aquinas, ibid., 10:75–77.
58. Earlier, in the sixth century, John Philoponus had already identified the heaven of the first day with the ninth sphere (see Duhem, *Le système*, 1913–1959, 2:496–497).
59. Here again, I follow Thomas, *Summa theologiae*, pt. 1, qu. 68, art. 4, 1967, 10:87–91: "Is there only one heaven" (Utrum sit unum caleum tantum).
60. In homily 4, Chrysostom insists (*Homilies on Genesis* [Hill], 1986, 56) that Scripture asserts unequivocally that the purpose of the firmament is to "keep one body of water from another." God then called this firmament heaven and therefore only one heaven, not many, can exist. Chrysostom does not, however, explicitly extend the firmament from the air to everything beyond, although he would clearly have meant it to include everything between the divided waters, a region that extended from the concave surface of the air to the convex surface of whatever celestial body (the fixed stars?) is in contact with the waters above. Galileo ([*De caelo*, qu. 1 (G)], 1977, 59) also cites Chrysostom as a supporter of a single heaven.

restrial elements distinct from the incorruptible celestial region.[61] If we take "heaven" (*caelum*) as "a particular sublime body, actually or potentially luminous, and by its nature indestructible," then, he explains that the term is used in Scripture in three ways:

> The first, completely luminous, is called the "empyrean heaven." The second, completely transparent, is called the "aqueous or crystalline heaven." The third, partly transparent partly luminous, is called the "sidereal heaven"; it is divided into eight spheres, viz. the sphere of the fixed stars and the seven spheres of the planets, which collectively can be referred to as eight heavens.[62]

Thomas did not identify any of these heavens as the firmament but appears to have associated the firmament with the eighth sphere of the fixed stars. If so, then Thomas would have made the firmament a part of the sidereal heaven. For many others, however, from the Middle Ages to the seventeenth century, the sphere of the fixed stars and the seven planetary spheres together were identified as the firmament. The latter was thus equivalent to what Thomas called the sidereal heaven.[63]

Some years before Thomas wrote his *Summa theologiae*, Robert Grosseteste had already taken this step as he grappled with the problem of the biblical firmament. In his *Hexaëmeron*, Grosseteste declares that many have carefully investigated the nature of the firmament and the number of heavens contained in it. By *firmamentum* Grosseteste understood a heaven that "is extended in thickness from the lowest wandering of the Moon up to [the region just above] the fixed stars where the superior waters are gathered."[64] Thus it is clear that for Grosseteste, the firmament is a vast region that includes the Moon and the fixed stars and everything in between. When, earlier in the *Hexaëmeron*, Grosseteste declared that by the name *firmamentum* he understood "the heaven in which the stars [*sidera*] are located," he intended to signify by *sidera* not only the fixed stars but also the planets.[65] Grosseteste's *firmamentum* is thus equivalent to Thomas's sidereal heaven.

Others adopted the same interpretation. In the early fourteenth century, Aegidius Romanus included within the firmament the entire region from the Moon to the fixed stars, which he regarded as a single, continuous orb,

61. In his earlier commentary on the *Sentences* (bk. 2, dist. 14, qu. 1, art. 2, 1929–1947, 2:350), Thomas denied that the firmament was composed of the four elements, assuming instead that it was made from a fifth body or essence, as described by Aristotle.
62. Thomas Aquinas, *Summa theologiae*, pt. 1, qu. 68, art. 4, 1967, 10:89. Bartholomew the Englishman used the same threefold division. Like Thomas, he did not specifically mention the term *firmamentum*. Bartholomew the Englishman, *De rerum proprietatibus*, bk. 8, ch. 2, 1601, 373.
63. John Philoponus had also assumed that Moses was describing a single heaven or firmament that embraced all the planets and stars (see Ch. 13, Sec. I).
64. My translation, from Grosseteste, *Hexaëmeron*, part. 3, ch. 6, 1, 1982, 106.
65. "Verumtamen, iudicio eius et aliorum expositorum verius intelligitur nomine firmamenti celum in quo locata sunt sidera." Grosseteste, ibid., ch. 3, 1, 103.

containing within it seven distinct and discontinuous cavities or channels through which the planets moved.[66] In the seventeenth century, versions of Aegidius's interpretation were adopted by Raphael Aversa, Mastrius and Bellutus, and Hurtado de Mendoza.[67]

In his metaphysical disputations of 1597, Francisco Suarez followed not only the general tradition of Bede, Strabo, and Grosseteste in assigning to the firmament both planets and fixed stars – Thomas's sidereal heaven – but he also included the air and, presumably, the fire above it, thus specifically adopting the opinion of Chrysostom. Suarez thought it probable that by *coelum* Moses intended all celestial bodies plus the air, observing that in Scripture air was frequently included in the term "heaven."[68] It follows that when God called the firmament *caelum*, air must have been included. Thus a tradition existed in which firmament was interpreted broadly, embracing not only the entire heavens between the Moon and the fixed stars but even extending down (through fire) to the air itself.

3. The firmament as the sphere of the fixed stars

The demands of Aristotelian–Ptolemaic astronomy and cosmology produced the final candidate for the *firmamentum*: the eighth sphere of the fixed stars. Among those who identified the fixed stars with the firmament were John of Sacrobosco, Michael Scot, Vincent of Beauvais, Thomas Aquinas, Campanus of Novara, Gregor Reisch, Christopher Clavius, Mastrius and Bellutus, and Giovanni Baptista Riccioli.[69] Michael Scot explains the iden-

66. Aegidius declares that the Sun and the Moon and all the planets are in the firmament, because "all the spheres of the planets make one body with the the eighth sphere [of the fixed stars], which [taken all together] is called the firmament. All such luminaries are in the firmament of the heaven." A few lines below, Aegidius adds: "it is appropriate that this whole region be called the firmament, because it was made between the waters and the waters" ("Ex hoc etiam magis concordamus cum Scriptura sacra dicente solem et lunam et omnes stellas esse in firmamento coeli, quia ex quo omnes sphaerae planetarum faciunt unum corpus cum octava sphaera, quod dicitur firmamentum, omnia huiusmodi luminaria sunt in firmamento coeli." Aegidius Romanus, *Opus Hexaemeron*, pt. 2, ch. 32, 1555, 49v, col. 1. And later he says, "Ideo congrue dicitur firmamentum esse factum inter aquas et aquas." Ibid., 49v, cols. 1–2.)
67. For further discussion of Aegidius's ideas and the various adaptations of his interpretations, see Chapter 13, Section III.9.
68. "Imo probabile satis est nomine coeli comprehendisse [i.e., Moses] omnia corporea usque ad aerem inclusive." Suarez, *Disputationes metaphysicae*, disp. 13, sec. 11, 1866, 1:448, col. 1, par. 26. Suarez cites (col. 2) the words "extendens coelum sicut pellem, qui tegis aquis superiora eius," from Psalm 103 (Vulgate), the relevance of which is hardly clear, since air is not mentioned. Without citing the texts, Suarez also says that Scripture mentions birds or a "flying creature of the heaven" (*volucrum coeli*), that is, of the air (in Psalm 8 of the Vulgate, we find the expression *volucres caeli* and in Psalm 8 according to the Hebrews *aves caeli*; see also note 49 of this chapter). He insists that *firmamentum* "is a certain part of the air, or includes air" (firmamentum vel aeris pars quaedam est, vel aerem includit).
69. Sacrobosco, *Sphere*, ch. 1, 1949, 77 (Latin); 118 (English); Michael Scot [*Sphere*, lec. 4], 1949, 282; Vincent of Beauvais, *Speculum naturale*, bk. 3, ch. 102, 1624, 1:230 ("Hoc caelum id est firmamentum est octava sphaera, quae et dicitur stellata"); for Thomas, see Campanus of Novara, *Theorica planetarum*, 1971, 156, line 205; 208, line 697 and 385–

tification of the eighth sphere with the firmament when he declares that "according to all the astronomers, the seven inferior orbs of the planets are moved contrary to the firmament [i.e. the eighth sphere of the fixed stars]; therefore it [the firmament] is not one continuous body."[70] It is in fact distinct from the seven planetary orbs. In this interpretation, *firmamentum* no longer represents a single heaven ranging from the lunar sphere to the fixed stars.

In the seventeenth century, Pedro Hurtado de Mendoza ([*De coelo*, disp. 3, sec. 3], 1615, 381, cols. 1–2, pars. 44–45) reviewed the two interpretations just described and argued that the conception of the firmament as the eighth sphere of the fixed stars, a view he attributes to Fathers Pererius and Molina, was essentially inconsistent with Scripture. According to Hurtado, Pererius and Molina held that the firmament, or eighth sphere, separates the ninth heaven, which they described as an icy sphere and identified with the watery heaven above the firmament, from the elementary waters below. Of the five arguments Hurtado marshaled against this position, three will be described here.

Not only is the identification of waters with heavens contrary to Scripture, but, because the waters that were divided must have had the same composition, the waters above should not be considered as different from the waters below.[71] Thus the former ought not to be taken as an icy, solid mass while the latter are identified with elementary, fluid waters. Moreover, because God placed the Sun and Moon in the firmament on the fourth day, and neither the Sun nor the Moon is located in the eighth sphere, we may infer that the firmament includes more than the eighth sphere. Finally, although the eighth sphere separates and divides the bodies on each side of its two surfaces, it cannot divide the waters from the waters, because its surfaces are not in direct contact with the inferior waters. Indeed, many other bodies – at the very least all seven planets, and perhaps fire and air as well – lie between the eighth sphere and those inferior waters. "Therefore," Hurtado concludes, "the eighth sphere does not divide the ninth sphere from the elementary waters."

386, n. 39; Gregor Reisch, *Margarita philosophica*, 1517, 245 ("et octavum quod celum stellatum sive firmamentum appellant") and 248; Clavius [*Sphere*, ch. 1], *Opera*, 1611, 3:11; Mastrius and Bellutus, *De coelo*, disp. 2, art. 2, 1727, 3:488, col. 1, par. 16 ("Whether above this heaven, called firmament [*coelum firmamentum vocatum*] because the stars firmly adhere to it, or [because] it is starred from the multitude of stars received in it, are assigned other heavens between this heaven and the empyrean"); and Riccioli, *Almagestum novum*, pars post., bk. 9, sec. 3, ch. 1, 1651, 273, col. 2.

Although Francisco Suarez included the planets, fixed stars, and even the air in his concept of the firmament (see this chapter, end of Sec. VI.2), he reports the opinion that "there are true waters above the firmament, that is, above the eighth and starry sphere" (quod sunt verae aquae super firmamentum, id est, super octavam et stellatam sphaeram). See Suarez, *Disputationes metaphysicae*, disp. 13, sec. 11, 1866, 1:440, col.1, par. 6.

70. "Sed inferiores orbes septem planetarum moventur contra firmamentum secundum omnes astrologos, ergo non est unum corpus continuum." Michael Scot [*Sphere*, lec. 4], 1949, 282.

71. As will be obvious, few treated the waters above as identical with those below.

What, then, is the firmament? For Hurtado it is "an aggregation of bodies interjected between terrestrial and celestial waters, which [the celestial waters] are contiguous to the empyrean heaven. This aggregation includes all the mobile heavens, fire, and air." Like Suarez, then, Hurtado thinks of the firmament as a heaven that ranges from the air below up through fire and into the celestial region, extending all the way up to and including the eighth sphere. Hurtado describes this interpretation as "the more common exposition."[72] Only by such an interpretation could the firmament be compatible with the scriptural demand that it divide, and therefore separate and be in direct contact with, the waters above and below. According to Hurtado, the firmament separates the waters above and below by means of "the last [or concave] surface of air that is contiguous to the sea" and "the last convex surface of the starry heaven contiguous to the superior waters." Thus the firmament must extend from the air to the fixed stars and embrace both corruptible and incorruptible matter.

For most of the period of this study, it seems that the firmament was most frequently assumed to embrace at least the seven planetary orbs and the sphere of the fixed stars,[73] and for some it also embraced the spheres of fire and air below the Moon. Those who included the spheres of fire and air were apparently prepared to effect a drastic departure from Aristotle and the overwhelming majority of his medieval followers: they stretched the meaning of "firmament" to embrace both incorruptible and corruptible parts.

Proper regard for the biblical firmament led inevitably to difficulties with the requirements of Aristotelian–Ptolemaic astronomy. For scriptural reasons, the firmament had to be simultaneously in contact with both the waters above and those below it. It thus had to range from the concave surface of the air to the convex surface of the eighth sphere. For astronomical reasons, however, the contrary motions of the planets with respect to the sphere of the fixed stars made it unfeasible to assign a single orb or heaven to the firmament. With the planetary orbs dissociated from the eighth sphere of the fixed stars, the term *firmamentum* seemed to apply best to that very eighth sphere, because the latter bordered on the crystalline sphere, which was identified in some manner with the biblical waters above the heaven. That the eighth sphere was not contiguous to the waters below the firmament seems to have been conveniently ignored. For some, very likely a

72. "Communior est haec expositio." Hurtado de Mendoza, *De coelo*, disp. 3, sec. 3, 1615, 381, col. 2, par. 45. Although Hurtado included fire and air in his concept of the firmament and seems to have shared this opinion with Suarez, he also assumed seven eccentric channels through which the planets moved, thus aligning himself with the tradition stemming from Aegidius. However, we shall see (Ch. 13, Sec. III.9) that whereas Hurtado believed the planets were self-moving in their deferent channels, Aegidius and others believed that they were carried by epicycles.

73. According to Christopher Scheiner, Hieronymus Vielmius understood "by the term 'firmament' that it contains the whole mass of celestial bodies" (Ait nomine Firmamenti totam caelestium corporum massam contineri). Scheiner [*Rosa Ursina*, bk. 4], 1630, 648, col. 2.

minority of scholastic authors, especially in the seventeenth century, the needs of astronomy and cosmology took precedence over a consistent interpretation of Scripture. But, as we have seen, and shall see again, a popular compromise subsumed the seven planetary spheres as diverse elements within a single, continuous, material sphere extending from the Moon to the sphere of fixed stars. Despite major obstacles, the urge to see in the celestial region a single orb, embracing, at the very least, all the planets and stars, was powerful.

VII. On the waters above the firmament: the crystalline heaven

From the time of the Church Fathers, the meaning and significance of the waters above the firmament were much debated. Because the account in Genesis spoke of waters above the firmament, Christian authors, following Augustine, were generally agreed on the necessity for a literal interpretation and were therefore committed to the existence of waters above the firmament.[74] Most scholastics were also certain that the suprafirmamental waters lay between the firmament and the empyrean, or last, heaven.[75] But what kind of waters? Thomas Aquinas insisted that they were material[76] but acknowledged that their nature was dependent on the composition of the firmament. As was his custom, he described the various possibilities.[77]

If the firmament signifies a sidereal heaven that is composed of the four elements, then the waters above the firmament could have the same nature as the ordinary element water. But if the sidereal heaven, or firmament, is not composed of the four elements, the waters above the firmament could not be identical with the element water. Under these circumstances, if the firmament divides the waters, the waters above the firmament must be different from the element water, for example, some kind of unformed matter.

On the assumption that the firmament is part of the atmosphere below the celestial region where clouds are formed, however, "then the waters that are above the firmament are the same as those that, when evaporated and taken up into the atmosphere, are the source of rain."

A question that frequently arose and was repeated by Thomas concerned the manner in which a fluid such as water could remain above the spherical firmament.[78] Two responses were frequently given, both cited by Thomas and attributed by him to Saint Basil. The first concerns the manner in which

74. For Augustine's assertion, see note 26 of this chapter.
75. See, for example, Richard of Middleton, *Sentences*, bk. 2, dist. 14, art. 1, qu. 1, 1591, 2:167, col. 1.
76. *Summa theologiae*, pt. 1, qu. 68, art. 2 ("Are there any waters above the firmament?"), 1967, 10:79.
77. What follows is drawn from ibid., 77–83.
78. Ibid., 77.

waters could remain stable on the convex surface of the firmament. Basil suggested that the outermost surface of the firmament may not be spherical. The fact that the concave surface of the firmament appears circular provides no warrant to infer that the convex surface is also circular; indeed it may be flat, just like the vaults of baths, which have a semicircular form in the interior but a flat surface on the roof.[79] Although repeated from time to time (for example, by Saint Ambrose, in his hexaemeral treatise) this explanation was not taken seriously.

The second response was. For here Thomas attributes to Basil, improperly it seems, the opinion that "the waters above the heavens are not necessarily fluid, but rather are crystallized around the heavens in a state similar to ice." This, says Thomas, "would be the crystalline heaven of some authors."[80]

Two contrasting opinions are discernible about the waters above the firmament: some thought of them as solid and hard; others considered them fluid.[81] This was perhaps the most significant issue concerning the crystalline orb. Since the hardness or fluidity of the latter forms part of the overall theme of the hardness or fluidity of all the heavens or orbs, the entire subject will be considered in Chapter 14 (especially Sec. IV).

VIII. The celestial luminaries created in the firmament on the fourth day

On the fourth day God said, "Let there be lights in the firmament of the heaven" and not only created the stars and planets but also began the process of adornment (*ornatus*).[82] According to Thomas, Moses explains a threefold purpose of the celestial luminaries: (1) to provide light to the earth; (2) to provide the change of seasons; and (3) to serve as signs for the weather,

79. For Basil, see *Exegetic Homilies*, homily 3, 1963, p. 42.
80. For Basil's version, see ibid., p. 43. In fact, Basil seems to deny that the suprafirmamental waters are hard like crystalline rock. Immediately after describing two interpretations of the term firmament, the second of which likened it to crystalline rock, Basil explains that "we compare the firmament to none of these things." Moreover, Basil explains that the term "has been assigned for a certain firm nature which is capable of supporting the fluid and unstable water" (ibid.). In the groupings that follow, I have placed Basil with those who argued for the fluidity of the waters above the firmament.
81. In the twelfth century, William of Conches avoided either opinion by denying outright the existence of waters above the firmament. He was especially annoyed with those who thought they were frozen, largely because he was convinced that congealed waters would be so heavy that "they must either descend to the earth by their natural heaviness or they must be moved. They cannot be moved, for motion cannot exist without heat. Thus, if they are moved, their movements generate heat in them, and if this happened they would be dissolved by that heat." William of Conches, *Glosae super Macrobii In somnium Scipionis*, translated in Helen Lemay, 1977, 229–230 (for the manuscripts used by Lemay see 235, n. 19).
82. For "adornment" and the other two distinctive phases of creation, see Section IV of this chapter.

which is important for various occupations.[83] But why, it was often asked, did God create them on the fourth day, after he had already allowed the production of plants on the third day, following the exposure of dry land? If plants depend for their growth on the Sun, why should they have been created before it? The answer: to reveal God's power to idolators and skeptics. The former would be humiliated to learn that plants could flourish without celestial bodies, which had been thought by many to be gods and essential to the production of plants, as Thomas reports; the latter, as Philo Judaeus (ca. 30 B.C.–after 40 A.D.) believed, would be confounded simply because reason would lead them to expect celestial bodies to be created before the plants that depended on them.[84]

With some exceptions, the nature and operation of those luminaries were not themes usually included in hexaemeral treatises and commentaries on the *Sentences*.[85] Such important matters were more appropriately considered elsewhere and are taken up later. On the assumption that the world was created, we shall now pursue certain questions that inevitably arose about that created world. Was it perfect? Was it finite or infinite? And if finite, what, if anything, might lie beyond it?

83. Thomas Aquinas, *Summa theologiae*, pt. 1, qu. 70, art. 2, 1967, 10:117.
84. Ibid., art. 1, 10:113, where Thomas cites Saint Basil's *Hexameron* (homily 5); for Philo, see Philo Judaeus, *De opificio mundi*, xiv.45, 1929, 1:35.
85. Henry of Langenstein (or Hesse) was a notable exception.

6

The finitude, shape, and place of the world

I. On the finitude of the world

The finitude of the world or cosmos was a cardinal principle of medieval natural philosophy and cosmology. Evidence for this fundamental tenet was not provided directly but was furnished in the manner of Aristotle: by demonstrating that no infinite body could exist and that the world was therefore necessarily finite. Much of the medieval literature on the possibility of an actual infinite was strictly of a logico-mathematical nature, consisting of imaginary, hypothetical examples. In coping with the pitfalls of the infinite, medieval scholastics split into two camps: the finitists, who, like Aristotle, assumed only a syncategorematic infinite – that is, a potential infinite in the manner of numbers, where for any given whole number, there is always another beyond it; and the infinitists, who believed that the existence of an actually infinite magnitude or quantity, a categorematic infinite, was not a contradictory concept and was therefore possible.[1]

The problem of an actual infinite as it bears upon the finitude of the world had two aspects, natural and theological. To argue against the infinity of the physical world, Aristotle sought to demonstrate by appeals to natural

1. See Duhem's lengthy discussion of the infinitely large in *Le Système*, 1913–1959, 7:89–157 and for the translation of it, see Duhem [Ariew], 1985, 73–131. The terms "finitist" and "infinitist" are Duhem's (*Le Système*, 7:151 and [Ariew], 1985, 125). Anneliese Maier also discussed the problem of the actual infinite and made some corrections in Duhem's account, in Maier, 1949, 196–215 ("Das Problem des aktuell Unendlichen"). Among the finitists we may mention Thomas Aquinas, Saint Bonaventure, Duns Scotus, Richard of Middleton, Walter Burley, Thomas Bradwardine, Jean Buridan, Albert of Saxony, and Johannes Versor; included among the infinitists are Gregory of Rimini, William of Ockham (see Maier, ibid., 206 and n. 100), John de Bassolis, Franciscus de Mayronis, Robert Holkot, Nicholas Bonetus, John Baconthorpe, John Major, Pedro Hurtado de Mendoza, and Roderigo de Arriaga.

   According to Nicholas Bonetus (Maier, 1949, 214), the actual infinite could be understood in two ways. In one way, an actual infinite is an infinite to which more can be added. Thus there might be an actual infinity of stones, but yet more stones could be added. This mode was encapsulated in the Latin phrase "quod non sint tot quin plures": "that there are not so many, but that there could be more." The second way assumes that nothing more can be added to the actual infinite, because there are no more things that can be added, a mode that is captured by the phrase "quod sint tot in actu quod plura non possint esse, quia omnia sunt actua posita": "that there are so many things in actuality that there could not be more, because all actual things have been posited."

philosophy and reason that an infinite body was impossible because of the many physical principles that it would violate. The second aspect, however, concerned Christian theology, in particular God's ability to create an actually infinite body or magnitude, which, in turn, involved the broader but vital question about God's absolute power.

The only actual existent infinites accepted during the Middle Ages were the infinite power of God and his infinite omnipresent immensity (see Ch. 9, Sec. II). Apart from these infinites and perhaps others associated with God, the infinitists did not believe in the existence of actually infinite magnitudes or bodies that were independent of God or created by God. What they sought to show was the possibility that God could create such infinites if he wished. Their goal was to demonstrate that the existence of an actually infinite body or magnitude or infinite multitude was an intelligible concept devoid of contradiction.[2] By logical analysis, they attempted to show the conditions under which an actual infinite could exist. In the process, they had to examine terms such as "greater," "smaller," "whole," and "part."[3] In his profound analysis, Gregory of Rimini even determined that "an infinite multitude can be part of another infinite multitude."[4] Numerous subtle illustrations were devised to construct imaginary actual infinites, frequently involving infinite divisibility and imagined aggregations of those parts. For example, God could make an infinite body in one hour by creating and conserving a one-foot body in every proportional part of the hour. "And because there are infinite proportional parts in this hour, it follows that at the end of the hour an infinity of one-foot bodies would exist from which an infinite body would be constituted."[5] A popular tactic was to weaken confidence in one or more of Aristotle's arguments against actual infinites by suggesting qualifications that would at least make Aristotle's arguments "nondemonstrative." Thus Marsilius of Inghen sought to counter Aristotle's argument that an actual infinite body would necessarily be composed of elements, or compounds of the elements, and that one of the constituent bodies would therefore have to be infinite, from which Aristotle concluded that it would destroy or corrupt the other elements or com-

2. Duhem explains ([Ariew], 1985, 109) that "objections against the possibility of the categorematic infinite were common in the schools. By various devices, one derived from its possibility conclusions of this kind: One can add something to infinity; there can be something greater than infinity; an infinity can be the multiple of another, etc. One calls these conclusions absurd and one deduces from them that the possibility of the categorematic infinite is contradictory."
3. Ibid., 110.
4. Ibid.
5. The example is from Buridan [*De caelo*, bk. 1, qu. 17], 1942, 79–80. See also Duhem [Ariew], 1985, 124. Buridan and Albert of Saxony objected to this argument because no last proportional part of an hour could be assigned and therefore no last one-foot body. As Duhem explains it, "When one divides to infinity, by any process whatever, there are no parts of which one can state that they are all the parts of this magnitude" (Duhem, ibid.). Thus the infinite does not become actualized, because not all the parts can be actualized. It is a syncategorematic infinite. For further details, see Duhem [Ariew], 1985, 119–126.

pounds.[6] To counter Aristotle, Marsilius adopts the principle that "a natural agent that has a certain power in itself does not act on any patient with its whole power."[7] Appealing to "experience," Marsilius argues that if iron were ignited and then extinguished in the sea, it would be doused no more quickly than if it were placed in the river Seine, and yet the sea or ocean has a much greater power than the Seine. The same argument holds for the sphere of fire, which exceeds by more than ten times the aggregate of the other three elements. If the fire acted by its total power on the other elements, it would destroy them, but we see that this does not occur. From these and other examples, Marsilius infers that Aristotle's argument about the immediate corruptive effect of an actual infinite on its component parts is invalid, thus lending further credence to the idea of an actual infinite body.[8] And yet toward the end of the question, Marsilius admits "that in fact there is no actually infinite body, and although this could not be demonstrated, nevertheless this agrees more with the senses, because everything we perceive is finite."[9]

Despite great interest in the possibility of an actual infinite, no one argued for the existence of an actually infinite body. The explanation appears obvious: to do so would have necessitated the abandonment of Aristotelian physics and cosmology. Indeed, Aristotle's arguments against an actually infinite body, and for a finite world, had many defenders. In the first book of *De caelo*, Aristotle emphasized the absurdities that would arise with respect to motion, as when he declared that "In general, where there is neither centre nor circumference [as would happen in an infinite body or world], and one cannot point to one direction as up and another as down, bodies have no place to serve as the goal of their motion. And if this is lacking, there cannot be movement; for movement must be either natural or unnatural and these terms are defined in relation to places, i.e. the one which is proper and the one which is alien to the body."[10] An infinite body must have an infinite weight, which is impossible, because such a body would move instantaneously. But impossible consequences follow even if it moves in a minimum of time. For then, the time of its motion would bear a finite ratio to the time of motion of a finite body. "Hence the infinite would have moved the same distance in the same time as the finite, which is impossible."[11] In the *Physics* (bk. 3, ch. 5), Aristotle provided other arguments against the existence of an actually infinite body, this time ex-

6. Aristotle, *Physics* 3.5.204b.11–205a.24.
7. "Suppono primo quod agens naturale quod habet in se aliquam virtutem non agit in quodlibet passum in quod agit secundum totam suam virtutem." Marsilius of Inghen [*Physics*, bk. 3, qu. 10], 1518, 45r, col. 2.
8. "Ex quibus sequitur quod consequentia Aristotelis non valet quod si esset aliquod simplex actu infinitum statim corrumperet omnia alia." Ibid.
9. "Secunda conclusio est quod de facto nullum est corpus actu infinitum et licet ista non possit demonstrari, tamen ipsa magis concordat sensui quia quodlibet est finitum quod sentimus." Ibid., 45v, col. 2.
10. Aristotle, *De caelo* 1.7.276a.8–11 [Guthrie].
11. Ibid., 1.6.273b.26–274a.13.

amining the material composition of the infinite, that is, whether it would be a compound body or composed of one or more elements or even homogeneous, and the place of the infinite. His varied arguments rejecting an actually infinite body formed the basis of almost every medieval and Renaissance scholastic discussion in which the existence of an actually infinite body was rejected.

Aristotle's defenders offered many arguments in his behalf, some of which are directly or indirectly traceable to the commentaries of Averroës. Drawing upon Averroës and others, Albertus Magnus, in the thirteenth century, presented seven arguments against an actual infinite,[12] which Bartholomew Amicus summarized and repeated in the seventeenth century.[13] In one of these arguments, three-dimensional finite bodies were assumed of necessity to have surface boundaries. But if the infinite is called a "body," it will have three dimensions and therefore be finite, which is contrary to the assumption that it is infinite. Another argument assumed that the infinite body was configured as a finite, concave heaven, the convex side of which extended to infinity. But this raises impossibilities. For the finite, concave part of the heaven revolves in a finite time. Therefore the convex part must also complete a revolution in a finite time. But the infinite cannot complete a revolution in a finite time, from which it follows that the convex surface is not infinite.

Few defended Aristotle more vigorously than Buridan, who considered the possibility of an actually infinite body in both his *Questions on De caelo* and his *Questions on the Physics*. In support of Aristotle's denial that a body that moved with a circular motion could be infinite, Buridan explains that a body moving with a circular motion would have a center (because of its circular motion) and not have a center (because an infinite body can have no center), thus yielding contradictory properties.[14] On the assumption that an infinite body moving with a circular motion would complete one circulation in a finite time, the absurd consequence follows that an infinite body traverses an infinite space or distance in a finite time.[15] Buridan also considered the consequences of rectilinear motion by an infinite body and concluded that such motion would be impossible, because the infinite body would have to move from one infinite place to another infinite place outside of it, thus requiring the existence of many other infinite places, which is absurd.[16] Johannes de Magistris added another twist by observing that if a

12. Albertus Magnus [*De caelo*, bk. 1, tract. 2, ch. 8], *Opera*, 1971, 53, col. 2–55, col. 2.
13. Amicus [*De caelo*, tract. 5, qu. 2, dubit. 2], 1626, 252, col. 1–253, col. 2.
14. Buridan [*De caelo*, bk. 1, qu. 15], 1942, 67. For similar arguments, see Albert of Saxony [*De celo*, bk. 1, qu. 7], 1518, 91r, col. 2–91v, col. 2, and Johannes de Magistris [*De celo*, bk. 1, qu. 3], 1490, p. 9, col. 1 [unfoliated; pages numbered from beginning of *De celo*].
15. Buridan, ibid., 69–70. In the fifteenth century, Johannes Versor [*De celo*, bk. 1, qu. 8], 1493, 6v, col. 1, seems to have accepted six arguments, which he says are derived from Aristotle, against the idea that a body moved with a circular motion could be actually infinite.
16. Buridan, ibid., qu. 16, pp. 74–75. For a similar argument, see Johannes de Magistris, *De celo*, bk. 1, qu. 3, 1490, p. 9, col. 1.

rectilinearly moving body could be infinite, the source of its motion would be either from itself or from outside. If the former, it would be a living animal, which is impossible; if the latter, the external mover would also have to be infinite, and there would be two infinites, mover and moved.[17]

Dilemmas associated with the actual infinite posed serious and perplexing theological problems. At the end of the sixteenth century, in their response to the question "whether by his divine power God could produce an actual infinite," the Coimbra Jesuits declared that "in this weighty and difficult controversy, the negative part is to be preferred [namely that God cannot create an actual infinite], both because it advances the better arguments and because the better-known philosophers have embraced it."[18] But a few lines later, they add an important qualification with the assertion that "Nevertheless, since the affirmative part [that God can create an actual infinite] also has its probability and not unworthy supporters, and especially because by no [single] argument can it be plainly refuted, we explain the arguments of each side, so that anyone can support any side he wishes."[19] From the fourteenth century on, each side enjoyed considerable support.[20]

Buridan reveals how the problem of a divinely created actual infinite posed serious theological problems. He explains that his argument about the rectilinear motion of an infinite body, described a bit earlier in this chapter, is relevant only to natural powers, not supernatural powers. As far as natural conditions apply, every body that moves with a rectilinear motion must be in a place, unless the supernatural power decrees otherwise. Now, it is a contradiction to assume that an infinite body could be in a place, because a place presupposes that an infinite body could be contained by another body. Without a place, it would seem that an infinite body could not move from place to place and therefore could not move rectilinearly. "But with respect to the divine power," Buridan explains ([*De caelo*, bk. 1, qu. 16], 1942, 75), "it was determined by the bishop of Paris and the university at Paris that it was an error to say that God could not move the whole world simultaneously with a rectilinear motion." With this reference to article 49 of the Condemnation of 1277,[21] Buridan allows that although

17. "Tertia ratio: si aliquid corpus recte motus esset infinitum, vel illud moveretur a se, et sic esset animatum, quod est impossibile; impossibile enim est infinitum esse animal; vel ab alio, et sic illud aliquid esset infinitum, et sic essent duo infinita, scilicet movens et quod movetur." Johannes de Magistris, ibid. Averroës is probably the source of de Magistris's argument (see Averroës [*De caelo*, bk. 1, comment. 72], 1562–1564, 5:49v, col. 1).
18. "In hac gravi et difficili controversia negativa pars praeferenda est, tum quia firmioribus nititur argumentis, tum quia eam melioiris notae Philosophi amplexi sunt." Conimbricenses [*Physics*, pt. 1, bk. 3, ch. 8, qu. 2, art. 3], 1602, col. 544.
19. "Nihilominus, quia affirmativa etiam pars suam habet probabilitatem et assertores non ignobiles, ac praesertim quia nullo argumento palam revincitur, utriusque partis rationes diluemus ut quam quisque voluerit tueri possit." Ibid.
20. In the thirteenth century, God's ability to create an actual infinite was overwhelmingly rejected.
21. "Quod Deus non possit celum motu recto. Et ratio est, quia tunc relinqueret vacuum." Denifle and Chatelain, 1889–1897, 1:546. On the condemnation, see Chapter 3, Section

the world is not in a place, because nothing exists beyond the world that might contain it, God, by his supernatural power, could cause it to move rectilinearly. The implication for an infinite body is obvious: despite the absence of a place, God could move an infinite body rectilinearly, just as he can move our finite world.

But could God create an infinite body or magnitude? In treating the question "Whether there is an infinite magnitude,"[22] Buridan was compelled to concede

> it must be believed on faith that beyond this world God could form and create other spheres and other worlds and absolutely create as many finite magnitudes as He wishes, so that for every finite thing created, He could create a greater [or larger] one that is twice as large, or ten times as large, or one hundred times as large, and so on through any proportion of one finite thing to another.[23]

But such reasoning cannot apply to an infinite magnitude, because, as Buridan explains,

> it is not necessary to believe that God could create an actually infinite magnitude, because when it has been created he could not create anything that is greater, since it is repugnant [or absurd] that there should be something greater than an actual infinite.[24]

Buridan thought it ridiculous to suppose that God would make something so great that he could not make it greater yet.[25] Such a situation would place limitations on God's power to make something greater if he chose to do it. Moreover, the power of an actual infinite would be equal to God's infinite power. For example, if God created an actually infinite fire, the latter "would be of infinite power and thus not of lesser power than God, which is not possible."[26] In this matter, Buridan may have reflected that, as a master of arts without theological training, he was vulnerable to criticism from theologians, and hence he may have thought it wise to label his argument a "persuasion" (*persuasio*) and to declare

III.2. It was declared an error to deny that God could move the world rectilinearly, even if a vacuum were left behind. For a detailed discussion of the impact of article 49, see Grant, 1979, 226–232.

22. "Utrum est aliqua magnitudo infinita." Buridan [*Physics*, bk. 3, qu. 15], 1509, 57r, col. 2–58r, col. 2.
23. "Primo igitur credendum est fide quod Deus posset ultra istum mundum formare et creare alias speras et alios mundos et omnino alias magnitudines finitas quantascunque vellet ita quod omni finita creata posset creare maiorem in duplo, in decuplo, in centuplo, et sic de omni alia proportione finiti ad finitum." Ibid., 57v, col. 1.
24. "Sed forte non oportet credere quod Deus posset creare magnitudinem actu infinitam quia illa creata non posset creare maiorem repugnat enim quod actu infinito sit aliquid maius." Ibid.
25. "Et tamen inconveniens est quod Deus possit facere creaturam potentie sue proportionatam sic quod non possit maiorem et perfectiorem facere." Ibid.
26. Ibid., cols. 1–2.

explicitly that with regard to "all the things that I say in this question, I submit to the determination of the lord theologians, and I wish to acquiesce in their determination."[27] Despite his evident uneasiness, Buridan held to his opinion.[28]

By contrast, infinitists painted quite a different picture. Pedro Hurtado de Mendoza considered the argument Buridan raised: if God produced an actual infinite, his omnipotence would thereafter be limited, because he could never again produce anything greater than an actual infinite.[29] Hurtado insists that the creation of an actual infinite would in no way limit God's omnipotence, because he could always annihilate such an infinite and re-create it or create another.[30] Indeed – and here Hurtado turns the tables on the finitists – unless God could create an actual infinite, his omnipotence would always be limited to the creation of finite creatures.[31] Thus, only if God has the power to create an actual infinite would his omnipotence really be saved.

In the sixteenth century, John Major had employed similar arguments in defense of the possibility of an actual infinite. Not only could God create an actual infinite, but it would in no way be equal in perfection or power to God or resist him in any way. Its inferiority to God is evident by the fact that God could corrupt and destroy it either instantaneously or successively.[32] Thus it posed no threat. But unlike Hurtado and almost all other infinitists, Major believed in an actual infinity of worlds. Few scholastics,

27. "Immo de omnibus que dicam in ista questione ego dimitto determinationem dominis theologis et acquiescere volo determinationi eorum." Ibid., 57v, col. 2.
28. In his *Questions on De caelo*, bk. 1, qu. 17, Buridan says much the same thing: "I believe that it is not possible even by the divine power of God that there be an infinite body with respect to magnitude, nor can there be an infinite effect with respect to perfection. Indeed, I believe that God could not make a body that is so great but that He could not make another greater body; nor could He make a thing so perfect that He could not make another thing more perfect." Buridan, *De caelo*, 1942, 79. For more on Buridan's ideas on God and the infinite, see Chapter 7, Section II.3.
29. "Prima ratio quam asserunt est: quia omnipotentia Dei est simpliciter infinita, ergo non potest producere infinitum in actu. Probatur consequentia: quia si semel potest producere infinitum, ergo etiam poterit actu producere quicquid est producibile. Sed tunc finiretur omnipotentia Dei, quia iam non valet quid ultra producere, ergo." Hurtado de Mendoza [*Physics*, disp. 13, sec. 2: "De infinito"], 1615, 315, col. 2.
30. For a similar opinion, see Arriaga [*Physics*, disp. 13, sec. 4: "De infinito"], 1632, 417, col. 2.
31. "Praeterea obiectum illud omnipotentiae si actu produceretur esset sine termino et fine quam ob causam eius productione non limitaretur omnipotentia Dei, neque exhauriretur, sed potius actu referretur ad suum adaequatum obiectum et cum illo adaequaretur actu. Quod posset annihilare et iterum reproducere. Immo retorqueo argumentum quia nisi posset fieri infinitum actu, semper praescriberetur terminus omnipotentiae Dei quia semper quod actu produceret esset finitum necessario quod si ex eo quod potest infinitum in potentia, est infinita; multo melius id erit ex eo quod infintum de facto producat." Hurtado de Mendoza, *Physics*, disp. 13, sec. 2: "De infinito," 1615, 315, col. 2.
32. "Ad primum, concedo quod Deus potest facere infinitum, sed nego quod illud esset infinite perfectionis vel infinite resistentie respectu Dei; potest enim a Deo subito vel successive corrumpi." Major [*Sentences*, bk. 1, dist. 37], 1519a, 93v, col. 1.

however, were bold enough to follow Major's path and opt for a real existent infinite. Most confined themselves to choosing between the possibility or impossibility of a divinely created actual infinite magnitude or body.

The possibility of an actual infinite challenged the medieval doctrine of God's absolute power, so strongly manifested in the Condemnation of 1277, to do or create anything he wishes. It was one thing to believe that, by his absolute and infinite power, God could do anything he pleased with finite magnitudes, bodies, and qualities. But, as we saw, the infinite was a puzzle. Was God's power constrained more by attributing to him the ability to create an actual infinite or by confining him solely to the production of finite things? In the absence of any overpowering natural or theological argument, scholastics can be found on both sides of this thorny issue. By the seventeenth century, however, we learn from Roderigo de Arriaga that "it is common among the moderns" to assume that God can create an actual infinite.[33]

But when we leave the realm of hypothetical arguments on the actual infinite and ask who among scholastic authors really believed in an actually created infinite magnitude or body, the numbers are few indeed. Apart from John Major and Thomas Compton-Carleton and perhaps a few others, almost all believed in a finite world.[34] And while some may have further held that an infinite imaginary space surrounded our finite world, they did not view that infinite space as a divine creation but only as the eternal immensity of God himself. With regard to the created world, however, scholastic natural philosophers from the Middle Ages to the Renaissance were far more likely to have agreed with Raphael Aversa that "it is absolutely certain, in accordance with Aristotle and all other philosophers and theologians, that the world and universe are truly finite and that no infinite body can be attributed [to them], but that all things are of an absolutely finite magnitude."[35]

II. On the shape of the world

Although the question of shape would have made little sense for an infinitely extended world, it was an important issue for a world of finite dimensions. Despite other possibilities, the world, or cosmos, was assumed to be spherical. As many as three aspects of sphericity were distinguishable: (1) the convex surface of the last, or highest, heaven; (2) the concave surface of

33. "In hanc sententiam inclinat P. Vasquez . . . et est communis inter modernos." Arriaga, *Physics*, disp. 13, sec. 4: "De infinito," 1632, 417, col. 2.
34. On Major, see Chapter 8, Section II.2; for Compton-Carleton, see Chapter 9, Section III.
35. "Certum tamen prorsus est, cum ipso Arist. et aliis omnibus philosophis ac theologis, mundum et universum esse vere finitum nullumque dari corpus infinitum, sed omnia esse finitae simpliciter magnitudinis." Aversa [*De caelo*, qu. 31, sec. 4], 1627, 9, col. 2–10, col. 1.

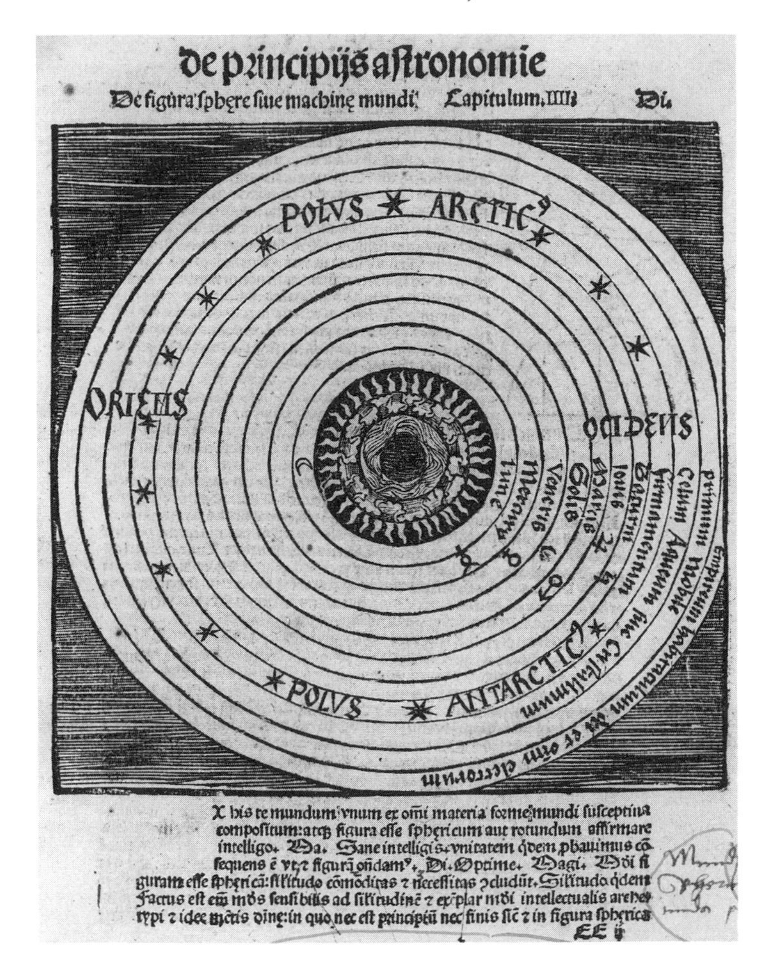

Figure 6. The finitude and shape of the world. (Gregor Reisch, *Margarita philosophica nova*, 1508, sig. EEii. Courtesy Lilly Library, Indiana University, Bloomington.)

the whole heaven, which was usually understood to be the concave surface of the lowest, or lunar, heaven; and (3) the sphericity of the heavens intermediate between the outermost and the innermost orb, each of which was assumed to have both a convex and a concave surface.[36] In discussing sphericity, I follow the medieval custom and employ the terms "sphere" and

36. "Pro questione primo notandum est quod questio potest intelligi de figura celi ultimi quoad superficiem eius convexam; vel de figura totalis celi quantum ad superficiem eius concavum; vel de figuris orbium intermediorum inter orbem supremum et infinitum quantum ad eorum superficies concavas et convexas." Albert of Saxony, *De celo*, bk. 2, qu. 5, 1518, 104v, col. 2. Pierre d'Ailly repeats essentially the same threefold division in D'Ailly, *14 Questions*, qu. 5, 1531, 153r.

"orb" to represent indifferently the Latin terms *spera* (or *sphaera* and *sphera*) and *orbis*.[37]

As with much else in cosmology, Aristotle and Averroës played a significant role in formulating the responses.[38] The explanation given by John of Sacrobosco in the first chapter of his treatise *On the Sphere* was, however, sufficiently brief and important to have exercised a considerable influence. Sacrobosco offered three reasons for assuming the sphericity or roundness of the heaven, or sky:

> likeness [*similitudo*], convenience [*commoditas*], and necessity [*necessitas*]. *Likeness*, because the sensible world is made in the likeness of the archetype, in which there is neither end nor beginning; wherefore, in likeness to it the sensible world has a round shape, in which beginning or end cannot be distinguished. *Convenience*, because of all isoperimetric bodies the sphere is the largest and of all shapes the round is most capacious. Wherefore, since the world is all-containing, this shape was useful and convenient for it. *Necessity*, because if the world were of other form than round – say trilateral, quadrilateral, or many-sided – it would follow that some space would be vacant and some body without a place, both of which are false, as is clear in the case of angles projecting and revolved.[39]

37. A useful distinction could be made between the terms *spera* and *orbis*, as Pierre d'Ailly (ibid.) observed when he declared that "properly speaking 'orbicular' and 'spherical' differ because orbicular [is a figure that] ought to be contained by two surfaces, namely concave and convex; the heaven is this way. Spherical, however, ought to be contained by a single surface, namely by a convex [surface] only. Nevertheless, sometimes one is used in place of the other, as in [what is] proposed [here]" (Sciendum est quod proprie loquendo orbiculare et sphaericum differunt quia orbiculare debet esse contentum duabus superficiebus, scilicet concava et convexa cuiusmodi est coelum. Sphaericum vero debet esse contentum unica superficie, scilicet convexa. Tamen quandoque ponitur unum pro alio, sicut in proposito). D'Ailly first makes this comparison earlier in the treatise (146v), when he declares that "an orb is called a spherical figure contained by a double surface in the middle of which is imagined a point." (Orbis vero dicitur figura sphaerica duplici superficie contenta in cuius medio imaginatus est punctus.) He goes on to apply this definition to celestial orbs, which "ought not to be called spheres, properly speaking; this is obvious because they are contained by a double surface, namely concave and convex" (Ex dictis sequuntur aliqua correlaria: primum orbes coelestes non debent dici sphaerae proprie loquendo patet quia continentur duplici superficie, scilicet concava et convexa).

    Obviously, d'Ailly chose to ignore the distinction, which was also repeated by Christopher Clavius [*Sphere*, ch. 1], *Opera*, 1611, 3:10, who explained: "Hoc namque differt orbis a sphaera, quod haec ad centrum usque tota sit solida unicaque tantum superficie, puta convexa exteriore concludatur; orbis autem non ita, sed duabus finiatur superficiebus: una exteriore et altera interiore, quales sunt omnes coeli." Although each heaven has both a concave and a convex surface and is, strictly speaking, an "orb" rather than a "sphere," most commentators followed d'Ailly and used the terms indifferently.

38. In his commentary on Genesis, Saint Augustine devoted a chapter to "The Shape of the Material Heaven," and seems to argue for its sphericity. Augustine reconciles two conflicting biblical passages, one in which the heaven is described as a stretched-out skin (Ps. 103.2) and another in which it is described as something suspended like a vault (Isa. 40.22). He concludes that "Our picture of heaven as a vault, even when taken in a literal sense, does not contradict the theory that heaven is a sphere." See Augustine, *Genesis*, bk. 2, ch. 9, 1982, 1:59–60.

39. Sacrobosco, *Sphere*, ch. 1, 1949, 120 (Latin text: 80–81). The argument from convenience was a popular one invoked by Roger Bacon, *De celestibus*, pt. 2, ch. 2, *Opera*, fasc. 4,

Commentators on Sacrobosco's *Sphere*, such as Michael Scot in the thirteenth century and Cecco d'Ascoli in the fourteenth, generally accepted and further elaborated on these three points.[40] But not all found his arguments convincing, although all accepted the sphericity of the world. In his commentary on the *Sphere*, Pierre d'Ailly found the third argument, the argument from necessity, wanting. He argued that even if all the celestial spheres moved and none rested, Sacrobosco had not proved that the convex surface of the last sphere is spherically shaped. For if it were oval in shape and moved around an axis that is the longest line connecting its farthest points, none of the dire consequences that Sacrobosco foresaw would occur, because an oval, like a sphere, has no outwardly projecting angles, and therefore nothing could project beyond the world. For if anything did, it could not be said to be in a place, and if it moved when the world rotated, it would abandon one location and occupy another, thus leaving behind an empty place of some kind, all of which is impossible in Aristotelian physics and cosmology.[41]

1913, 344; Cecco d'Ascoli [*Sphere*], 1949, 365; Albert of Saxony, *De celo*, bk. 2, qu. 5, 1518, 105r, col. 1; and Christopher Clavius (discussed next). As proof that among solid figures with equal perimeters the sphere had the greatest capacity or volume, Bacon and others cited the eighth proposition of the *De ysoperimetris*, an anonymous Greek mathematical treatise that was translated into Latin in the thirteenth century. (For the brief Latin passages from *De ysoperimetris* that speak of the circle and the sphere, see Clagett, 1964, 221; also 32 and 630–632.) Although omitting the proposition number, Albert of Saxony not only cites the treatise by name but attributes it to Archimedes (or, as the text has it, "Archimenides").

The ultimate, if not direct, source of the argument from necessity is Aristotle, *De caelo* 2.4.287a.11–23, where Aristotle argues that if, as he assumes, there is neither body, void, nor place beyond the outermost surface of the world, then that surface must be circular and not rectilinear. "For if it is bounded by straight lines, that will involve the existence of place, body, and void. A rectilinear body revolving in a circle will never occupy the same space, but owing to the change in position of the corners there will at one time be no body where there was body before, and there will be body again where now there is none" (Aristotle [Guthrie], 1960).

Relying on Alfraganus's *Differentie scientie astrorum*, Sacrobosco offered a fourth argument in behalf of cosmic sphericity by way of rejection of a flat sky. If the sky were flat, stars overhead should appear larger than when rising or setting. Thus the Sun should appear larger when overhead than when rising or setting. But the opposite is the case, since the Sun appears larger in the east or west than overhead. Alfraganus explains this phenomenon as a consequence of the refraction of the Sun's rays, caused by vapors that intervene between us and the Sun or other stars and planets. For the text of the Latin translation, see Alfraganus [Carmody], 1943, 5–6, and [Campani], 1910, 65–66. As might be expected, this argument was subsequently cited in discussions on the sphericity of the world, or, more specifically, on the sphericity of the concave surface of the heavens, as Albert of Saxony (*De celo*, bk. 2, qu. 5, 1518, 105r, col. 1) carefully noted.

40. For Michael Scot's brief discussion, which makes no mention of Sacrobosco's three specific points, see Michael Scot [*Sphere*, lec. 2], 1949, 257–259. Cecco d'Ascoli ([*Sphere*], 1949, 364–366) takes them up in the same order.

41. "Secunda conclusio est quod si ponamus quodlibet caelum moveri et nullum quiescens, adhuc ratio autoris non probat caelum quantum ad superficiem supremi caeli convexam esse sphaericae figurae. Patet conclusio quia si caelum poneretur ovalis figurae adhuc posset salvari eius motus sine hoc quod remaneret locus sine corpore aut corpus sine loco ponendo quod moveretur super axem quae est linea longissima in illo corpore et quod coni essent poli illius figurae. Et ita esset si caelum poneretur esse figurae compositae ex

In the late sixteenth and early seventeenth centuries, Christopher Clavius, who took issue with Sacrobosco on all three of his arguments for cosmic sphericity, agreed with d'Ailly concerning Sacrobosco's argument from "necessity" (*necessitas*). Clavius insists that it is valid only to the extent that it shows that the last heaven is "round in some way" ("coelum esse aliquo modo rotundum") and therefore not an angular body. But "round in some way" embraces not only spherical, but also oval, conical, or cylindrical. If this heaven were shaped in any of these forms, its rotation around its axis would occur in a manner similar to that of a globular or spherical body.[42] But Clavius argues that in fact the heavens must be spherical, because the inferior celestial bodies not only move with motions that are opposite to that of the first movable sphere (*primum mobile*), which represents the daily motion, but they also move on different poles. If these inferior heavens were not spherical, they would intersect and produce one of two impossible consequences: one heaven or round body would penetrate another, so that parts of two distinct bodies would occupy the same place; or, if they did not penetrate, one body would cause the other to divide and make room for it.[43]

D'Ailly also thought that Sacrobosco had failed to demonstrate the sphericity of the concave surface of the last heaven, which d'Ailly identifies as the concave surface of the lunar heaven. If it had an angular shape, the angular protuberances that extended below into the region of fire would leave behind a potentially empty space as the concave lunar surface moved around. But d'Ailly argues that the fluid fire below the concave surface of

duabus pyramidibus rotundis." D'Ailly, *14 Questions*, qu. 5, 1531, 153v. Albert of Saxony (*De celo*, bk. 2, qu. 5, 1518, 105r, col. 2) argued similarly, but much more briefly, that an oval or columnar shape for the convex surface of the world was not contrary to natural principles.

42. In the fourteenth century, Themon Judaeus appears to have raised the same objection against Sacrobosco (see Hugonnard-Roche, 1973, 119–120), as did Albert of Saxony (*De celo*, bk. 2, qu. 5, conclus. 1, 1518, 104v, col. 2). Aristotle seems to have sided with Sacrobosco and denied this argument on the grounds that if the celestial body "were of some other shape whose radii were unequal, that of a lentil or an egg for example," the revolution of such bodies would "involve the existence of place and void outside the revolution, because the whole does not occupy the same space throughout" (Aristotle, *De caelo*, 2.4.287a.20–22 [Guthrie], 1960). In a sense, both approaches are correct. The choice of the axis of rotation determines whether or not an oval-shaped body will occupy the same space.

43. Clavius [*Sphere*, ch. 1], *Opera*, 1611, 3:51. Medieval scholastics had made this point long before Clavius: for example, Themon Judaeus (see Hugonnard-Roche, 1973, 120); Albert of Saxony ([*De celo*, bk. 2, qu. 5, conclus. 4]), 1518, 104v, col. 2; d'Ailly, *14 Questions*, qu. 5, conclus. 5, 1531, 153v; and Versor, *De celo*, bk. 2, qu. 6, 20v, col. 2. Probably relying on Clavius, both Aversa, *De caelo*, qu. 31, sec. 4, 1627, 12, col. 2–13, col. 1, and Thomas Compton-Carleton [*De coelo*, disp. 4, sec. 3], 1649, 410 [mistakenly paginated 310], cols. 1–2, present substantially the same argument. However, whereas Compton-Carleton thought the convex surface of the outermost empyrean heaven could plausibly be assumed to be cubical (although he used the term *quadratum*), Aversa assumed it was spherical. The Coimbra Jesuits, citing Aristotle explicitly, also mention the void that would form if the last heaven were any shape other than spherical (Conimbricenses [*De coelo*, bk. 2, ch. 5, qu. 1, art. 4], 1598, 245). For more on the consequences associated with nested spheres, see Chapter 13, Section III.1–3.

the lunar heaven would immediately fill the potentially empty space "and thus no place would remain without a body."[44]

As for the argument from likeness (*similitudo*), Clavius insisted that it proves absolutely nothing. For the same argument that claims that the world was made round in imitation of the archetype so that no place on it would be either a beginning or an end applies equally well to humans and every other creature. Humans could also have been created round, so that they would have no beginning or end and thus be like the archetype. This has failed to occur, Clavius argues, because God chose not to manifest his perfection through one creature alone. He decided instead to make the whole world round because all creatures are contained within it and because the world manifests the goodness and perfection of God most efficaciously.[45]

Clavius accepts the argument from convenience (*commoditas*), but he denies that it is conclusive. Although the sphere is the most capacious of isoperimetric bodies, God could have employed another shape for the world that would be even larger than the present sphere and would still contain all the things that it now contains. But there are other reasons of convenience that God might have followed. Because Nature (*Natura*) always does what is best, it would have been appropriate to give the celestial body, which is the noblest of all bodies, the noblest shape, which is the sphere.[46]

The criticisms made by d'Ailly and Clavius were not intended to refute the case for sphericity, but to strengthen it. With an occasional special exception, virtually all scholastic authors between the thirteenth and the seventeenth century believed in the sphericity of the heavens and of all of their distinctive parts. In support of this conviction, they offered other arguments than those already described, some based on experience, others drawn from the principles of natural philosophy. Arguments from observational experience were common. Saint Bonaventure observed that from any part of the earth or oceans, the sky always seems to be equidistant from the earth, which could not occur unless the sky has an orbicular shape.[47] Aversa pointed to the circular motions of the Sun and planets. Thus the Sun is carried from east to west and again returns to the east as it completes its circular motion. The planets move in a similarly circular fashion. Therefore such motions are completed on spheres.[48] Observers on a ship moving

44. "Quarta conclusio est quod ratio autoris non demonstrat caelum esse sphaericum quantum ad superficiem concavum ultimi caeli, scilicet lunae. Patet conclusio quia dato quod poneretur esse angularis figurae cum moveretur aliquis conus, tunc ignis qui est fluxibilis subsequeretur. Et sic non remaneret locus sine corpore." D'Ailly, ibid., 153v. Albert of Saxony presents the same argument in *De celo*, bk. 2, qu. 5, 1518, 104v, col. 2.

45. Clavius, *Sphere*, ch. 1, *Opera*, 1611, 3:49. Clavius observes that God imitated sphericity or roundness in some things, as in tree trunks, branches, fruits, and in the extreme parts of animals.

46. Ibid.

47. Bonaventure [*Sentences*, bk. 2, dist. 14, art. 2, qu. 1], *Opera*, 1885, 2:341.

48. Aversa, *De caelo*, qu. 31, sec. 4, 1627, 12, col. 1. See also Conimbricenses, *De coelo*, bk. 2, ch. 4, qu. 1, art. 1, 1598, 239. Centuries earlier, Bonaventure (*Sentences*, bk. 2, dist. 14, art. 2, qu. 1, *Opera*, 1885, 2:341) made the same point in more general terms when

toward the horizon and witnessing, day after day, the rise of a star above the horizon at the same hour provide evidence that "the earth is everywhere spherical and so also are the heavens in their circulation around the earth."[49] According to the Coimbra Jesuits, terrestrial observers infer the sphericity of the heavens from the fact that the distance of the fixed stars from the earth seems never to vary, as well as from the observed paths of the fixed stars around the north pole, where those nearer the pole describe smaller circles than those farther away. Indeed, they also argued that since the shape of the planets was spherical, so also ought the shape of the heavens to be.[50]

Arguments in favor of the sphericity of the world were also based on the configuration of three of the the sublunar elements: water, air, and fire. Johannes Versor put it this way: "if water is spherical, it is necessary that air be spherical; and if air is spherical, it is necessary that fire be spherical, since every thing that touches in its entirety something spherical must be spherical with respect to the surface which touches it. And if fire is spherical, it is necessary that the heavens be spherical. But water is spherical. Therefore [every heaven] from the first to the last is spherical," and therefore so is the world.[51] The argument may be said to draw on experience because the underlying assumption is that the earth, around which the other three elements lie, is spherical. The determination of the earth's sphericity was based on various observations.[52] Also seemingly experiential was the claim that astronomical instruments, such as the astrolabe and the celestial globe, were circularly or spherically shaped to imitate the spherical shape of the heavens.[53]

Other supportive arguments relied heavily on natural philosophy, especially on the properties that Aristotle had assigned to the fifth element, or ether, which made up the celestial region. In a few syllogisms, Bonaventure invoked some of these. "To the simplest body, the simplest shape ought to be assigned; but the heaven is the simplest among bodies, since it

he declared: "Item, videmus sensibiliter, caelum moveri circulariter; sed motus circularis non competit nisi figurae circulari: ergo caelum est orbicularis figurae."

49. Aversa, ibid.
50. Conimbricenses, *De coelo*, bk. 2, ch. 4, qu. 1, art. 1, 1598, 239.
51. "Quarta ratio. Si aqua sit sperica oportet aerem esse spericum. Et si aer est spericus oportet ignem esse spericum cum omne tangens spericum aliquid secundum se totum sit spericum secundum illam superficiem que tangit ipsum. Et si ignis sit spericus, oportet celum esse spericum. Sed aqua est sperica. Ergo de primo ad ultimum celum est spericum." Versor, *De celo*, bk. 2, qu. 6, 1493, 20v, col. 2. Versor then offers a proof of the sphericity of water and follows this with a concluding statement that since the elements are spherically shaped, so is the whole world ("Unde patet quod elementa sunt sperice figure ex quibus sequitur quod totus mundus sit sperice figure"). Ibid., 21r, col. 1. The argument is drawn from Aristotle, *De caelo*, 2.4.287a.31–287b.4, and was fairly common. It also appears in Johannes de Magistris, *De celo*, bk. 2, qu. 2, 1490, 20 (counting from the beginning of *De celo*; or sig. K5r).
52. In his discussion, Aristotle includes the earth as the center and then declares that water is located around the earth, air around water, etc. (*De caelo*, 2.4.287a.31–287b.4). On the earth's sphericity, see Chapter 20, Section III.
53. See Albert of Saxony, *De celo*, bk. 2, qu. 5, 1518, 105r, col. 1, and Conimbricenses, *De coelo*, bk. 2, qu. 1, art. 1, 1598, 240.

is the most subtle; therefore the simplest body ought to have this shape. But this [simplest] shape is circular; therefore, etc."[54] Similarly, "to the most perfect body, the most perfect shape ought to be assigned; but the heaven is the most perfect of all [bodies], and the spherical shape is the most perfect of all shapes . . . ; therefore, etc."[55] And, finally, Bonaventure appeals to motion when he explains that "to the quickest speed there ought to be assigned the shape best suited for motion; but the heaven is the quickest body, [and], moreover, the shape best suited for motion is spherical; therefore, etc."[56]

Despite overwhelming support for a spherical cosmos, a few dubious or opposing voices were raised. One was that of Francesco Patrizi, who, although not a scholastic, seems to have had the traditional responses in mind. Patrizi doubted that any proper proofs had ever been given for belief in the theory of the rotundity of the celestial region,[57] but conceded that sound arguments had been given to show that it was not angular.[58] Patrizi's argument turns on the difference between the assumption of a world that is infinite and one that embraces traditional finite dimensions. As a defender and supporter of an infinite universe, Patrizi assumed that it could be round, because all lines drawn from its center would be of infinite length and therefore equal. Hence the infinite universe would necessarily be round.[59] But this does not apply to the traditional finite world, which, although probably curved, could be curved like an oval or a cone rather than like a sphere. Patrizi thus registered doubt but did not flatly reject a traditional spherical cosmos.

An unusual, and perhaps even bizarre, exception to the almost unanimous

54. "3. Item, simplicissimo corpori simplicissima debetur figura; sed caelum est simplicissimum inter corpora, cum sit subtilissimum: ergo simplicissimam debet habere figuram. Sed haec est figura circularis: ergo etc." Bonaventure, *Sentences*, bk. 2, dist. 14, art. 2, qu. 1, *Opera*, 1885, 2:341, col. 1. This argument is based on Aristotle, *De caelo* 2.4.286b.10–287a.5. Albert of Saxony [*De celo*, bk. 2, qu. 5], 1518, 104v, col. 2, repeats it.

55. "5. Item, corpori perfectissimo perfectissima debetur figura; sed caelum est perfectissimum omnium, et figura orbicularis est perfectissima omnium figurarum . . . ; ergo etc." Bonaventure, ibid., col. 2. The passage from Aristotle cited in the preceding note also applies here. The same argument appears in Albert of Saxony, ibid. In a similar vein, Versor (*De celo*, bk. 2, qu. 6, 1493, 20v, col. 2), declares that "the first figure [or shape] ought to be assigned to the first body, but the heaven is the first of all bodies, and the spherical shape is the first of all shapes; therefore the heaven is spherical."

56. "6. Item, corpori velocissimo debetur figura maxime ad motum idonea; sed caelum est velocissimum corpus, figura autem maxime apta ad motum est orbicularis: ergo, etc." Bonaventure, *Sentences*, bk. 2, dist. 14, art. 2, qu. 1, *Opera*, 1885, 2:341, col. 2. The argument is derived from Aristotle, *De caelo* 2.4.287a.24–30, and appears also in Albert of Saxony [*De celo*, bk. 2, qu. 5], 1518, 105r, col. 1.

57. "Nullae sunt ergo et astronomorum et philosophorum rationes quibus celi rotunditas adstruebatur." Patrizi, *Pancosmia*, bk. 10 ("De coeli rotunditate"), 1591, 87v, col. 2.

58. "Sed argumentum hoc sphaeristis istis validissimum, nihil concludit aliud quam coelum non esse angulatum. Non tamen illico est rotundum. Quia et ovale potest esse, et conicum, quod veterum quidam asserverunt." Ibid., col. 1.

59. Patrizi is, of course, mistaken to assume that an infinite world can have a center. Therefore his further assumption that all the infinite lines drawn from it would be equal is also unacceptable.

verdict in favor of sphericity was the suggestion that perhaps the convex surface of the last, or empyrean, heaven was square-shaped: that is, had the shape of a cube. As Compton-Carleton put it, "there is a difficulty about the convex part of the empyrean heaven."[60] Those who upheld the existence of an empyrean heaven, and most did, assumed that it surrounded the movable spheres but was itself at rest. Although the spherical shape was deemed the most perfect figure for the celestial heavens, Compton-Carleton, for example, suggests that in view of the conviction that the empyrean heaven is motionless, it would seem more appropriate that the convex surface be square-shaped, since the square is the most suitable shape for a body at rest.[61] He argued further that the substantial form of the empyrean heaven is not the most perfect of all things that could be created: therefore it had no need for the most perfect shape and need not be a sphere.[62] For those who rejected this strange mix of spherical and square surfaces, the most fundamental appeal would have been to a sense of symmetry and uniformity, namely that the surfaces of all the heavens should be of the same shape: spherical.[63]

The idea of the spherical shape of the universe did not die with the Ptolemaic system or with the birth of the Copernican. Tycho Brahe and Galileo envisioned a spherical cosmos. Even Kepler, who determined the elliptic orbits of the planets and thereby rejected the long-standing circular orbits of traditional astronomy, accepted an immobile sphere of fixed stars to enclose the planetary orbits (Kepler [*Epitome*], 1952, 854–855). Thus at least one distinguishably spherical convex surface was accepted by almost everyone, scholastic and nonscholastic. Few, if any, were so bold as to join Patrizi in doubting the sphericity of the heaven of the fixed stars, and therefore of the world itself, or of suggesting a square shape for the all-embracing empyrean heaven, as did Compton-Carleton. Otto von Guericke spoke accurately when he declared that "Ptolemaic, Tychonic, and Copernican astronomers assign only one sphere to all the fixed stars."[64]

60. "De coelo empyreo est difficultas quoad partem convexam." Compton-Carleton, *De coelo*, disp. 4, sec. 3, 1649, 410 (mistakenly paginated 310), col. 2.
61. In the *Timaeus* (55D-E), Plato assigned the cube as the shape of the basic unit of earth because he regarded the latter as the most immobile of the elements. Compton-Carleton, ibid., assumed that the concave surface was spherical because it was contiguous to the spherical tenth heaven, or *primum mobile*. At the close of the sixteenth century, some years before the appearance of Compton-Carleton's discussion, the Conimbricenses [*De coelo*, bk. 2, ch. 5, qu. 1, art. 4], 1598, 245, held the same opinion. The idea that a concave surface of the heaven, or firmament, can be spherical and its convex surface flat (and therefore cubical?), is found in Saint Basil's *Exegetic Homilies* (*Hexaemeron*), homily 3.4, 1963, 42.
62. "Sicut ergo forma substantialis coeli empyrei non est omnium perfectissima ex iis quae creati possint, ita nec est necessarium ut habeat figuram omnium perfectissimam." Compton-Carleton, *De coelo*, disp. 4, sec. 3, 1649, 410, col. 2.
63. In a rather obscure passage, this seems to be Aversa's response to those who conjectured that both the convex and concave surfaces of the empyrean heaven were square, but his reaction to the case involving a concave spherical surface with a square convex surface is unclear. See Aversa, *De caelo*, qu. 31, sec. 4, 1627, 13, col. 2.
64. "Astronomi quidem stellis fixis omnibus unam tantum sphaeram assignarunt, tam Pto-

For a host of reasons, good and bad, the idea that the world and all its distinctive constituent heavens, as well as its successively arranged elementary spheres, were spherical was rarely challenged in scholastic cosmology between the thirteenth and seventeenth centuries.

III. Are the outermost sphere, and the world itself, in a place?

Because the finitude and sphericity of the world were assumed by almost everyone, Aristotelian natural philosophers confronted a peculiar problem, which arose from the very nature of Aristotle's physics and cosmology. They had to determine whether the last, or outermost, sphere of the world is in a place.[65] Although this question was sometimes treated as equivalent to the question as to whether the whole world is in a place, queries about the place of the last sphere usually elicited different responses from queries about the place of the world. Indeed I shall treat the two questions separately, because they came to depend on the difference between external place (*locus*) and internal place (*ubi*), the latter associated with the concept of a three-dimensional void space in the sixteenth and seventeenth centuries.

1. Aristotle on the place of a body

To understand why this was a problem, it is essential to understand Aristotle's definition of the place of a body. In the fourth book of the *Physics*, where he considered the meaning of the concept of place, Aristotle concluded that any definition of the place of a body must meet the following conditions:[66]

1. The place of a thing must not constitute any part of that thing but serve only as its container.
2. The immediate place of a thing must be neither greater nor smaller than the thing contained.
3. Since a thing can depart from its place, the latter must be separable from what it contains.

lemaici quam Tychonici ut et Copernicani." Otto von Guericke, 1672, 223, col. 2. They do differ, says Guericke, in their estimation of the distances of the stars.

65. Scattered through his lengthy treatment of the doctrine of place, Duhem considered the problem of the place of the last sphere. See Duhem, *Le Système*, 1913–1959, 7:158–302, and vol. 10; for the English translation, Duhem [Ariew], 1985, 139–291.
66. In *Physics* 4.5.212b.27–28 [Hardie and Gaye], 1984, Aristotle explains that "not everything that is is in a place, but only movable body." Thus sphinxes and goat-stags, which are nonexistent, are "nowhere." Spiritual substances, such as celestial intelligences, though existent, were also without places. During the Middle Ages, spiritual substances were thought to occupy spaces in special ways characterized by the expression *ubi definitivum* (see Ch. 9, Sec. III); see also Grant, 1976b, 72–73. In my discussion of place and the place of the world, I draw, for the medieval part, upon Grant, 1981b.

4. A proper place is distinguishable with respect to "up" and "down," directions that are determined by the natural tendency of bodies to come to rest, some "up" and some "down."

With these criteria in mind, Aristotle presented four possible definitions of the place (*locus*, in the Latin translations) of a body: (1) the shape (or form) of the body itself; (2) the matter of the body; (3) the extension or dimension that lies between the surfaces of the containing body; and (4) the surface of the containing body. By process of elimination, Aristotle concludes that only the fourth definition – place as the surface of the containing body – meets the criteria. Indeed, Aristotle rendered an even more precise definition, characterizing place as "the boundary of the containing body at which it is in contact with the contained body."[67] Moreover that boundary, or innermost surface of the container, had also to be motionless,[68] a qualification that would present serious problems in the history of the exegesis of Aristotle's doctrine of place. When a body met these stringent criteria, it was said to be in a "proper place," that is, in a place that it alone occupied. By contrast, a place that contained more than one distinct body was characterized as a "common place."[69]

In the world as Aristotle envisioned it, then, the proper place of a body was always the innermost surface of another surrounding body. Because Aristotle's world was a material plenum, without the possibility of any existent vacua, Aristotle assumed that every body in the world that was capable of motion or change of size was somewhere, and therefore necessarily in a place. The outermost sphere of the world, however, was a notable exception. It could not be in a place, because Aristotle had argued that no body could possibly exist beyond our unique world.[70] Therefore no concave, or innermost, surface of any material body could surround the outermost sphere and serve as its place. Denial of a place to the last sphere was a consequence forced upon Aristotle in order to avoid an infinite regress of material places; for if the outermost sphere of the world were contained by another sphere, the latter, in turn, would require a containing sphere, and so on, ad infinitum, a process that would inevitably lead to the assumption of an infinite universe.[71] The anomaly of a last sphere that was not in a place apparently proved too much for Aristotle, who managed to find a kind of place for it by arguing that the last sphere was in a place indirectly by means of its parts, since "on the orb one part contains another."[72]

My "Catalog of Questions" (Appendix I), reveals that the possible place of the last sphere was one of the most widely discussed questions during

67. Aristotle, *Physics* 4.4.212a.5–7 [Hardie and Gaye], 1984.
68. Ibid., 212a.20–21.
69. Ibid., 4.2.209a.31–209b.1.
70. As we have already seen, in *De caelo* 1.9.278b.35–279a.18, Aristotle had argued that neither body, place, void, nor time existed beyond the world.
71. See Grant, 1978a, 272.
72. *Physics* 4.4.212b.12–14 [Hardie and Gaye], 1984.

the period covered by this study, serving not only as the subject matter of a question that specifically asked whether the last or outermost sphere is in a place (qu. 289) but also turning up in a question that asked "Whether every being is in a place" (qu. 283). Indeed, not a few characterized the question as difficult.[73]

2. *Responses to the problem*

a. The defense of Aristotle

The difficulty of the question is perhaps reflected in the range of responses. To begin with, authors had to decide whether or not the last sphere was in a place. A negative response meant that not every body in the world had a place; an affirmative reply compelled an author to fashion an explanation that could make sense of a place for a last sphere that had no material body surrounding it. Either way, the consequences posed a dilemma.

Some scholastic authors simply accepted Aristotle's position and assumed that the least problematic way of coping with the question was to deny that the outermost sphere could be in a place. Among those who took this path were Alexander of Aphrodisias (fl. 2nd–3rd c.), Avicenna, Johannes Canonicus, Roger Bacon, Jean Buridan, Albert of Saxony, and Marsilius of Inghen. Alexander's opinion was reported by Averroës, in the latter's *Commentary on the Physics*.[74] According to Averroës, Alexander ascribed to Aristotle the belief that the eighth sphere, or the outermost sphere of the fixed stars, is not in a place. Alexander explains that the eighth sphere does not move *per se* as a whole, because it does not change its place: that is, it does not leave one proper place for another, which is essential if the eighth sphere, or any body for that matter, can be said to have moved. Nor does the eighth sphere move *per se* by its parts, because it has no parts that are distinguishable from the whole that might serve as places.[75]

73. For example, Franciscus Toletus declared that "in this matter, there is special doubt, because it is of the greatest difficulty" (Dubium hoc praecipuum est in hac materia propter summam ipsius difficultatem). Toletus [*Physics* bk. 4, ch. 4, qu. 7], 1580, 121v, col. 1. Amicus explains that he introduced the question about the last sphere because that part of the universe posed great difficulty (Nos autem quaestionem instituimus de ultima sphaera, nam de ea parte universi est praecipua difficultas). Amicus, *Physics*, tract. 20, qu. 5, dubit. 2, art. 1, 1626–1629, 698, col. 1. See also Buridan, *Physics*, bk. 4, qu. 6, 1509, 72, col. 2, and John of Jandun [*Physics*, bk. 4, qu. 9], 1519, 64v, col. 1.
74. Averroës [*Physics*, bk. 4, comment. 43], 1562–1574, vol. 4, 143r, col. 1. Alexander's opinion was frequently repeated in the Middle Ages and Renaissance.
75. As justification for this claim, Alexander probably depended on Aristotle's *Physics* 4.4.211a.29–30 [Hardie and Gaye], 1984, where Aristotle explains that "when what surrounds . . . is not separate from the thing, but is in continuity with it, the thing is said to be in what surrounds it, not in the sense of in place, but as a part in a whole." Like all other spheres, the eighth is composed of homogeneous ether and therefore cannot have separable, distinguishable parts. For whatever part of the eighth sphere one takes that does not touch the convex surface of the sphere of Saturn below will be surrounded by identical ether that cannot function as its container and proper place.

Alexander denies as well that the outermost sphere can be said to be moved accidentally (*per accidens*) with respect to its parts. That could occur only if the whole were moved *per se*. Indeed, parts cannot be distinguished in the ethereal sphere either actually – because the parts cannot be made distinct – or in potentiality – because the parts cannot be distinguished even potentially. Alexander further argues that Aristotle held that a body moving with a circular motion is not in a place and does not move with a translatory motion.[76]

Finally, Alexander explains that it is not necessary that a body be in a place. After all, a place is not in the contained thing but in the container. The eighth sphere is neither in a place *per se* nor *per accidens*, and, because it is not in a place, Alexander concludes that it cannot be moved, because place is essential for motion.[77] In the fourteenth century, Buridan defended this position by insisting that although the question was difficult, it could be made much easier by taking place in its proper sense, namely as something that not only contains the located thing but is also distinct from it and contiguous to it.[78] From this standpoint, the last sphere cannot be in a place, since it contains all bodies and no body contains it.[79]

Alexander's opinion was taken to mean that the last heaven was not in a place *per se* or *per accidens*. Others, however, sought to show that the last sphere might be in a place in some other way than as in a proper place, that is, in a way that did not involve being surrounded by a material container in contact with it at every point. From the proposals, it is obvious that some, and probably many, felt uncomfortable in denying every sense of place to the last sphere. Their solutions formed an odd collection of ad hoc explanations that were cited by almost all who treated the question up through the seventeenth century. Although none of these seems to have won widespread support, they form a significant part of the literature on this important question, and a brief account of them must now be given.

76. The basis for this statement is Aristotle, ibid., 8.9.265a.32–265b.1, where we learn that in circular motion "any one point as much as any other is alike beginning, middle, and end, so that they are both always and never at a beginning and at an end (so that a sphere is in a way both in motion and at rest; for it continues to occupy the same place)."
77. Many presented much briefer accounts of Alexander's arguments. For example, in the sixteenth century, Toletus attributed to Alexander the view that "the last sphere could not be in a place in any way, neither *per se*, nor *per accidens*; nor could it be moved locally so that its motion could in no way be local" (Altera sententia fuit Alexander qui [ut Simplicius refert] dixit ultimam spheram nullo modo esse in loco, nec per se, nec per accidens, nec moveri localiter unde motum illum noluit localem esse). Toletus, *Physics*, bk. 4, ch. 4, qu. 7, 1580, 122r, col. 1.
78. After first declaring that "Ista questio reputata fuit difficillima et credo quod hoc fuit propter non distinguere equivocationem huius termini 'locus', dictum est enim prius quod uno modo dicitur locus proprie, scilicet continens locatum, divisum ab eo, et immediatum sibi." Buridan, *Physics*, bk. 4, qu. 6, 1509, 72r, col. 1.
79. "Et est prima conclusio quod capiendo locum proprie, scilicet pro continente, etc., ultima spera nec est in loco, nec habet locum qui sit locus ipsius quia supponimus nullum esse corpus continens ipsam sed ipsam omnia alia corpora continere." Ibid., 72v, col. 1.

b. The opinion of Avempace: place as the convex surface of an orb

Averroës not only served as the source of Alexander's argument, but he transmitted to the Latin West other opinions that formed the basis of virtually all subsequent discussions about the possible place of the last sphere.[80] Of these opinions, only one, attributed to Avempace (ibn Bajja, d. 1139), an Arab who lived in Spain,[81] assigned a place *per se* to the last sphere.[82] As the basis of his argument, Avempace insisted that a body that moves with a circular motion must be assigned a different kind of place than a body that moves with a rectilinear motion. A round body, or sphere, is not in a place by virtue of some external container, but by virtue of something that is internal. A celestial sphere is complete and perfect in itself because its circular surface is self-contained and it can neither increase nor diminish. Therefore no celestial sphere, including the last, needs an external container as its place. The fact that the celestial spheres are configured so that one celestial sphere seems to be contained by the concavity of the next upper, or external, sphere is merely accidental and quite unnecessary, since the essential and *per se* place of a sphere is internal. By contrast, bodies in the sublunar region require external places because they are compounded of the four elements and move with rectilinear motions, which are inherently imperfect and always capable of increase.

But what could serve as the internal place of a celestial sphere? For Avempace, the convex surface of the next inner sphere performed this function, so that, as Aegidius Romanus explained ([*Physics*], 1502a, 74r, col. 2), "if the eighth sphere is moved around the convex [surface] of the sphere of Saturn, the place of the eighth sphere is the convex [surface] of the sphere of Saturn."

By "internal" place, Avempace did not therefore mean internal to the sphere whose place we seek. His sense of internal place is rather "to be encompassed by." That is, the eighth sphere has as its place the convex surface of the orb of Saturn, which it encompasses but which is not an integral part of it. In a rather bizarre way, Avempace upheld Aristotle's criterion that the place must be distinct and separate from the body whose place it is. Despite the apparent tidiness of Avempace's solution and the widespread discussion of it, it won few if any supporters. For, as Thomas Aquinas rightly perceived ([*Physics*, bk. 4, lec. 7, par. 914], 1953, 202), Avempace's conception of place violated two fundamental properties of place as described by Aristotle, namely that place must be a container and that the container, or containing surface, must be equal to that which is located, or in place. Moreover, if Avempace thought he had succeeeded in

80. For these opinions and that of Averroës himself, see Averroës, *Physics*, bk. 4, comment. 43, 1562–1574, vol. 4, 141r, col. 1–143r, col. 1.
81. Averroës states that Avempace's opinion was also held by al-Farabi (ca. 870–950).
82. On Avempace, I rely heavily on Grant, 1981b, 75. The Latin text appears in Averroës, *Physics*, bk. 4, comment. 43, 1562–1574, vol. 4, 141v, col. 2–142r, col. 1.

assigning a place to the eighth sphere by resorting to the convex surface of the next lower sphere, he had only shifted the problem to the center of the cosmos. For, as Franciscus Toletus observed, by continuing this process "the place of the lunar heaven [or sphere] would be in [the sphere of] fire, fire in air, air in water, and water in earth, but earth would not have any place."[83]

c. The opinion of Averroës: the center of the world as the place of the last sphere

If Avempace's attempt to assign a place *per se* to the last sphere failed, Averroës' own effort to accord it a place *per accidens* managed to win some significant supporters of the stature of John of Jandun, Aegidius Romanus, Walter Burley, and William of Ockham.[84] Averroës agreed with Avempace that no container enveloped the last sphere and that even where containing surfaces did exist, as in the sequence of nested, concentric celestial spheres, the containing spheres were not places *per se* and were not required for the circular motion of any sphere.[85]

In order to assign a place *per accidens* to the last sphere, Averroës assumed that one of two meanings of *per accidens* was applicable. The relevant sense assumed that we could attribute to a whole what is ascribed to its part.[86] Using this interpretation, Averroës argued that the heaven is in a place *per accidens* because its part, the center of the world, or earth, is in a place *per se*. Thus did Averroës make the earth, or immobile center of the world, the basis of his interpretation. As support for his position, he appealed to the *Liber de motibus animalium localibus*, falsely ascribed to Aristotle, which declares that "those things that are moved *per se* need something at rest around which they are moved."[87]

Averroës seems to have conceived the universe as a wheel rotating around a fixed axle, where the latter represents the immobile earth while the rim of the wheel stands for the outermost celestial sphere. In this way, Averroës linked the outermost sphere, which is continually in motion, with the earth, which is always at rest in the center of the universe. He thus made a single entity of these two ordinarily disparate bodies and concluded that the last

83. "Nam coelum lunae non erit in loco nisi in igne, ignis in aere, aer in aqua, aqua in terra, iam terra non habebit locum ullum." Toletus, *Physics*, bk. 4, ch. 4, qu. 7, 1580, 121v, cols. 1–2.
84. For Averroës' own opinion, see Averroës, *Physics*, bk. 4, comment. 43, 1562–1574, vol. 4, 141r, col. 1–143r, col. 1. On Averroës, see Grant, 1981b, 75–76.
85. Averroës, ibid., comment. 45, 4:144r, col. 1.
86. Ibid. The other is when we attribute to an accident what we attribute to the thing possessing that accident (for example, when a color is said to be in a place because its subject is in a place).
87. "Dicendum est quod illa, que moventur per se, indigent aliquo quiescente circa quod moventur, ut declarat Aristoteles in libro *De motibus animalium localibus*." Ibid., comment. 43, 142v, col. 2. Although not by Aristotle, it was probably written by a member of his school. In the Aristotelian corpus, it bears the title *De motu animalium*.

sphere of the world is in a place *per accidens* because its center is in a place *per se*. According to Averroës, the fixity and permanence of the earth's place at the center of the world confers on the outermost sphere, which turns about it, an accidental place. And in the manner of Avempace, Averroës sought to determine the place of the last sphere by something that was not only physically distinct from the last sphere but was located somewhere within its concave surface. Despite the distinguished supporters mentioned earlier,[88] Averroës' interpretation was severely criticized. Because "the center [that is, the earth] is wholly extrinsic to the last sphere," Thomas Aquinas thought it "ridiculous to say that the last sphere is in place accidentally [simply] because the center is in a place."[89] Averroës' opinion was apparently equally ridiculous to Mastrius and Bellutus, who denied that a container could be in place by virtue of the thing it contains. The center of the world also fails to meet the basic requirement that a place must equal the located body and yet be outside of it.[90]

Other attacks were mounted against the idea that the motion of the last celestial sphere, or any celestial sphere, for that matter, requires a motionless physical center. Nicole Oresme insisted (*Le Livre du ciel*, bk. 2, ch. 8, 1968, 367) that although a rotating wheel lacked an immobile, physical center (at best the center was only an indivisible point that existed only in the imagination), it nevertheless rotated. And in an apparent allusion to Averroës, Oresme asks "the person who says there must be some motionless body in the center of a circularly moving body . . . what must be the size of this body, for we could not fix upon a quantity so small that a smaller one would not satisfy the demands of reason." It follows, therefore, that "the motion of the heavens does not require that the whole earth should rest." In the seventeenth century, Peter de Oña ([*Physics*, bk. 4, qu. 1, art. 6], 1598, 222v, col. 2) not only cited Thomas's arguments against Averroës but also observed that if the last sphere is in a place because it is moved circularly around the center, or earth, then so must all other celestial spheres be said to be in a place in the same manner. But since those orbs are not in a place in the manner described by Averroës, then neither is the last sphere.

Hierarchical considerations also played a role. Buridan thought it cosmically inappropriate to make the place of an incorruptible celestial sphere depend on the place of a mutable and inferior body, such as the earth.

88. Ockham [*Physics*, "De loco," qu. 79], 1984, 613–614, accepts it by noting that to say that "the heaven is in a place accidentally is [to say] nothing other than that the heaven contains the center around which it moves" and goes on to add that "this conclusion can be understood in another way, because if some body surrounds the heaven, then the heaven would be in a place *per se*." For Roger Bacon's defense of Averroës, see Duhem [Ariew], 1985, 145–147; for John of Jandun, see his *Physics*, bk. 4, qu. 9, 1519, 54r, col. 2–54v, col. 1.

89. "Cum igitur centrum sit omnino extrinsecum a sphaera ultima, ridiculum videtur dicere quod sphaera ultima sit in loco per accidens ex hoc quod centrum est in loco." Thomas Aquinas [*Physics*, bk. 4, lect. 7, par. 917], 1953, 203.

90. Mastrius and Bellutus [*Physics*, disp. 11, "De loco," qu. 4, art. 1], 1644, 825.

Indeed, the converse must hold: the order of inferior things must depend on superior things (Buridan [*Physics*, bk. 4, qu. 6, 4th conclus.], 1509, 72v, col. 1).

d. The opinion of Themistius: the place of the last sphere is the place of its parts

The last major explanation transmitted by Averroës was derived from Themistius (ca. 317–ca. 388), the Greek commentator, who developed Aristotle's idea that although bodies without containers cannot be in a place *per se*, they can be in a place accidentally by virtue of their parts.[91] According to Aristotle, the last heaven (*celum*),[92] which is not contained by any body and is therefore not anywhere as a whole, is nevertheless in a place, because its parts are in a place.[93] Since the last heaven has no container, Aristotle concluded that all things exist within this heaven and are arranged in such a way that "the earth is in water, and this in the air, and the air in the aether, and the aether in the heaven, but we cannot go on and say that the heaven is in anything else."[94]

Themistius based his interpretation on these conceptions. As Averroës explains it, Themistius insisted that "a celestial body is not in a place with respect to the whole, but with respect to its parts, namely with respect to the orbs which the greatest orb contains, according to the disposition [of these orbs] in the whole world."[95] For Themistius, then, the last sphere is in a place by virtue of its internal parts, which are the various spheres that lie within it. These cosmic parts are taken in descending order. That is, the last sphere, which (in this context) is the sphere of the fixed stars, is in a place because Saturn and all inferior spheres are in places. The containing place of Saturn is the concave surface of the last sphere, which lies directly above it. Similarly, the containing place of every other inferior orb is the concave surface of its encompassing neighbor immediately above. Despite the fact that the last sphere is not contained by any external surface, Themistius argues that it is nevertheless in a place, because all its inner parts are.

As with all of the interpretations concerning the place of the last sphere, there were weighty criticisms brought against this aspect of Themistius's explanation. Ockham dismissed it by appeal to the Commentator, Averroës, who had rejected it because Themistius insisted that the last, or eighth, sphere is moved only with respect to its parts but not with respect to its whole. In order for the whole to move, it would require an external place, since local motion is from place to place. Since Themistius rejected an

91. Aristotle, *Physics* 4.5.212a.31–212b.1. For Themistius, I have made significant use of Grant, 1981b, 77–79.
92. The Latin translations use *celum* for the Greek term *ouranos*, which can also mean "world."
93. Aristotle, *Physics*, 4.5.212b.7–10.
94. Ibid., 212b.20–22. In the Oxford translation of the *Physics*, Hardie and Gaye use "world" twice, for which I have substituted "heaven."
95. Averroës, *Physics*, bk. 4, comment. 43, 1562–1574, 4:141r, col. 2–141v, col. 1.

external place, he conveniently argued that the last sphere is moved only by virtue of the motion of its parts. But Ockham, following Averroës, rejects this, because, as he put it ([*Physics*, "De loco," qu. 79], 1984, *Opera philosophica*, 6:612), "it is impossible that the eighth sphere be in a place with respect to all its parts, but not [in place] with respect to the whole, since the whole is nothing other than [the sum of] its parts."

e. The second aspect of Themistius's opinion

During the Middle Ages, and prior to the translation of Themistius's paraphrase of Aristotle's *Physics*, Averroës' summary was assumed to be the full account of Themistius's description of the place of the last sphere. But after the translation of Themistius's *Physics* near the end of the fifteenth century, scholastics became aware of another aspect of his interpretation. Apparently he was not content to assign a place to the last sphere solely on the basis of its internal parts; he also introduced a bizarre sense of "containment," by identifying the convex surface of Saturn, which was in direct contact with the concave surface of the last sphere, as the "container" of the last sphere.[96] It was this aspect of Themistius's conception on which scholastics would thereafter focus, as in the sixteenth century, when Toletus explained that for Themistius "the last sphere, which according to Aristotle is the eighth, is not in place according to its convex surface, but according to its concave surface; and it is in the seventh sphere immediately next to it, which is Saturn. Thus the convex [surface] of the penultimate [sphere] is the place of the last sphere with respect to the [latter's] concave surface."[97] What Themistius did for the sphere of Saturn alone, Avempace, as we saw, did for all the celestial spheres.[98]

As he did with Avempace, Toletus identified the chief objection to Themistius's theory: the same thing cannot simultaneously be the located or contained body and also be the place of the container. But the seventh sphere (Saturn) is in the last sphere as in a real physical place; therefore the last sphere cannot be in the seventh sphere, which is the located thing, as if it were in a material place.[99]

f. The convex surface of the last sphere and the empyrean heaven as places of the last sphere

Averroës was not the sole source of opinions on the place of the last sphere. In the twelfth century, Gilbertus Porretanus, or Gilbert de la Porée (ca.

96. See Themistius, 1499, 44r, in Hermolaus Barbaro's Latin translation.
97. Toletus, *Physics*, bk. 4, ch. 4, qu. 7, 1580, 122r, col. 1. For a similar description, see Javelli [*Physics*, bk. 4, qu. 18], 1568, 1:588, col. 1, and Conimbricenses, *Physics*, pt. 2, bk. 4, ch. 5, qu. 2, 1602, col. 43.
98. Indeed, Avempace may have derived the basis of his interpretation from Themistius (see Duhem, *Le Système*, 1913–1959, 7:160, and Duhem, [Ariew], 1985, 141).
99. Toletus, *Physics*, bk. 4, ch. 4, qu. 7, 1580, 122r, col. 1.

1076–1154), argued that the place of the last sphere, or heaven, is its own convex surface.[100] More than a century later, Campanus of Novara invoked the immobile empyrean sphere as the true place of the last movable sphere (for the passage, see Ch. 15, Sec. II.2). But does this not merely pose the same question about the empyrean sphere? Is it in a place? Because it is the last sphere and is also immobile, and because place was always associated with things that move or are potentially able to move, Toletus ([*Physics*, bk. 4, qu. 7], 1580, 122r, col. 2) did not think it absurd to suppose that the empyrean sphere could exist without a place to contain it.

g. Internal space as the place of the last sphere and the world

No conclusive and definitive response was possible to the question as to whether or not the last sphere was in a place. But in the sixteenth and seventeenth centuries, radically different explanations were formulated which supplemented or replaced the idea of place as a container, or *locus*, whether the containment was considered external, as Aristotle defined it, or internal, as Themistius, Avempace, Averroës, and their various followers conceived it. One major concept assumed the existence of an internal space or place that bore no relationship to the internal superficial places of Themistius, Avempace, and Averroës. Another involved a space, whether void or imaginary or both, that was capable of receiving bodies without offering resistance to them. Let us examine the different types.

Buridan, Toletus, and Suarez were all concerned with the idea of internal space. Already in the fourteenth century, Buridan believed that space is nothing but the dimension of a body. To illustrate his meaning, he imagined a man located beyond the last sphere or heaven. If this man now raised his arm, Buridan explains that

> it would not be valid to say that he could not place or raise his arm there [simply] because no space exists into which he could extend his hand. *For I say that space is nothing but a dimension of body and your space the dimension of your body.* And before you raise your arm outside this [last] sphere nothing would be there; but after your arm has been raised, a space would be there, namely the dimension of your arm.[101]

100. In his treatise *Liber sex principiorum*, which was a metaphysical commentary on Aristotle's *Categories* (see Duhem [Ariew], 1985, 145); on Gilbert's life and work, see Gilson, 1955, 140–144, and 620, n. 72. Gilbert's opinion is discussed by the Conimbricenses, *Physics*, pt. 2, bk. 4, ch. 5, qu. 2, 1602, col. 42.
101. In the Latin text, I include a few lines preceding the beginning of the quotation just given: "Utrum si esset homo ultra speram ultimam ipse posset movere ultra illam sua membra, scilicet brachia. Et de hoc pono ultimam conclusionem quod homo sic posset movere membra quia nichil extrinsece ei resisteret nec valet dicere quod non posset illic brachium ponere vel elevare quia nullum esset ibi spacium in quo posset manum suam extendere. Dico enim quod spacium non est nisi dimensio corporis et spacium tuum dimensio corporis tui. Et antequam elevares brachium ultra illam speram nichil esset ibi; sed brachio elevato esset ibi spacium, scilicet dimensio brachii." Buridan, *Physics*, bk. 4, qu. 10, 1509, 77v, col. 1. On internal space, I draw upon Grant, 1981a, 122. For

Similarly, if God created a bean beyond the world, the magnitude of the bean would be its own space, because no separate space lies beyond the world for it to occupy. As with the bean, so with the whole world. For "the bean would not be created in an indivisible or a divisible space, because it would not be in a place or in any space, just as the whole world [*totalis mundus*] is not in a place or in any space, but, collectively speaking, all the magnitudes of the parts of the world are spaces."[102] Material bodies could not be in places or spaces beyond the world, because there are no such places or spaces. But any body placed out there would be in a space which is identical with the body's own dimensions. Thus the world itself is not in some separate place or space, but it is its own space or place, comprised of the individual spaces of all its parts. Although, as we saw, Buridan agreed with Aristotle that the last sphere is not in a separate place conceived as an external container, he nevertheless argued that not only is it in a space equal to its own dimensions but the world as a whole is in a place or space equal to its own dimensions.

In the sixteenth century, Toletus called the dimensions of a body "internal space" (*spatium intrinsecum* or *locus intrinsecus*), thus naming what Buridan had already clearly identified. The existence of internal space was undeniable, because body and space "imply each other as a mutual consequence. For if there is a body, there is a space; and if there is a true space, there is a body in it."[103] Among five properties Toletus assigned to place or space, the fourth and fifth are the most important. The fourth property asserts that "place contains the located thing," to which Toletus promptly adds the crucial qualification that "although place could be conveniently referred to a [containing] surface, that contains more truly and more perfectly which contains according to the whole corporeality [of the body, rather than just along a surface]."[104] Thus Toletus did not reject Aristotle's concept of place

more on Buridan, see Grant, 1963, 251–255. Albert of Saxony (*Physics*, bk. 1, qu. 9, 1518, 93v, col. 2) denied Buridan's interpretation by arguing that even without obstacles, no one could extend an arm beyond the extremity of the world, because no receptacle – that is, no place or space – could exist beyond the world to receive it. The argument about the extension of an arm beyond the sky is derived from the Stoics as transmitted by Simplicius in the latter's *Commentary on De caelo* in the translation from Greek to Latin by William of Moerbeke in 1271. For the passage in Moerbeke's translation, see Simplicius [*De celo*, bk. 1, comment. 96], 1540, 44v, col. 2.

102. "Dico quod illa faba nec crearetur in indivisibili nec in spacio divisibili quia non esset in loco, nec in aliquo spacio, sicut nec totalis mundus est modo in loco vel in aliquo spacio, sed collective loquendo omnes magnitudines partium mundi sunt omnia spacia." Buridan, *Physics*, bk. 3, qu. 15], 1509, 58r, col. 2.

103. "Esse autem hoc intrinsecum cuiusque corporis spatium negari non potest, ut videtur, quia invicem sese mutua consequentia inferant. Nam si corpus est, spatium est; et si verum spatium est, in eo corpus est." Toletus, *Physics*, bk. 4, ch. 5, qu. 8, 1580, 123r, col. 2.

104. "Quarta: Locum continere locatum. Nam hoc ad superficiem quamvis commode referri possit, tamen verius et perfectius continet quod secundum totam corpulentiam continet." Ibid., 123v, col. 2. Among examples to illustrate his meaning, Toletus offers the way in which light is contained in air and the soul in the body.

as an external containing surface, but he thought it better that a place involve the entire body, not just what surrounded its external surface.

Unless we assume the reality of internal place, Toletus envisioned circumstances in which a body would not be in a place, thus violating his fifth property, which asserts categorically that "every body is in a place."[105] A body must have an internal place, for otherwise we would be compelled to say that the heaven, or last sphere, is not in a place or that it is in a place by virtue of its center, alternatives which Toletus characterizes as "absurd and paradoxical and beyond the common sense of men." But if "as Aristotle rightly says, the heaven [or last sphere or the world itself] is not in a surrounding external place," it is, nevertheless, in a place or space that is internal to it.[106]

Toletus admits that the idea was not original with him but was "the very probable opinion of a certain modern." He was convinced that the concept of internal space, when added to Aristotle's correct but limited theory of external place, resolved the difficulties associated with the doctrine of place.[107] Although Toletus regarded the ontological status of internal space as unimportant – that is, whether it is really (*realiter*) distinct and separate from quantity or whether it is only formally (*formaliter*) distinct by reason alone – he seems to have agreed with John Philoponus, the sixth-century Greek commentator on Aristotle, that internal space is only a proper accident inhering in bodies rather than existing as a separable entity.[108] Only by reason, then, can we distinguish the internal space of a body from its quantity. Thus the last sphere, or the world itself, does not exist in a really separate void space, but its very dimensions constitute its own internal space.

105. "Quinta proprietas . . . quod omne corpus sit in loco." Ibid.

106. "Quia corpus quantum est nec in loco solum extrinseco, sed intrinseco, ne absolute cogamur dicere coelum non esse in loco, vel eius locum esse centrum, quae duo sane sunt absurda et paradoxa et praeter communem sensum hominum. Sed non est in loco extrinseco ambiente, ut recte in hoc ait Aristoteles, est tamen in loco et spatio intrinseco sibi." Ibid.

107. "At vero cuiusdam recentioris extat valde probabilis opinio que omnes quodammodo opiniones conciliat et omnes difficultates dissolvit." Ibid. For more on Toletus's ideas on internal space, see Grant, 1976a, 155–159. Although the name of the "modern" is unknown to me (Buridan is a candidate, because his *Questions on the Physics* was printed in 1509), the ultimate source for the idea of "internal space" is probably John Philoponus.

108. "Et ratione plereque Philoponi hoc ipsum spatium intrinsecum et quantitati proprium ostendunt, sed non separatum ut ipsa ponebat sed in rebus ipsis inhaerens tanquam proprium earum accidens." Toletus, ibid., 123r, col. 2–123v, col. 1. In fact, Philoponus expresses both opinions, in different works. In his *De aeternitate mundi contra Proclum*, written in 529, he held something akin to the ideas attributed to him by Toletus (see Grant, 1981a, 15). But in book 4 of his *Commentary on Aristotle's Physics*, he argued for the real existence of a separate void space that was, however, always occupied by the world and therefore never actually empty. This opinion was transmitted to the Middle Ages in Averroës, *Physics*, bk. 4, comment. 43, 1562–1574, vol. 4, 141r, col. 2. The first Greek edition of Philoponus's *Commentary on the Physics* was printed in 1535, followed by Latin translations published at Venice in 1554, 1558, and 1569. For further information, see Grant, 1978d, 557–558 and n. 30, which is reprinted in Grant, 1981c, XI, 557–558 and n. 30.

Perhaps it was because Toletus thought it unimportant whether internal space is really distinct from body or only formally so that Francisco Suarez, in his metaphysical disputations of 1597, mistakenly concluded that Toletus was a believer in a separate void space. In a lengthy discussion of the concept of *ubi*, Suarez rejected such a separately existing three-dimensional void space on purely Aristotelian grounds. For Aristotle had argued that if place were a three-dimensional empty space, it would be a body. But a body cannot receive another body, because "there would be two bodies in the same place" simultaneously.[109] On this basis, Suarez opposed the idea of a void place or space that was separate or distinct from the body that occupies it.[110]

As his own concept of *ubi*, Suarez assumed a formal and real mode of existence internal to a thing or body by virtue of which it is said to be here or there.[111] A body's *ubi*, or location, is in no way dependent on an external, circumscribing body but depends only on a material body and the cause which constitutes or preserves that body where it is. *Ubi* is an absolute in itself and wholly distinct from substance, quantity, and all corporeal accidents. As the best illustration of the properties of *ubi*, Suarez cites the last sphere, which for him was the empyrean heaven. The outermost sphere "has a true and real presence where it is and yet is not circumscribed by another body."[112] Although Suarez uses the term *ubi* where Toletus employed *locus intrinsecus*, the two interpretations are essentially the same.[113]

By its very definition, the concept of internal place (whether called *ubi* or *locus intrinsecus*) located the last sphere and the world itself in a place or space that was independent of any external place. No body need surround the last sphere or the world. Thus was solved the centuries-old problem as to whether the last sphere or heaven was in a place. Whether one assumed

109. Aristotle, *Physics*, 4.1.209a.5–7, 1984. For a discussion, see Grant, 1981a, 5–6 and Grant, 1978d, 551–552.

110. Suarez, *Disputationes metaphysicae*, disp. 51, sec. 1, 1866, 2:975, cols. 1–2, par. 12, where Suarez specifically mentions Toletus, and 978, col. 2–979, col. 1, par. 23. Suarez also attacked those who, while denying reality to such an empty space, thought it "something true, eternal, and immutable" and thus more than a mere figment of the imagination (hoc spatium, licet reale non sit, non tamen esse quid fictum per operationem intellectus, sed aliquid verum, aeternum et immutabile); ibid., 979, col. 1. A note by Suarez himself, or perhaps by his editors, cites Pedro Fonseca (1528–1599) (*Metaphysics*, V, 5, qu. 1, sec. 1) and the Conimbricenses (*Physics*, bk. 8, ch. 10, qu. 1, art. 4) as supporters of this approach.

111. I discuss Suarez's concept of *ubi* in Grant, 1976b, 78–79.

112. "Praeterea probatur haec pars exemplo ultimae sphaerae coelestis, quo confirmantur omnia quae diximus; nam illud corpus habet veram et realem praesentiam ibi ubi est, et tamen non habet aliud corpus quo circumscribitur; agimus enim de coelo empyreo, quo nullum est superius. Ergo hic modus praesentiae per se non requirit superficiem circumscribentem, neque ab ea pendet." Suarez, *Disputationes metaphysicae*, disp. 51, sec. 1, 1866, 2:978, col. 1, par. 21.

113. Suarez has a lengthy discussion on the advisability of using the terms *locus intrinsecus* and *locus extrinsecus* or, as he seems to favor, using *ubi* for *locus intrinsecus*. See ibid., sec. 2, 980, col. 2–982, col. 2. According to Suarez (980, col. 2, par. 4), Toletus employed *locus intrinsecus* and *locus extrinsecus*, equating the former with *ubi*.

a finite, spherical cosmos with no body or vacuum beyond it or accepted the popular late scholastic concept of a spherical, finite cosmos located in an infinite, imaginary space – however the latter may have been interpreted – the assumption of internal space or place guaranteed that the last sphere, or the world as a whole, would be in its own place.[114] Although scholastics continued to cope with the question in the traditional terms of an external place for the last sphere or world and invoked the various interpretations described in this chapter, the concept of internal place made all this superfluous for those who chose to adopt it.[115] Taken together with the concept of infinite, imaginary space, it marked a radical departure from the Aristotelian doctrine of place and forms part of the historical background in spatial concepts that led to Descartes, More, and Newton.

114. Although it was not made clear or obvious, the Conimbricenses seem to have assumed this (see Conimbricenses, *Physics*, pt. 2, bk. 4, ch. 5, qu. 2, art. 2, 1602, cols. 43–44).
115. Mastrius and Bellutus (*Physics*, disp. 11, qu. 4, art. 1, 1644, 826, col. 1) indicate that it was widely accepted when they declare that "the moderns [*recentiores*] say that the last heaven is in a place because it is in a space that it occupies and fills." Although the authors seem to reject this response, the identity of the *recentiores* is left unclear, but it probably refers to the scholastics, who held such ideas during the sixteenth and seventeenth centuries. Nonscholastic authors were unlikely to have considered the question at all.

7

The perfection of the world

Because God, a perfect being, had created the world, scholastics came to ask frequently, and rather naturally, whether the world itself was perfect. They fully recognized that the response depended upon one's definitions and conceptions of the terms "world" (*mundus*; sometimes *universum*) and "perfect" (*perfectus*).

I. Is the world perfect?

In the long history of imaginative discussions about the possible existence of other worlds, two kinds of worlds must be distinguished: world as cosmos and world as a single celestial body.[1] Our concern in this volume is with the world as cosmos, which Richard of Middleton defined in the thirteenth century as "the totality of creatures [spiritual, animate, and inanimate] contained within one surface which is contained by no other surface within this world and which includes the containing surface."[2] Like those of all medieval scholastic natural philosophers, Richard's definition was intended to encompass the world as Aristotle understood it. At the very least, the world, so conceived, included all celestial and terrestrial bodies, so that its containing surface was the convex surface of its last, or outermost, sphere, whether this was identified with the sphere of the fixed stars, a ninth transparent sphere, or the empyrean heaven.

Within this definition, no single celestial body (for example, the Moon or Sun) qualified as a world. The identification of world with a single celestial body did not happen until the Renaissance. Its occurrence involved "a shift that required the downfall of the long-cherished Aristotelian principle that there was a sharp distinction between the Earth and all other celestial bodies" (Dick, 1982, 2). Although a few scholastic authors also reduced, if not obliterated, the sharp distinction between the earth and the

1. I draw this distinction from Steven J. Dick's excellent study of the concept of a plurality of worlds (1982, 2).
2. "Respondeo vocando universum universitatem creaturarum infra unam superficiem contentarum quae a nulla alia superficiem continentur infra illam universitatem compraehendo etiam superficiem continentem." Richard of Middleton [*Sentences*, bk. 1, dist. 44, art. 1, qu. 4], 1591, 1:392, col. 2. For Duhem's French translation, see Duhem, *Le Système*, 1913–1959, 9:376; for an English version, see Duhem [Ariew], 1985, 452.

celestial bodies, the concept of a single celestial body as a world played virtually no role in the medieval scholastic tradition. For this obvious reason, it will play little role in this study.

Despite the general acceptance of Richard of Middleton's all-embracing definition of world, the scholastic penchant for hierarchical arrangements led some scholars to distinguish four different, but progressively more restrictive, senses of the concept of world. They were agreed that, in its broadest sense, world could be taken as the totality of all spiritual and corporeal things, thus including God, the intelligences, and all celestial and sublunar bodies; a second option omitted God but retained all the remaining entities, so that the world was constituted of the intelligences and all celestial and sublunar bodies;[3] as a third alternative, world could be restricted to celestial and sublunar bodies; and, finally, it might be confined solely to all the things in the sublunar region.[4]

On the basis of Aristotle's *Metaphysics* (5.16.1021b.12–1022a.2), "perfection" (*perfectio*) was usually taken in at least two ways: absolutely (*simpliciter*) and relatively (*secundum quid*). Something was absolutely perfect, or complete, when any additions made to it failed to cause an increase of its perfection, while something relatively perfect could be improved because it lacked some perfection.[5] An example of the former would be God, of the latter a physician, who, though he might be perfect as a physician, would be improvable in other ways.

Given the medieval penchant for hierarchical comparisons and ordering of things, it seemed self-evident, quite apart from any definitions of per-

3. I take this to represent Richard of Middleton's definition.
4. Among those who distinguished this fourfold sense of world are Nicole Oresme [*De celo*, bk. 1, qu. 3 ("Whether the world is perfect")], 1965, 40–42 (although Oresme distinguishes only three meanings of *mundus*, the third, which includes all celestial and terrestrial bodies, also mentions the sublunar region, which is dismissed as irrelevant to his discussion); Albert of Saxony [*De celo*, bk. 1, qu. 5 ("Utrum mundus sit perfectus")], 1518, 90r, col. 1; Galileo [*De caelo*, qu. 3 (E) ("On the Perfection and Unity of the Universe")], 1977, 47; and Bartholomew Amicus, *De caelo*, tract. 3, qu. 2, dubit. 1 ("An hic mundus sit perfectus"), 1626, 112, col. 1. John of Jandun [*De coelo*, bk. 1, qu. 6 ("An ipsum universum sit perfectum")], 1552, 5r, col. 2 and Jean Buridan [*De caelo*, bk. 1, qu. 12], 1942, 56, mention only the first and third definitions. But Buridan mentions some of the four definitions, and others, elsewhere in his *De caelo* (for a translation of the passage, see Chapter 2, Section III.2, of this study).
5. This twofold sense of perfection, based on Aristotle, is included in questions on the possible perfection of the world by John of Jandun, ibid.; Albert of Saxony, ibid., cols. 1–2; Galileo, ibid., 46–47; and Amicus, ibid. The terms *simpliciter* and *secundum quid* were used by all. Later in his discussion, Jandun (ibid., 5v, col. 1) explains that something is more perfect than another in two ways: "In one way intensively; in another way quantitatively and extensively." He goes on to declare that "the universe is more perfect than God extensively because it contains more perfections [pertaining to extension or quantity] than God does, but God is intensively more perfect because He has perfections in a nobler way than the whole universe because something [that] is in the whole universe distinctly and in a composite way is in God in a simple and nobler manner." In the seventeenth century, Melchior Cornaeus repeated essentially the same argument (Cornaeus [*De coelo*, disp. 1, qu. 2, dub. 2], 1657, 486).

fection, that these four progressively less inclusive interpretations of the concept "world" described entities that were also progressively less perfect.[6] For the most part, however, scholastics focused attention on the first and third definitions. In considering world as the totality of God and all spiritual and corporeal things, most would have conceded that world in this sense was absolutely perfect, because it contained every conceivable existing thing.[7] Nothing could be added to it.

Jean Buridan, however, seems to challenge this all-inclusive conception of world when he compares the perfection of God alone with that of God plus all other existing things. In his view, God plus all other things could not be better than God alone. The first conception – God plus all other things – is not some separate thing but a conglomeration of things that happens to include God. But God alone is a separate entity who, "according to faith and truth, is of infinite goodness and perfection, and nothing is better than infinite goodness nor more perfect than an infinite perfection, just as nothing could be greater than an infinite magnitude, if an infinite magnitude existed" (Buridan [*De caelo*, bk. 1, qu. 12], 1942, 55).

Usually world was taken as the totality of all natural bodies, both celestial and terrestrial, which is our third definition. In this sense, as Buridan observed, the world could not, strictly speaking, be a "perfect body" (*corpus perfectum*), because it was not a single body; it is nothing more than a collection or aggregation of many bodies. Only by allowing *corpus*, or "body," to represent, "or stand for" (*ad supponendum pro*), an aggregation of many bodies could the term *mundus*, or world, be considered a single, and therefore perfect, body (ibid., 56).

Taken in this improper though commonly used sense, Buridan agreed with the general opinion that the world could not be absolutely perfect because it could not be something better than God. It does, however, seem to have relative perfection, because not only is the corporeal world a perfect magnitude,[8] but all bodies are parts of it, and, since any whole is greater than any of its parts, the world must be more perfect than any of those parts. Therefore "no body is better or more perfect than the world," and "no body can be better or more perfect than the world by means of a natural power."[9]

But if the world was more perfect than any other single body, was it perfect in itself? Even the most casual observer could detect seeming im-

6. Albert of Saxony makes this explicit when he declares: "Accipiendo mundum primo modo est perfectior mundo accepto secundo modo; et mundus acceptus secundo modo est perfectior mundo accepto tertio modo; et sic consequenter." *De celo*, bk. 1, qu. 5, 1518, 90r, col. 1.
7. John of Jandun, *De coelo*, bk. 1, qu. 6, 1552, 5r, col. 1.
8. Aristotle argues this in *De caelo* 1.1.268a.29–268b.10.
9. Buridan, *De caelo*, bk. 1, qu. 12, 1942, 56. In the seventeenth century, Amicus (*De caelo*, tract. 3, qu. 2, dubit. 1, art. 2, 1626, 113, col. 1) held the same opinions, denying that the world is absolutely (*simpliciter*) perfect but allowing that it was relatively perfect (*secundum quid*).

perfections in nature. Were there not monsters, defects, and evil things in the world? Did not the existence of such things cast doubt on a perfect world?[10] Moreover, could our world be made more perfect than it is now? If so, then it must presently fall short of perfection. Most scholastics accepted the relative, or *secundum quid*, perfection of the world. In their view, monsters and seeming imperfections contribute to cosmic perfection, because, as Galileo expressed it, "the perfection of the universe consists in the variety of things, just as the ornament and perfection of a figure is found in the variety of its colors, some of which are less perfect than others."[11] Monsters, as well as evil, pernicious, and obnoxious things, serve a vital function. They provide the essential contrast with the better and more beautiful things of the world and thus make us more aware of them.[12] Although the sublunar region of the world suffered change and parts of it were always being generated or corrupted, this did not mean that it was constantly altering its state of perfection. Rather, as Nicole Oresme argued, generation in one part of the world was counterbalanced by corruption in another part.[13] Perfection was thus preserved.

By placing a stricter interpretation on the concept of perfection, Buridan was a rare exception to this point of view. He took seriously the condition that true perfection requires that the perfect thing be incapable of further improvement. The generation, corruption, and alteration constantly occurring in the terrestrial, though not the celestial, region of the world are not proportionate, nor are they always counterbalanced to preserve a perfect world. Indeed, the elements in the terrestrial realm need not and cannot maintain a perfect balance, since the celestial constellations and configura-

10. Most scholastics cited arguments – though not usually accepting them – about monsters and other defects as evidence of an imperfect world. For example, see, John of Jandun, *De coelo*, bk. 1, qu. 6, 1552, 51, cols. 1–2; Buridan, *De caelo*, bk. 1, qu. 12, 1942, 54; Nicole Oresme, *De celo*, bk. 1, qu. 3, 1965, 38; Albert of Saxony, *De celo*, bk. 1, qu. 5, 1518, 90r, col. 1; Galileo, *De caelo*, qu. 3 (E), 1977, 46–47; Conimbricenses [*De coelo*, bk. 1, qu. 1, art. 1], 1598, 8; Amicus, *De caelo*, tract. 3, qu. 2, dubit. 1, art. 2, 1626, 112, col. 2; and Bonae Spei [comment. 3, *De coelo*, disp. 1, dub. 2 ("An mundus sit perfectus et an dari possit perfectior")], 1652, 2, col. 1.
11. Galileo, *De caelo*, qu. 3 (E), 47. Raphael Aversa ([*De caelo*, qu. 31, sec. 5], 1627, 15, col. 1), Amicus (*De caelo*, tract. 3, qu. 2, dubit. 1, art. 4, 1626, 114, col. 2), Cornaeus (*De coelo*, disp. 1, qu. 2, dub. 1, 486), and Bonae Spei (comment. 3, *De coelo*, disp. 1, dub. 2, 1652, 2, cols. 1–2) said much the same thing. Amicus even employed the metaphor of a picture or painting. Long before Galileo, however, Richard of Middleton had used a similar metaphor (*Sentences*, bk. 1, dist. 44, art. 1, qu. 1, 1591, 1:390, col. 2), which he attributed to Saint Augustine in the *City of God*. Raphael Aversa proclaimed the perfection of the world more emphatically and in greater detail than did Galileo. Not only does the world include the perfections of all natural bodies and exceed in perfection any individual body, but it has the greatest fullness, variety, beauty, order, and proportion of all things. In this perfect world things are ordained for the utility of man, and man and all things are ordained for the utility of God. Aversa, *De caelo*, qu. 31, sec. 5, 1627, 14, col. 2–15, col. 1.
12. Thomas Aquinas insisted that a world without evil would not be as good as one with evil. It is better to have both good and evil than either one of them alone (*Sentences*, bk. 1, dist. 44, qu. 1, art. 2, 1929–1947, 1:1020).
13. Oresme, *De celo*, bk. 1, qu. 3, 1965, 52.

tions are not always in the most favorable positions to produce optimum effects in the sublunar region. The world must, therefore, be more or less perfect at one time than at another. Consequently it cannot be perfect *per se*. Thus when Buridan accepted the relative perfection of the world – taking world in an improper sense – he confined himself to the simple claim that the world is better than any and all other bodies.

II. Could God make our world more perfect?

If the world is only relatively perfect, it was almost inevitable that Christian scholastics should ask whether God could make, or could have made, something better than he made. Indeed, this very question was posed by Peter Lombard in book 1, distinction 44, of his *Sentences*.[14] It was thus easily transformed into one which asked whether God could make, or could have made, a better or more perfect world than the one he created. Peter noted that investigators have declared that God could not have made things better than he had, because if he could have done so and did not, this would indicate that he is invidious rather than the greatest good. Peter resolves the question by distinguishing between God and the things he created. Because of his wisdom (*sapientia*), God would have made everything properly the first time and therefore would not be able to – indeed would have no need to – better anything he made. Viewed from the standpoint of the created thing itself, however, there is room for improvement, since no created thing can be perfect. Therefore God could make any created thing better than he had made it. On the basis of Peter Lombard's response, scholastic natural philosophers and theologians built a considerable superstructure.

As will be evident in what follows, the possible betterment of our world could have two results: God could effect accidental changes that would improve the world but leave it essentially and recognizably the same; or he could produce substantial changes that would effectively destroy the old world and replace it with a new one. On the first alternative, our world would endure in recognizable form, despite all the changes. The second alternative, however, yields a succession of worlds where each successor world wholly replaces its immediate predecessor. God was free to follow either alternative at different times. In turning to the manner in which scholastic authors grappled with this problem, we shall see that most found these distinctions acceptable.

As was customary, they first asked about crucial terms. How was the term "better" (*melius*) to be understood? Most took it as equivalent to "goodness" (*bonitas*), which in turn had at least two meanings. The first, as Thomas Aquinas described it, was "essential goodness" (*bonitas essen-*

14. Peter Lombard, *Sentences*, vol. 1, pt. 2, 1971, 303–306.

tialis), which was equivalent to being (*ens*) itself. Thus for a man to possess life and reason was to be alive and therefore to possess "essential goodness." By contrast, "accidental goodness," the second usage, pertained to more transient things, such as health and knowledge.[15]

As applied to a world that was, despite its oneness, assumed to consist of an aggregation of parts, scholastics frequently distinguished three kinds of goodness. There was, first, the goodness that derived from the parts themselves; second, the goodness that was a function of the order or relation of those parts to each other; and third, the goodness that depended on the relationship of the order of the parts to the end or purpose of the world, which was always understood to be God himself. Could God make the world better with respect to these three kinds of goodnesses? Durandus Sancto Porciano (ca. 1270–1334), a Dominican, considered these problems in the same place as did most theologians who were his contemporaries and successors, namely in his commentary on book 1, distinction 44, of the *Sentences*.[16] Judging from frequent citations of his text well into the seventeenth century,[17] Durandus made a significant impact on those who wrote about the perfectibility of the world, perhaps because his discussion was extensive, thorough, and well-organized. On this problem, Durandus is a good guide.

1. Increasing the perfection of the world through its parts

To take the first type of goodness first: could God make separate parts of the world better, and thus in some sense improve the world? One seeming

15. Thomas Aquinas, *Sentences*, bk. 1, dist. 44, art. 1 ("An Deus possit facere aliquid melius quam facit"), 1929–1947, 1:1016. Omitting Thomas's examples, William of Ockham also distinguished "better" into essential and accidental goodness (see Ockham [*Sentences*, bk. 1, dist. 44], *Opera theologica*, 1979, 4:651).
16. Durandus de Sancto Porciano put it this way: "Sic dicendum est quod cum universum sit unum aggregatione continens plures partes habentes ordinem inter se, et ad finem. . . . Bonitas eius consistit in tribus, scilicet in bonitate partium, et ex bono ordine earum inter se, et ex bono ordine ad finem." See Durandus [*Sentences*, bk. 1, dist. 44, qu. 3 ("Utrum Deus possit facere universum melius quam fecerit")], 1571, 116v, col. 1, par. 10. The source of this aspect of the discussion is Aristotle's *Metaphysics* 12.10.1075a.11–12 [Ross], 1984, where Aristotle says that "We must consider also in which of two ways the nature of the universe contains the good or the highest good, whether as something separate and by itself, or as the order of the parts. Probably in both ways, as an army does." Aristotle goes on to declare that although things like fish, plants, and birds differ, they are yet related, because "the world is not such that one thing has nothing to do with another, but they are connected." Durandus's first sense of goodness is absent in Aristotle.
    Thomas Aquinas distinguished only the second and third of these, namely the mutual order of the parts of the world and the relationship of the whole world to its end or goal (*Sentences*, bk. 1, dist. 44, qu. 1, art. 2, 1929–1947, 1:1019). In this, Thomas followed Aristotle more closely than did Durandus. By contrast with Thomas, Richard of Middleton not only assumed the three types of goodness described by Durandus but added a fourth: the betterment of the end itself (*Sentences*, bk. 1, dist. 44, art. 1, qu. 1, 1591, 1:390, col. 1).
17. See Galileo, *De caelo*, qu. 3 (E), 1977, 45; Conimbricenses, *De coelo*, bk. 1, ch. 10, qu. 2, art. 2, 1598, 117, 119; Amicus, *De caelo*, tract. 3, qu. 1 ("De unitate mundi"), art. 3, 1626, 104, col. 2; and Bonae Spei, comment. 3, *De coelo*, disp. 1, dub. 3, 1652, 3, col. 2.

way to achieve this would be for God to add individual members to individual species, and thus improve the world quantitatively. Durandus rejects this approach, because, as any Aristotelian would have agreed, individuals of a species do not represent principal parts of the world; only species of things function in this manner.

If, then, God added new species or parts to the present world, would such an act make the world more perfect? Durandus is uncertain, because it is quite possible that God has already created all possible species in our world. If so, he could add no more, and the world cannot be made more perfect with respect to its parts. But if God has not produced all possible species or parts in our world, then by the creation of new species he could indeed make the universe more perfect.[18]

Most scholastic theologians would have opted for Durandus's second alternative and rejected the first. It was commonly assumed that God did not exhaust his creative powers in making our world, and therefore did not create all possible species or even the best form of a species.[19] Indeed, not only are there gaps in the order of goodness of species in our world, but there is an infinite gap or distance that lies between the most perfect created entity and God. The addition of new species of creatures to fill these gaps would thus occupy many presently empty degrees of goodness, or existence (recall that goodness equals being or existence). When the new species are

18. Durandus, *Sentences*, bk. 1, dist. 44, qu. 3, 1571, 116v, col. 1, par. 10.

19. Examples of this attitude can be cited from numerous scholastic authors. Thomas Aquinas made this assumption, as we see in his reply to an argument that denied God the power to make the world more perfect (the world contains every good thing; and since nothing can be better than every good thing, it follows that God could not make a better world). Thomas first insists that he is speaking about the real universe, not merely discussing the term *universum*. The real universe contains every actual and real good thing, but not every good thing that God can make. For the claim against a better world, see Thomas Aquinas, *Sentences*, bk. 1, dist. 44, qu. 1, art. 2, 1929–1947, 1:1018; for his reply ("Ad secundum"), 1:1020. In a similar vein, Aegidius Romanus declared that "the world embraces every good thing that actually exists. It does not however embrace everything that God could make. Thus we argue that after God there is nothing better than the world; [but] we do not argue, however, that the world cannot be improved" (Ad secundum dicendum quod universum aggregat omne bonum quod actu existit; non tamen aggregat omne quod Deus potest facere. Unde arguitur quod post Deum nihil est melius universo; non tamen habetur quod universum meliorari non possit). Aegidius Romanus [*Sentences*, bk. 1, dist. 44, qu. 2], 1521, 226v, col. 2.

Two examples can be cited from the seventeenth century. Aversa, who mistakenly assumed that Durandus favored the first alternative (indeed, so did Galileo, who also attributes the same opinion to Duns Scotus; see Galileo, *De caelo*, qu. 3 [E], 1977, 45), expressed this common opinion when he insisted (*De caelo*, qu. 31, sec. 5, 1627, 15, col. 2–16, col. 1): "it is not necessary that this world, this universe that God created, should contain all possible species of things. For surely even Durandus ought to acknowledge that God did not create all the species which he in fact [could have] created. But then all possible species of things could not be in this universe. Therefore it cannot be held that God was bound to create, or to have created, all possible species of things." Finally, Cornaeus explained that "God did not always make [or create] the best that he was able to, but could make better things into infinity" (Neque enim Deus facit semper optimum quod potest, sed in infinitum saepe meliora facere posset). *De coelo*, disp. 1, qu. 2, dub. 2, resp. 4, 1657, 486.

added to the old species, would the whole make a new world? Thomas Aquinas and others responded ambiguously, resting content to declare that such new additions would make the resultant world different from our present world in some respects but not in others; or, as Thomas put it, "it would not be absolutely the same, nor would it be absolutely different."[20] By not claiming the emergence of a new world, Thomas, and others who followed the same path, effectively denied that a new world would emerge from the mere addition of new species to old.

But this is a purely quantitative act. Durandus now asks whether God could improve the species of the world, and therefore the world itself, by intensively increasing the perfection of a species. The response depends on whether the intensification occurs with respect to substantial (or essential) goodness or only accidental goodness. In the former sense, God is unable to improve the world, because substantial perfection can be altered only by adding a new perfection to a species or by intensifying the perfections it already possesses.[21] But adding to, or even subtracting from, its perfections would alter the species to such an extent that the old substance would be destroyed and a new one created.[22] Nor can a preexisting perfection be intensified, because substantial forms cannot be intended or remitted; that is, essential perfection is not alterable.

Many others would also deny that the world could be bettered by adding to the essential or substantial goodness of its species or parts; they include Thomas Aquinas, Saint Bonaventure, Richard of Middleton, Aegidius Romanus, Galileo, and, as late as the seventeenth century, Melchior Cornaeus and Franciscus Bonae Spei.[23] To improve the essential goodness of species – that is, to change their essence – in our world would not perfect the world,

20. See *Sentences*, bk. 1, dist. 44, qu. 1, art. 2, 1929–1947, 1:1019. Thomas also says that after the new parts or species have been added to the world, the enlarged world bears a relationship to the "old" world "as a whole to a part" (Sed illud universum se haberet ad hoc sicut totum ad partem). Richard of Middleton adopted the same position, insisting that "if new parts were added to this universe, then the universe with its newly added parts would be related to the present universe as whole to part. Although whole and part are not absolutely different in number, they are nonetheless not absolutely the same; but they are partly the same, partly different" (*Sentences*, bk. 1, dist. 44, art. 1, qu. 1, 1591, 1:390, col. 2).
21. Durandus discusses this in the first question of book 1, distinction 44, of his *Sentences*, 1571 (see 115r, col. 2, par. 5).
22. As Durandus puts it: "Just as it cannot happen that a ternary can remain a ternary and have more or fewer units than three, so also it cannot happen that something that remains the same could have more or fewer essential perfections" (Unde sicut non potest fieri quod ternarius manens ternarius habeat plures vel pauciores unitates quam tres, sic non potest fieri quod res aliqua manens eadem habeat successive plures perfectiones essentiales). Durandus, *Sentences*, bk. 1, dist. 44, qu. 1, 115r, col. 2. For the number analogy, Durandus cites Aristotle, *Metaphysics*, book 8 (in modern editions, the reference is 8.3.1043b.35–1044a.1). Thomas Aquinas also included the same argument (*Sentences*, bk. 1, dist. 44, qu. 1, art. 1, 1929–1947, 1:1017).
23. Thomas Aquinas, *Sentences*, bk. 1, dist. 44, qu. 1, art. 1, 1929–1947, 1:1016–1017; Richard of Middleton, *Sentences*, bk. 1, qu. 44, art. 1, qu. 1, 1591, 1:390, col. 2; Aegidius Romanus, *Sentences*, bk. 1, dist. 44, 1521, 226r, col. 1; and Galileo, *De caelo*, qu. 3 (E), 1977, 46, par. 11. For the others, see the next note.

but would produce one that is wholly different. Our world would simply cease to exist.[24] It would be as if the attribute of rationality were added to the definition of a cow: the resultant creature would no longer be a cow but an entirely new species, namely human. If cows were deprived of the capacity to sense, they would live as plants.[25] Although God could do such things if he wished, he would not perfect our world but would instead produce another world. Indeed, Richard of Middleton argued that if God made essential changes in only some substances and left the others unchanged, the world as a whole would be transformed into something else, although with respect to its parts it could still be considered the same world.[26] But God can increase the intensity of the accidents, or accidental properties of species or parts of the world, and thereby increase the overall perfection of the world. Not only can God add new accidental perfections to any species without altering the essential nature of that species, but he can also intensify accidental perfections that a species already possesses. "God is therefore able to better many creatures accidentally."[27] By altering the accidental goodness of things in the world, God could indeed make the world better and thus improve its state of perfection.[28]

But to what extent does God have to make improvements in accidental perfections in order that we may say that the world is improved? Some argued that greater perfection could occur only if God improved *all* parts of the world proportionately, for otherwise the harmony among the parts of the world would be lost. Unless one assumes a great latitude in the proportional relationships among the parts of the world, Durandus sees no merit in this argument. After all, the state of innocence in which humanity was instituted made human bodies and souls accidentally better in those early stages of the world's existence than is the case today, and yet the

24. See Bonaventure [*Sentences*, bk. 1, dist. 44, art. 1, qu. 1], *Opera*, 1882, 1:782, col. 1; Cornaeus, *De coelo*, disp. 1, qu. 2, dub. 2 ("Potueritne mundus a Deo condi perfectior"), 1657, 486; and Bonae Spei, comment. 3, *De coelo*, disp. 1, dub. 3, 1652, 3, col. 2.
25. Thomas Aquinas, *Sentences*, bk. 1, dist. 44, qu. 1, art. 1, 1929–1947, 1:1017.
26. "Aut si aliquae remanerent et aliquae mutarentur in alias, adhuc non remaneret idem universum in numero secundum se totum, sed secundum partes." Richard of Middleton, *Sentences*, bk. 1, dist. 44, art. 1, qu. 1, 1591, 1:390, col. 2.
27. In speaking of "accidental goodness," Durandus explains: "Deus potest facere multas creaturas meliores quam sint. Cuius ratio est quia Deus potest facere quodcunque quod non repugnat naturae rei, sed multis rebus non repugnat habere plures perfectiones accidentales quam habeant; istas etiam quas habent possunt habere intensiori modo. Ergo Deus potest multas creaturas meliorare accidentaliter." *Sentences*, bk. 1, dist. 44, qu. 1, 1571, 115r, col. 2, par. 4.
28. Among those who accepted this opinion, I can mention Bonaventure, *Sentences*, bk. 1, dist. 44, art. 1, qu. 1, *Opera*, 1882, 1:782, col. 1; Aegidius Romanus, *Sentences*, bk. 1, dist. 44, 1521, 225v, col. 2; Thomas of Strasbourg [*Sentences*, bk. 1, dist. 44, art. 3], 1564, 117v, col. 1; and Gabriel Biel [*Sentences*, bk. 1, dist. 44, art. 2, pt. 3], 1973, 751. As an illustration of the improvement of an accidental property, Bonaventure offers a boy whom God could make into a giant. Although as a giant the boy would have much more substance and power than before, he would not have changed his species and become something else.

heavens, planets, stars, and elements were no better then than now. By implication, the proportion among the parts must have been better then than now. Or we might infer that there was no proportion among the parts then; or that there is none now. Durandus rejects these alternatives and concludes that "the goodness of the universe does not require that all its species be related in the same proportion with respect to accidental goodness."[29] If such a principle operated in the world and if an entire species, which is found in only one part of the world, were altered for better or worse, it would be necessary that all other species or parts of the world be similarly altered, in order to maintain the proper proportional relationships among all the parts. Durandus rejects this as absurd, although it was the position upheld by Thomas Aquinas.[30] Durandus chose to believe that the world could be improved not solely by a divine action that bettered all of its parts, but also if God improved only some of its parts, or only one of its parts.[31] Such an increase of perfection could be achieved only with respect to accidental goodness, because every created species or part could have more perfections conferred upon it or have those it possesses made more intense.[32]

Scholastics generally assumed that by increasing the accidental goodness, or perfection, of species either quantitatively (i.e., extensively) or intensively, or both, God could indeed make our world more perfect. Moreover, this more perfect world would remain the same as our present world because the essential properties of its many species would have been left unchanged. Indeed, there is still only one world, albeit somewhat altered. The question of whether God could make other distinct and

29. "Non ergo requirit bonitas universi quod omnes species eius semper se habeant in eadem proportione quo ad bonitates accidentales." Durandus, *Sentences*, bk. 1, dist. 44, qu. 3, 1571, 116v, col. 2, par. 12.

30. Thomas argued that God could not merely improve some species but would have to improve all of them. No harmonious order of goodness could result from a selective qualitative improvement of species while others remained unaltered. In this regard, one could compare the world to a cither. If all its chords were improved, it would produce a sweeter harmony; but if only some of them were improved, the result would be dissonance. Thomas Aquinas, *Sentences*, bk. 1, dist. 44, qu. 1, art. 2, 1929–1947, 1:1019. Galileo echoes Thomas's sentiments when he declares that "if any one of the things in the universe were better, it would destroy the proportion of the entire universe, just as if one string were made too loud, it would destroy the melody of the musical instrument." *De caelo*, qu. 3 [E], 1977, 46, par. 12.

Although a Dominican himself, Durandus does not mention Thomas. Without mention of Saint Thomas or Durandus, Thomas of Strasbourg (*Sentences*, bk. 1, dist. 44, art. 3, 1564, 117v, col. 1) sided with Durandus when he declared that if God had to improve all species accidentally before the whole world could be made more perfect, then it would follow that the Incarnation of Christ would have been detrimental to the world, since the other parts of the world were not also improved.

31. "Dicendum ergo quod Deus potest meliorare universum meliorando partes eius omnes, vel quasdam solum, vel unam tantum." Durandus, *Sentences*, bk. 1, dist. 44, qu. 3, 116v, col. 2, par. 13. Thomas of Strasbourg, *Sentences*, bk. 1, dist. 44, art. 3, conclus. 2, 1564, 117v, col. 1, says much the same thing.

32. Durandus, *Sentences*, bk. 1, dist. 44, qu. 1, 1571, 115r, col. 2, par. 4.

separate worlds that would exist simultaneously with ours and which might be identical with ours is a wholly different question, and it is considered in Chapter 8.

2. Increasing the perfection of the world through the order of its parts

Since the parts or species of the world cannot be improved with respect to their essential goodness, neither can any greater perfection be imposed on the order which the parts bear to each other.[33] By adding new species, however, God could make the order of the world more perfect. According to Bartholomew Amicus, God could make the order of the world more perfect simply by adding an accidental perfection to a single species. For if this were not true, then we would confront the following absurdity: "if a whole species found in one part of the world were accidentally perfected, either the order [of the parts or species of the world] would be destroyed, or all other species of the world would also [have to] be perfected."[34] Even the order that obtains between superior celestial bodies and inferior sublunar bodies could be improved with respect to accidental properties. For example, heavy and light bodies, which are governed by celestial influences, might be respectively conserved more efficaciously and perfectly in their natural "down" and "up" places.[35]

3. Increasing the perfection of the world by the improvement of its goal or purpose

The goodness of a world may also be judged by the relationship of the world to its goal, or ultimate objective, which is God. The world cannot be made more perfect by an intensification or augmentation of God's perfection because God is already perfect in every way.[36] But if the world

33. Ibid., 116v, col. 2, par. 14. Richard of Middleton offers the same rationale (*Sentences*, bk. 1, dist. 44, art. 1, qu. 1, 1591, 1:390, col. 1), as do Saint Bonaventure (*Sentences*, bk. 1, dist. 44, art. 1, qu. 3, *Opera*, 1882, 1:786, cols. 1–2), Thomas Aquinas (*Sentences*, bk. 1, dist. 44, art. 2, 1929–1947, 1:1020), and Thomas of Strasbourg (*Sentences*, bk. 1, dist. 44, art. 3, conclus. 3, 1564, 117v, col. 1).
34. "Confirmatur quia alioqui si tota species, quae reperiretur in una parte mundi perficeretur accidentaliter, vel destrueretur ordo, vel deberent omnes aliae species mundi perfici, quod utrumque est absurdum." This passage forms part of the fifth of six conclusions which Amicus included on the perfection of the world from the standpoint of its order. The fifth conclusion itself asserts (*De caelo*, tract. 3, qu. 2, dubit. 2, art. 4, 1626, 118, cols. 1–2) that "if one part [of the world] should be perfected, with the other parts remaining unchanged, the world could be made more perfect by reason of the [new] order" (Quinta conclusio: si una pars perficiatur, invariatis caeteris, potest universum reddi perfectior ratione ordinis).
35. Durandus, *Sentences*, bk. 1, dist. 44, qu. 3, 1571, 116v, col. 2, par. 15.
36. Most scholastics would have agreed with Thomas Aquinas, who declared (*Sentences*, bk. 1, dist. 44, art. 2, 1929–1947, 1:1020): "Similiter ordo qui est ad finem, potest considerari, vel ex parte ipsius finis; et sic non posset esse melior, ut scilicet in meliorem finem universum ordinaretur, sicut Deo nihil melius esse potest." For similar viewpoints, also

improves, whether by the increase of its species or the qualitative improvement of the species, it can then be said that the world has become better in relation to God, its ultimate end, because the more perfect it becomes, the more closely it approximates to God.[37]

In general, scholastic natural philosophers and theologians conceded that God could make our world more perfect. Indeed, if he chose, he could make better and better worlds, ad infinitum.[38] But could God have made a world of infinite perfection? Buridan denied the possibility, because he was convinced in general that God could not create an actual infinite, whether that infinite were an infinitely intense quality, an infinite body, or an infinite void space. God could not create a creature of infinite perfection, because such a creature would be as perfect as himself. Moreover, such an infinitely perfect thing could not be made more perfect and would constitute a limit to God's power, since not even God can create something greater or more perfect than an actual infinite.[39]

In response to the natural query as to why an omnipotent God with absolute, infinite power chose not to make a perfect world in the first instance, scholastics believed that God conferred on our world all the perfection that it needed and placed in it all the species that he thought it required. God was under no compulsion to create every possible species of being. Nor, indeed, did he have to produce the best version of what he did create. In accordance with the medieval doctrine of God's absolute power, most scholastics would have conceded that, by his omnipotence and inscrutable will, God could do as he pleased. There was no necessity to create every possible existent or to make everything the best that it could be.

These beliefs clashed with Plato's principle of plenitude, which, as Arthur Lovejoy characterized it in his famous book *The Great Chain of Being*, assumed that God could not help but create every possible existent being. To this fullness of the world was added the principle of continuity, drawn ultimately from Aristotle but also derivable from the principle of plenitude. The principle of continuity holds, in Lovejoy's words, that "If there is between two given natural species a theoretically possible intermediary type,

see Durandus, *Sentences*, bk. 1, dist. 44, qu. 3, 1571, 116v, par. 15; Richard of Middleton, *Sentences*, bk. 1, dist. 44, art. 1, qu. 1, 1591, 1:390, col. 1; and Cornaeus, *De coelo*, disp. 1, qu. 2, dub. 2, resp. 2, 1657, 486.

37. So Thomas Aquinas seems to argue in *Sentences*, bk. 1, dist. 44, art. 2, 1929–1947, 1:1020.
38. William Vorilong, in his commentary on the *Sentences*, bk. 1, dist. 44, declared so in the fifteenth century (see McColley and Miller, 1937, 388); also cited in Dick, 1982, 43. Toward the end of the sixteenth century, the Coimbra Jesuits argued in the affirmative when they discussed the question "Whether God could create more and more perfect worlds to infinity" (Utrum ne Deus alios atque alios mundos in infinitum perfectiores possit condere); see Conimbricenses, *De coelo*, bk. 1, ch. 10, qu. 3, art. 2, 1598, 121–123; for the enunciation of the question, see page 120.
39. See Buridan [*Physics*, bk. 3, qu. 15], 1509, 57v, col. 1. For a further discussion and for a list of those who did and did not believe that God could create an actual infinite, see Grant, 1981a, 128–129. For Buridan's ideas about the possibility of an actual infinite magnitude and for a general discussion of the problem of actual infinites, see this volume, Chapter 6, Section I.

that type must be realized – and so on *ad indefinitum*; otherwise there would be gaps in the universe, the creation would not be as 'full' as it might be, and this would imply the inadmissible consequence that its Source or Author was not 'good,' in the sense which that adjective has in the *Timaeus*."[40] If God had indeed created such a world, it would have contained every possible thing, and he would consequently be unable to make it better than he had initially made it. It is obvious, from what has already been said, that scholastic theologians and natural philosophers did not assume such a world. God had by no means exhausted his creative power with the creation of our world. By an act of will, he could make our world more perfect, either by adding new species or by improving some or all accidental properties of existent species. Scholastics were generally reluctant to place restrictions upon God's power or to argue that he had of necessity to do this or that.[41]

4. Does the world's perfection reside in its diversity or uniformity?

But if God conferred upon his created world all the perfection and harmony it required,[42] how had he chosen to manifest that perfection: by uniformity or diversity? In the widely known *De mundo*, falsely ascribed to Aristotle during the Middle Ages and Renaissance, the author insists that the creative power has brought harmony from opposites and seeming disharmony. He explains that

> a single harmony orders the composition of the whole – heaven and earth and the whole Universe – by the mingling of the most contrary principles. The dry mingling with the moist, the hot with the cold, the light with the heavy, the straight with the curved, all the earth, the sea, the ether, the Sun, the Moon, and the whole heaven are ordered by a single power extending through all, which has created the whole universe out of separate and different elements – air, earth, fire, and water – embracing them all in one spherical surface and forcing the most contrary natures in it to live in agreement with one another and thus contriving the permanence of the whole.[43]

40. Lovejoy, 1960, 58; see also page 59 and chapter 3, pp. 67–98.
41. For Lovejoy's criticism of Thomas Aquinas, see ibid., pages 73–81. Lovejoy sums up (p. 81) Thomas's attitude toward the principles of plenitude and continuity by claiming that "He employs both freely as premises . . . whenever they serve his purpose; but he evades their consequences by means of subtle but spurious or irrelevant distinctions when they seem to be on the point of leading him into the heresy of admitting the complete correspondence of the realms of the possible and the actual, with the cosmic determinism which this implies."
42. Except for the Gnostics, most authors, from Greco-Roman antiquity to the seventeenth century, assumed a world that was a harmonious perfection. For passages on the order and perfection of the world from Plutarch and Diogenes Laertius, see Heninger, 1974, 146–147. For Macrobius's belief that the purpose of the world soul is "to instill harmonious agreement in the whole world," see Macrobius, 1952, 192.
43. Aristotle, *De mundo* 5.396b.25–32 [Forster], 1984.

The doctrine of harmony, based on a concordance of opposites, found some significant support during the Middle Ages. In the late thirteenth century, Bernard of Verdun, a Franciscan astronomer, explained that "even the beauty and perfection of creatures consists more in variety, numerousness, and proportionality than in complete uniformity." The celestial region was no different, for "just as the beauty of the starry orb consists in the variety and numerousness of the stars, so [does the beauty] of the lower [celestial bodies or planets consist] in the variety and numerosity of their motions and orbs."[44]

It remained for Nicole Oresme to confront the issue by means of a debate between the muses Arithmetic, who represents rationality and uniformity in the world, and Geometry, who speaks for diversity. In his *Tractatus de commensurabilitate vel incommensurabilitate motuum celi*, Oresme has the two muses debate whether the celestial motions are commensurable, as Arithmetic argued, or incommensurable, as Geometry believed. In this debate Oresme presents numerous arguments for uniformity and diversity, but his sentiments are clearly with diversity, as represented by incommensurable celestial motions. In perhaps the most telling passage in favor of diversity, Geometry, on the assumption of an eternal world, explains that, if Arithmetic's opinion were true, commensurable celestial motions would produce "the same, or similar, motions and effects," which "would necessarily be repeated." Geometry, who expresses Oresme's opinion, counters the position of Arithmetic by observing that to avoid such boring sameness

> it seems more delightful and perfect – and also more appropriate to the deity – that the same event should not be repeated so often, but that [on the contrary] new and dissimilar configurations should emerge from previous ones and always produce different effects. In this way, the far-stretching sequence of ages, which Pythagoras knew as the golden chain, would not return in a circle, but would always proceed endlessly in a straight line. This could not happen, however, without some incommensurability [obtaining] in the celestial motions.[45]

It is more than likely that medieval and early modern natural philosophers would have followed Oresme and believed that God achieved the most agreeable degree of harmony and perfection for the world by means of diversity rather than sameness and exact repetition. For as Galileo declared, "the perfection of the universe consists in the variety of things."[46]

44. In Bernard's *Tractatus super totam astrologiam*, as translated in Grant, 1974, 523, col. 2.
45. Oresme, *De commensurabilitate*, 1971, 317. For a detailed discussion of the treatise and the debate, see Chapter 18, Section I.3.a.
46. Galileo, *De caelo*, qu. 3 (E), 1977, 47; for more on variety, see Section I of this chapter and note 11 (for others who were in agreement with Galileo).

8

The possibility of other worlds

The worlds that have been discussed thus far were the result of imagined substantial changes made by God that transformed one world into another, more perfect world. Indeed, some held that God could perfect worlds ad infinitum, always making the last more perfect than its immediate predecessor. Within this context, only one world would exist at any given time. At some point, God might choose to improve it sufficiently to transform it into a new and better world, a process that he could repeat indefinitely. Although Nicole Oresme observed that successive worlds need not be incrementally more perfect but could conceivably be of equal status,[1] discussions described in Chapter 7 were essentially theological in nature and had their locus in the *Sentences* of Peter Lombard. They were also hypothetical, and thus introduce us to a characteristic mode of scholastic disputation. No one believed that God had actually made a series of successively better worlds, or even one better world. Scholastics were primarily concerned with whether God *could* do such a thing and under what circumstances he might do it.

I. Plurality of worlds before 1277

If God could make successively better worlds, and all believed that he could, it seemed natural to inquire further whether he could also create one or more other worlds that would exist simultaneously with ours.[2] Here was a theme that not only was discussed in commentaries on the *Sentences* but was also central to Aristotelian natural philosophy, since in his *De caelo* (bk. 1, chs. 8 and 9), Aristotle considered the possibility of the existence of other worlds and rejected it. The possibility of simultaneously existing, indepen-

1. Here Oresme relies on Aristotle (*De caelo*, 1.10279b.12–16), who reports the opinions of Empedocles of Acragas and Heraclitus of Ephesus that the world is destroyed and reconstituted endlessly, and on Saint Jerome, who reports that Origen believed that God had destroyed and re-created our world innumerable times (Oresme, *Le Livre du ciel*, bk. 1, ch. 24, 1968, 167).
2. It was fairly common to distinguish the question on plurality of worlds into worlds that are successive and worlds that are simultaneous. See Buridan [*De caelo*, bk. 1, qu. 19], 1942, 88.

dent worlds was a theme in early Greek natural philosophy from the time of the atomists, Democritus and Leucippus. The nature of atomism as expounded by these two led inevitably to the assumption that an infinity of worlds existed.[3] It was against this tradition that Aristotle discussed the problem of other worlds and not only decided for the existence of a single, unique world but insisted on the *impossibility* of other worlds.[4]

The introduction of Aristotle's natural books (as we saw in Ch. 3, Sec. III) caused considerable intellectual turmoil at the University of Paris in the thirteenth century and resulted in the Condemnation of 1277. Of its 219 condemned articles, the thirty-fourth was central to the problem of a plurality of worlds, because it declared "That the first cause [God] could not make several worlds." Anyone in the diocese of Paris, and therefore also in the University of Paris, who thereafter argued that God could not make any world other than our own would, unless he recanted, automatically be excommunicated.

Before and after 1277, few in the Middle Ages held the opinion that God could not create other worlds if he wished. Among those who did, and whose names were cited in this connection in the sixteenth and seventeenth centuries, were two famous figures: Peter Abelard, who lived long before the condemnation of 1277, and John Wycliffe, who lived in the fourteenth century.[5] If, over the centuries, only two men were named as supporters of the opinion that God could not possibly create other worlds (neither of whom even lived in the 1260s and 1270s, when most of the 219 articles

3. For an excellent account of the history of the concept of a plurality of worlds, see Dick, 1982. For the Greeks, see chapter 1.
4. For Dick's description of Aristotle's views, see ibid., pages 12–19. On page 193, note 18, Dick cites Aristotle's minor references to other worlds in the *Physics* and *Metaphysics*. Duhem's account of Aristotle's ideas appear in *Le Système*, 1913–1959, 1:230–234 ("La pluralité des mondes") and in Duhem [Ariew], 1985, 431–435.
5. They are mentioned by the Conimbricenses [*De coelo*, bk. 1, ch. 9, qu. 1, art. 2], 1598, 114, where their names are spelled "Vitcleph" and "Abaylardus," and by Bartholomew Amicus [*De caelo*], 1626, 101, cols. 1–2, who spells their names "Petrus de Abellardo" and "Ioannes Vuicleffus." The Coimbra Jesuits declared of Wycliffe that "as he was stained by the poison of impiety, he also spewed forth, among other false dogmas, that God could not produce effects other than those that He [had already] made, and thus that He could not create more worlds because nature acts by necessity." To this they added that "Abelard follows the same [opinion]" (Vitcleph, ut erat impietatis veneno conspersus, inter alia falsa dogmata, illud quoque evomuit, non posse Deum alios effectus edere quam edit, atque adeo nec plures mundos creare propterea quod nature necessitate ageret. Idemque secutus fuit Abaylardus). If we judge on the basis of Amicus's summary, neither Abelard nor Wycliffe explicitly denied God's power to create other worlds. Rather, this inference was drawn from their belief, as Amicus expressed it, "that God can produce nothing except those things which He produced; nor can He make it in a better way than He did because He can produce nothing except what falls under His providence. But He made whatever falls under His providence. Therefore nothing can be made, except what has been made" (Dicebant Deum nihil producere praeter ea quae produxit, neque meliori modo quam produxit quia nihil potest producere nisi quod cadit sub sua providentia, at quicquid cadit sub providentiam factum est, ergo nihil potest produci praeter id quod factum est). Amicus goes on to say that the opinion held by Abelard and Wycliffe was condemned as an error at the Council of Constance (1414) because it denigrated God's infinite divine power and freedom. On Peter Abelard's attempt to limit God's power, see Courtenay, 1990, 44–53.

were presumably held, in order to be condemned), why was this article singled out for condemnation?

It was obviously not because scholastic authors held that God could not possibly create other worlds, but rather because some of them devised or repeated arguments that made it unfeasible or seemingly contradictory for God to create more than one world. Thus, even as they conceded God's power to create more worlds, certain authors prior to 1277, among whom Michael Scot, William of Auvergne, Thomas Aquinas, and Roger Bacon are most prominent, found reasons to limit God to the creation of a single world.[6] Thus, in the decades prior to 1277, Aristotle's forceful arguments against a plurality of worlds and in favor of a single, unique world seemingly triumphed over God's absolute power to create as many as he pleased. Aristotle had declared unequivocally that

> there is not, nor do the facts allow there to be, any bodily mass beyond the heaven. The world in its entirety is made up of the whole sum of available matter (for the matter appropriate to it is, as we saw, natural perceptible body), and we may conclude that there is not now a plurality of worlds, nor has there been, nor could there be.[7]

During the Middle Ages, and for most of the period of this study, the "worlds" in the problem of a plurality of worlds were assumed to be identical: that is, those worlds possessed the same elements, or species of matter, as are found in our world.[8] Although scholastics would allow that God, by his absolute power, could create different material worlds with different species, few pursued it further. Different worlds became a serious topic of discussion only in the sixteenth and seventeenth centuries.

In defense of a unique world, Aristotle's conception of motions on the earth played a significant role. Probably relying on gross observation, Ar-

6. After 1277, we may add John of Jandun to this list (see Duhem [Ariew], 1985, 461). Duhem discusses these and many more authors in what is unquestionably the lengthiest and most thorough, though somewhat repetitious, study of the medieval treatment of the plurality of worlds. See Duhem, *Le Système*, 1913–1959, 9:363–430, and brief summaries of the opinions of a number of individual authors taken up seriatim in volume 10. A brief section appears also in volume 8, pages 28–35, and scattered remarks are made on a number of authors discussed on pages 35–59 of that volume. For the translation of the section in volume 9, see Duhem [Ariew], 1985, 431–510. Although I shall cite Duhem, I shall also draw upon my own account in Grant, 1979, 217–226.
7. Aristotle, *De caelo* 1.9.279a.7–11 [Guthrie], 1960. Michael Scot's translation of Aristotle's *De caelo* along with the commentary on it by Averroës has been printed in the Junctas edition of the works of Aristotle. See Averroës [*De caelo*], 1562–1574, volume 5. For the passage that I have quoted, see book 1, text 98, folio 66r.
8. Largely because he was convinced that his theory of natural and violent (or constrained, or unnatural) motion necessarily applied to all matter, wherever it might be, Aristotle concluded that if other worlds existed they all "must be composed of the same bodies, being similar in nature" (*De caelo* 1.8.276a.31 [Guthrie]). He went on to explain (ibid., b.1–3) that "if the bodies of another world resemble our own in name only, and not in virtue of having the same form, then it would only be in name that the whole which they compose could be pronounced a world."

istotle divided terrestrial motion into natural motion and violent, or unnatural, motion. The former was apparent in the unimpeded fall of heavy bodies from heights. A stone, for example, is observed to fall in a straight line toward the center of the earth, which is said to coincide with the geometrical center of our spherical universe. Other bodies, fiery or airy or smoky ones, always seemed to rise toward the heavens. From such evidence, Aristotle and his medieval followers concluded that a heavy or earthy body, when unimpeded, moves naturally downward in a straight line toward the center of the earth, which was assumed to be the natural "place" of heavy bodies. Conversely, light bodies, when unimpeded, always move upward toward the concave surface of the lunar sphere, the innermost surface of the celestial region, which was regarded as the natural place of all light bodies.

A commentary on the *Sphere* of Sacrobosco ascribed to Michael Scot, probably composed between 1231 and 1236, already illustrates the most significant arguments against the existence of a plurality of worlds. One of the most widely used involved the concept of void space.[9] If several worlds existed, they would exist either in one place or in different places. Because it was axiomatic that two or more bodies could not exist in the same place simultaneously, it followed that they must occupy different places, which implied the existence of intervening space, a situation that would obtain even if two of the worlds were in contact at a single point. The intervening space must be either a plenum or a void. It cannot be a plenum, because such matter or body must be assumed to lie between the worlds, and therefore belongs to neither.[10] Nor could the space be void, since Aristotle had demonstrated the impossibility of void space in the fourth book of the *Physics*.[11]

In another argument against a plurality of worlds, Michael Scot focuses on their material relationship: the bodies, or more particularly the elements, of these worlds would be composed either of the same species of matter or of different species of matter. The latter alternative is dismissed. For if the elements of these worlds differed, they would share only the name "world"

9. All of the arguments cited here from Michael Scot's commentary on the *Sphere* of Sacrobosco appear in Michael Scot [*Sphere*, lec. 1], 1949, 252–254.
10. "Sed non est ibi corpus aliquod, quia illud esset extra omnem mundum, cum spatium illud sit extra omnem circumferentiam mundorum." Ibid., 252. Although Aristotle does not quite say this, Michael is here alluding to *De caelo* 1.9.278b.25–279a.8, where Aristotle argues that no body could exist beyond our heavens or world.
11. "Si vero non est ibi aliquod corpus replens spatium illud, erit ponere vacuum quod est impossibile in rerum natura, ut patet per Aristotelem *in 4. Physicorum*. Ergo impossibile est plures mundos esse." Michael Scot, *Sphere*, lec. 1, 1949, 252. Sometime between 1231 and 1236, when Michael Scot wrote his commentary on Sacrobosco's *Sphere*, William of Auvergne repeated much the same argument against a plurality of worlds, as did Roger Bacon decades later. See William of Auvergne, *De universo*, pt. 1 of pt. 1, chs. 13–14, 1674, 1:607–608; and Roger Bacon, *Opus majus*, 1928, 1:186. For a summary of Michael's argument, as well as the arguments by William and Bacon, see Duhem, *Le Système*, 1913–1959, 8:28–35. For a summary of Aristotle's arguments against void space, see Grant, 1981a, ch. 1, 5–8.

(*mundus*) and nothing else. Merely applying the term "world" to material conglomerations other than ours does not make them worlds. Hence it does not follow that several worlds exist, and Michael refuses to consider this position.[12]

On the assumption of identical worlds, Michael, in another argument, insists, with Aristotle, that the elements of these worlds are identical and have the same motions, so that "it would follow necessarily that the earth of another world would be moved toward the middle [or center] of this world, or conversely. And similarly, the fire of another world would be moved toward the upper part of this world, or conversely," all of which is impossible (Michael Scot [*Sphere*], 1949, 253). But if the earth of our world were moved naturally toward the center of another world, Michael infers that the immobility of the earth at the center of our world could not represent a natural state of rest, as it should by the principles of Aristotelian physics, but it would remain at rest there only by violent or unnatural means. Thus its alleged "natural" motion in our world is not natural at all. Indeed, in rising toward the center of another world, its natural motion would be upward rather than downward. As for fire, although its motion would be opposite to that of earthy bodies, fire would behave in the same manner. From all this, Michael draws the obvious inference: other worlds cannot exist.[13]

Michael, however, took cognizance of opponents who believed it possible for an omnipotent God to make other worlds, even an infinite number of them, and to create them from identical or diverse elements indifferently.[14] In response, Michael readily assents that by his absolute power God could, if he wished, create a plurality of worlds. But although God has the power to create many worlds, nature itself, as a caused entity, is incapable of receiving them, since it has not been endowed with a capacity to accommodate many worlds simultaneously.[15] Thomas Aquinas presented a similar

12. Michael should have been familiar with Aristotle's concept of a world, which is quoted in note 8 of this chapter.
13. Michael Scot, *Sphere*, lec. 1, 1949, 253. Duhem mistakenly believed that Michael had rejected a plurality of worlds solely on the basis of the impossibility of void space beyond the world, arguing that Michael, William of Auvergne, and Bacon "appear not to have bothered with the reasoning that the Stagirite so carefully developed, the reasoning concerning the movement of the earth toward its natural place. Michael Scot and William of Auvergne did not refer to it" (Duhem [Ariew], 1985, 446; also Duhem, *Le Système*, 1913–1959, 9:369).
14. Michael Scot, ibid. For this argument, I draw on Grant, 1979, 219. Since the authors embraced by the term *quidam* are not identified, it is not possible to determine the significance of Michael's remark. Prior to 1277, however, the substance of such discussions would probably have been meager and the impact minimal.
15. Ibid., 253–254. William of Auvergne similarly denied a plurality of worlds, attributing this to a deficiency on the part of other potential worlds rather than to God (*De universo*, pt. 1 of pt. 1, ch. 16, *Opera*, 1674, 1:611, col. 2). After summarizing Michael's arguments, Duhem declares (*Le Système*, 1913–1959, 9:365; [Ariew], 1985, 443) that in opposing to God's creative power an already "predetermined nature that puts limits and preconditions upon Him," Michael's God "is closer to the God of Averroës than the God of the Christians."

argument. He acknowledged that, by his absolute power, God could create other worlds. But God apparently refrains from doing so, because the best and most noble ends would not be served by additional worlds. For if such worlds were similar to ours, they would be superfluous; and if dissimilar, none would be perfect, since none could incorporate within itself the totality of natures of sensible bodies. Indeed, it would require a combination of all these separate worlds to make a perfect world, a state that could be more economically achieved by a single world. Better therefore to make a single, perfect world than many that are imperfect, and better also to assign goodness to a single world than to diminish that goodness by distributing it over many worlds.[16] Those who assume the existence of many worlds are like Democritus, who denied a guiding wisdom as the cause of our world and "who taught that this world, and an infinity of others happened from the clash of atoms."[17]

II. Plurality of worlds after 1277

Michael Scot, William of Auvergne, Thomas Aquinas, Roger Bacon,[18] and undoubtedly others, represent the pre-1277 interpretation of the possible existence of other worlds. The Condemnation of 1277 radically altered subsequent responses. "Before the decree," as Duhem rightly declared, "they [Parisian masters] accumulated reasons derived from Peripatetic physics in order to establish that the existence of several worlds is an impossibility; therefore they refused God the power to multiply worlds. They endeavored to prove that this refusal was not a limitation on God's creative omnipotence."[19]

With article 34 of the Condemnation of 1277, all this changed drastically. By declaring it an excommunicable error to deny that God could create more worlds, article 34 not only shifted the focus of attention from the naturally possible to the supernaturally possible, but also fostered a more critical examination of Aristotle's arguments in favor of a unique world.[20] Although article 34 demanded only that one concede that God could, if he wished, create other worlds, it did not require anyone to believe that he had in fact done so. With the exception of John Major, in the sixteenth century, and perhaps one or two others, no scholastic authors did so be-

16. Thomas Aquinas [*De caelo*, bk. 1, lec. 19, par. 197], 1952, 94–95, especially the first and third responses.
17. Thomas Aquinas, *Summa theologiae*, pt. 1, qu. 47, art. 3 ("Is there only one world?" [Gilby]), 1964–1976, 8:103.
18. For Bacon's views, see Duhem, *Le Système*, 1913–1959, 9:368–369; Duhem [Ariew], 1985, 444–446.
19. Duhem [Ariew], 1985, 455, from Duhem, *Le Système*, 1913–1959, 9:380.
20. "34. That the first cause could not make several worlds" (Quod prima causa non posset plures mundos facere). For the Latin, see Denifle and Chatelain, 1889–1897, 1:545; for the translation, Grant, 1974, 48.

lieve.[21] Nevertheless, article 34, and the general emphasis on God's absolute power that was fostered by the Condemnation of 1277, encouraged many not only to take seriously the possibility that God could create other worlds than our own but to assume that he had actually done so. From such an assumption, they sought to counter those arguments of Aristotle that had previously been accepted more or less routinely. Indeed, they sought nothing less than to make the possible existence of other worlds intelligible.

1. A plurality of simultaneous concentric and eccentric worlds

At least two kinds of distinct and simultaneous worlds were distinguished. The first, and least discussed, was one that assumed worlds within worlds, either concentric or eccentric. Concentric worlds were briefly mentioned and rejected by William of Auvergne and Roger Bacon (Duhem [Ariew], 1985, 443, 445–446). It remained for Nicole Oresme, however, to provide the most elaborate and imaginative scholastic discussion of concentric worlds.

In what sense could worlds be concentric? Oresme imagines a world within our world, specifically a world within our earth, although such worlds might also exist within the Moon or within a star.[22] Such worlds are imagined as replicas of our own, each consisting of a celestial and terrestrial region. The smallness of such worlds poses no serious problems, because size is relative. If our world were made "between now and tomorrow 100 or 1,000 times larger or smaller than it is at present, all its parts being enlarged or diminished proportionally, everything would appear tomorrow exactly as now, just as though nothing had been changed" (Oresme, *Le Livre du ciel*, bk. 1, ch. 24, 1968, 169). Because a world within the center of our earth would have everything proportionally smaller, Oresme assumes its viability.

If such a world occupied the center of our earth, however, the latter could no longer have a center and would consequently lose its natural place at the center of our world. Oresme denies this by explaining that "for the earth to be in its natural place, it is enough that the center of its weight should be the center of the world" (ibid.). Therefore the center of the concentric world within our earth also serves as the center and natural place of our surrounding earth.

Although the existence of such worlds is improbable, Oresme insists that concentric worlds are not impossible, because "the contrary cannot be proved by reason nor by evidence from experience" (ibid., 171).[23] Never-

21. Major is considered near the end of this chapter. Both Nicholas of Cusa and Giordano Bruno believed in the existence of other worlds, but whether either qualifies as a scholastic author is problematic. In his scholastic phase, Galileo did not accept the existence of other worlds.
22. For the entire discussion, see Oresme, *Le Livre du ciel*, bk. 1, ch. 24, 1968, 167–171.
23. See Section II.2 for Albert of Saxony's brief discussion of concentric worlds.

theless, it is apparent that Oresme did not take this version of a plurality of worlds seriously. His speculations about concentric worlds were, by his own admission, a mental exercise for amusement (*esbatement*).

Concentric worlds, and even eccentric worlds, were mentioned well into the seventeenth century. Bartholomew Amicus imagined that outside of our heaven – presumably the last heaven or sphere – God could make an earthy heaven, surrounded by a watery heaven, which, in turn, is surrounded by an airy heaven and then a fiery heaven, and so on upward or outward by adding the corresponding celestial spheres.[24] Within this configuration of worlds, Amicus says that Aristotle's objections to a plurality of worlds are irrelevant, because the earth of any encompassing concentric world would rest naturally in the middle of its own world and not incline toward the earth of our world or any other.

Amicus also distinguishes three kinds of plurality involving eccentric worlds. One is but the eccentric counterpart of concentric worlds. That is, one world is wrapped around another, with each successive unit in touch at all points with its immediately proximate neighboring worlds above and below it. A second kind of plurality of eccentric worlds postulated worlds that were distinct and separate from one another.[25] And, finally, Amicus imagined overlapping eccentric worlds, where each world incorporates some part of another world within itself.

Despite the potential interest in configurations of concentric and eccentric worlds, scholastic authors paid little attention to them – indeed, Amicus says no more than what has just been mentioned – and concentrated instead on the second conception of a plurality of worlds: simultaneously existing spherical worlds that are wholly independent of each other.

2. *A plurality of simultaneous worlds, each separate and distinct from the others*

The problem of a plurality of spherical worlds existing separately and simultaneously was undoubtedly one of the most popular cosmological questions discussed in the five centuries covered by this study. Its popularity continues undiminished to this day, although the conception of what a world is has altered radically, and the physics and astronomy that guide modern investigations bear little resemblance to their medieval counterparts.

From the outset, the meaning to be attached to world (*mundus*) or universe (*universum*) – the two terms were synonymous – was problematic. In Chapter 7, Section I, I described the different senses and definitions of "world"

24. Amicus, *De caelo*, tract. 3, qu. 1, art. 1, 1626, 101, col. 1. Whereas Oresme imagined his concentric world to lie within our earth, Amicus imagined his to lie beyond our world and to enclose it. In the fifteenth century, Johannes de Magistris also mentions concentric worlds of this kind (Johannes de Magistris [*De celo*, bk. 1, qu. 4], 1490, 9, col. 2).

25. Johannes de Magistris, ibid., simply says that the centers of eccentric worlds would lie outside one another.

that were rather commonplace. But there was an even larger definitional problem involved in the way that concept was understood. In considering the question "Whether God could make another world with this one remaining in existence," Thomas of Strasbourg detected the potential difficulty and addressed himself to its resolution by distinguishing two senses of the Latin term for "universe."[26] In the first way, the term *universum* can be taken according to its literal meaning, as a unity in which all things are enclosed under an outermost movable sphere (the *primum mobile*), which is turned and moved by a single prime mover (*primus motor*). On this Aristotelian approach, God could not make other worlds because the notion of world is effectively defined to exclude them. For if God made other worlds identical to ours, each would have its own outermost movable sphere and prime mover. Hence these worlds could not form a single, unified world, because world is defined as a unity, with only one outermost sphere and one prime mover. In this way, as Thomas puts it, "there would be a universe, and there would not be a universe, which implies a contradiction."[27]

The second definition, however, considers the notion of world with respect to the things that are represented or expressed by the term.[28] On this approach we can apply the term "world" to any collection of things, whether identical to or different from the things in our world. Thus God could make another collection of things that are identical with those in our world, and then we would wish to call that collection another world; or he could create a different and nobler collection of things, and that too would be considered a world. It is this second sense of world that Thomas found acceptable and which was probably in the minds of most other scholastic authors who gave the matter any thought.

That other worlds than ours could exist became a commonplace based on faith: by his absolute power to do as he pleased (short of a logical contradiction), God could make as many distinct and separate simultaneously existing worlds as he pleased. When scholastic natural philosophers assumed that God had indeed made such worlds, the worlds they imagined were identical replications of ours. Although many acknowledged that God could make worlds different from ours in every conceivable way, scholastic commentators concerned themselves overwhelmingly with the possible existence of identical worlds.[29]

26. Thomas of Strasbourg [*Sentences*, bk. 1, dist. 44, qu. 1, art. 4: "Utrum Deus possit facere aliud universum illo universo manente?"], 1564, 117v, col. 1.
27. "Uno modo quantum ad rationem nominis." Ibid. For d'Ailly's discussion of the term *universum*, see this volume, Chapter 1, Section II.2. For four meanings of the term *mundus*, see Chapter 7, Section I and note 4.
28. "Alio modo quantum ad realitatem vel realitates, expressam vel expressas, nomine universi." Thomas of Strasbourg, ibid.
29. This may have been determined by Aristotle's argument that if other worlds were possible at all, they would have to be of a kind identical with ours (*De caelo* 1.8.276a.32–276b.5.). He undoubtedly based this conclusion on his own determination that only four elements – earth, water, air, and fire – could exist, each of which moved with a single natural

On the assumption that God had made other worlds identical to ours, scholastic natural philosophers confronted a significant problem, one to which a number of them devoted a separate question. It was the problem raised by Michael Scot and described earlier in this chapter. If these worlds existed simultaneously, would the elements of one world tend to move toward the center and circumference of another world rather than to those in their own world? Would the earth of one world, or any part of it, seek the center of another?[30] For example, would a heavy, earthy particle in another world naturally seek the center of our world by first rising up in its own world and, upon reaching ours – however it might achieve that – fall toward its center? Heavy bodies would thus seem to possess two contrary natural motions, which Aristotle deemed impossible. No heavy body could rise. The same could be said of fire, which might rise in our world and, upon reaching another world, descend to its natural place between air and the lunar sphere. It would thus be capable of both rising and falling naturally. One and the same heavy or light body would thus be capable of two distinct contrary natural motions, a plain violation of Aristotle's dictum that a simple heavy or light body could have only one natural motion. Because no plausible alternative interpretation was readily available prior to 1277, Michael Scot, William of Auvergne, Roger Bacon, and others rejected the possibility of two or more worlds.

After 1277 and the condemnation of article 34, a plausible interpretation was formulated. One of the first to do so was Richard of Middleton, who inquired "Whether God could make another world" ([*Sentences*, bk. 1, dist. 44, qu. 4], 1591, 1:392, col. 1). Even if all the worlds are identical, as Aristotle assumed, Richard argues that neither the earth nor any of its parts would ascend in one world in order to reach the center of another world. Indeed, the earth of each world would remain at rest in its own world-center and any parts of it that might be removed would, if unimpeded, always tend to return to that same center. According to Richard, this was also "the opinion of Lord Stephen, bishop of Paris and doctor of sacred theology, who excommunicated those who dogmatized that God cannot make more worlds."[31]

motion to a specific natural place where it would naturally rest. For Aristotle, anything that could be truly called a world would have to be identical to ours and have the same four elements. If other worlds existed, they would have to possess the same structure as our world. John of Jandun was representative of medieval thought when he declared that "those who say that there are several worlds . . . say that they are all of the same species" (qui dicunt plures esse mundos, . . . dicunt omnes esse eiusdem speciei). John of Jandun [*De caelo*, bk. 1, qu. 24], 1552, 16v, col. 2.

30. This is question 66 ("If there were several worlds, whether the earth of one would be moved naturally to the middle [or center] of another") in the "Catalog of Questions," Appendix I.

31. "Et per hanc opinionem est sententia domini Stephani Episcopi Parisiensis et sacrae theologiae doctoris, qui excommunicavit dogmatizantes quod Deus non posset facere plures mundos." Richard of Middleton [*Sentences*, bk. 1, dist. 44, qu. 4], 1591, 1:392, col. 2–393, col. 1. The reference is to Etienne Tempier and article 34 of the Condemnation of 1277.

Moreover, if it were possible to remove the earth of another world and place it at the center of ours, that earth would remain at rest in the center of our world; and conversely, if the earth from our world were removed to the center of another world, it would remain at rest there with no inclination to move toward its former place. To reinforce the argument that each earth would remain in its new place, William of Ockham furnished an analogous argument for fire. If the fire of one world were removed to another world, it would remain there and not seek to return to its original world. Their behavior would be like that of two fires in our world, one moving toward the circumference of the heaven over Oxford, the other moving toward the heaven over Paris. If these masses of fire were switched, the fire now over Paris but formerly over Oxford would move directly upward toward that part of the celestial circumference over Paris, with no inclination to move back over Oxford.[32]

Many others also conceived each world as a self-contained, independent, closed system with its own proper center and circumference.[33] Aristotle's argument against a plurality of possible worlds, based on the idea that it was impossible for elements of one world to move to another world, which held sway in the thirteenth century prior to 1277, was thereafter usually abandoned.

Although many scholastics came to believe that Aristotle was wrong to deny the possibility of a plurality of worlds, certain others, while conceding that God could make other worlds supernaturally, continued to indicate, in a number of ways, that a multiplicity of worlds was not to be taken seriously. As a major strategy, they perfunctorily conceded that God could create other worlds supernaturally but then essentially proceeded to defend Aristotle. Thus Albert of Saxony confined his discussion to concentric and successive worlds, omitting and ignoring, perhaps deliberately, the most important category: distinct, nonoverlapping, spherical worlds. At first we are led to believe that Albert is attacking Aristotle when he argues that "against the plurality of concentric worlds, Aristotle's argument is not valid, namely [that argument] by which Aristotle holds that there cannot be several worlds because the earth of one would be moved naturally to the middle of another."[34] But Aristotle's argument was irrelevant because it applied only to distinct and identical, nonoverlapping spherical worlds and was in

32. Ockham [*Sentences*, bk. 1, qu. 44], 1979, 657; see also Dick, 1982, 34.
33. In this group, we find Godfrey of Fontaines (Duhem [Ariew], 1985, 451); Ramon Lull; Johannes Bassolis; William of Ockham; Walter Burley; Robert Holkot; William of Ware, or Varon (Duhem [Ariew], 1985, 457–458); Jean Buridan (*De caelo*, bk. 1, qu. 18, 1942, 86–87); Nicole Oresme (*Le Livre du ciel*, bk. 1, ch. 24, 1968, 173–175); Thomas of Strasbourg (*Sentences*, bk. 1, dist. 44, qu. 1, art. 4, 1564, 118r, col. 1); Albert of Saxony (Duhem [Ariew], 470); and Bartholomew Amicus (*De caelo*, tract. 3, qu. 1, art. 6, 1626, 107, col. 2).
34. See Albert of Saxony [*De celo*, bk. 1, qu. 11, conclus. 1], 1518, 95r, cols. 1–2, and Duhem [Ariew], 1985, 469–470.

no way intended for concentric worlds, whose centers would coincide by definition. Thus Albert cannot be included among the numerous scholastic authors who insisted that if other worlds like ours existed, the earth of each would remain naturally at rest in its own world.[35]

Albert also rejects concentric, as well as eccentric worlds, because each such world would require its own prime mover, and a plurality of prime movers cannot exist. Indeed, in a configuration of concentric or eccentric orbs, where each orb is a world, only the outermost orb or heaven could serve as the prime mover.[36]

Successive worlds are also naturally impossible, because one world would have to be destroyed and another come into being. But if a world is destroyed, it must be corruptible, including its celestial region. In Aristotelian cosmology and physics, as we shall see, the heavens are incorruptible and cannot be destroyed; therefore neither can the world of which they are a part.[37]

Albert's final conclusion is but a repetition of Aristotle's claim that no other world could possibly exist because all the matter in existence forms our world. The latter consists of five simple bodies, which includes the four elements and the compounds made from them, along with the celestial ether, or fifth element.

Although Albert of Saxony met the formal requirements that he concede that God could create other worlds of all kinds supernaturally, he chose to emphasize the natural impossibility that any such worlds could exist. He deliberately omitted from consideration simultaneously existing distinct worlds, perhaps because he did not wish to appear to oppose or reject the popular opinion that in simultaneously and distinctly existing worlds, the earth of one world would not move naturally or violently toward the center of another world. To have opposed this opinion and sided with Aristotle might have brought Albert under suspicion as one who, though upholding the letter of condemned article 34, was violating the spirit of it within the context of the Condemnation of 1277.

If Albert of Saxony obliquely avoided a potential difficulty, John of Jandun unequivocally supported Aristotle, although, as we shall see, he made the mandatory concession to God's power to create more worlds. Writing in the early fourteenth century, some years after the Condemnation of 1277, Jandun accepts without qualification Aristotle's argument about the earth of another world moving to the center of our world, an argument that defenders of a possible plurality of worlds were compelled to reject.[38] Jandun insists that the earth of one world would indeed move toward the center

35. Thus I here correct my assertion about Albert in Grant, 1979, 222, n. 36.
36. Albert of Saxony, *De celo*, bk. 1, qu. 11, conclus. 4, 1518, 95r, col. 2.
37. Ibid., conclus. 3.
38. For these and the subsequent arguments from Jandun, see John of Jandun, *De coelo*, bk. 1, qu. 24, 1552, 16v, col. 2. For Duhem's account, see Duhem [Ariew], 1985, 461–462.

of our world. But this is impossible; therefore, more than one world cannot exist. The impossibility derives from the common assumption that the other worlds would be identical to ours, that is, be of the same species as our world, in which event, according to Aristotle, it would follow, as we saw earlier in this chapter, that the earth of our world and the earth of any other world must have the same natural place, namely the center of our world. In order to reach the center of our world, the earth of another world would have to move through the circumference of its own world. To achieve, this, however, it must move naturally upward toward that circumference, which is impossible, because all heavy bodies of the same species move naturally downward.

But what if the earths of the two worlds were of different species: does it not follow that their natural places would differ, so that the earth of another would not seek to reach the center of our world? Jandun dismisses this suggestion as inappropriate because those who argue about a plurality of worlds always assume that the worlds are of the same species (see n. 29).

Jandun considers yet another possible objection to Aristotle, one that he describes as a "quibble" (*cavillatio*). For even if the worlds are identical, might it not be possible for the earth of another world to be sufficiently distant from the center of our earth and world that it would simply have no inclination to move toward our world, just as iron is not attracted to a magnet at any and all distances whatsoever? Aristotle answered this objection by denying that distance affects the form of an element or a body.[39] Natural bodies, wherever they are, tend to move to their one and only natural place. Distance cannot affect them. As for the iron and the magnet, the former does not move naturally toward the latter by its own inclination but only when its form is occasionally altered by the power of the magnet, which occurs only when that power can reach the iron. But heavy bodies do not fall to their natural place at the center of the world like iron moving toward a magnet. A magnet may exert its power over a distance, but the center of the world does not. Although heavy bodies may move in a passive manner when something else causes them to move, in a manner similar to the way a magnet causes iron to move, heavy bodies move toward their natural place in a more fundamental way by virtue of an internal principle of motion. Therefore they are not dependent on a motive power that varies with distance, as are the iron and magnet. Heavy bodies always seek their natural place, whatever their distance from the earth.

In another, rather murky, argument, John of Jandun rejects a plurality of worlds because if the other worlds are identical with ours, each would

39. For the basis of this argument, see Aristotle, *De caelo* 1.8.276b.22–26. Duhem's description is apt: "A body is heavy when it is by nature potential with respect to the center of the world, which is its natural place; whether it is near or far from the center of the world, it always has the potential to lodge there, and this potential cannot be spoken of in degrees; it can only be ended when the body is at the center of the world" (Duhem [Ariew], 1985, 433). Aristotle does not invoke the example of the attraction of a magnet for iron.

require immaterial movers (see Ch. 18, Sec. II.3) to produce its own perpetual celestial motions. Without a clear reason, Jandun rejects as impossible the existence of such additional immaterial movers. But he adds, significantly, that such limitations do not extend to the divine power, "because in Him infinite freedom and infinite power always exist to make other worlds. For although this [the creation of other worlds] cannot be derived from sensible things, which was the basis for Aristotle's argument, we must nevertheless believe this by assenting reverentially to the sacred doctors of the faith."[40]

Despite this weak concession to God's infinite power, John of Jandun reveals his sharp disagreement with most of his contemporaries and successors. He flatly denies the natural possibility of other worlds and gives arguments for so believing. By contrast, most of his scholastic colleagues allowed that other worlds were at least possible. They were prepared to deny that the earth of one world would seek the center of another world. Jean Buridan even characterized as "nondemonstrative" Aristotle's defense of the claim that heavy bodies in other worlds would move toward the center of ours ([*De caelo*, bk. 1, qu. 18], 1942, 86–87). This would only be true, countered Buridan, if the inclinations of heavy bodies depended solely on their common tendency to move toward a single center. But motions also depend on celestial bodies and God. Because every world would have its own celestial bodies and God's presence and control would be equal in all, the heavy, earthy bodies of a particular world would fall only to the center of their own world.

Of course, Buridan also emphasized God's power to make other worlds, declaring that "we hold on faith that just as God made this world, so could He also make another, or several others."[41] Buridan makes it quite evident, however, that God probably chose not to make those other worlds, because if he wished to create additional creatures of the kind that appear in our world, he could simply enlarge our world to double, or one hundred times, its present size.[42]

40. "Sed istud non praeiudicat potentiae divinae quia semper in ea servatur libertas infinita et potentia infinita faciendi plures mundos; quia licet non posset convinci a sensibilibus, a quibus accipitur ratio Aristotelis, tamen credendum est hoc firmiter sacris doctoribus fidei reverenter assentiendo." John of Jandun, *De coelo*, bk. 1, qu. 24, 1552, 16v, col. 2–17r, col. 1. For an English version of Duhem's French translation, see Duhem [Ariew], 1985, 461–462.

41. "Tenemus ex fide quod sicut deus fecit istum mundum, ita posset adhuc facere alium vel alios plures." Buridan, *De caelo*, bk. 1, qu. 18, 1942, 84.

42. This statement occurs in Buridan's *Questions on the Physics*, where he concedes to the theologians that God could make a space beyond our world and that one could not demonstrate the contrary. But he believes that no space or magnitude or other world exists beyond our world. For just as God could more simply and economically make our world larger to accommodate additional creatures, rather than make additional worlds, so also would it be in vain for him to create a space beyond our world that would serve no purpose. Buridan [*Physics*, bk. 3, qu. 15], 1509, 57v, col. 2. In his *Sentences*, William of Ockham adopted a radically different approach. He derived the possibility of other worlds from the conviction that "God could produce an infinite [number of] individuals

The question of other worlds produced more than just the anti-Aristotelian possibility that a plurality of simultaneous, identical worlds could exist as self-contained systems whose earths would remain at the center of their own worlds. It also challenged Aristotle's fundamental idea that each of the four elements had one absolutely determined natural place, a concept that was viable only if there was one possible world with a single center and circumference. With the possibility of more than one spherical world, and therefore more than one center and circumference, all centers would be equal and none unique, thus casting doubt on Aristotle's doctrine of natural place. It was Nicole Oresme who exploited its vulnerability in the course of a discussion of a plurality of worlds in his French commentary on Aristotle's *De caelo*, completed in 1377. The result was a significant departure from Aristotelian cosmology.

Whereas Aristotle had linked up, down, light, and heavy to an absolute sense of natural place, Oresme redefined their meanings in relative terms. He considered a body to be "heavy," and therefore "down," when it is surrounded by "light" bodies, which are assumed to be "up." Heavy and light, with their corresponding and interrelated concepts of up and down, were thus conceived independently of the natural places of bodies.[43] Oresme judged a body to be "heavy" and "down" when it is surrounded by light bodies, where the surrounding "light" bodies are conceived as "up." The independence of these relationships from Aristotle's natural places is made vividly apparent in a thought experiment. After imagining that a tile or copper pipe extends from the center of the earth to the heavens, presumably to the concave surface of the lunar sphere, Oresme insists that "if this tile were filled with fire except for a small amount of air at the very top, this air would drop down to the center of the earth for the reason that the less light body always descends beneath the lighter body. And if this tile were full of water save for a small quantity of air near the center of the earth, this air would mount up to the heavens, because by nature air always moves upward in water."[44]

of the same kind as those that now exist. Therefore God can produce as many individuals and more as are now produced. But He is not limited to producing them in this world. Therefore He could produce them outside of this world and make one [other] world of them, just as He made this world from the things that have already been produced." Ockham [*Sentences*, bk. 1, qu. 44], 1979, 655. Thus Buridan and Ockham both readily concede God's power to create additional creatures beyond those that presently exist in our world.

43. After distinguishing two senses of "up" and "down" (one with regard to us, as when we say that half the heavens lie "up," above us, and the other half "down," below us; and the other used with respect to heavy and light bodies, the latter sense being Oresme's sole concern here), Oresme declares that "up and down in this second usage indicate nothing more than the natural law concerning heavy and light bodies, which is that all the heavy bodies so far as possible are located in the middle of the light bodies without setting up for them any other motionless or natural place." Oresme, *Le Livre du ciel*, bk. 1, ch. 24, 1968, 173.

44. Ibid., bk. 1, ch. 4, 71. Although this example is relevant, it occurs early in the treatise. But in referring to it in book 1, chapter 24 (p. 173), Oresme explains that its purpose is

For Oresme, then, earth is heavy and down because it comes to rest naturally in the center of the lighter bodies that surround it. In fact, each heavy body at rest in a surrounding lighter element is a center of attraction for other heavy bodies. Although he fails to make the inference, Oresme has here the idea of a plurality of centers of attraction within our world, not only within each of many worlds. It is an idea that Leonardo da Vinci (1452–1519) pursued when he declared that "the moon is surrounded by its own elements: that is to say water, air, and fire; and thus is, of itself and by itself, suspended in that part of space, as our earth with its elements in this part of space; and that heavy bodies act in the midst of its elements just as other heavy bodies do in ours."[45] The idea of a "center of attraction," as Dick has observed (1982, 41), may have played a significant role in shifting interest from a plurality of worlds on the Aristotelian model to a plurality of earthlike planets within our cosmos, each of which would be conceived as a world in itself.

From the idea of a center of attraction, Oresme asked next what might occur if an earthy, or heavy, body was not surrounded by lighter bodies. He concluded that it could not be characterized as being either up or down. Therefore, if a vacuum existed between our world and another, a particle of earth from that world could not possibly move to the center of our world. For even if it could rise up and depart beyond the circumference of its own world – which, as we saw, was generally denied – it would enter the void between those worlds and come to rest. The latter result was inevitable because in a void, no lighter bodies exist to surround the heavy body that emerged from its world to seek the center of our world. Therefore it is neither up nor down and has no inclination to move in any direction. It must consequently remain at rest.[46] Indeed, Oresme insisted that the same result would obtain if God created a portion of earth and set it in the heavens where the stars are. Although he does not elaborate, Oresme's reasoning depends upon the ubiquitous belief, drawn from Aristotle, that, strictly speaking the ethereal substance of the stars is neither light nor heavy and thus offers no resistance to bodies.[47] Since the portion of earth is surrounded by a substance that is neither light nor heavy, it would have "no tendency

to show "how a portion of air could rise up naturally from the center of the earth to the heavens and could descend naturally from the heavens to the center of the earth."

45. Dick, 1982, 40, makes the point about a plurality of centers for Leonardo. The passage from Leonardo appears in Leonardo da Vinci [Richter], 1970, 2:130, sec. 902.

46. Without invoking a vacuum, Buridan argued similarly when he imagined that God destroys the heavens and all bodies but leaves a house standing, along with the air within it. In the midst of this air, a globe of earth is at rest. "This globe," Buridan declares, "would not be moved, because there is no reason why it would be moved toward one side rather than another, since one part of air would be no more up or down than another and some power would not reside in one than in the other [part] because [we have assumed that] order has been removed from the heaven itself." Buridan, *De caelo*, bk. 1, qu. 19, 1942, 86–87.

47. We shall see that "improperly speaking," properties such as lightness and heaviness, and numerous others, would be applied to the celestial ether.

whatsoever to be moved toward the center of our world." In effect, the ethereal substance shares similar properties with the vacuum: in neither would heavy bodies have any tendency to move toward the center of our world.[48]

Despite a conviction that "there never has been nor will there be more than one corporeal world," Nicole Oresme was motivated by article 34 of the Condemnation of 1277 to insist that "God can and could in His omnipotence make another world besides this one or several like or unlike it."[49] Here were the two beliefs that most scholastics shared: there was only one world, just as Aristotle and Genesis would have it, but God could, if he wished, make as many more as he pleased. This opinion was still accepted in the seventeenth century when Illuminatus Oddus declared that "There is in fact one universe, or five simple bodies, namely the heaven [that is, the celestial ether], fire, air, water and earth. No other than these, whether in species or number can be found, although this cannot be naturally demonstrated. With regard to the possible, another [world] can be produced by God both in species and in number."[50] But it was by examining the assumption that God could create other worlds that scholastics departed from Aristotle in significant ways. Whereas Aristotle had insisted that the existence of other worlds was impossible, scholastic authors argued that it was indeed possible, albeit by supernatural action. Moreover, they demonstrated that a plurality of worlds was physically intelligible. Each world would be a self-contained entity and have no effect on the others.

One scholastic author diverged radically from the opinions and approaches described thus far. John Major (1467/68–1550), a Scottish logician, theologian, natural philosopher, and Aristotelian commentator at the University of Paris, in a treatise on the "actual infinite" (*Propositum de infinito*, first published in 1506), proposed the existence of an infinity of worlds as one of a number of supporting claims for the existence of an actual infinite.[51] As his definition of world, Major assumed a collection of spheres, either eccentric or concentric, with all that was contained within them, which was

48. Oresme, *Le Livre du ciel*, bk. 1, ch. 24, 1968, 173.
49. Ibid., 177–179.
50. "De facto unum esse universum, seu quinque esse corpora simplicia huius mundi, scilicet caelum, ignem, aerem, aquam, et terram. Nec alia, sive specie sive numero, ab istis reperiri, licet hoc naturaliter demonstrari nequeat. De possibili vero posse alia, et specie et numero, a Deo produci." Oddus [*De coelo*, disp. 1, dub. 3], 1672, 5, col. 2. A world identical to ours, or which differs only accidentally, belongs to the same species as our world but differs "in number," since it is a distinct and separate world. However, a world that differs substantially from ours would be said to differ "in species."
51. Major revised and supplemented the *Propositum de infinito* in the 1510, 1519, and 1530 editions of his commentary on the *Sentences*. For a modern edition and French translation, which cites some of the later changes, see Major [Elie], 1938, 56–58, 60–62, 114. Duhem, *Etudes*, 1906–1913, 2: 92–94, and Dick, 1982, 38–39, provide summaries of Major's brief treatment. Duhem's account has been translated in Duhem [Ariew], 1985, 503–504. For a brief biographical sketch of Major with accompanying bibliography, see Lohr, 1988, 237–239.

essentially the medieval conception of a world.[52] Without providing much by way of sustained argument, Major declares that "naturally speaking" (*naturaliter loquendo*), he was unconvinced that there was only one such world.[53] Following Democritus, whom he cites, Major proclaims that "naturally speaking there are infinite worlds, [and] no argument can convince one of the opposite, [namely that the number of worlds is not infinite]."[54]

But what if someone should assume that all things, taken simultaneously, form only one world, which therefore embraces all the entities that we might otherwise call separate worlds? Because Aristotle did not dispute or consider such a question, Major dismisses this suggestion as a failure to understand Aristotle's texts.[55] Aristotle considered worlds that were assumed distinct and did not attempt to subsume them all under a single world. As for Aristotle's argument that the earth of one world would be moved toward the middle, or center, of another world, Major replies that this and any other arguments are easily refuted.[56]

Major's argument in favor of infinite worlds is an implicit one that seems to be based on the following strategy. He first insists that no one can give convincing reasons for supposing the existence of only one world. Therefore more than one must exist. But if more than one, how many? Since there is no good reason for any one particular number of worlds, it seems plausible to assume an infinite number. Thus did Major adopt unequivocally the old Democritean idea of an infinity of worlds. Because Major's opinion appeared in three subsequent editions of his commentary on the *Sentences*, it was probably read by other theologians. None, to my knowledge, joined him in support of this extraordinary, though anomalous, outlook.[57]

52. Judging by Major's statement a few lines below, these worlds could be either eccentric or concentric. Duhem, *Etudes*, 1906–1913, 2:93, n. 1, believed that the reference to "eccentric" worlds (*eccentrici*) was a mistake for "concentric." But there is no reason to assume a mistake, since earlier authors had spoken of eccentric and concentric worlds (for example, as we saw, Albert of Saxony). Moreover, Major speaks of "mundi eccentrici, fortassis concentrici" (Major [Elie], 1938, 62).
53. "Ad confirmationem videtur mihi mere naturaliter loquendo non potest convinci oppositum quod tantum est unus mundus, capiendo mundum, sicut utimur, pro aggregato spherarum et contentorum in eis." Major, ibid., 60.
54. "Naturaliter loquendo sunt infiniti mundi, nulla ratio convincit oppositum." Ibid., 56. In the fifteenth century, William Vorilong, a Franciscan theologian, declared in book 1, dist. 43, of his commentary on the *Sentences* that God could create an infinity of worlds (see McColley and Miller, 1938, 387–388), but he did not posit their actual existence. Occasionally scholastics mentioned infinite worlds almost in passing (for example, Jean de Ripa; see Grant, 1981a, 130).
55. "Si dicas omnia simul faciunt unum mundum, non intelligis propriam vocem. De hoc Aristoteles non disputasset." Major [Elie], 1938, 60–62.
56. "Ratio Aristotelis quod terra unius moveretur ad medium alterius facile diluitur et quelibet alia ratio ut apparet." These words follow immediately after Major declares that "Naturaliter loquendo infiniti mundi, nulla ratio convincit oppositum." Ibid., 56–58. From the refutations of this argument described earlier in this chapter, we know what Major had in mind.
57. Roderigo de Arriaga allowed that an infinite number of worlds was naturally possible, because there is nothing in nature to contradict this or to prevent it (Praeterea etiam

One might suppose that among the numerous scholastic discussions of a possible plurality of worlds, some would have considered the further possibility of the existence of life in those worlds. Thus far only one scholastic author is known to have raised the problem. In the fifteenth century, William Vorilong considered the existence of a second world and argued that our knowledge of it would be derived from "angelic revelation or by divine means." Moreover, the species of that world could differ from the species in our world. As for extraterrestrial life, William says that

> If it be inquired whether men exist on that world, and whether they have sinned as Adam sinned, I answer no for they would not exist in sin and did not spring from Adam. But it is shown that they would exist from the virtue of God, transported into that world, as Enoch and Elias [Helyas] in the earthly paradise. As to the question whether Christ by dying on this earth could redeem the inhabitants of another world, I answer that he is able to do this even if the worlds were infinite, but it would not be fitting for Him to go into another world that he must die again.[58]

Vorilong's brief discussion about extraterrestrial life was enunciated at approximately the same time as his great contemporary, Nicholas of Cusa (1401–1464), was proclaiming life in other parts of our cosmos – in the Sun and in the stars.[59] Although there is little to suggest that Vorilong believed in another inhabited world, his discussion of its possibility and his regard for some of the hypothetical consequences mark a significant departure from the usual concerns of scholastic authors on the existence of other worlds. It is appropriate, therefore, to characterize both Vorilong and Major as scholastics who adumbrated themes that later received detailed consideration outside of scholastic cosmology.

videntur naturaliter produci posse infiniti mundi quia nihil est in natura quod illis contradicat). Arriaga [*Physics*, disp. 13, sec. 3], 1632, 420, col. 1. But he does not indicate or suggest that they actually exist.

58. McColley and Miller, 1938, 388. In a note, McColley and Miller observe that in the sixteenth century "Philip Melancthon and others vehemently attacked the doctrine of a plurality of inhabited worlds for the reason that such a plurality was regarded as incompatible with the Atonement." Jesus Christ did not appear in other worlds to die and be resurrected in each.
59. For a brief and interesting summary of Cusa's views, see Dick, 1982, 40–42.

9

Extracosmic void space

Aristotle had not only denied the possible existence of other worlds, but he had also argued that no vacuum could exist beyond our world.[1] With perhaps a few minor exceptions, there was little serious discussion of the possibility of extracosmic void space prior to the Condemnation of 1277. When the problem did arise, Aristotle's rejection was usually adopted with little elaboration, as, for example, in the anonymous *Liber sex principiorum*, falsely ascribed to Gilbertus Porretanus;[2] in John of Sacrobosco's *Sphere*;[3] and in Robert Grosseteste's *Commentary on the Physics*.[4]

After 1277, however, the possibility of the existence of extracosmic space came to be discussed in two interrelated, though distinguishable, contexts. In the first, the primary concern was with the possible existence of void space that was independent of God but assumed to have been created by him before, during, or after the creation of the world. The possibility that God could create finite vacua at will was regularly conceded after 1277, although his ability to create an infinite vacuum was seriously questioned. In the second context, extracosmic void space was not assumed independent of God but was rather conceived as in some sense a property or attribute of his omnipresent immensity.

Before we consider these two aspects of void space, we should realize that the most powerful impetus in the Middle Ages for belief in the existence of a void space – indeed, an infinite void space – beyond our world came from the domain of theology by way of its concern for the nature of God's attributes. And just as book 1, distinction 44, of Peter Lombard's *Sentences*

1. After showing that body cannot exist beyond the world, Aristotle explains further that "This world is one, solitary and complete. It is clear in addition that there is neither place nor void nor time beyond the heaven; for (a) in all place there is a possibility of the presence of body, (b) void is defined as that which, although at present not containing body, can contain it, (c) time is the number of motion, and without natural body there cannot be motion. It is obvious then that there is neither place nor void nor time outside the heaven, since it has been demonstrated that there neither is nor can be body there" (*De caelo* 1.9.279a.10–17 [Guthrie], 1960). Aristotle's argument may be summarized as follows: since body cannot exist beyond the world, neither place nor vacuum could exist there, because "in all place there is a possibility of body" (no body, therefore no place) and because "void is defined as that which, although at present not containing body, can contain it" (no body, therefore no vacuum). My source for this summary is Grant, 1981, 105, although in this note I have changed from Stock's translation (Aristotle, 1984) to Guthrie's (Aristotle, 1960).
2. Minio-Paluello, 1966, 47, lines 8–9.
3. Sacrobosco [*Sphere*], 1949, 80–81 (Latin) and 120 (English).
4. Grosseteste [*Physics*, bk. 3], 1963, 58–59.

15 true?
1277
15
15

served as the locus of most discussions as to whether God could make the world better than he had made it,[5] so also did book 1, distinction 37, become the locus for discussions about extracosmic void space. For it was in this distinction that Peter Lombard described the ways in which God is said to be in things. Here theologians regularly discussed God's whereabouts – that is, the place where God was prior to the creation, the ways in which he could be said to be in a place, and whether God was in any way movable.[6] Although the *Sentences* was a natural locus for discussion of extracosmic space, some theologians in the sixteenth and seventeenth centuries found it equally convenient to treat infinite space in questions and commentaries on Aristotle's *Physics*, either the fourth or the eighth book.

I. Independent extracosmic void space

While conceding that God could create other worlds and/or a finite or infinite void space beyond our world, many scholastic authors denied the natural existence of such an empty space. Merely because God could create such entities did not mean that he had actually done so. Jean Buridan adopted this attitude when he cautioned that

> an infinite space existing supernaturally beyond the heavens or outside this world ought not to be assumed, because we ought not to posit things that are not apparent to us by sense, or [by] experience, or by natural reason, or by the authority of Sacred Scripture. But in none of these ways does it appear to us that there is an infinite space beyond the world. Nevertheless, it must be conceded that beyond this world God could create a corporeal space and any whatever corporeal substances it pleases Him to create. But we ought not to assume that this is so [just] because of this.[7]

If Buridan emphasized the improbability of an extracosmic void space, he nevertheless conceded the supernatural possibility of it. Just as with the possible plurality of worlds, most scholastic authors took a twofold approach: they followed Aristotle and denied that a void space was naturally possible but conceded that it was supernaturally possible. In a question "Whether the world is a finite or infinite magnitude," Albert of Saxony presents four brief imaginary situations in which God creates stones or

5. See Chapter 7, where book 1, distinction 44, is cited throughout.
6. See Grant, 1981a, 115. In the next two sections, on extracosmic void space, I draw heavily on my book *Much Ado about Nothing* (Grant, 1981a). Only a bare summary can be presented here.
7. For the Latin text, see Buridan [*De caelo*, bk. 1, qu. 17], 1942, 79. I have quoted my translation in Grant, 1976a, 150. Albert of Saxony [*De celo*, bk. 1, qu. 9], 1518, 93v, col. 2, expressed virtually the same sentiment. Gaietanus (or Caietanus) de Thienis [*Physics*, bk. 4], 1496, 28v, denied that Christians had to concede a vacuum beyond the world (nec oportet Christianos concedere vacuum extra celum).

worlds beyond our world that are located in a divisible, presumably void, space.[8] Indeed in the fourth, he imagines three worlds – ours and two others – any two of which touch at a point, from which he concludes that there must be spaces and distances that lie between the surfaces that are not in contact.[9] In the very next section of the question, however, Albert devotes five conclusions to demonstrate that nothing exists naturally beyond our world, neither body, place, vacuum, time, God, or intelligences.

Although the problems of other worlds and extracosmic void are to some degree linked, there is an important difference between them. In the former, God was always assumed capable of creating other worlds, and yet, except for John Major, scholastics did not believe in their actual existence. They were merely possibilities. With extracosmic void space, however, real existence was often assumed, occasionally on natural, but more often on supernatural, grounds. The possibility that God could create other worlds, coupled with Aristotle's definition of a vacuum as a place deprived of body but capable of receiving it, implied that void space might exist beyond our world.[10] No one presented the argument more succinctly and explicitly than Robert Holkot, an English Dominican friar (d. ca. 1349), who explains that if God could make another world, he could create it anywhere. Holkot then inquires whether anything now exists in the place where God could create this world. If something does exist there, then something, presumably a body, already exists beyond the world, contrary to Aristotle. But if nothing presently exists in the place where God could create that world, Holkot argues as follows: "Beyond the world, nothing exists; but beyond the world a body can exist [since God can create a world or a body there]; therefore a vacuum exists beyond the world, because a vacuum exists where a body can exist but does not. Therefore a vacuum is [there] now."[11] In this ar-

8. Encouraged to imagine worlds and bodies in all kinds of odd and bizarre situations, Albert of Saxony, ibid., hit upon one of the strangest when he declared that God could place a body as large as the world inside a millet seed and he could achieve this in the same manner as Christ is lodged in the host, that is, without any condensation, rarefaction, or penetration of bodies. Within that millet seed, God could create a space of 100 leagues, or 1,000, or however many are imaginable. A man inside that millet seed could traverse all those many leagues simply by walking from one extremity of the millet seed to the other. In his wildest flights of fancy and speculation, could Walt Whitman have had anything like this in mind when he penned the line "Every inch of space is a miracle"? *Leaves of Grass* (Ithaca, N.Y.: Cornell University Press, 1961), poem 8: "Poem of Perfect Miracles," 220.

9. "Quarto nam ponatur quod Deus extra hunc mundum formet duos alios mundos et illi tres mundi contingerent se sicut sphere imaginantur se contingere. Videtur quod inter illos tres mundos et puncta in quibus contingerent se esset spatium medium et distantia media. Aliter enim tangerent se secundum superficies et non secundum puncta, quod est falsum." Albert of Saxony, ibid., col. 1.

10. For Aristotle's definition of "void," see note 1 of this chapter.

11. Because of its importance, I cite the whole of Holkot's brief argument: "Praeterea, si Deus posset modo facere alium mundum ab isto, posset facere illum esse alicubi, sicut iste est modo, ita quod partes illius mundi distarent abinvicem extra mundum istum. Quero ergo quid est ibi modo: an aliquid an nihil. Si aliquid, ergo extra mundum de facto est aliquid. Si nihil, tunc arguitur sic: extra mundum nihil est, et extra mundum potest esse corpus; ergo extra mundum est vacuum, quia ubi potest esse corpus, et nullum

gument, Holkot seems to assert categorically, rather than hypothetically, the actual existence of an extracosmic vacuum. The mere possibility that God could create other worlds, which all Christians had to concede, was thus sufficient to infer the existence of a vacuum beyond the world. Perhaps it is a measure of the degree to which opinions and attitudes had changed since the thirteenth century that Walter Burley could declare that Christian theologians, and generally those who believed in the creation of the world, could hardly avoid the conclusion that a vacuum existed beyond our world, since they also conceded that God could create another world.[12]

But Holkot neglects to explain whether the extracosmic vacuum was a divine creation. Around 1354 or 1355, Jean de Ripa, a Franciscan theologian, argued, in book 1, distinction 37, of his commentary on the *Sentences*, that God could indeed create an actually infinite thing and therefore could have created – and perhaps did create – an actually infinite, imaginary void space.[13] But, although de Ripa thought that the divine immensity would fill any infinite void that God might create, he did not, and could not, identify this possible created void with God's immensity, as would others later, because "the infinity of a whole possible vacuum or imaginary place is immensely exceeded by the real and present divine immensity."[14]

Why did de Ripa arrive at this conclusion? Why did he not rather assume that God's infinite immensity was equal to and in some sense "coextensive" with infinite void space? Because, as de Ripa put it, if "the whole [infinite] imaginary place is equal to the divine immensity," God would, in some way, be "circumscribable by some imaginary thing outside Himself" and "He would be able to be present in some imaginary place that is equal to Himself." That is, if God's divine immensity is equal only to an infinite imaginary space, one could just as well declare that God is contained by the equal space as vice versa, an unacceptable consequence. De Ripa's distress was further intensified when he observed that if God's immensity and an infinite imaginary space were truly equal, then if, by natural or divine power, some creature also existed in that infinite imaginary void space, that creature might be as spatially, or "locally" (*localiter*), immense as God him-

est, ibi vacuum est. Ergo vacuum modo est." Holkot [*Sentences*, bk. 2, qu. 2], 1518, sig. bii, recto, col. 2. I have made minor changes in my translation in Grant, 1981a, 351, n. 130.

12. "Difficile tamen ut mihi videtur est vitare quin loquentes nostre legis et generantes mundum habeant ponere vacuum extra mundum quia ipsi dicunt quod sicut Deus creavit hunc mundum ita posset creare alium mundum." Burley [*Physics*], 1501, 89r, col. 1.
13. The term "imaginary" in the expressions "imaginary, infinite void space" and "imaginary infinite space" is explained in Section III of this chapter.
14. "Secunda conclusio: *Totius vacui possibilis seu situs ymaginarii infinitas immense exceditur ab immensitate reali et presentiali divina*" (de Ripa [*Sentences*, bk. 1, dist. 37], 1967, 235, lines 26–28). The translation is from Grant, 1981a, 133, where the text is cited on page 342, note 60. For a full discussion on de Ripa, see Grant, 1981a, 129–134. The quotation just given is from page 133. Elsewhere de Ripa says that "the divine immensity . . . immensely superexceeds every imaginary place" (ibid., 235, lines 35–40). Subsequent scholastic opinion on extracosmic imaginary infinite space would follow Bradwardine in assuming that God did not create infinite void space but rather that such a space has existed eternally in the form of his infinite immensity (see Sec. II of this chapter).

self and would thus be uncircumscribed by God. For de Ripa, this was impossible.

To avoid such impossible consequences, de Ripa assumed that God's immensity is greater than any infinite, imaginary space. The latter would always lie within God's infinite immensity. But de Ripa insists that anything contained within an infinite is infinitely exceeded by that infinite. Therefore God's immensity infinitely exceeds any infinite, imaginary void space.[15]

In this extraordinary discussion, de Ripa was led by an inexorable theological logic to distinguish two different kinds of infinites, one infinitely greater than the other. Because God is uncircumscribable and must circumscribe all things, he is a superinfinite, who is infinitely greater than any other infinite. Thus if God created an infinite, imaginary void, he would not only be omnipresent within it but would also infinitely exceed it.

In presenting and defending his position, de Ripa attacked an unnamed doctor (referred to only as *iste doctor*) – probably Thomas Bradwardine – whose opinions on the relations between God and infinite space had been formulated but a short time before. Because Bradwardine's approach triumphed over de Ripa's, it is essential that we understand his interpretation.

II. God-filled extracosmic void space

Although relevant ideas about God and extracosmic void space had been formulated in late antiquity, especially by the anonymous author of the *Asclepius* (or *De aeterno verbo*, as it was known in the Middle Ages), a Hermetic treatise, and by Saint Augustine,[16] the medieval phase in the history of this major theme was apparently given its first significant and noteworthy treatment by Thomas Bradwardine (ca. 1290–1349). In or around 1344, Bradwardine, an eminent mathematician, natural philosopher, and theologian at Oxford University who died as archbishop of Canterbury, wrote his *De causa Dei contra Pelagium* (*In Defense of God against the Pelagians*).[17] Within the context of a chapter titled "That God is not mutable in any way," Bradwardine inferred five corollaries that he judged consistent with God's immutability. Because of their importance, they deserve full citation:

1. First, that essentially and in presence, God is necessarily everywhere in the world and all its parts;
2. And also beyond the real world in a place, or in an imaginary infinite void.
3. And so truly can He be called immense and unlimited.

15. The last two paragraphs are drawn essentially from Grant, 1981a, 133.
16. For these two and others, see ibid., 112–115.
17. Here I draw specifically on ibid., 135–144.

4. And so a reply seems to emerge to the questions of the gentiles and heretics – "Where is your God?" And, where was God before the [creation of the] world?"
5. And it also seems obvious that a void can exist without body, but in no manner can it exist without God.[18]

The first two corollaries proclaim the ubiquity of God – that is, God's omnipresence within the world and beyond, in an infinite imaginary void space (*vacuo imaginario infinito*). As justification for God's ubiquity, Bradwardine invokes his infinite perfection and power,[19] which enabled him to be simultaneously in every part of the world when it was created and to have preceded the world in the very place where he created it. Prior to the creation, God's presence depended only on himself and not on the creation or any creature. Moreover, he must have been eternally in the place that the world would eventually occupy. For otherwise he would have had to arrive there from elsewhere, which is impossible, because any movement to the place of creation would constitute a mutation, and therefore an imperfection, in God's status.[20] But God could have created the world anywhere he pleased, so that we must assume an infinite number of void places in which God has existed eternally. Taken collectively, these places constitute an infinite imaginary void in which God exists everywhere.

As further support for God's ubiquity, Bradwardine declares that God is not confined to the place where he created the world, because "it is more perfect to be everywhere in some place, and in many places simultaneously, than in a unique place only."[21] Without need of any creature and eternally at rest, "God is, therefore, necessarily, eternally, infinitely everywhere in an imaginary infinite place, and so truly omnipresent, just as He can be called omnipotent."[22] Although Bradwardine did not make it explicit, it

18. Ibid., 135. The corollaries also appear in Grant, 1974, 556–557. The Latin text appears in Bradwardine, *De causa Dei*, 1618, 177, and is reprinted in Grant, 1981a, 344–345, n. 92. For a discussion of Bradwardine's ideas about God and space, see Koyré, 1949, 80–91. Koyré provides a French translation and Bradwardine's Latin text.
19. "Est ergo perfectionis et potentiae infinitae quod Deus necessario sit ubique." Bradwardine, *De causa Dei*, 1618, 180(E), and Grant, 1981a, 345, n. 99.
20. This opinion was shared by de Ripa, *Sentences*, bk. 1, dist. 37, 1967, 230–231, lines 30–47, and 234, lines 12–14. Nicole Oresme, *Le Livre du ciel*, bk. 2, ch. 2, 1968, 279, made much the same point when he declared that "if God made another world or several outside of this world of ours, it would be impossible that He not be in those worlds, and without moving Himself, because God cannot possibly be moved in any way whatsoever." In medieval physics and metaphysics, motion to and from anything was considered a sign of change and mutability. Because God was always assumed immutable, it was further assumed that he did not move from one place to another but occupied all positions simultaneously.
21. "Praeterea perfectius est esse in aliquo situ ubique, et in sitibus multis simul, quam in unico situ tantum." Bradwardine, *De causa Dei*, 1618, 178(E); Grant, 1981a, 346, n. 103; and Koyré, 1949, 89, n. 1. For the English translation, Grant, 1981a, 138, and Grant, 1974, 559.
22. Grant, 1974, 559; Bradwardine, *De causa Dei*, 1618, 178(E)–179(A); also see Grant, 1981a, 138, 346, n. 104.

also follows that the place of the world and the infinite void beyond are homogeneous, a point that Giordano Bruno would make much of in deriving the infinity of the universe.[23] From God's omnipresence in an infinite imaginary void place or space, the third corollary – that God is truly "immense and unlimited" – follows immediately. Two of the numerous quotations cited by Bradwardine in support of God's uncircumscribable, infinite immensity are noteworthy. The first was drawn from the *Book of the XXIV Philosophers*, an anonymous twelfth-century pseudo-Hermetic treatise, wherein the author declares that "God is an infinite sphere, whose center is everywhere and circumference nowhere."[24] The second came from Saint Augustine's *Confessions* (bk. 7, ch. 5), where Augustine sought to compare the creature to the creator by imagining the Lord embracing the world "in every part and penetrating it, but remaining everywhere infinite. It was like a sea, everywhere and in all directions spreading through immense space, simply an infinite sea. And it had in it a great sponge, which was finite, however, and this sponge was filled, of course, in every part with the immense sea." In representing the relationship between the finite world and God's infinite immensity by the figure of a sponge immersed in a boundless sea and penetrated by that sea, Augustine furnished a graphic model for those who came to believe that our world was immersed in an infinite void that was equated with God's immensity, which also penetrated the world.[25]

But if God is omnipresent in an infinite void space, does this imply that God is an actually extended magnitude? Not for Bradwardine or any subsequent scholastic author. Indeed, Bradwardine insists that God "is infinitely extended without extension and dimension. . . . He can be called immense since He is unmeasured; nor is He measurable by any measure; and He is unlimited because nothing surrounds Him fully as a limit; nor, indeed, can He be limited by anything, but [rather] He limits, contains, and surrounds all things."[26] Because Bradwardine identified infinite void space with God's immensity, and because God is "infinitely extended without extension and dimension," it follows that Bradwardine intended to deny extension to infinite void space.

In the final corollary, Bradwardine formulated the most fundamental property of infinite void space when he declared that "void can exist without body, but in no manner can it exist without God." Though devoid of body, void space was filled with God, or spirit. Bradwardine's inseparable asso-

23. See Greenberg, 1950, 49.
24. Although it was frequently cited between the twelfth and fourteenth centuries, Nicholas of Cusa may have been instrumental in disseminating the definition, which was often quoted in the seventeenth century by both scholastic and nonscholastic authors. For the significance and widespread use of this famous metaphor, see Grant, 1981a, 138–140. To those mentioned in Grant, 1981a, we may add John Case [*Physics*, bk. 8, ch. 10], 1599, 860.
25. Grant, 1981a, 140–141. Among those who used Augustine's figure, we may include the Coimbra Jesuits in the sixteenth century and Bonae Spei, Amicus, and Maignan in the seventeenth century.
26. Translated in Grant, 1981a, 141, from Bradwardine, *De causa Dei*, 1618, 179(A).

ciation of God and a dimensionless, extensionless void made his infinite vacuum radically different from that of the Greek atomists and Stoics, for whom vacuum was simply a three-dimensional space, devoid of body but not filled with spirit. But if Bradwardine's vacuum differed from that of the Stoics, his basic configuration of a spherical cosmos surrounded by an infinite void space was identical with theirs.[27]

Why did Bradwardine identify God's immensity with infinite void space, thus making the latter eternal and, in effect, a divine attribute? It was because Bradwardine was convinced that the world required a void space in which to be located upon creation. But such a world could not be an entity of eternal duration independent of God. In Bradwardine's opinion, such a precreation void space could have "no positive nature, for otherwise there would be a certain positive nature which is not God, nor from God . . . ; such a nature would be eternal with God, something no Christian could accept."[28] Indeed, article 201 of the Condemnation of 1277 specifically condemned the proposition that an independent precreation void could exist.[29] *A fortiori*, an infinite void space independent of God would be wholly unacceptable. Indeed, Bradwardine was one of those who also denied that God could create an actual infinite thing such as an infinite void. For these reasons, Bradwardine, and others who followed, arrived at the existence of an infinite void space by identifying it with God's infinite, extensionless immensity. Only in this way could they justify the existence of a precreation void that had to be assumed infinite in order that God might create the world anywhere he pleased.

Although Bradwardine's ideas were all but forgotten until the publication of his *De causa Dei* in 1618, it was he who seemingly first linked God and infinite void space and who considered that space as real. How could it be otherwise, if it was God's immensity? Despite some degree of opposition,[30] not only did many scholastics subsequently reaffirm the indissoluble bond between God and imaginary infinite void space, but some elaborated on aspects of it that Bradwardine had ignored, such as, for example, the manner in which God occupied the uncreated infinite void space that was his im-

27. On the Stoic conception, see Grant, 1981a, 106–108. The Stoic view was known from William of Moerbeke's 1271 translation from Greek into Latin of Simplicius's *Commentary on De caelo*.
28. Translation from Grant, 1981a, 111. For the Latin text, see Bradwardine, *De causa Dei*, 1618, 177, and Grant, 1981a, 326, n. 38.
29. For the translation and Latin text, see Grant, 1981a, 326, n. 37.
30. Although de Ripa disagreed with Bradwardine by denying that God's immensity was identical with an infinite imaginary void, he did accept the possible existence of such a void – and may even have thought of it as actual rather than merely possible. Duns Scotus, however, and most of his followers, denied that God had to be omnipresent by essence in an infinite vacuum in order that he might actually be present in every possible place in which he could create the world. Scotus insisted that God need not be present where he acts, since he can create anything he wishes anywhere at all merely by willing it. Since God can act anywhere at all by his will alone and therefore need not actually be present, an infinite extracosmic void is superfluous and unnecessary, and Scotus rejected it. For the discussion and references, see Grant, 1981a, 144–147.

mensity. Among those who discussed the existence of an imaginary, infinite void space, many identified it with God's immensity, including Nicole Oresme (fourteenth century); John Major, Michael de Palacio, the Coimbra Jesuits, and Francisco Suarez (sixteenth century); John Case, Pedro Hurtado de Mendoza, Bartholomew Amicus, Emanuel Maignan, Thomas Compton-Carleton, Franciscus Bonae Spei, Melchior Cornaeus, and Sigismundus Serbellonus (seventeenth century).[31]

III. The meaning of "imaginary" in the expression "imaginary infinite space"

Beginning in the sixteenth century, the discussions were often lengthier and more subtle. Scholastics became more aware of the need to define and qualify terms, concepts, and relationships. Perhaps the most important concept in need of explanation was the very meaning to be associated with the term "imaginary" in such expressions as "imaginary infinite space" or "imaginary, infinite void space." When Bradwardine and later scholastics used one or the other of these expressions, what did they mean? No single answer is possible, because a number of different meanings and usages emerged. The complexity of usage is further reflected in the two different expressions. Some authors speak explicitly of void space and others only of space, without indicating whether the space is void or not.

For some, especially those who rejected the actual existence of an infinite extracosmic void – for example, Aristotle, Averroës, Thomas Aquinas, and Pseudo–Siger of Brabant – the fact that we can conceive of an infinite empty space beyond our world is made possible by our imagination. They were in basic agreement with Aristotle, who, in explaining why certain people believed in the existence of the infinite, explained that in our thoughts, certain things appear to be inexhaustible or without end, for example, number, mathematical measures or magnitudes, "and what is outside the heaven."[32] Aristotle makes it clear (*Physics* 3.4.203b.25–29) that what lies "outside the heaven" is for some people an infinite void and place in which an infinite body or infinite worlds might exist. He seems also to imply that we have a strong intuitive sense that something must lie beyond. But for Aristotle, Averroës, Thomas, Pseudo-Siger, and Gabriel Vasquez (1551–

31. Oresme, Major, Suarez, the Coimbra Jesuits, Amicus, Maignan, and Bonae Spei are discussed in Grant, 1981a (on Bonae Spei, see n. 50 of this chapter). For the others, see Palacio [*Sentences*, bk. 1, dist. 37, disp. 2: "Num Deus est extra coelum"], 1574, 195v, col. 2; Hurtado de Mendoza [*Physics*, disp. 14, sec. 3], 1615, 323, col. 2; Compton-Carleton, [*Physics*, disp. 33, sec. 4], 1649, 337, col. 2; Cornaeus [*Physics*, bk. 4, disp. 3, qu. 1, sec. 2, dub. 9], 1657, 369–370; Serbellonus [*Physics*, disp. 5, qu. 1, art. 1], 1657, 803, col. 2; and Case, *Physics*, bk. 8, ch. 10, 1599, 868–869. Michael de Palacio was active in the second half of the sixteenth century at the University of Salamanca, where he was first student, then professor of philosophy (1545–1550) and subsequently professor of theology (1550–1554) (see Lohr, 1988, 297).
32. Aristotle *Physics* 3.4.203b.22–24. In this section, I rely on Grant, 1981a, 117–121 ("The Meanings of the Term 'Imaginary' in the Expression 'Imaginary Space' ").

1604), intuition was misleading: imaginary space was just that: imaginary, and nonexistent or fictitious.[33] Although he did not use the term "imaginary," that same intuition led Nicole Oresme to the opposite conclusion, when he insisted that "the human mind consents naturally, as it were, to the idea that beyond the heavens and outside the world, which is not infinite, there exists some space, whatever it may be, and we cannot easily conceive the contrary."[34]

A second opinion, held by Bradwardine but explicated more extensively and explicitly by Francisco Suarez, identified imaginary space with vacuum and contrasted it with real space, which exists wherever there is a body.[35] Imaginary space is converted to real space when occupied by body and is reconverted to imaginary space when vacated by body. In his lengthy discussion, Suarez, as did Bradwardine, equated God's infinite immensity with an infinite, imaginary void space. God does not need the real and positive space of bodies because he did not have to create the material bodies that constitute those real spaces. Without those bodies, all that would remain is imaginary space, or God's infinite immensity.

Almost all scholastic authors who identified infinite, imaginary space with God's infinite immensity felt compelled to deny extension or dimensionality to that space. To assign extension to imaginary space would have implied that God himself was an actually extended corporeal being. Although Benedict Spinoza, Henry More, and Isaac Newton took this momentous step, medieval and early modern scholastics, with the possible exception of Thomas Compton-Carleton (discussed later in this chapter), avoided it. To do this, they were compelled to grope for some means of describing a nondimensional space that, by its very association with God, had to be conceived as an existent something even though it lacked extension and could possess no positive attributes. Despite difficulties and perplexities, they were eventually led to describe imaginary space as some kind of negation. In this approach, Pedro Fonseca (1528–1599), a Jesuit who inspired the Coimbra Jesuits to publish their translations and commentaries on the works of Aristotle, and the Coimbra Jesuits themselves, who published the commentary on Aristotle's *Physics*, played significant roles.

The Coimbra Jesuits characterized imaginary space as an entity that was not only nondimensional but which also lacked real and positive properties and attributes. And yet it could not be fictional, a mere figment of the mind

33. In his *Disputationes metaphysicae*, Vasquez insisted that God was only in himself before the world, not in any imaginary space, which is absolutely nothing and impossible. As supporters of his opinion, Vasquez cites William of Auxerre, Saint Bonaventure, Duns Scotus, Thomas of Strasbourg, and others. Bartholomew Amicus cites Vasquez for the first of three different opinions on imaginary space. For the three opinions and references, see Grant, 1981a, 363, n. 90.

34. Oresme, *Le Livre du ciel*, bk. 1, ch. 24, 1968, 177. That Oresme understood this space to be void is evident from his characterization of it as "an empty incorporeal space quite different from any other plenum or corporeal space."

35. For Suarez, see Grant, 1981a, 155–156.

or reason, because it was something that could receive bodies.[36] Moreover, "God is actually in this imaginary space, not as in some real being but through His immensity, which, because the whole universality of the world cannot [accommodate it], must of necessity also exist in infinite spaces beyond the sky."[37] Although not a real and positive being, imaginary space has always existed. What then is it? It is a negation – Fonseca called it a "pure negation." But how could God exist in a privation, which is unreal and without properties? If God could exist in such a privation, he could also exist in darkness and other privations. The Coimbra Jesuits replied that space is different from all other negations because it can receive something positive, namely bodies. Therefore space is the only privation, or negation, in which God could exist.[38]

Up to this point, the scholastic authors whom we have considered either identified and equated imaginary space with vacuum (for example, Bradwardine, de Ripa, Oresme, and Suarez) or spoke only of imaginary space or place without explicitly introducing vacuum into their deliberations (Fonseca and the Coimbra Jesuits). Thus the latter group neither affirmed nor denied a possible equation between imaginary space and vacuum. By contrast, Bartholomew Amicus denied that imaginary space and vacuum could be the same, even though both were negations. Imaginary space is a negation because it can have no positive properties.[39] And since everything, for Amicus, had to be either positive or negative, it followed that imaginary space is a negative thing, even though we perceive space as if it were a positive quantity extended to infinity. What is imaginary space the negation of? For Amicus, it was the negation of a resistance for containing an extended thing. That is, bodies occupied imaginary space without seeming resistance from the latter. The negation that is space is not destroyed or negated when it receives body, but rather coexists with it. Space is a pure negation because it negates resistance to the reception of bodies, a property that it possesses eternally.

But why is vacuum characterized as a negation? What is it the negation of? Amicus explains that "a vacuum is called a negation of the body that fills it, since the vacuum is destroyed when the body that fills it arrives." By contrast, space is not destroyed by a body that fills it, "for space is conceived as a receptacle and container of body; but in receiving [body], it

36. Serbellonus, *Physics*, disp. 5, qu. 1, art. 1, 1657, 801, col. 2, explained that "it is not called imaginary because it is a mental fiction but because the intellect cannot directly apprehend it, so that it is necessary that the intellect apprehend it by means of the extension of any body in [that] immense [space]" (Nec dicitur imaginarium quasi mente confictum, sed quia intellectus non potest directe ipsum cognoscere. Ideo necesse est ut indirecte et admodum diffusi cuiusdam corporis in immensum ipsum apprehendat).
37. Conimbricenses [*Physics*, bk. 8, ch. 10, qu. 1, art. 3], 1602, col. 585. Also see Grant, 1981a, 163 (English) and 362, n. 79 (Latin), where, however, the column number differs because it was cited from the Cologne edition of 1602.
38. Grant, 1981a, 163. Serbellonus also characterized imaginary space as a negation or privation (Serbellonus, *Physics*, disp. 5, qu. 1, art. 1, 1657, 801, col. 2).
39. For the kinds of positive properties Amicus discussed, see Grant, 1981a, 166–167.

is not destroyed by what it has received, but is rather perfected by it."[40] Convinced by Aristotle's argument that no void could exist beyond our world, Amicus concluded that the world is surrounded by an infinite imaginary space, which he identified with the divine immensity.[41] Emanuel Maignan (1601–1676), a scholastic theologian and member of the Order of Minims, was in general agreement with Amicus and also emphasized that "imaginary space" must not be conceived as a separate container in which God's immensity exists, because God cannot be contained by anything else.[42]

Imaginary space posed a difficult problem to scholastics, because, although they assumed that God was omnipresent by his infinite immensity in an infinite, immobile space, they yet sought to avoid the language of quantity or extension in describing that space.[43] To employ such language would, in their judgment, transform God into a dimensionally extended or

40. Ibid., 169 (English), and 365, n. 109 (Latin).

41. Ibid., 1981a, 170–171.

42. Ibid., 177; based on Maignan, *Cursus philosophicus*, 1673, 244. Scholastics who identified God's immensity with infinite, imaginary space would probably have agreed with Maignan.

43. An important question that will not be discussed here is the relationship between God and imaginary space, that is, whether imaginary space or vacuum is God himself, by virtue of his infinite immensity, or whether in some sense God and infinite, imaginary space are distinct, even though the latter is said to be God's immensity. At the root of the difficulty and confusion lay certain statements by Saint Augustine, who declared that before heaven and earth existed, God was "in Himself" (*in seipso*). Some scholastics took this to signify that God was "in Himself" and therefore not in any place or space; therefore he could not be in an infinite, extracosmic void. Others, however, took God's infinite immensity to be God himself, or an aspect of the deity, such that God's infinite immensity was not only imaginary space but that God was "in Himself" when in that infinite, imaginary space. Although Augustine denied the existence of extracosmic void space, he conceded that if it did exist God would be omnipresent in it, simply because no reason could be adduced for confining God to our finite world (see Grant, 1981a, 113–114). Augustine's pronouncements on infinite space and God's relationship to it were ambiguous and confusing at best and were subsequently cited for and against the existence of an infinite void space.

Few scholastics explicitly considered this thorny problem. But Leonard Lessius and Otto von Guericke both insisted that imaginary space, which they equated with nothing and vacuum, is God himself (Grant, 1981a, 218–219). In this they took issue with Athanasius Kircher and, in effect, with Bradwardine as well, because both held that although a void could exist without body, it could not exist without God, thus seeming to imply that void and God are somehow distinct when, in truth, they are one and the same. But Compton-Carleton, *Physics*, disp. 33, sec. 1: "De spatio imaginario", 1649, 335, col. 2, cites Lessius by name and rejects his unqualified identification of imaginary space with God. After all, Compton-Carleton argues, among theologians a popular question is "Whether God exists in imaginary spaces beyond the heaven [or world], which would be an absolutely vain question if by imaginary space God Himself was understood. Indeed, if this were so, they [the theologians] would inquire whether God could be in Himself; but nobody doubts this. Therefore, whatever the case may be as to whether various properties of space could be united in God because of His immensity and eminently ubiquitous presence, God is not that [imaginary] space which the philosophers and theologians ask about in the proposed question [namely, whether God exists in imaginary spaces beyond the heaven]." Compton-Carleton concluded that God, by virtue of his immensity, existed in imaginary spaces beyond the world (Compton-Carleton, ibid., sec. 4, 337, col. 2).

quantified being, a move that would make of the deity an incorporeal extension. It would also make a positive being of that space. Indeed, the point of describing infinite space as "imaginary" was to distinguish it from what is real and positive in our ordinary experience, and to do this even though we may conceive imaginary space as a positive thing.[44] Because of the almost unavoidable tendency to conceive imaginary space in positive terms, scholastics occasionally were led to use semiquantitative language to describe it, as when Amicus characterizes imaginary space as "extending" to infinity; or when Maignan refers to it as "virtual extension" to contrast it with "formal extension." The latter applies only to corporeal quantities that have impenetrable parts, or parts that lie outside one another, which comes to the same thing. By contrast, "virtual extension" applies to things that have no parts, namely to "simple things" (*res simplices*), that is, spiritual substances, that do not occupy places the way bodies do. And yet all were agreed that spiritual substances must be somewhere. It was the effort to describe their presence here or there or everywhere that led to the use of extensional, or quantitative, terms. For how then, and in what sense, could God, as the uncreated divine immaterial substance, and angels, as created immaterial substances, occupy the places where they were?

Toward this end, Maignan formulated the concept of "virtual extension" (*extensio virtualis*), according to which a simple thing can be in a place "by its substance, so that the whole corresponds indivisibly to the whole extension of the place and to its particular parts."[45] Maignan's description of virtual extension appears to be equivalent to the medieval concept of *ubi definitivum*, which described the place of a spiritual substance in contrast to *ubi circumscriptivum*, which designated the place of a three-dimensional body. Because they were thought to have different degrees of intensive perfection, spiritual substances were said to occupy, or rather to be delimited by, imaginary places proportioned to their respective intensities. A finite spiritual substance could not only occupy the whole of its empty, finite imaginary place or space but the whole of the spiritual substance; for example, an angel or soul was said to be in every part, however small, of its finite place or *ubi definitivum*.[46]

Similarly, God, the only infinite, spiritual substance, not only occupies the whole of an infinite space or void, but he does this by being wholly and totally in each part, however small or large. Because God could be wholly and totally in every place of whatever size, it followed that his infinite immensity was not divisible. This "whole-in-every-part" doctrine, which appears already in Richard Fishacre's commentary on the *Sentences* (bk. 1, dist. 2, qu. 2), written around 1235,[47] proved the only way scholastics could assume that God's omnipresent immensity is coextensive with an

44. This is Amicus's opinion (Grant, 1981a, 172).
45. Grant, 1981a, 177.
46. See ibid., 130, and 343, n. 67.
47. See ibid., 143.

infinite space and also uphold his absolute indivisibility.[48] God's infinite immensity, which is coextensive with imaginary space, remains indivisible because he is wholly in every part of space that any body, or any of its parts, may occupy. Because God is wholly and indivisibly in every part of imaginary space, and because the space is in some sense a divine attribute (his immensity), both God and infinite, imaginary space are indivisible. The whole-in-every-part doctrine, which held sway for centuries, is, of course, a conception beyond the understanding, as Nicolas Malebranche (1638–1715) fully realized when, in 1688, he declared that "The immensity of God is His substance itself spread out everywhere, and all of it is present everywhere, filling all places without local extension, and this I submit is quite incomprehensible." But, "if you judge of the immensity of God by means of the idea of extension, you are giving God a corporeal extension," from which it would follow that "The substance of God will no longer be all of it wherever it is" – that is, the doctrine of the whole-in-every-part would no longer apply to an actually extended God.[49]

The ontological status of imaginary space was controversial and much discussed. In the seventeenth century, Franciscus Bonae Spei, a Reformed Carmelite, accepted the existence of an infinite extracosmic space that is God's infinite omnipresent immensity but refused to describe it as imaginary or as in any sense a negation. If extracosmic space is anything at all, it must exist in some sense. But if it has some degree, or sense, of existence, it cannot be a negation, because a negation signifies the denial of existence. In fact, Bonae Spei construed the "imaginary" in imaginary space as signifying nonexistence. Implicit in his discussion is a sense that imaginary space had been hypostatized into a separate existent capable of receiving bodies. No justification existed for assigning properties of any kind to an imaginary, nonexistent space. Thus Bonae Spei rejected the use of the term "imaginary" in association with an extracosmic infinite space identified with God's immensity and chose instead to call it "real space" (*spatium reale*). For what could be more real than God's immensity?[50]

Bonae Spei was an exception. Not only did the expression "imaginary space" appear frequently, but it was often characterized as a "pure negation" – to distinguish it from substance, accident, or privation. As God's immensity, however, "imaginary space" could hardly be a mere nothing or fiction. In fact, it was conceived as an existent entity that was neither a substance nor an accident and therefore was not an attribute of God. It was also homogeneous, immutable, continuous, indivisible, and capable of re-

48. Fishacre did not apply his idea to extracosmic space, but it proved serviceable in that connection.

49. Grant, 1981a, 222, cited from Malebranche, *Dialogues*, 1923, dialogue 8, pp. 212–213. For the French text, see Malebranche, *Entretiens*, 1965, 184–185.

50. For Bonae Spei, I have relied on Grant, 1981a, 178–180, where I refer to him as "Bona Spes."

ceiving bodies and coexisting with them. Because it was identified with God's immensity, however, scholastics, with at least one significant exception, avoided the attribution of real dimensionality to imaginary space. Such an attribution would have converted God to an extended being and made of him something akin to a body.

How close some came to that fateful step can be seen in the descriptive terminology applied to imaginary space. The Coimbra Jesuits likened it to the real and positive dimensions of body; Amicus conceived it in the mode of positive extension, and Maignan described it as "virtual extension." Scholastic ambivalence was probably best expressed by Bradwardine when he declared that God "is infinitely extended without extension."[51]

And yet at least one scholastic author seems to have believed both that infinite imaginary space is three-dimensional and that God's infinite immensity is omnipresent within it. In a discussion on "the true opinion concerning imaginary space,"[52] in 1649, Thomas Compton-Carleton, a Jesuit theologian, declared that "imaginary space is the indestructible negation of a real place."[53] Wherever a body exists, there we have a real place, or *ubicatio*. However, should the body be removed from that place, or *ubicatio*, the real place is destroyed, and its negation exists, namely imaginary space, which is indestructible.

Of five conditions that Compton-Carleton deemed essential for imaginary space, the first and most unusual is that it be infinitely extended three-dimensionally.[54] Although the attribution of three-dimensionality to imaginary space was momentous, Compton-Carleton says no more about this extraordinary claim. But linked to it is his later declaration that he sides with all those authors who assume that God exists by his immensity in infinite imaginary spaces beyond the heavens or the world.[55] For Compton-Carleton, then, imaginary space is an eternal, three-dimensional imaginary space that is identified with God's infinite immensity. Does this also imply that Compton-Carleton conceived of God as in some sense three-

51. These were the kinds of attributions that Bonae Spei rejected.
52. "Vera sententia de spatio imaginario." Compton-Carleton, *Physics*, disp. 33, sec. 2, 1649, 336, col. 1. Compton-Carleton's views were not included in Grant, 1981a.
53. "Dico itaque spatium imaginarium esse negationem non tollibilem ubicationis realis." Compton-Carleton, ibid.
54. "Notandum quae conditiones requirantur ad spatium imaginarium. Primo spatium imaginarium esse debet quid infinite suo modo diffusum, idque non versus unam tantum partem, sed omnes, ac proinde trine dimensum, seu extensum versus longitudinem, latitudinem, et profunditatem." Compton-Carleton, ibid. Properties 2 through 4 (ibid., cols. 1– 2), which were routinely accepted for imaginary space, are (2) immobility, (3) eternity, and (4) the capacity to receive bodies, though it cannot receive two impenetrable bodies in the same place simultaneously. The fifth, however, is also extraordinary because it denies homogeneity and proclaims instead that imaginary space has different, or heterogeneous, parts which correspond to real and positive places. In this, Compton-Carleton may stand alone.
55. "Dico itaque cum auctoribus secundae opinionae Deum esse in infinitis spatiis imaginariis extra coelum in immensum diffusis." Ibid., 337, col. 2.

dimensional? Not necessarily. If he assumed the whole-in-every-part doctrine as the manner in which God occupied any space, however small or large, he could have avoided the attribution of dimensionality to God. Unfortunately, Compton-Carleton indicates no awareness of any problem and is silent on the issue. We can be certain only that he assumed a universe in which our unique vacuumless, material, spherical cosmos is surrounded by a three-dimensional infinite, imaginary void space which is identified with God's infinite immensity. Except for God's immensity and the assumption that space contained heterogeneous parts, Compton-Carleton's universe was much the same as that proposed centuries earlier by the Stoics.[56] He thus marks a significant departure from his fellow scholastics, who assumed a nondimensional, God-filled imaginary space. Indeed, his views linking God's immensity with a three-dimensional infinite space precede by at least fifteen years the quite similar conceptions of More and Newton.

One noteworthy difference, however, separates Compton-Carleton from More and Newton. It was Henry More (1614–1687) who made the move that Compton-Carleton may have implied. For not only did More assume the existence of a three-dimensional infinite void space, but he boldly proclaimed that God was omnipresent in it as an actually extended three-dimensional being. More arrived at this momentous conclusion from his conviction that the whole-in-every-part doctrine, which scholastics had used to explain God's infinite omnipresence, was absurd, largely because it divided a whole into wholes instead of into parts. Convinced that everything, whether corporeal or incorporeal, is dimensionally extended, More unhesitatingly made of God a three-dimensional, incorporeal, extended being. Only by virtue of that infinite extension could he be omnipresent throughout an infinite void space. Indeed, More made infinite void space an attribute of God.[57] Newton, who adopted More's ideas about God and space, assumed the literal three-dimensional omnipresence of God in an infinite space that was God's property and, in effect, his immensity.[58]

In the historical developments described here, it was medieval and early modern scholastics who, for better or worse, introduced God into infinite space and thus effected a divinization of space.[59] Scholastic ideas about space and God form an integral part of the history of spatial conceptions between the late sixteenth and eighteenth centuries, the period of the Scientific Rev-

56. For further discussion of the Stoics, see Grant, 1981a, 106–108. Authors who also assumed a material cosmos surrounded by an infinite three-dimensional void were Hasdai Crescas (1340–1410) and Francesco Patrizi (1529–1597). On the former, see Grant, 1981a, 271, n. 33, and 321–322, n. 5, and for the latter, 199–206. Neither, however, related God to space. Patrizi exerted a considerable influence on Gassendi's views on infinite space (see Grant, ibid., 206–213).
57. Grant, ibid., 224, 227. More's definitive presentation appears in his *Enchiridion metaphysicum* of 1671.
58. Grant, ibid., 261; for a full discussion of Newton's views on space, see 240–255.
59. For earlier vague, though sometimes dazzling, metaphors from the patristic, cabbalistic, and Hermetic traditions, see Grant, ibid., 110–115.

olution. From the assumption that infinite space is God's immensity, scholastics derived most of the same properties of space as did nonscholastics, and did so before the latter. As God's immensity, space was almost always described as homogeneous, immutable, infinite, and capable of coexistence with bodies, which it received without offering resistance. In divinizing space, scholastics assigned to infinite imaginary space virtually all the properties, including even extension, that would be conferred on space during the course of the Scientific Revolution.[60]

60. For the basic content of this paragraph, see Grant, ibid., 262.

PART II

The celestial region

10

The incorruptibility of the celestial region

We have now investigated the finitude, shape, place, and perfection of the world as a whole and have also considered whether anything – bodies, worlds, voids and spaces – might lie beyond it. Now it is time to enter the world itself and examine its structure and operation. Following the overwhelming order of importance accorded by Aristotle and his scholastic followers to the celestial region over the terrestrial, I shall devote the rest of this study to the celestial region.[1]

Although the doctrine of the four terrestrial elements – earth, water, air, and fire – was abandoned by some as early as the sixteenth century, scholastic authors held firmly to it until the end of the seventeenth century. Did the four elements also play a role in the celestial region? Were one or more of them the real building blocks of the universe, both above and below the Moon, or was the celestial region from the Moon outward made of a special, separate substance, a fifth element?

The Middle Ages inherited two conflicting opinions, rooted in the texts of Plato's *Timaeus* and Aristotle's *De caelo*. The former argued, with little elaboration, that the four elements fill the subcelestial region and that the stars and planets are composed mostly of fire and to a much smaller extent of the other three elements,[2] while the latter developed an elaborate argument to demonstrate that the celestial region was not composed of any of the four elements but was constituted of an extraordinary substance, an ether or fifth element.

Although Plato's brief statement had been available in Latin since Chalcidius's partial translation in the fourth century, the *Timaeus* was always subsidiary to Aristotle's *De caelo* during the late Middle Ages.[3] However,

1. Although the final chapter of this volume (Ch. 20) is devoted to the earth, the emphasis there is on its cosmic relations (see Ch. 1, Sec. II.1 and n. 7, and the beginning of Ch. 20). In what follows, I rely heavily on Grant, 1991. Alterations have been made, however, in both order and content.
2. See Plato, *Timaeus*, 40A, in Plato [Cornford], 1957, 118, and Cornford's remarks on the same page. Solmsen, 1960, 291, indicates that for Plato the planets and stars were constituted solely of fire.
3. In Chalcidius's translation, the brief and cryptic passage reads: "Et diuini quidem generis ex parte maxima speciem ignis serena claritudine perpolibat, ut propter eximium splendorem nitoremque uidentibus esset uisurisque uenerabilis." See Chalcidius, *Commentary on Timaeus*, 1962, 33, lines 4–6. In Cornford's translation (Plato, 1957, 118), we find: "The

from late antiquity to the twelfth century, prior to the dominant role of Aristotle's *De caelo* from the thirteenth century onward, the opinion that the heavens were composed of one or more or all of the traditional four elements was widespread. Not only did Plato's *Timaeus* play a significant role in its dissemination, but so also did other works by Church Fathers and other figures from late antiquity, whose works were known in the Middle Ages.[4] Most authors who expressed opinions on the composition of the celestial region during the twelfth century accepted some version of the Platonic elemental heavens.[5]

With the introduction and fairly rapid acceptance of Aristotle's cosmology, natural philosophers assumed almost unanimously that the celestial region was composed of a special ether, or fifth element (*quinta essentia*), that differed radically from the four elements, which now were confined to the sublunar region. Richard of Middleton spoke for almost all scholastics when, in rejecting the opinion that the heaven is possessed of a fiery nature, he observed that this theory was not only "against the Philosopher [Aristotle] and the Commentator [Averroës] and against Avicenna and Algazali, but more than that it is against reason."[6] Richard insists that the heaven could not be of an elemental nature, because of the character of its natural motion and because of the nature of the elements and the compounds generated from those elements. By their very natures, elemental bodies are

form of the divine kind he made for the most part of fire, that it might be most bright and fair to see." From the preceding context, it is plain that Plato is speaking of the planets.

4. Among Church Fathers, Saint Augustine speaks of air and fire as material components of the celestial region (see Augustine, *Genesis*, bk. 3, chs. 3, 6, 1982, 1:77, 79, where air is mentioned; see ibid., bk. 2, ch. 3, 1:49–50, for fire; see also Duhem, *Le Système*, 1913–1959, 2:486. But Augustine also speaks of the firmament as possibly solid (see this volume, Ch. 14, Sec. V). Saint Basil's treatment is equally problematic, although Wallace-Hadrill, 1968, 20, declares categorically that for Basil "the firmament . . . must be composed of any of the four elements, singly or in combination." The evidence, however, is unclear, as can be seen in Basil, *Exegetic Homilies (Hexaemeron)*, homily 3.4, 1963, 43, where Basil says that we should not compare the firmament to a translucent stone, nor do "we . . . dare to say that the firmament is made either from one of the simple elements or from a mixture of them." For Basil, the firmament seems to be neither hard nor soft. Duhem, ibid., 2:480, observes that in describing the Aristotelian fifth etherial element, Basil did not repudiate it; neither, however, did he accept it. Among Latin authors, Chalcidius was one who ignored the Aristotelian ether, or fifth element, and opted for one of the four elements or some combination of them (see Chalcidius, *Commentary on Timaeus*, 1962, 71–76 and Stahl, 1962, 144). Macrobius, whose *Commentary on the Dream of Scipio* was widely known in the Middle Ages and Renaissance, describes Aristotle's opinion about an unchanging celestial region and also an interpretation that allowed for change. These were said to be differing interpretations among Platonists (see Macrobius [Stahl], 1952, 131–132). Thus Macrobius cannot be assigned to either group. Others who were committed to a rejection of the celestial ether because they believed in celestial corruptibility were John Philoponus and John Damascene (see note 10 of this chapter).
5. For Thierry of Chartres, see Crombie, 1959, 1:27–29, and Gilson, 1955, 146–147; for Adelard of Bath, see Thorndike, 1923–1958, 2:40; and for William of Conches, see Thorndike, 1923–1958, 2:56.
6. "Sed haec opinio est contra Philosophum et suum Commentatorem et contra Avicennam et Algazelem et quod maius est contra rationem." Richard of Middleton [*Sentences*, bk. 2, dist. 14, qu. 3: "Whether the firmament is of a fiery nature"], 1591, 2:169, col. 2.

capable of upward and downward rectilinear motion only, whereas celestial bodies move naturally with circular motion. Traditional ideas about the properties of celestial bodies also made the idea of a fiery celestial region seem impossible. For example, Saturn causes cooling and the Moon causes humidification, neither of which effects could be produced by fire. It is the Sun, not fire, that causes heat and dryness. For similar reasons, no other element or combination of elements can exist in the heavens. Indeed, the four elements possess contrary qualities, so that no one of them could preserve order and keep the elements together. They are constantly seeking to dissociate themselves. To preserve the sublunar world, the influence and governing power of a nonelemental body is required, and this task is provided by a fifth, celestial element.[7]

Aristotle had convinced medieval natural philosophers of the necessity of a special fifth element on the basis of striking behavioral differences between the terrestrial and celestial regions.[8] In *De caelo* (bk. 1, chs. 2 and 3), Aristotle contrasted the natural rectilinear motion of the four sublunar elements (earth, water, air, and fire) with the observed regular, and seemingly natural, circular motion of the planets and the fixed stars in the celestial region. The contrast between an incomplete, finite, rectilinear line and a closed, and therefore complete, circular line convinced Aristotle, if he needed convincing, that the circular figure was necessarily prior to the rectilinear figure. Because the four simple elemental bodies moved with natural rectilinear (upward and downward) motion, Aristotle concluded that the observed circular motion of the celestial bodies must of necessity be associated with a different kind of simple, elemental body: an ether, or fifth element.[9]

The most extraordinary feature of that ether – its incorruptibility – must now be described, as well as the manner in which that fundamental property affected the understanding of other celestial properties during the Middle Ages and well into the seventeenth century. As a sequel to the description of the ethereal properties, we shall examine the challenges to the Aristotelian ether that emerged among scholastic authors during the seventeenth century, in large part as a consequence of Tycho Brahe's astronomical achievements.

I. Aristotle on celestial incorruptibility

The most fundamental and striking property which Aristotle assigned to his celestial ether was incorruptibility. Prior to the introduction of Aristotle's physical works into the Latin West during the late twelfth and the early thirteenth century, the idea of celestial incorruptibility was prob-

7. Ibid., 169, col. 2–170, col. 1.
8. For the remainder of this paragraph, I follow Grant, 1983, 158.
9. Aristotle, *De caelo* 1.3.270b.20–26.

ably a minority opinion. As we have seen, it was not uncommon for scholars in late antiquity and the early Middle Ages to assume that the heavens were composed of one or more of the four elements. Since the elements were thought of as changeable entities, those who held that the whole world, including the heavens, was composed of one or more of them were committed, if only implicitly, to the idea of changeable or corruptible heavens.[10] The introduction of Latin translations of Aristotle's works during the twelfth and thirteenth centuries radically altered this tradition. A vital ingredient of Aristotle's "new" cosmology was the belief in celestial incorruptibility.

Aristotle distinguished two kinds of incorruptibility: one associated with eternality and *unchangeability*, the other linked with eternality and *changeability*. The cosmos, which for Aristotle was ungenerated, eternal, and indestructible, contained within itself both kinds of incorruptibility, one associated with each of two radically different parts into which he assumed the world was divided, namely the sublunar and celestial regions. Taken as a whole, the four elements of the sublunar, or terrestrial, region are as indestructible as the fifth element, or ether, of the celestial region. The totality of terrestrial matter that is composed of the four elements is constant and eternal, without beginning or end. Although the four elements are eternal, incorruptible, and indestructible as a whole, they are always changing, one into the other. Bits of fire are incessantly transformed into earth, and vice versa, while portions of air are always being transformed into water, and vice versa. Despite these changes, the earth retains its overall integrity, as do the other elements in their natural places. Bodies come and go, but the underlying static structure of the sublunar region remains constant. It is now indistinguishable from what it was in the past and from what it will be in the future.

Things are otherwise in the celestial region, where no part of the heaven could be transformed, or was transformable, into anything else. Apart from changes of position that arose as a consequence of regular circular motion, scholastic authors were agreed that substantial changes could not occur in the heavens. Our concern will be with this seemingly absolute sense of celestial incorruptibility.

10. Two who were explicit in their assumption of celestial corruptibility were John Philoponus, the sixth-century Greek Neoplatonic commentator on the works of Aristotle, and John Damascene. For selections on celestial corruptibility drawn from the works of Philoponus, see Philoponus [Böhm], 1967, 326–327, 329–331. Although the works of Philoponus, from which these selections were drawn, were unknown in the Middle Ages, some of the relevant passages had been quoted by his contemporary, Simplicius, who included them in his Greek commentary on Aristotle's *De caelo*, which was translated into Latin by William of Moerbeke in 1271. For a brief discussion, see Sambursky, 1962, 158–166, especially 164. John Damascene assumed the corruptibility of the heavens when he declared: "it is evident that the Sun, Moon, and stars are composite, and by their very nature subject to corruption." The passage is from Damascene's *De fide orthodoxa* as translated in John Damascene [bk. 2, ch. 7], 1958, 221.

II. The medieval defense of celestial incorruptibility

The scholastic defense of celestial incorruptibility included theory, observation, and an intuitive sense, expressed over many centuries, that the heavens, which were always associated with spiritual substances, must be incomparably superior to the terrestrial region.

1. Theory

According to Nicole Oresme, the cause of corruption is definable in two ways: (1) by means of a quality or substance that is contrary or opposite to another quality or substance; and (2) "by the absence of a conserving agent"; for example, the corruption (or destruction) of light occurs when the source of the light is removed.[11] Of the two, the first is more significant. The basis for the medieval belief in celestial incorruptibility depended on Aristotle's conception of generation and corruption, a conception that provided the rationale for a radical distinction between the celestial and terrestrial regions. Generation and corruption, the coming-to-be and passing away of things, or change in general, occurred in all substances consisting of matter possessed of a form or quality that was potentially replaceable by its contrary. For example, fire, which possessed the basic qualities of hotness and dryness, could be converted to earth, which possessed the qualities of dryness and coldness, if the hotness in fire were replaced by its contrary quality or form, coldness.[12] While one form was actualized in matter, its contrary was said to be in privation but potentially capable of replacing it. Eventually each potential form or quality would have to be realized; otherwise a form would remain unactualized, and nature would have produced it in vain. While one of a pair of contrary forms was actualized in matter, the other was absent and in privation, since two contrary forms could not exist simultaneously in one and the same body. Generation and corruption, and therefore all change, involved the possession of one, and the expulsion of another, of a pair of contrary forms or qualities. In asking the question "Are the heavens incorruptible?", Galileo, in his commentary on *De caelo*, explained the problem quite clearly as follows:

> [19]. ...every substance that is corruptible is such that it is composed of matter and form; and the form corrupts to the extent that it leaves the subject. And this always comes about from something else expelling it: for the subject (i.e., the matter) does not drive the form from itself, but always seeks to retain it; nor does the form leave of itself, since it seeks existence and by leaving it would be corrupted; therefore it must be expelled by another. But a form is

11. Oresme [*De celo*, bk. 1, qu. 10], 1965, 149 (Latin), 150 (English).
12. Aristotle described such changes in his *De generatione et corruptione* (bk. 2, chs. 3–4).

expelled when something else is induced into the subject where the form is, and which cannot coexist with it – for otherwise it would not be expelled; from this it is apparent that corruption must come about through the introduction of a contrary.

[20]. From this it is also understood that every substance that is corruptible must consist of matter and form, and that there cannot exist in matter another form having qualities contrary to the qualities of the form already existing in it. It is by these means that generation and corruption come about; for when one agent induces its qualities so that it may induce a form, it nececessarily expels the contraries, and when these are expelled the form is expelled also.[13]

Although the kind of change involving generation and corruption was assumed to occur incessantly in terrestrial bodies, it was denied for the ether that filled the celestial region beyond the concave surface of the lunar sphere.[14] Whatever one's opinion about the nature of the celestial ether – whether it was matter and form, matter alone, form alone, or something else (the "material" structure of the ether is discussed later in this chapter) – all who accepted its existence, and few denied it,[15] were agreed that it did not take on contrary forms and could not, therefore, suffer generation and corruption by natural means. Galileo encapsulated the distinction by explaining that "whatever undergoes corruption has a contrary, in the manner explained; and therefore . . . the heavens, since they lack contraries of this type, are incorruptible."[16] It followed that celestial matter was naturally incorruptible: that is, was not, as William of Ockham put it, corruptible by any "created agent" (*agens creatum*), although it was usually conceded that God, who had created the world supernaturally, could if he wished alter or destroy it supernaturally.[17] Arguments in defense of celestial incorruptibility depended heavily on the Aristotelian explanatory mechanism

13. Galileo [*De caelo*, qu. 4 (J)], 1977, 97–98. The full question extends over pages 93–102. Some duplication occurs in Galileo's next question, "Are the heavens composed of matter and form?", which I discuss in Chapter 12, Section II.1.
14. Aristotle, *De caelo* 1.3.270a.13–24.
15. In the Middle Ages, Robert Grosseteste was one who denied the existence of a fifth element. In his *De generatione stellarum*, he argued that the stars are composed of the four elements and are therefore corruptible. See Grosseteste, *De generatione stellarum*, 1912, 33, 35–36. I am grateful to Dr. Peter Sobol for calling Grosseteste's treatise to my attention.
16. Galileo, *De caelo*, qu. 4 (J), 1977, 98.
17. Ockham [*Sentences*, bk. 2, qu. 18], 1981, 401, used God's power to corrupt and destroy the heavens as a basis for distinguishing between absolute (*simpliciter*) and relative (*secundum quid*) celestial incorruptibility. Because God could destroy or change the celestial region by supernatural power, we must mean by celestiai incorruptibility that the heavens are not corruptible by any *created agent* (*agens creatum*). The heavens are therefore incorruptible only relatively, not absolutely. As a more straightforward example of concession to God's absolute power, Oresme, *De celo*, bk. 1, qu. 10, 1965, 152, declares that "according to faith one should hold that He [God] could annihilate the heaven without change in Himself due to His immensity, eternity and infinite power." Galileo, *De caelo*, qu. 4 (J), 1977, 101, spoke in a similar vein when he declared that "the heavens are corruptible in relation to God, just as are intelligences and rational souls and all creatures; but by their own nature they are incorruptible."

for change, which, as we saw, involved the presence or absence of contrary forms.

a. The absence of contrary forms in the heavens

In medieval Aristotelian cosmology, celestial matter could only have been assumed incorruptible because contrary forms did not exist there or were inoperative. The absence of contrary qualities was usually inferred from the universal belief that circular motion did not have a contrary motion, in contrast to rectilinear motion, which had contraries in the form of upward and downward motion. Once again, we can usefully call on Galileo, who explains that "the circular motion of the heavens is of such a nature that it can be perfect in itself, for it is always the same in the beginning, the middle, and the end." Since the body that possesses such a motion "can naturally be perfect," it therefore "lacks a contrary."[18] The celestial region was thus constituted of an ethereal substance that changed only with respect to position, as it moved with a perpetual, natural, uniform, circular motion. Change occurred only in the sublunar region, where things were always in the process of generation or corruption. Moreover, it was only in the domain below the Moon that matter could exist, since for Aristotle only things subject to generation and corruption could possess matter.[19]

But if the ethereal heavens lacked contraries, or opposite qualities, such as heaviness and lightness, rarity and density, hotness and coldness, and moistness and dryness, that were essential for the production of change in the sublunar realm, why did astrologers and natural philosophers describe Saturn as cold and dry; Mars as hot and dry; the Moon as cold and wet; and so on for the other planets? Indeed, unless we are shielded by clouds, the Sun makes us hot. How can we explain these properties if they do not really subsist in the ethereal bodies in question.[20]

Jean Buridan provided what was probably the standard medieval response: the heavenly region possesses those qualities only virtually, not formally. That is, the heavens can cause changes involving hotness and coldness in bodies below the Moon, but they do not possess any of those qualities formally, or *per se*. Hence those qualities do not operate in the celestial region.[21] Similar reasoning applied to secondary qualities that nor-

18. Galileo, ibid., 98. A few paragraphs earlier, Galileo had argued that rectilinear motion is not contrary to circular motion and concluded by means of a syllogism (ibid., 97) that "it is apparent that if the motion of the heavens has no contrary, neither do the heavens; but whatever has no contrary is incorruptible; therefore the heavens are incorruptible."
19. Aristotle, *Metaphysics* 8.5.1044b.26–28
20. Buridan [*De caelo*, bk. 1, qu. 9], 1942, 41.
21. Ibid., 43. This was also Buridan's response to those astrologers who spoke of certain signs of the zodiac as fiery, or airy. The attribution of such characteristics to zodiacal signs means no more than that those signs could produce the same properties in sublunar bodies as do fire and air. But the zodiacal signs themselves do not actually possess such qualities. For more on this theme, see Chapter 17, Section IV.4, and Chapter 19, Section V.

mally arose in the terrestrial zone from the activities of primary qualities. Thus colors arose from the opposite primary qualities black and white. Such pairs of secondary qualities as heaviness and lightness, and rarity and density, were derived below the Moon by naturally opposite upward and downward rectilinear motions.[22] Of the various pairs of qualities that might or might not exist in the heavens, scholastic natural philosophers chose to devote the most attention to the possible heaviness or lightness of the celestial ether and occasionally to consider rarity and density.

b. Do the contrary forms lightness and heaviness exist in the celestial ether?

In Aristotle's lexicon of natural philosophy, a "heavy" body was one that moved with a natural rectilinear motion toward the center of the world, and a "light" body was one that moved naturally away from the center.[23] On this basis, Aristotle concluded that "the body whose motion is circular cannot have either weight or lightness, for neither naturally nor unnaturally can it ever move towards or away from the centre."[24] Because the celestial ether moved with a natural circular motion, and therefore never approached or receded from the center of the world, it followed that it was neither light nor heavy, a judgment that won almost universal approval during the Middle Ages and retained considerable support during the sixteenth and seventeenth centuries.[25] Most of the arguments simply emphasized the accepted assumption that heavy and light bodies were carried toward or away from the center of the world. Indeed, as already mentioned, a heavy body was defined as one that moved naturally toward the center, and a light body was one that moved naturally away from the center, but the heavenly body moved in neither of these ways.

The lack of celestial heaviness and lightness is derivable in yet another way: from the absence of contrary qualities. Indeed, the four qualities – hotness, coldness, wetness, and dryness – are instrumental in conferring heaviness and lightness on bodies. But such pairs of contrary qualities are, as we have seen, absent from the heavens, which consequently can be neither heavy nor light.[26]

22. Ibid., 44. Buridan does not mention black and white as the source of other colors, but it is probably what he had in mind. For Aristotle's views on the formation of secondary colors from the primary colors black and white, see *On Colours*, 1.791a.1–792a.4.
23. Aristotle, *De caelo* 1.3.269b.1927. He discusses the problem at great length in book 4, chapters 1–4.
24. Aristotle, *De caelo* 1.3.269b.30–33 [Guthrie], 1960. John of Jandun repeats essentially the same thing in [*De caelo*, bk. 1, qu. 13], 1552, 10r, col. 2.
25. Among Aristotle's numerous supporters, see John of Jandun, ibid., cols. 1–2; Buridan, *De caelo*, bk. 1, qu. 9, 1942, 42; Versor [*De celo*, bk. 1, qu. 5], 1493, 4r, col. 2–4v, col. 1; Amicus [*De caelo*, tract 5, qu. 3, art. 2], 1626, 261, col. 1, for the statement of the problem and 262, cols. 1–2, for Amicus's opinion.
26. Johannes Versor, who includes this argument in his question as to whether the heavens are heavy and light, declares: "Secundo probatur quia gravitas et levitas sint qualitates

As was common in medieval questions, a hypothetical situation was also conceived. If the whole heaven moves neither upward nor downward, would a part of it do so? Thus Johannes Versor, and later Sigismundus Serbellonus, imagine that part of the heaven is removed and placed here below, presumably on the earth or near it. Would that separate, celestial part move rectilinearly upward, away from the center of the earth or world, to rejoin the whole heaven? Would the part not move toward the whole, just as a piece of earth removed from the whole earth would move naturally back toward the whole earth? Versor denies the analogy between these pieces of the heaven and the earth. Although the heavy piece of earth would move naturally toward the center of the whole earth, the piece of the heaven separated from its whole would move neither upward nor downward, because it cannot be moved with any but a circular motion.[27]

Serbellonus insists that if the piece of the heaven were located on the earth by some external agent, it would remain exactly where it was, because it has no tendency to move upward or downward. Despite its immobility, however, it could not be said to rest naturally on the earth, since it has no such natural inclination. By the same argument, that same piece of the heaven would lack any inclination to move if it were placed above the air or the water.[28]

Denial of heaviness and lightness to the celestial ether formed part of the overall tactic of associating with incorruptibility properties that were themselves virtually the opposite of those associated with corruptibility (though not in the sense of contraries). The fifth element, says Buridan, is prior to, and nobler than, heavy and light bodies. One can argue from the effects produced by these radically different kinds of bodies to the natures that produce those effects. Thus circular motion, which is an effect of the fifth body, or ether, is prior to and more perfect than rectilinear motion, which is the effect of the four elements. It follows that the celestial ether is nobler than the four elements and the bodies they form in the sublunar world. The celestial body is also prior to and nobler than the light and heavy bodies of the inferior or sublunar world, because it is more perfect than they are. As evidence for this claim, Buridan observes that the celestial body contains the inferior bodies and exceeds them in magnitude. The celestial ether is also more perfect because it dominates and governs the inferior part.[29]

communicate a quatuor qualitatibus primis activis et passivis. Sed celum denudatum est quatuor qualitatibus primis, ut postea probabitur. Ergo celum nullam habet gravitatem." Versor, *De celo*, bk. 1, qu. 5, 1493, 4v, col. 1.

27. Ibid.

28. Here is Serbellonus's response to the query about the removal of part of the heavens to the earth's surface: "Partem illam praecisam futuram indifferenter in qualibet parte spatii et ibi remansuram ubi collocaretur ab extrinseco agente. Non ascenderet itaque ex se ad coelum quia in se non habet principium talis motus; nec quiesceret naturaliter quia non habet qualitatem inclinantem ad talem quietem. Et eadem ratione qua existeret in terra existere posset in aere et supra aerem et ubicunque tandem ubi collocaretur ab agente." Serbellonus [*De caelo*, disp. 1, qu. 3, art. 3], 1663, 2:37, cols. 1–2.

29. Buridan, *De caelo*, bk. 1, qu. 9, 1942, 42.

c. Do the contrary forms rarity and density occur in the heavens?

Although the terms "rare" and "dense" seem to us relative, because "rare" can mean "less dense" and, conversely, "less rare" can be taken as roughly synonymous with "denser," they were not interpreted in this relativistic sense by Aristotle or his medieval followers. For them, density was equated with heaviness and hardness, whereas rarity was identified with lightness and softness. Thus density and rarity involved three pairs of opposite or contrary qualities.[30]

By analogy with light and heavy, the existence in the heavens of the properties rarity and density was usually denied. Buridan argued that secondary qualities, such as rarity and density, which are derived from primary qualities, in this case hotness and coldness, do not exist in the heavens in the same manner as they do in the sublunar region.[31] And yet the qualities rarity and density posed a much more serious problem than did lightness and heaviness, for which Aristotle had at least provided a plausible justification based on the differences between circular and rectilinear motion. Whereas scholastic natural philosophers routinely denied the existence of lightness and heaviness in the heavens, it was common for them to claim that some parts of the heavens were denser or rarer than others. Indeed, it was usually assumed that the planets and stars were denser parts of the celestial ether than were the invisible celestial orbs that carried them. We see the stars and planets because they are denser than the surrounding ether and reflect light to us. Albertus Magnus considered it necessary that the heavens vary in thickness, so that they could disseminate light to the terrestrial region and affect things in an appropriate manner.[32]

But it was not merely a matter of passive action. Averroës argued that "if two bodies are equal, and one is denser than another, it is necessary that the denser [body] exert a stronger action."[33] As an illustration, he pointed

30. See Aristotle, *Physics* 4.9.217b.11–19. Ironically, in his earlier *Categories*, ch. 8, 10a.16–24, Aristotle denied that "rare" and "dense" are qualities when he said: " 'Rare' and 'dense' and 'rough' and 'smooth' might be thought to signify a qualification; they seem, however, to be foreign to the classification of qualifications. It seems rather to be a certain position of the parts that each of them reveals. For a thing is dense because its parts are close together, rare because they are separated from one another, smooth because its parts lie somehow on a straight line, rough because some stick up above others." Aristotle [Ackrill], 1984.
31. Buridan, *De caelo*, bk. 1, qu. 9, 1942, 44. Although Buridan does not identify the primary qualities, he probably meant hotness and coldness. Peter of Abano (*Lucidator*, differ. 5, 1988, 288, lines 14–15) says that rarity and density are found in the celestial bodies in an equivocal sense.
32. "Et ideo necesse est caelum esse spissius et minus spissum, ut diversificetur suum instrumentum, quod est lumen, et ita per consequens diversimode moveat materiam ad diversas formas generatorum et corruptorum." Albertus Magnus [*De caelo*, bk. 2, tract. 1, ch. 2], *Opera*, 5, pt. 1, 1971, 107, col. 2. John of Jandun, *De coelo*, bk. 1, qu. 14, 1552, 10v, col. 2, observes that stars or planets are sufficiently dense to prevent us from seeing beyond them, whereas the rare parts of the heavens allow us to see beyond them, no doubt because they are transparent.
33. "Cum duo corpora fuerint aequalia et alterum eorum fuerit densius alio, necesse est ut

to the Sun, which he considered the greatest, most luminous, and densest of all the planets. The action on the orb carrying the Sun was therefore most powerful where the Sun itself was located. The Sun and other planets could, by virtue of their density, affect the air below more strongly than the other, much less dense, and therefore invisible, parts of the celestial ether.[34]

Many appear to have believed that celestial appearances and the nature of celestial effects on the sublunar realm required that some parts of the ether be denser than others. Indeed, by the mid-seventeenth century, Serbellonus reported ([*De caelo*, disp. 1, qu. 3, art. 2], 1663, 2:34, col. 1): "the common opinion [*communis opinio*] is that in the heavens there is true rarity and density and that light is reflected to us from the dense parts [but] the light that passes through the rare parts is not seen by us." Traditional denials of contrary qualities in the celestial region are, however, still heard in the seventeenth century, as when Raphael Aversa speaks of some "moderns" (aliqui recentiores) who denied true density and rarity in the heavens. They conceived it as "something like density and rarity, that is, a greater or smaller substance [that is, magnitude or volume] with respect to an equal quantity [of matter], but not in the same manner as in truly dense and rare inferior bodies."[35] In the terrestrial region, density arises from coldness and rarity from hotness; density and rarity, in turn, give rise to heaviness and lightness, hardness and softness, and solidity and fluidity, respectively. But these qualitative differences are not found in the heavens, and therefore neither are true density and rarity.[36]

At this point, an important distinction must be introduced. Celestial density and rarity could be conceived in two ways: statically or dynamically. In the former, already described, some parts of the ether are permanently

densius sit fortioris actionis." Averroës [*De caelo*, bk. 2, comment. 42], 1562–1574, 5:125v, col. 1. Averroës also declared that the qualities of rarity and density were in the heavens in a manner different than in elementary bodies. Most scholastics probably derived this idea from Averroës.

34. "Et, cum ita sit [a reference to the passage quoted in n. 33], pars orbis, in qua est stella, maioris actionis est in aerem vicinantem illi quam pars in qua non est stella. . . . Et ideo quia Sol est maximus stellarum et densior et magis luminosus contingit ut actio partis orbis, in qua est Solis, sit fortior." Averroës, ibid., who based his discussion on Aristotle, *De caelo* 2.7.289a.29–34. Guthrie, in a note to his translation of this passage in Aristotle's *De caelo*, comments that an explanation is provided in Aristotle's *Meteors* 1.3.341a.19 and then summarizes it as follows: "Perceptible heat is engendered by the Sun's motion because it is both rapid and near. The motion of the stars is rapid but distant, that of the Moon near but slow." See Aristotle, *De caelo* [Guthrie], 1960, 181, note c.

35. "Aliqui tamen recentiores docere maluerunt nequaquam in caelis esse veram densitatem et raritatem, sed potius aliquid instar densitatis et raritatis nempe maiorem aut minorem substantiam sub pari quantitate, non tamen eo modo quo in his corporibus inferioribus vere densis aut raris." Aversa [*De caelo*, qu. 34, sec. 2], 1627, 2:129, col. 2.

36. "Densitas et raritas apud nos inducunt gravitatem et levitatem faciunt corpora dura et mollia, solida et fluida; itemque ipsa densitas oritur ex frigore, raritas ex calore. Hae autem differentiae et qualitates non inveniuntur in caelis." Aversa, ibid. The Coimbra Jesuits (Conimbricenses, *De generatione et corruptione*, bk. 1, ch. 5, qu. 17, art. 1, 1606, 287) denied the existence of rarity and density, as well as all primary qualities, in the heavens.

denser than others: for example, planets and stars are aggregations of celestial ether and therefore denser than the invisible or transparent parts of the orbs that carry them. By contrast, dynamic rarity and density assumes that any particular part of the celestial ether can vary in density, being sometimes denser and sometimes rarer, as with sublunar bodies. During the Middle Ages, the dynamic interpretation was almost unanimously rejected, as when Oresme denied that celestial bodies were alterable "in density in the sense that one part of the heaven may be at times denser and at times rarer by a density of its own."[37]

Although, in theoretical terms, the perfect celestial ether should have been uniform and homogeneous, and therefore devoid of differences in density, scholastic natural philosophers, as we have seen, often found it necessary to acknowledge apparent differences between the visible celestial bodies and the transparent parts of the orbs that carried them.[38] But it was not until the seventeenth century, or perhaps the sixteenth, that some scholastics consciously distinguished between the two kinds of rarity and density. Both are encapsulated in Raphael Aversa's statement that "density and rarity can be found in celestial bodies; nevertheless it cannot be established in fact that parts of the heaven are at one time denser and at another rarer."[39] Although permanent differences of density and rarity could exist in the heavens, no one part of it could vary its density or rarity. As Pedro Hurtado de Mendoza explained, "[celestial] incorruptibility with greater and lesser rarity is not absurd, since all [such] dispositions lack a contrary."[40]

Aversa's conviction that density and rarity are found in celestial bodies was based on the definition that "density and rarity are nothing other than a body having more or less matter, or more or less substance, within a certain quantity [or magnitude or volume]."[41] In this sense, density and rarity appear in all bodies, whether terrestrial or celestial. But if the definition of quantity in the celestial region differs from that in the terrestrial region, then so will rarity and density, since the latter pair of contraries depends on the concept of quantity.[42] Thus did Aversa take cognizance of the prob-

37. Oresme, *De celo*, bk. 1, qu. 10, 1965, 153 (Latin), 154 (English). For Aegidius Romanus's rejection of the dynamic concept, see this volume, Chapter 13, Section III.9 and note 112.
38. Although Aristotle acknowledged qualitative differences within the celestial ether (see *Meteorology* 1.3.340b.6–10), he did not suggest that every celestial body was the denser part of its respective orb. See this volume, Chapter 17, Section I, and note 12.
39. "Pro resolutione dicendum est posse quidem in corporibus caelestibus inveniri densitatem et raritatem. Tamen de facto non constare alias esse caeli partes densiores, et alias rariores." Aversa, *De caelo*, qu. 34, sec. 2, 1627, 129, col. 2.
40. "Deinde non repugnat incorruptibilitas cum maiori et minori raritate; omnes enim dispositiones sunt sine contrario." Hurtado de Mendoza [*De caelo*, disp. 1, sec. 5], 1615, 366, col. 2.
41. Aversa adds that it is irrelevant whether or not the "matter" is true matter.
42. "Itaque si quantitas caelorum dicatur diversae rationis a nostra quantitate, sic consequenter raritas et densitas in sua ultima specifica ratione erit diversa. Si autem quantitas dicatur eiusdem prorsus rationis, similiter densitas et raritas erit eiusdem rationis." Aversa, *De caelo*, qu. 34, sec. 2, 1627, 129, col. 2.

lem and explicitly admit that rarity and density exist in the celestial region, something few Aristotelians did. But he himself adopted only the static kind of rarity and density (the first kind described earlier) and made no decision as to whether one and the same celestial part could vary its density.[43]

Some years later, Sigismundus Serbellonus went considerably beyond Aversa when grappling with the claim that if there is true quantity in the heavens, it would follow that true rarity and density also exist there, because rarity and density are properties of quantity.[44] Serbellonus rejects this argument with an unusual response. Rarity and density are not properties of any and every quantity but only those quantities that can be divided into parts that allow the penetration of minute corpuscles of another kind. But a solid body that is unmixed with foreign matter can be called only "dense," not "rare," since all its parts are united in a continuum. Serbellonus applies the same reasoning to a fluid body that preserves the greatest union or continuity of its parts and from which other kinds of matter are excluded. Such a fluid body could have no rarity but only density – a density, however, that cannot increase, just as with water.

How does all this apply to the celestial ether? Serbellonus interprets the celestial ether as a unified and largely continuous body, except in those places where it is divided by rising exhalations or by sublunar bodies or, indeed, even by planets. On this basis, Serbellonus considers the celestial ether to possess density over most of its vast extent but to possess rarity only at those places where different kinds of bodies penetrate – that is, where exhalations from sublunar bodies have risen or where the planets exist. Thus Serbellonus allowed for the occurrence of greater rarity in that part of the heavens where matter of a different kind could penetrate – and apparently it could penetrate – while simultaneously allowing that density existed in all the other homogeneous parts, but a density that was constant and incapable of increase.[45]

During the seventeenth century, scholastics gradually came to accept the existence of some form of rarity and density in the heavens, although they appear not to have embraced it fully. They achieved this by virtue of more detailed and self-conscious discussions of the meaning of rarity and density and how these opposites might be conceived in the celestial region. Although rarity and density were treated at great length in commentaries on book 4, chapter 9, of Aristotle's *Physics*, these discussions were concerned

43. But we shall see that Aversa thought it better to replace the idea of rarity and density in the heavens with the contrary properties opacity (*opacitas*) and diaphaneity (*diaphaneitas*).

44. "Obiectio prima: In coelis est vera quantitas; ergo etiam vera raritas esse potest et densitas. Consequentia probatur: quia raritas et densitas sunt proprietates quantitatis." Serbellonus, *De caelo*, disp. 1, qu. 3, art. 2, 1657–1663, 2:35, col. 2.

45. Ibid.; also see page 23, column 2, for a similar explanation. That Serbellonus could speak of sublunar matter and exhalations penetrating the celestial ether marks a radical departure from the Middle Ages and from Aristotle. Although it would appear that the intrusion of such foreign matter implies change in the heavens and therefore corruptibility, Serbellonus does not seem to have favored celestial corruptibility. For a discussion of celestial corruptibility, see Chapter 12, Section III.1.

almost exclusively with the terrestrial region. Not until the seventeenth century did scholastics confront the problem of rarity and density in the celestial region. This shift in emphasis was undoubtedly a direct consequence of Tycho Brahe's astronomical contributions regarding new stars and comets. Scholastics were on the defensive and had to cope with new discoveries and theories that demanded novel and more responsive explanations. Rarity and density in the celestial region was only one new challenge within the overall theme of celestial incorruptibility, to which we must now return.

d. Other arguments for celestial incorruptibility

Because of the absence of contraries in the celestial region, some scholastic authors, following Aristotle, inferred that celestial incorruptibility signified that matter itself was lacking in the heaven. The heaven cannot have matter, Buridan explains, because "every naturally generable or corruptible thing has matter." But "the heaven is not naturally generable or corruptible" and therefore cannot be said to have matter.[46] It was an argument intended to distinguish sharply between changeable terrestrial things and the unchangeable heaven. But, as we shall see later, it was challenged by other scholastics who also assumed celestial incorruptibility.

Another way of testing for celestial incorruptibility was to inquire whether the heavens could increase or decrease. According to Buridan,[47] Averroës said that increase, or augmentation, occurred in three ways: by nutrition, as with plants and animals; by rarefaction; and by adding things from outside, as in heaping up a pile of stones. None of these ways, however, applied to the celestial region. Nutrition requires substantial generation in which the nutrient is converted into the substance of the thing that ingests it. But such change was impossible in the heavens. Rarefaction depends on the primary qualities, hotness and coldness, which, as we saw, cannot exist in the heavens because they would cause generation and corruption. Nor, finally, can the heavens be augmented from outside, since the augmenting substances would need to be of a celestial nature. Where could such substances exist before they attached themselves to the heavens? If they were located some distance away, a rectilinear motion would be required for them to reach their goals. But the celestial substance lacks any natural capacity for rectilinear motion.[48]

46. "Sed dicendum est etiam quod caelum nec est generabile naturaliter nec corruptibile, et causa est quia non habet materiam, et omne naturaliter generabile vel corruptibile habet materiam." Buridan, *De caelo*, bk. 1, qu. 10, 1942, 46.
47. Ibid., 46–47.
48. In a question titled "Whether the world is perpetual," Albert of Saxony [*De celo*, bk. 1, qu. 17], 1518, 102r, col. 1, who frequently used specific arguments from Buridan's *De caelo*, repeats these arguments and much else from Buridan's question. Galileo, *De caelo*, qu. 4 (J), 1977, 100, also denied the augmentability of the heavens because of a lack of contraries in that region, arguing that "the heavens are not augmentable by any augmentation, strict or not; for both rarefaction and condensation and augmentations and

Potential problems arose from the universal conviction that the celestial region caused changes in the sublunar region. How could such changes be made without some part of the celestial region itself suffering change? How could the Sun and stars affect the earth with light and other influences without the assistance of the inferior celestial spheres? Was it not plausible to assume that the dissemination of influences from the heavenly bodies in the farthest reaches of the heavens could be multiplied only with the aid of the planets and spheres in the lowest parts of the heavens? The celestial recipients of these influences must surely undergo alteration of some kind. To solve this fundamental problem, Buridan and others distinguished a "proper" from an "improper" alteration. Only the latter could occur in the heavens, where the dissemination of light and other influences involved the multiplication and reception of qualities that not only lacked contraries but also met no resistance to their transmission. Such changes were "improper" – that is, not substantive – because they occurred without the involvement of contraries.

2. *Experience*

Of at least equal importance with the theoretical arguments just described was Aristotle's declaration that no changes in the celestial region had ever been observed or recorded.[49] In a typical repetition of this powerful defense of celestial incorruptibility, Buridan asserts that "from all the writings that we have from most ancient times, it does not appear that the heavens from that time to the present have been corrupted or worsened."[50] If the heavens were corruptible, some obvious physical sign of that corruptibility should have been detectable over so long a time span. Now, it might be thought that the observed alterations in the velocities of planets in the course of their regular movements might be just such a sign of celestial change. These, however, were merely apparent changes in velocity, attributable to the multitude of motions – some as yet unknown – that determined the ultimate positions of the planets.[51]

Aristotle's supporters also had to fend off a criticism based on astronomical experience, namely the well-known astronomical observations that revealed variations in the distance of the Moon and other planets from the earth. If these planets were sometimes nearer to and sometimes farther from the earth, did this not imply that at times they ascended rectilinearly and at other times descended rectilinearly? And if so, would these motions not signify heaviness and lightness in the planets? In response, Buridan, for

diminutions that are strict presuppose the actions of coldness and hotness, and these are contraries."

49. Aristotle, *De caelo* 1.3.270b.13–17.
50. "Secundum signum est quia per omnem memoriam quam per scripturas possumus habere ab antiquissimis, non apparet quod caelum ab isto tempore usque nunc sit corruptum vel peioratum" Buridan, *De caelo*, bk. 1, qu. 10, 1942, 46.
51. An argument found in Oresme, *De celo*, bk. 1, qu. 10, 1965, 151 (Latin), 152 (English).

example, explained that the variations in distance were the result of combinations of circular motions and in no way the consequence of rectilinear motions.[52]

The Sun's seemingly greater heat in the summer than in the winter seemed to imply an alteration in the Sun's power and therefore an inherent variability in the Sun itself. To preserve celestial incorruptibility, John of Jandun distinguished an equivocal use of the term "alteration" (*alteratio*), which enabled him to deny solar variability and conclude that the Sun's power was identical in summer and winter. The heat produced by the Sun did, however, vary in summer and winter but only because its rays were refracted more nearly at right angles in the summer than in the winter. These variations in refraction were caused by alterations in the media in the sublunar region and were therefore not real alterations of the Sun.[53] Although not based on direct experience, astrologers made similar claims about the planets when they assumed that the power of a planet varied as it occupied different signs of the zodiac. Scholastic natural philosophers, however, flatly denied real internal alterations in the power of either the planets or the zodiacal signs. As Buridan explained:

> a hot planet seems of greater power if it is in a hot sign than if it were in a cold sign because the sign and the planet can [then] simultaneously influence heating, and thus a great hotness arises here below. But if the hot planet is in a cold sign, the influence of the sign prevents the influence of the planet from acting, because the sign acts in a way contrary [to the planet]; [under these circumstances] the planet appears to possess little power.[54]

Between the thirteenth and fifteenth centuries, medieval natural philosophers developed an elaborate defense of celestial incorruptibility, which acquired the virtual status of a self-evident truth. Aristotle had facilitated the nearly unanimous acceptance of incorruptibility by denying celestial locations to shooting stars, comets, and similar phenomena, placing them instead below the Moon, in the upper atmosphere of the terrestrial region.[55] With such striking phenomena relegated to the sublunar realm and in the absence of challenges to the Aristotelian world view, it was easy to maintain both theoretically and empirically the unchangeability of the celestial region. Without any challenge to the Aristotelian system, celestial incorruptibility became an essential ingredient of world order.

52. Buridan, *De caelo*, bk. 1, qu. 9, 1942, 43 and John of Jandun, *De coelo*, bk. 1, qu. 13, 1552, 10r, col. 2. Jandun mentions eccentrics and epicycles (he uses the term *epicentrico* in conjunction with *eccentrico*; I take the former term to signify "epicycle") as the cause of the circular motion. We also saw earlier that the celestial ether lacked the capacity for rectilinear motion.
53. John of Jandun, ibid., qu. 17, 1552, 12r, col. 2–12v, col. 2.
54. Buridan, *De caelo*, bk. 1, qu. 10, 1942, 46 (for the astrologers' claim), and 48 (for Buridan's reply).
55. Aristotle, *Meteorology* 1.3.341a.33–35 and 1.4.342a.30–33, 1984.

By making the earth just another planet, Nicholas Copernicus, in 1543, hurled an implicit challenge to the concept of incorruptibility. As a planet, the heretofore imperfect earth was as perfect as other planets; or conversely, the previously perfect planets were now as imperfect as the earth. As profound as these implications were, they did not pose the most imminent threat to the medieval assumption of incorruptibility. The most immediate danger emanated from Tycho Brahe's direct observations of the new star of 1572 and, more importantly, the comet of 1577, followed by the telescopic observation of sunspots, associated most prominently with the name of Galileo.

III. Scholastic interpretations of celestial incorruptibility in the sixteenth and seventeenth centuries

Although the problem of celestial corruptibility or incorruptibility was always considered an important problem from the time of the Church Fathers to the seventeenth century, the doctors who judged this issue never condemned the contrary opinion. They were convinced that each alternative – corruptibility or incorruptibility – was consistent with faith and without danger of error. Unlike the fateful issue concerning the earth's centrality, which resulted in the condemnation of Galileo, celestial incorruptibility was never made a doctrinal issue but remained a straightforward philosophical problem.[56] An absence of doctrinal conflict and an apparent sense that celestial corruptibility would not of itself destroy Aristotelian cosmology were probably the factors that eventually enabled some scholastics to accept celestial corruptibility.

Most, however, remained faithful to the medieval tradition and defended celestial incorruptibility.[57] Their defense was a combination of old and new, the latter consisting largely of arguments devised to cope with the consequences of the well-attested new astronomical phenomena that had been witnessed in the 1570s, namely the new star of 1572 and the comet of 1577. Let us first examine the more important aspects of the old tradition.

56. On all this, see Amicus, *De caelo*, tract. 5, qu. 1, dubit. 1, art. 2, 1626, 232, cols. 1–2.

57. In commenting on *De caelo*, the Coimbra Jesuits declared that "by common consent, almost all the Peripatetic schools" (tota fere Peripaetica schola communi assensu) supported the incorruptibility of the heavens (Conimbricenses [*De coelo*, bk. 1, ch. 3, qu. 1, art. 2], 1598, 66). According to Franciscus de Oviedo [*De caelo*, contro. 1, punc. 2], 1640, 1:464, par. 15, "almost all philosophers, theologians, and interpreters of Scripture" (fere omnes philosophi, theologi, atque Scripturae interpretes) believed in celestial incorruptibility. By the seventeenth century, some scholastics had distinguished four senses of incorruptibility of which only the third and fourth, concerned with external and internal incorruptibility, are relevant. See Amicus, *De caelo*, tract. 5, qu. 1, dubit. 1, art. 1, 1626, 230, col. 2–231, col. 1, and Conimbricenses, ibid., 65. Although Serbellonus, *De caelo*, disp. 1, qu. 2, art. 3, 1663, 2:22, cols. 1–2, distinguishes only a physical and logical sense of incorruptibility, his distinctions closely resemble those of Amicus and the Coimbra Jesuits.

1. The traditional scholastic defense of celestial incorruptibility

a. The celestial substance

For many – including the youthful Galileo in his scholastic phase, the Coimbra Jesuits, Bartholomew Amicus, and Franciscus de Oviedo – the best evidence for celestial incorruptibility continued to be the experience of the ages, which revealed a remarkable constancy in the celestial region.[58] As Galileo explains ([*De caelo*, qu. 4 (J)], 1977, 99) "it has been found over all preceding centuries that no change whatever has taken place in the heavens. And this argument," he emphasizes, "has the greatest force: . . . since the motion of the heavens is of maximum velocity such that it could destroy even the most solid bodies, and since it has endured for such a long time, preserving the stars always at the same distance, opposition, and magnitude, from this argument it is most certain that the heavens are incorruptible." The argument from experience formed part of a traditional defense of celestial incorruptibility in the sixteenth and seventeenth centuries, by which time new arguments and refinements had been added to the medieval inheritance. Of major concern was whether celestial incorruptibility was the result of external or internal causes and whether it applied only to the celestial substance, or ether, alone or to celestial accidents as well.

According to Amicus, if the heavens were incorruptible by reason of external causes it meant that the heavens themselves were internally and essentially corruptible but were somehow preserved in a state of accidental incorruptibility by virtue of external causes. What kinds of external causes could preserve the heavens in a state of incorruptibility? Amicus distinguished three kinds: the removal of the external agent that was capable of causing the corruption of the heavens; the direct action of God, who could also preserve the heavens in a constant incorruptible state; or the addition of some quality which could make the heavens incorruptible, just as "the bodies of the blessed are said to become incorruptible by the gift of incorruptibility."[59] By contrast, a state of internal celestial incorruptibility would obtain if the heavens, or ether, consisted of matter or substance that had a predisposition for only one particular form but no capacity for another form. As we have already seen, without a capacity for another form – indeed a contrary form – generation and corruption could not occur.

The external and internal concepts of the sources of celestial incorruptibility were seemingly antithetical. Plato was associated with the first. He was said to believe that the heavens were inherently corruptible but had

58. See Conimbricenses, ibid., 66; Amicus ibid., art. 3, 232, col. 2; and Oviedo, ibid., 1:464, col. 1, par. 19.

59. Amicus's views on external and internal causes of celestial incorruptibility form part of a fourfold distinction of senses of incorruptibility. See Amicus, ibid., art. 1, 1626, 230, col. 2–231, col. 1. In similar fashion, angels were incorruptible by the grace of God, not by their own nature. For the similar views of Galileo and Oviedo, see note 82 of this chapter.

been made incorruptible by the will of God, who, in these circumstances, was considered an external force. Amicus opposed this interpretation because, in Aristotelian terms, it necessarily implied that the internally corruptible heavens could never actualize their alleged potential for change. He was convinced that nature would not allow the existence of a perpetually frustrated potential for change. In the interests of a harmonious world, nature would have made the celestial region internally incorruptible, and therefore without any capacity for change.

Sometime around 1648, Johannes Poncius (John Punch; 1599–1661), a Franciscan professor of philosophy and theology, also strongly supported an internal cause of celestial incorruptibility when he declared that the most common opinion among the Peripatetics and the theologians was that the heavens were incorruptible by their internal nature and could not be altered by any natural agent: not even by God in his ordained power.[60] The celestial region was intrinsically incorruptible, even if it was composed of matter that was identical with sublunar matter.[61] But if the matter of the heavenly region was identical with that of the sublunar region, did this not imply that celestial matter possessed contrary qualities and would therefore be corruptible? Poncius anticipated this objection by introducing into the heavens the concept of "conservative qualities" (*qualitates conservativas*), which, lacking contraries themselves, prevented terrestrial-type contrary qualities from operating, even if they existed in celestial matter.[62] As an internal cause, the operation of conservative qualities was tantamount to proclaiming an absence of contrary qualities in the heavens. Without contrary qualities, the conservative qualities possessed by celestial matter were capable of preserving the heavens because change could occur only by the operation of contrary forms or qualities. Sublunar substances, by contrast, are alterable because it is their inherent nature to require and receive contrary qualities.[63]

60. Poncius [*De coelo*, qu. 4: "Whether the heavens are corruptible"], 1672, 617, col. 1. For "ordained power" the Latin reads "potentiam ordinariam" but should presumably read "potentiam ordinatam." Although the work was first published in three volumes between 1642 and 1645, the section on *De coelo* was added to the editions published after 1648. Poncius, a Scotist, assisted Luke Wadding in the publication of the edition of the works of Duns Scotus (see Lohr, 1988, 362). God could, of course, alter the heavens by his absolute power.

61. This is the first conclusion of question 4 and reads: "The heavens could be incorruptible from their very internal nature, whether their matter were of the same species as that of sublunar matter or different" (Coeli possent esse incorruptibiles ex natura sua intrinseca, sive constarent materia eiusdem species cum sublunari, sive diversa). Poncius, ibid.

62. With reference to the heavens, Poncius declares: "Quia non obstante compositione ipsiu ex quacumque materia posset habere aliquas qualitates conservativas quae non haberent contrarias qualitates, neque enim in hoc est ulla prorsus repugnantia. Sed si haberet, non posset corrumpi naturaliter et hoc ob naturam suam intrinsecam. Ergo esse posset incorruptibile ex natura sua non obstante compositione eius ex quacumque materia, sive eiusdem, sive diversae speciei cum sublunaribus." Ibid.

63. In his questions on *De caelo*, Galileo, relied heavily on the doctrine of contraries, explaining first that in contrast to the upward and downward rectilinear motion of terrestrial bodies, celestial bodies moved only with a natural circular motion that lacked a contrary. Without a contrary motion to produce corruption, it followed that celestial bodies had to be

Special appeals based on perfection, harmony, and appropriateness were commonly invoked to demonstrate the intrinsic incorruptibility of the celestial region. Thus Amicus appealed to the perfection of the universe. If imperfect and corruptible things are part of a perfect universe, how much more ought incorruptible things to be part of it. But where could these be located? Because all other bodies – terrestrial bodies, that is – are imperfect and corruptible, the only logical choice for the location of incorruptibility is the heavens.[64] Another argument took its departure from the well-ordered perpetual generation and corruption of sublunar things.[65] Because every effect requires a proportionate cause, an invariant, incorruptible body is required to preserve the perpetual changes that occur in the determinate order of sublunar activity. Since the effect is perpetual, so also must the cause be perpetual and, indeed, incorruptible. This argument depends on the assumption – which Amicus makes – that experience teaches us that the order of the universe (*ordo universi*) in inferior things is determinate and invariable and cannot arise from the nature of inferior things themselves, which are variable and corruptible. Hence it ought to arise from an eternal and invariable principle, which must be the heavens.

As confirmation of this widely accepted principle, Amicus observes that the nature of a thing is proportionate to its goal or end. Now the goal of the heavens is to conserve the world perpetually by their continuous influence. But to act perpetually as a constant cause requires that the heavens be incorruptible.

Finally, that which is the measure of all corruptible things ought to be incorruptible by its very nature; otherwise, it would not be adequate to the task of preserving the order and perfection of the world. But God made the heavens to serve as a rule or model (*regula*) for all corruptible things and human actions.[66]

b. Celestial accidents or qualities

The incorruptibility discussed thus far is relevant only to the celestial substance, or ether. But scholastics were also concerned about the nature of celestial accidents, or qualities, that inhered in the celestial substance or were produced by it. Were such accidents also incorruptible? By distinguishing two kinds of accidental changes that had already been distinguished in the Middle Ages, Amicus argued that real accidental changes cannot occur in the heavens. In a separate question devoted to this problem, Amicus, as did

incorruptible (Galileo, *De caelo*, qu. 4 [J], 1977, 97). Because of the same lack of contraries in the heavens, Galileo, following Aristotle, also denies the possibility of augmentation in the heavens (ibid., 100).

64. Amicus, *De caelo*, tract. 5, qu. 1, dub. 1, art. 4, 1626, 235, cols. 1–2.
65. Ibid., col. 2.
66. Ibid.

Galileo earlier, divided accidents into "corruptive" and "perfective."[67] A "corruptive accidental change" (*mutatio accidentalis corruptiva*) is not possible in the heavens, because it requires the presence of contrary qualities, such as hotness and coldness, which produce substantial mutations, that is, cause the corruption of one substance and the generation of another, as in the transformation of a cold substance into a hot substance. Corruptive changes are possible only in sublunar bodies composed of the four elements.

By contrast, a "perfective accidental change" (*mutatio accidentalis perfectiva*) can occur regularly in the heavens, because it does not involve positive contrary qualities. With perfective accidents, changes occur from a privation, or absence, of a quality to a positive quality, or vice versa.[68] Such accidents appear in the transmission of both light and various celestial influences that of necessity pass through successive parts of the heavens because they cannot act at a distance. As examples of perfective accidents, Amicus mentions lunar eclipses and the continual generation and corruption of the Sun's light as the Sun is moved around the sky by its orb. Indeed, even the generation and corruption of new stars and sunspots are nothing more than accidental perfective mutations, because they are formed by the natural motions of bodies and epicycles.[69]

In the absence of contrary qualities, Amicus concludes that light and celestial influences produce qualitative changes that are like natural circular motions. Such changes are compatible with celestial incorruptibility. They can no more cause substantial corruption in the heavens than can the successive movements of angels between any two points. Thus did Amicus preserve the incorruptibility of the celestial region and also reveal one important defensive posture that many scholastics adopted in confronting the discoveries of such seemingly new celestial phenomena as new stars, comets,

67. "Whether [the heavens] are corruptible as to accidents," in Amicus, *De caelo*, tract. 5, qu. 2, dub. 2, arts. 1–4, 1626, 247, col. 2 -249, col. 1. For Galileo, see Galileo, *De caelo*, qu. 4 (J), 1977, 100–101.
68. Here is Amicus's basis for the distinction between perfective and corruptive accidents: "Secunda opinio esse capacem mutationis accidentalis, que est secundum accidentia perfectiva non autem quae est secundum accidentia corruptiva. Est communis fere omnium fundamentum sumitur ex fundamento opposito. Nam illa mutatio accidentalis debet poni in corpore incorruptibili que illi est proportionata, at non est sic que est secundum contraria. Sed mutatio secundum accidentia perfectiva est illi proportionata; at quae secundum accidentia corruptiva est illi contraria. Ergo perfectiva est in coelo; non autem corruptiva. Et haec opinio mihi placet." In Amicus, ibid. In support of this distinction, Amicus cites John of Jandun.

    In his questions on *De caelo*, Galileo, *De caelo*, qu. 4 (J), 1977, 100–101, declared that, along with "Simplicius, Averroës, and Saint Thomas," he believed that "alteration is twofold: one corruptive and the other perfective. The first is between contraries and involves corruption, and this has no place in the heavens; the other involves no contrariety and is found even in spiritual things – for which reason it is found also in the heavens." From the references given by Amicus and Galileo, we may conclude that the distinction between "perfective" and "corruptive" accidents was medieval, although the terminology may have been added in the sixteenth and seventeenth centuries.
69. Here Amicus favors the use of epicycles to explain the formation of new stars and comets. For further discussion, see Section III.3.a.

and sunspots: namely, to dismiss them as accidental perfective mutations. But what were these new discoveries that cast doubt on the existence of an eternally unalterable celestial region?

2. Celestial incorruptibility and the discoveries of Tycho and Galileo

Scholastics were aware that at least three instances of celestial change had been reported prior to the sixteenth century, one by Pliny, who mentions that Hipparchus had observed a new star that eventually disappeared, and two by Saint Augustine, who reported changes in the magnitude and course of Venus and a diminution in size of the star Spica in the constellation Virgo.[70] But these claims were insufficient to undermine Aristotle's defense of celestial incorruptibility. Those arguments were eroded only in the aftermath of the results achieved by Tycho Brahe, probably the greatest naked-eye-observational astronomer in history. Only when Tycho failed to detect parallax in his observations of the new star of 1572 and concluded that its lack of parallax signified that it was located in the sphere of the fixed stars above the planets were serious doubts entertained. The star's brief, but spectacular, two-year life span left a considerable impact, as it first emerged in brightness and then gradually dimmed and disappeared. Here indeed was an apparent case of celestial generation and corruption.

The comet of 1577 seemed to add the clinching argument against incorruptibility. Up to that time, Tycho had assumed, with all Aristotelians, that comets were generated beneath the Moon.[71] But his observations of the comet of 1577 convinced him that its parallax placed it in the celestial region, above the Moon in the region of Venus. With this knowledge, Tycho seemingly repudiated two fundamental Aristotelian and medieval beliefs: that comets were sublunar and that the heavens were immutable.[72] By 1611, Galileo, Christopher Scheiner, and Johann Fabricius, among oth-

70. See Pliny, *Natural History*, bk. 2, ch. 26, 1938–1963, and Augustine, *City of God*, bk. 21, ch. 8, 1948, 429, for the reference to changes in Venus. Amicus, *De caelo*, tract. 5, qu. 1, art. 7, 1626, 242, col. 1, provides the reference about Spica and cites book 21, chapter 10 of the *City of God*, which in Dods's translation contains no mention of Spica. For correct references to both, see Aversa, *De caelo*, qu. 33, sec. 2, 1627, 83, col. 1 for Augustine (only Venus is mentioned) and sec. 3, 83, col. 2–84, col. 1 for Pliny. Aversa adds a few more examples of alleged celestial corruptibility. Serbellonus, *De caelo*, disp. 1, qu. 2, art. 3, 1663, 2:23, col. 2, mentions Pliny and Hipparchus as reporters of new stars. Indeed, so did Tycho in his tract on the new star of 1572 (Thoren, 1990, 66).
71. See Hellman, 1944, 121.
72. Ibid., 130. Tycho wrote two treatises on the comet of 1577, one in Latin, the other in German. The former, *De mundi aetherei recentioribus phaenomenis*, was first published in 1588 but was begun immediately after the comet's disappearance in 1578. (Hellman describes the volume on 337–338.) The book's wider dissemination began only in the early seventeenth century. The German work, which was intended for a more general audience, "was probably written immediately after the disappearance of the comet in 1578 but first printed in 1922." Also see Hellman's discussion in *Dictionary of Scientific Biography*, 2:406–408. A third belief current in Tycho's time was also shattered, namely that in the solidity of the celestial orbs, discussed in Chapter 14 of this volume.

ers, had used the telescope to observe the variability of sunspots and add yet another impressive observational datum to the growing list of phenomena signifying celestial corruptibility and change.[73] Indeed, additional new stars were sighted in 1600 and 1604.[74]

How did scholastic natural philosophers react to these alleged changes in the celestial region? Did they accept the phenomena as genuinely celestial, or did they defend Aristotle's claim that such phenomena occurred only in the upper atmosphere just below the Moon? And even if they were genuine celestial events, were they really substantial changes, or only accidental? And, finally, did any scholastic natural philosophers accept the new changes and opt for celestial corruptibility? We must now seek answers to these vital questions.

3. *The scholastic reaction to the discoveries of Tycho and Galileo*

a. *Continued defense of celestial incorruptibility*

The celestial events reported as substantial changes in the heavens described in the preceding section were well known to scholastic authors of the seventeenth century but were dismissed for a variety of reasons.[75] The star allegedly seen by Hipparchus was not a star but was in the air itself. The changes in Venus reported by Saint Augustine were not real but probably caused by a diminution of rays or by planets that moved around Venus on epicycles or even by light reflected around such surrounding planets.[76] Among a number of explanations of sunspots, Amicus mentions one that assumes epicycles carrying around varied and irregularly shaped compressed, disklike figures.[77]

73. As Stillman Drake, 1957, 83, explains: "If blemishes could appear and disappear on the face of the sun itself, the incorruptibility and inalterability of the heavenly bodies was destroyed": unless, of course, the "blemishes" could be explained away.
74. Serbellonus mentions them. Serbellonus, *De caelo*, disp. 1, qu. 2, art. 3, 1663, 2:23, col. 2.
75. All were specifically mentioned by Amicus, who rejected them one by one. See Amicus, *De caelo*, tract. 5, qu. 1, dubit. 1, art. 7, 1626, 242, for the five arguments indicating celestial corruptibility, and page 246 for his rebuttals.
76. See ibid., 242, col. 1, for the description, and 246, col. 1, for Amicus's explanation. Here we have an instance in which Amicus invokes additional planets or stars to account for an alleged change in Venus. "It is a curious fact," Stillman Drake, 1957, 83, has observed, "that the conservative astronomers who for philosophical reasons had previously rejected Galileo's discovery of new moving stars in the heavens, now for philosophical reasons commenced to populate the sky with moving stars at a rate which made Galileo blush." Scheiner, for example, "accounted for sunspots by assuming a number of small planets to revolve about (or beneath) the sun and to obstruct our vision." Ibid.
77. Although some scholastics denied sunspots and others considered them consequences of a defect in the telescope, Bartholomew Amicus rejects these solutions. He thought it would indeed be amazing if all telescopes had the same deficiency, namely one that allowed the appearance of sunspots. In defending the validity of telescopic observations, Amicus may have had in mind Galileo, who challenged skeptics to construct an instrument that would reveal stars moving only around Jupiter but no other planet (see Drake, 1957, 73).

Amicus was one of those who refused to infer natural celestial corruptibility from the new celestial discoveries. Although he did not propose definitive explanations, Amicus presents four possible approaches that are representative of the way most scholastic authors chose to interpret the new discoveries. Their tactic was to deny the occurrence of substantial celestial changes.

The first way sought to preserve the Aristotelian interpretation by insisting that all the seemingly new phenomena are actually sublunar. To achieve this, they attributed the new phenomena to the physical effects of various external causes, such as an impure medium, the extreme distance of the objects, or the falsity of the instruments.[78] In general, the phenomena were explained as the result of tricks played by our senses. Scholastics thus exploited inevitable differences in observation, interpretation, and mode of argumentation among those who upheld the celestial location of the new discoveries. These differences were taken as a sign of uncertainty and weakness of argument. Indeed, Amicus believed that all comets after the time of Aristotle were sublunar, especially the one of 1618, which was below the Moon, since to the naked eye and to those viewing through a telescope (*tubum opticum*) it appeared weaker, duller, and more obscure than the light of the stars. Nor was its parallax sufficient to place it beyond the Moon. Indeed, those who placed the comet above the Moon did not agree on its altitude.[79]

A second approach invoked God's supernatural power.[80] One could readily concede that the newly observed phenomena were in the ethereal heavens in both substance and accidents, placed there by a supernatural power. Many reasons might have motivated God to create these celestial displays: as the sign of a great effect; to terrify mortals; or to demonstrate his dominion and power over his creatures. Because of their supernatural origin, the new celestial phenomena did not count as natural celestial alterations, and celestial incorruptibility remained inviolate.

Although explanation by miracle was given a strong boost when the Coimbra Jesuits adopted it at the end of the sixteenth century as the "more probable" (*verisimilior*) interpretation,[81] few scholastics chose to follow them

78. "Primum negando veritatem illarum apparentiarum, sed esse ludificationes sensuum ex variis causis ortas, scilicet medii impuritate, obiectorum distantia, instrumentorum falsitate." Amicus, *De caelo*, tract. 5, qu. 1, dubit. 1, art. 7, 1626, 242, col. 2. Serbellonus says much the same thing, even to the extent of following Amicus's precise word order from *ludificationes* to *falsitate*. Serbellonus, *De caelo*, disp. 1, qu. 2, art. 3, 1663, 2:23, col. 2, par. 44. Drake, 1957, 73, describes Galileo's attacks against those who sought to subvert his telescopic discoveries by challenging the validity of the instrument.
79. Amicus, ibid., 243, col. 1.
80. "Alterum responsionis genus admittit illas mutationes esse in coelo aethereo non solum secundum accidentia, sed etiam secundum substantiam, id est, at factas non virtute naturali, sed supernaturali ob finem Deo notum, sive ad significandum aliquem magnum effectum." Ibid, col. 2.
81. "Superest ergo ut verisimilior sit alia opinio omnium ultima quae asserit novam hanc stellam, non physica, sed supernaturali generatione a Deo fuisse procreatam." Conimbricenses, *De coelo*, bk. 1, ch. 3, qu. 1, art. 4, 1598, 71.

in the seventeenth century. Oviedo informed his readers that his concern for the question of celestial incorruptibility was confined to natural causes, since almost all scholars were agreed that God could corrupt the heavens if he wished.[82] Indeed some Church Fathers were thought to have denied incorruptibility to the heavens and to angels on the grounds that if God suspended his preservative powers, everything, including the heavens, would disappear.[83] Serbellonus thought it unphilosophical to appeal to the first cause if, without violence to reason, we could explain such phenomena by natural causes.[84]

The third approach was clearly the most important. It was a response against those who located some or all comets and new stars above the Moon.[85] To counter such claims, it was necessary to explain the emergence and disappearance of comets and new stars "without [the assumption of] a new generation and corruption of celestial substance."[86] The objective was thus to derive a new star from things that had always existed in the sky and were therefore ungenerated. Because visible stars remained at fixed distances from each other, they could not unite in any way to form a brighter, new star. Amicus reports a number of ways in which others had sought to attain this objective. One was to assume that a new star was nothing more than a concentration of reflected light (*splendor*) from an aggregation of planets and stars located in a particular part of the heaven.[87] A second explanation involved a union of small, ordinarily invisible stars (*stellae*), which upon coming together formed a sufficient mass to become visible.[88] A third way involved certain stars in the firmament, each of which is fixed in an epicycle that makes the star visible when the epicycle is turned toward us and invisible when it is turned away from us.[89] To all of these possibilities, Amicus himself raised fatal objections.[90]

82. Oviedo, *De caelo*, contro. 1, punc. 2, 1640, 464, col. 2, par. 20. Oviedo adds that not even angels are incorruptible by their nature, but only by the grace of God. Also in the context of celestial incorruptibility, Galileo, *De caelo*, qu. 4 (J), 1977, 96, observed that "God alone . . . is a completely necessary being," so that "angels and human souls are immortal by divine grace." For Amicus's similar opinion, see note 59 of this chapter.
83. See Compton-Carleton [*De caelo*, disp. 1, sec. 2], 1649, 397, col. 2.
84. "Nihilominus confugere ad primam causam in omnibus ex aliquorum auctoritate, non est satis philosophicum si in naturalibus causis sine violentia rationis sistere possumus." Serbellonus, *De caelo*, disp. 1, qu. 2, art. 3, 1663, 2:24, col. 1, par. 49.
85. "Tertium genus respondendi est aliorum qui neque omnes cometas dicit esse infra lunam, neque omnes supra lunam, sed aliquos infra aliquos supra." Amicus, *De caelo*, tract. 5, qu. 1, dubit. 1, art. 7, 1626, 244, col. 2.
86. "Sed ex iis qui ponunt aliquos esse supra lunam vario modo illam novitatem explicant sine nova generatione et corruptione celestis substantiae." Ibid.
87. Amicus, ibid., attributes this opinion to Christopher Clavius in the latter's *Commentary on the Sphere of Sacrobosco*.
88. Ibid., 245, col. 1.
89. Ibid., cols. 1–2.
90. Against the first position, Amicus notes that if new stars could be made from such reflected light it ought to happen more often, since the stars of the firmament always preserve the same relationships among themselves and since the planets also bear the same relationships to the fixed stars; therefore their reflected light will always be sent to the same parts of the sky, and new stars should form more frequently. Among a number of arguments

To account for the appearance of a new star, Amicus appealed to epicycles, which, although essential to Ptolemaic astronomy, had been rejected by many since Tycho Brahe had denied the existence of celestial spheres. In the medieval Aristotelian–Ptolemaic cosmological system of physical orbs, an epicycle was a small orb that was embedded between the convex and concave surfaces of a large deferent orb which surrounded the earth but was eccentric to it (that is, it had a geometrical center that was not the center of the earth). In this system, the planet was located within a concavity of the epicyclic orb, with the latter, in turn, carried around by the eccentric deferent (see Figure 8, in Ch. 13, Sec. II.3). As the most plausible explanation of the appearance of new stars, Amicus assumed the existence of numerous epicycles that were supposedly formed of denser, though still invisible, parts of celestial ether. When perchance three of these epicycles are aligned in such a way that their denser ethereal parts are clustered together, the Sun, which shines on all the stars, illuminates the ethereal cluster and makes it appear as a visible "new" star. As the epicycles move away from each other, the new star gradually fades. Although Amicus found this muddled explanation the most probable of the alternatives described – it had apparently been defended in a public debate on the comet of 1618 at the University of Ingoldstadt – he could not understand how the new star of 1572 in Cassiopeia could be a first-magnitude star during its first year and then jump to a third-magnitude star in the second year. Why, if the epicyclic motions were continuous, did they not produce a star of the second magnitude before they generated a star of the third magnitude? Amicus suggested yet other explanations for new stars and comets,[91] but like many scholastics would not commit himself to any single explanation to account for the new celestial phenomena. He remained content to present a series of possible explanations that depended on rearrangements or realignments of already existing bodies. Thus was celestial corruptibility denied.

But it was not just a matter of the rearrangement of preexisting bodies. Sigismundus Serbellonus appealed to the force of magnetism, which was used to account for a variety of effects. Magnetic force attracts bits and

against the second opinion, Amicus insists that such a union of stars could not occur in the orb of the fixed stars because the latter can have no motion relative to one another. Although this reason would also have sufficed against the third argument, Amicus denies the existence of stellar epicycles, on the grounds that their existence should cause new stars to occur more frequently, even if the motion of the epicycle was very slow, as was the case with the new star in Cassiopeia, which endured for two years. For these arguments, see Amicus, ibid. Serbellonus, *De caelo*, disp. 1, qu. 2, art. 3, 1663, 2:24, col. 1, posed a somewhat different objection when he asked how the new star in Cassiopeia could preserve the same distance from other stars if it was carried on an epicycle in a continuous revolution during the two years of its existence.

91. Amicus, ibid., 246, col. 1. In one he mistakenly attributes to Clavius the opinion that the new star of 1572 was not real but was simply an ordinary star that only appeared larger because of terrestrial exhalations that lie between us and the star, just as a coin placed in water appears larger because of refraction. Clavius, however, reported this as the opinion of others (see Clavius [*Sphere*, ch. 1], 1593, 208). Amicus's own opinion, as we shall see, upheld Tycho's judgment that the new star was really in the heavens.

pieces of celestial ether in the upper reaches of the heavens and brings them together to form opaque and solid bodies that can reflect the light of the Sun. The magnetic force is not constant, however, which presumably explained why such "new" celestial bodies are transient and why they can happen in different parts of the sky. Although such bodies have always appeared, Serbellonus did not conceive of them as "new" but thought of them rather as drawn from an already existing ether that could be temporarily rearranged by magnetic force.[92] Indeed, the formation of such bodies seemed to form part of the natural operation of the heavens and was not unusual or irregular.

Another tactic in defense of the traditional position was to draw empirical consequences from a hypothesis of corruptible heavens. On the assumption that fire would form part of corruptible heavens, Hurtado de Mendoza assumes that Moon and fire would oppose each other. The stronger of the two should consume all or part of the other, in the same manner in which fire consumes the part of the air nearest to it. But we observe nothing to indicate such a struggle.[93] Moreover, if the heavens are corruptible, they would be subject to corruption by some natural agent. Over thousands of years, that natural agent should have acted and corrupted the heavens sufficiently so that the heavens would have lost their ability to govern and perpetuate the world. Again, we observe nothing that would suggest this drastic consequence.[94]

b. Attempts to accommodate the new discoveries

Whereas Bartholomew Amicus used contemporary responses to vigorously defend traditional cosmology, Christopher Clavius (1538–1612) and Raphael Aversa (1589–1671) sought, however modestly, to adapt traditional theory to the new astronomical discoveries of the sixteenth and the early seventeenth century.

Clavius was one of the first scholastics to support Tycho. If the new star was no farther away than the atmosphere or air, it should have revealed different aspects. None had been observed. Nor indeed could the star be

92. Serbellonus, *De caelo*, disp. 1, qu. 2, art. 3, 1663, 2:24, cols. 1–2. The source of this celestial magnetic force is not explained.

93. "Itaque coeli sunt natura sua incorruptibiles probatur primo ex communi authorum consensu. Secundo, quia si luna esset corruptibilis, proculdubio damnum aliquod accepisset ab igne, cuius voracitas aut integram lunam aut bonam ipsius partem absumpsisset, sicut absumit partes viciniores aeris, licet sit summe humidus mutuoque pugnarent ignis et luna. Cumque ignis semper adhaerat lunae et sit potentior quam illa, illam vinceret; vel si est impotentior vinceretur a toto concavo lunae." Hurtado de Mendoza, *De caelo*, disp. 1, sec. 5, 1615, 367, col. 1. For a similar argument, see Oviedo, *De caelo*, contro. 1, punc. 2, 1640, 464, cols. 1–2, par. 19.

94. "Tertio quia si coelum est corruptibile, ergo ab aliquo agente naturali potest corrumpi, ergo illa potentia esset reducta in actum per tot annorum millia, atque ita coelum non esset aptum ad continentem et perpetuam mundi gubernationem." Hurtado de Mendoza, ibid., 367, cols. 1–2.

in any of the regular planetary orbs, because no astronomer had yet detected any motions that might indicate this. Clavius concluded that the star had to be in the most remote parts of the celestial region, that is, in the firmament, among the fixed stars.

What could have caused the new star? Clavius suggests that either it was created by God as a major portent or, if it was a natural celestial event, then one would also have to concede that comets could be created in the heavens just as they could be created in air. But if this was true, perhaps "the heaven is not a certain fifth element but a mutable body, although less corruptible than inferior [terrestrial] bodies."[95] Many philosophers of the ancient world had believed this, both before and after Plato and Aristotle. Indeed, some of the most eminent Church Fathers – Basil, Gregory of Nyssa, and Ambrose – taught the corruptibility of the heavens.

Just as we expect Clavius to opt for celestial corruptibility, he decides to withold his opinion ("meam enim sententiam in tanta re non interpono")[96] and rest content to have demonstrated that the new star is in the starry firmament. Nor did he wish to attempt answers to other difficult questions associated with the new star – what it portended or why it vanished after two years. Only God knows the answers to such questions. Thus did one of the most influential astronomers of the sixteenth and early seventeenth century decide in favor of the real existence of the new star but avoid a decision on the perplexing question as to whether it signified a corruptible celestial region.

Raphael Aversa sharply distinguished between new stars and comets: the former were celestial; the latter were phenomena of the upper atmosphere just below the Moon.[97] Thus Aversa agreed with Clavius that new stars were celestial phenomena. Acceptance of a celestial location for new stars was based on Aversa's high regard for astronomers like Tycho, who could find no parallax for the new star of 1572. Unlike Clavius, however, Aversa explicitly opted for the incorruptibility of the celestial region.

After rejecting the arguments of those who located the new star of 1572 in the upper atmosphere,[98] Aversa similarly opposed those who conceded that it was a celestial phenomenon but denied that it was newly generated in the sky. In the manner of Amicus, the defenders of this opinion insisted that the new star had been there all the time, but had remained hidden from our view until somehow rendered visible. In this explanation, the traditional assumption was made that the ether is not of equal density everywhere and that stars are denser accumulations of the ether. Some stars, however, were invisible because they lacked the requisite density to reflect sufficient light.

95. Clavius, *Sphere*, ch. 1, 1593, 211.
96. Ibid.
97. The arguments pro and con, with an ultimate resolution in favor of the sublunar nature of comets, appear in Aversa, *De caelo*, qu. 33, sec. 4, 1627, 91–100.
98. For Aversa's arguments against locating new stars in the sublunar region, see ibid., sec. 3, 85–86.

But somehow, in the course of celestial rotations, other, denser parts of the heavens were added to or aligned with a previously invisible star, rendering the whole aggregation visible. Conversely, when these parts were withdrawn from the new star it gradually disappeared from sight and resumed its previously invisible state.[99] Among his refutations of this explanation,[100] Aversa observes that if it were true and all celestial motions were uniform, a given denser portion of the sky ought to be augmented and diminished in the same way in every one of its revolutions. Hence a new star ought to come into being and pass away in every revolution.

Firmly rejecting the opinion that the new star of 1572 was a miracle,[101] Aversa was convinced that new stars were really new productions in the sky. In a final, unusual, and seemingly popular opinion, with which he appears to agree, Aversa describes how they might come into being.[102] By the time Aversa wrote in the 1620s, many had assumed that the formation and disappearance of new stars was caused by condensation and rarefaction of celestial matter. For those who assumed that the heavens were composed of fluid matter akin to air – as Tycho did and, by the 1620s, many others (see Ch. 14) – a greater condensation of the fluid at one point in the heavens could produce a new star, and a subsequent gradual rarefaction would cause the star to disappear. Some supporters of solid celestial spheres offered a similar explanation but found it more difficult to explain how condensation and rarefaction could occur in solid spheres. Aversa, who believed in the solidity of the celestial spheres, was aware of the difficulty and sought to avoid it by abandoning altogether the idea that various parts of the heavens were distinguishable in terms of rarity and density. It was better to conceive the heavens as opaque (*opacitas*) and transparent (*diaphaneitas*), because those qualitative properties are more appropriate to the heavens than the quantitative properties, rarity and density.[103] Aversa makes his point by com-

99. Ibid., 86, col. 2. According to Aversa, Franciscus Vallesius, who upheld this interpretation, invoked Genesis 2.1 and Ecclesiastes 3.14 to show that God would not create a new star. In the former, we learn that God completed the heaven and the earth; in the latter, we are told that all of his works are eternally preserved. It followed that God would not create a new star nor allow one to be generated.
100. Ibid., 86, col. 2–87, col. 2.
101. This is the seventh of eight opinions that Aversa describes (ibid., 89, col. 1). In the absence of any seemingly plausible natural explanation, the Coimbra Jesuits, as we saw, thought it reasonable to invoke a miracle. But Aversa insisted that the subsequent appearance of other new stars tended to weaken the case for a miracle. Moreover, when God performed the celestial miracles described in Scripture, the reason for them was made evident. But what was the divine purpose of the new star? No satisfactory response had been given, and Aversa opted to omit God from his explanation. Later in the seventeenth century, Franciscus Bonae Spei [*De coelo*, comment. 3, dub. 4], 1652, 10, col. 2, argued similarly against the miraculous explanation of the new celestial phenomena.
102. The eighth opinion in Aversa, ibid., 89, col. 2–90, col. 1. Because he mustered serious objections to all but the eighth opinion, it appears that Aversa favored it over the others, although he fails to make his support explicit.
103. Mastrius and Bellutus [*De coelo*, disp. 2, qu. 2, art. 3], 1727, 3:495, col. 1, par. 76, mention Aversa and seem to adopt his position. They insist that the heavens are every-

paring crystal and wood. Although crystal and glass are very dense, they are not opaque but transparent, whereas wood, which is more opaque than crystal or glass, is less dense than either. What makes the heavens transparent like air is their great density, or solidity. By analogy with crystal, celestial transparency is evidence of solidity rather than rarity.[104]

But Aversa thinks it inappropriate to say that stars are more solid than the heavens or that the heavens are more solid at some times than at others. Changes in the heavens cannot occur by mutations of substance or quantity but only by qualitative means. A qualitative change occurs when a diaphanous part of the heavens becomes somewhat opaque and can thus receive, and presumably reflect, sufficient light to become visible as a star (just as, presumably, the dense crystal becomes slightly opaque so that it reflects light). The new star will gradually disappear as its opacity diminishes and it ceases to reflect light.[105] But what could make a part of the heavens opaque and then diaphanous? Aversa admits that the cause of such effects – whether the result of something very powerful or a particular alignment of stars – is unknown, although he reports, with considerable skepticism, that some astrologers have attributed this process to certain planetary aspects.

Despite the concession that new stars were unequivocally celestial phenomena, Aversa denied that substantial or quantitative changes could occur in the celestial region. By rejecting the essentially quantitative concepts of rarity and density and replacing them with the presumably qualitative concepts of opacity and diaphaneity, Aversa believed that he had somehow preserved celestial incorruptibility. In effect, Aversa denied the existence of a quantitative sense of rarity and density in the heavens, but he nevertheless retained the idea of rarity and density by substituting for them the concepts

where of the same density but differ with respect to opacity and diaphaneity: "Dicimus astra omnia non necessario ex densiori materia constare quam sunt coeli; imo eiusdem densitatis, sed differre penes opacitatem et diaphaneitatem, ita ut astra sint partes coeli opacae, orbes vero sint omnino diaphani." Although, as we saw earlier in this chapter, Serbellonus believed "the common opinion is that in the heavens there is true rarity and density," his own view seems more in agreement with Aversa's interpretation (Serbellonus, *De caelo*, disp. 1, qu. 3, art. 2, 1663, 2:35, col. 1–36, col. 1).

104. Mastrius and Bellutus, ibid., 3:495, col. 1, par. 75, repeat the same example. They argue that what is opaque is dense, but what is dense may not be opaque, "as is obvious in a crystal, which is denser than wood but is nevertheless lucid and transparent" (quod opacum est densum est, non tamen e contra, ut patet in crystallo, quod densior [in place of *densivo*] est ligno et tamen perlucidum ac diaphanum est. Verum est tamen quod aliquando densum sumitur pro opaco).

105. In their joint work, Mastrius and Bellutus adopt the same opinion and even mention Aversa and Amicus. They explain that "with regard to the stars newly seen, we say that they have occurred from a certain accidental mutation made in the heavens from a certain concourse of stars unknown to us with respect to opacity and diaphaneity, so that the part that was previously transparent [*diaphana*] emerged opaque. Thus it could reflect light to us and be seen." And, on the basis of parallax, they, like Aversa, also placed the new star in the celestial region, at approximately the distance of the Sun but definitely not among the fixed stars. See ibid., qu. 3, 3:500, col. 1, par. 114.

of opacity and diaphaneity, which signified qualitative and accidental changes caused by the reflection (or lack of reflection) of celestial light.[106]

Although there were certainly other arguments in support of celestial incorruptibility, those that have been described here represent a good sampling of the arguments that scholastics believed most effective in the seventeenth century.[107] During the first three-quarters of that century, most scholastic authors continued to defend celestial incorruptibility. But by no means all did so, as we learn from Poncius, who, after mentioning certain modern astronomers – Tycho, Cornelius Gemma, Helisaeus Roëslin, Christopher Rothmann, and Kepler – who assumed that comets were generated anew in the heavens, declares that there were "even several Peripatetics" who agreed with them and inferred from this that the heavens are not ungenerable and incorruptible.[108] We shall return to this subject (Ch. 12, Sec. III.1) after we have examined scholastic ideas about celestial matter.

106. For more on Aversa's use of opacity and diaphaneity, see Chapter 17, Section IV.3.ii ("Averroës' Explanation").

107. In the 1590s, Galileo himself had defended celestial incorruptibility with quite similar but even more traditional arguments, arguments that were representative of the Jesuit theologians and natural philosophers at the Collegio Romano between 1570 and the early 1590s. See Galileo, *De caelo*, qu. 4 (J) ("Are the heavens incorruptible?"), 1977, 93–102. His relations with the Jesuits at the Collegio Romano are described in Wallace, 1981, 281, 308–309. For a list of the Jesuit authors on whom Galileo may have relied, see Galileo, ibid., 12–21. Of the group, Clavius is the best known.

108. "Alii recentiores Astronomi ut Tycho Brahe, . . . Cornelius Gemma, Helisaeus Roëslin, Christophorus Roshmannus [Rothmann], Keplerus, et alii, plures etiam Peripatetici, existimant cometas generari in caelo de novo atque adeo caelum non esse ingenerabile ac incorruptibile." Comments by Poncius, *Sentences*, bk. 2, dist. 14, qu. 3, 1639, vol. 6, pt. 2:738, col. 1.

11

Celestial perfection

Earlier we considered the question of the perfection of the world as a whole, touching only briefly on the perfection of the heavens. Two aspects of celestial perfection may be conveniently distinguished for consideration in this chapter: (1) differences in perfection within the celestial region; and (2) the comparison of the celestial region as a whole to the sublunar region, especially to the earth and living things. A third aspect, involving the governance of the terrestrial region by the celestial, is considered later in this volume.[1]

I. Intracelestial perfection

1. Are all celestial bodies in the same irreducible species?

The first aspect of celestial perfection essentially involves the following questions: if the whole celestial region is incorruptible, are all celestial bodies therefore equally perfect, or do they differ substantially?[2] And if they are all equal, can hierarchical gradations in perfection be distinguished in the heavens among the differently located celestial bodies? Potential substantive differences among celestial bodies was thus an important issue, as evidenced by the unusual question that emerged inquiring whether or not all the stars and planets belonged to the same irreducible, or, as it was often described, "most special," species (*species specialissima*).

Although both Jean Buridan and Nicole Oresme argued that the two basic opinions – that it was or was not the case that all celestial bodies belonged to the same species and therefore had the same nature – were both probable but that neither was demonstrable,[3] most chose to deny the claim

1. Until the seventeenth century, celestial incorruptibility implied unequivocally that the celestial region as a whole was more noble, and therefore more perfect, than the terrestrial part of the cosmos. The most significant consequence of this belief was the conviction that by virtue of their perfection and unchangeability, the heavens governed the incessant change of the terrestrial region.
2. This section is based largely on Grant, 1985c.
3. Buridan [*De caelo*, bk. 2, qu. 14], 1942, 184; Oresme [*De celo*, bk. 2, qu. 10], 1965, 608. Oresme also offered six criteria for judging whether things did or did not belong to the same species (ibid., 606). In his estimation, for celestial bodies to belong to the same species they must have an essential similarity, be equally noble, differ only accidentally, and be represented by the same name, concept and definition. Throughout the question, Oresme maintains an even hand and does not incline to either opinion. In this Buridan differs.

that all celestial bodies belonged to a single species. In this they opposed the opinion of Averroës, who believed that because they possessed the same essential properties, celestial bodies were like individuals in the same species. It was on this basis that Averroës was prepared to argue that if one planet was found to be spherical (for example, the Moon), it followed necessarily that all the others must also be spherical.[4] Averroës preferred this interpretation to Avicenna's, which classified each celestial body as the sole member of its own species and all the species as members of the same genus. For Averroës, this implied that as an individual species, each celestial body would have a distinct form that caused it to differ from all other celestial bodies. Therefore each celestial body must be composed of matter and form. But bodies composed of matter and form were subject to change, that is, to generation and corruption. Therefore celestial bodies would have been derived from something prior to themselves and would ultimately be transformed into something else, all of which Averroës, as a proper Aristotelian, deemed impossible.[5]

If all celestial bodies were members of the same species and were essentially identical, how could the planets differ in their motions and be seemingly diverse in other ways? These differences, Averroës argued, are like those found between the left and right side of the bodies of various species of animal and of human beings.[6] All differences are presumably of an accidental kind.[7]

Few scholastic natural philosophers were prepared to accept Averroës' analysis[8] or to remain relatively neutral, as did Buridan and Oresme. Indeed, by the late sixteenth century, the Coimbra Jesuits declared that "the greater and better part of philosophers" defended the contrary opinion, that the celestial bodies differed in species.[9] Galileo, who was a young contemporary

Despite his neutral stance, he seems to have believed that the celestial orbs and planets were not in the same ultimate or irreducible species. For although he allowed that all the orbs and stars were of the same nature and species with respect to the genus of substance and all belonged to the genus of simple bodies (ibid., 188), he also shows at some length (ibid., 189–190) that those bodies varied sufficiently among themselves so that they were different species within the same genus (some of his opinions are cited later in this chapter). Although Albert of Saxony [*De celo*, bk. 2, qu. 19], 1518, 113r, col. 2–114v, col. 1, drew most of his opinions from Buridan, he does not say that both opinions are equally probable and that neither is demonstrable. He utilizes Buridan's text to indicate his belief in the differences among celestial bodies.

4. See Averroës [*De caelo*, bk. 2, comment. 59], 1562–1574, 5:138r, col. 2.
5. See ibid., comment. 49, 5:131v, col. 2.
6. "Diversitas autem partium motuum non facit diversitatem formarum specificarum quoniam diversitas motuum est ex modo diversitatis animalium in parte dextra et sinistra et manifestum est quod istae diversitates inveniuntur in eadem specie eius, id est, in homine." Ibid.
7. For ten arguments defending the thesis of a single species for all celestial bodies, see Buridan, *De caelo*, bk. 2, qu. 14; Oresme, *De celo*, bk. 2, qu. 10, 1965, 608–620, offers eight.
8. One who did was John of Jandun [*De coelo*, bk. 2, qu. 13], 1552, 30v, col. 2–31v, col. 1.
9. "Nihilominus contrariam opinionem [that the celestial bodies belong to different species] veram arbitramur quam maior meliorque pars Philosophorum tuetur." Conimbricenses [*De coelo*, bk. 2, ch. 5, qu. 1, art. 2], 1598, 257.

of the Coimbra Jesuits, also strongly supported the theory of one species for every planet and its orb.[10] The many scholastic partisans of a distinct species, or *species specialissima*, for each celestial body, were convinced that differences among celestial bodies were more than accidental or superficial. One of the most common and basic distinctions was made between the Sun, which was self-luminous, and the Moon and all other planets, which were not. Indeed, most believed that the Sun was the source of light for all other planets.[11] Such a fundamental difference seemed to demand that the Sun be considered in a species different from the other planets. By detecting and assigning diverse natural properties to the planets and fixed stars – for example, some planets caused dryness and others wetness – and by assuming different powers for each planet depending on its position in its celestial path, astrologers and natural philosophers had provided powerful reasons for believing that celestial bodies differed in species.[12] Indeed, if there were no difference in species among the celestial orbs, there would be no reason for their arrangement in a certain order.[13] Moreover, if no substantial and specific differences existed among celestial orbs, then no differences should be assumed between the intelligences that move them. Thus any intelligence capable of moving one orb would be ipso facto capable of moving all the other orbs. Since intelligences are separate from the orbs they move and have no particular location, the same intelligence that moves Saturn could also move the orbs of the Sun and the Moon. Consequently, the assumption of one intelligence for each orb postulates many more intelligences than are required, thus rendering their existence vain and superfluous.[14]

Other arguments could be cited, but we have seen enough of the kinds

10. Galileo [*De caelo*, qu. 5 (K)], 1977, 144, 146. For a brief description, see Grant, 1983, 177–178.
11. Buridan, *De caelo*, bk. 2, qu. 14, 1942, 190, makes this point and cites a passage from the pseudo-Aristotelian *De proprietatibus elementorum*, where Aristotle is alleged to say that "the substance of the body of the Sun is different from the substance of the stellar [planetary] and lunar bodies. This is because the Sun has *lux* and *lumen* by itself, [whereas] the light of the Moon and the planets is derived from the Sun." In treating the same question ("Whether all stars belong to the same ultimate or most special species [*specie specialissime*]"), Albert of Saxony, *De celo*, bk. 2, qu. 19, 1518, 114r, cols. 1–2, who assumed that the planets and stars were in different ultimate species, cites and accepts as true the same passage from pseudo-Aristotle. Also arriving at the same conclusion were Johannes de Magistris [*De celo*, bk. 2, qu. 4], 1490, sig. K7r, and Pedro Hurtado de Mendoza [*De coelo*, disp. 1, sec. 3], 1615, 365, col. 2. However, as we shall see in Chapter 16, Section III, some scholastics argued for self-luminous planets.
12. Albert of Saxony, *De celo*, bk. 2, qu. 19, 1518, 114r, col. 2, and Buridan, *De caelo*, bk. 2, qu. 14, 1942, 190, emphasize the different properties and powers of the planets and fixed stars. On causing dryness and wetness and the property of the Moon for causing the tides, see Conimbricenses, *De coelo*, bk. 2, ch. 5, qu. 1, art. 2, 1598, 258. On attributing the powers of a planet to its motion and celestial position at any given time, see North, 1986, 56, 63.
13. "Quarto si orbes non different abinvicem specie, tunc non essent ordinati nisi accidentaliter unus super alium, quod tamen est falsum." Albert of Saxony, ibid., col. 1.
14. For this argument, see the second part of the fourth conclusion of Albert of Saxony, ibid, who probably derived it from Buridan, *De caelo*, bk. 2, qu. 14, 1942, 189.

that swayed most scholastics to assume significant differences among celestial bodies – planets, stars, and orbs.[15] With the widespread acceptance of significant differences, it seemed plausible to infer that various parts of the celestial region varied in degrees of perfection. With the great medieval emphasis on hierarchical differences in things of all kinds, both material and spiritual, the problem for natural philosophers was to determine the manner in which celestial perfection manifested itself in the known arrangement of the celestial bodies. As was so often the case, Aristotle's opinions served as the major points of departure for subsequent discussions.

2. *Aristotle on intracelestial degrees of perfection*

Although Aristotle wrote relatively little on the subject of intracelestial perfection, he was convinced that celestial bodies differed in degree of perfection. Unfortunately, the little he did say gave rise to conflicting interpretations.

In the first book of *De caelo*, Aristotle linked the perfection of the celestial region directly to its distance from the earth, or the sublunar region as a whole.[16] A celestial body was more perfect the farther removed it was from the earth.[17] Neoplatonists like Macrobius and the author of the pseudo-

15. For others, see Buridan, ibid., 186–188 (seven arguments) and his own opinions on 189–190; Oresme, *De celo*, bk. 2, qu. 10, 1965, 620–628; and Albert of Saxony, *De celo*, bk. 2, qu. 19, 1518, 113v, col. 2–114r, col. 2. For texts and summaries of the opinions of Thomas Aquinas, who believed that every individual celestial body was in a separate species, see Litt, 1963, 91–98.
16. "Thus the reasoning from all our premises goes to make us believe that there is some other body separate from those around us here, and of a higher nature in proportion as it is removed from the sublunary world." Aristotle, *De caelo*, 1.2.269b.13–18 [Guthrie], 1960. Here, in the thirteenth-century translation of Michael Scot, is one medieval version of this brief statement: "Et habens rationem potest ex omnibus predictis ratiocinari, quod aliud corpus est praeter ista corpora vicinantia et continentia nos, separatum ab eis, cuius natura nobilior est naturis eorum secundum suam remotionem ab eis et collocationem eius super ipsa." Michael's translation appears in Averroës, *De caelo*, bk. 1, text 16, 1562–1574, 5:11v, col. 2.
17. Averroës saw this as a higher-level analogy patterned on the relations among the four sublunar elements, where the nobler element occupies the higher location, as, for example, fire is higher, and therefore nobler, than earth. Averroës, *De caelo*, bk. 1, comment. 16, 1562–1574, 5:11v, col. 2. Because the earth and the Moon marked the beginning of the ascent of perfection for the inferior and superior regions, respectively, Averroës observes that in the *Liber de animalibus* Aristotle declared that "the nature of the Moon is similar to that of the earth" (Unde Arist. in *lib. de Animalibus* dicit quod natura lunae similis est naturae terrae); Averroës repeated this reference in *De caelo*, bk. 2, comment. 42, 127r, col. 1. Without mention of Averroës, Copernicus repeated this statement when he declared that "Aristotle says in a work on animals, the moon has the closest kinship with the earth." See Copernicus, *Revolutions*, bk. 1, ch. 10 [Rosen], 1978, 22. In a note (p. 360), Rosen, who gives Copernicus's source as Averroës' *De substantia orbis*, explains that Averroës and Copernicus were mistaken because "Aristotle's *Generation of Animals* (4.10.777b.24–27) regards the moon, not as akin to the earth, but as a 'second and lesser sun.' " In that very same passage, Aristotle also says that "the moon is a first principle because of its connexion with the sun and its participation in its light." *Generation of Animals* [Platt], 1984. Apparently with this same statement in mind, Averroës again refers to the *Liber de animalibus* in *De caelo*, bk. 2, comment. 32 (115v, col. 2) and com-

Aristotelian *De mundo* said virtually the same thing but changed their frame of reference from earth to first heaven. For them a body's degree of perfection was directly proportionate to its proximity to the first heaven, or God.[18] As the author of the *De mundo* expressed it,

> the heavenly body which is nighest to him most enjoys his power, and afterwards the next nearest, and so on successively until the regions wherein we dwell are reached. Wherefore the earth and the things upon the earth, being farthermost removed from the benefit which proceeds from God, seem feeble and incoherent and full of much confusion.[19]

In the second book of *De caelo*, Aristotle employed a different criterion for celestial perfection. Here the degree of perfection of a celestial body was determined by the number of motions it required to fulfill its cosmic objectives. As Aristotle explained, "the first heaven reaches it [i.e., its ultimate end or perfection] immediately by one movement, and the stars that are between the first heaven and the bodies farthest from it reach it indeed, but reach it through a number of movements."[20] In this scheme, the greatest perfection resides in the outermost sphere of the fixed stars, with its single daily motion, and the least perfection in the earth, which moves not at all.

From this arrangement, we expect that the degree of perfection between the two extremes corresponds to the ascending or descending order of the planets. Although Aristotle was aware of this expectation,[21] he explains (*De caelo* 2.12.291b.35–292a.3) that "the opposite is true, for the sun and moon perform simpler motions than some of the planets, although the planets are farther from the centre and nearer the primary body, as has in certain cases actually been seen."

Perfection could thus be interpreted according to distance from the earth, following the ascending order of the elements and planets; or it might be determined by the number of motions required to move a celestial body

ment. 42 (127r, col. 1), where he says that Aristotle linked the natures of Moon and earth because neither is self-luminous. Rosen assumes that the *Liber de animalibus* that Averroës attributed to Aristotle was the latter's *Generation of Animals*, but Averroës may have had some other work before him, perhaps not even by Aristotle.

18. For Macrobius and all Neoplatonists, the creative process was one in which there was a continual debasement and materialization as one moved farther "downward" from God, or the One. The celestial spheres were thus hierarchically ordered, those nearest the outermost reaches of the cosmos being more perfect than those nearer the earth. See Macrobius, *Dream of Scipio* [Stahl], 1952, 143–144.
19. See *De mundo* (*On the Universe*) 397b.26–32, in Aristotle [Forster], 1984. At least two Latin translations from the Greek original were made during the Middle Ages. For the Latin text of the passage, see Aristotle, *De mundo* [Lorimer], 1951, vol. 11, 1.2, 40 (anonymous translation) and 69 (translation of Nicholas Siculus).
20. Aristotle, *De caelo* 2.12.292b.22–25 [Guthrie], 1960.
21. When he declares (ibid. 291b.31–35): "considering that the primary body [i.e., the sphere of the fixed stars] has only one motion, it would seem natural for the nearest one to it to have a very small number, say two, and the next one three, or some similar proportionate arrangement."

around the sky, in which event the order of perfection could not follow the ascending order of the planets and orbs. These were the two major interpretations from Aristotle. Not only did they receive elaboration over the centuries, but new opinions were added. Other passages from the works of Aristotle not directly concerned with the problem of celestial nobility were also introduced and adapted to the literature on perfection.[22] Celestial perfection would eventually be compared to organic life, indeed even to the very lowest forms of it.

3. Medieval interpretations: Aquinas, Buridan, and Oresme

Although concern about celestial perfection became most intense during the course of the sixteenth and seventeenth centuries, there was certainly no lack of interest in that subject during the thirteenth and fourteenth centuries. Thomas Aquinas repeated, with apparent approval, Aristotle's declaration of ascending nobility. He added to it, however, a brief confirmation of the doctrine from Aristotle's *Physics* when he explained that "in the universe, containing bodies are related to contained bodies as form to matter and act to potency."[23] In subsequent centuries, this "container argument" was often cited in defense of ascending celestial nobility, as when Bartholomew Amicus invoked it by declaring that "as the whole is nobler than the part, so [also] is the container [nobler] than the contained, because it is related as whole to part."[24]

Although he considered the problem of celestial perfection difficult, Jean Buridan devoted a whole question to it in his *Questions on the Metaphysics*.[25] For Buridan, celestial perfection involved material celestial orbs, some of which carried planets, and an equal number of immaterial, separate intelligences, each of which moved one orb. Whatever level of perfection was assigned to the orb had also to be assigned to its intelligence. But how was perfection to be measured? Buridan felt compelled to eliminate direct methods for measuring the perfection of intelligences, because, as Aristotle argued in the twelfth book of the *Metaphysics*, each intelligence constitutes a separate species and no two of them are comparable, no more than a man

22. These were drawn primarily from Aristotle's *Physics* and *Metaphysics*.
23. Thomas Aquinas [*De caelo*, bk. 1, lec. 4], 1952, 23, par. 50. Although Thomas's reference to the fourth book of the *Physics* is to Aristotle's discussion on place, it should be emphasized that Aristotle does not consider whether the container, or place of a thing, is nobler than what is contained.
24. Amicus [*De caelo*, tract. 5, qu. 4, dubit. 2], 1626, 264, col. 2. As did Aquinas, Amicus also cites the fourth book of Aristotle's *Physics*. In the fifteenth century, Dominicus de Flandria [*Metaphysics*, bk. 12, qu. 7], 1523, sig. R3v, col. 2 (the work is unfoliated), following Thomas, accepts the container argument as upholding the ascending order of planetary nobility. Although he did not accept the ascending order of celestial nobility, Illuminatus Oddus, in the seventeenth century, presented it as the second of four arguments usually offered in support of ascending celestial nobility (see Oddus [*De coelo*, disp. 1, dub. 13], 1672, 35, col. 2).
25. "Whether the order of the celestial spheres in position is [also] their order in perfection." See Buridan [*Metaphysics*, bk. 12, qu. 12], 1518, 74r, col. 2–75r, col. 1.

could be compared to an ass. No increase or multiplication of qualities could enable an ass to equal or exceed a man.

Without a direct means of comparative measure, Buridan suggests that perfection be measured by effects. Because each intelligence was a motive power causing the motion of the sphere under its control, certain effects might be made the basis for comparison. Other things being equal, one might compare the perfection of motive intelligences on the basis of the perfection of the celestial bodies – that is, planets, stars, and orbs – that they move; or the magnitude of the mobiles, where the greater the magnitude, the greater the perfection of the intelligence that moves it; or one might compare the velocities of celestial bodies, where the greater the speed of the body, the greater is the perfection of its motive intelligence.[26]

To illustrate his meaning, Buridan compared the prime mover, or God, who moved the whole heavens, to the intelligence that moved the eighth sphere of the fixed stars in its motion of precession. Since the former completed one circulation in 1 day and the latter moved only 1 degree in 100 years, it followed that the ratio of their velocities, and therefore of the perfection of their respective motor powers, would be as 36,000 years to 1 day.[27]

The greater perfection of the prime mover was also measurable in other ways. It moved all the orbs, stars, and planets simultaneously, whereas every other intelligence moved only one partial sphere – that is, a single sphere within any complex of spheres that were together responsible for the motion of a single planet. Furthermore, the prime mover produced the quickest motion, since it moved the whole heaven in a day as compared to the mover of the Moon, which took a month; that of the Sun, which took a year, and so on.

If the prime mover is ignored, which of the motor intelligences that move only a single sphere, or at most a complex of spheres for a single planet, would be the most perfect? Would the order of perfection follow the ascending order of the planets and their orbs? Buridan acknowledges difficulties with this theory, the most serious of which concerned the Sun, which was always problematic in discussions about celestial perfection. On the face of it, the Sun seemed nobler than any other planet. Not only was it the greatest planet in magnitude, but it was generally thought to supply light to all, or most of, the other stars and planets. As the cause of generation in the sublunar world, it was the most patently active of all planets. Counting upward from the Moon, the nearest planet to earth, or downward from Saturn, the farthest planet from earth, the Sun was the fourth and middle

26. "Notandum est ergo quod rationabiliter ex multis arguitur maior perfectio potentie moventis primo ex maiori perfectione mobilis ceteris paribus; secundo ex maiori magnitudine mobilis ceteris paribus; tertio ex maiori velocitate motus ceteris paribus." Ibid., 74v, col. 1.

27. Ordinarily, God's infinite power was not comparable to the finite power of an intelligence. In this example, however, Buridan assumes that, in causing the daily rotation, God exerts only finite power. Ibid., col. 2.

planet and therefore, as Themon Judaeus put it, like a "wise king in the middle of his kingdom," or "like the heart in the middle of the body."[28] The fact that astronomers related the motions of the other planets to the Sun seemed to further reinforce the latter's premier role. And, finally, in an allusion to Aristotle's linkage of nobility to fewer motions, Buridan explains that by virtue of requiring fewer motions than Saturn and Jupiter, the intelligence that causes the Sun's motion must be nobler than those that move those two superior planets. And yet three other planets – Mars, Jupiter, and Saturn – had orbits beyond it. If nobility were a function of ascending order, Mars, Jupiter, and Saturn would be nobler than the Sun, despite all the other arguments in favor of the Sun's greater nobility.

Other discrepancies, based on widely received astrological properties, also threatened the theory of perfection associated with the ascending order of the planets. Thus Venus and Jupiter were called "fortunate" planets, while Mars and Saturn were "unfortunate." Similarly Jupiter and Venus were characterized as better and more benevolent planets than Saturn, from which it could also be inferred that their motor intelligences were more perfect than Saturn's. From this standpoint, order of position was plainly not associated with perfection.

Despite these seemingly weighty inconsistencies and discrepancies, Bur-

28. See Hugonnard-Roche, 1973, 75, and, for the translation, Grant [Lindberg], 1978a, 279; also Buridan's allusion to it in Buridan, *De caelo*, bk. 2, qu. 22, 1942, 226; also Albert of Saxony's mention of it in *De celo*, bk. 2, qu. 6, 1518, 105v, col. 1. The figure of "the king in the middle of his kingdom" was rather commonly cited in the sixteenth and seventeenth centuries, as, for example, by Galileo, *De caelo*, qu. 2 (H), 1977, 77, who also describes it as "the heart of all the planets" (*cor omnium planetarum*) and then justifies the Sun as "king" and "heart" by asserting that "the king lives in the center of the kingdom and the heart is in the center of animals, so that they can provide equally therefrom for all the people and for the members." For Galileo's later assessment of the Sun, as a Copernican, see his statement quoted at the end of Chapter 17, Section IV.2, in the present volume. The metaphor was used by the Coimbra Jesuits, *De coelo*, bk. 2, ch. 5, qu. 2, art. 2, 1598, 255, who cite Albumasar's *Great Introduction*, tract. 3, differ. 3, as their source, but I was unable to locate the passage in Albumasar, 1506 (for more on Albumasar's treatise, see below, note 35 of this chapter). It was also used by Amicus, *De caelo*, tract. 5, qu. 4, dubit. 2, 1626, 264, col. 2 (for his statement, see this chapter, Sec. I.4); and by Mastrius and Bellutus [*De coelo*, disp. 2, qu. 3], 1727, 3:499, col. 1, par. 108. Among the laudatory epithets that Kepler applied to the Sun in his *Mysterium cosmographicum* (1596) were "heart of the universe" (*cor mundi*) and "king" (*rex*). Kepler [Duncan], 1981, 76 (Latin), 201 (English).

The metaphor of the Sun as a king in the middle of his kingdom (*rex in medio regni*) may have derived from Haly Abenregel's *De iudiciis astrorum* (Hugonnard-Roche, ibid, 74, n. 4). The description of the Sun as "heart" (*cor*) appears as early as Macrobius, who, in his *Commentary on the Dream of Scipio* (1952, 169), characterized the Sun as "the heart of the sky," which was subsequently likened to "the heart in animals" or the "heart in the middle of the body," as mentioned earlier. Without reference to animals, Chalcidius [Waszink], 1962, 152, lines 1–2, says that the Sun is the heart and vitals of the whole world ("Ideoque solem cordis obtinere rationem et vitalia mundi totius in hoc igni posita esse dicunt"). Albertus Magnus [*De caelo*, bk. 2, tract. 3, ch. 5], *Opera*, 1971, 5, pt. 1:153, col. 1, compared the Sun to a heart when he declared that "the philosophers assimilated the whole orb [i.e. all the celestial orbs] to an animal in which the Sun is the principal part and holds the place of a heart. For this reason also, the orb of the Sun is placed by nature in the middle [position] of the orbs, just like the heart in an animal."

idan eventually adopted the ascending order of celestial perfection. He rightly observed that Aristotle had failed to provide the specific perfection with respect to which the spheres were to be ordered positionally. Fortunately, the great Arabian commentator Averroës had provided a proper response that took into account the apparent discrepancies just described. Although the prime mover, God, exceeded all intelligences with regard to properties associated with perfection, the perfections of other intelligences varied, exceeding in some and exceeded in others. We can see the same thing operating in terrestrial things, as, for example, in our knowledge that "a man is absolutely more perfect than a horse and yet a horse exceeds him in magnitude, speed, and strength. And so, although the three superior planets are absolutely nobler than the Sun, yet it is not absurd that the Sun should exceed them in some properties."[29]

Although Aristotle had not decided on the specific perfections by which the planetary spheres and their intelligences could be arranged in ascending order, Buridan believed that Thomas Aquinas had already proposed reasonable criteria for such an arrangement. For Thomas, an intelligence moving a superior sphere was more perfect than an intelligence moving an inferior sphere, a judgment he based on the container theory, which assumed that because a superior sphere contained an inferior sphere, the former must be prior to, and nobler than, the latter.

Buridan also resorted to Averroës' concept of comparison to establish the greater perfection of the planets and orbs farthest from the earth. For example, although lower spheres are quicker in their movements, superior spheres are greater in magnitude and have to traverse vastly greater distances on their circumferences. In this way, the superior sphere retains its overall absolute superiority, because, as Buridan expressed it, "having posited that an inferior is moved more quickly than a superior, nevertheless it is not moved so much more quickly than the superior sphere is greater [in magnitude]."[30] To further reinforce the concept of ascending perfection, Buridan believed that in governing the inferior world, the three superior planets, and the Sun as well, controlled the being and permanence of things, whereas the three planets below the Sun controlled the motions and mutations of

29. "Sed tunc ad ea que obiiciebantur de sole de venere de iove respondet Commentator [i.e., Averroës] quod diversa est habitudo primi motoris ad alios motores et aliorum motorum adinvicem. Primus enim motor, scilicet Deus, excedit omnes alios motores secundum omnes proprietates perfectionales que eis possunt attribui, licet non sic est de aliis intelligentiis adinvicem. Immo intelligentia simpliciter perfectior bene exceditur ab alia in aliquibus proprietatibus perfectionalibus, sicut etiam nos videmus in istis inferioribus. Homo enim est simpliciter perfectior equo et tamen equus excedit eum in magnitudine et in velocitate et in fortitudine. Et ita etiam quamvis tres planete superiores sint simpliciter nobiliores sole, tamen non est inconveniens quod sol excedat eos in aliquibus proprietatibus sicut arguebant rationes que de hoc fiebant et ita etiam diceretur de iove et venere ad saturnum." Buridan, *Metaphysics*, bk. 12, qu. 12, 1588, 75r, col. 1. For Duns Scotus's earlier attempt to compare the absolute perfection of different qualities, see this chapter, Section II.1.

30. Ibid., bk. 12, qu. 12, 74v, col. 2–75r, col. 1.

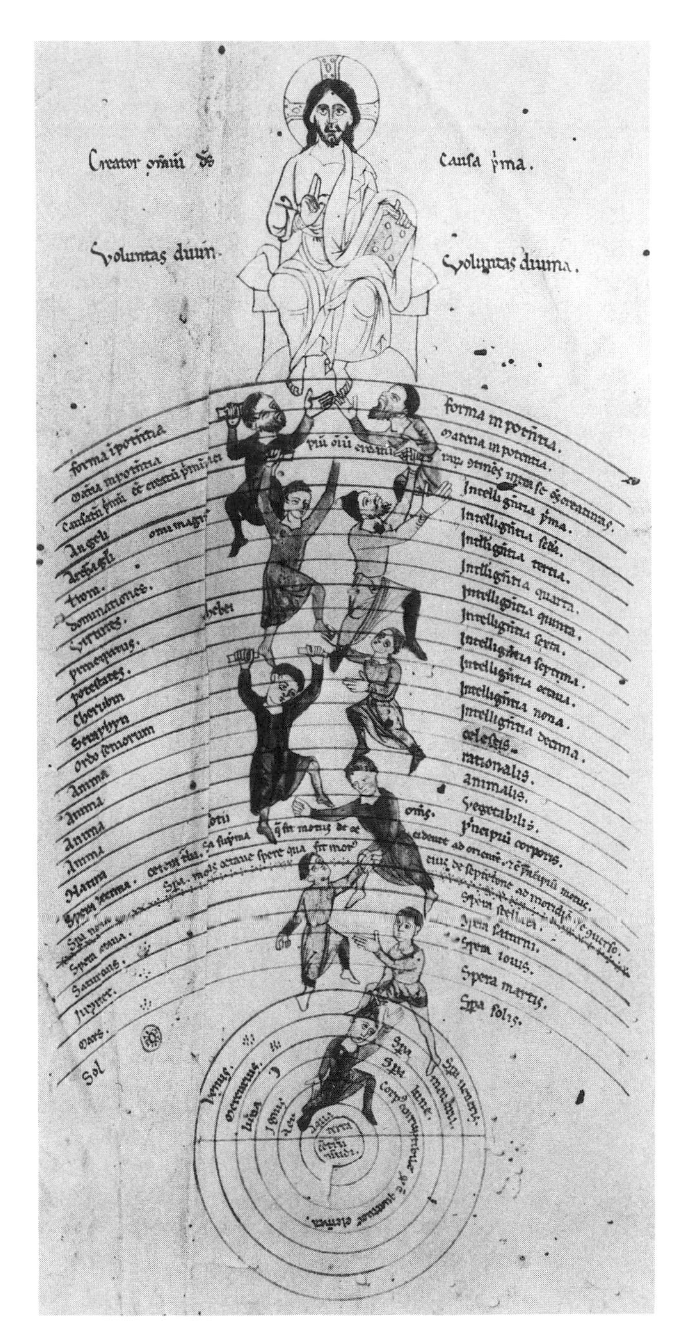

Figure 7. A twelfth-century hierarchy of being. Beginning with the earth as center, it ascends through the elements and planets, continuing on through the various levels of soul and ten levels of intelligences (the latter based on Avicenna's *Metaphysics*, but inverting his order), terminating in the "Creator of all things," depicted here as Christ enthroned. (Bibliothèque Nationale, MS. lat. 3236A, fol. 90r. Further details in Murdoch, 1984, 334–335.)

those same things.[31] Because "it is nobler to have a property concerning the being and permanence of things than [a property] concerning the motions and mutations of things,"[32] it followed, on the basis of this principle, that the three superior planets are nobler than the Sun and the three inferior planets.

But how did Buridan and others reconcile the principle of ascending nobility and perfection with Aristotle's order of perfection based on the number of motions required by a celestial body to complete a single revolution? Were the Sun and Moon, which used fewer motions than Saturn and Jupiter, nobler than the latter? If so, an ascending order of nobility for celestial bodies would be impossible. Buridan considered this problem when he inquired "Whether the Sun and Moon ought to be moved with fewer motions than other planets."[33] At the end of the question, Buridan proclaims the principle on which he based his solution. "It is not necessary universally," he declares, "that nobler things should have more or fewer actions; but [it is necessary] that in striving for the same end [or goal] nobler things have fewer actions." Since motions are actions, Buridan had the basis for a solution to the problem posed by Aristotle.

It required that the celestial region be divided into two parts. The first, and nobler, part embraced the sphere of the fixed stars and the three superior planets. Their goal and function was to control existence, the permanence and duration of things, and the overall well-being of the universe. To achieve these identical ends, the sphere of the fixed stars, which was the outermost and most noble body, performed only one motion, whereas the other three planets required more than one motion, although whether Saturn required fewer than Jupiter and Jupiter fewer than Mars is ignored.

By contrast, the Sun and the three inferior planets, Venus, Mercury, and the Moon, have a different goal, namely to control change in all things below the Moon. Buridan divides them into two groups with somewhat different objectives. The Moon is distinct from the other three and has as its cosmic mission the preparation of sublunar matter for the reception of the good influences of the celestial region. For this purpose, or, as Buridan put it, for this "inferior property," the Moon requires few motions. Of the other three planets, the Sun, as the most distant from the earth, has fewer

31. The same idea appears in Buridan, *De caelo*, bk. 2, qu. 21, 1942, 225. Albert of Saxony (in his questions on *De celo*, bk. 2, qu. 17, 1518, 92v, col. 1), who seems to have derived much of his material directly from Buridan's *De caelo*, repeats the same statement. In the fifteenth century, Dominicus de Flandria, *Metaphysics*, bk. 12, qu. 7, 1523, sig. R3v, col. 2, presented the same argument, except that he associated the Sun with the three lower planets as essentially a cause for mutations rather than preservation; the Coimbra Jesuits, *De coelo*, bk. 2, ch. 5, qu. 2, art. 3, 1598, 256, also linked the Sun with the lower planets.

32. "Quoniam nobilius est habere proprietatem super esse et permanentiam rerum quam super motus et mutationes earum." Buridan, *Metaphysics*, bk. 12, qu. 12, 1588, 75r, col. 1.

33. Buridan, *De caelo*, bk. 2, qu. 21, 1942, 223–225. Albert of Saxony, *De celo*, bk. 2, qu. 17, 1518, 92r, col. 2–92v, col. 2, treated the same question.

motions than Venus and Mercury; and perhaps Venus requires fewer motions than Mercury.

Even if astronomers had assigned more motions to Saturn than to Mars, or to Mercury than to Venus, Buridan would undoubtedly have preserved the ascending order of perfection by invoking properties that would have ranked Saturn as nobler than Mars and Venus as nobler than Mercury. To achieve this, he would have had only to invoke Averroës' principle that excess of absolute nobility did not preclude a less noble body from exceeding a more noble body in one or more of a range of qualities or properties. Thus even if Mars required fewer motions to achieve the same ends as Saturn, the latter would exceed Mars in the most important properties and preserve its greater absolute perfection.

Although the ascending order of the planets may have been the most popular theory of celestial perfection during the Middle Ages, some opted for the opinion Buridan rejected, namely that the Sun is the noblest and most perfect planet. An emphatic supporter was Nicole Oresme, who insisted that "the perfection of the heavenly spheres does not depend upon the order of their relative position as to whether one is higher than another."[34] Not only is the Sun, which lies in the middle position among the planets, "the most noble body in the heavens . . . more perfect than Saturn or Jupiter or Mars, which are all higher up," but "it is probable that Jupiter is more perfect than Saturn, and the Moon more so than Mercury." Thus not only did Oresme identify the Sun as the most noble and perfect planet, but he also accepted the consequence of his assumption by rejecting the ascending order of nobility. Although Oresme's *Le Livre du ciel et du monde* was probably unknown to early modern scholastic authors, versions of his ideas were widely adopted in the seventeenth century.

4. Continuation and extension of medieval interpretations in the sixteenth and seventeenth centuries

Whereas Nicole Oresme found the theory of the ascending order of nobility or perfection to be incompatible with the assumption of the Sun as the most perfect planet, the Coimbra Jesuits integrated the two concepts. To do so, however, they had to place a special interpretation on the three orbs that were usually assigned to the Sun in the Aristotelian–Ptolemaic system and whose individual motions interacted to produce the Sun's observed motion. Of these, the middle orb, in which the Sun was actually located, was not only assumed nobler than both of its companion solar spheres but also nobler than all other planetary orbs and planets. The other two solar orbs and all other celestial spheres were assumed progressively nobler according to their respective positions: that is, with the exception of the Sun's middle

34. See Oresme, *Le Livre du ciel*, bk. 2, ch. 22, 1968, 507.

orb, which was the most perfect of all orbs and planets, nobility or perfection varied directly with distance from the earth.

But if the Sun is the noblest planet, why was it placed fourth in the sequence of planets, rather than seventh and uppermost? According to the Conimbricenses, the Sun was wisely located in the middle so that it could diffuse its power, light, and heat in every direction, "as if from the middle of a kingdom. For if it were in the seventh sphere, its great distance from the earth would cause its light to heat inferior bodies much more weakly, and nearly all things would become hard from coldness. And if the Sun were in the first sphere, all things would burn up from its proximity."[35]

Retaining the two basic ideas of the Conimbricenses – the Sun as noblest planet, with the other planets and spheres arranged in ascending order of nobility – Bartholomew Amicus presented a more elaborate and interesting justification for his position.[36] In an effort to explain how the perfect order of the planets is preserved in the Ptolemaic system, the truth of which he assumes, Amicus presents four possible explanations, the last three of which he integrates into a final theory of his own.[37] First he presents the well-known arguments in favor of a hierarchical universe in which the lowest beings are linked with the highest by a series of graduated steps. Such a hierarchical system, which was always associated with Thomas Aquinas, could operate only if inferior beings are subject to superior beings. It was on this basis that astrologers assigned more perfect and nobler effects to planets and spheres in proportion to their distance from the earth. The container argument, which was considered a powerful support for the theory of ascending nobility, is also mentioned. Each containing sphere is superior to the sphere it contains.

35. Conimbricenses, *De coelo*, bk. 2, ch. 5, qu. 2, 255, in a question titled: "Whether as the celestial globes are higher, they are [also] of a nobler nature." As the source of this statement, the Conimbricenses correctly cite the Arabic astronomer and astrologer Albumasar (Abu Ma 'shar) (d. ca. 886), *Introductorium*, tract. 3, differ. 3, 1506. The text, which appears on 16v when the count is begun from the title page of this unfoliated edition, is as follows: "Ex his itaque patet quod si sol usque ad nonam speram sublimatus esset vel usque ad lunarem orbem humiliatus vel inde frigore vel hinc calore nimio mundum stare non posse. Quamobrem providus auctor omnium deus: solem tanquam universalem corporee vite vomitem in media mundi regione medium posuit." The *Introductorium* was twice translated into Latin in the twelfth century, the first time in a literal translation in 1133 by John of Seville, the second time in a somewhat abbreviated version by Hermann of Carinthia in 1140. John of Seville's translation was never published, whereas Hermann's was published three times, twice in Augsburg by Erhard Ratdolt (1489 and 1495) and once in Venice (1506), the edition used here. For information about the translations and editions, as well as the considerable influence which Albumasar had in the twelfth century, see Richard Lemay, 1962. On page 113, Lemay summarizes the relevant passage and, in a note, quotes the Latin text from John of Seville's translation. See also David Pingree, "Abu ma'shar al-Balkhi, Ja'far ibn Muhammad," in *Dictionary of Scientific Biography*, 1:35, col. 2.
36. Amicus, *De caelo*, tract. 5, qu. 4, dubit. 2, 1626, 264, col. 1–265, col. 1.
37. The first explanation, which he rejects, assigns the order of perfection to the pure free will of the creator, thus eliminating any preordained order of perfection. God, it seems, merely placed each sphere and planet where he pleased. If so, "each heaven attained that place which God gave to it." Ibid., 264, col. 1.

Despite this traditional array of arguments in favor of the ascending order of nobility, the most popular theory of perfection, Amicus expressed reservations about it. For "against this opinion stands the perfection of the Sun," which is greater than Jupiter and Saturn and yet is placed under them. Indeed, it is "placed in the middle, like a king in his kingdom and the heart in the body. Nor, indeed, is the excellence of the Sun over all other planets and the fixed stars to be doubted, since all receive their light, the most perfect of all qualities, from it" (Amicus [*De caelo*, tract. 5, qu. 4, dubit. 2], 1626, 264, col. 2).

Although convinced that the Sun was the most perfect planet and therefore superior to the planets above it, Amicus did not, as we shall see, abandon the theory of ascending nobility. He used it in conjunction with another theory – the third of the four interpretations he presents – which we must now examine. In this view, the order of the heavens is not maintained by nobility but by reason of symmetry (*commoditas*) and what is good and proper for the whole (*bonus universi*). The Sun has been placed in the middle of the planets so that it can diffuse its light to the stars and planets above and to the inferior planets and the sublunar region below. Preservation of the whole necesssitated a middle position for the Sun: if it were too low, "all things would burn up from the heat; and if it were placed too great a distance away, all things would freeze."[38] The Sun should also be located in the middle because it is the measure of other motions. Thus the three superior planets (Mars, Jupiter, and Saturn) agree in their epicyclic motion with the Sun, while the three inferior planets (Moon, Mercury, and Venus) agree in their deferent motions with the Sun.[39]

The stage was now set for Amicus to incorporate the second and third interpretations into a fourth, which would represent his own view. The order of the planets was primarily determined by symmetry (*commoditas*) and secondarily by nobility (*dignitas*). Symmetry and the common good of the whole took precedence over nobility whenever necessary, as, for example, when the Sun was placed as the middle planet. However, when preservation of the common good was not at issue, nobility always prevailed. Thus, with the exception of the Sun, all other planets and orbs,[40]

38. Ibid., col. 2. As holders of this opinion, Amicus cites Albumasar, "Ricc." (perhaps Richard of Middleton), "Saxony" (probably Albert of Saxony), and Christopher Clavius. As we saw, the Conimbricenses also adopted it.

39. "In medio Sol est constitutus quia est regula et mensura aliorum motuum diversa tamen ratione. Nam Mars, Iuppiter, et Saturnus ratione epicycli cum Sole in motu conveniunt. Luna vero Mercurius et Venus in orbibus deferentibus cum motu Solis conveniunt. Unde est in medio ut tres superiores planetas ab inferioribus segreget cum uniformitatem, cum utroque genere servet, sed diversa ratione." Amicus, *De caelo*, tract. 5, qu. 4, dubit. 2, 1626, 264, col. 2–265, col. 1. This argument, which Amicus attributes to Clavius, depends on the imposition of the earth's annual orbit on the motions of the outer planets as epicycles and on the inner planets as deferents. I am indebted to Danny Burton for this interpretation.

40. Amicus disagreed with those who distinguished the planets from their orbs; he argued that "the superior orbs are always nobler than the inferior contained [orbs], although the stars [i.e., planets] that are in the superior orbs are not nobler, as is obvious from the

from the Moon to the sphere of the fixed stars, were arranged in order of nobility, or perfection.

5. Challenge to the idea of intracelestial, hierarchical perfection

As is evident from the ideas of Nicole Oresme, the appearance of the Copernican heliocentric system was not an essential prerequisite for the abandonment of the theory of the ascending order of celestial perfection or for the subversion of its credibility. Nevertheless, the rejection of intracelestial perfection probably derived not from the evolution of theories debated in the Middle Ages but from the impact of the Copernican theory and Galileo's telescopic observations as they affected the status of the earth.

From the very nature of its fundamental assumptions, the Copernican theory was bound to have an impact on traditional ideas about celestial nobility and perfection. It made the earth a planet, and therefore as perfect or imperfect as the other planets.[41] Consequently, Copernicans had no good reason to perpetuate the traditional medieval belief that the earth was the least perfect body of the universe[42] and in no way comparable to the perfection of the celestial bodies. But it was Galileo's telescopic discoveries that furnished vital empirical support for what had previously been only an implication from the structure of the Copernican theory itself. Observations of celestial bodies led Galileo to conclude that the earth was no less perfect than any of the traditional planets. By insisting that all celestial bodies were as alterable and subject to change as the earth, Galileo had destroyed any basis for differential perfections and nobilities. Indeed, Galileo considered "the earth very noble and admirable precisely because of the diverse alterations, changes, generations, etc. that occur in it incessantly, . . . and I say the same of the moon, of Jupiter, and of all other world globes."[43] Because his name rarely occurs in discussions of celestial perfection, the effect of Galileo's radical ideas on scholastic authors, who retained the geocentric system, is uncertain.

Even before Galileo's *Dialogue* of 1632, but after his telescopic discoveries and his statement in *The Assayer* in 1623, seventeenth-century scholastics

Sun." Ibid., 265, col. 1. Amicus held that whether or not the planets and their orbs were of the same species, each planet and the orb that carried it were at the same level of nobility or perfection.

41. In *De revolutionibus orbium caelestium*, bk. 1, ch. 10, Copernicus declares that "I have no shame in asserting that this whole region engirdled by the moon, and the center of the earth, traverse this grand circle amid the rest of the planets in an annual revolution around the sun." Copernicus, *Revolutions* [Rosen], 1978, 20.
42. Distinguishing five degrees of perfection, Oresme declared that "the first and fifth are God and the earth respectively, and both are absolutely motionless, the one because of His great and infinite perfection and the other because of its very small degree of perfection." Oresme, *Le Livre du ciel*, bk. 2, ch. 22, 1968, 507.
43. Galileo, *Dialogue*, First Day [Drake], 1962, 58–59. In *The Assayer* of 1623, Galileo only hints at a repudiation of celestial perfection but does not actually discuss it. See, however, his remarks on the noblest shape of the sky in Drake and O'Malley, 1960, 279.

continued the medieval attacks against the idea that nobility or perfection increases with distance from the earth. Thus Raphael Aversa adopted an opinion that embraced both aspects of Oresme's attack.[44] Like many others, Aversa assumed that the Sun was the most noble and excellent of the planets. But unlike most of them, he refused to believe that the nobility of the other planets increased with their distance from the earth. The traditional astrological attributes assigned the planets made ascending nobility implausible. Saturn, which is the highest planet, is not nobler than Jupiter, nor is Mars, which is higher than Venus, nobler than the latter. Why should we think them nobler when both Saturn and Mars are considered harmful planets whereas Jupiter and Venus are assumed beneficial? Indeed, we could arrange the order of nobility among the planets on the basis of a variety of operations and properties, all of which would be arbitarary. With the exception of the Sun, Aversa concluded that "there does not appear any definite rule for [determining] the [nobility of] the other planets."[45]

In a manner similar to Aversa, Johannes Poncius, the seventeenth-century commentator on the works of Scotus, declared that no specific differences could be assigned between the various heavens, even though he considered it very probable that the Sun was the most noble of all the planets.[46] Another Scotistic commentator, Illuminatus Oddus, proclaimed with little discussion that although the heavens differ in number – that is, there are distinct orbs and planets – they do not differ in essential perfection.[47] Thus did Poncius, Oddus, and Aversa challenge the idea of intracelestial differential perfection. What had been almost unthinkable in the long medieval cosmological tradition had now come to pass. It derived, in large measure, from the contributions of Copernicus, Tycho Brahe, and Galileo.

II. Celestial bodies compared to animate and inanimate sublunar bodies

We must now consider the second aspect of the problem of celestial perfection and examine the manner in which medieval and early modern scholastics compared the incorruptible celestial bodies to corruptible animate and inanimate sublunar bodies.[48] On this question, scholastic authors diverged radically from Aristotle. Not only was the celestial region more perfect than the terrestrial region, in Aristotle's judgment, but it was also

44. Aversa [*De caelo*, qu. 33, sec. 8], 1627, 117, cols. 1–2.
45. "Aliorum vero astrorum non apparet certa regula." Ibid., col. 1.
46. Poncius [*Sentences*, bk. 2, dist. 14, qu. 2], 1639, 6, pt. 2:731, col. 1. Poncius was a supplementary commentator on this volume of Scotus's works. In the comment to which I refer here, his name is given as the commentator on page 730, column 1.
47. Oddus, *De coelo*, disp. 1, dub. 13, 1672, 36, col. 1.
48. In this section, I draw heavily upon Grant, 1985c, 152–162.

alive[49] – indeed, it was divine[50] – from which he inferred that it was more perfect than all living things, including man.[51] Despite the general acceptance of Aristotle's cosmology by natural philosophers in the Middle Ages, they eventually challenged all of these claims. Indeed, the divinity of the heavens never formed a part of medieval cosmology, because it was considered incompatible with the Christian faith.

Having denied divinity to the heavens, scholastics soon inquired whether they were even alive. Although Thomas Aquinas doubted their animation, he did not consider the problem relevant to the faith nor even an important question.[52] During the Middle Ages, most natural philosophers denied that the celestial region was animate, and by the sixteenth and seventeenth centuries virtually all scholastic authors repudiated the idea that celestial bodies were alive.[53] Thus when scholastics compared the heavens with the terrestrial region, they had in mind a comparison between an inanimate heavens and an inanimate sublunar region. In such a comparison, few if any would have denied the greater perfection of the celestial over the terrestrial.[54]

1. Two major opinions comparing the celestial region to living things

At some point in the history of the problem of celestial perfection, scholastics came to inquire whether inanimate celestial bodies were superior to living things on earth. Because none would deny that man, with his rational,

49. For Aristotle's belief that the celestial bodies are alive, see *De caelo* 2.2.285a.29–30 and 2.12.292a.20. The triumph of the theory of an inanimate celestial region during the Middle Ages to the end of the fourteenth century is described by Dales, 1980, 531–550. Whether the heavens are animate or inanimate was widely discussed.
50. *De caelo* 1.3.270b.1–14; 1.9.278b.15–17; *Metaphysics* 1074b.1–14; *On the Parts of Animals* 1.5.644b.23–26; also see the next note.
51. "For there are other things much more divine in their nature even than man, e.g., most conspicuously, the bodies of which the heavens are framed." Aristotle, *Nicomachean Ethics* 6.7.1141b.1–2 [Ross], 1984.
52. See Litt, 1963, 108–109; also Dales, 1980, 543–544.
53. After denying that celestial bodies are alive, Aversa declares that "this is the common opinion of the theologians and the philosophers." Aversa, *De caelo*, qu. 33, sec. 7, 1627, 110, col. 1; see also qu. 35, sec. 8, 196, col. 1. Among others denying life to the heavens, we may mention the Conimbricenses (*De coelo*, bk. 2, ch. 1, qu. 2, art. 2 ["it is concluded that the celestial orbs are not animated"], 1598, 166–169 [mistakenly paginated 145]), who provide a long list of those who rejected animation and also inform us that almost all scholastic theologians rejected animation; Franciscus Bonae Spei [*De caelo*, comment. 3, disp. 3, dub. 5], 1652, 11, col. 1–12, col. 2; and Hurtado de Mendoza, *De caelo*, disp. 1, sec. 4, 1615, 366, col. 2. As a glance at the "Catalog of Questions" in Appendix I reveals, the question "Whether the heavens are animated" (qu. 128) was one of the most widely discussed during the approximately five centuries relevant to this study. It is discussed further in Chapter 18 ("On Celestial Motions and Their Causes"), where it plays a role in the determination of the motive causes of the celestial orbs.
54. Oddus, *De caelo*, disp. 1, dub. 13, 1672, 36, col. 1, expressed the common sentiment when he declared: "Comparando vero coelum cum sublunaribus certissimum est apud omnes corpora coelestia esse omnibus inanimatis perfectiora."

spiritual, and intellective soul, is more perfect than any inanimate celestial body,[55] the comparison usually focused on living things that had a sensitive and/or vegetative soul. Indeed, the lowliest of living things – such as plants, flies, and wasps – were frequently compared to the heavens.

Two major opinions emerged, one associated with Thomas Aquinas, the other with Duns Scotus. Although Thomas does not appear to have considered the problem explicitly, he seems to have believed that the celestial region was more perfect than animals created by putrefaction.[56] The latter were generated solely by celestial power – especially that of the Sun – without the aid of semen. Creatures generated by putrefaction – for example, flies and gnats – were generated by the action of a celestial virtue on the four elements. Since Thomas believed that celestial bodies contained the properties of the elements in a more perfect and excellent manner, it is likely that he considered the heavens more perfect than animals generated by putrefaction.[57] Because Thomas believed that animals generated by semen required in addition the concurrent action of celestial virtue, it is possible that he also considered the heavens more perfect than all animals except man. Mastrius and Bellutus lend credence to this interpretation when they report that most Thomists believed that the heavens are nobler than sensitive and vegetative life.[58]

Nicole Oresme emphatically supported the Thomistic position when he declared that "it is not necessary that every body be animate in order to be nobler than a living one." By way of example, he explained that "a sapphire or an emerald is nobler than a fly or a dirty earthworm or several other living bodies with a material soul." Indeed, even "an intellectual soul is not as noble as the heavens, which are an eternal body." Only the "spiritual, eternal soul is nobler than the celestial body."[59] It was a view that still found occasional supporters in the sixteenth and seventeenth centuries.[60]

Duns Scotus formulated a radically different approach. Within the context of a discussion on whether any accident in the Eucharist could remain without a subject in which to inhere,[61] Scotus established certain criteria for comparing the absolute perfection of different properties. Thus he compared quantity and quality with respect to substance. On the assumption that substance is the most perfect of beings, and on the further assumption that, in absolute terms, quantity is more akin to substance than is quality, Scotus concluded that "quantity is more perfect than quality."

55. Ibid.
56. Aquinas's various statements on putrefaction have been collected and analyzed by Litt, 1963, 130–143.
57. Litt, ibid., 135–136.
58. Mastrius and Bellutus, *De coelo*, disp. 2, qu. 3, 1727, 3:499, col. 1, par. 109.
59. Oresme, *Le Livre du ciel*, bk. 2, ch. 22, 1968, 503.
60. From Aversa, *De caelo*, qu. 33, sec. 7, 1627, 113, col. 1, we learn that Piccolomini, in his *Liber de caelo*, ch. 23, "affirms that, with the exception of man, the heaven is nobler than animated bodies."
61. What follows is drawn from Duns Scotus [*Sentences*, bk. 4, dist. 12, qu. 2], *Opera omnia*, 1639, 8:731–732.

Moreover, he believed it possible to compare two different objects with respect to two different properties, if both properties were also possessed by God. Scotus assumes that the more necessary a property is, the closer it is to God. Incorruptibility is a necessary property, and is close to God, who is the most perfectly incorruptible thing. From this Scotus inferred that the incorruptible, physical heaven is closer to God than any corruptible thing. But not only is God an absolutely incorruptible entity; he is also an absolutely perfect intellectual creature. The nearest intellectual creature to God is an angel, after which follows the finite human intellectual nature, followed by all sensitive natures (animals), which approximate more closely to intellectual natures than do nonsensitive things (plants). In the order of things based on intellect, a fly is therefore nearer to God than are the heavens. But in the previously described order, based on incorruptibility and corruptibility, the heavens are closer in perfection to God than is the fly.

Relying upon Scotus, Illuminatus Oddus insisted that something is said to be more perfect than another thing with respect to some particular quality because it is nearer, and more similar to, the perfect being, God. But when two things are compared to God with respect to two different qualities possessed by God, that is said to be absolutely more perfect which has the greater perfection in the nobler of the two attributes. Thus, if we wish to compare the perfection of celestial bodies with a living thing that has a sensitive and vegetative soul, we cannot compare them with respect to degree of life, since celestial bodies are inanimate. But we might consider whether the incorruptibility of a celestial body is absolutely more perfect than the level of life in, say, a fly or wasp. Of these two qualities or attributes, *life* and *incorruptibility*, Oddus argues that "life in God has a more perfect nature than eternity and incorruptibility, for life is of the quiddity of God, [whereas] eternity is [but] an attribute. Therefore animate things, which are nearer to God in degree of life than the heavens, which are nearer [to God] in incorruptibility, will be absolutely more perfect [than the heavens]."[62] For Illuminatus Oddus, then, even a plant, which possesses only a vegetative soul, is more perfect than any celestial body.

If we may believe Raphael Aversa, who was not a Scotist, the Scotistic interpretations just described formed the common opinion around the time Aversa published his work in 1629. According to Aversa, those who accepted this opinion believed that the incorruptibility of the celestial region

> made the heaven more perfect and noble than all inferior bodies, even [those that are] animate. And by reason of its form, [the heaven] is simply and absolutely more perfect than all other inanimate bodies. But simply and absolutely, the form of the

62. Oddus, *De coelo*, disp. 1, dub. 13, 1672, 36, col. 2. Mastrius and Bellutus, *De coelo*, disp. 2, qu. 3, 1727, 3:500, col. 2, par. 119, say much the same thing about the comparison between the qualities of life and incorruptibility. They also conclude: "it must be conceded as absolutely true that a mouse, a flea, and a plant are absolutely nobler than the heavens" (ibid., 501, col. 1, par. 123).

heaven is more ignoble than a soul [*anima*], and the heaven itself is more imperfect than an animate body.[63]

Aversa further argues that incorruptibility favors living things, not the heavens, since "our soul is more immortal than the form and substance of the heaven."[64] In his judgment, the heaven does not operate by prudence and wisdom, as a living thing might, but simply as a natural cause.[65] To operate in this manner does not require that the heaven be "alive and essentially more excellent than all inferior substances." Aversa thus accepted the essentials of the Scotistic position, as indeed did his contemporary Thomas Compton-Carleton, who insisted that "the heaven, which is incorruptible, is relatively more perfect than a man, who is corruptible. Nevertheless, absolutely it [the heaven] is more imperfect than a man, because man is a principle of more perfect operations."[66]

For some, including Johannes Poncius, it was difficult to explain why "the heavens are more imperfect than any living thing."[67] Poncius found it implausible that plants, which possessed only a vegetative soul, were more perfect than the heavens. He therefore rejected this claim.[68] But he conceded that sensitive things, which included all animals, were more perfect than celestial bodies, because the former could imagine pleasure, whereas the latter could not. Poncius also drew a significant consequence. From the widely accepted principle, drawn perhaps from Duns Scotus, that only a more perfect thing can wholly or partially generate a less perfect thing,[69] he concludes that living things that are more perfect than celestial bodies could not be generated by celestial bodies.

2. *A big departure: The earth, with the life on it, is more perfect than the Sun*

What we have seen thus far is a break with an important aspect of the Aristotelian world view. Not only is man more perfect than the heavens

63. Aversa, *De caelo*, qu. 33, sec. 7, 1627, 112, col. 2. Among those who allegedly held this opinion, Aversa cites Thomas Aquinas, Ockham, Averroës, and the Coimbra Jesuits. As we saw earlier, it is not likely that Thomas held this opinion but that he considered the heavens nobler than sensitive and vegetative life.
64. "Et etiam quoad incorruptibilitatem, anima nostra est immortalis plusquam forma et substantia caeli." Aversa, ibid., 113, col. 1.
65. "Caelum deinde dicitur regere et gubernare haec inferiora esseque causa superior et universalis, non quasi per sapientiam et prudentiam res disponens, sed influendo et operando tanquam causa naturalis." Ibid., 114, col. 2.
66. Compton-Carleton [*De coelo*, disp. 3, sec. 2], 1649, 405, col. 2.
67. Poncius [*De coelo*, disp. 22, qu. ult., conclus. 3], 1672, 631, col. 2, par. 109.
68. Aversa, *De caelo*, qu. 33, sec. 7, 1627, 112, col. 2, also expressed doubts about plants being more perfect than celestial bodies.
69. "Anima quaecumque est perfectior formis coelestibus, ergo non possunt a corporibus coelestibus produci quia imperfectius non potest producere immediate perfectius." Poncius, *De coelo*, disp. 22, qu. ult., conclus. 3, 1672, 631, col. 2, par. 109. In his *Ordinatio*, bk. 1, dist. 7, qu. 1, n. 47, Scotus expresses the idea that it is necessary that the thing or principle that generates something be more perfect than the form or thing that it produces (Scotus, *Opera*, 1956, 4:127).

but so also are animals and, for some, even plants and insects. The celestial region was no longer the most excellent physical entity in the world. But it was still thought incomparably more perfect than the inanimate earth. It remained for someone to repudiate this traditional belief by uniting into one total being the living things on the earth with the earth itself. Giovanni Baptista Riccioli took this momentous step in his *New Almagest* (*Almagestum novum*) of 1651.

In the first of twenty-nine arguments for and against the cosmic centrality of the Sun,[70] Riccioli concludes with the electrifying statement that "the earth, with its living, and especially rational, animals is nobler than the Sun."[71] To justify his startling claim, Riccioli refers his readers to the first argument of the eighth chapter,[72] where at the outset he enunciates another startling proposition, namely that "the center of the universe is the most noble place in the world, for it is everywhere distant from the extremes and holds the middle position."[73] Riccioli adds, however, that the center of the world is the most noble place only in the natural order of things; in the supernatural order, "the center of the earth is the lowest and most wretched place."[74] But if, in the natural order, the center of the world is the most noble place, which body occupies it: the Sun or the earth? For Riccioli, of course, the earth occupies the center and does so because

> the earth ought not to be taken simply as one pure element of four or three elements; but [rather it ought to be taken] as one with living plants and animals, but especially with men for whom all the stars were made and [for whom they] are moved, as God attests in *Deuteronomy*. . . . It is the most excellent body of all the bodies of the world, if we judge by the magnitude of virtue and the dignity of the end [or goal], as is proper, rather than by the magnitude of the mass [of bodies].[75]

70. Riccioli, *Almagestum novum*, pars post., bk. 9, sec. 4, ch. 33, 1651, 469, col. 1. Riccioli titled this section "29 arguments in favor of the Sun's position in the center of the universe and [in favor] of the annual motion of the earth around the center of the universe simultaneously with the daily motion, and their solutions, furnished from chapters 8 to 18 inclusively."
71. "Tellus enim cum viventibus et animalibus praesertim rationalibus est nobilior sole." Ibid.
72. Ibid., 330, col. 2–331, col. 1.
73. "Non videtur dubitandum quin centrum universi sit locus nobilissimus in mundo; quippe qui aeque undique distat ab extremis et medium obtinet situm." Ibid., 330, col. 2.
74. "Dixi autem loquendo de ordine ac fine naturali, nam si de supernaturali centrum telluris est infimus ac miserrimus locus." Ibid., 331, col. 1. Although Riccioli speaks of the "center of the earth" and not the "center of the world," it is evident that he has equated the latter with the former.
75. "Tellus enim non debet sumi nude pro mero elemento uno ex quatuor, tribusve elementis, sed una cum plantis viventibus et animalibus, se praecipue cum hominibus, in quorum gratiam facta sunt et moventur sidera omnia, Deo id in *Deuteronomio* attestante. . . . Praestantissimum est corpus omnium mundi corporum, si magnitudinem virtutis ac dignitatem finis, ut par est, aestimemus potius quam magnitudinem molis." Ibid. If the comparison between earth and Sun were made on the basis of mass, the Sun, whose mass was assumed greater, would have been more excellent than the earth.

In these brief passages, Riccioli reveals extraordinary departures from the traditional scholastic interpretation of Aristotelian cosmology. We notice first the new emphasis on the importance of the center of the universe. Because it coincided with the earth's center, the world center in medieval cosmology had never been accorded much importance except as the alleged point around which the celestial orbs moved with uniform circular motion.[76] With the advent of the Copernican system, a dramatic change occurred. The center of the universe became the most important part of the cosmos, because it was usually assumed to coincide with the center of the Sun.[77] As Copernicus explained,

> At rest . . . in the middle of everything is the sun. For in this most beautiful temple, who would place this lamp in another or better position than that from which it can light up the whole thing at the same time? For, the sun is not inappropriately called by some people the lantern of the universe, its mind by others, and its ruler by still others. [Hermes] the Thrice Greatest labels it a visible god, and Sophocles' Electra, the all-seeing. Thus indeed, as though seated on a royal throne, the sun governs the family of planets revolving around it.[78]

Lofty praise of the Sun as the most noble planet was, as we have seen, partially based on its position as the middle, or fourth, planet among the seven. That praise was not unlike that which Copernicus heaped upon the Sun as the occupant of the center of a spherical universe around which all the other planets moved. Until Riccioli's *New Almagest*, Copernicus's glorification of the world's geometric center seems to have made little or no impact on scholastic cosmology. With the earth at the center of their world system, they could find little reason to extoll the virtues of a center that was judged immeasurably inferior to anything in the celestial region.

Riccioli changed all this. Among the arguments for and against the Copernican theory that involved a comparison of the conditions of the earth as compared to other planets but which did not involve motion,[79] Riccioli chose to battle the Copernicans on their own ground. To do this, he apparently decided to demonstrate that the earth was indeed superior to the Sun. In order to make the comparison plausible, Riccioli accepted the center

76. During the Middle Ages, the role of the geometric center of the world was further compromised by the widespread acceptance of eccentric and epicyclic spheres that moved around other centers. See Chapter 13, Section III.4–5, for further discussion.
77. Strictly speaking, Copernicus did not identify the cosmic center with the Sun's center but only said that "near the sun is the center of the universe." Copernicus, *Revolutions*, bk. 1, ch. 10 [Rosen], 1978, 20.
78. Ibid., 22.
79. This is the first of four general headings that Riccioli distinguished for organizing the arguments for or against removing the earth from the center of the universe. They involve a comparison of the earth with: "[1] the conditions of the planets, but not involving motion; or [2] the motion of the planets themselves; or [3] . . . other motions observed in the heavens; or, finally [4] . . . the motions or mutations observed in the sphere of the elements." *Almagestum novum*, pars post., bk. 9, sec. 4, ch. 8, 1651, 330, col. 2.

of the world as the noblest place in the natural order. In this, perhaps he was genuinely convinced that Copernicus had rightly judged the center of the universe as the most perfect place. If so, it was the best place only because the center would enable an appropriate body to receive things from other bodies and to communicate things to them in the most effective manner.[80]

With the center established as the noblest place in the universe, which of the two contending bodies – earth or Sun – was more fit to occupy it? We have seen that at least from the time of Duns Scotus, many scholastics had argued that living things on the earth were nobler and more perfect than the inanimate celestial planets, stars, and orbs. They were also agreed that the inanimate heavens were more perfect than the inanimate earth. But no one, it seems, conceived of the earth as a unity consisting of the physical globe *and* the living things in, on, and around it. Riccioli took this momentous step. In so doing, he elevated the earth from an inanimate entity inferior to the celestial region to the most noble body in the physical world. For although the comparison was between earth and Sun, Riccioli unambiguously declared the earth to be "the most excellent body of all the bodies of the world."

Though it was in vain, Riccioli's effort to make the earth worthy of location in the center of the universe by elevating its status over that of the Sun and the other celestial bodies is an important event in the history of scholastic Aristotelian cosmology. In his struggle with the Copernicans, Riccioli departed in two significant ways from traditional medieval cosmology. He made the center of the world the most noble place in the cosmos; and he made the earth more noble and perfect than any celestial body. The first change was apparently the result of Copernicus's emphasis on the center of the world as the only place worthy of serving as the Sun's location, the only place from which the Sun could properly exercise its dominant role in the universe. The center best served the operations and objectives of the world. Riccioli appears to have found those arguments compelling. The second departure, the greater nobility of the earth over the heavens, had, as we saw, a longer history. Ultimately it depended on

80. The crucial parts of Riccioli's analysis (ibid., 331, col. 1) depend on a syllogism that he formulates, parts of which he then refutes. The syllogism reads:

[Major premise:] To the most excellent of worldly bodies, the most excellent place ought to be assigned.
[Minor premise:] But the Sun, not the earth, is the most excellent body of the world and the most excellent place of the universe is the center.
[Conclusion:] Therefore the center of the universe ought to be assigned to the Sun, not to the earth.

Although Riccioli denies the first half of the minor premise and the conclusion, he concedes the major premise (and the second half of the minor premise) and argues that the most excellent body ought to occupy the most excellent place, which is the center of the world, not because of the "pure geometric excellence" of the center but "from a physical end and good which such a body ought to receive from others or to communicate to others." Being at the center enables the most noble body to realize these goals and objectives.

the widely held scholastic belief that the heavens were inanimate, though more perfect than the inanimate earth, and that living things were more perfect than the lifeless heavens. By making living things an integral part of the earth, Riccioli, or someone before him, if he was not the first, made the earth ipso facto more perfect than any celestial body. As the most noble and perfect body in the world, the earth, rather than the Sun, was more fit to occupy the center, the noblest of places.

12

On celestial matter: Can it exist in a changeless state?

That medieval scholastics regularly wondered whether the heavens, with all of the planets and stars, possessed matter may at first glance appear strange or even startling. Because the planets and stars are readily visible, and because visible effects were associated with matter,[1] it should have been obvious that some kind of matter must underlie the celestial appearances. Although most scholastics accepted the existence of such matter, others found it contrary to the principles of natural philosophy.

The question as to whether matter existed in the celestial region was thrust upon the Middle Ages by Aristotle's division of the world into radically different celestial and terrestrial (sublunar) regions. In Chapter 10, which concerned celestial incorruptibility, an issue that is intimately connected to the problem of celestial matter, we noted Aristotle's justification for this division: the existence of two radically different kinds of motion in the universe. One – finite, rectilinear, and therefore incomplete – was associated with elemental bodies and bodies compounded of those four elements; the other – circular and complete, without beginning or end – was the motion of celestial bodies alone. It followed for Aristotle that celestial bodies and the heavens in general consisted of a substance – or ether – different from any of the four elements. Incorruptibility, and therefore unchangeability, were its most fundamental properties, properties so amazing that they made the celestial substance radically different from bodies in the terrestrial region.[2]

We have also examined the theoretical basis for celestial incorruptibility: the absence of contrary qualities in the celestial region. In Galileo's words, "Whatever undergoes corruption has a contrary . . . ; and therefore . . . the heavens, since they lack contraries of this type, are incorruptible."[3] Moreover, Aristotle buttressed theory with the empirical claim that no changes in the heavens had ever been detected or recorded.

1. In *Metaphysics* 8.1.1042a.24–26, Aristotle says that "sensible substances all have matter." See Aristotle [Ross], 1984.
2. In *Metaphysics* 12.7.1074a.30–37, Aristotle emphatically denies that matter can exist in the heavens. For a brief discussion, see the beginning of Chapter 12 and Chapter 12, Section I.
3. Galileo [*De caelo*, qu. 4 (J)], 1977, 98.

During the Middle Ages, all were agreed with Aristotle and his commentator, Averroës, that generation and corruption were processes associated inherently with sublunar matter compounded of the four elements. Only the latter, and the bodies compounded of them, could undergo change from contrary qualities. Without contrary qualities, the celestial ether could not change and must therefore be incorruptible.

Did it follow that matter was something associated solely with terrestrial bodies, whose never-ending succession of contrary qualities produced incessant change? Did the absence of such contrary qualities in the heavens also imply an absence of matter as well? Or did the heavens include matter of some kind, and were they composed of matter and form? On this issue, like many others, Aristotle provided no clear guidance. Indeed, passages from his many works could readily be selected both to support and to oppose the idea of a celestial matter.[4] The medieval controversy over celestial matter involved two issues. The first and prior issue pitted those who denied its existence against those who affirmed it. Because most affirmed the existence of some kind of matter in the heavens, the secondary and more widely debated dispute concerned the nature and properties of an unchanging matter and the manner of its existence.[5]

I. That matter does not exist in the heavens

As Buridan and others were well aware, "philosophers are accustomed to use the term 'matter' in many ways,"[6] some of which clearly applied to the heavens. Thus, if matter is conceived as something composed of quantitative parts, the heavens must obviously possess matter, since they are composed of quantitative parts. The heavens would also possess matter if matter were defined as a substance that is the subject of motion or other accidents.[7] But these were not the senses in which those who rejected the existence of celestial matter – for example, Buridan, Godfrey of Fontaines, Peter Aureoli,

4. For an excellent illustration of how such juxtapositions were formed, see ibid., qu. 5 (K), 103–111 (for the way in which Averroës cited Aristotle's works against the existence of celestial matter); 111–112 (for those who cited Aristotle in favor of celestial matter); and 112–115 (for citations concerning the idea that celestial matter differs from, or is identical to, terrestrial matter). Inspection of similar questions by numerous other scholastic authors would reveal the same pattern.
5. In what follows, I rely heavily on Grant, 1983.
6. "Notandum est quod multis solent philosophi uti hoc nomine 'materia.' " Buridan [*De caelo*, bk. 1, qu. 11], 1942, 49. Oresme [*De celo*, bk. 1, qu. 11], 1965, 160–162, gives four different senses of "matter" and remarks that Aristotle speaks of matter in a variety of ways in the second book of his *Physics*.
7. Buridan, ibid. For the locus of these ideas, Buridan cites Aristotle's *Metaphysics*, bks. 7 and 8, respectively. Albert of Saxony [*De celo*, bk. 1, qu. 4], 1518, 89r, col. 1, whose text is occasionally an almost verbatim copy of Buridan's *De caelo*, gives virtually the same references to Aristotle.

and Albert of Saxony – considered the question.[8] As Buridan explained, for them "matter is called that from which a substance is composed with a substantial form inhering in it, which persists by itself, and which is called 'this something.' "[9]

By the time Buridan wrote, the opinions of Thomas Aquinas and Aegidius Romanus (which are described in Section II of this chapter) were taken as representative of two rival theories in favor of the existence of celestial matter. Buridan considered and rejected both[10] but admitted that it was difficult to refute the claim for the existence of celestial matter by demonstrating its opposite: namely, that matter does not exist in the heavens. Buridan agreed with Aristotle that something could not be made naturally from nothing. Consequently, the only proper way to define matter was by means of substantial transformations in which something, namely matter, remained constant as a body was generated and corrupted, that is, as the same matter acquired and lost successive forms. But like all his fellow scholastics, Buridan believed that such changes could not occur naturally in the celestial region, so that "it is absolutely in vain and without cogent reason that we should posit such matter in the heavens."[11]

Admitting, however, that he could not devise a formal demonstration for the nonexistence of celestial matter, Buridan sought to achieve the same result by invoking the widely used principles that nature does nothing in vain and that it is useless to "save the phenomena" with more when it can be done with less.[12] Rather than assume the existence of celestial matter, Buridan believed that all the phenomena could be saved by the assumption of a simple, uncomposed, celestial substance which, because it functions as a subject for an extended magnitude, must possess the property of extension. Indeed, this same celestial substance also serves as a subject for motion and other accidents.[13]

8. Galileo adds names to the list of those who rejected celestial matter, including Durandus de Sancto Porciano, Duns Scotus, Marsilius of Inghen, John of Jandun, "and all Averroists." Galileo, *De caelo*, qu. 5 (K), 1977, 105.
9. "Sed proprie loquendo materia vocatur ex qua cum forma substantiali sibi inhaerente componitur substantia per se subsistens quae dicitur 'hoc aliquid.' " Buridan, *De caelo*, bk. 1, qu. 11, 1942, 49. Hervaeus Natalis, *De materia celi*, qu. 3 ("Queritur utrum corpora omnia superiora et inferiora communicent in materia"), 1513, 38v, col. 1, put it in much the same way when he declared that speaking "more strictly, matter can be taken in another way, namely, as that which underlies a substantial form and is part of a composite subject. And now we shall speak about matter in this manner." (Alio modo potest accipi materia magis stricte illud, scilicet quod substernitur forme substantiali et est pars substantie composite. Et sic nunc loquimur de materia.)
10. Buridan, *De caelo*, bk. 1, qu. 11, 1942, 51–52. Albert of Saxony, *De celo*, bk. 1, qu. 4, 1518, 89r, cols. 1–2, also rejected the same two theories.
11. "Ideo frustra omnino et sine ratione cogente poneremus talem materiam in caelo." Buridan, ibid., 53.
12. Ibid., 52. Buridan attributed these ideas to Averroës, but the latter mentions only that nature does nothing in vain (see Averroës [*De caelo*, bk. 1, comment. 20], 1562–1574, 5:15r, col. 2). Saving the phenomena with the fewest possible assumptions is more akin to the principle of Ockham's razor.
13. Buridan, ibid. 52. Hervaeus Natalis, who wrote some years before Buridan, described

Peter Aureoli, a student of Duns Scotus and a trained theologian, had arrived at similar ideas in a much earlier discussion of the subject in his *Commentary on the Sentences*, where he also considered whether "the authorities of Sacred Scripture and the Catholic Doctors" thought it objectionable to assume that the heavens consist of a simple, incomposite substance.[14]

Aureoli first describes arguments he attributes to Aristotle and Averroës. These include their hostility to the conception of the heavens as a composite of matter and form and their conviction that the heavens have only fixed dimensions that belong to them as an inherent property.[15] He then presents his own opinions, of which the most striking is that the heavens not only consist of a simple, incomposite substance, but that this substance is neither matter nor form.[16] Aureoli also rejects the idea of the heavens as a form, or having a form, because form is something that determines and controls matter, accidents, and properties, and the heavens lack any mechanism with which to control their properties and perfections.[17] The explanation for this major difference between the terrestrial and celestial regions lies in the variability in size of animate and inanimate things in the former and, by contrast, the invariability of the figure and size of bodies in the latter. In animate, terrestrial bodies, sizes vary within certain limits that are controlled by the soul; in inanimate bodies, variations in the size of a thing are con-

(and then rejected) the opinion Buridan favored when he declared that supporters of this opinion say that "there is no potentiality for substantial being, namely for a substantial form. And this is the position of the Commentator [i.e. Averroës] in the *De substantia orbis* and of certain moderns who say that the heaven is a certain form spread out and extended by quantity, but which is supported by nothing" (Hervaeus Natalis, *De materia celi*, qu. 3, 1513, 38v, col. 2). This opinion was also reported by Peter of Abano in his *Lucidator dubitabilium astronomiae*, differ. 5, 1988, 288. As Peter puts it, the planets and stars are forms or dimensions that are not in matter, but are "spiritual bodies" (*corpora spiritualia*) or can be said to possess matter equivocally. Averroës had assumed indeterminate dimensions as an inherent property of the celestial body (see Averroës, *De substantia orbis* [Hyman and Walsh], 1973, 312). These dimensions, which were associated with the celestial ether's substantial form, were not divisible into determinate quantities, as were the indeterminate dimensions associated with prime matter in the terrestrial region.

14. "Articulus II: Utrum ponere caelum et naturam simplicem et non compositam ex materia et forma repugnat auctoritatibus sacrae Scripturae et Doctorum Catholicorum." Aureoli [*Sentences*, bk. 2, dist. 14, qu. 1, art. 2], 1595–1605, 2:189 col. 1.
15. Ibid., 186–87. If the dimensions of the heavens were as indeterminate as prime matter, the heavens would necessarily change their dimensions in the same manner as terrestrial bodies. Thus if the heavens were to be assumed incorruptible, Aristotle and Averroës, and all who followed them, had to assume invariant celestial dimensions. The heavens, or any part of them, are therefore not capable of increasing or decreasing their size by rarefaction or condensation or by addition or subtraction of any imaginable substance. See Aureoli, ibid., 187 col. 1 (D–F). On the difference between determinate and indeterminate dimensions or quantities, see Weisheipl [McMullin], 1963, 147–169.
16. "Caelum esse quantum non compositum, scilicet ex materia et forma, nec forma, nec materia, subiectum habens dimensiones tantum in actu suo." Aureoli, ibid., 189, col. 1 (A).
17. "De natura enim formae est quod sit in actu et determinet materiam et largiatur per modum exigentis et determinantis accidentia et proprietates. . . . Sed natura caeli non determinat proprietates suas et perfectiones postremas. Natura enim caeli in quantum huiusmodi non determinat sibi tantam quantitatem." Ibid., 188, col. 1 (E–F).

trolled by its form. If a cow, for example, were made as long as a serpent, it would cease to be a cow,[18] a possible catastrophe that is avoided by its soul. Similarly, if a proper upward motion were absent in fire, the latter would lack a form.[19]

By contrast, celestial matter and its bodies are invariant in size, shape, and motion and therefore require no forms or souls to regulate and determine those properties. Indeed, nothing inheres within celestial bodies that would enable them to determine their own properties and operations. Consequently, the heavens cannot possess a form, or forms.[20] As if to reinforce his argument, Aureoli observes that although the heavens are a finite entity, they are also eternal and permanent. But a finite thing lacks the nature to determine an eternally permanent thing. No intrinsic principle of the celestial substance – that is, no form or matter or combination thereof – could have conferred such properties on the heavens. Only an external power could have bestowed them, a power which Aureoli identifies with the celestial intelligences.[21]

If the heavens lack both form and matter, what kind of an entity could they be? On this crucial issue, Aureoli, who preceded Buridan, was somewhat more forthcoming. He judged the heavens to be an existing magnitude: a *quanta esse*, as he described it. "Just as matter is not understood except in relation to form," Aureoli explained, "so we cannot understand that the heavens have a definite [or fixed] quantity, figure, motion, and other properties unless [these properties] are [understood in] relation to an intelligence."[22] The *quanta esse*, or celestial magnitude, functions like a subject but has received its properties from an intelligence. Without matter and form, the two fundamental principles of all terrestrial things, the *quanta esse* and its properties fitted none of the traditional descriptions of the heavens.

18. "Unde anima quaelibet determinat figuram certam sui corporis sine qua non potest esse: facias enim bovem longum sicut serpentem, statim amittet esse bovis." Ibid., 188 col. 2 (A).
19. "Si ergo ignis non haberet ex se motum proprium eum, qui est sursum, qui est ei proprius, iam sequitur quod ignis non est forma." Ibid., (B).
20. "Sic in proposito cum caelum sit determinatae quantitatis in actu quia in eo non sunt dimensiones interminatae, cum etiam sit figurae rotundae et habeat motum circularem sine quibus impossibile est esse et talia non determinet sibi per naturam propriam, circumscripta anima. Patet quod caelum non est forma, cum forma quaelibet se ipsa, omni alio circumscripto, determinet sibi suas proprietates et operationes. Hanc rationem tangit Commentator *De substantia orbis*, tractatus 2." Ibid., (B–C).
21. "Hoc idem potest apparere de aeterna eius permanentia quam sibi non determinat, ut talis natura est. Nulla enim natura, quae habet dimensiones finitas, videtur sibi ex se determinare permanentiam aeternam. Sed hoc habet ex determinatione extrinseca, scilicet intelligentiae, quae largitur ei omnes huiusmodi perfectiones consequentes, non effective, sed exigitive et determinative tantummodo, ut dictum est saepe." Ibid., (C–D).
22. "Natura caeli est natura subiecti, et est esse quanta. Unde sicut materia non intelligitur nisi in analogia ad formam, sic non possumus intelligere caelum habere determinatam quantitatem, figuram, motus, et proprietates alias nisi in respectu ad intelligentiam, ut Commentator 2 *De caelo et mundo* dicit. Non ergo est forma quia forma non est in potentia ad suas postremas perfectiones; nec est materia quia materia est in potentia ad actum primum. Ergo est quasi medium, ut sic, ratio subiecti et essentia coniuncta." Ibid., (E–F).

With most of the traditional features of the celestial region rejected and virtually no really new ones available to replace them, Aureoli was left with an extended magnitude possessed of certain vital properties conferred by an otherwise undescribed intelligence. By offering so little information, Aureoli could present the positive features of his conception of the heavens in but a few lines.

Buridan, Aureoli, and others who rejected the existence of matter in the heavens were ultimately supporters of Averroës' position. And yet Averroës, despite his rejection of the idea that the celestial substance was a composite of matter and form,[23] and his insistence that the celestial substance had to be something simple and uncomposed, was prepared on occasion not only to call that simple substance a form, but also to call it matter, though clearly not matter in the ordinary sense, as described earlier by Buridan. In his commentary on *De caelo*, Averroës first concludes that "the celestial body does not have matter,"[24] arguing that even if matter existed in the incorruptible celestial region it would be superfluous, because it could never receive a new form (since there are no contrary forms) and thus could never change. Its potentiality would be forever frustrated and in vain, which is contrary to nature.

But later in the same commentary, Averroës speaks of the "matter of the form of a celestial body, which [matter] is actualized."[25] As "actualized" matter, it obviously could not be conceived as the prime matter underlying the four elements, which is a pure potentiality. Although celestial matter lacked any potentiality for substantial change (i.e., generation and corruption), or for qualitative or quantitative change and could not be stripped of its form, it did have a potentiality for place, as the celestial motions made evident.[26] Thus, with the celestial bodies clearly in mind, Averroës declared, in his commentary on the *Metaphysics*, that "eternal things, which are not generable but are moved with a translatory motion, have matter; not, however, generable matter, but [only] the matter of those things that are moved from place to place."[27]

23. In his *Commentary on De caelo*, Averroës declared that the heavens are not composed of matter and form as are the four simple elements, because "forms that are in matter are contraries, and if a form existed in matter without a contrary, then nature would act in vain, since no potentiality whatever could exist in this matter because potentiality occurs only when a form can separate from [its] matter." See Averroës, *De caelo*, bk. 1, comment. 20, 1562–1574, 5:15r, col. 1.
24. Ibid., bk. 1, comment. 21, 5:15v col. 2.
25. "Dicamus ergo quod ista natura neutra est, et non existens per se in actu, sed est materia formae corporis celestis, que est in actu." Ibid. bk. 1, comment. 95, 5:63v col. 2. Here Averroës speaks of matter and form.
26. "Et ideo in hac nulla potentia est qua denudari possit a sua forma et non habet nisi potentiam ad ubi." Ibid. 5:64r, col. 1.
27. "Omnia aeterna quae sunt non generabilia, sed moventur motu secundum translationem, habent materiam; sed non habent materiam generabilium sed materiam eorum que moventur de ubi in ubi." Averroës [*Metaphysics*, bk. 12, comment. 10], 1562–1574, 8:296v, col. 2–297r col. 1. For Aquinas's interpretation of these passages, see Wippel, 1981a, 286–287.

II. Two rival theories in support of the existence of celestial matter

1. Aquinas and Galileo: Celestial and terrestrial matter differ

Arrayed against Averroës and all who denied the existence of celestial matter were Thomas Aquinas and Aegidius Romanus, who, despite their agreement that the heavens are composed of matter and form,[28] disagreed radically as to the nature of that matter, Thomas assuming it radically different from, Aegidius identifying it with, terrestrial matter. These two differing interpretations by Thomas and Aegidius, along with the opposing opinion of Averroës, lay at the core of almost all discussions of the problem of celestial matter to the end of the seventeenth century. Because of their obvious importance, we shall examine them in some detail.

Although Thomas, and many who followed him, argued that the celestial region consisted of a composite of matter and form, the matter he had in mind "was of another kind than that of inferior [i.e., sublunar] bodies."[29] In the heavens, matter was in potentiality with respect to a perfectly actualized form that fulfilled all the possibilities of that matter, which therefore lacked a potentiality for any other forms.[30] Consequently, changes of substance, quality, or quantity could not occur in the heavens. The Sun, for example, was incapable of changing into anything else, nor did it come into being by the transformation of anything else.[31] By contrast, "the matter of the elements is in potentiality with respect to an incomplete form which cannot limit [or fulfill] the potentiality of the matter."[32]

28. Galileo, *De caelo*, qu. 5 (K), 1977, 111–112, cites numerous supporters of this general opinion, including "all the Arabs, with the single exception of Averroës," specifically mentioning Avempace (ibn Bajja) and Avicenna; Moses Maimonides, Saint Bonaventure, Thomas Aquinas "and likewise all Thomists," Aegidius Romanus (Giles of Rome), Albertus Magnus, Alessandro Achillini, and Julius Caesar Scaliger.
29. Thomas Aquinas [*De caelo*, bk. 1, lec. 6, par. 63], 1952, 31, declares: "quod materia caelestis corporis est alia et alterius rationis a materia inferiorum corporum." This passage, and some thirty-six others (drawn from a variety of Thomas's works) that reflect his views on the relationship of celestial and terrestrial matter, appear in Litt, 1963, ch. 3, pp. 54–80; for the quotation, see p. 79; for similar statements from Thomas's *Commentary on the Metaphysics*, and *Quaestio disputata de anima*, see Litt, ibid., 72, no. 22, and 77, no. 31, respectively. For a fine, brief account of Thomas's views, especially as they contrast with the opinions of Godfrey of Fontaines, see Wippel, 1981a, 285–291. Although he makes no mention of Thomas, Hervaeus Natalis, *De materia celi*, qu. 3, 1513, 38v, col. 2, adopted the same position, demonstrating it, however, not by positive arguments but by the fairly common medieval practice of refuting all of its alleged rivals – in this case, three other theories.
30. "Nam materia caelestium corporum est in potentia ad actum perfectum, idest ad formam quae complet totam possibilitatem materiae, ut iam non remaneat potentia ad alias formas. Materia autem elementorum est potentia ad formam incompletum, quae totam terminare non potest materiae potentiam." *De substantiis separatis* c. 8, n. 82, as cited by Litt, 1963, 79.
31. See Litt, 1963, 59; for Thomas's statement from his commentary on Boethius's *De trinitate*, see page 63.
32. For the Latin text, see note 30 in this chapter.

Despite the seeming total fulfillment of celestial matter by its single, unique, and permanent form, one aspect of potentiality remained. The uniform, circular motion of the planets and stars compelled Thomas, as it had Aristotle, to concede that a celestial body was in potentiality at least with respect to place,[33] even if not with respect to being. Although celestial and terrestrial matter shared a potentiality for place, they otherwise had nothing in common. Unlike the prime matter of the terrestrial region, which was a pure potentiality that could receive and lose all possible forms existing in that region, celestial matter was created with a single form so complete that it precluded the receipt of any other possible forms. Thus was the incorruptibility of the heavens preserved.

Whereas Thomas considered the problem of celestial matter briefly, in many treatises, Galileo, who was in essential agreement with Thomas, treated the problem at great length in only one treatise. In Galileo's questions on *De caelo*, at least four of the questions are relevant to the existence and nature of celestial matter.[34] As he subdivides the major opinions into a host of confirming arguments and objections, Galileo is the quintessential scholastic. Embedded within this rather heavy format are Galileo's own opinions, accompanied by objections to those opinions that are systematically answered at the end of each question. Of the great number of extant discussions on the problem of the possible existence and nature of celestial matter, Galileo's must rank as one of the most thorough. All but one of the opinions are described with scholastic fullness of detail and subtlety. And, as was characteristic of sixteenth- and seventeenth-century scholastics – but not of their medieval predecessors – Galileo cited a large number of authoritative sources: ancient, medieval, and Renaissance. Here we find specific citations to the relevant works of the supreme early authorities, Aristotle and Averroës, as well as to the works of lesser early figures such as Alexander of Aphrodisias, Simplicius, John Philoponus, and Avicenna. Numerous medieval arguments are cited from eminent scholastic authors such as Aquinas, John of Jandun, Saint Bonaventure, Albertus Magnus, Duns Scotus, Marsilius of Inghen, Durandus de Sancto Porciano, and especially Aegidius Romanus. From his contemporaries and predecessors of the fifteenth and sixteenth centuries, Galileo invoked the opinions and works of Alessandro Achillini, John Capreolus, Cajetan (Thomas de Vio), Julius Caesar Scaliger, and Marsilio Ficino.

From this impressive parade of authorities, we should not infer that Galileo had a deep familiarity with the vast literature on the problem of

33. See Litt, 1963, 56–57, especially note 3, which contains eight passages drawn from eight different treatises. On page 56, Litt also cites four passages in Aristotle's *Metaphysics* (8.1.1024b.6; 8.4.1044b.7; 9.8.1050b.21; 12.2.1069b.2).

34. These are questions no. 3 (I) ("Are the heavens one of the simple bodies or composed of them?"), Galileo, *De caelo*, 1977, 81–92; no. 4 (J) ("Are the heavens incorruptible?"), ibid., 93–102; no. 5 (K) ("Are the heavens composed of matter and form?"), ibid., 103–147; and no. 6 (L) ("Are the heavens animated?"), ibid., 148–58. The fifth question is the most relevant.

celestial matter. As William Wallace has demonstrated, Galileo drew heavily, and perhaps wholly, upon a few published treatises and a larger number of unpublished lectures, or *reportationes*, that had been produced by the Jesuit faculty of the Collegio Romano between the 1570s and 1590s.[35] Despite his heavy debt to the Jesuits of the Collegio Romano, the selection of the arguments and the final organization of the questions are probably Galileo's. If Galileo sincerely believed in the opinions that he adopted in these treatises, one can only conclude that he was a deeply committed Aristotelian scholastic when he wrote them, a picture that stands somewhat at variance with the Galileo of the other *Juvenilia*, the one who wrote the *De motu* and had clearly broken with Aristotle on the problem of motion in a vacuum.[36]

In the question "Are the heavens composed of matter and form?" Galileo, as we would expect in a scholastic treatise, considered both the negative and the affirmative positions. The champion for the negative side was Averroës, who, as we saw, had denied that the heavens were composed of matter and form.[37] Following a lengthy description of Averroës' position (Galileo [*De caelo*, qu. 5 (K)], 1977, 103–111), Galileo presented the case for the affirmative side, which constituted the majority opinion during the Middle Ages and the Renaissance, when its supporters included such famous authorities as Avicenna, Maimonides, Albertus Magnus, Aquinas, Aegidius, Bonaventure, Achillini, and Scaliger.[38] Within the majority group, debate centered on whether celestial matter differed from terrestrial matter.[39]

On the existence of celestial matter, Galileo – "following the common opinion of the peripatetics" – aligned himself with those who believed that "the heavens are composed of matter and form, whatever the matter may be" (ibid., 115). In support of the claim for the existence of celestial matter, Galileo presented numerous arguments, most of which were variations on a few basic medieval Aristotelian themes. For example, since our heaven is conceived as a particular entity, which he terms "this heaven," and all particular entities consist of matter and form, the heaven must also be composed of matter and form. A brief syllogism drawn from the *Metaphysics* constitutes a second argument: "Sensible substances contain matter; but a heavenly body is singular and sensible; therefore [it contains matter]."[40] In

35. See Wallace, 1981, 281, 309. For a list of the Jesuit authors on whom Galileo seems to have relied, see Galileo, *De caelo*, 1977, 12–21. Of this group, Christopher Clavius is the best known.
36. Whether Galileo believed the opinions he presents in the *Juvenilia* is difficult to determine. See my review of Wallace, 1981, in *Science* 214 (1981), 55–56.
37. As indicated earlier, Averroës occasionally spoke as if matter and form existed in the heavens, although not as a composite.
38. For these and other names, see Galileo, *De caelo*, qu. 5 (K), 1977, 111–112. Galileo declares that all Arabian authors with the exception of Averroës supported this opinion, as did all Thomists. An even larger list is furnished by Bartholomew Amicus [*De caelo*, tract. 4, qu. 1, dubit. 2, art. 2], 1626, 138, col. 2.
39. Galileo, ibid. 112. Galileo also identified a second major issue (ibid.) that turned on whether the form that is associated with celestial matter is also an intelligence.
40. Ibid., 116. The passage, where Aristotle says that "sensible substances all have matter," appears in the *Metaphysics* 8.1.1042a.24–26.

yet another argument, we are told that matter and form are the principles of natural things: "therefore, since the heavens are natural, they must be composed of matter and form" (ibid., 117).

By such arguments Galileo was convinced that matter of some kind must exist in the heavens. Like so many before him, he inquired next about the nature of that matter: Is it the same as, or different from, our terrestrial variety? He concluded that "the heavens are not composed of matter of the same kind as the matter of inferior bodies" (ibid., 132). To defend this position, Galileo found it necessary to attack in considerable detail the most important opposition thesis, namely Aegidius's well-known arguments in defense of the identity of celestial and terrestrial matter (ibid., 132–139). In the process, Galileo turned Aristotle's denial of celestial matter to his own advantage by insisting that it was not to be taken categorically. It was only Aristotle's way of denying that "there is in the heavens matter of the same kind as the matter of lower bodies" (ibid., 124; also 132). Galileo thus convinced himself that he was in agreement with Aristotle when he assumed that celestial matter was something quite different from terrestrial matter. If it existed in an incorruptible heaven, it had to be radically different – indeed, nothing less than incorruptible. Galileo, like all who adopted the same interpretation, was committed to a conception of celestial matter that made it function in ways that were largely the opposite of its terrestrial counterpart. The two matters were as radically different as the heavens from the earth.

The differences – and here Galileo relied on and agreed with Aegidius – derived from the causes of corruption, which were always explained in terms of matter, form, and privation (ibid., 134–135). Corruption occurs when matter that possesses a form seeks the contrary of that form. The privation of that contrary form is thus identified as the cause of the matter's corruption.[41] The privation itself, however, "arises from the fact that a form has a contrary" (ibid., 134). Unless that privation can be overcome, however, so that the matter can at some time possess that contrary form, it would be perpetually frustrated and therefore opposed to the operations of nature, which does nothing in vain.

If a form did not have a contrary, the matter that possessed that form could not be in a state of potentiality with respect to a contrary form. Therefore that matter could not be deprived of a contrary form, and privation would not, and could not, function as the cause of corruption.

Although such matter could not exist in the terrestrial region, Galileo declared that it did exist in the heavens. Celestial matter had only one form, which lacked a contrary. Without a contrary, privation could play no role, and celestial matter could have "no appetite for another form, for if it did, it would have an appetite to be deprived of its own existence" (ibid., 135).

41. "Hence the matter remains deprived, and is in potency to another form and therefore has an appetite for it; then corruption results." Ibid. 134.

Celestial matter was thus incorruptible, because it had no inclination for any other form. Nor did it have any inclination to destroy its only form, because that would indicate a desire for its own nonexistence, which is absurd (ibid., 141).

Galileo did not consider his arguments for the existence of an incorruptible celestial matter as demonstrative "but only as highly probable; because with the single exception of Averroës, it is that of practically all the peripatetics and because there is nothing that contradicts it, while there are many things in its favor" (ibid., 124).[42]

2. *Aegidius and Ockham: Celestial and terrestrial matter are identical*

Although Aegidius Romanus agreed with Thomas Aquinas that the celestial substance was a composite of matter and form, he differed from him by arguing that heavenly matter was essentially the same as terrestrial matter. In *De materia celi* (*On Celestial Matter*), a treatise devoted solely to the problem of the possible existence and nature of celestial matter, two questions form the basis of the work. In the first, Aegidius considers whether matter exists in the heavens[43] and, on the assumption that it does, poses the second: Is it essentially the same as terrestrial matter?[44]

To show that the celestial substance is composed of matter and form, Aegidius directly opposes the opinion of Averroës that the heavens are a simple, incomposite, indeterminate, and indivisible entity – that is, a form without matter. Aegidius stresses the evidence of our senses, declaring that we can observe that the heavens possess quantity and are therefore divisible.[45] But without matter, the heavens would be incapable of receiving the quantity necessary to make them divisible. Appealing again to sense,[46] Aegidius observes that one part of the heavens is thicker than another, as evidenced by the appearance of stars, which are visible because they are thickened celestial matter in contrast to celestial orbs, which are rarefied to the point of transparency.[47] Aegidius concludes that "we cannot save the

42. As we saw earlier, Averroës had Peripatetic followers in the Middle Ages. By the late sixteenth century, however, Galileo's claim may have been correct.
43. "Questio est utrum in celo sit materia vel sit celum corpus simplex, ut posuit Commentator." Aegidius Romanus, *De materia celi*, 1502b, 78r, col. 2. For bibliographical references to the life and works of Aegidius, see Wippel, 1981a, xi–xii.
44. "Queritur secundo: dato quod in celo sit materia utrum illa materia sit eadem per essentiam cum materia istorum inferiorum." Aegidius Romanus, ibid., 80v, col. 1.
45. "It is evident," Aegidius insists, "that the heavens have quantity, because to deny this men would have to hallucinate and deny the senses." Ibid., 79v, col. 1.
46. Contrary to Aegidius, Francisco Suarez, *Disputationes metaphysicae*, disp. 13, sec. 10, 1886, 1:436, col. 2, in the late sixteenth century, denied that one could demonstrate by any visible effects whether the heavens were composed of form alone, as Averroës, would have it, or of matter and form, as Thomas argued (and, of course, Aegidius, who is not mentioned).
47. To reinforce his position, Aegidius appeals to the second book of Aristotle's *On Generation and Corruption* (2.6.333a.21–24) where, according to Aegidius, Aristotle declared that 1

thickness and transparency in the heavens except by [the assumption of] matter."[48] From Aristotle's declaration that "everything that is perceptible is in matter" and that our entire world is compounded of matter,[49] Aegidius concludes that our heavens are "particular and sensible heavens" and necessarily imply a form in matter.[50] Convinced that matter must exist in the heavens, Aegidius, in the second question of his *De materia celi*, determines whether the matter of the celestial region is essentially the same as the matter of the inferior, or sublunar, region and concludes that they are indeed identical. According to Aegidius, none of the ancient doctors, saints, and philosophers whose works have reached us was of the opinion that matter exists in the heavens and therefore none of them believed that the heavens are composed of matter and form. Those among these venerable authorities who did assume the existence of matter in the heavens, however, did not think there was any difference between that matter and the matter here below, as "some masters [*magistri*] and modern doctors [*doctores*])" assert.[51]

Bolstered by a conviction that ancient authorities who considered the problem allowed that if matter did exist in the heavens, it would be identical with its terrestrial counterpart, Aegidius insisted that the heavens are not simple but are instead a composite entity made up of matter and form, a judgment that followed from the admitted quantity, thickness, transparency, and individuality of the celestial region.[52] But is the matter in this compound of matter and form a pure potentiality, or is it some kind of

pint of water equals 10 pints of air, a comparison that is possible only because air and water share matter – not form – in virtue of which one can say that water is "thicker" than air. The same kind of relationship obtains betweeen stars and celestial orbs. The ratio of 10 to 1 between water and air was merely a hypothetical example for Aristotle, although Aegidius invokes it as a fact.

48. "Ergo non poterimus salvare spissum et dyaphanum in celo, sed solum ex materia." Aegidius Romanus, *De materia celi*, 1502b, 79v, col. 2.
49. Aristotle, *De caelo*, 1.9.278a.8–16 [Guthrie], 1960.
50. For a fuller discussion, see Grant, 1983, 165–167.
51. "Dicendum quod in hac questione sic procedemus quod primo ostendemus quod nulli antiquorum doctorum nec philosophorum nec sanctorum de his qui pervenerunt ad nos secundum ea que vidimus fuerunt huius positionis: quod in celo esset materia et quod corpus celi circumscripta intelligentia esset compositum ex duabus substantiis, ex materia, scilicet, et forma, et tamen materia illa esset differens per essentiam ab ista, sicut aliqui magistri et moderni doctores posuerunt. Antiqui enim doctores vel negaverunt in celo esse materiam, ut posuit Commentator, vel si posuerunt ibi materiam dixerunt eam esse eandem cum materia istorum inferiorum." Aegidius Romanus, *De materia celi*, 1502b, 81r, col. 1. Among the saints who believed that celestial and terrestrial matter were identical, Aegidius mentions only Saint Augustine, while he cites Avicenna as representative of the philosophers (ibid., 81r, cols. 1–2). In his discussion of the question, William of Ockham, asserted the opposite when, after declaring that "there is matter in the heavens," he justified this claim by "the statements of the saints and doctors, who say that, in the beginning, God created matter from which the celestial bodies and other things were formed." Ockham [*Sentences*, bk. 2, qu. 18: "Utrum in caelo sit materia eiusdem rationis cum materia istorum inferiorum"], 1981, 5:399.
52. "Adhuc celum non est corpus simplex, ut posuit Commentator, quod probabamus et ex quantitate eius, et ex spissitudine, et ex dyaphanitate quas videmus in ipso et ex individuatione ipsius." Aegidius Romanus, ibid., 81v, col. 1. In this opinion, Aegidius says that he does not differ from other theologians (nec in hoc discordamus ab aliis theologis).

actuality? Aegidius argues for the former, insisting that if the matter were actualized it could not form a single essence with the actualized form, "because one thing is never formed essentially from two actualized things."[53] Celestial matter must, then, be "pure potentiality" (*pura potentia*), and therefore identical with the matter of inferior things.[54]

Aegidius offered three arguments to support his claim that celestial and terrestrial matter are essentially the same.[55] The first relies on a "principle of indifference," whereby Aegidius assumes that if celestial and terrestrial matter are both pure potentialities – as he believed – and if every form were stripped from those two matters, they would not differ in any way, "because there can be no distinction in pure potentiality." By their "indifference" (*per indifferentiam*), or lack of difference, the unity and identity of the two matters must be accepted.[56]

In the second defense of the identity of celestial and terrestrial matter, Aegidius attempts to compare celestial and terrestrial matter. He asks whether matter that serves as the subject of a higher form is not thereby more worthy, and therefore more actualized, than matter that is the subject of a less worthy form. Since all would readily agree that celestial forms are nobler than terrestrial forms, would it not follow that celestial matter is more actualized than terrestrial matter, and therefore different from it? Aegidius denies the very basis of the comparison by insisting that distinctions between matters that are pure potentiality could not arise from the hierarchical status of the forms which actualize those potentialities. If celestial and terrestrial matter are pure potentiality, no distinctions can be assigned between them on the basis of the greater or lesser nobility of the forms they may support.[57]

As his third defense of the idea that all matter is pure potentiality, Aegidius argues that as pure potentiality, matter is "a mean between being and noth-

53. "Nunquam ex duobus in actu sit unum per essentiam." Ibid.
54. "Eo ergo ipso quod ponimus ibi puram potentiam, oportet quod ponamus ibi unam materiam, et oportet quod illa materia sit eiusdem rationis cum materia istorum inferiorum." Ibid., col. 2.
55. Galileo summarizes seven arguments in favor of the identity of celestial and sublunar matter which he attributes to Aegidius (see Galileo, *De caelo*, qu. 5 (K): "Are the heavens composed of matter and form?", 1977, 113–115, pars. 42, 45, 47–49, 52, 56). For Galileo's rejection of these seven arguments, see pages 138–139. On pages 132–136, Galileo discusses other aspects of Aegidius's thoughts on this question. One may rightly infer that Aegidius's ideas were considered central in the debate on the relationship between celestial and terrestrial matter.
56. "Et quia materia celi est potentia pura et materia istorum inferiorum est potentia pura, si absolverentur ab omni forma, materia huius et illa non haberent per quid differrent quia in pura potentia non potest esse distinctio. Cum ergo dictum sit quod unitas materie et identitas eius accipienda sit per indifferentiam eo ipso quod materia celi absoluta ab omni forma non haberet per quid differret a materia istorum inferiorum sic absoluta, oportet quod sit una materia et eadem per essentiam et eiusdem rationis hic et ibi." Aegidius Romanus, *De materia celi*, 1502b, 82r, col. 1.
57. For a fuller discussion and documentation, see Grant, 1983, 168–169.

ing."[58] But it does not share in any of the properties of those extremes. "For it is not properly a being, because it is not something in act; nor is it absolutely nothing, because it is something in potentiality."[59] Or to put it another way, if matter "became something that had less being than pure potentiality, it would immediately become nothing, because it would be neither actual nor potential; but if it had more of being than pure potentiality, it would be necessary that it become an actualized thing."[60] Thus matter can become neither of these extremes without losing its unique status as pure potentiality. Because the argument applies to both celestial and terrestrial matter, it follows that both are pure potentiality and must always remain so.

For all these reasons, Aegidius concluded that not only does matter exist in the heavens but that as pure potentiality it is identical with the matter of the sublunar, or inferior, region. But, as Aegidius recognized, even if this were true, a major problem yet remained. If celestial and terrestrial matter are identical as pure potentiality capable of receiving forms, why do generation and corruption occur only in the sublunar region? In his response, Aegidius invokes the three fundamental principles of change: matter, form, and privation, the last-mentioned serving as the contrary to a form. Generation and corruption occur when all three are present. In the heavens, however, contraries or privations of a form are absent. Celestial forms lack any associated privation (*privatio admixta*) and are, consequently, perpetual and incorruptible:

If, therefore, you wish to assign a cause as to why these [sublunar] things are corruptible and not those [celestial] things, you should not assign this on the basis of the diversity of matter, because the matter is essentially the same here and there. But you should assign this [cause] based on the diversity of the form, because the [sublunar] form has a contrary and [therefore] has an associated privation; but that [celestial] form does not have a contrary, nor an associated privation. Thus these [sublunar] things are corruptible, not those [celestial] things.[61]

The celestial and terrestrial regions operate in contrary ways. Here below, a form with its associated privation is the cause of corruption, even though mat-

58. "Tertia via ad hoc idem sumitur ex eo quod materia est quid medium inter ens et nihil." Aegidius Romanus, *De materia celi*, 1502b, 82v, col. 1.
59. "Ponimus in terminis materia istorum inferiorum est potentia pura et ideo est media inter ens et nihil. Non enim est proprie ens quia non est quid in actu; nec est omnino nihil quia est quid in potentia." Ibid., col. 2.
60. "Et ut magis clare ostendamus propositum dicamus quod potentia pura non potest perficere in entitate nisi fiat ens in actu; nec potest in aliquo deficere in entitate nisi fiat nihil quia ex quo materia est pura. Si esset aliquid quod esset minus ens quam potentia pura illud statim esset nihil quia nec esset actus neque potentia; sed si plus habeat de entitate quam potentia pura oportet quod sit aliquid ens in actu." Ibid.
61. "Si ergo vis assignare causam quare illa sunt corruptibilia et non ista non assignes hoc ex diversitate materie quia eadem est materia per essentiam hic et ibi. Sed assignes hoc ex diversitate forme ut quia forma ista habet contrarium et habet privationem annexam; illa autem forma non habet contrarium nec privationem annexam. Ideo ista sunt corruptibilia, non illa." Ibid. 83r, col. 2. As the source of this opinion, Aegidius cites Avicenna's *Sufficientia*, book 1.

ter, as pure potentiality, must receive that contrary. In the heavens, however, the form cannot cause corruption because it lacks any contrary or privation. Thus the purely potential matter, which is capable of receiving another form, is never provided an opportunity to exercise that potentiality.[62]

Wielding his sharp, trusty razor, William of Ockham also upheld the identity of celestial and terrestrial matter,[63] though for reasons quite different from Aegidius's. Ockham frankly admitted that the identity of the two matters was not demonstrable but quickly added that neither was any other opinion on this issue.[64] However, by use of his razor and the concept of God's absolute power, Ockham hoped to make his case the most persuasive.

Although the incorruptibility of the heavens was routinely assumed, Ockham insisted that celestial incorruptibility was not absolute, because if it pleased God, he could corrupt and destroy the heavens. By celestial incorruptibility, then, we must mean that the heavens are not corruptible by any *created agent*. Thus the difference between terrestrial matter and celestial matter reduces to this: in the former, God and/or some created agent has the power to corrupt one form and generate another, whereas in the latter only God can corrupt or destroy the form. "Therefore," Ockham argues, "whether the matter is of the same kind or of a different kind does not affect its corruptibility or incorruptibility."[65] And even though the matter in each region is corruptible under different circumstances, the potentiality for corruption is always there. The "difference between the matter here and there," Ockham explains, is that "the matter here is in potentiality to other forms that can be produced by a natural, created agent and [also] to some which can only be created by God alone, as, for example, the intellective form. But the matter of the heaven is in potentiality to many forms, none of which can be produced or induced in that matter by a natural agent but only by God,"[66] who, if he wished, could introduce any terrestrial form,

62. "Dicemus itaque quod econtrario contingit de corruptione in istis inferioribus et de corruptione in supercelestibus. Nam hec forma si est causa corruptionis hoc est propter materiam; ibi autem est causa incorruptibilitatis propter formam. Totum tamen hoc contingit quia in istis inferioribus est admixta privatio; in supercelestibus vero est carentia privationis. Ideo illa sunt perpetua; hec autem corruptibilia." Ibid.

63. In his commentary on the *Sentences*, bk. 2, qu. 18: "Utrum in celo sit materia eiusdem rationis cum materia istorum inferiorium." See Ockham, *Sentences*, bk. 2, qu. 18, 1981, 5:395–409.

64. "Secundo dico quod in caelestibus et in istis inferioribus est materia eiusdem rationis omnino, licet haec pars non possit demonstrari sicut nec alia." Ibid., 400. Such declarations were not uncommon. For similar statements in the sixteenth century, see John Major [*Sentences*, bk. 2, dist. 12, qu. 3], 1519b, 65r, col. 2, and 66v, col. 1, and Suarez, *Disputationes metaphysicae*, disp. 13, sec. 10, 1866, 1:436, col. 2; for the seventeenth century, see Riccioli, *Almagestum novum*, pars post., bk. 9, sec. 1, ch. 5, qu. 2, 1651, 234, col. 2, where Riccioli also attributes this attitude to the Conimbricenses, Hurtado de Mendoza, and Roderigo de Arriaga.

65. "Ergo quod materia sit eiusdem rationis vel alterius nihil facit ad corruptibilitatem vel incorruptibilitatem." Ockham, *Sentences*, bk. 2, qu. 18, 1981, 5:401–402.

66. "Et in hoc est differentia inter materiam hic et ibi quod materia hic est in potentia ad alias formas quae possunt produci per agens naturale creatum et ad alias quae non possunt creari nisi a solo Deo, puta formae intellectivae. Sed materia celi est in potentia ad formas

say the form of fire or water, into celestial matter and thus subject the heavens to generation and corruption. In effect, the same substantial changes produced by natural agents in the terrestrial region could be caused in the celestial region by divine power.

But Ockham even imagines a situtation where the celestial matter would be acted upon naturally by a created agent. This might occur if God introduces the form of the element fire into celestial matter and if a quantity of water of greater active power than fire was sufficiently near. Under these circumstances, the water could naturally corrupt the form of fire and introduce the form of water into celestial matter.[67]

Thus the mere possibility that God could – although he almost certainly would not – effect the same substantial changes in the heavens that are caused by natural agents in sublunar things, led Ockham to conclude that "the matter in the heavens is the same kind as in inferior things." For why should we assume two different kinds of matter when one will do? "A plurality is never to be posited without necessity," Ockham insists. Two varieties of matter ought not to be introduced, "because all things which can be saved by the diversity of the nature of matter can be saved equally well, or better, by the unity of the nature [of matter]."[68] Because only one kind of matter exists in heaven and earth, and the matter in the latter is subject to generation and corruption, we may infer that for Ockham the heavens were also subject to generation and corruption, even though no potentially corruptive forms might ever appear there to cause an actual generation and corruption.

III. Celestial matter in the late sixteenth and the seventeenth century

1. *The focus of the debate: Thomas or Aegidius?*

The impact of these two rival thirteenth-century theories continued strong throughout the Middle Ages[69] and into the sixteenth and seventeenth cen-

multas quarum nulla per agens naturale potest produci nec induci in illa materia, sed solum a Deo possunt ista fieri." Ibid., 403.

67. "Nam materia caeli, ex quo est eiusdem rationis cum materia hic, est in potentia non tantum ad illas formas quae solum possunt causari a Deo et non ab agente creato, sed etiam ad formas quae possunt produci ab agente creato, puta ad formam ignis, aeris, etc. Posito igitur quod Deus in materiam caeli induceret formam ignis, sicut est possibile quia non includit contradictionem, si aqua tunc esset approximata et esset maioris virtutis in agendo quam ignis, corrumperet formam ignis et induceret formam aquae in materia quae primo erat sub forma caeli. Et ideo materia caeli simpliciter est in potentia passiva ad multas formas quas potest agens creatum producere." Ibid., 403–404.
68. "Sic igitur videtur mihi quod in caelo sit materia eiusdem rationis cum istis inferioribus. Et hoc, quia pluralitas nunquam est ponenda sine necessitate, sicut saepe dictum est. Nunc autem non apparet necessitas ponendi materiam alterius rationis hic et ibi quia omnia quae possunt salvari per diversitatem materiae secundum rationem possunt aeque bene vel melius salvari secundum identitatem rationis." Ibid., 404.
69. Nicole Oresme presents an anomalous situation. Although he seems to have accepted celestial matter as a composite of matter and form (Oresme, *De celo*, bk. 1, qu. 11, 1965,

turies. Most early modern scholastic authors rejected the opinion of Averroës and his followers and were agreed that the celestial region consisted of an ether that was composed of matter and form.[70] But within this majority group there was the same split between those who assumed the identity of celestial and terrestrial matter and those who assumed a radical difference.

The overall number of arguments for each opinion seems to have increased. Many of the traditional medieval arguments for the leading opinions were repeated and often elaborated. However, some elements were introduced that either played little role in the Middle Ages or were wholly new. In this regard, Tycho Brahe's rejection of hard, solid spheres and the gradual shift to the concept of fluid heavens played a role.

The case for a matterless heavens in the traditional sense ascribed to Averroës and his followers seems to have steadily lost support.[71] By the sixteenth and seventeenth centuries, additional arguments in favor of a heavens filled with a composite of matter and form were available. It was almost taken for granted that accidents such as quantity, along with rarity and density, that were common to terrestrial bodies also existed in the heavens.[72] Such properties were taken as evidence that matter must also

162–164), he rejects Aegidius's identification of celestial and terrestrial matter (164–166); rejects the idea that a completely perfect form is associated with celestial matter and negates the need of the latter for other forms (166–168); and rejects Thomas Aquinas's theory that celestial matter is of another nature than terrestrial matter (168). By both denying and affirming that celestial and terrestrial matter are identical, Oresme seems caught up in a contradiction.

70. Riccioli, *Almagestum novum*, pars post., bk. 9, sec. 1, ch. 5, 1651, 232–233, distinguishes three common opinions: that the heavens are a simple body not composed of matter and form; that the heavens are composed of matter and form; and a third opinion "that each of the first [two opinions] is probable" and attributes the third opinion to Duns Scotus [*Sentences*, bk. 2, dist. 14, qu. 1], Raphael Aversa [*De caelo*, qu. 33, sec. 5], 1627, 101, cols. 1–2, and Mastrius and Bellutus [*De coelo*, disp. 2, qu. 2, art. 1], 1727, 490, col. 2–492, col. 2. Riccioli mentions that the third opinion derives from the fact that one may argue from the authority of Aristotle that the heavens are not a composite of matter and form, whereas on the authority of theologians one may argue that they are. Scotus does this in the very question cited by Riccioli, except that Scotus substitutes philosophers for Aristotle. John Major, *Sentences*, bk. 2, dist. 12, qu. 3, 1519b, 65r, col. 2, was not unusual in declaring as the first of five conclusions that "although in the natural light it cannot be effectively proved that the heavens do not have matter and form, nevertheless, in the natural light the opposite [that the heavens do have matter and form] is more apparent."

71. A good barometer of its waning popularity is the list of supporters assigned by Riccioli to the two major opinions. His list for Averroës' opinion contains perhaps one seventeenth century figure ([Philip?] Faber), whereas he mentions approximately six seventeenth-century supporters for the rival theory that the heavens consist of matter and form. See Riccioli, *Almagestum novum*, pars post., bk. 9, sec. 1, ch. 5, 1651, 232, col. 2–233, col. 1. To Faber, we may also add Thomas Compton-Carleton [*De coelo*, disp. 3, sec. 3], 1649, 406, col. 2, who after declaring, "it is more probable that the empyrean heaven does not consist of matter and form but is a simple body" adds, in the next paragraph, "it seems more probable to me that all the heavens are not composed of matter and form but are similarly simple bodies." Although Compton-Carleton did not believe that it was impossible that matter could exist in the heavens, he thought it implausible, because whereas matter and form in terrestrial bodies were productive of continuous, obvious change, no such changes were visible in the heavens. He concludes, "we have no basis for asserting that in fact there is matter in them [i.e. the heavens]." Ibid., 407, col. 1.

72. See Aversa, *De caelo*, qu. 33, sec. 5, 1627, 104, col. 2; Conimbricenses [*De coelo*, bk. 1,

exist in the heavens. Suarez argued (*Disputationes metaphysicae*, disp. 13, sec. 10, 1866, 1:437, col. 1) that all mobile physical bodies are natural beings and fall under the domain of philosophy. He observes that in *De caelo*, book 1, chapter 2, Aristotle himself classifies the celestial bodies in the category of natural beings. But in the first two chapters of the second book of his *Physics*, Aristotle explains that nature is nothing but matter and form. Indeed, we do not even know another kind of physical nature. Suarez, therefore, concludes that celestial bodies, as natural, physical beings, must be composed of matter and form. A common scriptural argument invoked by Bartholomaeus Mastrius and Bonaventura Bellutus ([*De caelo*, disp. 2, qu. 2, art. 2], 1727, 3:492, col. 2, par. 57) emphasized that on the first day of creation God created all the matter from which the whole world, including both terrestrial and celestial components, was made; therefore, matter must exist in the heavens.

The interpretation that the heavens consist of matter and form easily triumphed over the matterless concept of Averroës and his followers. The real struggle concerned the nature of that matter: Was it identical with terrestrial matter, as Aegidius and his followers maintained, or was it radically different, as Thomas Aquinas and his supporters would have it? Eminent scholastics of the sixteenth and seventeenth centuries could be found on both sides of this issue. Thus Francisco Suarez (*Disputationes metaphysicae*, disp. 13, sec. 11, 1866, 1:440, col. 2), Raphael Aversa ([*De caelo*, qu. 33, sec. 6], 1627, 105, col. 2–109, col. 1), and Bartholomew Amicus ([*De caelo*, tract. 4, qu. 3, dubit. 1, art. 3], 1626, 146, col. 2) followed Thomas and differentiated between celestial and terrestrial matter, whereas Mastrius and Bellutus ([*De caelo*, disp. 2, qu. 2, art. 2], 1727, 3:492, col. 2, par. 57), as well as Giovanni Baptista Riccioli ([*Almagestum novum*, pars post., bk. 9, sec. 1, ch. 5, qu. 2], 1651, 235, col. 1), and Melchior Cornaeus ([*De caelo*, tract. 4, disp. 2, qu. 1], 1657, 488–489) assumed that they were identical.[73]

In this debate, the theory that preserved the popular Thomistic distinction between celestial and terrestrial matter not only perpetuated the traditional division between the celestial and sublunar realms, but also preserved the principle of celestial incorruptibility. The contributions of Tycho Brahe and Galileo during the late sixteenth and the first quarter of the seventeenth century seriously challenged both of these ideas. For scholastics who, where possible and feasible, were desirous of adapting and adjusting scholastic

ch. 2, qu. 4], 1598, 40; Suarez, *Disputationes metaphysicae*, disp. 13, sec. 10, 1886, 1:437, col. 1; and Major, *Sentences*, bk. 2, dist. 12, qu. 3, 1519b, 65v, col. 2. Aegidius himself had emphasized the existence of accidents in the heavens that were normally associated with matter and that were therefore good indicators that matter existed in the heavens. The kind of "rarity and density" that most scholastics had in mind was of the static (i.e., permanent) kind rather than the dynamic (or variable) type (see Ch. 10, Sec. II.1.c). Their sense of quantity, however, must have been of the terrestrial kind.

73. Riccioli expresses his choice this way: "Although we cannot know demonstratively and evidentally what the visible substance and nature of the heaven is, it is nevertheless more probable that it consists of matter that is identical with elementary matter."

cosmology to the latest scientific knowledge exemplified in the relevant work of Tycho and Galileo, the theory that identified celestial and terrestrial matter was the only hope of achieving a degree of accommodation with the emerging new cosmology.

For it was only by the assumption of a single cosmic matter that the rigid division between the celestial and terrestrial regions could be destroyed, along with its associated idea that the different, nobler, and more perfect celestial region was incorruptible. And yet in his version of the identity theory, Aegidius Romanus retained the rigid division of the celestial and terrestrial regions and the incorruptibility of the former. He could do this on the basis of the standard assumption that celestial forms had no contraries, whereas terrestrial forms did. Thus, although the matter was identical, the lack of contrary forms in the heavens prevented the kind of change that occurred in the terrestrial region.[74] A dramatic conceptual change occurred only when part or all of celestial and terrestrial matter was made identical in more substantial ways than in Aegidius's widely held theory, where the two matters were considered identical only as pure potentialities but otherwise radically different, the one incorruptible, the other corruptible. Significant changes followed upon the newly emerging cosmology, which was based on the consequences of the Copernican theory and the particular celestial discoveries of Tycho and Galileo, who revealed a dramatically different kind of heavens from the heavens as described in traditional medieval cosmology. The heavens that Tycho and Galileo described could bring forth comets, new stars, and variable sunspots. In this altered atmosphere of cosmological speculation, some scholastics assigned to celestial matter the same fundamental properties as one or more of the terrestrial elements. Despite centuries of assumed celestial superiority, the great change involved a terrestrialization of celestial matter, whereby it came to be conceived as corruptible.

2. *Scholastic repudiation of incorruptibility: the corruptibility of celestial matter*

The idea of a corruptible heavens was of ancient vintage and was well known during the Middle Ages, when it was most prominently associated with the name of John Philoponus, the sixth-century Christian Neoplatonist and critic of Aristotle.[75] Unfortunately, the relevant works of Philoponus were

74. For the details of Aegidius's theory, see Section II.2 of this chapter.
75. In his commentary on *De caelo*, Thomas Aquinas, *De caelo*, bk. 1, lec. 6, 1952, 29, remarks that Philoponus was one of those who assumed that the heavens were, by their nature, subject to generation and corruption. Thomas does not mention the source of his information, which was probably Simplicius's *Commentary on De caelo* (in the translation from Greek to Latin by William of Moerbeke in 1271). For Simplicius's remarks, see Simplicius [*De celo*, bk. 1, comment. 20], 1540, 11v, col. 2 and 12r, col. 1. For a discussion of Philoponus's opinion in favor of the identity of celestial and terrestrial matter, see Sam-

unknown until the sixteenth century, as were the details of his interpretation, which included an attack on Aristotle's ether, or fifth element, and the attribution of a fiery nature to the Sun and stars.[76] But, as we saw at the beginning of this chapter, the advent of Aristotle's natural philosophy rendered ideas about terrestrial elements in the heavens obsolete. Indeed, celestial corruptibility and the assignation of terrestrial elements to the heavens were discussed only for the purpose of refutation, as Hervaeus Natalis (ca. 1260–1323) did in his *De materia celi*.[77]

Almost four centuries after Hervaeus, during the second half of the seventeenth century, at least four scholastic authors – Giovanni Baptista Riccioli (1598–1671), Melchior Cornaeus (1598–1665), George de Rhodes (1597–1661), and Franciscus Bonae Spei (1617–1677), all Jesuits except Bonae Spei, who was a Carmelite – concluded that the heavens were corruptible. Of the four, Riccioli is by far the most significant and best known. In his *New Almagest* (*Almagestum novum*) of 1651 (bk. 9, sec. 1),[78] Riccioli considers "The Creation and Nature of Celestial Bodies," within which context – in chapters 5 and 6 – he discusses the problems relevant to our subject. The fifth chapter[79] is devoted to the nature of celestial matter – whether it is a simple or composite body and whether it is the same as or different from elemental matter. The decisions on these questions are relevant to the sixth chapter,[80] in which Riccioli specifically asks "Whether the heavens are generable and corruptible" (An caelum sit generabile et corruptibile).

After following the usual scholastic procedure and presenting the arguments pro and con for the various relevant questions, Riccioli concludes not only that the heavens are composed of matter and form – a popular opinion (as we saw) in the sixteenth and seventeenth centuries – but that celestial matter is probably the same as the matter of the sublunar region,[81] a thesis, as we noted, defended in the Middle Ages by Aegidius and Ockham.

From his assumption of the essential identity of celestial and terrestrial matter, Riccioli drew a radically different consequence: of the four elements

bursky, 1962, 154–166; for additional information, including a statement on celestial corruptibility by John Damascene, see Chapter 10, note 10.

76. Sambursky, 1962, 158.
77. For a summary of his critique, see Grant, 1983, 163–164. Of the four elements, Hervaeus thinks that only fire would be a plausible candidate as a celestial element. But it too fails to qualify because its naturally active qualities would consume the surrounding matter, nor could it account for the multiplicity of effects that the heavens were assumed to cause, such as coldness, wetness, and putrefaction.
78. Book 9 is in the second part (*pars secunda* or *pars posterior*) of the first volume, the only one of the proposed three volumes to be published.
79. Riccioli, *Almagestum novum*, pars post., bk. 9, sec. 1, ch. 5, qus. 1–3, 1651, 232–236.
80. Ibid., ch. 6, 237–238.
81. Riccioli presents his opinion in the following conclusion: "Licet non possit a nobis demonstrative atque evidenter sciri, quaenam sit caeli visibilis substantia et natura, probabilius tamen est illud constare ex materia eiusdem rationis cum elementari." Ibid., ch. 5, qu. 2, 235, col. 1.

(earth, fire, air, and water), the heavens must be composed of one or more of three of the elements (fire, air, and water).[82] Thus where Aegidius made prime matter the basis of the identity of celestial and terrestrial matter, Riccioli made elemental matter the basis – that is, matter that had already been actualized beyond the level of prime matter. It was thus essential that Riccioli identify the elements of which the heavens were composed. He concludes, for example: "It is more probable that the heaven in which the fixed stars are is watery; the heaven in which the planets are is fiery."[83]

Riccioli readily admitted that no genuine evidence or precise arguments could be offered in support of the claim that the heaven of the fixed stars was a congealed, watery solid and the heaven of the planets a fiery fluid.[84] Patristic authorities were, however, at hand. Some Fathers had held that the heaven consisted of elemental water and others that it was composed of elemental fire.[85] It therefore seemed a good compromise to identify the sphere of the fixed stars as the solid and watery sphere (both because the stars themselves remained fixed and unchanging and seemed to enclose the world and because the term *firmamentum* was used to describe the starry sphere) and to interpret the heaven through which the planets moved as a fiery fluid, since the paths of the planets varied.[86]

Riccioli's assumption of a fluid planetary heaven was not of itself a sufficient indication of a belief in celestial corruptibility,[87] but his assertion that these heavens consisted of two terrestrial elements was. In his chapter on the corruptibility or incorruptibility of the celestial region, which follows immediately after the chapter that identifies celestial and terrestrial matter, Riccioli declares the corruptibility of the celestial region. On the basis of his assumption that the heaven of the fixed stars is most probably watery and that the heaven of the planets is fiery, he infers "that from its very internal nature, the heavens have the capacity for generation and corrup-

82. By omitting any discussion of earth as a possible component of celestial matter, Riccioli indicates his rejection of it.
83. "Probabilius est caelum in quo sunt stellae fixae aqueum; caelum autem in quo sunt planetae igneum esse." Riccioli, *Almagestum novum*, pars post., bk. 9, sec. 1, ch. 5, qu. 3, 1651, 236, col. 1.
84. In chapter 7 (ibid., 238–244), Riccioli considers the question "Whether the heavens are solid or whether, indeed, some or all are fluid" (An caeli solidi sint, an vero fluidi omnes vel aliqui). At the end of the question, in a "unica conclusio," Riccioli declares that (ibid., 244, col. 1) "although it is scarcely evident mathematically or physically, it is much more probable that the heaven of the fixed stars is solid, that of the planets fluid." In Chapter 14, we shall consider medieval and early modern scholastic ideas about the hardness or fluidity of the celestial ether.
85. Ibid., ch. 5, qu. 1, 233, col. 2.
86. See ibid., qu. 3, 236, cols. 1–2.
87. A number of scholastics had argued that the fluidity or solidity of the heavens had no relevance to their corruptibility or incorruptibility. For example, see Amicus, *De caelo*, tract. 5, qu. 5, art. 1, 1626, 270, col. 2; Johannes Poncius, *De coelo*, disp. 22, qu. 5, 1672, 620, col. 1; and Franciscus de Oviedo [*De caelo*, contro. 1, punc. 2], 1640, 1:462, par. 2. Oviedo, indeed, believed that the heavens were both fluid and incorruptible. Ibid., 464, col. 1, par. 17.

tion."[88] But Riccioli informs us later that it was not only ideas from the Church Fathers and Scripture that led him to accept celestial corruptibility but also "the arguments derived from experience concerning spots and torches near the solar disk that were discovered by the telescope and from certain comets that have come into being and passed away above the Moon. These changes are more naturally explained by generation and corruption than by other more violent means or by nonviolent miracles."[89]

Perhaps because he was aware of how radically he had departed from traditional Aristotelian cosmology, Riccioli sought to salvage a remnant of celestial incorruptibility. Although, by their elemental nature, the heavens are intrinsically corruptible, they are not corruptible by any naturally created external agent. Thus, for Riccioli, the celestial region was "accidentally incorruptible" (*per accidens esse incorruptibile*), because no natural, external agent could corrupt it. This immunity from corruption by natural agents was perhaps a consequence of the heavens' great distance from the terrestrial region, which was external to them, or perhaps attributable to the great mass of the celestial region, or possibly the result of the distinctive nature of the primary qualities that God had placed in the heavens.[90]

Whatever the reason, Riccioli's concession to incorruptibility was of little consequence, as is evident when he likens celestial incorruptibility to that incorruptibility which applies to the whole earth and to the totality of air, each of which was really incorruptible as a totality even though its parts suffered continual change. Despite their overall incorruptibility, parts of the heavens were nevertheless capable of suffering corruption. In this the celestial region was just like the earth or air: it suffered change in its parts, but the whole endured unchanged.

Only with regard to the empyrean sphere did Riccioli accept the traditional opinion of incorruptibility. The empyrean sphere was not, however, a visible sphere, although it was required for the perfection of the universe and for the incorruptibility and eternal well-being of our bodies.[91]

Among the other three seventeenth-century scholastic authors who rejected celestial incorruptibility, Melchior Cornaeus did it in a single paragraph.[92] Because Cornaeus believed in the identity of celestial and sublunar matter, and because the latter is a principle of generation and corruption, he inferred that the heavens were corruptible.[93] Partial corruption in the

88. "Sequitur caelos hosce esse ab intrinseco et natura sua generationis et corruptionis capaces." Riccioli, *Almagestum novum*, pars post., bk. 9, sec. 1, ch. 6, 1651, 238, col. 1.
89. Ibid., 237, col. 2.
90. "Quia tamen sive propter distantiam ipsorum, sive ob ingentem molem, sive ob temperamentum insigne qualitatum secundarum cum primis quod Deus caelo indidit, non datur agens ullum naturale creatum quod possit caelos transmutare substantialiter; idcirco dixi per accidens esse incorruptibiles." Ibid., 238, col. 1.
91. Ibid.
92. Cornaeus, *De coelo*, disp. 2, qu. 1, dub. 3, 1657, 489.
93. In the next section, Cornaeus rejects the existence of a celestial ether, or fifth element, and suggests that fire is the most probable matter of the heavens. Ibid., dub. 4, 490–491.

heavens was also evident from the many new stars that had been reported from as far back as 125 B.C., including those of 1572, 1600, and 1604. Corruptibility was also implied by scriptural predictions of a Judgment Day in which the heavens will be destroyed. And finally Cornaeus argued that the Sun, the most beautiful part of the heavens, was corrupted almost daily by fires.

George de Rhodes went beyond Riccioli and argued for the fluidity of the entire heavens, including the sphere of the fixed stars.[94] Like Riccioli, however, he believed that although parts of the heavens were corruptible, the heavens, taken as a whole, were incorruptible.[95] And also like Riccioli, he judged the empyrean sphere to be absolutely incorruptible, whereas all other planets and spheres were corruptible. De Rhodes mentions Tycho, who, he explains, showed that the new star of 1572 was truly in the heavens. De Rhodes concludes that new stars could appear only in the celestial region, because they are newly generated there. Thus generation as well as corruption can occur in the heavens.[96]

Following a series of appeals to Scripture in favor of celestial corruptibility, Franciscus Bonae Spei grounds his belief in celestial corruptibility on "the various generations and corruptions in the heavens revealed by the most certain observations of the mathematicians" – that is, astronomers. Indeed, Bonae Spei insists that it is safer to accept the observations made with the telescope and other instruments than to follow the philosophers, who, "because of the very great distance [of the sky] and the weakness of their eyes, are easily deceived."[97]

In the seventeenth century, scholastic opinions about celestial incorruptibility changed rather dramatically from what they had been during the period between the Middle Ages and the end of the sixteenth century. Even if the majority of seventeenth-century scholastics retained the traditional opinion – and this is by no means certain – those like Riccioli, Cornaeus, de Rhodes, and Bonae Spei were prepared to abandon it and concede that

94. De Rhodes [*De coelo*, disp. 2, qu. 1, sec. 2], 1671, 278, col. 2–280, col. 1. De Rhodes specifically refutes the explanations that new stars are not "new" but have been in the heavens all the time but are seen only when they become sufficiently dense (a view attributed to Vallesius) and that new stars are produced by an accidental generation of opacity (a theory he rightly attributes to Aversa; see ibid., 279, col. 1). Since de Rhodes died in 1661 and his work was first published in 1671, the actual date of composition is unknown.

95. "Coelum licet corruptibile sit, nunquam tamen posse corrumpi totum; elementa enim tamersi sunt corruptibilia, semper tamen integra perstant sine ulla imminutione. Contingunt ergo in partibus coeli saepe mutationes." Ibid., 279, col. 2. For what it is worth, de Rhodes makes no mention of Riccioli in this section.

96. "Dico primo celum empyreum omnino incorruptibile; coelos autem sidereos esse corruptibiles." Ibid., 278, col. 2.

97. "Probatur secundo ratione variae contingerunt generationes et corruptiones in coelis, ut constat ex certissima mathematicorum observatione, quos utpote per tot saecula observationibus et tubis opticis aliisque instrumentis, certissime collimantes tutius est sequi quam philosophos, qui, propter longissimam distantiam et visus oculorum per se debilitatem, facile decipiuntur." Bonae Spei [*De coelo*, comment. 3, disp. 3, dub. 4], 1652, 10, col. 2.

substantial generation and corruption could and did occur in the celestial region. In answering the charge that Aristotle had declared the heavens to be immutable and incorruptible, Cornaeus even declared that

> if Aristotle were alive today and could see the alterations and conflagrations that we now perceive in the Sun, he would, without doubt, change his opinion and join us. Surely the same could be said about the planets, of which the Philosopher knew no more than seven. But in our time, through the works of the telescope, which was lacking to him, we know for an absolute certainty that there are more.[98]

Even some traditionalists like Aversa were prepared to break with Aristotle and allow that new stars and sunspots are celestial, rather than terrestrial, phenomena.

But why did scholastic Aristotelians yield on this seemingly important element in Aristotle's cosmology? On this, we can only speculate. Although many scholastics denied a celestial location to the new phenomena, others must have realized, as did Aversa, that astronomical data from the most respected astronomers of the day could not be ignored indefinitely. Thus the first breakthrough – to concede the celestial location of the new phenomena – was probably made rather painlessly, because it was still feasible, and even easy, to insist that such phenomena represented only accidental rather than substantial changes.

The eventual transition to the concept of celestial corruptibility was probably aided in no small measure by a widespread belief in the sixteenth and seventeenth centuries that Plato, Scripture, and many Church Fathers were agreed that the heavens were composed of one or more terrestrial elements and that the heavens were therefore capable of substantial change.[99] Indeed, as we saw, the eighth sphere of the fixed stars was often identified as the frozen or crystalline form of the scriptural waters above the firmament. Other Church Fathers had followed the Platonic idea that the heavens were made of the fourth and purest element, fire. Although such ideas were known during the late Middle Ages, Aristotle's conception of a fifth incorruptible element, or ether, had replaced Platonic and patristic interpretations.

In addition to the works containing ideas about the corruptibility of the heavens that had been available since the Middle Ages, others became available in the sixteenth century: for example, the works of Plato, Philoponus, and Saint Basil. In the developing anti-Aristotelian climate of the sixteenth

98. "Si Aristoteles hodie viveret et quas modo nos in sole alterationes et conflagrationes deprehendimus, videret absque dubio mutata sententia nobiscum faceret. Idem sane est de planetis quos Philosophus septenis plures non agnoscit. At nos hoc tempore opera telescopii (quo ille caruit) plures omnino esse certo scimus." Cornaeus [*De coelo*, disp. 2, qu. 3, dub. 3], 1657, 503.

99. Riccioli, *Almagestum novum*, pars. post., bk. 9, sec. 1, ch. 6, 1651, 237, col. 1–238, col. 1, discusses all three and provides a lengthy list of passages from the Bible and the Church Fathers.

century, celestial corruptibility was an opinion that became more difficult to ignore than during the Middle Ages. The dramatic celestial discoveries of the late sixteenth and the early seventeenth century provided the scientific basis for abandoning incorruptibility. Scholastics who found the celestial discoveries of Tycho and Galileo compelling could justify support for celestial corruptibility by direct appeal to Plato and, more significantly, to the Church Fathers. Or they may have found the astronomical arguments compelling only because of the corroborating statements of the Church Fathers.

Although I shall discuss in Chapter 14 the transition from the concept of solid heavens to that of fluid heavens, it is relevant to inquire at this point whether the gradual acceptance of the idea of fluid rather than solid heavens played a significant role in the abandonment of belief in celestial incorruptibility. Tycho's claim that the comet of 1577 was moving among the planets clearly implied the nonexistence of solid planetary spheres.[100] For those who accepted comets as supralunar, a gradual but inexorable shift toward the concept of fluid heavens began. But did the idea of fluid heavens imply corruptible heavens? At least one Jesuit scholastic, Antonio Rubio, in a work of 1615, believed that fluid heavens would have to be corruptible (presumably because of divisibility) and therefore rejected them.[101] But, as we have seen (n. 87 of this chapter) others – for example, Amicus, Poncius, and Oviedo – thought that the solidity or fluidity of the heavens was irrelevant to the issue of incorruptibility. Indeed, Oviedo believed that the heavens were both fluid and incorruptible. For some scholastics, then, fluidity alone did not necessarily entail divisibility. The matter of the heavens might be such that it was capable of receiving only a single form, or celestial matter might be incorruptible by virtue of its form, a form that adhered to its matter so firmly that another could not be received.[102] A seventeenth-century scholastic could therefore accept both fluidity and incorruptibility. Although the shift from solidity to fluidity was a significant change from the medieval tradition, it was not crucial for the issue of celestial incorruptibility.

IV. Some concluding observations about celestial matter and incorruptibility

From the various interpretations described earlier, it may be tempting to view Aegidius Romanus's identification of celestial and terrestrial matter as

100. Because the parallax of the comets placed them below the fixed stars, one could continue to believe, as did Riccioli, that the fixed stars were embedded in a solid sphere. (Donahue, 1972, 117, holds that the sphere of the fixed stars was the last element of the old cosmos to go.)
101. Donahue, 1972, 105.
102. Oviedo, *De caelo*, contro. 1, punc. 2, 1640, 1:462, col. 1, par. 2.

an anticipation of the final rejection in the seventeenth century of the traditional distinction between the heavens and the earth. Such a temptation should be resisted. Aegidius's identification of celestial and terrestrial matter as pure potentialities was made in the context of an Aristotelian universe in which the celestial region was assumed incorruptible and thus utterly superior to a continually changing terrestrial matter.

It was not Aegidius but John Philoponus who anticipated the seventeenth-century concept of a universal matter everywhere subject to change. During the Middle Ages only the idea was known, but not the works of Philoponus in which it was developed and justified. The dominance of Aristotle's physics and cosmology would have made such a bold idea unacceptable. About the time Philoponus's works became available in the sixteenth century, the Copernican theory, which made the earth just another planet, was at hand to dissolve the medieval distinction between the terrestrial and celestial regions. As the Copernican theory was gradually disseminated and adopted, it became inevitable that the earth and its sister planets would be conceived in terms of the same matter.

What was strongly implied in Copernicus's profound conceptualization was rendered graphically real by Galileo's telescopic observations some sixty-five years after the publication of *De revolutionibus* and only some fifteen to twenty years after Galileo had himself vigorously upheld the incorruptibility of the heavens and assumed the existence of a celestial matter that was radically different from its terrestrial counterpart. Beginning in 1610, however, with the publication of the *Sidereus nuncius* (*The Starry Messenger*), Galileo rendered the centuries-old tradition of celestial incorruptibility untenable.

Although Galileo did not explicitly declare the corruptibility of the heavens in the *Sidereus nuncius*, it was an obvious inference from the comparisons he made between the earth and the Moon, as when he observed that "the terrestrial roughnesses are far smaller than the lunar ones"[103] and that the earth "is movable and surpasses the Moon in brightness" and "is not the dump heap of the filth and dregs of the universe."[104] In his *Second Letter on Sunspots*, Galileo made the corruptibility of the heavens explicit,[105] and he also argued for celestial corruptibility at great length in his *Dialogue Concerning the Two Chief World Systems*,[106] where, as evidence for corruptibility, he cited comets, new stars, sunspots, and the Moon's rough surface.[107] All the planets and stars were declared alterable by Galileo, even though the Moon and other planets might differ from the earth because their similar,

103. Galileo, *Sidereus nuncius* [Van Helden], 1989, 51.
104. Ibid., 57.
105. Galileo, *Letters on Sunspots* (1613) in Drake, 1957, 118. *The Second Letter on Sunspots* is dated August 14, 1612, and was printed in 1613.
106. Galileo, *Dialogue*, First Day [Drake], 1962, 41–100.
107. Ibid. 51, 72, 100. Galileo did not, however, believe that the Moon's matter was like that of our earth, and from this he inferred that if plants and animals existed on the Moon, they would be unlike those on earth.

if not identical, matter was affected differently by the Sun and other environmental conditions.

Despite possible differences in lunar and earthly matter or even between the earth and any other planet, Galileo wholly rejected the traditional conception of the earth's inferiority with respect to celestial bodies, an inferiority that was based on the alleged immutability of the heavens as contrasted to the alterability of terrestrial bodies. Through Sagredo, his spokesman in the *Dialogue*, Galileo posed a question, the answer to which would have been assumed as self-evident in the Middle Ages. Why should immutability be more noble than mutability? "For my part," Sagredo replies,

> I consider the earth very noble and admirable precisely because of the diverse alterations, changes, generations, etc. that occur in it incessantly. If, not being subject to any changes, it were a vast desert of sand or a mountain of jasper, or if at the time of the flood the waters which covered it had frozen, and it had remained an enormous globe of ice where nothing was ever born or ever altered or changed, I should deem it a useless lump in the universe, devoid of activity and, in a word, superfluous and essentially nonexistent. This is exactly the difference between a living animal and a dead one; and I say the same of the moon, of Jupiter, and of all other world globes.[108]

These words were written nearly one hundred years after the publication of Copernicus's *De revolutionibus*. Despite the repudiation of celestial immutability by Tycho in the late sixteenth century and by Kepler in the second decade of the seventeenth century, Galileo thought it necessary to devote most of the discussion of the first day of his *Dialogue* to that same theme. There his interpretations of his own telescopic observations reduced the issue of celestial incorruptibility to a nullity. After Galileo, scholastics, as we have seen, were much less likely to assign incorruptibility as a property of celestial matter.

108. Ibid., 58–59.

13

The mobile celestial orbs: concentrics, eccentrics, and epicycles

As we saw earlier (Ch. 1, Sec. II.2 and n. 12), the term *caelum*, which signifies heaven or heavens, was used to designate the entire celestial region and occasionally even the spheres of the elements below. Indeed, Thomas Aquinas conceived all seven planetary spheres plus the sphere of the fixed stars as part of a single "sidereal heaven." The customary usage of the term *caelum* was, however, more restricted and was intended to designate the sphere or spheres that carried a single planet around, as when Galileo declared: "we maintain that there are ten movable heavens [*caelos mobiles*], and beyond these, that there is an eleventh, immovable heaven."[1] Although in this chapter occasions will arise where the term "heaven" *(caelum)* will be applied to a single orb or to the spheres of a single planet, I shall more regularly use the terms "sphere(s)" or "orb(s)."[2] The term "heavens," or "heaven," will continue to be used to designate the celestial region as a whole.

I. One heaven (or sphere) or many?

Despite the multiplicity of planets and stars, it was not immediately obvious to Aristotelian natural philosophers that the heavens were filled with independent orbs to which those celestial bodies were in some way attached and which functioned to carry them around in their orbits. In his *Metaphysics* (12.8.1074a.30–37 [Ross], 1984), Aristotle declares that "there is but one heaven." For if more than one existed, matter would necessarily exist in the heavens, because "all things that are many in number have matter." But matter cannot exist in the heavens. Moreover, "that which is moved always and continuously is one alone; therefore there is one heaven alone." But in the same *Metaphysics* (12.8.1073b.17–1074a.16), immediately pre-

1. Galileo [*De caelo*, qu.1 (G)], 1977, 63. The Latin is from his *Juvenilia*, in Galileo, *Opere* [Favaro], 1891–1909, 1:41.
2. On the largely ignored distinction between the terms "orb" and "sphere," see Chapter 6, note 37.

ceding his declaration that the heavens are one, Aristotle describes and expands upon the famous astronomical systems of Eudoxus and Callippus. Not only did he increase the 33 spheres of Callippus to 55, but he also conceived of the spheres as physical bodies, rather than as convenient mathematical constructions in the manner of Eudoxus and Callippus. In *De caelo*, Aristotle declares emphatically that each planet is carried around in its own sphere.[3] Was there a conflict here? Would the existence of spheres divide the celestial region into a multiplicity of heavens? And would a series of distinct spheres imply divisible, rather than indivisible, heavens? Both possibilities were contrary to basic Aristotelian principles.

In the sixth century, John Philoponus, an important Greek commentator on the works of Aristotle but also a Christian, wrote a significant commentary on the six days of creation. In it, he assumed that in Genesis Moses was describing a single heaven, or firmament, which included all the planets and stars. Although astronomers were free to save the appearances by imagining as many orbs and motions as they wished, they had never demonstrated the existence of such entities. Indeed they did not themselves agree on the number of these extra orbs and motions.[4] In his commentary on *De caelo*, Averroës conceived of the totality of celestial orbs as a single heaven. Indeed, he thought of it as a single animal, whose partial orbs were like its members: that is, the motions of the partial orbs were like the motions of the members of an animal. It was because of this oneness that the whole heaven could be moved with a single daily motion.[5]

During the Middle Ages, scholastics frequently inquired whether the heaven was one, continuous, indivisible body or a divisible entity comprised of a series of distinct spheres, including one sphere for all the fixed stars and at least one sphere for each planet. Robert Grosseteste considered the question on the oneness or multiplicity of the firmament, which for him embraced everything from the Moon to the fixed stars, difficult.[6] He confessed an inability to answer it, asserting that "no one can declare anything certain about the number of heavens or their motions or movers or na-

3. "But the upper bodies [i.e., the planets] are carried each one in its sphere." Aristotle, *De caelo* 2.7.289a.30 [Guthrie], 1960. In his thirteenth-century translation, Michael Scot rendered this passage as "stellae autem, quae sunt in orbe superiori, procedunt in suo orbe." For Michael's translation in the edition of Averroës' commentaries, see Averroës [*De caelo*, bk. 2, text 42], 1562–1574, 5:124v, col. 2.
4. Duhem, *Le Système*, 1913–1959, 2:111–112, 498–499, describes Philoponus's interpretations as he found them in the latter's Greek text of *De opificio mundi libri VII* edited in 1897 by Walter Reichardt. As his source, Duhem gives book 3, chapter 3, pages 113–116. Toward the end of the fifteenth century, in Italy, Giovanni Gioviano Pontano (1429–1503) rejected the existence of physical orbs and assumed that the orbs of the astronomers were simply convenient devices to aid the understanding. He believed that the planets were self-moving. See Trinkaus, 1985, 455. Even before Tycho Brahe's discoveries, the Jesuit Robert Bellarmine rejected the existence of orbs in favor of a fluid celestial medium (see Ch. 14, Sec. VIII).
5. See Averroës, *De caelo*, bk. 2, comment. 42, 1562–1574, 5:125v, col. 1.
6. Grosseteste asks: "Que sit autem huius firmamenti natura, et quot sint celi contenti in hoc uno celo quod dicitur firmamentum." *Hexaëmeron*, part. 3, ch. 6, 1, 1982, 106, lines 1–2.

tures."[7] Indeed, we cannot even be certain that we know the number of planets, since other planets might exist that are invisible to us but which might play a role in the generation of sublunar things.[8] In arriving at a judgment about the oneness or multiplicity of the heavens, scholastic natural philosophers frequently had to consider whether the celestial orbs were *continuous* or *contiguous*, and the latter question occasionally involved opinions about whether the celestial region was composed of a fluid substance or a series of hard celestial spheres. Indeed, in the seventeenth century, when "solid" sphere was synonymous with "hard" sphere (see Ch. 14, Sec. VII), Roderigo de Arriaga declared that for those who assumed a fluid heaven, the problem of the number of spheres was easy to resolve. They would "assume only one heaven [conceived as a single sphere] through which the planets and stars move as [do] fish in the sea, or birds in the air. The difficulty lies in the case in which the heavens are solid."[9]

At the outset, we must draw attention to an apparent ambiguity in the medieval interpretation of "one heaven(s)," which was understood as synonymous with the whole celestial region. Some conceived of such a single heaven, or heavens, as devoid of orbs, while others thought of it as subdivided into orbs that were either continuous or contiguous. By contrast, some, perhaps most, assumed a celestial region that embraced numerous separate heavens in the form of independent orbs that were regarded as contiguous to, and therefore distinct from, one another.

Those who believed in a single, physical heaven without orbs[10] understood by this the entire celestial region, which embraced all of the planets and stars and for some even extended down to the region of the air. In this system, the planets, which are not attached to orbs, are moved either by themselves or by an intelligence. Although John Damascene and John Chrysostom may have adopted such an interpretation, and Robert Grosseteste at least entertained it,[11] an orbless heaven found few supporters during the

7. "Nullus potest de numero celorum aut eorum motibus aut motoribus aut ipsorum naturis aliquid certum profiteri." Ibid., part. 3, ch. 8, 3, 109, lines 7–9. Confessions of ignorance about the firmament were not unusual. Durandus de Sancto Porciano [*Sentences*, bk. 2, dist. 14], 1571, 155v, col. 2, reports that Bede identified the firmament with the sidereal heaven (*coelum sydereum*) that divides the waters from the waters, while some thought the firmament was air, others that it was water and yet others that it was fire. "What these are and for what they were created," asks Durandus, "are known only to the one who created [them]" (Quales autem sunt, et ad quid conditae, ille solus novit qui condidit).
8. "Sed unde scietur quod non sint plures stelle erratice nobis invisibiles, generacioni tamen in inferiori mundo necessarie et utiles?" *Hexaëmeron*, part. 3, ch. 8, 3, 108, lines 24–25. The existence of such planets could only be known by revelation ("Unde igitur sciri posset, nisi a divina revelacione, an non sint plurime huiusmodi stelle invisibiles nobis, quarum quelibet suum habeat celum movens ipsam ad profectum generacionis in mundo inferiori?"). Ibid., lines 27–30.
9. "Nova hic celebris occurrit de caelorum numero difficultas. Qui caelos fluidos dicunt facile se ab hac quaestione expediunt: ponunt enim unum tantum caelum per quod planetae astraque discurrunt, sicut pisces in mari aut aves in aere. Difficultas est casu quo caeli sint solidi." Arriaga [*De caelo*, disp. 1, sec. 4], 1632, 504, col. 1.
10. Because of its special nature, the empyrean heaven was always treated as a separate entity.
11. Although John Damascene assumed that the firmament embraced all seven planets (he

Middle Ages but came into vogue following upon Tycho Brahe's famous rejection of hard, celestial orbs in favor of a fluid, or soft, celestial region.[12]

The concept of a single heaven divided into spheres was, however, not uncommon. Thus Thomas Aquinas assigned the seven planets and the fixed stars to a single heaven, the "sidereal heaven," distinguishing eight spheres, one for each planet and the fixed stars.[13] Whether Thomas assumed a continuous heaven is uncertain, but Aegidius Romanus leaves no doubt that he himself viewed the seven planets and the fixed stars as part of one continuous heaven that nevertheless contained deferents and eccentrics that were contiguous and therefore discontinuous. Aegidius thus insisted on a heaven that was continuous in one sense but discontinuous in another. (How he presented this important and rather influential concept is described in Section III.9.) More common, however, was Christopher Clavius's interpretation of a separate firmament of fixed stars, under which there are seven separate planetary spheres.[14]

Until the seventeenth century, when the effects of Tycho's authoritative repudiation of the existence of solid celestial spheres began seriously to affect scholastic thought, almost all scholastic authors assumed the existence of real, physical spheres in the heavens.[15] The most fundamental reason for the postulation of celestial orbs derived from the long-observed diversity of planetary motions. Because planets were thought incapable of self-movement and therefore unable to move through the celestial ether "like a bird through air or a fish through water," as the popular phrase went,[16]

speaks of "the seven zones of the firmament" [septem zonis firmamenti], where each zone contains a planet), he appears not to have assumed a division into orbs but rather into "zones." He states, "they say that there are seven zones of the heavens, one better than the other" (Septem vero zonas aiunt esse caeli, aliam alia excelsiorem). See John Damascene, *De fide orthodoxa*, 1955, 80, 82. Whether Damascene included the fixed stars in the firmament is unclear, but Thomas Aquinas seems to have thought so (see Ch. 5, Sec. VI.2).

12. For example, Francisco de Oviedo ([*De caelo*, contro. 1, punc. 4], 1640, 471, col. 2), who rejected the existence of hard orbs, held that a single, overall heaven could be conceived as consisting of several subheavens, "because part of a heaven through which the Moon moves can be called the 'heaven of the Moon,' and another part through which the Sun is moved can be called the 'heaven of the Sun,' and the same for the other parts [of the heaven] through which the other planets are moved."
13. See Thomas Aquinas, *Summa theologiae*, pt. 1, qu. 68, art. 4, 10:89. In the seventeenth century, Amicus included all of these in the firmament. After arguing that "the solidity that we proved applies to the firmament, probably also applies to all the heavens of the planets," he declares that in Scripture "the term firmament not only signifies the heaven of the fixed stars but also [those] of the wandering stars." Amicus [*De caelo*, tract. 5, qu. 5, art. 3], 1626, 279, col. 2. Amicus accepted the existence of celestial orbs.
14. Clavius [*Sphere*, ch. 1], *Opera*, 1611, 3:23.
15. As we saw in note 4, Pontano denied their existence. But Pontano does not qualify as a scholastic author (see Trinkaus, 1985, 449).
16. Buridan declares that a planet can be imagined to be self-moving "by dividing the orb itself, just as a bird flies through air or a fish swims in water, or even as a man walks in air." See Buridan [*De caelo*, bk. 2, qu. 18], 1942, 210–211. Many others used either the bird or fish analogy. For example, see Aegidius Romanus's discussion of eccentrics and epicycles, where he declares that it is unreasonable to say that "the planets are moved in

they had to be carried around by gigantic orbs in which they were embedded.[17] On the further assumption that every motion attributed to a planet required its own unique orb, it followed that every orb moved with a single, unique motion, in a single direction, but that different orbs could move in diverse directions. The motion of a single orb in a single direction could not account for the astronomical phenomena.

The need to account for the celestial phenomena was the reason why celestial orbs were deemed essential. They had been part of the Greco-Arabic legacy of treatises that were translated into Latin in the twelfth and thirteenth centuries. More particularly, discussions about celestial orbs were embedded in the extensive Aristotelian–Ptolemaic body of astronomical and cosmological literature. No serious opposition to their acceptance appeared. Despite the assumption of a multiplicity of orbs, however, some scholastic authors conceived of the celestial region as one continuous body, because they assumed that the surfaces between successive spheres were continuous.[18] Others denied such continuity and argued for the contiguity of the surfaces, and therefore for the distinctiveness of each sphere. Still others assumed a single, continuous heaven with diverse planetary channels or cavities that functioned as deferent or eccentric orbs. These interpretations depend on Aristotle's definitions of the continuity and contiguity of surfaces. However, before defining these terms and examining the manner in which they were applied to the celestial orbs, we shall first describe the kinds of spheres about which medieval cosmologists were concerned.

II. Concentric versus eccentric orbs

1. Aristotle's system of concentric spheres

Between approximately 1160 and 1250, two rival cosmological systems entered western Europe and vied for acceptance. The first was derived from the works of Aristotle, where it was assumed that the stars and planets were carried around on, or in, concentric or homocentric spheres. As concentric spheres, they shared a common center, which was both the geometric center of the world and the center of the earth. In this system, which he derived from his predecessors Eudoxus and Callippus, and which he describes all

orbs as fish are moved in water" (Aegidius Romanus, *Opus Hexaemeron*, 1555, 49r, col. 2). Duhem translates the section in *Le Système*, 1913–1959, 4:114. One or the other of the two descriptions appears in Conimbricenses [*De coelo*, bk. 2, ch. 5, qu. 1, art. 1], 1598, 246; Clavius [*Sphere*, ch. 4], 1593, 515; Aversa [*De caelo*, qu. 32, sec. 6], 1627, 66, col. 1; Amicus [*De caelo*, tract. 5, qu. 4, dubit. 3, art. 1], 1626, 266, col. 1; Arriaga, *De caelo*, disp. 1, sec. 3, subsec. 1, 1632, 499, col. 2; and Oviedo, *De caelo*, contro. 1, punc. 4, 1640, 1:471, col. 2. For Robert Bellarmine's use of the metaphor, see Chapter 14, Section VIII and note 76.

17. How those orbs were thought to move is considered in Chapter 18.
18. For example, Saint Bonaventure [*Sentences*, bk. 2, dist. 14, pt. 2, art. 1, qu. 1], *Opera*, 2:352, col. 1.

too briefly in book 12, chapter 8, of the *Metaphysics*,[19] Aristotle assumes the existence of 55 spheres. On the basis of his study of Eudoxus and Callippus, Aristotle assigned 33 concentric spheres to account for the motions of the seven planets. Of these 33 spheres, 4 each were assigned to Saturn and Jupiter and 5 each to Mars, Venus, Mercury, the Sun, and the Moon. Then Aristotle declares (ibid., 1073b.38–1074a.5):

> It is necessary, if all the spheres put together are going to account for the observed phenomena, that for each of the planetary bodies there should be other counteracting [literally "unrolling"] spheres, one fewer in number [than those postulated by Callippus for each set] and restoring to the same function each time as regards position the first sphere of the planetary body situated below; for only thus is it possible for the whole system to produce the revolution of the planets.[20]

Thus did Aristotle assign 22 additional "unrolling" spheres, for a total of 55. D. R. Dicks (1970, 200–201) has described the relationship between the regular and unrolling spheres of Saturn as follows:

> Thus for the four spheres of Saturn A, B, C, D, a counteracting sphere D′ is postulated, placed inside D (the sphere nearest the earth and carrying the planet on its equator) and rotating round the same poles and with the same speed as D but in the opposite direction; so that the motions of D and D′ effectually cancel each other out, and any point on D will appear to move only according to the motion of C. Inside D′ a second counteracting sphere C′ is placed, which performs the same function for C as D′ does for D; and inside C′ is a third counteracting sphere B′ which similarly cancels out the motion of B. The net result is that the only motion left is that of the outermost sphere of the set, representing the diurnal rotation, so that the spheres of Jupiter (the next planet down) can now carry out their own revolutions as if those of Saturn did not exist. In the same manner, Jupiter's counteracting spheres clear the way for those of Mars, and so on (the number of counteracting spheres in each case being one less than the original number of spheres in each set) down to the Moon which, being the last of the planetary bodies (i.e. nearest the earth), needs, according to Aristotle, no counteracting spheres.

This is indeed the extent of Aristotle's "detailed" account of the number and interrelationships of the concentric spheres that carry the planets around the sky. Not only is it exceedingly brief, to the point of obscurity, but Aristotle even left scholars in confusion over the actual number of spheres, suggesting that if one subtracts certain motions from the Sun and Moon, the total number of spheres would be 49 (1074a.12–14). Some have argued that 55 is required, while others reject 55 and argue for 49 or 61.[21] Because

19. Indeed, this is the only place among the currently extant works of Aristotle in which he provides any details about his system of concentric spheres.
20. The translation is by D. R. Dicks, 1970, 200.
21. Dicks accepts 55 (1970, 202–203), while Hanson (1973, 69) argues for 49 or 61, with the latter preferable.

Aristotle fails to mention or further explicate the unrolling spheres anywhere else in his works, a close examination of what the real physical relations of these spheres might be yields little but frustration and confusion.[22] It is with good reason that Dicks (1970, 203) has suggested that Aristotle may have considered it only "an interesting speculation, but one that would not stand up to close scrutiny."

If Aristotle was spare in the description of his system of concentric spheres, his Greek commentators – Sosigenes, Simplicius, and Philoponus, for example – were more lavish and made some effort to explain Aristotle's meaning and intent. From the thirteenth to the seventeenth century, Aristotle's system and number of concentric orbs were rarely discussed in questions on the *Metaphysics*, although they did receive consideration, if only summarily, in section-by-section commentaries on the *Metaphysics*.[23] Not only were commentaries on the *Metaphysics* rarer than questions on the *Metaphysics*, but the difficulty and obscurity of Aristotle's purely concentric system tended to discourage full-blown discussions. Indeed, it was supplemented, if not largely supplanted, by a second system that also reached the Latin West in the great wave of Greco-Arabic translations of the twelfth and thirteenth centuries.

2. *Ptolemy's system of eccentric spheres*

In the second system derived ultimately from Claudius Ptolemy's *Hypotheses of the Planets*, the planets were assumed to be carried around by a system of material eccentric and epicyclic spheres, whose centers were geometric points that did not coincide with the earth's center.[24] Despite its momentous role in medieval cosmology, the *Hypotheses of the Planets* was not translated into Latin during the Middle Ages. In some as yet unknown manner, however, its fundamental ideas reached western Europe, probably in works translated from Arabic. The precise treatise, or treatises, in which these ideas were embedded have yet to be identified. Although works attributed to Alhazen (ibn al-Haytham), Alfraganus (al-Fargani), and perhaps Thabit ibn Qurra included descriptions of material eccentric and epicyclic spheres,

22. Dicks (1970, 203) declares that "if the heavens really operated in this manner, with the counteracting spheres effecting their respective cancellations of planetary motions, how did astronomers ever manage to make the obervations that lay behind the original Eudoxan scheme with its planetary loops and retrogradations?"
23. Jean Buridan's *Questions on the Metaphysics* lacks such a discussion, while Thomas Aquinas's *Commentary on the Metaphysics* includes one (Thomas Aquinas [*Metaphysics*, bk. 12, lesson 10], 1961, 2:904–907).
24. In this section, I rely on Grant, 1987b, 189–195. Duhem, *Le Système*, 1913–1959, vols. 3–4, describes the controversy between the defenders of concentric orbs and those who sided with Ptolemy and assumed eccentric and epicyclic orbs. Duhem chose to treat this important topic by taking up a series of individual authors in chronological order. The concentric–eccentric controversy was but one of a number of themes discussed for each author. Sections of varying lengths are thus isolated within descriptions of individual authors and are nowhere adequately summarized or synthesized.

the basic scholastic version of eccentrics and epicycles described in this chapter finds no counterpart in the Latin translations of these Arabic treatises.[25]

References to epicycles and eccentrics appear in widely used thirteenth-century works like Sacrobosco's *On the Sphere* and in the anonymous *Theorica planetarum*, although neither author implies or suggests that they might be real, material, solid orbs.[26] If Roger Bacon (ca. 1219–ca. 1292) was not the first to mention material eccentrics and epicycles in the Latin West, he may well have been the first scholastic natural philosopher to have presented a serious evaluation of their cosmological utility.[27] After some hesitation and ambivalence, Bacon rejected physical eccentrics and epicycles and opted for Aristotle's system of concentric spheres. Ironically, it was his description of the system of eccentrics and epicycles that was most widely adopted by medieval natural philosophers and which still found defenders well into the seventeenth century. Although many scholastics ignored the topic of eccentrics and epicycles, those who gave it more than a cursory glance included, in the thirteenth century, Albertus Magnus and Duns Scotus; in the fourteenth century Aegidius Romanus, John of Jandun, Jean Buridan, and Albert of Saxony; in the fifteenth century Pierre d'Ailly,[28] Cecco d'Ascoli, Johannes de Magistris, Johannes Versor, and Thomas Bricot; in the sixteenth century John Major, the Coimbra Jesuits, and Christopher Clavius; and in the seventeenth century Bartholomew Amicus. Bacon's account, perhaps

25. Although it has been said (Pedersen [Lindberg], 1978, 321 and then 319) that the machinery of the material spheres suggested by Ptolemy's *Hypotheses of the Planets* was presented to Western astronomers in two brief cosmological treatises by Thabit ibn Qurra, namely Thabit's *De hiis quae indigent antequam legatur Almagesti* and *De quantitatibus stellarum et planetarum et proportio terrae*, an examination of these treatises reveals nothing relevant for the problem of material or physical eccentric and epicyclic spheres (for the texts, see Thabit ibn Qurra [Carmody], 1960, 131–139, 145–148). What Thabit may have passed on to the West on the subject of material eccentric spheres will be seen later in this chapter.

26. For Sacrobosco's discussion, see Sacrobosco, *Sphere*, 1949, 113–114 (Latin), 140–141 (English). For the *Theorica*, see Olaf Pedersen's translation in Grant, 1974, 452–465 (eccentrics are defined on page 452). In speaking of "real, material, solid orbs," I deliberately refrain from signifying whether those orbs were conceived as hard or soft, a problem that will be considered later.

27. Sometime in the 1260s, Bacon presented almost identical accounts in his *Opus tertium* and in the second book of his *Communia naturalium*, or the *De celestibus*, as I shall refer to it. In the latter, Bacon added a significant chapter on whether eccentrics and epicycles were consistent with Aristotelian cosmology and followed this with a lengthy description of the Ptolemaic system. See Bacon, *Opus tertium*, 1909, where Bacon's discussion on eccentrics and epicycles extends over pages 99–137. For the *De celestibus*, see Bacon, *Opera*, fasc. 4, 1913, 419–456. A. C. Crombie and J. D. North believe the *Opus tertium* was written sometime between 1266 and 1268 (see their article on Bacon in *Dictionary of Scientific Biography*, 1970–1980, 1:378) and suggest that Bacon may have written his *Communia naturalium* in this same period. By contrast, Lindberg [Bacon], 1983, xxxii, who accepts 1267 as the date of composition for the *Opus tertium*, believes that no firm date can be attached to the *Communia naturalium* and concludes: "all that can be said is that it represents the early stages of the broadening of Bacon's outlook, the usual guess placing it in the early 1260s."

28. If d'Ailly's treatment was perhaps less extensive than Bacon's, it was nevertheless lengthier than most others.

the most important medieval description of eccentrics and epicycles, as these were understood by natural philosophers, forms the basis of this discussion.

Because Bacon and subsequent authors regularly assumed at least three eccentric orbs for each planet,[29] I shall frequently refer to the "modern" theory as the "three-orb system," but I shall also refer to it as the "Aristotelian–Ptolemaic system," since Aristotle's concentric spheres were assigned a significant role within the system of eccentrics.

3. The system of eccentrics: Roger Bacon

After demonstrating the impossibility of any system comprised of eccentric orbs in which the eccentricity is due to an eccentric convex surface, an eccentric concave surface, or the eccentricity of both surfaces, Bacon introduces another interpretation – "a certain conception of the moderns," as he put it[30] – in which the external surfaces of each planetary orb are concentric but which contain at least three eccentric orbs. To illustrate the system, Bacon describes the motions of the Sun and Moon. Because the Sun has only an eccentric orb, it will be useful to follow his account of the Moon, which has both an eccentric and an epicyclic orb.

In the diagram (Fig. 8),[31] let *T* be the center of the earth and world and also the center of the lunar orb. The entire sphere of the Moon lies between the convex circumference *ADBC* and the concave circumference *OQKP*, which are both concentric to *T*. Between these two circumferences, three orbs are distinguished (namely a', b', and c') by assigning another center, *V*, toward the Moon's *aux*, or apogee. Around *V* as center are two circumferences, *AGFE* and *HNKM*, which signify the surfaces that enclose the lunar deferent and form the eccentric orb b'. Surrounding the eccentric orb or deferent is the outermost orb, a', lying between surfaces *ADBC* and *AGFE*; and surrounded by the eccentric orb is the innermost orb, c', lying between the concave surface *HNKM* and the convex surface *OQKP*. Between the surfaces of the middle, or eccentric, orb is a concavity that contains a spherical epicycle. The latter may be conceived in two ways: either as a solid globe, which Bacon calls a "convex sphere" (*spericum convexum*), which resembles a ball (*pila*) because it lacks a concave surface; or as a ring with two surfaces, one convex (*KLFI*), the other concave (*RYSΘ*), where the

29. See Duhem, *Le Système*, 1913–1959, 4:112.
30. "De quadam ymaginatione modernorum." See Bacon, *Opus tertium*, 1909, 125 (and 125–134 for the exposition and critique of the modern theory) and Bacon, *De celestibus*, fasc. 4, 1913, 438 (and 437–443 for the modern theory). Although the "modern" theory seems ultimately derived from Ptolemy's *Hypotheses of the Planets*, Bacon's immediately preceding discussion (*Opus tertium*, 119–125; *De celestibus*, 433–437), concerning the impossibility of a total planetary orb being composed of eccentric orbs where one or both of the external surfaces is eccentric, may have derived from an earlier attempt to materialize eccentrics on the basis of Ptolemy's description in the *Almagest*.
31. The figure appears in Bacon, *Opus tertium*, 1909, 129; the relevant text is on pages 128–131. The figure was used in Grant, 1987b, 191. To identify the three distinct orbs, I have added the letters a', b', c'.

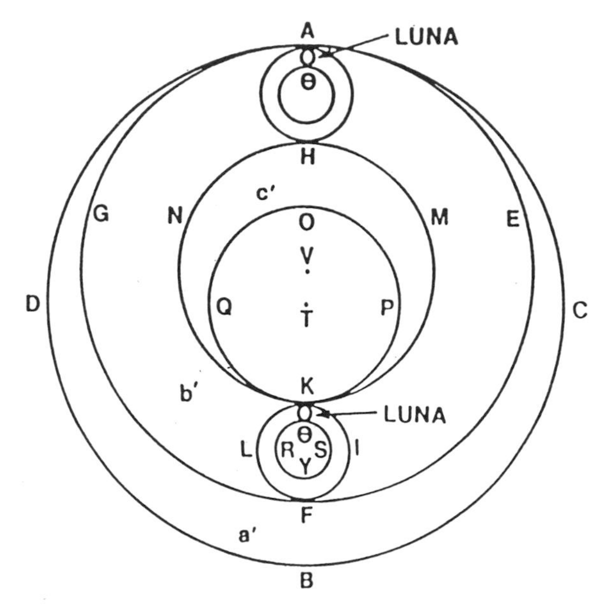

Figure 8. Representation of the Moon's concentric, eccentric, and epicyclic orbs as described in Roger Bacon's *Opus tertium*. (Diagram from Bacon, *Opus tertium*, 1909, 129.)

central core belongs wholly to the eccentric orb and forms no part of the epicyclic sphere itself. The Moon (*luna* in Figure 8), or planet, is a solid spherical figure which has only a convex surface and is located within a concavity of the epicyclic orb. The eccentric sphere is assumed to move around its center, *V*, carrying the epicycle with it; the epicyclic sphere, in turn, has its own simultaneous motion and carries the planet with it. I shall first describe medieval concern for eccentrics and then consider epicycles.

When extended to all the planets, it was this system that was widely adopted during the late Middle Ages. Even those who did not accept the three-orb system believed it saved the astronomical appearances better than did the systems of concentric spheres proposed in Aristotle's *Metaphysics* (12.8.1073b.11–1074a.14) and in al-Bitrūjī's more technical *De motibus celorum*.[32] Not even Averroës' strong support for Aristotle's concentric astronomy and cosmology[33] could entice medieval natural philosophers from the conclusion that Ptolemaic eccentric orbs were superior for saving the astronomical phenomena. Indeed, only a few unambiguous defenders of Aristotle's purely concentric cosmology can be identified.[34] But during the

32. The Latin text appears in Bitrūjī [Carmody], 1952.
33. Averroës' defense of Aristotle was made in his middle commentaries on *De caelo* and the *Metaphysics*. The relevant passages have been collected and analyzed by Carmody, 1952, 556–586.
34. For example, William of Auvergne, Alexander of Hales, and Saint Bonaventure. See Duhem, *Le Système*, 1913–1959, 3:404.

thirteenth, and even into the fourteenth, century, there were scholastic natural philosophers who refused to embrace, without qualification, a world constructed almost wholly of eccentric orbs. In a number of places in his writings, Thomas Aquinas was either noncommittal (*De trinitate*) or rejected eccentric orbs (*Commentary on the Metaphysics*). In his last treatise (*De caelo*), he argued that the existence of eccentrics and epicycles was undemonstrated, because even though they are useful for saving the astronomical appearances, they might not be physically real.[35] Although John of Jandun was convinced that the three-orb system could save the astronomical appearances – indeed, he proclaimed that he knew of no argument that could repudiate it – he rested content in the end to proclaim it as merely "possible."[36]

Such reservations and hesitations were, however, more and more the exception by the end of the thirteenth and the early fourteenth century. Scholastics increasingly came to assume the existence of real, material eccentric orbs, as did, for example, Albertus Magnus,[37] Duns Scotus, Aegidius Romanus (ca. 1245–1316), and Durandus de Sancto Porciano (d. 1334).

4. *The system of eccentrics: Pierre d'Ailly*

By the end of the fourteenth century, some 130 years after Bacon's account, the three-orb system had received its definitive scholastic form. No one expressed it better than Pierre d'Ailly, who presented as detailed an account as could be expected from a natural philosopher who was not a technical astronomer. Keeping in mind Figure 8, it will be useful to sketch d'Ailly's description of the three-orb system as he presented it in his *14 Questions on the Sphere of Sacrobosco*.[38] Not only does he employ a technical terminology that is largely absent in Bacon, but the objections he raises and the solutions he proposes were representative of the way material eccentrics and epicycles were interpreted by most natural philosophers from the late Middle Ages to the end of the sixteenth century.

According to d'Ailly, the heavens are made up of a combination of concentric and eccentric orbs. The totality of every sphere or orb (*orbis totalis*) is concentric and includes within it all other orbs necessary to produce the position of the planet. Within the concentric surfaces of each planetary orb are the eccentric orbs. Each eccentric orb or sphere, usually described as a "partial orb" (*orbis partialis*),[39] contains the center of the world as well as

35. For these passages, see Litt, 1963, 348, 350–352. Thomas's position is akin to that of Moses Maimonides (see Maimonides, *Guide* [pt. 2, ch. 24], 1963, 2:322–327).
36. John of Jandun [*Metaphysics*, bk. 12, qu. 20: "Whether a plurality of eccentric orbs and epicycles is really in celestial bodies"], 1553, 141r, col. 1–142r, col. 1.
37. Although Albertus Magnus accepted the existence of eccentrics, his arrangement of them differed from Bacon's popularly accepted description.
38. Pierre d'Ailly devoted the thirteenth of his fourteen questions on the *Sphere* of Sacrobosco to the question "Whether it is necessary to assume eccentric and epicyclic circles to save the appearances in planetary motions." See d'Ailly, *14 Questions*, qu. 13, 1531, 163v–164v.
39. Although the expression *orbis partialis* was rather common, d'Ailly did not use it. Clavius, *Sphere*, ch. 4, 1593, 502, however, did use and define it, saying that a whole sphere was

its own proper center that is eccentric with respect to the center of the world.

Like Bacon, d'Ailly distinguished three types of eccentric orbs but divided them into two classes: one, called *eccentricus simpliciter*, has the same center for both its concave and convex surfaces; the other, designated *eccentricus secundum quid*, has the center of the world as the center of one surface and a point outside the center of the world as the center of the other surface. The former surface is concentric, the latter eccentric. Thus the eccentric surface of an eccentric orb *secundum quid* may be either convex or concave, yielding two different types of eccentric orbs for a total of three.[40] Because it has two eccentric surfaces, an *eccentricus simpliciter* will always be of uniform thickness and is called the "deferent" orb because it carries the planet. The deferent orb is divided into four equidistant points: the *aux*, which is most distant from the center of the world; the opposite of the *aux*, which is the point on the deferent nearest to the center of the world; and the two opposite points, located between the *aux* and opposite of the *aux*, are called the mean distances (*longitudines mediae*).[41] By contrast, the *eccentricus secundum quid*, with one surface concentric and the other eccentric, is thicker in one part and thinner in another.[42] When eccentric orbs are moved, the thin part of one moves with the thick part of another and conversely.[43]

composed of partial orbs (*orbes partiales*). Philip Melanchthon (1550, 52v) also spoke of the geometers who fashion three partial orbs for the Sun ("Solis tres partiales orbes").

40. Without employing the terms *eccentricus simpliciter* and *eccentricus secundum quid*, Albert of Saxony ([*De celo*, bk. 2, qu. 7: "Whether for saving the appearances of the planetary motions, it is necessary to assume eccentric orbs and epicycles"], 1518, 106v, col. 1) divided eccentric orbs into the same three types as did Bacon and d'Ailly. Because d'Ailly's arguments are similar to, and even follow the order of, Albert's, it is not unreasonable to suppose that d'Ailly may have used Albert's question as one of his chief sources. Among authors who wrote after d'Ailly, Johannes de Magistris [*De celo*, bk. 2, qu. 3], 1490, 21, col. 1, expressed the same threefold distinction when he said, "there is a certain eccentric with respect to each surface [i.e., its convex and concave surfaces are eccentric]; another [kind] is eccentric with respect to only one surface [i.e., either to the convex or concave surface]" (Quidam est eccentricus secundum utramque superficiem, alius est eccentricus secundum unam tantum). Thomas Bricot repeats the same idea (Bricot [*De celo*, bk. 2], 1486, 21v, col. 2), as does John Major ([*Sentences*, bk. 2, dist. 14, qu. 4], 1519, 75v, cols. 1–2). In the late sixteenth century, Clavius spoke of the same two classes of eccentrics and used the same terminology as d'Ailly, as did Aversa, Amicus, Mastrius and Bellutus, and Cornaeus in the seventeenth century. See Clavius, *Sphere*, ch. 4, 1593, 499; Aversa, *De caelo*, qu. 32, sec. 5, 1627, 58, col. 2; Amicus, *De caelo*, tract. 5, qu. 4, dubit. 3, art. 3, 1626, 267, col. 1; Mastrius and Bellutus [*De coelo*, disp. 2, qu. 1, art. 2], 1727, 3:488, col. 2, par. 22; and Cornaeus [*De coelo*, disp. 2, qu. 2, dub. 3], 1657, 494–495. As we shall see, of this group only Cornaeus rejected eccentrics and epicycles unequivocally.
41. The same four points are mentioned by Major [*De celo*, bk. 2, qu. 2], 1526, 15.
42. Although the diversity of thickness in eccentrics *secundum quid* was obvious, it was explicitly mentioned by some: for example, Clavius, *Sphere*, ch. 4, 1593, 499; Aversa, *De caelo*, qu. 32, sec. 5, 1627, 58, col. 2–59, col. 1; Amicus, *De caelo*, tract. 5, qu. 4, dubit. 3, art. 2, 1626, 266, col. 2; and Cornaeus, *De caelo*, disp. 2, qu. 2, dub. 3, 1657, 495.
43. See d'Ailly, *14 Questions*, qu. 13, 1531, 163v. Duns Scotus agreed when he declared that "the thicker part of one lies against the thinner part of another, and conversely" (Semper enim spissior pars unius est contra partem minus spissam alterius, et e converso). Duns Scotus [*Sentences*, bk. 2, dist. 14, qu. 2], *Opera*, 1639, 6, pt. 2:732. Major, *Sentences*, bk.

With these definitions and concepts, d'Ailly describes next the three-orb scheme in a manner that differed little from Bacon's earlier description of the theory of the "moderns."[44] According to d'Ailly, astronomers imagine three eccentric orbs as constituting the whole sphere of a planet. Two of these orbs are eccentrics *secundum quid*, that is eccentric with respect to one surface only. One of them is the outermost orb and the other the innermost orb. As eccentrics *secundum quid*, the outermost orb is eccentric only with respect to its concave surface, while the innermost orb is eccentric only with respect to its convex surface. Between these two orbs lies the third, which is *eccentricus simpliciter*, because both of its surfaces are eccentric. Indeed, the middle eccentric is constituted of the concave surface of the outermost eccentric sphere and the convex surface of the innermost sphere. The middle orb is called the "deferent orb" and carries the planet itself. With the exception of the outermost and innermost orbs of the world, each orb was conceived as a ring-like figure that contained other ring-like orbs and was contained by other ring-like orbs.[45] Raphael Aversa provides further details when he explains that

> the planet itself, which is carried in such an orb [i.e., in the eccentric deferent], now arrives over the thicker part, now over the thinner part of the extreme lower orb. And so for half of its path it recedes a bit from the earth, and in the other half it draws nearer [to the earth]. And it also takes more time in the thicker half than in the thinner half, because that part of its circuit is greater. Thus it seems to turn more slowly then. [Finally], when it is in the summit of the thicker part, it is said to be in the *aux*; and when it is in the thinner part it is said to be in the opposite of the *aux*. The *aux* is called apogee; [the] opposite [of the *aux* is called] perigee.[46]

5. *Epicycles*

Although the deferent orb carries the planet, it does so by means of an epicycle, thought of as either a solid globe or a ring (and depicted in the

2, dist. 14, qu. 4, 1519b, 75v, cols. 1–2, seems to say much the same thing, when he declares that "the differences of thickness are so arranged that the thicker part of one always corresponds to the thinner part of another" (Sunt difformes spissitudinis sic ordinati ut semper parti spissiori unius pars tenuior alterius respondeat). Aversa, *De caelo*, qu. 32, sec. 5, 1627, 58, col. 2, and Mastrius and Bellutus, *De coelo*, disp. 2, qu. 1, art. 2, 1727, 3:488, col. 2, par. 22, also held the same opinion.

44. With this proviso: d'Ailly's range of technical terminology was much greater.
45. In the seventeenth century, Aversa gave a thorough summary of the three-orb system along the same lines and with much the same terminology in *De caelo*, qu. 32, sec. 5, 1627, 58, col. 2–60, col. 2. For d'Ailly's description of an orb as a ring-like figure with a convex and concave surface, see my discussion in Chapter 6, note 37.
46. "Unde sit ut ipsum astra, quod in tali orbe defertur, modo incedat super partem crassiorem, modo super tenuiorem orbis extremi inferioris. Et sic pro medietate sui circuitus paulatim altius a terra recedat; pro altera medietate proprius accedat. Atque etiam plus temporis conficiat in medietate crassiori quam tenuiori quia maior est illa pars sui circuitus. Et ita pro tunc videatur tardius gyrare. Atque dum est in summo partis crassioris dicitur esse in auge; dum est in imo partis subtilioris dicitur esse in opposito augis. In auge dicitur apogaeum; in opposito perigaeum." Aversa, ibid., 59, col. 1.

diagram as the circle *KLFI*). D'Ailly describes the epicycle as "a small circle on the surface of the deferent orb that does not contain within itself the center of the world; and the body of the planet is imagined to be in it. And this epicycle is assumed to be contiguous, and not continuous, with the eccentric deferent because it is moved with a motion other than the motion of the eccentric deferent."[47] Like the eccentric deferent, the epicycle has four equidistant points: the *aux* of the epicycle; the opposite of the *aux* of the epicycle; and two points equidistant from the *aux* and the opposite of the *aux* called "stations" (*stationes*), one of which marks the point at which the planet begins its retrograde motion (point *I* of the epicycle in Figure 8), the other of which marks the point where it begins its direct motion (point *L* in Figure 8).[48]

Like eccentric orbs, epicyclic orbs represent the various observed motions and dispositions of the planets, which, according to d'Ailly, are of three types: direct motion, retrogradation, and station. When a planet is in the *aux* of its epicycle, its motion is said to be direct and quickest, because the direction of its motion on the epicycle is the same as that of the eccentric deferent. But when the planet is in the opposite of the *aux* of the epicycle, its motion is retrograde and slower, because it now moves in a direction opposite to that of the eccentric deferent. Should the planet arrive at one of the points of station, it would move neither with the deferent nor contrary to it, so that its speed will seem neither to increase nor decrease and it will appear stationary.[49]

6. The great compromise: the three-orb system

Although d'Ailly's three-orb system incorporates the three types of eccentric orbs that he, Bacon, and many others distinguished, the enclosing surfaces

47. D'Ailly, *14 Questions*, qu. 13, 1531, 163v–164r. For the position of the epicycle, see Figure 8. Sacrobosco defines an epicycle as "a small circle on whose circumference is carried the body of the planets, and the center of the epicycle is always carried along the circumference of the deferent." Sacrobosco, *Sphere*, ch. 4, 1949, 141 (Latin text, 114). Major, *Sentences*, bk. 2, dist. 14, qu. 4, 1519b, 75v, col. 2, also says that the epicyle is contiguous, rather than continuous, with its deferent, "because it is moved with another motion than the motion of the deferent" (quia movetur alio motu quam motu deferentis).
48. D'Ailly, ibid., 164r. Brief definitions of the points of station appear in Sacrobosco's *Sphere* (ch. 4, 1949, 141) and in the *Theorica planetarum* (for both, see Grant, 1974, 450 and 461, respectively). Major, *Sentences*, bk. 2, dist. 14, qu. 4, 1519b, 75v, col. 2, identifies the same four points. However, the four points on the epicycle are in fact not equidistant, as d'Ailly and Major assert. Although the first and second stations must always be equidistant from the true apogee (*aux*) of the epicycle, the points of station always fall nearer to the perigee of the epicycle (opposite of the *aux*) than to the apogee. It follows therefore that the four points on the epicycle mentioned by d'Ailly cannot be equidistant. See Campanus of Novara, *Theorica planetarum*, 1971, 225–227, 231, 313. I am grateful to my late colleague, Victor E. Thoren, for bringing this to my attention,
49. In the thirteenth century, Bartholomew the Englishman provided an earlier descriptive version of eccentrics and epicycles (bk. 8; 1601, 398–399). Although he speaks of equant, deferent, and epicyclic circles and also of direct and retrograde motion, as well as of station, Bartholomew's account lacks the more sophisticated terminological distinctions found in d'Ailly's treatise.

– namely the surface of the outermost eccentric orb and the concave surface of the innermost eccentric orb – had to be concentric. This was, of course, a primary feature of the Aristotelian–Ptolemaic compromise that saved the geocentric system. Because a concentric orb has the geometric center of the world as its center, d'Ailly explains that "the first movable sphere [the *primum mobile*, sometimes equated with the sphere of the fixed stars] is a concentric orb, and generally every total orb is concentric, where 'total orb' [*orbis totalis*] is taken as the aggregate of all the orbs required to save the total motion of a planet."[50] In this manner, Aristotle's cosmology of concentric spheres was saved, even though, in violation of his physical principles, eccentric orbs with centers other than the center of the world formed the basis of the compromise system.

Here, then, was the compromise that produced the three-orb system in which three partial eccentric orbs are encompassed within two concentric surfaces that together form a single concentric orb (see Figure 8). The latter was then perceived as representing the total motion of the planet. The three-orb system was forced upon Aristotelian natural philosophers because Aristotle's straightforward concentric system could not account for the astronomical phenomena. Thomas Bricot observed that there are numerous irregularities in the celestial motions, but not all require the assumption of eccentrics or epicycles.[51] But one irregularity that definitely requires an eccentric is the variation in a planet's distance from the earth.[52] It was the most fundamental reason why almost all scholastic natural philosophers felt compelled to accept the compromise and abandon Aristotle's purely concentric system. They consoled themselves with the thought that the total orb (*orbis totalis*), which contained the three eccentrics, was concentric and that in some sense one could still speak of a concentric world system. The difficulty in all this was the fact that although the earth remained the center of the concentric orbs, it could not function as the center of the eccentric orbs. Because most scholastics could see no plausible alternative, it was an anomaly they had to accept. That is why, although d'Ailly mentions rival interpretations which rejected epicycles and eccentrics, he regards the Pto-

50. "Unde orbis concentricus dicitur orbis sub utraque eius superficie continens centrum mundi et habens eius centrum cum centro mundi. Isto modo primum mobile est orbis concentricus et generaliter quilibet orbis totalis est concentricus et ibi capitur orbis totalis pro aggregato ex omnibus orbibus requisitis ad salvandum motum totalem unius planetae." D'Ailly, *14 Questions*, qu. 13, 1531, 163v.

51. He describes five such irregularities, or difformities (*difformitates planetarum*), and declares that "such irregularities can be saved by the assumption of several motions in the same celestial body without the assumption of eccentrics and epicycles" (Et tales difformitates possunt salvari per positionem plurium motuum in eodem corpore celesti sine positione eccentricorum vel epiciclorum). Bricot, *De caelo*, bk. 2, 1486, 29r, col. 2. Among the irregularities, Bricot includes a planet's proper motion, which is contrary to its daily motion; movement along the zodiac from tropic to tropic; and variations in latitude with respect to the ecliptic.

52. "Sed est alia difformitas in appropinquatione planete ad nos et in elongatione eius a nobis. Et talis non potest salvari sine eccentricis, sed bene posset salvari sine epiciclis." Ibid.

lemaic theory of eccentrics and epicycles as "more common" (*est magis communis*) and unambiguously adopts it.

Until the seventeenth century, most scholastic authors assumed the truth of the three-orb compromise. In that century, however, Tycho Brahe's analysis of the comet of 1577, which indicated the nonexistence of solid, hard spheres of any kind, began to make its influence felt. The gradual acceptance of a fluid celestial region led even some scholastic authors to abandon the idea of eccentric and epicyclic spheres. But the traditional opinion continued to hold the allegiance of some important scholastics in the late sixteenth and the seventeenth century, including Clavius, Amicus, Mastrius and Bellutus, and perhaps Aversa.[53] Of this, however, more will be said when we consider scholastic opinions on the hardness or fluidity of the celestial region. Now, however, we must examine the physical and cosmological consequences that followed from the assumption that eccentrics and epicycles were real, material spheres.

III. Cosmological problems with eccentrics and epicycles

By enclosing each set of eccentric planetary orbs within concave and convex surfaces that were themselves concentric with respect to the earth's center, Ptolemy himself had seemingly made a strong gesture toward reconciling his own cosmology with that of Aristotle. In both systems, the earth's center was the center of motion for each total planetary orb. Natural philosophers could thus continue to believe that the fundamental structure of Aristotle's system was preserved: the external surfaces of every planetary sphere were concentric with the earth. Ptolemy sharply diverged from Aristotle, however, by his assumption that within the external concentric surfaces that comprised each total planetary sphere or orb were three or more partial eccentric orbs with centers other than that of the earth. In violation of Aristotle's dictum that all celestial spheres move with uniform motion around the earth as center, Ptolemy assumed the motion of all his eccentric spheres to be around points other than the earth's center. Although most scholastics recognized that eccentric orbs and epicycles explained planetary variations in distance and latitude that went unaccounted for in Aristotle's system of concentric spheres, they were also aware of a number of significant problems in the received system of eccentrics and epicycles that were potentially subversive of Aristotelian cosmology and physics. As defenders of the three-orb system, they sought to explain these anomalous

53. Clavius, *Sphere*, ch. 4, 1593, 525; Amicus, *De caelo*, tract. 5, qu. 4, dubit. 3, art. 3, 1626, 267, col. 1; Mastrius and Bellutus, *De coelo*, disp. 2, qu. 1, art. 2, 1727, 3:488, col. 2. Aversa, *De caelo*, qu. 32, sec. 6–7, 1627, 66, col. 2–74, col. 1, seems to think of his eccentrics and epicycles as cavities or channels in a firm, material celestial region. With some reservation, we may classify him as a qualified supporter of eccentrics and epicycles. Clavius's lengthy discussion of eccentrics and epicycles may be the most detailed and significant of all scholastic accounts (see Clavius, *Sphere*, ch. 4, 1593, 499–525).

situations and to reconcile eccentrics and epicycles with Aristotelian cosmology and physics.

In what follows, I shall describe the manner in which medieval natural philosophers coped with material eccentrics and epicycles within the framework of Aristotelian cosmology and physics. An early account – probably one of the first – of the problems inherent in a system of eccentrics and epicycles appears in Vincent of Beauvais' *Speculum naturale*, where, sometime around 1244,[54] in a chapter titled "Whether there is any space between the spheres of the planets,"[55] Vincent asked whether there is any body or void space between spheres or whether the spheres mutually touch.[56] Citing an author whom he calls "Avenalpetras," who is probably the Arabian astronomer al-Biṭrūjī,[57] Vincent relates that Avenalpetras assumed that the celestial orbs are in contact and, further, denied that there are planetary eccentrics, epicycles, or elevations and depressions of the planets. Should such things exist in the heavens, they would clearly imply that the fifth body, or celestial ether, is divisible. For if a planet were elevated or depressed – that is, varied its distance from the earth – it would follow that the supposedly indivisible fifth element, or ether, would in fact be divided. Under such circumstances, either the ether would fill the places left vacant by the planet as it moved higher or lower, or the places from whence the planet withdrew would remain void. Since neither of these alternatives is naturally possible, Avenalpetras rejected the existence of eccentrics, epicycles, and planetary variations in distance.[58]

Some fifteen or twenty years later, many of the cosmological objections that were raised against the three-orb system of material eccentrics and epicycles also appear, though not always clearly expressed, in Roger Bacon's treatises of the 1260s. Perhaps the best statement of them and their incom-

54. For this approximate date, see William Wallace's article on Vincent in *Dictionary of Scientific Biography*, 1970–1980, 14:34–35.
55. "Utrum aliquod spacium sit inter sphaeras planetarum." Vincent of Beauvais, *Speculum naturale*, bk. 3, ch. 104, 1624, 1:cols. 230–231.
56. "Utrum inter sphaeras sit corpus aliquod vel spacium illud sit vacuum; an sphaerae contingant se invicem." Ibid., col. 230.
57. For other instances of Vincent's mention of "Avenalpetras," see ibid., cols. 226 (ch. 97) and 228 (ch. 100). In a discussion of Albertus Magnus's *Summa de creaturis*, Gilson (1955, 281) identifies the name "Anavelpetra" with the famous Arabian astronomer al-Biṭrūjī (Gilson has "al-Bitrogi"). Is "Anavelpetra" the same as "Avenalpetras"? But in his commentary on *De caelo*, Albertus refers to al-Biṭrūjī as "Alpetrauz" or "Alpetragius," but not Anavelpetra or Avenalpetras (see Albertus Magnus, *De caelo*, 1971, 276, under "Al-Biṭrūjī"). Vincent, however, says that Avenalpetras rejects eccentrics and epicycles, as indeed did al-Biṭrūjī (see Biṭrūjī, 1952, 11). Thus the identification of al-Biṭrūjī with Avenalpetras gains plausibility. In his *Lucidator*, differ. 6, 1988, 315, line 1, Peter of Abano refers to al-Biṭrūjī as "Avempetras."
58. "Et dicit Avenalpetras quod ille sese contingunt. Sed ipse ponit quod non sunt eccentrici, nec epicycli, nec elevationes, nec depressiones planetarum, quod dicit ea ratione: quia secundum naturam quintum corpus non est divisibile. Si autem esset elevatio ac depressio planetarum tunc oporteret quod corpus illud divideretur ac succederent stellae elevatae ac depressae in locum suum; aut quod locus unde recederet stella vacuus remaneret. Quorum utrumque est impossibile secundum naturam." Vincent of Beauvais, *Speculum naturale*, bk. 3, ch. 104, 1624, 1:col. 230.

patability with Aristotelian cosmology was presented by Cecco d'Ascoli (1269–1327), who, in his *De eccentricis et epicyclis*, where he upheld the existence of eccentrics and epicycles, summarized the major impossible consequences that opponents of eccentrics and epicycles believed would result from their existence. Cecco explains that "if there were eccentrics and epicycles, then rarefaction or condensation would occur, which is impossible by the first [book] of [Aristotle's *De*] *celo et mundo*; or a vacuum would occur, which is impossible, as is said in the fourth [book] of [Aristotle's] *Physics*; or there would be a separation of the spheres, which is impossible, as is obvious in the second [book] of *De celo et mundo*; or there would be a penetration of bodies, which is false, as is obvious in the fourth [book] of the *Physics*."[59]

1. *Vacua and condensation and rarefaction in the heavens*

In Pierre d'Ailly's language, these possibilities apply only to eccentric orbs that are *secundum quid*, that is, eccentric orbs that have one eccentric surface and one concentric surface. Bacon, d'Ailly, and many others well into the seventeenth century considered such eccentric orbs to be of unequal thickness because the points of apogee and perigee were unequally distant from the center of the world.[60] The absurdities described by Cecco d'Ascoli derived from the rotations of such orbs. For example, if we assume that eccentric orb c' (see Figure 8) has rotated 180 degrees, it follows that the thickest and thinnest parts of it will have exchanged places. Because the thickest part of c' will occupy more space than the thinnest part, it must make a space for itself by pushing away some of the surrounding matter that now occupies the place it must enter; *or* it must occupy the same place with that matter. If it displaces the matter presently in that place, the displaced matter, in turn, must find a place for itself and therefore must divide adjacent celestial matter. Within the set of planetary orbs that contains orb c', matter must condense somewhere, so that the thicker part of c' can occupy a greater place. To accommodate the rotations of such orbs, celestial matter must be conceived as divisible and condensible, both of which pos-

59. "Si esset ponere excentricos et epiciclos, tunc modo esset rarefactio aut condensatio, quod est impossibile, ut patet primo *Celi et mundi*; aut vacuum, et hoc est impossibile, ut dicitur 4° *Phisicorum*; aut scissio sperarum, quod est impossibile, ut patet 2° *Celi et mundi*; aut corporum penetratio, quod est falsum, ut patet 4° *Phisicorum*." Cecco d'Ascoli, *De eccentricis* [Boffito], 1906–1907, 161. In her edition of Peter of Abano's *Lucidator dubitabilium astronomiae*, Graziella Federici Vescovini has reedited this work of Cecco's as Appendix II (Peter of Abano, *Lucidator*, 1988, 383–394). Both editions were made from the same manuscript and are virtually identical. Cecco's Latin text, as cited in this note, appears on page 384, lines 19–24. Roger Bacon also insisted that "it is impossible to assume an eccentric orb of any planet, because then it would be necessary that the celestial body be divisible; or that two bodies be in the same place; or that a vacuum exist." Bacon, *Opus tertium*, 1909, 119; Bacon, *De celestibus*, pt. 5, ch. 13, *Opera*, fasc. 4, 1913, 433–434.
60. For example, in Figure 8, a' is thickest between points FB, where F is the point of perigee, and thinnest at point A, the point of apogee; whereas eccentric sphere c' is thickest at OH, where H is the point of apogee, and thinnest at point K, the point of perigee.

sibilities were ruled out by Aristotle and most of his followers. At the other side of the eccentric orb, the thinnest part will be unable to fill the space formerly occupied by the thickest part. In these circumstances, either a vacuum will form, or, in order to prevent a vacuum, matter adjacent to orb *c′* must instantaneously fill any empty spaces. Under these circumstances, either void spaces exist in the heavens, or celestial matter is divisible and capable of rarefaction to fill a potential vacuum. In the Aristotelian physical world, neither alternative was acceptable.

2. *Are the celestial spheres continuous or contiguous?*

The problems for Aristotelian cosmology that Cecco d'Ascoli raised about eccentrics were primarily about the relationships between the external surfaces of any two successive orbs – that is, between the convex surface of a contained sphere and the concave surface of its containing sphere. In Aristotelian terms, there are three possibilities: (1) the surfaces are continuous, that is, they coincide; (2) the surfaces are contiguous, that is, they are distinct but in direct contact at every point; or (3) they are wholly or partially distinct and without contact. Aristotle distinguished two kinds of contact. In one way, things that are in succession and touch are said to be "contiguous." Thus each of two distinct surfaces in contact at all points would be successive and contiguous and have the same shape.[61] But if those two surfaces became one and the same, they would be considered "continuous." Indeed, "continuity would be impossible if these extremities are two." To emphasize his point, Aristotle further explained that "if there is continuity there is necessarily contact, but if there is contact, that alone does not imply continuity; for the extremities of things may be together without necessarily being one."[62] Aristotle used the concept of continuity in his definition of the "place" of a thing, which he defined as "the boundary [or inner surface] of the containing body at which it is in contact with the contained body." Instead of two surfaces, however, the surface of the container and the surface of the thing contained formed only one surface and were therefore continuous. Or, as Aristotle expressed it: "place is coincident with the thing, for boundaries are coincident with the bounded."[63]

Perhaps the first one to apply the concept of contiguity to celestial orbs was Ptolemy, who, after presenting the distances from the earth of the successive planetary spheres in his *Hypotheses of the Planets*, declares that

61. In *De caelo* 2.4.287a.6–7, Aristotle [Guthrie], 1960, declares that "what is contiguous to the spherical is spherical."
62. Aristotle, *Physics* 5.3.227a.9–12, 21–23 [Hardie and Gaye], 1984.
63. Ibid., 4.4.212a.5–6 and 212a.30. Aristotle's doctrine of place was as applicable to nested celestial spheres as it was to terrestrial objects, as is evident from his denial of the existence of places beyond the world (*De caelo* 1.9.279a.8–15), thereby implying the existence of places everywhere within the world.

If (the universe is constructed) according to our description of it, there is no space between the greatest and least distances (of adjacent spheres), and the sizes of the surfaces that separate one sphere from another do not differ from the amounts we mentioned. This arrangement is most plausible, for it is not conceivable that there be in Nature a vacuum, or any meaningless and useless thing. The distances of the spheres that we have mentioned are in agreement with our hypotheses. But if there is space or emptiness between the (spheres), then it is clear that the distances cannot be smaller, at any rate, than those mentioned.[64]

Campanus of Novara repeated the substance of Ptolemy's position, as we shall see in the next paragraph. But when justification of contiguity or continuity was required, scholastics turned to Aristotle. Christopher Clavius reveals how Aristotle's definitions of contiguity and continuity were applied to celestial orbs. If a line were drawn from the center of the world and intersected with the ninth and tenth heavens or spheres, Clavius argues that the point on the ninth sphere and its immediate neighbor on the tenth sphere would be two distinct points in the mind but one and the same point in actuality. Despite the oneness, and therefore the continuity, of the successive points on the surfaces of the two successive orbs, Clavius was aware that the ninth and tenth spheres had different motions and therefore insists that the convex surface of the ninth orb and the concave surface of the tenth orb were contiguous rather than continuous. Indeed, Clavius insists not only that the successive and immediate surfaces are contiguous but that nothing can lie between them, for otherwise an infinite process would result. For example, if a globe were assumed in air, nothing could lie between the convex surface of the globe and the concave surface of the air surrounding the globe. For if something, say body *a*, could intervene between the two surfaces, then something else would have to intervene between the convex surface of the air and the concave surface of body *a*, as well as between the convex surface of body *a* and the concave surface of the air; and so on ad infinitum. In order to allow for distinct and even contrary motions and for the motion of a superior orb to be communicated to an inferior orb, Clavius settled for successive celestial surfaces that were one and the same in reality but conceptually distinct. Moreover, he chose to characterize them as contiguous without the possibility of intervening matter (Clavius [*Sphere*, ch. 1], *Opera*, 1611, 42).

Clavius has made an important point, one that was probably implied in medieval discussions: if the convex surface of an orb is continuous with or contiguous to the concave surface of the next successive superior orb, those two surfaces will be equidistant from the center of the earth. Whether successive spherical surfaces were continuous or contiguous was thus of no consequence with regard to the measurement of planetary distances. In his widely used *Theorica planetarum*, Campanus of Novara declared that the

64. From Ptolemy [Goldstein], 1967, 8, col. 1. See also Ptolemy [Toomer], 1984, 420, n. 4.

convex surface of one planetary sphere was exactly equal to the distance of the concave surface of the next-highest celestial sphere. As Campanus expressed it, "the highest point of the lower [sphere] coincides with the lowest point of the higher."[65] In this brief passage, did Campanus consider the relevant surfaces continuous or contiguous? We cannot say with any certainty, since either alternative is compatible with the equality of the distances of those surfaces. Also compatible with continuity or contiguity is Campanus's conviction that only by a fusion of these two celestial surfaces could waste space be avoided, either in the form of a vacuum or as some kind of separate matter distinct from the orbs themselves. The Aristotelian definitions of contiguity and continuity are compatible with the virtual or actual fusion, respectively, of two successive surfaces, so that not only are their distances equal but nothing can lie between them. Without discussion of the alternatives, or even any realization of the issues involved or any awareness of the choice they had made, most scholastic authors unknowingly opted for one or the other alternative, confident that they had neither fallen into difficulty about planetary distances nor allowed matter or void to intervene between successive spherical surfaces. Only a few, including Clavius, were sufficiently knowledgeable to articulate the issues.

Even if the distinction between continuity and contiguity of successive celestial surfaces could be ignored with respect to planetary distances and intervening matter, it was of crucial importance with regard to the diversity of celestial motions, as Roger Bacon recognized when he declared that continuous surfaces would cause those orbs to "be moved with equal velocity, even with the same motions, which is contrary to experience."[66] Although Bacon's contemporary, Robertus Anglicus, thought he could reconcile continuity of celestial surfaces with diversity of celestial motions,[67] Bacon, and probably most other natural philosophers and astronomers, including Richard of Middleton, Nicole Oresme, Albert of Saxony, John Major, and Christopher Clavius, believed that the obvious facts of astronomy required a denial of continuity and the assumption of contiguity.[68] The

65. "Per hoc enim sequitur quod supremum inferioris sit infimum superioris sue." Campanus of Novara, *Theorica planetarum*, 1971, 331; see also 331–337 and 53–55. From the Latin text, we observe that Campanus uses no term for "coincides." It was usually assumed in medieval Islamic and Latin astronomy that the distance from the earth of the convex surface of one planetary sphere was equal to the distance from the earth of the concave surface of the next sphere (see the tables of distances and dimensions in ibid., 356–363). In effect, since the two distances were identical, so were the surfaces. Although the distances of the innermost and outermost circular surfaces were fixed, the distances of the planets varied within their respective epicycles.
66. Bacon, *Opus tertium*, 1909, 123; Bacon, *De celestibus*, pt. 5, ch. 13, *Opera*, fasc. 4, 1913, 436.
67. In his commentary on the *Sphere* of Sacrobosco, written around 1271, Robertus avoided the major problem with continuous orbs by assuming that the outer surfaces of celestial orbs were immobile, with only their middle parts, which he likened to a fluid, being capable of motion. Under these conditions, each orb could move independently of the others. See Sacrobosco, *Sphere*, 1949, 147 (Latin) and 202–203 (English).
68. Richard of Middleton [*Sentences*, bk. 2, dist. 14, art. 3, qu. 1], 1591, 2:184, col. 1; Oresme,

different directions in which planets were carried by their orbs or the diverse speeds at which they rotated made it obvious that successive celestial surfaces could not form a single unified, continuous surface but had to be distinct entities, in contact at every point: that is, contiguous. Although Campanus of Novara failed to express the distinction between continuity and contiguity, the diversely directed motions of the spheres makes it almost mandatory to attribute to him an assumption of contiguity.[69] For around five centuries, scholastics who assumed the existence of celestial orbs probably assumed that the surfaces of successive orbs were contiguous rather than continuous.

There was thus little reason to believe that extraneous matter or the dreaded void could intervene between successive surfaces of celestial orbs solely because they were contiguous or continuous. A more serious source of fear that intervening matter or void might intrude between surfaces arose for a quite different reason, namely from the idea that an eccentric material orb that possessed one concentric surface and one eccentric surface was not only of unequal thickness but was ovoid in shape. If so, then Aristotle may have furnished the basis for this mistaken notion when, in demonstrating the sphericity of the heavens, he declared that if the world were not spherical but "lentiform, or oviform, in every case we should have to admit space and void outside the moving body, because the whole body would not always occupy the same room" (Aristotle, *De caelo* 2.4.287a.12–24 [Stocks], 1984).

In commenting on this passage, Nicole Oresme distinguished different circumstances under which oval-shaped planetary orbs would or would not produce the impossible consequences described earlier. He argues ([*Le Livre du ciel*, bk. 2, ch. 10], 1968, 391) that "if the planetary spheres were oval in shape, being moved in a manner different from the sovereign [or last] heaven and on different axes . . . , either there would have to be an empty place or penetration in the heavens – that is, one heaven would pierce through the other – or there would have to be condensation or compression, all of which are impossible in nature."

Judging from certain responses, some scholastics seem to have believed that eccentric orbs of uneven thickness would indeed produce the alleged

Le Livre du ciel, bk. 2, ch. 9, 1968, 385; Albert of Saxony, *De celo*, bk. 1, qu. 4, 1518, 89v, col. 1; and Major, *Sentences*, bk. 2, dist. 14, qu. 4], 1519b, 75v, col. 1. In his *De multiplicationes specierum*, Bacon declares unequivocally that "the spheres are contiguous and possess distinct surfaces" (Bacon [Lindberg], 1983, 119). In the late sixteenth century, Clavius insisted that the celestial orbs are contiguous, with each superior orb including its immediately inferior orb "and there is no medium between one and the other, as in the peels of an onion, where everywhere we see the upper [peel] surround a lower [peel]" (Sunt autem omnes orbes coelestes contigui prorsus et immediati inter se, ita ut semper superior inferiorem includat, nihilque inter unum atque alterum sit medium non secus ac in tunicis caeparum videmus superiorem undique circundare inferiorem). Clavius, *Sphere*, ch. 1, *Opera*, 1611, 3:10.

69. As do Benjamin and Toomer (see Campanus of Novara, *Theorica planetarum*, 1971, 412, n. 47), and Aiton, 1981, 90.

impossibilities were it not for an otherwise unexplained synchronization of motions. On the assumption that eccentric orbs move uniformly, d'Ailly, for example, held that when the thickest part of one eccentric is moved toward its opposite side, another eccentric orb moves uniformly in the opposite direction. When the two eccentric orbs, say *a* and *b*, have simultaneously moved 180 degrees, the thickest part of orb *a* will have come to occupy the place formerly occupied by the thickest part of orb *b*; and similarly, the thinnest part of orb *a* will also have come to occupy the place formerly occupied by the thinnest part of orb *b*.[70] In this manner a balance is always maintained, and the dreaded impossibilities are perpetually avoided. How and why such synchronization of orbs should occur is nowhere explained, but the idea was already presented early in the fourteenth century by Cecco d'Ascoli and repeated long after d'Ailly by Georg Peurbach in the fifteenth century and Christopher Clavius in the sixteenth and seventeenth centuries.[71]

3. *The rejection of continuity and contiguity: the assumption of matter between two orbs*

The existence of material eccentrics was made to seem viable in yet another way: by the assumption of intervening matter between the orbs, which implied the rejection of spheres that were either continuous or contiguous. In a brief though important passage, Vincent of Beauvais describes such an opinion when he reports:

Some say that the spheres do not mutually touch and that a body of the same nature lies between them. Indeed [that body] is divisible by the spheres but is not transmutable into another species, and in it eccentrics and planets are elevated and depressed [i.e., vary their distances from us]. Nevertheless, some ancients say that

70. See d'Ailly, *14 Questions*, qu. 13, 1531, 163v, for the argument and 164v for d'Ailly's brief response.
71. Between 1322 and 1327, Cecco d'Ascoli, *De eccentricis* [Boffito], 1906, 166, described the same mechanism for synchronizing the motions of eccentric orbs by what he called "proportional motions" (*proportionales motus*). Although Cecco, who defended the existence of eccentrics and epicycles, thought the idea of "proportional motions" was a good idea, he denied that Ptolemy had it in mind. Peurbach, *Theoricae*, 1987, 10, used the same expression, "proportional motions," when he declared that "the deferent orbs of the apogee of the Sun move by their own proportional motions, so that the narrower part of the superior is always above the wider part of the inferior, and go around equally fast." Some two hundred years after d'Ailly, Clavius approvingly presented the same explanation. See Clavius, *Sphere*, ch. 4, 1593, 521 (for a description of the impossibilities) and 523 (for the response). The idea may be traceable to Bernard of Verdun, at the end of the thirteenth century. After mentioning the usual charge that eccentrics would produce the dreaded impossibilities, Bernard explains that the "different parts" of eccentric orbs "succeed themselves continually in the points or places of the farther and nearer distance" – that is, in the points of apogee and perigee – "that are imagined in the convexity of the surrounding orb." The translation is mine from Grant, 1974, 523 (I have here replaced "longitude" with "distance"). Bernard did not use an expression comparable to "proportional motion(s)."

eccentrics and planets traverse the body that lies between the spheres but yet do not divide it. This occurs because of the formality of these bodies, just as light [*lumen*] traverses through the air [and does not divide it].[72]

Whereas Vincent of Beauvais was content to report the opinions of others using the anonymous phrase "some say," Albertus Magnus not only provided a much clearer and more complete account in his commentary on *De caelo*,[73] but he unhesitatingly adopted the concept of a divisible matter intervening between the planetary orbs. Like Bacon, Albertus seems to have considered eccentrics as ovoid in shape and was therefore convinced that if eccentrics are nested one within the other their motions would cause gaps between their surfaces.[74] Because Albertus believed that these gaps or spaces could not be void, one of two alternatives must occur: either the various eccentric spheres would rarefy and condense, to prevent formation of a vacuum; or another body must intervene between any two successive eccentric spheres.

To refute the first alternative, Albertus argues that if the circles or spheres rarefied and condensed, their shapes would vary. Hence their motions would be essentially unknowable, and the data derived from those motions would be false.[75] If, however, we say that the circles, that is, spheres, themselves are sometimes in direct contact because no medium intervenes between their surfaces, then, when the surfaces of these eccentric orbs do separate because of their shapes, a new body that did not exist there before cannot be generated to fill the void which must inevitably occur. That no other body could have existed there before the separation of the surfaces is obvious, because if such a body did exist there before separation, then two bodies would have occupied the same place, which is impossible.[76]

72. "Quidam etiam dicunt quod sphaerae non contingunt se et corpus eiusdem naturae est inter eas. Divisibile quidem a sphaeris sed non transmutabile in speciem aliam et in illo elevantur et deprimuntur eccentrici et planetae. Quidam tamen antiqui dixerunt quod eccentrici et stellae transeunt per corpus quod est inter sphaeras et tamen ipsum non dividunt. Et hoc contingit propter formalitatem ipsorum corporum, sicut et lumen transit per aerem." Vincent of Beauvais, *Speculum naturale*, bk. 3, ch. 104, 1624, 1:col. 231.

73. Albertus Magnus, *De caelo*, bk. 1, tract. 1, ch. 11, *Opera*, 1971, vol. 5, pt. 1:29–30. Weisheipl (1980, 27) believes that Albertus wrote all of his Aristotelian paraphrases, including that on *De caelo*, between 1250 and 1270. Thus his account may have been written after Vincent's report.

74. Because Albertus assigned only one eccentric to each planet, we may infer that he was not reporting Bacon's "modern" three-orb system.

75. "Si enim nos diceremus, quod ipsi circuli rarificantur et inspissantur, tunc non semper tenerent figuram eandem, et sic non posset sciri motus eorum, quod constat omnibus illis esse falsum, qui sciunt canones motus excentricorum." Albertus Magnus, *De caelo*, bk. 1, tract. 1, ch. 2, *Opera*, 1971, 5, pt. 1:30, col. 1.

76. "Et si nos diceremus, quod aliquando ita se tangunt circuli, quod nihil est medium ipsorum, tunc oporteret nos concedere, quod tunc intercideret vacuum inter eos, quando per motum distant a tali situ coniunctionis, quia quando seiunguntur, non generatur ibi novum corpus, quod prius non fuit. Prius autem nullum potuit esse medium, si omnino et in omni parte se tangendo impleverunt, quia si fuisset ibi tunc aliud corpus, fuissent duo corpora in eodem loco, et hoc est impossibile." Ibid.

"Because of this," Albertus concludes, "I say that they [the successive spheres] never touch but that intervals [or gaps between the spheres] in some particular place are sometimes greater and sometimes smaller, and that a rare or dense body existing between the circles [or spheres] fills them." Albertus adopts this interpretation and identifies it with "the opinion of Thebit, a wise philosopher, in a book which he composed on the motion of the spheres."[77] Thus every sphere is separated from its immediate neighbors by a certain kind of celestial matter that is capable of rarefaction and condensation.

That Albertus adopted such an opinion – and it seems that he did – is quite astonishing. It marked a radical departure from Aristotle's cosmology. Indeed, Albertus distinguished the matter that lies between orbs from the ether, or fifth element, that composes the rest of the celestial region. Unlike ether, which is incorruptible, indivisible, and therefore suffers no rarefaction or condensation, the intervening matter can rarefy and condense and must therefore be divisible. Although the creator of the celestial orbs made some parts permanently rarer and some parts permanently denser than other parts,[78] he made "the intervening body contractable and extendable so that

77. "Et haec est sententia Thebit philosophi sapientis in libro, quem composuit de motu sphaerarum." Ibid. Hossfeld, the editor of Albertus's *De caelo*, identifies his source as Thabit's (or Thebit's) *De motu octavae sphaerae* (see Hossfeld's note to line 29 on page 30). In Carmody's edition of this treatise (see Thabit ibn Qurra [Carmody], 1960, 102–107), there is no such passage. Hossfeld's claim, however, is based on a single manuscript, Paris, Bibliothèque Nationale, fonds latin, MS. 7195, 140vb–143va, which is but one of a number of manuscripts that Carmody used for his edition. If such a passage existed in Thabit's *De motu octavae spherae*, it is unlikely that all traces of it would have disappeared from Carmody's edition. Indeed, the theory of intervening matter between the celestial orbs does not appear in any other of Thabit's works that were translated into Latin and which have been edited by Carmody.

However, according to Bernard Goldstein (1986, 277–278), Albertus, in an unpublished part of the fourth book of his *Sentences*, cited "a passage from Thabit's lost treatise, *Libro de excentricitate orbium*," in which "we are told that there is a subtle matter that fills the space between the spheres. This matter is uniform, transparent, and subject to division [i.e., fluid], but not to alteration." Moses Maimonides also describes a similar interpretation and attributes it to Thabit (*Guide*, pt. 2, ch. 24, 1963, 2:325). But it is probably from Maimonides, rather than Thabit (B. Goldstein, ibid., 278), that Levi ben Gerson derived the idea that the celestial region "consists of planetary shells separated by fluid layers with certain properties that allow us to compute their thicknesses" (ibid., 273). In the Hebrew text of his *Astronomy*, Levi sought to compute these thicknesses, but the calculations were apparently not incorporated into the Latin translations of his treatise (ibid., 285). Precisely what Thabit may have had in mind is thus left vague. On Maimonides and Thabit, see also Duhem's brief discussion, *Le Système*, 1913–1959, 2:118–119.

The ultimate source of the theory of separate matter lying between successive orbs may have been Ptolemy himself, when, after describing the contiguity and contact of successive surfaces of orbs in his *Hypotheses of the Planets*, he allows that "if there is space or emptiness between the (spheres), then it is clear that the distances [between successive surfaces of successive planetary orbs] cannot be smaller, at any rate, than those mentioned." See Ptolemy [Goldstein], 1967, 8, col. 1 (for the full passage, see Sec. III.2 of this chapter).

78. See the distinction between static and dynamic rarity and density, in Chapter 10, Section II.1.c. Albertus Magnus denies dynamic rarity in the regular celestial ether from which the orbs are composed but allows it for the special matter that lies between orbal surfaces.

it should fill what lies between the spheres."[79] Thus did Albertus abandon the important Aristotelian concept of celestial homogeneity and assume instead the existence of two different kinds of eternal celestial substances: one divisible, and therefore changeable; the other indivisible and unchangeable. Moreover, in his interpretation, the celestial orbs were no longer in direct contact.

Few chose to follow Albertus's radical theory.[80] Convinced that eccentrics were essential to account for the astronomical phenomena, Albertus was obviously prepared to abandon certain important Aristotelian concepts in favor of a system that would save the phenomena and also preserve a viable cosmology. As we shall see, others proceeded in a similar fashion.

4. If eccentrics exist, can the earth lie at the center of the world?

Although the potential impossibilities just described were probably considered the most serious cosmological difficulties for scholastic authors, a number of other objections appeared rather regularly. D'Ailly reports that some questioned whether, if eccentrics existed, the earth could lie at the center of the universe.[81] This objection was apparently based on the assumption that *all* celestial orbs are eccentric. If so, the earth could not be their center, by definition. But d'Ailly and others replied that the earth lies at the center of the "total orb," that is, it lies at the center of all the concentric surfaces that serve as boundaries for each set of planetary orbs.[82] As Albert of Saxony explained, the absence of the earth from the center of eccentric orbs posed no problem, because eccentrics are included within the totality of planetary

79. Albertus Magnus, *De caelo*, bk. 1, tract. 1, ch. 2, 1971, 5, pt. 1:30. Albertus adds that this is also the opinion of Avicenna (in the *De caelo et mundo* of the latter's *Sufficientia*) and Averroës (in the *De substantia orbis*, which Albertus cites as *Liber de essentia orbis*).

80. One who did was Cecco d'Ascoli, who declared that "orbs are neither continuous nor contiguous, but there is an intervening body between them, which, according to Thebit and Albertus, is capable of being compressed." For the Latin text, see Cecco d'Ascoli, *Sphere*, 1949, 353. In the seventeenth century, Giovanni Baptista Riccioli cited this very passage as evidence that Cecco believed in the fluidity of the heavens (see Riccioli, *Almagestum novum*, pars post., bk. 9, sec. 1, ch. 7, 1651, 239, col. 2). In his *De eccentricis et epicyclis*, Cecco makes no mention of bodies intervening between successive orbs. Although he seems not to have adopted it, Thomas Aquinas mentions the theory without reference to Albertus when he explains (in his commentary on Boethius's *De trinitate*) that supporters of eccentrics and epicycles believe that this opinion avoids the dilemma that two bodies might have to occupy the same place and that the substance of the spheres could be divided. Thomas describes the intervening matter as "another substance, which lies between the spheres and which, like air, is divisible and without thickness, although [unlike air] it is incorruptible." For the Latin text, see Litt, 1963, 348. Robertus Anglicus mentions "another opinion" (alia tamen opinio) in which matter is assumed between orbs. See Robertus Anglicus, *Sphere*, lec. 1, 1949, 147–148 (Latin); 203 (English).

81. D'Ailly, *14 Questions*, qu. 13, 1531, 163v.

82. Ibid., 164v. Paul of Venice accepted the same argument and also used the expression "total orb of the planets" (*totalis orbis planetarum*); see Paul of Venice, *Liber celi*, 1476, 31, col. 2 (the last two lines; because the work is unfoliated and is provided with few signatures, the page numbers have been determined by counting from the beginning of the *Liber celi et mundi*).

orbs, the external surfaces of which are concentric.[83] Here, of course, was the fundamental feature of the compromise between Ptolemaic astronomy and Aristotelian cosmology. With the extreme surfaces of every planetary sphere assumed concentric with respect to the earth, the eccentric orbs contained within those concentric surfaces could possess their own centers without any adverse effect on Aristotelian cosmology.

5. *Eccentrics and the problem of a plurality of centers*

But even if the earth could serve as a center of the universe, the existence of at least one other center for the eccentric orbs would involve at least two different centers for celestial bodies. If two such centers existed, could a heavy body move naturally downward to its natural place, when the latter is defined as a unique center, coincident with the earth's center, that functions as a *terminus ad quem*? According to d'Ailly, some denied that a heavy body could reach its natural place at the center of the world, arguing that a heavy falling body would either have to move to both centers simultaneously or, because it could not choose between them, would not move at all.[84] D'Ailly responded that despite the different centers, every heavy body would nonetheless move toward the center of the world, because the latter is the center of the "total orb," that is, the center of all the concentric surfaces which enclose all the eccentric orbs.[85] In a similar vein, Johannes Versor argued that a plurality of eccentric and concentric centers would not render meaningless the idea of a unique, absolute "down" location. "Down" in the universe, Versor explains, was usually taken "in relation to the whole heaven, or in relation to a whole orb, but not with respect to partial circles [or orbs]."[86] But the heaven, "is concentric with respect to each of its extremal surfaces; and the same holds for any orb, even though a partial

83. Albert of Saxony, *De celo*, bk. 2, qu. 7, 1518, 106r (mistakenly foliated 107), col. 2, for the objection; 106v, col. 2, for the reply.
84. D'Ailly, *14 Questions*, qu. 13, 1531, 163v, who reports this argument, speaks of only two centers, one for the world and the other for all eccentric orbs. But eccentric orbs had many centers, because differences in their planetary eccentricities precluded a common center. Paul of Venice, *Liber celi*, 1476, 31, also spoke of two centers, but Albert of Saxony spoke of "many centers" (plura centra) (*De celo*, bk. 2, qu. 7, 1518, 106r, col. 2, for the objection; 106v, col. 2, for Albert's reply).
85. D'Ailly, ibid., 164v. Paul of Venice offered the same solution (*Liber celi*, 1476, 31), explaining that the earth is "in the middle of the total orb of the planets, because the orb is totally concentric to the world" (tamen est [i.e., the earth] in medio totalis orbis planetarum eo quod orbis totaliter est concentricus mundo). Albert of Saxony, *De celo*, bk. 2, qu. 7, 1518, 106r, col. 2, presented the same objection with much the same response on 106v, col. 2.
86. Versor [*De celo*, bk. 2, qu. 9: "Whether eccentric, concentric, and epicyclic circles (i.e., orbs) are to be assumed in the heavens to save the appearances of the planetary motions"], 1493, 22v, col. 1, for the objection, and 23r, col. 1, for Versor's reply. Here we see the common distinction that most natural philosophers drew between "the whole orb" (*orbis totalis*, or as Versor put it, *orbis integer*), which embraces three or more eccentric orbs, and a "partial circle" or "orb" (*circulus* [or *orbis*] *partialis*), which refers to only one of the constituent eccentric orbs of a "whole orb."

surface of one part of the orb has a center distinct from the center of the world."

Although the concentric surfaces of each planetary sphere enabled the earth to retain its cosmic centrality, and although the earth remained the natural place of heavy bodies, the defenders of solid eccentrics and epicycles had made a significant departure from Aristotelian cosmology: they allowed celestial bodies to move around more than one center. Eccentric orbs were assumed to move around their own centers rather than around the earth as center. To accept the three-orb system as truly representative of the physical cosmos was to admit that, contrary to Aristotle, Averroës, and Maimonides, planetary spheres could rotate around geometric points other than the center of the earth, that is, other than the geometric center of the universe. Most scholastics passed over this significant shift with little or no comment, but a few, like Cecco d'Ascoli, Nicole Oresme, and Jean Buridan met the issue head-on. Because he believed that the celestial orbs were not all of the same nature and that celestial bodies differed in matter and form and in their motions, Cecco insisted: "it is therefore not absurd that they [the planets] should have different and immobile centers."[87] Oresme flatly declared: "whether Averroës likes it or not, we must admit that they [the heavenly bodies] move around various centers, as stated many times before; and this is the truth" ([*Le Livre du ciel*, bk. 2, ch. 16], 1968, 463). Buridan firmly stated: "The Commentator [Averroës] speaks improperly when he says that the spheres are located by a [common] center; . . . modern astronomers [*astrologi*] do not concede that all celestial spheres have the same center; indeed, they assume eccentrics and epicycles" ([*De caelo*, bk. 2, qu. 14], 1942, 191, lines 19–23).

6. *Would planets move with rectilinear motion if eccentrics and epicycles existed?*

Because a key purpose of eccentrics and epicycles was to account for changes in planetary distances from the earth, it was alleged that eccentrics and epicycles would cause planets to ascend and descend rectilinearly as they alternately approached and withdrew from the earth. Following Albumasar, Bacon held that motion is threefold: namely, from the center of the world (*media*); toward the center; and around the center.[88] Celestial bodies move only around the center of the world, that is around the earth. For if a planet moved around another center, it would sometimes be nearer the earth and sometimes farther away. But if a celestial body varied its distance from the

87. In replying to an argument against eccentrics, Cecco declares: "Si dicatur omnes orbes esse eiusdem nature, quod est falsum, ut dicit Albertus in libro *Celi et mundi*, et quia corpora celestia diversa sunt in forma et materia, et in motu diversa erunt: non ergo erit inconveniens quod habeant diversa centra et inmobilia." Cecco d'Ascoli, *De eccentricis*, 1906, 167; see also the edition in Peter of Abano, *Lucidator*, 1988, 393.

88. Bacon, *De celestibus*, pt. 5, ch. 17, *Opera*, fasc. 4, 1913, 444, lines 10–11, cites Albumasar's *De conjunccionibus* as his source.

earth in this manner, it could do so only by a rectilinear motion or by a motion compounded of rectilinear and circular motion. To move toward or away from the earth rectilinearly, a celestial body would have to be either heavy or light, or compounded of both; or it might have an entirely different nature. But rectilinear motion toward or away from the earth and out of a circular orbit would involve a celestial body in violent action, which was contrary to the nature of the celestial ether.[89]

The usual response was to deny that variations in planetary distances were the result of rectilinear motion. D'Ailly argued[90] that upward and downward motion could happen only where generation and corruption occurred, namely in the terrestrial region.[91] Paul of Venice met the same objection by a different argument (*Liber celi*, 1476, 31, col. 2). To qualify as rectilinear ascent and descent, motions must be measured along a radius of the world. Such measurements were therefore not applicable to circular motion, from which it followed that the motion of planets on circular eccentrics and epicycles did not qualify as rectilinear.

7. *The problem with epicycles*

It would appear that the acceptance of eccentric orbs also implied a commitment to epicyclic orbs. But at least one scholastic, Jean Buridan, accepted the former but not the latter. Epicycles posed a special problem, because of the Moon's observed behavior. Since the Moon always shows the same face to us, Aristotle had argued that it cannot be said to rotate or revolve. On the assumption that all planets are alike in their basic properties, he inferred (*De caelo* 2.8.290a.25–27) from the Moon's behavior that no planets rotated around their own axes. Aristotle's denial of rotation to the Moon and other planets played a significant role in arguments about the reality of material epicycles.

Although the fundamental problem about epicycles is traceable to Roger Bacon in the thirteenth century,[92] it was Jean Buridan and Albert of Saxony in the fourteenth century who described the two approaches available to natural philosophers.[93] Buridan discussed the issue in a question on

89. "But there is no violence in the heavens, as Aristotle says in the book *On the Heaven and the World* and in the eighth [book] of the *Physics*; and it is obvious that nothing perpetual is violent." Bacon, ibid., 444, lines 26–29.
90. D'Ailly, *14 Questions*, qu. 13, 1531, 163v.
91. Ibid., 164v. Albert of Saxony had earlier presented the same objection and resolution (see his *De celo*, bk. 2, qu. 7, 1518, 106r, col. 2, for the objection, and 106v, col. 2, for the response).
92. Bacon, *Opus tertium*, 1909, 130–131; the *De celestibus* omits this section. Duhem, *Le Système*, 1913–1959, 3:436, conjectures that Bacon may have been the first to propose this objection to the existence of solid epicycles. Since Bacon speaks as if others had already proposed the criticism, this seems unlikely.
93. See also Chapter 17, Section IV.3a.iii, for a further discussion of these ideas in connection with the Moon and its spots.

"Whether epicycles are to be assumed in celestial bodies."[94] He based his opinion on the behavior of the "man in the Moon," that is, the spot on the lunar surface that had the appearance of a man whose feet always point toward – or lie at – the bottom of the Moon. Buridan argued that if the Moon had an epicycle, the man's feet should sometimes appear in, or point toward, the upper part of the lunar disk. Thus if the man's feet are at the bottom of the lunar disk when the Moon is in the *aux*, or apogee, of the epicycle, the feet ought to be in the upper part of the lunar disk when the Moon reaches the opposite of the *aux*, or perigee, of the epicycle. But such an occurrence is never observed. The feet always remain at the bottom of the lunar disk, thus calling for the rejection of an epicycle.[95] Buridan suggests a way to account for this phenomenon and retain the epicycle. We would have to assume that "just as this epicycle is moved around its proper center, so also is the body of the Moon moved around its proper center in a motion contrary to that of the epicycle and with an equal speed" (Grant, 1974, 526). Only in this way will the upper part of the man always appear in the upper part of the lunar disk.

Assuming with Aristotle that all planets possess the same fundamental properties, Buridan infers that if the Moon has a proper rotatory motion, all the other planets should also possess that same motion. In agreement with Aristotle, however, he was convinced that no planet could rotate around its own center. Planets not only move from one position to another; they also cause transmutations in sublunar bodies. Consequently, if planets rotated around their own centers, the rotations ought to affect the way in which they cause sublunar effects. That is, each planet ought to produce differential effects; otherwise its rotatory motion would be superfluous. Taking the Sun as exemplar, Buridan argues that it does not produce such differential effects, probably because it is a uniform, homogeneous body whose upper and lower parts are identical. Any rotatory motion by the Sun around its own center would therefore be superfluous, because no sublunar changes would result. "But if the Sun does not have such a motion, it does not seem reasonable that the Moon should have it, since the Sun is much nobler than the Moon."[96]

94. Buridan [*Metaphysics*, bk. 12, qu. 10], 1518, 73r- 73v. The quotations are from my translation of this question in Grant, 1974, 524–526.

95. Buridan is only partially correct. If the Moon were carried on an epicycle but lacked rotatory motion, the appearance of the Moon would indeed change. But the change would not be as Buridan describes it. The "man in the Moon" would not turn upside down from apogee to perigee, but rather a terrestrial observer would see the man in the Moon for awhile and then not see him. For this interpretation, I am indebted to my student, Mr. James Voelkel, and to an anonymous reader.

96. "Et si sol non habeat talem motum nec videtur rationabile quod luna habeat, cum sol sit multo nobilior quam luna." Buridan, *Metaphysics*, bk. 12, qu. 10, 1518, 73v, col. 1. Buridan believed that planets were also unlikely to have proper motions, because each such motion would require a special mover. We would then have to assume "as many intelligences as there are stars in the sky, because each star would require a special mover for its special motion." But "Aristotle did not assign [or concede] such a multitude [of motions and intelligences]." Ibid. (Latin) and Grant, 1974, 526 (English).

Buridan then derives the following consequence: if the Moon does not have a proper motion around its own center, it cannot have an epicycle. For if the Moon had an epicycle but lacked a proper motion, the head and feet of the man in the Moon would change positions every time the epicycle's apogee and perigee rotated 180 degrees. Because no such change is observed, Buridan concludes that the Moon can have no epicycle, from which he generalizes that "if an epicycle is not posited in the orb of the Moon, it ought not to be posited in the orb of the other planets, since all the reasons which apply to the other planets should also apply to the Moon" (Grant, 1974, 525). Thus did Buridan conclude that "all appearances can be saved by eccentrics [alone] without epicycles" (ibid., 526).

One response to Buridan was to allow that a particular planet might indeed behave differently from its sister planets. Albert of Saxony adopted just such a strategy. After describing the problem much as Buridan had,[97] Albert assumes that the Moon possesses a proper motion around its own center in a direction that is contrary to the motion of its epicycle. As for those who say that "other planets do not have proper motions around their proper centers, therefore the Moon does not," Albert counters, without elaboration and perhaps with Buridan in mind, that the Moon's nature differs from that of the other planets because the Moon's upper and lower parts can affect sublunar things differentially. Its proper motion around its own center is, therefore, not superfluous but brings the lower part of the Moon to the upper part and the upper to the lower. Because the Moon's proper motion is contrary to the motion of its epicycle, we do not observe these continuous and regular turnings of the spot in the Moon.[98]

Albert's interpretation prevailed and was repeated with the same arguments by Christopher Clavius in the numerous editions of his commentary on the *Sphere* of Sacrobosco that appeared in the late sixteenth and early seventeenth century.[99] Although both Albert and Buridan sought to save the observed behavior of the spot in the Moon, they did so in radically different ways. Whereas Buridan insisted on the uniformity of planetary behavior and properties, Albert permitted divergence. Buridan sought for consistency: either all planets rotated around proper centers, or none did; either all planets moved on epicycles, or none did. The astronomical ap-

97. Albert of Saxony, *De celo*, bk. 2, qu. 7, 1518, 106r, col. 2 (the fifth principal argument).
98. Although Albert of Saxony admitted that he had often seen a black spot in the Moon, he denied that it resembled a man.
99. In agreement with Albert's position were d'Ailly (*14 Questions*, qu. 13, 1531, 163v, for the objection, and 164v for the response) and Paul of Venice, *Liber celi*, 31, col. 2, for the objection and 32, col. 1, for the response. Paul argued that because the Moon has "diversity in its parts," it requires a proper motion, whereas the other planets lack diversity and need no proper motions. Without invoking diversity, Bernard of Verdun (Grant, 1974, 523–524) retained the lunar epicycle and also assumed that the Moon somehow turns, or is turned, so that "the spot always appears to us in the same shape [or form]." Although this interpretation was quite traditional by the time Clavius wrote, he attributes it to Jean Fernel (1497–1558). See Clavius, *Sphere*, ch. 4, 1593, 522, for the objection, and 525 for his response.

pearances could be saved only if planetary homogeneity and uniformity were preserved. By contrast, Albert of Saxony thought it more important to save the appearances than to preserve the uniformity of planetary behavior. In Albert's scheme, it was not necessary that all planets should move on epicycles (the Sun did not). Nor, as we saw, was it necessary that either *all* planets or *no* planets move around their own centers. If the phenomena could be saved by assuming that some planets really moved around their own centers and others did not, Albert was satisfied.[100]

In the seventeenth century, Melchior Cornaeus agreed with Buridan and rejected the existence of a lunar epicycle. As we shall see later (Ch. 14, Sec. VIII.1.b.iii), however, the context of his discussion and the reasons for his decision were radically different.

8. Summary of differences with Aristotle

Although a few other arguments were sometimes cited for and against eccentric and epicyclic spheres,[101] those mentioned here were unquestionably the most important for cosmology. Despite the widespread conviction that eccentrics and epicycles saved the astronomical phenomena and that Aristotelian concentric astronomy did not, scholastic natural philosophers were also aware that those same eccentrics and epicycles appeared to violate important aspects of Aristotelian cosmology. In order to save the astronomical phenomena and avoid alleged cosmological impossibilities, some, and in a number of instances many, natural philosophers made significant departures from Aristotelian cosmological principles. Among the most significant were the assumptions that (1) eccentric celestial orbs move with circular motion around centers other than the earth; (2) that the Moon and all other planets have proper motions around their own centers in a direction opposite to that of their epicycles; (3) that successive orbs are not in direct contact and the space between those orbs is occupied by a celestial substance that is divisible, though incorruptible; and, finally, (4) that celestial bodies, and therefore the celestial substance, need not be homogeneous.[102]

Scholastic commentators were aware that Aristotle's description of his

100. Nonetheless, one may ponder why Albert did not infer from the lunar rotation the rotation of all planets around their respective axes. Perhaps he thought, as did Paul of Venice later (see note 99 to this chapter), that they lacked the Moon's diversity and therefore did not require axial rotation as they were carried by their respective epicycles.

101. For example, Bacon argued that although the surface of an eccentric sphere is spherical, the sphere itself is nonuniform, as is evident from its varying thickness. Natural philosophers, however, insist that celestial bodies must be simple and homogeneous, and therefore invariant with respect to thickness. This is but another aspect of the homogeneity argument. See Bacon, *Opus tertium*, 1909, 133 (for a few additional arguments, see 132–137); Bacon, *De celestibus*, pt. 5, ch. 15, *Opera*, fasc. 4, 1913, 440 (and 439–443 for the same additional arguments).

102. As is evident by the assumption that the substance between orbs differs from that of the orbs themselves; that the planets have different basic properties; and, as we saw in the preceding note, that one and the same sphere may vary in thickness.

system of concentric spheres as he described it in the twelfth book of his *Metaphysics* was inadequate. They sometimes reacted to that description in strange ways. Thus in the seventeenth century, a Scotistic commentator, perhaps Hugo Cavellus (1571–1626), offered a lengthy analysis of Aristotle's concentric spheres, explaining how Aristotle arrived at 55 and how one could reduce this to 47, which he mistakenly believed was Aristotle's final total.[103] Our commentator declares that a single, uniformly moving, concentric sphere could not properly represent the motion of a planet, because planets alter their speeds and seem to change directions, moving sometimes directly and sometimes retrogressively. To take these anomalies into account, Ptolemy and other astronomers utilized deferents (presumably eccentrics) and epicycles. But Aristotle also recognized that a single planet had more than one motion and accounted for this by building on the systems of Eudoxus and Callippus: that is, he assigned a plurality of concentric spheres to account for the motion of each planet.[104] In describing how Aristotle did this, our commentator makes no further mention of eccentrics or epicycles. Indeed, it is as if he had equated the two systems simply because both assigned a plurality of spheres – not just one – to account for the motion of each planet. In light of this, we are not surprised that, despite writing in the seventeenth century, our commentator finds no reason to mention Copernicus or the Copernican system. Aristotle's system is analyzed as if the Ptolemaic and Copernican systems had never existed.

9. On the physical nature of eccentrics

Despite a rather large number of authors who considered the suitability of eccentrics for astronomy and cosmology, relatively few ventured opinions about the nature of such spheres, especially the deferent orbs. Were they hollow and void, or filled with some rare or dense substance? Or were they solid? Were the spherical epicycles carried around as immobile bodies within their deferent orbs? Or did they move through the orb itself? Were the planets carried within an epicycle, or were they self-moved? As we shall see, most of those who did consider the internal nature of the orbs did so in the late sixteenth and the seventeenth century. Few scholastics troubled to describe the nature of the spheres themselves, especially the deferent orb, which carried a spherical epicycle.

103. See Cavellus, *Metaphysics*, bk. 12, summa 2, ch. 4, 1639, 4:448–450. Aristotle offered 49 as an alternative to 55.

104. "Notandum etiam quod ultra motum diurnum, qui est motus primi mobilis, deprehensum est plures esse lationes planetarum secundum instrumenta mathematica, puta astrolabium et quadrantem, etc. Et etiam per rationem quatenus motus planetae apparet quandoque velocior, quandoque tardior; et planeta quandoque videtur directus, quandoque retrogradus, quandoque stationarius statione prima, vel secunda: quod non potest esse secundum motum sphaerae, cum ille sit omnino uniformis. Et ideo ad salvandum hos diversos motus Ptolemaeus et alij periti Astrologi investigaverunt circulos planetarum praeter sphaeras, scilicet deferentem et epicyclum, etc. Et ideo bene ait Philosophus quod astrorum errantium plures sunt lationes quam una." Ibid., 448, col. 1.

One of the few who did was Aegidius Romanus, whose discussion in his *Hexaemeron* appears to have had some influence in the seventeenth century, when not only was it cited explicitly by Mastrius and Bellutus but its key ideas were adopted by others.[105] Convinced that eccentrics and epicycles existed in the heavens and that only they could save the astronomical phenomena,[106] Aegidius provides a brief physical interpretation for the eccentric system. He assumes that the celestial region, from the concavity of the lunar orb to the fixed stars, or eighth sphere, was one single, continuous body or orb. But just as a man or a lion is said to be one body but yet contains within itself things that are discontinuous – such as, for example, marrow in a bone, or blood in the veins – so also does the continuous single orb embracing the region from the Moon to the sphere of the fixed stars contain discontinuities within it.[107] Those discontinuities are represented by the seven eccentric or deferent planetary orbs, which are embedded discontinuously within the mass of continuous matter that comprises the single orb that stretches from the Moon to the eighth sphere of the fixed stars.

Thus did Aegidius attempt to integrate unity – the unity of a single orb from the Moon to the fixed stars – and diversity – the diversity of the seven eccentric deferents, one for each of the seven planets. As he put it: "because of the unity of the whole body, there is one sphere and one heaven; and because of the diversity of deferents and eccentrics, we can [also] say that there are many spheres and many heavens."[108] To ensure that his readers were in no doubt about his conception, Aegidius, as we saw, reinforced his description with vivid analogies.

Because he described the mode of existence of an eccentric deferent embedded within the continuous substance of the single orb that contained

105. For Aegidius's account, see *Opus Hexaemeron*, pt. 2, ch. 32, 1555, 49r, col. 1–54r, col. 1. The quotations are drawn largely from 49v, cols. 1–2 (Aegidius repeats his major ideas on 53v, col. 1). Duhem, *Le Système*, 1913–1959, 4:110–119, gives a summary account. For the statement about Aegidius by Mastrius and Bellutus, see note 112 of this chapter.

106. Aegidius, ibid., pt. 1, ch. 16, 1555, 15v, col. 2, says "Advertendum etiam quod nos dicimus esse in caelo eccentricos et epicyclos." Apart from saving the astronomical phenomena, Aegidius opposed the idea that the planets could move themselves "just as fish are moved in water" (sicut pisces moventur in aquis). The planets are not self-moved, because "then there would be a division [*scissio*] of the orb, or there would be a vacuum, or two bodies would be in the same place" (quia tunc esset scissio orbis, vel esset vacuum, vel essent duo corpora in eodem). Ibid., pt. 2, ch. 32, 49r, col. 2. These same arguments also served to attack the existence of eccentrics, as can be seen in Cecco d'Ascoli's description of them (see the end of Section III and all of Section III.1 of this chapter).

107. "Advertendum autem quod unum animal, ut puta unus homo vel unus leo, dicitur esse unum corpus, non tamen omnia quae sunt in ipso sunt continua, ut medulla non est continua ossi, sed contigua; et sanguis non est continuus venae, sed contiguus. Sic potest a globo lunari usque ad octavam sphaeram; includendo ipsam sphaeram octavam dici unum corpus propter continuationem totius. Non tamen omnia quae sunt in eo sunt continua quia deferentes sive eccentrici non sunt continui cum huiusmodi coelo." Aegidius, ibid., pt. 2, ch. 32, 49v, col. 2.

108. "Propter unitatem totius corporis est una sphaera et unum coelum; propter diversitatem autem deferentium et eccentricorum possunt dici multae sphaerae et coeli multi." Ibid.

and surrounded it as being "like marrow in a bone," Aegidius implies that the overall heaven is hard and solid, like a bone, and that the eccentric cavities are filled with a soft or fluid material, akin to the marrow of a bone. The other analogy of blood in the veins conveys a similar relationship. Aegidius asserts that the soft matter in each deferent is contiguous with respect to its immediate surrounding surfaces. Thus each eccentric deferent is really a hollow cavity filled with a soft substance. Within each deferent orb is a spherical epicycle carried around by its deferent. The epicycle in turn is discontinuous, or contiguous, with respect to the eccentric. Because each of them is discontinuous, the eccentric and epicycle each has its own proper motion by which the planet effects its retrograde or direct motion.[109]

The relationship of the soft matter in the eccentric deferent to the matter surrounding it is, however, left unclear. Are the two matters the same or different? The analogies indicate differences, but such an inference would imply two unchangeable substances, or one unchangeable and one changeable, as Albertus Magnus had assumed earlier. Indeed, Aegidius may have thought of them as the same substance, with differing densities. Once again resorting to an analogy with animals, Aegidius observes that "just as in animals, all parts are not equally dense, because the bone is denser than flesh, so also in the heavens, all parts are not equally dense, because a star [or planet] is denser than an orb."[110] But whereas the rare part of an animal is more changeable than a denser part, this is untrue for the celestial region, which is unalterable and incorruptible and where every part always remains at the same level of rarity or density.[111]

Aegidius's ideas about a single heaven between the Moon and fixed stars, containing within itself eccentric deferents as hollow cavities, were adopted by Raphael Aversa and discussed, and perhaps adopted, by Mastrius and Bellutus, who describe them as "zones" (*zonae*) or rings (*anuli*).[112] They assumed a solid, single, presumably hard, starry heaven in which all the fixed stars and planets are embedded. Since the planets do not move themselves, each is carried around in its zone by an epicycle. According to Aversa,

109. Sed habent [that is, the deferents or eccentrics] suos proprios motus per quos deferuntur planetae et epicycli, ubi sunt fixi planetae non sunt continui ipsis circulis deferentibus. Et inde est quod habent suos motus proprios per quos dicitur planeta retrogradus vel directus." Ibid.

110. "Advertendum etiam quod sicut in animali omnes partes non sunt aeque densae quia os est densius carne, sic in huiusmodi coelo omnes partes non sunt aeque densae quia stella est densior orbe." Ibid.

111. "Differunt tamen haec in animalibus et in coelo quia in animalibus partes magis rarae sunt magis passibiles, sed in coelo ita est impassibiles et ita inalterabilis et se mota ab omni peregrina impressione pars rara, sicut et densa." Ibid.

112. Aversa, *De caelo*, qu. 32, sec. 7, 1627, 72; Mastrius and Bellutus, *De coelo*, disp. 2, qu. 1, art. 2, 1727, 3:489, cols. 1–2, par. 28. Mastrius and Bellutus also mention that Aegidius seems to teach this opinion in his *Hexaemeron*, pt. 2, ch. 33 (Duhem, *Le Système*, 1913–1959, 4:111, correctly cites chapters 32 and 36). The term "zones" (*zonae*) may derive from John Damascene, who, as we saw earlier (note 11 of this chapter), conceived of the firmament as a single body divided into seven zones. On page 490, column 2, Mastrius and Bellutus offer quite a different opinion (see this volume, Ch. 14, n. 134).

this single heaven is firm and solid, turning by itself from east to west with the daily motion as it carries all the celestial bodies with it. Within this single heaven, the seven zones, or bands, are discontinuous with the rest of the heaven and are actually cavities within it.[113] Although Aversa denies that these cavities are void, he does not indicate what fills them.[114]

Whereas Aegidius, Aversa, and Mastrius and Bellutus agreed that the planet was carried around by an epicycle, Pedro Hurtado de Mendoza, according to Aversa,[115] assumed that the planets are self-moved rather than carried around by an epicycle. Like Aegidius and those who followed him, Hurtado assumed only one overall, presumably solid and hard, heaven within which are channels that function as deferent orbs. Inside each deferent orb is a planet, which is assumed to move by its own effort through its orb, the interior of which is assumed to be either void or filled with fluid. Aversa rejected the interpretation, because it has self-moving planets and because he found a heaven that is part solid and part fluid objectionable.[116]

The Coimbra Jesuits and Francisco de Oviedo also rejected the concept of eccentric zones with self-moving planets. Citing Hurtado as a proponent of it, Oviedo argues[117] that if the planets moved by themselves in these celestial channels or cavities, almost the entire heaven would be hollowed out, because the Sun and planets do not move in a single track over which they pass endlessly; rather they move over a broad band of the sky. The Sun, for example, does not always move over the same path, but moves from the equinoctial circle to each solstice. If Hurtado's interpretation were true, the Sun would require a hollow cavity that extended from one solstice to the other. The same reasoning would apply to planets like Mercury and Venus. Moreover, what fills the spaces in the cavities that are at any given moment unoccupied by the planet? Oviedo thinks they would be void and that, as a consequence, we would be unable to see the stars, since the species that enable us to see them would be untransmittable through celestial vacua.

Oviedo, who seems to have assumed the fluidity of the region of the fixed stars and planets and to have treated the region as a single heaven, denied the existence of eccentrics and channels. Indeed he denied the existence of any other heavens. For if the single heaven is really fluid, each planet must "traverse the whole space of the heaven freely, just as a fish in

113. Aversa mentions that some hold that Mercury and Venus do not have total orbs but move around the Sun by means of epicycles (Imo plures iam sunt, qui conveniunt in hoc, ut Venus et Mercurius non habeant alios proprios orbes et caelos totales, sed moveantur solum per epiciclos circa solem). Aversa, *De caelo*, qu. 32, sec. 7, 1627, 72.

114. In the third of five opinions on the fluidity or solidity of the heavens, Riccioli, *Almagestum novum*, pars post., bk. 9, sec. 1, ch. 7, 1651, 239, col. 2, describes a theory similar to Aversa's. Like Aversa, he attributes it to Hurtado de Mendoza (in the latter's [*De caelo*, disp. 2, sec. 1], 1615; I have not found it there).

115. Aversa, *De caelo*, qu. 32, sec. 6, 1627, 65, col. 2. Oviedo also attributed this interpretation to Hurtado (see Oviedo, *De caelo*, contro. 1, punc. 4, 1640, 471, col. 1), as did Amicus (*De caelo*, tract. 5, qu. 5, art. 5, 1626, 284, col. 2).

116. Aversa, ibid., sec. 7, 69, col. 1–73, col. 2.

117. Oviedo, *De caelo*, contro. 1, punc. 4, 1640, 1:471, cols. 1–2.

a river traverses the whole space of the waters up and down, once and again."[118]

The Coimbra Jesuits mentioned and rejected this interpretation even before Hurtado de Mendoza wrote.[119] If the celestial orbs are really channels in the sky through which the planets move by themselves, the substance that fills those channels must either be the same as the rest of the heaven or be of a sublunary nature. Traditional Aristotelian arguments are invoked to reject both alternatives. Thus if the substance filling the interiors is of a celestial nature, then, when the planet moves from one place to another within the interior of its orb, either another body succeeds it in the place it just vacated, or not. If not, then a vacuum would exist in nature, which is absurd; if celestial matter succeeds into any place vacated by the planet, it could only do so by the processes of rarefaction and condensation, which cannot occur in celestial matter. But if the interior matter is of a sublunary nature, the proponents of such a theory would have assumed something corruptible in the celestial region, which is also absurd.

10. On the assumed physical reality of eccentrics and epicycles

Whatever the reason for the seeming reluctance to speculate on the nature of the orbs and their inner structure, it was not because of any doubts about the physical reality of eccentric and epicyclic orbs.

From discussions of the three-orb system, it is obvious that those who accepted the "new" system believed in the physical reality of material eccentrics and epicycles. The controversy in the Latin West was not between those who argued for a system that merely saved the appearances regardless of physical reality and those who insisted that any astronomical system must not only save the phenomena but also represent physical reality.[120]

118. "Potest enim quodlibet astrum libere totum spatium caeli percurrere, sicuti potest piscis in flumine totum aquarum spatium sursum et deorsum semel et iterum pertransire." Ibid., 472, col. 2.

119. See Conimbricenses, *De coelo*, bk. 2, ch. 8, qu. 1, art. 1, 1598, 324. In arguing for a fluid heaven and against solid orbs, George de Rhodes opposed the opinions of both Hurtado de Mendoza and Aversa. Without elaboration, he was convinced that the vast cavities which both assumed in the heavens implied the existence of a vacuum and also would result in collisions. De Rhodes [*De coelo*, disp. 2, qu. 1, sec. 2], 1671, 280, col. 2.

120. A few did adopt the first alternative, including Maimonides and Thomas Aquinas. Sympathetic to Aristotelian cosmology but aware that it could not save certain crucial astronomical phenomena and also disturbed by the cosmological dilemmas inherent in any system of solid eccentrics, they argued that the phenomena might perhaps be saved in ways that had not yet been understood or, as Maimonides put it (for Thomas, see Sec. II.3 and note 35 of this chapter), "the deity alone fully knows the true reality, the nature, the substance, the form, the motions, and the causes of the heavens" (Maimonides, *Guide*, pt. 2, ch. 24, 1963, 2:327). Although what Aristotle says about the sublunar region "is in accord with reason," Maimonides believes that the heavens are too far away and too noble for us to grasp anything "but a small measure of what is mathematical" (ibid., 326). Few in the Middle Ages shared the cosmological uncertainty exhibited by Aquinas and Maimonides. One who did, but went even further, was Henry

Rather the dispute involved a decision as to which system of cosmic spheres best represented physical reality – a purely concentric system or a mixture of concentric and eccentric spheres. Repeated invocations of dire physical consequences that might or might not follow from one or the other of the two rival systems serve only to confirm that medieval natural philosophers were arguing about the structure of cosmic reality, not about convenient and arbitrary arrangements of geometric figures that might save the astronomical appearances. Clavius expressed traditional scholastic sentiments when he explained ([*Sphere*, ch. 4], 1593, 525) that eccentrics and epicycles are not monstrous and absurd things but were adopted by astronomers for good reasons. Just because eccentrics have a diversity of centers and some eccentrics vary in thickness should not cast doubt on them. After all, parts of the Moon vary in density, as indicated by its spots. Indeed, different parts of the heavens, not just the Moon, differ in density. Why, then, should eccentrics and epicycles be rejected because of variations in thickness or because of a diversity of centers? Not until the end of the sixteenth century, after the appearance of the new star of 1572 and the comet of 1577, was the physical existence of eccentrics and epicycles seriously challenged.[121]

IV. On the number and order of the mobile heavenly orbs

1. On the order of the heavens

Earlier in this chapter (Sec. II.6), we emphasized the great compromise which produced the almost universally accepted union of Aristotelian and Ptolemaic ideas about the relationship of the celestial spheres. The Aristotelian–Ptolemaic fusion, which Ptolemy had already made in his *Hypotheses of the Planets*, depended on a distinction between the concept of a "total orb" (*orbis totalis*) and a "partial orb" (*orbis partialis*), to use medieval terminology. The total orb was a concentric orb whose center is the center of the earth, whereas a partial orb was an eccentric orb, that is, an orb whose center is a geometric point lying outside the center of the world. The concentric total orb, whose concave and convex surfaces have the earth's center as their center, is composed of at least three partial orbs (see Figure 8), one of which, the eccentric deferent, carries an epicycle in which a planet is embedded. Thus were the concentric orbs of Aristotle fused with the eccentric orbs of Ptolemy.[122]

of Langenstein, or Henry of Hesse (1325–1397), who was convinced that eccentric and epicyclic orbs were imaginary and were not to be found in the heavens. He devised a system that was "a curious hybrid of homocentric astronomy and an Arabic innovation introduced into Ptolemaic astronomy by Thabit ibn Qurra" (Steneck, 1976, 70). For more on Thabit's innovation and Henry's own system, see Steneck, 69–72.

121. Indeed, even before these two celestial phenomena appeared, Robert Bellarmine, sometime between 1570 and 1572, had already assumed a fluid heavens and rejected the physical existence of eccentrics and epicycles. See Chapter 14, note 75.

122. On the continuity or contiguity of these orbs, see Section III.2 of this chapter.

From the "Catalog of Questions" (Appendix I), we learn that the most popular question in medieval cosmology (qu. 97) concerned the number of celestial spheres, "Whether there are eight or nine, or more or less." Scholastic natural philosophers faced a curious problem: should they count total orbs or only partial orbs? In fact, one could count either, as long as the sum was not represented as the total number of celestial orbs. Such a move would have been redundant, since each total concentric orb was formed from at least three partial eccentric orbs. Without the latter, there would be no concentric orb. Pierre d'Ailly makes all this quite clear when he declares that a celestial sphere is "the aggregate of all the orbs needed to save all the appearances concerning any planetary motion. In this way three eccentrics with an epicycle and with the body of the planet are said to be only one sphere and this is how we speak about spheres in what is proposed here."[123]

There were thus two basic ways to count orbs in the Aristotelian–Ptolemaic compromise system. Duns Scotus, for example, assigned 5 eccentric orbs to Mercury and 3 to every other planet, for a total of 23 eccentric, mobile orbs. To this, he added 1 orb for the eighth sphere of the fixed stars and 1 for the ninth, or crystalline, heaven, for a total of 25 orbs.[124] Of these 25 orbs, at least 23 are eccentric.[125] But if Scotus had counted only concentric, or total, orbs, he would have had only 9 orbs. Thus we may attribute to Scotus a total number of partial, eccentric orbs, say 25, or a total number of concentric orbs, say 9. But we may not speak of their sum, or 34 cosmic orbs, because each concentric orb is but the sum total of its eccentric orbs.

Fortunately, between the thirteenth and seventeenth centuries most scholastic natural philosophers left little doubt about their intentions: when inquiring about the number of celestial spheres in the universe, they counted only concentric orbs, although well aware that the latter were constituted

123. "Sed tertio modo dicitur aliqua sphaera una quia est aggregatum ex omnibus orbibus requisitis ad salvandum omnia illa quae apparent circa motum alicuius planetae. Et isto modo tres eccentrici cum epiciclo et corpore planetae non dicuntur nisi una sphaera et ita loquendum est de sphaeris in proposito." D'Ailly, *14 Questions*, qu. 2, 1531, 149r. This is the third of three ways in which d'Ailly believes the term *sphaera* can be used. In the second way, each eccentric orb of the aggregate is counted separately, as are also the epicycles. The first mode includes any spherical part of the heavens that is not separated from the whole of it, a definition that also includes spherical celestial bodies, such as stars (ibid., 148v–149r).

124. Duns Scotus, *Sentences*, bk. 2, dist. 14, qu. 2, *Opera*, 1639, 6, pt. 2:733: "Saltem caeli mobiles circundantes terram erunt vigintiquinque, scilicet vigintitres planetarum et praeter hoc caelum octavaum et caelum nonum." In his widely used *Theoricae novae planetarum*, Peurbach assigned 24 orbs to the seven planets, allocating 3 to all except the Moon, which had 4, and Mercury, which had 5 (see Peurbach, *Theoricae*, 1987, 9–27). Albertus Magnus, *Metaphysics*, bk. 2, tract. 2, ch. 24, *Opera*, 1964, 16, pt. 2:514, col. 1, describes Ptolemy's system as one of three proposed by "moderns." Albertus assigned 49 orbs (he says 50, but is mistaken in his count) to Ptolemy's system, attributing 3 to the eighth sphere of the fixed stars; 2 to the Sun; 5 to the Moon; 7 to Mercury; and 8 each to Venus, Mars, Jupiter, and Saturn. Albertus's source for this strange total is a mystery.

125. The eighth sphere of the fixed stars is concentric and perhaps also the ninth sphere.

of eccentric, or partial, orbs. The concentric orbs of the Middle Ages and Renaissance differed from the 55 or 49 distinguished by Aristotle in *Metaphysics* 12.8.1073b.2–1074a.14. The 7 orbs assigned by Aristotle to Saturn, for example, were all concentric (though some turned on different axes) and were not counted or conceived as a single orb. By contrast, the 3 partial orbs assigned to Saturn in the three-orb, Aristotelian–Ptolemaic compromise were all eccentric but could also be interpreted as forming a single concentric orb, because the outermost and innermost surfaces were concentric with the earth's center. Schematic representations of the celestial orbs made during the Middle Ages and Renaissance were not drawn from Aristotle's cosmological system but from the concentric orbs in the three-orb compromise.[126]

If it was not Aristotle's system of concentric orbs that was incorporated into the three-orb compromise, why, then, do I designate the compromise "Aristotelian–Ptolemaic"? Since the compromise was ultimately the work of Ptolemy, would it not be more accurate to call it the Ptolemaic compromise? Indeed it would. But the term Aristotelian–Ptolemaic seems more appropriate, because it emphasizes the most essential feature of Aristotle's celestial system: the concentricity of each planetary orb with the earth's center. It was that concentricity that enabled Aristotelian natural philosophers to embrace the system. Within the three-orb system, the order and number of the concentric heavens or spheres varied throughout the period of our study. All were agreed on the existence of seven planets and the fixed stars, and some included planets and fixed stars as part of the firmament.[127] Here, then, was the basic core of the celestial heavens.

Disagreement arose, however, on the order of the planets. Because ancient and medieval astronomers could find no parallax for planets other than the Sun and Moon, there was no way to determine the order of the three superior planets. Nevertheless, Saturn was assumed farthest from the earth, with Jupiter next, followed by Mars, an order that was based on the time it took each of them to complete its sidereal period: thirty years for Saturn, twelve for Jupiter, and two for Mars. Almost all astronomers and natural philosophers agreed on this. The order of Sun, Venus, and Mercury posed quite different problems. Without detectable parallaxes, and because Venus and Mercury always remained in the vicinity of the Sun, the order of these three planets was not determinable except on the basis of arbitrary, nonastronomical reasons. The Moon, however, was universally assumed to be the closest planet to earth.

Because of the various combinations possible between the Sun, Venus,

126. For an 11-orb version, see the diagram in this chapter from Peter Apian's *Cosmographicus liber* (Figure 9) and also see Reisch, *Margarita philosophica*, 1517, 244; an 8-orb version appears in Paris, Bibliothèque Nationale, fonds latin, MS. 6280, 20r (12th century), which is reproduced in Grant, 1978a, 276.

127. One who did not was Clavius, who equated the firmament solely with the sphere of the fixed stars. See Clavius, *Sphere*, ch. 1, *Opera*, 1611, 3:23.

and Mercury, a number of different planetary orders were commonly proposed. Albert of Saxony, for example, mentions four different arrangements that had at least some support. Moving from the outermost to the innermost planet, one interpretation placed the sphere of Saturn first, then those of Jupiter, Mars, Venus, Mercury, Sun, and Moon. In another interpretation the sphere of Venus was placed above the Sun and the sphere of Mercury below it. A third interpretation reversed the positions of Mercury and Venus with respect to the Sun. A fourth interpretation placed Venus and Mercury below the Sun, because it was thought more reasonable and elegant that the Sun should be in the middle of the planets, "like a king in the middle of his kingdom in order that the Sun should exercise its influence equally above and below."[128]

Of these opinions, Ptolemy mentions the two which locate Venus and Mercury either above or below the Sun (*Almagest*, bk. 9, ch. 1), thus ignoring the two in which Venus was placed above the Sun and Mercury below, and vice versa. As his own choice, Ptolemy selected the order which placed Venus and Mercury below the Sun, arguing that this conveniently separated three suprasolar planets, which could be any angular distance from the Sun, from the two subsolar planets which could not.[129] Although in the *Almagest* Ptolemy chose not to specify whether it was Mercury or Venus that lay directly below the Sun, in his later *Hypotheses of the Planets* he locates Venus right below the Sun and Mercury below Venus, just above the Moon.[130] Thus Ptolemy favored an order of planets in which the Sun was the middle, or fourth, planet, whether counting downward from Saturn above or upward from the Moon below. Although this order may have been the most popular, medieval justifications of it rarely mentioned Ptolemy's argument but rather emphasized the importance of the Sun's centrality, usually citing the popular metaphor of the king in the middle of his kingdom.[131]

128. "De ordine autem talium orbium septem planetarum quidam posuerunt primo spheram Saturni, deinde spheram Jovis, deinde spheram Martis, deinde speram Veneris, deinde spheram Mercurii, deinde spheram solis et ultimam spheram lune. Ita quod illi posuerunt Venerem et Mercurium supra solem. Alii autem posuerunt Venerem supra solem et Mercurium infra; alii autem econverso. Sed quicquid de hoc sic rationabilius esse videtur quod tres planete sint supra solem et tres infra et sol in medio tanquam rex in regni medio ad finem quod supra et infra possit equaliter influere et illuminare." Albert of Saxony, *De celo*, bk. 2, qu. 6, 1518, 105v, col. 1. Melanchthon (1550, 51v) described this order as "the oldest and common opinion" (Sequamur igitur vetustissimam et communem sententiam, quae medium locum Soli tribuit, sitque hic ordo: Saturnus, Iupiter, Mars, Sole, Venus, Mercurius, Luna). For the widespread use of "the king in the middle of his kingdom" as a metaphor for the Sun, see Chapter 11, note 28.
129. In book 9, chapter 1, Ptolemy speaks of the five planets, thus excluding the Sun and Moon. Perhaps he chose to exclude the Moon from the subsolar company of Venus and Mercury because the Moon could be any angular distance from the Sun. The Moon, of course, had to be, and was always, counted among the planets below the Sun.
130. Van Helden, 1985, 20–23, has a detailed discussion.
131. See Chapter 11, Section I.3 and note 28. Although Copernicus emphasized the Sun's centrality in quite another way, by having all the planets revolve around it, the Sun's importance as the middle planet was heavily emphasized in scholastic cosmology.

Despite certain criticisms of Ptolemy's order of the planets,[132] most followed it,[133] although they could have chosen another opinion (the first mentioned in my discussion of Albert of Saxony), one that placed the Sun below Venus and Mercury (that is, Venus–Mercury–Sun; or, with the positions of Venus and Mercury reversed, Mercury–Venus–Sun), leaving only the Moon below it. Clavius called this the "Egyptian system," as it was known traditionally, and cited both Plato and Aristotle as supporters.[134]

A trace of another kind of planetary "order" should be mentioned, one that probably derives from *The Marriage of Mercury and Philology* of Martianus Capella, who may have composed it sometime between 410 and 439. Martianus was probably the first extant Latin author to have adopted an arrangement of the planets wherein Mercury and Venus are assumed to orbit around the Sun, rather than the earth, for which reason Copernicus mentioned Martianus as a precursor.[135] By proposing Sun-centered orbits for Mercury and Venus, Martianus could not have accepted a single, fixed order of the planets but was committed to a variable order, such that when Mercury and Venus are above the Sun with respect to the earth (that is, farthest from the earth) their descending order is Venus–Mercury–Sun, but when they are below the Sun (or nearest the earth) their descending order is Sun–Mercury–Venus (Eastwood, 1982, 147). Because Mercury is closest to the earth when the two inferior planets are above the Sun and Venus is

132. Albert of Saxony, *De celo*, bk. 2, qu. 6, 1518, 105v, col. 1, reports a counterargument based on lunar eclipses. If the Moon, which lies between us and the Sun, can eclipse the Sun, why do not Venus and Mercury also eclipse it? Some explain this by the greater transparency of Venus and Mercury, which permits the Sun's rays to penetrate them. The lesser transparency of the Moon blocks the Sun's rays and creates an eclipse. A second explanation is based on the principle that the closer an opaque body is to us, the greater the eclipse it can cause as compared to a body, or bodies, that are more remote. But Venus and Mercury are so far away from us that the part of the Sun that they eclipse is not visible to us. In his *Hypotheses of the Planets*, Ptolemy suggests that the smallness of Venus and Mercury may explain why so few of their transits across the Sun have been observed (see Van Helden, 1985, 21).

133. For example, Clavius, *Sphere*, ch. 1, *Opera*, 1611, 3:43. Additional supporters are mentioned in Chapter 11, Section I.3 and note 28 of this volume.

134. For Plato, Clavius mentions the *Timaeus*, and for Aristotle he cites *De caelo*, book 2, chapter 12, and *Meteorology*, book 1, chapter 4, in neither of which does Aristotle present an order of planets (see Clavius, *Sphere*, ch. 1, *Opera*, 1611, 3:42). Indeed, Aristotle does not even mention an order of the planets in his famous discussion on the number of spheres in *Metaphysics* 12.8.1073b.18–1074a.14. However, Clavius also mentions Aristotle's *De mundo*, suggesting it might have been falsely attributed to Aristotle, as indeed it was. In the pseudo-Aristotelian *De mundo* 392a.20–30 (see Aristotle [Forster], 1984), the author does indeed present the Egyptian order of the planets, with Mercury, Venus, and Sun in descending order.

135. After rejecting Vitruvius, Chalcidius, and Macrobius as Latin sources for Sun-centered planetary motion, Eastwood (1982, 146, n. 1) argues that Martianus was the first Latin author to propose heliocentric orbits for Mercury and Venus, a theory that Martianus may have derived ultimately from Theon of Smyrna (fl. ca. 130; Eastwood, ibid., 151). For Copernicus's citation of Martianus, see Copernicus, *Revolutions*, bk. 1, ch. 10 [Rosen], 1978, 20. Eastwood also shows (1992, 233, 256) that no evidence exists in support of the claim that Heraclides of Pontus (4th c. A.D.) had previously proposed heliocentric orbits for Mercury and Venus.

closest when they are below the Sun, it follows that no fixed order exists, but that Mercury, Venus, and the Sun vary their order with respect to the earth. Although the concentric orbits that Martianus clearly attributed to Mercury and Venus would be altered by the ninth century – indeed, they would usually be depicted as intersecting – the varied order of Mercury, Venus, and the Sun remained.[136] By the seventeenth century, this variable arrangment of the three planets was known as the "Capellan" system and was taken as the basis of Tycho Brahe's own Sun-centered system involving five planets. The Capellan system had numerous supporters, many of whom were Jesuits (Schofield, 1981, 172–183).

The varied order of the inferior planets would appear again in the fourteenth century in Jean Buridan's *Questions on De caelo*. In Buridan's discussion, however, the orbits of Venus and Mercury are neither concentric nor intersecting, but eccentric and epicyclic. Observing that the inferior planets traverse their orbits in less time than the superior planets, Buridan asks why Mercury, Venus, and the Sun seem to complete their orbits in the same time, something no other planets do. In response to his own question, Buridan declares:

> some reply that this occurs because these three planets are fixed in the same sphere, although they have different epicycles and eccentrics within it. And this is probable because, as the Commentator [i.e., Averroës] says, many ancients assumed that Venus and Mercury were above the Sun, [while] others [assumed that] they were below the Sun. [Now] this could be [true], because when they are in the *auxes* of their eccentrics and epicycles, they are higher than [or above] the Sun; and [when they are] in the opposite of the *auxes*, they are lower than [or below] the Sun.[137]

Buridan's description of the variable order of Mercury and Venus with respect to the Sun seems equivalent to assigning them Sun-centered orbits within the frame of the Ptolemaic system of eccentrics and epicycles. To achieve these orbits, someone had boldly proposed that Mercury, Venus, and the Sun be encompassed within a single sphere, although each planet would have its own eccentric deferent and epicycle. Buridan characterized this arrangement as "probable" (*probabile*) because it reconciled the differences of opinion mentioned by Averroës, namely that some ancients placed Mercury and Venus above the Sun while others located them below the Sun.[138] The interpretation Buridan describes saves both alternatives.

If his report for a variable order for the Sun, Mercury, and Venus was

136. Eastwood, 1982, 149–155, describes the transformation of Martianus's text.
137. "Aliqui respondent quod hoc est quia illi tres planetae fixi sunt in eadem sphaera, licet in ea habeant diversos epiciclos et diversos eccentricos. Et hoc est probabile quia sicut dicit Commentator, multi antiqui posuerunt Venerem et Mercurium supra solem, alii autem infra solem; quod poterat esse quia quando erant in augibus eccentricorum et epiciclorum suorum, tunc erant altius quam sol, et in opposito augium erant bassius." Buridan, *De caelo*, bk. 2, qu. 20, 1942, 220.
138. Buridan does not say where Averroës made this remark.

derived ultimately from Martianus Capella, Buridan gives no indication of it. By contrast, Capella's name is the only one mentioned by Copernicus, although he also alludes to "certain other Latin writers."[139] Is it possible (though perhaps unlikely) that Copernicus had in mind one or more Latin writers of the late Middle Ages, perhaps Buridan himself? And who did Buridan have in mind when he speaks of "some" (*aliqui*) who placed the Sun at the center of the orbits of Mercury and Venus and encompassed the three planets within a single eccentric–epicyclic sphere? Is it plausible to suppose that Buridan and Copernicus might have included one or more of the same individuals? These are questions to which answers seem unlikely.

Indeed, we cannot even propose a definitive reply to the question of whether Buridan was himself a supporter of Sun-centered orbits for Mercury and Venus. For although he calls that interpretation "probable" (*probabile*), Buridan introduces an alternative opinion by observing that Ptolemy adopted the fixed order of Sun–Venus–Mercury, where the Sun is farthest from the earth and Mercury nearest, and each planet has its own sphere. To explain how these three planets complete their independent revolutions in the same time, we would have to assume, says Buridan, "a similar ratio of moving intelligences to moved spheres."[140] Which of the two alternatives Buridan favored is thus left unclear.[141]

What emerges from Buridan's discussion that is of considerable significance is the fact that he reported sympathetically a limited heliocentric system involving Mercury and Venus, encompassed within a system of eccentrics and epicycles. Since Copernicus used Ptolemaic eccentrics and epicycles, Buridan seems thus far to have left us the first extant, unequivocal description of limited heliocentric orbits within a system of eccentrics and epicycles.[142] Echoes of it were still heard in the seventeenth century.[143]

139. Copernicus, *Revolutions*, bk. 1, ch. 10 [Rosen], 1978, 20. Among "other Latin writers," Rosen suggests that Copernicus may have included Vitruvius, *Architecture*, IX, 6 (Copernicus, ibid., 358). Although Eastwood has eliminated Vitruvius, Macrobius, and Chalcidius as real believers in Sun-centered planetary motion, it does not follow that Copernicus and others would have viewed earlier, potentially relevant Latin authors in the same light. In his *Almagestum novum*, Riccioli considered the Egyptians as the inventors of the Capellan system and, in addition to Capella, also named Vitruvius, Macrobius, and Bede as its supporters (see Schofield, 1981, 173 and 347, n. 23).

140. Here are Buridan's words on Ptolemy: "Dicitur tamen quod Ptolomeus geometrice invenit sphaeram solis esse supra sphaeram Veneris et sphaeram Veneris supra sphaeram Mercurii. Et tunc causa propter quam illae sphaerae sic aequali tempore perficerent suas circulationes, esset similis proportio intelligentiarum moventium ad sphaeras motas." Buridan, *De caelo*, bk. 2, qu. 20, 1942, 220, lines 28–33. On the relationships of intelligences to the spheres they move, see Chapter 18, Section II.

141. Although in his *Questions on the Metaphysics* (bk. 12, qu. 10, 1518, 73r–74r), Buridan concludes that all astronomical appearances can be saved by eccentrics alone without epicycles (for my translation, see Grant, 1974, 524–526), he speaks of both eccentrics and epicycles in the heliocentric argument about Venus and Mercury. Nevertheless, he calls the latter argument "probable" and raises no objections to the inclusion of epicycles. I am ignorant of the order of composition of Buridan's *Metaphysics* and his *De caelo*.

142. Albert of Saxony considered the same question (*De celo*, bk. 2, qu. 16, 1518, 111v, col. 2–112r, col. 2) as did Buridan and included most of what Buridan discussed. It is

2. The number of orbs

a. Do orbs exist beyond the eight orbs of the planets and fixed stars?

Whatever the order chosen, all were agreed that the planets and fixed stars together accounted for at least eight concentric orbs or heavens. Cosmologists and astronomers were, however, soon compelled to decide whether any spheres existed beyond the fixed stars and if so, whether they were mobile or immobile. With respect to mobile orbs, responses usually depended on the number of motions assigned to the sphere of the fixed stars. Throughout the Middle Ages, at least three motions were attributed to it: (1) a daily motion from east to west; (2) a precession of the equinoxes of 1 degree in 100 years, producing a complete revolution of the starry sphere in 36,000 years; and (3) a progressive and regressive motion of the stars known as "access and recess," or "trepidation," a theory proposed by the ninth-century Arab astronomer Thabit ibn Qurra.[144] Although Thabit's trepidation theory was intended as a substitute theory for the precession of the equinoxes, not as an additional motion, many natural philosophers treated them as two distinct motions, as we shall see.

If we accept the principle that every motion requires its own separate sphere, a sphere would have to be added for every such motion. But this did not signify that a sphere could not move with multiple motions. To understand this, we must realize that medieval natural philosophers distinguished between the "proper" motion of a sphere and those motions that were imposed upon it externally from the motions of superior spheres. A sphere could have only one proper motion, usually characterized as a simple motion. Thus the eighth sphere of the fixed stars might have the motion of precession as its proper motion. But it also had a daily motion and a motion of trepidation. The sources of these two motions had to be sought in orbs that were independent of, and distinct from, the eighth sphere. On the universal principle that "no sphere is ever moved with the motion of an inferior sphere but is moved with the motion of a superior sphere," as Pierre d'Ailly put it,[145] the other two motions of the eighth sphere were

noteworthy, however, that he makes no mention of the Sun-centered orbits of Mercury and Venus.

143. For Amicus's discussion, see Chapter 14, Section VIII.2.c.

144. The theory of trepidation arose from discrepancies in the observation of precession (see Campanus of Novara, *Theorica planetarum*, 1971, 378–379, and Dreyer, 1953, 276–277). Ordinarily, either precession or trepidation should have been employed, but some scholastics assigned both motions to the stars. Albertus Magnus, *Metaphysics*, bk. 2, tract. 2, ch. 24, *Opera*, 1964, 16, pt. 2:514, col. 1, attributes all three motions to the eighth sphere and mentions Thabit, or Thebit, as the discoverer of trepidation, which he calls "the motion of accession and recession" (motus accessionis et recessionis), or "progression and regression."

145. This is the third of four assumptions, where d'Ailly says: "Tertio supponitur quod aliqua sphaera nunquam movetur ad motum sphaerae inferioris sed bene ad motum sphaerae superioris." D'Ailly, *14 Questions*, qu. 2, 1531, 149r.

quite naturally attributed to superior spheres: one motion to a ninth sphere, and the other to a tenth sphere. Thus with three motions assigned to the eighth sphere, it was usual to add two spheres; if four motions were assigned to the sphere of the fixed stars, three additional orbs were customarily added. The spheres that allegedly existed beyond the eighth sphere were assumed to be devoid of celestial bodies and therefore wholly transparent and invisible.

For those who assumed only an east-to-west daily motion for the stars, no additional orb was necessary; eight movable orbs sufficed.[146] Others, for example, Peter of Abano, argued for nine spheres. Peter assumed two motions for the sphere of the fixed stars: one, the daily motion, he assigned to the eighth sphere; the other, the precession of the equinoxes, he attributed to the ninth sphere.[147] In a somewhat different arrangement, Illuminatus Oddus ([*De coelo*, disp. 1, dub. 14], 1672, 41, col. 2) assumed nine mobile heavens, equating the ninth with the *primum mobile* and crystalline orb. But Albert of Saxony, Roger Bacon, Themon Judaeus, and Pierre d'Ailly, among others, attributed all three motions to the fixed stars and therefore assumed the existence of ten mobile orbs.[148] Thus Albert assigned the daily motion to a tenth orb, the *primum mobile*; a motion of trepidation to the ninth sphere; and the motion of precession to the eighth or starry sphere.[149]

Clavius characterized the theory of ten mobile orbs as "the most celebrated that has appeared in the schools of astronomy to this day"[150] but did not himself adopt it, while the Coimbra Jesuits defended it as their opinion, observing that not only astronomers, but also many Peripatetic philosophers had embraced it.[151] Both Clavius and the Conimbricenses mention that a

146. Albert of Saxony, *De celo*, bk. 2, qu. 6, 1518, 105v, col. 1, explains: "Ulterius sciendum est quia aliqui philosophi non perceperunt octavam spheram moveri pluribus motibus sed unico, scilicet ab oriente in occidentem, dixerunt spheram octavam esse ultimam et nullam esse ultra." D'Ailly, *14 Questions*, qu. 2, 1531, 149r, mentions the opinion of those who insist on the existence of only eight orbs. They say that the eighth orb moves with only one motion (the daily motion) and that there is no need to assume a ninth orb. D'Ailly rejects this opinion and observes that astronomers (*astrologi*) deny the attribution of only a single motion to the eighth sphere. Amicus, *De caelo*, tract. 4, qu. 6, dubit. 2, 1626, 185, col. 2, also mentions the eight-orb interpretation and observes that its proponents reject the motion of trepidation.

147. See Peter of Abano, *Lucidator*, differ. 3, 1988, p. 217, lines 21–22 through p. 218, line 2. Peter specifically argues against those who assume ten spheres.

148. Albert of Saxony, *De celo*, bk. 2, qu. 6, 1518, 106r [incorrectly foliated as 107], col. 1; d'Ailly, *14 Questions*, qu. 2, 1531, 149r; and Bacon, *De celestibus*, pt. 4, ch. 3, *Opera*, fasc. 4, 1913, 388; ibid., pt. 5, ch. 18, 447, 449; ibid., ch. 19, 455; for Themon Judaeus, see Hugonnard-Roche, 1973, 105.

149. Although d'Ailly assigned motion of precession as the proper motion of the eighth sphere, he does not specify the motions of the ninth and tenth orbs. It was usual, however, to assign the daily motion to the outermost moving orb, or *primum mobile*. It is therefore likely that d'Ailly followed Albert of Saxony and assigned the daily motion to the tenth orb and trepidation to the ninth orb. Clavius reports a different arrangement (see the next paragraph).

150. "Hic igitur denarius numerus orbium coelestium in scholis astronomorum celeberrimus ad hanc usque diem extitit." Clavius, *Sphere*, ch. 1, *Opera*, 1611, 3:23.

151. "Haec igitur sententia de denario coelestium sphaerarum numero nobis probatur quam

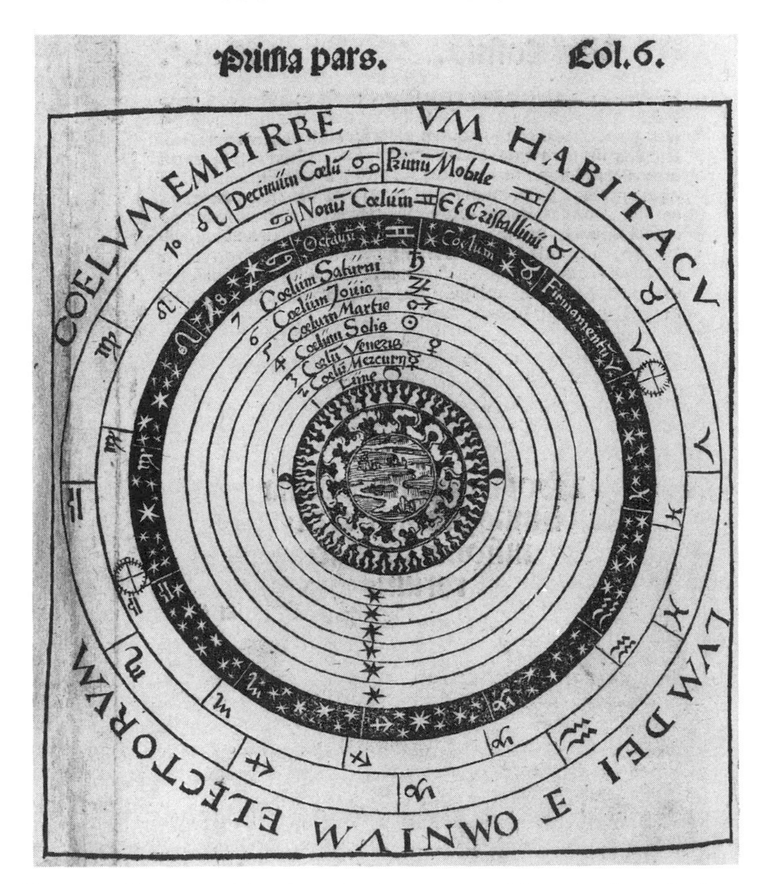

Figure 9. The movable celestial spheres, ranged in order from the lunar orb to the "first movable heaven" (*primum mobile*). Encompassing the whole is the immobile empyrean heaven, "dwelling place of God and all the elect." (Peter Apian, *Cosmographicus liber* (1524), col. 6. Courtesy Lilly Library, Indiana University, Bloomington.)

ten-orb scheme was accepted in a similar form by such great astronomers as Albategnius (al-Battani),[152] Thabit ibn Qurra, King Alfonso of Spain, Georg Peurbach, and Regiomontanus, all of whom applied these three motions to the eighth, ninth, and tenth spheres.[153] Thus they assigned the daily motion from east to west to the tenth sphere, which carried with it the eighth sphere of the fixed stars and all the inferior planetary orbs; to the ninth heaven they attributed a west-to-east motion that carried with it

non solum Astronomi, quorum observatio et experientia hac in re fidem meretur, sed etiam multi e Peripatetica schola nobiles Philosophi amplectuntur." Conimbricenses, *De coelo*, bk. 2, ch. 5, qu. 1, art. 1, 1598, 247.

152. Only the Coimbra Jesuits mention Albategnius.

153. The same interpretation is also reported, though not accepted, by Amicus, *De caelo*, tract. 4, qu. 6, dubit. 2, 1626, 185, col. 2–186, col. 1.

the firmament, or sphere of fixed stars, and all the inferior planetary orbs; and finally, to the eighth orb of the fixed stars they assigned the motion of trepidation.[154] The functions assigned to the eighth to tenth orbs could vary, as when some assigned the motion of precession to the ninth orb rather than to the eighth orb.[155]

Clavius moved from the ten-orb system to one with eleven mobile orbs. In this, he was influenced by Copernicus, whom he called "a most learned man and most praiseworthy" (vir longe doctissimus, omnique laude dignissimus). According to Clavius, Copernicus assigned four, not three, motions to the fixed stars.[156] In apparently following Copernicus, Clavius assumed three mobile orbs beyond the eighth sphere of the fixed stars and assigned the four motions to orbs eight to eleven. To the eleventh orb, he assigned the daily motion from east to west; to the tenth orb, a motion from north to south and south to north; to the ninth orb, he assigned "a certain unequal libration from east to west and west to east"; and to the eighth sphere of the fixed stars, a proper motion from west to east, which appears to represent the precession of the equinoxes.[157]

Although Aristotle had assigned multiple orbs to each planet (but only one to the fixed stars), he devised a system in which the orbs of one planet could not effect the orbs of another. Moreover, since the planet itself was carried by only one of the multiple orbs assigned to it, the remaining orbs

154. Since both texts are substantially similar, I shall cite only that of Clavius: "Post Ptolemaeum deinde, annis interiectis MCXL fere, Tebith, Alphonsus Hispanorum rex Anno Domini MCCL, Georgius deinde Peurbachius et Ioannes de Regiomonte insignes astronomi, deprehenderunt quidem in stellis fixis duos motus praedictos, sed eas praeterea observarunt tertio quodam motu, quem accessus et recessus dixerunt, ut paulo post declarabitur agitari. Quare cum corpus simplex unico tantum motu ferri sit aptum, ut volunt Philosophi non potest nonum coelum esse primum mobile, sed supra ipsum erit aliud statuendum coelum quod sit primum mobile. Ita enim fiet ut decimum hoc coelum motu diurno quem habet proprium ab oriente in occidentem, secum trahat omnes coelos inferiores atque adeo Firmamentum quoque cum stellis fixis spacio 24 horarum; nonum deinde coelum circumvehat suo proprio motu quem obtinuit ab occidente in orientem et Firmamentum et reliquos omnes coelos infra ipsum; octavum denique coelum, seu Firmamentum, in quo stellae fixae existunt, moveatur tanquam proprio motu, accessu illo et recessu, quem praefati astronomi repererunt." Clavius, *Sphere*, ch. 1, *Opera*, 1611, 3:23. The Conimbricenses declare that the motion of access and recess is also called trepidation ("Et denique firmamentum motu sibi proprio moveatur, accessu illo et recessu, quem titubationis, seu trepidationis, motum vocant"). Conimbricenses, *De coelo*, bk. 2, ch. 5, qu. 1, art. 1, 1598, 247.

155. As reported by Amicus, *De caelo*, tract. 4, qu. 6, dubit. 2, 1626, 185, col. 2.
156. "Nostra denique tempestate Nicolaus Copernicus, vir longe doctissimus, omnique laude dignissimus, non solum tres in stellis fixis motus observavit, sed quatuor." Clavius, *Sphere*, ch. 1, *Opera*, 1611, 3:23.
157. "Nam ad motum undecimi coeli, seu primi mobilis, moventur omnia astra ab ortu in occasum; et ad motum decimi coeli a septentrione in austrum et ab austro in septentrionem per 24 minuta sub coluro solstitiorum; ad motum vero noni coeli habent librationem quandam inaequalem ab ortu in occasum et ab occasu in ortum sub ecliptica decimae sphaerae per minuta 140; motu denique proprio octavi orbis stellae fixae circumvehuntur ab occasu in ortum." Ibid., 23–24. Amicus, *De caelo*, tract. 4, qu. 6, dubit. 2, art. 4, 1626, 188, col. 2, rejects the eleven-orb system, believing that an eight-orb system is more probable.

carried no star or planet. With this in mind, Clavius posed an interesting objection against himself. He explains that in the twelfth book of his *Metaphysics*, Aristotle insisted that every motion of an orb directly represents the motion of a star or planet, or, as the Conimbricenses explain it ([*De coelo*, bk. 2, ch. 5, qu. 1, art. 3], 1598, 250), Aristotle argued that a celestial motion exists for the sake of a planet or star. Since no star or planet exists in the ninth through eleventh orbs, would this not signify a vain and superfluous existence? Clavius replies that although no star or planet exists in those orbs, the motion of each such orb, as we have seen, exercises a direct influence on a planet or star that exists in another heaven. The Coimbra Jesuits argued further (ibid., 251) that orbs do not exist solely to cause the motion of planets or stars, as is evidenced by the common assumption that all parts of an orb – even those parts that are distant from the celestial body it carries – exercise an influence on inferior bodies below the Moon.

Clavius was apparently satisfied with his response to the objection but, like all of his predecessors and contemporaries, ignored the fact that the extra orbs simultaneously affected a given celestial body or bodies. Thus the three or four orbs (i.e., orbs nine to eleven or twelve) assigned to represent each of three or four motions assigned to the eighth sphere of the fixed stars act continuously and simultaneously on the eighth orb. The fixed stars would thus be subject to three or four simultaneous motions, which is what was supposed to be avoided by invoking one orb for each motion of any celestial body. If one were merely saving the phenomena geometrically, the number of spheres assigned to represent the total number of motions would be irrelevant and without physical consequences. But for Clavius and other astronomers and natural philosophers, the orbs were assumed physically real, and all motions were real motions. In the end, the fixed stars would be subject to contrary motions, since the orbs above would act simultaneously on the orb of the fixed stars. Such problems were apparently ignored. Indeed, once the multiple motions of a planet or the stars were assigned to independent orbs, the problem was considered resolved.

At least one medieval natural philosopher opposed the assumption of a ninth sphere beyond the eighth. Nicole Oresme, noting that astronomers had determined that the eighth orb of the fixed stars had a movement composed of several different motions, for which reason they assumed the existence of a ninth sphere, thought such a move superfluous. Unfortunately, Oresme rests content merely to inform his readers that in an earlier work he had explained how two different, simultaneous motions could be assigned to the starry orb without invoking a starless ninth orb.[158]

But if we are ignorant of Oresme's way of avoiding a ninth orb, there were apparently other suggestions for obviating the need for additional orbs beyond the eighth. Albert of Saxony reports one in which two intelligences are assigned to the eighth orb, one to move it from east to west, the other

158. Oresme, *Le livre du ciel*, 1968, 488–491. Oresme does not name the earlier treatise.

to move it from west to east, a solution that Albert rejects because it violates Aristotle's dictum that to one and the same orb only one motive intelligence can be assigned.[159] One might also argue for the existence of only eight orbs by assuming that the eighth orb does not itself move with a plurality of motions – three, to be precise. Thus two of the motions most frequently assigned to the eighth orb of the fixed stars might be assigned to the earth, namely a west-to-east motion, which would account for the daily motion of the heavens from east to west, and the progressive and regressive motion of trepidation, which Thabit ibn Qurra had discovered.[160] In this scheme, only the motion of precession is assigned to the eighth sphere.[161] Because there is no direct evidence to demonstrate that the earth moves in these ways, Albert expresses reservations about this interesting opinion, although he then tantalizingly suggests, without elaboration, that one might devise a way to avoid the difficulties.[162]

b. The theological heavens: the firmament and the crystalline orb

The sacred text of Genesis 1.6 demanded that waters of some kind be assumed to lie above and beyond the heaven of the firmament, or fixed stars, thus giving rise to two theological heavens or orbs: the firmament and the crystalline orb. Earlier in this study (Ch. 5, Secs. VI–VII), we considered some of the essential features of these two orbs. We saw that the firmament (*firmamentum*) was given a number of interpretations. A few associated it with the air beneath the heavens; others identified it with the region between, and including, the Moon and the fixed stars; while some limited it solely to the eighth orb of the fixed stars. Either of the last two interpretations was compatible with the three-orb compromise system. For even if one counted all the planets and stars as comprising the firmament, the planets were still conceived as subdivided into seven orbs, while the fixed stars were embedded wholly in an eighth sphere; or one could simply assign the term "firmament" solely to the eighth sphere of the fixed stars. On either interpretation, the end result was identical. The waters above the firmament, however, required a somewhat more complex interpretation before they could be assimilated into the secular cosmology of the Aristotelian–Ptolemaic compromise system.

159. Albert of Saxony, *De celo*, bk. 2, qu. 6, 1518, 105v, col. 2–106 [mistakenly foliated 107]r, col. 1.
160. "Aliter potest sustineri quod non essent nisi octo orbes et quod octava sphera non moveretur pluribus motibus sed quod ipsa apparet moveri pluribus motibus est ex eo quod terra movetur ab occidente in orientem et per unum alium motum terre possit salvare apparentia motus accessus et recessus octave sphere que invenit Thebit." Ibid., 106 [mistakenly foliated 107]r, col. 1.
161. For further discussion of the earth's possible axial rotation, see Chapter 20, Section V.
162. "Sed istud non videtur esse omnino tutum quia non apparet prima facie quid terram sic moveret. Nihilominus forte qui niteretur in defensionem huius opinionis posset excogitare faciliter modum hoc evadendi et plura alia dictam opinionem multum colorantia." Albert of Saxony, *De celo*, bk. 2, qu. 6, 1518, 107 [really 106]r, col. 1.

Although some Greeks thought of the celestial region as composed of the four elements, which therefore included water, most would have denied the existence of a large mass of water beyond the fixed stars. Indeed, Aristotle denied the possible existence of water beyond the concave surface of the lunar sphere.[163] Biblical exegesis, however, demanded that the waters above the firmament be conceived as real, although their precise nature was open to debate.

Relatively early in the history of Christianity, those waters were conceived as crystalline, a term which, as we saw earlier, was sometimes thought of as applying to fluid waters and sometimes to waters that were congealed and hard like a crystal. The latter gained support from Ezekiel 1.22, which speaks of an awesome crystal stretched like a vault over the heads of the animals of the firmament.[164] Thus for Saint Jerome and Bede the waters above the firmament were conceived as crystal-like, which signified hardness, whereas for Saints Basil, Gregory of Nyssa, and Ambrose they were fluid. Whether fluid or hard, however, during the early Middle Ages, say from the fifth to the mid-twelfth century, the crystalline orb was usually located above the sidereal heaven, or firmament of fixed stars, and below the empyrean heaven.[165]

Whether two or three in number, the starless and planetless orbs were assumed to be not only material, physical entities but also transparent and invisible. It was therefore easy to identify one or all of them with the waters above the firmament, or the crystalline sphere, as those waters were often described. Some considered the identification of waters with one or more orbs above the firmament, or above the eighth sphere of the fixed stars, essential on theological grounds,[166] or at least viewed it as a sphere that was named "crystalline" by theologians.[167] Because of their "clarity and transparency," Clavius identified the ninth through eleventh orbs with the crystalline sphere.[168] For the same reason, as well as for the freezing power that they allegedly have, the Coimbra Jesuits identified the ninth and tenth mobile orbs with the watery, or glacial, heaven, above the firmament, which is usually characterized by one common name: "crystalline."[169] Occasionally

163. See Duhem, *Le Système*, 1913–1959, 2:488.
164. "Et similitudo super capita animalium firmamenti, quasi aspectus crystallis horribilis, et extenti super capita eorum desuper." See Bible (Vulgate), 1965; also Campanus of Novara, *Theorica planetarum*, 1971, 393–394, n. 54.
165. Thomas Aquinas [*Sentences*, bk. 2, dist. 14, qu. 1, art. 4], 1929–1947, 2:354, cites Rabanus Maurus (ca. 776–856) as one who assumed seven heavens: empyrean (*empyreum*), crystalline (*chrystallinum*), sidereal (*sidereum*), fiery (*igneum*), olympian (*olympium*), ethereal (*aethereum*), and airy (*aereum*).
166. See Campanus of Novara, *Theorica planetarum*, 1971, 183, and Clavius, *Sphere*, ch. 1, *Opera*, 1611, 3:24.
167. Michael Scot declares (*Sphere*, 1949, 283): "Secundum celum dicitur nona sphera que a theologis dicitur cristallinum."
168. Clavius, *Sphere*, ch. 1, *Opera*, 1611, 3:24.
169. "Nonum et decimum, quos, ut alibi retulimus, theologorum nonnulli significari putant in sacris literis nomine aquarum cum Geneseos 1., dicunt Deum aquas ab aquis interposito firmamento secrevisse, et in Psalmo 148, cum dicuntur aquae esse super coelos *aquae,*

there was hesitation in such identifications. Convinced on theological grounds that there was a crystalline sphere, Campanus of Novara was initially uncertain about identifying it with the ninth orb, though he eventually did so.[170] Peter of Abano went further and insisted that those who theologized more spheres by "assuming a crystalline, or aqueous [sphere] and an empyrean, or fiery [sphere]" did so on the basis of revelation, not reason or experience, on which he based his work.[171]

The concept of a ninth orb probably entered western Europe within the corpus of Greco-Arabic astronomy and cosmology that was translated from Arabic into Latin. Thus Roger Bacon held that in some translations of Aristotle's *Metaphysics* a ninth orb is assumed,[172] as it was in works by Thebit and Al-Bitrūjī (Alpetragius).[173] The introduction of a ninth orb was made within a secular and astronomical context, having no connection with theology. A ninth orb was therefore sometimes, and perhaps even often, mentioned without any reference to the waters above the firmament or the crystalline sphere. It was the first astronomical sphere beyond the sphere of the fixed stars and was usually assigned one of the motions associated with the latter. John of Sacrobosco, for example, says no more about the ninth sphere than that it is the "first mobile orb" (*primum mobile*), or the "last heaven."[174] Other thirteenth- and fourteenth-century authors – for example, Robertus Anglicus and Cecco d'Ascoli – also mention it without any reference to its "crystalline" character or its theological connection.

Precisely who linked the biblical account with a ninth and starless orb may never be known. As early as the sixth century, John Philoponus identified the ninth sphere,[175] not with the waters above the heaven, but with the heaven created on the first day, which he described as a transparent, starless orb that surrounds the firmament created on the second day.[176] But

quae super coelos sunt, laudent nomen Domini. Neque incongrue hi duo orbes aquarum nomine designari possunt propterea quod cum nullae in iis stellae fulgeant, sed admodum translucidi et perspicui sint aquarum referunt similitudinem tum quia perfrigerandi vim habere creduntur. Quare a quibusdam coelum aqueum, sive glaciale, ab aliis Christallinum uno communi nomine appellantur." Conimbricenses, *De coelo*, bk. 2, ch. 5, qu. 1, art. 5, 1598, 252.

170. He was uncertain in the *Theorica planetarum* but not in *Tractatus de sphera*. See Campanus of Novara, *Theorica planetarum*, 1971, 183 and 393, n. 53.
171. "Propter secundum sciendum quod theologizantes plures speras, sive celos, figurant, super octavam, quidem, ponentes cristallinum sive aqueum, et empireum seu igneum, de quibus in presentarium nihil, cum potius que ipsorum, revelatione quam ratione sciantur, aut experientia, quibus hoc opus nititur sistere." Peter of Abano, *Lucidator*, differ. 3, 1988, 214, lines 4–8.
172. Nowhere in the works of Aristotle is there even a vague hint of a ninth orb beyond the fixed stars.
173. Bacon, *De celestibus*, pt. 4, ch. 2, *Opera*, fasc. 4, 1913, 387–388.
174. See Sacrobosco, *Sphere*, ch. 1, 1949, 118–119.
175. He attributed the discovery of a ninth sphere to Hipparchus and Ptolemy. In the *Hypotheses*, Ptolemy seems to have assumed a ninth sphere, whose function was to move the starry sphere but which was not connected with the precession of the equinoxes. See Ptolemy, *Hypotheses*, 1907, 122, 125, and Pedersen, 1974, 249, n. 7.
176. For a summary of Philoponus's views in his *De opificio mundi*, see Duhem, *Le Système*, 1913–1959, 2:496–501.

it was not the heaven created on the first day with which Christian natural philosophers and theologians eventually identified the ninth orb – and even tenth and eleventh orbs. It was rather with the waters above the firmament, which they often described as a crystalline orb.

In these brief and general descriptions and allusions to a ninth sphere, we learn of certain significant features. During the thirteenth century, some authors – Robertus Anglicus and Michael Scot, for example – upheld the existence of a ninth orb, despite a degree of uncertainty. Robertus admits that he was aware of no authority who had demonstrated the necessity for more than eight orbs. Nevertheless, because two motions were associated with each of the eight planetary orbs, as well as with the eighth starry orb, he thought it reasonable to assume that a ninth orb existed which possessed only a single, simple motion – presumably the daily east-to-west motion. But the ninth orb exercised a cosmic influence not only by motion but also by light, which was evenly distributed over the orb, in contrast to planet-bearing orbs where light was concentrated around and near the celestial body.[177] Michael Scot offered very nearly the same argument.[178]

At some point, however, perhaps during the initial influx of Greco-Arabic astronomical and cosmological literature, the starless, transparent ninth orb was linked with the biblical waters above the firmament and thereafter identified with the crystalline orb. The two are already joined in Vincent of Beauvais' *Speculum naturale*, probably composed over the period 1244–1254.[179] Although, as we saw, not everyone explicitly made the connection when discussing the ninth orb, many did, so that throughout the period of this study the ninth sphere was frequently equated with a crystalline orb of biblical origin.

We have now considered all the mobile orbs in medieval cosmology but have yet to examine the final theological sphere, the immobile, empyrean heaven. Because the latter was widely discussed and always problematic, a separate chapter (Ch. 15) will be devoted to it.

177. Robertus Anglicus, *Sphere*, lec. 1, 1949, 148 (Latin), 203 (English).
178. Michael's discussion is embedded in a question format. See Michael Scot, *Sphere*, lec. 2, 1949, 259–260.
179. In his lengthy discussion on the crystalline sphere (bk. 3, chs. 90–100, cols. 221–229), Vincent focuses mostly on its theological aspects, but in chapters 97 and 100, columns 226 and 228, respectively, he links it explicitly with an astronomical ninth heaven.

14

Are the heavens composed of hard orbs or a fluid substance?

Prior to the impact of Tycho Brahe's astronomical research, scholastic authors found no reason to devote even a single question to consider whether the celestial orbs might be hard or soft. Tycho, however, had made the question virtually unavoidable. The issues he raised challenged the very existence of eccentric and epicyclic orbs and inevitably posed questions about the hardness or softness of the celestial ether. Although it was Tycho who first made the ether's hardness or softness an issue central to cosmology, the problem had a long, but vague and even muted, history. Because medieval scholastic natural philosophers rarely discussed the matter directly or in useful detail, information about the hardness or fluidity of orbs must be gleaned indirectly from discussions in other contexts.

I. Modern interpretations of medieval orbs

A widely held opinion today is that scholastic authors thought the celestial orbs were solid, where "solid" is taken as synonymous with hard or rigid.[1] Here the image is one of transparent glass or crystalline globes. Hardly in contention as to popularity with the first opinion today is a second, which assumes that medieval thinkers faithfully adhered to Aristotle's dicta about the celestial ether. Thus the orbs or spheres could be neither solid nor fluid because Aristotle had argued that contrary qualities such as hardness and softness, density and rarity, and so on, were inapplicable to the incorruptible, celestial ether of which they were composed. Nicholas Jardine observes (1982, 175) that to pose a question about the hardness or softness of celestial spheres would have been considered a "category mistake." Hardness and softness are qualitative opposites found only in terrestrial matter. Since pairs of opposite qualities are the source of all terrestrial change, they must of necessity be absent from the celestial region, where change is im-

1. For a lengthy list of scholars who hold this opinion, see Grant, 1987a, on which I rely for much of what follows. See 153.

possible. Thus to inquire about the possible hardness or softness of celestial orbs is to ask an irrelevant question.

Perhaps in order to avoid posing such questions, William Donahue supposes that the Peripatetics conceived of the heavens and its orbs as "immaterial or quasi-material" entities, whatever these terms may signify.[2] A similar position is adopted by Edward Rosen, who denies (1985, 13) that Aristotle argued for the existence of solid celestial orbs. Because Aristotle denied the existence of corruptible terrestrial matter in the incorruptible and unchanging ethereal heavens, it followed, Rosen argued, that his celestial ether was not material and therefore could be neither solid nor hard, as Pierre Duhem had claimed. Indeed, by similar reasoning (though Rosen does not draw the inference), it could not be fluid either. From a narrow, strict-constructionist standpoint, one might defend such an approach. In reality, however, this interpretation misunderstands Aristotle's intent, and, if applied to the Middle Ages, would distort medieval opinion. It will be well to eliminate this second interpretation before proceeding.

Although Aristotle may have denied that alterable matter like that on earth could exist in the heavens, his ether may be construed as a fifth kind of substance, or element – a *quinta essentia*, as many commentators would call it – with properties, as we have seen, radically different from those of the four sublunar elements. Whatever Aristotle may have thought about the properties of the celestial ether, there is no doubt that in *De caelo* he assumed the corporeality, and therefore physicality, of the heavenly orbs.[3] As nonspiritual, corporeal, and therefore three-dimensional physical entities composed of ether, celestial orbs had to be something akin to hard or soft – even though Aristotle himself was committed to a formal denial of contrary qualities such as hardness and softness, hotness and coldness, and rarity and density. In the course of discussions on the celestial orbs, one would sooner or later find it necessary to speak of their physical nature, despite Aristotle's strictures. Were they hard or soft? If one or the other, then could they also be said to be, in some sense, dense or rare? But Aristotle seems to have precluded such analyses. Perhaps this is why he chose to ignore the physical nature of celestial spheres and why he offered no helpful clues as to how one might speak about them.

Indeed, this may well explain why his medieval scholastic commentators also chose to ignore the problem. But just as many scholastic authors ignored

2. See Donahue, 1975, 251, 256–259, 275.
3. In *De caelo* 2.12.293a.8, Aristotle declares that "the last sphere moves round embedded in a number of spheres, and each sphere is corporeal." Aristotle [Guthrie], 1960. In the Latin translation of Aristotle's *De caelo* that accompanies Averroës' commentary, the Latin translation of Aristotle's second phrase is "et omnis orbis eorum est corpus." See Averroës [*De caelo*, bk. 2, text 70], 1562–1574, 5:70r, col. 1. In his *Hexaëmeron*, Robert Grosseteste reports: "John Damascene also implies in his book of *Sentences* that the existence of an immaterial body, that which is called a fifth body among the wise men of the Greeks, is impossible," but he declares that Aristotle and his followers did assume the existence of a "fifth body" in addition to the four elements. Grosseteste, *Hexaëmeron*, part. 3, ch. 6, 1, 1982, 106. John may have had in mind the passage from Aristotle just cited.

Aristotle's famous dictum that neither place, nor void, nor time could exist beyond the outermost sphere of the physical world and began to inquire what indeed might lie beyond, so, to a lesser extent, did some of those same scholastic authors reveal an opinion or judgment, usually indirectly, about the hardness or softness of the celestial orbs, which they all assumed to be physical bodies.

To my knowledge, no medieval natural philosopher rested content to depict the celestial orbs as immaterial entities devoid of physical properties. This remains true, despite the fact that many denied in the abstract that the celestial ether could possess terrestrial attributes such as hotness and coldness, or rarity and density. Only when confronted with specific problems about the spheres themselves – that is, about their arrangement and the relationships between successive surfaces – do scholastic natural philosophers speak in a quite different vein and reveal, perhaps inadvertently, a concern about real, physical spheres. Indeed, we have already seen that numerous medieval discussions about possible physical problems that might affect eccentric orbs – for example, whether vacua can occur between successive celestial surfaces or whether two orbs can overlap and occupy the same place simultaneously – provide ample evidence that the spheres were conceived as physical bodies. I am aware of no instance in which physical considerations were dismissed because the celestial orbs were deemed immaterial or quasi-material. Despite Aristotle's well-known attitude, those orbs were judged to be physical, and it was therefore difficult to avoid the attribution of some physical properties to them. Although the attribution of contrary qualities to Aristotle's ethereal orbs is a "category mistake" within Aristotelian cosmology, some of Aristotle's legions of commentators often found it unavoidable to attribute terrestrial qualities to celestial bodies.

With the second interpretation eliminated from further consideration, we shall now attempt to determine whether, during the late Middle Ages, the celestial orbs were conceived as hard and rigid or fluid and soft, or perhaps some combination of these properties. The problem of the hardness or softness of the celestial region is rather complex. The fact that a natural philosopher may have assumed the existence of celestial orbs does not permit us to make any inference as to whether he thought them hard or soft. Ptolemy himself may have aided the confusion by his apparent assumption in the *Almagest* that planets move about in fluid media, a view he seems to have abandoned in the *Hypotheses of the Planets*.[4] Moreover, there was dis-

4. Duhem, *Le Système*, 1913–1959, 2:479, explains that in the *Almagest* Ptolemy regarded the heaven as a fluid in which the stars moved freely but that he abandoned this idea in his later *Hypotheses*. In his translation, Taliaferro has Ptolemy say that the planets "can all penetrate and shine through absolutely all the fluid media" (Ptolemy, *Almagest*, bk. 13, ch. 2 [Taliaferro], 1952, 429). In Toomer's translation (Ptolemy, *Almagest*, bk. 13, ch. 2, 1984, 601), however, Ptolemy says that the nature of the heavens "is such as to afford no hindrance, but of a kind to yield and give way to the natural motions of each part, even if [the motions] are opposed to one another." Is a substance that yields and gives way to the motions of other parts also a fluid substance or merely elastic?

agreement about the nature of the different traditional divisions of the heavens, which often influenced the properties that were assigned. By the seventeenth century, Giovanni Baptista Riccioli distinguished five different interpretations concerning the hardness or softness of the heavens.[5] There were those who believed that all the heavens were wholly solid – that is, hard – while others thought them wholly fluid. The last three opinions assumed a partly solid and partly fluid celestial region. Thus Riccioli reports a third opinion wherein the fluid part apparently consisted of ring-like channels filled with a subtle or tenuous air-like substance. The surfaces and everything else were presumably hard. A fourth opinion assumed that the extremities of the celestial region were solid, namely the heaven, or sphere, of the fixed stars and the heaven, or sphere, of the Moon. Everything between these two extremes was of a fluid nature. The fifth and final opinion, which Riccioli says is "now the most celebrated" (nunc celeberrima opinio) – indeed it was a direct legacy from Tycho Brahe – assumed a solid sphere for the fixed stars with the planetary heavens being of a fluid nature.

Among the numerous partisans of these opinions that Riccioli mentions, few are from the late Middle Ages.[6] Indeed, Riccioli includes only the names of Thomas Aquinas, who, he says, took no position on the fluidity or solidity of the heavens; Michael Scot and Cecco d'Ascoli, who assumed that the heavens were fluid; and Saint Bonaventure, who assumed that the heavens were partly fluid, partly solid.[7] Although, as we shall see, other names may be added, the paucity of medieval names is probably no oversight but reflects the fact that few medieval scholars expressed opinions on this interesting problem, which only became a major issue in the late sixteenth and the seventeenth century. It was during this later period that the expression "solid sphere" became virtually synonymous with "hard sphere" or "rigid sphere." From the seventeenth century on, the association of "solid" with "hard" was applied retrospectively to the Middle Ages and became fixed in the subsequent literature of the history of astronomy and cosmology. Thus it is that when modern scholars speak of "solid" orbs in the Middle Ages, they usually mean orbs that are hard and rigid.[8] But was this the medieval understanding of a celestial orb? To ascertain whether the description of an orb as solid also implied its hardness or rigidity, we must

5. See Riccioli, *Almagestum novum*, pars post., bk. 9, sec. 1, ch. 7, 1651, 238, col. 2–240, col. 2.
6. Most are from the ancient period (Greeks and Romans) and from the sixteenth and seventeenth centuries, which would have been the modern period for Riccioli.
7. Riccioli, *Almagestum novum*, pars post., bk. 9, sec. 1, ch. 7, 1651, 239, cols. 1–2. Bonaventure's position is unclear. Although he accepts the existence of orbs, he also insists that they are continuous, subtle, and rare like water. Indeed, they have no terminating surfaces as do solid bodies (see Bonaventure [*Sentences*, bk. 2, dist. 14, pt. 2, art. 1, qu. 1], *Opera*, 2:352, cols. 1–2). Bonaventure's heavens seem more fluid than solid. Nevertheless, Amicus also attributed to Bonaventure heavens that were a mixture of fluid and solid (Amicus [*De caelo*, tract. 5, qu. 5, art. 2], 1626, 275, col. 1).
8. For example, Jardine, 1982, 175, assumes that solidity is the opposite of fluidity and is therefore equivalent to hardness.

examine the meanings that were assigned to the term *solidum* during the Middle Ages.

II. The meaning of the term *solidum* in the Middle Ages

At the beginning of his famous thirteenth-century treatise *On the Sphere* (*De spera*), John of Sacrobosco defined a sphere by citing Euclid and Theodosius, both of whom considered it a solid body.[9] As a consequence, it became usual, in commentaries on Sacrobosco's *Sphere*, to inquire about the nature of a sphere and occasionally to ask about the sense in which a sphere was a solid. In a commentary on the *Sphere* ascribed to Michael Scot, we learn that the term *solidum* is spoken of in three ways:

> in one way, it is the same as hard, just like earth; in another way solid is the same as continuous, and thus the elements and supercelestial bodies are called solid; in a third way it is like a three-dimensional thing, and thus it is the same as a body. Therefore it is not superfluous to say that *a sphere is a solid body.*[10]

Although this significant passage poses serious problems, it is striking that Michael – for convenience, let us assume that he was the author – invokes earth to exemplify the meaning of a *hard solid*, but mentions celestial bodies[11] (and the elements) to illustrate the meaning of a *continuous solid*. Does this signify that Michael thought of the celestial bodies as continuous but not hard? This may depend on whether the term "elements" (*elementa*), in the second sense of solid, includes or excludes earth. Was Michael, in effect, dividing the elements into hard (earth) and soft (water, air, and fire), with only the latter assumed continuous? If so, the celestial bodies would also be continuous *and soft*, just like water, air, and fire.

Another possibility suggests that Michael had something else in mind, namely to signify that solid bodies possessed all three attributes: hardness, continuity, and tridimensionality. This interpretation seems less plausible, because, in a sentence immediately following the one proclaiming his threefold sense of the term *solidum*, Michael provides a clue that he may have intended three quite distinct senses of solid rather than to suggest that a solid possesses all three attributes. For he says:

> It is also known that a surface [*superficies*] is threefold: it is plane, as in a wall; it is concave, as in a tub; [and it is] convex, as in a mountain. And it is by such a surface

9. See Sacrobosco, *Sphere*, 1949, 76–77 (Latin) and 118 (English).
10. "Item nota quod solidum dicitur tribus modis: uno modo est idem quod durum, sicut terra; alio modo solidum idem quod continuum, et sic elementa et corpora supercelestia solida dicuntur; tertio modo, id est, quod trina dimensio, et sic idem est quod corpus, unde non est ibi negatio, *Spera est corpus solidum.*" Ibid., 256.
11. By the expression *corpora supercelestia*, I assume that Michael means all celestial bodies – that is, both planets and spheres.

[i.e., the convex surface] that a round solid is contained, because it includes everything within itself, leaving nothing outside.

Is there a parallel between the respective threefold senses of the terms "solid" and "surface"? Since Michael obviously intended three senses of the term surface, may we also infer that he intended three distinct senses of the term solid? If so, then he may also have intended that celestial bodies be conceived of as continuous but not hard, or at least not necessarily hard.

An examination of a discussion by Robertus Anglicus, in the latter's *Commentary on the "Sphere" of Sacrobosco*, written around 1271, lends support to this possibility. In a passage that he may have drawn, and perhaps even copied, from Michael Scot, Robertus describes the same three senses of the terms *solidum* and *superficies*.[12] He assumes the existence of nine celestial orbs and also proclaims the immutability of the material from which they are composed. The orbs are distinguished as being larger and smaller orbs by means of "greater and smaller intelligible [i.e., imaginary] circles." That is, an orb, or sphere, is the space that is cut off between two such circles and is the place where each planet is carried.[13] Robertus illustrates the arrangement of the nine celestial orbs by imagining nine wheels of such sizes that they can be arranged concentrically. These nine nested wheels are assumed to be in the air and to move around the same center. Robertus now explains that the quantity or volume of air between any two wheels is like a celestial orb which carries around the planet that lies within it. By choosing air enclosed by wheels as his analogy with celestial orbs, Robertus leaves the impression that he conceived of the celestial orbs as somehow fluid in nature – or at least fluid in their interiors, if not in their surfaces.

This interpretation gains credibility when Robertus later considers whether the celestial spheres are continuous (*continue*) or contiguous (*contigue*) and decides that they are continuous, which means that the convex surface of one sphere is identical with the concave surface of the next-superior orb.[14] But if the successive surfaces of successive orbs are continuous, Robertus acknowledges the following problem:

since the orbs are moved by contrary motions, . . . then one and the same [surface] would be moved by contrary motions, which is impossible. Also, it would then follow that, if one orb were moved by some motion, all the other orbs would be moved by the same motion, which, nevertheless, we know to be impossible.[15]

12. For the Latin passages, see Robertus Anglicus, *Sphere*, lec. 1, 1949, 145 (and 200 for Thorndike's translation). Robertus substitutes the expression *corpora celestia* for Michael Scot's *corpora supercelestia*. Aiton, 1981, 90, mentions Robertus's discussion of the term "solid" and correctly explains that "The Earth was solid in the first sense, while the celestial bodies were solid in the second and third senses, but were not necessarily hard."
13. Robertus uses the terms *spera* and *orbis* interchangeably.
14. "Ad primam questionem dicendum quod omnes orbes novem sunt continui." Robertus Anglicus, *Sphere*, lec. 1, 1949, 147. If the surfaces were distinct, the orbs would be contiguous. This discussion appears in ibid., 146–147 (Latin), 202–203 (English).
15. Ibid., 202. These arguments have already been described.

In replying to this difficulty, Robertus indicates that orbs are fluid when he says:

We suppose the outer edge of the orb immobile and the middle of the orb to be moved, just as we see that the center of water is moved, yet at its sides the water is still. And it seems much more likely that this can be done in the orbs, which are much simpler than water. Nor, as is now clear, need all orbs be moved when one orb moves, although they are continuous, just as it is not necessary that, when a part of the water is moved, all the water should be moved, although the water is continuous.[16]

Whatever Robertus may have meant by these examples, the fact that he used water to illustrate the continuity and motion of celestial orbs suggests that he thought of those orbs as continuous and fluid rather than as continuous and hard.[17] Moreover, his description is of great interest, for he seems to say that the surface of an orb can be assumed to be immobile while the part toward its center is in motion. Thus the planet itself is somehow carried by the fluid part of the orb lying within its immobile surfaces. But what is the nature of an orb's immobile surfaces? The water analogy indicates fluid surfaces, since the latter are in no way differentiated from their mobile content.

By adopting an approach in which only the inner part of an orb was assumed to move while its outermost surface lay immobile, Robertus avoided the seemingly impossible dilemma that would have resulted from an assumption of the continuity of the celestial orbs. For on that assumption, two separate but successive and immediately proximate orbs would necessarily move in the same direction, because the convex surface of the inner sphere would be continuous – that is, identical – with the concave surface of the next-superior orb. Despite the assumption of continuity, however, Robertus could now declare that although the convex surface of the planetary orb of Mars and the concave surface of the sphere of Jupiter, for example, were one and the same, those two planetary spheres could, nonetheless, move in different directions, because only the middling parts of each sphere or orb – not the surface itself – would move.

Although the evidence is stronger for Robertus Anglicus, both he and Michael Scot appear to have thought of the celestial orbs as soft rather than hard. Whatever the merits of that claim, however, one strong inference

16. Ibid., 202–203.
17. Bonaventure seems to have held the same opinion (see note 7 of this chapter). Aegidius Romanus may have had Robertus in mind when he attacked, as a "fatuous" opinion, the idea that "just as water that remains continuous can be moved according to one part to one place and according to another part to another place; thus the orbs should be able to do the same thing. But this cannot occur without the division of the water. Hence this cannot be applied to the orbs without division of the orbs." Aegidius Romanus, *Opus Hexaemeron*, pt. 2, ch. 32, 1555, 49r, col. 2. Aegidius regarded the heavens as solid overall, with softness confined to the channels marking out the eccentric deferents.

may be drawn from the discussions of our two thirteenth-century commentators on the *Sphere* of Sacrobosco: when the term *solidum*, or any of its variants, occur in the context of a discussion on celestial orbs, they may, in the absence of any other decisive criteria, refer to either hard or soft spheres. The modern interpretation, which always equates solid with hard, is unwarranted.

But a move toward equating solid with hard may already have been under way in the fourteenth century. We may infer this from Pierre d'Ailly, who, by the end of the fourteenth or early fifteenth century, said that *solidum* is taken in three ways, the first of which assumes that it is "firm or hard, just like iron or stone, and this is the common usage."[18] In the course of the sixteenth century, and certainly by the seventeenth century, the earlier ambivalence vanishes: fluid then was regularly opposed to solid, with the latter unequivocally equated with hard, as when Johannes Poncius declared: "some say that the heaven is a continuous body and fluid, like air . . . ; and other moderns think that the firmament is a solid [i.e., hard] body."[19] For Michael Scot and Robertus Anglicus, and perhaps for other natural philosophers during the thirteenth to fifteenth centuries, the concept of a solid body may have entailed a fluid state. But during that very period the identification of hard with solid was well under way.

III. The three major positions

From my summary of Riccioli's five opinions on the fluidity or hardness of the heavens, it is evident that these opinions reduce to three alternatives: (1) all are hard; (2) all are fluid, or soft; and (3) some are hard and some fluid.[20] Most scholastics appear to have adopted the third, or mixed, interpretation, assuming some orbs to be hard and others soft. A prime example is Aegidius Romanus, who assumed a hard overall orb from the Moon to the sphere of the fixed stars but conceived of the eccentric deferents as channels filled with a soft substance. Thus far the authors whom we have sampled have spoken only in generalities. We must now examine medieval

18. "Est advertendum quod tripliciter solet capi solidum. Primo modo prout tantum valet sicut firmum vel durum sicut ferrum vel lapis et sic eo utuntur vulgares." D'Ailly [*14 Questions*, qu.1], 1531, 148r.

19. Poncius wrote these remarks in his commentary on book 2 of Scotus's *Sentences*. See Poncius [*Sentences*, bk. 2, dist. 14, qu. 2] 1639, vol. 6, pt. 2:727, col. 1. Poncius observes that the solidity – that is, hardness – of the heavens is an opinion "more common" to Peripatetics (727, col. 2). Although Amicus accepted solid and hard celestial orbs, he reports that "in bk. 1, ch. 1 of his *Almagest*, it seems that Ptolemy assumes that solid and hard bodies are mutually distinct" (Videtur Ptolomaei I, Almeg. c.1 . . . supponit esse corpora solida et dura inter se distincta). See Amicus, *De caelo*, tract. 5, qu. 5, art. 2, 1626, 275, col. 2.

20. Among authors who distinguish these three positions are Amicus, ibid., 275, col. 1 and Roderigo de Arriaga [*De caelo*, disp. 1, sec. 3, subsec. 1], 1632, 499, cols. 1–2. Both present four different opinions.

opinions about the orbs or heavens themselves. Because so few considered the various parts of the heavens systematically, information must be gleaned from a variety of sources, often from discussions of this or that particular heaven or of the heavens as a whole. Therefore, I shall now describe what medieval natural philosophers thought about the hardness or softness of the two major celestial subdivisions, namely the crystalline heaven and the firmament, the latter, as we saw (Ch. 5, Sec. VI.2), often including the planetary orbs, which were frequently conceived as subdivisions of the firmament.

IV. The crystalline orb

In its theological aspect, the crystalline sphere developed from commentaries on Genesis 1.7, which spoke of a division of waters between those above the firmament and those below. From the time of the Church Fathers, the meaning and significance of the waters above the firmament were, as we have seen, much debated. Because the biblical text spoke of waters above the firmament, Christian authors, following Saint Augustine, were generally agreed on the necessity for a literal interpretation of this particular text and were therefore committed to the existence of waters of some kind above the firmament (see Ch. 5, Sec. VI).[21] All else was seemingly arguable. Indeed, the debate hinged on the interpretation placed on the terms "waters" (*aquae*) and "firmament" (*firmamentum*), the latter largely determining the meaning of the former.[22]

From the time of the Church Fathers to the end of the Middle Ages, a variety of interpretations of the waters above the firmament were proposed. The interpreters divide essentially into two groups: those who thought of the waters as solid and hard and those who considered them fluid. Among the former we may include Saint Jerome and Bede, the latter likening the waters to "the firmity of a crystalline stone" (*cristallini lapidis quanta firmitas*).[23] In his famous and widely used *Sentences*, composed in the twelfth century, Peter Lombard indicates an awareness of Bede's opinion when he

21. One who was not was William of Conches (fl. 1120–1149), who, in his *Philosophia mundi*, denied that waters could exist above the firmament and insisted that the scriptural passage in which this is asserted must be interpreted allegorically. See Lemay, 1977, 231.

22. To see how the meaning of "firmament" determined the meaning of the "waters" above that firmament, and to obtain an excellent sampling of the different interpretations placed on both of these terms, see Thomas Aquinas, *Summa theologiae*, pt. 1, qu. 68, arts. 1 ("Was the firmament made on the second day?"); 2 ("Are there any waters above the firmament?"); and 3 ("Does the firmament separate some waters from others?"), 1964–1976, 10:71–77, 77–83, and 83–87, respectively. Aquinas explains (79): "we maintain that these waters are material. Just what they are must be explained in different ways depending on various theories about the firmament." He then proceeds to offer a number of interpretations (for some of them, see Ch. 5, Sec. VI).

23. For the relevant passages from Jerome and Bede, see Campanus of Novara, *Theorica planetarum*, 1971, 393–394, n. 54.

declares that the waters above the heaven (i.e., firmament) are not in a vapory form but are suspended by icy solidity (*glaciali soliditate*) to prevent their fall.[24] Here we find the conviction that in liquid form the waters would surely flow downward; only if they were frozen or congealed would they remain suspended above the firmament.

Those who assumed that the waters above the firmament were fluid formed the larger group during the Middle Ages and included Ambrose, John Damascene, Alexander of Hales, Robert Grosseteste, Richard of Middleton, Saint Bonaventure, Vincent of Beauvais, and an anonymous author of a French encyclopedia written around 1400.[25] Within this group, some, like Richard of Middleton and Saint Bonaventure, provide little or no description of the other alternative, namely that the supraheavenly waters might be hard like a crystal. They were agreed, however, that although these waters were unlike ordinary elemental water, they did share with it a few important properties, namely, transparency (*perspicuitas*) and coldness (*frigiditas*), and for Richard also the property of wetness (*humidum*). Vincent of Beauvais, who assumed that the waters were immutable, believed they were luminous (*luminosum*), transparent (*perspicuum*), and subtle (*subtile*).[26]

Although terms like "crystalline" and "icy solidity" seem to imply hardness, they could be interpreted otherwise. In commenting on Peter Lombard's passage on the icy solidity of the waters above the firmament, Bonaventure insists that the sense of *solidity* that implies that those waters are heavy and held in position above the firmament by violence is contrary to the order of the universe. We should rather understand that "those waters compare with icy solidity [*glaciali soliditati*] not because of heaviness but because of continuity and stability because they do not ebb or flow, nor do they descend downward."[27] Bartholomew the Englishman is even more explicit when he explains that the waters above the heaven (firmament) are called "crystalline, not because they are hard [*durum*] like a crystal but

24. See Peter Lombard [*Sentences*, bk. 2, dist. 14, ch. 4, par. 1], 1971, 396.
25. Saint Ambrose, *Hexameron*, 3rd homily (bk. 2, the 2nd day), ch. 3, 1961, 51–59; John Damascene, *The Orthodox Faith*, bk. 2, ch. 9 [Chase], 1958, 37:224; Alexander of Hales, *Summa theologica*, inquis. 3, tract. 2, qu. 2, tit. 2, ch. 5 ("Qualiter illae aquae dicantur caelum crystallinum"), 1928, 2:341; Grosseteste, *Hexaëmeron*, part. 3, ch. 3, 4, 1982, 104; Richard of Middleton [*Sentences*, bk. 2, dist. 14, qu. 1 ("Utrum coelum crystallinum dictum sit naturae aquae")], 1591, 2:167–168; Bonaventure, *Sentences*, bk. 2, dist. 14, art. 1, qu. 1 ("Utrum caelum crystallinum sit natura aquae"), *Opera*, 2:335–338; Vincent of Beauvais, *Speculum naturale*, chs. 90–100, 1624, cols. 221–229, especially col. 224; for the statement in the French encyclopedia, see Hyatte and Ponchard-Hyatte, 1985, 11.
26. Richard is silent, whereas Bonaventure mentions only that Bede believed that the waters above the heavens rested, and were sustained, by virtue of their solidity: "Et ibidem aquae illae quiescunt et sustentantur vel sua soliditate, sicut videtur Beda dicere, vel sua subtilitate, vel etiam Dei virtute, quae sic ordinavit." Bonaventure, ibid., 337, col. 2. For Vincent, see *Speculum naturale*, bk. 3, ch. 95, 1624, 1:col. 224. Vincent also describes the alternative opinion that the waters are congealed like a crystal and rejects it because he can find no cause that would congeal the waters; ibid., col. 221. Grosseteste described the opinion that the waters above the firmament were like a hard, crystalline stone (*cristallus lapis*) and also rejected it (*Hexaëmeron*, part. 3, ch. 3, 4, 1982, 104).
27. Bonaventure, *Sentences*, bk. 2, dist. 14, pt. 1, art. 3, qu. 2, dub. 1, *Opera*, 2:350, col. 1.

because they are uniformly luminous and transparent. Moreover, it [i. e., the crystalline heaven] is called watery insofar as water is moved by virtue of its subtlety and mobility."[28] A similar description is provided by the author of a French encyclopedia, who, around 1400, declared that

> Others call it [the ninth sphere] the "crystalline sphere" or the "crystalline heaven" [or sky], not because it is of hard and solid material like crystal, but for its luminosity and its great transparency and uniformity. And it is also the heaven [or sky] that theologians call "watery," not because there are waters such as those which are here below, rather they are light [*soubtilles*] waters of a noble nature similar to the heaven [or sky] in clarity and luminosity.[29]

Vincent of Beauvais declares that the waters above the firmament should not be understood as the element water but rather as a kind of diffuse matter (*materia confusionis*) that has within itself "hotness, dryness, wetness, and coldness, luminosity, darkness, transparency, and opacity."[30] Hardness is not even mentioned.

For many, if not most, of those who considered the suprafirmamental waters "crystalline," the latter term did not signify the hardness of the waters but rather their immutability, transparency and luminosity. When medieval authors spoke of the crystalline sphere, they usually had in mind those properties of a crystal such as luminosity, transparency, and even a quasi immutability, rather than hardness.

V. The firmament and the planetary orbs

In Chapter 5, we saw the variety of meanings that attached to the term "firmament" (*firmamentum*), the heaven that God created on the second day (Genesis 1.6–8) to divide the waters above from the waters below and in which he placed the luminaries on the fourth day (Genesis 1.14–19) to divide day and night. Because of the explicit biblical assertion that the luminaries were in the firmament ("Fiant luminaria in firmamento caeli"), it was not

28. "Et ideo in summo dicitur crystallinum non quia durum sicut crystallus, sed quia uniformiter est luminosum et perspicuum. Aqueum autem dicitur quemadmodum aqua ex sua subtilitate et mobilitate movetur." Bartholomew the Englishman, *De rerum proprietatibus*, bk. 8, ch. 3 ("De coelo aqueo sive crystallino"), 1601, 379. As we saw, Vincent of Beauvais held a similar view about the crystalline sphere.
29. Hyatte and Ponsard-Hyatte, 1985, 11. I am grateful to Prof. Reginald Hyatte, who not only called my attention to the reference but also translated the relevant passage.
30. "Nos itaque dicimus quod ubi dicitur *Deus divisit per firmamentum aquas ab aquis*, non sumitur aqua pro elemento, sed pro materia confusionis, quae habet in se calidum et siccum, humidum et frigidum, luminosum et tenebrosum, perspicuum et opacum." Vincent of Beauvais, *Speculum naturale*, bk. 3, ch. 95, 1624, 1:col. 224. Later, in ch. 95, col. 225, Vincent explains that "this heaven [the crystalline] is not crystalline with respect to every property of a crystal, but according to the property of transparency." (Dicimus quod coelum illud non est chrystallinum secundum omnem proprietatem chrystalli, sed secundum proprietatem perspicui.)

unusual for natural philosophers to extend the firmament to include not only the eighth sphere of the fixed stars but also the planetary orbs. Now we must determine whether that firmament was conceived as hard, or fluid, or some combination of these contrary properties.

Arriving at an opinion was not often easy, as two of the greatest Church Fathers, Augustine and Basil, illustrate. Despite his observation (in his commentary on Genesis), that too much subtlety and learning had been expended on explicating the nature of the firmament,[31] Augustine advised those who analyzed the meaning of the firmament to "bear in mind that the term 'firmament' does not compel us to imagine a stationary heaven: we may understand this name as given to indicate not that it is motionless but that it is solid and that it constitutes an impassable boundary between the waters above and the waters below."[32] Without choosing between them, Augustine thus explained the "firmity" of the firmament in two ways: it is firm either because it is motionless or because it is solid and prevents the passage of waters from above or below. He gives no indication that those who assumed a "solid" and impenetrable firmament meant also to signify that it was hard. However, elsewhere in the same commentary on Genesis, Augustine speaks of air and fire as material constitutents of the heavens, thereby suggesting soft and fluid heavens.[33]

Saint Basil suffered from similar equivocation, which he exhibited over the span of a few lines of his *Hexaemeron*, where he wrote:

> Not a firm and solid nature, which has weight and resistance, it is not this that the word "firmament" means. In that case the earth would more legitimately be considered deserving of such a name. But, because the nature of superincumbent substances is light and rare and imperceptible, He called this firmament, in comparison with those very light substances which are incapable of perception by the senses.[34]

With his denial of solidity to the firmament, Basil goes on to deny as well that it could be composed of the simple elements or of any combination of them.[35] Despite his uncertainty about its composition, Basil seems to incline toward a fluid firmament.

31. Augustine proclaimed that he himself had "no further time to go into these questions and discuss them, nor should they have time whom I wish to see instructed for their own salvation and for what is necessary and useful in the Church." Augustine, *Genesis*, bk. 2, ch. 8, 1982, 1:60–61.
32. Ibid.
33. See Chapter 10, note 4.
34. Basil, *Exegetic Homilies (Hexaemeron)*, homily 3, 1963, 47. Earlier in homily 3 (p. 43), Basil had denied that the firmament could be compared to water that is "like either frozen water or some such material which takes its origin from the percolation of moisture, such as is the crystalline rock which men say is remade by the excessive coagulation of the water, or as is the element of mica which is formed in mines."
35. See Chapter 10, note 4. But in homily 3, Basil seems to equate the heavens with the firmament when he declares (ibid., 49–50): "we have observed in many places that the visible region is called the heavens due to the density and continuity of the air which clearly comes within our vision and which has a claim to the name of heavens from the

A solid, hard firmament had at least one important defender in late antiquity. Approximately two hundred years after Saint Basil, John Philoponus considered the nature of the firmament in his commentary on the six days of creation. Defending the account of Moses in Genesis, Philoponus insisted that the latter had given a better explanation of the firmament than either Plato or Aristotle. Whereas Plato had assumed a heaven composed of the four elements and Aristotle had invoked the existence of a fifth corporeal element, the ether, Moses, by contrast, had assumed that the firmament was formed in the midst of the waters. Because the substance of the firmament is transparent and water and air are the only two elements that possess this property, Moses assumed that the transparent heavens are formed of these two elements, though composed perhaps more of water than air. The term *firmamentum*, implying solidity, also suggests that these two elements were transformed from their natural fluid state to a solid, and presumably hard, body.[36]

Despite the seeming ambivalence or inconsistency of Augustine and Basil, and Philoponus's explicit support of solidity, most Christian authors and Latin Encyclopedists during late antiquity and the early Middle Ages, and even into the thirteenth century, probably thought of the heavens as fiery or elemental in nature, and therefore fluid. In this category, Christopher Scheiner included Gregory of Nyssa, Chalcidius, Isidore of Seville, John of Damascus, Peter Damian, Hugh of Saint Victor, Peter Lombard, Alexander of Hales, and Bonaventure,[37] to which we may add Macrobius, Michael Scot, Robertus Anglicus, and perhaps Peter of Abano.[38] A notable exception was Robert Grosseteste, who described the heavens not only as the most subtle of all bodies, but, relying on the words of Job 37.18, also characterized it as the most solid of bodies, like fused metal.[39] Others undoubtedly offered no opinion at all. For some of these individuals, Scheiner inferred belief in a fluid heavens because they assumed that the Sun, or in certain instances even all the planets, were fiery bodies.[40]

word 'seen,' namely, where the Scripture says: 'The birds of the heavens,' and again, 'the flying creatures below the firmament of the heavens.' "

36. Here I follow Duhem, *Le Système*, 1913–1959, 2:499–500.
37. Scheiner, *Rosa Ursina*, bk. 4, 1630, 627–635. Scheiner devoted the fourth and final book of his *Rosa Ursina* to the themes of fluidity and corruptibility of the heavens by citing passages from numerous authors who, in his judgment, had expressed explicit or implicit opinions. Scheiner withheld his own opinion, declaring at the end of this lengthy section, that he would give his opinion at another time and in another place. Ibid., 773, col. 1.
38. For Michael Scot and Robertus Anglicus, see Section II of this chapter; for Peter of Abano, see Duhem, *Le Système*, 1913–1959, 4:253, where Duhem, in discussing Peter's *Lucidator*, says that in the latter Peter assumes that each planet moves within the medium of a fluid substance which constitutes its heaven. In her edition of the *Lucidator*, Vescovini Federici seems to arrive at the same conclusion about Peter's acceptance of self-moved planets, but she makes no claims about a fluid medium (*Lucidator*, 1988, 269–270). Indeed, Peter makes no mention of a fluid medium.
39. Est itaque celum corpus primum . . . quia subtilissimum; et tamen, ut dicit Iob, *solidissimum quasi ere fusum*." Grosseteste, *Hexaëmeron*, part. 3, ch. 16, 2, 1982, 117. For other references to this frequently cited passage from Job, see note 51, this chapter.
40. In this group, Scheiner (*Rosa Ursina*) includes Saint Basil (627, col. 2); Gregory of Nyssa

During the late Middle Ages, most authors were vague and noncommital, despite the fact that the very name *firmamentum*, with its implications of strength, power, stability, and even of solidity and hardness, seemed to invite an explanation and thus to provide an occasion for the expression of opinions about its possible hardness or softness. Few, however, chose to avail themselves of an opportunity to explain why, in Genesis, the term *firmamentum* was used for the heaven created on the second day.[41] Two who did were Vincent of Beauvais and Campanus of Novara. Vincent declared that the term was used because that heaven is ungenerated and incorruptible rather than because it is immobile with respect to place. It is indeed indissoluble, because it lies beyond the action and passion of contraries.[42] Campanus explains that the firmament is so called because "its motion always seems to be firm and uniform and because the fixed stars seem to be firmed in it." [43] By describing the fixed stars as being "firmed" in the firmament, Campanus is perhaps implying the existence of a hard firmament. The brevity of his discussion makes judgment uncertain. Nowhere, however, do Vincent or Campanus explicitly associate solidity or hardness with the term *firmamentum*. Centuries later, Bartholomew Amicus also argued that the term implied firmness and solidity.[44] For Amicus, however, and all other scholastics of the seventeenth century, solidity signified hardness. By contrast, his seventeenth-century contemporary Roderigo de Arriaga insisted that nothing could properly be deduced from the term *firmamentum* about the solidity or hardness of the heaven called by that name.[45] Just because this particular heaven is called "firm" (*firmus*) does not warrant an inference of hardness, because the heavens could be "firm, stable, and incorruptible, even if they are not hard."[46]

Our information, such as it is, derives from brief statements in a variety of contexts. During the late Middle Ages, the hardness or softness of the celestial orbs was not judged a significant topic, as evidenced by the fact

(628, col. 1); Isidore of Seville (634, col. 1); John Chrysostom (628, col. 1); Peter Damian (629, col. 1); Hugh of Saint Victor (631, col. 2); Peter Lombard (629, col. 2); Alexander of Hales (629, col. 2); Bonaventure (629, col. 2), and others.

41. Although Bonaventure, for example, discussed the firmament in a few questions in his *Sentences*, bk. 2 (*Opera*, 2:338–341, 351–352), he nowhere considers why the term *firmamentum* was used to describe the one or more heavens embraced by it. The same may be said about Richard of Middleton in the latter's second book of his *Sentences*.
42. "Nos autem dicimus ad primum quod firmamentum dicitur a firmitate naturae quia non generatur, nec corrumpitur et non ab immobilitate secundum locum. . . . Et propter hoc dicitur firmamentum quia indissolubilis est concensus ille cum extractus sit extra actionem et passionem contrariorum." Vincent of Beauvais, *Speculum naturale*, bk. 3, ch. 102, 1624, 1:col. 230.
43. Of the firmament, which he identifies with the eighth sphere, Campanus says: "Et dicitur firmamentum quoniam ipsius motus semper videtur esse firmus et uniformis et quia stellae fixe videntur firmari." These words appear in Campanus, *Sphere*, 1531, 196r.
44. For more on this point, see Section VIII.2 of this chapter.
45. "Vides ergo ex nomine *firmamenti* nihil plane de soliditate deduci." Arriaga, *De caelo*, disp. 1, sec. 3, subsec. 4, 1632, 502, col. 2, par. 37.
46. "Ubi caeli dicuntur firmi, nihil de soliditate contineri; sunt enim firmi, stabiles, et incorruptibiles, etiamsi non duri." Ibid., par. 38.

that scholastics did not see fit to devote a *questio* to that theme. Why did they ignore this question, which became so important in the seventeenth century? Perhaps because Augustine thought a Christian's interpretation of the meaning of "firmament" was relatively unimportant,[47] an attitude that was reinforced by Aristotle himself, who ignored the issue and provided no guidance.

The hypothesis of fluid heavens, which went largely unchallenged prior to the thirteenth century, came to have a rival after the introduction and dissemination of Aristotelian–Ptolemaic astronomy and cosmology in the thirteenth century. Whereas previously the idea of fluid or soft heavens was overwhelmingly dominant, the existence of orbs and their possible hardness now emerged as an opposition hypothesis.

Richard of Middleton illustrates the change that had occurred. During the course of a discussion on whether the planets, or heavens, form one continuous body, Richard describes the two rival interpretations, without choosing between them.[48] The first opinion assumes that the heavens are of a fluid nature, which Richard identifies with Aristotle's fifth element, or ether. In this fluid theory, no distinction is made between orbs on the basis of different forms nor on the basis of any discontinuity of their surfaces, as, for example, one stone is distinguished from another. Celestial orbs differ only because of their diverse motions. But these diverse motions do not produce discontinuiuty in the fluid medium through which the planets can move readily and easily. Indeed, Richard may mean here that the orbs and planets are themselves fluid parts of the overall ether and that celestial motions consist of parts of the ether moving in different directions, as we can observe in water, "when its different parts move to different positions." Under these circumstances, the water does not lose its form.[49]

Others, however, present a second opinion in which "the fifth body consists of celestial solids that are not divisible. Thus [in] Job 37 it is said that the heavens, which are most solid, are made of metal. And the Philosopher, in the second book of *De caelo*, proves that the stars [planets] are not moved from one part of an orb to another part, because the orb would be divided, which he assumes absurd. And also to many, it seems that if the orbs were fluid and [therefore] divisible by nature, they should appear corruptible." Because of the contrary motions of the orbs, those who adopted this interpretation denied the continuity of the eccentric orbs and not only assumed their contiguity with the surfaces that surrounded them but also assumed that the surface of an epicycle was contiguous with the

47. Judging from his discussion of the firmament in the *Summa theologiae*, pt. 1, qu. 68, art. 1, 1967, 10:71–77, Thomas Aquinas was certainly one of these. Grosseteste thought it would be tedious and prolix to present a detailed analysis of the nature of the firmament (see Grosseteste, *Hexaëmeron*, part. 3, ch. 6, 1, 1982, 106).
48. Richard of Middleton, *Sentences*, bk. 2, dist. 14, art. 3, qu. 1, 1591, 2:184, col. 1.
49. It is possible that Richard is reporting a version of the opinion of Robertus Anglicus, as described in Section II of this chapter.

surface of its eccentric deferent. Indeed, the planet itself is also contiguous with the concave and convex surfaces of the epicycle that surrounds it.

Richard was probably one of the first in the Latin West to present the case for heavens composed of solid and hard orbs. Although he gives no clear indication as to which of the two alternatives he preferred, it is likely that the hard-orb hypothesis was itself derived from some earlier account.[50] The justification from Job 37.18 was the most explicit biblical support available for belief in hard orbs and would be frequently cited – especially in the seventeenth century – in defense of hard orbs and against those who believed in fluid heavens.[51] In this passage, Elihu asks Job whether, like God, he could fabricate the heavens as if they were made of molten metal,[52] thus implying that God had made the heavens hard like metal.

Despite the general absence of detailed and explicit discussions of the hardness or fluidity of the celestial orbs in the Middle Ages, some scholastic authors give evidence of having assumed the hardness of the spheres. During the course of the fourteenth century, Themon Judaeus and Henry of Hesse, and in the late fifteenth century Hartmann Schedel, explicitly argued for hard spheres, while in the fourteenth century Nicole Oresme did so indirectly, as perhaps did Pierre d'Ailly in the early fifteenth century. Let us now consider the manner in which these few individuals indicated their preference for hard orbs.

In his questions on the *Meteorology*, Themon debated whether the sky or heavens are of a fiery nature and rejected the possibility, arguing that if the heavens had an elemental nature, they would be like earth and water rather than fire. This is because "a heaven [i.e., orb] is a hard [*durum*] body without capacity for flowing."[53] But "fire in matter proper to it is not hard and lacking in the capacity to flow, as is obvious by experience, as we see in flames. [Experience] also shows [that it is quite otherwise with] water, ice,

50. Perhaps from Grosseteste, who, in the first half of the thirteenth century, had linked Job 37.18 with a hard heavens (see Section V, this chapter).
51. Besides Richard, others who cited it were the Conimbricenses [*De coelo*, bk. 1, ch. 3, qu. 1, art. 4], 1598, 70; Hurtado de Mendoza [*De coelo*, disp. 1, sec. 5], 1615, 366, col. 2; Amicus, *De caelo*, tract. 5, qu. 5, art. 2, 1626, 273, col. 1 and art. 3, 278, col. 1; Aversa, *De caelo*, qu. 32, sec. 6, 1627, 67; Arriaga, *De caelo*, disp. 1, sec. 3, subsec. 4, 1632, 502, col. 1; Poncius, *Sentences*, bk. 2, dist. 14, qu. 2, 1639, vol. 6, pt. 2:727, col. 2–728, col. 1; Compton-Carleton [*De coelo*, disp. 1, sec. 3], 1649, 399, col. 1; Riccioli, *Almagestum novum*, pars post., bk. 9, sec. 1, ch. 7, 1651, 240, col. 2; Bonae Spei [comment. 3, *De coelo*, disp. 3, dub. 7], 1652, 14, col. 1; Cornaeus [*De coelo*, disp. 2, qu. 2,], 1657, 499; and Oddus [*De coelo*, disp. 1, dub. 12], 1672, 35, col. 1 (Oddus gives the correct text, but the wrong reference, citing Job 3.32).
52. In the Vulgate, the text reads: "Tu forsitan cum eo fabricatus es caelos, qui solidissimi quasi aere fusi sunt." The Douay-Challoner translation of the Vulgate (ed. John P. O'Connell, Chicago: Catholic Press, 1950) renders "qui solidissimi quasi aere fusi sunt" as "which are most strong, as if they were of molten brass." A recent translation from the Hebrew text provides a more graphic version to describe the rigid heavens: "Can you beat out the vault of the / skies as he does, / hard as a mirror of cast metal?" *New English Bible*, 1976.
53. See Themon Judaeus [*Meteorology*, bk. 1, qu. 3 ("Utrum coelum sit nature ignis")], 1518, 157v, col. 2.

and earth. For earth [and water] can be made hard and even transparent [*perspicua*], as is obvious from glass and ice."[54] Since fire cannot be hardened, it cannot be the material from which the celestial orbs are composed.

Around 1390, in a commentary on Genesis (*Lecturae super Genesim*), Henry of Hesse presented an unusual interpretation of the celestial orbs. According to Steneck (1976, 61–62), Henry argued that the firmament created on the second day was comprised of "a series of concentric shells or spheres that stretch from the region of the Moon to the region of the fixed stars. They are clear, firm, impenetrable, and have thickness. . . . In fact, the image of glass globes spinning on fixed axes around the central earth, so commonly used to describe the medieval conception of the celestial orbs, seems to fit quite nicely the discussion in the *Lecturae*." The various orbs had congealed like water or lead. Henry rejected Aristotle's celestial ether, or fifth element, and insisted that the heavenly region was composed of matter similar to that of the earth. Moreover, he further argued that on the fourth day of creation the planets were formed from mixtures of elemental matter that rose up through the various hard celestial orbs. Because Henry believed that the movement of such relatively coarse matter through the hard, celestial orbs was physically impossible, and since he was not prepared to abandon his interpretation, he chose to explain it by miraculous intervention. Hartmann Schedel conveyed his conviction of a hard firmament in his *Liber chronicarum* of 1493, by a simple declaration that God "made [the firmament] solid, out of water congealed like a crystal."[55]

Few were as explicit as Themon Judaeus, Henry of Hesse, and Hartmann Schedel. Only indirectly and occasionally can we infer the apparent opinions of others about the hardness of the spheres. In this connection, Nicole Oresme's discussion in his French translation of, and commentary on, Aristotle's *De caelo* is of interest. Here Oresme describes the surfaces of all celestial spheres as perfectly polished and smooth.[56] Because no vacua can exist between any two celestial surfaces,

54. Here is the full text of Themon's second conclusion: "Si celum esset nature elementalis potius esset nature aque vel terre quam ignis. Probatur conclusio: quia celum est corpus durum influxibile, alias fieret permixtio astrorum et stellarum nimium irregularis propter divisionem eius. Sed ignis in propria sibi materia non est durus et influxibilis, ut patet per experientiam, videmus enim de flammis. De aqua autem et glacie et terra patet. Terra enim potest indurari etiam perspicua fieri, ut patet de vitro et glaciebus. Ergo celum potius esset nature aque vel terre, quod est propositum." Ibid.

55. Hartmann Schedel, *Liber chronicarum*, pt. 1 [Rosen], 1493. Although the book is unpaginated, see "On the Work of the Second Day," lines 4–5. On the opposite page, the Latin text reads: "Ex aquis congelatis in modum cristalli solidavit et in eo fixa sidera." Schedel seems to have conflated the crystalline sphere with the firmament. In note 34 of this chapter, we saw that when Basil also mentioned crystalline rock and the firmament in the same passage, he sought to dissociate the two.

56. "The primary and sovereign heaven . . . does not push nor pull the heavenly sphere which is immediately under it. In addition, this concavity or concave area is very completely and perfectly polished, planed, and smoothed so that it could not be more so . . . and without any roughness or denticulation." Oresme, *Le Livre du ciel*, bk. 2, ch. 5, 1968, 317.

it follows necessarily that the concave surface of the sovereign heaven and the convex surface of the second or next heaven below must be absolutely spherical, with no roughness or humps, and that these heavens must move one inside the other without any friction. Rather, the passage of one surface above the other must be as smooth, as gentle, and as effortless as possible. The same holds for the second and third heavens and thus through all of them in descending order down to the concave surface of the lunar sphere, which is concentric with the earth and with the heavenly body which contains or comprises or is composed of all the partial heavens; otherwise, all this body taken together would be thicker in one part than in another, which is neither probable nor reasonable. Therefore the concave surface of the lunar heaven must be perfectly spherical.[57]

The perfect sphericity of the concave lunar surface causes the convex surface of the sphere of elemental fire to assume a perfectly spherical shape. Ordinarily, none of the imperfect four elements could assume a perfectly spherical shape. The convex surface of fire, which Oresme describes as "perfectly polished and spherical," is, however, an exception. But "this is not due to the element of fire itself, but to the concave surface of the lunar sphere which contains the fire and which is perfectly spherical, being everywhere in contact with the fire without intermediate plenum or vacuum."[58] On the basis of Oresme's discussion, it seems reasonable to assume that he judged the celestial surfaces to be hard. Otherwise the lunar concavity could not have shaped the outermost surface of grosser fire into a perfectly spherical surface.

The inclusion of Oresme among those who probably thought of the spheres as hard is based not on his description of the orbs as "perfectly polished" but on his assumption that the lunar concavity shaped the contiguous, convex surface of the perfect, fiery sphere below. Similarly, one ought not to include an anonymous fourteenth-century author among proponents of hard spheres simply because of an assertion that "celestial bodies do not rub together in their local motions because they are highly polished. Nor is there any friction between them that could generate heat."[59]

With Pierre d'Ailly, we complete our list of those who directly or indirectly indicated a world of hard celestial orbs. D'Ailly reveals his belief in hard orbs in the context of a conclusion in which he informs his readers that Sacrobosco had demonstrated that the intermediary spheres of the

57. Ibid., ch. 9, 387.

58. Ibid., ch. 11, 399.

59. "Dicendum est quod corpora celestia in suis motibus localibus non confricantur quia sunt corpora politissima. Nec inter ipsa sit aliqua confricatio talis ex qua possit gigni calor." See "Compendium of Six Books," Bibliothèque Nationale, fonds Latin, MS. 6752, 214v. For a description of the contents of the treatise, see Thorndike, 1923–1958, 3:568–584. Sometime between 1570 and 1572, Robert Bellarmine, the future Cardinal Bellarmine who played a significant role in the Galileo affair, rejected hard, polished, contiguous orbs by arguing that such orbs had no tendrils, braces, or glue to enable them to cling to each other so that one orb could drag another with it (see Lerner, 1989, 268, n. 35).

celestial region are spherical with respect to both their concave and convex surfaces. This is true, d'Ailly continues, only if we assume that "the heavens are not breakable [*frangibile*], [not] fluid [*fluxibile*], [not] augmentable, or diminishable" and that there can be no penetration of dimensions or existence of a vacuum.[60] By denying fluidity and breakability, d'Ailly implies the existence of hard celestial orbs.

VI. On the difficulties of determining whether natural philosophers assumed hard or fluid orbs in the late Middle Ages

From what has been described thus far about the problem of the hardness or softness of celestial orbs, we perceive a gradual shift: the widespread assumption of fluid heavens and orbs in the thirteenth century was yielding to an assumption of their hardness in the fourteenth century. Richard of Middleton described both theories but refused to choose between them. Shortly after, Aegidius Romanus proposed a combinatory theory in which the overall heaven was hard but the eccentric deferents were fluid (see Ch. 13, Sec.III.9). As the fourteenth-century progressed, Themon Judaeus and Henry of Hesse explicitly opted for hard orbs, and Nicole Oresme did so implicitly. Explicit defenders of fluid orbs or heavens have yet to turn up in the late fourteenth and fifteenth centuries. By the late sixteenth century, the hardness theory had become explicit and widespread. Tycho Brahe acknowledged it as the major opinion he had to destroy. Even as the existence of hard orbs was losing support in the seventeenth century, it remained the rival theory to the concept of a fluid heaven.

Although in retrospect the hardness or softness of the celestial orbs appears to us an important cosmological problem, our medieval predecessors seem not to have shared that judgment. They were quite content to discuss the celestial orbs without any explicit, or even implicit, indication as to their hardness or softness. The issue rarely surfaced. This reluctance was not confined to Latin scholastics but is equally apparent in authors who make up the Greco-Arabic tradition. Aristotle himself never saw fit to raise the problem. Nor did Ptolemy in the *Almagest*, not even in the first book, where he presents a modest amount of information about the nature of the physical world.[61] Even when an author specifically discussed the interre-

60. "Quinta conclusio est quod ratio autoris bene demonstrat caelum esse sphaericum quantum ad superficies tam concavas quam convexas sphaerarum intermediarum. Patet conclusio supposito quod caelum non sit frangibile, fluxibile, augmentabile, nec diminuibile; supposito etiam quod non possit esse penetratio dimensionum nec vacuum." D'Ailly, *14 Questions*, qu. 5, 1531, 153v.

61. During the seventeenth century, it was not uncommon for scholastic authors to classify earlier authors as proponents of hard or fluid heavens. These identifications were sometimes arbitrary and without foundation, as when Amicus insisted (*De caelo*, tract. 5, qu. 5, art. 2, 1626, 275, col. 2) that in the first chapter of book 1 of his *Almagest*, Ptolemy

lationships of celestial spheres, rarely did he see fit to interject an explicit judgment on their hardness or fluidity, as is evident from al- Bitrūjī's *De motibus celorum*, which Michael Scot translated from Arabic to Latin in 1217. Here Bitrūjī explains:

It is well known by all men that the whole heavens are composed of mutually different spheres and that one touches another in perfect contact. And because one [sphere] is moved inside another, there is a finiteness of the rotation and an equality of surfaces. And these [orbs] are continuous with each other because no other body lies between them.

And it is known that the concave surface of a higher [orb] is the place of the [orb] next below it and between them there is neither a plenum of another extraneous body, nor is there a vacuum, but one [orb] touches the other [orb] with its whole surface.[62]

In this passage, Bitrūjī gives no indication of his opinion on the hardness or softness of the celestial orbs. But if the orbs are contiguous, as is likely (see Ch. 13, Sec. III.2), or even if they are assumed continuous, does this alone provide a clue about his opinion concerning the hardness or softness of the surfaces of the celestial orbs? It does not. If an author simply opted for continuity or contiguity without providing any additional clues about hardness or softness, we would have no good independent reasons for assuming either hard or soft orbs. In Chapter 13 (Sec. III.2), we saw that Campanus of Novara probably assumed contiguous orbs in order to avoid the possibility of the overlap of two successive orbs or the possibility of void space between them. Because these are impossibilities in Aristotelian natural philosophy, the modern editors of Campanus's *Theorica planetarum* infer that Campanus "supposes the spheres to be solid,"[63] by which, it seems, they mean hard. If so, the inference is unwarranted. Even if the spheres were composed of a single fluid substance, the same objection would obtain: the overlapping parts of the two spheres could not possibly occupy the same place simultaneously. Thus Campanus's spheres may indeed be solid, but whether that solidity is to be associated with hard or soft surfaces and orbs cannot be determined.

Whereas Campanus and most others provided no useful clues for deciding the issue, Robertus Anglicus, who really seems to have assumed the continuity of successive surfaces, presents explicit information that enables us to conclude that he considered the orbs fluid. In order to permit the different

"assumes that the [celestial heavens] are solid and hard bodies and mutually distinct" (videtur Ptolomaei I *Almeg.*c.1, ubi supponit esse corpora solida et dura inter se distincta).

62. My translation from Bitrūjī, *De motibus celorum* [Carmody], 1952, 82. The words "there is a finiteness of the rotation" are an uncertain rendition of the Latin "ideo ipse est in fine rotationis." For an English translation based on Arabic and Hebrew versions of Bitrūjī's treatise, see Bitrūjī, *On the Principles of Astronomy*, 1971, 65–66.
63. Campanus of Novara, *Theorica planetarum*, 1971, 412, n. 47.

orbs to have independent motions and to have continuous and identical outer surfaces, Robertus assumed that the outer surfaces of celestial orbs were immobile but that their inner parts, which he likened to a fluid, were capable of motion.[64]

The use of the contiguity theory of distinct and separate celestial surfaces that are everywhere in perfect contact to avoid the possibility of void space or extraneous matter between them may seem at first glance to indicate that the surfaces are hard. If the convex surface of the planetary sphere of Mars were contiguous to the concave surface of Jupiter's orb, and if those two distinct, touching surfaces moved with different speeds and perhaps even in different directions, one might argue that they would retain their separateness and move independently of each other only because their surfaces were hard. But because of the attributes traditionally assigned to the celestial ether, the conditions just described do not by themselves warrant the inference of hard spheres. They are equally compatible with soft and fluid spheres. After all, not only is the celestial ether unalterable and incorruptible, but it was usually assumed to be more subtle than air and fire.[65] With such properties, the ethereal orbs could be contiguous but fluid. Moreover, because of their presumed unalterable and incorruptible natures, fluid orbs could be in contact at every point and also retain their perfect spherical shapes. Perfect, incorruptible, contiguous fluid surfaces, no less than hard surfaces, could be polished and move without resistance or friction. If I earlier attributed to Oresme an implicit belief in hard spheres, it was not because of his assumption of polished surfaces moving without resistance, but rather because of his declaration that the lunar concavity shaped the contiguous, convex surface of the fiery sphere below it.

Because the celestial ether was traditionally assigned properties that made it appear rare and fluid-like, it might seem that, in the absence of an explicit assertion in favor of hard spheres, or indirect independent evidence indicating rigid spheres, it would be more plausible to assume that a medieval author considered the celestial orbs solid and soft rather than solid and hard. Such a judgment would also be unwarranted and misleading. In the absence of reasonably compelling evidence, it is wiser to draw no inferences.

64. See Chapter 13, Section III.2 and note 67.
65. Aristotle nowhere says this, but it seems to follow from his ordering of the four elements (earth, water, air, and fire), the latter three of which become rarer and more subtle as their distance from the earth increases. Since the celestial ether extends beyond fire, it should exceed the latter in rarity and subtlety. In the strict sense, of course, Aristotle denied that contrary qualities such as rarity and density could be applied to the ether. But because physical attributions were sometimes unavoidable, the celestial ether was likely to be considered purer, rarer, and more subtle than any other substance. Even planets, which were usually conceived as aggregations of ether sufficiently dense to reflect light and become visible to us, were hardly thought of as "dense" in the sense in which that term might be applied to any of the four elements or bodies compounded of them.

VII. When did "solid orb" become synonymous with "hard orb"?

Because his world system required an intersection between the orbits of Mars and the Sun, which would have been impossible if hard spheres existed, Tycho Brahe used his knowledge that the comet of 1577 was moving in the celestial region beyond the Moon to deny the existence of solid celestial orbs and to suggest instead that the heavenly region was composed of a fluid substance.[66] The solid celestial orbs whose existence Tycho denied, were, of course, of the hard and rigid variety. In 1588, he explained that he "first showed and clearly established that by the motions of comets they [the heavens] are fluid and that the celestial mechanism is not a hard and impervious body filled with various real orbs, as has been believed by many up to this point, but that it is very fluid and simple, with the orbits of the planets free, and without the efforts and revolutions of any real spheres."[67] From Tycho's assertions that "many" or "very many" contemporaries believed that the heavens were composed of hard, celestial orbs, we may infer that this was the commonly held opinion of his day. Tycho's influence was so great that seventeenth-century astronomers and natural philosophers who mentioned solid spheres, whether or not they agreed with him, did so with the understanding that they were hard and rigid.

Almost from the first formulation of Tycho's radical interpretation, scholastic natural philosophers found themselves divided. As a direct reflection of that division of opinion, scholastic authors introduced a new question into their commentaries on Aristotle's *De caelo*, one that was unknown to the Middle Ages. They asked whether the heavens are solid or fluid. Although some sided with Tycho and assumed fluid heavens without spheres, whereas others defended the existence of hard spheres and still others argued for a combination of hard and soft, almost all were agreed that a solid sphere signified a hard sphere.[68] Giovanni Baptista Riccioli underscores the pow-

66. On this, see Thoren, 1979, 53–67.
67. "Ubi per Cometarum motus prius ostensum et liquido comprobatum fuerit, ipsam Coeli machinam non esse durum et impervium corpus varijs orbibus realibus confertum, ut hactenus a plerisque creditum est, sed liquidissimum et simplicissimum, circuitibusque Planetarum liberis, et absque ullarum realium Sphaerarum opera aut circumvectione." Tycho Brahe, *De mundi aetherei*, 1922, 4:59. On page 222 of the same work, Tycho says much the same thing, emphasizing that "very many modern philosophers . . . distinguished the heavens into various orbs made of hard and impervious matter" (et recentiores etiam Philosophos quamplurimos, qui Coelum ex dura et impervia materia Orbibus varijs distinctum); for a similar statement, see also page 223. Tycho's *De mundi aetherei* was reprinted in 1603 and 1610 and was thus widely known. For these and other references to Tycho, I am indebted to my late colleague, Professor Victor Thoren.
68. For example, see Arriaga, *De caelo*, disp. 1, sec. 3, 1632, 499, col. 1–504, col. 1 ("Whether the heavens are incorruptible and solid"); Serbellonus [*De caelo*, disp. 1, qu. 2, art. 4], 1663, 2:25, col. 1–28, col. 1 ("Whether the celestial bodies are solid or liquid"); Thomas Compton-Carleton, *De coelo*, disp. 1, sec. 3, 1649, 398, col. 2–399, col. 2 ("Whether the heavens are solid or fluid"); and Cornaeus, *De caelo*, disp. 2, qu. 2, dub. 3, 1657, 494–500 ("Whether the heavens are hard and solid").

erful association of solidity with hardness when he explains that although the term *soliditas* means three-dimensional, it also "has associated with it hardness [as] opposed to softness, as we say that marble is solid, and metal, as long as it does not liquefy, and even ice before it melts."[69] Here was a significant departure from the Middle Ages, when, as we saw, the term *solidum* could signify either hardness or softness.

But when did a solid orb come to imply a hard and rigid orb? From what has been said already, the indissoluble nexus between solid and hard probably did not occur during the Middle Ages. Although Themon Judaeus, Henry of Hesse, and Hartmann Schedel made explicit commitments to hard orbs, and a few others implied the hardness of the celestial orbs, most offered no opinion. Natural philosophers who commented on Aristotle's *De caelo* and theologians who commented on Peter Lombard's *Sentences* either did not think about the problem at all or, if they did, felt no compulsion to discuss it. Astronomers were no different. Thus ibn al-Haytham (Alhazen) and Georg Peurbach, whose treatises played instrumental roles in disseminating knowledge about solid spheres during the late Middle Ages and the Renaissance, undoubtedly spoke about real, material orbs, but neither gives any indication whether those orbs are rigid or fluid.[70] Although it is likely that Copernicus believed in the existence of solid spheres, it does not follow that he, "like every other astronomer of his time, envisioned planetary models to be composed of non-intersecting, rigid spheres."[71] Nothing that Copernicus said or implied in the *De revolutionibus* enables us to decide with any confidence whether he assumed hard or fluid spheres. Copernicus fits the pattern of the Middle Ages, when explicit opinions about the rigidity or fluidity of the orbs were rarely presented.

Approximately one-half century before Copernicus published the *De revolutionibus*, the same pattern is revealed in a widely read sixteenth-century encyclopedia. In the *Margarita philosophica*, Gregor Reisch describes a sphere as "round and solid" and characterizes it as "a solid body contained by a single surface."[72] In a rather strange depiction of the empyrean sphere,

69. "Controversia igitur est de soliditate presse sumpta quae praeter trinam dimensionem habet adiunctam duritiem mollitiei oppositam, quomodo solida dicimus esse marmora et metalla quamdiu non liquescunt et ipsa glacies antequam dissolvatur." Riccioli, *Almagestum novum*, bk. 9, sec. 1, ch. 7, 1651, 238, col. 2.

70. Swerdlow, 1976, who assumes that solid orbs or spheres were rigid, implies that in both the Latin translation of ibn al-Haytham's "On the Shape of the Universe," which Swerdlow claims introduced "solid sphere planetary models into medieval European astronomy" (117), and in Peurbach's *Theoricae novae planetarum*, the solid spheres were assumed hard and rigid (see 109–110, 116–117). I have examined both treatises and encountered only the usual silence on the issue of hardness or softness. For ibn al-Haytham's treatise, see ibn al-Haytham, 1942, 285–312; for Peurbach, I have used the edition included in d'Ailly's *Spherae tractatus* (Venice, 1531). As we saw earlier, Campanus of Novara also ignored the issue.
71. Swerdlow, 1976, 108–109. In the seventeenth century, Cornaeus (*De caelo*, disp. 2, qu. 2, dub. 3, 1657, 496) included Copernicus among those who believed in fluid heavens.
72. "Sphera est tale rotundum et solidum quod ab arcu semicirculi circumducto describitur.

Reisch says that it may also be called "crystalline because it is a nonconcave, transparent, and lucid solid body [*corpus solidum*] in which the subtle bodies of the blessed move without resistance or penetration of dimensions."[73] Despite an attribution of transparency and lucidity to this crystalline solid body, Reisch neglects to inform his readers whether it is hard or soft.

The powerful bond that was forged between hardness and solidity in the seventeenth century was apparently not achieved by the time of Copernicus. Because Tycho claimed that many of his contemporaries believed in hard spheres, it is not unreasonable to assume that the firm connection between solidity and hardness became explicit and commonplace during the period between the emergence of Copernicus's *De revolutionibus* and Tycho's publication of his cometary researches in the late 1580s. How this occurred and who might have been instrumental in its development are unknown, and I shall not pursue this further except to suggest that Clavius does not seem to have been the disseminator of hard and solid celestial spheres, as has been argued.[74]

Indeed, despite Tycho's assertion about his contemporaries, another hypothesis may be equally plausible: that the firm connection occurred after Tycho's publication of 1588 and because of it, not between the time of Copernicus and Tycho. By showing that the paths of comets and the intersection of the orbits of the Sun and Mars made hard orbs an untenable assumption and virtually compelled acceptance of an orbless, fluid heavens, Tycho himself forced the issue. To Tycho and his followers, the denial of fluidity left only one meaningful option: hard orbs. In the debate that followed, the word "solid" came somehow to be inextricably linked with hardness, which, in turn, came to be the major qualifier of the term "orb." Whereas once the word "solid" signified any fluid or hard body that was of a continuous nature – that is, without vacua – later the idea of solid came to be opposed to permeability and penetrability and was instead linked to hardness. The divorce of solidity from fluidity and its nexus with hardness was not a necessary development but may have occurred as a consequence of Tycho's contributions and influence. But the connection may not have taken hold until the seventeenth century. In any event, Clavius seems to have ignored it in the numerous late sixteenth- and early seventeenth-century editions of his commentary on Sacrobosco's treatise. In these edi-

Vel sphera est corpus solidum unica superficie contentum." Reisch, *Margarita philosophica*, bk. 7, ch. 3, 1517, 242; see pp. vi–vii for a brief summary of Reisch's life.

73. "Dicitur etiam . . . crystallinum eo quod sit corpus solidum non concavum transparens et lucidum in quo tamen corpora beatorum subtilia sine resistentia aut dimensionum penetratione ambulabunt." Ibid., ch. 8, 247.

74. Donahue, 1975, 263. Although Clavius assumed material and physical orbs and even says that "the world is indeed a solid sphere" (Mundus siquidem est sphaera solida; Clavius [*Sphere*, ch. 1], *Opera*, 1611, 3:9), to my knowledge nowhere in his *Commentary on the Sphere of Sacrobosco* does he explicitly characterize them as hard nor even imply it. Indeed, like most of his predecessors, he does not raise the issue. By "solid sphere" Clavius meant only that the world is a plenum.

tions, he proceeded like most of his medieval predecessors: he defended the existence of solid orbs but found no reason to indicate whether they were hard or soft.

Whether orbs were hard or fluid, however, was much less important than whether planet-bearing orbs existed at all as opposed to a celestial region devoid of orbs but filled entirely with fluid substances. The new celestial discoveries – especially comets and planetary satellites – were not only destined to affect which alternative was chosen but even prompted solutions that incorporated both.

VIII. The scholastic reaction to Tycho Brahe: hard orbs or fluid heavens, or both, in the late sixteenth and the seventeenth century?

In Chapter 10, we considered at some length the nature of celestial incorruptibility. Tycho Brahe's investigations of the new star of 1572 and the comet of 1577 directly challenged that long-held, powerful Aristotelian concept. Not only did Tycho repudiate solid and hard orbs, but he replaced them with a fluid material through which the planets moved. Dramatic celestial changes of the kind represented by new stars and comets convinced Tycho and many others that not only were the heavens composed of a fluid, rather than hard, substance but that changes did indeed occur in the heavens, which could no longer be conceived as incorruptible.

Although it was Tycho who first gave scientific arguments for rejecting hard orbs in favor of fluid heavens, the fluid-heavens hypothesis had been popular in the Middle Ages prior to the introduction of the Aristotelian ether and the gradual emergence of hard orbs. It reemerged in the sixteenth century at the very time when the hard-orb theory was at its height. Between the years 1570 and 1572, Robert Bellarmine (1542–1621) emphatically rejected hard orbs – indeed orbs of any kind[75] – and insisted that celestial bodies moved freely through a fluid medium "like birds in the air and fish in the sea."[76]

Another early opponent of hard orbs and advocate of fluid heavens was

75. In his Louvain Lectures of between 1570 and 1572 (unpublished until 1984), Bellarmine declared that "such complex and extraordinary structures as epicycles and eccentrics are dreamed up so that that even the astrologers are reticient to speak about them." Bellarmine, 1984, 22 (English), 23 (Latin). Although Bellarmine used the term *astrologi*, he clearly means astronomers.

76. Bellarmine explains: "Si assere velimus coelum sydereum non esse nisi unum, et illud igneum, vel aereum: quod saepius conformius scripturis esse diximus: necessario iam dicere debemus, stellas non moveri ad motum coeli, sed motu proprio sicut aves per aerem, et pisces per aquam." Ibid., 19 and 38, n. 88; also quoted in Baldini, 1984, 301. See also Lerner, 1989, 268. The brief translation is mine. Defenders of orbs, whether hard or soft, found it difficult to believe that celestial bodies could be self-moved "like birds in the air and fish in the sea." See Chapter 13, note 16, where other users of one or both of these metaphors are also mentioned.

Francesco Patrizi, who, as early as 1591, insisted that the heavens were neither hard nor solid. However, it seems unlikely that Patrizi was influenced by Tycho, since he relied not on astronomical arguments but on metaphysical and general cosmological claims. For Patrizi, the basic building blocks of our universe are things like light, heat, and space, none of which are hard or solid.[77] But if nothing in the heavens is hard or solid, then the heavens could not support stars or planets that are fixed in it like knots.[78] Patrizi concludes that "all Philosophers and Astronomers err who teach that the planets [*stellae*] are fixed in the heavens like knots in tables."[79]

Belief in the existence of solid spheres was nevertheless common in the late sixteenth and the early seventeenth century. By 1630, however, if not earlier, it seems to have lost its dominance, lingering on as a minority opinion.[80] In his *Cursus philosophicus* of 1632, Roderigo de Arriaga explained that just a few years earlier celestial incorruptibility and hard solidity "were absolutely beyond controversy."[81] By the time his book appeared, fluid and corruptible heavens had largely replaced the two previously entrenched concepts and had done so because "of the diligent observations of certain mathematicians and astronomers, which [observations] were discovered with the aid of new and excellent instruments, especially the telescope. Thus did some [individuals] begin to wholly invert the structure of the heavens."[82] In one important sense Arriaga was typical of seventeenth-century scholastic authors. Most were well aware of the arguments by Tycho, Galileo, and Kepler on the nature of the heavens. As the century moved on, a scholastic literature developed which incorporated the arguments of these great figures of the new science and cosmology. Scholastic authors could thus learn indirectly of these arguments from members of their own group or directly from the works of these three astronomers. In one or both of these ways, most became aware of the crucial observations and arguments. As they did, they gradually abandoned hard orbs for fluid heavens, so that by 1672,

77. "Si vero ex nostris fundamentis, e spacio, ex lumine, ex calore, ex fluore, tamquam ex communibus rerum omnium elementis, celum dixerimus esse compositum verum equidem dicemus, sed non tanta posse esse duritie ac soliditate ut tam vehementi rotatu, non dispergatur. Si vero coelum, quod Chaldaei docuerunt solum est lumen, lumini nulla est soliditas, nulla durities." Patrizi, *Pancosmia*, 1591, 89, col. 2.

78. "Si nihil in coelo est durum, si nulla est in eo soliditas, nullam profecto fixionem, vel nodorum, vel stellarum potuit suscipere." Ibid.

79. "Toto ergo errarunt coelo et Philosophi et Astronomi omnes, qui stellas coelo fixas, uti nodos tabulis esse docuerunt." Ibid.

80. Donahue, 1975, 273, declares that "by the end of the 1620s the debate over the fluidity of the heavens was very nearly concluded." The estimate seems reasonable. Thoren declares (1990, 254) that "in the second half of the sixteenth century at least, intellectuals in general and Tycho Brahe in particular believed that something real existed in the heavens to carry the planets through their appointed rounds." By 1587, however, Tycho came to reject the existence of hard orbs. Thoren, ibid., 258.

81. "Utrumque ante aliquot annos omnino extra controversiam fuerat." Arriaga, *De caelo*, disp. 1, sec. 3, 1632, 499, col. 1.

82. "Propter quorumdam mathematicorum et astronomorum diligentes observationes quas, novis exquisitisque instrumentis adiuti, invenerunt, et praecipue tubi optici subsidio, caelorum structura penitus a nonnullis inverti coepit." Ibid.

when George de Rhodes published his discussion, he could say of the planetary heaven, "no one now denies the fluidity of the heaven of the planets."[83]

Because hard orbs had been regularly linked with incorruptibility, it seemed natural to associate fluid heavens with corruptibility. Some scholastics, however, found these rigid pairings unwarranted, perhaps because some wished to assert fluidity and nonetheless retain the concept of celestial incorruptibility. "Crystal, stone, wood, etc. are solid bodies," observes Arriaga, "but are not incorruptible, and some substance might be easily permeable and yet not be corrupted."[84] He was further convinced that "some experiences can be adduced for proving the fluidity of the heavens which do not thereby prove their corruptibility. Contrarily, other experiences can be adduced to prove that the heavens are corruptible which cannot show that they are fluid. Thus it is necessary to distinguish between them."[85]

Among late sixteenth- and seventeenth-century scholastic natural philosophers whose works play a significant role in this study, opinion was divided on the issue of fluidity or solidity. Those who defended the existence of solid orbs were the Coimbra Jesuits, Bartholomew Amicus, and Thomas Compton-Carleton, while the defenders of fluid heavens included Pedro Hurtado de Mendoza, Roderigo de Arriaga, Francisco de Oviedo, Giovanni Baptista Riccioli, Franciscus Bonae Spei, Melchior Cornaeus, Sigismundus Serbellonus, and George de Rhodes. Of the eight defenders of fluidity, seven published after 1632, thus strengthening the view that by 1630 most scholastics had abandoned hard and solid orbs in favor of fluid heavens.[86] Of the authors listed as supporters of fluidity, some held a third opinion, which envisioned heavens that were partly hard and partly fluid. In this group we may place Hurtado de Mendoza, Aversa, Riccioli, and Serbellonus.[87] They, and others who shared this interpretation, may be appro-

83. "Prima ergo pars de fluiditate coeli planetarum a nemine nunc negatur." De Rhodes [*De coelo*, bk. 2, disp. 2, qu. 1, sec. 2, pt. 2], 1671, 280, col. 1.

84. The full statement reads: "Primo suppono non esse idem corpus esse liquidum et esse corruptibile; neque e contrario idem esse corpus solidum et incorruptibile: nam crystallus, lapis, lignum, etc., sunt corpora solida et non sunt incorruptibilia; et potest esse aliqua substantia facile permeabilis, licet non possit corrumpi." Arriaga, *De caelo*, disp. 1, sec. 3, subsec. 1, 1632, 499, col. 1.

85. "Hinc suppono secundo aliquas experientias adduci ad probandum caelos esse fluidos quae non propterea quidquam probant de eius corruptibilitate; alias vero e contrario ad probandum eos esse corruptibiles quae non ostendunt illos esse fluidos, unde eas oportet valde inter se distinguere." Ibid. Amicus, *De caelo*, tract. 5, qu. 5, art. 1, 1626, 272, col. 1 also agreed with Arriaga's position.

86. The seventeenth century scholastic authors were not chosen for this study by virtue of the opinions they held but largely because they seemed to include relevant discussions and were reasonably well distributed through the century.

87. Riccioli, for example, assumed it more probable that the heaven of the fixed stars was solid and the heaven of the planets fluid. Thus we read, in the enunciation of his final and sole conclusion, "it is more probable, although hardly evident mathematically or physically, that the heaven of the fixed stars is solid and that of the planets fluid" (Probabilius multo est, licet nondum mathematice aut physice evidens, caelum fixarum solidum

priately listed with the fluid theorists, because the assumption of fluidity in the heavens, whether for all or part of it, marks a strong departure from what was taken as the major opposition theory in the late sixteenth century, namely heavens conceived as totally solid and hard.

What did terms like "fluid" and "liquid" mean to opponents of hard orbs? In responding to the question "Whether the heavens are fluid," Roderigo de Arriaga explained that the fluid he had in mind need not be a "watery liquid" (*liquor aqueus*), for "it suffices if they [the heavens] are easily permeable, much like our air, which is, nevertheless, not called absolutely fluid."[88] Thus the heavens could range from a liquid to a gas and still be categorized as a fluid. The meaning of fluidity was apparently extended in this manner to avoid the charge that watery, liquid heavens would fall down upon us in the form of rain. A vaporized fluid, analogous to air, was more readily conceived to remain in its celestial location high above us.[89]

1. *Scholastic arguments for fluid heavens*

Arguments for or against fluid heavens came from a variety of sources. Some were drawn from Scripture and the Church Fathers and were largely appeals to authority; others were derived from Aristotelian physics and cosmology and the scholastic additions thereto. But the most significant and most challenging drew on, or were responses to, the new discoveries, or the "new phenomena" (*nova phaenomena*) as they were sometimes called,[90] associated most prominently with the names of Brahe and Galileo. We shall have occasion to consider all of these types.

a. Scripture

Just as the passage from Job 37.18 served as the most important biblical support for hard orbs, so did Isaiah 51.6 serve to uphold fluid heavens with the words "quia caeli sicut fumus liquescent" (because the heavens appear as smoke).[91] Indeed, a number of scriptural quotations were arrayed on each side of the controversy and largely offset each other. One and the same

esse, planetarum autem fluidum). Riccioli, *Almagestum novum*, pars post., bk. 9, sec. 1, ch. 7, 1651, 244, col. 1.

88. "Tertio suppono, cum quaerimus an caeli sint fluidi non quaeri a nobis an sint quasi quidam liquor aqueus, qui facile labitur; sufficit enim si sint facile permeabiles ad modum quo est noster aer, qui tamen non vocatur absolute liquor." Arriaga, *De caelo*, disp. 1, sec. 3, subsec. 1, 1632, 499, col. 1.
89. Illuminatus Oddus argues this way (*De coelo*, disp. 1, dub. 12, 1672, 35, col. 1).
90. The expression *nova phaenomena* probably meant "new celestial appearances." For its use, see Amicus, *De caelo*, tract. 5, qu. 5, art. 2, 1626, 273, col. 1 and de Rhodes, *De coelo*, bk. 2, disp. 2, qu. 1, sec. 2, pt. 2, 1671, 280, col. 1.
91. The Latin is from the Vulgate.

author might even present scriptural passages in support of each side.[92] Nevertheless, such passages were invoked because scriptural authority was still deemed important. Moreover, because each side could muster biblical support, it was not unusual for scholastic authors to show that the biblical passages cited by their opponents were irrelevant or inappropriate. Thus although Amicus ([*De caelo*, tract. 5, qu. 5, art. 3], 1626, 278, col. 1) cites the passage from Job in his argument in favor of solid orbs, which he believed the true opinion, he also cites an argument, in the section where he presents the case for fluid orbs, to show that the same passage was irrelevant to the case for solid, hard spheres.[93] In this argument, we are told that the Job passage does not really attribute hardness to the heavens and that the words are not those of God or Job but of Elihu, whose utterances are not accepted as true.[94] Indeed, Melchior Cornaeus rejected the relevance of the passage because the words were those of Elihu, an unlearned man, whom God subsequently denounces for uttering ignorant opinions (in Job 38.2).[95] George de Rhodes went much further and simply denied the relevance of the argument from Job, as well as three other biblical passages. Scriptural texts do not signify that the firmament is a hard, solid body, de Rhodes insists, but only that it is "a body that has a constant and perpetual state" ([*De caelo*, bk. 2, disp. 2, qu. 1, sec. 2], 1672, 280, col. 2). De Rhodes even denied the relevance for either side of Isaiah 51.6. "When it is said," he argued, "that the heavens [literally] pass away [*liquescent*]," this only signifies that they will be changed into a better state.[96]

Because they could be assembled for either side, biblical passages could not play a crucial role in the debate over hardness and softness, as they did in the Copernican controversy, where all relevant scriptural passages upheld one side of the dispute, namely the traditional interpretation of the Sun revolving around a stationary earth. Moreover, the Church never intervened in the issue of the hardness or fluidity of the heavens. Although scriptural

92. For Riccioli's scriptural citations in behalf of hardness, see his *Almagestum novum*, pars post., bk. 9, sec. 1, ch. 7, 1651, 240, col. 2–242, col. 1; for his citations in favor of fluidity, see ibid., 242, col. 1–244, col. 1; for Amicus's citations of biblical passages in favor of fluidity, see his *De caelo*, tract. 5, qu. 5, art. 2, 1626, 272, col. 2–275, col. 2 and in favor of solid hardness see 275, col. 2–278, col. 1. Arriaga cites the Job and Isaiah passages in *De caelo*, disp. 1, sec. 3, subsec. 4, 1632, 503, col. 1, par. 39.
93. However, this was not his real opinion, as will be seen shortly.
94. Amicus, *De caelo*, tract. 5, qu. 5, art. 2, 273, col. 1. De Rhodes, *De coelo*, 1672, bk. 2, disp. 2, qu. 1, sec. 2, pt. 2, 280, col. 1, argues that these are only the words of Job's friend and also asserts that the words about the metallic solidity of the heavens are meant to apply to an immense extent of air.
95. Cornaeus, *De coelo*, disp. 2, qu. 2, dub. 3, 1657, 499. Franciscus Bonae Spei, *De coelo*, comment. 3, disp. 3, dub. 7, 1652, 14, col. 1, had, a few years earlier, used the same argument and the same appeal to Job 38.2. Without specifically citing Job 38.2, Serbellonus, *De caelo*, disp. 1, qu. 2, art. 4, 1663, 2:25, col. 2, repeated the same argument as Cornaeus and Bonae Spei.
96. "Cum dicitur quod coeli liquescent, significatur tantum quod mutabuntur in meliorem statum." De Rhodes, *De coelo*, bk. 2, disp. 2, qu. 1, sec. 2, pt. 2, 1671, 280, col. 2. The relevance of this passage rested wholly on the term "liquescent," which, in the context, did not even mean "liquify" but rather signified "pass away" or "melt away."

passages could be cited for each side of the controversy, thus effectively eliminating them as a critical factor in the ultimate outcome, individual authors could still be powerfully persuaded by their personal interpretations of relevant scriptural texts, as we shall see when we examine Amicus's defense of hard heavens.

b. The new discoveries

The most significant arguments in favor of fluid heavens were based upon the observational achievements of Tycho Brahe and Galileo Galilei, the former relying on the naked eye, the latter on the recently invented telescope. The cumulative impact of their obervations and the inferences drawn from them transformed cosmology. Comets that Aristotle had characterized as sublunar phenomena were now placed by Tycho in the vicinity of Mars in the celestial region. Tycho had also identified the new star of 1572 as a genuinely novel celestial phenomenon and thus challenged the traditional opinion of celestial incorruptibility. With his telescope, Galileo added to these the satellites of Jupiter[97] and a picture of celestial bodies that had irregularities, especially the Moon, whose mountains and valleys made it akin to the earth.[98] Galileo proclaimed "the earth very noble and admirable precisely because of the diverse alterations, changes, generations, etc. that occur in it incessantly, . . . and I say the same of the Moon, of Jupiter, and of all other world globes" (*Two Chief World Systems* [Drake], 1962, 58–59). Also significant for the debate about hard or fluid orbs were Galileo's discoveries of the phases of Venus and sunspots.[99]

i. Comets. Of the new discoveries, Tycho Brahe's determination of the celestial nature of comets was perhaps the most dramatic event in turning scholastic opinion from hard orbs to fluid heavens. To appreciate the momentous challenge that Tycho's achievements posed to Aristotelian cometary theory, and therefore to Aristotelian cosmology, it is necessary to describe briefly Aristotle's theory of comet formation as expressed in his *Meteorology*. At the beginning of the latter treatise, Aristotle declares that the region of the world with which meteorology is concerned is "nearest to the motion of the stars" (1.1.338b.20–22 [Webster], 1984), by which he meant the region of air and fire just below the Moon. The kinds of phenomena that occur in this region include comets, meteors, and the Milky Way. Comets are thus not celestial phenomena but occur in the upper

97. Galileo also mistakenly identified the rings of Saturn as satellites. See Drake's article in *Dictionary of Scientific Biography*, 1970–1980, 5:241, col. 2. As a consequence, scholastics occasionally mentioned Saturn's "satellites."

98. Galileo described the lunar irregularities and the satellites of Jupiter in *The Starry Messenger* (*Sidereus nuncius*) of 1610. For translations, see Galileo [Drake], 1957, 21–58, and Galileo [Van Helden], 1989.

99. See Galileo's *Letters on Sunspots* (1613) in Galileo [Drake], 1957, 87–144. Galileo discovered the phases of Venus in 1610, after he had written the *Starry Messenger*.

atmosphere between earth and Moon. In the fourth chapter, Aristotle says (ibid., 1.4.341b.7–25) the upper region is comprised of two kinds of exhalations:

One kind is rather of the nature of vapour, the other of the nature of a windy exhalation. That which rises from the moisture contained in the earth and on its surface is vapour, while that rising from the earth itself, which is dry, is like smoke. Of these, the windy exhalation, being warm, rises above the moister vapour, which is heavy and sinks below the other. Hence the world surrounding the earth is ordered as follows. First below the circular motion comes the warm and dry element, which we call fire, for there is no word fully adequate to every state of the smoky evaporation; but we must use this terminology since this element is the most inflammable of all bodies. Below this comes air. We must think of what we just called fire as being spread round the terrestrial sphere on the outside like a kind of fuel, so that a little motion often makes it burst into flame just as smoke does; for flame is the ebullition of a dry exhalation. So whenever the circular motion stirs this stuff up in any way, it catches fire at the point at which it is most inflammable. The result differs according to the disposition and quantity of fuel.

From this physical arrangement of the upper atmosphere, Aristotle explains (ibid., 1.4.342a.16–30) the formation of various meteoric occurrences, including comets:

When the phenomenon is formed in the upper region it is due to the combustion of the exhalation. When it takes place at a lower level it is due to the ejection of the exhalation by the condensing and cooling of the moister exhalation; for this latter as it condenses and inclines downward contracts, and thrusts out the hot element and causes it to be thrown downwards. . . . So the material cause of all these phenomena is the exhalation, the efficient cause sometimes the upper motion, sometimes the contraction and condensation of the air. Further all these things happen below the moon.

Aristotle assumed that the dry and warm, or fiery, exhalation and a great part of the air below it is carried circularly around the earth by virtue of the circular celestial revolution. In the process of being carried around, and under the right conditions, parts of the dry and warm exhalation or the upper air might ignite. "We may say, then," Aristotle continues (ibid. 1.7.344a.15–21),

that a comet is formed when the upper motion introduces into a condensation of this kind a fiery principle not of such excessive strength as to burn up much of the material quickly, nor so weak as soon to be extinguished, but stronger and capable of burning up much material, and when exhalation of the right consistency rises from below and meets it. The kind of comet varies according to the shape which the exhalation happens to take.[100]

During the Middle Ages, most theories about comet formation remained close to Aristotle's account. If they diverged, it was not with respect to the sublunar location of comets. By placing the comet of 1577 in the celestial region and doing so on the basis of careful observation of the comet's parallax, Tycho changed forever the debate about comets, as is readily apparent by a glance at Riccioli's lengthy section on comets in his *Almagestum novum* of 1651,[101] where he summarized virtually all the relevant arguments with respect to the formation, substance, location, and distance of comets.[102] Opinions on the location of comets ranged from below the Moon, to above the Moon, and to some comets below and some above. Theories about the matter from which comets were formed varied from the sublunar elements in various manifestations to celestial matter either by means of condensation, by the alteration of parts of the heavens, and even by matter flowing from the Sun and planets themselves.[103] Toward the end of what was surely one of the lengthiest and most detailed studies of comets in the seventeenth century, Riccioli arrived at certain cautious conclusions that conceded only the probability, but not the certainty, of supralunar comets. Because he was not yet convinced that there had been any absolute demonstration that any comets had occurred above the Moon,[104] Riccioli concluded in favor of the probability that some comets occurred above the Moon and some below. History, he acknowledged, could furnish no information to help determine cometary locations.[105] It was not only that comets moved across the heavens in ways that made the existence of hard orbs difficult to defend – comets were thus frequently invoked in support of fluid heavens, as Melchior Cornaeus argued[106] – but the typical theory of comet formation also made solid, hard orbs seem impossible. Thus Sigismundus Serbellonus, who agreed with Aristotle that sublunar exhalations could produce comets,[107] was con-

100. See also Jervis, 1985, 11–13.
101. Riccioli devoted the eighth book to comets and new stars.
102. In *Almagestum novum*, pars post., bk. 8, sec. 1, ch. 23, 1651, 117, col. 2–120, col. 1, Riccioli cites the opinions of others on the place, parallax, and distances of comets from the earth.
103. Ibid., ch. 13, 57, col. 2–58, col. 2.
104. Under the heading "Conclusiones de distantia et loco cometarum," the second conclusion reads: "Nullus adhuc cometarum demonstratus est absolute fuisse supra lunam, sed ex hypothesi tantum probabili quidem, sed tamen incerta." Ibid., ch. 23, 119, col. 1.
105. Fourth conclusion: "Probabile est aliquos cometas fuisse supra lunam, aliquos vero infra, etiam ex illis de quorum loco ex nuda historia nihil constat." Ibid., col. 2.
106. He asserts that "the comet, which we saw in 1618, was, according to the common opinion of astronomers, in the heaven itself. Therefore the heaven is not hard, but permeable and fluid, like air" (Cometes ille quem anno 1618 vidimus communi astronomorum consensu intra ipsum coelum fuit. Ergo coelum non est durum, sed permeabile et liquidum ad instar aeris). Cornaeus, *De coelo*, disp. 2, qu. 2, dub. 3, 1657, 499.
107. In opposition to Aristotle, however, Serbellonus also believed that comets could be celestial and that they could be produced by effluences given off by planets. In short, they could also be produced by celestial matter.

vinced that comets demonstrate fluid heavens. Not only are comets visible below the Moon, but they are also seen above the Sun, Mars, and Saturn and have existed as far away as the region immediately below the firmament of the fixed stars. Because comets develop from exhalations given off by the earth, it follows that "if the heavens were solid, they [comets] could not be seen above any heaven, but all would be below the Moon, which is contrary to the common observation of the astronomers. Therefore the heavens are fluid, so that they could be penetrated by these exhalations."[108]

Riccioli used the same argument to deny hardness to the heaven of the Moon, although he assumed that the sphere of fixed stars was hard because it seemed the only way to preserve the distances between the stars and to avoid the assignation of a separate mover for each star.[109]

In one of ten arguments in support of fluidity, Riccioli declares that the oblique and free trajectories of comets above the Moon, which astronomers have demonstrated,[110] would be incompatible with solid eccentric, concentric, and epicyclic heavens. He implies that in a world of solid orbs, epicycles would also be required to carry comets. But since comets appear only occasionally, where would the matter come from to form a special epicycle for the occasion, and from whence would a place appear to accommodate it? Because no such special adjustments seemed possible or plausible, Riccioli concludes that the free trajectory of comets demonstrates the fluidity of the heavens, as Kepler, drawing on Tycho, argued in the fourth book of his *Epitome of Astronomy*.[111]

108. "Cum igitur cometae oriantur ab exhalationibus, aut a terra elevatis aut a planetis profluentibus, si coeli solidi essent, non possent videri supra coelum aliquod, sed omnes essent infra lunam, quod est contra communem astrologorum observationem. Fluidi ergo sunt coeli, ut permeari possint ab ipsis exhalationibus." Serbellonus, *De caelo*, disp. 1, qu. 2, art. 4, 1663, 2:25, col. 2. In my summary earlier, I omitted Serbellonus's assertion about the role of planetary exhalations in comet formation. Not only was this another incompatible addition to Aristotle's comet theory (since Aristotle denied the existence of comets in the celestial region, no planet could have given off exhalations or participated in comet formation), but it makes what Serbellonus says inconsistent. For despite the inability of earthly exhalations to penetrate beyond the lunar orb, comets could form from exhalations given off by planets themselves, even though their movements in a heaven of hard orbs would pose serious problems. The formation of comets from matter given off by planets was a serious new theory in the seventeenth century (see Riccioli, *Almagestum novum*, pars post., bk. 8, ch. 13, 1651, 57, col. 2–58, col. 2).

109. Ibid., bk. 9, sec. 1, ch. 7, 244, col. 1.

110. Although Riccioli believed in fluid heavens, the arguments presented here are drawn from a variety of sources. They are not necessarily his own. Indeed, a few paragraphs earlier we saw that he thought it only probable, *and not demonstrated*, that comets occur above the Moon.

111. My interpretation is based on the following passage: "Quartum argumentum sumitur a multiplici ac vago motu, seu libera et obliqua traiectione cometarum illorum, quos Tycho et alii censentur demonstrasse genitos et motos supra lunam fuisse. . . . Iam si celi essent solidi, tota eorum moles cessisset in eccentricos, concentricos, et epicyclos planetarum, nec superesset materia aut locus pro epicyclis cometarum. Idcirco ex traiectionibus cometarum fluiditatem celi demonstratum a Tychone putarunt Keplerus in *Epitome Astronomie*, lib. 4, pag. 422." Riccioli, *Almagestum novum*, pars post., bk. 9, sec. 1, ch. 7, 1651, 242, col. 2–243, col. 1.

ii. New stars. We saw earlier how scholastics of the sixteenth and seventeenth centuries coped with the problem of new stars in their efforts to retain or abandon the traditional concept of celestial incorruptibility.[112] Few linked new stars to the problem of hard or fluid heavens, perhaps because many scholastics either denied that new stars were real celestial phenomena or, if they recognized them as genuine celestial occurrences, explained them as some configuration of already existing bodies. But George de Rhodes, who assumed that new stars were wholly new phenomena, found it easier to imagine such events occurring in fluid heavens. Because astronomers judged the distance of new stars to be the same as those of the fixed stars, he found it difficult to envision how, if the orb of the fixed stars were hard, a new star could suddenly appear.[113]

iii. New discoveries concerning the Moon, Mercury, Venus, Mars, and the satellites of Jupiter and Saturn. Other discoveries were, however, more directly relevant. Melchior Cornaeus reveals the manner in which one or more of these discoveries could be applied to the debate on the hardness or fluidity of celestial orbs. In the fourteenth century, as we saw earlier (Ch. 13, Sec. III.7), Jean Buridan and Albert of Saxony took opposite positions on whether or not the Moon had an epicycle. Albert assumed not only that the Moon was carried around by an epicycle but that the Moon rotated around its own center in the same period as the epicycle but in a contrary direction. Only in this way, he insisted, would we always see the same face of the Moon.

Cornaeus rejected this argument and the existence of a lunar epicycle, the interior of which would have to be either void or filled with matter. If the Moon rotated around its own center as it was being carried by its epicycle, the huge lunar mountains, which are greater and higher than those on earth and which make the Moon's surface rough and uneven, would necessarily sweep from one place to the next. As the Moon rotated, each irregular lunar prominence would be carried from one place to another and either move into a void or leave one behind, which is impossible;[114] or, if the Moon's epicycle is filled with matter, the peaks of the lunar mountains would have to penetrate that matter, that is, occupy the same space with it, which is also impossible. Insofar as his argument applies to void space, Cornaeus has here drawn on Aristotle's demonstration for a spherically shaped world, which was popularized by Sacrobosco in the Middle Ages. A world that was not spherically shaped but had protruding angles that revolved in a circle would, as Aristotle put it, "never occupy the same

112. See Chapter 10, Section III.3.a, b; also Chapter 12, Section III.1.
113. "Secundum est de novis illis stellis, quas nuper dixi apparisse saepius in coelo eandem habere distantiam a terra quam habent reliquae stellae firmamenti, quod etiam cum soliditate coelorum stare non potest." De Rhodes, *De coelo*, bk. 2, disp. 2, qu. 1, sec. 2, pt. 2, 1671, 280, cols. 1–2.
114. In a briefer paragraph, Riccioli seems to say the same thing; Riccioli, *Almagestum novum*, pars post., bk. 9, sec. 1, ch. 7, 1651, 243, col. 2, par. 21.

space, but owing to the change in position of the corners there will at one time be no body where there was body before, and there will be body again where now there is none."[115] Because a lunar epicycle, whether void or filled with matter, seems unable to account for the rotation of a Moon that has an irregular surface, Cornaeus rejected epicycles and assumed that the Moon moves through a fluid medium.

For scholastic authors like Cornaeus, who were compelled to reject Copernicus's ecclesiastically condemned heliocentric system but who accepted Tycho's geoheliocentric cosmos,[116] hard, solid orbs were virtually impossible. In Tycho's scheme, which departs radically from Aristotelian cosmology, a number of planetary motions are centered on bodies other than the earth. Cornaeus mentions Mercury and Venus, which move around the Sun as center and are therefore sometimes above and sometimes below it, a state of affairs that was based on Tycho's geoheliocentric system and Galileo's discovery of the phases of Venus;[117] the intersection of the orbits of Mars and the Sun, so that Mars is sometimes below the Sun, and sometimes above it;[118] and finally the four satellites of Jupiter are also sometimes above Jupiter and sometimes below it, and sometimes ahead of it and sometimes behind it. And yet all these subsystems also move around the earth. It would be impossible, says Cornaeus, for these celestial bodies to be fixed in solid, hard heavens.[119] On the assumption that these arguments were obvious to his readers, Cornaeus offers no further elaboration.

Years earlier, however, Bartholomew Amicus, who rejected fluid heavens and was a supporter of hard planetary orbs, mentioned (*De caelo*, tract. 5, qu. 5, art. 2, 1626, 273, cols. 1–2) all of the same phenomena and briefly explained why partisans of fluid heavens thought the theory was compatible with that concerning the various celestial subsystems that Tycho and Galileo had identified. The phases of Venus could not occur, he explained, unless the heavens were fluid, for otherwise Venus, in circling the Sun (rather than the earth, in an otherwise geocentric universe) and moving above and below

115. Ironically, Aristotle was arguing for a spherical world, whereas Cornaeus was arguing in behalf of an irregular Moon. For further discussion, see Chapter 6, Section II and note 39.
116. Tycho assumed that the Sun moved around an immobile earth at the center of the world – just as in the Aristotelian–Ptolemaic system – but that all the other planets moved around the Sun as the center of their orbits. For a description of Tycho's world system, see Thoren, 1990, ch. 8, 236–264.
117. The phases of Venus are explicitly mentioned by Amicus, *De caelo*, tract. 5, qu. 5, art. 2, 1626, 273, cols. 1–2.
118. Tycho Brahe made this an integral part of his geoheliocentric world view in opposition to Copernicus's heliocentric system. See Thoren, 1990, 254.
119. Cornaeus, *De coelo*, disp. 2, qu. 2, dub. 3, 1657, 499. Although Cornaeus does not use the term "hard" (*durum*), it is clearly implied. Riccioli, who also mentions the satellites of Saturn, explains that the satellites of Jupiter and Saturn make it unfeasible to admit solid, hard orbs, because the latter would impede the motions of the satellites. Like Cornaeus, Riccioli does not explicitly mention hard orbs – he speaks only of "the solidity of the heavens" (*soliditas celi*) – but they are surely the subject of his discussion. For, as we saw earlier, "solid" and "hard" were inextricably linked in the latter part of the sixteenth and the seventeenth century.

it, would smash through any hard solar heaven or orb in which the Sun was somehow fixed. To reinforce the argument, Amicus mentions Tycho's claims about an intersection of the orbits of Mars and the Sun, a situation that would make it impossible for hard orbs to exist. The satellites of Jupiter and Saturn (these were in fact the rings of Saturn, mistaken for satellites) would similarly smash through any hard orbs associated with these planets.[120]

c. Other arguments

In the preceding section we saw that hard orbs were suspect because of the various subsystems that had emerged from the work of Tycho and Galileo. It was commonly assumed that Mercury and Venus, and even Mars, moved around the Sun as center and that the satellites of Jupiter moved around the latter as center; even Saturn was alleged to be the center of two satellites that perpetually circled it. With so many centers of motion other than the earth, the existence of planetary orbs appeared untenable, because it was assumed that as the circling bodies moved above and below the body at the center, the orb supporting the latter would be smashed, thus disrupting the normal movements of the heavens. Riccioli took the matter a step further by presenting an argument which declared it vain to multiply so many real and solid planetary orbs and their motions. Such a variety of motions created a mutual danger of collision and obstruction between the planets and orbs and also unnecessarily caused the imagination to grow weary in the contemplation of so many allegedly real and solid epicycles, eccentrics, concentrics, and epicyclic eccentrics.

As the climax of the argument, Riccioli invokes the doctrine of simplicity, arguing that it was unlikely that the Divine Wisdom would create a vast and complex machinery of orbs to carry around a single planet like Saturn when he could have done it so easily by the use of a motive intelligence. Hard orbs appear even more incongruous when one realizes that a planet is like a point with respect to the orb that carries it – indeed it bears a smaller ratio to its orb than any drop of water to the ocean. Why construct a vast orb to carry a small planet? The implication is obvious: the Divine Wisdom would have rejected hard orbs and resorted to the simpler expedient of fluid heavens.[121]

120. Scholastics usually presented the best-case scenario for the opposing viewpoint, as Amicus does in this paragraph. De Rhodes, *De coelo*, bk. 2, disp. 2, qu. 1, sec. 2, pt. 2, 1671, 280, col. 1, also cites the four satellites of Jupiter and the two attributed to Saturn as evidence that "the stars [or satellites] are moved in fluid heavens, like birds in air with an angel moving them through a liquid space" (ut in caelo fluido moveantur stellae, ut aves in aere movente illas Angelo per spatium liquidum). For Amicus's arguments in favor of solid, hard orbs, see Section 2.6 of this chapter.

121. Here is the relevant text: "Tertium argumentum. Frustra multiplicantur tot orbes reales ac solidi planetarum et motus eorum. Immo non solum frustra, sed cum periculo mutuae collisionis et impedimenti spectata tanta varietate motuum vel certe absque necessitate cogimur imaginationem defatigare in tot realibus ac solidis epicyclis, eccentricis, con-

One of the main reasons for the introduction of epicyclic orbs in the ancient world was to account for variations of planetary distances from the earth. In the seventeenth century, Cornaeus rejected those orbs and accounted for variations in planetary distance within the context of a fluid medium. "If a planet sometimes approaches the earth and sometimes recedes from it," he declares, "it [the planet] is not fixed in a solid body, but it is necessary that it be in a liquid and permeable body, so that at times it can approach and at times withdraw."[122]

Cornaeus raises an obvious objection against his own position: eccentrics and epicycles can also account for variations in distance. Why, then, reject hard orbs in favor of a fluid medium? In response Cornaeus invokes traditional arguments that had been raised against eccentrics and epicycles since the thirteenth century and were neatly summarized by Cecco d'Ascoli (see Ch. 13, Sec. III). Taking as his illustration the Sun, which has no epicycle but only an eccentric, Cornaeus argues that as the Sun moves from its farthest point from the earth to its closest point, it would have to move through its eccentric.[123] But if it does so, there must either be a vacuum for it to move through or, if matter exists in the eccentric, the Sun would have to penetrate that matter, either by dividing it or occupying the same place with it.[124] Cornaeus then takes up the case of a planet with an epi-

centricis, eccentricis epicyclis. . . . Denique incongruum videtur Divinae Sapientiae, ut propter motuum unius planetae, puta Saturni, qui facillime a se vel ab intelligentia moveri potest, moveatur tanta et tam vasta machina quanta est totum caelum cuiusque planetae, qui comparatus ad suum caelum non est nisi instar puncti et minor est quam sit gutta respectu oceani." Riccioli, *Almagestum novum*, pars post., bk. 9, sec. 1, ch. 7, 1651, 242, col. 2, par. 15.

Simplicity arguments were usually invoked for any advantage they might provide to bolster one or another side of an argument. Because Riccioli sought to present a thorough case for each side, he also felt an obligation, perhaps, to defend against simplicity arguments that were proposed against hard orbs (ibid., pars post., bk. 9, sec. 4, ch. 33, 467, col. 2). Why would God make a world in which huge orbs had to travel at enormous speeds and perhaps generate great resistances to those speeds? He could surely have achieved the same results in a much simpler way. But these were irrelevant problems. If the celestial orbs could endure such speeds, the speeds would pose no serious problems. Moreover, God, or the motive intelligences that move the spheres, would have no difficulty in overcoming such potential resistances, however large they might be. Nor indeed would our senses suffer ill effects from these great speeds, since they are regulated by celestial intelligences.

122. "Si planeta aliquando appropinquat terrae, aliqando vero ab eadem recedit, ergo non est in corpore solido infixus, sed necesse est ut sit in liquido et permeabili, ut aliquando possit accedere et recedere." Cornaeus, *De coelo*, disp. 2, qu. 2, dub. 3, 1657, 497. As evidence for distance variations, Cornaeus invoked the telescope (*tubum opticum*), by means of which one could project an image of the Sun in such a way that the Sun's diameter would vary in size, thus indicating that its distance from the earth varied. Distance variations were also detectable from the observation of eclipses.

123. I am not certain why Cornaeus assumes that the Sun would move through its eccentric rather than being carried by it.

124. "Excentrici et epicycli non expediunt nondum. 1: quia ut sol ex opposito Augis ex *O* veniet in *I* ad Augem, debet se necessario commovere per suum excentricum. Quomodo autem hoc sit vel sine vacuo, vel sine penetratione?" Cornaeus, *De coelo*, disp. 2, qu. 2, dub. 3, 1657, 497, does not provide all the details but my description seems to represent his intent. The letters *O* and *I* are references to a figure on page 494.

cycle.[125] Now it is the epicycle which is assumed to move through its eccentric deferent. Under these circumstances, the inside of the eccentric will be either a plenum or a void. If void, we would have an enormous empty space in the universe, which Cornaeus denies; moreover, motion would occur in this void, which is also denied.

Should the space within the eccentric deferent be a plenum, the matter will be either solid or fluid.[126] If solid, or hard, the planet and the hard matter within the eccentric deferent must interpenetrate, because as hard bodies neither will yield to the other. But if the matter within the eccentric is fluid, one ought to say that the whole heaven is fluid, not just a part. For if the eccentric deferent is fluid and the two eccentric orbs that enclose it – namely the eccentric that surrounds the eccentric deferent and the one that is enclosed by it – are hard and solid, how will the outermost eccentric communicate a motion to the inner eccentric if the two solid eccentrics are separated by a fluid orb?[127]

Cornaeus includes yet another important, though brief, argument in favor of fluidity when he asks how a vast body like the heavens could be solid and hard and yet be moved with such rapidity without suffering a violent disruption of its parts and without fire arising from its intense motion.[128]

Roderigo de Arriaga thought the heavens were more likely to be fluid than solid. Neither reason nor authority suggested abandonment of the fluid hypothesis. Indeed, fluid heavens seemed to save diverse celestial phenomena better than an assumption of solidity.[129]

2. *Scholastic arguments for hard spheres*

a. *The heavens conceived as a combination of hard and fluid orbs*

A combination of hard and soft orbs may be traced back at least to Aegidius Romanus in the early fourteenth century (see Ch. 13, Sec. III.9). For Aegidius, the eccentric deferent was like "marrow in a bone" or "blood in the veins," thus associating a soft, or fluid, material with one that is much harder. A similar opinion, in which a solid heavens is divided into seven

125. Ibid., 498.
126. Ibid.
127. Here Cornaeus assumes that motion is transmitted from orb to orb. When they specifically considered the motions of the celestial spheres, scholastic natural philosophers assigned a separate mover to each orb, so that no orb depended on another for its motive power. For a full discussion, see below, Chapter 18, Section II.
128. "Vix cogitari potest quomodo corpus tam vastum et solidum tanta rapiditate moveri possit sine partium violentia disruptione, ac sine incendio ex nimia agitatione orto." Cornaeus, *De coelo*, disp. 2, qu. 2, dub. 3, 1657, 500. As part of his defense of the earth's daily rotation, Copernicus explains that "Ptolemy has no cause, then, to fear that the earth and everything earthly will be disrupted by a rotation created through nature's handiwork. . . . But why does he not feel this apprehension even more for the universe, whose motion must be the swifter, the bigger the heavens are than the earth?" Copernicus, *Revolutions*, bk. 1, ch. 8 [Rosen], 1978, 15.
129. See Arriaga, *De caelo*, disp. 1, sec. 3, subsec. 4, 1632, 503, col. 1, par. 41.

zones or channels, was adopted in the seventeenth century by Hurtado de Mendoza and Aversa, and was at least described by Mastrius and Bellutus. But they cannot be said unequivocally to have combined hard and fluid parts, because they leave it unclear as to whether the hollow interior of the eccentric deferent of each planet was filled with fluid matter. For those who assumed it was, the hard and fluid interpretation is of interest because it represents an attempt to assume a degree of fluidity while simultaneously retaining hard planetary orbs. But the most popular version that combined hardness and fluidity was adopted by some of those who abandoned all planetary orbs except the sphere of the fixed stars. Riccioli, who was perhaps the most prominent of this group, assumed that the sphere of the fixed stars was hard, whereas the region of the planets was a fiery fluid. This idea was, as he put it, "the most celebrated contemporary opinion," supported, according to Riccioli, by the likes of Oviedo, Arriaga, and probably Mastrius and Bellutus.[130] Although he recognized that this assumption was neither mathematically nor physically evident, Riccioli thought it was more probable than any other.[131] It seemed best suited to reconcile the numerous opinions of the Church Fathers and the theologians (that is, the scholastic doctors of the Middle Ages). Fluidity in the region of the planets was not only consonant with observations of modern astronomers but required the least degree of violence and the smallest number of motions and devices. It was most compatible with the new discoveries, that is "the phenomena of comets, of Mars, Venus, and Mercury, . . . and the motions of the satellites of Saturn and Jupiter and of sunspots."[132]

As for the assumption of a hard orb for the fixed stars, Riccioli thought it the best explanation to account for the unchanging distances between the stars themselves and also the best means of avoiding the needless multiplication of movers for the many stars.[133]

The association of a solid, hard heaven of the fixed stars with a fluid heavens in which the planets are moved directly by intelligences without the need of orbs is also reported by Mastrius and Bellutus, who characterize

130. "Et nunc celeberrima opinio." Riccioli, *Almagestum novum*, pars post., bk. 9, sec. 1, ch. 7, 1651, 240, col. 1. For the names of its supporters, see 240, cols. 1–2. This "most celebrated" opinion was the fifth of those that Riccioli categorized under the fluidity of the heavens.

131. "Probabilius multo est, licet nondum mathematice aut physice evidens, caelum fixarum solidum esse, planetarum autem fluidum." Ibid., 244, col. 1.

132. "Constat id argumentis et responsionibus utrimque hactenus adductis: hac enim distinctione tum probabilitatis ab evidentia tum caeli fixarum a caeli planetarum concilientur plurime opiniones patrum ac doctorum inter se et cum astronomorum recentiorum observationibus minorique violentia aut multiplicitate motuum ac machinarum; minori quoque periculo repugnantiae physice inter motus tam varios planetarum explicantur phenomena cometarum, Martis, Veneris ac Mercurii . . . et motus comitum Saturni et Iovis et macularum solarium." Ibid. This was essentially Arriaga's opinion (see Section VIII.1.c of this chapter).

133. "Si sphaera fixarum solida ponatur, promptius redditur ratio cui servent perpetuo eamdem inter se distantiam; nec multiplicandi erunt innumerabiles motores fixarum." Riccioli, *Almagestum novum*, pars post., bk. 9, sec. 1, ch. 7, 1651, 244, col. 1.

it as a "very probable [*valde probabilis*] opinion enunciated by certain contemporaries."[134] Of the opinions they report, this may have been the one they judged most plausible. Although Amicus was a defender of hard and solid heavens, he reported this opinion as acceptable to moderns because it was compatible with scriptural statements about the solidity of the firmament of the fixed stars (probably Job 37.18, which he cites later) and also seemed to account for the new discoveries which indicated that planets moved by their own motions through a fluid or airy medium.[135]

b. The case for solid, hard spheres

Despite the inexorable, if gradual, abandonment of hard orbs in favor of fluid heavens, a system of hard orbs had its defenders. One of the most prominent was Bartholomew Amicus, some of whose arguments were subsequently repeated by Riccioli, despite the latter's defense of fluid planetary heavens. With Amicus, scriptural arguments played a significant role, especially Job 37.18, which was traditionally invoked in favor of solid orbs. Although God reproved Elihu's discourse, Amicus interpreted this as a moral rejection only, not one that pertained to natural things. On the contrary, God seems to accept Elihu's statement that the heavens are hard.[136] The very name *firmamentum*, which applies to the heaven of the fixed stars, implies firmness and solidity. Moreover, a solid body is required to divide the waters from the waters, since a liquid body has no proper boundaries. Without a solid, hard firmament to play this role, the waters would mix with the things around them.[137] But the term *firmamentum* does not apply only to the sphere of the fixed stars but also to all the other heavens and planets. After all, in Genesis 1.14–17, which Amicus cites, God placed in

134. "Nota vero quod valde probabilis est etiam quorundam recentiorum sententia ponens coelum stellarum solidum in quo fixa existant astra ad motum coeli mobilia, deinde coelum aliud fluidum in quo planetae moveantur ab intelligentiis." Mastrius and Bellutus [*De coelo*, disp. 2, qu. 1, art. 2], 1727, 3:490, col. 2, par. 38. Earlier we saw that Mastrius and Bellutus described another opinion in which the heavens were divided into seven zones or channels (see Ch. 13, Sec. III.9 and n. 112). The opinion they favored is left uncertain.

135. "Secunda opinio est dicentium caelos planetarios esse ex materia liquida quia facilius motus et apparentiae planetarum salvantur caelum; vero stellarum fixarum esse solidum in quo stellae sunt fixe ut nodi in tabula, unde non per se sed ad motum orbium moventur. Hanc significant dicentes stellas in firmamento fixas esse; planetas vero per aerem vagari vario motu. . . .

    Hec tamen opinio potest a recentioribus accipi quia ex asserentibus liquiditatem celorum est omnium probabilissimam satisfacit scripturae tribuenti firmamento soliditatem, quod est celum stellatum fixarum, et satisfaciunt novis apparentiis dum planetas moveri per se propriis motibus per orbem afferunt." Amicus, *De caelo*, tract. 5, qu. 5, art. 2, 1626, 274, col. 2. In this report, Amicus seems to envision the planets as moving through a fluid medium that is located within an orb.

136. After citing the passage, Amicus says: "Neque obstat quod illud sit dictum ab Elihu cuius discursus fuit a Deo reprobatus c.38, nam ibi fuit reprobatus discursus moralis quo Job sanctitatem innocentiam accusabat. Sed quoad naturalia ibi ducta a nullo sapiente reprobantur, sed potius recipiuntur." Amicus, ibid., art. 3, 278, col. 1.

137. Ibid., 278, cols. 1–2.

that very firmament the luminaries he created on the fourth day. Planets and stars are all part of the firmament.[138]

Reason also tells us that solid, hard orbs are needed to carry the planets perpetually at great velocities. Without them, the celestial bodies would self-destruct and fail to preserve the order and constancy of the heavens.[139] Amicus demonstrates this with three brief arguments: (1) If the planets were not embedded in solid orbs, one intelligence would be required for each planet and each fixed star; so great would be the number of intelligences required that a needless multiplication of entities would result.[140] (2) A mover for each planet would be a much less effective way of preserving uniformity of celestial motion, just as it would be if all the stars and planets were moved only by a single mover. For in the latter situation, the intelligence, or angelic intellect, would, because of its finitude, be less able to attend to individual stars and planets.[141] Finally, (3) because angels move bodies only when they are actually in touch with them, it follows that if angels moved the stars and planets directly, each angel would have to be moved right along with its own celestial body in order to assist it as it moves along, a situation that is avoided if the planets are embedded in solid orbs.[142]

138. "Secunda conclusio soliditatem, quam probavimus convenire firmamento, probabile est convenire omnibus caelis etiam planetarum." Ibid., 279, col. 2.

139. "Secundo probo ratione: nam orbes sunt ordinati ad deferenda sydera perpetuo et ordinate. At hic finis exigit soliditatem orbium quia alioquin per tantam velocitatem motus facile corpora dissparentur atque adeo non posset servari tanta constantia et ordinatio motuum et corporum motorum." Ibid.

140. "Tum quia si sydera pe se et non infixa orbibus moverentur magnus exigeretur intelligentiarum numerus ad tot stellas presertim fixas ordinate movendas unde sine necessitate sufficienti multiplicarentur entia." Ibid. Riccioli's eighth argument in favor of solid orbs is similar: "Si non concedantur orbes solidi, oportebit ad movenda corpora celestia multiplicare innumerabiles intelligentias, tot nimirum, inquit Tannerus, quot sunt stellae fixae, quotque maculae solis." Riccioli, *Almagestum novum*, pars post., bk. 9, sec. 1, ch. 7, 1651, 241, col. 2.

141. "Tum quia ex motibus factis a tot diversis motoribus non posset servari tanta uniformitas quia non possent semper attendere ad servandam tantam vel tantam velocitatem, et alias circumstantias, ex quibus pendet uniformitas motus. Et hoc idem probat si omnes stellae moverentur ab eodem, nam licet intellectus Angelicus comprehendat naturas rerum corporalium, adhuc tamen ob sui finitatem minus potest attendere ad singula, dum pluribus intendit." Amicus, *De caelo*, tract. 5, qu. 5, art. 3, 1626, 280, col. 1.

142. "Tum quia cum Angeli non moveant corpora nisi sint sibi praesentia, necessario oporteret ipsos moveri cum ipsis stellis, assistendo illis ut moveant, quae omnia vitantur si moveri dicamus in orbibius infixas." Ibid. Amicus probably derived the argument about the need for angels to be present where they act from Thomas Aquinas, who denied action at a distance even for spiritual creatures, including God (see Grant, 1981a, 146). In his category of arguments in defense of hard orbs, Riccioli, *Almagestum novum*, pars post., bk. 9, sec. 1, ch. 7, 1651, 241, col. 2, presents the same argument. As a counterargument, he declares that an intelligence need not travel around with the body it moves but could rather remain in a particular place and impress enough impetus into a star or planet to carry it around for one revolution. When the point at which the impetus had been injected arrives again at the same place, the intelligence impresses the same quantity of impetus to carry the body around for another revolution, and so on. But Riccioli also insists that no difficulties would arise if an intelligence moved around with the celestial

If the celestial substance were really fluid, the enormous velocities of the gigantic celestial bodies that move through it would seem of necessity to produce a loud noise, especially at the point of impact.[143] Although Amicus fails to draw the inference, it is obvious: because we hear no such sound, the heavens cannot be of a fluid nature.[144] On a more positive note, Amicus declares that solid, interconnected, and interrelated orbs confer a greater nobility and system on the heavens than would be the case with stars and planets moving through a fluid medium as fish move through the sea.[145]

Amicus also argued that the attribution of liquidity to the firmament was contrary to common sense. But though he was a staunch advocate of solid orbs, Amicus conceded that solidity was originally an unnatural state for the heavens. When God created the heaven on the first day, he apparently produced a fluid, watery heaven. The true nature of the firmament was thus fluid. On the second day, however, God, in dividing the heaven of the first day, made the solid firmament, which is really the fluid heaven of the first day made hard and solid. The solidity of the firmament is therefore an accidental property of the heavens, secondary to its true and original fluid nature.[146] Amicus observes that there are those who believe that the heavens are naturally fluid and remained that way and there are others who hold that the heavens were originally fluid but were made unnaturally solid and hard. Because the fluidity of the heavens seems natural in both theories, Amicus concludes "that from authority, from the motions of new stars, and from similar things, which [Christopher] Scheiner reports, it is sufficiently probable that the heavens are fluid. But I do not follow this [opinion],

body to which it is assigned. For more on Riccioli's ideas on the application of impressed forces to celestial bodies, see Chapter 18, Section II.6.a and note 226.

143. "Conf. quia mirum est ex tanta velocitate motuum corporum liquidorum tam ingentium non gigni sonum adeo ingentem ut ad nos perveniat, nam sonus gignitur ex collisione corporum ad acrem, id est, corpus liquidum." Amicus, ibid.

144. Riccioli, *Almagestum novum*, pars post., bk. 9, sec. 1, ch. 7, 1651, 241, cols. 1–2, describes the same argument mentioning that the sounds should be akin to those hissing or whistling sounds that emanate from stones launched in the air by slinging machines. He also cites counterarguments from Tycho Brahe, Christoph Rothmann, and Franciscus de Oviedo. Thus Rothmann denied that such sounds could reach our ears, because of the great distances and the rarity of the celestial ether. Oviedo's response was predicated on the familiar analogy between the movement of fish in water and planets in the heavens. Just as there is no sound in the water itself when fish swim through it, so also there is no sound in the fluid heavens as the planets move through them. Moreover, defenders of hard orbs ought to be asked why fire and air do not produce audible sounds as a result of the circulation of the lunar heaven.

145. "Conf. secundo nam quo corpora sunt superiora eo magis sunt nobiliora et maiori quodam artificio ornata. At hoc artificium magis apparet ponendo multos orbes tum mobiles inter se connexos et ordinate motos . . . quam si ponatur unum liquidum per quod stellae discurrant ut pisces per mare." Amicus, *De caelo*, tract. 5, qu. 5, art. 3, 1626, 280, col. 1.

146. "Firmamentum secundo die productum sola soliditate differt ab eodem producto initio, sed soliditas, cum sit accidens, non variat naturam rerum, ergo neque naturam firmamenti. Si prius erat liquidum ex natura, similiter erit natura liquidum sub soliditate. Hec autem variatio in caelo facta est ob bonum universi." Ibid., art. 4, 281, col. 1.

nor do I retreat from ancient opinion without an urgent reason and [also] because solidity [and hardness] conform more to Scripture to which every human intelligence is subjected."[147] Although Amicus thought the fluidity of the heavens improbable on scriptural grounds, he conceded that the opinions drawn from scriptural texts and Church Fathers were not so clear or obvious in support of the solidity of the heavens. Moreover many learned contemporary theologians, philosophers, and astronomers (he calls them "mathematicians") thought they were fluid. For these reasons, Amicus declares that despite the improbability of fluid heavens, it was by no means rash to uphold this theory.[148]

Giovanni Baptista Riccioli represents another significant seventeenth-century source for arguments favoring celestial solidity and hardness. Although he rejected heavens filled with solid orbs, Riccioli, like many other scholastic authors, sought to present a balanced account in the dispute over fluid or solid heavens. Just as he did in his section defending fluid heavens, he compiled arguments in favor of total or partial solidity and hardness and often included a common rebuttal of each argument. Riccioli reiterated what was probably obvious to everyone by his day: solidity signifies not only three-dimensionality but also has an associated meaning of hardness as opposed to softness.[149] As he did in his presentation favoring fluid heavens, Riccioli cites arguments from authority in support of celestial solidity, drawing upon Holy Scripture (including Job 37.18, the most frequently mentioned), the Church Fathers,[150] and Aristotle. Thus we are told that in *De caelo* (bk. 2, ch. 7) Aristotle declares that a heaven and the star or planet that is part of it are made of the same material. But a star or planet is a solid body; therefore, so is the heaven or orb that carries it.[151]

Leaving authority and moving on to more substantive arguments, Riccioli

147. "Ex quibus puto satis probabile esse caelos esse fluidos ex auctoritate, et motibus novarum stellarum et similibus, quae affert Scheiner. Sed eam non sequor, ne recedam ab antiquata opinione sine ratione urgente et quia soliditas est magis conformis scripturae cui omnis humana intelligentia subdidebet." Ibid., 282, col. 1.

148. "Ego vero in hac diversitate opinionem asserentium caelum esse liquidum existimo esse quidem improbabile, non tamen temerariam. Nam scripturae loca et Patrum testimonia non ita clare soliditatem caelorum exprimunt, ut interpretationem non admittant ut patet ex iis quae adversarii adducunt. Idque confirmo nam nostre aetate multi sunt ex Theologis, Philosophis, et Mathematicis, multae eruditionis, qui liquiditatem caelo convenire nituntur probare quos non est aequum temeritatis censura notari." Ibid., art. 3, 281, col. 1.

149. Riccioli, *Almagestum novum*, pars post., bk. 9, sec. 1, ch. 7, 1651, 240, col. 2–242, col. 1, for the arguments, and 238, col. 2, for the linkage between solidity and hardness.

150. Ibid., 240, col. 2–241, col. 1.

151. "Tertium argumentum est Aristotelis lib.2 *De celo*, cap.7, ubi ait congruum esse rationi ut caelum sit ex eodem corpore cuius est sidus; sidus autem quodlibet esse solidum, ergo et celi corpus ex quo est." Ibid., 241, col. 1. For the passage in Aristotle, see *De caelo* 2.7.289a.11–12. In the counterargument, Riccioli concedes the identity of the matter of the planet and its orb or heaven but insists that they differ in other ways, for otherwise, it would follow that because a planet is luminous and opaque, so also would the whole heaven be luminous and opaque. But this is false. *Almagestum novum*, pars post., bk. 9, sec. 1, ch. 7, 1651, 241, col. 1.

cites one that was intended to show a fatal flaw in the theory of fluid heavens. On the assumption that the planets move through fluid heavens that could have no void spaces and in which bodies cannot penetrate one another, Riccioli explains that a planet moving through such a fluid would cause either the whole fluid, or a part of it, to undulate. This would occur by the impact of the planet's forward motion, which would condense the fluid in front of it. As this occurs, the matter behind the planet would necessarily rarefy in order to prevent formation of a vacuum in the places that the planet has just vacated. But condensation and rarefaction are alien to the heavens and are signs of corruptibility. Fluid heavens would thus produce impossible consequences.[152]

Riccioli also reports an argument in which hard orbs are judged better for explaining the occurrence of a plurality of simultaneous, and even contrary, motions. Indeed, it is impossible for the same body to move with several motions simultaneously unless it achieves this with a motion of its own (*per se*) in one direction while the one or more remaining motions are produced by the motion of the solid body to which it is affixed or in which it is embedded.[153]

c. The new discoveries and solid orbs

To defend solid, hard orbs, it was essential to deny that the new discoveries implied fluid heavens and/or to deny or cast doubt on the new discoveries

152. "Quartum argumentum: Si celum in quo moventur sidera esset fluidum et non admittatur vacuum aut penetratio corporum, sequeretur ad motum sideris vel totum fluctuare celum, vel partem celi a sidere impulsam condensari, partem vero a sidere destitutam rarefieri. At condensatio et rarefactio repugnant caelo et sunt indicium corruptibilitatis." Ibid. To this argument, Riccioli presents a number of replies. Some argue that in the celestial region mutual penetration of bodies is possible, or that rarefaction and condensation are possible in incorruptible heavens; and some insist that the heavens are corruptible and that therefore condensation and rarefaction can occur. During the entire period covered by this study, fear of the consequences that bodies and celestial matter might interpenetrate and that celestial matter might rarefy and condense frequently compelled defenders of celestial orbs to explain how such dire consequences could be avoided (see Ch. 13, Sec. III, especially 1–3 and n. 59).

153. "Sextum argumentum indicatum ab Aristotele et inculcatum a Pereiro sumitur a multiplicitate motuum. Impossibile enim est idem corpus moveri pluribus motibus et quidem contrariis, quomodo constat moveri planetas, immo et fixas, nisi uno motu per se moveantur in unam plagam, reliquis autem moveantur ad motum corporis solidi, cui sint affixa aut insidentia." Ibid., 241, col. 2. As a counterargument, Riccioli declares that even if another body were required to account for all the motions of a single planet, it would not have to be solid and hard. After all, fish are carried downstream by a rapidly moving river while they simultaneously attempt to move upstream in the opposite direction. But Riccioli reports further that Clavius had countered this argument by observing that if the planets were really moved freely through a fluid medium like fish and birds, their motions would be just as indeterminate and uncertain. It would be as if they did not know that they were being moved in a fluid medium by an intelligence. Hence we could have no certain knowledge of planetary motions. ("Aliter hoc argumentum proponit Clavius in cap. 4, sphere pag. 449, ait enim si moverentur ut pisces et aves in fluido, liberum ac nimis vagum fore planetarum motum et sic nullam certam fore scientiam de ipsorum motibus, quasi vero nequeant ab intelligentia moveri in fluido, servatis tamen legibus motuum." Ibid.)

themselves. Amicus did both. Following Tanner, Amicus argues that the new phenomena indicate that certain planets – presumably Mercury and Venus – do not have proper and distinct orbs that surround the earth but are carried in epicycles around the Sun and actually lie within the Sun's orb. Thus did Amicus, and others, accept that part of Tycho's system which made the Sun, rather than the earth, the center of the orbits of Mercury and Venus. Indeed, the result was a variation of traditional geocentric cosmology that resembles the medieval opinion reported in the fourteenth century by Jean Buridan.[154] And just as Buridan thought this configuration probable, so did Amicus assume that the epicycles – he makes no mention of eccentrics – which carry Mercury and Venus around the Sun do so without any penetration of one orb by another and without any crashing of orbs.[155]

IX. The diversity of opinion

What do all these diverse and often conflicting opinions signify? Probably an inability to determine convincingly the operational structure of the heavens. Numerous opinions and variations on those opinions were inevitable, because scholastic natural philosophers did not and could not know the true nature of the heavens. There was much room for disagreement about the hardness or fluidity of the celestial region. Even some of those who adopted a fluid medium for the planetary region opted for a hard orb for the fixed stars, not only because it seemed more economical to have one mover for all the stars rather than one mover for each star, but perhaps also because it seemed more fitting that the whole of the cosmos be enclosed by a firm, hard surface to keep it all together and prevent its dissipation into the region beyond.[156] Indeed, Galileo himself seems to have thought it best to enclose

154. See Chapter 13, Section IV.1, for a discussion and an English translation of Buridan's text (n. 137 for the Latin text).

155. "Ad tertium ductum ex novis phaenomenis, resp. Tannerus disp. 6, q. 3, d. 3, par. 6, ex horum planetarum motibus id solum consequi non habere proprios et distinctos orbes terram ambientes, sed solum epicyclos in orbe solari solem ambientes per quos planetae circa solem feruntur sine ulla vel penetratione, vel fractura orbis, quod stat cum caeli soliditate et incorruptibilitate, sed his obstat communis sensus philosophorum et theologorum, qui illis planetis tribuunt proprios orbes mobiles circa centrum mundi. Sed hoc non putant absurdum." Amicus, *De caelo*, tract. 5, qu. 5, art. 5, 1626, 285, col. 1. Amicus did not agree with those philosophers and theologians who insisted that all orbs must have the earth as center. The new astronomical observations showed that Mercury and Venus moved around the Sun in proper epicycles. ("At cum observatum sit per novas observationes illos planetas non moveri circa terram, sed circa solem; ideo negandi sunt illis proprii orbes et ob id conceduntur proprii epicycli." Ibid.) Serbellonus, *De caelo*, disp. 1, qu. 2, art. 4, 1663, 2:26, cols. 1–2, reports that not only are Venus and Mercury contained in the solar orb in proper epicycles but so also is Mars. He rejects the existence of solid eccentrics and epicycles because division, penetration, and fracture of orbs would inevitably occur.

156. According to Van Helden (1985, 63), "Kepler continued to believe," to the very end of his life, "that the fixed stars were arranged in a spherical shell."

the fluid heavens between two hard surfaces, with the planets and stars distributed at various altitudes between them.[157]

In a certain sense, the controversy had reached a stalemate. For although solid orbs seemed incapable of explaining the occurrence of celestial comets or new stars, partisans of solid orbs could, as we have seen, either deny that such phenomena were celestial or invoke normally invisible celestial bodies that are carried in epicycles within larger spherical complexes and that somehow cluster together to produce a visible body. On this approach, change is accidental rather than substantial and involves a mere rearrangement of previously existing bodies. As for the various subsystems with centers other than the earth, these were usually explained by the assumption of epicycles for the satellites themselves or for Mercury, Venus, and even Mars. Sunspots were also explained in a similar manner. Although in retrospect such constructions seem of an ad hoc nature, they enabled a steadily diminishing group of scholastics to hang on to the chief elements of the old system despite the ovewhelming challenge that confronted them.

Within the broad categories of hard and soft, a wide variety of interpretations was formulated, with no decisive way to choose among them. The seeming advantage of the theory of hard and solid eccentrics and epicycles was that the planet was carried around within a hard epicycle. Although there were grave problems about the place of the planet within the epicycle itself and how the planet would relate to either a vacuum or some kind of matter within the epicycle, the planet was at least fixed within the epicycle. But if each planet seemed to be carried within its hard epicycle, the problem of motion was certainly not solved; rather, it was removed one step to the orb itself. What caused the orbs themselves to move is a problem that I shall consider later (Ch. 18).

And yet the fluid theory of the heavens emerged as the most appropriate interpretation of the new celestial phenomena. It triumphed because it seemed more congruent with those phenomena and made fewer incredulous demands on reason. Fluidity did not, however, triumph because of any overwhelming and certain arguments. Indeed, those who abandoned hard orbs in favor of fluid heavens had to confront the problem of planetary motion directly. What enabled a planet to move in its orbit like fish in the water or birds in the air, as the popular analogies expressed it? For those who not only assumed fluid planetary spheres but also a fluid zone for the fixed stars, there was the additional problem of assigning a motive cause to each of the more than one thousand visible stars. For no longer could they rely on a single hard orb to carry around the fixed stars that had been previously imagined as fixed in their hard sphere like knots in a piece of wood. Although he recognized that if the firmament were solid, only one

157. Galileo, *Dialogue*, Third Day [Drake], 1962, 325–326. Although the words are put into the mouth of Simplicio, they seem to represent Galileo's opinion. As early as March 23, 1615 (in a letter to Monsignor Dini), Galileo had already rejected the real existence of "solid, material, and distinct orbs" (see Finocchiaro, 1989, 61–62).

mover would be required to carry all the stars simultaneously, Melchior Cornaeus preferred to believe that God did not create hard orbs but rather assigned one angel to each star, of which there were more than a thousand. After all, God was not destitute of angels, and a star was not so small that it did not deserve its own motive angel.[158]

Thus, where Kepler had proposed a causal, physical mechanism based on magnetic forces to account for the motions of orb-free planets in his *Astronomia nova* (1609) and *Epitome astronomiae Copenicanae* (1617–1620),[159] scholastic natural philosophers, who assumed, as did Cornaeus, fluid heavens, relied heavily upon angels or intelligences as celestial movers. But unlike Cornaeus, as we shall see in Chapter 18, some associated impressed forces with angelic movers and thus tended to make the latter more mechanical than spiritual. In so doing, they joined Kepler and others in a quest for more impersonal forces to explain the motion of orb-less celestial bodies.

Until the theory of universal gravitation in Newton's *Principia* settled the matter once and for all, there were only ad hoc theories, which were no more convincing than the causal explanations invoked for celestial orbs. Nonetheless, the cumulative evidence inclined strongly and suggestively toward fluidity, and more and more scholastics embraced it as the most plausible alternative.

158. Cornaeus, *De coelo*, disp. 2, qu. 2, dub. 3, 1657, 500, first raises an objection against himself: "Si firmamentum non est solidum, ergo singulis astris assignandus est angelus motor, qui per liquidum conducat et certo itinera dirigat. Atqui si firmamentum statuamus solidum, unus sufficiet pro omnibus" and then replies: "Concedo sequel. Neque tam parva res est stella ut angeli custodiam non mereatur, neque tam inops angelorum est Deus ut pro omnibus et singulis stellis non sit ei sufficiens eorum copia." We shall learn more in Chapter 18 about causal factors in celestial motion.

159. Kepler relied on two forces. According to Koyré (1973, 323), he assumed a rotation of the Sun, which "sends out into space (in the plane of the ecliptic) a motive whirlpool which carries the planets round and impresses on them a circular motion round the Sun; at the same time the planetary magnets, in accordance with a mechanism which has been fully described above, causes the planets to approach and recede from the Sun. As a result of being subjected to this two-fold influence, the planets do not describe circles in the sky, but describe ellipses having the Sun at one of their foci."

15

The immobile orb of the cosmos: the empyrean heaven

Thus far we have considered only mobile orbs, all of which had astronomical functions and at least two of which – the eighth (the firmament) and ninth (the crystalline orb) – also had biblical sanction.[1] In the chapter on creation, we saw that in commentaries on Genesis the heaven, or orb, created on the first day was often called the "empyrean,"[2] even though the latter is not explicitly mentioned anywhere in the Bible. Its existence was derived from faith, not rational argument. Inferences about it fashioned an invisible, immobile orb that enclosed the world, a place that was conceived by many as the "first and highest heaven, the place of angels, the region and dwelling place of blessed men"[3] or as "the dwelling place of God and the elect."[4] If the number of mobile orbs is n, the empyrean orb was always numbered $n + 1$. In the most popular, ten-orb mobile system, it was the eleventh and final orb.[5]

I. Features and properties of the empyrean heaven

According to Thomas Aquinas, the empyrean heaven had been recognized much earlier by Venerable Bede and Walafrid Strabo.[6] Without using the

1. The "crystalline orb" is biblical only when it is interpreted as the congealed form of the waters above the firmament. As we saw earlier (Ch. 13, Sec. IV.2.b), the tenth orb might also be included within the concept of a crystalline orb.
2. For a description of some of its salient features, see Chapter 6, Sections II and III.2.f and g.
3. "Coelum empyroeum est primum et summum coelum, locus angelorum, regio et habitaculum hominum beatorum." Bartholomew the Englishman, *De rerum proprietatibus*, bk. 8, ch. 4 ("De coelo empyreo"), 1601, 379–380. Although God is everywhere, Bartholomew says that he is especially in the empyrean heaven.
4. "Coelum empirreum habitaculum Dei et omnium electorum" is the description found in the figure of the world that appears on column 6 of Peter Apian's *Cosmographicus liber* (Landshut: Johann Weyssenburger for Petrus Apianus, 1524), reproduced in this volume as Figure 9.
5. For this eleven orb system, see Figure 9.
6. *Summa theologiae*, pt. 1, qu. 66, art. 3, 1967, 40. Although Saint Basil's role was inconsequential, Thomas included him as a third person involved in the development of the

word *empyreum*, Venerable Bede had indeed, in the eighth century, distinguished an immobile heaven created on the first day from the mobile, observable celestial bodies created later.[7] Without specifying immobility, Alcuin did much the same thing in the ninth century.[8]

But the empyrean sphere did not emerge as a distinct entity called the *caelum empyreum* until the twelfth century, when Anselm of Laon, Peter Lombard, Hugh of Saint Victor, and Roland Bandinelli (the future Pope Alexander III) furnished brief descriptions that gained acceptance in subsequent scholastic literature. According to a description by Anselm of Laon (d. 1117) in the *Glossa ordinaria*,[9] the empyrean sphere was understood as "fiery or intellectual, which is so called not by virtue of its burning [or heat] [*ardor*] but from its brilliance [*splendor*], since it is immediately filled with angels."[10] Not only did Peter Lombard, in his famous *Sentences*, quote the words of Anselm, but he also identified the empyrean sphere with the invisible heaven created on the first day and thus distinguished it from the heaven created on the second day and made visible on the fourth day.[11] Indeed, Peter believed that the empyrean heaven was created simultaneously with the angels and all corporeal things.[12]

During the thirteenth century, all the great theologians – William of Auvergne, Alexander of Hales, Albertus Magnus, Saint Bonaventure, Duns Scotus, Richard of Middleton, and Thomas Aquinas – had come to accept the empyrean heaven. By then it had become the dwelling place of God and the angels, as well as the abode of the blessed.[13] Although conceived

concept of an empyrean sphere. For a brief but useful account of the empyrean sphere as it developed in the late Middle Ages, see the article on it in the *Dictionnaire de théologie catholique*, vol. 2, pt. 2, cols. 2503–2508. The earlier history is described by Maurach, 1968. An important collection of Latin descriptions of the empyrean sphere from Bede to Thomas Aquinas appears in Litt, 1963, 255–258, n. 1. Duhem included a brief account in *Le Système*, 1913–1959, 7:197–202, of which pages 197–200 have been translated in Čapek, 1976, 43–45. See also Grant, 1978a, 275–276. For an account of the fate of the empyrean heaven in the first half of the seventeenth century, see Donahue, 1972, 223–259.

7. Bede [*Genesis*], 1967, 4, lines 35–41. The term *empyreum* had already been applied to the extramundane region by Martianus Capella in the fifth century. In Campanus of Novara, *Theorica planetarum*, 1971, 393, n. 52, Benjamin and Toomer cite the Latin text from Dick's edition of *The Marriage of Mercury and Philology* (*De nuptiis Philologiae et Mercurii libri VIII*), bk. 2, line 200, p. 76.
8. For the passage, see Litt, 1963, 257.
9. During the Middle Ages, the *Glossa* was falsely ascribed to Walafrid Strabo (this attribution is repeated by Benjamin and Toomer (see Campanus of Novara, *Theorica planetarum*, 1971, 393, n. 52). On the assignation of the *Glossa ordinaria* to Anselm, see Thomas Aquinas, *Summa theologiae*, 1967, 10:40, n. 5.
10. My translation is from the Latin text quoted in Campanus of Novara, ibid. The translation first appeared in Grant, 1978a, 275–276. This assertion was frequently repeated (see, for example, Vincent of Beauvais, *Speculum naturale*, bk. 3, ch. 88, 1624, col. 220).
11. The brief biblical description of a seemingly distinct heaven created on each of the first two days formed the basis for belief in an empyrean heaven. Because the heaven of the second day was clearly intended as the firmament, only the vaguely described heaven of the first day remained as a viable candidate for the empyrean.
12. Peter Lombard [*Sentences*, bk. 2, dist. 2, ch. 5], 1971, 340; see also Litt, 1963, 256–257, n. 1.
13. During the first half of the thirteenth century, there was apparently an attempt to identify

to be a body,[14] the empyrean heaven was the most subtle of all bodies and contained within itself the purest light in the universe. Albertus Magnus envisioned it as formed from fire, the most noble of the simple elemental bodies, but Thomas Aquinas denied it any connection with the four elements and insisted rather that it was composed of pure ether, the fifth element in Aristotle's cosmology.[15] Despite the empyrean's purity and splendor, it was invisible[16] and its effects were imperceptible. Like the other celestial spheres, the empyrean heaven was thought incorruptible; unlike them, however, it was always assumed immobile.

In the fourteenth century, Thomas of Strasbourg (fl. 1345) ([*Sentences*, bk. 2, dist. 2, qu. 2], 1564, 134r, col. 1–134v, col. 1) gathered much that had been elaborated in the preceding century and presented a thorough discussion of the properties of the empyrean sphere, to which he assigned four basic attributes. It is the most lucid, or light-filled, sphere. As the first and noblest of celestial bodies, the empyrean ought to possess the noblest corporeal quality, which is light. Moreover, as the noblest of bodies, it should be independent of all other celestial bodies. Because the latter receive their light solely from the Sun, the empyrean should receive its light directly from God. Although it was filled with light, the empyrean sphere transmitted no light to the celestial and terrestrial regions below. Empyrean light was effectively blocked by the nontransparent nature of the eighth sphere of the fixed stars.

By contrast – and this is its second property – the empyrean sphere was transparent and rarefied, made so to enhance the pleasure of the blessed, so that each inhabitant could see friends in the same state.

Its third property was incorruptibility, which was inferred from Aristotle's general attribution of incorruptibility to the other planets. Because the empyrean was nobler than all other celestial bodies, it must also be incorruptible.

Finally, Thomas of Strasbourg argues that the empyrean heaven must be immobile. Immobility is the only appropriate state for the blessed, who are themselves in a perfect state of rest. Although Aristotle denied the existence of immobile spheres, Thomas declares that this has no validity for Christians, "who assume a certain body above the first movable body [*primum mobile*] itself, [a body] that is absolutely independent of the first movable sphere, as is the empyrean heaven itself." The empyrean heaven is not,

the empyrean heaven with one of the astronomical orbs, namely the ninth, or crystalline, sphere. In 1241 and then officially in 1244, the bishop of Paris condemned those who located the Blessed Virgin and the glorified soul in the ninth, or crystalline sphere, instead of in the empyrean heaven, or tenth sphere. Indeed, Michael Scot explains that the empyrean heaven was introduced to serve as a moving cause prior to the first movable sphere, or *primum mobile*. For all this, see Vescovini Federici's discussion in Peter of Abano, *Lucidator*, 1988, 200–201. It appears that the ecclesiastical authorities wished to have a place for the blessed that was distinct from any sphere that had an astronomical function.

14. See Alexander of Hales, *Summa theologica*, inquis. 3, tract. 2, qu. 2, tit. 1, memb. 1, art. 2, 1928, 2:329 and *Dictionnaire de théologie catholique*, vol. 2, pt. 2, col. 2506.
15. *Dictionnaire de théologie catholique*, vol. 2, pt. 2, cols. 2506–2507.
16. See Chapter 16, Section I and note 7.

however, inherently immobile, because, like any celestial sphere, God could move it if he wished. But the empyrean heaven is called immobile, because universal order does not require its motion and because God does not subject it to the power of any other creature that could move it.

II. Arguments for and against an immobile sphere

Although an immobile empyrean orb was widely accepted on theological, rather than cosmological, grounds, its alleged immobility clashed with Aristotelian natural philosophy, which cast serious doubt on such a possibility. It was, therefore, not unusual for scholastics to inquire whether an immobile orb could exist beyond all the mobile orbs, as Jean Buridan makes evident in his discussion of the question "Whether a resting or immobile heaven should be assumed above the heavens that move."[17]

1. The arguments against

First, Aristotle held that every natural body has a natural local motion, so that to every simple natural body, some natural simple motion must be assigned. On this basis, the empyrean heaven could not be immobile. Second, the empyrean heaven ought to be nobler than the heavens that are moved, because the former is above them, contains them, and confers powers on them. But it cannot be nobler, because motion is nobler and prior to rest. Therefore bodies that are moved naturally are nobler than bodies at rest, which is why the immobile earth is the most ignoble of bodies. We are thus confronted with a contradiction: the empyrean heaven would be both nobler and not nobler than the celestial spheres that are in motion. Finally, although the empyrean heaven is immobile, it could be moved by God. Therefore it has the potentiality for motion. But if it remains forever immobile, it would fail to realize that potentiality and would be perpetually frustrated. "Thus," the argument concludes, "it would be absurd that the heaven should never be moved."[18]

Despite these Aristotelian arguments, and "because of the statements of the theologians" (propter dicta theologorum), Buridan found it necessary to defend the existence of the empyrean sphere. After all, Aristotle assumed many things contrary to the Catholic faith, because he sought only to derive arguments from sensation and experience. Therefore one need not believe

17. "Utrum sit ponendum caelum quiescens sive non motum supra caelos motos." Buridan [*De caelo*, bk. 2, qu. 6], 1942, 149. The empyrean heaven was a customary topic in commentaries on the second book of Peter Lombard's *Sentences* but also turns up in questions on *De caelo* and in commentaries on the *Sphere* of Sacrobosco.
18. For the three arguments, see Buridan, ibid., 151–152.

Aristotle when he departs from Scripture,[19] and Buridan subsequently dismisses the arguments just described. As the noblest body, the empyrean sphere does not require action or motion. And although it could indeed be moved supernaturally by God, the empyrean sphere has no inclination – that is, no potentiality – for motion and therefore would not be eternally frustrated by its failure to achieve it.

Despite his seeming defense of the empyrean sphere on theological grounds, Buridan's commitment to it is doubtful. At the end of the question, he adds a section for those who wish to defend the Aristotelian position and promptly offers guidance by refuting five arguments in favor of the empyrean sphere that he had previously presented.[20] One of the favorable arguments – the fourth – held that the earth was divided into two halves by the equatorial circle. The half toward the antarctic pole is uninhabitable. The other half, toward the arctic pole, is divided into two quarters, one of which is habitable, the other uninhabitable. Since the earth is ruled by the heavens, it is essential that the heavens arrange things so that one of these two northern quarters is habitable and the other uninhabitable and covered with water. But this arrangement cannot be caused by mobile heavens,

> because the same parts of the heaven and the same planets [and stars] are turned uniformly over this quarter and over the other. Therefore it is necessary that this be caused by a resting heaven, one part of which – the one that is over us – has influence and dominion over the well-being of animals and plants, and the other has more dominion over the gathering of waters.[21]

In rejecting this favorable argument, Buridan suggests that the habitability of the earth might have been ordained from all eternity by God and be in no way dependent on the empyrean sphere. Indeed, in the very next question, he explains how the earth's overall topography could have been preserved eternally and does so without alluding to the empyrean sphere.[22] And in a later question of the same treatise, when he considers the possibility of the earth's axial rotation, Buridan again rejects the empyrean sphere. To determine if the earth could rotate axially, Buridan asks whether rest or motion is nobler. Rest is nobler, he argues, when a body comes to rest following upon a motion toward its natural place. That is, the rest acquired after the motion of a body toward its natural place is superior to the motion that brought it to its natural place. By contrast, motion is nobler than rest for those bodies that are always in their natural places and which have no other goal than to move in their natural places with their natural motions.

19. "Et potest responderi ad rationes Aristotelis, quod ipse multa posuit contra veritatem catholicam, quia nihil voluit ponere nisi posset deduci ex rationibus ortum habentibus ex sensatis et expertis; ideo non oportet in multis credere Aristoteli, scilicet ubi dissonat sacrae scripturae." Ibid., 152.
20. Ibid., 153.
21. Ibid., 150–151.
22. Question 7, "Whether the whole earth is habitable." Ibid., 154–160.

This, of course, describes the behavior of celestial bodies, each of which revolves with uniform circular motion in its natural place. Since motion is nobler than rest for celestial orbs, it would follow by implication (Buridan does not explicitly mention the empyrean sphere) that an immobile empyrean heaven would be less noble than the mobile celestial spheres that it surrounds and contains.[23] Because no one who accepted the empyrean sphere would have relegated it to a status more ignoble than that of the mobile celestial spheres contained concentrically within it, we may plausibly conclude that Buridan's support for the empyrean sphere was at best dubious.[24]

If Buridan's opposition to the empyrean sphere was ambivalent, Albert of Saxony's was not. Without naming the empyrean sphere, Albert flatly rejects the existence of an immobile heaven.[25] His three arguments against it are the same as those reported by Buridan. But whereas Buridan accepted the immobile empyrean sphere by reason of faith[26] while simultaneously furnishing reasons for rejecting it on natural grounds, Albert of Saxony ignores theology and, on the basis of the three Aristotelian arguments, rejects the existence of an immobile sphere, thus becoming one of the few who did.

Was the existence of the empyrean heaven a matter of doctrine and faith? An error condemned by the bishop of Paris in 1244 suggests that it was. That error, according to Federici Vescovini, is "the thesis that the glorified soul and the Blessed Virgin are not in the empyrean heaven with the angels but in the aqueous, or crystalline, heaven and above the firmament, in fact in the ninth sphere."[27] In Paris, at least, the empyrean heaven seems to have had Church sanction. A long tradition of general acceptance throughout Christendom may, in any event, have conferred upon the empyrean heaven a kind of quasi-doctrinal status. If Albert of Saxony violated church tradition or doctrine, there is no evidence that he suffered any penalty or adverse reaction.

2. *In defense of an empyrean sphere*

No one can doubt that scholastics overwhelmingly accepted the existence of an empyrean sphere. And yet, because it was an invisible entity without detectable effects, belief in its existence was based largely on conjecture,

23. These arguments appear in question 22, "tertia persuasio" ibid., 228, for the arguments in favor of the greater nobility of rest, and 230, lines 25–31, for the conditions under which motion is nobler. For more on the comparative nobility of motion and rest, see this volume, Chapter 20, Section V.2.b.
24. As we shall see in Section IV of this chapter, Amicus used the rest–motion and noble–ignoble dichotomies to defend the existence of an immobile empyrean heaven.
25. Albert of Saxony [*De caelo*, bk. 2, qu. 8], 1518, 107r, col. 1–107v, col. 2.
26. Buridan says that he accepts a resting, or immobile, heaven that lies beyond the mobile heavens "because we assume an empyrean heaven there on faith" (Arguitur quod sic, quia ex fide nos ponamus ibi caelum empyreum). Buridan, *De caelo*, bk. 2, qu. 6, 1942, 149.
27. See Peter of Abano, *Lucidator*, 1988, 200.

exegesis, and authority.[28] Thomas of Strasbourg argued that the existence of the empyrean heaven, which he identified with the tenth orb, could not be shown necessary by reason[29] but only by a probable argument.[30] Campanus of Novara spoke for many when, in his *Theorica planetarum*, he declared that "whether there is anything, such as another sphere, beyond the convex surface of this [ninth] sphere, we cannot know by the compulsion of rational argument. However, we are informed by faith, and in agreement with the holy teachers of the church we reverently confess that beyond it is the empyrean heaven in which is the dwelling place of good spirits."[31] Indeed, for some the empyrean heaven had no scriptural basis whatever.[32] Its widespread acceptance was based on theological authorities who had conceived it as a place for spiritual beings and who were then gradually committed to consider further the nature and properties of that supreme heaven, arriving at those that we have been examining.

Although some adopted the empyrean heaven solely by reason of faith, arguments based on natural philosophy were also formulated for its existence. Without mentioning it by name, preferring to characterize it only as an immobile orb, Pierre d'Ailly, a staunch supporter of the empyrean heaven, discussed it at some length in his *14 Questions on the Sphere of Sacrobosco*, where, in the second question, he gives three arguments as to why an immobile sphere must exist beyond the ten mobile spheres (149v). The first opinion conceives of the empyrean heaven as the place and container of the world. It assumes that a mobile sphere must change place either as a whole or with respect to its parts, from which it follows that the mobile sphere is in a place. The place of that mobile sphere must surround and contain it. Moreover, that surrounding place must also be immobile, a consequence which d'Ailly does not actually establish. But the idea that all movable things ought to have an ultimate immobile place had earlier mo-

28. In his *Sentences* [bk. 2, dist. 2, qu. 2, art. 1], 1929–1947, 2:71, Thomas Aquinas explains that "the empyrean heaven cannot be investigated by reason because we know about the heavens either by sight or by motion. The empyrean heaven, however, is subject to neither motion nor sight . . . but is held by authority."
29. As the first of two conclusions in his *Sentences*, bk. 2, dist. 2, qu. 2, 1564, 133v, col. 1, Thomas of Strasbourg declares: "quod caelum decimum esse, quod sancti appellant empyreum, non potest ostendi necessaria ratione."
30. Thomas of Strasbourg's second conclusion (ibid., 133v, col. 2) asserts "quod caelum empyreum esse potest aliquo modo declarari probabili persuasione."
31. Campanus of Novara, *Theorica planetarum*, 1971, 183. I have made one change in the translation by Benjamin and Toomer. Following "rational argument" they add "alone," which implies that rational argument, along with other methods, is used to arrive at the existence of the empyrean sphere. In fact, Campanus, like Thomas Aquinas before him, was arguing that rational argument was useless in determining the existence of that special sphere.
32. *Dictionnaire de théologie catholique*, vol. 2, pt. 2, cols. 2505, 2508. Among those who found no scriptural basis for the empyrean sphere were Thomas Aquinas, Durandus de Sancto Porciano (ca. 1275–1334), and Cajetan (Thomas de Vio; 1468–1534). For Dante, it had no real existence in space but only in the Divine Mind (see *Convivio*, II.iii.II). For this reference and for a thorough set of citations to Dante's treatment of the empyrean heaven, see Toynbee, 1968, 181 ("Cielo empireo").

tivated Campanus of Novara to declare (1971, 183) that "The empyrean's convex surface has nothing beyond it. For it is the highest of all bodily things, and the farthest removed from the common center of the spheres, namely the center of the earth; hence it is the common and most general 'place' for all things which have position, in that it contains everything and is itself contained by nothing."[33] Those who assumed this argument – and all who believed in the existence of an empyrean orb did – agreed with Aristotle that our world was surrounded by an ultimate convex surface but departed from him by the assumption of its immobility.[34]

D'Ailly's second argument is based on Aristotle's claim (*De caelo*, bk. 2, ch. 2) that the heavens possess absolute differences in directions, which he identified as right and left, front and back, and above and below. D'Ailly insists that mobile spheres could not exhibit such directions if they turned, because that direction which is now right would become left and the part that is up would become down. Only in an immobile sphere can such absolute directions and differences of position be found.

The third argument concerns possible celestial influences. Could an immobile empyrean heaven influence terrestrial change and perhaps even be essential for that purpose? If so, it would not only have a theological role but an important and vital cosmic role as well. Because of the potential significance of such an empyrean heaven, we shall consider this problem, including d'Ailly's third argument, at some length.

III. Can the empyrean heaven influence the terrestrial region?

Until the seventeenth century, when some raised doubts about its existence,[35] controversy about the empyrean sphere focused not on its existence but on its function and influence. Although in a subsequent chapter (Ch. 19), we shall address the crucial theme of celestial influence as it pertained to all the mobile orbs, in this chapter we shall consider the subject of celestial influences with reference only to the immobile, empyrean heaven.

Throughout the five centuries embraced by this study, ambivalence best characterizes scholastic attitudes about alleged influences of the empyrean heaven. As an invisible, immobile entity whose essential raison d'etre was purely theological and which lacked explicit biblical justification, the em-

33. This passage is also translated in Duhem [Ariew], 1985, 175. Dante also says that the empyrean heaven "contains all bodies and is contained by none." See Toynbee, 1968, 181, for references. In Chapters 8 and 9, we saw the kinds of things that were conceived to exist, or possibly exist, beyond the world or its last convex surface.
34. On Aristotle's discussion of the place of the last sphere, see Chapter 6, Section III. Duhem considers the empyrean heaven within a general treatment of the medieval doctrine of place. *Le Système*, 1913–59, 7:197–202.
35. For example, Thomas White, who rejected it, and Nicolaus Caussin, who questioned its existence. Donahue, 1972, 253, 251.

pyrean sphere seemed remote and uninvolved with the governance of the world. Uncertainties about it were also attributable to the fact that "the saints," as Saint Bonaventure expressed it, "speak little of this heaven because it is hidden from our senses, and the philosophers [say] even less."[36]

From the thirteenth century onward, however, discernible efforts were made to attribute some effects to the empyrean heaven and thus involve it in cosmic operations. Scholastics began to inquire whether the empyrean heaven could influence inferior things. There was nonetheless some reluctance to involve the empyrean sphere in regular cosmic operations. Bartholomew the Englishman, for example, denied that the empyrean heaven was needed for the continuation of generation among inferior things.[37] One of its functions was, rather, to complete the body of the universe by serving, along with the earth, as one of a pair of immobile, bodily extremes. Whereas the earth was opaque (and presumably the heaviest body in the universe), the empyrean was the most luminous, most subtle, and least heavy body. In his commentary on Sacrobosco's *Sphere*, Michael Scot describes the empyrean sphere as immobile and uniformly filled with light, a light that can extend its influence only to the inferior heavens, though not uniformly, because those less perfect heavens are incapable of receiving the empyrean light in a perfect way.[38]

Not all were so cautious or negative. Indeed, the anonymous thirteenth-century author of the *Summa philosophiae*, falsely attributed to Robert Grosseteste, made the empyrean sphere the most noble and powerful body in the universe. "The empyrean heaven," he declared, "is the original principle of rest of all natural things." Moreover, it was the empyrean heaven, not the center of the earth, with respect to which all the mobile spheres originally moved. On this basis, he proclaims that "the immobility of the empyrean heaven is more the universal cause of every transmutation of generable and corruptible things than the *primum mobile* and the other inferior spheres, just as a primary cause is more a cause than a secondary cause."[39]

Although most scholastics assigned a causal role to the empyrean heaven,

36. "Dicendum, quod quamvis Sancti parum loquantur de hoc caelo, quia latet nostros sensus, et philosophi adhuc minus." Bonaventure [*Sentences*, bk. 2, dist. 2, art. 1, qu. 1], *Opera*, 1885, 2:71, col. 2.

37. Bartholomew the Englishman, *De rerum proprietatibus*, bk. 8, ch. 4, 1601, 380.

38. "Et primum celum a theologis dicitur empyreum non ab ardore sed a splendore et est uniformiter plenum lumine et immobile, non uniformiter tamen influit lumen suum in inferioribus celis eo quod actio agentis non recipitur in passum per modum ipsius agentis sed etiam per modum patientis, ut dicitur in libro *De substantia orbis*." Michael Scot [*Sphere*, lec. 4], 1949, 283.

39. In chapter 3, tractatus 15, the anonymous author of the *Summa philosophiae* treats of the empyrean sphere (see Grosseteste, *Summa philosophiae*, 1912, 545–548). The Latin texts of the two translations just cited are respectively: "Caelumque empyreum totius quietis rerum naturalium originale esse principium" (ibid., 546) and "Quietem caeli empyrei magis esse causam universaliter omnis transmutationis generabilium et corruptibilium, quam primum mobile ceteraeque sphaerae inferiores, sicut causa prima magis est causa quam causae secundariae" (ibid., 547). For a French translation of the second passage (and considerably more), see Duhem, *Le Système*, 1913–1959, 7:199–200.

few were prepared to go so far as to make it "the universal cause of every transmutation." In the question, "Whether the empyrean heaven influences inferior things" (Utrum caelum empyreum influat in haec inferiora), Saint Bonaventure presents five arguments in favor of influence and five against.[40] He concludes that "any of these opinions is sufficiently probable. But which is more true is not clearly apparent."[41] Neither reason nor authority can make obvious why the empyrean heaven has to exercise any influence on sublunar things, "since the influence of the other inferior [celestial] bodies seems to suffice." Nevertheless, Bonaventure inclined toward acceptance of an empyrean influence on inferior things, justifying his decision by observing that among all bodies, the empyrean heaven was the first created and has the greatest size (*moles*) and power (*virtus*). Its size enables it to locate all bodies by surrounding and containing them; its power, or influence, enables it to animate and conserve things, although it may achieve this through the other celestial spheres.[42]

Despite a similar initial ambivalence, Thomas Aquinas eventually arrived at the same conclusion. Because he accepted Aristotle's principle that only bodies in motion could affect other bodies, Thomas at first (in his commentary on the *Sentences*) denies that an immobile body like the empyrean sphere could influence other bodies.[43] But later, in his *Quodlibetal Questions*, he concedes that, upon further reflection, the empyrean heaven does indeed influence inferior bodies.[44] The world order demands that corporeal things be governed by spiritual things and inferior bodies be ruled by superior bodies. It would be absurd if the empyrean heaven did not influence inferior bodies, for then it would form no part of the universe.

To explain how the empyrean sphere influences inferior bodies, Thomas employs a hierarchy of descending perfections based on rest as most perfect, followed by uniform motion, and finally difformity – that is, nonuniformity – of motion. By its rest, the empyrean heaven causes the first movable sphere to move with a uniform motion and thus produce the daily motion as it moves from east to west; the first movable sphere, in turn, causes all the spheres below it to move with motions that vary with the distances of their planets from the earth. As the planets vary their distances, they cause generation and corruption in sublunar bodies. Thus the empyrean sphere is the direct cause of uniform motion and therefore the direct cause of permanence in the universe, and through the uniform motion of the first

40. *Sentences*, bk. 2, dist. 2, art. 1, qu. 2, *Opera*, 1885, 2:73–75.

41. "Quaelibet harum opinionum satis probabilis est; quae autem sit magis vera, non plane apparet." Ibid., 74, col. 2.

42. Ibid., 74, col. 2–75, col. 1.

43. "Respondeo dicendum quod caelum empyreum nullam habet influentiam super alia corpora, quae rationabiliter poni possit." Thomas Aquinas, *Sentences*, bk. 2, dist. 2, qu. 2, art. 3, 1929–1947, 2:76.

44. Quodlibetum 6, qu. 11, art. 19: "Whether the empyrean heaven exercises influence over other bodies." See Thomas Aquinas, *Quaestiones quodlibetales*, 1949, 130; also Litt, 1963, 260–261. Thomas wrote his *Sentences* during 1252–1256 and his sixth quodlibetal question in 1272 (see Weisheipl, 1974, 358–359, 367).

movable sphere, the empyrean sphere is also the indirect cause of the generation and corruption caused by all the celestial spheres below the first movable sphere.

Like Bonaventure and Thomas Aquinas, most theologians expressed opinions on the empyrean heaven. Without providing much detail, Richard of Middleton ([*Sentences*, bk. 2, dist. 2, art. 3, qu. 3], 1591, 2:44–45), who unhesitatingly accepted the influence of the empyrean on inferior things, suggests that the immobile empyrean sphere moderates the influences exerted by the mobile celestial spheres. The empyrean sphere was ordained for the use of man both in the future life, when it would presumably serve as the abode of the elect, and in the present life, when it influences inferior things by its power. Also convinced of an empyrean influence on inferior things, Thomas of Strasbourg sought to convince his readers of its necessity, although he fails to describe the form of that influence.[45] Recognizing that the chief obstacle to the concept of influence stemming from an immobile empyrean heaven was the widely accepted idea that only a body in motion can influence another body,[46] Thomas counters by invoking the Joshua miracle, which, as we shall see, was frequently cited as evidence that even when at rest the celestial region could influence inferior things. Although the Sun had been commanded to halt for a day, celestial influences continued to affect the earth, "for otherwise, those living here below would have been dead."[47] Similarly, a magnet at rest attracts iron.[48] Thus a body at rest is capable of exerting influence on distant things. Thomas of Strasbourg therefore insists that not only can the immobile empyrean heaven influence inferior things but that it can do so directly, without the mediation of celestial motions. That this does not actually happen is a consequence of the nature of inferior things, which can only receive empyrean influences indirectly through the mediation of the celestial motions.

D'Ailly's third argument, alluded to earlier, seemed to offer empirical evidence in favor of an immobile heaven. In its various versions, it was frequently repeated in the sixteenth and seventeenth centuries. Citing differences in fruits and customs and in many other things that one finds on parts of the earth that lie on the same latitude, or, as d'Ailly expressed it,

45. Thomas of Strasbourg, *Sentences*, bk. 2, dist. 14, qu. 1: "Whether the heavens are the cause of inferior things" ("An caelum est causa horum inferiorum"), 1564, 156v, col. 2–158v, col. 2; for the discussion of the empyrean heaven, see 158r, col. 2–158v, col. 2.

46. As Thomas of Strasbourg explains (ibid., 158r, col. 2): "quidam venerabilis doctor in suo scripto dist. 2 secundi libri Sententiarum quia eo ipso quod celum empyreum non movetur non potest sibi competere aliqua realis influentia in cetera corpora quia corpus non agit nisi per motum."

47. "Ad primum dicendum quod nec illa consequentia valet nec eius probacio quia tempore Iosue sol non movebatur per spacium unius diei et tamen habuit realem influentiam in ista inferiora alias ista viventia hic inferius fuissent mortua." Ibid., 158v, col. 1. Among those who cited the Joshua miracle as evidence that the immobile heavens could influence inferior things were Hervaeus Natalis in the thirteenth century, Nicole Oresme in the fourteenth, and the Coimbra Jesuits in the sixteenth.

48. "Magnes non motus trahit ferrum." Ibid.

that lie "between east and west [equi]distant from the poles," d'Ailly insists that these differences cannot be explained by the circular motions of mobile spheres, because, as the same configurations of celestial bodies swept over the same latitude, all the inhabitants and all the fruits and vegetation along that latitude would be exposed to the same celestial influences. Under these circumstances, samenesses rather than differences ought to be found among members of a given species on the same latitude. But an immobile sphere like the empyrean could produce the differences we see. It could act differentially on various parts of the earth and, by virtue of its immobility, do so constantly and permanently.[49] As evidence of such differential power, d'Ailly repeats a common belief that all stars have greater power in the east than in any other part of the heaven.[50]

IV. Concepts of the empyrean orb in the late sixteenth and the seventeenth century

The range of ideas about the empyrean heaven that were predominant in the late Middle Ages remained prevalent in the late sixteenth and the seventeenth century, although some new interpretations and attributes were added. Because it failed to manifest its existence in any of the usual ways, there was a continuing sense that the existence of the empyrean heaven was incapable of rational proof.[51] Although a rigorous demonstration was unattainable, certain kinds of evidence served in its stead. In one of the most thorough and detailed treatments of the empyrean heaven, Bartholomew Amicus presents the kinds of reasons that were deemed persuasive. By the seventeenth century, Amicus could appeal for acceptance of the empyrean heaven to the authority of the Church Fathers, the traditional acceptance of the scholastics, and the common acceptance of the Church.[52] With such weighty credentials, few if any scholastic authors were inclined to deny the existence of this immobile orb. Indeed, Amicus also assumed, along with many others, that when Genesis declares that God created heaven and earth, this signified either the empyrean heaven only or the latter along with all the other heavens. Moreover a number of biblical passages speak of this heaven as the domicile of God and the blessed, and the "heaven" in question was usually assumed to be the empyrean.[53] It was probably with such

49. We saw, in Section II.1 of this chapter, that Buridan rejected this argument.
50. Mastrius and Bellutus [*De coelo*, disp. 2, qu. 7, art. 1], 1727, 3:510, col. 2, par. 189–511, col. 1, par. 190, repeat this in the seventeenth century.
51. See Donahue's translation of a relevant passage (1972, 255) from p. 507 of the *De caelo* section of Roderigo de Arriaga's *Cursus philosophicus*.
52. "Prima concl.: Certum dari coelum empyreum. Primo patet ex auctoritate Patrum et Scholasticorum et communi assensu Ecclesiae." Amicus [*De caelo*, tract. 4, qu. 6, dubit. 5, art. 2], 1626, 194, col. 1. This is the first of six conclusions that represent Amicus's own opinions and extend over pages 193, col. 2–197, col. 2.
53. Amicus cites Psalms 67, 102, 113, and Deuteronomy 10.

thoughts in mind that Otto von Guericke cited Jacques du Bois, an ecclesiastic of Leyden, as insisting that anyone "who denies that there is an empyrean [heaven] above the visible heavens does not believe the sacred words nor does he believe the foundations of Christianity" (*Experimenta nova*, 1672, 49).[54]

The empyrean was also a perennial problem for Aristotelians because it confronted them with the problem of whether motion or rest was nobler. Was an immobile celestial body, even one that surrounded all of the moving spheres, more noble than the moving spheres it contained? The kind of argument that Buridan formulated found its counterparts in early modern scholasticism. Amicus ([*De caelo*, tract. 4, qu. 6, dubit. 5], 1626, 194, col. 1), for example, mentions an objection to the existence of the empyrean heaven, namely that mobile bodies are nobler than immobile bodies, as is obvious from the earth, which is immobile and deemed the most ignoble of all bodies. If the empyrean heaven is said to be the noblest of all spheres, it cannot, therefore, be immobile. Amicus resolves the dilemma (ibid., 197, col. 2) by emphasizing the virtue of immutability. The empyrean heaven is not immobile because it lacks the capacity for motion but is immobile by divine decision, since it does not require motion to realize its own ends. Amicus further denies that an immobile body in its proper place is more ignoble than a mobile body in its proper place. Indeed, immobility is a perfection, as is obvious from the fact that God is immutable and therefore immobile. But motion is an imperfect thing, because it is a path to perfection and not the perfection itself. The peculiar demands of the empyrean heaven fostered an ambivalent attitude toward the contraries rest and motion. Whereas Buridan exalted motion over rest, which he compared to the ignoble earth, Amicus linked rest with divine immutability and exalted immobility over motion.[55]

The invisibility of the empyrean heaven was also puzzling. If it really existed, why was it not visible, since it is said to be the most lucid body? Most responses emphasized the great distance of the empyrean heaven and the thickness of the inferior heavens that were interposed between it and us. For such reasons, the brilliant splendor of the empyrean was imperceptible to human observers.[56] Pedro Hurtado de Mendoza appealed to the divine when he suggested that God had made a certain opaque curtain to conceal the empyrean heaven from unworthy eyes in a manner similar to the silken cover used to conceal the Holy of Holies.[57]

54. Also in the seventeenth century, George de Rhodes insisted that the empyrean heaven must not be denied by any Catholic. De Rhodes, *De coelo*, bk. 2, disp. 2, qu. 1, sec. 2, pt. 3, 1671, 281, col. 1.
55. Nicole Oresme also considered rest nobler than motion (see Oresme [*De spera*, qu. 8], 1966a, 161). Scholastic authors sometimes assigned greater nobility to rest or motion solely on the basis of the argument they sought to demonstrate.
56. See Amicus, *De caelo*, tract. 4, qu. 6, dubit. 5, art. 4, 1626, 197, col. 2.
57. Hurtado de Mendoza [*De coelo*, disp. 2, sec. 5], 1615, 375, col. 1, par. 63. In his discussion of the empyrean heaven, von Guericke, *Experimenta nova*, 1672, 49, cols. 1–2, repeats

The central issue about the empyrean heaven was not, however, its existence, which almost all scholastics assumed, but rather its alleged effects on inferior parts of the universe, ranging from the orb immediately below it all the way to the earth itself. Late sixteenth- and seventeenth-century scholastics adoped a variety of opinions, ranging from a denial of causal efficacy – Raphael Aversa, Mastrius and Bellutus, and Illuminatus Oddus – to the acceptance of direct influence on terrestrial events – Christopher Clavius, the Coimbra Jesuits, and Bartholomew Amicus. On balance, however, there was a shift of opinion toward a denial of terrestrial influence to the immobile empyrean heaven.

1. Does the empyrean heaven cause terrestrial effects?

In a manner similar to Thomas Aquinas, the Coimbra Jesuits, for example, argued that if the empyrean heaven were to be considered part of the universe, as they believed, influences of some kind had to flow from it, for otherwise it would not be part of the world, and, according to Mastrius and Bellutus and others, its existence would be in vain. Believers in an active empyrean heaven also argued, as Raphael Aversa informs us ([*De caelo*, qu. 35, sec. 5], 1627, 180, col. 1), that because the substance of the empyrean heaven is nobler than that of all other orbs below it, it must surely influence terrestrial change and events just as they do. Otherwise it would be more ignoble, rather than more noble and efficacious, than the orbs below it. To counter this argument and deny causal efficacy to the empyrean heaven, Mastrius and Bellutus argue that the empyrean sphere's perfect substance does not imply a more perfect operation,[58] since a perfect substance need not exercise influence on anything below it. Despite a lack of causal efficacy on inferior things, the empyrean sphere would not have been created in vain, because it serves as the abode of the blessed and produces a light appropriate to that region.[59]

But what kind of influences, if any, could the empyrean heaven exert on the regions below, especially the earth? We saw earlier how Pierre d'Ailly responded to this fundamental problem, and we shall now see that in the

this opinion when he declares: "Some state that the farthest part of the empyrean is solid and is darkened from [or by?] the inferior [or nearest] part as if by a certain curtain of thick and opaque matter, lest that light and [that] heaven be perceived [or seen] by unworthy mortals here in the world; just as once among the Jews, the Holy of Holies was concealed with a certain cover."

58. Mastrius and Bellutus, *De coelo*, disp. 2, qu. 7, art. 1, 1727, 3:510, col. 2, par. 189–511, col. 1, par. 190.

59. Illuminatus Oddus [*De coelo*, disp. 1, dub. 24], 1672, 74, col. 1, accepts the same argument – indeed he even mentions Mastrius and Bellutus (71, col. 2) – when he declares that "perfection in a substance does not always imply a more perfect operation. And this is obvious in angels, who, although they are most perfect, can elicit no substantial action. Thus it suffices that it [the empyrean heaven] should serve as a dwelling place for the blessed so that it cannot be said to be in vain." Aversa says much the same thing (*De caelo*, qu. 35, sec. 5, 1627, 180, cols. 1–2).

late sixteenth and the seventeenth century the response was divided: some denied causal efficacy to the empyrean, while others upheld it. Let us examine the latter position first.

a. The affirmative arguments

For those who believed in the causal efficacy of the empyrean heaven, the problem was to identify the direct effects that it caused. Because its sublunar effects were hidden, the Conimbricenses, for example, professed ignorance as to the manner in which the empyrean heaven acted on inferior things.[60] Others, however, were more forthcoming. In support of an active empyrean heaven that helped govern the sublunar region, some, like Bartholomew Amicus, invoked proof "by natural reasons" (*rationibus naturalibus*), a kind of proof that had by then become rather common and supplemented traditional appeals to theology and Scripture. Such "proofs" involved the identification of discernible terrestrial, or sublunar, effects that were attributable to the empyrean sphere and could thus reinforce belief in its real existence. The most dramatic evidence lay in the diversity of effects observed in different parts of the earth and even on the same parallel of latitude. For instance, men of a given region have the same inclinations, whereas those of another have different inclinations; and the trees of one region are alike, whereas those of another are different. In common examples, it was claimed that although the fastest horses are bred in Hungary, on the forty-seventh parallel, elsewhere on the same parallel they are not produced at all; apes (*simiae*) are generated in Mauretania but not elsewhere on the same parallel. Such diversity along the same parallels ought to be ascribed to the heavens, since it appears unlikely that they arose because of differences in the earth itself. But are the heavens that can explain these diversities mobile or immobile? Mobility cannot explain them, because mobile heavens sweeping uniformly and regularly over the same parallel would produce uniform, rather than diverse, terrestrial effects in every part of the earth under that parallel. The cause of these differences ought therefore to be attributed to the empyrean sphere, which, because of its immobility, could focus rays from every one of its parts to the corresponding region of earth beneath.[61]

To produce diverse effects, however, different powers had to be assumed to radiate from different parts of the empyrean heaven. But if, as most

60. See Conimbricenses [*De caelo*, bk. 2, ch. 3, qu. 2, art. 2], 1598, 196.
61. The substance of this argument with a variety of examples appears in Amicus, *De caelo*, tract. 4, qu. 6, dubit. 5, art. 2, 1626, 195, cols. 1–2 and tract. 6, qu. 3, 353, cols. 1–2; Clavius [*Sphere*, ch. 1], *Opera*, 1611, 3:24; Aversa, *De caelo*, qu. 35, sec. 5, 1627, 180, col. 2, who mentions Clavius as his source for the argument; and Mastrius and Bellutus, *De caelo*, disp. 2, qu. 7, art. 1, 1727, 3:510, col. 2, par. 189. Clavius, Mastrius and Bellutus, and Illuminatus Oddus, *De coelo*, disp. 1, dub. 24, 1672, 73, col. 2, mention the examples of Hungary and Mauretania. Although the same examples are sometimes described differently, their object is to show that only the immobile empyrean heaven could be the cause of such diversity.

believed, the empyrean heaven is homogeneous, its impact should be the same all over. How, then, could such a homogeneous body radiate different influences from its different parts? Recognizing a potentially troublesome problem, Amicus, who defended the causal efficacy of the empyrean heaven, denied true homogeneity to it, arguing that "just as diversity in inferior things belongs to the perfection and pleasure of sight, so in the empyrean heaven," that is, just as the perfection of sight is partially constituted from the diversity of things seen, so also is the perfection of the empyrean sphere partially constituted from the diversity of inferior things that it helps produce by its unhomogeneous nature. Amicus was not alone in attributing heterogeneity to the empyrean heaven.[62]

b. The negative arguments

The sense that the empyrean heaven was not created for the purpose of influencing sublunar things remained strong. Those who adopted this attitude had to reply to the argument that if the empyrean heaven exercised no influence on the physical things it contained, then the most perfect substance God had created – and it was routinely so described – was created in vain. This charge was easily countered by insisting that God's primary purpose in the creation of the empyrean heaven was to provide a domicile for the bodies and spirits of the blessed. Its function was supernatural, rather than natural.[63] And of course it was always thought to function as the container of the world.[64] But even as the most perfect of things, it was not essential that it engage in more perfect operations, such as influencing terrestrial things more perfectly than did the mobile celestial bodies it contained. To illustrate the point, Illuminatus Oddus noted that although angels are the most perfect of things, they do not cause substantial actions.[65]

62. Thomas Compton-Carleton [*De coelo*, disp. 2, sec. 4], 1649, 404, col. 2, denied homogeneity to the empyrean sphere when he declared: "it seems to me that a variety of effects that are perceived in different parts of the earth are poured forth from the diverse, heterogeneous parts of the empyrean heaven." Melchior Cornaeus [*De coelo*, tract. 4, disp. 2, qu. 1, dub. 5], 1657, 491, also seems to have considered the empyrean heaven heterogeneous (*quasi heterogeneas*) when he divided it into three parts, the lowest of which is solid and serves as a foundation for blessed bodies; the middle is fluid and respirable, like air; the highest part is also solid and encloses the respirable, airlike substance and serves as the roof of the celestial structure. As the source of this threefold division of the empyrean sphere, Cornaeus cites Lessius's *De summo bono*, ch. 8.
63. Aversa, *De caelo*, qu. 35, sec. 5, 1627, 180, col. 1, declares that "iam probatur non esse de facto ullam operationem tribuendam celo empireo in haec inferiora corpora quia caelum empireum per se et proprie conditum et ordinatum fuit ut esset domicilium Beatorum."
64. Oddus, *De coelo*, disp. 1, dub. 24, 1672, 73, cols. 1–2, explains that although the empyrean heaven lacks the capacity for action, it serves to contain the bodies of the blessed and the blessed spirits and is the boundary for the physical universe ("Empireum quamvis omnis actio ab eo removeatur, sed satis est deservire ad continendum intra se corpora beatorum et domicilium esse beatorum spirituum, et machinam hanc universi terminare").
65. "Nam ut saepe dictum est, perfectio in substantia non semper arguit perfectiorem operationem. Et patet in Angelis, qui licet perfectissimi sint, nullam tamen possunt actionem substantialem elicere. Unde sufficit ut beatis ad inhabitandum inserviat, ne otiosum dicatur." Ibid., 74, col. 1.

The argument that dramatically characterized the empyrean heaven as heterogeneous so that its influences could explain the diverse customs, plants, and animals that lay along a given parallel of latitude also came under attack. The idea of describing the empyrean heaven as heterogeneous must have struck Raphael Aversa as contrary to one's expectations about the most perfect of all celestial spheres, which he regarded as the same in all its parts.[66] Rather than ascribe heterogeneity to the empyrean heaven, Aversa assigned the cause of terrestrial diversity along a given parallel of latitude to the earth itself, which possessed varied dispositions, different seeds of things, and a multiplicity of mixtures of water and air within the bowels of the earth. These various dispositions influence events on the earth's surface and can cause diversity.[67] Indeed, even celestial bodies moving uniformly over the earth's surface might cause differential effects. Oddus emphasizes the disparate relationships that distinct parts of the earth may have with celestial bodies. Conjunctions from the celestial region may affect different terrestrial areas differentially. "If efficient and material causes vary," Oddus concludes, "it is little wonder that some effects are generated in one area and not in another."[68]

Some assumed that if the empyrean heaven affects the sublunar region, it does so by somehow transmitting effects to the other celestial spheres below. But Aversa wonders what it is that the empyrean transmits to those inferior celestial orbs. After all, the latter possess their own proper qualities, attributes, and causal powers and are not in need of anything from a superior, immobile sphere. In sum, the empyrean has nothing to confer on the mobile spheres below.

If the transmission of physical influences by the empyrean heaven was highly controversial, the transmission of light was more readily conceded, because of the alleged brilliance and splendor of the empyrean heaven. But this particular influence is discussed later, in the chapters on celestial light (Ch. 16) and the nature of celestial effects on the terrestrial region (Ch. 19), where light is distinguished as one of three basic modes of influence.

2. *The status of the empyrean heaven*

From the thirteenth to the seventeenth century, few scholastic theologians denied the existence of the empyrean heaven, and several, as we saw, even

66. "Nam sine dubio hoc caelum [i.e., the empyrean] est totum homogeneum et eiusdem rationis in omnibus suis partibus." Aversa, *De caelo*, qu. 35, sec. 5, 1627, 181, col. 1. This argument and the others in this and the next paragraph are drawn from the sixth of seven arguments Aversa musters to deny claims for an active empyrean heaven. For all seven arguments, see ibid., 179, col. 1–181, col. 1; the sixth appears on 180, col. 2–181, col. 1.

67. Oddus, *De coelo*, disp. 1, dub. 24, 1672, 73, col. 2, says much the same thing. The earth's powers and dispositions vary from region to region and thus may plausibly explain different effects in different regions.

68. Ibid.

went so far as to insist that rejection of the empyrean heaven was tantamount to a rejection of Christianity. All accepted its theological function as the abode of the blessed, but disagreement arose when it came to assigning a cosmological role to the empyrean sphere. It was one thing to concede its existence, and quite another to believe that it influenced the physical world. The absence of discernible effects, its acknowledged invisibility, and its patently theological nature made arguments about the cosmological role of the empyrean heaven inconclusive. Whereas thirteenth-century theologians and natural philosophers were prepared to attribute some influence on the physical world to the empyrean heaven, many denied it that capability by the seventeenth century. In the numerous commentaries or *questiones* on *De caelo* written by scholastic theologians in the seventeenth century, either as independent treatises or as part of a larger *cursus philosophicus*, many – like Sigismundus Serbellonus, Francisco de Oviedo, and Johannes Poncius – omitted serious consideration of the empyrean heaven.[69] For them it may no longer have seemed appropriate to include such a blatantly theological heaven in a cosmological treatise.

As we saw, however, serious discussions continued. Among those who still thought it worthy of discussion in a cosmological treatise, as many seemed to reject its influence as proclaimed it. When account is taken of those who ignored the subject and those who opposed any empyrean influences, we may conclude that the number of scholastic theologians who believed that the empyrean heaven influenced celestial and/or terrestrial physical operations was relatively small by the middle of the seventeenth century. Copernican cosmology may have played a significant role in producing this result. But not even Copernican cosmology and the Newtonian Scientific Revolution could completely cause the disappearance of defenders of an empyrean influence. Well into the eighteenth century, Juan Hidalgo, a Spanish Augustinian Hermit, who wrote a *cursus philosophicus* "according to the thought of Blessed Aegidius Romanus," vigorously repeated his master's thirteenth-century defense of empyrean physical influence (1737, 2:73, col. 2–76, col. 1). There are no surprises in Hidalgo's discussion, much of which is directed specifically against Aversa. As the most noble of all bodies, the empyrean heaven should influence inferior things, not by contact but by its operation. Although the empyrean heaven is immobile, its influence is disseminated by the motions of the inferior celestial spheres. As a part – indeed, the noblest part – of the universe, the empyrean heaven must exercise an influence on other parts. The philosophers have mistakenly accounted for the behavior of inferior things by assigning all the causes to the mobile heavens. But they should attribute some influences and causal changes to the empyrean sphere, which was discovered by revelation and ought to be properly fitted into the cosmic picture. Hidalgo's readers would find no mention of the great figures who were responsible for the triumph

69. Others, like Pedro Hurtado de Mendoza, treated it only briefly.

of the Copernican revolution. Absent are the names of Copernicus, Tycho, Galileo, Kepler, Newton, and all the others. Although his body was present in the eighteenth century, Hidalgo's mind, filled with the thoughts of Aegidius, lived only in the thirteenth century. Perhaps there were a few more like Hidalgo in the eighteenth century, who defended the empyrean heaven as an active influence in the physical world. Already in the seventeenth century, however, most of their scholastic predecessors had eliminated it from their cosmology.

16

Celestial light

As a theme and metaphor, light has been perhaps the most pervasive and ubiquitous topic in Western science, philosophy, theology, literature, poetry, and art. Its chief significance for cosmology, however, lies in its relationship to celestial bodies.

I. The sources

Light as it was treated in the tradition of geometric optics and in visual theory, especially in that of the medieval perspectivists, plays little role in what follows. Although the connections between light and theology are more fruitful, they are not our major source of information for the relations between light and celestial bodies and hence are mentioned only briefly here at the beginning. The most fundamental cosmological link between light and theology derives from the first chapter of Genesis, which declares that God created light on the first day and that he created the celestial luminaries to light heaven and earth on the fourth day. We would, therefore, expect to find discussions of light in commentaries on the creation account of Genesis. During the Middle Ages, commentaries on the six days of creation were rarely made in special hexaemeral treatises, but they were usually made in commentaries on the *Sentences* (*Sententiae*) of Peter Lombard, which, written around 1150, was the most famous theological textbook of the Middle Ages and a required text for comment by all theological students. There, in book 2, distinction 13, Peter discussed the first of the six days of creation and asked whether light was corporeal or spiritual; why God found it necessary to create the Sun if he had already created light on the first day; and the meaning of day and the distinction between day and night (Peter Lombard, *Sentences*, bk. 2, dist. 13, ch. 2, 1971, 389–391). In book 2, distinction 14, Peter briefly mentions the creation of Sun, Moon, and stars on the fourth day and asks about their utility (ibid., chs. 9–10, 398–399).

For nearly five centuries, scholastic commentators on the *Sentences* routinely included questions on the nature of light. Most of the questions, however, discussed aspects of light that were only tangentially, if at all, relevant to light as it affected celestial bodies. Saint Bonaventure, for example, inquired whether the light of the first day was corporeal or spiritual; how light made day and night; whether light is a body or the form of a body; whether light as the quality of a body is a substantial or accidental form; whether the light that emanates from a body is itself a body; and whether the light that emanates from a body is a substantial or accidental form.[1] Bonaventure's responses reveal no particular concern for the relationship of light to celestial bodies.[2] Thomas Aquinas and others posed similar questions in their commentaries.[3] Although the responses to these questions occasionally contained material relevant for cosmology,[4] most were concerned with the nature of pure light, with little regard for the role of celestial bodies.

Comments on the fourth day of creation produced similar results. Here the questions are about the "luminaries," or visible heavenly bodies created on the fourth day. The comments are usually about their planetary natures, their motions, and how they compare to terrestrial bodies, but rarely about their relationship to light.[5] Certain questions about the empyrean heaven concerned light. One that was usually posed, in commenting on the second distinction of the second book of the *Sentences*, asked "Whether the empyrean heaven is luminous." Although there was general agreement that the immobile, invisible, empyrean heaven was the location of purest light, this light was often assumed to be superior to the light that was diffused throughout the physical cosmos.[6] Indeed, the empyrean was deemed to be invisible to us because of opacity on its concave side, as if it were covered by a curtain (*cortina*).[7]

Theological works – hexaemeral treatises, commentaries on the *Sentences*,

1. Bonaventure [*Sentences*, bk. 2, dist. 13], *Opera*, 1885, 2:311–329.
2. Thomas of Strasbourg reacted similarly. In his single but lengthy question on light, titled "Whether light is a real form" (An lumen sit forma realis), Thomas considers four subthemes, of which only the first – whether celestial light is the same as terrestrial light – is relevant, and this is treated briefly in less than half a column. The rest of the question is given over to whether light is a real or intentional form; whether it is produced from the potency of a medium; and whether two intentional forms can be simultaneously in the same part of a medium. See Thomas of Strasbourg [*Sentences*, bk. 2, dist. 13, qu. 1], 1564, 155r, col. 2–156v, col. 1.
3. See Thomas Aquinas [*Sentences*, bk. 2, dist. 13], 1929–1947, 2:328–340. The same applies to Aquinas's *Summa theologiae*, which contains a series of questions on the six days of creation (pt. 1, qu. 67, arts. 1–4, 1967, 10:52–91).
4. As when Richard of Middleton, in his commentary on the *Sentences*, bk. 2, dist. 13, art. 1, qu. 4, 1591, 2:159, col. 1, considered whether all light is of the same species and concluded that celestial light differs from terrestrial light, since the latter "is like a certain participation in celestial light, or an imitation of it." Thus light from terrestrial fires differs from the light we receive from the celestial region.
5. See Bonaventure, *Sentences*, bk. 2, dist. 14, pts. 1–2, *Opera*, 1885, 2:335–365. The same may be said for Thomas Aquinas (see *Sentences*, bk. 2, dist. 14, 1929–1947, 2:346–360 and *Summa theologiae*, pt. 1, qu. 70, arts. 1–3, 1967, 10:106–125).
6. Thomas Compton-Carleton [*De coelo*, disp. 2, sec. 3], 1649, 403, col. 2, made this distinction. He regarded empyrean light as "a higher light than the light of the other heavens and of the sublunar [region]."
7. Mastrius and Bellutus [*De coelo*, disp. 2, qu. 5], 1727, 3:506, col. 1, par. 157. Hurtado de Mendoza [*De coelo*, disp. 2, sec. 5], 1615, 375, col. 1, par. 63, says the same thing. Arriaga [*De caelo*, disp. 1, sec. 6], 1632, 507, col. 2, says that although the empyrean heaven is most lucid, it is commonly thought that it has an opaque cover on our side of it. Although, he adds, this claim cannot be demonstrated, neither can it be rejected.

and summas of theology – are therefore not the best source of information about scholastic conceptions of light in relationship to celestial bodies, although one can learn about important aspects of light itself: for example, that it was not a body but was either a substantial form of a luminous body – that is, the form that makes the Sun a luminous body[8] – or the accidental quality of a luminous body, so that light could be a quality derived from the substantial form of the Sun,[9] or indeed both.[10] One can also learn that opinion was divided about the transmission of light, some assuming an instantaneous transmission[11] while others argued that it was disseminated successively, in very short but still temporal intervals.[12] The most helpful source of discussion on celestial light is questions on *De caelo*. But even here, the physical aspects of celestial light were accorded only modest attention (see the "Catalog of Questions," Appendix I, qus. 226–234). The most widely discussed question (qu. 231), and easily the most important, however, was whether all stars and planets receive their light from the Sun or are self-luminous. Along with a few other aspects of celestial light, the source of light for celestial bodies is the focus of our attention here.

II. *Lux* and *lumen*

In connection with light sources, scholastics followed Avicenna and distinguished two aspects of light, to which they applied different terms. The term *lux* was associated with light as the luminous quality of a self-luminous body, such as, for example, the body of the Sun. The light from a luminous body that emanated into a surrounding medium, such as the Sun's rays, was characterized by the term *lumen* (see Bacon [Lindberg], 1983, 365, n. 10). The latter was thus capable of lighting bodies that were not self-luminous. As Bartholomew the Englishman expressed it: "*Lumen* is a certain emission, or irradiation, by the substance of *lux*. *Lux* is the original substance

8. Bonaventure has been cited as upholding this position (see Steneck, 1976, 47 and Wallace in Thomas Aquinas, *Summa theologiae*, 1967, 10:61, n. e). As we shall see, however, Bonaventure seems to uphold both positions (see n. 10 of this chapter).
9. This was the most popular view and was held, for example, by Thomas Aquinas (see *Summa theologiae*, pt. 1, qu. 67, art. 3, 1967, 10:61); Richard of Middleton (*Sentences*, bk. 2, dist. 13, art. 1, qu. 3), 1591, 158; and Henry of Hesse (Steneck, 1976, 47).
10. Bonaventure held that both opinions were founded on some truth. It is a substantial form because insofar as other bodies participate more or less in *lux*, so do they have greater or lesser being; but insofar as light is perceptible and is capable of augmentation and diminution it is an accidental property. See *Sentences*, bk. 2, dist. 13, *Opera*, 1885, 2:321, col. 1.
11. This was the opinion of Aristotle (*De anima* 2.7.418b.21–26), who was followed by Richard of Middleton, *Sentences*, bk. 2, dist. 13, art. 2, qu. 2, 1591, 2:162, col. 1. Among those who adopted this opinion in nontheological works were Robert Grosseteste (see Lindberg, 1976, 97) and Witelo (Lindberg's translation in Grant, 1974, 395).
12. This opinion was incorporated into a theological work by Henry of Hesse (see Steneck, 1976, 49). Of those who adopted this opinion, Alhazen was the chief authority and Roger Bacon one of his leading followers (see Grant [Lindberg], 1974, 396–397, where Bacon also describes Alhazen's opinion).

from which *lumen* arises" (Grant [Eastwood], 1974, 383). Although medieval authors may have "ignored the distinction or employed it haphazardly" (Bacon [Lindberg], 1983, 365, n. 10), it was repeated many times well into the seventeenth century.

Another way to characterize the problem of light sources is to view it as one in which natural philosophers sought to determine whether this or that celestial body, or group of celestial bodies, was made visible to us by *lux* or *lumen*.[13]

III. Are the stars and planets self-luminous, or do they receive their light from the Sun?

The role of the Sun lay at the heart of the problem. Was it the ultimate, unique source of celestial light for all planets and stars? If so, it might follow that all visible celestial bodies lacked light, or *lux*, of their own and derived their illumination from the *lumen* of the Sun. According to Albert of Saxony, and many other scholastic authors, authorities were divided on this issue. Aristotle and Averroës, who both assumed the Sun as the sole source of light,[14] were opposed to Macrobius and Avicenna, who "concede that the Moon has light from the Sun" but assume "that all the other planets [and stars] [*astra*] are self-luminous."[15] Lunar eclipses and the waxing and waning

13. Mastrius and Bellutus, *De coelo*, disp. 2, qu. 5, 1727, 3:505, col. 1, par. 151, who give much the same definitions, pose their question in the form: "What are *lux* and *lumen*, and how do they occur in the heavens?" (Quid sint lux et lumen et quibus conveniant in celis?). On p. 506, cols. 1–2, pars. 157–162, they discuss the theme we shall pursue in the next paragraph.

    The term *splendor* was also used. In the thirteenth century, Thomas Aquinas used it (along with *lux*, *lumen*, and *radius*) to indicate reflected rays that reach a smooth and polished surface, such as water and silver, and are then projected again (see *Sentences*, bk. 2, dist. 13, qu. 1, art. 3, 1929–1947, 2:332). An anonymous fourteenth-century author also employed *splendor* and defined the three terms as follows: *lux* is found in a lucid body, such as the Sun; *lumen* occurs in a transparent medium, such as air; and *splendor* occurs in a body that has light by reflection. "Thus indeed we say that *lux* is in the Sun, *splendor* in the other planets, and *lumen* is in the air." (Dicimus quod lux est in sole, splendor in aliis planetis et lumen in aere.) See "Compendium of Six Books," Bibliothèque Nationale, fonds Latin, MS. 6752, 214v. Although the term *splendor* was used to characterize light in the empyrean region, it was rarely used to describe the light of planets. For further discussion of these terms, see Conimbricenses [*De coelo*, bk. 2, ch. 7, qu. 9, art. 2], 1598, 300–301.
14. Plato, *Timaeus*, 39B, 1957, 115, also appears to have assumed that the Sun was the sole source of light for all planets and stars.
15. "In hoc enim Aristoteles cum Averroi contrariantur Avicenne Macrobio and pluribus aliis. Aristoteles enim in libro *De proprietatibus elementorum* vult dicere quod omnes stelle habent lumen a sole, sicut luna et quod non sunt de se lucide quia apparet sibi quod consimili modo debeat esse de luna et aliis astris. Sed Avicenna, cum suis sociis, licet concedat lunam habere lumen a sole, ponit tamen istam conclusionem: quod omnia alia astra habent lumen a se." Albert of Saxony [*De celo*, bk. 2, qu. 20], 1518, 115r, col. 2. See also Oresme [*De celo*, bk. 2, qu. 11], 1965, 640, who adds to Macrobius and Avicenna the names of Heraclitus and Tulius (Cicero).

    Macrobius, in his *Commentary on the Dream of Scipio*, bk. 1, ch. 20 (1952, 168) adopts

of the Moon every month were sufficient to convince all that the Moon had its light from the Sun. During the Middle Ages most natural philosophers seem to have sided with Aristotle and Averroës and argued for the Sun as the sole source of light for the stars and planets. But the answer was by no means obvious. Nicole Oresme and Albert of Saxony insisted that neither side of the argument was demonstrable,[16] although both, for different reasons, thought one side more acceptable than the other. Because of his "love of Aristotle, the Prince of Philosophers," Albert rejects six arguments attributed to Avicenna and defends Aristotle's opinion that all planets and fixed stars receive their light from the Sun.[17] By contrast, Oresme declares (near the end of the question) that the self-luminosity option was more probable.[18] Each side had a sweeping analogical argument based on two assumed truths: that the Sun shines by itself and that the Moon receives its light from the Sun. Those who thought the planets and stars were self-luminous pointed to the self-luminosity of the Sun. If the latter was self-luminous, then, on the assumption that all celestial bodies are in the same species, all other celestial bodies should be self-luminous.[19] By the same token, those who viewed the Sun as sole source of celestial light reasoned that if the Moon received all its light from the Sun, then so also should all other celestial bodies receive their light from the Sun. As we shall now see, many representative arguments for these two opposing positions were of this nature or were based on hierarchical or metaphysical considerations.

this interpretation from Cicero, who, according to Macrobius, allowed that the planets receive light from the Sun, but insisted that they "also have their own light – that is, with the exception of the Moon, which, as we have repeatedly noted, is devoid of light." Macrobius was associated with this position well into the late sixteenth and the seventeenth century, when he (and Avicenna) were mentioned by the Conimbricenses, *De coelo*, bk. 2, ch. 7, qu. 4, 1598, 303, and by Riccioli, *Almagestum novum*, pars prior, bk. 6, ch. 2, 1651, 395, col. 1, who gives the references as book 1, chapter 19, but quotes the correct passage.

16. Oresme, ibid. and Albert of Saxony, ibid. Albert likens each of these arguments to a *problema neutrum*, that is, an argument in which two alternatives are either equally probable or equally incapable of demonstration ("Breviter ista dubitatio utrum omnia astra preter lunam et solem habeant lumen suum a sole est quasi unum probleuma neutrum, sic quod rationes que fiunt pro una parte possunt solui faciliter sicut rationes adducte pro alia"); on *problemata neutra* see Maier, 1949, 199.
17. "Et ideo ob amorem Aristotelis, principis philosophorum, solvam rationes sex iam factas pro opinione Avicenne contra Aristotelem tenendo cum Aristotele quod omnes stelle preter lunam et solem, sive sint planete sive stelle fixe, habent lumina sua a sole." Albert of Saxony *De celo*, bk. 2, qu. 20, 1518, 115r, col. 2. Some of these rejected opinions are described below in Section III.3b of this chapter; see also Section III.3a, for Albert's important qualification at the end of the question, where he allows that planets may have a weak light of their own.
18. Oresme, *De celo*, bk. 2, qu. 11, 1965, 652.
19. Ibid., 644 (seventh argument). We saw earlier that most scholastics denied that all celestial bodies belonged to the same ultimate species. Those who used this argument would have to adopt the single-species argument and then explain how all the diverse celestial bodies could belong to a single species.

1. The Sun as sole source of celestial light

In his discussion of the Moon, Averroës provided one rather popular model of the Sun's mode of activity. The Moon derives its light from the Sun and is therefore not self-luminous. But "it has been demonstrated," Averroës explains, "that if the moon acquires the power of lighting up from the sun, it is not from reflection. That has been proven by Avenatha [that is, by Abraham ben Meir ibn Ezra] in an interesting treatise."[20] "If it illuminates," he continues, "it is by becoming a luminous body itself. The sun renders it luminescent first, then the light emanates from it in the same way that it emanates from the other stars; that is, an infinite multitude of rays is issued from each point of the moon."[21] Duhem likened this process to florescence and explained how Averroës used it to account for the dark spots on the Moon ([Ariew], 1985, 482): "When the light of the sun predisposes and excites them, the various parts of the moon become luminescent; but they do not all become luminescent in the same way." The parts that become least luminescent show up as dark spots.

Albertus Magnus played a significant role in developing arguments in defense of the Sun as the unique source of celestial light. Not only did his arguments exert an influence in the Middle Ages – Albert of Saxony seems to have adopted his key ideas – but they continued to do so into the seventeenth century. Albertus's treatment of the subject thus provides a convenient point of departure for our discussion of the Sun as the sole source of celestial light.[22]

That the Sun was the unique source of celestial light was evident to Albertus by a metaphysical argument that was frequently repeated well into the seventeenth century. Something that exists in many things at once must exist primarily and fundamentally in one of them, which functions as the cause of all the rest. If light is an example of this principle, as Albertus believed, then one thing must be the source and cause of light in everything else. The Sun was the obvious candidate.[23]

20. Oresme, who accepted this argument (*Le Livre du ciel*, bk. 2, ch. 16, 1968, 457), explains that reflection is not the mode of diffusion of the Moon's light. For if the Moon produced reflection like a mirror, "the sun would appear in only a small portion of that part of the moon which seems lighted to us, and at times it would appear in no part at all. . . . It would be exactly as though we were looking at the sun in a mirror or in the water; we do not see it from every position from which we can see the mirror, nor from every angle, but only from a certain position and at a certain distance, and from another distance we see it in another place." All of this follows, says Oresme, from the law of reflection.
21. The translation is from Duhem [Ariew], 1985, 481. For the Latin text, see Averroës [*De caelo*, bk. 2, comment. 49], 1562–1574, 5:131r, col. 2–131v, col. 1.
22. Albertus Magnus [*De caelo*, bk. 2, tract. 3, ch. 6 ("A digression explaining how all the stars are illuminated by the Sun")], *Opera*, 1971, 5, pt. 1:153, col. 1–155, col. 1. See also pages 29 and 107, where Albertus discusses the dissemination of celestial light.
23. "Omne quod est in multis secundum unam rationem, primo est in uno aliquo quod est causa omnium illorum, sicut omnium calidorum causa est ignis. Lumen ergo, quod est multiplicatum in caelo et multis modis est in luminibus caeli, oportet, quod primo sit in

Those who assumed this principle had to explain how the planets could appear visibly different and yet receive their light only from a single source, the Sun. Albertus coped with this problem by assuming that planets differ in their ability to receive the Sun's light and that these differences were based upon a principle of nobility: the greater the nobility of a planet, the greater its capacity to receive solar light.[24] In each of the noblest and purest celestial bodies, light enters the surface on the side facing the Sun and moves immediately to the opposite surface, thus filling the entire body with light, just like the light that burns in a candle.[25]

But the planets affect the light they receive in different ways. Thus Jupiter's light has never diminished in purity from its first reception. By contrast, Mars, as a less noble body, alters the incoming light to a reddish color, as does the star Aldebaran, whereas Venus converts its light to a pale hue and Saturn tends to obscure its light.[26] Although these planets affect the light they receive in different ways, the Sun's light penetrates them all and fills their interiors. Only the Moon is an exception. Because it is of a terrestrial nature, as Aristotle declared,[27] the light received by the Moon does not fill its entire interior but only penetrates partway. For this reason, the Moon, the least noble of the planets, appears only partially illuminated and exhibits phases.[28]

uno, quod est causa multitudinis huius. Ab illo autem quod est causa luminis, in omnibus causatur lumen. Ergo oportet esse unum, a quo recipiatur lumen in omnibus quae lucent. Hoc autem nihil ita convenienter ponitur sicut sol. Ergo a sole lumen est in omnibus stellis." Ibid., 153, col. 2, lines 83–93. In the fifteenth century, Johannes Versor and Thomas Bricot repeat and accept essentially the same argument (see Versor [*De celo*, bk. 2, qu. 11], 1493, 23v, col. 2, and Bricot [*De celo*, bk. 2], 1486, 22r, col. 1 ["Dubitatur quinto utrum astrum aliud a sole lumen suum recipit a sole"]), as did Johannes Velcurio in the sixteenth century (Velcurio, [*Physics*], 1554, 76, col. 2). In the seventeenth century, Aversa ([*De caelo*, qu. 35, sec. 3], 1627, 171, col. 2) and Amicus ([*De caelo*, tract. 6, qu. 5, dubit. 1, art. 3], 1626, 359, col. 2) explicitly reject the argument. Because something is first in a genus does not signify, Amicus insists, that it is the cause of the other, lesser things in that same genus. As the noblest member of the genus, it only sets a standard for nobility, just as with colors, white is noblest but is not the effective cause of other colors.

24. For the same assumptions, see Versor, ibid., 23v, col. 1, and Bricot, ibid.
25. "Sed tamen in omnibus his tanta est pervietas, quod stella recipiens lumen ex parte, qua vertitur ad solem, statim impletur ipso secundum totam superficiem et per omnia interiora eius, sicut si sit lumen accensum in candela." Albertus Magnus, *De caelo*, bk. 2, tract. 3, ch. 6, 5 pt. 1:154, cols. 1–2, lines 43–47. Without mention of a candle, Versor adopts the same position (ibid.).
26. Albertus Magnus, ibid., 154, col. 1. It is obvious that Albertus has assigned a lesser nobility to Saturn than to Jupiter, and thus implies, perhaps unwittingly, a rejection of the concept that nobility was a direct function of distance from the earth.
27. In *De caelo*, bk. 2, comment. 42, 1562–1574, 5:127r, col. 1, Averroës mentions that in his *Liber de animalibus* Aristotle declares that "the nature of the Moon is similar to that of the nature of the earth because of the obscurity [or darkness] which is in it [that is, in the Moon]." For a similar statement by Averroës, see his *De caelo*, comment. 49, 131v, col. 1. See Chapter 17, Section IV.3.a for further discussion.
28. If the Moon were completely penetrated, as are the stars and other planets, it would always appear fully illuminated as long as it received some sunlight, however little. Earlier in his commentary, Albertus emphasized (*De caelo*, bk. 2, tract. 1, ch. 2, *Opera*, 1971, 5,

The solar light that penetrates the planets is not reflected to us but is rather embodied or incorporated into them (*incorporatum est lumen stellis*). The light we see in a planet is "embodied light," which lights up the planet from its center to its circumference. Without sunlight, however, embodied light cannot be retained continuously. It is not like the light of a candle, which burns continuously as long as it has matter to kindle it. When the matter is used up, the candle is corrupted and extinguished. If the planets depended on some analogous matter to maintain their light, they would be corrupted upon the exhaustion of that matter. But planets and stars are incorruptible and in this aspect are not analogous to candles. Rather, they are like spherical vessels, which light up whenever a ray of light touches them.

The Moon is a partial exception. Albert of Saxony, whose opinions on this are strikingly similar to those of Albertus Magnus, who may have been his source, held that, like the other planets, the Moon not only received light directly from the Sun but embodied or incorporated that light into its transparent body. Indeed, Albert believed that the Moon was transparent at the surface and perhaps throughout its magnitude, although not as clear as Venus and Mercury. Because of its large size, however, sunlight was unable to penetrate and fill the whole of it. Consequently the side of the Moon turned away from the Sun is illuminated, albeit not as intensely as the side turned toward the Sun.[29] When we gaze upon the Moon, however, the light "we see is not only the light of the Sun reflected on the body of the Moon but [is also] the light of the Sun embodied and incorporated [*imbibitum et incorporatum*] into the Moon." And like Albertus Magnus, Albert of Saxony explains the variation of certain properties and powers in

pt. 1:107, cols. 1–2) thickness of celestial bodies as the major criterion for the reception of light. There he declared that "the cause of light in stars is their thickness, and in some the light [*lumen*] is received in their depths, and in others the light is diffused on the surface. And those in which the light is received in the depths are made luminaries, like stars that light like candles; and those in which the light is diffused on the surface are made bright like milk, which is called the Milky Way . . . because there the orb is thicker. . . . And so it is necessary that the heaven be thicker and less thick so that its instrument, which is light, can be diversified and move in different ways to different forms of generated and corrupted things." In the later discussion that I described, Albertus ignores "thickness" and stresses degree of transparency, which seems associated with a body's nobility. Indeed, he even regards as false the claim that "in order to [produce] light a pure transparent thing requires only thickness" (Et quod dicitur, quod diaphanum purum ad hoc, ut luceat, non indiget, nisi quod sit spissum, falsum est). Ibid., bk. 2, tract. 3, ch. 6, 154, col. 2. Noteworthy is Albertus's apparent location of the Milky Way in the celestial region among the stars, rather than in the upper reaches of the sublunar region, as Aristotle had insisted. Oresme seems to have adopted the same interpretation (see note 44 in this chapter).

29. "Unde dico quod lumen solis incorporatur in luna ita quod luna est corpus perspicuum et transparens, saltem circa superficiem eius et forte per totum, licet propter magnitudinem corporis lunaris lumen solis non possit totum corpus lunare penetrare sic quod eque intense appareat lumen in parte lune versa a sole sicut in parte lune versa ad solem." Albert of Saxony, *De celo*, bk. 2, qu. 20, 1518, 115r, col. 1. Thus "thickness" – in the guise of size or magnitude, rather than density – seems to have played a role for Albert of Saxony.

different planets and stars by reason "of the different natures of those stars in which light is incorporated and embodied [*incorporatur et imbibitur*]."[30]

The conception of the planets as transparent bodies able to retain the light of the Sun – the sole source of light in the celestial region – and to disseminate it in proportion to their transparency may have become the most widely adopted explanation of celestial light during the Middle Ages.[31] It continued to have its supporters well into the seventeenth century.[32] Because of the medieval emphasis on nobility and hierarchy, it seemed natural to inquire why the Sun, as the source of celestial light, was not placed above all the planets and stars, so that its light could descend and illuminate all celestial bodies. Why did it occupy the middle, or fourth, position among the planets and light up both superior and inferior planets, in violation of the principle that inferior bodies cannot influence or affect superior bodies? "Wise nature" (*sapiens* or *sagax natura*), responded Albertus Magnus, decreed that the planet providing life-giving light and heat should be in the middle, so that it would be neither too far away, and thus leave bodies too cold, nor too close and cause bodies to become too hot.[33]

One problem, however, seems to have escaped explicit discussion. Why, during a solar eclipse, should the side of the Moon turned toward us be completely darkened if the Moon is a transparent body? Would it not be plausible to assume that sunlight striking the far side of the Moon would produce illumination sufficiently strong to be noticeable on the earth side of the Moon? Perhaps it was with this in mind that an anonymous author of a fourteenth-century compendium of natural philosophy adopted a radically different position, arguing that all planets are opaque and dense bodies. As a dense and opaque body, the Moon can cause an eclipse of the Sun when interposed between us and the Sun.[34] On the very next page, however, in seeking to explain the spots on the Moon, he declares, inconsistently,

30. The Latin text of the first of these two passages is: "Sic ergo lumen lune quod videmus non est solum lumen solis reflexum super corpus lune sed lumen solis lune imbibitum et incorporatum." Albert of Saxony, ibid. The second reads: "Lumen unius astri causat calorem et lumen alterius frigiditatem in istis inferioribus. Hoc est propter diversas naturas ipsorum astrorum quibus illud lumen incorporatur et imbibitur." Ibid., 115v, col. 1.
31. For Oresme's description and acceptance of this interpretation, see *Le Livre du ciel*, bk. 2, ch. 16, 1968, 457–459, and Chapter 17, Section IV.3a.ii, of this volume. In his description of this opinion, Duhem [Ariew], 1985, 482, says that "the property they attributed to the moon is not unlike what we call fluorescence."
32. For example, see George de Rhodes [*De coelo*, bk. 2, disp. 2, qu. 1, sec. 3], 1671, 284, col. 2 ("Dico secundo astra omnia, tum fixa, tum errantia, nullam videntur habere lucem propriam, sed ea omnia lluminari a luce solis").
33. Albertus Magnus, *De caelo*, bk. 2, tract. 3, ch. 6, *Opera*, 1971, 5, pt. 1:154, col. 2. For the probable source of Albertus's opinion and for others who held it, see Chapter 11, Section I.4 and note 35. For more on the central role of the Sun, see Chapter 11, Section I.3 and especially note 28.
34. "Compendium of Six Books," Bibliothèque Nationale, fonds latin, MS. 6752, 226v. Because the Sun is greater than the earth, our author observes that the Moon cannot totally eclipse the Sun. For a discussion of MS. BN 6752 and a description of its contents, see Thorndike, 1923–1958, 3:568–584. For the Latin text, see Chapter 17, note 140.

that the Moon is not equally dense. Indeed, some parts of it are so rare that they cannot retain any light and are therefore observed as dark spots.[35] But even in allowing rarity, which was usually associated with transparency, our author assumes that such rarity cannot hold light. Despite his assumption of differing densities within the Moon, he insists that the Moon cannot hold sunlight but can only reflect it from its surface. Presumably this was also true of all other planets. If so, the anonymous author had departed drastically from the more widely accepted interpretations of Albertus Magnus and Albert of Saxony.[36]

In a surprising penultimate sentence to his question on solar light and its reception by the planets, Albert of Saxony declares that the Moon and every other star or planet have a weak self-luminous light that does not originate in the Sun. He infers this striking conclusion from observation of a quarter Moon in clear air. Instead of seeing only the quarter Moon, we see the entire face of the Moon, although the part turned toward the Sun is more lucid than that which is turned away from the Sun. If there were no light within the Moon itself, we ought to see only the quarter of the Moon, because there are no solar rays hitting the part that is turned away from the Sun.[37] In his explanation of this phenomenon, Albert concedes that "the Moon and every other planet have a weak and remiss light within themselves. But that light which is noteworthy and which illuminates us perceptibly is from the Sun."[38] In this way, Albert of Saxony conceded some element of truth to the Avicennan position even as he supported the more traditional opinion of the Sun as sole source of light in the celestial region.[39]

Long before Albert of Saxony's acknowledgment of a second source of planetary light, Richard of Middleton had distinguished two kinds of celestial light in all the planets except the Sun. One kind is natural to the planet itself but is weak and tenuous. Here we recognize the natural planetary light which Albert of Saxony also identified. The second light is not at all natural to the planets, because it is received from the Sun.[40] From the

35. "Compendium of Six Books," MS. BN 6752, 227r. For the Latin text, see this volume, Chapter 17, note 140.
36. In his discussion of lunar spots, he mentions Albertus Magnus and was probably familiar with his *De caelo*.
37. At the outset of the question, Albert says the following: "Probatur primo de luna quod habeat lumen ex se. Nam aere existente sereno et luna existente in prima quadra videtur totale corpus lune, licet illa pars que est versa ad solem videatur lucidior quam alia pars versa a sole." *De caelo*, bk. 2, qu. 20, 1518, 114v, col. 1.
38. "Ad primum bene conceditur quod luna et quodlibet aliud astrum habet lumen debile et remissum ex se. Sed quod habet lumen notabile, quod nos notabiliter illuminat, hoc es a sole." Ibid., 115v, col. 1.
39. In contrast to Albert, Thomas Bricot, *De celo*, bk. 2, 1486, 22r, col. 1, rejects the idea that all planets except the Moon need not receive their light from the Sun.
40. "Dicendum quod in luna et in qualibet alia stella a sole, duplex est lux: una sibi naturalis, quae debilis est et tenuis; alia non sibi naturalis quam a sole recipit. Prima non est eiusdem speciei cum luce solis; secunda tamen est eandem speciem reducitur cum luce solis." Richard of Middleton, *Sentences*, bk. 2, dist. 13, art. 1, qu. 4, 1591, 2:159, col. 2.

thirteenth century, at least two kinds of planetary light were distinguished. Although little was said about the light of the fixed stars, it eventually emerged as a third distinct, though by no means weak, source of light.

2. That the celestial bodies have some or all of their light from themselves

To arrive at a representative selection of arguments that the planets and stars, with the exception of the Moon, are partly or wholly self-luminous, we can do no better than draw upon Nicole Oresme and Albert of Saxony, two of the leading scholastic authors of the fourteenth century.[41] Both were convinced, as we saw earlier, that except with reference to the Moon, no demonstration could determine whether the Sun was the sole source of light among the planets and stars.[42] Most noteworthy, perhaps, is the emphasis (especially by Oresme) on arguments that appeal to the strong medieval sense of cosmic hierarchy and assume that things farther away from the earth are "better" and "nobler" than those nearer to it.

The fixed stars ought to be self-illuminating because they are part of the *primum mobile*, or eighth sphere, which, according to Aristotle, is the most noble heaven.[43] Since light is the most noble quality and the fixed stars the most noble bodies in the most noble heaven, the fixed stars ought to shine by themselves.[44]

Turning to the planets, Oresme explains that Saturn, Jupiter, and Mars ought to be self-luminous because they are superior to, or higher than, the Sun. This argument trades on an oft-repeated concept of cosmic hierarchy that superiors act on inferiors and not the contrary. Thus the Sun cannot act on the three superior planets and therefore furnishes no light to them.[45]

Comets, Oresme observes, are self-luminous, shining by themselves at night "in the shadow of the earth" in the elemental region of the world just below the Moon. Since comets can shine independently in this manner,

41. In the thirteenth century, Bartholomew the Englishman, *De proprietatibus rerum*, bk. 8, ch. 30, 1601, 414, declared that the Moon "has a substantial darkness, since it does not have light from itself, as do the other planets" (Habet enim substantialem obscuritatem, cum non habeat sicut alii planetae lumen a seipsa).
42. Although Albert believed that either side of the issue was demonstrable, he sided with Aristotle and adopted much that Albertus Magnus espoused.
43. The *primum mobile* means the first movable sphere and was always the outermost moving sphere in the cosmos. Thus it could be any sphere from the eighth to the tenth. Since Aristotle identified the eighth sphere, or sphere of fixed stars, with the *primum mobile*, Oresme merely repeats the identification.
44. Oresme, *De celo*, bk. 2, qu. 11, 1965, 640–642. In the third argument, Oresme identifies the nature of the Milky Way with that of the eighth sphere and argues that the Milky Way must also shine by itself. Oresme seems to have located the Milky Way in the celestial region, as did Albertus Magnus (see this chapter, note 28), which marks a sharp departure from Aristotle (see Chapter 14, Section VIII.1.b.i and Chapter 19, Section III.2 and note 26, for Aristotle's views).
45. Oresme, ibid., 642. Albert of Saxony's second argument in defense of the Avicennan position is basically the same (see his *De caelo*, bk. 2, qu. 20, 1518, 115r, col. 2). For Roger Bacon as a possible exception, see Chapter 19, note 5.

"so much more [should] the stars of the heaven, which are more noble and ought to shine by themselves perpetually."[46]

Another argument trades on the nature of sunlight, which was assumed to produce heat and dryness. Now, if the Sun's light is the sole source of planetary light, reflected light from the planets should also produce heat and dryness. But astronomers and experience reveal that "some light from the stars produces cold, and the light from the moon produces cold and moisture."[47] The inference is obvious: the planets and stars can produce such effects only by their own light, the effects of which differ from sunlight. Not to be ignored is the fundamental medieval argument that if the Sun shines by itself, the other planets ought to do likewise, because all are members of the same species.[48]

An argument by Oresme about the Moon resembles the one cited earlier from Albert of Saxony, except that where Albert speaks of a phase of the Moon, Oresme appeals to a lunar eclipse, which, despite the exclusion of solar light, manifests colors that are really lights, lights that could only come from the Moon itself.[49]

Supporters of the Avicennan and Macrobian position argued that if all the planets had their light from the Sun and possessed none of their own, planets ought to exhibit variations in light – that is, they ought to undergo phases – just as the Moon does, and some of them ought to cause eclipses of the Sun (Mercury and Venus) or be eclipsed by the earth (Mars, Jupiter, and Saturn) when it is interposed between them and the Sun. But no such phenomena are observed.[50]

46. Oresme, ibid. Indeed, Oresme pushes this point even farther in the next argument, when he explains that in the sublunar world things like fire and the scales of certain fish are self-luminous; therefore, so much more ought the perfection of self-luminosity to exist in the far more excellent celestial bodies.
47. Ibid., 644. The "experience" (or "experiences") that would reveal information of this sort is unmentioned. Albert of Saxony presents a briefer version in *De celo*, bk. 2, qu. 20, 1518, 115r, col. 1.
48. This argument would have had little force, because few believed that all the planets belonged to the same species.
49. Oresme, *De celo*, bk. 2, qu. 11, 1965, 644.
50. Albert of Saxony's fourth argument (*De celo*, bk. 2, qu. 20, 1518, 115r, col. 2). Oresme offers a similar argument in ibid., 646, but confines himself to Mercury and Venus. In the repudiation of this argument, it was noted that no eclipses of Mars, Jupiter, and Saturn could occur when the earth is interposed between them and the Sun, because the earth's shadow does not extend to the planets above the Sun. As his authority for this, Albert invokes the *Theorica planetarum* of Campanus of Novara, who is alleged to have said that the earth's shadow does not extend beyond Mercury (Albert, ibid., 115v, col. 1). But even if it did, Mercury remains so close to the Sun that if the earth's shadow did reach it, the earth's diameter would not fit between Mercury and the Sun and therefore could not cause an eclipse.

    Actually, the data Campanus provides indicate that the earth's shadow would extend beyond Mercury and reach as far as the sphere of Venus. Thus the apex of the earth's shadow extends 866,536 4/11 miles from the earth's surface to the apex of its cone (Campanus, *Theorica planetarum*, 1971, 149). In summary tables for Mercury and Venus, the concave surface of the sphere of Mercury is 209,198 miles and its convex surface 579,320 miles (ibid., 358). Obviously, the earth's shadow will extend beyond this and

Finally, the lights of the planets differ from one another: Mars is like the light of fire, and Saturn is like that of white lead. But if they all have their light from the Sun, they ought to be of similar appearance.[51]

3. Seventeenth-century scholastic interpretations and the new discoveries

We must emphasize that the seemingly most popular explanation of planetary light – the one adopted by Albertus Magnus and Albert of Saxony – did not explain the visible light of the planets in terms of the direct reflection of solar light from their allegedly opaque bodies. Rather, it assumed planets and stars that were transparent and could therefore receive solar light throughout the extent of their bodies. They were visible to us because they were completely filled with the Sun's light. During the Middle Ages, this interpretation was opposed by one attributed to Avicenna, which, as we saw, assumed self-luminosity for every planet and star except the Moon. Both of these rival opinions were encapsulated by Nicholas Copernicus, who speaks of those who "do not admit that these heavenly bodies have any opacity like the moon's. On the contrary, these shine either with their own light or with the sunlight absorbed throughout their bodies."[52]

into the orb of Venus, the concave surface of which is 579,320 miles and its convex surface 3,892,866 miles (ibid., 359).

In the seventeenth century, Melchior Cornaeus (*De coelo*, tract. 4, disp. 2, qu. 4, dub. 9, 1657, 512) explained that the fixed stars, which receive their light from the Sun, were not eclipsed by the earth's shadow because the latter culminated in the apex of a cone and never reached the firmament. Thomas Compton-Carleton, *De coelo*, disp. 2, sec. 3, 1649, 403, col. 1, says much the same thing as Albert of Saxony, denying that Mercury and Venus could eclipse the Sun and that the earth can eclipse the superior planets. He explains that Venus is only a hundredth part of the Sun (in volume?) and Mercury is even smaller; and since the Sun is 166 times greater than the earth (presumably in volume), the earth's shadow cannot reach the superior planets. Drawing upon "Albatenius, Thebit, and other astronomers," Galileo ([*De caelo*, qu. 2 (H)], 1977, 79), explains that "the visual diameter of the sun has a tenfold ratio to the visual diameter of Venus," so that geometrically "the visual diameter of the sun . . . has a hundredfold ratio to the visual circle of Venus." Van Helden, 1985, 71, says this relationship is found in Ptolemy's *Planetary Hypotheses*.

51. Albert of Saxony's sixth argument (*De celo*, bk. 2, qu. 20, 1518, 115r, col. 2). Earlier in this section, we saw that Albertus Magnus replied to this argument by insisting that the planets had different capabilities for receiving the Sun's light and that these differences produced different colors and appearances.

52. Copernicus, *Revolutions*, bk. 1, ch. 10 [Rosen], 1978, 19. Copernicus adds (ibid.) that these planets "do not eclipse the sun, because it rarely happens that they interfere with our view of the sun, since they generally deviate in latitude. Besides, they are tiny bodies in comparison with the sun. Venus, although bigger than Mercury, can occult barely a hundredth of the sun. So says Al-Battani of Raqqa, who thinks that the sun's diameter is ten times larger [than Venus's], and therefore so minute a speck is not easily descried in the most brilliant light." In this passage, Copernicus epitomizes widely held medieval views, especially that Venus "can occult barely a hundredth of the sun." Copernicus immediately rejects this interpretation. That Venus could only occult a hundredth of the Sun was repeated by Compton-Carleton, *De coelo*, disp. 2, sec. 3, 1649, 403, col. 1.

As we saw, however, Albert of Saxony had already conceded that solar light was not the only source of planetary and stellar illumination: to a very small extent, the Moon, and by inference the rest of the planets, seem to be weakly self-luminous. Unwittingly, Albert adumbrated an interpretation that would become popular in the sixteenth and seventeenth centuries. The Coimbra Jesuits bore witness to this when, at the end of the sixteenth century, they declared: "the more common assertion of the astronomers is that both the fixed stars and the planets receive their light from the Sun but nevertheless possess some light by themselves."[53] Among scholastic supporters of this interpretation in the seventeenth century were Bartholomew Amicus, Thomas Compton-Carleton, and Melchior Cornaeus.[54] All would have agreed, however, that the Sun provides the greater part of planetary and stellar light.[55] Their interpretation assumed that most of the light for celestial bodies came from the Sun but that all of these bodies also had some additional source of light from within themselves.

Another rather widely held opinion was opposed to the one described by the Coimbra Jesuits. The rival interpretation either explicitly or by implication denied any degree of self-luminosity to the planets, including the Moon. All planetary light derived directly from the Sun and was subsequently disseminated by reflection. Within this group, the planets could be viewed as entirely opaque bodies or as materially different bodies that were unequally lucid and possessed different degrees of purity. Johannes Velcurio was in the latter group,[56] while Raphael Aversa and George de Rhodes are identified with the former.[57]

Let us now consider seventeenth-century opinions on the Moon, the other planets, and the fixed stars, in that order.

53. "Statuenda tamen est hec assertio quam communior astrologorum consensus approbat: tam stellae fixae quam planetae lumen a sole mutuantur; ita tamen ut aliquid ex se lucis possideant." Conimbricenses, *De coelo*, bk. 2, ch. 7, qu. 4, art. 1, 1598, 303.
54. In the first of five conclusions about the source of stellar and planetary light, Amicus declares (*De caelo*, tract. 6, qu. 5, dubit. 1, art. 3, 1626, 358, col. 1) that "the stars [and planets] have some light from themselves." Compton-Carleton (*De coelo*, disp. 2, sec. 3, 1649, 403, col. 1) assumes that "Neither the Moon, nor the other planets, nor even the fixed stars, receive all their light from the Sun" (Nec luna aut alij planetae, nec stellae etiam fixae omnem lucem suam accipiunt a sole). In his reply to the question "Whether stars receive light from the Sun" (An stellae accipiant lumen a sole), Cornaeus, *De coelo*, tract. 4, disp. 2, qu. 4, dub. 9, 1657, 512, responds: "Yes, but nevertheless they also have some from themselves" (Ita, sed tamen aliquod etiam ex sese habent).
55. Compton-Carleton, ibid., declares that "the stars [and planets] receive the greatest part of their light from the Sun. This is manifestly proved from eclipses of the Moon and also from the fact that planets nearer to the Sun are illuminated more and more intensely, as if they were ablaze, as is obvious in Mars and Venus."
56. Velcurio declares (*Physics*, 1554, 76, col. 2–77, col. 1): "Ergo omne lumen syderum refertur ad solem. . . . Ergo sol est causa efficiens luminis in omnibus stellis. Et per consequens omnes stellae recipiunt lumen suum a sole." He goes on to explain: "Quod autem non omnia astra sunt aeque lucida, neque eiusdem vel paris luminositatis, nihil impedit haec nam culpa non solis est, sed materia in stella quarum quo quaeque est purior et nobilior. . . . Quo autem quaeque spissior est. . . . Et quo iusto rarior est."
57. Despite divergence on important points, Aversa and de Rhodes were influenced by Galileo.

a. The Moon

The problem of the Moon's source of light continued to loom large in the seventeenth century. The problem was essentially the same as it was for Albert of Saxony: how to explain light or colors in areas of the Moon in which solar rays were absent and which should have been totally dark and invisible. Raphael Aversa enunciated the problem succinctly: How can we explain a certain whiteness in that part of the Moon when there is no Sun shining on it, and how, during the course of an eclipse, do we account for a certain ruddy color?[58] Although Albert of Saxony and others in the Middle Ages had discerned lights in the darkened areas of the Moon, Aversa was more than likely summarizing Galileo's distinction in the *Sidereus nuncius* between two different kinds of lunar light – one in the normal course of the changing dark and light areas of the lunar surface and the other detected during lunar eclipses.[59] Aversa explains that "there is doubt from whence and how the Moon has this light, whether from itself or from the same Sun; and if from the Sun, whether it is directly from the Sun or from another body illuminated by the Sun."[60]

Among responses to such questions was the Albertus Magnus-Albert of Saxony medieval explanation in which the Moon is conceived as somewhat transparent rather than absolutely opaque. The Moon is thus assumed pervious to solar rays and capable of receiving a weak light that penetrates beneath its surface.[61] Aversa rejects this interpretation, insisting that the

58. Amicus describes a similar phenomenon when he proposes the Moon as evidence that the planets had some light of their own. Although the Moon seems to receive all its light from the Sun, it nevertheless seems to have some light of its own, because it becomes reddish during an eclipse, which could only occur if it had some light within itself (Primo patet ex luna quae videtur totum a sole lumen recipere et tamen habet aliquid luminis ex se, ut patet in ipsius eclypsi, cum tunc rubeat, qui rubor non posset esse nisi aliquid luminis in se construaret. Ergo idem dicendum de alijs astris). Amicus, *De caelo*, tract. 6, qu. 5, dubit. 1, art. 3, 1626, 358, col. 1. Thomas Compton-Carleton repeats the same argument (*De coelo*, disp. 2, sec. 3, 1649, 403, col. 1). Amicus's opinion is similar to Albert of Saxony's in behalf of a small amount of weak light in the Moon itself. Indeed, Amicus draws on a number of arguments which Albert of Saxony presented in defense of the self-luminous nature of celestial bodies.
59. Galileo, *Starry Messenger* (*Sidereus nuncius*) in Drake, 1957, 42–43 and Galileo, *Sidereus nuncius* [Van Helden], 1989, 54–55.
60. The full text follows: "At vero quia visu ipso cernimus partem quoque lunae avertam a sole quando luna non est plena et videmus illam albicantem; imo et in eclipsi conspicimus adhuc lunam fusco quodam lumine rubescentem. Dubium est unde et quomodo luna hoc lumen habeat: an a se vel ab eodem sole; et si a sole, an ab illo immediate vel ab alio corpore illuminato a sole." Aversa, *De caelo*, qu. 35, sec. 2, 1627, 168, col. 2. Mastrius and Bellutus asked similar questions (*De coelo*, disp. 2, qu. 5, 1727, 3:506, col. 1, par. 157) about all the celestial bodies, including the Moon: "In another part of what is sought, we ask in what celestial bodies does light [*lux*] occur: whether all the heavens and celestial bodies [*astra*] have light, or only the celestial bodies. And of these [i.e., the celestial bodies] whether all, or only some have light" (In altera parte quaesiti quaeritur quibus corporibus coelestibus conveniat lux: an scilicet omnes coeli et astra habeant lucem an solum astra; et istorum an omnia, an vero quaedam tantum).
61. "Primus dicendi modus est lunam non esse penitus opacam et imperviam lumini, sed esse aliquantum diaphanam et ita etiam in profunditate recipere immediate debile saltem

Moon is perfectly opaque and in no way transparent. Solar rays are incapable of penetrating below the lunar surface.[62] By contrast, Aversa's contemporary, Bartholomew Amicus, accepted the opinions of Albertus Magnus and Albert of Saxony – citing both in the course of his rather lengthy discussion – and assumed that the Moon (as well as all the other planets) had light of its own which it receives from the Sun and "incorporates" (*incorporatio*) into itself, in the manner described by his medieval predecessors.[63] The illumination of the Moon could not be explained solely in terms of reflection, since only a small part of the Moon would be seen as illuminated, namely that observed part where the angle of incidence equals the angle of reflection, as is obvious with mirrors. If the Moon diffused light solely by means of reflection, only a small part of the earth would be illuminated. Amicus concludes that the Moon receives solar light into its very depths, or, as he put it earlier, "incorporates" the light of the Sun into itself and then diffuses it to the earth.[64]

If Aversa found that the Moon was not diaphanous and did not receive solar light within itself, he was equally convinced that it lacked an innate light of its own. Such a light should be visible throughout the Moon's circuit around the earth, which is manifestly not so.[65] Nor does it receive additional weak light from Venus, as some, including Tycho Brahe, believed.[66]

lumen a sole. Ita opinati sunt illi, qui superiori quaest., sect. 2, dicebant lunam non esse perfecte opacam, sed esse aliquantum diaphanam et ita perviam radijs eiusdem solis." Aversa, *De caelo*, qu. 35, sec. 2, 1627, 168, col. 2. This light, which is ultimately from the Sun, should be distinguished from the weak, self-luminous lunar light that Albert of Saxony mentions.

62. "Verum satis ibi ostensum est lunam esse potius ita perfecte opacam ut transitum non praebeat in sua profunditate radijs solis." Ibid.
63. As he so frequently does, Amicus, *De caelo*, tract. 6, qu. 5, dubit. 1, art. 3, 1626, 360, col. 1, first poses a question in the name of an imaginary opponent or critic: "You ask [literally "say"] how does the Moon receive light from the Sun"? (Dices quomodo luna recipiat lumen a sole?) "I reply," says Amicus (literally "Let it be replied") that "it receives it by incorporation to the extent of a one-third part of itself" (Respondetur recipere illud per incorporationem, usque ad tertiam eius partem).
64. "Quia si non per incorporationem esset per reflexionem. Unde quando luna esset ad latus nostri aspectus, non appararet illuminata nisi parva pars lunae, scilicet pars a qua angulus incidentiae et angulus reflexionis ad nostrum aspectum essent aequales. Tum quia si esset solum corpus reflectens non posset illuminare nisi tantam partem terrae quanta est in se, ut patet ex reflexione speculorum. Et tamen apparet totam terram illuminari que est maior quam luna est. Ergo dicendum est recipi in profunditate et inde diffundi lumen." Ibid., 360, cols. 1–2.
65. Aversa, *De caelo*, qu. 35, sec. 2, 1627, 169, col. 1.
66. These opinions were all briefly mentioned by Galileo in the *Sidereus nuncius*. In four or five lines Galileo describes them all (Galileo, *Sidereus nuncius* [Van Helden], 1989, 54). Thus in speaking of the white light that shines even in the darkened parts of the Moon, Galileo comments: "Some have said that it is the intrinsic and natural brightness of the Moon herself; others that it is imparted to it by Venus, or by all the stars; and yet others have said that it is imparted by the Sun who penetrates the Moon's vast mass with his rays. But such inventions are refuted with little difficulty and demonstrated to be false." Galileo rejected the Venus explanation as "childish," saying "For who is so ignorant as not to know that near conjunction and within the sextile aspect it is entirely impossible for the part of the Moon turned away from the Sun to be seen from Venus?" See also

As a scholastic aware of current opinion, Aversa also considers what he calls Galileo's "very new mode of explanation" (valde novus dicendi modus est Galilaei). Although the Moon receives almost all of its light from the Sun, Galileo argued, in the *Sidereus nuncius*, that reflections of sunlight from the earth to the Moon augmented the Moon's light and affected otherwise darkened areas of the lunar surface, a phenomenon Galileo called "secondary light." Galileo's reasoning, Aversa explains, is simply that if the smaller Moon can illuminate the much bigger earth, it follows that the earth, which receives more solar light than the Moon, should reflect some of it to the lunar surface. Aversa finds the consequences of this belief unacceptable, since it inverts the order of the universe,[67] that is, instead of the heavens sending down light and other influences to affect the terrestrial region, as everybody believed, Galileo's explanation reverses the process and has the earth, and therefore the terrestrial region, affecting and influencing the nobler celestial region. Aversa expressed further displeasure at Galileo's inclusion of the earth among the planets, a move that implied that the planets are like our earth, an association that was repugnant to defenders of traditional cosmology. "The earth," Aversa insists, "is an unpolished, uneven, and rough body, [whereas] the Moon is just like every other planet: it is very pure and very polished, more like our metals and gems."[68]

From whence, then, does the Moon's secondary light derive? It comes from the surrounding parts of the heavens, seemingly from sunlight reflected to the Moon from other celestial bodies. Obviously this excludes the earth, which for Aversa is not a planet. In a curious sense, Aversa accepted an argument resembling Galileo's, but instead of the earth as the Moon's source of reflected sunlight, he vaguely invoked other neighboring celestial bodies. Mastrius and Bellutus adopted the same explanation and even mentioned Aversa and three other seventeenth-century scholastic authors (Licetus, Scheiner, and Tanner) who also accepted it. Speaking explicitly of "secondary light" (*lumen secundarium*), they trace its source to "the nearer parts of the heaven illuminated by the Sun." To counter Galileo, they argued that if solar light were reflected from the earth and reached the second region of the air on its way toward the Moon, it would heat the intervening air.[69] This does not occur. Moreover, solar rays reflected from

the translation in Galileo, *Starry Messenger* [Drake], 1957, 42–43. For references to Tycho, see Galileo, *Sidereus nuncius* [Van Helden], 1989, 54, n. 52. As we shall see, Aversa seems to have adopted as his explanation a version of the interpretation that attributed the cause of the Moon's extra light to "all the stars."

67. "Verum de se satis incredibile apparet immitti et diffundi lumen a terra ad caelum, ad caelestia corpora, quod plane est invertere ordinem universi." Aversa, *De caelo*, qu. 35, sec. 2, 1627, 170, col. 1.

68. "Nec unquam obtinebit Galilaeus, ut terra credatur unum ex astris et ullum astrum credatur ut terra. Terra est corpus rude, asperum, et ruidum [in place of *ruuidum*]; luna, sicut et omne aliud astrum, est corpus summe tersum et perpolitum plusquam apud nos metalla et gemmae." Ibid.

69. After citing Galileo's explanation of secondary light, Mastrius and Bellutus declare (*De coelo*, disp. 2, qu. 5, 1727, 3:506, col. 2, par. 159): "Sed probabilior dicendi modus est

the earth would not reach the Moon because they are too weak to reach beyond the second region of the air.[70]

b. The other planets

Due in large measure to Galileo's observations with the telescope, the range of explanations about the source of celestial light was considerably expanded in the seventeenth century. By his discovery of the phases of Venus in 1610,[71] Galileo had shown that another planet besides the Moon revealed a continual variation in the amounts of lightness and darkness on its surface and did so as a consequence of its position in relationship to the Sun. It was not only convincing evidence that Venus revolved around the Sun rather than the earth but that the brightness of Venus was produced by the Sun's light. Since Galileo had shown that darkness lay over that part of the Venusian surface where the Sun failed to shine, it was implausible to suppose that Venus had its own independent source of light. Without solar light, it would lie immersed in darkness. On this approach, the planets were best interpreted as opaque rather than transparent bodies. Moreover, if Venus's brightness was derived from solar light, it seemed a further reasonable inference that solar light was also the cause of the varying degrees of color and brightness in the other planets. A powerful reinforcement to the phases of Venus was Galileo's slightly earlier observation and description of the Moons, or satellites, of Jupiter.[72] With these powerful claims, scholastic natural philosophers in the seventeenth century faced a significant challenge to traditional interpretations about the planetary sources of light.

Of the serious opinions offered as explanations for the source or sources of planetary light, none denied the role of the Sun. In general, three opinions

quem afferunt Aversa, quaest. 35, sect. 2; Licetus *lib. de Lap. Bon.*, cap. 50; Scheiner disp. 37; Tannerus I disp. 6, quaest. 7 et alii, lumen illud esse quoddam lumen secundarium proveniens a partibus celi propinquioribus a sole illuminatis. . . . Tum quia radii solares per reflexionem a terra non transcendunt secundam aeris regionem, aliter haec esset calida." The second, or middle, region of air is the location of clouds and is always cold (see Ch. 20 [immediately preceding Sec. I] for d'Ailly's description of the sublunar region).

70. "Videmus enim reflexionem radiorum solis a terra non pertingere secundam aeris regionem, nam reflexio non diffunditur ad quamcunque distantiam propter imbecillitatem specierum." Mastrius and Bellutus, ibid., 507, col. 1, par. 162. In the preceding sentence (also see note 69), Mastrius and Bellutus say that if sunlight reflected from the earth were sent toward the Moon it would heat the air of the "second region." They also say that the reflected rays are too weak to reach the Moon. Would those rays also be too weak to heat the air?

71. In his *First Letter on Sunspots* to Mark Welser, dated May 4, 1612, Galileo expresses astonishment that his opponent Apelles (Christopher Scheiner) had not yet heard of his method for determining whether "Venus and Mercury revolve about the sun or between the earth and the sun." Galileo explains that this method was "discovered by me about two years ago and communicated to so many people that by now it has become notorious. This is the fact that Venus changes shape precisely as does the moon." Galileo, *Letters on Sunspots* (1613) in Drake, 1957, 93. The discovery was made in 1610 but after publication of *The Starry Messenger* in March of that year.

72. In his *Sidereus nuncius* of March 1610.

emerged. The first assumed that the Sun is the sole source of planetary light; this opinion has three subdivisions. In one, the planets are assumed opaque and their light is but a reflection of solar light. In another, the planets are assumed partly or wholly transparent and, instead of reflecting solar light, the latter penetrates the diaphanous planetary matter and fills it with light, thus rendering the planet visible because it is always filled with light. In the third opinion, for which Campanus of Novara apparently qualifies, the Sun is assumed the source of planetary light, but no judgment is made about the opacity or transparency of the planets.[73]

The second opinion attributes the primary source of planetary light (*lumen*) to the Sun but also assumes that each planet has within itself a source of light (*lux*). The third opinion is the now familiar one, derived from Macrobius and Avicenna, that only the Moon receives light from the Sun but all other planets are self-luminous. Although this opinion was frequently repeated in the Middle Ages, few, if any, accepted it. In the seventeenth century, Aversa includes it in his discussion, even mentioning Macrobius and Avicenna by name, and cites a few of its alleged supporters (Molina, Lucillus Philalthaeus, and Scaliger).

Those who assumed that the Sun was the sole source of planetary light and that it shone on opaque planets could appeal for support to Galileo's telescopic observations, as did Raphael Aversa. Citing Galileo's *Letter on Sunspots*, Aversa invokes Galileo's telescopic observations of the phases of Venus and concludes that "according to this observation it is clearly demonstrated that Venus does not have native light but is illuminated by solar light."[74] A paragraph later, Aversa introduces the satellites, or Moons, of Jupiter and mentions that one or another of them is always eclipsed by the shadow of Jupiter when the latter is interposed between its satellite and the Sun. Aversa agrees that Jupiter and its satellites receive their light from the Sun. Indeed, he suggests that this may also explain the behavior of the two "satellites" allegedly observed around Saturn.[75] On the basis of such powerful evidence, Aversa seems to have accepted the idea that all the planets receive their light from the Sun and have none of their own.

In this Aversa may have been unusual. Almost all other seventeenth-century scholastic natural philosophers accepted the Sun as the primary source of light for the planets but also assumed that the latter were to some

73. I interpret Campanus, *Theorica planetarum*, 1971, 148, lines 79–88 (translation on p. 149) in this manner.
74. After citing "Galilaeus in Epist. 3, de Maculis Solis," Aversa (*De caelo*, qu. 35, sec. 3, 1627, 172, col. 2) goes on to describe the opinion of Galileo and likeminded supporters: "Docent planetas illuminari a sole quia iam per telescopium certis observationibus deprehensum est Venerem mutari ac variari ad instar lunae et non in suo orbe plenam sed vere corniculatam apparere ita ut qua parte versa est ad solem notabili lumine fulgeat et in oppositum cornua vertat. Iuxta hanc ergo obervationem perspicue demonstratur Venerem non suo nativo lumine, sed solari illustratione lucere."
75. The rings of Saturn were initially interpreted as satellites.

extent self-luminous.[76] Upon inspection of the arguments, which are occasionally ambiguous, the light peculiar to a planet is either solar light that is received within a diaphanous planet or a kind of light that is left unidentified. In those instances where the light of the planet is transformed solar light, Albertus Magnus and/or Albert of Saxony is probably the ultimate source. For example, Johannes Poncius, the seventeenth-century Scotistic commentator, declares that "planets receive their light principally from the Sun,"[77] offering as reasons the varying brightness of planets in proportion to their distance from the Sun and the middle position of the Sun with respect to the other planets, which made it easier to communicate light to them. Indeed, if the Moon is self-luminous, it ought to appear lucid during a lunar eclipse.

Poncius then qualifies his statement that the planets receive their light "principally" (*principaliter*) from the Sun, "because it is not improbable that they have some light from themselves." As evidence of this, he mentions only the Moon and gives a common instance: the Moon reveals some light during its eclipse. By the time Poncius wrote, the lunar light seen during an eclipse and even light seen in the darkened portions of the lunar surface during its normal phases were usually explained as the result of solar light reflected to the Moon from one or more other celestial bodies, and even the earth (as Galileo argued). That was as far as Poncius would go with claims for self-luminosity. Indeed, he rejected a general claim for it. He did this by posing a common objection against his position that the Sun is the principal source of planetary light: what of the frequently mentioned argument that even when the earth is interposed between the Sun and other planets, the latter are not eclipsed but continue to shine brightly? Is this not evidence of self-luminosity? On this piece of evidence, Poncius gives the usual negative response: the earth's shadow reaches only to Venus and therefore cannot cause an eclipse of Saturn, Jupiter, and Mars, which continue to shine from the reception of solar light. And although the earth's shadow can reach Venus and Mercury, those two planets do not recede sufficiently from the Sun to enable the earth to be interposed between them.[78]

Mastrius and Bellutus also argued for some sense of self-luminosity, commencing their discussion with an assertion that the other planets (*astra*) are probably "illuminated not only by the Sun but also have a proper greater light, [just] as [does] the whiteness of the Moon."[79] Like most scholastics,

76. We saw earlier that Amicus, Compton-Carleton, and Cornaeus held such an opinion. To them we may add Poncius and also Mastrius and Bellutus.

77. "Planetae suam lucem principaliter a sole participant." Poncius [*De coelo*, disp. 22, qu. 8], 1672, 625, col. 1.

78. Ibid. The same opinion was held in the Middle Ages (Campanus of Novara, Albert of Saxony) and by others in the seventeenth century (Cornaeus and Compton-Carleton). See note 50, this chapter.

79. "Tandem de aliis astris dicendum probabiliter videtur, non solum a sole illuminari, sed

they insisted that the Sun was the primary source of planetary light, appealing, as did Aversa, to the phases of Venus, Mercury, and even Mars to buttress their claim.[80] But what about self-luminosity? They give two reasons for accepting it. First, the planets (*astra*) must be diaphanous, for otherwise they would not even be seen. For their second reason they mention the different colors of the planets, a phenomenon that indicates "a difference of light, and yet if they were illuminated by the Sun, the light in all of them ought to be of the same kind."[81] Taken together, these two reasons strongly suggest that Mastrius and Bellutus had in mind two manifestations of light: sunlight and another kind of light that is peculiar to each planet. Both are explicable in terms of the medieval interpretations associated with Albertus Magnus and Albert of Saxony. The diaphanous nature of each planet enables it to receive the Sun's light throughout its body. Indeed, sunlight fills it. It is able to shine because it then diffuses its sunlight. Each planet absorbs light in its own way and thus alters it. These alterations of the light are manifested by the different colors of the planets.

On this interpretation, the "proper light" of each planet is really solar light received in the interior of each planet and altered by it. In the final analysis, there is only one kind of light, and it manifests itself in different and unequal ways.[82] This interpretation gains further credibility in the following paean to the Sun offered by Mastrius and Bellutus (*De coelo*, disp. 2, qu. 5, 1727, 3:506, col. 2, par. 160):

> From these things it is obvious how the Sun is the first measure in the genus of lights and the measure of others and the source of light. For it exceeds the others and communicates light to all. It can be spoken of as if it were the only light because by comparison to it the other [lights] can be thought of as shadows.

Although Mastrius and Bellutus believed that solar light was the only kind of light, it could take two forms. In effect, they postulated two kinds of light, pure sunlight and sunlight that is altered within the planet itself. Earlier on, Bartholomew Amicus held a similar opinion, believing that planets had some light of their own in addition to solar light. But Amicus was even vaguer than Mastrius and Bellutus.

etiam propriam lucem habere majorem quam sit albicatio illa lunae." Mastrius and Bellutus, *De coelo*, disp. 2, qu. 5, 1727, 3:506, col. 2, par. 160.

80. Ibid. According to Mastrius and Bellutus, someone named Fontana claimed to have observed horns of light on Mars, which resembled those on the Moon.

81. The text for the first reason is: "Secunda pars quod etiam sint astra ex se ipsis lucida potest suaderi . . . ex hoc: quod omnia alia astra sunt diaphana, ut dicemus, ergo si lucida non essent, non viderentur, vel saltem non tam lucida conspicerentur, sicut nec coeli cernuntur propter diaphaneitatem;" and the text for the second reason: "Quia planetae . . . apparent diversum coloris quod indicat in luce differre et tamen si a sole tantum illuminarentur deberet lumen in omnibus esse eiusdem rationis." Ibid.

82. Mastrius and Bellutus, ibid., insist that the light in each planet is not equally intense. Since the planets are in different species, equal intensities are not essential (An vero haec lux sit aeque in omnibus astris intensa, negative respondendum quia nulla apparet necessitas hujus aequalitatis, maxime quia sunt specie distincta).

In the course of six arguments intended to illustrate that planets (*astra*) have their own light, Amicus provides no real clue as to the nature of that light, although it is clear that he regarded the Sun as the primary source of planetary light.[83] The kinds of arguments he presents are unhelpful in determining the nature of the proper light allegedly possessed by a planet. Thus in his fifth argument, Amicus uses an analogy with the Sun. Since the Sun is lucid, the other planets should also be lucid, because they are of the same generic, though not specific, nature. By their single generic nature they should all have light of their own, but by their different specific natures the degree and intensity of light varies in each.[84] Amicus does not say whether this generic light, which is specifically adapted to each planet, is the Sun's light. That planetary light differs from sunlight is conveyed by Amicus in his third argument. Because planets (*astra*) exercise different effects on inferior, sublunar things, they cannot act only by means of sunlight. For if the planets received only sunlight, it would be identical in each of them, and they would all produce the same effects.[85] However, since the effects differ, so must the powers of the planets. Therefore each planet must have its own proper light.

But then, as if to nullify this argument, Amicus poses a difficulty: different sublunar effects may occur not only because of light but also from other powers of the planets. Consequently, one cannot be certain that light does differ in each planet, because the different effects may be caused by other planetary powers. But if no other power operates except light, then Amicus concedes that the light must be received in a different way in different planets.[86] Here Amicus seems to acknowledge that the "proper light" of each planet is simply sunlight differently adapted by the specific natures of each planet. He does not seem to conceive of a "different" kind of light in each planet. There can be no doubt, however, that Amicus conceived of the Moon as a partially transparent planet capable of receiving and retaining solar light in the transparent part and that he considered the remaining planets as totally transparent and capable of receiving solar light throughout their bodies.[87]

83. In a fourth conclusion, Amicus says that "Besides a proper light, all planets receive light from the Sun" (Quarta conclusio: omnia astra praeter lumen proprium recipiunt lumen a sole). Amicus, *De caelo*, tract. 6, qu. 5, dubit. 1, art. 3, 1626, 360, col. 2.
84. Ibid., 359, cols. 1–2.
85. This argument seems to conflict with the fifth argument just described. In that argument, all planets should be lucid in the same way as the Sun, because they are all generically the same. Nevertheless, the light will vary from planet to planet because the planets differ specifically and each will alter the generic light in its own unique way. In the third argument, however, Amicus says that if the planets receive only sunlight, they would all exercise the same effects. But why should this happen? Would their specific differences not cause each planet to adapt the light to its own specific nature, so that each planet will cause different effects?
86. Amicus, *De caelo*, tract. 6, qu. 5, dubit. 1, art. 3, 1626, 358, col. 2.
87. After describing the earth as the most opaque of all bodies, Amicus declares (ibid., art. 4, 362, col. 1): "Ita luna est minus diaphana quam alii planetae. Unde habet in se tam

Not all who believed that sunlight was both the ultimate source of planetary self-luminosity and also the source of ordinary planetary light conceived of planets as wholly or partly transparent. An exception was Sigismundus Serbellonus, who argued that planets and stars possess a source of light within themselves that is independent of the Sun[88] and also insisted that they are illuminated by the Sun.[89] Arguing from a common type of analogy favorable to his position, Serbellonus concluded that all stars and planets are self-luminous, because they are solid, opaque, and composed of the same substance. Since they are all alike in these and other vital properties, he assumes that they belong to the same species. From this, it seemed a reasonable inference that if the Sun had the property of light (*lux*), so also should the planets and stars.[90] Since they shared essential properties, we may plausibly infer that Serbellonus, who is silent on this, assumed that the light possessed by each planet and star was identical in its properties to the light of the Sun.

Despite the mysterious and often obscure manner in which the "proper light" of the planets is described and defended, it seems likely that the proper light of a planet was usually thought of as sunlight that the planet had somehow transformed below its surface. Most of these scholastics would also have conceived of the planets as partly or wholly transparent bodies capable of absorbing, retaining, and diffusing sunlight, in a manner similar to that described by Albertus Magnus and Albert of Saxony.

Serbellonus was a notable exception, because he assumed solid, opaque planets and stars. He insisted that the light of the Sun reaches all the planets and even the fixed stars. Because planets, including the Sun and stars, are opaque and solid bodies, they reflect the Sun's light and are illuminated. As opaque bodies, the planets and the fixed stars do not receive solar light into their depths but only at the surface, where it is reflected. Thus Serbellonus rejected the idea of transparent planets that could receive solar light into their interiors.[91] The planets were bright because of reflected light and also, presumably, because of their own proper light, which, as we saw, is probably identical with sunlight. But how could planets produce their own proper light if the Sun's light did not penetrate below their opaque surfaces? If Serbellonus retained the two kinds of light, namely direct sunlight and sunlight transformed into a planet's proper light, he did not explain how

varias partes, ut macula appareat et lumen non recipiat secundum totam profunditatem sicut recipiunt alii planetae."

88. "Stellae et planetae omnes habent lumen ex se." Serbellonus [*De caelo*, disp. 1, qu. 5, art. 1], 1663, 2:46, col. 1, par. 3.

89. "Omnia sydera illuminari etiam a sole." Ibid., col. 2, par. 7. I have interpreted the term *sydera* to be equivalent to *stellae et planetae*, in note 88.

90. After declaring that "Stellae et planetae omnes habent lumen ex se" (see n. 88), Serbellonus justifies his assertion: "quia sunt eiusdem substantiae, soliditatis et opacitatis; magis enim et minus non variant speciem. Ergo non est maior ratio cur lux sit proprietas solis et non aliorum syderum." Serbellonus, ibid., 2:46, col. 1, par. 3.

91. "Dico igitur lumen solis recipi in sola superficie, sive planetarum, sive stellarum fixarum." Ibid., 2:47, col. 1, par. 10.

the former was transformed into the latter. The puzzle might have been resolved if Serbellonus had explained how the Sun, as an opaque body, radiates light to the other planets and fixed stars and how its light is distributed. Does the light of the Sun lie only on its surface or deep within? On the basis of Serbellonus's discussion, we may assume that whatever solution he proposed for the Sun would apply to every other celestial body, each of which is essentially the same as the Sun.

Few, if any, scholastics took seriously the medieval Avicennan-Macrobian theory of self-luminosity, which held that only the Moon received light from the Sun but that all other planets were self-luminous. The best that could be said for this theory in the Middle Ages is that Nicole Oresme and Albert of Saxony thought neither it nor its rival (that all planets receive their light from the Sun) was demonstrable. Although Oresme thought the self-luminosity theory more probable, it found little support and only occasional mention.[92]

c. *The fixed stars*

With regard to the planets, most scholastics assumed that solar light was the single source of planetary illumination but was manifested in different ways because of the diverse natures of the planets themselves. But what about the fixed stars? Did the Sun also illuminate them, or did they produce their own light independently of the Sun?

In turning to this topic we must, at the outset, address a problem of terminology. The term *planeta* (plural: *planetae*) was always used to signify a planet, never a fixed star, whereas the expression *stella fixa* (plural: *stellae fixae*) was the most common expression for a fixed star.[93] But three terms caused considerable confusion, namely *aster* (plural: *astra*); *sidus* (plural: *sidera* or *sydera*), and *stella* (plural: *stellae*). At some point in the historical evolution of these terms, they seem to have had more or less specific meanings. According to Macrobius (*Commentary on the Dream of Scipio*, bk. 1, ch. 14, 1952, 147), the term *stella* was used for the five planets as well as for those fixed stars that Ptolemy had not included in his forty-eight constellations. Thus *stella* could be used for five of the planets (presumably not for the Sun and Moon, however) and for all fixed stars not in a constellation recognized by Ptolemy. But the terms *aster* and *sidera* (*sidus*, but Macrobius

92. In the late fifteenth or early sixteenth century, Bricot, *De celo*, bk. 2, 1486, 22r, col. 1, who seems to have assumed that the Sun was the source of light for the planets, also mentions a challenge to the latter interpretation when he reports the opinion of those who argue that only the Moon receives light from the Sun, whereas the other planets have their own proper light independently of the Sun. He does not mention Avicenna or Macrobius. In the seventeenth century, Arriaga thought that it was doubtful that planets (*planetae*) derived their light from themselves (An planetae habeant a se aliquam lucem plane dubium). Arriaga, *De caelo*, disp. 1, sec. 6, 1632, 507, col. 2.

93. Occasionally we see an expression such as *astra fixa*, which appears in Mastrius and Bellutus, *De coelo*, disp. 2, qu. 2, art. 3, 1727, 3:495, col. 2, par. 78. Some version of the adjective *fixus, fixa, fixum* almost invariably signified a fixed star.

actually uses the plural form) were to be applied only to the fixed stars, with *aster* indicating a single star and *sidera* a constellation of them, such as Aries or Taurus. Although Macrobius's distinctions were repeated, albeit with some occasional changes and distortions,[94] they gradually dissolved, and the three terms came to be used indifferently for planets and fixed stars. The term *planeta* was frequently replaced by one of our three terms, especially *stella*. For example, when Melchior Cornaeus (*De coelo*, tract. 4, disp. 2, qu. 4, dub. 9, 1657, 512) discussed "An stellae accipiant lumen a sole," he was actually asking "Whether the planets receive light from the Sun" and was in no manner concerned with the fixed stars. Here, as in most instances, *stella* signifies planet rather than star. Amicus, who usually used the term *stella* for planet, introduces at one point the term *sydera* and consciously uses it to represent both planets and fixed stars.[95] In the course of two successive sentences, Johannes Velcurio describes Jupiter and Mars, respectively, as *sydus Iovis* and *stella Martis*. And within a few lines, he again speaks of the *lumen syderum* as derived from the Sun and also proclaims that all the *astra* are equally lucid. Velcurio used these three terms indifferently for the planets and the term *stellae fixae* for the fixed stars.[96] Although the context often determines the meaning of these terms, occasions arise where an author may say something about *stellae* and leave it to the reader to determine whether he is referring to planets or fixed stars alone, or both.

With this in mind, it appears that few in the Middle Ages inquired about the source of light for the fixed stars, although it may appear they do, because of these terminological problems just described. Thus when Oresme discusses at some length the question "Consequenter queritur utrum omnes stelle habeant lumen a sole vel alique ex se," which the translator renders as "Consequently, it is sought whether all the stars receive their light from the Sun [or whether] some stars [produce light] in themselves," inspection of the whole question reveals that Oresme is not speaking of "stars," in

94. According to Stahl (Macrobius, *Commentary on the Dream of Scipio*, 1952, 147, n. 41), Isidore of Seville and Honorius of Autun were influenced by Macrobius's distinctions. In an interesting passage, Vincent of Beauvais, *Speculum naturale*, bk. 15, ch. 16, 1624, col. 1102, claiming to draw upon Isidore of Seville, says that a *stella* is any single star; *sydera* are many stars, such as the Pleiades; and *astra* are great stars, such as Orion and Lucifer. He goes on to say that "authors confuse these names and use *astra* for *stellae* and *stellae* for *sydera*." Vincent seems to identify all three terms with the fixed stars. Immediately after, however, he declares that "Stars [*stellae*] do not have proper light but are said to be illuminated by the Sun, just like the Moon." Here *stellae* seems to signify the planets.

95. Because of its obvious relevance to our discussion, I translate the passage: "Of stars [*syderum*], some always preserve the same distance between them, as do the 'stars of the firmament' [*sydera firmamenti*]. Some [however] do not preserve the same distance, as [for example] the planets [*planetae*]. The stars [*sydera*] of the first kind are moved in the same way, because if [they were moved] in a different way, they would not preserve the same distance. But it is not so with the other [stars, or *sydera*, namely the planets] because they must necessarily be moved with different motions, because otherwise the difference of distance and nearness could not be caused." Amicus, *De caelo*, tract. 5, qu. 7, dubit. 2, art. 1, 1626, 338, col. 1.

96. See Velcurio, *Physics*, 1554, 76, col. 2 and 77, col. 1.

the sense of "fixed stars," but only of planets.[97] Indeed, the term *planeta* occurs only once in the question, in the expression "stelle sive planete que sunt sub sole scilicet Venus et Mercurius," which the translator correctly renders as "the stars or planets which are under the sun, such as Venus or Mercury."[98] The term *stella* is thus a synonym for *planeta* and turns up in such expressions as *stella Saturni* (the star, or planet, of Saturn).[99] Oresme's question, which at first glance seems directly pertinent to our present concern about the source of the light of the fixed stars, is in fact irrelevant.[100] This difficulty was already inherent in thirteenth-century discussions about the luminosity of celestial bodies. Thus Bartholomew the Englishman held that the *stellae* are essentially self-luminous but required supplemental light from the Sun. However, in declaring that all *stellae* except the Moon have their own proper light, Bartholomew means to include only the planets under the term *stellae*.[101] For the most part, then, *stella* means planet rather than fixed star, usage which continues into the seventeenth century.

Inspection of medieval discussions about the light sources of celestial bodies reveals that a considerable number of scholastics discussed the light source of the planets but that few included the fixed stars. Interest in the light source or sources of the fixed stars becomes manifest in the late sixteenth century, when the Coimbra Jesuits declare that the more common opinion of astronomers was that both the fixed stars and the planets receive their light from the Sun,[102] an opinion that was echoed some seventy years later by George de Rhodes, who declared that "all the stars, both fixed and errant, seem to have no proper light but are all illuminated by the light [*lux*] of the Sun."[103] De Rhodes insisted that the fixed stars were not so far away that the Sun's light could not reach them, as some argued. In a slight modification of that opinion, Pedro Hurtado de Mendoza also assumed that the sidereal heaven had its light from the Sun but allowed that each star might have a very small amount within itself.[104]

97. Oresme, *De celo*, bk. 2, qu. 11, 1965, 637 (Latin), 638 (English).
98. Ibid., 645 (Latin), 646 (English).
99. Oresme also uses the term *astra*, as in the expression "Septimo sol lucet ex se, ergo et alia astra." But *astra* is simply a synonym for *stella*, and both mean planet. Kren translates both *stelle* and *aster* as "star" or "stars," undoubtedly because they both subsume the term "planet." But in the context of Oresme's discussion, they can only mean planet.
100. Indeed, we have already considered Oresme's important question with respect to the Moon and the planets.
101. "Unde omnes stellae habent lumen proprium praeter lunam. Et quamvis stellae ex se sint luminosae, ad consummationem tamen suae luminositatis recipiunt complementum ab ipso Sole." Bartholomew the Englishman, *De rerum proprietatibus*, bk. 8, ch. 33, 1601, 420.
102. For the passage, see note 53 of this chapter.
103. "Astra omnia, tum fixa, tum errantia, nullam videntur habere lucem propriam, sed ea omnia illuminari a luce solis." De Rhodes, *De coelo*, bk. 2, disp. 2, qu. 1, sec. 3, 1671, 284, col. 2.
104. "Coelum sydereum nullam aut exiguam habere nativam lucem." And some lines below, he declares: "Non propterea nego aliquid nativae lucis syderibus, sed illud dico esse perexiguum collatum cum lumine quod a sole mutuantur." Hurtado de Mendoza, *De coelo*, disp. 2, sec. 5, 1615, 375, col. 1, par. 64.

He based his argument on Genesis 1.14–15, where, after God had already created the firmament, he is said to have created the luminaries in the firmament to provide light for it. Thus the firmament is seen only by virtue of the luminaries, or the Sun and Moon. If it had its own native light, it would have been seen before the creation of the luminaries. Nevertheless, the stars may have a very small amount of light, which is, however, insufficient to make them visible to us without the Sun. Thomas Compton-Carleton was less tentative in his attribution of some light to the fixed stars when he said that "neither the Moon, nor the other planets, nor even the fixed stars receive all their light from the Sun,"[105] thus implying that the fixed stars receive part of their light from some other source or from themselves.[106]

But scholastic authors were hardly of one mind, and some were convinced that the fixed stars were self-luminous. In 1627, a few years before Galileo likened the fixed stars to "so many Suns" (*Dialogue*, Third Day [Drake], 1962, 327), thus implying their self-luminosity, Aversa left no doubt as to his interpretation: the fixed stars probably are not illuminated by the Sun but are self-luminous.[107] The planets are related to the Sun, which lies in their midst and with respect to which they move. Therefore it is proper that they should receive their light from the Sun. But the fixed stars have no such relationship to the Sun and derive no light from it; rather they have their own proper light. Moreover, by contrast to the opaque bodies of the planets, which receive light only on their surfaces, the fixed stars are transparent bodies that are suffused with light throughout their depth.[108] Thus the transparency that Albertus Magnus, Albert of Saxony, and a number of early modern scholastics attributed to the planets, Aversa assigns only to the fixed stars. And whereas Albertus and Albert filled the transparent planetary bodies with light from another

105. "Nec luna, aut alij planetae, nec stellae etiam fixae omnem lucem suam accipiunt a sole." Compton-Carleton, *De coelo*, disp. 2, sec. 3, 1649, 403, col. 1.
106. Compton-Carleton offers no further elaboration.
107. In the late sixteenth century, Giordano Bruno (1548–1600), and perhaps even Nicholas of Cusa (1401–1464) in the fifteenth century, had assumed that all stars were Suns. See Dick, 1982, 108 for Bruno, and ibid., 40, for Cusa, who assumed that all celestial bodies were like the earth, which he characterized as a star that was self-luminous. By implication, then, all stars should be self-luminous.
108. "Deinde haec omnia nullatenus procedunt de stellis fixis atque adeo nullum prorsus indicium suppetit ut lumen suum dicantur a sole recipere. Et quidem verisimile videri poterit bene quidem planetas, non autem stellas fixas illuminari a sole quia planetae agnoscunt pro suo principe solem et in medio illum continent et in suo motu habent respectum ad solem bene ergo putari debent ab illo et per illum lucere. At stellae fixae per se propriam agunt aciem et non habent huiusmodi respectum ad solem. Censeri ergo potius debent non illius beneficio sed propria virtute lucere. Et iuxta hoc planetae quidem dici debebunt corpora opaca atque in solo externo ambitu lumine illinita. Stellae autem fixae esse corpora perspicua et in tota sua mole ac profunditate eodem suo lumine perfusa. Iuxta id quod dicebamus quaestione superiori sect. 2. Et hoc etiam iuuare potest ad rationem reddendam cur maxime stellae fixae scintillent et sol non solum scintillare sed et veluti ebullire cernatur." Aversa, *De caelo*, qu. 35, sec. 3, 1627, 173, col. 2–174, col. 1.

body, the Sun, Aversa assumes that every fixed star is transparent and possesses its own source of light.

Giovanni Baptista Riccioli considered the opinions of Galileo, Kepler, and Descartes, all of whom assumed that the fixed stars possessed their own light and received none from the Sun, as far more probable than any other interpretation. In addition to his acceptance of their reasons, Riccioli thought it was more appropriate "to the majesty of the Divine Creator that there not be a single light for the stars but that a multitude should light in the manner of the Sun. Nor do they [the fixed stars] require another source of light other than God, the Father of all lights." To strengthen his argument, Riccioli quotes from Baruch 3.34–35, where it is said that "The stars shone at their appointed stations and rejoiced; he [God] called them and they answered, 'We are here!' Joyfully they shone for their Maker."[109] The stars were thus capable of shining with their own light and had no need of the Sun.[110]

Johannes Poncius, and perhaps Roderigo de Arriaga, also joined the ranks of those scholastics who regarded the fixed stars as essentially self-luminous bodies. Poncius couched his statement in a manner analogous to the way in which he identified the source of planetary light. Just as the Sun is principally (*principaliter*) the source of planetary light, so also do "the fixed stars have their light principally from themselves."[111] Poncius describes this opinion as "more common."[112] He could find no experience that would lead us to believe that the fixed stars receive their light from the Sun. But he invokes one experience that indicates that the fixed stars are self-luminous, namely the fact that they are visible at noon from the deepest wells.[113]

109. *New English Bible*, 1976, p. 179 (of the Apocrypha).
110. Indeed, Riccioli (*Almagestum novum*, pars post., bk. 6, ch. 2, 1651, 395, col. 2) thought it natural to ask which of these two lights was stronger, that of the Sun or that of the fixed stars. He concluded that they might have light of equal strength (ibid., 396, col. 1).
111. "Stellae fixae habent lucem suam principaliter a seipsis." Poncius, *De coelo*, disp. 22, qu. 8, 1672, 625, col. 1, par. 74.
112. "Haec est communior." Ibid.
113. "Et non constat ulla experientia quod stellae illae mutuent a sole suam lucem; imo potius suffragatur experientia opposito, nam in ipso meridie ex altissimis puteis stellae fixae videri possunt." Ibid. It is on the basis of this same argument that Arriaga may be classified among those who believed that the fixed stars are self-luminous. He thought it probable that "the light of these stars [*stellae*] is independent of the Sun." Although the passage, which follows, may also be interpreted to apply to the superior planets instead of the fixed stars – the term *stellae* is equally applicable to both – the fixed stars seems more appropriate in light of Arriaga's example, which is the same as that of Poncius. Here is the relevant passage: "An planetae habeant a se aliquam lucem plane dubium. Probabile est habere quia si vera est sententia docens etiam in meridie videri ex profundis puteis stellas, cum ipsae sint supra solem et tunc a sole non respiciantur ne illuminentur (suppono enim solem esse opacum a tergo, ut possit ad nos melius lucem mittere), necessarium est ut illa lux stellarum sit independens a sole." Arriaga, *De caelo*, disp. 1, sec. 6, 1632, 507, col. 2. For Compton-Carleton's use of the well argument, see note 124.

By the time Poncius published his relevant work, in 1642 and 1643,[114] it is likely that the assumption of the fixed stars as self-luminous rather than illuminated by the Sun was indeed "more common." If it was not more common among scholastic authors, it was nevertheless widely accepted. How was it that the Sun, which had previously been thought to be the primary light of the world on which other bodies depended for their illumination, came to be perceived as no more light-giving than a fixed star and perhaps even less so?

The explanation may lie in the newly developed ideas about a vastly expanded universe that had its roots in the Copernican theory. Because no stellar parallax could be detected from the earth's orbital motion, an enormous spatial gap had to be assumed between Saturn and the fixed stars.[115] In the Copernican scheme, the universe was of enormous size, perhaps even unmeasurable, and the fixed stars were a vast distance from the Sun. But it was not merely a matter of distance. Tycho Brahe, an opponent of the Copernican system, showed that if the Copernican scheme were true, a third-magnitude star would have a diameter 200 times greater than that of the Sun, which he thought absurd.[116] It would seem an odd universe in which a relatively small Sun could illuminate huge stars so very far away.[117]

But scholastics were not Copernicans, and those who followed Tycho may have known that he had not only assumed a smaller distance of the fixed stars than was traditional but also reduced their size.[118] What reason would scholastic natural philosophers have had to assume that the fixed stars were self-luminous and independent of the Sun? Perhaps Galileo's analysis of his telescopic observations impressed them. After all, Galileo, and later Kepler, had declared that the stars were themselves Suns.[119] As Suns, the fixed stars would provide their own light.

No compelling argument or overwhelming piece of evidence promoted and encouraged the partial scholastic acceptance of self-luminous fixed stars. Indeed, during the Middle Ages the question about the light source of the fixed stars was rarely, if ever, raised. Those to whom the question did occur probably assumed that the Sun was the source of all stellar light. Not until the late sixteenth and the seventeenth century did the issue acquire a measure of prominence, and this solely because of revolutionary new concepts and

114. Although I have used the 1672 edition, the first edition of 3 volumes appeared in 1642–1643 (see Lohr, 1988, 362).
115. See Van Helden, 1985, 46–48.
116. Ibid., 51–52.
117. Otto von Guericke emphasized the enormous distance between the Sun and the nearest fixed star as a major factor in denying that sunlight could light the stars, even though it lit the planets. Guericke, 1672, 229, col. 1.
118. Van Helden, 1985, 51.
119. Guericke (1672, 230, col. 1) cites both Galileo and Kepler as among those who identified the stars as Suns. For the former he cites the *Sidereus nuncius* (1610), for the latter, *Dissertatio cum Nuncio sidereo* (1610).

observations associated primarily with the names of Copernicus, Brahe, and Galileo.

IV. Is the light of the stars and planets of the same species?

From the time of Richard of Middleton in the thirteenth century, those who attributed a certain degree of weak self-luminosity to the planets and who also assumed the Sun as the primary source of planetary light had seemingly recognized at least two different species of light in the heavens.[120] Although this might have seemed a logical consequence based on the distinction of two different lights, it was probably a minority opinion.

As with almost all scholastic issues, arguments for another interpretation could usually be formulated with some plausibility within the Aristotelian system. In this case, however, it was probably a strong desire to treat nature in the simplest terms and therefore to subsume all light under a single species. Thomas Compton-Carleton, for example, insisted that, with the exception of the light in the empyrean heaven, which differs from the light of the physical world, there is only one species of light, not only in the heavens but also in the sublunar world.[121] The light for all these bodies is derived ultimately from the Sun. But what of the claim that the planets have their own light, independently of the Sun? Does this not make it differ in species from the light of the Sun? In a strange reply, Compton-Carleton explains that the natural light of the planet is intensified and supplemented by solar light, which could not occur if they were really distinct in species,[122] a point that Amicus, who also argued that celestial light does not differ specifically, elaborated some years before when he remarked that an internal light could be intensified by an external light only if they were of the same species. To be intensified to a more intense degree implies that the thing intensified and the thing intensifying are the same kind of thing – that is, in the same species.[123]

120. For example, Richard of Middleton emphatically distinguished two species of light. See note 40, this chapter.

121. The Coimbra Jesuits (*De coelo*, bk. 2, ch. 7, qu. 9, art. 2, 1598, 301) adopted a similar attitude, holding that all light produced by the forces of nature belonged to the same species. After expressing some doubts, they even included the lights associated with "the glorious bodies," whose light was said to be as lucid as that of the Sun and therefore in the same species.

122. "Quoad lucem autem aliorum coelorum probatur eam specie non distingui. Recipiunt quippe a sole lucem, sicut aer et res omnes sublunares. Ergo non est cur lux illa sit diversae rationis a luce solis. Dices primo secundum nos singuli caeli et astra habent aliquid lucis ex se, ergo illa saltem erit diversae speciei secundum diversitatem substantiae a qua oritur. Negatur tamen consequentia illa enim lux intenditur a luce solis quod tamen fieri non posset si esset specie adaequate distincta." Compton-Carleton, *De coelo*, disp. 2, sec. 3, 1649, 403, col. 2.

123. "Secundo quia lumen intrinsecum intenditur ab extrinseco, ergo sunt eiusdem speciei, nam intensio non fit nisi per gradus qualitatis unius speciei." Amicus, *De caelo*, tract. 6, qu. 5, dubit. 2, 1626, 363, col. 1.

But what about the seeming variety of lights in the heavens, namely the light of the Sun, the planets, and the fixed stars: Are they not different? In denying any differences, Compton-Carleton points to the rainbow, which, although it contains different colors, is produced by one kind of light. Differences in light are only apparent and arise from differences in the properties of bodies, especially density and transparency. Moreover, the light of the Sun does not destroy the light of the planets and fixed stars, although it may seem to do so, because their light is not seen during the day. But if you descend into a deep well, where the Sun's light cannot reach, you will perceive the stars clearly, as at night. What makes the stars and planets invisible during the day is the principle that if a sensible thing, say the Sun's light, exceeds by a great amount a lesser sensible thing, say the light of individual planets and fixed stars, the greater will drastically interfere with the smaller.[124]

V. Celestial light as a mix of old and new

Despite their contributions, however, most scholastic arguments about the sources of light were, as we have seen, not technical but general, analogical, and scriptural. Raphael Aversa, for example, did not argue that the fixed stars were Suns. Indeed, he assumed they were totally different: whereas the Sun was an opaque body, the fixed stars were transparent and self-luminous. For thinkers who followed a centuries-long tradition about the hierarchy of the heavens, it may well have seemed perfectly plausible for the higher, and therefore nobler, fixed stars to have their own light rather than be dependent on light from a planet so much closer to the earth and therefore presumably less perfect. But Aversa joined Galileo in the belief that the fixed stars were self-luminous and not lighted by the Sun. In this decision, he was probably influenced by the controversies and debates concerning the new concepts and observations that had emerged since the end of the sixteenth century.

What applies to the fixed stars is equally applicable to the entire range of problems about celestial light and its manifestation in all celestial bodies. Many, though by no means all, scholastic natural philosophers in the seventeenth century mention the names of Copernicus, Brahe, Galileo, and Kepler. Directly or indirectly, they were aware of the new and dramatic

124. "Dices secundo apparet alterius quasi rationis lux in stellis ac veluti fulva. Contra etiam in iride apparet lux diverse rationis et tamen non est, sed prorsus eiusdem. Hoc ergo solum provenit ex diversa dispositione corporis in quo recipitur secundum diversam temperiem, densitatis, diaphaneitatis, etc. Nec lux solis destruit lucem planetarum aut stellarum firmamenti quod interdiu non appareant, sed hoc ex eo provenit: quod maius sensibile, praesertim si valde magnus sit excessus, impediat minus. Unde si quis medio die descenderet in profundum puteum quo lux solis non pertingeret aeque clare, ut aiunt, perciperet stellas, ac nocte." Compton-Carleton, *De coelo*, disp. 2, sec. 3, 1649, 403, col. 2–404, col. 1.

claims made in the Copernican and Tychonic systems and knew about Galileo's telescopic discoveries. Despite their commitment to Aristotelian cosmology on traditional and religious grounds, scholastics sometimes adopted ideas from the new science and incorporated them into their own cosmic scheme without great difficulty. As a consequence, scholastic cosmology encompassed both traditional and new elements, even though the latter were rather poorly integrated into the Aristotelian system. It is difficult to imagine how a hierarchically based Aristotelian cosmology could easily reconcile a planetary system which had the Sun, the fourth planet from the earth, as the primary source of all planetary light and could also assume self-luminosity for the fixed stars. The relationship between the fixed stars and the Sun was thus complex. To add to the difficulties, Galileo had called the fixed stars Suns. Did this imply that the Sun was therefore a fixed star? For Copernicans, who assumed an immobile Sun, the Sun could indeed be conceived as a fixed star, but not for Aristotelians, who assumed a Sun that moved around a stationary earth. For the most part, scholastics who thought the fixed stars self-luminous avoided such questions and their implications. They coped with questions about light in the usual ad hoc manner. Although light was perhaps the most important and spectacular attribute of the heavens,[125] there were other properties and powers that marked out the heavens as special. It is now time to describe them.

125. Among the visible qualities, Aversa, *De caelo*, qu. 34, sec. 3, 1627, 133, col. 2, called it "the most noted."

17

The properties and qualities of celestial bodies, and the dimensions of the world

"One may find it surprising," wrote Friedrich Solmsen in his informative study of Aristotle's physical system of the world, "that Aristotle does not say more about the nature of the heavenly bodies."[1] On this subject, Aristotle's commentators had no choice but to find their own way.

I. The celestial ether

Because planets, stars, and orbs were assumed to be constituted of a special celestial ether, I have had occasion to mention it at different points in this study. It is now time for a more systematic examination of this extraordinary substance.[2]

With the notable exception of Robert Grosseteste, and perhaps a few others during the Middle Ages, the celestial region was assumed to be composed of ether, rather than of fire or some combination of the four elements.[3] In his popular encyclopedia, Bartholomew the Englishman described Aristotle's ether as "something beyond the lunar globe that is of a separate nature from the nature of the inferior elements. Thus the ether is neither heavy nor light, neither rare nor dense, nor is it divisible by the penetration of another body. No corruption or alteration, universally or particularly, affects the ethereal nature, which would happen to it if its

1. Solmsen, 1960, 316, n. 50.
2. For Solmsen's description of Aristotle's conception of the ether, or "first body," as the latter called it, and the role it played in his system, see ibid., 287–309.
3. Grosseteste, *De generatione stellarum*, 1912, 33, says that "a star does not possess the nature of a fifth essence" (Stella autem non est de natura quintae essentiae). Indeed he also argues (ibid.) that stars are not only composites of matter and form but also composed of elements, by which he clearly means elements that do not differ from our terrestrial elements. Grosseteste was but following the earlier Platonic tradition characteristic of the early Middle Ages, which became popular again in the sixteenth and seventeenth centuries when the observation of seemingly real changes in the sky undermined the notion of celestial incorruptibility and consequently also of the ether that was allegedly incorruptible.

origin or composition were drawn from the elements."[4] At approximately the same time, John of Sacrobosco declared in his famous *Sphere* that "Around the elementary region revolves with continuous circular motion the ethereal, which is lucid and immune from all variation in its immutable essence. And it is called 'Fifth Essence' by the philosophers."[5] Elaborating on these few sentences, Christopher Clavius reveals that the ether was understood in much the same way nearly four centuries later.[6] He distinguished five major properties, the first of which is that the ethereal region encloses the elementary region as its container and is therefore its place. As the place of the elementary region, philosophers consider the ethereal zone more excellent, because it is removed from the incessantly changing region of the elements and also because it exists among the divine movers of the orbs that enjoy the best life. The second of its properties is light, which is much nobler than that of the elements. As the third property of the celestial ether, Clavius mentions its capacity for avoiding change: it cannot be altered, or diminished, or increased, or generated or corrupted, all of which attributes are the opposite of those in the four elements. Its fourth property is its continuous circular motion, which is the cause of continuous generation and corruption and stands in contrast to the natural rectilinear motion of the inferior, terrestrial region, which is not perpetual but always comes to an end. As its fifth and final property, Clavius observes that philosophers call the ethereal region a fifth essence. For centuries, scholastics had thought of the celestial ether in virtually the same terms.

Up to this point, we have had occasion to consider the celestial ether in a number of different contexts. In Chapter 10, we examined its incorruptible nature, and in Chapter 12 considered whether or not it was perceived as matter (for the most part, it was) and whether that incorruptible celestial substance was a composite of matter and form, and whether celestial matter was different from or identical to corruptible terrestrial matter. And in Chapter 14, we pondered the fundamental question as to the hardness or fluidity of the ethereal orbs. In these primarily metaphysical discussions, the level of discourse was mostly abstract and general. We must now approach the ether as a substance comprised of a large number of seemingly different bodies: planets, stars, and orbs. Indeed, one may even ask whether the waters above the firmament, which form the crystalline or ninth orb, could be formed of the same matter as the firmament, which separates the waters forming the crystalline orb from the waters below.[7] If the ether is

4. "Quicquid enim supra lunarem globum est, naturae est separatae a natura inferiorum elementorum. Unde aether neque est grave neque est leve; neque rarum, neque densum; neque per alterius corporis penetrationem divisibile. Naturam enim aethereum nulla ingreditur corruptio vel alteratio universaliter vel particulariter, quod ei accideret si ex elementis compositionem aut originem contraxisset." Bartholomew the Englishman, *De rerum proprietatibus*, bk. 8, ch. 5 ("De aethere"), 1601, 381.
5. Sacrobosco [*Sphere*, ch. 1], 1949, 119.
6. What follows appears in Clavius [*Sphere*, ch. 1], *Opera*, 1611, 3:20.
7. At first glance, this question seems to cry out for a response, but to my knowledge it was

a single substance, why do many of the celestial bodies appear to differ from one another?

If the medieval followers of Aristotle had taken his statements about the ether in *De caelo* literally, or as Bartholomew the Englishman and Clavius understood them, they would have been compelled to conclude that all celestial bodies and the ether as a whole possessed the same properties. For, on the assumption of ethereal homogeneity, it follows that the ether and all the celestial bodies within it are identical in appearance and power and in all distinguishable qualities. And yet Aristotle himself cast doubt upon this interpretation when, in his *Meteorologica* (1.3.340b.6–10), he indicated that the ether was not uniform in quality, especially in those parts bordering on the terrestrial region. Gross observation, moreover, made it apparent that the celestial bodies did indeed differ: the Sun from the Moon, and the two latter bodies from the other five planets, which, in turn, seemed to vary from each other. For a very long time, astrologers had forecast their predictions and assessments on the basis of assumed differences in the powers and appearances of the planets and stars. Aristotle himself recognized such differences. The author of the *De proprietatibus elementorum*, falsely attributed to Aristotle during the Middle Ages, insisted that the Moon and other planets diverged from the Sun and that the planetary orbs differed from the planets they carried, because the former, although they receive light, are not illuminated as are the planets, an indication that their substances varied. Indeed, the anonymous author was prepared to distinguish three distinct celestial substances.[8] How, then, was this apparent conflict between Aristotle's theory of a uniform, homogeneous, and incorruptible ether reconciled with observed differences among the planets, stars, and even the orbs themselves?

We must first recognize that what I have characterized as an "apparent conflict" was rarely made explicit during the Middle Ages. Aristotle failed

ignored. It seems implausible to suppose that the waters above the firmament could be made of the same substance as the firmament itself, that is, that both could be composed of an identical celestial ether. The firmament, after all, was created to form a barrier between the waters above and below. Thus no single celestial ether could have formed both the crystalline sphere and the firmament, whether the latter was interpreted as the eighth sphere of the fixed stars only or as comprised of the fixed stars and planets.

8. Here is the complete text of this unusual passage: "Iam ergo apparet et est manifestum quod substantia corporis stellarum et lunae est alia a substantia corporis Solis, sicut candela quando opponitur ei speculum. Illuminatur enim speculum a substantia candelae et est alia a substantia speculi. Iam ergo apparet nunc quod illud quod diximus et quod ostensum est quod substantia corporis stellarum et lunae est alia a substantia corporis Solis. Et similiter est substantia corporis orbis absque substantia corporis stellarum et lunae, quod est quia substantia corporis orbis ex substantia corporis stellarum et lune recipit lumen et non illustratur splendore sicut illustratur corpora stellarum et similiter oportet quod substantia sit corporis orbis alia a substantia corporis Solis. Iam ergo apparet per ista accidentia et has inquisitiones quod corpora eorum sunt diversarum substantiarum et quod orbis cum eis quae sunt in eo sunt tria elementa, scilicet tres substantie." Averroës, *Opera*, 1562–1574, 7:208r, col. 2–208v, col. 1, where the treatise is titled *De causis libellus proprietatum elementorum Aristoteli ascriptus*, although it was usually cited as *De proprietatibus elementorum*.

to identify it as a problem and offered little help in its resolution. Scholastic natural philosophers devoted no questions to it. And yet the problem was implicitly acknowledged, because it was discussed in an indirect manner. Celestial properties have two distinct aspects, visible and invisible. The first of these concerns such visually detectable variations as brightness and color among stars and planets or the radical differences between the Sun and the Moon; the second applies to latent and invisible properties and powers that were assumed to inhere in some sense in each celestial body and which were thought to produce changes in the terrestrial region. Although visible and invisible celestial properties were rarely, if at all, distinguished, I shall devote a separate chapter (Ch. 19) to the influence of the celestial region on the terrestrial, which is largely concerned with latent properties and influences.

In the broadest sense, any scholastic natural philosopher who gazed skyward on any clear evening could see a panorama of wandering planets and fixed stars, between which there were vast tracts of darkness. As we have already seen, Aristotle had assumed that this entire celestial region is filled with a fifth element, or incorruptible, pure, transparent ether. The ether exists in the form of gigantic but invisible spheres, within which are embedded the visible stars and planets. As the spheres turn with circular motion, they carry around all the visible celestial bodies. But the concept of a pure, homogeneous, transparent ether posed a monumentally difficult problem, one that Aristotle ignored but Alexander of Aphrodisias did not, as we learn from Simplicius, who reports that Alexander asked how celestial differences could occur in the simple celestial ether. If celestial bodies differed in density or rarity or with respect to color and other properties, how could such differences occur in an element in which no changes were thought possible because no contrary properties could exist therein?[9]

Alexander's response, with which Simplicius concurred, was destined to have some influence on medieval natural philosophers. They were agreed that contraries, such as white and black or hotness and coldness, could coexist if they were not in the same subject, that is, not in the same body. The same principle applies to the celestial region. For example, the Sun might possess hotness, but never coldness, and, simultaneously, Saturn might possess coldness but never hotness. Under these circumstances, hotness and coldness are present in the heavens simultaneously, but because each member of the pair is in a different body, they cannot cause generation or corruption. Similarly, if density exists in one part of the heavens, or in

9. See Simplicius [*De celo*, bk. 2, comment. 28], 1540, 69v, col. 2–70r, col. 2, especially 70r, col. 1, where Simplicius says: "Dubitat bene Alexander quomodo simplici existente quinta dicta substantia circulariter mobilis corporis tanta apparet differentia corporis astrorum ad celeste. Si autem differunt totaliter spissitudinibus aut raritatibus aut secundum colores aut secundum quasdam alias tales species, quomodo simplicia dicuntur aut quodmodo impassibilia siquidem passiones secundum has fiunt differentias et sunt differentie passionum." Simplicius's work was translated into Latin by William of Moerbeke in 1271. Also see Guthrie's summary of the main points in Aristotle, *De caelo* [Guthrie], 1960, 176–177.

one celestial body, and rarity exists simultaneously in another, they do not oppose each other, nor is one transmuted into the other.[10] Thus rarity and density, and hotness and coldness, and other pairs of contraries, could seemingly exist in the heavens as long as no pair was embodied in the same celestial subject. As we saw earlier, numerous scholastic natural philosophers accepted some version of this interpretation, frequently accepting it for rarity and density but seldom if at all for such opposites as hotness and coldness or wetness and dryness.[11] But they were overwhelmingly agreed that pairs of contrary qualities could not be involved in the same subject, for then change would indeed have to occur. With the idea fairly well accepted that differences of rarity and density could exist in the celestial ether (though not in one and the same specific body), many in the Middle Ages and Renaissance went a step farther and assumed that the regions of high ethereal density were celestial bodies. This was not an idea that drew inspiration from Aristotle, who was content merely to assume that each star was composed of the same substance as the orb that carried it.[12] In the later interpretations, a celestial body was thus a relatively small region of highly concentrated ether. Albert of Saxony made this plain when he declared that "stars [or planets: *astra*] are denser parts of their orbs," and others agreed.[13] By the seventeenth century, Mastrius and Bellutus report: "It is

10. "Que autem in scientia contrariorum rationes albi et nigri, calidi et frigidi, veluti distantes in indistanti existentes possunt coexistere invicem, immo et subsistunt. Sed neque in scientia album ad id quod in materia est nigrum oppugnat, neque corrumpunt se invicem, neque generantur ex invicem quia non sunt nata in eodem fieri subiecto. Sic igitur etiam in celo spissum, si forte et hic rarum sit neque oppugnant adinvicem neque transmutantur in invicem quia non sunt nata fieri in eodem subiecto cum sint alterius nature existentia. Sed neque in celo spissum ad id quod ibi rarum oppugnat, neque statio ad motum quia et subiecta differentia secundum naturam sunt. Verbi gratia, poli et equinoctialis circulus. Hi quidem ad stationem apti nati sunt, hic autem ad motum. Et sol, si forte, et astra, que videntur spissa." Simplicius, ibid., 70r, col. 2.
11. For details and illustrations, see above, Chapter 10, Section II.1.c.
12. See *De caelo*, 2.7.289a.12–14 [Guthrie], 1960, where Aristotle concludes that "The most logical and consistent hypothesis is to make each star consist of the body in which it moves, since we have maintained that there is a body whose nature it is to move in a circle." And yet we must observe that Jean Buridan ([*De caelo*, bk. 2, qu. 9], 1942, 41) attributed to Aristotle the opinion that "a star [or planet] is the denser part of its orb" (unde Aristoteles ponit quod stella est densior pars sui orbis) as did John Major when he declared that "according to him [Aristotle] a star is the denser part of its orb" (Major [*De celo*, bk. 2, qu. 6], 1526, sig. iiii verso, col. 1). As his source, Major cites the pseudo-Aristotelian *De proprietatibus elementorum*, which, in the version used here (see n. 8), lacks such a statement.
13. "Ymaginandum est ergo quod astra sunt densiores partes suorum orbium." Albert of Saxony [*De celo*, bk. 2, qu. 20], 1518, 114v, col. 2. In the thirteenth century, Roger Bacon and Albertus Magnus described the star as the denser part of its orb. For Bacon, see Section II and note 32, this chapter; for Albertus Magnus see his *De caelo*, bk. 2, tract. 3, ch. 4, 1971, 149, col. 2, where he argues that "a star [or planet] is like the worthier part of the orb . . . and that the greater density is in the star itself" (Et ideo dicendum videtur, quod stella est quasi dignior pars orbis. . . . Et quod densitas maior est in ipsa stella).

    Albert of Saxony reports a second opinion (ibid.), in which the planets and stars are conceived as holes in their respective heavens or orbs. In this interpretation, the outermost sphere of the world is assumed completely filled with light. Below this light-filled out-

commonly said that the stars are denser parts of the heaven, [with] the remaining parts rarer."[14]

A denser part of the ether was probably thought sufficiently dense to reflect light and thereby make the planet or star visible to us, an idea that appears in the thirteenth century when Robertus Anglicus explained that "a star which is in an orb does not differ from the orb except in greater and less aggregation of light, and in greater and less density, wherefore a star is an aggregation of light in an orb with a greater density than the orb."[15]

Despite the alleged common acceptance of the opinion that a star is a denser aggregation of ether, Mastrius and Bellutus, like Raphael Aversa before them, rejected this interpretation. Aversa, as we saw (Ch. 10, Sec. III.3.b) denied the existence of the quantitative opposites rarity and density in the heavens and replaced them with the qualitative opposites opacity (*opacitas*) and diaphaneity (*diaphaneitas*), or transparency, which he used to demonstrate the solidity of the celestial spheres. Mastrius and Bellutus adopted Aversa's arguments and declared that "all the stars [or planets] are not necessarily constituted from denser matter than are the heavens. Indeed, [they are] of the same density but differ with respect to opacity and diaphaneity, so that the stars [or planets] are parts of the opaque heaven, while the orbs are absolutely diaphanous."[16] The shift to opacity and diaphaneity instead of density and rarity reflects the idea that what is transparent (for example, crystal) may be denser than what is opaque (for example, wood). One could thus defend the solidity and hardness of the celestial orbs by describing them as transparent, while the planets and stars could be described as opaque and visible because they reflect light. On this approach, a star or

ermost orb lie all the planetary orbs. Mars, for example (Albert provides no examples), is merely a hole in the orb of Mars through which we see the light of the outermost orb. Thus Mars is not truly a celestial body but is simply a hole or opening in a celestial orb through which the brilliant light of the outermost sphere perpetually shines. Albert rejects this theory because it cannot account for the variations in lunar light or the differences in the colors of the planets.

14. "Communiter dicitur stellas esse partes coeli densiores, reliquas vero partes esse rariores." Mastrius and Bellutus [*De coelo*, disp. 2, qu. 2, art. 3], 1727, 3:495, col. 1, par. 74. In agreement were the Conimbricenses [*De coelo*, bk. 2, ch. 7, qu. 1, art. 1], 1598, 194; Illuminatus Oddus [*De coelo*, disp. 1, dub. 10], 1672, 29, col. 1; and Bartholomew Amicus [*De caelo*, tract. 4, qu. 7, dubit. 3, art. 2], 1626, 208, col. 1. Riccioli, *Almagestum novum*, pars prior, bk. 6, ch. 2, 1651, 397, col. 2, attributes this opinion to Galileo, to Francisco Suarez and to "many others." Galileo, however, does not seem to adopt this position himself but has Sagredo attribute it to Simplicio (see Galileo *Dialogue*, First Day [Drake], 1962, 43).
15. See Robertus's commentary on the *Sphere* of Sacrobosco in Robertus Anglicus [*Sphere*, lec. 2], 1949, 206; Latin (151). Bartholomew the Englishman, *De rerum proprietatibus*, bk. 8, ch. 33, 1601, 417, quoting someone he calls "Alphra" (for Alfraganus?) defines a star as "the light aggregated in its orb" (Secundum *Alphra*, stella est lux aggregata in suo orbe). As it stands, this statement indicates that the star is a concentration of light and says nothing about density, as did Robertus Anglicus. Perhaps density is understood or implied.
16. Mastrius and Bellutus, *De coelo*, disp. 2, qu. 2, art. 3, 1727, 3:495, col. 1, par. 76.

planet may be no denser than the rest of the orb, since density is not necessarily directly proportional to opacity.

Thus from Aristotle's theoretically assumed uniform, unchanging density of the heavens, scholastics came to assume both rarity and density in the heavens, assigning rarity to the entire orb, except for the planet or star that it carried, which they assumed to be the densest part of the orb. Because of anomalies in the relationship between transparency and density, where a denser object could be more transparent than a less dense object, some scholastics, in the seventeenth century, would abandon the opposites density and rarity in favor of opacity and diaphaneity (i.e., transparency).

II. Whether all celestial orbs and bodies belong to the same species

Whether the celestial ether was to be assumed homogeneous and of the same nature throughout its vast extent seems at first glance a troubling problem in Aristotelian cosmology. The extent to which this was a problem is manifested by the frequency with which scholastics considered some form of the question "Do planets, stars, and orbs belong to the same ultimate species?"[17] If the ether was absolutely uniform and homogeneous, one would expect that all celestial bodies and orbs would be assumed in some fundamental sense identical and therefore members of the same irreducible species. And yet, as we saw in our earlier examination of this theme (Ch. 11, Sec. I.1, "Are all celestial bodies in the same irreducible species?"), most natural philosophers opposed the opinion of Averroës and decided that one species could not account for the great diversity of bodies, orbs, and motions in the heavens. Albeit in that earlier discussion major emphasis was on the sameness of or differences among the visible celestial bodies themselves, namely stars and planets, natural philosophers also made other comparisons. Did the planetary orbs differ from one another? Did stars and planets differ from the orbs that carried them? Indeed, natural philosophers even asked whether the three or more partial orbs of a single planetary orb differed – that is, whether the two or more eccentric orbs and the single concentric orb of a total planetary orb were sufficiently distinct to be assigned to separate species.[18]

17. See "Catalog of Questions," qus. 91, 92, 94, 95, Appendix I.
18. Amicus discussed all of these questions in four *dubitationes* in *De caelo*, tract. 4, qu. 7, 1626, in the following order: (1) "An coeli totales inter se specie distinguantur" (dubit. 1: 202, col. 1–205, col. 2), that is, whether the total heavens differ in species from each other; for example, does the total planetary orb of Mars differ from the total planetary orb of Jupiter; (2) "De distinctione partium unius orbis qua totum coelum circumdant" (dubit. 2: 205, col. 2–206, col. 1), which concerns the eccentric and concentric parts of the whole sphere; (3) "An astra differunt specie a suis orbibus" (dubit. 3: 206, col. 2–208, col. 2), namely whether the stars or planets differ in species from their orbs; and (4) "An omnia astra inter se specie distinguantur" (dubit. 4: 208, col. 2–210, col. 2), which

Although there were defenders of the view that all celestial bodies belong to the same species, and a few were convinced that neither opinion was demonstrable,[19] most scholastic natural philosophers opted for diversity and differences of species (for the arguments, see Ch. 11, Sec. I.1). The denial that all celestial bodies belong to the same species was often based on obvious visual differences between planets and on the differences between their effects, especially the Sun and Moon. Thus Albert of Saxony appeals to the different properties and natural powers that planets and fixed stars possess. This is apparent by contrasting the Sun with the other planets. Whereas the Sun provides its own light, the planets and Moon get their light from the Sun. Obviously the Sun possesses radically different properties from the other planets. But an ultimate, irreducible species ought to have similar properties and accidents.[20] As additional evidence, Albert points to the diverse manner in which planets and stars influence contrary things and species in the sublunar world. Indeed, the astrologers proceed on this assumption. No single ultimate species could embrace such diversity.

Centuries later, George de Rhodes declared (*De coelo*, bk. 2, disp. 2, qu. 1, sec. 3, 1671, 284, col. 1) that the planets (*astra*), and presumably the fixed stars, differ in species from the invisible heavens in which they are embedded, and further observed that they differ from each other. These differences were explained by the diversity of properties which reside in the planets and not in the surrounding heavens. For example, density and opacity are properties of the planets but not of the surrounding heavens, as are also many occult properties. Thus certain planets have occult properties that can arouse storms and hail; the Sun and Moon, in different aspects, cause a marvelous diversity in things; the polestar governs the needle that is attracted

is the more traditional question as to whether the stars and planets differ from each other in species.

19. For Buridan and Oresme, see Chapter 11, Section I.1. In his discussion of whether the whole or partial orbs of one planetary sphere differ from another in species, Amicus, ibid., dubit. 1, art. 2, 205, col. 1, cites as a third opinion the view that neither of the first two interpretations is demonstrable and that each is probable.

20. "Quantum ad tertium sit prima conclusio: sol est alterius speciei a luna et aliis stellis. Istam conclusionem ponit et probat Aristoteles in libro *De proprietatibus elementorum* et dicit sic: dico autem quod substantia corporis solaris est alia a substantia corporum stellarum et lune quod est quia sol de se habet lumen et lucem. Lumen autem lune et stellarum est acquisitum a sole, ita quod Aristoteles fundat se super hoc quod substantie eiusdem speciei specialissime debent habere accidentia propria et naturalia consimilia. Et si habeant dissimilia hoc est propter extranea agentia et concurrentia ad corruptionem vel propter diversas dispositiones sue materie aut suorum agentium talia autem non habet locum in celo." Albert of Saxony, *De celo*, bk. 2, qu. 19, 1518, 114r, cols. 1–2. In the printed edition of the *De proprietatibus elementorum* that appears in Averroës, *Opera*, 1562–1574, vol. 7, I found no such statement. In the next paragraph, Albert again (ibid., 114r, col. 2) appeals to its authority when he quotes the following line from it: "and similarly, the substance of the body of an orb is other than the substance of the body of the stars." For the Latin text and the context of this line, see note 34 in this chapter. Amicus, *De caelo*, tract. 4, qu. 7, dubit. 3, 1626, 207, col. 1, quoted essentially the same line ("quod substantia corporis orbis est alia a substantia corporis stellarum"). But Amicus may have derived it directly from Albert of Saxony, whom he cites frequently, rather than from the other text.

by a magnet. They also say that Saturn is hot lead; that Jupiter is clear and liquidy; that Mars is bloody; Mercury vigorous, and so on.

But there was also debate on purely theoretical grounds concerning whether the orbs that made up one planetary sphere, or "total orb," belonged to the same species as the orbs that made up every other "total orb" or planetary sphere.[21] In the thirteenth century Michael Scot and Roger Bacon denied this and argued for separate species. Michael insisted that just as the four terrestrial elements are in the same genus but differ with respect to species, since earth differs from water and water from air, and so on, so also the aggregate of celestial bodies, both orbs and planets, belongs to the same genus, but each body is in a different species in accordance with its distinct nature.[22] Roger Bacon argued that to have a multitude of individuals in a species was a sign of imperfection and seems to imply that this is characteristic of the terrestrial realm. But in the perfect celestial region it is necessary that only one individual be in a species.[23] Uniqueness is thus a greater perfection than multiplicity. Moreover, as Thomas Aquinas observed, since each individual in the celestial region is incorruptible, a multiplicity of individuals in a species is not essential for the preservation of a celestial species, as it is for the corruptible members of a terrestrial species.[24]

A further comparison between celestial and terrestrial species concerned the place of the members of a species. According to Bacon, things that belong to the same species seek the same natural place. Thus all parts of the earth descend toward the center of the world. When any stone or piece of earth is removed from below another piece of earth, the latter descends naturally. The same may be said for quantities of water. The reverse occurs with elemental fire: if a fiery body that lies above other fire is removed, the unremoved fire rises naturally to replace the fire that has been removed. Because they are all members of the species fire, all bits of fire behave similarly.

Bacon then applies the same reasoning to the celestial orbs. He implicitly assumes that the celestial orbs are neither light nor heavy but can act as if they are either one or the other, depending on circumstances. If all orbs belong to the same species and are nested one within another, the removal of a sphere nearer to the earth, say the fourth orb, would result in the immediate descent of the next-superior orb, say the fifth, to replace it. Or, conversely, the removal of the fourth orb might cause the third orb to rise naturally to fill the vacated space. But if celestial orbs could naturally rise and descend in this manner because they all belonged to the same species, they would be capable of natural rectilinear motion, which is impossible according to Aristotelian physics. It follows that the celestial orbs do not belong to the same species.[25]

21. On "total" and "partial" orbs, see Chapter 13, Section II.4.
22. Michael Scot [*Sphere*, lec. 4], 1949, 284.
23. Bacon, *De celestibus*, pt. 4, ch. 4, *Opera*, fasc. 4, 1913, 393.
24. See Litt, 1963, 96.
25. Because I have interpreted Bacon freely and expansively and even added examples, I

Thus not only did celestial bodies differ, but so also did one celestial orb differ from another. But what about the one orb with a multiplicity of stars: the orb of the fixed stars? Do the many fixed stars – according to Ptolemy, there were 1,022 of them – in that single orb differ in species? Albertus Magnus argued that "stars [or planets] that are in different orbs differ in species, but those that are in one orb have no such difference."[26] Thus all the fixed stars are in the same species but a different species from each of the planets. How would Albertus have reacted to the almost universal assumption that the fixed stars differed in power[27] and might therefore be specifically distinct from each other? Under these circumstances, each star was perhaps a member of a unique species. Albert of Saxony seems to have adopted this opinion. In any event, he denied that the stars of the eighth sphere could be in the same ultimate species (*species specialissime*).[28]

But what of the relations between an orb and the star or planet that it carried: Were the orb and the planet of the same species? Aristotle, as we saw earlier, identified the planet and its orb as being essentially the same body (see note 12, this chapter). Nevertheless, as with the relations between planet and stars alone, three possible answers won support: yes, no, and an opinion that neither the affirmative nor negative thesis could be demonstrated.[29]

Despite the various subdivisions of the question, most scholastics found it untenable to assume that celestial bodies and the orbs that carried them around could both be subsumed under a single species. They were more likely to agree with Johannes de Magistris that "all orbs and all stars [and planets] differ mutually with respect to an ultimate [not further reducible] species." For Magistris and many of his colleagues, the fact that orbs and stars have different properties and that celestial bodies differ from one another – after all, one planet (the Sun) is self-luminous, and the others are not; some produce heat, others cold – as do the celestial orbs themselves, was sufficient basis for assigning each planet, star, and orb to a different species.[30] An even better basis for assigning them all to different species

present the full text: "Item ea que sunt ejusdem speciei appetunt, eundem locum specie, ut partes terre omnes descendunt ad centrum, et remoto inferiori superius naturaliter descendit, et sic de aqua in spera sua, quia graves sunt in speris suis, ignis vero quia solum levis est in spera sua, ideo parte superiori remota pars inferior ascendit. Cum ergo sic est propter idemptitatem speciei, tunc si orbes essent ejusdem speciei, remoto inferiori superior descenderet, vel amoto superiore inferior natus eset ascendere. Ergo nati erunt movere motu recto: quod est inpossibile." Bacon, *De celestibus*, pt. 4, ch. 4, *Opera*, fasc. 4, 1913, 393–394. Bacon compared other aspects of the four elements to the alleged behavior of the ethereal orbs in order to determine whether they belong to the same species, which he denied.

26. "Et quod quaeritur de differentia stellarum inter se, dicendum videtur, quod stellae, quae sunt in diversis orbibus, differunt specie, sed quae sunt in orbe uno, non tantam habent differentiam." Albertus Magnus, *De caelo*, bk. 2, tract. 3, ch. 4, 1971, 150, col. 1.
27. Thomas Aquinas assumed this (see Litt, 1963, 98 and n. 4).
28. For the passage and reference, see below note 34, this chapter.
29. Amicus discusses the three opinions in *De caelo*, tract. 4, qu. 7, dubit. 3, 1626, 206, col. 2–208, col. 2.
30. "Omnes orbes et omnia astra a se invicem differunt specie specialissima. Patet quia habent proprietates specie differentes. Ideo differunt specie. Antecedens patet quia aliquod astrum

was, as John Major would have it, the fact that each planet, and therefore its orbs, had different motions.[31] Indeed, Roger Bacon used the assumption that the ether of a planet is denser than the ether of the rest of the orb to show that star and orb were not of the same species. A star or planet is a locus of action and requires a high concentration of the celestial ether. It is precisely the high concentration of celestial substance that makes a star a star; and it is the dispersion, or rarity, of the same substance that makes up the rest of the orb.[32] Bacon argued in a similar manner in the immediately preceding paragraph with respect to the generative power associated with the celestial region. The orb and its star should not be said to possess generative powers differing only in degree so that they may be included in the same species. The star or planet is but a small body by comparison to the half of the sphere that accompanies the star over a particular region or dwelling. On this basis, the entire half of the sphere that lies over a particular region should possess far more power to influence that region than the small star or planet. And yet the other half, from which the star or planet is absent, also passes over that same region. But, Bacon implies, the half without the star or planet does not affect the region without the presence of the star, from which Bacon concludes that the star and the orb are not of the same species.[33]

As evidence that the substances of planets and stars belong to different ultimate species than do the substances of the orbs that carry them, Albert of Saxony pointed to the fact that the Sun illuminated stars and planets but not their orbs. Indeed, Albert reinforced his position by observing that if the eighth sphere of the fixed stars were of the same irreducible species as the many stars it carried, all of those stars would have to be members of the same species. But they cannot belong to the same species, because, as we saw earlier, they have different natural powers and exercise effects on different things in the sublunar world.[34]

From the thirteenth and into the seventeenth century, this was probably the more popular opinion. In the seventeenth century, Amicus thought it more probable that the stars or planets (*astra*) differ in specific nature from their orbs.[35] And although Riccioli conceded the identity of the matter of

est per se luminosum, sicut sol; alia autem non. Etiam quoddam est calefactivum, aliud est frigefactivum, etc." Johannes de Magistris [*De celo*, bk. 2, qu. 4], 1490, sig. k7r.

31. After announcing that any heavens, or total planetary orbs, are in different species, as are any two planets, Major, *De celo*, bk. 2, qu. 6, 1526, sig. iiii recto and verso, says: "Probatur habent alios et alios motus. Ergo habent formas specie differentes."
32. Bacon, *De celestibus*, pt. 4, ch. 7, *Opera*, fasc. 4, 1913, 402.
33. Ibid.
34. Albert considers the issue in his *De celo*, bk. 2, qu. 19, 1518, 114r, col. 2, where he says, in a fourth conclusion: "Quantum ad quartum sit ista conclusio: quod orbes sunt alterius speciei ab astris que sunt in eis. Istam conclusionem ponit et probat Aristoteles in libro *De proprietatibus elementorum* dicens et similiter substantia corporis orbis est alia substantia corporis stellarum. Quod patet quia stelle illustrantur a sole et non orbes. Et confirmatur hoc: si orbis octavus esset eiusdem speciei specialissime cum stellis suis, tunc omnes ille stelle essent eiusdem speciei cuius oppositum dixit conclusio immediate precedens."
35. Amicus, *De caelo*, tract. 4, qu. 7, dubit. 3, 1626, 207, cols. 1–2. As an authoritative opinion

a planet and its orb or heaven, he insisted that they differ in other ways, for otherwise, it would follow that because a planet is luminous and opaque, so also would the whole heaven be luminous and opaque, which is false.[36]

The most basic argument in favor of including an orb and its planet in the same species was that a simple, inanimate body such as an ethereal orb is homogeneous with respect to all its parts. Since stars and planets are denser parts of their respective orbs in the same manner as knots in a table, it follows that orb and planet are in the same species. Although he considered it a difficult question, Albertus Magnus upheld this opinion when he insisted that "no part differs in species from its whole." For Albertus, the star or planet is the worthier part of its orb, but both are in the same species.[37]

In view of the properties which Aristotle attributed to the celestial substance, especially incorruptibility and homogeneity, it would have seemed prima facie that the prevalent view from the thirteenth to the seventeenth century would have placed all celestial bodies and orbs in the same ultimate species. But we have seen that it was quite otherwise. Scholastics emphasized diversity and found no plausible way to include them within a single species.

Diversity, rather than homogeneity, was the major mode of celestial existence. Celestial variety was apparent by an array of visible and invisible qualities. The former were assumed to exist in celestial bodies only virtually rather than actually.[38] But both visible and invisible qualities were an inherent part of the heaven and affected its behavior and powers. Among the qualities that pertained to the celestial region, we shall consider first the distances and dimensions of its major celestial parts, the planets and orbs.

III. The dimensions of the world and its celestial bodies

In my examination of the continuity or contiguity of the celestial orbs (Ch. 13, Sec. III.2), I depicted the concept of nested spheres, which was so characteristic of medieval cosmology. Nothing has yet been said, however, about the dimensions of those spheres or the magnitudes of the planets and

in support of different natures, Amicus cites (ibid., 207, col. 1) the *De proprietatibus elementorum*.

36. Riccioli, *Almagestum novum*, pars prior, bk. 4, ch. 19, 1651, 241, col. 1.

37. "Sed difficilius est, quod quaeritur, utrum stella ab aliis partibus orbis differat specie, quia nos videmus, quod nulla pars specie differt a suo toto; inferius autem probabitur, quod stella est pars orbis, in quo est. Et ideo dicendum videtur, quod stella est quasi dignior pars orbis." Albertus Magnus, *De caelo*, bk. 2, tract. 3, ch. 4, 1971, 149, col. 2. Amicus (*De caelo*, tract. 4, qu. 7, dubit. 3, 1626, 206, col. 2), who seems to oppose this opinion, cites a number of its supporters, including John Major, John of Jandun, and Albertus Magnus.

38. That is, the celestial body could cause effects associated with a certain property without actually possessing that property. See Chapter 10, Section II.1.a, where Buridan is the focus. Albert of Saxony expresses the same opinion in *De celo*, bk. 1, qu. 2, 1518, 87v, col. 1, when he says that "some parts of the heavens are hot or cold virtually [*virtualiter*] but not formally [*formaliter*]." For the same sentiment by Themon Judaeus, see his *Meteorology*, bk. 1, qu. 3, 1518, 157v, col. 2–158r, col. 1.

fixed stars. In the domain of cosmic dimensions, Aristotle provided little guidance. He neglected to discuss the size of the world or the distances of planets and stars, or even the relative sizes of these bodies. Perhaps this explains the virtual absence of such discussions in the *questiones* and commentaries on Aristotle's *De caelo*. Even Roger Bacon, who is one of the few scholastics to treat cosmic dimensions, did so not in his questions on *De caelo*[39] but in his famous *Opus majus*. Despite the absence of dimensions in scholastic natural philosophy, I shall briefly summarize the primary dimensions as these were generally understood in the Middle Ages.

The ultimate source of cosmic dimensions and the concept of nested spheres was Ptolemy's *Planetary Hypotheses*,[40] which was unavailable in medieval Europe. The essential measurements and the overall scheme entered western Europe embedded in al-Farghani's *Differentie scientie astrorum*, or, as it was known in Gerard of Cremona's translation of 1175, *Liber de aggregationibus scientie stellarum*,[41] and in two brief treatises by Thabit ibn-Qurra.[42] Al-Farghani's cosmic dimensions proved the most popular and became the standard version in the Latin West (Van Helden, 1985, 34).

Despite the use of somewhat varied values, most estimates were sufficiently close so that we may ignore the differences. Thus Bacon, in his *Opus majus* (1928, 1:248–259), drew his values directly from al-Farghani, while Campanus of Novara arrived at somewhat different results because he used a more precise value for the earth's radius (3,245 5/11 miles instead of the al-Farghani-Bacon value of 3,250 miles) and added the planetary diameters to the thicknesses of the spheres. Because of the large number of manuscript copies of both Campanus's *Theorica planetarum* and Robertus Anglicus's 1271 commentary on the *Sphere* of Sacrobosco, which included Campanus's values without acknowledgment, I shall cite Campanus's values as properly representative of medieval notions of the dimensions and distances of celestial bodies.[43]

Although the earth was not considered a planet until Copernicus made it so in his heliocentric system of 1543, the earth was the essential standard of measure for cosmic dimensions. Using a value of 56 2/3 miles for the

39. I have equated Bacon's *De celestibus*, which forms part of his *Communia naturalium*, with a typical questions on *De caelo*. See Bacon's *Opera*, fasc. 4, 1913.

40. Van Helden, 1985, 26–27, declares that "The Ptolemaic System of cosmic dimensions was an ingenious mélange of philosophical tenets, geometric demonstrations with spuriously accurate parameters, planetary theories, and naked-eye estimates. It was a speculative by-product of the first complete system of mathematical astronomy. . . . The resulting schemes of sizes and distances was, therefore, as plausible as it was ingenious, and it came to occupy an important place in cosmological thought among Moslem astronomers and then the Christian schoolmen."

41. Van Helden, 1985, 33. John of Seville translated the *Differentie* in 1137. Van Helden (ibid., 29) explains that al-Farghani did not know the *Planetary Hypotheses* but came up with very similar values from Ptolemy's *Almagest*.

42. Van Helden, 35. The full titles are *De hiis que indigent expositione antequam legatur Almagesti* and *De quantitatibus stellarum et planetarum et proportio terre*. The texts appear in Thabit ibn Qurra, 1960.

43. For all this, see Van Helden, ibid., 34.

measurement of 1 degree north or south along a meridian of the earth's surface, Campanus gives 20,400 miles as the earth's circumference.[44] Campanus accepted al-Farghani's value for a mile as equal to 4,000 cubits, but not only is the cubit undefined, but 4,000 cubits does not correspond "to any 'mile' actually in use in western Europe during any period."[45] Nor are we helped much by knowing that "the Arabic 'mile' ('mil') employed by al-Ma'mun's measuring teams was considerably longer than the 'mile' ('miliare') of the West in Campanus's time."[46] Despite the fact that Campanus and others based their measurements on the mile, we are ignorant of its real length. It follows, however, that a terrestrial circumference of 20,400 miles in the West represented a smaller earth than did its Arabic counterpart and that all subsequent measurements based on that value would yield a smaller cosmos than that measured by the longer mile of the Arabs.

To arrive at a value for the earth's radius, Campanus first derived its diameter, namely 6,490 10/11 miles (i.e., 20,400 × 7/22), and then its radius, 3,245 5/11 miles. With a value for the earth's radius, all the measurements could be calculated. In giving their distance measurements, Campanus and all medieval astronomers and natural philosophers assumed that the celestial orbs were nested one within another and in direct contact (see Ch. 13, Sec. III.2), so that the convex surface of the Moon was assumed identical with the concave surface of the sphere of Mercury; the convex surface of Mercury was identical with, or exactly the same distance from the earth as, the concave surface of the sphere of Venus; and generally, the convex surface of a sphere was assumed to be exactly the same distance from the earth as the concave surface of the very next planetary sphere.

Following the conclusion of his *Theory of the Planets*, Campanus of Novara gathers into tabular form valuable data on each planet from the treatise itself. All the data for a given planet are assembled together, and the order of presentation begins with the Moon and moves upward to Saturn. Although in Table 1 I also follow the planetary order from Moon to Saturn – thus omitting the sphere of the fixed stars and the invisible, planetless and starless ninth and tenth spheres[47] – instead of listing all the data for each planet independently of the others, I have consolidated the

44. Both values are from al-Farghani. See Alfraganus [Carmody], 1943, 13–14; Alfraganus [Campani], 1910, 89; and Campanus of Novara, *Theorica planetarum*, 1971, 146 (Latin), 147 (English).
45. See the note by Benjamin and Toomer in Campanus of Novara, *Theorica planetarum*, 1971, 25, n. 2. For the appearance of this value in Campanus's text, see 146, line 53. Al-Farghani assumed " 'black cubits' of 24 fingerbreadths each" (Van Helden, 1985, 30–31). Unfortunately, Campanus says nothing comparable.
46. Benjamin and Toomer in Campanus of Novara, ibid., 381, n. 17. Al-Ma'mun was caliph during the period 813 to 833, when the measurements were made.
47. "Concerning the ninth and tenth heaven," Bacon declared (*Opus majus*, pt. 4 ["Mathematics"], 1928, 251) that "nothing can be known by means of instruments belonging to the senses in the matter of their altitude, thickness, and size; also concerning the thickness of the eighth heaven, as we are able to do in regard to the other heavens, since all these matters are hidden from our sense, and therefore in regard to them certification of their magnitudes, altitudes, and thicknesses is lacking."

Table 1. *Campanus of Novara's data on planets and spheres*

Planet	Concave surface of sphere	Convex surface of sphere
Distances of concave and convex planetary surfaces from center of earth (miles)		
Moon	107,936 20/33	209,198 13/33
Mercury	209,198 13/33	579,320 28/33
Venus	579,320 560/660[a]	3,892,866 550/660
Sun	3,892,866 550/660	4,268,629 110/660
Mars	4,268,629 110/660	35,352,075 420/660
Jupiter	32,352,075 420/660	52,544,702 280/660
Saturn	52,544,702 280/660	73,387,747 100/660
	Thicknesses of planetary spheres (miles)	
Moon	101,261 26/33	
Mercury	370,122 5/11	
Venus	3,313,545 650/660	
Sun	375,762 220/660	
Mars	28,083,446 310/660	
Jupiter	20,192,626 520/660	
Saturn	20,843,044 480/660[b]	
	Diameters of planets (miles)	
Moon	1,896 26/33	
Mercury	230 26/33	
Venus	2,884 560/660	
Sun	35,700	
Mars	7,572 480/660	
Jupiter	29,641 540/660	
Saturn	29,209 60/660	

data for all the planets under the following headings: the distances of the concave and convex surfaces of the planetary spheres;[48] the thickness of each total planetary sphere; the diameters of the planets; and finally their circumferences.

Although Campanus includes data on other aspects of planetary dimensions,[49] the information presented in Table 1 is sufficient to convey

48. The distance for the concave surface of each planetary sphere represents the near edge of that planet at its least distance from the center of the earth; the distance for the convex surface of that same sphere represents the far edge of the planet at its greatest distance from the center of the earth.
49. He provides mile measurements for the circumferences of the convex surfaces of the spheres and for the deferents and epicycles of the planets. Although these may be spurious interpolations, the tables also include data for the mean daily motion of a planet (in miles) on its epicycle or eccentric, known as the *dieta* of a planet. On their spuriousness, see Benjamin and Toomer in Campanus of Novara, *Theorica planetarum*, 1971, 28–29.

Table 1 (*cont.*)

Distances of concave and convex planetary surfaces from center of earth (miles)		
Planet	Concave surface of sphere	Convex surface of sphere
	Circumferences of planets (miles)	
Moon	5,961 11/33[c]	
Mercury	725 11/33	
Venus	9,095 63/660	
Sun	112,200	
Mars	23,800	
Jupiter	93,160	
Saturn	91,800	

[a]Although the distance of the convex surface of Mercury is listed as 579,320 28/33 and that of the concave surface of Venus as 579,320 560/660, the distances are equal, since 28/33 = 560/660. Why different fractions were used is a puzzle.

[b]Drawing on al-Farghani, Bacon (*Opus majus*, pt. 4 ["Mathematics"], 250–251) gives the following values (in miles) for the thicknesses of the spheres: Moon, 99,504; Mercury, 334,209; Venus, 3,097,250; Sun, 325,000; Mars, 24,882,000; Jupiter, 17,969,250; Saturn, 18,541,250. These values are reasonably close to those given by Campanus.

[c]This value is a correction, by Benjamin and Toomer, for 5,958 32/33 as given by Campanus (Campanus of Novara, *Theorica planetarum*, 1971, 399, n. 67).

something of the relationships among the planets with respect to their distance, size, magnitude, and the thickness of their spheres. By setting the concave surface of a planetary sphere, say Jupiter's, exactly equal to the convex surface of the planet that it immediately encloses and touches, namely Mars, Campanus and all astronomers and cosmologists made certain that there were no void spaces or extraneous matter between any celestial orbs.[50] These widely accepted dimensions and distances bear little relationship to modern values. Of the planets, the most distorted figures are associated with the Sun. The distance measurements varied from an apogee of approximately 3,900,000 to 4,200,000 miles, in contrast to the modern value of approximately 93,000,000 miles. The medieval value of 35,000 miles for the diameter of the Sun must be contrasted with the modern value of 864,000 miles. A glance at the table reveals anomalous measurements that seem to make the cosmos an entity of idiosyncratic elements rather than a beautifully crafted, harmonious world of an infinite creator. No patterns seem to emerge. The thicknesses of the planetary

50. Earlier, we saw that Albertus Magnus assumed the existence of corruptible matter between the surfaces of orbs. Although interesting, and perhaps reflective of an earlier Arab opinion, few adopted it.

spheres seem to defy understanding. Venus, for example, is approximately nine times thicker than the planets immediately above and below it (the Sun and the Moon). With the superior planets, we observe an enormous increase in thickness over the Sun and inferior planets. For what reason is the planetary sphere of Mars thicker than those of Jupiter and Saturn above? The diameters, and therefore the sizes, of the planets also seem to vary in strange ways. Since there is an absence of discussion of these seeming anomalies, we have no way of determining whether scholastic authors were concerned about them.

From the center of the earth to the convex surface of Saturn, an approximate distance of 73,000,000 miles was assumed.[51] The last visible component of the universe was the eighth sphere of the fixed stars. Following the pattern previously described, the concave surface of the sphere of the fixed stars was assumed continuous with the convex surface of the sphere of Saturn, from which we may infer that Campanus of Novara and almost everyone else would have assumed that the fixed stars were at least 73,000,000 miles from the center of the earth. But were they all scattered over the concave surface of the eighth sphere at the same distance from the earth's center, or were they distributed at different distances within an eighth sphere, or firmament, that had some degree of thickness in the manner of the planetary spheres? Or perhaps, as Moses Maimonides suggested, each of the many fixed stars has its own sphere with the same poles, so that each star undergoes the same motion,[52] an opinion that could be interpreted to mean that the stars are at different distances.[53] To better appreciate the problem, we shall first consider one important aspect of the fixed stars that was universally accepted in the Middle Ages.

From Ptolemy's *Almagest*, medieval natural philosophers accepted the division of the fixed stars into six different magnitudes, or orders, of brightness. Before the introduction of photometric methods in 1903, the classification into magnitudes was rather arbitrary.[54] On the apparent assumption that all the stars were on the same surface and therefore equidistant from the center of the earth, some Arab astronomers associated greater size with greater brightness. First-magnitude stars were therefore assumed physically larger than second-magnitude stars, and so on. According to Roger Bacon, who cites as his authorities Ptolemy and Thabit ibn Qurra, there are 15 stars of the first magnitude, each of which is 107 times the size of the earth;

51. Following al-Farghani, Bacon estimated the distance of the sphere of the fixed stars as 65,357,500 miles. Bacon, *Opus majus*, pt. 4 ["Mathematics"], 1928, 249.
52. See Maimonides, *Guide*, pt. 2, ch. 11, 1963, 2:274.
53. Peter of Abano, *Lucidator*, differ. 3, 1988, 215, so interpreted Maimonides' statement when he declared: "Quidam vero, ut Raby Moyses, tot asserit fore speras quot etiam stelle, ut unaqueque suo tantum contineatur celo, quod movebatur, quoniam stellarum quedam videntur maiores tamquam minus distantes, alie vero minores ceu remotiores, nonnulle autem medie tamquam mediocriter elongate." The medieval Latin title of Maimonides' *Guide of the Perplexed* is *Dux seu Director dubitantium aut perplexorum in tres libros divisus.*
54. See Pedersen, 1974, 259.

45 stars of the second magnitude, each of which is 90 times the size of the earth; 208 stars of the third magnitude, each of which is 72 times the size of the earth; 474 stars of the fourth magnitude, each of which is 54 times the size of the earth; 217 stars of the fifth magnitude, each of which is 36 times the size of the earth; and 62 stars of the sixth magnitude, each of which is 18 times the size of the earth.[55] Although Bacon's total count mistakenly falls one shy of 1,022 – Ptolemy included 63 stars in the sixth magnitude, rather than 62 – he sought to follow the Ptolemaic tradition as he had received it from Ptolemy and Arabic astronomers, especially Alfraganus and Thabit ibn Qurra.[56] Nearly four hundred years later, Mastrius and Bellutus repeated the same magnitudes, the same number of stars in each magnitude, and the same comparisons of the stars in each magnitude with the size of the earth.[57]

Even if Mastrius and Bellutus and most of their scholastic colleagues in the seventeenth century repeated the old values, the Copernican system had stirred up controversies about the sizes of the planets and fixed stars. New values for the diameters of fixed stars were derived or assumed, and new values for their comparative sizes were computed. Tycho Brahe, Galileo, and Kepler all entered into this new field. In arguing against the Copernican system, Tycho showed that in the latter, a fixed star of the third magnitude would have a diameter approximately the size of the earth's orbit around the Sun (Van Helden, 1985, 51). Galileo rejected Tycho's claims and argued that the Sun would be smaller than any visible fixed star (ibid., 74), while Kepler argued that the fixed stars are many times smaller than the Sun (ibid., 89). For the most part, scholastic natural philosophers did not enter these controversies, which were focused on the Copernican system. They

55 In his comparison of the size of the earth with those of the fixed stars, it is unclear what dimension Bacon is measuring. In his *Opus majus*, pt. 4 ["Mathematics"], 1928, 254–257, Bacon compares the diameters of planets; by multiplying each diameter by 3 1/7, the Archimedian value for pi, he also compares the circumferences of planets; and by multiplying the diameter and circumference of each planet, compares their respective surface areas; and by cubing diameters, he compares the volumes of the planets. Peter of Abano, *Lucidator*, differ. 1, 1988, 149, lines 26–31, through 150, line 2, accepts 1,022 fixed stars divided into six magnitudes but presents a strange and radically different set of dimensions for the six magnitudes. First-magnitude stars are 6 times the quantity of sixth-magnitude stars; 5 times the quantity of fifth-magnitude stars; 4 times the quantity of fourth-magnitude stars; and so on.

56. Perhaps because his treatise was titled *Theorica planetarum*, Campanus had little to say about the fixed stars (see also Benjamin and Toomer's comment, Campanus of Novara, *Theorica planetarum*, 1971, 435, n. 72).

57. Mastrius and Bellutus, *De coelo*, disp. 2, qu. 2, art. 3, 1727, 3:495, col. 2, par. 78. In the manner of Bacon, Mastrius and Bellutus err in the number of stars they include in the sixth magnitude, giving 73 instead of 63. George de Rhodes also repeats the number of stars in each of the six magnitudes, but his numbers are even less accurate (see *De coelo*, bk. 2, disp. 2, qu. 1, sec. 4, 1671, 286, col. 2). Christopher Clavius and Raphael Aversa correctly give the number of stars in each of the six magnitudes (Clavius, *Sphere*, ch. 1, *Opera*, 1611, 3:73; Aversa [*De caelo*, qu. 32, sec. 2], 1627, 2:43, col. 2). In his famous early sixteenth-century encyclopedia *Margarita philosophica*, bk. 7, ch. 20, 1517, 255, Gregor Reisch repeated the number of fixed stars in each of the six magnitudes and their comparison to the size of the earth.

tended rather to repeat the old values based on the traditional geocentric system they inherited from the Middle Ages.

In that system, even stars of the sixth magnitude, the smallest visible stars in the heavens, were assumed larger than the largest planet, the Sun. With their acceptance of the division of the fixed stars into six magnitudes graded according to size, from the largest in the first magnitude to the smallest in the sixth, medieval astronomers and natural philosophers seem to have accepted the idea that all stars of whatever magnitude were fixed in their sphere at an equal distance from the earth, for only on this assumption is the division into six magnitudes plausible. A first-magnitude star can be assumed 107 times greater than the earth, and a second-magnitude star 90 times greater only if it is assumed that they are the same distance from the center of the earth. If not, and without additional information, comparisons to the earth's size would have been meaningless.

Albertus Magnus posed the question directly ([*De caelo*, bk. 2, tract. 3, ch. 11], 1971, 167, cols. 1–2): Are differences in magnitude the result of fixed stars that are of the same size but located at different distances from the earth? Or are the stars really of different sizes but located on the same celestial surface, equidistant from the earth's center? Albertus was one of the few who raised this question, and his response may be representative of those who did.

Against the first alternative, Albertus observes that if the six magnitudes of stars were all of the same size but at different distances, six heavens or orbs would have to be assumed, one orb for each magnitude of stars. But this is absurd, because if six such orbs existed we would expect them to turn with different speeds, just as the planetary spheres do in the daily motion. That is, the latter are all carried around together in the course of a day, but the spheres themselves are moved with different velocities. But observation of all the fixed stars, both the greater and the smaller, shows that they are moved with a single velocity. And this must occur solely because they are on the surface of one sphere. Because the fixed stars seemed to move with the same velocity, Albertus, and undoubtedly most others who considered the problem, located them on a single spherical surface equidistant from the earth. On this basis, they were distinguishable into at least six different sizes, or magnitudes.

What Albertus Magnus did not consider, however, was the possibility that the fixed stars were distributed at different distances from the earth within the same single sphere, rather than spread over six spheres in order of magnitude. Galileo himself, if not others before him, suggested this interpretation. Galileo, however, includes the planets along with the fixed stars in this unique sphere. We learn his views from the *Dialogue Concerning the Two Chief World Systems*, where he has Salviati ask: "Now what shall we do, Simplicio, with the fixed stars? Do we want to sprinkle them through the immense abyss of the universe, at various distances from any predetermined point, or place them on a spherical surface extending around a center of their own so that each of them will be the same distance from that center?"[58]

To this Simplicio is made to reply: "I had rather take a middle course, and assign to them an orb described around a definite center and included between two spherical surfaces – a very distant concave one, and another closer and convex, between which are placed at various altitudes the innumerable host of stars. This might be called the universal sphere, containing within it the spheres of the planets which we have already designated."[59]

That Galileo found Simplicio's suggestion acceptable seems evident when Galileo has Salviati then say: "Well, Simplicio, what we have been doing all this while is arranging the world bodies according to the Copernican distribution, and this has now been done by your own hand." Despite his rejection of the hard orbs of Ptolemaic astronomy, Galileo appears to have accepted the existence of at least one sphere whose two circular bounding surfaces enclose the entire celestial region and therefore all celestial bodies, including planets and fixed stars, the latter, like the planets, located at different distances from Sun or earth. The heaven is thus a single, gigantic, fluid sphere, apparently bounded by hard convex and concave surfaces.

Galileo's idea that the fixed stars are distributed at varying distances from any previously determined point of reference, whether earth or Sun, was approved by many, including Otto von Guericke.[60] Some scholastics were also receptive, such as, for example, Athanasius Kircher,[61] while others, for example, Mastrius and Bellutus, were prepared to take a neutral stance and thereby cast a degree of doubt on the traditional interpretation of the stars' equidistance from the earth.[62]

58. Galileo, *Dialogue*, Third Day [Drake], 1962, 325–326.
59. Ibid., 326. At first glance, it seems that Galileo should have described the distant surface as convex (rather than concave) and the nearer surface as concave (rather than convex). That is, the most distant enclosing surface of the world would be convex, and the nearest surface (for an Aristotelian like Simplicio) would be the concave surface of the lunar orb. On this approach, we are obviously including the thicknesses of the farthest and nearest *orbs* and distributing the planets and stars between the extreme surfaces of those orbs. Galileo, however, wishes to exclude the thicknesses of the two most extreme orbs, namely the nearest (that of the Moon) and the farthest, and to distribute the planets and stars between the concave surface of the outermost sphere and the convex surface of the nearest sphere. Drake has faithfully and correctly translated Galileo's Italian text, which reads: "cioè una altissima concava e l'altra inferiore e convessa." Galileo [Favaro], *Opere*, 1891–1909, 7:353.
60. Curiously, von Guericke mistakenly attributes to Galileo the view that the fixed stars are scattered on a spherical surface that is equidistant from a center. Von Guericke leaves no doubt that he believes the fixed stars are at varying distances: "Etenim non credo (*cum Galilaeo in Dialogo 3. Systematis Cosmici*) stellas esse sparsas in sphaerica superficie distante aequaliter a centro, sed existimo distantias earum adeo varias esse, ut aliae aliis, bis, ter, quater, etc., a nobis remotiores sint." Guericke, *Experimenta nova*, 1672, 224, col. 1. Guericke offers a number of proofs (224–225).
61. According to the quotations cited by Guericke, ibid., 224, cols. 1–2.
62. "Whether indeed all the fixed [stars] are of equal distance from the earth – whether one is above, another below in the depth of the heaven – nothing certain can be stated" (An vero omnia fixa sint equalis distantiae a terra, an vero unum superius, alterum inferius

The idea of a diversity of distances among the fixed stars cast doubt on the widely accepted medieval value of 73 million miles as the distance for all the fixed stars and therefore the distance of the outer limit of the physical and visible world.[63] During the Middle Ages Campanus's dimensions, or those presented by al-Farghani, which were quite similar, were widely accepted. Indeed, in the last, posthumously published edition of his commentary on the *Sphere* of Sacrobosco in 1611, Christopher Clavius published a table of dimensions the values of which were largely the same as those of al-Farghani.[64] The advent of Tycho Brahe's system, moreover, did not produce significant dimensional changes and caused little anxiety to traditional scholastics.[65] The Copernican system was quite another matter.

Although Copernicus did not compute absolute distances of the planets, the value in medieval miles for Saturn in the Copernican system is approximately 39 million, an astonishingly small value.[66] Thus the distance from the Sun to Saturn in the Copernican cosmos was only a little more than half the distance from the earth to the convex orb of Saturn in the medieval cosmos. In sharp contrast with Campanus and the medieval tradition, Copernicus also allowed gaps "between the greatest heliocentric distance of one planet and the least distance of the next one. This meant that the nesting spheres procedure, with the Moon as the starting point, was no longer possible."[67] The medieval tradition of nested spheres was thus necessarily abandoned by Copernicus and his followers.

But if Copernicans reduced the size of the medieval Ptolemaic planetary cosmos by nearly half, they made the distance to the fixed stars immensely greater than in the medieval system. Because they could detect no stellar parallax as the earth moved in its vast orbit around the Sun, Copernicans were compelled to assume that the fixed stars were a vast distance away from the earth. Copernicus, who did not calculate the distance of the fixed stars, assumed that the universe was immense and even like an infinite magnitude, although, as Koyré reminds us, it was finite because the sphere of the fixed stars served to contain and limit it.[68] According to Otto von

in profunditate celi sint collocata, nihil certum statui potest). Mastrius and Bellutus, *De coelo*, disp. 2, qu. 2, art. 3, 1727, 3:495, col. 2, par. 77.

63. The invisible and bodiless heavens beyond the eighth sphere of the fixed stars do not count in arriving at the size of the world.
64. Clavius, *Sphere*, ch. 1, *Opera*, 1611, 3:100–102, and Van Helden, 1985, 53.
65. Van Helden, 1985, 53.
66. Ibid., 47 (also see the table of distances for the Copernican system on 46), reckons Saturn's greatest distance from the earth to be 12,252 e.r. (earth radii), a figure which assumes that both Saturn and the earth are at aphelion and that Saturn is in conjunction. The 12,152 e.r. is composed of 11,073 e.r. for Saturn's distance from the Sun at aphelion and 1,179 e.r. for the earth's distance from the true Sun at aphelion. By multiplying 12,252 by 3,245.5, the earth's radius according to Campanus of Novara (actually 3,245 5/11), we obtain 39,763,866 miles for Saturn's distance from the earth.
67. Van Helden, ibid., 44.
68. In *Revolutions*, bk. 1, ch. 10 [Rosen], 1978, 20, Copernicus declares that "the size of the universe is so great that the distance earth–Sun is imperceptible in relation to the sphere of the fixed stars." Indeed in the same chapter (ibid., 21), Copernicus explains that

Guericke, Riccioli attributed values for the distance of the fixed stars to Copernicus and a number of Copernicans. In earth radii, Copernicus assumed 47,439,800; Galileo 49,832,416; Bullialdus 60,227,920; and Kepler 142,746,428.[69] Even the smallest of these numbers, assigned to Copernicus himself, would, if the earth radii were multiplied by 3,245 5/11, the medieval value in miles for the radius of the earth, yield a figure of 153,963,688,669 miles. We would thus have a heliocentric universe of nearly 154 billion miles as compared to the medieval geocentric world of 73 million miles. Although 73 million miles was a widely accepted approximation for the distance of the fixed stars, one anomalous medieval estimation was truly astonishing. Levi ben Gerson (1288–1344) calculated the fixed stars at a distance of 159,651,513,380,944 earth radii, which, when multiplied by 3,245 5/11, the miles in an earth radius, yields a number that dwarfs those provided by Riccioli.[70]

Only telescopic observations would produce reliable values for the distance of the planets and fixed stars and for the diameters of the planets. But that story has been brilliantly told elsewhere.[71] Let us now briefly consider the remaining properties and qualities of the celestial bodies, taking in order the fixed stars, the Sun, the Moon, and the remaining planets.

IV. The properties and qualities of the stars and planets

1. The fixed stars

a. On the number of stars

Because the empyrean heaven and its alleged terrestrial influences have been discussed in Chapter 15, we shall commence our examination of celestial properties with the sphere of the fixed stars, or the eighth sphere, as it was generally known. On the authority of the *Almagest* (bks. 7 and 8), where

although the earth revolves around the Sun, we fail to detect any "backward and forward arcs" of the fixed stars because of the "immense height" of those stars, "which makes even the sphere of the annual motion, or its reflection, vanish from before our eyes." Earlier, in bk. 1, ch. 6, Copernicus declares (13) that "the heavens are immense by comparison with the earth and present the aspect of an infinite magnitude." See also Koyré, 1973, 65.

69. Guericke, *Experimenta nova*, 1672, 224. Copernicus did not calculate the distance of the stars.
70. Van Helden, 1985, 40, cited from Bernard Goldstein, *The Astronomical Tables of Levi ben Gerson* (Hamden, Conn.: Archon Books, 1974), 28–29.
71. See Van Helden, 1985. On 160, Van Helden explains that "The Ptolemaic cosmic dimensions were not challenged until the sixteenth century, when the Copernican and Tychonic schemes of the cosmos dictated a new set of relative distances. Copernicus and Tycho did not, however, make significant changes in the solar distance, so the absolute distances in their systems were of the same order of magnitude as the Ptolemaic distances. While Copernicus had to make the sphere of the fixed stars very large indeed, the sphere of Saturn and everything in it actually shrunk. Only the telescope could change this."

Ptolemy provided longitudes, latitudes, and magnitudes for 1,022 fixed stars, it was commonly assumed that the totality of visible stars was 1,022, all embedded in the eighth sphere.[72] Many, if not most, Christians were convinced that quite a few more stars existed than were numbered by Ptolemy. Already in late antiquity, Saint Augustine emphasized the enormity of the number in the course of a discussion about the Lord's promise to Abraham that his descendants would be increased like the countless multitude of stars (Gen. 15.5).[73] Inspection of the visible sky, however, hardly revealed a countless multitude of stars. To render the Lord's promise to Abraham plausible, Augustine insists that "it is not to be believed that all of them can be seen. For the more keenly one observes them, the more does he see. So that it is to be supposed some remain concealed from the keenest observers." For Augustine, the authority of the book of Genesis "condemns those like Aratus and Eudoxus, or any others who boast that they have found out and written down the complete number of the stars." Even Aristotle seemed to buttress this opinion in *De mundo*, a work falsely ascribed to him but in which one could read that "the multitude of fixed stars is innumerable to men."[74]

Augustine's attitude was adopted by at least two authors in the thirteenth century. Robert Grosseteste asked whether there might exist planets and fixed stars that are invisible to us. Although such knowledge could only be known by divine revelation, Grosseteste insists that if invisible celestial bodies existed, they would also affect generation and, presumably, corruption in the inferior, terrestrial world.[75] In his *Summa philosophiae*, Pseudo-Grosseteste mentions that the divine word bears witness that it is impossible to specify a number for the fixed stars and promptly quotes from Genesis

72. For a discussion of Ptolemy's methodology in constructing the tables, see Pedersen, 1974, 249–260.
73. Augustine, *City of God*, bk. 16, ch. 23, 1948, 2:135.
74. "Multitudo quidem igitur fixarum innumerabilis est hominibus" is the text in an anonymous Latin translation edited by W. L. Lorimer (see Aristotle, 1951, 25, lines 21–22). In another translation by Nicholas of Sicily, we find (ibid., 55, lines 17–18): "Igitur non errantium multitudo inscrutabilis hominibus est." In the Oxford translation by E. S. Forster, see 392a.16–17 in Aristotle [Barnes], 1984. The *De mundo* was written in Greek sometime between 50 B.C. and A.D. 140. Aversa, *De caelo*, qu. 32, sec. 2, 1627, 43, col. 1, not only cites this line from the *De mundo* (he writes: "Atque Aristoteles, *libro de Mundo ad Alexandrem*, cap.2 ait numerum syderum inerrantium iniri non posse") but also says that the same sentiment is implied in Aristotle, *De caelo*, bk. 2, text 61, and in *Meteorology*, bk. 1, ch. 8. In the *De caelo* passage (2.12.292a.10–14), Aristotle asks: "what can be the reason why the primary motion should include such a multitude of stars that their whole array seems to be beyond counting?" (Aristotle [Guthrie], 1960). I have found no relevant passage in the *Meteorology*, although it is also mentioned in the passage quoted from Mastrius and Bellutus, toward the end of this section (IV.1.a).
75. "Dicunt enim philosophi galaxiam esse ex stellis minutis fixis nobis invisibilibus. Unde igitur sciri posset, nisi a divina revelacione, an non sint plurime huiusmodi stelle invisibiles nobis, quarum quelibet suum habeat celum movens ipsam ad profectum generacionis in mundo inferiori? Stelle enim que galaxiam constituunt, licet indistinguibiles sint secundum visum, non carent effectu generacionis et profectus in mundo inferiori." Grosseteste, *Hexaëmeron*, part. 3, ch. 8, 3, 1982, 108.

15.5.[76] Nevertheless, ancient wise men proved the existence of 1,022 stars. Thus it was that the 1,022 fixed stars came to be viewed not as the total number of stars in existence but only the number visible to us.[77] Johannes Velcurio adopted much the same approach in the sixteenth century when he declared that the fixed stars are innumerable but that astronomers had observed 1,022 stars in the firmament.[78]

Augustine's prescient intuition that there are many more stars than meet the eye would become reality in the seventeenth century when Galileo turned his telescope to the heavens and published his discoveries in the *Starry Messenger* (*Sidereus nuncius*). The scholastic reaction to the discovery of numerous previously hidden stars is well illustrated by Mastrius and Bellutus, who declared that

> the fixed stars commonly are innumerable and judged almost infinite, as Aristotle says in *Meteorology*, bk. 1, ch. 8, and as is had from Sacred Scripture, Genesis 15, where God says to Abraham, "Look up into the sky and count the stars if you can"; [and in] Jeremiah 33, "Just as the stars of the heaven and the sands of the sea cannot

76. Pseudo-Grosseteste, *Summa philosophiae*, tract. 15, ch. 6, 1912, 551. In ch. 7, 554, Pseudo-Grosseteste declares that the stars in the heavens are innumerable and that many are hidden from us, although some occasionally appear.
77. Bacon, *De celestibus*, pt. 4, ch. 5, *Opera*, fasc. 4, 1913, 395, specifies 1,022 fixed stars whose magnitudes can be measured by instruments and says of all other stars that "there is no number possible by human consideration in this life" (Manifestum est autem per Aristotelem in libro *Secretorum* et per Ptholomeum in octavo *Almagesti*, et per Alfraganum in libro suo *de Motibus Celorum* quod stelle fixe, quarum quantitas potest comprehendi per instrumenta in locis note habitacionis, sunt 1022. De aliis vero non est numerus possibilis in hac vita per consideracionem humanum). Centuries later, in a quite different context, de Rhodes, *De coelo*, bk. 2, disp. 2, qu. 1, sec. 4, 1671, 286, col. 1, remarks that although the stars are innumerable, "those which can be distinctly known and designated without [use of] a telescope [*tubo*] are 1,022, divided into 48 constellations by astronomers, and into many more by modern astronomers" (Dico primo, tametsi stellae . . . esse videantur innumerabiles, tamen illae quae notari et designari distincte sine tubo possunt, non sunt nisi mille viginti duae divisae ab Astronomis in constellationes quadraginta octo et a recentioribus vero astronomis notatae sunt multo plures).

    Long before de Rhodes, Clavius had cautioned that the words about Abraham in Scripture should not be interpreted as indicating an infinite number of stars (Non sunt ergo accipienda verba illa Scripturae in hoc sensu, ut dicamus infinitas stellas esse). Clavius, *Sphere*, ch. 1, *Opera*, 1611, 3:74. Clavius believed that no more than 1,022 stars existed in the six magnitudes distributed among the 48 traditional constellations, but he conceded (ibid.) that "Scripture speaks of all the stars that are in the heavens, even those that are smaller than those which are contained in the six differences [i.e., in the six different magnitudes], which are perhaps innumerable" (Scripturam loqui de omnibus stellis, quae in coelo sunt, etiam de illis, quae minores sunt, quam quae in sex differentijs continentur, quae fortasse innumerabiles sunt). Ibid. Thus Clavius allowed that if the smallest stars are considered, which are smaller than sixth-magnitude stars, then perhaps the stars are innumerable. But he declared himself persuaded that no more than 1,022 stars existed in the first six magnitudes of fixed stars (Mihi certe facile persuadeo non esse plures in sex dictis differentijs contentas quam 1022). Ibid.
78. "Stellae fixae numerari possunt, cum sint innumerabiles. Tamen Mathematicorum observarunt praecipue mille et vigintiduas stellas in firmamento, ut in circulo Zodiaco et in alijs septentrionalibus meridionalibusque syderibus." Velcurio [*Physics*], 1554, 77r. Reisch, *Margarita philosophica*, bk. 7, ch. 20, 1517, 254, also adopted Augustine's interpretation and even cites *City of God*, book 16.

be numbered, so will the seed of David be multiplied." This ought not to be understood about the stars that are visible, for these are numbered by astronomers, as we said, but [it should be understood] about all things [or celestial bodies] that exist in the heavens, for very many are not seen by us, which is because modern astronomers, using bigger and better instruments, have discovered new stars that were unknown to the ancients. For Tycho Brahe, in bk. 1 of his *Progymnasmata*, chs. 2 and 3, found thirteen stars in the constellation of Cassiopeia, [and] Galileo, in his *Sidereus nuncius*... notes 80 stars in the belt of Orion [that were not known before]; indeed, he found 500 stars in different places, and the nebulous stars, such as the Milky Way, are nothing other than an aggregation of minute stars. Therefore in speaking of fixed stars, those which are more visible and notable of appearance, especially in the wintertime, when more of them appear because of the purity of air from exhalations and the greater depression of the Sun below the horizon, have been numbered by astronomers as no more than 1,022 distributed in six classes of magnitudes.[79]

Brahe's naked-eye observations and Galileo's telescopic discoveries of previously invisible stars did not fall on a totally unprepared scholarly community.[80] Biblical pronouncements that in God's universe the fixed stars were beyond count – indeed, Roger Bacon would not hesitate to use the term "infinite" (*stellae aliae infinitae*)[81] – were interpreted to mean that a very large number of stars must lie hidden from our sight. This was Augustine's understanding of Genesis 15.5, and it was probably his enormous prestige that impelled medieval and early modern scholastic authors to make the same assumption.[82]

b. *The shape of the fixed stars*

By contrast with the rare and completely transparent orb that carried it, each fixed star was assumed to represent a dense accumulation of celestial

79. Mastrius and Bellutus, *De coelo*, disp. 2, qu. 2, art. 3, 1727, 3:495, col. 2, par. 78. The *Pyrogymnasmata* is Tycho's *Astronomiae instauratae progymnasmata* (Prague, 1602; Frankfurt, 1610). Aversa had earlier adopted a similar position in an even fuller account (*De caelo*, qu. 32, sec. 2, 1627, 43, col. 2–44, col. 2), which probably influenced Mastrius and Bellutus, who frequently cite Aversa.
80. For Galileo's description of his discoveries of previously unseen stars, see Galileo, *Sidereus nuncius* [Van Helden], 1989, 59–63 and Galileo, *Starry Messenger* [Drake], 1957, 47–50.
81. In his *Opus majus*, pt. 4 ["Mathematics"], 1928, 258, Bacon declares that in addition to the stars that belong to the six different magnitudes, there are "other stars in infinite number, the size of which cannot be ascertained by instruments, and yet they are known by sight, and therefore have sensible size with respect to the heavens, like the part with respect to the whole" (Deinde sunt stellae aliae infinitae, quarum quantitas non potest sciri per instrumenta, et tamen sunt visu notabiles, et ideo habent quantitatem sensibilem respectu coeli, sicut pars respectu totius). For the Latin text, see Bacon, *Opus majus* [Bridges], 1900, 1:236.
82. In using such terms as "beyond count" or "infinite," it is unlikely that scholastics meant an "actual infinite," although they probably intended a "potential infinite" in the Aristotelian sense – that is, however many stars might be counted, there were always more to count.

ether. But how were these concentrations of ether shaped? Most discussions about the shape of *stellae*, or "stars," were really about the planets rather than the fixed stars, since, as we saw, the term *stella*, without *fixa*, usually signified planet rather than fixed star.[83] But the term *stellae* could also signify both planets and fixed stars. Albert of Saxony applied the term to both entities in explaining the first of three ways by which he thought a sphere or orb could be defined. A sphere could be an unseparated part of a celestial orb, such as, for example, a star (*stella*). Since many fixed stars exist, then, if each fixed star is a sphere, it follows that very many celestial spheres would exist.[84] Unfortunately, few scholastics expressed opinions on this subject, although it is probable that, like the planets, the fixed stars were perceived as spherical in shape. As substances composed of the celestial ether, which naturally assumed a spherical shape, Aristotle had assumed (*De caelo* 2.8.290a.7–10) that the fixed stars were also spherical.

But there were other reasons for assuming spherical stars. As we have seen, the observed sphericity of the Moon and Sun was extended to all celestial bodies, including the fixed stars. By analogical reasoning with the Sun and Moon and by strong traditional prejudices in favor of sphericity as the most desirable of all shapes and the most suited for the incorruptible celestial ether, it was regularly assumed that each fixed star was spherical.[85] Telescopic observations eventually forced a different interpretation. With his telescope, Galileo found that "the fixed stars are not seen bounded by circular outlines but rather as pulsating all around with certain bright rays."[86] Thus "the rays and fringes surrounding their images hid any sharp outline," until eventually most astronomers concluded "that, seen from the Earth,

83. Although the five authors cited in question 127 ("Whether the planets [*astra*] are spherical in shape") of the "Catalog of Questions" (Appendix I), use the terms *astra* and *stella*, their concern is with the shape of the planets, not of the fixed stars. Thus Melchior Cornaeus, *De coelo*, tract. 4, disp. 1, qu. 4, dub. 9, 1657, 512, considers whether *stellae* receive their light from the Sun ("An stellae accipiant lumen a Sole") but speaks only of the Moon, Mercury, and Venus.

84. "Est una distinctio et sit ista quod sphera vel orbis dicitur unus tripliciter: uno modo quia est quedam pars celi spherica non separata a toto nec suppositaliter existens. Illo modo stella bene dicitur una sphera et sic secundum talem intellectum loquendo valde multe sunt sphere celestes." Albert of Saxony, *De celo*, bk. 2, qu. 6, 1518, 105v, col. 2. The second and third ways concern orbs that are eccentric or concentric with respect to the center of the world.

85. For the special attitude toward the sphere, see Chapter 6, Section II. In a section "On the Fixed Stars" ("De stellis fixis"), Bartholomew the Englishman (*De rerum proprietatibus*, bk. 8, ch. 33, 1601, 418) appears to assume the sphericity of the fixed stars, although there is some ambiguity that suggests he may be speaking about the planets.

86. *The Sidereal* or *Starry Messenger* in Galileo, *Sidereus nuncius* [Van Helden], 1989, 58; see also Galileo, *Starry Messenger* [Drake], 1957, 47. In his *Letters on Sunspots*, published in 1613 but actually written in 1612 (see Sec. IV.2.a of this chapter), after the *Starry Messenger*, Galileo twice argued that the fixed stars are round. In the first letter, he declares that "stars, whether fixed or wandering, are seen always to keep the same shape, which is spherical" (Drake, 1957, 100) and in his third letter, he remarks that "the telescope shows us the shapes of all the stars, fixed as well as planets to be quite round" (ibid., 137). I have no idea why Galileo altered his opinion and whether it was his final judgment. Neither Van Helden nor Drake mentions the discrepancy.

the fixed stars appear as dimensionless points."[87] In this way, did stars "lose" their round shapes.

c. On the "fixity" of the fixed stars

Throughout the Middle Ages and into the seventeenth century, the fixed stars, as their name so aptly implies, were always assumed to be embedded in the eighth sphere. And like the planets, they were thought incapable of self-motion and were therefore assumed to be carried around by the eighth sphere itself. Indeed, since the fixed stars moved with at least two, and perhaps even three distinct motions, two additional spheres, the ninth and tenth, which carried no planets or stars, were added. In this way, one stellar motion was assigned to each of the eighth, ninth, and tenth spheres. When hard planetary orbs were gradually replaced in the seventeenth century by a fluid heaven and the planets were perceived to be orbless bodies that were either self-moved or moved by some external force or intelligence, the fixed stars continued to be perceived as bodies incapable of self-motion, embedded in an orb of some thickness that carried them around. The disposition of the fixed stars thus varied little from what it was during the Middle Ages. The issue that came to a focus in the seventeenth century, however, was whether all of the fixed stars were located on the same surface, at the same distance from the earth, and therefore in a hard orb no thicker than the thickest star, or whether they were dispersed at various distances from the earth and thus encased in an orb of enormous thickness. In Section III of this chapter, we saw that the latter alternative gained support, although its status in the scholastic community is uncertain because relatively few discussed it.

Also from our discussion in Section III, we learned that visible, fixed stars were traditionally assumed larger than the planets and earth and that ever since the days of Ptolemy the stars had been divided into six magnitudes, ranging from the largest to the smallest. Smaller, invisible stars, however, were also assumed to exist. Indeed, as a consequence of Galileo's telescopic observations, the whitish, cloudlike Milky Way was seen to be composed of clusters of small stars that were individually indiscernible to the naked eye.[88] That they were perceived as stars in the celestial region marked a radical departure from the medieval Aristotelian tradition, which denied the Milky Way celestial status and located it in the upper reaches of the sublunar region.[89] During the Middle Ages, then, the Milky Way was

87. Van Helden, 1985, 89. Galileo may have started the trend away from the attribution of sphericity to each fixed star, but whether he retained that opinion is unclear (see this chapter, note 86).

88. See Galileo, *Sidereus nuncius* [Van Helden], 1989, 62, and Galileo, *Starry Messenger* [Drake], 1957, 49; also Mastrius and Bellutus, *De coelo*, disp. 2, qu. 2, art. 3, 1727, 3:496, col. 2, par. 87.

89. For Aristotle's account, see *Meteorologica*, bk. 1, ch. 8. Albertus Magnus, Nicole Oresme, and Henry of Langenstein appear to represent exceptions to the medieval tradition of

not thought to be composed of stars but was rather described as having been formed by the motions of stars and planets that caused alterations in the upper airy and fiery regions.

d. The fixed stars as Suns

Although few considered the problem, prior to the late sixteenth century stars were usually assumed to receive their light from the Sun. Giordano Bruno broke with this tradition by interpreting the stars as Suns, and therefore self-luminous bodies (Dick, 1982, 108), an idea that became popular in the seventeenth and eighteenth centuries and numbered among its supporters Galileo and Descartes, as well as Henry More, Christiaan Huygens, Richard Bentley, William Derham, Immanuel Kant, and, among scholastics, Athanasius Kircher.[90] But here again, few scholastics discussed the issue of the fixed stars as Suns. As Aristotelians and Catholics opposed to the Copernican system, scholastics were unlikely to identify fixed stars as Suns that were the centers of Copernican-type solar systems.

e. The twinkling of stars

Twinkling might have seemed an inherent property of fixed stars as contrasted with the planets, but Aristotle had denied the objective reality of the phenomenon itself. In *De caelo* (2.8.290a.7–24 [Guthrie], 1960), he declares that "our sight, when used at long range, becomes weak and unsteady" and goes on to explain that

> this is possibly the reason also why the fixed stars appear to twinkle but the planets do not. The planets are near, so that our vision reaches them with its powers unimpaired; but in reaching to the fixed stars it is extended too far, and the distance causes it to waver. Thus its trembling makes it seem as if the motion were the stars' – the effect is the same whether it is our sight or its object that moves.

Although Aristotle's explanation was challenged, his denial of the objective reality of the phenomenon in the stars was accepted. Thus Albertus Magnus ([*De caelo*, bk. 2, tract. 3, ch. 8], 1971, 159, col. 2) attributed the twinkling or shaking motion to the great distance of the stars and to the small angle under which we see each fixed star. The shaking motion occurs in our eyes, where the apex of the visual pyramid terminates. This apex is made to tremble or shake by a visual spirit that is activated by objects that are very

locating the Milky Way in the sublunar region (see Ch. 16, n. 28, for Albertus; Ch. 16, n. 44, for Oresme; and Steneck, 1976, 86, for Henry of Langenstein and Albertus Magnus).

90. See Dick, 1982, 52, 108, 116, 129, 133, 148, 152, and 168. In the seventeenth and eighteenth centuries, most nonscholastics considered the fixed stars to be Suns. Except for Kircher, all were Copernicans. But even Kircher assumed that planets encircled the starry Suns (Dick, 116).

far away. Thomas Aquinas offered a similar explanation. The motion, or shaking, actually occurs in our eyes, but it seems as if the fixed stars are vibrating, just as those who sail near the shore think that they are at rest and the mountains and the earth are moved.[91]

In the seventeenth century, scholastics not only attacked Aristotle's explanation, but some also attributed scintillation, or twinkling, to the planets as well as the fixed stars. Thus George de Rhodes (*De coelo*, bk. 2, disp. 2, qu. 1, sec. 3, 1671, 285, col. 1) rejected Aristotle's distance argument when he observed that Mercury, though much nearer than Saturn, twinkled more than the latter.[92] Others found that stars equidistant from the earth twinkle in different ways and concluded that the interior of the eye could not cause such differences.[93]

For these reasons, some sought the cause not in the wavering motion in the eye but in the external medium. According to de Rhodes, Christopher Scheiner explained the phenomenon as arising from continuously agitated air between us and the stars, just as if there were smoke between our eye and an object.[94] According to Serbellonus ([*De caelo*], disp. 1, qu. 5, art. 1, 1662, 2:47, col. 2), the interposed vapors cause various refractions, which then produce the vibrations and twinkling. Since vapors congregate more fully nearer the horizon than toward the vertex, the stars near the horizon twinkle more than those near the vertex.[95] As the cause of twinkling, the external medium had serious drawbacks: if intervening vapors are the cause of twinkling, then "the planets ought to sparkle in the same way as the fixed stars" and "the stars ought not to twinkle in a clear [or fair] sky."[96]

Raphael Aversa reports ([*De caelo*], qu. 34, sec. 3, 1627, 134, col. 1) that Tycho Brahe assigned the cause of stellar twinkling to the fluid medium through which he assumed all the stars moved like birds through air or fish in water. As they moved through their fluid medium, the stars themselves seemed to vibrate or twinkle. Aversa inquires (ibid.) why the planets, which

91. Thomas Aquinas [*De caelo*, bk. 2, lec. 12], 1952, 202, par. 405. The Coimbra Jesuits repeated Thomas's analogy (Conimbricenses, *De coelo*, bk. 2, ch. 8, qu. 1, art. 1], 1598, 326).
92. Mastrius and Bellutus, *De coelo*, disp. 2, qu. 5, 1727, 3:507, col. 1, par. 165, repeat the same argument against Aristotle's position, as does Aversa, *De caelo*, qu. 31, sec. 9, 1627, 34, col. 1. Serbellonus [*De caelo*, disp. 1, qu. 5, art. 1], 1663, 2:47, col. 2, says that he himself observed Mars twinkle, or vibrate, more than once and that Simplicius observed that Mercury twinkled (Scintillant aliquando etiam planetae et de Mercurio notavit etiam Simplicius; de Marte non semel egomet observavi).
93. "Quia stellae aeque distantes diversimode scintillant." Mastrius and Bellutus, ibid.
94. De Rhodes (*De coelo*, bk. 2, disp. 2, qu. 1, sec. 3, 1671, 285, col. 1) cites Scheiner's *In Disquisitionibus mathematicis*.
95. Amicus, *De caelo*, tract. 5, qu. 7, dubit. 3, art. 2, 1626, 344, col. 2, attributes the twinkling partly to the intervening vapors and partly to the species of the object that are received in the sensorium and activate the mobile spirits, which then make the object appear to move.
96. "Sed neque videtur ratio illa satisfacere quia deberent eodem modo micare planete ut stellae fixae si vapores interiecti causae essent scintillationis. Sed neque scintillare deberent stellae sereno coelo." De Rhodes, *De coelo*, bk. 2, disp. 2, qu. 1, sec. 3, 1671, 285, col. 1.

move through the same fluid medium, do not also twinkle. Indeed, Aversa reports yet another hypothesis (ibid., 134, col. 1–135, col. 1) in which it is assumed that each fixed star rotates around its own center and gives the appearance of twinkling. By contrast, the planets do not rotate around a center, as is evident from the Moon, which always reveals the same face. Aversa rebuts this claim by invoking the planets, some of which are seen to twinkle by the naked eye (Mercury, for example) and others recently viewed through a telescope (*tubo optico*). Since some planets twinkle and yet do not rotate around their own centers, stellar rotation cannot explain the same phenomenon in the fixed stars.

Aversa offers his own opinion in which he seeks to explain the twinkling phenomenon in both the planets and the fixed stars. Most planets and all the fixed stars are partly opaque and also have light diffused through their bodies. As the heavens turn, this combination, or admixture, of opacity and lucidity produces a twinkling, or tremulous, effect. Because the Moon is a completely opaque celestial body, it fails to twinkle, as do many, if not all, of the others.[97]

Because no crucial evidence – not even from the telescope – was forthcoming to determine the cause of twinkling or vibrating stars, theories and opinions multiplied. Indeed, de Rhodes ([*De coelo*], bk. 2, disp. 2, qu. 1, sec. 3, 1671, 285, col. 2) even reports that Galileo and others attributed twinkling to the angular shape of the fixed stars.

2. *The Sun*

With the exception of an occasional encyclopedia article or hexaemeral treatise, no *questio* was devoted to the Sun, which was rarely singled out for separate, descriptive treatment.[98] With the advent of the Copernican system in the sixteenth century and the discovery of sunspots in the second decade of the seventeenth century, the Sun became a focus of attention, as we see in works by Riccioli and Otto von Guericke and, of course, in Galileo's *Letters on Sunspots*. Medieval natural philosophers did, however, comment here and there on the role and position of the Sun, and from such

97. Aversa, *De caelo*, qu. 34, sec. 3, 1627, 135, col. 1. De Rhodes (ibid., 285, cols. 1–2) cites what may be the same explanation and attributes it to Aquilonius, book 5, proposition 81. Mastrius and Bellutus (*De coelo*, disp. 2, qu. 5, 1727, 3:507, col. 1, par. 166) also cite "Aquilonius, bk. 5, proposition 81," but the view they attribute to him is quite different. In their version, which they consider probable, each star twinkles because it also turns around a proper center. Since these rotatory speeds differ, one star twinkles more or less than another. In this version, the stars not only move with the sphere in which they are all embedded, but each star also rotates around its own center. Immediately after this description, Mastrius and Bellutus describe Aversa's opinion, mentioning him by name and declaring his interpretation no less probable than their own.

98. For encyclopedia articles, see Bartholomew the Englishman, *De rerum proprietatibus*, bk. 8, ch. 29, 1601, 405–409, and Vincent of Beauvais, *Speculum naturale*, bk. 15, chs. 2–6, 1624, 1:cols. 1094–1097. For a hexaemeral treatise, see Grosseteste, *Hexäemeron*, part. 5, ch. 20, 1-ch. 21, 3, 1982, 179–181.

isolated remarks some sense of their ideas about that great luminary can be determined.

a. The position of the Sun

Of the seven planets, the Sun was judged the largest and occupied the fourth position, counting upward from the Moon or downward from Saturn. It was thus in the middle position among the planets. Because it was recognized as the primary light source of the cosmos, the Sun was usually viewed as the most important of the planets, even though it failed to meet the primary Aristotelian criterion for the most perfect planet: it was not the farthest planet from the earth, a fact that also vitiated the otherwise desirable feature of possessing the fewest motions of all the planets. But, as we saw in Chapter 11, Section I.3, a combination of features made it easy to treat the Sun as the most fundamental and vital of all planets: it was the fourth and middle of the seven planets, and it radiated its light in all directions, which made the middle position seem natural and appropriate for it. Medieval natural philosophers exalted the Sun's middle position by two fundamental metaphors: the Sun was like a "wise king in the middle of his kingdom," or "like the heart in the middle of the body."[99] As Macrobius observed, however, the Sun was in the middle by number only and not by space. For it is closer to the Moon, the lowest planet, than to Saturn, the highest.[100] The middle position of the Sun was essential for the preservation of the universe. Robert Grosseteste expressed a common opinion when he declared that if the Sun were elevated to the region of the fixed stars, the elements and compounds around the earth would be destroyed, presumably because the earth's heat would be greatly reduced. Similarly, if the Sun descended to the region of the Moon, its proximity to the elements would also destroy them.[101] Albertus Magnus saw further significance in the Sun's central location ([*De caelo*, bk. 2, tract. 3, ch. 15], 1971, 178, col. 1). Nature ordained this arrangement so that the Sun could first receive the powers of other planets above and below it and subsequently transmit them to the Moon by means of its light. Because of its motion and nearness to the elements directly below, the Moon, in turn, transmits these powers to those

99. These are the words of Themon Judaeus. For the names of others who used one or both of these metaphors, see Chapter 11, note 28.

100. "The region of the sun is not found to be in the middle since it is farther removed from the top than the bottom is from it. . . . Saturn, the highest planet, passes through the whole zodiac in thirty years, the sun, the middle one, in a year, and the moon, the lowest, in less than a month. The difference between the sun's orbit and Saturn's is as the difference between thirty and one, but the difference between the moon's orbit and the sun's is as the difference between twelve and one." Macrobius, *Commentary on Dream of Scipio*, 1952, 166.

101. "Si autem sol ascenderet ad circulum stellarum fixarum, destruerentur elementa et elementata; et rursum si descenderet ad circulum lune, destruerentur eciam." Grosseteste, *Hexäemeron*, part. 5, ch. 21, 2, 1982, 180, lines 20–22. Clavius provided similar reasons to justify the Sun's middle position (*Sphere*, ch. 1, *Opera*, 1611, 3:45).

elements, for which reason philosophers call the Moon "queen of the heavens" (*regina caeli*).

In the event that the Sun's role in maintaining and preserving life may have escaped anyone, Aristotle proclaimed its momentous power by observing that generation ("coming-to-be") and corruption ("passing away") of all sublunar things were totally dependent on the Sun.[102] By comparison, the status of the other planets was modest, if not insignificant. It was in the approaches to, and the withdrawals from, the earth that the Sun caused generation and corruption. On its approach, it caused the waters surrounding the earth to evaporate, but when it receded from the earth those waters condensed and fell.[103]

As anyone could observe, the Sun not only provided light to sustain life, but also heat. Aristotle briefly explained how the Sun performed this vital function. Surprisingly, it was not because the Sun was a hot body, for, as we saw earlier, neither the Sun nor any other planetary body was deemed naturally hot.[104] The Sun provided heat to the world because of its motion. For, as Aristotle explained (*Meteorology* 1.3.341a.19–24), "motion is able to dissolve and inflame the air; indeed moving bodies are actually found to melt. Now the Sun's motion alone is sufficient to account for the origin of warmth and heat. For a motion that is to have this effect must be rapid and near, and that of the stars is rapid but distant, while that of the moon is near but slow, whereas the sun's motion combines both conditions in a sufficient degree." Because it is not inherently hot, the Sun is also not a fiery body. Its color is given as white or yellow.[105]

Scholastics found no reason to challenge Aristotle's description of the Sun's power and properties, although they did occasionally supplement his ideas. Macrobius, whose *Commentary on the Dream of Scipio* was widely known during the Middle Ages, called the Sun the "regulator" of the other planets, "because it controls the departing and returning of each planet through a fixed allotment of space." That is, the Sun determines the direct and retrograde motions of the planets and the transitional stationary points between these directional changes. "In this way," says Macrobius, "the sun's power and influence direct the movements of the other planets over their appointed paths."[106] In an obvious elaboration on the same theme, Bartholomew the Englishman says that of the planets, only the Sun and Moon lack retrogradation, although each is moved on an epicycle. He goes

102. *On Generation and Corruption*, 2.10.336b.16–24. In *Physics* 2.2.194b.13–14 [Hardie and Gaye], 1984, Aristotle again emphasized the Sun's role when he declared that "man is begotten by man and by the sun as well."
103. Aristotle, *Meteorology*, bk. 1, ch. 9.
104. In *Meteorology*, 1.3.341a.15–16, Aristotle says that "we may now explain how it [i.e., heat] can be produced by the heavenly bodies which are not themselves naturally hot." Aristotle [Webster], 1984.
105. See *Meteorology* 1.3.341a.37, and *On Colours* 1.791a.3 and 3.793a.14.
106. Macrobius, *Commentary on Dream of Scipio*, 1952, 169. Somewhat earlier, Pliny, Cleomedes, and Theon of Smyrna had similar ideas (ibid., n. 4).

on to explain that some think that the reason why the Sun lacks retrogradation is "because solar rays are the cause of this retrogradation. For the power of the solar rays sometimes repels them [i.e., the planets] and makes them retrograde, and sometimes attracts them; and thus forces them to remain as if [they are] immovable."[107] With the ability to cause all planetary motion, the Sun's powers were substantially augmented. The advent of the Copernican system only enhanced the Sun's unique role. No longer was it merely the middle planet, but it now lay at the very center of the cosmos. Galileo viewed the Sun as the heart of the planetary system and assigned to it a rotatory motion, which, in turn, caused all the planets to move around. "And just as if the motion of the heart should cease in an animal, all other motions of its members would also cease, so if the rotation of the Sun were to stop, the rotations of all the planets would stop too."[108]

b. Sunspots

Until their discovery by telescopic observation in the seventeenth century, there is no literature on sunspots, although reports of them were recorded long before.[109] In 1611, as many as five separate observers – Scheiner, Galileo, Fabricius, Hariot, and Passignani[110] – discovered them and set the stage for a significant controversy about their true nature. In this instance, a scholastic natural philosopher, Christopher Scheiner (1573–1650), was himself a codiscoverer of sunspots. More than that, it was Scheiner's version that was printed first, in 1612, albeit under the pseudonym "Apelles."[111]

107. "Non enim sol neque luna retrogradiuntur, quamvis in epicyclo moveantur. Et hoc ideo ut dicunt aliqui quia radii solares sunt causa istius retrogradationis. Nam virtus radiorum solarium aliquando illos repellit et facit retrogrados, quandoque eos attrahit. Et sic eos quasi stare cogit, ut dicit Alphra." Bartholomew the Englishman, *De rerum proprietatibus*, bk. 8, ch. 22, 1601, 399. "Alphra" may be Alfraganus. See Alfraganus [Campani], 1910, ch. 24, 154, and [Carmody], 1943, ch. 24, 41. Alfraganus talks about the rising and setting of the planets with respect to the Sun and retrogradation. But he says nothing about the Sun's rays attracting and repelling the planets and thereby causing retrogradations, stations, and direct motions.

108. Galileo, *Letter to the Grand Duchess Christina*, in Galileo [Drake], 1957, 213. For Galileo's judgment of the Sun in his scholastic, or pre-Copernican, days, see Chapter 11, note 28.

109. In his *Letters on Sunspots* (1613), Galileo ([Drake], 1957, 117) cites "the *Annals of French History* by Pithoeus, printed at Paris in 1588, on page 62, where (in the *Life of Charlemagne*) one reads that for eight days together the people of France saw a black spot in the solar disk. . . . This was believed to be Mercury, then in conjunction with the sun, but this is too gross an error. . . . Therefore this phenomenon was definitely one of those very large and very dark spots, of which another may be encountered in the future." Sufficiently large sunspots are indeed visible to the naked eye. Aversa (*De caelo*, qu. 34, sec. 9, 1627, 155, col. 1) reports substantially the same account and adds an even earlier one in the time of Cleomedes, when nebulous sunspots are reported in what Aversa calls "bk. 2 of *De mundo*, not far from the beginning." I have found no such reference if, by *De mundo*, Aversa means the work of that name falsely attributed to Aristotle.

110. See William R. Shea's article "Scheiner, Christoph," in *Dictionary of Scientific Biography*, 1970–1980, 12:151, and also Shea, 1970, 501, n. 21.

111. The book was published by Scheiner's friend, Mark Welser of Augsburg, under the title

Scheiner's letters on sunspots incited Galileo's first two letters on sunspots, which were also sent to Mark Welser. In response to Galileo's criticism, Scheiner published an expanded version of his letters during the summer of 1612,[112] which Galileo answered with a third letter to Welser in 1612. It was these three letters, published by Prince Federico Cesi in Rome in 1613, that constituted Galileo's famous *Letters on Sunspots*.[113] The newly discovered phenomenon of sunspots became yet another battleground in the ongoing struggle between traditional Aristotelian cosmology and the new science. Scheiner sought to defend the incorruptibility of the heavens and shaped his interpretation toward that end.[114] Thus he denied that the spots were on the Sun and assumed instead that each spot was a star circling around the Sun and near it. To buttress his case, he even argued that these spots, or stars, had phases. In this way, Scheiner could claim that the stars, or spots, had always been there and had not come into being. Celestial incorruptibility was thus preserved.

Some years later, Mastrius and Bellutus adopted essentially the same interpretation. After rejecting the explanation proposed by Aversa (to be described shortly), they identify the spots as small stars, followers and attendants of the "Prince of Planets." Despite the seeming obscurity and darkness of these spots, they are not devoid of light but simply receive less light from the Sun. They appear dark because "when a smaller luminous object is conjoined with a luminous object, it is seen not as luminous, but as something dark."[115]

By contrast, Galileo "affirmed that the spots were contiguous to the surface of the sun, that their properties were analogous to those of clouds, and that they were carried around by the rotation of the solar body."[116] As something analogous to clouds, the sunspots seemed to Galileo to be changeable entities.[117] Since he rejected the incorruptibility of the heavens, such changes posed no problems for Galileo. Indeed, in his final letter on sunspots, Galileo mentions that "there are not a few Peripatetics on this side

Tres epistolae de Maculis Solaribus scriptae ad Marcum Velserium (Augsburg, 1612). It has been reprinted in the *Edizione nazionale* of Galileo's collected works.

112. It bore the title *De maculis solaribus et stellis circa Jovem errantibus accuratior disquisitio* (Augsburg, 1612).

113. Partly translated by Stillman Drake with the title *History and Demonstrations Concerning Sunspots and their Phenomena* and printed in Drake, 1957, 87–144.

114. See Shea, 1970, 500, 502. For these highlights of the debate, I rely on Shea.

115. "Diximus has maculas esse stellulas quasdam veluti Solis asseclas perpetuo planetarum principem comitantes . . . nam quamvis subobscurae videantur non est illis lumen aliquod denegandum. Quando enim minus luminosum cum majori luminoso conjungitur, non ut luminosum sed ut aliquod obscuram conspicitur." Mastrius and Bellutus, *De coelo*, disp. 2, qu. 2, art. 3, 1727, 3:497, col. 2, par. 92. Oddus says that he follows Mastrius and Bellutus, and indeed he does so, almost verbatim (see Oddus, *De coelo*, disp. 1, dub. 15, 1672, 45, col. 1).

116. Shea, 1970, 501. Galileo used a combination of mathematics and observation in his letters. For the details, see Shea, 1970, 498–519.

117. "It should be mentioned," Galileo declares in his *Second Letter on Sunspots*, "that the spots are not completely fixed and motionless on the face of the sun, but continually change in shape, collect together, and disperse" (Galileo [Drake], 1957, 109).

of the Alps who go about philosophizing without any desire to learn the truth about the causes of things, for they deny these new discoveries or jest about them, saying that they are illusions." They defend celestial incorruptibility, "which perhaps Aristotle himself would abandon in our age." And they do so in the manner of Apelles, "save that where he puts a single star for each spot, these fellows make the spots a congeries of many minute stars which gather together in greater or smaller numbers to form spots of irregular and varying shapes" (Galileo, *Letters on Sunspots* [1613], Third Letter, in Drake, 1957, 140–141).

But for those Peripatetics who identified sunspots as stars, Galileo saw that another essential feature of Peripatetic philosophy would crumble (ibid., 141–142). They could not be fixed stars, since they continually changed position with respect to each other. "Hence anyone who wished to maintain that the spots were a congeries of minute stars would have to introduce into the sky innumerable movements, tumultuous, uneven, and without any regularity. But this does not harmonize with any plausible philosophy. And to what purpose would it be done? To keep the heavens free from even the tiniest alteration of material." In order to defend celestial incorruptibility, some Peripatetics were thus prepared to accept a lesser disaster and permit certain celestial bodies, the sunspot-stars, to move in the most random and unpredictable manner, thus abandoning a cherished and traditional notion that the celestial region is a place of predictable, if not sublime, regularity.

As was so often the case, scholastic natural philosophers were themselves not of the same mind. They were not all committed to sunspots as stars. Raphael Aversa, who accepted some, though by no means all, of Galileo's ideas on sunspots, is a significant exception. On the basic phenomena, most scholastics, including Aversa, were probably in agreement. Thus, near the beginning of his discussion of solar spots, Aversa declares that

> These spots of the Sun are not like the spots on the Moon but very dissimilar. They appear as certain clouds; they are very small in comparison to the Sun; they are not fixed and stable; they are not always the same; they do not last long. Some depart, others arrive. Now there are more, now fewer, now a single one, now none. They have various and irregular shapes: some are round, others verge toward the vertical. They do not always retain the same shape with which they begin. They change magnitude and state; they increase or decrease. Sometimes they seem to be congregated and thickened, sometimes they seem to be separated and dispersed. Sometimes many are conjoined into one, sometimes one is dissolved into many.[118]

Aversa admitted that the subject of sunspots was obscure and perplexing, posing a special difficulty, "because if these spots were in the Sun or in the heavens, as they appear [to be], they seem to cause great damage to the

118. Aversa, *De caelo*, qu. 34, sec. 9, 1627, 155, col. 1. There is more, which I omit.

stability and incorruptibility of the heavens, since they seem everywhere to be generated anew and to be destroyed in a brief time."[119] A number of moves were made to interpret the spots without abandoning celestial incorruptibility. Some of these had already been mentioned and rejected by Scheiner, namely that sunspots were an illusion of the eyes (they had been observed by several distinct individuals), that defective telescopic lenses had somehow produced them (but eight telescopes had produced the same results), and that the spots were disturbances in the terrestrial atmosphere (among a number of reasons, Scheiner argued that no cloud or vapor could follow the Sun's small diameter during daylight hours).[120] Aversa repeated some of these arguments but rejected them. He also repudiated interpretations that assumed that the spots are elemental vapors that rise up from below, partly attracted by the Sun itself. These vapors remain in the upper atmosphere below the heavens, but directly beneath the Sun. On this interpretation, the spots are clouds of vapors in the upper terrestrial region. In another interpretation, the lack of any sensible parallax between the Sun and the spots suggested that the vapors rise into the heavens themselves, ascending quite near to the Sun. But Aversa counters that although terrestrial vapors may rise to the highest point of the terrestrial region, they cannot break into the heavens.[121] Aversa next turns to Galileo's view that the spots are really vapors, but celestial rather than terrestrial. Indeed, Galileo assumes that sunspots are formed from vapors given off by the Sun itself (Aversa [*De caelo*, qu. 34, sec. 9], 1627, 157, col. 2), an interpretation Aversa rejects because if the vapors or exhalations came from the Sun, they should be as lucid as the Sun itself and therefore unable to obscure it. Moreover, to assume such solar vapors or fumes implies heavens and celestial bodies that are fluid and therefore changeable, a consequence that Aversa considers unthinkable. Once again the assumption of celestial incorruptibility serves to decide an argument.

After rejecting a few more opinions,[122] Aversa presents his own interpretation,[123] declaring that the spots are "truly and really" (*vere et realiter*) in the Sun itself. Because the same spots do not return and they are always varied and different, Aversa concludes (ibid., 160, col. 1) that "the spots

119. "Sed adhuc res tota satis obscura et perplexa remanet. Et praecipue difficultatem affert: quia si hae maculae sint in sole sive in caelo prout apparent, magnum praeiudicium afferre videntur stabilitati et incorruptibilitati caelorum, cum passim de novo gigni et brevi spatio deleri videantur." Ibid., 156, col. 1.
120. Shea, 1970, 499.
121. "Et magis adhuc errant elevando terrestres vapores intra regionem caelestem. Hi enim esto ad summum ascendere possint in supremam elementaris mundi regionem, prorsus non poterunt irrumpere confinia caeli." Aversa, *De caelo*, qu. 34, sec. 9, 1627, 157, col. 2.
122. Ibid., 158–159. One opinion considered the spots as hard planets that are always in the heavens and another conceived them as "certain denser and opaque parts of the heavens which are revolved around the Sun itself by a certain epicycle, just as . . . Mercury and Venus are revolved around the Sun" (ibid., 159, col. 1).
123. Scheiner quoted the whole of Aversa's opinion verbatim in his *Rosa Ursina*, 1630, 744, cols. 1–2.

cannot be constituted in the Sun as its firm and perpetual properties." The spots "are formed anew in the Sun and last a short time and then disappear, and others again succeed them."

But how can this happen? Aversa suggests two ways. The first assumes that the Sun is a fluid, rather than solid, body, which is held together as if it were contained in a vessel. Aversa imagines the fluid to be like molten metal which contains within itself certain opaque and dark parts, which from time to time rise to the external face and are subsequently reduced in size as they change their shapes. All this is compatible with the Sun's rotation in its own proper place. As his second mode, Aversa assumes that celestial causes can produce in the Sun a certain opacity that impedes light and which is capable of enduring for a certain time, after which it is destroyed, a process that continues perpetually.

Do these changes of the opaque parts of the Sun signify that mutations occur in the heavens? Aversa denies this by invoking a fairly standard response: the changes in opacity just described are only accidental, not substantial. Despite the obvious acceptance of changes within the Sun, Aversa, like many other scholastic natural philosophers, sought to avoid abandoning the principle of incorruptible heavens. To do this, he simply treated the changes as if they were accidental, and therefore of no consequence.

As already mentioned, Mastrius and Bellutus criticized Aversa's explanation. They could not reasonably account for his assumption that some parts of the Sun itself were rendered more opaque than other parts and that the transformation into opaque parts is to be characterized as merely an accidental mutation. Moreover, why should these spots move parallel to the Sun's diameter from the eastern part of the Sun to the western part, rather than the converse? And why, at some time, do these spots not exist simultaneously at the extreme, upper surface of the Sun? That is, why are they not found over the whole surface of the Sun, rather than only around the Sun's equatorial diameter? And why does this accidental mutation not occur everywhere on the Sun, instead of only around the diameter?

Because the nature of sunspots was simply unknown, all explanations were subject to easy challenge. For Aristotelians, however, committed to celestial incorruptibility, the discovery of sunspots posed serious difficulties and seemed to signify another blow against the traditional cosmos. To preserve incorruptibility, they were compelled to resort to explanations that denied change altogether or that allowed for distinctions between substantial and accidental changes. In the latter event, scholastic authors could come fairly close to acceptance of Galileo's views on sunspots. But whereas Galileo saw the spots as clear evidence of change and mutability, scholastics devised explanations that retained as strong a sense of incorruptibility as was feasible, even where they assumed a fluid Sun.

The Sun also had other properties, mostly of an astrological kind, which played a significant role in medieval ideas about the influence of the celestial region on the terrestrial. (These are described in Chapter 19.) Now, however, we must turn our attention to the Moon.

3. The Moon

The Moon was perhaps the most interesting of all celestial bodies. Unlike the other planets, its proximity to earth made some of its surface details visible to the naked eye. Those details suggested properties that seemed to conflict with Aristotelian generalizations about the celestial region as a whole. In our earlier examination of the Moon, in Chapter 16, where we considered only the lunar reception of solar light and whether the Moon possessed its own proper light, no such conflicts arose. Now, however, we must investigate medieval and early modern interpretations of the Moon's most striking visible feature: its spots (singular: *macula*; plural: *maculae*), which were sometimes called "the man in the Moon" or the "face" in the Moon.[124]

a. The lunar spots

In his splendid summary of medieval opinions about the lunar spots, Pierre Duhem remarks that even before Galileo revealed the Sun's spots, Aristotle's physics "was already confronted by a perpetual contradiction with the spot on the Moon. It was impossible to observe the spot on the Moon without thinking that it denotes a certain heterogeneity in the structure of the Moon, a certain irregularity incompatible with the geometric purity of celestial essence as defined by Peripatetic philosophy; it requires that one consider the moon as a body comparable to our earth."[125]

In his *Generation of Animals*, Aristotle associated the Moon with elemental fire. It was Averroës, however, who linked the Moon with the earth and attributed this linkage to Aristotle, allegedly in the latter's *Liber de animalibus* (*Book on Animals*).[126] Prior to the thirteenth century and under Platonic influence, the Moon was indeed conceived as formed from sublunar elements (Duhem [Ariew], 1985, 482). In a work falsely ascribed to Venerable

124. Buridan ([*Metaphysics*, bk. 12, qu. 10], 1518, 73r, col. 2) referred to it as the figure of a man (*imago hominis*) and Plutarch as a human face, as is evident by the very title of his famous treatise "Concerning the Face Which Appears in the Orb of the Moon" (see Plutarch, *Moralia*, 920, 1957, 37)
125. Duhem [Ariew], 1985, 479. Duhem has a separate section devoted to lunar spots (see Duhem, *Le Système*, 1913–1959, 9:409–430, and the translation by Ariew, 479–497).
126. Averroës [*De caelo*, bk. 2, comment. 42], 1562–1574, 5:127r, col. 1, says that "Aristotle, in his book *On Animals*, says that the nature of the Moon is similar to the nature of the earth because of the darkness which is in it. Therefore the luminous part of orbs is similar to the nature of fire." Presumably Aristotle is referring here to the dark spots on the Moon. Averroës repeats the same sentiment in ibid., comment. 49, 131v, col. 1. Duhem says ([Ariew], 1985, 480) that he could find no such reference in Aristotle's *History of Animals*. Because Averroës said it was so, however, this passage became influential.

Bede, the lunar spot was explained as a consequence of an inadequate mixture of the earth with the other three elements. Thus the improperly mixed earth prevents the transmission of light and causes the spot (ibid., 482–483).

By the latter part of the thirteenth century, there was a strong tradition for linking the Moon with the terrestrial region. But as Aristotelian natural philosophy became entrenched in the university curriculum and won universal support, the Moon was conceived as constituted of the same incorruptible ether, or fifth element, as all other celestial bodies. Within this framework, what was the lunar spot, and how could it be fitted into an Aristotelian context? As Duhem put it (ibid., 483): "The Aristotelian doctrine about celestial essence did not seem reconcilable with the existence of the spot," nor with the Moon as an earthy, elemental body.

i. Aversa's summary account: nine interpretations. By the time Raphael Aversa published his account of the lunar spots in 1627, he was able to present nine interpretations, including his own, that had been proposed over the centuries ([*De caelo*, qu. 34, sec. 8], 1627, 151, col. 2–154, col. 2). There are those who follow Plutarch and deny that spots are really on the Moon, suggesting instead that they are either an illusion or the result of vapors interposed between our eyes and the Moon. But Aversa observes that "the constant and perpetual appearance of the spot to all men in every place and time shows that it is truly in the Moon itself." In chapter 9 of the second book of his *Natural History*, Pliny offers a variation on this theme by attributing the spot to terrestrial vapors rising up to the Moon and adhering to its surface. Aversa rejects this suggestion because he deems it impossible that terrestrial vapors can penetrate into the celestial region. Aversa also rejects as untenable the suggestion that the spot is really a reflection of our earth and seas from the mirrorlike surface of the Moon.[127] At full Moon, the half of the earth that faces the full Moon is dark and therefore could not be reflected.

Almost inevitably, one response invoked bodies lying between the Moon and Sun, bodies that allegedly cast shadows on the lunar surface and thus formed the spot. The lunar spot is, however, constant and always appears the same. Intervening bodies would therefore have to maintain the same positions with respect to the Sun and the Moon, an unlikely possibility.

All of the explanations thus far appealed to phenomena external to the Moon. Aversa insists, however, that the spots are in, or on, the Moon

127. In this connection, Aversa mentions (*De caelo*, qu. 34, sec. 8, 1627, 152, col. 1) Plutarch as one who repeats this idea. Plutarch (*Moralia*, 920–921, 1957, 41) does indeed repeat an opinion he ascribes to Clearchus, declaring: "The man, you see, asserts that what is called the face consists of mirrored likenesses, that is images of the great ocean reflected in the moon." Plutarch speaks only of oceans, not land. The interpretation was also repeated by the Conimbricenses, *De coelo*, bk. 2, ch. 7, qu. 4, art. 3 (mistakenly given as 2), 1598, 306.

itself.[128] As his fifth interpretation, he reports the opinion of those who viewed the lunar spots as "a certain mixture of fire and air," in which "the fire is the more lucid [component] and the air the more obscure [and dark component]."[129] The airy parts comprise the lunar spots. Aversa found this wholly unsatisfactory. In this interpretation, celestial bodies are constituted of terrestrial elements. Thus if the Moon were partly made of fire, the latter should always be seen independently of the Sun. Or if the fire is illuminated by the Sun, so also should the air be, in which event the spots would disappear.

As his sixth opinion, Aversa presents an interpretation based on a Pythagorean idea that the Moon is like another earth. Thus it has unequal parts, including high mountains, valleys, waters and seas. And "by means of such inequalities of its parts, it makes various shadows on itself, just as on our earth shadows are made alongside mountains and valleys. These shadows are the spots on the Moon." Aversa observes that, in his *Sidereus nuncius*, Galileo also assumed that the Moon was like our earth and suggested that the dark spots on the Moon are seas and the lighter parts are land areas.[130] The suggestion that there could be land areas, seas, and air on a celestial body like the Moon as on our earth Aversa regarded as utterly absurd.[131] Such talk might be tolerated in poets, but was simply unworthy in philosophers.[132]

Ironically, the seventh interpretation is ascribed to a poet: Dante.[133] In *Paradiso*, Canto 2, Dante suggests that the lunar spots are caused by differences of density in the Moon, but Beatrice suggests instead that they are produced by differences in the intelligences that move the heavens, as well as by differences in the celestial bodies themselves. As the lowest sphere, farthest removed from the empyrean heaven, the Moon's intelligence and its bodily quality are the most inferior of all celestial bodies. Hence it receives the Sun's light unevenly and thus reveals spots.[134]

128. "Statuendum ergo est vere maculas esse in ipsa luna." Aversa, *De caelo*, qu. 34, sec. 8, 1627, 152, col. 2.
129. "Quinto. Aliqui apud eundem Plutarchum dixerunt lunam in se esse quandam ignis aerisque mixturam et ignem quidem esse lucidiorem, aerem vero subobscurum." Ibid.
130. See Galileo, *Sidereus nuncius* [Van Helden], 1989, 43 and Galileo, *Starry Messenger* [Drake], 1957, 34. In 1610, in his *Conversation with Galileo's Sidereal Messenger*, Kepler adopted Galileo's interpretation of the lunar spots when he conceded to Galileo: "You have proved your point completely. I admit that the spots are seas, I admit that the bright areas are land" (see Kepler [Rosen], 1967, 109, n. 276).
131. Within a few lines of the conclusion of his *Sidereal Messenger*, Galileo declares that "not only the Earth but also the Moon has its surrounding vaporous orb. And we can accordingly make the same judgment about the remaining planets." Galileo, *Sidereus nuncius* [Van Helden], 1989, 86; see also Galileo, *Starry Messenger* [Drake], 1957, 58.
132. "Nil absurdius de caelestibus corporibus dici potest quam fingere ibi terras et maria et aerem. Fuissent haec quidem in poetis toleranda, sed philosophis prorsus indigna." Aversa, *De caelo*, qu. 34, sec. 8, 1627, 153, col. 1.
133. Aversa remarks (ibid., cols. 1–2) that it is no more discordant to admit poets into the chorus of philosophers than it is to have philosophers enter into the realm of poets.
134. Aversa does not express it quite this way but simply reports that the intelligences affecting the heavens act upon its various parts in different ways. My interpretation is drawn

ii. Averroës' explanation the most popular in Middle Ages. As an eighth interpretation, Aversa summarizes the opinion of Averroës.[135] In Chapter 16, Section III.1, we noted Averroes' explanation of the lunar spot as a function of the uneven luminescence the Moon receives from the Sun. The lunar spots appear where the luminescence is weakest. To account for this, Averroës and other Peripatetic commentators conceived of the Moon as composed of rare and dense parts. There was, however, disagreement as to whether the lucid parts were dense or rare. In his summary account of the various interpretations of lunar spots, Aversa says that Averroës and Albertus Magnus associated greater density with the darker areas and greater rarity with the most lucid parts.[136] But others – Aversa mentions Richard of Middleton, Aegidius Romanus, the Coimbra Jesuits, and Caesar Cremoninus (ca. 1550–1631) – argued that the dark spots are rarer and, by implication, that the lucid parts are denser.[137] Although many scholastics denied that contrary qualities such as rarity and density could exist in the heavens as they do in the terrestrial region, they allowed that the Moon must in some sense vary in density to produce lunar spots. Aversa, however, found this unsatisfactory. Since "lunar spots are nothing other than less lucid parts," Aversa suggests that the terms "denser" (*densior*) and "rarer" (*rarior*) be replaced by "opacity" (*opacitas*) and "transparency" (*diaphaneitas*) – that is, by the concept of a greater or lesser aptitude for receiving and diffusing light. The dark lunar spots are either more diaphanous parts of the Moon, and thus less suitable for reflecting light, or they are the more opaque parts, less suitable for receiving light.[138] Thus the Moon ought not to be conceived as dense and rare but rather as composed of a variety of qualitative parts that are more or less capable of receiving light and which we might better designate as opaque or diaphanous, depending on whether we associate the dark spots with opacity or diaphaneity.[139]

from *The Comedy of Dante Alighieri the Florentine: Cantica III, Il Paradiso*, tr. Dorothy L. Sayers and Barbara Reynolds (Baltimore: Penguin, 1962), 63, 72.

135. In *Le Système du monde*, Duhem's account of the lunar spots is confined almost exclusively to Averroës' opinion and those who adopted it.

136. "Octavo: Iam Averroes 2. *Caeli*, Com. 49, et alij Peripatetici communiter maculas lunae referunt ad diversitatem accidentalem partium eius penes raritatem et densitatem, quod scilicet aliae partes lunae sint rariores, aliae densiores. Et sic aliae magis aliae minus lucidae. Averroes significat et Albertus in *lib. de Quatuor Coevis*, q. 4, art. 2, affirmat partes maculosas seu minus lucidas esse partes densiores et ideo minus aptas ut perfundantur lumine." Aversa, *De caelo*, qu. 34, sec. 8, 1627, 153, col. 2.

137. For the Coimbra Jesuits, see Conimbricenses, *De coelo*, bk. 2, ch. 7, qu. 4, art. 3 [mistakenly given as art. 2], 1598, 306. In support of this opinion, Duhem cites Aegidius Romanus and also John of Jandun (Duhem [Ariew], 1985, 487–488). We may add the anonymous fourteenth-century author who, as we saw (Ch. 16. Sec. III.1), associated the rarity of the Moon with its dark spots; also Amicus, *De caelo*, tract. 6, qu. 5, dubit. 3, art. 3, 1626, 367, col. 1 ("Quinta conclusio: partes luminosae lunares sunt ex densitate; obscuriores vero ex raritate").

138. "Itaque partes maculosae in luna vel sunt partes magis diaphanae et ita minus aptae ad reflectendum lumen; vel magis opacae ea opacitate quae reddit corpus minus aptum ad recipiendum in ipsa prima superficie lumen." Aversa, *De caelo*, qu. 34, sec. 8, 1627, 154, col. 1.

139. On Aversa's discussion of "opacity" (*opacitas*) and "transparency" (*diaphaneitas*), see Chapter 10, Section III.3.b.

Not everyone thought of the Moon as a mix of density and rarity or of opacity and diaphaneity. Some, perhaps only a few, thought of it as completely opaque and dense, as an anonymous fourteenth-century author seems to have characterized the Moon and other planets.[140] Although Copernicus reports two opposing interpretations about the planets, namely (1) that they are all opaque and (2) that whether opaque or translucent, they "shine either with their own light or with the sunlight absorbed throughout their bodies" (Copernicus, *Revolutions*, bk. 1, ch. 10 [Rosen], 1978, 18–19), he exempts the Moon, which was also considered opaque in the second interpretation. This may even represent his own opinion. But it was probably a minority view among scholastics, because, as Nicole Oresme explains (*Le Livre du ciel*, bk. 2, ch. 16, 1968, 459), if the Moon "were a non-transparent and dark body like iron or steel, it would reflect the sun's light like a mirror, which fact we have already shown to be not true" (see Ch. 16, n. 20). The Moon is, rather, "a transparent and clear body such as crystal or glass, at least in those parts near its surface." Such crystals are dark, but if they are thin enough the Sun's light will pass through. Because of the Moon's great size, however, the Sun's rays cannot pass all the way through. Only if the Moon were homogeneously transparent in the parts where the Sun's rays hit would it be uniformly illuminated. The occurrence of lunar spots, however, is sufficient indication, argues Oresme, that the Moon is not uniformly illuminated.

iii. Does the moon have a proper motion? The man in the moon and the lunar epicycle. Observation of the lunar spots, or "the man in the Moon," had long ago revealed that the Moon always displays the same face to us, with the clear implication that the Moon lacks a proper rotatory motion.[141] But how can the Moon always exhibit the same face to us if it is carried around by an epicycle, as is assumed in astronomy? For when the Moon is in the *aux*, or apogee, of the epicycle, it would present a different image than when it is in the opposite of the *aux*, or perigee. As Buridan explained: "It would follow that in that spot [*macula*] of the Moon which appears as if it were an image of a man whose feet always appear to be below [or toward the bottom], the feet would sometimes appear above [in the upper part of the Moon]," which is contrary to experience.[142] The feet always remain at the bottom of the lunar disk.

Among those who assumed a transparent, rather than opaque, Moon,

140. "Unde quia luna est corpus opacum et densum sicut quilibet planeta." "Compendium of Six Books," Bibliothèque Nationale, fonds latin, MS. 6752, 226v. On the very next page, however, our anonymous author seems to retract this assessment when he says that the Moon is not equally dense: "Ad hoc dicendum quod luna non est equaliter densa. Unde alique partes eius sunt ita rare quod non retinent lumen." Ibid., 227r. For a more detailed discussion, see Chapter 16, Section III.1.

141. For the basic structure of this section, I draw on Grant, 1989, 241–242, and repeat something of what was said in Chapter 13, Section III.7 ("The Problem with Epicycles").

142. Translated in Grant, 1974, 525, col. 2, from Buridan's *Metaphysics*, bk. 12, qu. 10, 1518, 73r, col. 2. For the error in Buridan's reasoning, see Chapter 13, note 95.

one of the most unusual explanations was provided by Bernard of Verdun in the late thirteenth century. Bernard conceded that if the Moon were carried by an epicycle, the same part of the Moon would not always face the earth. How, then, does it happen that the Moon does indeed show us the same face? This might occur because the combination of orbs that moves the Moon may be such as to keep the same face toward us. But perhaps the same part of the Moon does not always face the earth and yet appears to do so. Here Bernard assumes (Grant, 1974, 524, col. 1) that the Moon is "a solid spherical body of great transparency which extends to every part of its depth and in the interior of it there is a spot of spherical shape." However the Moon may revolve in its orbit and whatever lunar surface confronts us, the spot will always appear to us with the same spherical shape. Thus did Bernard conjure up a fully transparent Moon with a dark, spherical spot that bore little resemblance to the unspherical lunar spots that observers actually saw.

To retain the epicycle and also remain faithful to observation – that is, to account for the perpetual appearance of the same lunar face – most scholastic natural philosophers assumed that the Moon moved with a rotatory motion contrary to that of its epicycle and with an equal speed. Thus would the Moon always display the same face. Here was a significant departure from Aristotle (*De caelo* 2.8.290a.25–27), who had assumed that the Moon lacked any rotatory motion because it always revealed the same face. At least one planet was not a merely passive body carried around by its epicycle but had a proper motion of its own. By analogy with the Moon, some scholastics – for example, Paul of Venice – inferred that the remaining planets also had rotatory motions.[143] Albert of Saxony, however, resisted this temptation and insisted that, of all the planets, only the Moon rotated, because it differed from the other planets. In Albert's approach, planets were divisible into those that had a capacity for self-rotation and those that did not. Moroever, Albert thought of the Moon as composed of nonhomogeneous parts that were capable of affecting the sublunar world in different ways.

Such a radical distinction between planets was no part of Aristotelian cosmology. In defense of that cosmology, Jean Buridan took a stand opposite to Albert's and denied to the Moon both a proper motion and an epicycle. Buridan recognized that if the Moon were carried around by an epicycle, it also had to have a proper rotatory motion "contrary to that of the epicycle and with an equal speed."[144] However, "by a parity of reasoning," if the Moon had a special rotatory motion, "the same would seem to apply to the other planets and stars, since every planet [*stella*] is a spherical body like the Moon. No reason can be offered for the Moon's motion which ought not to apply to any other planet [*stella*]."

143. Paul of Venice, *Liber celi*, 1476, 31, col. 2, explains that "there seems no reason why the Moon ought to have a proper motion more than the other planets" (Deinde non videtur ratio quare luna plus debeat habere motum proprium quam alii planete).

144. Buridan, *Metaphysics*, bk. 12, qu. 10, as translated in Grant, 1974, 526, col. 1.

Buridan thus assumed with Aristotle that all planets possess the same basic properties, so that what applies to the Moon must also apply to the other planets.[145] Like Aristotle, however, he was also convinced that no planet could rotate around its own center. Planets not only change position, but they also cause transmutations in sublunar bodies. Consequently, if planets rotated around their own centers, the rotations ought to affect the way in which they cause sublunar effects. Unless the rotatory motion of each planet contributed something to its total effect on sublunar things, rotatory motion would be superfluous. Taking the Sun as exemplar, Buridan insists that because it is a homogeneous body, its proper rotation would cause no effects on the terrestrial region. All the faces that the Sun might show to the sublunar world would be identical, and therefore no changes would emerge from its rotation. "But if the Sun does not have such a motion, it does not seem reasonable that the Moon should have it, since the Sun is much nobler than the Moon."[146] But Buridan conveniently ignores the fact that, by virtue of its dark spots, the Moon's surface is not homogeneous and therefore might well produce a differential sublunar effect if it had a proper motion.

As we saw earlier, Buridan boldly took the next step and denied an epicycle to the Moon.[147] Thus whereas Albert of Saxony assigned to the Moon both an epicycle and a proper rotatory motion, Buridan denied both. These opinions and options remained fundamental well into the seventeenth century. But it was Albert of Saxony's interpretation that was adopted by most scholastics, even into the seventeenth century.[148] Aversa offered two solutions. One is virtually the same as that proposed by Albert of Saxony, namely that the Moon turns on its epicycle and simultaneously and in the same period undergoes a proper rotatory motion in the opposite direction.[149] It was apparently the one Aversa preferred.

By the seventeenth century, however, new suggestions were proposed. Aversa reports that those who abandoned epicycles and orbs and assumed the heavens were fluid no longer had to worry about the lunar epicycle. With fluid heavens, the Moon, like all the other planets, was assumed to be self-moved around the earth in its monthly period and required no rotatory motion.[150] Otto von Guericke, who abandoned epicycles and orbs,

145. Here I draw on Grant, 1987b, 203–204.
146. "Et si sol non habeat talem motum, nec videtur rationabile quod luna habeat, cum sol sit multo nobilior quam luna." Buridan, *Metaphysics*, bk. 12, qu. 10, 1518, 73v, col. 1.
147. See Chapter 13, Section III.7.
148. The Coimbra Jesuits adopted it. See Conimbricenses, *De coelo*, bk. 2, ch. 7, qu. 4, art. 3 (mistakenly given as 2), 1598, 307.
149. "Melius Peripatetici ac etiam astronomi respondent ex hac uniformi apparentia lunae colligi lunam ipsam per se in proprio centro revolvi velut in oppositum ac volvitur suus epiciclus et eodem tempore cum illo perficere suam revolutionem, ut ita semper ad nos conversam gerat eandem suam faciem non obstante motu epicicli." Aversa, *De caelo*, qu. 34, sec. 8, 1627, 154, col. 2.
150. "Hanc difficultatem statim superant illi qui ablata omni distinctione epiciclorum et orbium faciunt caelum fluidum et dicunt omne astrum per se solutum moveri." Ibid., col. 1.

explained the phenomenon by an attractive force emanating from the earth which holds the Moon in its orbit so that we always see the same face.[151]

b. Other lunar properties

Despite Galileo's interpretation of the Moon as earthlike and composed of valleys and mountains and an envelope of surrounding airy vapors, scholastics avoided an interpretation of the Moon that would make of it some mixture of the four elements. Giovanni Baptista Riccioli thought it probable that the Moon had a celestial form rather than a form that was a mixture of our terrestrial elements.[152] Moreover, scholastics (for example, Amicus) insisted that the lunar surface is smooth, without elevations and depressions as on earth.[153] For if there were mountains, there would either be a vacuum between the prominences, which is impossible in Aristotelian physics, or a body would fill the intervening space. If a body should fill the intervening space, it would have to do so perfectly, in order to produce a perfectly spherical surface. Otherwise a vacuum would occur, which is impossible. Amicus concludes that the lunar surface is perfectly spherical, a conclusion with which Nicole Oresme would probably have agreed when he called the Moon "a perfectly polished spherical body" (*Le Livre du ciel*, bk. 2, ch. 16, 1968, 459).

Of all the properties or attributes assigned to the Moon during the Middle Ages, wetness was perhaps most intimately associated with it. As Pierre d'Ailly put it: "The Moon is cold and wet and the mother of waters."[154] The Moon's association with humidity and similar attributes was relevant to its capacity for producing effects on earth (see Ch. 19).

4. The other planets

In the course of our study of the fixed stars, the Sun, and the Moon, we have had occasion to notice some of the essential properties of the remaining planets: Mercury, Venus, Mars, Jupiter, and Saturn. Most assumed that the fixed stars and all of the planets, including Sun and Moon, were spherical. Occasionally a modest qualification was made. Nicole Oresme, for example, insisted that neither the Moon nor any planet was perfectly spherical, "because the earth is not spherical but would appear so to anyone seeing it

151. Guericke, *Experimenta nova*, 1672, 149, col. 1 and 178, col. 1–179, col. 1.
152. "Ego vero probabilius censeo lunam non habere formam mixtam ex elementis nostris actu aut virtute constantem; neque ex ipsis genitam aut in ea resolubilem, sed habere formam coelestem et omnino diversam a formis elementorum nostrativum et mistorum ex his genitorum." Riccioli, *Almagestum novum*, pars prior, bk. 4, ch. 2, 1651, 187, col. 1.
153. "Prima conclusio: lunae corpus est secundum superficiem aequalis et laevis sine partibus prominentibus et depressis quales sunt partes terrae montanae et imae. Nam inter illas partes spatium contentum vel esset vacuum, et repugnat naturae, vel plenum corpore." Amicus, *De caelo*, tract. 6, qu. 5, dubit. 3, art. 3, 1626, 365, col. 1.
154. From d'Ailly's *Ymago Mundi* as translated in Grant, 1974, 632, n. 5.

from the same distance that we see the moon," which is obvious "because of the earth's shadow in a lunar eclipse." He regarded claims "that there are mountains and valleys on the moon's surface which we cannot see because of its great distance from us" as "pure guesswork." Despite an absence of "perfect sphericity," Oresme thought it reasonable to believe that "the luminous bodies in the heavens are of noble and perfect figure and that such a figure is a sphere" (*Le Livre du ciel*, bk. 2, ch. 20, 1968, 497).

Usually the planets were assumed to be transparent rather than opaque and consequently capable of receiving light from the Sun into the depths of their volumes from whence it was diffused to terrestrial observers. A few thought the planets were capable of self-illumination with or without the supplemental light of the Sun. Although many disagreed with Aristotle and assumed that the Moon had a proper rotatory motion, they were reluctant to extend this property to all the planets and generally assumed that planets lacked rotatory motions. Apart from the obvious fact that the Sun and Moon were readily observable and some of their effects directly perceptible, one other major difference distinguished the two great luminaries from the other five planets: the Sun and Moon lacked retrograde motions and stations, which were common to the other five.

Such qualities were rarely discussed in one place under a rubric mentioning the planets. In *questiones* and commentaries on Aristotle's *De caelo*, one or more of these attributes might be mentioned in relation to some other issue or problem or only incidentally. The anonymous author of the "Compendium of Six Books," whom I have cited in a number of contexts, devotes the sixth and final part of his treatise to the celestial region but apparently did not think it sufficiently important to include a separate section on the planets and their properties.

There are, however, treatises, in which planetary properties are described. Thus, in the thirteenth century, Bartholomew the Englishman and Vincent of Beauvais devoted separate sections to the planets in their respective encyclopedias, as did Pierre d'Ailly in his famous *Ymago mundi*, which served Christopher Columbus as a primary source of knowledge about the cosmos and the earth. This information was about the visible and invisible properties of the planets and derived largely from Ptolemy's astrological treatise the *Tetrabiblos*, or *Quadripartitum*, as it was known in its medieval Latin translation. The astrological tradition reached late medieval Europe directly from the works of Latin encyclopedists such as Macrobius, Martianus Capella, and Isidore of Seville and from Arabic astronomical and astrological sources.[155] In these treatises the planets were regularly assigned properties that were supposed to characterize the nature of their terrestrial influences. A typical list appears in d'Ailly's *Ymago mundi*, where he declares (Grant, 1974, 632, col. 2) that

155. For these sources, see R. Lemay, 1962.

> Saturn is naturally cold and dry in its effect [on other things]; it is pale and of an evil disposition. Jupiter is hot and wet, clear and pure, thus tempering the maliciousness of Saturn. Mars is hot and dry, fiery and radiant, thereby harmful and provoking to war. . . . Venus is hot and wet, most splendid amongst the stars, and always companion to the Sun, called Lucifer when it precedes the Sun [as morning star] and Vesper when it follows the Sun [as an evening star]. Mercury is radiant and keeps pace with the Sun, being never more distant than 24 degrees. Thus it is rarely perceptible.

Drawing primarily upon Ptolemy (probably indirectly), Macrobius, Martianus Capella, Isidore of Seville, the Venerable Bede, and Mesahalla (Misael, as it appears in the text), Bartholomew the Englishman elaborates on a similar, and often identical, set of properties, taking up each planet separately in five successive chapters.[156] Bartholomew's account is little more than an embellishment of what we have already read in d'Ailly's later account. Jupiter, for example, is "benevolent, hot and wet, diurnal and masculine, and temperate in its qualities. In color, it is silvery, bright, clear, and smooth."[157] When joined to the good planets, it "causes good and useful effects" (bonas et utiles facit impressiones) on the elements below. Among these are certain positive effects in the human body, where Jupiter helps bring about beauty, specifically beautiful coloring, eyes, teeth, and hair. Indeed, Jupiter also produces a round beard, which Bartholomew must have thought beautiful, or at least Ptolemy did, since Bartholomew cites Ptolemy as his source.

Because it was astrologically important, Bartholomew also saw fit to include Jupiter's relations to the signs of the zodiac. Sagittarius and Pisces are its houses, but it rules in Cancer and ends its rule in Capricorn. Astrologer's have the highest opinion of Jupiter, whose influences are all positive. According to them, "it signifies wisdom and reason and is truthfulness. Hence when it appears in the ascendant, it signifies . . . reverence, honesty, faith, and discipline." Moreover, Jupiter strengthens the goodness of all the signs when it is in them, except the twelfth, "where they [the astrologers] say that Jupiter signifies servitude, poverty, and sadness for quadrupeds and sorrow for the family and slaves" (Bartholomew the Englishman, *De rerum proprietatibus*, bk. 8, ch. 24, 1601, 402).

Bartholomew's other planetary descriptions are in the same vein and need not be described further. Almost all astrologically motivated descriptions were similar. Within this congeries of astrological properties, we see the

156. Bk. 8, chs. 23 (Saturn), 24 (Jupiter), 25 (Mars), 26 (Venus), and 27 (Mercury), in Bartholomew the Englishman, *De rerum proprietatibus*, 1601, 400–405.

157. "Iupiter . . . planeta est benevolus, calidus et humidus, diurnus et masculinus; in suis qualitatibus temperatus; in colore est argenteus, candidus, clarus, et blandus." Ibid., bk. 8, ch. 24, 401. These properties, as well as those for the other planets, derive ultimately from Ptolemy's *Tetrabiblos*. This will be evident by examining the list of subentries under the name of each planet in the index to Ptolemy's *Tetrabiblos* in Ptolemy, *Tetrabiblos* [Robbins], 1948.

standard application of the four Aristotelian primary qualities – hotness, coldness, dryness, and wetness. Although Saturn is said to be "naturally cold and dry," we recall from earlier discussions that the planet Saturn itself would not have been thought cold and dry. Indeed, the attributes assigned to the planets and heavens from terrestrial elements, compounds, and qualities were assumed not to inhere in the celestial bodies, but the latter were deemed capable of producing such attributes in the sublunar region. Once again, in the *Ymago mundi*, d'Ailly furnishes a concise and illuminating passage on this important theme:

The heaven [or sky] is not of the nature of the four elements, nor does it have any of their qualities, since it is not generable or corruptible; nor is it called hot except virtually [*virtualiter*], since by its power [*virtute*], it makes [things] hot. Nor is it properly colored, for it is clear [and lucid]; nor is it properly light or heavy; soft or hard; rare or dense. Only improperly is it called hard because it is unbreakable and impenetrable; and only improperly is it called dense or thick because a star is said to be the densest part of its sphere. (Grant, 1974, 632, n. 5, col. 2.)

Despite the attempt of d'Ailly, and most other scholastics, to preserve this Aristotelian distinction between real and virtual qualities, the former relevant only to the terrestrial region, the latter to the celestial, most found themselves speaking of these attributes – especially density and rarity – as if they were real.[158]

One attribute, however, life itself, was not distinguished into real and virtual. If life was somehow a property of the celestial ether and its celestial bodies, it had to be real. But could the celestial region be somehow alive? And if so, in what sense?

V. Are the heavens alive?

Between the thirteenth and the seventeenth century, one of the most popular two or three questions in scholastic *questiones* on Aristotle's *De caelo* was whether the heavens are alive (see "Catalog of Questions," Appendix I, qu. 128). Interest in the animation of the heavens was no doubt promoted by the fact that, in the *De caelo* itself, Aristotle had unequivocally assumed that the heavens were alive.[159] Plato's similar belief added further substance to this idea,[160] which had widespread support among the pagan Greeks.

158. Indeed, in his Louvain Lectures, presented between October 1570 and Easter 1572, the future Cardinal Bellarmine characterized as ridiculous the notion that the Sun is warm "virtually." See Lerner, 1989, 267. I am grateful to W. G. L. Randles for calling my attention to Lerner's important article.

159. In *De caelo* 2.12.292a.19–22, Aristotle asserts ([Guthrie], 207, 1960): "The fact is that we are inclined to think of the stars as mere bodies or units, occurring in a certain order but completely lifeless; whereas we ought to think of them as partaking of life and initiative." He also proclaims their animation in ibid., 2.2.285a.29–30, when he says

Christians, however, viewed the problem quite differently. During the early centuries of Christianity, eminent churchmen such as Saints Basil, John Damascene, and Jerome, as well as the Christian convert and Aristotelian commentator John Philoponus, denied life of any kind to the heavens.[161] Indeed, we are told that Origen's assumption of living celestial bodies was condemned at the fifth synod of the Church in Constantinople in 553.[162] Saint Augustine, however, accepted an animated heavens in his *De immortalitate animae*, only to waver in later life, confessing in the *Retractiones*: "I do not affirm that it is false that the world is animate, but I do not understand it to be true."[163] Because Augustine found no solution to this vexing problem in either reason or Scripture, his enormous prestige could not readily be invoked to decide the issue. Christian opposition to animated heavens was probably based on the fear that "if the heavens were admitted to be alive or in any sense divine, they would be the objects of idolatrous worship."[164] Although a few Christians like Origen and Tatian assigned life to celestial bodies, most in late antiquity and the early Middle Ages whose writings were influential in the later Middle Ages adopted one of three positions: they denied life to the heavens and celestial bodies, were ambivalent about it, or ignored the problem altogether.[165]

Beginning with the second half of the twelfth century, the works of Aristotle and his Arabian and Jewish commentators began to enter Europe. By the second quarter of the thirteenth century, the relevant works of three of the most notable commentators – Avicenna, Averroës, and Moses Maimonides – were known to Christian scholastics in Latin translation. All three of these influential authorities upheld Aristotle's opinion about animated heavens. Depite the weight of their opinions, they persuaded few

that "the heaven is alive and contains a principle of motion." Although Aristotle was generally thought to have assumed that the orbs and planets were alive, there are other passages in *De caelo* where Aristotle has been interpreted as denying life to the celestial bodies. On this point, see Wolfson, 1973, 23.

160. See Plato, *Timaeus* 32C–33B, where Plato explains why, although the world is alive, it does not need eyes or ears, air to breathe, or hands or feet. Indeed, the world is not merely alive but "a blessed God" (34B). See Plato [Cornford], 1957, 54–55, 58.

161. For references, see Dales, 1980, 533. Dominicus de Flandria placed Saint Jerome with Origen as one who attributed life to the celestial bodies (see this chapter, n. 195). Indeed, so did Galileo ([*De caelo*, qu. 6 (L)], 1977, 150).

162. Oviedo [*De caelo*, contro. 1, punc. 1], 1640, 460, col. 2, and Aversa, *De caelo*, qu. 33, sec. 7, 1627, 111, col. 2, cite this condemnation, both attributing it to a passage in Nicephorus's *Historiae Ecclesiasticae*, bk. 17, ch. 28, which they quote as follows: "Si quis dicit caelum et solem et lunam et stellas et aquas quae supra caelos sunt, animantes [Oviedo has *animales*] quasdam esse et materiales virtutes anathema sit." In addition to Origen, Dales, 1980, 532, n. 9, also mentions Tatian as a Christian who assumed animate heavens.

163. Cited from Dales, ibid.

164. Dales, ibid., 534.

165. I arrive at this conclusion from the evidence presented by Dales, 1980, who, however, declares: "the assumption that the heavenly bodies were either alive and intelligent, or moved by beings who were, was by far the most widely held view" (535), a situation that began to change only in the mid-eleventh century.

Christian scholastics to assume the existence of living orbs and planets.[166] And yet, judging by the frequency with which it was discussed, the issue did not fade away during the five centuries covered by this study; indeed, it was one of the most widely discussed questions. The reason for this may lie with the motion of the celestial bodies. What could cause these gigantic orbs to rotate and carry the planets around? Since humans and animals are living things and capable of self-motion, it seemed natural to inquire whether the heavens themselves might not be self-moving entities, and therefore in some sense alive.

1. Two senses of life

In what sense might planets be alive? During the Middle Ages, the attribution of life to celestial bodies could be taken as analogous to human life as a whole or only to its highest level, intelligence. The most common interpretations of life associated it with soul, as described in Aristotle's *De anima* (bk. 2, chs. 1–4). Although the soul was conceived as a single entity and was often identified with the form of a body, Aristotle distinguished three possible levels within it. The lowest is the nutritive, or vegetable, soul, which controls growth and decay in all plants and animals but is the only level attained by plants. The second level, the sensitive, or animal, soul, exists in all animals and is responsible for perception and certain offshoots of the latter, namely imagination and memory, as well as the appetitive faculty, which prompts motion. Finally, the third level concerns reason, which is peculiar to man.[167] In these three levels, the higher always encompasses the lower. Thus to possess a rational soul implies possession of the two lower levels, sensitive and nutritive.

Life could also be associated with an intellective soul alone, either as integrated with each celestial body or as a separate substance that might exist independently of celestial bodies but associated with them. In his *Metaphysics* (bk. 12, ch. 8), Aristotle himself had appeared to assign one intelligence, or "unmoved mover," to each celestial orb, assuming either 49 or 55 orbs. During the Middle Ages, these Aristotelian intelligences were usually conceived as the causes of celestial motion and sometimes, if not often, identified with angels.[168] Although there were those who denied the identification between intelligences and angels,[169] both were thought to

166. Adelard of Bath, in the mid-twelfth century, believed that the celestial bodies are alive (ibid., 536). Although he may have been influenced by Aristotle, and perhaps Avicenna, both Averroës and Maimonides wrote their relevant works approximately twenty-five to fifty years after the death of Adelard.
167. For a brief, summary account, see Ross, 1949, 129–131.
168. Oviedo, *De caelo*, contro. 1, punc. 1, 1640, 461, col. 1, considers intelligences and angels identical when, after mentioning intelligences, he says: "which we now call angels" (seu intelligentias, quas nos modo angelos vocamus).
169. For example, Albertus Magnus and his student, Theodoric of Freiberg. See Weisheipl, 1961, 307.

possess only the highest level of soul, namely the intellectual property of reason. Avicenna distinguished between souls and separated substances, more particularly intelligences, which he thought of as angels (Wolfson, 1973, 50): the former (soul) was "the perfection and form" of a celestial body, while the latter (separated substance or intelligence) was wholly distinct from the body whose intelligence it was.[170] Both of these interpretations are subsumed in Grosseteste's complaint that philosophers "write contrary things about these matters. . . . [They] tried to prove that the heavens were animated, and some of them thought all the heavens were animated by one soul, others that each was animated by its own. Some thought the heavens were moved by a soul which was not united to them in an individual unity, but by an intelligence or intelligences not unible to the body in a personal union."[171] The relevant question for us, however, is whether possession of an intellective soul was thought to confer life upon a celestial body.

2. The theological reaction to the idea of living celestial bodies: the Condemnation of 1277

Although Saint Augustine was ambivalent about the animation of celestial bodies, he allowed for its possibility. Indeed, he did not consider it a vital problem, because there was nothing in Scripture that denied or condemned the idea that celestial bodies might be alive. Thomas Aquinas adopted the same attitude (Litt, 1963, 108–109; Dales, 1980, 543–544), as did Robert Grosseteste in his *Hexaëmeron.*[172] Despite the seeming neutrality of Scripture on this issue, as attested by such eminent figures as Augustine, Thomas Aquinas, and Grosseteste, theological fears were apparently aroused when two of the greatest arts masters of the thirteenth century, Siger of Brabant and Boethius of Dacia, appear to have accepted Aristotle's judgment that the heavens are alive.[173] For Siger, every celestial body possessed an intellective soul which was united to it – indeed, it was the cause of its motion – but was yet ontologically distinct.[174] Siger, however, denied that the intellective soul conferred on its celestial body the full life of an inherent soul that functioned as the substantial form of a body. That is, it was not much like the threefold soul that Aristotle described in *De anima*, since it

170. Avicenna [*De philosophia prima (Metaphysics)*, tract. 9, ch. 4], 1508, 105r, col. 1. The Latin text is also cited in Aegidius Romanus, *Errores philosophorum*, 1944, 31, n. 76.
171. Translated by Dales, 1980, 540; for Latin, see Grosseteste, *Hexäemeron*, part. 3, ch. 6, 1–ch. 7, 1, 1982, 106–107.
172. For the translated passage, see Dales, 1980, 540–541. However, a few chapters later Grosseteste declares that "the heavenly lights are only bodies and not living beings" (541–542).
173. On Siger, see Hissette, 1977, 131–132; for Boethius, see ibid., 69–70, where Hissette attributes the ninety-fourth article of the Condemnation of 1277 as perhaps deriving from Boethius of Dacia.
174. For this reason, Siger did not consider the intellective soul as the substantial form of the celestial body with which it was uniquely associated.

lacked the vegetative aspect of plants, the sensitive soul of animals, and the intellective soul of man. This quite limited attribution of life to celestial bodies would have its followers up into the seventeenth century.

Only occasionally did any scholastic author, such as, for example, Peter Aureoli, attribute a degree of life to celestial bodies that exceeded what was assigned to them by Siger of Brabant and Boethius of Dacia. However, since both Siger and Boethius were popular teachers, they probably exercised an influence on other arts masters. Worried about the consequences of such a belief for Christian doctrine, some conservative theologians in Paris during the decade of the 1270s came to view acceptance of animated celestial bodies as potentially dangerous. In his *Errors of the Philosophers* (*Errores philosophorum*), Aegidius Romanus (1944, 31) listed among Avicenna's errors "The subject of the animation of the heavens."[175] The bishop of Paris, Etienne (or Stephen) Tempier, and the theologians who compiled the list of 219 errors at Paris in 1277 must have considered the animation of the heavens a dangerous idea, because they condemned it in at least five different articles: for example, in article 92 ("That celestial bodies are moved by an internal principle, which is soul; and that they are moved by a soul and by an appetitive power [that is, by force of desire] just as an animal; for just as an animal is moved by desire, so also is the sky"), and article 94 ("That there are two eternal principles, namely the body of the sky and its soul").[176] Because there are at least five condemned articles concerning animated celestial bodies, we have no way of knowing which one Richard of Middleton had in mind when he declared that "to assume that celestial bodies are animated is an article condemned by Stephen, a certain bishop of Paris and doctor of theology."[177]

Richard himself opposed the idea on grounds that would be commonly cited. Celestial bodies could not be considered alive because they lacked vegetative, sensitive, and rational levels of soul. Thus they required no nourishment, were not increased in size, and did not generate bodies similar to themselves. Therefore they could have no vegetative, or nutrient, soul.[178] Nor could they have a sensitive, or animal, soul, since they lack organs for sensation. Indeed, their lack of a vegetative soul implies a lack of a higher-

175. In the preceding paragraph, we saw that Avicenna thought celestial bodies were alive because each had a soul as its form (quod anima cuiusque caeli est eius perfectio et eius forma). Avicenna, *De philosophia prima (Metaphysics)*, tract. 9, ch. 4, 1508, 105r, col. 1.

176. Cited from Grant, 1974, 49. See also articles 95, 102, and 110. Dales, 1980, 545–546, translates all but article 92. For further discussion of these articles, see Hissette, 1977, 130–133 (for art. 92); 69–70 (for art. 94); 67–69 (for art. 95); 136 (for art. 102); 194–195 (for art. 110), where the numbers of the articles differ and are 73 (92), 32 (94), 31 (95), 75 (102), 119 (110).

177. "Item ponere corpora caelestia animata est articulus excommunicatus a domino Stephano quodam Parisensis episcopo et sacrae theologiae doctore." Richard of Middleton [*Sentences*, bk. 2, dist. 14, art. 3, qu. 4], 1591, 2:187, col. 1. John Major also refers to one of the same condemned Parisian articles (for the text, see Ch. 18, n. 197 of this volume).

178. "Non habent enim animam vegetativam quia nec nutriuntur neque augentur nec sibi simile de se generant. Generativa autem nutritiva et augmentativa partes sunt potentiales animae vegetativae." Richard, ibid., 2:187, col. 2.

order sensitive soul, because the latter presupposes the former. Finally, they are also devoid of an intellective, or rational, soul, because the latter can only affect a body by means of nutritional and sensitive powers, which are lacking in celestial bodies. It follows, Richard concludes, that "celestial bodies are absolutely inanimate."[179] Thus did Richard and most scholastic natural philosophers deny life to the celestial bodies in the Aristotelian sense of an indissoluble, integrated union of body and soul. Nor did they see analogues to the three levels of soul Aristotle had distinguished. John of Jandun, Nicole Oresme, Galileo, and Franciscus Bonae Spei, for example, agreed with Richard's position.[180] Thus Oresme (*Le Livre du ciel*, bk. 2, ch. 5, 1968, 315) attacks Averroës, who held that the intellective soul was entirely within the heavens itself. Oresme asks "how an intellective soul could dwell in a living body without the sensitive soul also being present"; nor indeed could the sensitive soul exist naturally without the vegetative soul. In his sixteenth-century questions on *De caelo*, Galileo similarly ([qu. 6 (L)], 1977, 149–150) rejects the existence of a vegetative and sensitive soul incorporated within celestial bodies.[181] Bonae Spei (*De coelo*, comment. 3, disp. 3, dub. 5, 1652, 12, cols. 1–2) not only rejected the vegetative, sensitive, and rational souls but also describes a fourth kind of soul, the locomotive, which was held to signify life on the assumption that celestial bodies were self-moving entities (see next section).

3. The intellective, or rational, soul

It was the intellective, or rational, soul, however, that would pose the greatest difficulty. Two major interpretations are distinguishable concerning the intellective soul alone. Could an intellective soul, or an intelligence, be incorporated into, or fully "inform," a celestial body without the presence of the vegetative and animal souls? That is, could a purely spiritual substance, like an intelligence, be so intimately related to its celestial body that it bears a relationship to it as form to matter? Or is the intelligence related to its celestial body as an external mover to the body it moves.[182] On the first alternative, the intimate, integrated relationship between spiritual intelligence and celestial body clearly implies that the celestial body is a living thing in the sense of a thinking being. Few scholastics adopted this interpretation, for much the same reason offered by Richard of Middleton,

179. "Nec habent animam intellectivam quia nullum corpus potest informari anima intellectiva nisi habente vim delectandi, vegetandi, et sensificandi, quia suas operationes intelligibiles per corpus non exercet nisi mediantibus aliquibus viribus sensitivis. Cum ergo, ut dicit Philosophus 2 *De anima*, non sit anima praeter praedictas sequitur quod corpora celestia simpliciter sunt inanimata." Ibid.
180. Although Jandun rejected the attribution of three levels of soul to celestial bodies, he did accept some level of life for them (see this section [V.3] of this chapter, and n. 191).
181. For much the same response from John of Jandun, see his *De coelo*, bk. 2, qu. 4, 1551, 25r, col. 2.
182. Litt, 1963, 108, says that Thomas Aquinas confronted these two alternatives.

namely that the intellective soul could not inform a celestial body in the absence of the vegetative and animal souls, a judgment in which Thomas Aquinas, Nicole Oresme, and Galileo concurred.

An exception was Peter Aureoli, who appears to have assumed living celestial bodies. Aureoli argues that life can only be attributed to celestial bodies if we follow Aristotle and his Commentator, Averroës, and assume that an intelligence is united to its celestial body as a natural form, so that the relationship of the intelligence and the celestial body is like that of the oneness of matter and form in the usual sense.[183] He explicitly denies that life can be associated with celestial bodies if intelligences are only associated with them as external movers. But even where celestial body and intelligence are united as matter and form, Aureoli was only prepared to call the heaven an animal equivocally (quod aequivoce animal dici potest). Indeed, he spends the rest of the article presenting four propositions which focus largely on the differences between real animals and the equivocal manner in which "animal" is applied to the heavens[184] and concludes that "celestial bodies are animated, and are [therefore] animals, to some extent (indeed, one animal), but equivocally with animals here [on earth]."[185]

But what of the second alternative: that the intellective soul is an intelligence capable of separate existence from the celestial body with which it is associated? All who accepted this interpretation thought of the intelligence, or intellective soul, as the cause of the celestial body's motion. Indeed, as a separate substance distinct from its celestial body, the intellective soul, or rational soul, or intelligence or angel as it was also called, was like an external mover in its relationship to the mobile or body it moved. Under these circumstances, did scholastic authors consider celestial bodies alive? Peter Aureoli, as we just saw, denied life to a celestial body that was assumed to have only an external motive intelligence.

In his *Summa theologiae*, Thomas Aquinas adopts a different attitude in grappling with the question "Are the luminous heavenly bodies living?"[186] Indeed he presents an excellent recapitulation of what has been said earlier, in Section V.2 of this chapter. He argues first (*Summa theologiae*, pt. 1, qu. 70, art. 3, 1967, 10:121) that "there is no possibility for the soul of a heavenly body to engage in the activities of a nutritive soul – nourishment, growth, generation – for these do not fit in with a body incorruptible by its nature." Nor are the operations of a sentient, or sensitive, soul appropriate for it, "since all the senses are based on that of touch, the scope of which is elemental qualities. Furthermore, all the organs of the sensory powers must

183. Aureoli [*Sentences*, bk. 2, dist. 14, qu. 3, art. 1], 1596–1605, 2:200, col. 1.
184. This extends over 200, col. 1–201, col. 2.
185. "Ad quaestionem dico quod concedi potest quod corpora celestia sint aliqualiter animata et animalia (imo unum animal), sed aequivoce cum animalibus quae sunt hic." Ibid., 201, col. 2. Franciscus Bonae Spei [*De coelo*, comment. 3, disp. 3, dub. 5], 1652, 11, col. 2 cites Aureoli's belief that the heavens are animated.
186. "Utrum luminaria caeli sint animata." Thomas Aquinas, *Summa theologiae*, pt. 1, qu. 70, art. 3, 1967, 10:119.

be composed of elements mixed in proper proportion and such elements are by their natures foreign to the heavenly bodies" (ibid., 123).

We are thus left with the intellective soul that is divorced from sensation and imagination but which is concerned with understanding (*intelligere*) and motion (*movere*). Understanding, or intellectual activity, requires no body for its activity, except where the latter provides images to it. But we already saw that celestial bodies cannot engage in sentient activities. Hence "intellective activity would not be the point of a soul being united to a heavenly body" (ibid.). This leaves only motion as the reason why an intellective soul would be joined to a celestial body. But how is it joined to the body it moves? Not as a form is joined to a body, but rather externally, as "a contact of power" (per contactum virtutis), that is, "the way in which a mover is conjoined to the object itself."[187] Since the intellective soul is not a body, it cannot be in contact with the object it moves; but the physical object that is moved can, in some sense, be in contact with its intellective mover.[188]

Thomas assumes that "if the celestial bodies are alive, they have an intellect without the senses." He concludes that "the heavenly bodies are not 'living' in the way plants and animals are, but in an equivocal sense. Thus between those who hold that the heavenly bodies are alive and those who deny it, there is not real but merely verbal disagreement."[189] For Aquinas, then, if a heaven or celestial body can be called alive in any sense, it would be by virtue of an intelligence associated with it as an external mover. The intelligence was clearly conceived as alive. But was the celestial body alive? It is on this crucial point that Thomas is ambivalent. For despite his conviction that an intellective soul moves each celestial body in the manner of an external mover, he was uncertain as to whether such a relationship conferred life upon the celestial body.

Johannes de Magistris was tentative in a somewhat different manner. After making a number of distinctions, Magistris concedes that the heavens could only be alive if they possessed souls, or intelligences, that at most assisted their motions but were not fully integrated with the celestial bodies.[190] Whereas Thomas Aquinas assumed that this was indeed the re-

187. Ibid., 123.

188. For a summary of these arguments, see Wolfson, 1973, 47–51, especially 48.

189. The first citation is my translation of "Corpora igitur caelestia, si sunt animata, habent intellectum sine sensu" in Thomas Aquinas, *De caelo*, bk. 2, lec. 13, 1952, 207, par. 418; for the second passage, see Thomas Aquinas, *Summa theologiae*, pt. 1, qu. 70, art. 3, 1967, 10:123–125. For an excellent summary of Thomas's opinion, see Wolfson, 1973, 47–51. Litt, 1963, 108–109, emphasizes the doubt which Thomas felt on this issue.

190. Magistris's arguments are not clear. But he first draws two distinctions about the heavens and two about the meaning of soul, or life. Thus one could conceive of the heavens as a circularly moved sphere separated from any moving intelligence or intelligences; or, one might think of the heavens as a single integrated whole embracing both the spherically moved celestial bodies and motive intelligences. Similarly, two interpretations of the soul are possible: one in which the soul does not form a unified whole with the body whose soul it is but merely assists or aids that body externally; another which assumes that the soul inheres in the body whose soul it is and forms a unity with it. With these

lationship between intellective soul and celestial body and was nonetheless uncertain whether this conferred life on the latter, Johannes de Magistris was convinced that if the relationship between intellective soul and celestial body was as just described, then the celestial body would indeed be alive. Thus Thomas was convinced of their relationship but uncertain about life, whereas Magistris was uncertain about the relationship but convinced that if it did exist so did life in the celestial body.

Most scholastics, however, were more emphatic in their opinions, largely deciding against the attribution of life. Those who did assume animation usually did so in special senses. For example, in the early fourteenth century John of Jandun rejected the attribution of a three-level soul to celestial bodies but seems to have followed Averroës and assumed that they were alive at some level because they were nobler than real living things.[191]

From the fourteenth century on, however, scholastics overwhelmingly opposed the attribution of life to celestial bodies. In what was perhaps a disagreement with Thomas Aquinas, Nicole Oresme explains that "Just because an angel moves a body, it is not necessary that it should give it life nor that it be in the body by union and information or otherwise, save by voluntary application or appropriation, as in the case of a man moving a boat on which he stands" (*Le Livre du ciel*, bk. 2, ch. 5, 1968, 319). Thus just as a man may row a boat and move it without being an intimate or integral part of the boat, so an intelligence or angel may cause the motion of a celestial body but be no part of the celestial body it moves. Therefore the latter need not be alive.[192] Thus did Oresme conclude that he had shown "by natural reason, contrary to Aristotle's statements, that the heavens are not a living nor an animated body." As further evidence that an intelligence need not be integral to the celestial body it moves, Oresme declares that "if we assume the heavens to be moved by intelligences, it is unnecessary that each one should be everywhere within or in every part of the particular heaven it moves." At creation, God may have placed special motive powers and resistances into the celestial bodies in such a perfectly harmonious manner that the heavens move without violence and without the need for any

two distinctions made for the heavens and the souls, Magistris now declares that in the first sense of heavens, celestial bodies are not alive; indeed, they would be moved by an external mover, in the first sense of soul just described. The remainder of the argument is unclear, but Magistris seems to allow the conditional animation of the heavens, taken in the second sense. For the argument, see Magistris, *De celo*, bk. 2, 1490, sig. k3 recto, col. 2.

191. John of Jandun, *De coelo*, bk. 2, qu. 4, 1552, 25r, col. 2–25v, col. 1. For more on Jandun's attitude, see the latter part of this section (V.3) and note 221. For Averroës' thoughts, see Wolfson, 1973, 42–44. In support of his contention that celestial bodies are animate, Averroës cites an argument from Alexander of Aphrodisias that "it is impossible that the noblest of animate beings should be inanimate" (Wolfson, ibid., 43).

192. Amicus gives a similar example when he says that just as an orb might be called alive because it is moved with the assistance of something external, so also could a ship be characterized as alive because it is moved with the assistance of rowers (Dices secundo si ex assistentia orbis dicitur animatum, eadem ratione dicetur navis ex assistentia remigum). Amicus, *De caelo*, tract. 4, qu. 4, dubit. 2, art. 3, 1626, 171, col. 2.

further application of power, either externally or internally. As Oresme expressed it in a famous metaphor, "the situation is much like that of a man making a clock and letting it run and continue its own motion by itself."[193] Thus did God produce the regular motion of the heavens without having to incorporate intelligences, or angels, into the celestial bodies themselves. What Oresme may have understood by celestial motive powers and resistances is, however, a puzzle, since a few lines later he suggests that "as soon as God had created the heavens, He ordained and deputed angels who should move the heavens and who will move them as long as it shall please Him."[194] But if God placed motive powers *within* celestial bodies, why does he also need angels as celestial movers, angels who, as we saw earlier, are external to the celestial bodies they are alleged to move? Although an answer to this question may elude us, we may conclude that insofar as Oresme assumed angels as external celestial movers, he was following the tradition of Thomas Aquinas; but whereas Thomas conceived the celestial bodies as equivocally alive, Oresme flatly denies their animation.

The scholastic attitudes that have been described here were maintained in the seventeenth century. The two fundamental positions arrived at in the Middle Ages were maintained: virtually all denied that the heavens were informed by souls, and therefore really alive. Differences arose, however, as to whether to concede some degree of life, as Thomas Aquinas was thought to have done, solely because a living intelligence, or angel, was the external instrument that moved each orb in its circular path. In the seventeenth century, Mastrius and Bellutus upheld the negative position, while Bartholomew Amicus supported the affirmative. Both sides claimed that their interpretation was the most common.

Mastrius and Bellutus painted a picture of nearly solid opposition to any attribution of life to celestial bodies when they declared that "the form of the heavens does not belong to the genus of living things but of nonliving things, both according to truth and according to Aristotle; and it is also derived from Scotus, bk. 2, dist. 14, qu. 1. And it is common among both the Fathers and scholastic doctors and among Peripatetics, which would be too long [a list] to enumerate, but it can be seen in Amicus and Aversa."[195]

193. Oresme, *Le Livre du ciel*, bk. 2, ch. 2, 1968, 289. On the basis of this passage, Lynn White, Jr. (1978, 239) considered Oresme "the first to foreshadow the Deist concept of the clockmaker God." Earlier, Oresme cites a brief passage from Cicero's *De natura deorum*, II. 38, where Cicero speaks about "the absolutely regular movement of a clock" (ibid., 283). Because his clock was mechanical and Cicero's was not, Oresme is the more plausible forerunner of the Deist concept.

194. Oresme, ibid., 289. See also Dales, 1980, 548, who cites part of this paragraph.

195. *De coelo*, disp. 2, qu. 2, art. 2, 1727, 3:493, col. 2, par. 62. According to Dominicus de Flandria ([*Metaphysics*], 1523, sig. R5v, col. 1), Origen and Saint Jerome were said to have assumed animated celestial bodies; Basil and John Damascene assumed they were lifeless; while Saint Augustine remained undecided. On the basis of what is reported in the next paragraph, I conclude that Mastrius and Bellutus were wrong to include Amicus in this list. Thomas Compton-Carleton [*De coelo*, disp. 1, sec. 1], 1649, 397, cols. 1–2, also proclaimed that a lifeless heavens was the common opinion among the Fathers of the Church and the theologians and went on to name recent theologians who shared

The discussion of one of the "scholastic doctors" and "Peripatetics," Raphael Aversa, was important and influential. Indeed, it is probably from Aversa that Mastrius and Bellutus derived their statement about the widespread acceptance of the lifelessness of celestial bodies. For Aversa emphatically declared that "the common opinion of the theologians and philosophers" is that "the form of celestial bodies is absolutely not a soul and that celestial bodies are not animated or living."[196] To support his claim, Aversa explains why none of the traditionally accepted three levels of soul – vegetative, sensitive [i.e., animal], and rational – is appropriate to the heavens.[197] The operations of the vegetable, or nutritive, soul – namely generation, nutrition, and augmentation – simply do not appear in the heavens. A heaven, or celestial body, does not generate another heaven similar to it as a man generates a man and a plant a plant.[198] Generally, the activities of a vegetative soul can only occur where there is generation and corruption, activities that are lacking in the ungenerable and incorruptible heavens. A sensitive soul is also lacking in the heavens because no external or internal senses exist there. In order to have sensation, something must act on the senses. But such actions are alien to celestial bodies. Moreover, animals experiencing sensation are generable and corruptible, which activities require a vegetative soul that is lacking in celestial bodies.[199]

What about the existence of an intellective soul in the heavens? In his reply Aversa provides useful insights into the elusive distinctions that were made among angels, souls, and forms. He distinguishes three possibilities: (1) the intellective soul is an intelligence which moves the heaven; (2) it is a soul that belongs to the same species as our human soul; or (3) it is of another species altogether.[200]

that opinion, including the Coimbra Jesuits, Hurtado de Mendoza, Arriaga, and Oviedo. Despite the inclusion by Mastrius and Bellutus of Aristotle among those who denied life to the heavens, Aristotle was generally acknowledged to have assumed animated heavens.

196. "Dicendum tamen omnino est formam corporum caelestium non esse animam, corpora caelestia non esse animata, nec esse viventia. Haec est communis Theologorum et Philosophorum sententia." Aversa, *De caelo*, qu. 33, sec. 7, 1627, 110, col. 1.

197. A few scholastic authors, before and after Aversa, similarly sought to show why the three levels of soul were absent from the heavens. For example, the Conimbricenses, *De coelo*, bk. 2, ch. 1, qu. 1, art. 2, 1598, 167; Oviedo, *De caelo*, contro. 1, punc. 1, 1640, 460, col. 2–461, col. 1; and Bonae Spei, comment. 3, *De coelo*, disp. 3, dub. 5, 1652, 11, col. 2–12, col. 1. Some, like Galileo, *De caelo*, qu. 6 (L), 1977, 149–150, only covered the vegetative and sensitive souls.

198. "Tertio specialiter in caelo non esse animam vegetativam quia operationes propriae huius animae sunt vitalis generatio, nutritio, et augmentatio; sed in caelo nulla ex his operationibus invenitur. Non utique unum corpus caeleste generat vitaliter aliud sibi simile eo pacto quo homo hominem et planta plantam." Aversa, *De caelo*, qu. 33, sec. 7, 1627, 110, col. 2.

199. I cite only the final part of the argument: "Tum demum, animalia omnia iuxta naturalem suam constitutionem debent esse generabilia et corruptibilia et indigent operationibus vegetativis. Anima ergo sensitiva corporibus caelestibus omnino repugnat." Ibid., 111, col. 1.

200. "Quinto specialiter in caelo non esse animam intellectivam. Quia vel haec esset eadem

Aversa denies the first possibility by invoking angels, which he seems to conceive as synonymous with intelligences, as celestial movers. An angel differs from a form and a soul. It is a total and complete substance, whereas without matter a form is incomplete. But if an angel is whole and complete, it is also external to the celestial body it moves and is therefore a *motor assistens*, moving its celestial body just as a sailor in a ship (*nauta in navi*) is separate from the ship but nonetheless moves it – presumably, by steering and guiding it.[201] Thus an angel is not like a "composing and informing form" (*forma componens et informans*), in the manner of a soul in a body, but an external entity that moves the body with which it is associated.[202]

Can an intellective soul exist in the heavens that is of the same species as our human souls? Here Aversa invokes arguments that are by now familiar. Without simultaneously existing vegetative and sensitive souls, which occur in human souls but not in celestial orbs, an intellective soul that is like a human soul could not exist in the heavens. For a humanlike soul to exist in the heavens, the latter would require a variety of parts and organs in the manner of a human body, a diversity that simply does not exist in the heavens.[203]

But could another kind of soul exist in the heavens? For example, could an intellective soul exist that lacks a sensitive soul? In replying to this third possibility, Aversa straightaway denies it. First, the soul is appropriately and adequately divided into the three levels which Aristotle distinguished, namely vegetative, sensitive, and rational. A soul that lacks sentiency would thus be anomalous. But perhaps more significantly, Aversa declares that a soul is united to a body not for the sake of the body but for its own sake.[204]

ipsa intelligentia quae caelum movet; vel esset quaedam alia anima eiusdem speciei cum anima nostra; vel cuiusdam alterius speciei." Ibid.

201. Mastrius and Bellutus, *De coelo*, disp. 2, qu. 2, art. 2, 1727, 3:494, col. 1, par. 67, also used the sailor–ship analogy to express the relationship between an external mover and the celestial body it moves, but they argued that there is a greater and more powerful union between an intelligence and the celestial body it moves than between a sailor and the ship he moves. In their view, a ship is not said to be animated because of the sailor's assistance as a mover, but a celestial body may be called animated by virtue of the assistance it receives from its externally located intelligence. Mastrius and Bellutus emphasized that they did not speak of true animation in the sense of an "informing" soul but only of "animation by assistance" (Ad Aristotelem dicimus non loqui de vera animatione, quae fit per substantialem informationem, sed de animatione per assistentiam, quatenus intelligentia perpetuo et necessario in ipsius sententia assistit celo illudque movet unde maior est unio intelligentiae cum celo quam nautae cum navi).

202. "Non primum quia substantia angelica est per se totalis atque completa; forma vero est quaedam pars incompleta nec aliter fieri potest compositum per se unum ex materia et forma. Angelus ergo se habet ad caelum tanquam motor assistens, sicut nauta in navi, non tanquam forma componens et informans, sicut anima in corpore." Aversa, *De caelo*, qu. 33, sec. 7, 1627, 111, col. 1.

203. "Neque anima caeli potest esse eiusdem speciei cum anima nostra. Tum quia haec simul est sensitiva et vegetativa, quale non est caelum. Tum quia requirit tantam membrorum et organorum varietatem talemque corporis humani figuram, quae nullatenus est in caelo." Ibid.

204. Here he cites Thomas Aquinas, *Summa theologiae*, pt. 1, qu. 70, art. 3.

Thus an intellective soul that lacked a sensitive aspect or level would not even require a body for its operation. Without a body, its activity would be purely spiritual, confined to intellection, or understanding.[205]

Despite the condemnation of animated heavens in 1277, most scholastics ignored the religious implications of the issue. Aversa was one of the few who did not, choosing rather to draw out some potential religious consequences for faith. Despite his awareness of Duns Scotus's assumption that reason could not demonstrate that the heavens were lifeless and Thomas Aquinas's argument that the issue of celestial animation did not pertain to faith,[206] Aversa believed that the nonanimation of the celestial bodies could be demonstrated by reason and was upheld by faith.[207] As if to ridicule the idea of living celestial bodies, Aversa infers theologically absurd consequences from such an assumption. For example, if celestial bodies are living things with intellective souls, they would possess free will and could thus do either good or evil; they would therefore be candidates for merits and demerits, rewards and punishments, from which it followed that they could be blessed with the saints or damned with the impious. Celestial bodies that sin would be sent to hell, just as were the wicked angels. By contrast, celestial bodies that are meritorious should be adored, just as are the bodies of saints on earth.[208]

Although Franciscus de Oviedo was not mentioned by Mastrius and Bellutus (the three published their relevant works in the same year, 1640), he was also one of those "scholastic doctors" and "Peripatetics" who denied life to the heavens. Oviedo mustered three philosophical arguments against the claim for celestial life (*De caelo*, contro. 1, punc. 1, 1640, 460, col. 2, par. 5), the first two of which differ from those advanced by Aversa. In the first, he observes that the heavens are so vast in magnitude that they could not consist of living forms, because life requires the nearness of its

205. Although Aversa omits mention of it, the same argument would apply to an intellective soul devoid of a vegetative level.

206. "Unde immerito Scotus in 2. d. 14, q. 1 in fine, dixit solum credi caelum esse inanimatum; non vero ratione probari quia alioqui nulla apparet conditio in caelis repugnans animae. Ex adverso D. Thomas, 2 *Contra Gentiles*, cap. 70, indicavit non pertinere ad doctrinam fidei si caelum dicatur animatum anima intellectiva vel inanimatum." Aversa, *De caelo*, qu. 33, sec. 7, 1627, 112, col. 1. Bonae Spei also cites Scotus and Thomas (comment. 3, *De coelo*, disp. 3, dub. 5, 1652, 11, cols. 1–2).

207. "Ex his patet hanc veritatem quod corpora caelestia non sint animata et ratione et fide probari." Aversa, ibid. In support of his own opinion that faith upholds an unanimated heavens, Aversa mentions Saint Bonaventure, Capreolus, Hervaeus Natalis, and the Coimbra Jesuits.

208. Aversa, ibid., 111, col. 2–112, col. 1. The Coimbra Jesuits had earlier presented similar arguments (*De coelo*, bk. 2, ch. 1, qu. 1, art. 2, 1598, 167–168), as did Bonae Spei (comment. 3, *De coelo*, disp. 3, dub. 5, 1652, 12, col. 1) later in his attack against the existence of a rational soul in the heavens. See also, Major, who, after explaining that John Damascene held that the celestial bodies are inanimate and insensible, declares that "If the heavens were animated, then similarly they could be just and blessed and consequently to be adored [or worshiped], the opposite of which is said in Scriptures" ([*Sentences*, bk. 2, dist. 12], 1519b, 65v, col. 1).

parts, so that parts separated by some distance can support and sustain each other.[209] As his second argument, Oviedo contrasts the homogeneity and organization of the parts of the heavens with the lack of organization and homogeneity in living things, observing that a lack of organization (*deorganizatio*) and homogeneity are essential for the proper functioning and operation of living things.[210] In the third argument, Oviedo emphasizes that celestial bodies lack the three basic levels of life: the vegetative, sensitive, and rational. But, as was not uncommon, Oviedo, following suggestions made by Aristotle in *De anima* (bk. 2, ch. 3),[211] also considered a fourth level of soul, namely a level associated with a body's motion.[212] Although motion is an aspect of life, and the celestial bodies do move, one cannot infer from this alone that celestial bodies are alive. Indeed, Oviedo insists that "even if they were self-moved, this would not confer life upon them, because the locomotive power is not vital, nor is progressive motion in living things formally a vital action per se."[213]

209. "Primam peto, ex ipsorum vastissima magnitudine, quae non recte componitur cum viventium formis quae partium approximationem exigunt, ut vicissim foveri possint et haec in aliam influere." Oviedo, *De caelo*, contro. 1, punc. 1, 1640, 460, col. 2.

210. "Secundam depromo ex ipsorum partibus, quas homogeneas esse cernimus nulla deorganizatione elaboratas. Deorganizatio autem necessario est requisita in viventium formis ad suarum operationum functiones." Ibid. The Coimbra Jesuits had earlier used a similar argument (Conimbricenses, *De coelo*, bk. 2, ch. 1, qu. 1, art. 2, 1598, 167), as did Compton-Carleton somewhat later (*De coelo*, disp. 1, sec. 1, 1649, 397, col. 2). Although making the same point, Benedictus Hesse (*Physics*, bk. 8, qu. 17, 1984, 735) used contrary language. For him "every animated thing is organized; the heaven is not organized; therefore it is not alive." Thus where Oviedo says animated things are "deorganized" (*deorganizatio*), Benedictus Hesse says they are "organized" (*organizatio*). Mastrius and Bellutus (*De coelo*, disp. 2, qu. 2, art. 2, 1727, 3:495, col. 1, par. 73) adopted an approach contrary to Oviedo's. They argued that the heavens consist of parts that are heterogeneous – not homogeneous – but that this provides no warrant to infer that the heavens form an organic body. The heterogeneity of the heavens derives from the differences in species among the various parts of the heavens (ibid., art. 4, 498, col. 1–499, col. 1, where Mastrius and Bellutus consider "In what manner the celestial bodies differ among themselves" [Quo pacto coelestia corpora inter se differans]).

211. In *De anima* 2.3.414a.30–31, Aristotle enumerates the various psychic powers that living things possess when he declares that "those we have mentioned are the nutritive, the appetitive, the sensory, the locomotive, and the power of thinking" ([Smith], 1984). Thus was locomotion projected into medieval discussions about the possibility of animated celestial bodies.

212. John of Jandun, *De coelo*, bk. 2, qu. 4, 1552, 25r, col. 2, distinguished the same four levels of soul when he declared that if a heaven is animated with an inhering soul, the latter would be "animated either with a sensitive soul, or a vegetative [soul], or an intellective [soul], or a motive soul, beyond which four subdivisions there are no further souls" (Si celum esset animatum primo modo [that is, as an inherent or informing soul], vel esset animatum anima sensitiva, vel vegetativa, vel intellectiva, vel motiva, per sufficientem divisionem quia non sunt plures animae, ut patet secundo *De anima*).

213. "Sit tertia ratio nullum reperiri rationis vestigium suadens haec corpora vita aliqua vegetativa, sensitiva, aut rationali gaudere, quod sic probo. Si aliquod fundamentum vitae in his corporibus posset reperiri esset motus localis. . . . Non autem hanc inferre constabit ex infra dicendis ubi probabo caelos non moveri a se; deinde etiamsi a se moverentur, non ex eo virtute vitali essent donandi quia virtus locomotiva vitalis non est, neque motus progressivus adhuc in viventibus per se formaliter est actio vitalis." Oviedo, *De caelo*, contro. 1, punc. 1, 460, col. 2–461, col. 1.

Indeed, Oviedo argues that however important locomotion is, or any other power, philosophers have not considered motion as an intrinsic aspect of life but have always understood life to consist of the vegetative soul, "which does not differ formally from a principle of increasing by ingesting something"; the sensitive soul, "which consists in the principle of eliciting sensations"; and the rational soul, "which is formally the principle of understanding." Whatever might be the importance of locomotion, it does not rank with these three absolutely essential principles of life.[214]

But these three levels are not observed in celestial bodies. The latter do not become lean from lack of food or become fat from an excess of it. A nutritive power would prove unbeneficial to the heavens, because it is only appropriate to corruptible things. Nor is there any need for a sensitive power, because the heavens do not exhibit grief or joy. And, finally, the heavens do not possess subtle reason by means of which they could compose complex things or resolve a serious difficulty.[215]

If Oviedo was prepared to reject the existence of the three levels of soul in the heavens and also to deny that locomotion of celestial bodies was a sign of life, Franciscus Bonae Spei (*De coelo*, comment. 3, disp. 3, dub. 5, 1652, 12, cols. 1–2) chose to believe that although the three traditional aspects of soul were absent from the heavenly bodies, each of the latter possessed an internal "locomotive soul" (*anima locomotiva*) which enabled it to be self-moving.[216] After all, Aristotle had argued that things that are self-moving are alive. In a less than clear and cogent argument, Bonae Spei insisted that "things that could move themselves locally above and below [that is, upward and downward] in their natural centers [or with respect to their natural centers] are alive, or animated."[217] Since the heavens are also entities that are capable of moving themselves above and below and are therefore self-moving, they must possess a locomotive soul and also be alive.[218]

To reinforce his position, Bonae Spei likens the celestial bodies and their

214. "Neque Philosophi vitam aliquam cognoverunt praeter vegetativam, quae non differt formaliter a principio se augendi per intus sumptionem; et sensitivam, quae in principio sensationes eliciendi consistit; et rationalem, quae dicit formaliter intelligendi principium. Ac proinde omne illud cui non convenit aliquod ex his tribus principiis omnino vitae omnis expers est, quantumvis potentia locomotiva, aliave quacumque virtute polleat." Ibid, 461, col. 1.

215. Ibid. Although the content of the arguments was somewhat different, the Conimbricenses, *De coelo*, bk. 2, ch. 1, qu. 1, art. 2, 1598, 167, also showed that a soul could not be joined to a material heaven with respect to its vegetative, sentient, and rational levels nor because of local motion.

216. "Dico . . . verisimilius est secundum philosophiam naturalem praecise coelos esse animatos anima locomotiva distincta a vegetativa, sensitiva, et rationali." Bonae Spei, comment. 3, *De coelo*, disp. 3, dub. 5, 1652, 12, col. 1.

217. "Illa quae in centro suo naturali movent se localiter supra et infra sunt viventia, sive animata; atqui coeli sunt tales, ergo, etc." Ibid.

218. Bonae Spei does not explain the sense in which a celestial body moves "above and below" or upward and downward. Perhaps he was thinking of the variation in distance of the planets, which was sometimes depicted as resulting from an up and down motion with respect to their centers.

motions to those of a bird. Surely no one would deny life and soul to a bird, which we judge to fly by itself because we cannot perceive an external mover.[219] Similarly, we cannot perceive external movers for the heavenly bodies and must assume they are self-moved. Moreover, one must either concede that a progressive motion caused by an internal principle is not an action stemming from a living thing or assert that it is. If it is not, then, a fortiori, the sensations and "vegetations" (that is, growth and nourishment) of ordinary bodies do not indicate life in those bodies. But if one concedes that the local motion of the heavens arises from an internal principle, then it must follow that they are also alive and have a soul, a "locomotive soul."[220] Although Bonae Spei's surprising opinion found little support in scholastic circles, it is noteworthy because it assigned a degree of life to the celestial bodies by virtue of their motion, which was caused by an internal principle. The customary approach was to attribute some sense of life to the celestial bodies through the action of external movers, namely the intelligences that were usually alleged to move them. Although Thomas Aquinas was ambivalent, it was not unusual to assume that this was also his opinion. Indeed, it also seems to have been the opinion of John of Jandun, who denied the existence of vegetative, sensitive, and intellective souls in the celestial bodies but allowed that a celestial body could be alive by virtue of an external mover.[221]

A strong advocate of the Thomistic view was Bartholomew Amicus, who informs us that many philosophers and moderns, especially Peripatetics, "say that the heavens are animated, for although they deny that they are animated by an informing soul, they do not absolutely deny that a thing could be called animated by means of an assisting intelligence."[222] Thus Amicus was prepared to argue that "the aggregate of an orb and an intelligence is, according to Aristotle and truth, an animated [or living] body."[223]

219. "Nec certe ullus est qui aviculae vitam et animam negaret si in ipsa solum volatum aliunde certus quod ab extrinseco non moveretur deprehenderet." Bonae Spei, comment. 3, *De coelo*, disp. 3, dub. 5, 1652, 12, col. 1.

220. "Vel dicendum motum progressivum ab intrinseco qua talem non esse actum vitalem, quod dici non potest alioqui idem dici a fortiori deberet de sensationibus, vegetationibus, etc.; vel dicendum coelorum motum localem ab intrinseco pariter esse actum vitalem, ac per consequens animam in ipsis praesupponere." Ibid., col. 2.

221. "Tertio dicendum quod coelum est animatum anima movente secundo modo animati si accipiatur cum motore." John of Jandun, *De coelo*, bk. 2, qu. 4, 1552, 25r, col. 2. Jandun had defined the "second way" that a celestial body could be animated in terms of an external mover ("Alio modo animatum dicitur anima non inhaerente corpori nec constituta per subiectum, quae est principium dans sibi esse, sed solum animatum anima non inhaerente corpori nec constituta per subjectum, quae est principium motus et coniungitur ei secundum motum quia est principium immediatum et determinatum motus"). Ibid.

222. "Communiter tam philosophi quam recentiores, praecipue Peripatetici, dicunt coelum esse animatum, nam quod negant esse animatum per animam informantem, non negant (altered from "negat") simpliciter dici animatum per intelligentiam assistentem." Amicus, *De caelo*, tract. 4, qu. 4, dubit. 2, art. 3, 1626, 170, col. 1.

223. "Prima conclusio: coelum ut dicit aggregatum ex orbe et intelligentia est secundum Aristotelem et veritatem corpus animatum." Ibid.

It is not likely, however, that Amicus thought of celestial bodies as fully alive. His ideas were probably more in keeping with those of Dominicus de Flandria, who, in the fifteenth century, adopted the thoughts and very nearly the words of Thomas Aquinas when de Flandria declared that "celestial bodies are not animated as are plants and animals, but equivocally. Thus among those who assume that they are animated and those who say they are not, little or no substantive difference is found, except in the words."[224] That Amicus probably held similar thoughts is suggested by the fact that Dominicus assumed, as did Amicus, that "the heavens are animated with an assisting soul and are not animated with an informing soul which would be an essential part of the whole celestial body." By "assisting soul" (*anima assistente*), or intelligence, Dominicus did not mean a soul distinguishable into vegetative, animal, and rational aspects, as Aristotle defined soul in the second book of *De anima*, but only intended a spiritual substance that was distinct from, and external to, the body.[225] Indeed, Dominicus explains that "the spiritual substance is united to the celestial body as a mover to a mobile."[226]

From the thirteenth and into the seventeenth century, the question about the animation of the celestial bodies drew at least five distinguishable responses. One, enunciated by Peter Aureoli in the fourteenth century, accorded the highest level of life to celestial bodies by viewing the relationship between an intelligence and its celestial body as a unity and oneness in the manner of matter and form in terrestrial bodies. A second opinion, held by both Benedictus Hesse and John Major, denied both an informed intellective soul and life to celestial bodies.[227] Neither, however, reveals whether he thought an external intelligence was associated with each celestial body as an "assisting mover."

For the most part, however, scholastics across the centuries opted for one of the three remaining interpretations, although some authors were sufficiently ambiguous as to defy easy classification. There were those who – and we may count this as a third opinion – denied a genuine informing soul to the heavens and assumed instead an external assisting intelligence that caused the circular motion of each celestial body. They were, however,

224. "Sic igitur patet quod corpora celestia non sunt animata sicut plante et animalia, sed equivoce. Unde inter ponentes ea esse animata et ponentes esse non animata parva vel nulla differentia reperitur in re, sed in voce tantum." De Flandria, *Metaphysics*, bk. 12, qu. 7, 1523, sig. R5v, col. 2.

225. "Ex predictis ergo colligitur quod celum est animatum anima assistente et non est animatum anima informante, que sit pars essentialis totius corporis celestis. Et quando dicitur quod corpus celeste est animatum anima assistente non accipitur ibi anima secundum quod diffinitur a Philosopho in secundo *De anima*. Sed accipitur ibi anima pro substantia spirituali." Ibid.

226. "Et substantia spiritualis unitur corpori celesti tanquam motor mobili." Ibid.

227. See Benedictus Hesse, *Physics*, bk. 8, qu. 17, 1984, 735–737, and Major, *Sentences*, bk. 2, dist. 12, 1519b, 65v, col. 1. Major does not assign a separate question to the animation of the heavens but considers it as the fifth conclusion in a discussion of whether the heavens consist of matter and form. There he declares that "the heavens do not have an intelligence or intellectual nature as their form."

ambivalent about the attribution of any level of life to the heavenly bodies. Sharing this opinion were, as we have seen, Thomas Aquinas and Dominicus de Flandria. Within a fourth category, we may include those who were more positive and, because of the external intelligence associated with each celestial body as the cause of its motion, allowed a certain level of animation to all heavenly bodies but did not consider them fully living things. Siger of Brabant, Boethius of Dacia, John of Jandun, Johannes de Magistris, Bartholomew Amicus, and Mastrius and Bellutus seem to belong to this group.[228] As a fifth opinion, numerous authors, such as Michael Scot, Richard of Middleton, Nicole Oresme, Galileo, the Coimbra Jesuits, Raphael Aversa, Roderigo de Arriaga, Francisco de Oviedo, and Thomas Compton-Carleton, assumed that each celestial body was moved by an external assisting intelligence, or angel, but they saw no reason to attribute any level of life to the celestial body itself.[229] Indeed, Arriaga argued that even if celestial orbs or bodies were moved intrinsically, this was insufficient to confer life upon them.[230]

4. Heavens not really animated

On the basis of all that has been said in this chapter, we may plausibly conclude that the heavens were not gradually deanimated during the Middle Ages, largely because they were never really animated in the first instance.[231]

228. Because of Thomas's ambivalence, many placed him in this group, although others assigned him to the following, or fifth, category.
229. For Richard of Middleton, Oresme, and Oviedo, see the earlier part of this section. Although Michael Scot (*Sphere*, lec. 5, 1949, 289) does not mention intelligences or angels, he holds that the mover of a celestial body is external to the body and is the entity that has a desire for motion. The celestial body itself is wholly inanimate. Galileo asserts (*De caelo*, qu. 6 [L], 1977, 154) that "apart from intelligences no other souls are constituent in the heavens" and that these intelligences are external "assisting forms" (155). In Galileo's lengthy discussion of the possible animation of the heavens, there is no indication that he assigned any level of life to the heavens. Arriaga also assumed that the heavens are moved by an assisting intelligence, or angel ("Respondeo caelos non a se sed ab extrinseco, ab intelligentia assistente, id est, ab angelo moveri"). Arriaga, disp. 1, sec. 2, 1632, 498, col. 2. That Oviedo, *De caelo*, contro. 1, punc. 1, 1640, 462, col. 1, considered angels as celestial movers is evident by his statement that "De facto caelos ab Angelis moveri mihi persuadet authoritas Patrum et Scholasticorum, qui caelestes motus Angelis attribuunt." Compton-Carleton, who believed that intelligences moved the heavens (*De coelo*, disp. 4, sec. 3, 1649, 409, col. 2), denied any trace of life to the heavens (ibid., disp. 1, sec. 1, 397, col. 2): "Nullum enim vitae vestigium in coelis deprehendimus. Unde gratis et sine ullo fundamento vitam iis quis tribuerit."
230. "Non enim sufficit ad vitam virtus producendi aliquid ab intrinseco." Arriaga, ibid.
231. Dales, 1980, leaves the impression that there was a gradual falling away from the conception of an animated heaven to one that is deanimated but that the process of celestial deanimation was incomplete during the Middle Ages. Dales does not make the distinction between fully living celestial bodies and the weak form of life associated with an external intelligence or angel that has been emphasized here. Only the latter played a significant role in medieval deliberations. Moreover, the special level of life associated with an intelligence, which was not really comparable to life in bodies informed by a soul with vegetative, sentient, and rational levels, was not usually transferred to the celestial body that it moved.

Almost without exception, scholastic natural philosophers between the thirteenth and the seventeenth century did not attribute life to celestial bodies themselves, although they probably all assumed that the intelligences that moved those bodies were alive. When natural philosophers asked whether the heavens were alive, they were really inquiring whether an external, spiritual intelligence or angel, which had neither vegetative, sentient, nor rational levels of activity but was itself alive in some sense, could somehow confer life upon the physical, celestial body that it moved. A few thought it could, but most did not.

Our attention in the fifth section of this chapter was focused on the ways in which life might have been attributed to the celestial orbs and intelligences associated with them. Our concern was for life as an attribute or property of heavenly bodies and beings. In Section II of Chapter 18, we shall be concerned once more with orbs, intelligences, and angels. Although we shall again have occasion to deal briefly with the animation of the heavens in Chapter 18, Section II (especially in II.4 and II.5.a), the primary emphasis there will be on intelligences and angels as possible motive powers of the orbs and planets.

18

On celestial motions and their causes

In its various manifestations, celestial motion was the most frequently discussed theme in medieval cosmology. A glance at Section XIII in the "Catalog of Questions" (Appendix I) reveals that at least 73 questions about the heavens were devoted to the nature and causes of the motions of the orbs, a number that far exceeds all other categories of questions.[1] The questions ranged considerably, embracing the prime mover, the naturalness of circular motion and its regularity and uniformity; whether distinct and separate intelligences or internal forms or natures move the planets; whether planets are moved with contrary motions and whether the motions of planets and orbs weaken over time; and many others. As the "Catalog" indicates, scholastics emphasized certain topics over others, and it is on these that we shall focus here.

The treatment of celestial motion may be conveniently divided into two broad categories, the kinematics and dynamics of motion. In the former, motion is viewed as a phenomenon in space and time, without reference to the forces or entities that may produce and sustain it. By contrast, the dynamics of motion focuses on causes and therefore considers anything relevant to the production and preservation of a body's motion. Although kinematic and dynamic aspects of motion are not mutually exclusive, we shall adhere to this convenient division as much as is feasible.[2]

I. The kinematics of celestial motion

1. Uniformity and regularity

a. Definitions

In *De caelo* (bk. 2, ch. 6), Aristotle argues that the primary motion of the first, or outermost, heaven – the medieval *primum mobile* – is uniform and

1. Only the number of questions about the sublunar region approximates it.
2. The distinction between kinematic and dynamic motion was explicitly made at Merton College, Oxford, in the fourteenth century (see Clagett, 1959, 205–209).

circular.[3] He does not, however, extend the claim for uniformity to the inferior planets, because the resultant motion of each of them derives from a combination of other motions. Thus Aristotle acknowledged the celestial appearances by recognizing that, as Averroës would put it many centuries later, "the orbs that are under it [that is, under the starry orb, or outermost heaven] are sometimes quicker and sometimes slower, . . . and their motions appear diverse," because "there is not one motion but several, which are united to move the planet."[4] Medieval natural philosophers sought to differentiate between the seemingly uniform and the seemingly irregular types of motion and in the process even distinguished between "uniform" and "regular" motion. Much effort was expended on defining terms such as *uniformitas* and *difformitas*; *regularitas* and *irregularitas*; and *simplex* and *compositus*.

Although, up to this point, I may have used the terms "uniform" and "regular" synonymously, some natural philosophers distinguished sharply between them.[5] Albert of Saxony insisted that it is not absurd that certain motions could be uniform but not regular and that some might be regular but not uniform.[6] Albert measured uniformity of motion with respect to the parts of a body and regularity of motion with respect to time.[7] Thus "a motion is said to be uniform by which one part of some mobile moves just as quickly as another part [of that same mobile]." For example, although a stone descends with an accelerated motion, every part of that stone is moving with the same speed at any particular point in its descent.[8] Opposed to uniform motion is "difform" motion, where one part moves more

3. In Michael Scot's thirteenth-century translation, the uniform motion of the first heaven is described as "equalis sine diversitate."
4. "Et quia in orbibus, qui sunt sub isto, aliquando est velocitas et aliquando tarditas . . . quoniam . . . non est unus motus, sed plures qui congregantur ad movendum stellam." Averroës [*De caelo*, bk. 2, comment. 35], 1562–1574, 5:118, col. 2.
5. Versor [*De celo*, bk. 2, qu. 8], 1493, 22r, col. 1, explains that "some assume a difference between a uniform motion and a regular motion," thus implying that some did not make this distinction. Pierre d'Ailly was one who did not, but, as we shall see, Buridan and Albert of Saxony did.
6. "Ulterius sciendum est quod non est inconveniens aliquem motum esse uniformem et non esse regularem. . . . Similiter non est inconveniens aliquem motum esse regularem et tamen non esse uniformem." Albert of Saxony [*De celo*, bk. 2, qu. 13], 1518, 110r, col. 2. Thomas Bricot said the same thing ([*De celo*, bk. 2], 1486, 18r, col. 1).
7. Nicole Oresme [*De celo*, bk. 2, qu. 8], 1965, 580–584, assumed the same distinction. By contrast, Pierre d'Ailly, *14 Questions*, qu. 3, 1531, 150v, ignored it or found it unacceptable. He chose to take the terms "regular" and "uniform" as synonymous, as evidenced by such expressions as "uniform or regular" (uniformis quo ad tempus seu regularis) or "regular or uniform" (regularis seu uniformis).
8. "Sciendum est quod differentia est inter motum regularem et uniformem. Nam uniformitas motus attenditur quantum ad partes mobilis, ita quod ille motus dicitur uniformis quo movetur aliquod mobile cuius una pars movetur ita velociter sicut alia, sicut si lapis aliquis descenderet non obstante quod ille motus in fine esset velocior quam in principio." Albert of Saxony [*De celo*, bk. 2, qu. 13], 1518, 110r, col. 1. Versor, *De celo*, bk. 2, qu. 8, 1493, 22r, col. 1, gives essentially the same definition. D'Ailly, ibid., gives a similar definition, but also adds a second definition of uniform motion with respect to time, which is the counterpart of the definition for regularity just given.

quickly than another, as is the case with a revolving wheel. The parts nearer the axle are not moved as quickly as those nearer the circumference, although they all complete their circulations in the same time.[9] Thus a sphere – and therefore a celestial sphere – does not move with uniform motion.

Regularity of motion, by contrast, is measured only with respect to time (*ex parte temporis*). The body moves with the same speed in every part of time. It follows that irregularity of motion occurs when a body moves more quickly at one time and more slowly in another.[10] In the seventeenth century, Bartholomew Amicus ([*De caelo*, tract. 5, qu. 6, dub. 6, art. 1], 1626, 314, col. 1) adopted not only Albert's definitions of uniform and regular motion but also his examples.

From these definitions, Amicus concluded that "uniformity can be found without regularity, as was said about a falling stone; and regularity without uniformity, as is obvious from the motion of a wheel."[11] As Albert of Saxony explained, at every point of the descent of an accelerated falling body, all parts of it are moving with equal speed and are thus moving uniformly. But they are not moving regularly, because the velocity of the body alters by virtue of its acceleration: it will be greater at the end of its descent than at the beginning.[12]

Additional regularities and irregularities were also distinguished. Buridan, for example, described irregularities "on the part of the mobile" (*ex parte*

9. "Motus autem dicitur difformis cuius una pars movetur velocius et alia tardius, sicut esset motus rote: partes enim eius circa axem non movetur ita velociter sicut partes circa circumferentiam, licet bene ille partes eque velociter circuant." Albert of Saxony, ibid. Oresme, *De celo*, bk. 2, qu. 8, 1965, 580, gives essentially the same definitions, as does Versor, *De celo*, bk. 2, qu. 8, 1493, 22r, col. 1.
10. "Regularitas autem motus attenditur ex parte temporis ita quod motus ille dicitur regularis quando ipsum mobile movetur eque velociter in una parte temporis sicut in alia. Sed ille motus dicitur irregularis quo movetur aliquod mobile quod in una parte movetur velocius et in alia tardius." Albert of Saxony, ibid., bk. 2, qu. 13, 1518, 110, col. 2. Buridan [*De caelo*, bk. 2, qu. 11], 1942, 173, takes up this case as the third kind of distinguishable irregular motion. It is irregularity "from the standpoint of velocity" (*ex parte velocitatis*); for Buridan's other two kinds of irregularities (*ex parte mobilis* and *ex parte spatii*, see the paragraph following the next one). In the fifteenth century, Bricot presented the same three distinctions of irregular motion (Bricot, *De celo*, bk. 2, 1486, 18r, col. 1). Buridan was probably his ultimate source. Oresme gives the same definition as Albert (Oresme, *De celo*, bk. 2, qu. 8, 1965, 582).
11. "Secundo notandum: uniformitatem reperiri posse sine regularitate, ut dictum est de lapidis descensu; et regularitatem sine uniformitate, ut patet motu rotae." Amicus [*De caelo*, tract. 5, qu. 6, dubit. 6, art. 1], 1626, 314, col. 1. Amicus also holds (ibid.) that both uniformity and regularity can be found in the same body, as, for example, in a heavy body moving downward through a medium that resists it equally at every point from beginning to end. Under these circumstances, the body will traverse equal distances in equal parts of time (Et denique posset utranque reperiri in aliquo motu, ut in gravi descendente per medium a principio ad finem aequaliter resistens. Unde illud grave in aequalibus partibus temporis aequalia pertranseat spatia).
12. "Ulterius sciendum est quod non est inconveniens aliquem motum esse uniformem et non esse regularem. Patet de motu gravis deorsum in medio uniformi quod movetur uniformiter quia una eius pars movetur ita velociter sicut alia et tamen non movetur regulariter quia movetur in fine velocius quam in principio." Albert of Saxony, *De celo*, bk. 2, qu. 13, 1518, 110r, col. 2. Oresme, *De celo*, bk. 2, qu. 8, 1965, 582, gives the same example.

mobilis) and "on the part of space" (*ex parte spatii*), irregularities that Amicus described as "external." A mobile is irregular when its shape is uneven or irregular. But the heavens – indeed, each heaven or orb – is spherical and thus as regular as is possible.[13] Albert of Saxony observes further that the celestial mobiles – that is, the planets and orbs – are also immutable, and therefore no irregularity can arise from them.[14] Although the assumption of eccentric orbs in the heavens seems to pose a problem, since they are nonuniform with respect to the center of the world, Buridan avoids this potential dilemma by resorting to the concept of a "total orb" (*totalis sphaera*). Eccentrics are contained within a total orb, whose outer and inner surfaces are concentric. The total orb is thus regular, although it contains eccentric "partial orbs" that are irregular with respect to the center of the world.[15]

Irregularity with respect to space would occur if the body were moved through a twisting or angular space. But this cannot apply to celestial motion, because "celestial bodies are not moved through any spaces distinct from themselves."[16] Here Buridan means that every celestial orb always rotates in the same place and never moves through a space that lies outside itself. Hence there can be no irregularity with respect to celestial space, because no such space exists.

Not all of these definitions were employed, but two that have not yet been mentioned played an important role: the definitions of simple and compound motion. According to Buridan, a circular simple motion (*motus circularis simplex*) is that of a single, continuous mobile or orb that is moved by a single mover around one set of poles, with motion around any other poles excluded. A compound (*compositum*) motion applies to a single mobile within which two or more simple motions are distinguishable. The motion of that single mobile is the resultant of a plurality of simple motions, each of which moves around a different pole and has its own mover. In compound motion, then, we find a plurality of simple motions, and therefore a plurality of orbs, poles, and movers.[17] The observable planetary motions

13. Buridan, *De caelo*, bk. 2, qu. 11, 1942, 173. Bricot repeats the same description (*De celo*, bk. 2, 1486, 18r, col. 1), as does Amicus, *De caelo*, tract. 5, qu. 6, dubit. 6, art. 1, 1626, 314, col. 2.
14. "Irregularitas motus non potest venire nisi propter mutationem mobilis aut motoris. Modo mobile puta celum est inalterabile." Albert of Saxony, *De celo*, bk. 2, qu. 13, 1518, 110v, col. 1.
15. Buridan, *De caelo*, bk. 2, qu. 11, 1942, 173. Buridan does not use the term "orbis partialis" or an equivalent, but it is what he had in mind. For a discussion of total and partial orbs, see Chapter 13, Section II.4 (on d'Ailly).
16. Buridan, ibid. Bricot gives the same definition (*De celo*, bk. 2, 1486, 18r, col. 1). Amicus mentions "irregularity with respect to space" (*per spatium irregulare*), which is a space, or path, that is devious or twisting. But he dismisses such talk, because the heavens move with circular motion (Amicus, *De caelo*, tract. 5, qu. 6, dubit. 6, art. 1, 1626, 314, col. 2).
17. "Et tunc solet distingui quod in caelo motus localis potest dici simplex vel compositus; et vocatur motus circularis simplex, quia est unius mobilis continui et super eosdem polos et ab uno motore, circumscripto omni motu praterquam super illos polos. Sed motus

are the result of compound motion, whereas simple motion is unobservable and is therefore arrived at by reason.[18]

b. Application of definitions

Conclusions about celestial uniformity and regularity were based on the application of most of these definitions to celestial motions. The daily motion of the heavens taken as a whole is not uniform, because different parts move with different speeds: the parts of the heavens around the celestial equator move more quickly than do the parts around the poles of the world. This occurs because in equal times, the lineal distances traversed by parts near a larger circumference are much greater than the distances described by the parts near the poles or near a smaller circumference.[19] Despite the lack of a uniform daily motion, Bartholomew Amicus emphasized that the daily motion was regular (*regularis*), because all of its parts described equal angles around the axis of the world.[20]

The motion of the *primum mobile*, or outermost movable sphere, sometimes identified with the sphere of the fixed stars, was characterized as regular because no change in velocity occurs.[21] Every change in velocity, whether from an increase or decrease, that is, from intension or remission, to use medieval parlance, must occur at the beginning of a motion (as in violent motion), at the end of a motion (as in natural motion), or in the middle of a motion (as in projectile motion or the motion of animals, which often move laterally, as birds do). But in the circular motion of the *primum mobile*, there is no beginning, middle, or end and therefore no change in

compositus dicitur congregatio in eodem mobili plurium talium motuum simplicium super diversos polos a diversis motoribus." Buridan, ibid., 174. With a few minor variations, Albert of Saxony offers a virtually verbatim version of Buridan's account (see Albert of Saxony, *De celo*, bk. 2, qu. 13, 1518, 110r, col. 2). It is more likely that Buridan was Albert's source than vice versa. Bricot presents similar definitions (*De celo*, bk. 2, 1486, 18r, col. 2). In the seventeenth century, Amicus adopted the same two definitions, probably deriving them from Albert of Saxony's *De celo*, which he cites and with which he seems to have been familiar.

18. "Et ille motus compositus in ipsis planetis est ille qui nobis apparet; simplices autem non apparent nobis distincte ad invicem nisi per ratiocinationem." Buridan, ibid. Albert of Saxony, ibid., has the same passage almost verbatim.
19. "Sit prima conclusio: ista motus ipsius celi non est uniformis probatur ex eo quod non omnes partes celi moventur eque velociter. Partes enim celi circa equinoctialem moventur velocius quam partes celi circa polos ex eo: quod in equalibus partibus temporis maius spatium lineale describunt, puta maiorem circumferentiam circa axem mundi." Albert of Saxony, ibid., 110v, col. 1.
20. "Etiam si motus caeli non sit uniformis, adhuc tamen est regularis, nam partes illius motus semper in aequalibus temporibus describunt aequales angulos circa axem mundi." Amicus, *De caelo*, tract. 5, qu. 6, dubit. 6, art. 3, 1626, 316, cols. 1–2.
21. Oresme declares (*Le Livre du ciel*, bk. 2, ch. 13, 1968, 413) that "the highest heaven has simple regular motion, but each heaven beneath this primary one moves with compound motion, combining several simple regular movements." In modern physics and cosmology, a body moving with circular motion would be assumed to change its direction, and therefore its velocity, at every instant. For scholastics, however, velocity in rectilinear or circular motion was straightforwardly equivalent to speed.

velocity.[22] Moreover, if irregularity did occur in the *primum mobile*, it would have to arise either from the body of the *primum mobile* itself or from its mover. Because it is simple, ungenerated (although created), and incorruptible, the *primum mobile* cannot change and become irregular. It follows that since a mover is more noble and excellent than that which it moves, the mover of the *primum mobile*, whatever it may be, must also be absolutely immutable. Hence the *primum mobile* is not susceptible to irregular motion.[23]

The definition of simple motion makes it obvious that no simple celestial motion is irregular; or, to put it positively, every simple motion is regular. Thus no part of the heavens which moves with a simple motion – and this encompasses all individual orbs – is moved faster at one time than at another.[24]

The motion of individual planets is irregular, because the position of any one of them at any time is the result of two or more simple motions around different poles, which produces a compound motion. As Oresme expressed it: "A motion composed of many motions can be irregular, as could be demonstrated geometrically, so that if something is moved in two motions each of which is regular, and these take place around diverse centers or on diverse poles, it is necessary that the [resultant motion] take place irregularly, as is clear in the theory of the sun."[25] The planets move with more than two simple motions, and the resultant compound motion causes the planets to move sometimes more quickly and sometimes more slowly. Indeed, unlike the Sun, the other planetary motions are sometimes retrograde, sometimes stationary, and sometimes direct.[26]

22. See Versor, *De celo*, bk. 2, qu. 8, 1493, 22r, cols. 1–2, and Amicus, *De caelo*, tract. 5, qu. 6, dubit. 6, art. 3, 1626, 315, col. 1. Buridan, *De caelo*, bk. 2, qu. 11, 1942, 174, applies this reasoning to all of the planets as well, not just the *primum mobile*; Albert of Saxony, *De celo*, bk. 2, qu. 13, 1518, 110v, col. 1, does the same.
23. In a second argument for the regularity of the first heaven, Versor (ibid., col. 2) declares: "Si in illo motu esset irregularitas vel illa irregularitas proveniret ex parte mobilis vel motoris, vel utriusque. Non mobilis quia primum celum est simplex, ingenitum, incorruptibile et omnino intransmutabile et ergo semper est in eadem dispositione ad susceptionem motus et ad obediendum motori. Et cum motor sit prestantior mobili, sequitur quod motor primi mobilis, scilicet celi, est simplex et omnino intransmutabilis. Ergo semper est in eadem dispositione et virtute ad movendum. Ergo ex parte sui non est irregularitas in motu celi." Versor offers four arguments in favor of the regularity of the *primum mobile*.
24. As Buridan expressed it, "Dicendum est quod nullus motus simplex in caelo est isto modo irregularis, ita quod nec caelum nec aliquod astrum nec aliqua pars caeli movetur velocius uno tempore quam alio motu aliquo simplici, scilicet circumscriptis aliis motibus." Buridan, *De caelo*, bk. 2, qu. 11, 1942, 174. See also Amicus, *De caelo*, tract. 5, qu. 6, dubit. 6, art. 3, 1626, 316, col. 1.
25. Oresme, De celo, bk. 2, qu. 8, 1965, 584 (English), 583 (Latin). In his later French commentary, *Le Livre du ciel*, bk. 2, ch. 13, 1968, 413, Oresme gives a detailed example of the Sun's proper and daily motions. Earlier Buridan, ibid., 175, had given the same example.
26. See Albert of Saxony, *De celo*, bk. 2, qu. 13, 1518, 110v, col. 1 (conclus. 4). Buridan, ibid., says similar things, explaining that the Sun undergoes two simultaneous motions, a daily motion and an annual motion between the Tropics of Cancer and Capricorn. The other planets have even more motions and undergo even greater difformities of motion

If medieval scholastic natural philosophers had confined themselves to the data of astronomical observations, they would have had no reason to assume regularity of celestial motion. The observations would have driven them to the conclusion that celestial motions are irregular. For Buridan and Albert of Saxony, these apparent irregularities were the result of compound motions. But appearances could hardly serve as conclusive grounds for accepting a fundamentally irregular system of planetary motion. Traditional Aristotelian metaphysical principles concerning the world were based on certain assumptions about underlying realities. The most fundamental of these proclaimed the incorruptibility of the celestial ether along with its future eternity. This assumption alone would have driven medieval cosmologists to belief in an underlying celestial regularity. An incorruptible ethereal substance of which the planets and the surrounding medium were composed, and which moved – by whatever means – with circular motion, could not alter its speed. To do so would have implied change, and therefore corruptibility. Hence the traversal of equal distances in equal times had to be assumed at a basic level. That level lay with the constituent orbs of the system, each of which was assumed to move with a "simple circular motion" (*motus circularis simplex*). Each individual orb moved with regularity and uniformity. The observed planetary positions, which reveal changes in velocity and direction (from progressive to retrograde motion), result from the interaction of two or more simple motions. Ironically, the resultant and observed irregular compound motions were perceived as somehow less real than the simple motions that produced them.

Whatever the appearances, scholastic authors treated each planet as if it moved with uniform speed. It therefore seemed natural to inquire whether all planets moved with the same uniform speed. All knew that the planets completed their respective periodic revolutions in different times.[27] But, as Buridan expressed it ([*De caelo*, bk. 2, qu. 20], 1942, 220), "although the Moon completes a circulation more quickly than the Sun, it does not follow that it [the Moon] is moved more quickly because the path of the Sun's sphere is much greater." Despite the disparity in their circular paths, the Sun might move as quickly as the Moon. Or it might move with the same velocity as the Moon, "if we assumed that the ratio of the Sun's sphere to the Moon's sphere in magnitude is as the ratio of a year to a month; and if it were a greater ratio, it would be moved more quickly [than the Moon]."

How might the disparity between the periodic revolutions of superior and inferior planets be explained? If the speeds of a superior and an inferior planet are equal, then their circles, or paths, must be unequal; and if the

(Et adhuc inveniuntur maiores difformitates in aliis planetis, quia moventur pluribus motibus quam sol). See also Oresme, *Le Livre du ciel*, bk. 2, ch. 13, 1968, 413.

27. Mercury, Venus, and the Sun, however, were observed to complete their periodic revolutions in the same time. The periods of the other planets differed. See Buridan, *De caelo*, bk. 2, qu. 20, 1942, 220.

circles are equal, one planet or orb must move faster than the other (ibid., 221). If the latter is the case, then two possible causes may produce differences in velocity: the power or force that moves one planet or orb is greater than that of another; or, if the movers are equal in power, the resistance to them must vary to produce unequal speeds. As a faithful Aristotelian, Buridan immediately rejects celestial resistances or impediments. "I believe," he concludes, "that a greater velocity exists in the heavens because of the greater perfection of a mover or from the smallness of the mobile." Buridan hastens to add that even in the absence of celestial resistance, the movers will not move any mobile, or planet, with an unlimited, or infinite, velocity. Despite the lack of resistance, the movers cannot cause unlimited speeds, because they are themselves only of finite power.

The ratio of a motive power, or intelligence, to its mobile, or planet, determines the speed of that planet. For example, if "we assume that mobile [or planet] *A* is 100 times greater than *B* and the two move with equal speeds, then the power [or force] moving *A* would have to be 100 times more powerful than the power that moves *B*" (ibid., 222). But if *A* is moved with twice the speed of *B*, then the power that moves *A* must be twice as great as the power that moves *B*.

Although Buridan devoted a question to determine whether the inferior spheres of planets are moved more quickly than the superior spheres, he could not, nor could anyone else, offer any absolute response.[28] Despite observational knowledge of the planetary periods, it was not possible for Buridan and his fellow scholastics to determine the real speeds of the planetary orbs. Whether the orb that carried Mars moved more quickly, less quickly, or at the same speed, as the orb that carried Venus could not be determined without arbitrary assumptions about the size of orbits, speeds, motive powers, and even magnitudes.

In the sixteenth and seventeenth centuries, some abandoned much if not all of the basic terminology of regular and uniform motion, while others retained it. The Coimbra Jesuits, for example, seem to have abandoned the earlier language, although they did employ a few similar terms. They inquire whether the celestial motions are "uniform and equable" (*uniformes et aequabiles*), thus signifying that these two terms are equivalent. Although they ignore the distinction between simple and compound motion, they did differentiate between "equality of distance" (*aequalitas spatii*) and "equality of time" (*aequalitas temporis*). The former concept is equivalent to the medieval idea of uniformity and difformity, the latter to that of regularity and irregularity. As an illustration of the first, they use the familiar example of an irregularity with respect to the whole heavens, presumably the daily motion, where those parts nearer the poles traverse a smaller distance in

28. Albert of Saxony treated the same question as did Buridan and in a similar manner. See Albert's *De celo*, bk. 2, qu. 16, 1518, 111v, col. 2–112r, col. 2.

the same time as parts farther away. But the overall motion of the heavens is uniform, because all the parts complete their motion in the same time, that is in an "equality of time."[29]

By contrast, Bartholomew Amicus, as we have already seen, retained virtually all of the terminology, probably because he drew most of it, approvingly, from Albert of Saxony's *De caelo*.

c. The effect of the triumph of fluid heavens over hard orbs

The distinction between simple and compound motion presupposed acceptance of the existence of celestial orbs, which, by the late sixteenth century, at the latest, were assumed by most astronomers and natural philosophers to be hard and rigid. By approximately 1630, as we have already seen (Chapter 14, Sec. VIII) many and probably most scholastic natural philosophers followed Tycho Brahe and switched from hard orbs to a fluid medium devoid of orbs, although some continued to assume a hard orb for the fixed stars. With the abandonment of celestial orbs, the motion of the planets could no longer be attributed to orbs, which disappeared from the heavens.

As a consequence, the rationale for distinguishing between simple and compound celestial motion vanished. The irregular observed planetary motions, with their retrogradations, direct motions and stations, could no longer be conceived as the consequence of a combination of simple regular motions made by numerous hard, celestial orbs. As most scholastics were aware, the Copernican system allowed for a direct and reasonable explanation for these irregularities. They were the consequences of an earth in annual motion, passing and being passed by other planets. Although the Copernican option was not available to Catholic natural philosophers, those scholastic natural philosophers who assumed fluid heavens would have been constrained to reject the traditional distinction between simple and compound motion, as did Melchior Cornaeus, who, in a brief paragraph, mentions the numerous irregularities of planetary motion and concludes that

29. After asserting that one of two things to be explained is whether the celestial motions are uniform and equable, the Conimbricenses continue ([*De coelo*, bk. 2, ch. 6, qu. 2, art. 2], 1598, 288): "Sciendum est posse nos expendere vel aequalitatem spatii, quod a mobili eiusve partibus decurritur, vel aequalitatem temporis, quod in tali spatio peragrando consumitur. Si igitur secundum priorem considerationem loquamur motus coelestis quoad partes ipsius coeli non est aequabilis quandoquidem aequali tempore aliae partes minus, aliae maius spatium peragrant, ut primo superioris articuli argumento ostendebatur." To illustrate the circumstances under which "equality of distance" is not realized, the Conimbricenses point to an example (287) in which parts of a heaven traverse unequal distances in the same time, as when parts nearer the poles traverse smaller circles than do parts farther from the poles. On "equality of time," they say (288): "Si autem sermo sit de motu coeli secundum posteriorem notionem, de qua potissimum philosophi et astrologi loqui consueuerunt, dicendum erit esse illum prorsus uniformem et aequabilem."

the planets are not carried around with a "simple and equable motion, nor can they be."[30]

2. Contrary motions

The widely accepted observation that each planet had two simultaneous motions in opposite directions, namely the daily east-to-west motion and the periodic west-to-east motion, posed a perplexing dilemma for medieval natural philosophers, especially since Aristotle had denied the possibility of two such contrary motions in one and the same body.

A contrariety of forms or dispositions was deemed impossible if they existed in one and the same thing simultaneously, although they might exist in one and the same thing successively. From this standpoint, Jean Buridan argued ([*De caelo*, bk. 1, qu. 8], 1942, 38) that a circular motion could not be contrary to a rectilinear motion, because both motions could exist simultaneously in the same body, as is evident when those who play with globes move one rectilinearly from one terminus to another even as the globe is continually rotating. But circular motions around the same poles in opposite directions are contrary motions, although circular motions in opposite directions around different poles did not qualify as contrary motions.

All were agreed that the celestial orbs could not move with contrary motions around the same poles (see Sec. II.7). But what about contrary motions that are successive? Was it possible that an orb could move in one direction for a period of time and then move in the opposite direction for the next period of time? This suggestion was rejected, because if the celestial motions are assumed eternal into the future, the contrary motion of a present spherical motion could never occur: for why should an orb moving in one direction suddenly reverse that direction?

Could motions to contrary places occur, as happens in rectilinear motions when a body moves between upward and downward directions? Buridan also denied this possibility, as did his fellow scholastics. Celestial motions do not move from one contrary place to another. Indeed, a celestial orb does not move from its present place into another place; it always rotates in the same place.[31]

30. "Ergo semper eodem simplici et aequabili motu non feruntur, neque ferri possunt." Cornaeus [*De coelo*, tract. 4, disp. 1, qu. 4, dub. 3], 1657, 504. In accepting fluid heavens rather than hard orbs, Cornaeus (508) denies that eccentrics and epicycles are physically real but allows that they could be imagined, as Christopher Scheiner had suggested in his *Rosa Ursina*.

31. Buridan includes all these arguments. For similar ones, see Johannes de Magistris [*De celo*, bk. 1, qu. 2], 6, col. 1. In his discussion, Bricot, *De celo*, bk. 1, 5r, col. 1, assigns three requisite conditions for the occurrence of contraries in local motions: (1) that the motion occur from contrary place to contrary place; (2) that the opposite motions occur along the same shortest line or path; and (3) if one of the contrary motions is natural to something, the other contrary motion must be violent. Bricot goes on to argue that none of these conditions applies to celestial motions.

Although we have thus far considered only the kinematic aspects of contrary motion, later in this chapter (Sec. II.7), we shall see how contrary motions became intertwined with the theories of al-Bitrūjī, who linked conceptions of force and the dragging of inferior orbs by superior orbs.

3. On the commensurability or incommensurability of the celestial motions and the Great Year

The relationship between celestial motions was hardly confined to those of a contrary nature. A number of medieval and early modern authors found occasion to inquire about the commensurability or incommensurability of those motions.[32] Interest in this theme goes back to classical antiquity, when Greek and Roman authors believed that the celestial motions are commensurable, that is, related by rational ratios. Commensurability seemed to follow from the widespread belief in the uniformity of nature and especially in the regular and uniform repetition of celestial configurations and events. One of the most important of these events was the Great, or Perfect, Year, which Cicero defined as follows: "On the diverse motions of the planets the mathematicians have based what they call the Great Year, which is completed when the sun, moon, and five planets having all finished their courses have returned to the same positions relative to one another. The length of this period is hotly debated, but it must necessarily be a fixed and definite time."[33] During antiquity and the Middle Ages, numerous periods were proposed for the Great Year, the most popular being 36,000 and 49,000 years. The former was derived from Ptolemy's *Almagest*, based on a value of precession of the equinoxes of 1 degree in 100 years.[34]

The temptation to draw deterministic inferences from these repetitions proved irresistible, especially to Stoic authors, who concluded that the Great Year entailed an exact and identical substantive and sequential repetition of all celestial configurations and terrestrial events. In their judgment, Socrates and Plato, and every other individual, would return in each Great Year and do precisely what they had done in every preceding Great Year. Some found this deterministic doctrine of individual return attractive, especially when it was linked with astrology.

For opponents of this doctrine – for obvious reasons, Christians were among its severest critics, with Saint Augustine as the most famous of

32. In what follows, I largely follow my own essay, "The Concept of Celestial Commensurability and Incommensurability from Antiquity to the Sixteenth Century," chapter 3 in Oresme, *De commensurabilitate*, 1971. Brief remarks about Amicus and Riccioli have been added.
33. The passage is from Cicero's *On the Nature of the Gods* (Cicero [Rackham], 1933, 173); see also Oresme, *De commensurabilitate*, 1971, 103.
34. Ptolemy gives these values in *Almagest*, bk. 7, ch. 2 [Toomer], 1984, 328.

them[35] – the best way to subvert the doctrine of individual return was to undermine the doctrine of the Great Year. One powerful method of achieving this goal was to suggest and, if possible, to demonstrate that some or all of the alleged celestial motions were really incommensurable. Indeed, one had only to show that any two planetary motions were incommensurable. Then, if all the planets were assumed to start from some particular configuration, they could never again enter into the same relationship in the same places.

This argument, however, was not applied in the ancient world and during the centuries of the early Middle Ages. Although the doctrine of the Great Year and celestial commensurability seem to have played little or no role during the early Middle Ages, they must have entered Europe during the late twelfth or the thirteenth century, as is evident from one of the articles condemned in 1277. The sixth article declares "That when all the celestial bodies have returned to the same point – which will happen in 36,000 years – the same effects now in operation will be repeated."[36]

Not long after, Duns Scotus (ca. 1266–1308) may have been the first to use the incommensurability argument to reject precise cyclical returns when he declared that "This opinion [i.e., exact return] can also be disproved with respect to its cause, for if it could be proven that some celestial motion was incommensurable to another, . . . then, I say, it follows that all the motions will never return to the same place" (Oresme, *De commensurabilitate*, 1971, 119). As an example of incommensurable motions, Scotus assumes that two bodies are moved with equal speeds on a square. One body moves on the side of the square, the other body moves on the diagonal of the same square, and Scotus infers that they would never return to the same positions they held at the outset. Scotus admits, however, that a "great discussion" would be required to determine if such incommensurability actually exists.

In 1343, Johannes de Muris completed his *Quadripartitum numerorum*, which included a section on the commensurability or incommensurability of celestial motions.[37] De Muris was one of the first Latin scholars to enter into a mathematical discussion of commensurability, imagining various

35. In the *City of God*, bk. 12, ch. 13, Augustine speaks of philosophers who have introduced "cycles of time, in which there should be a constant renewal and repetition of the order of nature; and they have therefore asserted that these cycles will ceaselessly recur, one passing away and another coming, though they are not agreed as to whether one permanent world shall pass through all these cycles, or whether the world shall at fixed intervals die out, and be renewed." Augustine [Dods], 1948, 1:498. Later, in the same chapter (499), Augustine warns against interpreting Solomon's famous remark that "there is no new thing under the sun" as support for cycles "in which, according to those philosophers, the same periods and events of time are repeated; as if, for example, the Philosopher Plato, having taught in the school at Athens which is called the Academy, so, numberless ages before, at long but certain intervals, this same Plato, and the same school, and the same disciples existed, and so also are to be repeated during the countless cycles that are yet to be." Augustine calls upon Christians to reject this doctrine.
36. See Oresme, *De commensurabilitate*, 1971, 109–110.
37. For a summary of the relevant propositions in de Muris's *Quadripartitum*, see Oresme, ibid., 86–97.

scenarios where two or three bodies traveling with commensurable velocities move simultaneously on circles. In purely kinematic terms, de Muris calculates the number of conjunctions they would enter into and the number of days between conjunctions, and so on.

De Muris devotes a brief section to incommensurable motions. He assumes two concentric but incommensurable circles or circumferences, which are related as the diagonal and side of a square, where the sides are presumably taken as unity. If two bodies, one on each circle, are assumed in conjunction and begin to move with commensurable motions, they will never again through all eternity conjunct in the same point, because, despite the commensurability of their motions, they traverse incommensurable distances.

Johannes de Muris, like most of those interested in celestial incommensurability (except for Nicole Oresme), was little concerned as to whether the planetary motions were commensurable or incommensurable. He knew that after astronomers predict a conjunction, they use their senses to determine whether the conjunction has occurred.[38] Long before de Muris wrote, Averroës had explained (Oresme, ibid., 108) why technical astronomers found it unprofitable to argue about the commensurability or incommensurability of celestial motions. Whatever the judgment, it was ostensibly irrelevant, because astronomers were well aware that their observations and data were only approximate. What might not count as a conjunction, in the strict and precise mathematics of incommensurability, might well be visually perceived as a conjunction. If so, it would be treated as a real conjunction.

And long after de Muris, Bartholomew Amicus, who considered the problem of celestial incommensurability in the seventeenth century, argued that we cannot measure the celestial motions with precise accuracy. Indeed, we cannot even know the relationship between the velocities of two bodies near us; how much less, then, can we know such relationships between the velocities of celestial bodies.[39]

a. *Nicole Oresme*

Among those who considered the problems of celestial commensurability and incommensurability during the medieval and early modern periods, the

38. In chapter 24 of his *Quadripartitum*, de Muris declares: "But I am unconcerned whether it is ever this [way] or that [i.e., whether the celestial motions are commensurable or incommensurable]. For an astronomer, however, it suffices that he can predict conjunctions of planets and stars in such a way that the senses are incapable of showing it otherwise." See Oresme, ibid., 371. To my original translation, I have added the word "celestial" within the square brackets.

39. Amicus, *De caelo*, tract. 5, qu. 6, dubit. 10, art. 3, 1626, 333, col. 2, offers this as the first of four conclusions in which he asserts his own opinion. "Prima conclusio non potest a viatoribus certe sciri scientifico mensura adaequata caelestium motuum quia difficilius cognoscitur mensura quantitatis caelorum et motuum caelestium velocitatis quam duorum sensibilium nobis propinquorum sed horum mensura et motuum velocitas praecisa non potest certo cognosci; ergo neque caelestium."

interest was at best modest and usually peripheral. An exception, perhaps the only one, is Nicole Oresme, who devoted two complete treatises to the subject: the *Ad pauca respicientes* and the later *Tractatus de commensurabilitate vel incommensurabilitate motuum celi*. In his original and unusual *De proportionibus proportionum*, he presented the mathematical foundations for his belief in the probable incommensurability of celestial motions. The importance of this theme for Oresme is further underscored by his use of it in a number of other works.[40]

The mathematical theorems on which Oresme based his conclusion that the celestial motions are probably incommensurable were first formulated in his *De proportionibus proportionum*, or *On Ratios of Ratios*, composed in the 1350s.[41] Here Oresme developed ideas that first appeared in rudimentary form in Thomas Bradwardine's *Tractatus de proportionibus*. For Oresme, a "ratio of ratios" (*proportio proportionum*) always involved two rational or irrational ratios that are related by a third ratio, which could be either rational or irrational (what we would call an "exponent"). Thus in the relationship $A/B = (C/D)^{p/q}$, the exponent p/q could be either rational or irrational. The exponent itself is the ratio of ratios. A number of the propositions of the *De proportionibus proportionum* attempt to determine whether the ratio of ratios is rational or irrational. Thus if the ratio of ratios p/q is rational, the ratios A/B and C/D, which could be rational or irrational (or one rational and the other irrational), would be considered commensurable and represent a "rational ratio of ratios." Similarly, if p/q is irrational, then A/B and C/D form an "irrational ratio of ratios." In the context of Oresme's proportionality theory, the ratios 3/1 and 27/1 form a rational ratio of ratios, because $27/1 = (3/1)^{3/1}$; but 3/1 and 6/1 form an irrational ratio of ratios, because $6/1 \neq (3/1)^{p/q}$, where p/q is rational. In this case, p/q is irrational, although Oresme could not express the relationship in either symbolic or rhetorical terms.

In the tenth proposition of the third chapter of the *De proportionibus*, Oresme constructed a demonstration to show that any two given, unknown ratios were more likely to be incommensurable than commensurable. The demonstration involved probability considerations. Oresme's "proof" is by way of an illustration. For any given sequence of ratios, $n/1$, where $n = 2, 3, 4 \ldots$, Oresme shows that more irrational than rational ratios of ratios can be formed. As an example, he takes 100 rational ratios from 2/1 to 101/1 and by taking them two at a time shows that 9,900 possible ratios of ratios can be formed. Because he was only interested in ratios of greater inequality, where the numerator is greater than the denominator, only half of the total, or 4,950, are relevant. Of these 4,950 possible ratios of ratios, Oresme shows that only 25 can be rational, with the remaining 4,925 irrational.[42]

40. We find some trace of relevant discussions in his *De spera*, *De celo*, *Questiones super geometriam Euclidis*, *Quodlibeta*, and in his final work, *Le Livre du ciel*.
41. This section is drawn primarily from Grant, 1988, 35–38.
42. The 25 are formed from the following geometric series: $(2/1)^n$, where n = 1, 2, 3, 4, 5,

From these relationships, Oresme concluded that in a set of 100 rational ratios, from 2/1 to 101/1, the ratio of irrational to rational ratios of ratios is 4,925/25, or 197/1. By extrapolation, he further argued that if more and more rational ratios were taken, say 200 or 300, and so on, the disparity between irrational and rational ratios of ratios would increase, from which Oresme inferred the existence of many more irrational than rational ratios. That there could be more of one type of ratio of ratios than another was made plausible for Oresme by analogy with perfect and cube numbers, because "however many numbers are taken in a series, the number of perfect or cube numbers is much less than other numbers and as more numbers are taken in the series the greater is the ratio of non-cube to cube numbers or non-perfect to perfect numbers. Thus if there were some number and such information as to what it is or how great it is, and whether it is large or small, were wholly unknown, . . . it will be likely that such an unknown number would not be a cube number."[43] From this example and by analogy, Oresme infers that ratios of ratios exhibit the same characteristics, which he expressed at the conclusion of the tenth proposition:

And so it is clear that with two proposed unknown ratios – whether they are rational or not – it is probable that they are incommensurable. . . . Therefore, if many [unknown ratios] are proposed it is [even] more probable that any [one of them you choose] would be incommensurable to any other [that you might choose]. . . . Now the more there are, the more one must believe that any [one of them you might choose] is incommensurable to any other [you might choose], for if it is probable that one proposed ratio of ratios is irrational, it is more probable when many are proposed that any one [you might select at random] would be irrational.[44]

In the fourth chapter, Oresme applies the ratio of ratios to physical magnitudes. The relationships are now transformed into the following general type: $\frac{F_2}{R_2} = \left(\frac{F_1}{R_1}\right)^{\frac{v_2}{v_1}}$, where F is a force applied to a resistance, R, and v_2/v_1 is the ratio of velocities that arises from the two force-resistance ratios. Oresme's examples in chapter 4 concern terrestrial motions, where, if $F > R$, and F is applied to R, a velocity is produced. Oresme then extends the range of these relationships to the celestial orbs. But how, we might ask, can a ratio F/R be applied to celestial orbs, where there are no forces or resistances? Oresme suggests that such a ratio "ought not to be called a ratio of force to resistance except by analogy, because an intelligence moves

6. Taking these six ratios two at a time, we obtain $(6 \times 5)/2 = 15$ possible rational ratios of ratios; $(3/1)^n$, where $n = 1, 2, 3, 4$, from which six rational ratios of ratios can be formed; and from each of the following, one rational ratio of ratios can be formed: $(5/1)^n$, $(6/1)^n$, $(7/1)^n$, and $(10/1)^n$, where in each instance $n = 1, 2$. See Oresme, *De prop. prop.*, 1966b, 41.

43. Ibid., 249–251.

44. Ibid., 253–255. Some of the words added within square brackets are additions to my original translation.

by will alone and with no other force, effort, or difficulty, and the heavens do not resist it, as I believe were the opinions of Aristotle and Averroes."[45] The "analogy" was probably fairly strong, and many scholastics probably thought of an intelligence or angel as a force and of the orb it moved as, in some sense, a resistance. Amicus was probably typical when he spoke of a celestial mobile having the same constant ratio to its mover.[46]

What Oresme hinted at here, he formalized in the final proposition of the *De proportionibus* (ch. 4, prop. 7) when he took the momentous step of applying his mathematical conclusions to celestial motions. Thus did Oresme conclude that "When two motions of celestial bodies have been proposed, it is probable that they would be incommensurable, and most probable that any celestial motion of the heaven [that you might choose] would be incommensurable to the motion of any other [celestial] sphere [that you might choose]" (Oresme, *De prop. prop.*, 1966b, 305 [translation altered]).

Oresme even suggests four propositions to which his ideas are applicable, one of which asserts that "If two planets, with respect to longitude and latitude, should be conjuncted once in a point, they will never again be conjuncted [in that same point]."[47] In addition to these four propositions, Oresme declares that he can demonstrate "many other no less beautiful propositions from the same principle." Moreover, "many errors about philosophy and faith could be attacked by the use of these [propositions], as [for example], that [error] about the Great Year which some assert to be 36,000 years, saying that celestial bodies were in an original state and then return [to it in 36,000 years] and that past aspects are arranged again as of old" (Oresme, *De prop. prop.*, 1966b, 307). Here, then, Oresme thought his conclusion about celestial incommensurability could destroy the theory of the Great Year, a concept which, as we saw, was condemned in 1277.

But in two later treatises on celestial motions, Oresme did not apply the probability considerations based on ratios of ratios of force to resistance. In his two treatises devoted solely to the commensurability and incommensurability of celestial motions, the *Ad pauca respicientes* and a much expanded version of it, the later *Treatise on the Commensurability or Incommensurability of the Celestial Motions* (*Tractatus de commensurabilitate vel incommensurabilitate motuum celi*), he employed purely kinematic theorems.[48] The commensurabilities and incommensurabilities are derived from relationships of distances traversed, times, and velocities.[49] Force and resistance play no

45. Ibid., 293. In his final work, *Le Livre du ciel*, Oresme spoke vaguely of forces and resistances implanted in celestial orbs by God. On this see Section II.4.d of this chapter.
46. "In motu caeli non est medium resistens sicut est in motu recto, sed tantum ibi esse mobile, quod semper servat eandem proportionem cum motore, nam semper est similium partium, figurae, et naturae." Amicus, *De caelo*, tract. 5, qu. 6, dubit. 6, art. 3, 1626, 315, col. 2.
47. Oresme, *De prop. prop.*, 1966b, 307. For clarification, I have added the bracketed material.
48. The *Ad pauca respicientes* is not described here.
49. According to Jan Von Plato, 1981, 190, "a few of Oresme's theorems on the consequences

role, and probability considerations are ignored. Oresme is content to determine whether two or three mobiles will conjunct once, whether they will conjunct again or have ever conjuncted before, and similar questions. But he does not ask what, if any, force–resistance relationships produced those velocities and whether they are more likely to be incommensurable than commensurable. Indeed, the proposition cited two paragraphs earlier from the *De proportionibus* – that "If two planets, with respect to longitude and latitude, should be conjuncted once in a point, they will never again be conjuncted [in that same point]" – which Oresme proposed as an example of the kind of proposition to which his probability doctrine of ratios of ratios would apply, was included in both of these treatises on celestial commensurability and incommensurability.[50] But in neither of these instances is there any trace of the application of ratios of force and resistance.[51] The relationships are purely kinematic.

The two treatises differ in their orientation. The *Ad pauca respicientes*, the earlier one, includes a supposition in which the relationship between any two randomly chosen quantities from among many quantities is assumed to be probably incommensurable (see n. 51). This is applied to celestial motions in only one proposition. The later *Treatise on Commensurability*, however, includes no such assumption, probably because its format excluded such a possibility. The entire treatise is intended as an objective attempt to determine whether the celestial motions are commensurable or incommensurable. Thus Oresme could not assert the probable incommensurability of celestial motions, although he refers to it once in the great debate that concludes the treatise.[52] For a better appreciation of its structure and content, especially the crucial debate at the conclusion of the work, let us briefly examine its organization.

The treatise is divided into a prologue and three parts. Part 1 consists of

of the possible incommensurability of the revolution times of bodies in uniform circular motion have their counterpart in the modern theory of ergodic dynamical systems."

50. In the *Ad pauca respicientes*, see pt. 1, prop. 4 (Oresme, *De prop. prop.*, 1966b, 395) and in the *Treatise on Commensurability*, pt. 2, prop. 1 (Oresme, *De commensurabilitate*, 1971, 249).

51. Nevertheless, in pt. 1 of his *Ad pauca respicientes*, Oresme (*De prop. prop.*, 1966b, 385–387) includes the following probability assumption, as supposition 2: "If many quantities are proposed and their ratios are unknown, it is possible, doubtful, and probable that any [one of them] would be incommensurable to any other [of them you might choose]." Oresme invokes this supposition only once, in pt. 2, prop. 17 (ibid., 423), which announces: "It is probable that in any instant the celestial bodies are related in such a way that they were never so related in the past, nor will be so related at any time in the future; nor was there, nor will there be, a similar configuration or disposition through all eternity." The demonstration depends on the probability that some of the quantities relevant to the motions and positions of planets – namely, "circles, latitudes, distances, eccentricities and many motions and diversities" – are probably incommensurable, as the second supposition proclaims. But this is in no way linked to the incommensurabilities and irrationalities associated with the doctrine of ratios of ratios and its application to ratios of force and resistance.

52. It is introduced as one more argument in favor of celestial incommensurability, following a series of arguments in favor of celestial commensurability.

twenty-five propositions concerned with commensurable motions, in which bodies are assumed to move with commensurable speeds on concentric circles;[53] part 2 contains twelve propositions, in each of which the motions are incommensurable. In these two parts, Oresme sought to derive various consequences from the motions of two or more bodies whose speeds are first assumed to be commensurable and then incommensurable. He wished to show how bodies would relate to one another if their motions were assumed commensurable, and how they would relate if their motions were assumed incommensurable. In both the *Ad pauca respicientes* and the *Treatise on Commensurability*, Oresme was concerned with exact punctual relations of bodies moving with circular motion. Thus he ignored aspects near or around a point.[54] The punctual character of Oresme's approach is illustrated by his definition of conjunction, which is said to occur "when the centers of any mobiles are on the same line drawn from the center [of the world]" (Oresme, *De commensurabilitate*, 1971, 179).

In a significant passage, Oresme (ibid.) defines commensurability and incommensurability and emphasizes the purely kinematic character of his treatise:

> I take the commensurability and incommensurability of circular motions in terms of the magnitude of the angles described around the center or centers, or in terms of the circulations, which is the same thing. Thus, things are moved commensurably when, in equal times, they describe commensurable angles around the center, or when they complete their circulations in commensurable times. Circulations are incommensurable when they are completed in incommensurable times, and when, in equal times, incommensurable angles are described around the center. Accordingly, conjunctions, oppositions, aspects, and all the motions ascribed to the heavens by astronomers are to be measured in this way, since a ratio of velocities varies as a ratio of the circular lines described by the mobiles. Whether or not it ought to be taken in this way is not relevant to what is proposed here.

Oresme thus acknowledges that his definitions may differ from the way astronomers actually proceed. In most propositions, he arbitrarily applies kinematic consequences to astronomical aspects, with conjunction as the paradigm case.

As if to reestablish a strong link between his kinematic approach and nature, Oresme goes on to explain (ibid.) that

> incommensurability can be found in every kind of continuous thing, and in all instances in which continuity is imaginable, either extensively or intensively. For a magnitude can be incommensurable to a magnitude, an angle to an angle, a motion to a motion, a speed to a speed, a time to a time, a ratio to a ratio, a degree to a degree, and a voice to a voice, and so on for any similar things.

53. As a minor exception to this statement, the circles are eccentric in proposition 20.
54. See Oresme, *De comensurabilitate*, 1971, 8 and 178, lines 45–49.

In the twelve propositions of part 2, Oresme demonstrates the consequences of celestial incommensurability but does so without using the theorem from the *De proportionibus proportionum*. Instead, he usually assumes incommensurable velocities – and occasionally mixes in a commensurable motion – of two or more bodies moving on concentric circles and describes their possible relationships. He shows that if any two celestial motions are incommensurable, the precise positional relationships of the bodies with those motions would be impossible to determine. If two or three celestial bodies enter into any astronomical aspect, say conjunction or opposition, they could never again enter into that same relationship in the same points. Knowledge of precise past relationships and future dispositions would be impossible. Celestial events, such as conjunctions, oppositions, and eclipses, would necessarily be unique and nonrepetitive.

A reader who had finished the first two parts would have no inkling whether celestial motions were commensurable or incommensurable, or both. This judgment was supposed to be made in part 3, the concluding section, where Oresme presents a debate presided over by Apollo and involving as protagonists Arithmetic and Geometry, each of whom cites numerous classical sources in her own support. Arithmetic, who presents her case first, argues passionately for the commensurability of celestial motions, while Geometry does the same for incommensurability.[55] The debate occurs within the framework of Oresme's dream and terminates without a judgment. For just as Apollo is to render a decision as to whether the celestial motions are commensurable or incommensurable, Oresme awakens.

As a vehicle, the dream was a convenient literary device for avoiding a definite decision. But it also served an ulterior motive. Not only has Oresme's rude awakening robbed us of a unique opportunity to acquire a profound truth, but it has made us dependent on the appeals of Arithmetic and Geometry for whatever insight we may attain on this important problem. Readers are now aware that at best they must formulate their own judgments on the basis of the two orations, which incorporate the only kinds of arguments and appeals that humans can know and understand. Let us briefly describe some of them.

In behalf of commensurability, Arithmetic emphasizes the enormous utility of rational ratios, which produce pleasure rather than the offensive effects of irrational ratios. It is rational ratios that produce the harmonious music of the celestial spheres. By far the most telling argument is Arithmetic's claim that unless the celestial motions are commensurable, astronomical predictions and tables, as well as knowledge of future events – thus did Arithmetic align herself with astrology – would be impossible.[56]

Geometry represents magnitude in general, and therefore encompasses both rational and irrational relationships. She concedes that rational ratios

55. This paragraph is drawn from Oresme, ibid., 6, and Grant, 1988, 37.
56. Here I draw on Oresme, *De commensurabilitate*, 1971, 68–69.

do indeed produce beauty in the world. But if they were united with irrational ratios, a much richer variety of effects would be produced. Indeed, Geometry combines the two and thereby adds much more splendor to the heavens. If Arithmetic did produce the music of the spheres, it would be monotonous, since it would be based solely on fixed, rational ratios. Only by adding Geometry's infinite variation could celestial music be made interesting.[57] Indeed, if the heavenly motions were commensurable, conjunctions and other astronomical aspects could only occur in a certain finite number of places in the sky, which, as a consequence, would appear to be preferred over other places. Would it not be better that such events be capable of occurring anywhere in the sky?

Arithmetic's arguments had presupposed the possibility of acquiring exact knowledge. Geometry denies the possibility, insisting that we must rest content with approximations. Exact knowledge is not even desirable, since not only would it discourage further observations, but the precise knowledge of future events that it might provide would make us like the immortal gods, a repugnant thought.[58]

To avoid these difficulties, Geometry urges that we assume celestial incommensurability. As her final argument, she reminds Arithmetic and Apollo (321) that a mathematical demonstration elsewhere – clearly a reference to his *De proportionibus proportionum* – has shown that "when any two unknown magnitudes have been designated, it is more probable that they are incommensurable than commensurable, just as it is more probable that any unknown [number] proposed from a multitude of numbers would be non-perfect rather than perfect. Consequently, with regard to any two motions whose ratio is unknown to us, it is more probable that that ratio is irrational than rational."

Oresme understood that sense perception could not determine the commensurability or incommensurability of the celestial motions. This is made clear at the outset of the dialogue in part 3, when Apollo informs him (285) that our senses cannot achieve exactness. "For if an imperceptible excess – even a part smaller than a thousandth – could destroy an equality and alter a ratio from rational to irrational, how will you be able to know a punctual [or exact] ratio of motions or celestial magnitudes?" Apollo convinces Oresme that "the judgment of the senses cannot attain exactness. But [even] if the senses could attain such exact knowledge, one could not know whether he had judged rightly, since an insensible [or undetectable] addition or subtraction could alter a ratio but would not change the judgment. Therefore, I do not vainly presume that the aforesaid problem is solvable by mathematical demonstration."

Oresme did not believe that he had demonstrated the incommensurability

57. Geometry, however, appears skeptical of the very notion of music from the celestial spheres.

58. Because Oresme has staged a debate among ancient pagans, he speaks of "gods" and includes no biblical references.

of the celestial motions. But there is good reason to believe that he favored incommensurability over commensurability, largely because he had elsewhere shown the *mathematical probability* that some, and perhaps most, celestial motions are incommensurable. Moreover, the structure of part 3 of the *Treatise on Commensurability* favors Geometry's arguments. Incommensurability provides us with a more varied world and also guarantees that we shall not attain exact knowledge, especially about the celestial motions. Finally, as we saw, Geometry alone appeals to a mathematical demonstration, one that is probable but, under the circumstances, the only kind attainable. Amid all the rhetoric and appeals to authority, that demonstration was the most solid foundation on which to base a judgment.

By advocating the probability of celestial incommensurability, Oresme sought to persuade others that celestial effects were inherently unpredictable. As a dedicated foe of judicial astrology, he hoped to weaken its foundations and strike a blow at the astrologers, who had aroused his deep concern by virtue of their considerable influence on the king of France, Charles V. He was annoyed by their pretentious claims of punctual exactness, claims that they could never fulfill. His treatises on celestial incommensurability were partly intended to deflate the astrologers, as well as to emphasize our inability to acquire exact knowledge.[59]

Declarations of originality in medieval treatises are relatively rare. Oresme makes such a claim for his *Treatise on Commensurability* when he informs us that "if another has set out the more fundamental principles [or elements found in this book], I have yet to see them." Moreover, he also tells us that he "did not release this little book without [first] submitting it for correction to the Fellows and Masters of the most sacred University of Paris."[60] Perhaps because of this, and the fact that he wrote on the subject in many places, a number of Parisian scholastics utilized some of his conclusions, including Henry of Hesse, who mentions Oresme by name; Pierre d'Ailly, who plagiarized most of part 3 of the *Treatise on Commensurability*;[61] Marsilius of Inghen; and Jean Gerson (1363–1429). To this group, we may add the Bolognese physician and astrologer John de Fundis, who, in 1451, wrote a hostile commentary on Oresme's *Ad pauca respicientes*, the only critique known thus far on either of Oresme's two major treatises on celestial incommensurability. If Oresme exercised any influence, it was either through the authors just mentioned or by reason of the dissemination of his *De proportionibus proportionum* and *Ad pauca respicientes*, which were twice published together in the first decade of the sixteenth century, once in Venice in 1505, and again in Paris, around 1510.[62]

59. In the *Ad pauca respicientes*, pt. 2, prop. 19, Oresme (*De prop. prop.*, 1966b, 427) provides reasons on the basis of which one should properly conclude "that astrology is vain."
60. Both statements occur in Oresme, *De commensurabilitate*, 1971, 175.
61. D'Ailly incorporated these passages in his *Tractatus contra astronomos* (in Oresme, ibid., 130–131).
62. In both editions, Oresme's works are accompanied by other treatises, mostly on proportionality. The *De proportionibus proportionum* and *Ad pauca respicientes* are printed as if

b. Amicus and Riccioli

Because of the difficulty of the subject, celestial incommensurability was never widely discussed. Few traces of it are found in the seventeenth century. Among those who considered the subject were Bartholomew Amicus and Giovanni Baptista Riccioli. Of the two accounts, Amicus's is the more substantial.[63] Although Oresme is never mentioned – only al-Battani, Pliny, and Ptolemy are cited – most of Amicus's ideas have counterparts in his works, such as, for example, Amicus's description of a basic theorem for commensurability and one for incommensurability. With regard to the former, he says that if two bodies moving commensurably are now in conjunction, they will be conjoined again in the same place.[64] But if two bodies are now in point *A* and are moving incommensurably, they will never again conjunct in the same place, nor indeed will they conjunct in any other point which is separated from the point of conjunction by a distance that is commensurable to the whole circle.[65]

Amicus describes three opinions as to whether the celestial motions are representable by rational or irrational ratios. All seem to have counterparts in Oresme's work. Those who favor rational ratios argue that God "has made all things in number, weight, and measure," which signifies rationality.[66] Moreover, just as the celestial orbs are the most perfect bodies and are therefore assigned the most perfect shape, the sphere, so also are rational ratios nobler than irrational.[67] Indeed an irrational ratio is rather a "dispro-

they comprised a single work, bearing almost identical titles: *Tractatus proportionum nicholai oren* (the Paris edition varies slightly by substituting "horen" for "oren"). For further information, see Oresme, *De prop. prop.*, 1966b, 130–131.

63. In the sixteenth century, Girolamo Cardano devoted 6 of 233 propositions in his *Opus novum de proportionibus* to mobiles moving in circles. Although counterparts to all of his propositions can be found in Oresme's *Treatise on Commensurability*, Cardano nowhere applies his propositions to celestial motions. See Oresme, *De commensurabilitate*, 1971, 142–160.
64. "Si duo mota commensurabiliter sunt nunc coniuncta, ipsa in eodem puncto coniungentur." Amicus, *De caelo*, tract. 5, qu. 6, dubit. 10, art. 1, 1626, 332, col. 2. For Oresme's version, see *Treatise on Commensurability*, pt. 1, prop. 4 in Oresme, ibid., 192 (Latin).
65. "Quod si duo incommensurabiliter mota sunt nunc coniuncta in puncto A, numquam alias in eodem coniungentur, nec in alio distante ab hoc per partem suo toti commensurabilem." Amicus, ibid., 333, col. 1. Amicus has here described propositions 1 and 2 of part 2 of Oresme's *Treatise on Commensurability*. The first proposition (Oresme, ibid., 249) declares that "If two mobiles have moved with incommensurable velocities, and are now in conjunction, they will never conjunct in that same point at other times." The second proposition adds the second part of Amicus's enunciation: "If two mobiles are now in conjunction, they will never conjunct at other times in any point separated from their present point of conjunction by a part of the circle commensurable to the whole circle." Ibid., 251.
66. This famous passage is from the Book of Wisdom 11.21. Oresme puts these words into Arithmetic's oration (see Oresme, ibid., 295 and 340–341, n. 19). For the rest of Amicus's arguments cited in this paragraph, see Amicus, ibid., 333, cols. 1–2.
67. Arithmetic says (Oresme, ibid., 291): "Therefore, just as the more perfect figure is appropriate for the celestial orbs, so [also] is the more noble ratio best suited for their motions, so that there is no lack of physical beauty to these bodies."

portion."[68] According to writers on "perspective," or optics, things that delight the sight are related by rational ratios. The harmonies that produce pleasure in tastes and odors and the harmonies that delight the ear are all based on rational ratios.[69] Finally, and most importantly, Amicus presents the common argument that without rational ratios we could not predict future conjunctions and aspects and therefore could not know future effects. The science of astrology would be in jeopardy.[70]

The second opinion that Amicus describes represents those who hold that irrational ratios better represent the relationships of the celestial region.[71] The perfection of the universe consists in variety; therefore it is best if the heavens embrace all kinds of ratios, including those that are irrational.[72] To reinforce this point, Amicus emphasizes that when magnitudes are unknown, it is more probable that they are mutually incommensurable.[73] He also repeats another Oresmeian idea, namely that if the celestial motions are commensurable, conjunctions would only occur at a small number of points. But this is false, for why should certain parts of the ecliptic be deprived of conjunctions of Sun and Moon? With the assumption of incommensurable motions, however, conjunctions could occur in any part of the ecliptic.[74]

The third opinion represents Amicus's own view. We wayfarers (*viatores*) cannot know the ratios of celestial motions, because a very small mutation will change one kind of ratio into another, say a rational ratio into an irrational ratio.[75] Indeed, since we cannot know things near our senses, how much less likely is it that we can know the velocities of celestial motions. If the celestial ratios were irrational, not only would we be unable to know the positions of future aspects, but we could not represent such positions by numbers or fractions of numbers with respect to a whole circle. Astronomers could not use rational, fractional parts of circles as the basis for a set of precise measures.

68. "Nam irrationalis est potius disproportio." Arithmetic says much the same thing (ibid., 290–292, lines 90–91), even using the term *disproportio*.
69. For Oresme, see his *De commensurabilitate*, 1971, 293.
70. Oresme says similar things in ibid., 305.
71. These arguments appear in Amicus, *De caelo*, tract. 5, qu. 6, dubit. 10, art. 2, 1626, 333, col. 2.
72. In the great debate in Oresme's *Treatise on Commensurability*, Geometry makes this point (Oresme, *De commensurabilitate*, 1971, 311).
73. This is, of course, Oresme's fundamental assumption, which he could claim to have derived from his probability theorem in the *De proportionibus* (see Section I.3.a of this chapter).
74. See Oresme, *De commensurabilitate*, 1971, 319 for this same point.
75. "Tertia opinio est dicentium a viatoribus non posse agnosci qualis sit proportio motuum caelestium quia per insensibilem partem potest fieri mutatio unius generis proportionis in aliud, scilicet rationalis in irrationalis, et mihi placet." Amicus, *De caelo*, tract. 5, qu. 6, dubit. 10, art. 2, 1626, 333, col. 2. Much earlier, in his *Treatise on Commensurability*, pt. 3, Oresme expressed the same sentiment (ibid., 285). In a feigned rebuke to humans by Apollo, Oresme has the latter declare that we are ignorant of "ratios relating to things of the world," largely because "an imperceptible excess – even a part smaller than a thousandth – could destroy an equality and alter a ratio from rational to irrational."

Despite our inability to know the exact ratios of motions, Amicus believes that we can preserve astrological, and presumably astronomical, science. We may distinguish two aspects of astrology ([*De caelo*, tract. 5, qu. 6, dubit. 10, art. 3], 1626, 334, col. 1, conclus. 4), theoretical and judicial.[76] The theoretical part relies on approximate, not exact, measures, just as some other sciences do, such as the science of weights and measures and the musical consonances. The judicial aspect, which treats of powers arising from conjunctions and aspects, also depends on a broader concept of measure. The efficacy of these powers depends on a broadly conceived sense of a conjunction, one that may range over some degrees of longitude, not one that is punctually exact. It is like the power of a medicine, the strength of which is not precisely known, any more than is the exact degree of a sickness.

Amicus seems to have conceded that both rational and irrational ratios exist among the celestial motions. But we cannot know for certain whether any particular ratio of celestial magnitudes is rational or irrational. Thus it is essential to accept astronomical aspects as falling within sufficiently broad limits so that we can determine where they occurred in the past and where they will occur in the future.

Riccioli adopts essentially the same attitude. He explains that astronomy has not yet been able to determine "whether the periods of celestial motions are mutually commensurable and consist of rational ratios, or whether indeed they consist of irrational ratios."[77] The probability that they might be incommensurable is never mentioned. We learn more about Riccioli's attitude when he takes issue with Kepler, who, in his *Epitome*, argued that the ratios of the periodic times of the planets are irrational and thus partake of infinity, that is, indefiniteness, which, for Kepler, meant an absence of beauty. Because of their indefinite nature, Kepler denied that irrational ratios could be the work of a creator Mind.[78] Riccioli granted that such ratios might be irrational, and therefore incapable of precise expression: any expression of them would tend toward infinity, that is, involve an infinite process. But Kepler was mistaken in his interpretation of infinity as lacking in beauty. After all, the Divine Immensity and Eternity have every perfection of the infinite and do not lack beauty. There is also no evidence to indicate that there is a lack of beauty in celestial motions that encompass both kinds of ratios. Finally, Riccioli finds it unacceptable that Kepler makes

76. Although Amicus uses only the term *astrologica*, and not *astronomica*, it is likely that he meant the theoretical side to apply to both astronomy and astrology. The judicial side, as we shall see, seems relevant only to astrology.

77. "An autem periodi motuum celestium sint commensurabiles invicem et ex rationalibus proportionibus constantes, an vero ex irrationalibus nondum adeo profecit Astronomia, ut certo controversiam hanc dirimere liceat satisque fuerit opiniones indicasse." Riccioli, *Almagestum novum*, pars post., bk. 9, sec. 2, ch. 6, 1651, 269, col. 2.

78. Riccioli cites book 4, page 512, of Kepler's *Epitome*. For a translation of the passage, see Kepler, *Epitome* [Wallis], 1952, 894. For the Latin quotation, see Riccioli's *Almagestum novum*, pars post., bk. 9, sec. 2, ch. 6, 1651, 270, col. 1.

God the author of rational ratios but denies that he is the author of irrational ratios.[79]

Christopher Clavius found occasion to mention briefly the Great Year and celestial incommensurability. He reports that belief in a Platonic year of 49,000 years is widespread. "But," says Clavius, "they seem to assert this rashly, since according to many the motions of the heavens are mutually incommensurable, so that it cannot happen that all the stars enter into the same positions and order that they now have or once had."[80] Thus was Oresme's intricate doctrine of the probable incommensurability of the celestial motions encapsulated in a sentence or two.

c. *The impact of the doctrine of celestial incommensurability*

No one from the medieval and early modern periods has yet been identified who can be said to have treated the subject of celestial commensurability or incommensurability in any manner comparable to that of Oresme, whether in extent, depth, or sophistication. The accounts of Amicus and Riccioli pale by comparison. Both refused to adopt one side or the other and allowed that both types of ratios were seemingly relevant. The evidence was inconclusive. We are unable to arrive at precise or exact measures and observations. For Riccioli (*Almagestum novum*, 1651, pars post., bk. 9, sec. 2, ch. 6, 269–270) the beauty of the world arises from both kinds of ratios.

Oresme was, of course, aware of such arguments and attitudes. One is even tempted to say that he invented most of them. On numerous occasions, Oresme acknowledges that our senses are weak and therefore we cannot achieve precise measurements. But his arguments had nothing to do with measurements and whether we can empirically detect whether a ratio is commensurable or incommensurable. Indeed, we cannot. Oresme was not interested in approximations but in propositions or theorems based on exact punctual relationships. Although we cannot detect these punctual relationships observationally, we can draw consequences from reasoning mathematically about them. If the celestial motions are probably incommensurable, as Oresme believed on the basis of his own mathematical

79. In a continuing response to Kepler, Riccioli declares (ibid., col. 2), at the conclusion of his section on whether the celestial motions are rational or irrational: "Sed neque evidens est proportiones, quae sunt nobis ineffabiles, aut saltem quia in infinitum tendunt esse expertes omnis pulchritudinis, cum Divina Immensitas et Aeternitas habeat perfectionem omnem infiniti, nec tamen sua pulchritudine careat. Sicut neque evidens est non esse pulchrius in motibus caelestibus esse utriusque generis proportiones. Denique non placet quod Deum authorem facit proportionum effabilium, negat autem ineffabilium." Kepler used the term *ineffabiles* for "irrational," as he explains when he says "ineffabiles, irrationales vulgo." See Kepler, ibid., where Wallis includes the Latin phrase.

80. "Sed temere hoc assere videntur, cum enim secundum plerosque motus coelorum sint inter se incommensurabiles fieri non potest, ut unquam omnia sidera eundem situm et ordinem quem nunc habent aut olim habuerunt obtinere possint." Clavius [*Sphere*, ch. 1], *Opera*, 1611, 3:29. In the first edition of 1570, Clavius makes no mention of incommensurability (see Oresme, *De commensurabilitate*, 1971, 134–135, n. 123).

demonstrations, we can derive a set of consequences that may be revealing, and even startling, about the "real" celestial aspects that we can never actually detect. Incommensurability meant that precise relationships could never be known, not only because our senses are weak, but more importantly because of the nature of mathematics or by virtue of the very structure of the universe itself, either of which reason guarantees that astronomical aspects never repeat. Predictions of future celestial configurations are therefore not possible, nor is the determination of exact past relationships.

The implications of this doctrine were potentially far-reaching.[81] In a world where celestial incommensurability is probable, both astronomy and astrology become inherently inexact sciences. Nature, or mathematics, decrees their inexactness. If Jupiter and Mars, for example, could only conjunct in a given point through all eternity, any other alleged conjunctions at that same point would be merely approximate and would produce slightly different effects. The Great Year, as we saw, would be impossible, because all the planets can never enter into the same disposition in which they were 36,000, or 49,000, years before.

Oresme even used the doctrine of celestial incommensurability to refute Aristotle's fundamental concept that whatever had a beginning must have an end and that what has no end cannot have had a beginning. He had only to use the incommensurability doctrine to determine one or more unique celestial events, which terminated one cosmic condition that had existed from all past eternity and which immediately thereafter began a new cosmic condition that would last through all of future eternity.[82]

Unfortunately, none of these interesting consequences was really discussed in subsequent centuries. In a strong sense, Oresme's incommensurability doctrine was unique; it had neither predecessors nor successors. All that survived of Oresme's ideas was the declaration of the probability of celestial incommensurability. By the sixteenth and seventeenth centuries, few even knew what this assertion signified and why it was proclaimed. Not only were the supporting details ignored or unknown, but the question as to whether the celestial motions were commensurable or incommensurable became one of whether they were rational or irrational, and the latter, in turn, was usually resolved by the assertion that our senses were incapable of determining whether celestial motions were truly represented by the one or the other, or both. Because of this, astronomers and astrologers were content with approximations that made the question of commensurability or incommensurability essentially irrelevant. What entered into general discussion, as we find it in Amicus's *De caelo*, were bits and pieces of the doctrine that were probably derived ultimately from Oresme's works, though without benefit of any context or realization of the source.

81. In this and the next paragraph, I rely on Grant, 1978c, 112–113.
82. For a few examples, see ibid., 113.

II. The dynamics of celestial motion

1. Aristotle on internal and external celestial movers

In examining the possible animation of the heavenly bodies, we had occasion to consider motive intelligences and their role as causal agents in celestial motion (Ch. 17, Sec. V). We must now investigate more systematically not only motive intelligences but the entire range of medieval and early modern scholastic ideas about the causes of celestial motion. As was so often the case in cosmology, Aristotle played a central role in shaping ideas about celestial movers, although his statements in the *Physics*, *De caelo*, and the *Metaphysics* left subsequent readers and commentators with seemingly conflicting advice. They found Aristotle advocating internal movers in some places and external movers in other places. Thus in *De caelo* (1.2.269a.5–7), Aristotle seems clearly to favor an internal principle of motion when he asserts that the fifth element in the heavens, or the celestial ether, is a "simple body naturally so constituted as to move in a circle in virtue of its own nature" and when he explains later in the same treatise (2.1.284a.14–15) that the motion of the heaven "involves no effort, for the reason that it needs no external force of compulsion, constraining it and preventing it from following a different motion which is natural to it."[83]

But in the *Physics* (bk. 8, ch. 6) and the *Metaphysics* (bk. 12, ch. 8), Aristotle presented a different picture of celestial motion, one in which only external movers were operative. In the *Metaphysics*, where he considers the astronomical theories of Eudoxus and Callippus, Aristotle (as we saw in Ch. 13, Sec. I), arrived at a total of 55 spheres to account for the motion of all the celestial bodies. Although his system of concentric spheres had few supporters in the Middle Ages, his description in the *Metaphysics* of the manner in which those spheres were moved was destined to have considerable influence. For here, in contrast to *De caelo*, Aristotle speaks of external, spiritual movers, one for each orb.[84] In fact, he speaks of them as "immovable movers," of which the one associated with the outermost sphere of the fixed stars is called the "first immovable mover." These movers are transcendent, or at least separate and distinct from the orbs with which they are associated. They are also without magnitude, and therefore lack parts and divisibility. Although the immovable movers are all identical, the first is immobile both essentially and accidentally, whereas all the other immovable movers are immovable essentially but not accidentally.[85] The accidental motion of all immovable movers but the first is, presumably, an

83. For the translations, see Aristotle [Guthrie], 1960. See also Wolfson, 1973b, 23. Aristotle seems to have assumed that sublunar bodies are also intrinsically capable of natural self-movement. If unimpeded, a heavy, earthy body will, by its nature, tend to fall toward the center of the world.
84. Here I follow Wolfson, 1973a, 1–11.
85. See Wolfson, 1973a, 7, and Stewart, 1973, 544–545.

accidental motion caused by the "circular motion of the outer sphere, which is moved by the first mover" (Stewart, 1973, 539). Whether this distinction between the "first unmoved mover" and all the others led Aristotle to assume a hierarchical distinction between them is a moot point.[86] If Aristotle did distinguish between them, he differentiated only the first unmoved mover from all the rest. The unmoved movers of the second sphere, or fourth sphere, or fortieth sphere had identical natures. As we shall see, however, Avicenna, Averroës, and others invoked criteria to distinguish one from another.

In the *Metaphysics* (bk. 12, ch. 7), Aristotle discusses the first immovable mover, or prime mover, as it would be known in the Middle Ages. The first heaven, or outermost sphere of the fixed stars, moves with an incessant, uniform motion that had no beginning and will have no end. The first immovable mover is the cause of that motion, but, as its name clearly implies, it does not itself undergo motion of any kind. How, then, does it cause motion without being in motion? "It produces motion by being loved," was Aristotle's response (*Metaphysics* 12.7.1072b.3–4). But what loves the prime, or first, mover? Does each physical, celestial orb love it? Or is it rather the unmoved mover associated with each orb? Although Aristotle is unclear on this point, the latter interpretation seems the more plausible. For although Aristotle thought of the celestial ether, from which the orbs, planets, and stars were made, as in some sense alive, and even divine, he gives no indication that ethereal bodies were capable of the intellective action of loving. Such an activity would have been more appropriate to the unmoved mover associated with each sphere, since Aristotle attributes to them a capacity for thought. Indeed, their sole activity is thinking about their own thoughts. For this reason, Aristotle may have assumed that the immaterial, unmoved mover associated with each sphere was capable of loving and that each orb moves because its unmoved mover loves the first immovable mover, which Aristotle conceived as God.[87]

86. In Aristotle's judgment (*Metaphysics* 12.8.1073a.36–1073b.3), the unmoved movers, are "substances which are of the same number as the movements of the stars [i.e., planets], and in their nature eternal, and in themselves unmovable, and without magnitude." The bracketed qualification is mine. Among these substances, he adds, "one . . . is first and another second according to the same order as the movements of the stars" (Aristotle [Ross], 1984). Thus they have an order, but Aristotle does not here indicate whether they differ. Wolfson, 1973a, 17, says that "there is no distinction at all in the immaterial movers themselves; the distinction between them is only a distinction in their relation to things outside themselves – a distinction of external relation which, as we have shown, does not affect their nature." By contrast, Stewart, 1973, 545, believes that "there seems to be in Aristotle a hierarchy of being implicit in the doctrine of the unmoved mover. The first unmoved mover is prior in that it is unmoved both essentially and accidentally, whereas the planetary movers are only unmoved essentially. The first unmoved mover is also prior in the sense that the planetary movers receive additional motion from the movement of the outer heaven, which is moved by the first unmoved mover."

87. Ross, 1949, 181, and Kelly, 1964, 7, adopt this interpretation. But it is problematic, since the first unmoved mover thinks only its own thoughts – indeed, its thoughts are of itself – and is presumably unaware of anything outside of itself. If this is also an attribute of the other unmoved movers – recall that Aristotle assumed them identical – then they too

As God, the prime mover enjoys the best kind of life. Taking no cognizance of things external to itself, the prime mover lives the fullest life by thinking of itself, as the most worthy object of thought. "We say therefore," Aristotle concludes (*Metaphysics* 12.7.1072b.29–30 [Ross]), "that God is a living being, eternal, most good, so that life and duration continuous and eternal belong to God; for this is God." And, as if to emphasize the uniqueness of the prime mover, Aristotle assumes that it is located at the circumference of the universe, because "the things nearest the mover are those whose motion is quickest, and in this case it is the motion of the circumference that is the quickest: therefore the mover occupies the circumference" (*Physics* 8.10.267b.7–8).[88]

The concept of a prime mover played a significant role in Aristotle's physics and cosmology and is the theme of the eighth book of the *Physics*. In the sixth chapter of that book, Aristotle explains that up to this point he had "established the fact that everything that is in motion is moved by something, and that the mover is either unmoved or in motion, and that, if it is in motion, it is moved at each stage either by itself or by something else; and so we proceeded to the position that of things that are moved, the principle of things that are in motion is that which moves itself, and the principle of the whole series is the unmoved."[89] But if everything that is moved is moved by another, "we have a series that must come to an end, and a point will be reached at which motion is imparted by something that is unmoved" (ibid., 8.10.267b.1–2). The series of movers terminates with the first immovable mover, thus avoiding a potentially infinite sequence of movers and moved things. Versions of this argument would be frequently repeated.

Aristotle seems to have bequeathed two different explanations for the cause of celestial motion: one internal, the other external. In truth, in accordance with Thomas Aquinas's interpretation, Aristotle was also the source of a third account that fused the first two.[90] According to Thomas, who probably drew his interpretation from Avicenna, who, in turn derived it from a long Neoplatonic tradition, Aristotle distinguished two immaterial substances associated with each celestial orb: a soul and a separate intelligence.[91] The former is an integral part of its orb, whereas the latter is distinct

would be unaware of any other mover and thus could hardly be expected to love something outside themselves, not even the first unmoved mover. Thus a version of the first interpretation may have been intended. Each celestial sphere is somehow capable of loving its own unmoved mover and moves around and around as a consequence of that love.

88. Aristotle [Hardie and Gaye], 1984. Ross, 1949, 180–181, regards this statement as "an incautious expression." He believes that "Aristotle's genuine view is that the prime mover is not in space" and cites *De caelo* 1.9.279a.18, where Aristotle declares that "there is neither place, nor void, nor time, outside the heaven. Hence whatever is there, is of such a nature as not to occupy any place" (Aristotle [Stocks], 1984).

89. *Physics* [Hardie and Gaye] 8.6.259a.29–259b.1. See also *Physics* 8.5.256a.13–21.

90. Mastrius and Bellutus ([*De coelo*, disp. 2, qu. 4, art. 3], 1727, 3:503, col. 1) present the same three opinions.

91. In the *Treatise on Separate Substances*, Thomas says ([Lescoe], 1963, 46): "Thus, according

from its orb, though associated with it in some sense. It follows that as each orb moves around with uniform circular motion, its soul also moves around with it. Hence the soul of a celestial orb is necessarily in motion. But the soul of an orb is not the direct cause of the orb's motion. Motion arises because of the soul's intellectual desire and love for the separate intelligence that is also associated with the same orb. The direct cause of motion is therefore the separate intelligence, which causes the soul to love and desire it so that the soul will move its orb around and around.[92] Thus does the separate intelligence act as a final, and perhaps even as an efficient, cause of the motion of the orb and its soul. In apparent agreement with Thomas, a modern commentator on Aristotle suggests (Ross, 1949, 181) that "we must probably think of each heavenly sphere as a unity of soul and body desiring and loving its corresponding 'intelligence.' "[93] On this interpretation of Aristotle, which Thomas himself did not accept, the soul functions as an internal mover and the separate intelligence as the external mover. In Thomas's own interpretation, each intelligence is an unmoved mover, and the first of these is accorded no special status, which seems to agree with Aristotle's intent.

2. *Medieval concepts of the prime mover*

Scholastics usually chose one or another of these three interpretations. But whether the cause of celestial motion was conceived as internal or external, or a combination of the two, a major problem concerned the ultimate source of celestial motions. Was it God, or the prime mover, as God was often called in the context of cosmological discussions? If so, how did God cause the celestial motions?

Whereas Aristotle was ambiguous about the relationship of the prime mover to other movers, Christians were unequivocal: as God, the prime mover was in no way to be confused with any other celestial movers,

to the position of Aristotle, between us and the highest God, there exists only a twofold order of intellectual substances, namely, the separate substances which are the ends of the heavenly motions; and the souls of the spheres, which move through appetite and desire" (Sic igitur secundum Aristotelis positionem inter nos et summum Deum non ponitur nisi duplex ordo intellectualium substantiarum, scilicet substantiae separatae quae sunt fines coelestium motuum, et animae orbium quae sunt moventes per appetitum et desiderium). See also Weisheipl's excellent account (1961, 320). Gilson, 1955, 196, has a succinct description of Avicenna's interpretation.

92. Duhem, *Le Système*, 1913–1959, 8:324, describes precisely this relationship as the cause of celestial motion when he declares that "Greek and Arab Neoplatonists exercised the genius of their minds on the relation which unites the intelligence to the orb that it moves: to each orb, they will conjoin a soul that will be its immediate mover. This soul will know, admire, love the separate intelligence and desire to become like it; and this eternal desire will maintain the perpetuity of celestial motion."

93. The difficulty with Thomas Aquinas's interpretation, and David Ross's as well, is that in the sections of the *Metaphysics* where Aristotle considers the intelligences, or separate substances, he makes no mention of the souls of orbs and, consequently, neither have I in my brief description of Aristotle's views. Both Albertus Magnus and Robert Kilwardby adopted substantially the same interpretation as did Thomas (Weisheipl, 1961, 320).

whether external or internal; indeed, God was the creator of these other celestial movers. Of these two radically distinct entities – the prime mover and all other celestial movers – we shall consider the prime mover first. Because Aristotle discussed the unmoved mover only in his *Metaphysics* and *Physics* – and not in *De caelo* – questions about the prime mover in medieval cosmology usually appear in questions on those two treatises.[94]

a. *The existence and attributes of the prime mover*

The existence of a prime mover was most frequently demonstrated by the necessity to deny an infinite regress of cause–effect relationships. It was always assumed that such relationships could not proceed indefinitely and had to terminate with a prime mover. The denial of infinite regress could be applied to proofs of a prime mover in different ways. For Mastrius and Bellutus the existence of the prime mover was demonstrable in two ways: (1) by physical means, using motion as the illustration of infinite process; and (2) by metaphysical means, using causality and production, where it is assumed that causes cannot proceed *in infinitum* and must eventually come to a first cause that is itself uncaused.[95]

Marsilius of Inghen also employed the denial of an infinite regress to demonstrate the existence of a prime mover. We must arrive at some being which, if it is moved, is moved by itself. Marsilius then shows that the prime mover must be independent of all other things. For we can always ask whether something is dependent or independent of any other thing. If it is independent, it is the prime mover. If not, we can ask the same question of another being. But this process cannot proceed to infinity. It must stop at something that is independent of everything else, which is, Marsilius insists, the prime mover.[96]

There was almost universal agreement with Aristotle that a being had to exist – a first, or prime, mover – who provided the starting point for all cosmic motions of the celestial and sublunar regions. For most scholastics, this seemed to follow obviously from the Aristotelian principle, as expressed by Peter de Oña, that "every thing that is moved, is moved

94. For Aristotle's discussions, see *Metaphysics*, book 12, and *Physics*, book 8, and for the medieval questions on those discussions, see the "Catalog of Questions," Appendix I, qus. 152–165.
95. "Duplici medio demonstrari potest existentia primi motoris physico et metaphysico. Medium physicum est motus, quo usus est Ar. 8. *Phys.* et 12 *Metaphys*; metaphysicum est causalitas et productio quo demonstravit 2 *Met.* 2 in nullo genere cause posse in infinitum progredi, sed ad unum primum deveniendum esse." Mastrius and Bellutus [*Physics*, disp. 15, qu. 8, art. 3], 1727, 2:380, col. 1. Mastrius and Bellutus seem to hold that the metaphysical proof is stronger than the physical.
96. "Tunc ponitur ista conclusio quod aliquis est primus motor probatur quia in moventibus et motis essentialiter ordinatis non est processus in infinitum, ut patet septimo huius. Igitur est devenire ad primum motorem, qui si movetur, movetur a seipso et non ab alio." Marsilius of Inghen [*Physics*, bk. 8, qu. 8], 1518, 84v, col. 2. In the next, or second conclusion, Marsilius shows that the prime mover must be independent.

by another, a process that cannot go to infinity [from which one may infer the existence of a prime mover that is not moved by anything else]."[97] Thus if *A* moves *B* and *B* moves *C* and *C* moves *D*, the process cannot proceed infinitely, because there are no infinite processes in nature. And yet it appears that a body is put into motion by something that is itself already in motion. To terminate this potentially infinite process, something had to exist which was itself absolutely immobile and which could somehow also cause celestial orbs and bodies to move.[98] Although he expressed it differently, Marsilius of Inghen said much the same thing. A celestial mover could only be the prime mover if it had "primacy of causality" (*de primevitate causalitatis*) over all other celestial movers and independence from them. The latter condition presupposes that the other movers can cause motion only by virtue of the prime mover, which also has more influence than any of the others.[99]

As used in medieval scholasticism, the term prime mover (*primus motor* or *primum movens*) was equated with God,[100] and therefore had the customary attributes assigned to the deity, among which were immobility, immutability, infinite power or strength, indivisibility, and absence of magnitude.[101]

97. "Omne quod movetur, ab alio movetur: sed non est processus in infinitum, ergo." Oña [*Physics*, bk. 8, qu. 1, art. 8], 1598, 376v, col. 1. Mastrius and Bellutus, *Physics*, disp. 15, qu. 8, art. 3, 1727, 2:380, col. 1, mention and discuss the same principle.
98. Although we shall not enter into further subtleties associated with the concept that "every thing that is moved is moved by another," occasional exceptions were noted. Thus William of Ockham reports, with seeming approval, that the Commentator, Averroës, insisted that something could be moved by another in two ways, the first of which involved a mover that was numerically distinct from the thing moved. In this sense the principle "omne quod movetur ab alio movetur" always holds. In a second way, however, something could be moved by another thing that was in a distinct place or location, and in this sense the principle could fail. As his example, Ockham mentions that although the first sphere is moved by something that is really distinct from it, namely the prime mover, the latter has no distinct place because as God, it is not locatable and therefore has no place (Ad primum istorum respondet Commentator quod moveri ab alio potest esse dupliciter: vel ab alio numero distincto vel ab alio distincto situ. Primo modo omne quod movetur ab alio movetur. Secundo modo, non oportet: licet enim prima sphaera movetur ab aliquo realiter distincto, ut a primo motore, ille tamen non distinguitur situ ab isto, quia non habet situm). Ockham, *Brevis Summa libri Physicorum*, bk. 8, ch. 2, 1984, 125–126. Ockham's example had little or no impact on the principle "every thing that is moved is moved by another."
99. In response to the question as to what is meant by "primacy" when we use the term "prime mover" (*primus motor*), Marsilius explains: "Quantum ad primum notandum quod quando dicitur motorem celi esse primum hoc intelligitur de primevitate causalitatis et independentie in ordine essentialium moventium que sic se habent quod posterius non movet nisi in virtute primi moventis et influentis plusquam faciat movens posterius." Marsilius of Inghen, *Physics*, bk. 8, qu. 9, 1518, 85v, col. 1.
100. In a question on the prime mover, Buridan [*Physics*, bk. 8, qu. 11], 1509, 119r, col. 1, for example, asks "Whether the prime mover, namely God, is of infinite power [strength]" (Utrum primus motor, scilicet Deus, sit infinite vigoris). See also Peter de Oña, *Physics*, bk. 8, qu. 1, art. 8, 1598, 377r, col. 1. Marsilius of Inghen, *Physics*, bk. 8, qu. 8, 1518, 84v, col. 2, explains that the term *primus motor* is equivalent to the expressions *prima causa* and *ultimus finis*.
101. In a single paragraph, Chrysostom Javelli includes all of these attributes when he demonstrates [*Physics*, bk. 8, tract. 4, ch. 3], 1568, 1:201, col. 2, that "motor primus est omnino immobilis et incorporeus et penitus indivisibilis. Et arguitur sic: nullam habet

All of these attributes were discussed, sometimes in separate questions (see "Catalog of Questions," in Appendix I).[102] Immobility was among the most widely considered, because it was essential that the prime mover be able to cause motion without itself being in motion.[103] By analogy with natural examples, scholastics found it easy to assume the existence of such an entity. They observed that a magnet causes iron to move toward it but is itself unmoved; color causes sight, which is a kind of motion, but the color itself does not move; and even if the Sun were at rest, it would heat terrestrial, or inferior, things.[104] Indeed, if a prime mover were moved by something else, it would not be a prime mover.[105] Therefore it must either be moved by itself or not moved at all. But it cannot move itself, because it would then both be a cause of motion and be in motion – that is, it would be simultaneously in a state of actuality and potentiality.[106] Whatever qualifies as a prime mover, however, must be devoid of potentiality. Hence the prime mover cannot move itself.[107] Moreover, Buridan argues that the first thing in every genus ought to be the most simple. In the genus of things causing motion, this absolutely simple first thing is what causes things to move but is not itself moved.[108] During the Middle Ages, motion was divided into three types: natural, violent, and voluntary motion. Because all ordinary natural and violent terrestrial motion comes to rest and celestial

magnitudinem, ergo etc. Probatur antecedens: quoniam consequentia est nota, non habet magnitudinem infinitam quoniam illa non datur, ut probatum est in tertio physico[rum]; nec finitam quoniam probatum est in capite praecedenti non posse esse in magnitudine finita potentiam infinitam. Sed motor primus immobilis est potentiae infinitae cum sit movens tempore infinito, ut dictum est supra. Ergo constat ipsum esse penitus sine magnitudine et incorporeum, etc. Haec de tertio capite sufficiant." In earlier chapters, Javelli shows that the prime mover is one (200, col. 1) and of infinite power (200, col. 1–201, col. 2).

102. Not all scholastics included a discussion on the prime mover in their questions or commentaries on the *Physics*. Among those who did not are Thomas Compton-Carleton ([*Physics*], 1649) and Pedro Hurtado de Mendoza ([*Physics*], 1615), who includes a section on the First Cause but does not really treat of the prime mover.
103. From our sample list of authors, at least nine, from Pseudo-Siger of Brabant in the thirteenth century to Peter de Oña in the seventeenth, asked whether the prime mover is absolutely immobile (see the "Catalog of Questions" in Appendix I, qu. 159).
104. Buridan, *Physics*, bk. 8, qu. 6, 1509, 114v, col. 2, offers these very examples when he says: "Color movet visum absque hoc quod moveatur; et magnes quiescens attrahit ferrum; et si sol quiesceret adhuc calefaceret ista inferiora." John Major [*Physics*, bk. 8], 1526, sig. kiii recto, col. 1, presents the same examples, along with a Latin text that is very nearly identical with Buridan's.
105. See Buridan's "second way" (*secunda via*). Buridan, ibid.
106. As a continuous cause of motion, the prime mover would be considered in a state of actuality. If it could also cause its own motion, it could only do so because it has the potentiality for motion. Therefore at some point, it would be simultaneously in a state of actuality and potentiality.
107. Albert of Saxony [*Physics*, bk. 8, qu. 8], 1518, 81r, col. 1. For a much more detailed argument that makes the same point, see Buridan, *Physics*, bk. 8, qu. 6, 1509, 114v, col. 2–115r, col. 1.
108. Buridan, ibid. With a few different examples, Albert of Saxony makes essentially the same argument in *Physics*, ibid.

motion does not, but continues on without end, celestial motion could be neither violent nor natural in the ordinary Aristotelian sense.[109] It was assumed to be voluntary motion, that is, motion produced by an act of will. Within this context, a popular and perhaps convincing medieval argument was developed, pointing toward a prime mover. It involved three seemingly exhaustive possible relationships between a voluntary mover and a thing that is moved. Thus Buridan and Albert of Saxony argued that there is (1) something that causes motion and is itself in motion, just as (2) there is something that is in motion but does not itself cause motion; therefore (3) something must exist that causes motion but is not itself in motion, which would be the prime mover.[110]

Now a mover, or something that causes motion, is naturally prior to, and more perfect than, something that is only moved – that is, only in motion. Buridan argues further that something that is prior to something else ought to be separable from it more readily than a posterior thing is separable from the thing prior to it. Therefore an immobile and immutable mover that is incapable of being acted on and which meets all the conditions just described ought to exist. But this is the prime mover.[111]

Albert of Saxony invoked the daily motion in support of a prime mover who ought to cause things to move perpetually and uniformly. The daily motion was an example of a perpetual and uniform motion from which one could infer an immobile prime mover. Variation in motion, explains Chrysostom Javelli, who agrees with Albert's argument, "proceeds from variation in the mover." Without detectable variation in the daily motion, Albert and Javelli inferred the immobility of the prime mover. Moreover, if the prime mover were itself in motion, it would bear continually differing relationships to terrestrial things and cause different effects than we now observe. Because no such effects are detected, we may infer a prime mover that causes perpetual and uniform motion.[112]

109. This is Hugo Cavellus's argument (*Metaphysics*, bk. 12, qu. 8, 1639, 4:830, col. 2), but it is applicable to the Middle Ages.

110. Buridan, *Physics*, bk. 8, qu. 6, 1509, 115r, col. 1; Albert of Saxony, *Physics*, bk. 8, qu. 8, 1518, 81r, col. 1. In the seventeenth century, Cavellus assumed the same tripartite division (ibid.).

111. Buridan, ibid.

112. This is Albert's fourth conclusion in favor of a prime mover (see *Physics*, bk. 8, qu. 8, 1518, 81r, col. 1). For Javelli's discussion, see *Physics*, bk. 8, tract. 4, ch. 1, 1568, 1:200, col. 1, where Javelli seeks to prove "quod ad sempiternitatem unius motus requiritur immobilitas omnino ipsius motoris" and offers the following as proof: "Secundum probatur ex regularitate et uniformitate quam videmus in primo motu qui est motus diurnus et arguitur sic: motus primus est maxime regularis; ergo motor primus est omnino immobilis. Probatur consequentia quoniam ad sensum videmus irregularitatem motus procedere a varietate motoris. Nam quia moventia inferiora non semper sunt in eadem dispositione quia fatigantur in movendo, ideo in inferioribus non datur motus unus continuatus regularis quia igitur motor primus causat motum maxime regularem et aeternum, signum est quod semper eodem modo se habet; ex consequenti est omnino immobilis et invariabilis." Although Buridan omits mention of the daily motion, he gives essentially the same argument (Buridan, ibid., cols. 1–2). Buridan and Albert offer

b. Is there more than one prime mover?

Convinced of the existence of at least one prime mover, scholastics also inquired whether there might be more than one. In the sense that God is the prime mover, there could, of course, not be more than one, although the question was posed. It was, however, possible that other unmoved movers might exist that functioned as prime movers. The central problem was to define or explicate the imaginable ways in which a plurality of prime movers might exist. Chrysostom Javelli raised the question ([*Physics*, bk. 8, tract. 4, ch. 3], 1568, 1:201, col. 2) because the first heaven, or the outermost physical orb of the cosmos, is the first moved thing and seems to have a double mover, namely an intelligence that is conjoined to the heaven itself and also a separate mover that is distinct from the first heaven. Could these be two prime movers? Javelli, and most scholastics, distinguished between an intelligence that is somehow conjoined or associated with the first, or outermost, physical orb[113] and a mover that is completely separate from that orb. According to Javelli, the conjoined intelligence is not really a prime mover, because it is subordinated to the separated mover, by which we must understand God as prime mover.

Marsilius of Inghen developed a much more elaborate approach concerning the possibility of a plurality of prime movers and, as expected, denies it.[114] He distinguished three ways in which a plurality of prime movers could be imagined: (1) if the mobile celestial bodies were all equally prime – that is, if there are no differentiations among any of the celestial bodies to be moved – and completely independent movers were then applied to any one of them; (2) to one of these equally prime mobiles, different prime movers are applied simultaneously; and (3) to the same, or another, equally prime mobile, different equally prime movers are applied successively (Marsilius of Inghen [*Physics*, bk. 8, qu. 9], 1518, 85v, cols. 1–2). Marsilius then subdivides each of these three ways. For example, the first way can be subdivided into two other ways. In one way, we can imagine a plurality of worlds in each of which there is one prime mobile (for example, the sphere of the fixed stars) to which one prime mover is applied. As a second subdivision, we can also imagine only one world in which all of the celestial orbs are completely independent of each other and of exactly equal status – that is, they are equally prime mobiles. It follows that the mover

a few more arguments, which need not be included here. For a radically different set of arguments on the very same question, see Benedictus Hesse [*Physics*, bk. 8, qu. 16], 1984, 730–735, who emphasizes terminological and logical distinctions.

113. Authors varied in their conception of the first physical heaven. It could have been the tenth, ninth, or eighth spheres. The tenth and ninth spheres would have represented long-term motions of the fixed stars (precession and/or trepidation), whereas the eighth sphere represented the daily motion.

114. "Utrum plures sint motores primi." Marsilius of Inghen, *Physics*, bk. 8, qu. 9, 1518, 85r, col. 2–86r, col. 2.

of each of these equally prime mobiles is a prime mover and that there are as many prime movers as there are celestial orbs.[115]

In a series of six conclusions, Marsilius rejects all arguments for a plurality of prime movers. Although he allows for the possibility of a plurality of worlds, Marsilius, in the first conclusion, denies that this implies a plurality of prime movers, one for each world. In fact, he argues that a plurality of worlds is impossible, and therefore so is a plurality of prime movers.[116] Marsilius further argues (in the second conclusion) that the celestial orbs are not equal, because they differ in location (*situs*), place (*locus*), and position (*positio*). Since no two celestial orbs are equal, their movers cannot be equally prime. Only one prime mover can be supreme.

Following the sixth and final conclusion, Marsilius declares: "it is obvious that a plurality of prime movers does not exist in any of the ways in which a plurality of prime movers is imaginable. Indeed, there is only one prime mover. But the unity of the prime mover is proved, not only from what has been said, but from the order of the universe, from its state of efficient and final causes, [and] from the unity and perfection of the world" (ibid., 86r, cols. 1–2).

3. Medieval concepts of the other celestial movers

During the late Middle Ages, to inquire whether circular motion was natural to a celestial orb was to ask, in effect, how a celestial orb was moved, that is, what caused its motion.[117] To argue that its motion was really natural – that is, occurred by the very nature of a celestial orb – was in effect to reject other causes of motion, whether internal (impressed forces or souls) or external (intelligences or angels). The most significant alternative to natural celestial motion arising from the very nature of the orb was voluntary motion effected by agents capable of willing, such as intelligences, angels,

115. "Primus modus potest subdividi secundum quod dupliciter possunt imaginari plura mobilia eque primo. Uno modo quod sint plures mundi in quorum quodlibet sit unum mobile primum et cuilibet illorum mobilium priorum sit unus motor primus applicatus. Alio modo quod in isto mundo sint plura mobilia eque primo ita quod sint plures orbes celi et quilibet illorum sit independens ab alio. Et sic cuilibet motor potest dici motor primus." Ibid., 85v, col. 2.

116. Marsilius explains (ibid.) that "if there were a plurality of worlds, the earth of one would descend to the earth of another; and there would be a plurality of times and a plurality of gods, and between these worlds there would be a vacuum. All these are impossible." In light of the general acceptance of the possibility of a plurality of worlds after the Condemnation of 1277 and therefore the rejection of the argument that "the earth of one [world] would descend to the earth of another" (see Ch. 8, Sec. II.2), Marsilius's rejection of that possibility and his acceptance of the movement of an earth from one possible world to another are surprising. Perhaps his position on these issues was conditioned by his immediate goal of denying the existence of more than one prime mover.

117. In the "Catalog of Questions" in Appendix I, questions 173–175 specifically raise some of these issues, but questions on the role of intelligences and angels frequently involved issues about the causes of motion of celestial orbs.

and souls.[118] But there was at least one other conceptual possibility: something might be impressed into an orb that enabled it to move but that was not inherently natural to it, such as, for example, an impressed force, or "impetus," as it was often called. We must now examine these different motive powers and see how they were thought to operate.

Most scholastic natural philosophers and theologians were agreed that one angel or intelligence was associated with each orb, or with each planet (see Sec. II.4.e). No single angel was capable of moving all the heavens, because of the latter's enormous magnitude and variety of motions.[119] They were also in general agreement that God did not cause – although he could if he wished – the celestial motions directly, as an efficient cause, but rather had assigned this task to an intermediate or secondary cause of his own creation.[120] But what secondary cause? Had God chosen to move them by some distinct and separate entity, say an intelligence or angel? Or had he rather chosen to move the celestial orbs by some form, or soul, or natural innate force and thus allow them to be self-moving? A number of scholastics considered this issue in the form of a question that asked whether the celestial orbs are moved by intelligences or by some internal or innate form or power, perhaps even a soul, although, after the Condemnation of 1277, this last option, which implied fully living orbs, was exercised by few, if any.[121] Most chose one or the other of the two options: either separate,

118. One could also argue that celestial motion was violent, but no one did. In a series of four questions about the heavens, Michael Scot asks first, "Whether celestial motion is natural, violent, or voluntary" (Primo, utrum motus celi sit naturalis vel violentus vel voluntarius). Michael Scot [*Sphere*, lec. 5], 1949, 288.

119. "Quia Angeli virtus est limitata et finita, non videtur autem omnes celos posse esse adaequatam sphaeram activitatis unius Angeli propter ingentem magnitudinem; insuper propter motuum varietatem." Mastrius and Bellutus, *De coelo*, disp. 2, qu. 4, art. 4, 1727, 3:504, col. 2, par. 147, offer a brief discussion.

120. Raphael Aversa [*De caelo*, qu. 34, sec. 7], 1627, 150, col. 1, put it directly: "It is nevertheless certain that God could indeed move both the first movable sphere [*primum mobile*] and every other heaven [or orb]. In fact, he does not move them immediately by himself" (Certum tamen esse debet Deum potuisse quidem per se immediate movere tam primum mobile quam omne aliud caelum. De facto tamen non movere se solo immediate). Compton-Carleton, *De coelo*, disp. 4, sec. 3, 1649, 409, col. 2, asserts that although P. Lessius and some others hold that God alone moves the celestial orbs, the common opinion is that they are moved by intelligences (Quaeres primo, utrum moveantur coeli ab Intelligentijs? P. Lessius cum alijs nonnullis dicit moveri eos a solo Deo. Communis tamen sententia affirmat moveri ab Intelligentijs). Riccioli, *Almagestum novum*, pars post., bk. 9, sec. 2, ch. 1, 1651, 248, col. 1, also characterized it as the "common opinion." The common opinion had been just as common in the Middle Ages, when it was held by Thomas Aquinas and numerous others (for example, Ulrich of Strasbourg; see Duhem, *Le système*, 1913–1959, 8:39).

It was contrary to faith to allow that any other thing or entity, except God, could create anything. Thus Thomas Aquinas [*Sentences*, dist. 14, qu. 1, art. 3], 1929–1947, 2:353, describes an opinion, probably Avicenna's, in which God, as first cause, creates an intelligence, which in turn creates the soul of a celestial orb, from which soul the substance of the orb itself is produced. Thomas repudiates this opinion, because "our faith, which assumes that only God is the creator of things, does not suffer it" (Hoc autem fides nostra non patitur, quae solum Deum rerum creatorem ponit).

121. As we saw, Bonaventure [*Sentences*, bk. 2, dist. 14, pt. 1, art. 3, qu. 2], *Opera*, 1885, 2:348, cols. 1–2, rejected souls as possible celestial movers.

external movers or some internal power, though probably not a soul. A few, however, thought the alternatives might be equally probable, a position that Saint Bonaventure favored when he asked "Whether celestial motion occurs by [means] of a proper form or by an intelligence."[122] During the Middle Ages a celestial mover was conceived in a variety of ways: as an intelligence, an angel, a form, or a soul. Some of these terms were used interchangeably. In Greek and Arabic Neoplatonism, soul and intelligence were quite distinct, the former conjoined to its orb, the latter associated with that orb as a separate and distinct entity. The identification of angels and intelligences had already been made by Avicenna, al-Ghazali, and Moses Maimonides and was soon widely adopted in the Christian West, although with significant exceptions.[123] Occasionally the term "intelligence" was equated with an internal form. Whatever their differences, these entities shared certain fundamental properties: all were divinely created, spiritual, immaterial, devoid of magnitude, indivisible, and incorruptible, although the divine will could choose to destroy them. Of these entities, intelligences and angels were generally categorized as external movers, while souls and forms were usually perceived as internal movers conjoined to their bodies.[124] However, when intelligences were equated with forms, they were assumed to function as internal movers.

John of Jandun illustrates both usages for the term "intelligence" when he asks "Whether an intelligence that moves an orb is conjoined to the orb essentially as the substantial form of its matter."[125] In his lengthy discussion, Jandun, who describes the question as "unusual" (*inconsueta*), distinguishes two opinions: the first is that an intelligence that moves its orb is united to it as part of its very being, so that it would be false to say that it is essentially separated from its orb.[126] The second opinion, which is not explicitly as-

122. Ibid., 348, col. 1–349, col. 2. In addition to the eleven other individuals who discussed some form of this question (see the "Catalog of Questions," Appendix I, qu. 197), we may add Franciscus de Marchia, whose opinions Duhem describes (*Le Système*, 1913–1959, 8:325–328). Clavius, although he grappled with numerous cosmological issues in his famous *Commentary on the Sphere of Sacrobsoco*, appears to have ignored celestial movers, except for an occasional mention of intelligences, usually with reference to someone else's discussion. He offers no distinct causal account of the celestial motions. Perhaps this is only a reflection of the fact that Clavius was more astronomer than natural philosopher.

123. Duhem, ibid., 6:30. Duhem (8:324) holds that Muslim, Jewish, and Christian philosophers, who received the concept of celestial souls and intelligences from pagan sources, adapted these entities to monotheism by identifying angels with intelligences and calling the latter by the former term. Although the identification of angel with intelligence was quite common, I have observed no analogous equation of angel with celestial soul.

124. In this regard, see Thomas Aquinas, *De caelo*, bk. 2, lec. 3, par. 3, 1952, 156.

125. "Utrum intelligentia movens orbem sit coniuncta orbi secundum esse, ut forma substantialis suae materiae." John of Jandun [*Metaphysics*, bk. 12, qu. 13], 1553, 134r, col. 2.

126. "Notandum quod ista quaestio est inconsueta et procedam sic: quod primo tangam duas opiniones cum suis rationibus; secundo solvam rationes secundum tenentes has opiniones et quilibet tunc eligat quae sibi videbitur probabilior. Prima est opinio unius et est quod intelligentia movens orbem effectivae est unita orbi in esse existentiae ita quod est falsum ipsam dicere esse separatam in esse existentiae ab orbe." Ibid., 134v, col. 1.

serted but is clearly understood and which Jandun thought more probable, holds that an intelligence is wholly separate from its orb.[127]

That Jandun thought the theory of a separate intelligence was only more probable, rather than reasonably or definitely certain, is not difficult to understand. Some of those who defended the first opinion and assumed that an intelligence is united to its orb seem to have qualified their understanding of "united" or "conjoined" to make it more palatable to those who argued for a complete separation between intelligence and orb. In one interpretation, for example, they insisted that the intelligence and its orb were not one with respect to their being but only with respect to "appropriation" (*secundum appropriationem*), that is with respect to a particular relationship, namely that between a mover (the intelligence) and a thing that is moved (the celestial orb). Strictly speaking, the relationship is between a mover and the thing it moves, not between a form and its matter. Supporters of this interpretation viewed the relationship as analogous to that between a sailor and his ship. The sailor and the ship do not form one unified thing but, in some important sense, the sailor causes the ship to move.[128]

Another argument which sought to further weaken any sense of essential or organic union between intelligence and orb focused on the distribution within the orb of the power (*virtus*) of the intelligence. Ordinarily, forms which perfect and complete some extended thing or object were thought of as spread uniformly throughout its magnitude. The relationship between intelligence and orb, however, is quite different. The former is related to the latter as a point is united to its line. Although a point is united to a line essentially, it is not extended throughout the line. Here the essentiality is retained between intelligence and orb, even as the dissemination of the power of the intelligence is relegated to a minute part of the orb.[129] The partisans of this opinion thought of intelligences as internal movers but weakened the sense of unity between intelligence and orb. For the most part, however, intelligences were unambiguously conceived as external movers, as I shall now emphasize.

4. External movers: intelligences (or angels)

Following Avicenna, the terms "intelligence" (*intelligentia*) and "angel" (*angelus*) were usually taken as synonymous where the term intelligence was

127. Jandun says (ibid., 135r, col. 2) that "those who hold the second opinion, which seems more probable and seems to be the intention of Aristotle and the Commentator, solve the arguments of the other [or first] opinion" (Sed isti qui tenent secundam opinionem, quae videtur esse probabilior et de intentione Aristotelis et Commentatoris, solvunt rationes alterius opinionis).

128. For the Latin passage, see ibid., 134v, col. 2. Amicus, *De caelo*, tract. 4, qu. 4, dubit. 1, art. 1, 1626, 163, col. 1 and art. 4, 167, col. 1, also mentions the sailor–ship analogy.

129. John of Jandun, ibid., 135r, col. 1.

applied to an angel that was associated with a celestial orb.[130] Celestial intelligences thus formed a species, or subset, within the genus of angels. Although the terms angel and intelligence could be used interchangeably, the *questiones* literature reveals a preference for intelligence.[131] By the seventeenth century, it was the "common opinion" that external intelligences, or angels, caused the motion of the planets and stars, and not some form or internal mover.[132] Of the two options, then, that which assumed intelligences as separate, external movers clearly predominated. Whatever the term employed, intelligences or angels were usually conceived as spiritual substances created by the First Intelligence (God) even before the world itself was created. The properties attributed to them were usually those that Aristotle had attributed to the prime mover and the other unmoved movers, namely immobility, indivisibility, and absence of magnitude. Although angels possessed considerable power, they could not exceed the finite powers conferred on them at the Creation. An angel could move bodies, indeed, it could move a rare and light celestial orb, but it did not follow, for example, that it could move the heavy earth or create a vacuum.[133] Its powers were great, but limited and puny in comparison with those of God. Although for Aristotle the immaterial intelligences were limited to the number of celestial motions, where it is understood that each motion is effected by a celestial orb, Christians assumed an enormous number of immaterial, spiritual substances. Although many identified planetary intelligences with angels, most angels were assumed to enjoy separate existences and to be completely independent of bodies of any kind.[134]

130. On Avicenna, see Weisheipl, 1961, 303. Albertus Magnus was a notable exception, rejecting the identification of intelligences and angels (Weisheipl, 1960, 323, and 1961, 307). Albertus explains that the ancient Peripatetics did not discuss angels but that among moderns certain Arabs – for example, Avicenna and Algazel – and certain Jews – for example, Isaac Israeli and Moses Maimonides – had discussed them. "Moreover, they agree that intelligences are substances which the common people call angels" (Concorditer autem isti dicunt quod intelligencie sunt substancie quas uulgus angelos uocat et dicunt). Weisheipl, 1960, 323. From the thirteenth to the seventeenth century, most considered intelligences and angels identical, as did Richard of Middleton ([*Sentences*, bk. 2, dist. 14, art. 1, qu. 6], 1591, 2:173) and Giovanni Baptista Riccioli (*Almagestum novum*, pars post., bk. 9, sec. 2, ch. 1, 1651, 248, col. 1). In the *Convivio*, Dante remarks that the common people refer to the intelligences as angels (see Kelly, 1964, 24).

131. In the "Catalog of Questions" in Appendix I, see questions 195–202, and 206, where *intelligentia* was preferred over *angelus* or its variants. Among users of the term "angel," we can include Cornaeus, *De coelo*, tract. 4, disp. 1, qu. 4, dub. 2, 1657, 504, who says that "the common opinion of the ancients, the philosophers, and the Fathers held that each orb was turned by an angel," and Oresme (*Le Livre du ciel*, bk. 2, ch. 2, 1968 289), who uses the terms "intelligence" and "angel" interchangeably within the same paragraph. Bonaventure (*Sentences*, bk. 2, dist. 14, pt. 1, art. 3, qu. 2, *Opera*, 1885, 2:349, col. 1) also used them synonymously when he declared that "another position is that God moves the heaven by means of a created intelligence, or an angel" (Alia vero positio est, quod Deus movet caelum mediante Intelligentia creata, sive mediante Angelo). Riccioli also used the terms interchangeably (Riccioli, ibid.).

132. For more on the "common opinion," see note 120 of this chapter.

133. Suarez, *Disputationes metaphysicae*, disp. 35, sec. 6, 1866, 2:476, col. 2, par. 27.

134. See Thomas Aquinas, 1963, 45.

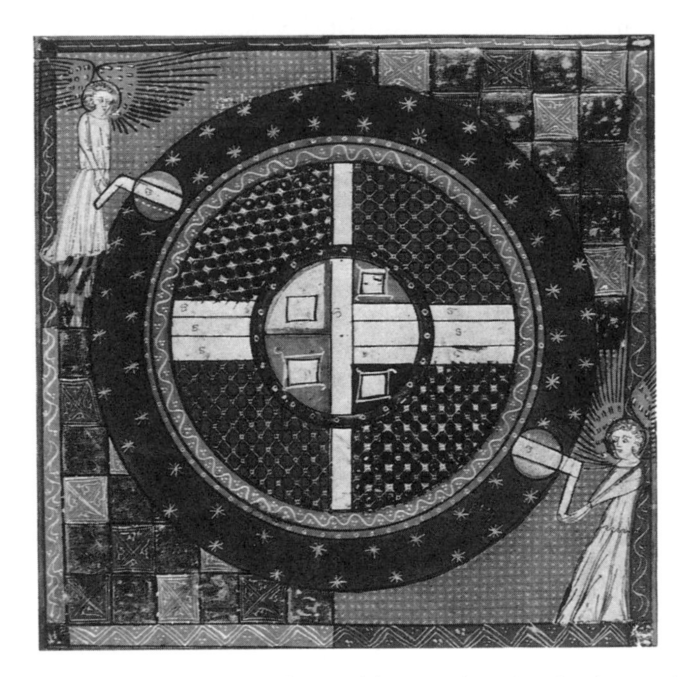

Figure 10. Angels cranking the world to produce its circular motion. (British Library, Harl. MS. 4940, fol. 28. For another illustration of angels cranking the world, see Murdoch, 1984, 336.)

Perhaps the most crucial question about celestial intelligences and angels concerns the manner in which they were thought to move the celestial orbs. We saw that in the Neoplatonic tradition, as reported by Thomas Aquinas and others (see Sec. II.1 and n. 92 of this chapter), an intelligence causes the soul of its orb to love it to such an extent that the soul causes the body to move. Under these conditions, the intelligence operates more as a final cause than an efficient cause. But while this opinion was frequently reported, it was rarely, if ever, held by scholastic authors.

a. The Condemnation of 1277

In the Condemnation of 1277 at Paris, article 212 condemned the opinion "That an intelligence moves a heaven by [its] will alone."[135] Why did the bishop of Paris, who ordered the condemnation, find this article objectionable? It is likely that he was influenced by a treatise titled *Liber de intelligentiis*, which has been attributed to Witelo, the great Polish optical writer who

135. "Quod intelligentia sola voluntate movet celum." Denifle and Chatelain, 1889–97, 1:555.

lived in the thirteenth century.[136] The author of the *Liber de intelligentiis* explains that only God can move by will alone, for only God is pure act so that his action is identical with his will. But the action of any creature is a mixture of act and potentiality and cannot, therefore, be identical with its will. It follows that an intelligence cannot move an orb simply by willing it but must have some contact with, or penetration of, the orb's body.[137] Hence it cannot will the motion of an orb from a distance. Perhaps the article was also condemned because the assumption that an angel moves an orb by its will alone seems too similar to the way the human soul moves the body whose substantial form it is. The implied analogy suggested that just as the soul and its body are alive, so also are the intelligence and its orb alive. In other articles condemned in 1277, the bishop of Paris had repudiated the idea of living celestial orbs.[138]

Because it was always understood that the will is guided by an intellect, scholastics may generally have interpreted the condemned article to include the intellect as well as the will. Hervaeus Natalis makes this abundantly clear by repeatedly coupling the two terms in his analysis of angels as celestial movers.[139] Richard of Middleton explicitly links the angelic intellect and will when he explains how a heaven is moved by an intelligence: "a certain potency [or power] of this intelligence directed by the intellect and commanded by the will brings forth a certain power of the heaven from potency to act."[140]

When he condemned this article, the bishop of Paris, who was an ideological foe of Thomas Aquinas, may have had the latter in mind. For Thomas had indeed allowed for the possibility that an intelligence could move a celestial orb by its will alone and may even have assumed it. His fellow Dominican, Bernard de Trille, unambiguously proclaimed that angels move the orbs by will alone (Duhem, *Le Système*, 1913–1959, 6:46).

Among Parisian scholastics who wrote treatises after the Condemnation of 1277, some seem to have accepted and acknowledged the condemned article without mentioning it, while others mentioned it and addressed the problem directly. Among the former was Richard of Middleton, who,

136. I follow Duhem's summary of the *Liber de intelligentiis* in *Le Système*, 1913–1959, 5:367 and 6:30. The text has been edited by Clemens Baeumker, *Witelo, ein Philosoph und Naturforscher des XIII. Jahrhunderts*, Beiträge zur Geschichte der Philosophie des Mittelalters, vol. 3, pt. 2 (Münster, 1908).

137. See Duhem, ibid., 6:30, and, for a translation of the Latin text, 5:367; also Hissette, 1977, 134.

138. See Hissette, 1977, 134, and also condemned articles 92 and 213, in Denifle and Chatelain, 1889–1897, 1:548 and 555, respectively, and Hissette, 130–133. For other reasons why the article may have been condemned, see the discussion on Hervaeus Natalis that follows.

139. Hervaeus Natalis [*Sentences*, bk. 2, dist. 14, qu. 1], 1647, 243, col. 2–246, col. 2. Most scholastics did the same. Suarez, however, remarks (*Disputatio metaphysicae*, disp. 35, sec. 21, 1866, 2:474, col. 2) that some authors insist that although the intellect directs the will, the latter is the real motive force.

140. I give the full quotation from Richard shortly.

perhaps around 1293 or 1294, wrote a brief account,[141] wherein he explains that

> a heaven is moved by an intelligence in this sense: a certain potency [or power] of this intelligence, directed by the intellect and commanded by the will, brings forth a certain power of the heaven from potency to act, by which [power] the heaven is determined for a definite motion commanded by the will of the intelligence; and by this power it moves the heaven itself. These things are so arranged [or ordained] for moving the heaven that an angel under God is the principal mover of the heaven. The power [*virtus*] caused in the heaven by this angel in the aforesaid manner is as if it were the instrumental mover.[142]

Within this paragraph, Richard embodies the new approach, which sought to avoid entanglement with the condemned article. The angel's will does not directly and immediately move its heaven. Instead, the angel actualizes a "force" or "power," a *virtus*, which causes the orb or heaven to move. As long as the intelligence or angel keeps this power, or instrument, actualized, "just so long will the heaven be moved; but when it ceases to cause the aforesaid influence [or power] and to conserve it in the heaven, the heaven will rest."[143]

Those in the second group provided more details, and even a name, for the *virtus* invoked by Richard of Middleton. Two of the most important of these authors were Godfrey of Fontaines (ca. 1250–ca. 1306), who considered the problem in his fifth *quodlibet*, and Hervaeus Natalis (d. 1323) of the Dominican Order, who treated it in his *Commentary on the Sentences*.[144] Because Godfrey wrote his account around 1288, probably before Hervaeus wrote his, it will be advisable to consider Godfrey's views first. Godfrey warns that one ought not to assume that an angel can move a heaven by its will alone, because this proposition was included among the condemned articles, by which he undoubtedly means the articles condemned at Paris in 1277.[145] It was condemned, Godfrey explains, because an angel was said

141. According to Gilson, 1955, 347, Richard "completed the Fourth Book of his *Commentary on the Sentences* soon after 1294." If he wrote the four books sequentially, the second may have been completed around 1293 or 1294.

142. "Dico ergo quod celum movetur ab intelligentia per hoc: quod aliqua potentia ipsius intelligentiae intellectu dirigente et voluntate imperante aliquam virtutem educit de potentia celi ad actum qua coelum determinatur ad motum determinatum imperatum ab intelligentiae voluntate; et per illam virtutem ipsum coelum movet se. Sunt ergo ista sic ordinata in movendo coelum quod angelus sub Deo est celi principalis motor. Virtus causata in celo ab ipso angelo modo praedicto est quasi instrumentalis motor." Richard of Middleton, *Sentences*, bk. 2, dist. 14, art. 1, qu. 6, 1591, 2:173, col. 2.

143. "Et quandiu haec virtus continuabitur in coelo per ipsius intelligentiae potentiam tamdiu celum movebitur et cum desinet praedictam influentiam causare et conservare in celo celum quiscet." Ibid.

144. Both authors were treated by Duhem in *Le Système*, 1913–1959, 6:46–51, where he translated sections of their arguments.

145. "Nec tamen dicetur angelus movere caelum sola voluntate, quod est reprobatum inter articulos condemnatos." Godfrey of Fontaines, *Quodlibet* V, qu. 6, 1914, 22–23.

to move anything with complete freedom of choice, not according to some determination and relationship which it has to the thing that it moves. Indeed, an angel cannot be said to move an orb or heaven by its will alone, because a good angel is a minister of God and moves only what has been determined for it to move, whereas a bad angel cannot move any bodies without God's permission. Thus an angel's mere decision to move a body does not signify that the body will move. As Godfrey explains, "To order that something happen does not signify that it will happen."[146]

But if an angel does not move by will alone, how does it move an orb or heaven? Godfrey suggests that, in addition to intellect and will, we should assume that an angel possesses a third entity, which he calls a "motive force" (*virtus motiva*) and which he describes as "something spiritual, simple, indivisible, non-material, and non-local." In Godfrey's view, the intellect and will bring the *virtus motiva* into play in order to move the body which has been willed to move. Godfrey describes the *virtus motiva* as a third executive power or force (*tertia potentia executiva*). It is called into activity by the intellect and will, which cannot by themselves effect or complete the acts that they will. Godfrey explains that "such a proximate and immediate motive force is assumed . . . because to grasp and desire a motion or to move a thing or to order that some body be moved does not suffice, but a third executive power [or force] is required which by its action [effects] what has been grasped or commanded."[147]

Although Hervaeus's account bears some resemblance to Godfrey's, there is no evidence that the former was directly influenced by the latter. Hervaeus explains that it was Aristotle who asserted that intelligences move the orbs by intellect and will alone, but that the intellect and will of an intelligence operate by the very nature, or essence, of the intelligence. Therefore, despite the sense of choice and decision that intellect and will seem to imply, an intelligence would be predetermined and have no capacity to choose whether or not it wishes to move its orb.[148] Hervaeus further criticizes those who believe that intelligences operate by intellect and will alone by observing that if these two essential properties formed the essence of an intelligence, it could will anything at all without limit, since it would be absolutely perfect. But such perfection belongs only to God.[149]

What is Hervaeus's solution? The key lies in his interpretation of the condemned article, which he describes as "a certain article, which, as is

146. "Et quia imperare quod aliquid fiat non est facere illud." Ibid., 24.
147. "Sed ponetur talis virtus proxime et immediate motiva . . . , quia scilicet apprehendere et appetere motum vel rem movere vel imperare quod corpus aliquod moveatur non sufficit, sed requiritur tertia potentia executiva per eius actum sic apprehensi et imperati; et per hoc patet ad argumenta." Ibid.
148. As Hervaeus put it (*Sentences*, bk. 2, dist. 14, qu. 1, art. 3, 1647, 244, col. 2), "according to this path, any whatever intelligence is more determined by its very nature for moving its orb than fire is for heating [things]" (Et secundum hanc viam unaquaeque intelligentia ex natura sua erat magis determinata ad movendum orbem suum quam sit ignis ad calefaciendum).
149. Ibid., 244, col. 2–245, col. 1.

said, condemns as erroneous to posit that angels move the orbs by will alone, so that they could move bodies at the pleasure [or command] of their will."[150] Hervaeus thus understands the objection of those who condemned the article to be that angels or intelligences could move any body whatever, without limitation of size, merely by willing it. Thus angels would have infinite capacities equal to God's. Hervaeus of course rejects such an interpretation and suggests that intelligences can only exercise their wills to move things that are proportioned to their finite power. Even if an angel should wish to do so, it cannot move bodies larger than the finite maximum for which it has been created.

Just as Godfrey of Fontaines did, Hervaeus distinguishes between the will of an intelligence and its capacity to execute that will. The message of the condemned article is that an angel or intelligence cannot move by its will alone, if by this we understand a capacity to move anything indifferently.[151] Hervaeus explains that "Besides the will which the angel possesses, a definite relationship [or proportion] exists between the mobile body and the will of the mover, insofar as the will of the mover is a finite motive force [*virtus motiva finita*]."[152]

By adding a *virtus motiva* to the intellect and will, Godfrey and Hervaeus may have begun a tradition that continued into the seventeenth century. Not only does an intelligence require an intellect and will, but the intellect and will can only be exercised by means of a third spiritual entity, which is conceived as a "finite motive force," a *virtus motiva finita*, and which Godfrey called an "executive power." The intellect and will of an intelligence are necessary conditions for an angel to move an orb, but they are not sufficient. To achieve motion an angelic will must also relate as a finite motive force to the body it wishes to move. And for motion to occur, the angel's finite motive force must be sufficient to move the body it wills to move. In all this, we must understand that the *virtus motiva* is in the angel, not in the body or orb that it moves. The *virtus motiva* is not an impressed force which the angel transmits to the orb in order to move it. Neither Godfrey nor Hervaeus seem to have had any knowledge of impressed forces or the impetus theory, which received its full development somewhat later in the fourteenth century.[153]

150. "Sciendum quod est quidam articulus, qui ut dicitur, condemnat sicut erroneum ponere Angelos movere orbes per solam voluntatem, ita quod ad nutum pro voluntate possint movere corpora." Ibid., 245, col. 2.
151. "Et sic potest intelligi articulus ut sit sensus: quod angelus per solam voluntatem non movet, ita quod ad nutum subsit sibi quod oportet indifferenter movere." Ibid., 246, col. 1.
152. "Quia praeter velle, quod est ex parte angeli, requiritur determinata proportio inter corpus mobile et ipsum velle moventis, secundum quod velle moventis est virtus motiva finita." Ibid.
153. In a section titled "On How the Intelligences Move the Heaven or Planets" (De modo quo intelligentiae caelum vel sidera movens), Riccioli, *Almagestum novum*, pars post., bk. 9, sec. 2, ch. 1, 1651, 250, col. 1, says that it is customary to ask "Whether intelligences, or, universally, angels, move the heavens by intellect and will alone or by a

By the late sixteenth century, at the latest, this changed. The assumption of a third power, or an executive power (*potentia executiva*), to carry out the wishes of the intellect and will of an angel was by then fairly common. The tradition of Godfrey of Fontaines and Hervaeus Natalis, and perhaps other thirteenth- and fourteenth-century authors, had taken hold. In the late sixteenth and the seventeenth century, however, they would ask whether that executive force, or *virtus motiva*, transferred or transmitted an impressed force to the physical orb itself. With this question we are dealing with two forces, one external to the celestial orb (the angel or intelligence), the other internal to it (the impressed force). It will be more appropriate to treat this later development in a subsequent section of this chapter (Sec. II.6).

Although normally independent of bodies, intelligences were capable of uniting with bodies for special operations: for example, with the celestial orbs to cause their motion. In this capacity, intelligences were said to "assist" (*assistere*) the celestial orbs to move. The expression "assisting intelligence," which was apparently absent from medieval usage, appears to have come into vogue during the sixteenth and seventeenth centuries.[154] But precisely what "assisting" was intended to convey is unclear. At least three different interpretations have some plausibility. Were intelligences assisting entities because, as God's creations, they were surrogates given the duty and power to move the celestial orbs and could therefore be viewed as God's assistants and helpers? Or did intelligences assist the orbs because they were their sole movers? Or, finally, were the orbs perhaps moved by two complementary powers, one internal, the other external, where the intelligence, or external power, was only a partial mover and therefore could be conceived as assisting the internal force?

In a one-sentence declaration that "the heavens are not moved by themselves; nor [are they moved] by an internal form; nor directly by God alone; but they are moved by assisting intelligences or angels," Giorgio Polacco seems to lend support to the first opinion that the intelligences are God's assistants.[155] By denying that God moves them alone and following that with a declaration that assisting intelligences or angels move the heavens, Polacco links God with the intelligences and implies that the role of the latter is to assist the former. Bartholomew Amicus is one of a number of scholastics who provide arguments for believing that the second interpre-

power distinct from each [of them]" (Quaeri solet an intelligentiae celos, vel universim angeli, corpora moveant solo intellectu ac voluntate, an vero per potentiam ab utraque illa distinctam). Among medieval authors who assumed that such a power was necessary, Riccioli lists Duns Scotus, Henry of Ghent, Franciscus Mayronnis, Peter Aureoli, and one "Godophredus," probably Godfrey of Fontaines.

154. Among those who used the term *assistere* or *assistentes* are Galileo, *De caelo*, qu. 6 (L), 1977, 155; Aversa, *De caelo*, qu. 34, sec. 7, 1627, 147, cols. 1–2 and 150, col. 2; Riccioli, ibid., 248, col. 1; and Giorgio Polacco, *Anticopernicus Catholicus*, 1644, 63.

155. "Coeli nec a se ipsis; seu forma sua intrinseca; nec immediate a solo Deo; sed ab intelligentiis, seu angelis, assistentibus moventur." Polacco, ibid.

tation may have been largely intended. In one place, Amicus declares that an "assisting form" (*forma assistens*) is something that gives motion and operation to a subject without serving as its actual form.[156] Here "assisting" seems to imply that the intelligence supplies the full motive power. Amicus offers a similar description later when he describes four different opinions as to whether the cause of celestial motions is internal or external. As the second opinion, which Amicus characterizes as the most common among theologians and philosophers of the Peripatetic school, the celestial orbs are said to move by assisting intelligences, where once again each intelligence seems to be the total immediate cause of its own orb's motion.[157] But evidence in favor of the second opinion is also compatible with the first interpretation. The intelligences may move the orbs wholly by their own efforts, but they may also have been perceived as doing God's work and therefore performing as assistants.

Amicus also provides evidence in favor of the third interpretation. In that opinion, a celestial orb is assumed to be moved partly by a proper, internal form and partly by an external assisting intelligence.[158] Because it was not the sole mover, the intelligence in this situation could be construed as only assisting the motion of the orb. Whatever "assisting intelligence" was intended to convey, Raphael Aversa was convinced that it was essential that every heaven or orb be moved by one. Among one of eight arguments in favor of intelligences as opposed to internal forces or forms, Aversa insists that the latter could not account for the various activities of planets and orbs. For not only do orbs and planets move in one direction rather than another and toward one pole or another, but the first movable sphere (*primum mobile*) moves toward the west, while the planets move from west to east. Moreover, they all move around different poles, and their speeds and times differ. Aversa concludes that such differences of behavior could not be produced by an internal nature and by some power or force of the heaven. Rather we should assume that "every heaven is moved by an assisting intelligence which has been divinely ordained to move in a certain direction

156. "Tertio notandum formam assistentem dici quae dat subiecto motum et operationem absque eo quod illud informet." Amicus, *De caelo*, tract. 4, qu. 4, dubit. 1, art. 1, 1626, 162, col. 2.

157. "Secunda opinio est dicentium moveri ab intelligentiis orbibus assistentibus et nullo modo a caeli forma active; ex ordinatione primae causae ad mundi conservationem est communis omnium theologorum et philosophorum ex schola peripatetica." Ibid., tract. 5, qu. 6, dubit. 2, art. 2, 293, col. 2. Riccioli, who also presented four different interpretations of the cause of celestial motions, lends support to this interpretation in the last of them when he identifies the common opinion on celestial movers, explaining that the "heavens and planets are moved by intelligences or by angels as assisting and immediately effective causes." As did Amicus, Riccioli also described this interpretation as the "most common" (Quarta opinio: eaque communissima est caelum et sidera moveri ab intelligentiis seu ab angelis tamquam assistentibus et immediatis causis effectivis). Riccioli, *Almagestum novum*, pars post., bk. 9, sec. 2, ch. 1, 1651, 248, col. 1.

158. "Tertia opinio est dicentium hunc motum partim esse a propria forma, partim ab intelligentia assistente." Amicus, ibid., 294, col. 1.

and with a certain speed."[159] Although Aversa chose to ignore the possibility that God might have achieved the same results by programming each orb with some kind of internal power, rather than assigning it an already programmed external intelligence, others had long before considered that option, as we shall see (Sec. II.5.b, c)

b. The location of an intelligence with respect to its orb

If an intelligence, or angel, is a separate, spiritual substance rather than an informing form, where is it located with respect to the orb it moves? Does "separate" mean outside the orb? Or is the intelligence located within the orb but nevertheless completely distinct from it? Few scholastics posed such questions, and therefore they rarely provided explicit responses. Amicus, for example, insisted that an intelligence was united to its whole orb only by the contact of force or power, but not by informing the orb as an inherent part of it.[160] Aversa thought that it ought to have a real presence in its orb, but not in the whole orb. Its immediate presence should be felt only where it could truly "assist" (*assistere*) the orb. In Aversa's opinion, the fastest-moving part of the orb was most in need of assistance, so he suggested that an assisting intelligence might be located in that part of the orb that lies in the middle, between its poles.[161] Under these circumstances, does an intelligence always remain immobile, in a fixed space, as every part of the rotating orb passes by? Or does it assist only a particular fixed part of the heaven, and thus turn around with the orb while remaining in that fixed place?[162] Although Aversa offers no opinion on these alternatives, he pro-

159. "Septimo ulterius, si spectemus naturam ipsius caeli, nulla potest esse maior ratio cur potius caelum moveatur ab hac parte versus illam et non e contra, vel versus polum aliamve quamvis partem; cur primum mobile ob hoc ortu in hunc occasum, planetae vero ab occasu in ortum; cur primum mobile super illos polos, planetae super alios. Item cur motus fiat in tanto vel tanto tempore; cur aliud caelum moveatur velocius, aliud tardius. Non utique potest omnis huiusmodi differentia determinari ex intrinseca natura et virtute caeli. Oportet ergo fateri omne caelum moveri ab intelligentia assistente quae iuxta divinum praescriptum movet tali certa parte et tanto certo tempore." Aversa, *De caelo*, qu. 34, sec. 7, 1627, 149, col. 1.

160. "Intelligentia unitur orbi toti tantum per contactum virtutis, ergo non per informationem quia informatio supponit indistantiam formalem inter extrema." Amicus, *De caelo*, tract. 4, qu. 4, dubit. 1, art. 3, 1626, 166, col. 1.

161. "Debetque unaquaeque Intelligentia per realem praesentiam insidere suo orbi quem movet. Et quia dici non debet esse praesens toti caelo, assistere debet immediate alicui certae parti. . . . In qua vero parte assistat Intelligentia verisimile est assistere parti mediae inter polos cuius partis motus velocissimus est." Aversa, *De caelo*, qu. 34, sec. 7, 1627, 150, col. 2. Aversa seems to have in mind a sphere whose fastest-moving part would be the circumference of its equatorial circle, or, as he puts it, the "middle part between the poles" (parti mediae inter polos).

162. "Et vel ipsa Intelligentia semper manet immota in quodam certo spatio, verbi gratia in nostro oriente aut alio, in quo proinde successive fiat praesens diversis caeli partibus quae in gyrum ibi succedunt. Vel assistit semper certae cuipiam parti caeli atque adeo semper cum illa circumgyrat." Ibid., 150, col. 2–151, col. 1.

vides a relatively rare discussion on the problem of the location of intelligences with respect to their orbs.[163]

Some years later, Riccioli wrote in a similar vein. What distinguishes Riccioli's account is that it was presented as a separate question – "Whether and where the motive intelligences rest" – within a chapter of his *Almagestum novum* (pars post., bk. 9, sec. 2, ch. 1, 1651, 250, col. 2). Although he does not mention Aversa, Riccioli seems to agree with him, insisting that while it is fitting that an intelligence lead each planet (not an orb, however; see Sec. II.4.e), the intelligences "are not everywhere nor can they produce an impetus or motion in a body that is enormously distant. . . . Their place is limited and circumscribed, and their force is finite."[164] Mastrius and Bellutus adopted a similar stance when they declared that "not even an angel is diffused through the whole orb as in a place, but only in a part, namely that which is moved most quickly, as is the case for the part near the equator."[165] Indeed, it will be somewhere in the epicycle on which the planet is carried.

Riccioli, who rejected planetary orbs (see Sec. II.4.e) did, however, assume the existence of a single orb for the fixed stars. Where did the intelligence, or intelligences, reside within or in relation to this orb? Adopting a rather skeptical tone as to whether one or more intelligences cause the circular motion of the starry orb, Riccioli suggests locations for each scenario. He explains that "if there is only one [intelligence that moves the sphere of the fixed stars, it will be in] whatever point of the heaven on the equator is east with respect to some horizon. It will be necessary to determine some particular region, say Palestine, in the eastern horizon of which the angel is." But perhaps one angel is insufficient to move so vast an orb as that of the fixed stars. If so, then several intelligences might be arranged around the orb. It appears that the angel, or angels, are within the orb of the fixed stars itself. At the termination of his question, Riccioli invokes Job 38.33 by way of acknowledging his, and everyone's, feeble understanding of such matters.[166]

Those who assumed that celestial intelligences are separate movers of celestial orbs, or planets (see Sec. II.4.e), and who further held that a separate intelligence was somehow within the boundaries of the orb it moved, did not usually hold that an intelligence was diffused throughout the extent of its physical orb, as is evident from the views of de Oña, Aversa, and Riccioli described in the preceding paragraphs, to which we may add that of Sig-

163. Ibid., 151, col. 1, attributes the first alternative to Aristotle.

164. Riccioli buttresses his argument by appeal to Scripture and the Church Fathers.

165. "Ex quo patet ad 3: non enim angelus est per totum orbem diffusus tamquam in loco, sed uni tantum parti, illi scilicet quae velocius movetur qualis est pars aequatori propinqua. Et si zonis planetarum unus angelus tantum assistit erit in illa parte in qua immersus reperitur epiciclus." Mastrius and Bellutus, *De coelo*, disp. 2, qu. 4, art. 4, 1727, 3:504, col. 2, par. 148.

166. In the New English Bible, the verse reads: "Did you proclaim the rules that govern the heavens, or determine the laws of nature on earth?"

ismundus Serbellonus, who located the angel within a part of its planet or sphere but assumed that its power radiated throughout the sphere.[167] In the fourteenth century, Nicole Oresme held a similar opinion when he insisted (*Le Livre du ciel*, bk. 2, ch. 2, 1968, 289) that an intelligence need not "be everywhere within or in every part of the particular heaven it moves." Indeed, as we saw, John of Jandun indicated that even those who assumed intelligences as internal movers denied that they were "spread" or "diffused" homogeneously throughout the orb itself. An intelligence could thus be located somewhere within the physical bounds of the orb but was not assumed to occupy every part of it. If we recall that most scholastics conceived of intelligences as angels, the assumption that the intelligence, or angel, is located within the boundary of its orb but not in any particular place within the orb fits well with traditional medieval ideas about the nature of angels.

During the Middle Ages, spiritual substances – that is, angels or souls – were assumed to occupy places or spaces in ways that were both analogous to, and different from, the ways in which material bodies occupy places.[168] According to Peter Lombard, who was instrumental in forging this distinction, a body or corporeal substance occupies a place that is coextensive with its length, depth, and width. Peter called this a body's *ubi circumscriptivum*. By contrast, a spiritual substance, an angel or soul, despite its lack of dimension and magnitude, was assumed to occupy a place, but was not necessarily coextensive with it. The place merely delimited or set bounds to the location of the angel or soul. In fact, it might be better to express this as Oresme did (*Le Livre du ciel*, bk. 2, ch. 2, 1968, 289): "for each angel there is a certain amount of space which it cannot exceed nor increase by the power with which it was endowed at its creation or by its nature, not without divine miracle, although its power can indeed be diminished."[169] It was essential to place limits to the locations of angels, because no created thing can be everywhere and everything must be somewhere. This way of occupying a place was characterized by the expression *ubi definitivum*.

Although few, if any, to my knowledge, explained the location of intelligences in celestial bodies by this distinction, it was apparently applied. An angel or intelligence need not have a fixed location within the boundaries of the physical orb but could be anywhere within, ranging from the occupation of some particular point or locale to the occupation of the full

167. "Dico intelligentiam movere immediate omnes partes sphaerae, etiam si uni tantum assistat per suam entitatem quia omnibus assistit per suam virtutem cum omnes sint intra sphaeram suae activitatis, quemadmodum Sol illuminat totam suam sphaeram aeris." Serbellonus [*De caelo*, disp. 1, qu. 4, art. 3], 1663, 2:44, col. 1, par. 35.

168. Here I follow Grant, 1981a, 129–130 and 342–343, nn. 66–67.

169. Amicus, *De caelo*, tract. 5, qu. 6, dubit. 6, art. 3, 1626, 315, col. 2, says that intelligences cause motion by the arrangement of the "author of nature," whose will they never transgress because of their uprightness (Resp. secundo: intelligentias movere ex ordinatione auctoris naturae, cuius voluntatem, ob rectitudinem quam habent, nunquam transgrediuntur).

dimension of the orb. Most, however, seem to have restricted intelligences to some part of an orb or planet.

c. Are intelligences immobile, or do they move with their orbs?

At first glance, it would appear that separate movers, or intelligences, should be carried around by the orbs in which they are located. This seems to follow from the assumption that each intelligence is located somewhere within the termini of its orb. And yet, as we saw, Aristotle described all intelligences as "unmoved movers." How could an intelligence be carried around as its orb turns and also be an unmoved mover? That is, how could it be simultaneously in motion and immobile? Few confronted this question directly. But once Aristotle's unmoved movers became identified with Islamic and Judeo-Christian angels, the question had no clear-cut solution. Angels were capable of motion. Did this also hold for angels that were celestial movers? No definitive answer emerged.

Some would probably have denied motion to the intelligence of an orb, largely because Aristotle had described them all as unmoved movers, even though the first unmoved mover, or prime mover, was identified with God, who was always sharply distinguished from all other unmoved movers of his own creation. Although Aristotle made no explicit and unambiguous assertions about hierarchical differences between unmoved movers, the unique, infinitely powerful, and omnipresent God of the Christian faith could not be equated with each of a plurality of identical, or perhaps similar, independent unmoved movers. Indeed, distinctions between unmoved movers had already been made in the Aristotelian commentary literature of the Islamic world by Avicenna and Averroës, both of whom called the external unmoved movers "intelligences" (Wolfson, 1973a, 11–12), a term that would thereafter be commonly employed in the Latin West.

Despite the lesser status of all celestial movers other than the prime mover, some opted for their immobility.[170] Peter de Oña ([*Physics*, bk. 8, qu. 1, art. 9], 1598, 379r, col. 2) argued that an assisting intelligence does not inform an orb – it is completely separate from it – and therefore does not move around with the orb. The common view was that an intelligence caused celestial motion by being desired, either as an object of love by the soul of the orb or by the orb itself, but was not itself in motion. There is, however, something troubling about intelligences being both angels and unmoved movers. For angels were capable of self-motion and could traverse distances. They were not naturally immobile. Francisco Suarez took cognizance of this anomaly when he observed that Scripture, and many theologians, assumed the motion of angels but that Aristotle had proclaimed the immobility of celestial intelligences. To cope with this seeming con-

170. We saw earlier (in Section II.4.b of this chapter) that Aversa asked whether or not an intelligence moved with its orb but failed to provide an answer.

tradiction, Suarez distinguished between separate substances that move celestial orbs and those that do not. The former are immobile, the latter are not. If motive intelligences that cause the motion of celestial orbs were mobile, they might – presumably because they have will and intellect – depart from their respective orbs. But this chaotic possibility is foreclosed, not by the union of intelligence and orb or the "informing" of the latter by the former, but by the accidental nature of the relationship of a natural mover to its mobile. Despite their distinct identities, they are perpetually joined in a binding relationship.[171]

But there were dissenters. Oresme (*Le Livre du ciel*, bk. 2, ch. 2, 1968, 287–289) was among those who insisted that celestial intelligences do move with their orbs, emphasizing his disagreement with Aristotle and Averroës on this point. In the seventeenth century, Serbellonus repeats this sentiment when he declares that "the motive intelligences of a planet are moved with their motions [that is, with the motions of the planet],"[172] an opinion that Riccioli seemed to share when he declared that the angel that caused the Sun's motion – he rejected hard, planetary orbs – also moved around with the Sun.[173] Because of the conflict between Aristotle's concept of unmoved movers and the Judeo-Christian concept of angels, it is difficult to determine on any a priori basis how a given author might have resolved this difficult issue.

d. Do intelligences use energy in moving celestial orbs?

In the medieval view, by functioning as an object of desire and love an intelligence somehow causes its orb to move. In a not readily explicable manner, an intelligence was thought to exert some kind of force on the orb, whether as a final or efficient cause, or both. If intelligences exerted something identifiable as a force to move the celestial orbs in their circular paths, it was not unreasonable to inquire whether an intelligence could, in some sense, become fatigued or tired from its efforts. After showing that the heavens were indestructible and ungenerated, Aristotle himself set the stage for discussion when he confirmed that not only does the heavenly region not suffer from the maladies of a mortal body, but "its motion involves no effort, for the reason that it needs no external force of compulsion, constraining it and preventing it from following a different motion which is natural to it."[174] Averroës' elaboration on this passage was widely

171. See Suarez, *Disputationes metaphysicae*, disp. 35, sec. 3, 1866, 2:455, col. 2–456, col. 1, pars. 49–50. In his disputation 35, "On Created Immaterial Substance" ("De immateriali substantia creata"), sec. 3, 2:424, col. 2–477, col. 1, Suarez has an illuminating discussion of "The Attributes of Intelligences."

172. "Intelligentias motrices planetarum moveri ad motum ipsorum." Serbellonus, *De caelo*, disp. 1, qu. 4, art. 3, 1663, 2:43, col. 2, par. 33.

173. For the text, see note 190 of this chapter.

174. *On the Heavens* 2.1.284a.14–16 in Aristotle [Guthrie], 1960. For the Latin versions of these words from both Greek and Arabic sources, see Maier, 1958, 215.

accepted ([*De caelo*, bk. 2, comment. 3], 1562–1574, 5:96v, col. 2). The heavens do not suffer from fatigue or effort, because the labor and effort required to move them differs from the labor and effort of animals. The latter have contrary motions, in the sense that the soul may seek to go in one direction but the material, or heavy, part of the animal seeks to move to a contrary part. This tension of opposites, which requires effort and produces fatigue, is absent in the heavens. As John of Jandun explained it ([*De coelo*, bk. 2, qu. 2], 1552, 24r, col. 2), a celestial heaven, or orb, does not have any tendency to move with a motion other than circular. It lacks any inclination to move up or down, because it is neither heavy nor light. Hence no effort or labor is required to move it.

In Michael Scot's thirteenth-century translation of *De caelo* from Arabic to Latin, the Latin expression for "effortless," or "without labor," is *non fatigatur*.[175] But what does a term like *fatigatio*, "fatigue" or "effort," really mean in the context of celestial motion? In inquiring "Whether or not a heaven is moved with any fatigue," Jean Buridan – and others – saw the need to define the meaning and usage of that term.[176] Because active powers resist and purely passive powers do not, Buridan concluded that only the former can suffer fatigue. Moreover, a concept like fatigue or effort is not applicable to inanimate objects.[177] To say that an ax is tired because it no longer cuts as well as it did, or that earth is disturbed because it is no longer as fertile as it was, is nothing but an improper mode of speaking.[178] Such terms are also inapplicable to certain animate beings, such as plants, because the latter are incapable of cognition. Buridan is thus led to define *fatigatio* as "the diminution of a cognitive, active power because of its action over a long period."[179] But this would not include such instances as a man who is tired or troubled because he is weak from old age or illness or because

175. For the Latin passage, see Averroës, *De caelo*, bk. 2, text 3, 1562–1574, 5:96v, col. 2, K–L.

176. Indeed, Buridan, *De caelo*, bk. 2, qu. 1, 1942, 130, considered the terms *fatigatio, vexatio, labor*, and *poena* as synonymous. Among others who discussed the definition of *fatigatio* were Albert of Saxony, *De celo*, bk. 2, qu. 9, 1518, 107v, col. 2 and Amicus, *De caelo*, tract. 5, qu. 6, dubit. 7, art. 1, 1626, 311, cols. 1–2, who used the term *defatigatio*, as indicated in the very title of the question "An motus caelorum sit cum defatigatione."

177. "Postea videtur mihi quod adhuc fatigatio vel labor non solent proprie attribui inanimatis." Buridan, ibid., 131.

178. Albert of Saxony, *De celo*, bk. 2, qu. 9, 1518, 107v, col. 2, observes that a stone that falls or a fire that burns ought not to be characterized as "fatigued." Noncognitive active powers are incapable of fatigue. Bricot, *De celo*, bk. 2, 1486, 16v, col. 2, says the same thing.

179. "Sed nunc oportet videre quid est fatigatio vel vexatio. Et apparet mihi quod fatigatio est diminutio virtutis activae cognoscitivae, propter longam eius actionem." Buridan, *De caelo*, bk. 2, qu. 1, 1942, 131. Buridan probably influenced the definitions offered by Albert of Saxony, ibid., and the fifteenth-century scholastic Johannes de Magistris (*De celo*, bk. 2, qu. 1, 1490, sig. k3r, col. 1), who both define the term *fatigatio* in almost the same words. For similar versions, see Versor, *De celo*, bk. 2, qu. 1, 1493, 16v, col. 1; Paul of Venice, *Liber celi*, 1476, 32, cols. 1–2; and Amicus, *De caelo*, tract. 5, qu. 6, dubit. 7, art. 1, 1626, 311, col. 1. Amicus, who wrote one of the lengthiest questions on whether the heavens suffer fatigue from their motions, attributes the definition to Albert, whom he mentions at least five times in his discussion.

of a deficiency in some part of his body. It would, however, include those instances where the weakness arose because a man walked around too much or was unable to sleep.[180] In cases involving sensitive powers in a cognitive being, the parts engaged in laborious effort lose certain spirits, and even humors, which nature had provided for such efforts. By rest, the parts recuperate.[181]

Since intelligences are cognitive active powers and act as the motive forces for the celestial orbs, it would follow that any diminution of that power in any of them would produce some kind of "fatigue." However, this fails to occur, for one of two reasons: either because the motive intelligences possess an intensive, infinite potential for keeping the celestial orbs in uniform motion or because by their very natures they are "fatigueless forces," so that the level of effort and power never varies.[182]

Although all were agreed that the celestial intelligences did not become fatigued, they disagreed over the manner in which they preserved their level of power and how they were to be characterized. Was each intelligence able to cause the motion of its orb because it possessed an infinitely intensive power? Or could it perform its task with a finite power? In part, the problem involved the relationship between the prime mover, or God, and the created intelligences.

To resolve the problem, however, one had first to explain a seemingly anomalous situation, namely how all the celestial motions could be of finite velocity in the absence of resistance. In medieval physics, where velocity was conceived as arising from the application of a force to a resistance, a velocity would become infinitely great as the resistance became infinitely small or zero (Maier, 1958, 207). Thus even a finite motive power, when applied to a body moving without resistance, should produce an infinite velocity. But, as was evident, the celestial bodies move with finite speeds. How could this be? The solution lay in a vital distinction: celestial movers, or intelligences, were deemed voluntary agents operating by intellect and will (and also capable, some thought, of generating a kind of motive force), in contrast to natural motive powers, which acted naturally or mechanically. With the exception of Oresme, the force–resistance model for representing the velocity of terrestrial motions was judged inapplicable to the heavens.

180. "Unde non dicimus hominem fatigatum vel vexatum si sit impotens ex senectute vel ex aegritudine vel ex defectu membri, sed si fiat impotens vel minus potens quia nimis ambulavit vel nimis vigilavit." Buridan, ibid.

181. Albert of Saxony offers much the same argument (*De celo*, bk. 2, qu. 9, 1518, 107v, col. 2) and probably drew it from Buridan, whom he relied on rather frequently. De Magistris (*De celo*, bk. 2, qu. 1, 1490, sig. k3r, col. 1), Versor (*De celo*, bk. 2, qu. 1, 1493, 16v, col. 1), and Bricot (*De celo*, bk. 2, 1486, 16v, col. 2) present similar arguments.

182. See Maier, 1958, 216. I have relied heavily on Maier's splendid account of the *vires infatigabiles*. Another possible cause of the diminution of celestial motion might arise from external resistances opposed to the rotating orbs. This suggestion was routinely rejected, because the celestial ether in which the orbs were immersed and from which they were made was assumed incorruptible and unchangeable. Therefore no resistance could arise to vary the motion of the rotating orbs.

In earlier treatises, Oresme had also accepted the prevailing interpretation,[183] although he seemed willing to speak of forces and resistances in the heavens analogically. In his last and perhaps greatest work, *Le Livre du ciel et du monde* (bk. 2, ch. 2, 1968, 289), Oresme seems to move beyond analogy to argue that God had initially implanted forces and resistances in each celestial orb but goes on to say that "these powers and resistances are different in nature and in substance from any sensible thing or quality here below." Unfortunately, Oresme does not explain how the force and resistance of an orb relate to the intelligence, or angel, which is also a motive power. Is it the implanted force–resistance relationship that moves the orb, or is it the intelligence? Or are force and intelligence one and the same?[184] Although the language of ratios of force and resistance was thought inapplicable to the celestial motions, some scholastics thought analogically in terms of force and resistance relationships, usually implicitly, though occasionally explicitly.[185]

All were agreed, however, that the prime mover possessed infinite, intensive power and that he caused motion both as a final and as an efficient cause. As the means of producing celestial motions, the prime mover created intelligences that were voluntary agents capable of operating by intellect and will. Latin scholastics were aware that Aristotle and Averroës had treated the prime mover and the other intelligences as virtual equals, at least in the sense of assuming that they were all eternal.[186] The equation of God with the intelligences that he himself had created was unacceptable to Christians. Hence they turned to other solutions.

Scholastics commonly assumed that the longer a given uniform motion endured, the greater the power that would have to be expended by its motive force, which implied that an infinite force would be required to sustain an eternal motion (Maier, 1958, 192). Obviously, an eternal motion could only occur if the world was eternal. In the syncategorematic sense, as we saw earlier (Ch. 4), time and motion and the world as a whole were often considered eternal, or at least assumed to be so for the sake of nu-

183. In the *De proportionibus*, ch. 4, prop. 6, Oresme declares (*De prop. prop.*, 1966b, 293) that the relationship of an intelligence to the orb it moves should be described as one of force and resistance only by analogy, "because an intelligence moves by will alone, and with no other force, effort, or difficulty." Although many had abandoned the idea that intelligences move by the will alone and introduced an executive power that angels used to produce a *virtus motiva*, as we saw, Oresme was still operating here with the old principle condemned in 1277.

184. A few lines after the preceding quotation (ibid., 289), Oresme suggests angels as movers when he declares that "possibly, as soon as God had created the heavens, He ordained and deputed angels who should move the heavens and who will move them as long as it shall please Him."

185. All those who spoke of an angel using its executive power to generate a *virtus motiva* to move its orb were probably implicitly thinking of a force–resistance relationship. But Amicus, *De caelo*, tract. 5, qu. 6, dubit. 6, art. 3, 1626, 315, col. 2, was somewhat explicit when he declared that movers and mobiles in the celestial region have the same constant ratio (*proportio*). See note 46 of this chapter for Amicus's text.

186. See Maier, 1958, 193.

merous arguments. Moreover, we also saw that many in the Middle Ages, especially Thomas Aquinas, thought that the creation of the world was compatible with its possible eternity. Thus infinites of various kinds were frequently assumed.

Of these, a few were relevant to celestial motions. First there is a temporal infinite, which presupposes the eternity of the world, and therefore the eternity of celestial motions. On the assumption of an infinite, eternal time, however, a fundamental problem arose: how could a finite intelligence move its orb with a uniform, finite celestial motion over a period of infinite time? Could this be achieved only by means of an infinite motive force, or could it be done with a finite motive power? Whatever the appropriate response to these questions, natural philosophers and theologians were agreed that, in celestial intelligences, the prime mover was the immediate source of all motive power, whether finite or infinite. Both options won some support. Let us first consider the implications that follow from the assumption that intelligences possess infinite motive power.

Because it is a creature of intellect and will, an intelligence could choose to apply its infinite motive power in one of two ways: it could either (1) apply it all at once and produce an instantaneous or infinite motion of a celestial orb; or (2) utilize it gradually and produce a finite motion over an infinite time. In the latter situation, the intelligence's total motive power is like an infinite source of energy that is gradually consumed over an infinite time as it produces a uniform, finite motion at every instant. Maier explains (1958, 194) that "such a *virtus motiva* can consume its energy of motion little by little in an infinitely long time in such a way that the effect at any moment is a finite one. Just as Sortes [i.e., Socrates] can, by choice, apply the energy standing at his disposal and use it either in labor of short but intensive duration or in a less intensive way and lasting longer, so also can the intelligences do this." For the heavens, the intelligences have obviously chosen to produce finite, uniform motions and thus to expend their infinite supply of energy bit by equal bit, through all eternity.

Not all were prepared to concede that a finite entity, such as an intelligence, could include within itself an infinite force.[187] The infinite duration of celestial motions did not warrant the inference that the intelligences that caused those motions require infinite motive powers to maintain them eternally. Only the prime mover is an intensively infinite being. All other celestial intelligences, or movers, are only capable of finite intensity. Despite the finitude of their respective motive powers, the intelligences do not exhaust themselves in causing the unceasing, eternal motion of their orbs. Exhaustion or depletion of energy fails to occur, because the prime mover constantly replaces it. According to Maier (1958, 195), in this explanation "the operation of the celestial movers consists in this, that they transform

187. For this interpretation, see ibid., 195–196. As representatives of this group, Maier includes Buridan and William of Alnwick (195–206).

energy, which is not their own but which is furnished to them successively from outside [by the prime mover], into motion."

In both approaches, the uniformity and finitude of celestial motions were explained and reconciled with the infinites of duration and power.

e. The rejection of hard planetary orbs did not cause the abandonment of separate intelligences

In his *Epitome of Copernican Astronomy*, composed between 1618 and 1621, Kepler considered the problem of intelligences as celestial movers. Because he followed Tycho Brahe and rejected hard, solid orbs, Kepler inquired whether intelligences might yet play a role in causing planetary motion in an orbless world. In the fourth book of the *Epitome*, he initially declares: "If there are no solid spheres, then there will seem to be all the more need of intelligences in order to regulate the movements of the heavens, although the intelligences are not gods. For they can be angels or some other rational creature, can they not?" The question for Kepler, then, was whether intelligences, or angels, or mind – he used these terms interchangeably – could move planets, rather than orbs, directly. Kepler concluded that they could not. He sought to convince his readers with a number of related, and rather unimpressive, arguments. Thus he insisted that if mind could produce planetary orbits, those orbits would be perfect circles. But, as he himself had shown, the orbits were not circular but elliptical. Since right-thinking intelligences would not willingly choose elliptical over circular orbits, Kepler concluded that intelligences did not direct planetary motions. His remaining arguments challenge the ability of intelligences to draw correct planetary orbits in the fluid ether.[188] "It is not possible," Kepler insisted (*Epitome*, 1952, 892), "for the planetary globe to be carried around by an intelligence alone." For Kepler, "the natural powers which are implanted in the planetary bodies can enable the planet to be transported from place to place" (ibid.).

Scholastics who continued to accept the existence of hard orbs – for example, the Coimbra Jesuits, Amicus, Compton-Carleton, and Serbellonus – continued to regard intelligences as celestial movers. After 1630, however, many if not most scholastics – including Pedro Hurtado de Mendoza, Roderigo de Arriaga, Francisco de Oviedo, Giovanni Baptista Riccioli, Franciscus Bonae Spei, Melchior Cornaeus, and George de Rhodes – abandoned hard, planetary orbs in favor of the fluid heavens advocated by Brahe and Kepler, although some of them – Riccioli and Hurtado de Mendoza, for example (see Ch. 14, Sec. VIII and n. 87) – also assumed that the fluid ether was surrounded by a hard orb of fixed stars. All of them, how-

188. For Kepler's arguments, see the *Epitome*, 1952, 891–893. Riccioli, *Almagestum novum*, pars post., bk. 9, sec. 2, ch. 1, 1651, 249, cols. 1–2, who summarized and repudiated Kepler's arguments against intelligences, observed that the circle is not the only perfect figure. Indeed, the ellipse is also perfect, as are many other figures.

ever, regarded the planets as unattached bodies immersed in a very rare, fluid ether. How did these seemingly free-moving planets traverse their paths? Were intelligences involved?

Many scholastics continued to invoke intelligences as celestial movers, only they now associated them with the planets, rather than the orbs. Having rejected hard planetary orbs, Arriaga confronted the question of why the denser planets did not fall down through the rarer ethereal medium through which they moved. Why did they not fall toward the center of the world? Because, as he put it, "they are sustained by the angel by which they are moved around."[189] Riccioli makes much the same argument when he declares that "Sacred Scripture sufficiently confirms that an angel which moves the Sun also moves around itself and accompanies the Sun, which it moves."[190]

5. *Internal movers*

a. *Souls and the possible animation of the heavens*

In medieval tradition, souls of orbs were conceived very differently from intelligences. The latter were distinct and separate from the orbs they moved, even though they might have been located "within" them in some sense. Most, though not all, of those who imagined souls as celestial movers conceived of them as organically integrated into, or "conjoined" (*conjuncta*) to, their respective orbs and therefore not separate substances.[191] For them the soul of a celestial orb was as integrated into its orb as the human soul was into the human body. The soul of an orb functioned as the form, and the material orb functioned as its body. If celestial orbs were composites of body and soul, did this not imply that the celestial orbs were alive? It did indeed, which is why most scholastic authors shied away from the assumption of souls as celestial movers.[192]

But what about intelligences? Were they essential forms of their orbs, so that they might also be construed as forming a living thing with their physical orbs? An anonymous fourteenth-century author represented me-

189. First Arriaga (*De caelo*, disp. 1, sec. 4, subsec. 2, 1632, 505, col. 2) asserts the question: "Dices illi planetae sunt solidi et densiores, ergo sunt graviores quam caeli; ergo cadent quia cum caeli fluidi sint non possunt eis resistere." He replies that "eos [the planets] sustenari ab angelo a quo circumaguntur."

190. "Hoc posito si ad litteram accipiatur, ut accipi potest, Sacra Scriptura, satis hinc potest confirmari angelum, qui Solem movet, ipsum quoque circuire ac Solem quem movet concomitari." Riccioli, *Almagestum novum*, pars post., bk. 9, sec. 2, ch. 1, 1651, 250, col. 2. A few lines later, Riccioli mentions that Arriaga holds the same opinion, referring to the same passage cited in the preceding note.

191. According to Weisheipl, 1974, 28, Thomas Aquinas must be classified as an exception, because he assumed that the soul of a sphere was a separate substance and therefore regarded it of little consequence "whether the sphere was moved by a soul inherent in the body or by a distinct substance, separate from matter, moving the sphere through its efficient causality." Also see note 194 in this chapter.

192. For more on the animation of the heavens, see above, Chapter 17, Section V.

dieval opinion when he denied this, because "an intelligence is not able to be the form of any body, and therefore the heaven is not animated, because an intelligence is not conjoined to the heaven with respect to itself [*secundum se*] but only with respect to its operation, as is said in the *De substantia orbis*." Because an intelligence is not conjoined essentially to the matter of the orb, the author goes on to show that "an intelligence is not a form or soul of the heaven, and consequently the heaven is not animated except in this sense, that the heaven has operations as if it were an animated body, and this is because of the intelligences conjoined to it, not with respect to essence, but with respect to operation." Thus intelligences, or angels – our author seems to take them as synonymous – are conjoined to orbs but not as their forms or souls. For "a separated soul and an angel differ in this: that a soul is essentially unitable to a body as a perfection to its perfectible thing; an angel, however, is not unitable. For although we read that angels assume bodies, they are nevertheless never united to them with respect to essence but only with respect to appearance or operation."[193]

Even better reasons were formulated for rejecting souls for celestial orbs. In his questions on *De caelo*, Galileo provides an excellent summation of medieval ideas on this important subject (qu. 6 [L], 1977, 149). If each celestial orb was informed by a real soul, the latter, according to Aristotle and all natural philosophers, would have to consist of three levels of activity: the vegetative; the sentient or sensitive; and the intellective. But a celestial orb cannot have a vegetative soul, because it would then require food, and this would make it corruptible and therefore not eternal. Nor can it have a sensitive soul, because it would require a vegetative soul and a sense of touch, neither of which exists in the heavens.

This leaves only the intellective soul. Can an intellective soul alone animate the heavens? Although some have believed this possible by assuming that the intellective soul is an intelligence and the form of the heavens, Galileo rejects this and denies that intelligences can inform the orbs. Intelligences exist apart from bodies and move them as efficient causes. Galileo concludes (ibid., 154) that "apart from intelligences no other souls are constituent in the heavens."

The essential and organic union of genuine souls with physical orbs would have made living things of the orbs. Although Thomas Aquinas was indifferent to this problem and perhaps ambivalent about it,[194] the medieval

193. My translation from "Compendium of Six Books," BN Latin MS. 6752, 231r. The anonymous author does not cite Aristotle's *Metaphysics*, which attributes celestial motion to intelligences that are distinct from the spheres they move. The *De substantia orbis* was a widely known work by Averroës.

194. Thomas (in his *Summa contra gentiles*, bk. 3, ch. 23, and in *De caelo*, bk. 2, lec. 3, 1952, 156, par. 3) thought it important only that an orb be moved by an intellectual substance. Whether that intellectual substance was an informed soul or a separated substance was irrelevant. Since the former would have implied a living orb and, on certain interpretations, even the latter might have been so construed, we may reasonably infer that Thomas was either unable to arrive at a judgment or thought it unimportant to do so. For the translated passage, see Dales, 1980, 543 and note 46; see also Litt, 1963, 109,

tradition after Thomas was overwhelmingly against the attribution of life to the heavens.[195] This was perhaps partially a consequence of the Condemnation of 1277 by the bishop of Paris, who condemned at least four articles that proclaimed the existence of souls in the heavens; that souls were equivalent to intelligences; and that intellective souls could cause celestial motions.[196] In the sixteenth century, John Major mentions that the assumption of celestial animation was condemned by Stephen, the bishop of Paris.[197] Because he agreed with the bishop of Paris, Major assumed that "the heaven does not have an intelligence or intellectual nature as its form."[198]

Even before the Condemnation of 1277, Saint Bonaventure had rejected as "erroneous" – that is, theologically erroneous – the idea that the heavens had souls and was therefore a "great animal" (*magnum animal*). The position Bonaventure attacked was the one often attributed to Aristotle in which it was assumed that each celestial orb had a soul that was directed by a mediating intelligence.[199] In Bonaventure's judgment, an inhering soul was tantamount to the assumption of a living orb. Not only are Catholic doctors against this opinion – Bonaventure reminds his readers that John Damascene had rightly declared that "the heavens are inanimate and insensible"[200] – but Bonaventure observes that the philosophers also find it wanting, because an intellectual substance, such as a soul, could not be united to a body without vegetable and sentient functions. If the celestial orbs possessed such functions, however, they would be corruptible bodies, which is contrary to the universally accepted doctrine that they are incorruptible.

b. Forms or nature

If souls had few adherents, other kinds of internal movers won significant support during the Middle Ages. Already in the thirteenth century, John Blund and Robert Kilwardby argued for the natural, intrinsic capability of the heavenly orbs for self-motion. Sometime around 1200, when the works of Aristotle began entering western Europe, Blund denied that souls move the heavens. The firmament and other heavenly bodies are moved by an

who observes that even if the celestial orbs are alive, their life is uniquely intellective and in no way vegetative or sensitive.

195. Most of the Church Fathers – for example, Saints Basil, John Damascene, and Jerome – rejected animate heavens, as did John Philoponus, who was not a Church Father. A notable exception was Saint Augustine, who expressed uncertainty. See Dales, 1980, 533. Hissette (1977, 68) observes, of the much-debated question about the animation of the celestial spheres, that many medieval theologians were inclined to reject it.
196. For translations, see Dales, 1980, 545–546.
197. "Secundo ponere corpora celestia animata est articulus excommunicatus a domino Stephano episcopo Parisiensi." Major [*Sentences*, bk. 2, dist. 12], 1519b, 65v, col. 1.
198. "Quinta conclusio: celum non habet intelligentiam vel naturam intellectualem tanquam formam eius." Ibid., 65r, col. 2.
199. If Aristotle held such an opinion, he never made it explicit. See Section II.1 of this chapter and especially notes 91–93.
200. Bonaventure, *Sentences*, bk. 2, dist. 14, pt. 1, art. 3, qu. 2, *Opera*, 2:348, col. 2.

innate nature – he makes no mention of angels, intelligences, or forms – which causes it to move with circular motion.[201] "To Blund," Dales informs us (1980, 538), "the forms of the heavenly bodies are totally immanent in matter and not connected through a hierarchy of intelligences to God Himself, as they were in both Neoplatonic and Aristotelian thought."

In 1271, in response to a question as to whether angels move the celestial bodies, Robert Kilwardby expressed an opinion quite similar to that of Blund.[202] The celestial orbs move by their own spontaneous, natural inclinations in the same manner as heavy and light terrestrial bodies move downward and upward by their own natural tendencies. By attributing celestial motions to innate natural tendencies, Kilwardby was led to deny that God is the immediate mover of the heavens and that angels are celestial movers. As Weisheipl explains Kilwardby's interpretation (1961, 316), "To each planet and orb God gave an innate natural inclination to move in a particular way in rotational motion; to each he accorded an innate order, regularity and direction without the need of a distinct agency like a soul, an angel or Himself here and now producing the motion."

c. *Impetus or impressed force*

Where Blund and Kilwardby were satisfied with the postulation of a vague, innate capacity conferred on celestial orbs to enable them to move in regular, uniform, circular motions, Jean Buridan, in the fourteenth century, applied his well-developed and quantified impetus theory to celestial motion.[203] In his *Questions on the Physics* (bk. 8, qu. 12), following an elaborate description of the properties and behavior of impetus, or impressed force, in the pro-

201. "Dicimus quod firmamentum movetur a natura, non ab anima, et alia supercelestia; et ille motus naturalis est propter perfectionem habendam in inferioribus." Blund [Callas and Hunt], 1970, 4, par. 10. See Dales, 1980, 538, where Blund's brief paragraph is translated, and Weisheipl, 1961, 317.

202. In 1271, John of Vercelli, master general of the Dominican Order, sent a list of forty-three questions to three Dominican masters in theology: Albertus Magnus, Thomas Aquinas, and Robert Kilwardby. As Weisheipl observes, the first five questions are of great interest to historians and philosophers of science. They are, in Weisheipl's translation:

1. Does God move any physical body immediately?
2. Are all things which are moved naturally, moved under the angels' ministry moving the celestial bodies?
3. Are angels the movers of celestial bodies?
4. Is it infallibly demonstrated according to anyone that angels are the movers of celestial bodies?
5. Assuming that God is not the immediate mover of those bodies, is it infallibly demonstrated that angels are the movers of celestial bodies?

See Weisheipl, 1961, 286–287; for his discussion of Kilwardby's views, 315–317.

203. For excellent accounts of Buridan's impetus theory, see Clagett, 1959, 505–540; Maier, 1951, "Die Impetustheorie," 113–314; and, in Maier, 1964, "Die naturphilosophische Bedeutung der scholastischen Impetustheorie," 353–379, which has been translated by Steven Sargent under the title "The Significance of the Theory of Impetus for Scholastic Natural Philosophy," in Maier, 1982, 76–102.

jectile motion of terrestrial bodies, Buridan prepares the stage for the application of impetus to celestial motion by proclaiming that the Bible does not specify that intelligences move the celestial bodies. Since the Bible provides no textual support for intelligences as celestial movers, Buridan feels free to dispense with them and to suggest instead that

> God, when He created the world, moved each of the celestial orbs as He pleased, and in moving them He impressed in them impetuses which moved them without his having to move them any more except by the method of general influence whereby he concurs as a co-agent in all things which take place. . . . And these impetuses which He impressed in the celestial bodies were not decreased nor corrupted afterwards, because there was no inclination of the celestial bodies for other movements. Nor was there resistance which would be corruptive or repressive of that impetus.[204]

Buridan's impetus was thus a permanent quantity which, in the absence of external resistances or contrary tendencies, would move each celestial orb with the same velocity forever. Although Buridan applied his brilliant impetus theory to celestial motion, we must not suppose that he was seeking to produce a single, unified mechanics for the terrestrial and celestial regions. As evidence of this we may point to his retention of Aristotle's sharp dichotomy between the two regions.[205]

Although Buridan's *Questions on De caelo* was not published until 1942, his *Questions on the Physics*, which included the detailed version of his impetus theory, was printed in Paris in 1509, 1515, and 1516.[206] Despite these early sixteenth-century printings, Buridan's version of celestial impetus seems to have exercised little direct influence on the subsequent history of scholastic natural philosophy. In the literature consulted for this study, it is not cited. Instead, Buridan's contemporary, Albert of Saxony, came to represent fourteenth-century impetus theory as it was applied to celestial motion. Ironically, Albert probably derived his version from Buridan.

Albert of Saxony's brief discussion of celestial impetus occurs in his questions on the *Physics*, book 8, question 13, the final question of the treatise, usually cited as the *ultima questio*. That Albert's version, rather than Buridan's, became historically important is probably attributable to the fact that Albert's treatise was published not only in Paris in 1516 and 1518, but

204. Translated by Clagett, 1959, 536, par. 6; reprinted in Grant, 1974, 277. Because of the theological implications of his theory, Buridan was sufficiently uneasy to conclude his brief account of the cause of celestial motion by emphasizing its tentative nature "so that I might seek from the theological masters what they might teach me in these matters as to how these things take place" (Clagett, ibid., 277–278). In his *De caelo*, bk. 2, qu. 12, 1942, 180–181, Buridan says essentially the same thing in fewer words but makes no mention of the theologians. See Clagett, ibid., 561, par. 7, for a translation of the passage, and Grant, 1974, 282, where it is reprinted.
205. See Clagett, 1959, 525.
206. See Lohr, 1971, 168–169.

also in Padua (1493) and Venice (1500, 1504, and 1516) (Sarnowsky, 1989, 450). By the late sixteenth and the seventeenth century, Italy had emerged as a far more significant source of scholastic natural philosophy than Paris. Hence Albert's Italian publications (including his questions on *De caelo*) were widely read and cited, while Buridan's remained virtually unknown.

In that final question of the eighth book of his *Physics* (1518, 83v, cols. 1–2), Albert of Saxony presents a brief version of Buridan's celestial impetus theory, although he does not employ the term *impetus*, but uses equivalents such as *virtus motiva* (twice), *virtus impressa* (twice), and *qualitas motiva*.[207] The question itself concerns projectile motion in general, with Albert asking, "By what is a projectile moved upward after its separation from the projector?"[208] After describing three opinions, Albert declares: "There is another opinion which for now I think more true, [namely] that a projector impresses into a projectile a certain motive power which is a certain innate quality, unless some impediment occurs in the opposite direction toward which the projector projects."[209] After two illustrations showing that a denser body can receive more impressed force than one less dense and that a longer lance can be hurled a greater distance than a shorter lance because the former is capable of receiving more impressed force, Albert applies the concept of impressed force to the heavens by observing that if this opinion is correct, "it would not be necessary to assume as many intelligences as there are orbs. Thus it could be said that the First Cause created the celestial orbs and impressed one such motive quality on each of them, which moves the orb. Nor is this power [or force] corrupted there, because such an orb is not inclined toward any opposite motion, etc."[210]

On the basis of this discussion, Albert of Saxony was frequently and rightly included among those who denied the need of motive intelligences and who explained the celestial motions by means of an internal force; he was also, but less appropriately, included among those who assigned the cause of celestial motions to the internal forms of orbs without mention of

207. It appears that Albert relied on an earlier version of Buridan's *Physics*, composed around 1350, in which Buridan himself had not yet developed his technical terminology but used terms like *qualitas* and *virtus motiva*, much the same as Albert did. Buridan's final version, called the *ultima lectura* and represented by the 1509 Paris edition, was probably completed after 1355 but before Buridan's death around 1358. It was in this final version that Buridan employed the term *impetus*. For the details, see Sarnowsky, 1989, 50–51.

208. "Ultimo quaeritur a qua moveatur proiectum sursum post separationem illius a qua proicitur." Albert of Saxony, *Physics*, bk. 8, qu. 13, 1518, 83r, col. 2.

209. "Alia est opinio quam pro nunc reputo veriorem quod proiiciens imprimit proiecto quandam virtutem motivam que est quedam qualitas que innata est movere nisi fiat impedimentum aliunde ad eandem differentiam positionis ad quam proiiciens proiicit." Ibid., 83v, cols. 1–2.

210. "Iuxta istam opinionem potest dici quod non esset necesse ponere tot intelligentias quot sunt orbes. Unde diceretur quod prima causa creavit orbes celestes et cuilibet eorum impressit unam talem qualitatem motivam que illum orbem taliter movet. Nec illa virtus ibi corrumpitur propter hoc quod talia orbis non est inclinatus ad motum oppositum, etc." Ibid., col. 2.

motive forces.[211] Aversa managed to attribute both to Albert when he placed him among those who thought that "the heavens are moved by themselves and by an internal power [or force] proper to their form, just as animals are moved among us and heavy bodies are moved downward and light bodies upward."[212]

By the late sixteenth and the seventeenth century, Albert's terminology was supplemented by the term *impetus*. Although he did not accept it, Riccioli gives a good description of celestial impetus theory, suggesting that God could make a body with impetus already impressed in it. If such an impetus were not gradually corrupted by a contrary impetus produced by heaviness, it should endure forever. Because lightness and heaviness are absent from the heavens, the impetus would not be destroyed and would indeed function as a proper form. Once in orbit, the celestial orb would move around perpetually.[213]

Early modern scholastic natural philosophers largely avoided celestial impetus or internal movers and opted instead for external movers in the form of intelligences or angels. The extent to which this interpretation was entrenched in the seventeenth century is revealed by Thomas Compton-Carleton, who declared that this "common" opinion "has come into such use among all, so that it is almost a crime to deny it."[214]

A few, however, thought it unnecessary to appeal to God or angels as movers. Indeed, they thought it as likely, if not more likely, that the celestial orbs were moved by an internal form or power. Franciscus Bonae Spei was one of them. While conceding that common authority favored angels as the cause of celestial motions, Bonae Spei (*De coelo*, comment. 3, disp. 3, dub. 6, 1652, 13, cols. 1–2) countered that reason made it far more probable that the heavens were moved internally. If angels really moved the heavens, why could we not say that God or angels moved a stone toward the center of the world; or, since we can observe that animals are self-moved, why do we not also say that the heavens are similarly self-moved by internal

211. See Conimbricenses, *De caelo*, bk. 2, ch. 5, qu. 5, art. 2, 1598, 266; Bonae Spei [comment. 3, *De coelo*, disp. 3, dub. 6], 1652, 13, col. 1, par. 85; and Mastrius and Bellutus, *De coelo*, disp. 2, qu. 4, art. 3, 1727, 3:503, col. 1; Aversa, *De caelo*, qu. 34, sec. 7, 1627, 147, col. 1; and Riccioli, *Almagestum novum*, pars post., bk. 9, sec. 2, ch. 1, 247, col. 1.

212. "Ioannes Maior 2. *Caeli* cap.5, Albertus de Saxonia 8. *Physics*, quaest. ultima, volunt caelum moveri per se et ab intrinseco virtute propriae suae formae, sicut apud nos moventur animalia et sicut moventur gravia deorsum, levia sursum." Aversa, ibid. Although I have been unable to locate Major's discussion, he was usually cited along with Albert as a proponent of internal force or form.

213. Riccioli, *Almagestum novum*, pars post., bk. 9, sec. 2, ch. 1, 1651, 247, col. 1. As advocates of this theory, Riccioli, who did not accept it, cites Major and Albert of Saxony, the latter in the last question (*quaestio ultima*) of the last book of his *Physics* (that is, question 13; see n. 208).

214. "Communis tamen sententia affirmat moveri ab intelligentijs quod ita iam invaluit apud omnes, ut pene nefas sit id inficiari cui proinde ob tot tamque doctorum hominum auctoritatem subscribo omnesque constanter asserunt non posse motum illum provenire ab intrinseco." Compton-Carleton, *De coelo*, disp. 4, sec. 3, 1649, 409, col. 2.

means?[215] Even Scripture favors the internal hypothesis. After all, Joshua did not order angels to cease moving the Sun and Moon but directly commanded the Sun and Moon to cease their motions.[216] While Bonae Spei did not argue with any strong sense of commitment for internal movers, he seems to have preferred them over external movers. He was, in any event, prepared to deny the need to explain the cause of celestial motion by appeal to God and angels.

In a similar manner, Thomas Compton-Carleton, although he accepted the testimony of the doctors of the Church and the many others who denied that motion can arise from an internal cause, admitted that he failed to see the force of the argument against internal movers. After two arguments in which an internal cause of celestial motion is shown to be no less tenable than an external cause, Compton-Carleton says that the solution to the problem lies in deciding whether or not the motion of the heaven is natural. If celestial motion is derived from an internal cause, the motion must be natural; but if not, then the cause would seem to be "preternatural" to the heaven, where, in the manner of fire in its own sphere, an orb, or heaven, neither seeks nor rejects motion.[217]

Franciscus de Oviedo (*De caelo*, contro. 1, punc. 1, 1640, 461, col. 2) thought that explanations of the cause of celestial motions could only be probable. There were no compelling arguments. God could indeed have created celestial bodies that are internally self-moved. Because the Fathers and doctors of the Church leaned toward external movers, Oviedo accepts their traditional position.

Despite the widespread acceptance of external movers in the form of intelligences or angels, there was, even among supporters of the common opinion, such as Compton-Carleton and Oviedo, a sense that the question was difficult and that the deciding factor might be the weighty, traditional opinions of the Fathers and doctors of the Church. To have assumed the existence of internal movers would not have been judged outrageous in the late sixteenth and the seventeenth century. From the previous discussion, however, we may conclude that internal forces, such as forms or impetus, were more popular in the thirteenth and fourteenth centuries than in the sixteenth and seventeenth.

215. "Confirmatur primo: alioqui non est cur similiter non dicam lapidem in centrum a Deo aut Angelis moveri. Item motum progressivum animalium ab illis et non ab animabus intrinsecis esse. Si enim propter experientiam motus illis tanquam ab intrinseco sese moventibus tribuatur, quidni et coelis? Bonae Spei, comment. 3, *De coelo*, disp. 3, dub. 6, 13, col. 1, par. 86.

216. "Confirmatur secundo quia scriptura magis nostrae sententiae favet quam oppositae qui Iosue primo datur soli et lunae imperium ne moveant se, non autem Angelis ne solem et lunam moveant, quod alioqui fieret." Ibid., par. 87.

217. "Et per haec patet solutio ad id, quod hic quaeri solet utrum scilicet motus coeli dicendus sit illi naturalis, necne? Si enim sit ab intrinseco non est dubium quin sit caelo naturalis; sin minus, dicendus potius illi videtur praeternaturalis, sicut et motus ignis in sua sphaera, ubi nec motum petit, nec respuit." Compton-Carleton, *De coelo*, disp. 4, sec. 3, 1649, 409, col. 2–410 (incorrectly paginated as 310), col. 1.

6. External and internal motive forces that act simultaneously

a. *Acceptance of angels and impressed force: the Coimbra Jesuits and Riccioli*

Now that we have devoted separate sections to external and internal movers, we are ready to examine their combined use. To qualify for discussion in this section, an author must have reported or adopted explanations of celestial motion by means of (1) intelligences or angels *and* (2) an additional force that is caused by an intelligence and somehow impressed in a celestial orb. We shall see that the post-1277 interpretations of Godfrey of Fontaines and Hervaeus Natalis, and probably others, were developed further. Godfrey and Hervaeus both rejected the condemned idea that angels move celestial orbs by will (and intellect) alone and introduced a third force – an "executive power" (*potentia executiva*) that carried out the wish of the angel by applying the force – the *virtus motiva* – needed to move the orb. Around 1320, Franciscus de Marchia (ca. 1290-after 1344) assumed that an angel moved its orb by impressing a certain power (*virtute . . . impressa*) into it.[218] Thus rather than have the motive power operate from within the angel, de Marchia conceives of the motive force as something impressed from the angel into the orb. By the late sixteenth century, the Coimbra Jesuits and Francisco Suarez, and perhaps others before them, expressed opinions about the possibility of an impressed force, or impetus, being transmitted by an angel to its celestial orb.

Among those who adopted the concept combining angels and impressed force were the Coimbra Jesuits, in their commentary on *De caelo* first published in 1592, where they clearly imply that the idea was accepted earlier by others;[219] and Giovanni Baptista Riccioli, who, in turn mentions others who also adopted this approach in the seventeenth century.

The discussion by the Coimbra Jesuits occurs in the context of a question as to whether the celestial motions are uniform. That the angel associated with each heaven impresses force into that heaven or orb is assumed. On the basis of such an assumption, the question is whether celestial motions can be uniform. The response hinges on the way in which the angel impresses its force. If it moves its heaven in the same manner as a javelin or wheel, then uniform motion seems unlikely. The continuation of such motions requires that the force or energy involved in pushing the javelin or wheel be constantly renewed or the motion will cease. By analogy, the impressed force in the orb will be continually diminished, and the orb would come to a halt unless the impressed force is restored. If the force is impressed

218. See de Marchia's commentary on the *Sentences*, bk. 4, qu. 1, in Schneider, 1991, 50, lines 3–6 and 52, lines 87–89; also see ibid., 236–237.

219. With respect to the idea of an angel impressing a force on a heaven, the Conimbricenses say (*De coelo*, bk. 2, ch. 6, qu. 2, art. 3, 1598, 289) "responderi solet a quibusdam productionem impulsus." No names are provided.

in an instant, it will soon diminish, and another instantaneous increment of force will have to be impressed. As a consequence, alternate diminutions and increments would occur, and the motion could not be uniform. Similarly, if the angel impresses force over some temporal interval and in successive instants, the impressed force will increase over every moment of that time interval, and the speed of the orb will continually increase. Here again, there will be no uniform motion.[220]

Two solutions are proposed. The first holds that the impressed force which the angel transmits to its heaven or orb is sufficient to move the orb at its requisite speed. In this solution, the impressed force remains perpetually constant, never diminishing or increasing.[221] This solution is akin to Buridan's (described earlier in this chapter), where, instead of employing an angel, God impresses the totality of impressed force, or impetus, on each orb at the beginning of the world, after which its motion is forever uniform because the impressed force remains constant.

This response was found wanting[222] because it assumed that the impressed force was an invariable quantity. The Conimbricenses were apparently convinced that the impressed force "is a quality with a disappearing nature" (sit qualitas suapte natura evanescens), just as are the impressed forces that push an arrow or a stone upward.[223] But if the celestial motions are uniform, how does an angel maintain a constant level of impressed force to produce a uniform motion for its own orb? "Some think more correctly," say the Conimbricenses,

> that the impulse which the angel impresses continually in any whatever part of time is successively diminished and that it is simultaneously restored by the same angel in a perennial and continuous influx, so that as much as is lost on the one hand is restored on the other. And thus the impulse [or impressed force] is perpetually conserved in the celestial spheres with the same intention and degree.[224]

220. "Secundo, impulsus quem Angelus coelo imprimit, suapte natura evanescit, quemadmodum et is qui nostratibus rebus imprimitur, ut iaculo vel rotae. Igitur ne coelum interquiescat oportebit eiusmodi impulsum assidue redintegrari. Haec autem redintegratio vel fit in instanti, vel in tempore et successive. Si in instanti, certe immediate post illud minuetur impulsus, atque adeo motus coeli remittetur; si in tempore et successive, aliquod ei incrementum accedet ex continua productione sicque motus celerior evadet. Non potest ergo coelum aequabiliter agi." Ibid., art. 1, 287.
221. "Impulsus perseverat secundum omnia instantia ac totum tempus, quod a principio motus coelestis hucusque defluxit et neque ulla nova impulsus pars advenit, neque ulla abit. Sed idem impulsus integer sub eodem gradu et intentione perpetuo efficitur et conservatur." Ibid., art. 3, 289.
222. "Verum haec responsio non satisfacit." Ibid.
223. In this question, the Coimbra Jesuits bring together the two basic medieval versions of impetus theory: temporary versus permanent impetus.
224. "Quare rectius alii occurrendum censent, videlicet impulsum quem Angelus coelo imprimit continenter in qualibet parte temporis successive diminui simulque perenni et continuo influxu ab eodem Angelo refici, ita ut quantum ex una parte amittur, tantundem ex alia resarciatur. Atque ita perpetuo conservari in sphaeris coelestibus impulsum sub eadem intentione et gradu." Conimbricenses, *De coelo*, bk. 2, ch. 6, qu. 2, art. 3, 1598, 289–290.

Thus did the Coimbra Jesuits favor the notion of a temporary, impressed angelic force. By continually impressing increments of force into its orb to replace what was just lost – indeed, we have to suppose that the replacement of impressed force occurs simultaneously with the loss – the level of force is kept constant, and every celestial motion could be uniform.

In a briefer treatment, Riccioli assumes that angels possess a force or power distinct from the intellect and will.[225] Following in the footsteps of Godfrey of Fontaines, whom he mentions, and many others, Riccioli calls it an "executive power" (*potentia exequutiva*). It is the means by which an angel impresses a certain "translative quality," or *impetus*, into a mobile, or celestial body, from which motion follows immediately.[226] The combination of angels and impressed force, or impetus, was thus a reasonably popular theory in the sixteenth and seventeenth centuries, by which time the more vague thirteenth-century *virtus motiva*, or *virtus impressa*, was transformed into the more precise fourteenth-century *impetus*, which, as we noted, had received its fullest development from Buridan.

b. Rejection of the combination of angels and impetus: Suarez and Serbellonus

In his *Disputationes metaphysicae*, first published in 1597, Francisco Suarez rejects the idea that angels impress forces into the orbs to cause their motions. In this, he may have been reacting to the Coimbra Jesuits. At the outset, Suarez, without mentioning names, assumes the probability of earlier opinions that are traceable to Godfrey of Fontaines and Hervaeus Natalis. Thus he thought it more probable that an angel has a proper "motive power" or "force" (*virtus motiva*) that is distinct from the intellect and will.[227] Indeed, Suarez even calls this force an "executive power" (*potentia executiva*), thus using the same expression as Godfrey of Fontaines more than 300 years earlier.[228]

But where Godfrey and Hervaeus spoke only of a motive force that was used by the executive force to implement the will of the angel or intelligence, Suarez speaks of impressed forces transmitted by the angel into the body

225. He characterizes it as the "more correct" position and says that it was adopted by Duns Scotus, Henry of Ghent, Franciscus Mayronnis, Peter Aureoli, Godfrey of Fontaines, the Conimbricenses, Suarez, and Amicus. Riccioli, *Almagestum novum*, pars post., bk. 9, sec. 2, ch. 1, 1651, 250, col. 1.

226. "Suppono autem ex dictis alibi a nobis in tractatu de angelis eos imprimere qualitatem quandam translativam mobilis a loco ad locum, quae vocatur impetus et ex qua sequitur immediate motus." Riccioli, ibid. He adds that Molina, Vasquez, and Tanner also teach the same doctrine.

227. "Unde etiam ad se movendum est mihi probabilius Angelum habere propriam virtutem motivam distinctam a voluntate." Suarez, *Disputationes metaphysicae*, disp. 35, sec. 6, 1866, 2:475, col. 1, par. 22. Earlier in the same paragraph, Suarez explains that a faculty exists in intelligences that differs from the intellect and will. (Haec autem ratio eumdem locum habet in intelligentiis; est ergo in eis etiam haec facultas diversa ab intellectu et voluntate).

228. Ibid., cols. 1–2.

or orb. Indeed, Suarez uses the term *impetus* and thinks of it as something impressed into the heaven or orb but distinct from the motion it produces.[229] Thus he, or his source, was influenced by the traditional impetus theory as it was developed in the fourteenth century.

Suarez explains that some feel that just as "a man impresses an impulse when he moves any body, so also does an angel, because local motion does not seem otherwise possible."[230] The impressed impetus is characterized as a "certain quality" that is ordained to produce local motion. The impetus produces motion when the mobile body is separated from its mover. Suarez admits that one could easily defend this mode of speaking.

However, because it seemed unnecessary and superfluous, Suarez could see no good reason to assume an impressed-force theory of celestial motion. Why should an angel impress impetus into something in order to move it when the angel could achieve the same result by the direct application of its own instrumental motive force (*virtus motiva*), which it can exercise by its continual presence in or on the orb or body? Hence the angel need not impress impetus into an orb to enable the latter to move when it is separated from its original mover, in this case the angel itself. An angel can always be directly present to the body it moves. It requires no surrogate force to act in its behalf. Impulses or impressed forces are in no way necessary for motion; otherwise God would be unable to move a body without first impressing a force into it, which is absurd.[231]

When an angel moves itself, it does not impress an impulse or impetus into its own substance. It moves itself only by making a motion. Why, therefore, should we, without good reason, multiply entities to explain the way angels move orbs?[232] Moreover, when an angel moves, it is surely unnecessary that it should first wish to impress impetus and then wish to move a body. It is only necessary that it wish to move the body. And for this purpose it has sufficient force and can produce the desired effect without the aid of anything else. Drawing on Ockham's razor, Suarez declares that

229. "Rursus vero inquiri potest circa hanc facultatem, an intelligentia ad movendum corpus imprimat in illo impetum distinctum ab ipso motu." Ibid., col. 2, par. 24. The Conimbricenses did not use the term *impetus*, preferring *impulsus* almost exclusively.

230. "Aliqui enim sentiunt, sicut homo imprimit impulsum dum movet aliquod corpus, ita etiam Angelum quia motus localis non videtur aliter posse fieri." Ibid.

231. "Fortasse tamen non est necessarium ut Angelus, quando immediate movet corpus imprimat ei impulsum, sed immediate ipsum motum, et nihil aliud. Ratio est, quia impulsus solum ponitur ut sit instrumentum moventis quando separatur a mobili; ergo quando movens immediate adest, et in se habet sufficientem vim motivam, non indiget impulsu tanquam instrumento separato quo moveat. Nec vero talis impulsus est de necessitate motus localis; alias nec Deus ipse posset movere corpus nisi imprimendo illi impulsum, quod videtur ridiculum et gratis dictum. Cur enim non poterit Deus efficere in corpore solum localem absque alio novo effectu?" Ibid., 2:475, col. 2–476, col. 1, par. 25.

232. "Nulla est ergo ratio vel necessitas imprimendi talem impulsum. Sicut Angelus movendo se non imprimit suae substantiae impulsum, sed motum; quid enim necesse est multiplicare entia aut instrumenta sine causa?" Ibid., 2:476, col. 1, par. 25.

"it is vain to do with more what can be done by less."[233] Not only can angels not impress impetus into bodies, but they cannot impress any qualities into bodies.[234]

Although Suarez thought the impressed-force explanation defensible, he did reject it. Unfortunately, Suarez mentions no names in his entire discussion, so that we cannot identify other discussants nor learn of any supporters of the theory combining angels and impressed force. But whatever Suarez's source for this topic, it bears unmistakable links with the tradition of impetus theory deriving from the fourteenth century.

Some seventy years after the appearance of Suarez's discussion, Sigismundus Serbellonus adopted much the same interpretation but offers a clearer exposition and reveals an interesting adaptation to the new cosmology. Serbellonus distinguishes two ways in which an intelligence could move a mobile. It could do so directly or by means of a force which it impresses into the mobile. To move a body the first way, the intelligence communicates motion directly and successively to the body it moves. The motion arises immediately from a relationship between mover and moved. As we might surmise, the second way includes an extra step. Initially, the intelligence immediately moves the mobile, but in doing so it also transmits to it an impressed force, an *impulsus*, which then takes over and continues the motion of the celestial body. The impressed force is capable of moving its mobile even if the intelligence itself remains immobile.[235]

With the two options described, Serbellonus indicates his preference. Although there is some ambiguity, he seems to opt for the simpler explanation, which is that the angel moves its planet directly, without impressing any force into it.[236] His argument is much the same as Suarez's. Why should an angel employ an instrument, such as impressed force, when it can achieve the same results by moving something directly and immediately?[237] More-

233. My interpretation of: "Denique declaratur, quia cum Angelus movet, non est necesse ut velit imprimere impetum, sed tantum ut velit movere. Nam ad hoc habet sufficientem virtutem et ille effectus motionis localis est de se factibilis absque alio; ergo tunc efficiet Angelus solum motum absque alia priori qualitate vel impulsu; ergo verisimile est semper sic movere, quia frustra faciet per plura quod potest facere per pauciora." Ibid.

234. "Atque hoc etiam modo generalius defenditur, Angelos nullam qualitatem per se posse corporibus imprimere." Ibid.

235. Here is the text on which this paragraph is based: "Dicend. 1. Intelligentiam posse movere coelum mediante motu solum et mediante impulsu impresso. Explicatur. Tunc corpus movetur per motum solum quando movens est ipsi immediatum et successive recipit ab ipso motionem; per impulsum vero cum imprimitur illi motus virtute cuius moveri potest etiamsi motor desistat a motione per aliquod tempus. Unde motus simpliciter in primo sensu consurgit ex complexo mobilis et moventis in actu; secundo impulsus vero fit quidem per complexum ex mobili et motore in principio impressionis, at sequitur per se in mobili." Serbellonus, *De caelo*, disp. 1, qu. 4, art. 3, 1663, 2:43, col. 1, par. 29.

236. In the latter part of his argument, Serbellonus tries to answer a few other objections to the assumption of impressed force, but it seems that he thought it best to eliminate impressed forces.

237. "If therefore an impressed force [*impulsus*] is unnecessary for local motion, why is it

over, an angel would move a sphere with the same power with which it moves itself. Since it causes itself to move without impressing any force into itself, why can it not also move a sphere or mobile by itself without resort to an impressed force?[238]

But what celestial bodies do the angels or intelligences move? As one who believed in the fluidity of the planetary heavens – he insisted that the planets are not affixed to orbs[239] – and for whom only the firmament of the fixed stars was solid and hard, Serbellonus associated the motive intelligences with the planets rather than with orbs. He declares that there are "as many motive intelligences of celestial bodies as there are such distinct bodies." Because there are so many different motions among the planets, no single intelligence could move them. Hence one intelligence is assigned to each planet. Indeed, not only does one intelligence assist each planet, but we must assume that at least one intelligence moves the four satellites of Jupiter, and perhaps one intelligence moves each satellite; the same must be said of the two alleged satellites of Saturn. The satellites of Jupiter, and presumably also of Saturn, are too far away to be moved by the same intelligences that move Jupiter and Saturn directly.[240] As was often stressed, the range of influence of intelligences or angels was limited. Hence more intelligences were needed.

If angels or intelligences caused the planetary motions, how did they do it? Did they move by intellect and will or by some third force or power? When Serbellonus published his second volume in 1673, with its important section on the heavens, almost four hundred years had elapsed since the Condemnation of 1277, when the idea that an intelligence could move a heaven by will alone was denounced. What was Serbellonus's response to this perennial problem? A brief introductory sentence at the beginning of his lengthy discussion on "How Intelligences Move Celestial Bodies" tells it all. "We do not inquire," he commences, "about the quality of motive power in intelligences, namely, whether it is distinct from the intellect and will or, indeed, results from each. This [problem] pertains to the theolo-

assumed in angels who are intimately present [in their celestial bodies] and have sufficient power to push [the bodies in which they are present]?" (Si ergo impulsus non est necessarius ad motum localem, non est cur ponatur in angelis cum sint intime praesentes et habeant sufficientem virtutem impellendi). Serbellonus, *De caelo*, disp. 1, qu. 4, art. 3, 1663, 2:43, col. 1, par. 29.

238. "Deinde angelus per eandem virtutem movet sphaeram per quam se ipsum movere potest. Sed se movere potest sine impulsu impresso in se ipso; ergo etiam sphaeram ita movere poterit." Ibid.

239. "Deinde cum planetae, per nos, non sint affixi propriis orbibus. . . . " Ibid., art. 2, 2:42, col. 1, par. 23.

240. "Non solum autem cuilibet planetae propria intelligentia assistit, sed etiam cuicunque alteri syderi, quod in firmamento non sit: sic quatuor satellites Iovis propriam intelligentiam motricem habere debent quia vario ac multiplici motu circa Iovem feruntur et ratione distinctionis et distantiae, quam ab ipso servant, non recte sub dominio eiusdem intelligentiae constituuntur. Hoc idem dicendum de duobus Saturni comitibus." Ibid., par. 25. As noted earlier, the rings of Saturn were for a time thought to be two satellites like those of Jupiter.

gians,"[241] thus suggesting that the article condemned in 1277 continued to play a role. Otherwise, Serbellonus would have felt no need to explain that he was omitting any judgment about the means by which an angel moved its planet or orb.

Having decided that an angel moved its planet by itself and not by an impressed force, Serbellonus chose to delve no further. It sufficed that he had chosen between the angel as direct cause of motion or as indirect cause by virtue of its ability to impress a force into bodies. The rest – whether an angel moved itself and its orbs by intellect and will or by something else – was best left to the theologians. We see here how intelligences and angels had become little more than names for motive forces that were used to explain the efficient causes of celestial motion. Indeed, as we shall see in Section II.8 of this chapter, Serbellonus, though perhaps more emphatic than most, belongs to a long tradition going back to the thirteenth century.

c. The influence of al-Bitrūjī: Mastrius and Bellutus

The impressed force discussed in the preceding sections was supposedly introduced into every orb or planet by an angel or intelligence. Each orb or planet received a quantity of impressed force, or impetus, from the single angel assigned to it. There was, however, another important version of impressed force that we must now consider and for which we turn to Mastrius and Bellutus, who incorporated quite a different concept of impressed force into their explanation of the causal factors in celestial motion. In a lengthy question on the local motion of the heavens, with numerous subdivisions, Mastrius and Bellutus had much to say about celestial movers. Although they concede that God could have created internally self-moving celestial bodies, they declare instead that "the heavens are moved externally, not internally,"[242] by an "incorporeal and spiritual" intelligence. Moreover, they further assume that "angels move by a power formally distinct from the intellect and will," which they describe as an "executive power" (*potentia executiva*),[243] thereby associating themselves with all those scholastics – ranging from Godfrey of Fontaines to Francisco Suarez – who saw fit to describe this centuries-old causal explanation of celestial motions.

In the world of hard planetary spheres which formed the frame of their cosmology, Mastrius and Bellutus assumed that the outermost sphere of the world, or the *primum mobile*, was moved by an intelligence. In their

241. "Non quaerimus de qualitate virtutis motivae in intelligentiis, an scilicet sit distincta ab intellectu et voluntate; an vero ex utraque resultet. Pertinet enim hoc ad Theologos." Ibid., art. 3, 43, col. 1, par. 28.

242. "Dicimus, primo, celos ab extrinseco moveri, non ab intrinseco." Mastrius and Bellutus, *De coelo*, disp. 2, qu. 4, art. 3, 1727, 3:503, col. 1, par. 138.

243. "Angelos movere per potentiam formaliter distinctam ab intellectu et voluntate." Mastrius and Bellutus, ibid., art. 4, 1727, 3:504, col. 2, par. 148. A few lines later they explain that the "executive power" (*potentia executiva*) transmits the will of the angel.

view, it was the only intelligence associated with the celestial orbs.[244] The manner in which this unique celestial intelligence moved its orb, the *primum mobile*, was not by the will and intellect alone but by a third entity, an "executive force," which transmitted the will of the angel to the orb in the form of a motive force. In their interpretation of Aristotle, Mastrius and Bellutus viewed the *primum mobile* as an aggregate of orb and intelligence, where the latter moves the orb by virtue of its love of the prime mover. Thus the prime mover, which is common to all motions, may be regarded as the final cause and the intelligence as an efficient cause in producing the motion of the *primum mobile*, which is also "the first [of all motions and also] the cause, regulator, and measure of other motions."[245]

But what about all the other enclosed spheres? Was each of them also moved by an intelligence, as Aristotle and most other scholastics had believed? By implication, Mastrius and Bellutus seem to deny this, since they speak of only one celestial intelligence, which they associate solely with the *primum mobile*. Did this one intelligence, therefore, move all the orbs of the celestial region? No.[246] The influence and power of a single angel was deemed woefully inadequate to supply operating power to the entire celestial region.[247] The motions of the celestial region were too complex and its magnitude too great to be served by a single intelligence.

According to the Coimbra Jesuits, this was precisely the erroneous theory held by Girolamo Fracastoro in his *Homocentrica, sive de stellis*, published in 1538. Fracastoro believed that a single intelligence could move all celestial orbs. But, in an approach that Mastrius and Bellutus would adopt, the Coimbra Jesuits assert ([*De caelo*, bk. 2, ch. 5, qu. 8, art. 1], 1598, 277), that one intelligence – whether actually in the outermost sphere or only conjoined to it – could not simultaneously produce the variety of differing, and perhaps even oppositely directed, motions of so many orbs. A single intelligence would be incapable of producing the totality of impulses or forces required for all the celestial motions. Indeed, because of the inadequacy of a single intelligence or angel, the transmission of its motive force

244. In their assumption that a single intelligence or angel is the ultimate cause of all celestial motions, Mastrius and Bellutus, like Girolamo Fracastoro before them, appear to have repudiated the Aristotelian rule that every orb has its own angel or intelligence, so that there are as many celestial intelligences as there are orbs.

245. Here is the relevant text for this passage: "Dicimus . . . Aristotelem in texto 43 assignasse numerum intelligentiarum quarum singulae sunt particulares et proprii motores orbium. Praeter istas autem assignavit primum motorem communem omnibus, de quo locutus fuerat in texto 36, dum dixit movere ut amatum et desideratum. In texto vero 38, accepit primum mobile pro aggregato ex orbe et intelligentia, nam supra dixerat primum motorem movere ut amatum, ergo nequit intelligi de primo orbe tantum, non enim orbis movetur ad amorem, sed de orbe et intelligentia. Dicitur autem specialiter et appropriate primum mobile movere, quamvis ut sic concurrat ad alios orbes quia motus primi mobilis est primus, causa, regula, et mensura aliorum motuum." Mastrius and Bellutus, *De coelo*, disp. 2, qu. 4, art. 4, 1727, 3:504, col. 1, par. 146.

246. "Idcirco non sufficit unus angelus ad omnes motus celorum exercendos." Ibid., 505, col. 1, par. 150.

247. "Quod quamvis nullam reperiat contrarietatem in medio, tamen determinatam sphaeram habet suae actionis." Ibid.

(*vis motiva*) by means of its executive power would produce disorder and confusion in the inferior orbs.[248]

If the unique intelligence associated only with the *primum mobile* is incapable of moving all the orbs, and there are no other intelligences, what could cause the motion of all the orbs enclosed within the *primum mobile*? Mastrius and Bellutus reply: the *primum mobile* itself. Just because the *primum mobile* is moved by an external intelligence does not preclude it from being a mover. They envisioned the following process: acting as an efficient cause, the single intelligence directly causes the primary motion of the *primum mobile*. As the latter moves in its east-to-west daily circular motion, it transmits impulses, or impetuses, to the solid, hard, contiguous celestial orbs that it surrounds and contains. As the impetuses, or impressed forces, are transmitted downward from one orb to another, the force is weakened, and the orbs cannot complete a full circulation in twenty-four hours. Consequently, the daily paths of the planets are not perfect circles, but incomplete arcs of circles that form spiral lines.[249]

These spiral lines result from the daily east-to-west motion and what appears to be a periodic or sidereal west-to-east motion. In reality, Mastrius and Bellutus argue, the planets move only in an east-to-west direction, and not at all from west to east. However, the velocities of the planets in their east-to-west daily motions are such that they traverse slightly less than the full circle attained in twenty-four hours by the daily motion of the sphere of the fixed stars. Hence every day each planet completes slightly less than a complete circulation, and its path actually traces out an imaginary spiral line. As it completes its daily spiral – that is, as it falls slightly short of completing a full circular path each day – every planet gives the appearance of moving backward from west to east.[250]

Mastrius and Bellutus fashioned a causal explanation of celestial motion

248. "Namque licet non oporteat Angelum in toto supremo globo esse, ut eum moveat, sed satis sit ei coniungi in certo situ, fieri tamen non poterit ut ab eo situ dispertiat impulsum ad praedictos motus omnium sphaerarum et errantium syderum, cum plerique eorum motuum sint ad partes oppositas, vel quasi oppositas, et alii majorem, alii minorem impulsum desiderent pro magnitudine temporis et spatii quod conficiunt. Sane quidem si per intermedia corpora impulsus ille descenderet susque deque omnia turbaret, nec tam efficeret motum quam tumultum. Addimus etiam nec vim motivam unius Angeli quae haud dubie limitata est videri sufficere ad producendum simul tantum impulsum quantum omnes ii motus requirunt." Conimbricenses, *De coelo*, bk. 2, ch. 5, qu. 8, art. 1, 1598, 277. Mastrius and Bellutus mention only Aristotle, not Fracastoro. It seems that they only shared with Fracastoro the idea of a single intelligence operating on the celestial orbs.

249. "Sic primum mobile movet coelos inferiores ab ortu in occasum et quia virtus in corpore distanti minus efficaciter diffunditur, hinc impetus a primo mobili impressus non causat in coelis inferioribus perfectum circulationem, sed cum aliquali retardatione propriae intelligentae impellunt orbes ad latera, ex quibus virtutibus resultat motus quidam modificatus ab ortu in occasum per lineas spirales." Mastrius and Bellutus, *De coelo*, disp. 2, qu. 4, art. 1, 1727, 3:502, col. 2, par. 134. Fracastoro expressed a similar judgment when he assumed that an outer orb can communicate motion to an inner orb. It was not necessary to associate an intelligence or angel with each orb or planet.

250. For the description of spiral motion, see Mastrius and Bellutus, ibid., 501, col. 2–502, col. 1, par. 130.

from three separate components: (1) a single intelligence that moves the outermost mobile sphere, or *primum mobile*, an idea ultimately derived from Aristotle, whom they mention, and perhaps also from the *Homocentrica* (1538) of Girolamo Fracastoro, whom they do not mention; (2) an impressed force, or impetus, that originates from the daily motion of the *primum mobile* but which is successively diminished as it is transmitted downward from orb to orb; this concept seems ultimately traceable to al-Bitrūjī's *De motibus celorum*, as translated by Michael Scot in the early thirteenth century, and perhaps also to Fracastoro; and (3) spiral paths of the planets, produced by the successively diminishing impressed forces, an idea probably derived from al-Bitrūjī.

As real paths of planets, spiral lines can be traced back to Plato and were mentioned in the Middle Ages by Averroës, al-Bitrūjī, Albertus Magnus, Sacrobosco, and Oresme.[251] By invoking spiral lines as a consequence of a gradual dissipation of an impressed force as it passes from orb to orb, Mastrius and Bellutus seem to have causally derived the two basic planetary motions, east-to-west and west-to-east, from the single east-to-west motion of the *primum mobile*, a motion caused ultimately by the single intelligence associated with the *primum mobile*.

The idea that impressed force could be transmitted "inward" or "downward" from the *primum mobile* to each of the inferior planetary orbs was probably derived ultimately from al-Bitrūjī's *De motibus celorum*, of which printed editions had been available since 1531;[252] or it was derived from Fracastoro's *Homocentrica*, available since 1538, where Fracastoro seems to acknowledge a debt to al-Bitrūjī (whom he calls "Albateticus"). A glance at al-Bitrūjī's account will reveal how strong is the probability that Mastrius and Bellutus drew directly from it:

We say that the highest sphere always moves about two fixed poles from east to west, completing one revolution in a day and a night. It causes the motion of the universe, and its motion is swifter than all the motions below it, for all those spheres lose some of its motion. The amount of this loss varies with their distance from the highest spheres [sphere?] which causes their motion. Each of these lower spheres desires to imitate the highest sphere and trails along according to the amount of power which it retains from the highest sphere, but its form is maintained by its proper motion. Thus each of them moves with this other motion about its poles, trailing behind the motion of the highest sphere and combining with it, imitating the highest sphere. The spheres below the highest sphere differ in the speed with which they trail the highest sphere, and this varies according to their distance from

251. For a brief description of the generation of spiral lines, see Oresme, *De comensurabilitate*, 1971, 31–33, especially note 49 with its figure. Al-Bitrūjī describes spiral motion in chapter 9 of his *De motibus celorum*. See al-Bitrūjī [Carmody], 1952, 40–41, 51–54, 98 (21), and al-Bitrūjī [Goldstein], 1971, 85–86, par. 72; 101, par. 106.

252. See al-Bitrūjī [Carmody], 51 (Carmody's summary of ch. 8), 92–94 (all of ch. 8); al-Bitrūjī [Goldstein], 76–79, pars. 57–62 (all of ch. 1).

the highest sphere since the power for their motion depends on the distance from the prime mover, the source of all power. The motion of the highest simple sphere is simple and invariable, always maintaining the same speed. The spheres which are closer to the simple sphere move more swiftly and powerfully, for the source (of their motion) is the highest sphere. But those bodies which are farther away are less powerful, and their motion is slower, for motion depends on the amount of power. This statement – that motion depends on the amount of power – is one on which everyone will agree.[253]

The most frequently used Latin expression for "power" or "force" in Michael Scot's translation of al-Bitrūjī's *De motibus celorum* is *virtus*.[254] The term *impetus* is absent, because it came into common use only in the latter part of the fourteenth century, when, as we saw earlier, Buridan even associated it with the celestial orbs (though he assumed that God had implanted such forces at the creation). Its use by Suarez, Mastrius and Bellutus, the Coimbra Jesuits, Amicus, and Serbellonus, in the late sixteenth and the seventeenth century indicates that this significant term was still in vogue.[255]

7. Al-Bitrūjī and the desire to avoid contrary motions

Al-Bitrūjī's explanation of celestial motion and the acceptance of some version of it by Alessandro Achillini, Mastrius and Bellutus, and others, was motivated in part by a desire to avoid assigning contrary motions to one and the same celestial body. In the usual medieval discussions of celestial motions, it was assumed that one and the same planet was continually undergoing simultaneous contrary motions. For not only did every planet have a daily motion from east to west in twenty-four hours, but at the same time it was slowly moving in an opposite west-to-east direction to complete a periodic motion through the zodiac: Saturn in thirty years, Jupiter in twelve, and so on. From the standpoint of Aristotelian physical theory, one and the same body could not undergo simultaneous contrary motions.[256] "It seems absolutely impossible in any way," Raphael Aversa insisted, "that any mobile be truly carried with contrary motions, unless it

253. Goldstein's translation from al-Bitrūjī, 1971, 76–77, par. 57 (I have inserted the bracketed word). I have used Goldstein's translation from the Arabic and Hebrew because it corresponds reasonably closely to the Latin translation in al-Bitrūjī [Carmody], 1952, 92 (1–3). According to Dreyer (1953, 298), Fracastoro "assumes that an outer sphere may communicate its motion to an inner one, while an inner one does not influence an outer one, and he is therefore able to let the Primum Mobile communicate its daily rotation to all the planets without having with Eudoxus to assume one sphere for each planet to produce the daily rotation."
254. See al-Bitrūjī [Carmody], 1952, 46.
255. For the Coimbra Jesuits, see *De coelo*, bk. 2, ch. 5, qu. 8, art. 1, 1598, 276; for Amicus, *De caelo*, tract. 5, qu. 6, dubit. 1, art. 1, 1626, 289, col. 2.
256. In *Physics* 8.8.264a.28–29, Aristotle declares that "it is impossible for a thing to undergo simultaneously two contrary motions." Aristotle [Hardie and Gaye], 1984.

could be replicated in many places."[257] To avoid these difficulties, Aristotle, with his concentric orbs, and Ptolemy, with his concentrics and eccentrics, and many other astronomers and cosmologists assumed that every real astronomical motion is produced by only one orb. The daily motion had its own orb, and the periodic motion of each planet was also represented by a single orb.

Despite these efforts, it was obvious that the position of one and the same planet was nonetheless represented by at least two oppositely directed orbs. Thus it still seemed as if cosmology was in violation of Aristotle's dictum against contrary motions. According to Clavius, critics of contrary celestial motions, apparently arguing by analogy with upward and downward rectilinear motion on earth, insisted that one of those motions would have to be violent. But violent motions had to terminate and could not be perpetual. Hence some celestial motions would be of finite duration, rather than perpetual, an unacceptable consequence. Moreover, it was the nature of a violent motion to weaken over time, which violated the general consensus that the heavens always move with the same, uniform speed. Finally, contrary motions are superfluous. Why assume contrary motions when the same results can be achieved by a single, unidirectional motion?

Although many natural philosophers and astronomers ignored this dilemma, it remained a perennial underlying problem in medieval and early modern scholastic cosmology. Al-Bitrūjī's (or Alpetragius's) solution was well known. He denied contrary motions and assumed that all celestial motions were from east to west but that the planets moved with lesser velocities than the daily motion. In the passage from al-Bitrūjī quoted earlier, the latter explains that "The spheres below the highest sphere differ in the speed with which they trail the highest sphere, and this varies according to their distance from the highest sphere since the power for their motion depends on the distance from the prime mover, the source of all power."

By adopting some version of al-Bitrūjī's theory of concentric spheres with its exclusive east-to-west motions, a number of medieval and early modern authors sought to eliminate contrary motions. They devised various causal accounts to explain the motions. Mastrius and Bellutus, as we saw, assigned one intelligence for the *primum mobile* and assumed that the inferior orbs were moved by means of successively transmitted impressed forces. In lieu of impressed forces, they might have assigned one intelligence to each of the inferior orbs but may not have wished to invoke intelligences needlessly.

Christopher Clavius thought al-Bitrūjī's theory sufficiently important to provide a lengthy description of its rationale, along with a refutation.[258]

257. "Verum quocumque modo apparet penitus impossibile idem mobile ferri vere contariis motibus nisi replicetur in pluribus locis." Aversa, *De caelo*, qu. 34, sec. 6, 1627, 145, col. 1.

258. "Alpetragius [i.e., al-Bitrūjī] and Achillini, with other authors, embraced this opinion because in no way could they imagine that one and the same celestial body could be

Clavius sought to defend traditional Ptolemaic astronomy and cosmology, with its eccentrics and deferents turning in different (and often opposing) directions. To do this, he had somehow to eliminate or sidestep the problem of contrary motions. Clavius assumes the traditional opinion that the *primum mobile* moves from east to west to account for the daily motions and that each of the inferior spheres from Saturn to the Moon has its own proper motion from west to east. As the *primum mobile* moves in its daily east-to-west motion, it drags the inferior planetary orbs with it, so that they also have a daily east-to-west motion. Therefore each planetary orb has a proper motion to the east and also a simultaneous daily motion to the west. But are these not contrary motions?

Clavius denies this. The two simultaneous oppositely directed motions of a planetary orb are not contraries, because of a distinction Clavius makes between an *absolute* (*simpliciter*) motion and one that is *relative* (*secundum quid*). His crucial argument is made by means of an analogy involving a ship moving at maximum speed from east to west (surrogate for the *primum mobile*), while its captain walks at a slower pace from the bow of the ship to its stern, or from west to east (surrogate for a planet's proper motion):

> With this assumed, do you not see that the captain is moved from east to west absolutely because the motion of the ship is much quicker than the proper motion toward the contrary direction by which he is moved, and because of this he always withdraws more from the east and approaches the west. However, is he not simultaneously moved relative [*secundum quid*] to the east, that is, [is he not moved] toward the eastern parts of the ship, but not absolutely? Do you not also see that if the ship remained unmoved, the captain would then be moved absolutely from west to east, since it [the ship] always approaches more to the east and withdraws from the west? Finally, do you not see that the same thing would happen if the captain moved with a quicker motion than the ship? In this manner, therefore, must it be understood that the inferior heavens are moved from west to east under the zodiac of the *primum mobile*.[259]

These motions fail to qualify as contrary because at every instant the proper eastward motion of a planetary orb is not equal to its daily westward motion

carried with two motions: from east to west and west to east." Translation from Clavius, *Sphere*, ch. 1, *Opera*, 1611, 3:25. For the full discussion, see 25–28.

259. "Moveatur navis aliqua ab oriente in occidentem maxima celeritate; nauclerus autem eodem tempore, gradu admodum tardo perambulet navim a prora in puppim. Quo posito, nonne vides nauclerum simpliciter quidem moveri ab oriente in occidentem eo quod ad motum navis celerius multo quam proprio motu in contrariam partem moveatur et ob id semper magis ab oriente recedat occidenti vero appropinquet? Simul tamen secundum quid moveri ad orientem, id est, ad partes orientales navis, non autem simpliciter? Nonne etiam vides, si navis immota consisteret nauclerum simpliciter tunc moveri ab occidente in orientem, cum semper magis ad orientem accederet et ab occidentem recederet? Nonne denique idem contingere conspicis, si nauclerus citatiori motu incederet quam navis? Ita igitur intelligendum est coelos inferiores moveri sub Zodiaco primi mobilis ab occidente in orientem." Ibid., 28.

caused by the dragging effect of the *primum mobile*. They would be contrary only if, after referring their motions to the same fixed point, one motion advanced as far toward that point as the other motion withdrew from it. But this does not occur, because from any fixed point and for any interval of time less than a day, a planet's westward daily motion will be greater than its proper eastward motion during the same interval.[260] Raphael Aversa set much laxer conditions for contrary motions, which he nevertheless rejected. No fixed point was necessary to establish opposite motions; one could simply allow that two opposite motions were oblique to each other, moving around different poles and in different paths. Yet even with these far less stringent conditions, Aversa thought it "impossible that the same mobile be carried with different motions in different directions."[261]

Although Clavius acknowledges that these two motions are indeed commonly spoken of as "contrary motions," he quickly adds that this is so only because the opposing terms "east" and "west" are used.[262] And because they are not really contrary motions, Clavius says that neither of them will be a violent motion. But even if one of them were violent, it could still be perpetual and retain its power at the same level, because the motive cause that produces its regular motion is itself perpetual and capable of acting without loss of energy or effort.[263] The motive cause, presumably the prime mover, will supply the necessary energy required to maintain every celestial motion at its appropriate speed.

Aversa had a more plausible solution. There can be only one motion for any single heaven or orb, but that single motion may arise from several causes. One cause might push it in one direction, another cause in another direction, presumably even in the opposite direction. As a result of the interaction of these competing causes, an orb would tend toward one direction or another. But it would move in only one direction and never in contrary or opposite directions at the same time.[264]

260. "Ex hac porro declaratione et exemplis adductis, perspicuum relinquitur duos praedictos coelorum motus quorum unus est ab oriente in occidentem, alter ab occidente in orientem, non esse contrarios cum non simpliciter ad terminos contrarios, puta ad orientem et ad occidentem fiant, ut explicavimus. Contrarii namque motus referri debent ad unum idemque punctum fixum, ut videlicet uno motu ad illud punctum accedatur et ab alio ab eodem recedatur, quod in motibus coelorum minime fieri diximus." Ibid., 29.

261. Here is the full text: "Addunt aliqui duos dictos motus unius caeli non esse vere contrarios quia unus non tendit directe adversus alterum, sed in obliquum, nempe super diversos polos et per diversam lineam. At similiter impossibile apparet idem mobile ferri diversis motibus, id est ad diversas partes." Aversa, *De caelo*, qu. 34, sec. 6, 1627, 145, cols. 1–2.

262. "Dicuntur tamen isti duo motus communi loquendi modo contrarii et oppositi ratione terminorum contrariorum, puta orientis et occidentis." Clavius, *Sphere*, ch. 1, *Opera*, 1611, 3:29.

263. "Quare ad primam rationem Alpetragii et Achillini respondendum est illos motus non esse contrarios . . . et ob id neutrum esse violentum. Adde, non sequi etiamsi concederemus alterum illorum esse quodammodo violentum illum non fore perpetuum atque debilitari posse, cum causa eius motiva sit perpetua et infatigabilis. Illud enim violentum solum dicitur non posse esse perpetuum quod causam fatigabilem et non perpetuam habet. Hoc enim simpliciter et per se violentum dicitur." Ibid., 26.

264. "Res ergo ita intelligenda et explicanda est quod vere et realiter unus sit motus unius-

8. Did angels and intelligences function as mechanical forces?

We have now seen enough to conclude that medieval and early modern explanations of the causes of celestial motions were unusually varied. Within the most basic division into external and internal movers, there was a rich variety of explanations and suggestions to account for the regularities and irregularities of celestial motion: intelligences, angels (the two were usually but not always synonymous), forms, souls, and natural inclinations or tendencies of the orbs themselves. Even the abandonment of hard orbs did not compel scholastic natural philosophers to reject their array of explanations. They could simply attribute the same intelligences and motive forces to the individual planets. It would take Isaac Newton's theory of universal gravitation to sweep them from the sky.

Because angels or intelligences were easily the most widely invoked causes of celestial motion, we shall conclude this chapter with a brief assessment of the manner in which they were perceived. We have seen that although they were not assumed to be fully living beings – they lacked nutritive and sentient levels of activity – their exercise of will and intellect were thought to confer some level of life on these bodiless, immaterial creatures. Since medieval celestial intelligences and angels were the direct descendants of Aristotle's unmoved movers, the language of love – the love of an orb for its intelligence – associated with Aristotle's unmoved movers reinforced the strong sense that intelligences were alive. Scholastic thinkers had little difficulty in assigning life to immaterial, dimensionless, spiritual entities such as intelligences and angels. After all, God also possessed these properties and a host of others – omnipresence, omnipotence, infinite immensity, and so forth. Surely, one would have to consider God as a living being. Thus whatever else may be thought about celestial intelligences, all were agreed that they were alive in some important sense.

Despite the attribution of life, however, intelligences and angels associated as movers of celestial orbs seem to have performed more as mechanical forces than as living, spiritual entities. With an emphasis on souls, angels, and intelligences, we might suppose that scholastics perceived their world as primarily an organic whole rather than as a largely mechanically operating system. In truth, intelligences were made to operate far more as impersonal, efficient causes than as love-inspiring, spiritual, final causes. Indeed, the language of love was largely ignored or abandoned, because it was superfluous and even misleading. Each intelligence was assumed to move its own sphere as a direct, efficient cause, not because of its love for something else or because of something else's love for it.

When Jean Buridan and his followers suggested that God might have chosen to move the celestial orbs by impetuses impressed in them at the

cuiusque caeli et orbis, tamen proveniens a pluribus causis quarum una impellit mobile versus unam partem, alia versus aliam. Et ita prodit motus quidam certo modo attemperatus et modificatus, ut mobile semper ad unam simpliciter partem tendat, sicut in unico semper loco reperitur." Aversa, *De caelo*, qu. 34, sec. 6, 1627, 145, col. 2.

Creation rather than by external intelligences, he was merely substituting one impersonal force for another. Despite Buridan's concern about the reception of his suggestion by the theologians, there seems to have been little or no reaction. Indeed, Albert of Saxony also adopted this interpretation and passed it on to scholastics in the sixteenth and seventeenth centuries.

Although many medieval scholastics may have agreed with Moses Maimonides that "t'is love that makes the world go round," the love involved had little or nothing to do with emotion or spirit. The love Maimonides alludes to was an impersonal force, just as were the intelligences, angels, impetuses, and souls. It was probably so even with Aristotle himself. The metaphysical language associated with celestial motion masks a genuine attempt to understand the causes of celestial motion in purely physical terms. Indeed, by denouncing those who thought the heavens were alive because they were said to have a soul and also those who thought that angels could move their orbs by will alone, the authors of the Condemnation of 1277 would seem to have strengthened this tendency.

What Francis Carmody said (1952, 46) about Averroës (ibn Rushd) is applicable to the whole range of medieval thought about celestial movers: "One may well think of desire for perfection and assimilation through love as metaphysical interpretations; but there is evidence even in Ibn Rushd that these terms, based on Aristotle's principles, had lost the intervening implications and become, in due process of semantic reduction, no more than the vocabulary of physical thought."

19

The influence of the celestial region on the terrestrial

If celestial motion and its causes represented the category with the greatest number of cosmological questions, the second largest category was easily the influence of the celestial region on the terrestrial. It is tempting to ascribe this keen interest to the role of astrology, since astrological influences constituted a formidable subset of the total range of effects that were thought to be exercised by the celestial region on the terrestrial. If by astrology, however, we mean the prediction of natural events and human behavior on the basis of knowledge of alleged powers inherent in individual celestial bodies and their positions in the heavens, as well as their manifold configurations and interrelationships, then astrology plays very little role in scholastic natural philosophy. Commentaries and questions on Aristotelian treatises rarely contained specifically astrological discussions or predictions. Such discussions would have been deemed irrelevant, for which reason they will be inconspicuous in this account.

I. Celestial actions and human free will

As composite entities, human beings were comprised of a material body and an immaterial, spiritual soul under which the human will was subsumed. From the early centuries of Christianity, Church authorities vigorously opposed the pagan astrological tradition that assumed almost axiomatically that celestial bodies influenced both the human body and its immaterial soul and will. Since human free will, which enabled Christians to choose to sin or not to sin, was a part of the intellective soul, Christian theologians vigorously denounced those who claimed that celestial influences were determinative of human actions and that free will was an illusion. The Christian position was that the heavens could not act directly on the free will because the latter was part of the intellective, or rational, soul, which was a spiritual entity. The rule for Christians was that corporeal agents could not directly affect a spiritual entity.[1] John Major reflected the centuries-

1. Amicus sums up these rules ([*De caelo*, tract. 6, qu. 8, dubit. 7, art. 3 (incorrectly given as 4)], 1626, 443, col. 1): "Tertio sequitur caelum neque agere directe in liberum arbitrium patet ex dictis quia est pars animae intellectivae (ex 3 *De anima* 42) atque adeo spiritualis. Et ideo non subdita directe agenti corporeo, ut patet ex dictis."

long Christian attitude when he concluded that "the heaven has no influence on the rational soul."[2] Although the practice of and belief in astrology were widespread during the Middle Ages,[3] most Christians were careful to avoid compromising free will and the rational soul. On the face of it, at least, they would have supported John Major's conclusion.[4]

Because of the inability of celestial influences to affect the rational soul, what follows in this chapter concerns only the effect of celestial bodies on material bodies of the terrestrial region.

II. The general claim for celestial influence

Knowledge about the impact of the celestial region on the terrestrial was deemed of vital importance because it was thought that the former, which was incorruptible and unchanging, was the ultimate source of physical change in the latter, which was subjected to incessant generation and corruption. That sublunar things might in some way affect and cause changes in the celestial region was never seriously entertained.[5] The changeable and ignoble bodies of the terrestrial region were judged incapable of affecting the incorruptible heavenly bodies.

That celestial bodies exerted a vital and even controlling influence over material things in the terrestrial region, which included everything below the concave surface of the innermost lunar sphere, was universally accepted.[6] Themon Judaeus was representative of his age when he declared that "every natural power of this inferior, sensible world is governed by the heaven."[7]

2. "Quarta conclusio: celum in animam rationalem non habet influentiam ullam." Major [*Sentences*, bk. 2, dist. 14, qu. 8], 1519b, 79v, col. 1.
3. Although Nicole Oresme accepted the reality of celestial influences on the terrestrial realm, he was an ardent foe of astrological prognostication. (For more, see Ch. 18, Sec. I.3.a.)
4. Although Giovanni Pontano assumed, in at least one place in his *De rebus coelestibus*, that celestial bodies affect the human soul and mind, "elsewhere in this treatise and in other writings, he says that astrology deals only with the corporeal and that the mind and soul are independent of stellar influences." Trinkaus, 1985, 453. For Dante's more conventional opinions, see North, 1986, 82–85.
5. Roger Bacon is a qualified exception. In his *De multiplicatione specierum*, pt. 1, ch. 5, he argued (Bacon [Lindberg], 1983, 73) that "a species is produced in celestial bodies not only by other celestial bodies, but also by terrestrial things." Since we see things by means of species of the eye, "when we see the stars the species of our eye must be generated in the heaven, just as in the elemental spheres" (ibid., 75). The very act of sight causes a terrestrial thing to affect the heavens. But Bacon denied that these visual species could destroy one thing in the heavens and generate another and thus did not challenge the doctrine of celestial incorruptibility. The visual species that reached the stars did not alter the heavens. Rather, they assimilated celestial nature to terrestrial "for the sake of well-being and greater unity of the universe and to meet the needs of sense, especially sight, the species of which comes to the stars and to which the species of the stars come in order to produce vision."
6. In this chapter, I draw heavily on Grant, 1987c. Amicus, *De caelo*, tract. 6, qu. 1, art. 1, 1626, 348, col. 1, says that all Peripatetics and theologians believed that the heavens act on inferior things. He also mentions numerous philosophers and Church Fathers who supported that opinion.
7. "Tertia conclusio: omnis virtus naturalis huius mundi inferioris sensibilis gubernatur a celo." Themon Judaeus [*Meteorology*, bk. 1, qu. 1], 1518, 155v, col. 2–156r, col. 1. Similarly, Major, *Sentences*, bk. 2, dist. 14, qu. 7, 1519b, 78r, col. 2, declared that the inferior

Such a claim required little by way of proof, according to Thomas of Strasbourg, because "we daily perceive many real changes that we can reduce to no other agent than the heaven itself."[8] Mastrius and Bellutus distinguished four kinds of effects that seemed to embrace virtually the whole range of possibilities producible by celestial bodies: (1) accidents, by which they intended principally the primary qualities of hotness, coldness, wetness, and dryness; (2) inanimate bodies, comprising elements; bodies compounded of elements, or "mixed bodies," and the latter's two subdivisions, "imperfect mixed bodies," such as rain, clouds, and snow, and "perfect mixed bodies," such as metals, stones, and gems; (3) animated bodies having souls to the vegetable and sensitive levels; and (4) animated bodies having rational souls.[9] To these, we might add the lowest form of animated body, one that is spontaneously generated from putrefaction and is therefore lower on the ladder of nature than a body with a vegetable soul.

The ubiquitous and pervasive medieval principle of celestial dominance over terrestrial matter derived its authority and status from Aristotle and Ptolemy, although a variety of other authors – Latin, Greek, and Arabic – had reinforced that belief in a number of physical and astrological works that were available in western Europe by the end of the twelfth century.[10]

III. The basis of belief in celestial causes of terrestrial change

1. The theory of celestial causation

It was Aristotle above all others who provided the intellectual basis for the conviction that the heavenly region excelled the terrestrial. In a series of arguments in *De caelo* (bk. 1, chs. 2 and 3), Aristotle contrasts the natural, uniform, and eternal circular motions of the heavenly bodies with the natural, nonuniform, and finite rectilinear motions of the four elementary

world must be contiguous to the superior motions so that our sublunar world can be governed from there.

8. After enunciating his first conclusion ("Prima est: quod celum habet actionem realem in ista inferiora"), Thomas of Strasbourg [*Sentences*, bk. 2, dist. 14, qu. 1, art. 2], 1564, 157v, col. 1, declares: "Ista conclusio non indiget multa probatione quia quotidie multas reales alterationes percipimus quas in nullum aliud agens reducere possumus quam in ipsum celum."
9. "Ad quatuor genera omnes effectus producibiles a coelestibus corporibus reducere possumus, claritatis gratia: primum genus accidentia constituent; secundum corpora quaedam inanimata, ut elementa, mixta, etc.; tertium omnia animata vegetabili anima, ac sensitiva; quartum tandem, quae constant anima rationali." Mastrius and Bellutus [*De coelo*, disp. 2, qu. 7, art. 2], 1727: 3:511, col. 1, par. 191. In par. 195 (511, col. 2), they further divide mixed bodies into "imperfect" and "perfect."
10. Among these, the Arabian author Albumasar, or Abu Ma'shar Ja'far ben Muhammad ben 'Umar al-Balkhi, was perhaps the most influential. His *Introductorium in astronomiam* was heavily astrological but drew much from Aristotle's natural philosophy. Without realizing it, many Latin authors of the twelfth century first met elements of Aristotle's natural philosophy in Albumasar's *Introductorium*, which was translated in 1133 by John of Seville and in 1140 by Hermann of Carinthia. See R. Lemay, 1962, and North, 1987, 7–9.

terrestrial bodies and concludes that each type of motion must be associated with radically different kinds of simple bodies. By virtue of uniform circular motions that have neither beginning nor end, the celestial bodies are assumed to be eternal and incorruptible. All terrestrial bodies, by contrast, suffer incessant generation and corruption. Because he was convinced that the celestial substance was incorruptible, Aristotle assumed that it was a divine ether, declaring its divinity in a number of places.[11] It therefore seemed fitting that the incorruptible, divine, celestial substance should exercise an influence over the behavior of the corruptible and ever-changing sublunar bodies.

Using Aristotle's basic cosmology of four sublunar elements and a fifth, radically different celestial element, or ether, Claudius Ptolemy (fl. 2nd c. A.D.), in his *Tetrabiblos* (or *Quadripartitum*, as it was known during the Latin Middle Ages), assumed that

> a certain power emanating from the eternal ethereal substance is dispersed through and permeates the whole region about the earth, which throughout is subject to change, since, of the primary sublunar elements, fire and air are encompassed and changed by the motions in the ether, and in turn encompass and change all else, earth and water and the plants and animals therein.[12]

That such a power emanated from the heavens and affected the earth was made virtually self-evident to Ptolemy and almost everyone else by the behavior of the Sun and the Moon.[13] By analogy with, and extrapolation from, these two most prominent celestial luminaries, the other planets and stars were also assumed to cause a never-ending succession of terrestrial effects. Because celestial bodies possessed different powers and had different positions, their effects also varied. Depending on a complex set of relationships, planets and stars could cause either beneficial or harmful effects. By a judicious combination of observation and theory, it was even possible to know when and where these effects would occur.[14] Thus did Ptolemy transform Aristotle's vaguely described celestial influences, focused, as we

11. *De caelo* 1.3.270b.1–14; *Metaphysics* 1074b.1–14; *De partibus animalium* 1.5.644b.23–26; *Nicomachean Ethics* 6.7.1141b.1–2; and *De mundo* 2.392a.8–9. Although the last-named treatise was not by Aristotle, it was always so regarded in the Middle Ages. See also Solmsen, 1960, 289–290.
12. Ptolemy, *Tetrabiblos*, bk. 1, ch. 2 [Robbins], 1948, 5–7. For a brief summary of its content, see Thorndike, 1923–1958, 1:110–116.
13. "For the sun, together with the ambient, is always in some way affecting everything on the earth, not only by the changes that accompany the seasons of the year to bring about the generation of animals, the productiveness of plants, the flowing of waters and the changes of bodies, but also by its daily revolutions furnishing heat, moisture, dryness, and cold in regular order and in correspondence with its positions relative to the zenith. The moon, too, as the heavenly body nearest the earth, bestows her effluence most abundantly upon mundane things, for most of them, animate or inanimate, are sympathetic to her and change in company with her; the rivers increase and diminish their streams with her light, the seas turn their own tides with her rising and setting, and plants and animals in whole or in some part wax and wane with her." Ibid., 7.
14. Ibid., 11–13.

shall see, largely on the Sun, into a solid foundation for judicial and horoscopic astrology.[15]

The ideas of Aristotle and Ptolemy concerning the role of celestial influences on terrestrial affairs had an enormous impact. Because Aristotle's works formed the core of the university curriculum during the Middle Ages, his ideas on celestial influences were frequently discussed in commentaries on his works, in commentaries on the *Sentences* of Peter Lombard, or in the form of separate questions in treatises titled *De materia celi*. Rarely did scholastics devote a distinct, separate treatise to the subject. An exception was Thomas Aquinas, who, in response to a question from a soldier, sought to explain the manner in which the celestial region influenced terrestrial bodies.[16]

Thomas begins his analysis with the elements. Many actions of an element are caused directly by the nature of the element itself. "A stone, for example, is moved towards the center [of the earth] according to the property of earth dominant in it. Metals also have the power of cooling according to the property of water."[17] But the manifold activities of a body do not always follow from the nature of its primary constituent element, as, for example, in the case of the magnet's attraction for iron and the ability of certain medicines to purge humors. Such actions constituted occult phenomena, and their causes had to be sought in the behavior of superior agents, of which two kinds are distinguishable: (1) the heavenly bodies, and (2) "separate," or "separated," spiritual substances, a category that embraced angels, demons, and celestial intelligences.[18]

A superior agent can act on an inferior body in one of two ways: it can either impart to it some power or form that enables the inferior body to perform a certain action, or it imparts no form or power to it but by its own motion causes the inferior body to move, just "as a carpenter uses a saw for cutting." As illustrations of the second type of occult phenomena, Thomas mentions the power of the Moon to cause the ebb and flow of the tides, which it accomplishes by its own movements that somehow physically agitate the water.[19] Separate substances achieve similar effects by curing sick people through contact with a saint's relics. Here again, no form is

15. See also North, 1987, 7–8.
16. Thomas Aquinas, *Letter* [McAllister], 1939. McAllister omits the Latin text but gives an English translation from volume 24 of the Vivès edition of Aquinas's *Opera omnia* (1871–1880); for a later edition, see *Sancti Thomae de Aquino Opera omnia iussu Leonis XIII P.M. edita*, vol. 43 (Rome, Editori di San Tommaso, 1976), 183–186. McAllister (15) says the letter was probably written sometime between 1269 and 1272 and, since Thomas died in 1274, reflects his mature thoughts on this important subject. The authenticity of the treatise has never been challenged.
17. Thomas Aquinas, *Letter* [McAllister], 1939, 20. The translation extends over pages 20 to 30.
18. Ibid. 21, 79–80. For Thomas, God is the creator of all separate substances. Whether Thomas also included God, who is, of course, a spiritual substance, in the class of separate substances is irrelevant to our discussion, but Thomas seems clearly to imply this in his *Treatise on Separate Substances* [Lescoe], 1963, 108.
19. North (1987, 8) describes a similar idea by Albumasar, from whom Thomas may have derived it.

implanted in the relics, but the divine power nevertheless uses them to perform the miraculous cures (Thomas Aquinas [McAllister], 1939, 22).

Species of bodies or objects that have had forms implanted in them by a superior agent act in a constant manner. Thus *every* magnet attracts iron, and rhubarb *always* purges a certain humor. By contrast, species of inferior bodies in which forms or powers are not intrinsic act irregularly. In such instances, the superior agent chooses certain members of a species and confers the power on them alone. Thus "not every bone nor all relics of the saints heal upon touch, but those of some at some times, . . . nor does all water ebb and flow according to the movement of the moon" (ibid.).

But how are the two types of superior agents – that is, the celestial bodies or orbs and the separate substances – related? As was customary during the Middle Ages, Thomas assumes that separate substances are superior to corporeal celestial bodies. Not only do the former exist apart from matter, but Thomas assumed that they are unmoved,[20] whereas the latter exert their power by their movements. Indeed, the separate substances operate through celestial bodies to affect the inferior bodies of the terrestrial region.[21] Thomas now had the basis for a grand cosmic hierarchy, one that was popular in the Middle Ages. At the top are separate substances, followed by the celestial bodies, which are controlled by the separate substances. The inferior bodies, in turn, are organized "in such a way . . . that some are less perfect and closer to matter, while others, however, are more perfect and closer to superior agents."[22] In this scheme, the forms of the elements were the most imperfect, while bodies that approached greater uniformity of composition "became in some way or other like to heavenly bodies. . . . And therefore the greater the uniformity of mixture which such bodies approach, so much the more noble a form do they receive from God. Such is the human body, which, enjoying a most uniform composition, as the excellence of touch in men indicates, has a most noble form, namely a rational soul."[23] Thomas's hierarchical order was arranged according to the nobility of specific forms, all of which (except, of course, the soul) derived from superior agents associated with the celestial region.

Hierarchy was thus the essential reason why celestial bodies and substances affected the behavior of terrestrial bodies. It was fitting and proper that what is more noble and more perfect should influence and guide what is less noble and less perfect. With perhaps an occasional exception (possibly

20. As we saw in Ch. 18, Sec. II.4.c, scholastics differed on this point, some assuming that separate substances are immobile, others that they are mobile and move with their orbs.
21. Thomas Aquinas, *Letter* [McAllister], 1939, 25. As John of Jandun put it ([*De coelo*, bk. 1, qu. 1], 1552, 2r, col. 2): "The Commentator [Averroës] says in the first book of this [*De caelo*] that the heaven is the tie [or link: *ligamentum*] between abstracted [that is, separated] substances and inferior things."
22. Thomas Aquinas, ibid., 26.
23. Ibid. Thomas placed inanimate bodies, such as stones, metals, and minerals, above simple elemental bodies. The former have not only the powers of the elements but also other virtues derived from specific forms that were imparted to them by superior agents.

Roger Bacon), Saint Bonaventure spoke for virtually all scholastics of the Middle Ages when he declared:

The reason why superior things act on inferior things . . . is because they are nobler bodies and excel in power, just as they excel with respect to location. And since the order of the universe [*ordo universitatis*] is that the more powerful and superior should influence the less powerful and inferior, it is appropriate for the order of the universe that the celestial luminaries should influence the elements and elemental bodies.[24]

When the creator of the world made the celestial bodies incorruptible, he ordained that they should rule over corruptible and inferior things and for the attainment of this goal assigned appropriate powers to each celestial luminary.

2. *The empirical basis for celestial causation*

Thus far I have emphasized metaphysical and even intuitive appeals and arguments in support of celestial causal superiority. But empirical evidence was also invoked. Indeed, beginning with Aristotle there was a steady accumulation of experiences that were believed to demonstrate the reality of celestial causation in the terrestrial world. The radical contrast between celestial and terrestrial bodies, the former incorruptible and the latter continually undergoing change, was itself founded upon observation. "For in the whole range of times past, so far as our inherited records reach," declared Aristotle, "no change appears to have taken place either in the whole scheme of the outermost heaven or in any of its proper parts,"[25] a claim he supported by locating meteors, comets, shooting stars, and the Milky Way in the fiery sphere just below the Moon.[26] By contrast, incessant changes in the sublunar region were readily observable,[27] a fact that confirmed the superiority of the celestial region to the terrestrial.

The most graphic evidence of specific celestial causation, however, was provided by the Sun. Once again Aristotle pointed the way. Numerous observations of the Sun's annual motion in the ecliptic prompted him to

24. Saint Bonaventure [*Sentences*, bk. 2, dist. 14, pt. 2, art. 2, qu. 2 ("Utrum diversa luminaria diversas habeant impressiones super corporalia")], *Opera*, 1885, 2:360, col. 2. Richard of Middleton includes a similar paragraph in *Sentences*, bk. 2, dist. 14, art. 3, qu. 5, 1591, 2:188, col. 2. For medieval views on celestial perfection, see Grant, 1985c. Because visual rays had to be multiplied or extended all the way to the celestial region for us to see celestial bodies, Roger Bacon was convinced that terrestrial influences could affect the celestial region, but he denies that this would detract from the nobility of the heavens. See Bacon [Lindberg], 1983, 72–75 (text and translation), lxi–lxii (Lindberg's description); also see note 5, this chapter.
25. Aristotle, *De caelo* 1.3.270b.13–17 [Stocks], 1984.
26. See the first book of Aristotle's *Meteorology*, especially 1.1.338b.21–25, 1.3.341a.32–35; 1.4.342a.30–31; and 1.8.346b.10–15.
27. Aristotle, *De generatione et corruptione* 1.1.314b.13–15.

declare that "there are facts of observation in manifest agreement with our theories. Thus we see that coming-to-be occurs as the sun approaches and decay [occurs] as it retreats."[28] The Sun also produces the rains that make life on earth possible (*Meteorology* 2.2.354b.24–33). Indeed, the Sun's circular motion produces the cycle of the seasons, and therefore the things associated with those seasons (*De generatione et corruptione* 2.11.338b.1–5). Aristotle also assigned a role to the Sun in human generation when he declared that "man is begotten by man and by the sun as well" (*Physics* 2.2.194b.13–14 [Hardie and Gaye]). Although the Sun's actions on the earth were more noticeable than those of any other celestial body, Aristotle held that the totality of celestial motions was the ultimate source of change for terrestrial bodies (*Meteorology* 1.2.339a.20–33).

During the period under investigation here, approximately 1200 to 1700, a considerable variety of other "experiences" was offered as clear evidence that the heavens influenced the inferior region.[29] As the most obvious source of celestial influence, the Sun and Moon were most frequently cited. According to the Coimbra Jesuits, the Sun's motion was assumed to cause the four seasons of the year along with the two equinoxes and two solstices, thus producing coldness and hotness, which in turn made the generation and destruction of things possible.[30] Lesser activities of the Sun were not ignored, such as, for example, its power to cause heliotropism in flowers and to make the cock crow daily before sunrise.[31] In the fourteenth century, Themon Judaeus cited a number of such experiences (*per experientias*).[32] After the Sun has risen, its light causes heat and also makes seeds and fruits grow; the setting of the Sun at night causes the temperature to drop and also produces wind and rain.

The Moon manifests its terrestrial influence by the well-known associ-

28. Ibid., 2.10.336b.16–19 [Joachim], 1984.
29. Although many scholastics provided lists of experiences, the examples presented here have been drawn from Themon Judaeus, *Meteorology*; the Conimbricenses, *De coelo*; Amicus, *De caelo*; and de Rhodes, *De coelo*.
30. Conimbricenses [*De coelo*, bk. 2, ch. 3, qu. 1, art. 2], 1598, 191. Amicus also included this one in his lengthy list of experiences (*De caelo*, tract. 6, qu. 1, art. 1, 1626, 348, col. 2). The tractate on the influences of the heavens on inferior things ("Tractatio sexta: De influxibus caelorum") constitutes the longest tract in Amicus's extensive commentary on *De caelo*.
31. Conimbricenses, ibid. The experience of heliotropism was included by Amicus, ibid., 349, col. 2, while the cock-crowing example was repeated by de Rhodes [*De coelo*, disp. 2, bk. 2, qu. 3, sec. 1], 1671, 297, col. 1.
32. Themon Judaeus, *Meteorology*, bk. 1, qu. 1, 1518, 155v, col. 2. Themon was at the University of Paris during the 1350s and perhaps 1360s. In a brief preface to the work just cited, George Lokert, its editor, ranked Themon, Albert of Saxony, and Jean Buridan as a famous triumvirate of eminent men in the celebrated school of Paris. The passage is quoted by Hugonnard-Roche, 1973, 11. These three scholars were known to Lokert because each had had one or more *questiones* published on Aristotle's natural books. Because none of his questions on Aristotle's natural books was published until this century, Oresme, who was more brilliant than any of the three, may have been unknown to Lokert.

ation of its motion with the tides. Indeed, the association of the Moon with liquid prompted Themon to mention that humors in animals increased when the Moon waxed and decreased when it waned. Physicians regularly linked the Moon's motion with the critical days of an illness, which suggested to Themon that "this could not occur unless the Moon did act powerfully on these inferior things."[33]

But the Sun and Moon were only the most obvious celestial agents. The other planets also caused a great variety of terrestrial activities. Toward the very end of the sixteenth century, the Coimbra Jesuits described powerful correspondences that were assumed to obtain between the planets and terrestrial objects. The planets corresponded to the different parts of the human body: the Sun to the heart; Mars to the gallbladder; Jupiter to the liver; Mercury to the mouth and tongue; Saturn to the head; and so on.[34] Also mentioned are the famous Shakespearean "ages of man" linked to the seven planets, a linkage that was probably already ancient when Ptolemy included them in his *Tetrabiblos*.[35] The Moon corresponds to infancy; Mercury to boyhood; Venus to adolescence; the Sun to youth; Mars to the virile age; Jupiter to old age; and Saturn to very old age or senility. Because gold cannot produce gold, nor a gem produce a gem, George de Rhodes (*De coelo*, bk. 2, disp. 2, qu. 3, sec. 1, 1671, 297, col. 1), in the seventeenth century, inferred that only the heavens could cause metals and gems.

The ancient world was the source of this and most similar beliefs. Although Aristotle was not among them, some already believed that each planet influenced a particular species of metal. The following correspondences, which de Rhodes and many others repeated, were already established: gold was related, or corresponded, to the Sun; silver to the Moon; iron and steel to Mars; lead to Saturn; tin to Jupiter; mercury to Mercury; copper or bronze to Venus.[36] Another famous correspondence related the planets and bodily humors, for it was assumed that Mars causes yellow bile, Saturn melancholic humor, the Moon generates phlegm, and the Sun and Jupiter create blood.[37] Many also assumed that the planets dominated the four sublunar elements, where planet and element paired off according to similar qualities. Thus Saturn dominated earth, the Moon water, Mercury air, and Mars fire. Indeed, these four planet-element pairings were believed to de-

33. Themon Judaeus, ibid. See also Conimbricenses, *De coelo*, bk. 2, ch. 3, qu. 1, art. 2, 1598, 191; Amicus, *De caelo*, tract. 6, qu. 1, art. 1, 1626, 348, col. 2, and de Rhodes, *De coelo*, disp. 2, bk. 2, qu. 3, sec. 1, 1671, 297, col. 1, for similar and nearly identical examples of lunar efficacy.
34. Conimbricenses, ibid., and Amicus, ibid., 349, col. 1.
35. See Ptolemy, *Tetrabiblos*, bk. 4, ch. 10 [Robbins], 1948, 441–47. Earlier in the treatise, Ptolemy had proposed four ages of mankind (bk. 1, ch. 10, 59–61).
36. De Rhodes, *De coelo*, disp. 2, bk. 2, qu. 3, sec. 1, 1671, 297, col. 1; see also Themon Judaeus, *Meteorology*, bk. 1, qu. 1, 1518, 155v, col. 1 and Amicus, *De caelo*, tract. 6, qu. 1, art. 1, 1626, 348, col. 2–349, col. 1.
37. Conimbricenses, *De coelo*, bk. 2, ch. 3, qu. 1, art. 2, 1598, 191, and Amicus, ibid., 349, col. 1.

termine human complexions: the Moon and water produce a phlegmatic complexion; Mercury and air a bloody complexion; Mars and fire a choleric complexion; and Saturn and earth a melancholic complexion.[38]

Finally, it was generally assumed that celestial virtues were the cause of magnetism. So closely were these related that together they were thought capable of producing perpetual motion. Themon Judaeus reports that the author of the *De magnete* (probably Peter Peregrinus, who wrote a *Letter on the Magnet*) declared that if a piece of iron were placed at each of the two poles of a properly mounted spherical magnet, the magnetic sphere would revolve perpetually and synchronously with the heavens, because each part of the magnet would correspond to a similar part of the heavens. Thus each part of the heavens influences its corresponding part on the magnet, "which would not happen," says Themon, "unless every part of the heavens influenced that stone [or magnet] here below."[39]

The phenomena just described were characterized as "experiences" and were thought sufficiently corroborative to convince anyone of the pervasive role played by celestial bodies in terrestrial affairs. Among such experiences, Themon even included the belief of astrologers that human fortunes and misfortunes depended on the heavens.[40]

We have now examined the theoretical and empirical basis for belief in the powerful role of celestial causation in terrestrial affairs. But we must keep in mind that if Aristotle and medieval natural philosophers fully accepted a powerful role for celestial control over terrestrial activities, that role was not total and complete. Aristotle and his medieval followers also accorded to ordinary terrestrial bodies composed of matter and form the ability to develop their own potentialities and to affect other bodies without the aid of the heavens. As Thomas Aquinas expressed it, "A stone . . . is moved towards the center [of the earth] according to the property of earth dominant in it" (*Letter* [McAllister], 1939, 20). In its fall to earth, the stone could affect other things with which it came in contact. All of this did not seem to involve celestial bodies. How could this be? If terrestrial bodies had inherent powers of development and also had the power to affect other bodies, then what role did the heavens play?

A frequently cited statement from Aristotle may serve to explain this apparent dilemma. In the *Physics* (2.2.194b.13–14 [Hardie and Gaye]), as we saw earlier in this section, Aristotle explained that "man is begotten by man and by the sun as well." Often using some version of this phrase, medieval natural philosophers signified that both terrestrial and celestial powers were involved in terrestrial activities. A man and a woman were able to produce offspring naturally, since they possessed the requisite causal powers. But although those powers were necessary, they were apparently

38. Amicus, ibid.
39. Themon Judaeus, *Meteorology*, bk. 1, qu. 1, 1518, 155v, col. 2. Citing only Agricola, Amicus, ibid., provides a similar account of the magnet and perpetual motion.
40. Themon Judaeus, ibid.

not sufficient. Human offspring could not be generated or survive unless the Sun performed its role in the universal scheme of things, that is, unless the Sun produced the seasons and the heat that enabled life to survive. At the very least, celestial bodies were essential as concomitant causes that worked in harmony with the causes that operated within terrestrial bodies. Problems arose, however, in apportioning roles to celestial and terrestrial causes. Because the celestial region was acknowledged as nobler than the terrestrial realm, there was often a tendency to overemphasize the causative role of the celestial region, to the detriment of terrestrial causation. Astrologers, of course, did so regularly. But others did so as well, since it had become customary to heap praise on the heavens and their powers and to declare that all things are governed by the heavens. A reaction to this tendency is already detectable in the Condemnation of 1277, as we shall see.

IV. Can celestial bodies generate living things?

Because scholastic natural philosophers were agreed that celestial bodies were nobler than inanimate terrestrial bodies,[41] they had no difficulty in assuming that celestial bodies played a significant, if not essential, role in causing generation and corruption among inanimate terrestrial objects.[42] However, no broad agreement developed as to whether those same celestial bodies could also produce living things in the terrestrial region. A major obstacle was the powerful Christian belief that any living thing, however lowly, was nobler than any inanimate celestial body. Mastrius and Bellutus took it "as absolutely true that a fly, a flea and a plant are absolutely nobler than the heavens,"[43] a view that was contrary to Aristotle's opinion. On the assumption that a principal cause cannot be less noble than its effect, it would require some ingenuity to argue that celestial bodies were capable of generating animate beings nobler than themselves.

Neverthless, scholastics were reluctant to abandon the idea that celestial

41. As Aversa expressed it ([*De caelo*, qu. 35, sec. 8], 1627, 194, col. 1): "Corpora enim haec inanimata sunt absque dubio ignobiliora corporibus caelestibus."

42. In his *Orthodox Faith*, bk. 2, ch. 7, 1958, 219, John Damascene was one of the few Christians who denied the causal efficacy of celestial bodies on terrestrial bodies when he declared that "the stars do not cause anything to happen, whether it be the production of things that are made, or events, or the destruction of things that are destroyed." Although Thomas Aquinas cites this passage ([*Sentences*, bk. 2, dist. 15, qu. 1, art. 2], 1929–1947, 2:370, 371), he sought to explain it away by insisting that Damascene was denying that celestial bodies could influence inferior bodies solely by themselves, without any help from God. Thus Damascene was essentially denying a form of idolatry (Ad secundum dicendum, quod Damascenus intendit negare a corporibus caelestibus illam causalitatem quae idolatriam inducebat, ut patet ex praedictis). Ibid., 372.

43. "Quare concedendum ut absolute verum muscam, pulicem et plantam esse celo nobiliores simpliciter." Mastrius and Bellutus, *De coelo*, disp. 2, qu. 3, 1727, 3:501, col. 1, par. 123. For more on the greater nobility of even the lowest animate beings over celestial bodies, see Chapter 11, Section II.

bodies played a role in the generation of some, or all, living things. Aristotle proved a source of support. Aristotelian natural philosophers recognized at least two kinds of living things that were the result of a generative and/or natural process: those that were born by means of seed, which scholastics called "perfect animals," and those living things that were spontaneously generated from decaying matter, as were a number of insects, or from within by secretions from the organs of animals.[44] These spontaneously generated and secreted creatures of the second category were described as "imperfect animals."[45] At least some imperfect animals, say a frog, were thought capable of coming-to-be either by spontaneous generation or by means of seed. A question that naturally arose was whether a frog that was spontaneously generated by means of putrefaction was of the same species as a frog that was generated by means of seed.[46] Because he could find no way to distinguish between them, Johannes Poncius assumed that they were no less members of the same species than were two men of the human species.[47] The most plausible place to involve the heavens – especially the Sun – in the generation of living things was in the spontaneous generation

44. In his *Generation of Animals* and *History of Animals*, Aristotle developed an elaborate ladder of nature, based upon various reproductive modes associated with different levels of vital heat (see Ross, 1949, 116–117). At one point Aristotle says (*History of Animals*, 5.1.539a.23–25 [Thompson], 1984) that "with animals some spring from parent animals according to their kind, whilst others grow spontaneously and not from kindred stock; and of these some come from putrefying earth or vegetable matter, as is the case with a number of insects, while others are spontaneously generated in the inside of animals out of the secretions of their several organs."

45. Johannes Poncius, in his supplemental commentary on the *Sentences* of Duns Scotus, makes this common distinction when he says: "Moreover, here I call those animals perfect which always require the generative action of other animals by means of the effusion of seed; those animals which are sometimes generated without a similar mode of propagation – for example, frogs, wasps, flies, and similar things – are, for this reason, called imperfect animals" (Voco autem hic . . . animalia perfecta illa quae semper requirunt actionem generativam aliorum animalium mediante effusione seminis; ad differentiam aliorum animalium, quae absque simili propagatione quandoque generantur, ut sunt ranae, vespae, muscae, et similia, quae propterea animalia imperfecta nuncupantur). See Poncius [*Sentences*, bk. 2, dist. 14, qu. 3], 1639, vol. 6, pt. 2:742, col. 1.

46. "Here you seek whether animals made directly from putrid matter without generated seed – for example, wasps [and] frogs – belong to the same species with animals made from generated seed – for example, wasps and frogs" (Quaeres hic obiter an animalia, verbi gratia, vespae, ranae, ex putri materia absque semine generata sint eiusdem speciei cum animalibus, verbi gratia, cum vespis et ranis, ex semine generatis). Poncius, [*De coelo*, disp. 22, qu. ult. (qu. 10)], 1672, 633, col. 1, par. 116.

47. "Secondly, this is proved [as follows]: because there is no apparent reason for making a specific distinction between them and [because] all other judgments indicate that they belong to the same species no less than do two men. Therefore, they belong to the same species" (Probatur secundo: quia nulla apparet ratio distinctionis specificae inter illa; et omnia alia iudicia sunt quod sint eiusdem speciei non minus quam duo homines. Ergo sunt eiusdem species). Ibid. In support of his position, Poncius cites Scotus, Thomas Aquinas, Richard of Middleton, Ockham, and John of Jandun. We may add Michael Scot ([*Sphere*, lec. 9], 1949, 314), who cites with seeming approval the claim by Averroës, in the latter's commentary on book 10 of Aristotle's *Metaphysics*, "that there is no difference between things generated from seed by propagation and from things made by putrefaction" (Item *10 methe.* dicit commentator quod nulla est diversitas inter generata ex semine per propagationem et ex se per putrefactionem).

of the lowest forms of life. But even with the higher forms, indeed the very highest, the human species, Aristotle had attributed a role to the Sun when he declared that "man is begotten by man and by the sun as well,"[48] signifying that without the Sun the conditions of human life could not be sustained. Thus were scholastics encouraged to suggest ways in which the heavens might be involved in the generation of some, or all, living things.

Thomas Aquinas distinguished sharply between the generation of perfect and imperfect animals. The former were not generated solely by the universal power (*universalis virtus*) of the Sun but also required a seed within which a particular power (*virtus particularis*) resided. But for imperfect animals, namely those generated by means of putrefaction in the absence of any seed, the power of the heavens suffices.[49] Indeed, in causing the putrefaction that results in an imperfect animal – say, a fly or a frog or a flea, to name commonly used examples – it seems that the heavens produce not only their bodies but also their souls. The heavens are therefore the direct cause of life in these imperfect creatures. Although Thomas seems to have had no difficulty with this concept, others would later find objectionable the idea that the heavens could confer life, even on spontaneously generated creatures. Pedro Hurtado de Mendoza, for example, conceded that while the Sun could produce plants and minerals, it was unable to generate living things from putrefaction, because the Sun is less perfect than those living things.[50] Most scholastics, however, would find a role for the celestial bodies, especially the Sun, in the generation of living things that were considered more perfect than the Sun. At the very least, many felt that an explanation or justification was in order.

One of these was Durandus de Sancto Porciano. Durandus shows that others had been thinking about this problem when he reports an opinion in which it was assumed that, although the heaven is not a living thing, it can nonetheless generate living things because it is moved by a living thing, namely, an intelligence. Through the power of its living intelligence, therefore, an orb could produce a living thing in the terrestrial world.[51] Durandus rejects this opinion because he finds it implausible that a heaven could receive anything from its intelligence, whether it was the power of acting or the power to produce anything.[52]

But like most others, Durandus was nevertheless convinced that some-

48. *Physics*, 2.2.194b.13–14 [Hardie and Gaye], 1984.
49. For twenty-eight statements from the works of Thomas Aquinas, see Litt, 1963, 130–133.
50. "Plantas et mineralia a Sole produci ostendi, disput. 9 *Physic.* sect. 4, animantia vero etiam ex putrefactione genita non fiunt a Sole, qui est illis imperfectior, perfectio enim obiecti arguit maiorem perfectionem quam quaelibet actio solaris." Hurtado de Mendoza, *De coelo*, disp. 2, sec. 6, 1615, 375, col. 2, par. 67.
51. "Est intelligendum quod quidam dicunt quod coelum, licet non sit vivens, quia tamen movetur a substantia vivente, scilicet ab intelligentia, virtute eius potest aliquod vivens generare." Durandus [*Sentences*, bk. 2, dist. 15, qu. 1], 1571, 157v, col. 1, par. 6.
52. After giving a few reasons, Durandus concludes: "Quod non acquirit virtutem agendi, non acquirit virtutem aliquid producendi." Ibid.

how the heavens could produce life or play a role in its generation. He proposed a theory that was repeated during the sixteenth and seventeenth centuries. Durandus likened the heavens, or any individual heaven or orb, to a seed. Although a seed is not itself a living thing, it is derived from a living thing and can produce a living thing. As something analogous to a seed, a heaven, or orb, is not alive, yet it possesses the power to produce life, which it does not receive from its intelligence. It receives this power directly from God, who initially conferred upon the heavens the seed of all generable and corruptible things. God produces this power in the heavens, and once there it functions just as does the semen of a horse, which generates a power in the horse to produce a horse.[53]

The Coimbra Jesuits, who accepted Durandus's theory, added further qualifications and refinements. Some had objected that the seed to generate a new living thing ought somehow to be conjoined to the thing that is generated. And yet if the heavens are like seeds, they are far removed from the terrestrial things that they are alleged to help generate. To counter this apparent difficulty, the Coimbra Jesuits point out that a seed requires two things: (1) that it not be separate from the thing that it generates; and (2) that it should have within itself a generative power transmitted by the thing that generated it. Only the second condition is met. But that is sufficient to enable the heavens to function as if they were seeds, since it is not necessary that things that are called similar be similar in every respect.[54]

Durandus and the Coimbra Jesuits ignored a glaring difficulty. If a horse, for example, can produce the semen that will produce another horse, is it not superfluous to invoke the heavens? In his seventeenth-century summary of the theory of Durandus and Conimbricenses, Raphael Aversa supplied an answer when he explained that they and others assigned to the heavens the role of a "disposing cause," or the role of an "instrument of God" (*instrumentum Dei*) rather than a direct, generating cause. The principal cause is the animal itself. Thus a frog is the principal cause in the generation of another frog.[55] But the heavens are a disposing cause that makes the generation of the frog possible.[56]

53. "Ideo dicendum quod sicut in semine, quod non est vivens, est virtus generativa rei viventis, pro eo quod semen decisum est a vivente. Sic in coelo, licet non sit vivens, est virtus generativa rerum viventium, non quia motum ab intelligentia, nec quia virtus sit ei influxa ab aliqua intelligentia creata, sed quia productum est a Deo tanquam quoddam semen omnium generabilium et corruptibilium. Et sic coelum agit in virtute Dei ad productionem viventium, sicut semen equi in virtute equi generantis ad producendum equum." Ibid., par. 7.
54. "Respondemus coelum non vocari a nobis semen, sed veluti semen quod sit affectum virtute quasi seminaria viventium. Nimirum ad rationem seminis duo (ut nunc cetera omittamus) requiruntur. Alterum, ut non sit disiunctum a re quae generatur; alterum, ut habeat in se virtutem generativam a genitore tranfusam. Nos ergo non ob priorem, sed ob posteriorem conditionem dicimus corpus coeleste esse quasi semen. Non enim ea quae similia vocantur in omnibus similia esse oportet." Conimbricenses, *De coelo*, bk. 2, ch. 3, qu. 6, art. 2, 1598, 212.
55. Although a frog was considered an "imperfect" animal, it was generally believed that it could be produced from seed in a normal reproductive mode, as well as generated spon-

The "seed" (*semen*) theory espoused by Durandus and the Coimbra Jesuits met opposition from Pedro Hurtado de Mendoza and Raphael Aversa in the seventeenth century. Hurtado declares his disagreement with Durandus and the Conimbricenses, "who assert that the heavens are sprinkled as if with seeds endowed with the power to generate these animals." He offers little detail but says that they have assumed their seed theory "without any basis" and that the theory "is improbable, if the talk is about a proper seed." Thus Hurtado rejects the analogy. In his view, Durandus and the Conimbricenses used the term "seed" improperly.[57]

Aversa argued that animals that procreate by seed need only the seed and not a concomitant external influence from the heavens. The production of an animal does not arise from something external to the animal, as happens when a fire generates a fire. No additional instrument (*instrumentum*) is required for producing a new animal.[58] Indeed, Aversa found superfluous the idea of the heavens as an "instrument of God." If God is the universal cause of all forms and powers but also finds it necessary to act in the production of each effect, then the heavens are no more an "instrument of God" than any other secondary cause. With such an interpretation, all secondary causes would seem to be superfluous, because they could not be genuine principal causes able to act in accordance with the powers conferred upon them.[59]

taneously without seed. Poncius listed frogs among animals that are sometimes generated without seed (see notes 45–46 of this chapter).

56. "Propter hoc conati sunt aliqui explicare caelum esse solum causam disponentem ad generationem talium viventium, causam vero propriam et principalem esse adhuc alia similia viventia, ita ut in generatione ranae, verbi gratia, causa principalis sit adhuc alia rana. Quia, inquiunt, omnes dispositiones a caelo inductae sunt instrumentum debitum ac proportionatum alteri ranae ad producendum sibi similem ranam." Aversa, *De caelo*, qu. 35, sec. 8, 1627, 195, col. 1. Poncius *Sentences*, bk. 2, dist. 14, qu. 3, 1639, vol. 6, pt. 2:742, col. 2, describes a similar theory in which he says that "the principal productive cause of a frog is another frog, in virtue of which the heaven produces this frog" (Alii dicunt causam principalem productivam animalium imperfectorum esse alia similia animalia, verbi gratia, causam principalem productivam ranae esse aliam ranam, in virtute cuius caelum producit illam ranam).

57. "Neque assentior Durando et P. Conimb. asserentibus coelum esse respersum quasi seminaria virtute gignendi haec animalia, tum quia sine fundamento ponitur, tum quia si intelligatur de proprio semine est improbabilis." Hurtado de Mendoza, *De coelo*, disp. 2, sec. 6, 1615, 375, col. 2, par. 67.

58. "Ac praeterea modus proportionatus et debitus generandi sibi simile in animalibus non est per modum extrinsecum, quo verbi gratia ignis generat ignem; sed per modum vitalis propagationis. Igitur in illo casu nullum est instrumentum proportionatum et debitum alteri animali ad gignendum illud quod de novo producitur." Aversa, *De caelo*, qu. 35, sec. 8, 1627, 195, col. 2.

59. "Ideo rursus Durandus, . . . Conimbricenses, . . . et Ruvius, . . . Fonseca, . . . Suarez, . . . et quidam alii asserunt caelum operari potius tanquam instrumentum Dei a quo vere accepit omnem suam virtutem. Et ita Deum esse causam principalem illorum animalium. Verum si hoc dicatur inquantum Deus est causa universalis dans omnibus rebus suas formas et virtutes et actualiter quoque concurrens ad omnem operationem; sic non magis dici debebit caelum in tali operatione agere ut instrumentum Dei quam omnia alia causa secunda in omni alia operatione. At hoc non toliit quominus causa secunda in suo genere dicatur vere causa principalis ac debeat esse vel aequalis vel maioris perfectionis cum suo effectu." Ibid., 196, col. 2.

Did this imply that Aversa denied a role to the heavens? Not at all. Aversa found common cause with Durandus and others, who had identified, as he believed, a "formative power" (*virtus formativa*) in the heavens which played a role in the formation of living things.[60] Whether the heavens act as an instrument of God or an instrument of an intelligence or even as a principal cause, Aversa was convinced that they acted "by a peculiar power and faculty that they have both for forming the bodies of such living things, with their organs and parts, and for producing and bringing forth a soul from the potency of matter." This power is appropriately called "a formative and animative power" (*virtus formativa et animativa*).[61]

But did this not raise the old dilemma: How can something less noble generate something more noble? that is, how could the less noble heavens generate more noble living things? Aversa explains that the heavens are not necessarily more ignoble than all living bodies. Although the rational soul and a man are nobler than any heaven,[62] plants, which operate solely with a vegetative soul, may not be absolutely nobler than celestial bodies.[63] Indeed, Aversa regarded the heavens as the principal cause of all living things that were generated without natural means, which presumably encompassed all things that Aristotle, and many others, had regarded as produced by spontaneous generation.[64] Aversa leaves unclear whether the heavens are also the "principal cause" (*causa principalis*) of living things that are generated from seeds.

In his 1672 Aristotelian commentary from the Scotist point of view, Johannes Poncius took issue with Aversa. He first enunciates an important conclusion:

> Celestial bodies act together for the production of all living things, even man, by disposing matter in this or that manner in order to receive souls. However, they

60. "Caeterum Durandus et alii plures tribuunt caelo peculiarem virtutem formativam ad gignenda talia animalia inditam illi a Deo, veluti virtutem seminalem." Ibid., 197, col. 1.
61. "Verum, sive caelum in tali operatione agat ut instrumentum Dei aut Intelligentiae, sive ut causa principalis, agit utique per peculiarem virtutem et facultatem quam habet, tum ad efformanda corpora talium viventium, cum suis organis et membris, tum ad producendam ac educendam de potentia materiae animam. Nam utique debet exercere talem operationem per virtutem aptam et proportionatam. Atque haec virtus recte dicetur virtus formativa et animativa et ad instar virtutis seminalis in suo genere." Ibid., cols. 1–2.
62. "At sine dubio comparando formam caeli cum anima rationali et celum cum homine, manifeste quidem apparet valde nobiliorem gradum esse animae et hominis." Ibid., qu. 33, sec. 7, 113, col. 1.
63. "Sed fortasse necesse non erit dicere corpora celestia esse simpliciter ignobiliora omnibus corporibus animatis. Quinimo si comparatio fiat cum plantis, quae sola vegetativa anima vivunt, non sunt sane adeo insignes operationes huius animae, ut debeant simpliciter praeferri operationibus caeli." Ibid., 112, col. 2.
64. "Dici ergo tandem poterit caelum ipsum esse sufficientem causam principalem talium viventium iuxta id quod notavimus q. 33, sect. 7 non debere necessario caelum dici essentialiter ignobilius omnibus corporibus viventibus. Sic enim optime caelum dicetur vera causa principalis illorum viventium que absque propagatione naturaliter generantur." Ibid., qu. 35, sec. 8, 197, col. 1. For citations from question 33, section 7, see notes 62–63.

do not immediately achieve the production of any soul – neither as a principal cause [*causa principalis*] nor as an instrumental [*instrumentalis*] [cause] – [not] even of imperfect animals.[65]

In this conclusion, Poncius makes two points. First he insists that celestial bodies are capable of producing all living things, even man. As evidence, he cites Aristotle's oft-quoted declaration that "man is begotten by man and by the sun as well."[66] Poncius would have agreed with Thomas Compton-Carleton, who described the heaven as a "universal cause" (*causa universalis*) because it influences every sublunar effect,[67] an expression that Thomas Aquinas had already used in the thirteenth century with much the same signification.[68] Even reason supports this conclusion, since one can reason that someone born under one constellation would have different inclinations and tendencies than if born under another constellation. Some constellations are even thought to be helpful and others harmful.[69]

Poncius's second point – that the heavens cannot directly produce a soul – conflicted with an opinion shared by a number of scholastic authors, especially Aversa, who was among those "who thought that the souls of imperfect animals, namely of those [animals] that are produced without the intervention of seed, are produced by the heavens from putrid matter, as [with] frogs, wasps, fleas, and similar things."[70] Because he remained con-

65. "Corpora coelestia concurrunt ad productionem viventium omnium, etiam hominum, disponendo materiam tali vel tali modo in ordine ad receptionem animarum. Non tamen attingunt immediate productionem ullius animae, etiam animalium imperfectiorum, neque tanquam causa principalis, neque tanquam instrumentalis." Poncius, *De coelo*, disp. 22, qu. ult. (qu. 10), conclus. 3, 1672, 631, col. 2, par. 108. Alluding more than likely to both inanimate and animate things, Thomas Compton-Carleton, *De coelo*, disp. 2, sec. 4, 1649, 404, col. 1, declared that "the heaven is a principal cause with respect to effects that are more imperfect than it; [but] it is an instrumental cause with respect to effects that are more perfect than it" (Est autem coelum causa principalis respectu effectuum qui sunt ipso imperfectiores; instrumentalis respectu effectuum ipso perfectiorum).
66. Aristotle makes this remark in *Physics*, 2.2.194b.14 [Hardie and Gaye]. Poncius, ibid., expresses the saying as "Quod sol et homo generet hominem" which can be translated literally as "the Sun and a man generate a man." Michael Scot, *Sphere*, lec. 9, 1949, 314, cites Aristotle's *On Generation and Corruption*, book 2, as the source of this remark. See also Hurtado de Mendoza, *De coelo*, disp. 2, sec. 4, 1615, 374, col. 2, par. 57. Although Compton-Carleton (ibid.) cited the usual text (Aristotle, *Physics*, bk. 2, text 26), he altered this frequently cited phrase to read: "the Sun and a horse generate a horse" (sol et equus generant equum). He also attributed to Aristotle (in the latter's *De generatione*, bk. 1, text 55) the opinion that the Sun is "the author and parent of sublunar things."
67. "Quaeritur ad quos in particulari effectus concurrant caeli et astra? Universim affirmare videntur auctores coelum influere in omnes omnino effectus rerum sublunarium et propterea illud vocant *causam universalem*." Compton-Carleton, ibid.
68. Thomas used *causa universalis* as equivalent to the expressions *natura universalis* and *agens universale*. Litt, 1963, 154–166, has gathered fifteen examples of these usages.
69. Poncius (*De coelo*, disp. 22, qu. ult. [qu. 10], 1672, 631, col. 2, par. 108) offers this information in support of the first part of his conclusion, namely that celestial bodies produce all living things.
70. Poncius (ibid.) includes Aversa, Fonseca, Marsilius [of Inghen?], and Suarez with those "qui putant animas animalium imperfectiorum, eorum scilicet, quae producuntur absque interventu seminis ex putri materia, ut sunt ranae, vespae, pulices, et similia, produci a coelis." It is likely that Thomas Aquinas also believed this.

vinced that any soul is more perfect than celestial bodies, Poncius assumed that "[souls] cannot be produced by celestial bodies because what is more imperfect cannot directly produce something more perfect."[71]

And yet Poncius was puzzled about the assumption that the heavens are more imperfect than any living thing and found it difficult to explain. Life at the vegetative level hardly seemed more perfect than the celestial bodies.[72] Plants probably seemed less vital than spontaneously generated fleas or wasps. But Poncius did not abandon his belief that all living things were more perfect than any celestial body.

Despite differences of opinion, scholastic natural philosophers assigned a significant role to the celestial bodies in producing living things, even as they conceded that most, if not all, living things were more noble and perfect than the heavens. We must now consider the ways in which celestial bodies were thought to effect sublunar bodies, both inanimate and animate.

V. The instrumentalities of celestial action

With the causes of so many effects and experiences attributed directly to celestial bodies, it was natural to inquire how those effects were transmitted to the sublunar region. Three instrumentalities of celestial action were often identified: (1) motion (*motus*); (2) light (*lumen*); and (3) influence (*influentia*).[73] Scholastics were unanimous in the belief that these instrumentalities could operate only if the celestial and terrestrial regions were joined in a manner that made it impossible for vacua to exist between them, for otherwise, as Themon Judaeus explained, "the heaven could not act on this world . . . ,because action cannot occur through a vacuum."[74] Only because

71. "Probatur secunda pars: quia anima quaecumque est perfectior formis coelestibus, ergo non possunt a corporibus coelestibus produci quia imperfectius non potest producere immediate perfectius." Ibid., par. 109.
72. "Rationem autem assignare cur coeli sint imperfectiores quolibet vivente, est mihi valde difficile; nec sane occurrit ulla quae comprehendat viventia vita vegetativa tantum." Ibid.
73. Among those who accepted this threefold division were Themon Judaeus, *Meteorology*, bk. 1, qu. 1, 1518, 155v, col. 1; Albert of Saxony [*De celo*, bk. 2, qu. 12], 1518, 109v, col. 1; Johannes de Magistris [*Meteorology*, bk. 1, qu. 2], 1490, 8, col. 2 (unpaginated; I have counted from the beginning of the *Meteorology*); Paul of Venice [*Liber celi*], 1476, sig. g3r, col. 2–g3v, col. 1; and John Case [*Physics*, bk. 8, ch. 6], 1599, 825. The same threefold division was still in vogue during the seventeenth century; see, for example, Amicus, who treated each at great length in *De caelo*, tract. 6, qus. 5 (light), 6 (motion), and 7 (influences or occult qualities), 1626, 356, col. 1–398, col. 2; Mastrius and Bellutus, *De coelo*, disp. 2, qu. 2, art. 2, 1727, 3:494, col. 1, par. 65 (where they substitute *illuminatio* for *lumen*) and disp. 2, qu. 7, art. 4, 517, col. 1, par. 226; and Illuminatus Oddus [*De coelo*, disp. 1, dub. 29], 1672, 100–108. Oresme's position is problematic: he accepted it in one work, and in another he both accepted and subsequently rejected it. Thus he may have been one of the few who did not accept this threefold division of the transmission of celestial effects (see Sec. V.3 of this chapter).
74. The essentials of Themon's brief argument follow: "Necesse est istum mundum esse contiguum et immediatum celo. Probatur quia alias oporteret poni vacuum inter celum et mundum istum et tunc celum non posset agere in mundum istum . . . quia actio non fieret per vacuum. Patet etiam ex hoc: quia agens debet esse immediatum passo; et celum

the cosmos was a continuum were motion, light, and influence able to cause the variety of effects that was deemed indispensable for the preservation of inanimate and animate things in the terrestrial region.

The three instrumentalities caused effects largely by producing the four primary qualities – hotness, coldness, wetness, and dryness – that generated the four elements.[75] Speaking for most scholastic natural philosophers, Bartholomew Amicus declared: "No one denies that all elementary qualities are produced by the heavens, but, because they can be produced directly, or *per se*, and indirectly, or *per accidens*, the debate concerns the manner in which they are produced."[76] Although the three instrumentalities were the means by which celestial causes operated, most scholastics, probably following a long astrological tradition, attributed to each planet the power to produce one or more primary qualities. Thus the Sun was capable of causing hotness in sublunar things and the Moon wetness. Astronomers and astrologers were agreed that "just as some planets make things hot, so others make things cold."[77] Indeed, Amicus ([*De caelo*, tract. 6, qu. 5, dubit. 5, art. 2], 1626, 371, col. 2) traced coldness to Saturn, wetness to the Moon, dryness to Mars, and hotness to the Sun. Although Amicus and other scholastics traced the causes of the four primary qualities to different planets, they did not mean to attribute these qualities to the planets themselves. The Sun itself was not a hot body, nor was the Moon a cold body, and so on. The Sun was capable of causing heat in other bodies but did not itself possess the attribute hotness. The medieval mode of expression was to call the Sun *virtually* hot but not *actually* hot. This explains the usual terminology, such as "frigefactive," and "calefactive." The Sun, for example, was calefactive, that is, capable of causing heat in other bodies, but was not

est agens et illa inferiora patiuntur ab illo." Themon Judaeus, *Meteorology*, bk. 1, qu. 1, 1518, 156r, col. 1. De Magistris also insisted that the heaven and inferior regions must be continuous in order to avoid a vacuum (see his *Meteorology*, bk. 1, qu. 2, 1490, 8, col. 2). The denial of vacuum at the points of contact between the celestial and terrestrial regions was but a special case of the universally held medieval dictum that "Nature abhors a vacuum." See Grant, 1981a, 67–100. Although medieval natural philosophers frequently discussed imaginary vacua, not one of them, to my knowledge, accepted the real existence of void space anywhere within the cosmos.

75. Aristotle explains the generation of the four elements from the four primary qualities in his *On Generation and Corruption*.

76. "Nemo negat quod producantur a caelo omnes qualitates elementares. Sed quia possunt produci directe, seu per se, et indirecte, seu per accidens, discrimen est de modo producendi." Amicus, *De caelo*, tract. 6, qu. 8, dubit. 4, 1626, 418, col. 2. In offering his own opinion, Amicus declares that "these four qualities ought to be intended by nature, *per se* and directly, by the actions of agents that are ordained for the preservation of the universe. Among these agents are the heavens, without the influence of which the status of the universe could not be preserved." Ibid., dubit. 4, art. 2, 420, cols. 1–2. Mastrius and Bellutus held that the planets are the direct cause of coldness, wetness, hotness, and dryness (*De coelo*, disp. 2, qu. 7, art. 2, 1727, 3:511, cols. 1–2, pars. 192–193).

77. "Unde sicut aliquae stellae calefaciunt, ita alique frigefaciunt; et omnes astronomi dicunt hoc." John of Jandun, *De coelo*, bk. 2, qu. 12, 1552, 30r, col. 2. We must recall that although the Sun could cause heat in other things, it was not itself hot; nor was the Moon wet, and so on. The planets did not possess the primary qualities that they were able to cause in other things.

itself hot. Saturn was a frigefactive body, but was not itself a cold planet. Followers of Aristotle had no choice but to make such a distinction, because they were convinced that the four primary qualities – hotness and coldness, and wetness and dryness – could not exist in the heavens as pairs of opposites. For if they did, they would inevitably cause changes and alterations in that region, the occurrence of which Aristotle, and all his followers, denied.[78]

Our concern, however, is not with the qualitative properties assigned to this or that planet but with the three instrumentalities that were believed to produce the multiplicity of effects that governed and preserved the world.

1. Motion

Of the three types of celestial causes, the category of motion was probably considered the most fundamental for generation and corruption. Although Themon Judaeus had attributed to Aristotle the position that celestial motion is the primary motion and therefore causes and rules all other motions,[79] it was Aristotle's great Arabian commentator who made a much more sweeping claim for the dominion of celestial motion over terrestrial change. In his widely known *De substantia orbis*, Averroës declared that

> the Giver of the continuation of motion is [also] the Giver of celestial motion [*dator motus coeli*]. For if not, motion would be destroyed, and if motion were destroyed, so would the heaven itself [be destroyed]. Indeed, the heaven exists because of its motion; and if celestial motion were destroyed, the motion of all inferior beings would be destroyed and so also would the world. From this [then], it is verified that the Giver of the continuation of motion is the Giver of existence to all other beings.[80]

For Averroës, then, not only were terrestrial bodies influenced and affected by celestial motion but their very existence depended on it. For much of the thirteenth century, Averroës' attitude was influential. Sometime around 1271, Robertus Anglicus explained that the sky moved continuously because without celestial motion nothing could be moved here below, thus implying that the inferior region could not exist without the heavenly motion.[81] Despite some evidence to the contrary, Thomas Aquinas was also identified,

78. For earlier discussions, see Chapter 17, Section IV.4, and Chapter 10, Section II.1.a.
79. Themon cites book 8 of the *Physics*, where Aristotle derives the unmoved mover and establishes the superiority of circular motion over rectilinear and therefore of celestial motion over terrestrial. See Themon Judaeus, *Meteorology*, bk. 1, qu. 1, 1518, 156r, col. 1.
80. See Averroës, *De substantia orbis*, 1562–1574, 9:10v, col. 1. The passage has also been translated by Duhem in *Le Système*, 1913–1959, 6:59–60.
81. This is one of four reasons given by Robertus to explain why the sky moves. Robertus Anglicus [*Sphere*, lec. 3], 1949, 154 (Latin), 208 (English).

rightly it seems, with the Averroist position.[82] According to Johannes Versor in the fifteenth century, Thomas held that a cessation of celestial motion would be followed by a cessation of terrestrial, or inferior, motion.[83] At the end of the sixteenth century, the Coimbra Jesuits attributed the same opinion to Thomas, even as they disagreed with him on this important issue.[84]

During the next two centuries, supporters of the Averroës–Aquinas position were much in evidence: John of Jandun in the fourteenth century and Johannes Versor and Johannes de Magistris in the fifteenth.[85] In his commentary on *De substantia orbis*, John of Jandun fully supported this opinion of Averroës. Jandun explains that the destruction of all inferior things would follow upon the destruction of celestial motion, not only because all inferior things depend for their existence – that is, their coming-to-be – on celestial motion, but also because they depend on it for their very preservation. With reference to the second book of Aristotle's *De generatione et corruptione*,

82. The "contrary evidence" appears in Aquinas's letter on the occult works of nature, where he seems to have assumed that the actions of an element that derived from its nature were independent of celestial motions or influences (see Sec. III.1 of this chapter). However, relevant passages from his other works support the interpretation of Versor, the Coimbra Jesuits, and Amicus, as a scanning of a limited sampling of nine passages gathered by Litt, 1963, 146–147, reveals. Although Amicus includes Aquinas among those who adopted the Averroist position, he does say that some believe that Thomas changed his opinion (Amicus, *De caelo*, tract. 6, qu. 8, dubit. 2, art. 2, 1626, 407, col. 1).
83. "Sed secundum sanctum Thomam est dicendum quod cessante motu celi cessarent omnes motus inferiores." Johannes Versor [*De celo*, bk. 2, qu. 5], 1493, 20r, col. 1. Perhaps Versor had in mind Aquinas's statement in *De potentia* qu. 5, art. 8, where Thomas declares that "with the cessation of celestial motion . . . there will be no action by which matter is changed, [a process] which [normally] follows generation and corruption" (Et ideo cessante motu coeli . . . non autem erit actio per quam transmutatur materia, quam sequitur generatio et corruptio). My translation from Thomas Aquinas, *Quaestiones disputatae*, 1925, 2:200, col. 1. Another statement supporting Versor's interpretation occurs in Aquinas's [*De caelo*, bk. 2, lec. 4], 1952, 166, par. 342, where Aquinas declares: "it is better to say that, with the motion of the heaven ceasing, every motion of inferior bodies would cease, just as Simplicius says, because the powers of inferior bodies are like material and instrumental things with respect to celestial powers, so that those [material and instrumental powers] do not move [i.e., do not cause motion] unless they have been moved [by something else, namely by celestial powers]." In this passage, Thomas is presenting his own opinion, not Aristotle's. Since Thomas's *De caelo* was apparently his last work, we may conclude that we have here his final opinion on this topic.
84. The Conimbricenses, *De coelo*, bk. 2, ch. 3, qu. 4, art. 2, 1598, 202, explain: "therefore Saint Thomas thought that if the heaven ceased its motion, neither the heaven itself nor sublunar bodies could produce anything from prior actions. . . . This," they continue, "is the opinion of Saint Thomas, which, although it is defended in the Peripatetic school with great probability and has neither few nor ignoble defenders, has nevertheless proved unsatisfactory for us."
85. Nicole Oresme might appear to belong to this group when he declares (*Le Livre du ciel*, bk. 2, ch. 22, 1968, 511) that "the position and grouping of the heavenly bodies, the number, speed, quality, and diversity or difference of their motions are all arranged expediently and have for their principal purpose the generation and preservation of terrestrial bodies." (North, 1986, 93, offers his own translation of this passage.) However, as we shall see shortly, in our examination of Oresme's use of the Joshua miracle, Oresme qualified this in the very same treatise and allowed that preservation of terrestrial bodies did not wholly depend on celestial motions.

Jandun explains that although the Sun's motion around the ecliptic causes generation and corruption in the lower world, the daily motion of all celestial bodies causes the eternal continuity and preservation of inferior bodies. With such total dependence on celestial motion, the inferior, or sublunar, part of the cosmos would obviously be destroyed if the celestial motions ceased.[86] For Jandun, as for all who adopted this position, the only exception to the total dominance of celestial bodies over terrestrial bodies was human actions. "Celestial bodies," Jandun explained, "do not have the power for causing the intellect to understand or not to understand and for necessitating the will to choose or not to choose, or to will or not to will."[87]

If the celestial region affected the terrestrial region by motion, some scholastics, especially those who followed the Averroist position, inquired whether more than one motion is required to cause generation and corruption. In his treatment of this question, Johannes Versor presented a series of six conclusions drawn from Aristotle's *De caelo* and *De generatione et corruptione*.[88] Versor used the first five to establish the existence of the four elements (earth, air, water, and fire) and their need to undergo continual generation and corruption.[89] In the sixth conclusion,[90] however, he argues that a plurality of celestial motions is necessary to cause terrestrial generations and corruptions. However, the heavens do not only cause generations and corruptions but also conserve things. No one thing in the universe, with its single circular motion, can perform both activities simultaneously.

Permanence and continuity of existence are the province of the "prime

86. "Notandum est quod dicit [i.e., Averroës] si destrueretur motus coeli, et tunc destruerentur omnia inferiora. Et hoc ideo quia coelum non est solum causa in esse istis inferioribus, sed etiam in conservari. Nam, sicut apparet secundo *De generatione* et Commentator assumit secundo *Coeli et mundi*, motus solis in obliquo circulo est causa generationis et corruptionis; motus autem diurnus est causa continuitatis et perpetuitatis et conservationis aeterne in re naturali. Quare patet coelum esse causam istorum inferiorum non solum in esse sed etiam in conservari; sed destructa causa conservativa, destruitur effectus; quare destructo motu coeli, destruerentur omnia ista inferiora." John of Jandun, *De substantia orbis*, 1552, 49r, col. 2.

87. "Corpora coelestia non habent virtutem ad causandum intellectum ad intelligendum vel non intelligendum et ad necessitandum voluntatem ad eligendum vel non eligendum, vel volendum vel non volendum." For this opinion, see John of Jandun, *De coelo*, bk. 1, qu. 1, 1552, 2v, col. 1.

88. Versor, *De celo*, bk. 2, qu. 5 ("Whether, in order to save generation and corruption in inferior things, it is necessary that the heaven be moved by a plurality of motions"), 1493, 19v. A somewhat briefer presentation of the same six conclusions, called "conditionals" (*conditionales*), was also given by de Magistris, *De celo*, bk. 2, qu. 2 ("Whether a plurality of spheres and their motions must be necessarily assumed because of generations and corruptions"), sig. k4v, col. 1. Since Versor and de Magistris were contemporaries in Paris during the second half of the fifteenth century, it is not implausible to assume that one derived the six conclusions or conditionals from the other or else that both drew on an earlier source. Indeed, early in the fourteenth century, John of Jandun considered the same problem ("Whether in the heavens, a plurality of motions ought to be assumed for the different parts [of the heavens]") in five "consequences" (*consequentiae*). John of Jandun, *De coelo*, bk. 2, qu. 7, 1552, 27r, cols. 1–2.

89. The first five are based on Aristotle, *De caelo* 2.3.286a.3–286b.9, and *De generatione et corruptione*, ch. 3.

90. The sixth conclusion was drawn from Aristotle's *De generatione* 2.10.336a.15–336b.15.

motion" (*primus motus*), that is, the daily east-to-west motion of the spheres of the fixed stars and planets. But the variety of generations and corruptions is caused by the approach and withdrawal of the Sun in its annual west-to-east motion in the ecliptic (the oblique circle). The situation is even more complex, however, because generations require a different source of motion than do corruptions. Hence a plurality of celestial motions is essential to maintain the daily operation of the universe.

In view of his strong emphasis on the necessity of a plurality of celestial motions for terrestrial generations and corruptions, we are not surprised to learn, later in the same question, that Versor selects motion as the most fundamental of the three instrumentalities ([*De celo*, bk. 2, qu. 5], 1493, 20r, col. 1). Indeed, light and influence are both dependent on motion for diffusion and multiplication to inferior things. Hence they could not themselves cause the motions of inferior things. Celestial motion was thus the supreme and sole cause of terrestrial motion and change.

a. *Celestial motion and the generation of heat in terrestrial things*

If the celestial region caused effects in the terrestrial zone, what effects did it transmit, and how did it transmit them? What mechanism could convert celestial motions to terrestrial effects? One persistent claim was that celestial motions produced heat in the inferior world.

We can scarcely doubt the belief of most scholastics that, directly or indirectly, the motions of the celestial orbs played a dominant, if not decisive, role in terrestrial change. This was so because nearly everyone believed that, in some sense, celestial motions caused the two pairs of fundamental qualitative opposites that Aristotle had identified as the ultimate mechanisms for all terrestrial change, namely hotness and coldness and wetness and dryness.[91] Of these qualities, hotness was easily the most active and important, and it was the quality most discussed. For Aristotle had declared in *De caelo* (bk. 2, ch. 7) that as the planets rotate, they cause heat and light in the air below. Because each planet is carried around by an orb, the planets themselves and their orbs do not produce friction in the celestial ether and do not catch fire, but somehow they manage to heat the air below the Moon. Indeed, the Sun is a direct cause of heat, because it is near the air and its motion is sufficiently rapid to cause the air to heat up.[92] According

91. Thus Jean Buridan explains ([*De caelo*, bk. 2, qu. 17], 1942, 207, lines 26–30) that what follows is true and may be conceded, namely that "the local motion of the heaven causes in inferior things both hotness and coldness, and wetness and dryness, not [however] properly and by itself [i.e., not directly]; but [indirectly through] the Sun and the planets, which cause hotness and coldness, wetness and dryness in inferior things." Buridan's opinion conflicts directly with that which he attributes to Averroës, where the latter denies the capacity of individual planets and orbs to cause heat in inferior bodies but attributes that to the entire heaven, taken as a single thing.

92. See Aristotle, *Meteorology* 1.3.341a.19–28 [Webster], 1984, where Aristotle explains why the Sun is the ideal planet to heat the air. "For a motion that is to have this effect must

to a usual interpretation of Aristotle, the latter taught that "all planets heat inferior things by their motions because they [the planetary motions] drag the sphere of fire, and the upper part of the air, with them." These motions cause parts of the air and fire to separate and collide, setting them on fire.[93] Thus Aristotle had attributed the cause of heat in inferior things to both the Sun and other planets. For obvious reasons, it was the Sun that received the most attention, making it plausible to ask how the Sun's motion could heat the air when the Sun lies beyond the Moon in the celestial region and the air lies below the fire, or is perhaps even intermingled with the fire, in the sublunar region. In order for the motion of the Sun, or its orb, to produce heat in the sublunar region – even the sublunar region immediately below the concave surface of the lunar sphere – it would have to generate the heat by means of friction, "whereby," as Simon de Tunstede explained, "a body capable of causing heat is rubbed against another body."[94] But how could the Sun produce heat by friction when it was not even in contact with the airy region below?

Here indeed was a difficult problem, as Jean Buridan recognized. He readily conceded that it was manifest to our senses that motion causes heat and can make some things hot. After all, if we run we become hot; if we vigorously rub two hard bodies together, we also produce heat; and if we blow on a fire we intensify the heat of the fire. But the difficulty lies in determining what local motion is and the particular mode by which it causes something to heat. According to Buridan, there are many different opinions on these issues.[95]

But Buridan himself could find no persuasive argument that celestial motions *directly* caused heat in inferior things. They achieved this *indirectly* by agents, that is, by means of some property or innate capacity that was distinct from motion.[96] If it were only circular motion that caused heat,

be rapid and near, and that of the stars is rapid but distant, while that of the moon is near but slow, whereas the sun's motion combines both conditions in a sufficient degree."

93. This appears to be Amicus's summary (*De caelo*, tract. 6, qu. 6, dubit. 4, 1626, 387, col. 2) of Aristotle, *Meteorology*, book 1, chapter 3 (340b.4–11), although Amicus cites book 1, chapter 4.

94. "In ista quaestione supponendi sunt modi per quos calor potest produci a lumine vel motu. . . . Calor generatur a motu ex confricatione corporis calefactibilis ad aliquod aliud corpus." See Simon de Tunstede (Pseudo-Scotus) [*Meteorology*, bk. 1, qu. 12], 1639, 3:28, col. 2.

95. "Dicendum est breviter quod ad sensum est manifestum de multis quod motus calefacit vel est aliquando causa caloris; ut si curramus, et si corpora dura fricentur simul fortiter, . . . ; et sufflans in ignem acuit calorem ignis. Et sic quaestio est difficilis, propter quam causam vel per quem modum motus localis sic calefaciat; et sunt de hoc multae et diversae opiniones." Buridan, *De caelo*, bk. 2, qu. 16, 1942, 200. He also considers whether the sphere of the Sun heats inferior things more than do other spheres: "Ista quaestio est satis dubia, quia valde dubium est quae res sit motus localis, et quo modo motus localis agat calditatem" (ibid., qu. 17, 207).

96. As we shall see, Hervaeus Natalis held that motion was not even the active principle of the heavens. That distinction belonged to a mysterious, and not further defined, quality of motion, which remained constant whether or not the heavens moved. Both Buridan

then each sphere ought to cause as much heat as any other sphere. But we must rightly concede "that the motion of the Sun's sphere causes more hotness in inferior things than the motion of other spheres, because it [i.e., the Sun's sphere] adapts [or accommodates] for us the planet that has the greatest capacity for making heat, namely the Sun itself."[97]

However, Buridan reports that others – Averroës and Peter of Auvergne, for example – sought to indicate ways by which one might link celestial motion directly to the production of heat in inferior things. Averroës conceived of the multiplicity of planetary orbs and the starry orb as forming one single heaven that was akin to an animal. In this scheme, the multiplicity of orbs of the single heaven were analogous to the various members of the animal.[98] This one great body had as its concave surface the innermost surface of the lunar sphere. Since it was a single body, capable of acting as a single body, the heaven could affect the region below the Moon's concave surface because it was in direct contact with it and hence could act, as Buridan summarized it, "not only by its ultimate [concave] surface, but also by its total depth." Thus the Sun, and all other orbs of the whole heaven, could act on the inferior region, but none of them, including the Sun, could affect anything independently. Thus the Sun did not need to be in direct contact with the air in order to heat it.[99] As a member of the single body of the heaven, its effect, and that of all the other orbs, was felt at the surface of that body. Averroës' idea of conceiving of the multiplicity of orbs as a single heaven was perhaps motivated by the near-futility of attempting to explain the transmission of celestial effects from one orb to another orb in Aristotle's concentric system.[100]

Peter of Auvergne, who completed the unfinished *De caelo* of Thomas Aquinas, suggested another means by which the Sun could generate heat in inferior bodies by means of motion, although, like Averroës, he did not assume direct contact with any solar sphere. Peter suggests that by its motion the Sun radiates a certain insensible quality which is multiplied all the way to the region of the air, where it is capable of causing the air, and perhaps

and Hervaeus seem to have shared the idea that celestial motion did not act directly on terrestrial things. What else they may have shared is unclear.

97. "Et sic bene concederetur quod motus sphaerae solis magis agit in ista inferiora caliditatem quam motus aliarum sphaerarum, quia applicat nobis planetam summe calefactivum, scilicet ipsum solem. Haec igitur communiter sunt concessa." Buridan, *De caelo*, bk. 2, qu. 17, 1942, 207.
98. See Averroës [*De caelo*, bk. 2, comment. 42], 1562–1574, 5:125v, cols. 1–2. See also this volume, Chapter 13, Section I.
99. See Buridan, *De caelo*, bk. 2, qu. 17, 1942, 208.
100. In establishing his concentric system, Aristotle had inserted extra spheres ("unrolling spheres") that were designed to prevent the set of orbs associated with one planet from affecting the orbs associated with another planet (see Ch. 13, Sec. II.1). Thus effects could not be transmitted down through the orbs toward the sublunar region. But even if scholastics ignored Aristotle's scheme, they could not have devised any method for transmitting effects from one orb to another. The same may be said for those who adopted the Aristotelian–Ptolemaic compromise of eccentric and epicyclic orbs.

other inferior things, to become heated.[101] Thus did Peter seek to avoid the difficulties associated with attempts to explain the Sun's generation of heat by any motion of its own that placed it in direct contact with things that were far removed from it.

Buridan himself offers no definitive solution but suggests that something like the opinions of Averroës and Peter of Auvergne might prove reasonable.[102] He was emphatic on only one thing: celestial motions did not cause heat in inferior things directly, but only indirectly, in some unspecified manner, perhaps in some way described by Averroës or Peter of Auvergne.

One argument that seems obvious in retrospect concerns the most plausible place where the heavens might cause heat in the uppermost regions of the sphere of fire, namely the interface between the concave surface of the celestial region – which is equivalent to the concave surface of the lunar sphere – and the convex surface of the sphere of fire. Themon Judaeus posed this possibility, only to reject it on the grounds that no friction could exist between two such highly polished surfaces. Thus celestial motion could not generate any mutual friction that would produce heat.[103]

The possibility that heat might arise from the celestial orbs themselves as they rotated was occasionally raised but quickly rejected. An anonymous fourteenth-century author described the celestial orbs as so highly polished that they were incapable of causing heat by friction, which is the way heat is generated.[104] In the same vein, Sigismundus Serbellonus reminds us ([*De caelo*, disp. 1, qu. 4, art. 4], 1663, 2:45, col. 2) that the celestial ether is of the greatest subtlety and fluidity and thus does not lend itself to friction. Moreover, no friction could arise without resistance, which is absent in the heavens. Despite the fact that the anonymous author just mentioned con-

101. Peter resorts to an analogy with "a certain fish," probably an electric eel, which has the capacity to stun, for which reason it is even called *stupor*. When this fish is taken in a net, it shocks the hands of the fisherman who holds it. This capacity to stun is not innate in the fish but is received from outside (we are not told from where it comes) in the form of a certain quality that enables the fish to stun those who touch it. My source for Peter's opinion is Buridan, *De caelo*, bk. 2, qu. 17, 1942, 208.

102. After describing the interpretations of Averroës and Peter of Auvergne, Buridan says: "You should speak up about these [interpretations] as you think best" (Dicatis ergo de istis sicut videtur vobis bonum). Ibid.

103. "De tertio, scilicet qualiter celum ignem calefacit. Sit prima conclusio: quod hoc non fit per confricationem celi cum igne. Probatur: quia celum et ignis non confricantur quia sunt politissima corpora in superficiebus se tangentibus. Sed illa non fricantur, ut patet per predicta." Themon Judaeus, *Meteorology*, bk. 1, qu. 6, 1518, 161r, col. 1.

104. The anonymous author first explains that the claim that motion always causes heat when the parts in motion are in contact is not universally true (Hec autem opinio universaliter non valet quia motus localis non semper est causa caloris, ut puta quando nulla fit confricatio partium rei mote). Anonymous ("Compendium of Six Books"), 214r. In a vacuum, a body in motion would not cause heat because no air exists with which it could be in contact to cause friction and generate heat. But even where there is contact between moving bodies, heat might not be generated, as in the celestial bodies, which, though in contact, are too highly polished to produce friction (Ad propositum igitur redeuntes dicendum est quod corpora celestia in suis motibus localibus non confricantur quia sunt corpora politissima. Nec inter ipsa sit aliqua confricatio talis ex qua possit gigni calor prout predicta opinio imaginabatur). Ibid., 214v.

sidered the orbs hard, whereas Serbellonus conceived them as fluid, both denied the possibility of friction between rotating celestial orbs.

But there were those who argued that some celestial motions could produce heat in the region of fire down to the upper region of air.[105] Simon de Tunstede held strictly to the idea that heat is generated by friction between a body capable of being heated and some other body. After eliminating light and other spiritual influences, Simon argued ([*Meteorology*, bk. 1. qu. 12, conclus. 1], 1639, 28, col. 2) that celestial spheres could, by motion alone, cause heat in inferior things, but their range could extend no farther than the upper region of air. He obviously assumed that friction occurred at the interface between the heavens and the upper region of fire. The fire generated there was able to fall downward to the upper reaches of the region of air and cause it to become, or remain, hot.[106] Amicus favored a similar interpretation, in which the circular motion of the lunar sphere, in contact with the fiery region below, produces and conserves heat by contact with the inferior matter directly below. Indeed, the Moon is the collection point, so to speak, for the motions of all the celestial orbs.[107]

Scholastic opinion was thus divided. Some assumed that celestial motion directly produced heat, while others assigned an indirect role to motion, which somehow produced a quality that enabled inferior bodies to become hot. Motion, however, was not the only possible cause of heat. Light seemed an even more obvious source, and we shall determine later in this chapter (Section V.2) whether opinions about the role of light in the production of heat were as diverse as those about motion. But first we must examine the extent to which natural philosophers made the terrestrial region dependent on the celestial region as the source of all, or most, change.

b. The degree of dependence of terrestrial change on celestial motion

From the thirteenth to the seventeenth century, there was a current of scholastic opinion that followed Averroës and made sublunar motion (and therefore change) totally dependent on celestial motion. If the latter ceased, so would the former. But already in the late thirteenth century this opinion was challenged by a straightforward denial of its claim. In the Condemnation of 1277, the bishop of Paris and his advisers clearly had it in mind when in article 156 they condemned the opinion that "if the heaven should stand [still], fire would not act on tow [or flax], because nature would cease

105. On the three regions into which the sphere of air was divided, see the passage from d'Ailly's *Ymago mundi* that I quote near the beginning of Chapter 20.

106. Simon explained this by assuming that the strong impetus (*impetus*) which fire had for rising carried it upward with sufficient momentum so that it was reflected from the concave surface of the heavens, thus descending to the upper region of the air, which it caused to become hot.

107. Amicus, *De caelo*, tract. 6, qu. 6, dubit. 4, 1626, 390 (incorrectly given as 382), cols. 1–2.

to operate."[108] By the condemnation of article 156 and a few others, the bishop of Paris left no doubt of his distaste for the idea that terrestrial actions were totally dependent on celestial motions and, by implication, perhaps independent of God's actions.[109] As a consequence, one had to concede, at least in the diocese of Paris, that if the celestial motions ceased, fire could indeed act by its own power and burn flax.

The impact of article 156 is already manifest during the last years of the thirteenth century. In a discussion of the third day of creation in his commentary on the *Sentences*, Richard of Middleton argued that the four elements created on the third day were not made by the power of the heaven from celestial matter created on the first day but were separately created by God.[110] He then inquired whether the elements could operate without celestial influence.[111] Of the three opinions Richard distinguished for this question, the second asserts that if the heavens exert no influence on the elements, the latter would cease to exist. This opinion is "false and dangerous," Richard declares "because it seems to favor those who say that prime matter was produced by God through the mediation of the celestial body."[112] Richard identifies Moses Maimonides as a supporter of this condemned opinion. According to Richard, Moses held that "just as if the heart should cease from its motion for the blink of an eye, a man would die and his motion and powers would be destroyed, so also if the celestial motions should rest through the point of an hour [that is, for a moment], the whole world would disappear and all the things in it would be destroyed." Moses said all this, Richard continues, "believing that the heaven does not influence anything except by motion."[113]

But it was precisely such an opinion, Richard insists, that the bishop of Paris condemned when he threatened with excommunication those who held that if the heavens ceased to move the elements could not operate and

108. The original version of article 156 appeared as "Quod si celum staret, ignis in stupam non ageret, quia Deus non esset." Denifle and Chatelain, 1889–1897, 1:552. I have translated the text as emended by Hissette, 1977, 142. Using the alternative readings supplied by Denifle and Chatelain, Hissette changed "Deus non esset" to "natura deesset."

109. One of the "others" is article 186, which states: "That the sky never rests, because the generation of lower things, which is the end purpose of celestial motion, ought not to cease; another reason is because the heaven has its being and power from its mover, which things are preserved by its motion. Whence if its motion should cease, its existence would cease." Cited from Grant, 1974, 50.

110. Richard of Middleton, *Sentences*, bk. 2, dist. 14, art. 2, qu. 5, 1591, 2:181.

111. "Utrum elementa possent aliquid operari si caelum non influeret in ipsa." Ibid., 182, col. 1. Richard's arguments appear in question 6 on pages 182–183.

112. This is surely a reference to article 38 of the Condemnation of 1277, which denounces the claim "that God could not have made prime matter without the mediation of a celestial body." From Grant, 1974, 48.

113. Richard's reference is to "Rabbi Moses, ch. 67." I have not found the reference in Maimonides' *Guide for the Perplexed*, but Amicus (*De caelo*, tract. 6, qu. 3, art. 3, 1626, 354, col. 1) says "Rabbi Moyses" held that "the heaven is in the universe as the heart in an animal, [so that] resting for an hour all things would cease to exist." No reference is given.

therefore fire would not burn tow, a clear allusion to article 156. Thus did Richard reject any attempt to make the sublunar elements and the bodies compounded of them totally dependent on the celestial motions. He believed that "although the elements could not do all the things they could do with the influence of the heavens, they could nevertheless operate n some ways by a natural operation." They could do this because God had created the elements independently of the celestial bodies. Under the influence of condemned article 156, Richard rejected a necessary nexus between heaven and earth whereby the latter was totally dependent on the former. The sublunar region was now accorded some capacity for independent activity.

c. Hypothetical consequences of the cessation of celestial motion: elemental and mixed bodies

In the manner of Richard of Middleton, other scholastic authors imagined a terrestrial world in which all celestial motions had ceased by God's command. With the heavenly bodies assumed stationary, many believed that the regular powers of those same bodies would continue to function as before, except that the powers would now affect only those portions of the earth over which the bodies remained stationary.

Under these circumstances, Hervaeus Natalis, a Dominican follower of Thomas Aquinas (though he differed with his master on this issue) insisted that terrestrial change was not just a matter of a stationary Sun heating or not heating elements. Elements had powers of their own. Fire, for example, had sufficient power to act on its own, and the recipients of its power possessed a capacity for receiving its action – all without the motion of the celestial region. In fact, Hervaeus held that motion was not even the active principle of the heavens. That distinction belonged to its quality, which remained constant whether or not the heavens moved.[114] Hervaeus conceded, however, as would others, that without celestial motion the diversity of change would be considerably diminished, because every stationary star and planet could exert its power only over a limited part of the earth.

Up to this point, the terrestrial bodies under discussion have been of an elemental nature – that is, earth, water, air, and fire. Many were agreed that such bodies would continue to change despite the cessation of all celestial motion. But what about mixed bodies – that is, what about bodies compounded of at least two elements? Would they be capable of change if celestial motions ceased? Sensing that he had gone as far as he could in downplaying the power of celestial motion, Hervaeus explains (*De materia celi*, qu.7, in *Quolibeta Hervei*, 1513, 50r, col. 1) that if the celestial motions were not to be wholly superfluous – for he had already shown that they

114. Hervaeus discusses the problem of celestial influence in his *De materia celi*, questions 7 and 8 (see Hervaeus Natalis, *Quolibet Hervei*, 1513, 47v, col. 1–49r, col. 1 (qu. 7), 49r, col. 1–51v, col. 1 (qu. 8); for his denial of celestial motion as an active principle, see 48v, col. 1.

were unnecessary for the continued activity of elemental bodies – it was essential that they cause the generation of mixed bodies. If they were also unnecessary for the generation and corruption of mixed bodies, it would follow that the motion of the sky is superfluous, an unacceptable consequence. Celestial motion must therefore be essential for the generation and corruption of at least some mixed bodies. In this category Hervaeus included all mixed bodies derived from putrefaction, a process that depended exclusively on the heavenly movements and which produced all metals, minerals, and some living things. For putrefaction to occur, however, different parts of the sky had to pass over one and the same place and successively transmit rectilinear rays to that place. With the cessation of celestial motion the production of all such mixed bodies would therefore immediately cease.

But if new mixed bodies were no longer generable upon the cessation of celestial motion, what would happen to mixed bodies already in existence? They would certainly not disappear at the instant when the motions ceased. At the very least, the forms of the elements that constitute each mixed body would continue to exist for a time. The continued existence of mixed bodies seemed apparent to Hervaeus from his conviction, commonly held, that the heavens exercised a preservative power over all inferior things. As evidence of this, he invoked the miracle of Joshua (Joshua 10.10–13) and assumed that when Joshua commanded the Sun to stand still, all other celestial bodies also came to a halt. Although the Sun's cessation of motion was a true miracle, "the mixed bodies that remained have not been attributed to any miracle by any doctor."[115] Hervaeus took this as proof that mixed bodies would continue to exist by natural means after the heavenly motions ceased. Their existence would eventually terminate, however, because mixed bodies cannot endure forever, as is evident with living bodies.

During the fourteenth century, it was not unusual, especially in commentaries on *De caelo*, to inquire whether terrestrial elements and bodies could act independently if the celestial motions ceased; or, alternatively, whether a plurality of celestial motions was required for the occurrence of generation and corruption in inferior bodies. In 1377, one hundred years after the great Parisian condemnation, Nicole Oresme considered the latter question in his brilliant *Le Livre du ciel et du monde.*[116] Oresme straightaway denies that a plurality of motions is necessary for sublunar generation to occur. Rather he insists that

> if the heavens were at rest, change and growth would still exist, because if fire were at the present moment applied to a matter which it heated and burned, it is unrea-

115. As proof that mixed bodies would not disappear instantaneously if all heavenly motions ceased, Hervaeus declares: "Huius enim probabile signum potest accipi ex hoc quod accidit tempore Iosue quando sol stetit. Probabile enim videtur quod tunc omnis motus corporum celestium cessaverit. Et licet statio solis attribuatur miraculo, tamen mixta tunc remansisse non attribuitur miraculo ab aliquo doctore." Ibid., 50r, col. 2.

116. Oresme, *Le Livre du ciel*, bk. 2, ch. 8, 1968, 375–377. As was rather common, Oresme grappled with the problem in the following form: "If generation exists, there must be many motions in the heavens."

sonable to suppose that it would stop heating or burning even should celestial motions be stopped. To say the contrary is to support an article condemned at Paris.[117]

At this juncture, and following upon an obvious reference to article 156, Oresme introduces the miracle of Joshua at the battle of Gibeon as a counterinstance to the claim that the inferior region is totally dependent on celestial motions. When Joshua commanded the Sun and Moon to cease their motions, Oresme assumes that all other celestial motions also came to a halt. While all were at rest, however, "generation and destruction did not cease because during the period of cessation the enemies of Gibeon were killed." While this corruption was going on at Gibeon, generation was taking place elsewhere, for during this very time Hercules was said to have been conceived by Jupiter and Alcmena.[118]

Jean Buridan and Albert of Saxony agreed with Oresme that if the celestial motions ceased, generation and corruption would continue indefinitely. They offered quite similar explanations. Buridan, for example, believed that the region of earth lying under the Sun would convert water to air and fire, whereas the opposite side of the earth, where coldness dominated, would continually convert fire and air into water. While water would diminish by evaporation on that side of the earth subject to the steady, invariant rays of the motionless Sun, it would continually increase in that region of the earth perpetually deprived of sunshine. "Now," as Buridan explains ([*De caelo*, bk. 2, qu. 10], 1942, 171),

it is always natural that where water is higher, it is moved to a lower place. And so this water is continually moved around the earth to a place under the Sun; and this air or fire, generated under the Sun, is also moved to the opposite side. And thus the water that comes under the Sun is always converted into air; and the air coming to the opposite side is converted into water.

Under a motionless sky, and under the conditions described, such generations and corruptions could continue forever. Albert of Saxony ([*De celo*, bk. 2, qu. 12, conclus. 3], 1518, 109v, col. 2) held that even if the heavens ceased their motion, the rays of the Sun as well as rays of influence would

117. In a much earlier work, his unpublished *Quodlibeta*, probably composed in the 1350s, Oresme had, without allusion to article 156, argued that generation and corruption would continue even if the sky stopped moving. See Thorndike, 1923–1958, 3:414. Thomas of Strasbourg (*Sentences*, bk. 2, dist. 14, qu. 1, art. 2, 1564, 157v, col. 2) specifically cites the text of article 156 against those who insist that celestial motions are essential for terrestrial changes. ("Nec valet, si dicitur, quod hoc facit praesupponendo motum celi quia articulus Parisiensis dicit 'dicere quod ignis non possit comburere stupam sibi approximatam, stante motu caeli, error.' ") Later, on the same page, Thomas indicates that the condemned article played a role in the position he adopted.

118. Oresme explains (*Le Livre du ciel*, bk. 2, ch. 8, 1968, 377) that although this story is a fable, he mentions it only because "it is probable that the memory of this marvelous night dwelled among the pagans up to the time when Hercules was reputed a god and deified by them, and they thought or imagined that he had been conceived that night."

continue to radiate to earth and cause changes. Obviously, such generations and corruptions would differ from what we normally observe, because the influences of the immobile Sun and planets would no longer be diversely applied to the different regions of the earth.

If some medieval scholastics were convinced that generations and corruptions could continue if all celestial motions ceased, it was, *a fortiori*, even more plausible for them to assume the persistence of change if the sky had but a single, unique motion which ceased. As Buridan explained it, "I also say that if there were only one celestial motion, there would yet be generations and corruptions, etc., because there could be no fewer [generations and corruptions with a single celestial motion] than if the whole heaven should rest."[119] But as Albert of Saxony reminded his readers, a single celestial motion could not produce the usual variety of daily generations and corruptions. Only a multiplicity of celestial agents could generate such a diversity of effects. The daily motion and the oblique motions of Sun and planets in the zodiac are required. For "the Sun and the other planets sometimes approach us and sometimes withdraw from us, thereby making a diversity of times for us, namely winter and summer, and [making] the diversity of generations and corruptions in inferior things."[120]

By the late sixteenth and the seventeenth century, the anti-Aristotelian and anti-Averroist position, as represented by the Coimbra Jesuits, Aversa, Amicus, and Poncius, seems to have triumphed among scholastic authors, who continued to cite article 156 and the Joshua argument against total terrestrial dependence on celestial motions.[121] As their medieval predecessors did, early modern scholastics usually distinguished (sometimes only implicitly) those effects that were solely attributable to the celestial motions (such as the seasons, or night and day) and those for which they were a necessary but not a sufficient cause. With the cessation of celestial motions, the seasons would cease immediately, as would day and night. One would experience only one season, or only day, or only night. But in the numerous cause–effect relationships where the heaven was a concurrent cause, actions ought not to cease instantaneously or abruptly. It did not seem plausible to Aversa (*De caelo*, qu. 35, sec. 9, 1627, 200, col. 1) that a man taking a walk should become immobile as soon as the heavens ceased their motions, or

119. "Dico etiam quod si esset solus unus motus caelestis, adhuc essent generationes et corruptiones etc., quia non minus essent quam si totum caelum quiesceret." Buridan, *De caelo*, bk. 2, qu. 10, 1942, 171. Albert of Saxony repeats this opinion (*De celo*, bk. 2, qu. 12, conclus. 4, 1518, 109v, col. 2).

120. Albert of Saxony, ibid., conclus. 5–6. Earlier in the fourteenth century, John of Jandun made much the same point (*De coelo*, bk. 2, qu. 7, 1552, 27r, cols. 1–2).

121. However, only the Conimbricenses allude to article 156, when they speak of "the consensus of the Parisian doctors, who condemned by one of their articles the opinion of those who believed that if the celestial motions ceased, tow [or flax] could not be burned by fire" (Conimbricenses, *De coelo*, bk. 2, ch. 3, qu. 4, art. 2, 1598, 203). For the Joshua argument, see ibid.; Aversa, *De caelo*, qu. 35, sec. 9, 1627, 200, col. 1; Amicus, *De caelo*, tract. 6, qu. 8, dubit. 2, art. 2, 1626, 408, col. 2; and Poncius, *De coelo*, disp. 22, qu. 9, 1672, 630, col. 1, par. 100.

that a stone that was falling from a height should cease falling in the middle of its descent, or that a fire should stop burning wood; it seemed paradoxical to him to assume that such activitics would instantly terminate upon the cessation of the celestial motions. Actions that really depended on the heavens would cease their activities only gradually.[122]

Amicus went farther (*De caelo*, tract. 6, qu. 8, dubit. 2, art. 3 [corrected from 2], 1626, 409, col. 2–410, col. 1). Even during the occurrence of celestial influences in winter, when the weather is cold, we see that fire continues to produce heat. In fact, Amicus considerably reduced the dominance of the heavens when he argued that although the actions of some inferior causes depend on the heavens, others do not. Although the production of some living things and things that arise from putrefaction, and also the generation of metals, depend on celestial influences and motions, many actions do not.

Despite a wide-ranging attempt to attribute more causal activities to terrestrial bodies and to reduce the previous overwhelming dominance of the celestial region with motions and other causal influences, early modern scholastic natural philosophers were undoubtedly in agreement with their medieval predecessors that our terrestrial region is largely governed from the heavens and that if the influences and motions of the heavens were to cease for a brief time, "the status of this world would be disorderly [indeed]."[123]

d. A dissenting voice

The ideas described here marked a considerable departure from Aristotle and Averroës, who held that terrestrial generation and corruption were wholly dependent on the celestial motions. We saw that Thomas Aquinas, Robertus Anglicus, John of Jandun, Johannes Versor, and Johannes de Magistris, among others, upheld this judgment. Most of those who sided with the Averroist approach would probably not have denied the Joshua account as a counterinstance to their position. But they would have insisted that this was an abnormal state of affairs, an instance of God's direct intervention in the regular activities of the physical world. Indeed, this was a way of neutralizing the impact of article 156. Marsilius of Inghen, who spent most of his academic life as a master of arts before becoming a theological master near the end of his life, took precisely this approach.

In his commentary on the *Sentences* of Peter Lombard, Marsilius considered the question "Whether the firmament dividing the waters from the

122. Amicus says this (ibid., 410, col. 1).

123. Aversa (*De caelo*, qu. 35, sec. 9, 1627, 200, col. 2) expressed it this way: "Just as an animal is preserved and functions by [means of] life and by the motion of its heart, so is this whole inferior world preserved and operated by the influence and motion of the heaven. But it is sufficient [to know] that with the cessation of the influence and motion of the heaven for a brief period, the status of this world would be disorderly and that the world is truly disposed and governed by the influence and motion of the heavens."

waters is, by its motion, the cause of generation."[124] Those who deny this insist that if the heavens were stopped or removed completely, the Sun would yet illuminate and heat inferior things. Generation and alteration would thus continue. This is confirmed by an article of Paris that declares it an error "to say that when the heaven has ceased its motion fire could not burn tow next to it."[125] In his response, Marsilius distinguishes two ways to approach the problem: natural and supernatural. With the heavens assumed motionless, he concedes that God could, presumably by his absolute power, cause the fire to burn tow. But God could do this only if he decides to deviate from the "accustomed course of nature" (*solitus cursus nature*). As long as the accustomed, or common, course of nature obtains, not even God could cause the fire to burn tow.

Although Marsilius's approach effectively neutralized article 156,[126] the latter appears nevertheless to have produced a reaction against the defenders of total terrestrial dependence on celestial motion. Richard of Middleton, Hervaeus Natalis, Nicole Oresme, Jean Buridan, Albert of Saxony, and others insisted that terrestrial elements and ordinary compound bodies would continue to undergo change even if all the celestial motions ceased. A popular counterinstance to the Aristotelian position was the biblical account of the Joshua miracle. Article 156 and the Joshua miracle were common ingredients in the attempt to uphold the natural ability of terrestrial bodies to suffer change in the absence of celestial motion. By the late sixteenth and the seventeenth century, the anti-Aristotelian position seems to have triumphed among scholastic authors, who, as we saw, continued to cite article 156 and the Joshua argument.

In part, the controversy concerned the relationship between God and the celestial bodies. To make all terrestrial change dependent on celestial motion was to concede too much power and dominance to the celestial region. As Richard of Middleton recognized, it made it appear as if the celestial substance had somehow created the terrestrial elements. The condemnation of article 156 in 1277 was probably a response to those who followed Averroës' strongly deterministic interpretation in *De substantia orbis*. The continued operation of the sublunar world depended only on God, not on celestial motions. As if to confirm this, many scholastics restricted the influence of celestial motions by assigning varying degrees of independent action to terrestrial bodies. At the same time, however, they acknowledged that without regular celestial motions the world as we know it would be impossible.

124. Marsilius of Inghen [*Sentences*, bk. 2, dist. 14, qu. 10], 1501, 241v, col. 2.

125. "Confirmatur per articulum Parisiensis dicente quod dicere quod ignis non possit comburere stupam approximatam cessante motu celi: error." Ibid., 242r, col. 1.

126. It neutralized it because one could no longer make the categorical claim that with the heavens motionless, fire could not burn tow. For, if he chose to intervene, God could make the tow burn supernaturally. Since we cannot know whether or not God would intervene, we cannot infer that tow will not burn if the heavens are motionless.

2. *Light*

That the Sun's light produced terrestrial daylight was too obvious to merit discussion. Attention was concentrated on the relationship between celestial light and heat. Most natural philosophers assumed that light, or *lumen*, from a celestial source was a cause of hotness. Disagreement arose over the manner in which light causes hotness[127] and even from which particular celestial bodies. Jean Buridan, and other natural philosophers, took it as obvious to the senses that the Sun's light caused heat. As further evidence of this, one could produce heat with burning mirrors and even focus the Sun's rays so intensely that their heat would burn things.[128] The common assumption that all planets receive their light from the Sun made it appear reasonable to infer that if the Sun's light is heat producing, so also is the light of the other planets.[129]

a. Light and its relationship to hotness and coldness

However, if *all* celestial bodies emit heat-producing light, how could this be reconciled with another common assumption that some planets cause coldness? Was there a problem here? Could one and the same planet cause hotness and coldness? At first glance, it might have been tempting to reject such a claim and to argue that some planets produced coldness and were frigefactive, in contrast to others that produced hotness and were calefactive. The responses to this seeming dilemma were varied, but surprisingly, some would allow one and the same planet to cause both hotness and coldness.

Medieval interpreters of Averroës, perhaps with reference to his discussion in *De caelo* (Averroës [*De caelo*, comment. 42], 1562–1574, 5:126v, col. 1), sought to explain coldness as a lack of hotness. According to John of Jandun, Averroës argued that celestial bodies did not cause hotness and coldness but only hotness. As far as Averroës was concerned, "no celestial body causes coldness, but all cause hotness and this in different ways, because some make more heat and some less, as is appropriate to every

127. In a question that asks "Whether light generates hotness" (An lumen generat caliditatem), John of Jandun explains: "there is no doubt that light makes hotness, but there is diversity [of opinion] concerning the manner in which it causes the hotness" (Sciendum est de quaestione quod lumen calefacit non est dubium; sed de modo secundum quem calefacit est diversitas). John of Jandun, *De coelo*, bk. 2, qu. 10, 1552, 29r, col. 1.

128. See Buridan, *De caelo*, bk. 2, qu. 15, 1942, 193. This is the first of four brief conclusions which Buridan advances in his question on "Whether, by their light, celestial bodies are productive of heat." Of the four, only the first two are directly relevant to light; the remaining two are mentioned later in this chapter in the section on "influence" (Sec. V.3). Albert of Saxony, *De celo*, bk. 2, qu. 21, 1518, 115v, col. 2, presents the same four conclusions in the same order. It would appear that he derived them directly from Buridan. Amicus, in turn, drew directly upon Albert and repeated the same conclusions (see Amicus, *De caelo*, tract. 6, qu. 5, dubit. 5, 1626, 370, col. 1, under *tertia opinio*).

129. This is Buridan's second conclusion (Buridan, ibid., 193–194). It is repeated by Amicus, ibid., art. 2, 372, col. 2.

element and compound."[130] Because all planets received solar light, they could produce heat. But since planets varied in the degree of light they received, they also differed in the intensity of heat they could generate. Coldness was thus a lesser degree of hotness, or no hotness at all. On this heat continuum, coldness was not readily determinable but it could be perceived if the level of heat was very low or nonexistent. As an anonymous fourteenth-century author put it, "such planets do not cause coldness, except in the sense that they cause less heat."[131] As evidence that heat was the sole operating quality and that cold was merely its perceptible absence, it was observed that when the Moon is full the nights are warmer than at other times, which could occur only because there was more light at that time.[132]

In Averroës' approach, if the amount of light a planet receives varied sufficiently, it might sometimes generate sufficient heat to cause hotness, but at other times its quantity of light would be insufficient to produce perceptible heat. Supporters of the theory could then say it causes coldness. Consequently, one and the same planet could be said to cause both hotness and coldness at different times. If at full moon there was more heat than at, say, new moon, when the Moon was invisible because no light reached it, would this qualify as an instance of the production of hotness (at full moon) and coldness (at new moon)? Only if the absence of light at new moon was deemed to cause coldness at that time.

Although their intention may have been otherwise, those who adopted the theory attributed to Averroës were assuming, in effect, that light could produce both hotness and coldness, albeit the latter only indirectly by virtue of an absence, or insufficiency, of the former. Opposition was inevitable. For most scholastics, hotness and coldness were equal members of a pair of primary contrary qualities.[133] Coldness could not be the mere absence of heat. One thing, such as light, could not produce both and could not generate the other merely by its absence. Thus was the question occasionally proposed, usually to be formally rejected, whether light, in addition to causing hotness, could also produce coldness. Could some stars emit light that was heat-producing and other stars – cold stars – radiate light that was cold-producing? Most Aristotelian natural philosophers would have found

130. Jandun goes on to explain that "perhaps the reason that he [Averroës] induced this is because motion and light produce hotness; but all bodies are moved with a local motion and with quickness, which is required for motion; and also all celestial bodies that cause hotness have light, although some more and some less." John of Jandun, *De coelo*, bk. 2, qu. 12, 1552, 30r, col. 1. In this question, Jandun asks "Whether some of the celestial bodies make inferior things cold." Buridan mentions the same opinion of Averroës, when he declares (*De caelo*, bk. 2, qu. 15, 1942, 195), that "Averroës speaks as if all the elements are naturally hot, which is false concerning earth and water."

131. "Vel dicunt aliter, scilicet, quod tales stelle non frigefaciunt nisi ad istum sensum quod minus calefaciunt." Anonymous ("Compendium of Six Books"), 215r.

132. "Et ad robur sue opinionis allegant Aristoteles dicentem *Libro de animalibus* quod noctes in plenilunio sunt calidiores quam in alio tempore, quod non accidit nisi ratione maioris luminis ipsius lune." Ibid.

133. Of the two, hotness was considered the active quality.

such a suggestion unacceptable.[134] A single instrumentality such as light could not cause both members of a pair of contraries such as hotness and coldness.[135] Because, as Buridan expressed it ([*De caelo*, bk. 2, qu. 15], 1942, 194), "this world is governed by the heaven in all its dispositions," and since both earth and water are basically cold elements, it followed that "virtually cold" celestial bodies must exist in order to produce the coldness in these elements.[136] One and the same celestial body could have powers other than light. It could also be frigefactive.[137] Thus some celestial bodies cause hotness, others coldness, and some bodies are basically hot (fire and air) and others cold (water and earth).

An explanation reported by an anonymous author in the fourteenth century conflicted with both Averroës and Buridan. It not only treated hotness and coldness as distinct qualities but allowed that both could issue from one and the same planet. This was achieved by assuming that any illuminated planet could cause heat. But if it also causes coldness, this is attributable to an invisible quality called "influence" (*influentia*).[138] A planet was thus capable of possessing two attributes, light and influence, simultaneously, each performing a different specific function.

b. How light produces heat

Numerous explanations were proposed to describe how light, or *lumen* from the planets, which was not thought of as either hot or a body, produced heat. Buridan describes at least four, but rejects them all.[139] They are of interest because they show the range of responses which confronted natural philosophers. Of the two that I shall describe here, the first assumes that

134. John of Jandun was a rare exception. In his *De coelo*, bk. 2, qu. 12, 1552, 30r, col. 2, he sought to explain that generic light could not produce both hotness and coldness, but the light of specific planets was somehow different. Thus although "light as light" cannot produce coldness, the specific, individualized light of this or that planet could do so: for example, the light of Saturn.
135. Amicus offers this explanation when he declares (*De caelo*, tract. 6, qu. 5, dubit. 5, art. 2, 1626, 374, col. 1): "Secundo patet nulla astra esse frigefactiva per suum lumen quia eadem causa non est per unum instrumentum causa immediata contrariorum unde in illis est ponenda alia virtus per quam frigefaciant distincta a lumine et motu." Albert of Saxony (*De celo*, bk. 2, qu. 21, 1518, 115v, col. 2, conclus. 3), citing astrologers as his authority, argued that some planets cause coldness in inferior things – but not by their light.
136. See Buridan's third conclusion in *De caelo*, bk. 2, qu. 15, 1942, 194.
137. "Quarta conclusio est, quod necesse est stellas habere virtutes alias a suis luminibus, quia habent virtutes frigefactivas, et per lumina sua non frigefacerent, imo potius calefacerent." Ibid.
138. "Ad hoc respondent aliqui dicentes quod omnis planeta seu stella inquantum illuminat calefacit. Sed si frigefacit, hoc est ratione alterius qualitatis invisibilis, quam non percipimus, que dicitur influentia." Anonymous ("Compendium of Six Books"), 214v–215r. For a detailed treatment of "influence," see Section V.3 of this chapter.
139. They are given in response to a doubt, the third, which Buridan raised against his own four conclusions. The third doubt "concerns the mode by which light makes things hot" (Tertia dubitatio est de modo per quem lumen calefacit). Buridan, *De caelo*, bk. 2, qu. 15, 1942, 194.

as solar light passes through the sphere of fire it incorporates the heat of fire within itself, just as light passing through a colored glass incorporates the color of the glass within itself. Buridan judges this response absurd, because light is not a body. Moreover, if light did incorporate heat within itself in the sphere of fire, it would lose it shortly afterward as it absorbed coldness in the middle region of the sphere of air, which was generally conceded to be a frigid zone.[140]

The second explanation assumed that light rays refracted and intersected in the sphere of air. As a consequence of the activity of these light rays, the air was parted in many places and thus rarefied and made hot.[141] Indeed, the consequent motion of the parts also made the air hot. Against this opinion, Buridan argues that the greatest amount of solar light and the greatest amount of heat occur in the summer. Thus the air should be most agitated during the summer. But, counters Buridan, the contrary is true, because the air is most serene and tranquil at that time of year. An even more telling counterargument, from Buridan's standpoint, concerned light rays, to which he denied the status of bodies and which therefore could not divide the air. But even if they could, it is impossible that the totality of air should be everywhere divided.

As his solution to the problem, Buridan argues that it is a natural property of light to heat something that is heatable, as long as that thing's resistance to being heated is not stronger than the activity of the light.[142] Although light causes heat, it will not do so in all situations. To produce heat, (1) the light must be sufficiently intense; (2) the resistance to it must not be too great; and (3) the subject or thing that is to be heated must have the capacity to absorb the heat – in scholastic parlance, it must be of a heatable

140. Ibid., 195–196. Albert of Saxony repeats the same idea in *De celo*, bk. 2, qu. 21, 1518, 115v, col. 2–116r, col. 1, as does Amicus, *De caelo*, tract. 6, qu. 5, dubit. 5, 1626, 370, cols. 1–2, who derived this argument from Albert. Achillini, in his *Liber de orbibus*, 1545, 58r, col. 1, mentions that the middle region of air is cold. In this criticism, Buridan could have had Albertus Magnus in mind, since Albertus formulated a similar explanation, which is described later in this section (also see n. 151).

141. Buridan, ibid., 196, lines 7–18. Buridan cites Aristotle, *De caelo* 2.7.289a.20–22, as support for the claim that celestial bodies cause heat and light by causing friction in the air. Buridan disagrees with Aristotle.

142. "Naturalis proprietas luminis est calefacere subiectum suum si sit calefactibile, et quod resistentia non sit fortior quam activitas luminis." Buridan, ibid., 197. In this regard, Buridan says (ibid.) that light is like many other natural properties, such as, for example, "heaviness" (*gravitas*), which has the natural property of moving downward the body in which it exists. It will do this if the body is not already down and if the resistance which the body encounters is not greater than the power of the heaviness. Similarly for the property of hotness, which causes the subject on which it acts to rarefy, if that subject is rarefiable, and so on (Sicut enim naturalis proprietas gravitatis est movere deorsum subiectum in quo est, si non sit deorsum et non sit resistentia excedens virtutem gravitatis; et sicut etiam naturalis proprietas caliditatis est rarefacere subiectum suum, si sit bene dispositum ad raritatem). Albert of Saxony's opinion is the same as Buridan's. It is the final opinion Albert describes, and he introduces it with the words "there is another opinion which I repute to be true." *De celo*, bk. 2, qu. 21, 1518, 116r, col. 1. He also repeats Buridan's arguments as they are described in my next paragraph.

nature.[143] Or, as Themon Judaeus expressed it, the light from a luminous body "can cause heat in a body susceptible of heat."[144] But it cannot cause heat in a body that is not susceptible. Indeed, some luminous bodies can illuminate a medium, or body, and yet not heat it. The Sun, for example, illuminates other celestial bodies but does not heat them, because celestial bodies are not heatable. Indeed, the middle region of the air can be illuminated by planets, but it is not heated.[145]

But why does light cause heat? Buridan dismisses the question as irrelevant. Light causes heat because it is a natural property of light to do this in a thing that is capable of being heated.[146] It is the same with all natural properties. If we ask why "heaviness" (*gravitas*) moves its body downward, the answer, says Buridan, is that it is a natural property of heaviness to do precisely that. Whereas some sought to provide a mechanism by means of which light generated heat, Buridan found this a vain enterprise.

Others, however – Albertus Magnus, for example, and perhaps most scholastics – sought to provide a cause for the production of heat by light. Albertus thought of light as a "proper form of planets" (*propria forma stellarum*) by means of which the matter of generable things is brought into existence.[147] The motion of the first movable heaven, and of the heavens as a whole, radiated a kind of general light into generable things. This light (*lumen*) was akin to a kind of life that existed in all natural things. It produced heat by dissolving matter. But Albertus distinguished another kind of light that was associated with specific celestial bodies, just as the light of Jupiter differs from the light of Saturn.

Since it was understood that every planet was carried by an invisible orb, Albertus asked whether both orb and planet radiated light. Because of their invisibility, orbs could not shine, from which Albertus concluded that they did not radiate light; and if they did not radiate light, they could not cause heat. Thus it was only the visible celestial bodies that emitted light and that were potentially capable of producing heat. Of these, however, only the

143. "Sed etiam sciendum est, quod ad hoc quod lumen calefaciat, tamen non apparet notabiliter effectus eius nisi lumen sit bene intensum, et quod non sit nimia resistentia: et cum hoc etiam, quod subiectum sit naturae calefactibilis." Buridan, ibid., 197–198.

144. As the first of three conclusions on light as a cause of heat, Themon declares (*Meteorology*, bk. 1, qu. 8, 1518, 162v, col. 1): "Tunc sit prima conclusio: quod omne corpus luminosam mediate lumine suo causat vel potest causare calorem in medio susceptivo caloris."

145. "Tertia conclusio: quod non omne luminosum omne medium quod illuminat calefacit." As one of his examples, Themon says: "Similiter corpus celeste illuminatur a sole et tamen non calefit quia celum non est calefactibile; et similiter de media regione propter virtutes aliarum stellarum vel influentias potest illuminari et non calefieri." Ibid. The "middle region" probably refers to the sphere of air.

146. "Ita non oportet quaerere propter quid lumen calefacit, nisi quia naturalis proprietas luminis est calefacere ipsum calefactibile." Buridan, *De caelo*, bk. 2, qu. 15, 1942, 197, lines 29–31.

147. "Si autem quaeratur causa, quare lumen ita facit calorem, cum secundum dicta Peripateticorum neque lumen sit corpus calidum neque radius eius: videtur esse dicendum, quod lumen est propria forma stellarum et corporis caelestis, qua universaliter movetur materia generabilium ad esse." Albertus Magnus [*De caelo*, bk. 2, tract. 3, ch. 3], 1971, 5:147, col. 2.

Sun radiated light that was strong enough to generate heat.[148] The Sun alone was capable of generating heat, because of its greater magnitude, thickness, purity, and so on.[149] The Sun's rays were thus suitable for generating heat, whereas the rays of the other planets were not.

In Albertus's interpretation, the rays of the Sun do not move directly to the sublunar region and generate heat. They acquire their heat-making power from the fiery region through which they must pass. Albertus explains that the Sun's light moves fire "just as the sphere of the fixed stars has to move earth, and the sphere of the Moon has to move water, and the five other planets have to move air."[150] Albertus seems to accept the interpretation of "some philosophers" (*quidam philosophi*) that the Sun's rays do not themselves induce heat, but that the heat of its rays is derived from the fire which is moved by the Sun's rays as they pass through it. Solar rays cause the fire to move in the same manner as iron is moved by a magnet.[151] The Sun's rays thus "attract" or "draw" heat given off by the fire which is agitated by the movement of those rays, a process that strongly suggests that Albertus conceived of light as a body.[152] After the now-heated solar rays move beyond the sphere of fire on their way toward earth, they produce heat by refraction when many rays are brought to a focus at a point.[153] Why Albertus chose to invoke refraction as the basic mode of heat production is unclear. A better account was provided by Themon Judaeus,

148. "Si autem queratur, utrum totus orbis hoc modo sit causa caloris vel saltem omnes stellae, dicimus, quod non, quia licet totus orbis sit de natura corporis lucidi, sicut dicit Avicenna, non tamen totus orbis micat. Et ideo pars, quae non micat, non emittit lumen, et ideo nullo istorum modorum est causa caloris. Sed et inter stellas solus sol emittit radios fortes, et ideo solus sol hoc modo est causa caloris." Ibid., col. 1.

149. Albertus presents five reasons (ibid.) for this, which when taken collectively reveal what a perfect heat-causing body the Sun really is. The first four reasons describe properties that enable the Sun, and no other planet, to be a heat-generating body. These properties are its magnitude, which far exceeds that of the other planets and enables it to cause stronger effects; its thickness (*spissitudo*), or density, which exceeds that of all the other planets and enables the Sun to produce more friction and therefore cause more heat; its light, which is denser than in other planets and therefore able to emit stronger rays and better able to burn things in the sublunar region; and its greater purity and subtlety than the other bodies. For the fifth reason, see the next note.

150. "Quinta et ultima causa est, quia sol in natura sui luminis habet movere ignem, sicut et sphaera stellarum fixarum habet movere terram et sphaera lunae habet movere aquam et sphaerae quinque aliorum planetarum habent movere aerem." Ibid. Albertus adds a sixth cause (ibid., cols. 1–2), which he attributes to Avicenna, namely that the nearness of the Sun to the sublunar region gives it an advantage over the other planets (although Venus and Mercury are nearer, they are never far from the Sun and are moved with it).

151. "Et ideo etiam quidam philosophi dixerunt, quod radius solis in eo quod radius, non habet inducere calorem, sed calor est a proprietate ignis, qui movetur a radiis eius sicut ferrum a magnete et adducitur cum radiis eius." Ibid. Precisely how the solar rays–fire relationship was meant to operate in the manner of the iron–magnet relationship is left unclear.

152. As we saw, Buridan denied that light was a body. For Buridan's rejection of the view adopted by Albertus, see note 140 of this chapter.

153. "Reflexio autem est causa caloris, eo quod in reflexione multi radii diriguntur ad punctum unum, ubi propter multiplicatum calorem aut calet locus aut in toto incenditur." Albertus Magnus, *De caelo*, bk. 2, tract. 3, ch. 3, 147, col. 2. Although Albertus used the term *reflexio*, he seems to mean refraction.

who argued that, other things being equal, incident rays produce more intense light and heat than do reflected or refracted rays.[154] But when an aggregation of light rays is reflected or refracted, those rays are able to produce a more intense heat than an incident ray, as is evident from what happens when rays are focused with concave burning glasses.[155] Nevertheless, Albertus Magnus conceived a means by which light, which is not hot by nature, could become hot by drawing heat to itself on its way to earth.[156]

Some were prepared to deny that light itself could cause heat and instead attributed the generation of heat to bodies that were illuminated by light. Buridan ([*De caelo*, bk. 2, qu. 15], 1942, 196) rejected this explanation because heat was not a form in these bodies. For example, water is composed of prime matter and the substantial form of water, neither of which was capable of causing heat. Thus water illuminated by light might become hot, but it does not possess hotness. Hot water is not composed of water plus hotness but is only water. Thus it will not cause heat in other things simply because it has been illuminated.

Whereas Albertus Magnus chose to explain in concrete terms how light causes heat, Buridan chose to avoid such explanations. Instead he relied on the notion that light causes heat because it is its nature to do so and then described the conditions under which it would naturally act to produce heat. Both approaches were influential in the seventeenth century in one and the same person, Bartholomew Amicus. Buridan's ideas were influential because Albert of Saxony chose to make most of Buridan's questions his own, and Amicus, in turn, adopted Buridan's concepts and arguments from Albert of Saxony. But Amicus was also familiar with Albertus Magnus's arguments in the latter's commentary on *De caelo* and cites Albertus by name a number of times. Thus not only did Amicus, in his lengthy discussion, repeat Buridan's idea that it is no more necessary to seek the manner in which light produces heat than it is to ask why heaviness (*gravitas*) is the cause of falling, but he also accepted Albertus's arguments about the role of the Sun in producing heat and the way solar rays acquire heat in the region of fire.[157] Either Amicus was unaware of any conflict, or he found it quite plausible to deny the necessity of a causal explanation for the generation of heat from light while, at the same time, accepting Albertus's causal explanation of the way in which the Sun's rays generate heat.

154. "Secunda conclusio: quod omne luminosum fortius agit lumen et calorem per radium incendentem quam per reflexum radium vel refractum, ceteris paribus." Themon Judaeus, *Meteorology*, bk. 1, qu. 8, 1518, 162v, col. 1. Themon goes on to explain why this occurs and compares the behavior of the rays to light rays striking a concave burning glass.
155. Ibid.
156. Amicus, *De caelo*, tract. 6, qu. 5, dubit. 5, art. 2, 1626, 373, col. 1, who briefly touches on Albertus's idea that light causes heat by agitating the fire in the fiery region under the Moon, thought Albertus failed to satisfactorily explain how light could become hot. Later in the same section (art. 3, 375, col. 2–376, col. 1), however, Amicus repeats virtually all of Albertus's arguments that I have described here.
157. Amicus, ibid., 373, col. 1 and 375, col. 2–376, col. 1.

Amicus, however, thought it important to inquire why light produces more heat at some times, and less heat at other times.[158] In certain subjects, more light seems to be accompanied by less heat and less light with more heat. For example, in the second, or middle, region of air, there is less heat than in the inferior region, even though the former region is nearer to the Sun and should have more light. Amicus admits that although we cannot explain such phenomena, we ought to accept them, because they are experiences we cannot deny.[159] In the sixteenth century, John Major argued that "other things being equal, where there is more light, there is more heat."[160] Things, however, were not always equal, since Major conceded that there is more direct light in the middle region of air than in the lower region of air nearer to us. But the greater quantity of light in the middle region of air, which is naturally cold, fails to produce any heat there. From the early thirteenth to the late seventeenth century, scholastic natural philosophers were agreed that, directly or indirectly, light caused heat. Because light was not considered naturally hot, few, if any, argued that it could cause heat directly by moving down from luminous celestial bodies, especially the Sun, to the earth. Light had to be altered somehow and somewhere so that it could then cause heat in inferior things. Some – for example, Albertus and Amicus – thought light was heated in the region of fire when its rays divided the fire and caused heat to be produced by friction. In some manner akin to the way iron moves to a magnet, heat was drawn to light rays, which then became hot and could cause heat in terrestrial things.

Before light rays could do this, however, they had to pass through the middle region of air, which, in accordance with Aristotle's statements in the *Meteorology*, was always assumed to be frigid. Buridan could thus counter that just as light incorporates hotness in the region of fire, so it would absorb coldness as it passes through the middle region of air.[161] Obviously, in Buridan's judgment, light could not transmit heat to the earth in this way. Indeed, he could find no plausible explanations to account for the manner in which light transmitted heat from planets to terrestrial objects. Convinced that celestial light did cause heat in terrestrial objects, but unable to determine any causal mechanism by which this was achieved, Buridan concluded that no explanation was required. It was enough for us to know that, by its very nature, celestial light caused heat in terrestrial objects.

When light rays reached the terrestrial region, their capacity for causing

158. "Quarta conclusio non est quaerendus modus quo lumen producit calorem, sed cur aliquando producitur maior, aliquando minor." Ibid., 373, col. 1. In taking up the first part of this conclusion, Amicus uses the analogy about heaviness.

159. Just as Aristotle explained that people continued to accept the actuality of motion despite Zeno's paradoxes that were designed to deny its possibility. For the texts, see Amicus, ibid., 373 (mistakenly paginated 383), col. 2.

160. "Respondetur concedendo ubi est plus de lumine plus est de calore ceteris paribus." Major, *Sentences*, bk. 2, dist. 14, qu. 7, 1519b, 78r, col. 1.

161. "Iterum, sicut incorporaret sibi caliditatem in sphaera ignis, ita incorporaret sibi frigiditatem transeundo per mediam regionem aeris quae semper est frigida, et sic frigefaceret." Buridan, *De caelo*, bk. 2, qu. 15, 1942, 195–196.

heat was usually explained in terms of geometric optics. Heat was produced by incident, reflected, and refracted light rays,[162] which Buridan would probably have explained by the nature of light.

3. Influence

Those terrestrial effects that were inexplicable by celestial motion or light were usually attributed to "influences." Celestial causes described as influences were invisible, as contrasted with the other two celestial causative agents, motion and light. Buridan warned that we ought not to deny such influences simply because they are invisible. After all, we do not perceive the force that is disseminated from a magnet to the iron it attracts, and yet we can see from the effect that it must be of considerable power.[163] Amicus explained that because such an influence is not directly detectable by the senses – we know it only by its effects – as are celestial motion and light, it is said to be "occult" and "is called by the term 'influx' [*influxus*], or 'influence' [*influentia*]."[164]

In the manner of its diffusion, celestial influence was most nearly akin to celestial light and even to heat.[165] But there were significant differences. Not only was light visible and influences invisible, but, more importantly, influences could penetrate solid, opaque bodies where light could not.[166] One of the most striking illustrations of influence, cited by supporters of the influence theory, was the production of metals in the bowels of the earth, where light could not penetrate.[167] Indeed, it was because of the

162. Simon de Tunstede (Pseudo-Scotus), *Meteorology*, bk. 1, qu. 12, 1639, 3:28, col. 2, describes the way that light produces heat: "Whence heat is generated from light from the coming together of opposite rays, which are the incident ray and the reflex, or refracted, ray" (Unde calor generatur a lumine ex concursu radiorum oppositorum, qui sunt radius incidens et reflexus, seu refractas).

163. "Et non debet hoc negari ex eo quod illam influentiam non percipimus sensibiliter quia etiam non percipimus illam quae de magnete multiplicatur per medium usque ad ferrum, quae tamen est magnae virtutis." Buridan, *De caelo*, bk. 4, qu. 2, 1942, 250.

164. "Hic autem influxus caeli perspicue fit per motum et lumen, si vero non apparet sensui, dicitur virtus occulta et communi nomine 'influxus,' seu 'influentia,' dicitur." Amicus, *De caelo*, tract. 6, qu. 7, art. 1, 1626, 391, col. 1 (mistakenly paginated 381). On page 397, column 2, Amicus says that we know influences by their effects.

165. Themon Judaeus described *influentia* as "a certain quality or power diffused through the whole world, just as the species of heat or light is multiplied" (Dico quod est quaedam qualitas sive virtus diffusa per totum mundum sicut multiplicatur species caloris vel luminis). *Meteorology*, bk. 1, qu. 1, 1518, 155v, col. 2.

166. This is the second of three ways in which Themon Judaeus (ibid.) distinguished influence from light: "Secundo differunt quia talis influentia non impeditur saltem totaliter si aliquod corpus interponatur sed transit per corpora opaca et densa per que non potest transire lumen sicut apparet de magnete quia unus movet alium superius in vase natantem si sub vase bene denso teneatur." Amicus, *De caelo*, tract. 6, qu. 7, art. 2, 1626, 393, col. 1, mentions Themon by name and cites the three ways.

167. See Themon Judaeus, ibid., and Amicus, ibid., 394, col. 2. Major, *Sentences*, bk. 2, dist. 14, qu. 7, 1519b, 78r, col. 2, and the Conimbricenses, *De coelo*, bk. 2, ch. 3, qu. 3, art. 2, 1598, 199, also assumed the existence of influences that generated metals in the earth's interior. In his *Questions on the Meteorology* (*Questiones Meteororum*), Nicole Oresme says the same thing. For the Latin text, see Oresme, *De causis mirabilium* [Hansen], 1985, 45–

inability of light to penetrate to the earth's interior that some scholastics inferred the existence of invisible influences. Following alchemical tradition, John Major ([*Sentences*, bk. 2, dist. 14, qu. 7], 1519b, 78r, col. 2) linked the influence of particular planets to the formation of specific metals. Thus influences from Saturn produced tin, those from Jupiter lead, from Mars iron, and so on, in the descending order of the planets, for gold, copper, mercury, and silver. Rather than vague influences, perhaps it would be more appropriate to assume that heat, which is caused by celestial light or motion, reaches the interior of the earth and causes metals and other things. On this approach, celestial influences are superfluous to explain changes deep within the earth. Amicus rejects this argument. Heat produced on or above the earth's surface could not reach deep enough into the earth to cause the required effects. The density, opacity, and frigidity of the earth would prevent this.[168]

Where a terrestrial effect was not plausibly attributable to celestial light or motion, many found it reasonable to assign its cause to some kind of influence radiating from the heavens, always assuming that this influence was greater on sublunar objects nearer the heavens than farther from them.[169] The concept of influence served as a convenient explanatory device for those terrestrial effects that were not adequately explicable by celestial motion or light. A good example was the magnetism by which the poles of a magnet attract iron. Light could not explain this phenomenon, because magnetic attraction occurred even in a dense fog, where light could not penetrate.[170] Amicus also argued that because celestial light is from the Sun and is all of one and the same species, it could not produce contrary effects in the terrestrial region. If one wished to argue that light was not of one species but varied from planet to planet, then, in Amicus's view, this was tantamount to an admission of influences, because these different lights would cause different effects. But to explain the way in which any one of these lights penetrated to the bowels of the earth posed a formidable problem.[171] In the thirteenth century, Richard of Middleton considered the charge that if light is the same in species for all celestial bodies, then the latter could not cause diverse effects in the sublunar region.[172] Richard replies

46, n. 30. But, in his *Contra divinatores horoscopios*, as we shall see, Oresme denied the existence of celestial influences.

168. "Nam hic calor non pervenit ad tantam distantiam praesertim resistente densitate, opacitate, et frigiditate terrae." Amicus, ibid.

169. See for example, Oresme, *Contra divinatores horoscopios* [Caroti], 1977, 228.

170. Amicus, *De caelo*, tract. 6, qu. 7, art. 2, 1626, 394, col. 2.

171. "Et iam patet ex dictis non posse salvari omnia etc. Nam lumen productum a sole in omnibus astris est eiusdem speciei, quare ex illo non possunt oriri tam multi effectus. Illud autem quod est in singulis astris: si eiusdem speciei procedit idem argumentum; si diversae iam virtualiter admittitur influentias et cum maiori difficultate, ut patet ex penetratione usque ad viscera terre." Ibid., art. 4, 397, col. 2.

172. "Item actio corporum superiorum in ista inferiora est per lucem. Sed omnia luminaria communicant in luce. Ergo in istis inferioribus non causant effectus diversos in specie." Richard of Middleton, *Sentences*, bk. 2, dist. 14, art. 3, qu. 5, 1591, 2:188, col. 1.

that celestial bodies do not affect inferior things by light alone but also by motion and by influences that are radiated by the same substantial forms of the celestial bodies that emit light rays. Indeed, influences radiating from the Moon were responsible for the ocean tides.[173] In his interpretation, Amicus gives the impression that influences might radiate in some manner from the planets themselves. Since Amicus, and everyone else, assumed that light radiated from planets, Amicus inferred that influences also radiated from each planet. Long before, Richard of Middleton had explicitly assigned the source of light and influences to the substantial forms of the planets.[174] Themon Judaeus differed sharply. Not only did he differentiate between the terrestrial effects of light and influence, but he also distinguished the sources from which they arose: light came from the planets, but influences were caused by the orbs, that is, the starless or planet-free parts of the heavens.[175] Unfortunately, the specific source of celestial influences was a subject that attracted little attention.

Despite the arguments in favor of celestial influences, there were a few scholastic dissenters who denied their existence and who believed that all terrestrial effects traceable to the celestial region could be accounted for solely by motion and light. Perhaps the most eminent of this group was Nicole Oresme, although in his case the situation is confused. In his questions on Aristotle's *Meteorology*, Oresme accepts the three instrumentalities. He seems to do so again in his *Contra divinatores horoscopios*, where, at one point, he declares unequivocally: "I assume that celestial bodies act on this inferior [region] by light, motion, and influence"[176] and proceeds to discuss instances of celestial influence on the terrestrial region. But later in the same work, Oresme rejects influences, arguing that light and motion were sufficient to account for "saving the action or production in these inferior things and their diversity." Indeed, Oresme emphasizes that the heaven acts "by its light and motion and not by other unknown qualities, which are called influences, etc."[177] In defense of his position, Oresme invokes Av-

173. "Ad quintum dicendum quod corpora superiora non alterant ista inferiora per influentiam luminis tantum, sed etiam per motum et per influentia immissas a suis formis substantialibus cum radiis luminosis per quam influentiam causat luna mirabiles aestus oceani, hoc est fluxum eius et refluxum." Ibid., 189, col. 1. Oresme also said this in his *Questions on the Meteorology* (*Questiones Meteororum*; see Oresme, *De causis mirabilium* [Hansen], 1985, 46, n. 30).
174. For the text, see note 173.
175. For the text, see Themon Judaeus, *Meteorology*, bk. 1, qu. 1, 1518, 155v, col. 2, and Grant, 1987c, 11, n. 42.
176. "Suppono quod corpora celestia agunt hic inferius per lumen, motum, et influentiam." Oresme, *Contra divinatores horoscopios* [Caroti], 1977, 228.
177. Here is the full passage: "Tunc ponuntur conclusiones: Prima: quod ad salvandum actionem seu productionem in istis inferioribus et diversitatem ipsorum non oportet ponere in celo qualitates seu influentiam aliam quam lucem et motum . . . ita quod celum quidquid agit per suam formam seu essentiam per lucem suam et per motum suum agit et non per alias qualitates ignotas, que dicuntur influentie etc." Ibid., 274. Hansen assumed (Oresme, *De causis mirabilium* [Hansen], 1985, 45) that although Oresme initially accepted the threefold division in his *Meteorology*, he later rejected influence in his *Contra*

erroës, the Commentator, who argued that a diversity of proportions between qualities and elements "is the cause of the diversity of inferior bodies composed of the four elements; therefore that diversity is not attributable to the diversity of the superior [celestial] bodies."[178] Terrestrial elements and compounds act upon each other to produce the diversity we observe. Apart from the effects of celestial motions and celestial light, no separate, superfluous influences need be invoked. Oresme concludes that "no other cause must be sought than that which produces hotness or coldness, etc.; but that is light and motion, etc."[179] In at least one place in one work, Oresme clearly rejected the existence and operation of celestial influences and did so because of a conviction that all terrestrial effects could be saved by celestial motions and light alone.[180]

Other scholastic authors also rejected the role of celestial influences. In his *Liber de orbibus*, Alessandro Achillini thought that heat, rather than influences, could produce metals in the bowels of the earth.[181] If we are to believe the Coimbra Jesuits ([*De coelo*, bk. 2, ch. 3, qu. 3, art. 1], 1598, 196–197), those who rejected celestial influences did so because they were reluctant to attribute a significant causative role to a force that was imperceptible to any of our senses. From a positive standpoint, they believed (197) that light, which could generate heat, was, along with the effects of motion, sufficient to explain all terrestrial effects that could be attributed to the heavens.[182]

Despite some defections, most scholastic authors accepted a role for celestial influences. Why were celestial influences widely accepted as a third causative agent in the heavens? Probably because they seemed the most plausible and intelligible way of extending celestial action into the bowels of the earth. It was a far greater strain on credulity to attribute that function to light, or to the heat that light might produce, than to assume that an invisible celestial influence could penetrate opaque bodies.

divinatores horoscopios. But Hansen seems to have missed Oresme's earlier acceptance of influence in the same *Contra divinatores*.

178. "Secundo, quia secundum Commentatorem in secundo *De generatione* et super *De celo* diversitas proportionum qualitatum et elementorum est causa diversitatis istorum inferiorum compositorum ex quattuor elementis, igitur non propter diversitatem superiorum." Oresme, *Contra divinatores horoscopios* [Caroti], 1977, 277.

179. "Ergo non est querenda alia causa quam illa que facit ad calorem vel frigus etc., sed illa est lux et motus etc., ergo etc." Ibid.

180. No definitive judgment can be rendered until the order of Oresme's relevant works is determined and until we can properly reconcile the two conflicting viewpoints in his *Contra divinatores*. Oresme's attitude in this work is, however, consistent with the position he adopted in his final work, *Le Livre du ciel et du monde*, where, by means of his "Joshua argument" (described in Sec. V.1.c of this chapter), he sought to make the terrestrial region much less dependent on the celestial.

181. "Ad secundum: pars terrae exterior passa a celestibus partem illi continuam alterat in visceribus terrae latentem aptam ut in metalla convertatur per antiperistasim, quoque calor in visceribus terrae fortificatur, etc." Achillini, *Liber de orbibus*, 1545, 58r, col. 2.

182. Among those who rejected celestial influences in favor of motion and light alone, the Coimbra Jesuits, *De coelo*, bk. 2, ch. 3, qu. 3, art. 1, 1598, 196, mention Averroës, Avicenna, Avenazra (Abraham ben Ezra?), Pico della Mirandola, and George Agricola.

Perhaps as important as any reason is the utility of the concept of invisible influences. Occult phenomena could be attributed to such radiations. Magnetic attraction was a typical example. The lodestone was an earthy body, but it had unusual attractive properties that could not be explained in ordinary Aristotelian terms.[183] Invoking a celestial influence seemed a reasonable way of resolving that problem and similar ones. Moreover, invisible forces and powers were so common in medieval thought that adding yet another, in the form of a celestial influence, caused little intellectual discomfort.

VI. Universal nature and the preservation of the world

If the purpose of celestial motions, light, and influence was to cause, directly or indirectly, the great variety of changes that occur regularly in the terrestrial region, there was yet another kind of action, indeed an influence, that was occasionally invoked to explain how the world as a whole was preserved and its order maintained.

In the ordinary course of events, a body behaves in accordance with its "particular nature" (*natura particularis* or *specialis*).[184] The latter embraced the basic properties that made a body behave in its characteristic manner. Thus a heavy body moves downward to its natural place, while a light body moves upward to its natural place. But matter is also influenced by a "universal cause" (*causa universalis*) or "universal nature" (*natura universalis*), which originates in the celestial region, from whence it is diffused throughout the sublunar realm. Also described as a "celestial" or "supercelestial agent" (*agens celeste* or *superceleste*), or a "universal agent" (*agens universale*), or even a "celestial force" (*virtus caelestis*),[185] this powerful influence was contrasted with the "particular nature."

In the material plenum of the Aristotelian cosmos, the universal nature, which was not Aristotle's concept, was conceived as the vigilant guardian of material continuity, a universal regulative power.[186] Its most essential characteristic was an inexorable tendency to preserve and maintain the continuity of the universal plenum.[187] Or, to put it in other terms, its major objective was to prevent formation of vacua within the world. Whenever formation of a vacuum was imminent and danger of separation threatened

183. To my knowledge, Aristotle does not explain the attractive power of magnets.
184. I rely here on Grant, 1981a, 69.
185. Walter Burley used the first three expressions in his *Questiones circa libros Physicorum*, bk. 4, qu. 6 ("Utrum vacuum possit esse"), 65v, col. 2–66r, col. 1. For the Latin text, see also Grant, 1981a, 314, n. 107.
186. Among the descriptions that Bacon and Burley offered for the universal nature, we also find *virtus regitiva universi*. See Bacon, *Liber primus, Communia naturalium*, in *Opera*, fasc. 3, 1911, 222, lines 25–26 and Burley, ibid., 65v, col. 2. Centuries later, Amicus [*Physics*, tract. 21 ("De vacuo")], 1626–1629, 2:745, col. 2, used the expression *virtus regitivae totius universi*.
187. Bacon, ibid., 220, lines 11–15.

cosmic continuity, the omnipresent universal nature would cause bodies to act contrary to their particular natural tendencies: heavy bodies would rise and light bodies fall. To preserve the continuity of the universe, the heavens would even transform their natural circular motion to rectilinear motion. If fire were suddenly to descend from its natural place below the concave surface of the lunar sphere, John Dumbleton believed that the heavens would immediately move down in a straight line to prevent formation of the dreaded vacuum.[188] Whether the universal nature was distinct from bodies and acted independently from them when necessary or whether it was an inherent property of matter is left unclear.[189] Also uncertain, though likely, is the derivation of the universal cause from the pseudo-Aristotelian *Book on Causes* (*Liber de causis*), which had been translated from Arabic to Latin in the twelfth century.[190]

What is clear, however, is that by the end of the fourteenth century the concept of a universal nature was widely accepted and most frequently employed to explain nature's abhorrence of a vacuum.[191] Indeed, it was sometimes encapsulated, without further elaboration, in the phrase "Nature abhors a vacuum." The popularity of this important concept continued on into the fifteenth, sixteenth, and seventeenth centuries, when Achillini, Suarez, Aversa, Amicus, and Compton-Carleton, among others, appealed to it.[192] In a brief but thorough account, Suarez ([*Disputationes metaphysicae*, disp. 18, sec. 2], 1866, 1:614, col. 1, par. 40) explained that God intervenes with the universal nature to prevent the destruction of the order of the universe, to prevent matter from remaining without form, and to make certain that matter is not reduced to nothing. To prevent the latter situation, water, for example, is made to move upward to fill any potential vacuum. But Suarez was uncertain whether God performed these acts by his will alone or delegated the task to an intelligence.

Whereas Suarez described the universal cause in terms of what it was intended to prevent, Amicus put it more positively in explaining why it

188. The passage is from Dumbleton's *Summa logicae et philosophicae naturalis*, pt. 6, ch. 3, and is described in Duhem, *Le Système*, 1913–1959, 8:162–163. For more on this, see Grant, 1981a, 304, n. 19.

189. For further discussion, see Grant, 1973, 330–331, n. 6 (where I argue that Bacon and Burley seem to have conceived it as something external), and 334, n. 13 (where Pseudo-Aegidius Romanus seems to assume that God confers on things the ability to be joined and connected when necessity demands it).

190. For the evidence, see Grant, 1973, 330–331, n. 6.

191. During the thirteenth century, Roger Bacon, Pseudo-Grosseteste, Johannes de Quidort, and Thomas Aquinas used the concept. Litt, 1963, 154–166, cites fifteen occasions where Thomas opposes the notion of a universal cause to that of a particular cause and five instances where Thomas used the expression *natura universalis*. Among those who accepted it in the fourteenth century were Aegidius Romanus, John of Jandun, Franciscus Mayronnis, Graziadei d'Ascoli, Johannes Canonicus, and Henry of Hesse. For references, see Duhem, *Le Système*, 1913–1959, 8:152–158, 160–161, and 166–168. To Duhem's list, we may add Walter Burley.

192. See Achillini, *Liber de orbibus*, 1545, 58v, col. 1 (and 59r); Amicus, *De caelo*, tract. 6, qu. 8, dubit. 2, art. 2, 1626, 407, col. 1; Aversa, *De caelo*, qu. 35, sec. 9, 1627, 201, col. 1.

was called a "universal cause." In the first place, it was not so called because of any influence it exercised on the actions of other particular causes but rather "because its action is extended to more effects than that of a particular cause; secondly, [because] it is ordained for the universal good; and thirdly, because, in some way, it helps other agents in their actions, since many effects require the cooperation of the heavens, as is obvious from the generation of plants and similar things."[193] Others, however, like Thomas Compton-Carleton, viewed the "universal cause" as something that influenced every sublunar effect but seemingly ignored its role as an instrument for the maintenance of universal order.[194]

From the thirteenth and into the seventeenth century, the concept of a universal nature or cause, or whatever name was employed, served a wide range of purposes that seemed to vary from author to author. Most, however, emphasized its role as preserver of the cosmos, a function it performed by preventing formation of vacua within the material continuum of our world. In the broadest sense, however, the universal cause had a rather strange purpose. It was God's way of preserving the world against contingencies that only he could cause.

193. "Dices caelum est causa universalis. Respondetur dici causam universalem, non quia influat in actiones aliorum agentium particularium, sed quia eius actio extenditur ad plures effectus quam causa particularis; secundo, ad universale bonum ordinatur; tertio [Amicus has "3"] quia iuuat aliquo modo alia agentia in actionibus suis, cum multi effectus exigant concursum celorum, ut patet ex generatione plantarum et similium." Amicus, ibid., col. 2.

194. See note 67 of this chapter for the relevant text.

20

The earth and its cosmic relations: size, centrality, shape, and immobility

In his questions on *De caelo*, Albert of Saxony did something quite unusual, if not extraordinary. He presented an overall plan for the organization of his questions by dividing his treatise into three parts, which were meant to embrace the four books of Aristotle's *De caelo*. In the first book, Albert devotes questions to the whole world, while in the second he treats the nobler part of that world, namely the heavens, to which he adds the earth, "insofar as it is the center of the heavens and of the whole world." And in the third, or final, book, Albert considers the parts of the world that are less noble than the heavens, namely the heavy and light bodies that make up the four elements.[1] These three themes – the world as a whole, and its two fundamental parts, the nobler celestial region and the less noble elemental region – form the basis of Albert's questions on *De caelo*.

To this point, I have followed Albert's outline, which he believed, rightly perhaps, was what Aristotle himself had in mind. We have thus far considered the world as a whole (Chs. 4–9) and then treated its nobler part, namely the heavens (Chs. 10–19). But we have not yet completed Albert's vision of the celestial region, since he also added to it the earth, "insofar as it is the center of the heavens and of the whole world." In this category, Albert included four questions, the final four of the second book (bk. 2, qus. 23–26), wherein he asks: first, whether the earth is in the center of the world; second, whether it always rests in the middle of the world; third, whether it is spherical; and fourth, whether the whole of it is habitable. Although the earth was not considered a planet, its location near or at the center made it a vital part of the cosmic network. Of Albert's four themes (centrality, immobility, sphericity, and habitability), habitability will be treated only in passing, largely replaced by the size of the earth, a subject

1. "Postquam tractavi circa primum huius quasdam dubitationes concernentes totalem mundum. Et similiter tractavi circa secundum huius dubitationes concernentes partem nobiliorem mundi, scilicet celum et cum hoc terram prout est centrum celi et totalis mundi. Nunc volo tractare dubia concernentia gravia et levia, sicut sunt elementa que sunt partes mundi totalis minus nobiles quam sit celum." Albert of Saxony [*De celo*, bk. 3, qu. 1], 1518, 119v (incorrectly given as 117v), col. 2.

that Albert ignored (this shift of emphasis is reflected in the title of this chapter). The primary aim of this chapter, then, is to concentrate on the sorts of questions about the earth that had cosmic relevance, rather than to treat of the earth's topography, geography, flora, and fauna, or to describe the generation and behavior of the four elements and the compounds (minerals and metals) formed from them.[2]

Before proceeding to these questions about the earth, I shall provide a brief sublunar context for the earth by describing the broad features of the three elements that were above it: fire, air, and water. I can do no better than quote from Pierre d'Ailly's *Image of the World* (*Ymago mundi*), one of the most popular treatises of the late Middle Ages, a work that was printed early and exercised great influence in the sixteenth and seventeenth centuries.[3] "After treating to some extent of the heaven," d'Ailly explains, "we must now consider the things that lie under it":

Immediately after [or below] the sphere of the moon, the philosophers place the sphere of fire, which is the most pure there and invisible because of its rarity. Just as water is clearer than earth and air than water, so this fire is rarer and clearer than air, and so is the heaven [or sky] rarer or clearer than fire, except for the stars, which are thicker [or denser] parts of the sky so that the stars are lucid and visible. Afterwards is the sphere of air, which encloses water and earth. This is divided into three regions, one of which is the outermost (next to fire) where there is no wind, rain, or thunder, nor any phenomenon of this kind, and where certain mountains, such as Olympus, are said to reach. Aristotle says that starry comets appear and are made there and that the sphere of fire and this supreme region of air with its comets are moved simultaneously with the heaven [or sky] from east to west. The middle region [of air], however, is where the clouds are and where various phenomena occur, since it is always cold. The other [and third] region is the lowest, where the birds and beasts dwell. Then follow water *and* earth, for water does not surround the whole earth, but it leaves a part of it uncovered for the habitation of animals. Since one part of the earth is less heavy and weighty than another, it is, therefore, higher and more elevated from the center of the world. The remainder [of the earth], except for islands, is wholly covered by waters according to the common opinion of philosophers. Therefore, the earth, as the heaviest element, is in the center or middle of the world, so that the earth's center is the center of its gravity; or, according to some, the center of gravity of the earth and also of water is the center of the world. And although there are mountains and valleys on the earth, for which reason it is not perfectly round, it approximates very nearly to roundness. Thus it is that

2. Thus I shall not treat the third and final part of Albert's threefold division on the light and heavy elements of the sublunar realm, which formed the substance of books 3 and 4 of Aristotle's *De caelo*. Although some might wish to incorporate the sublunar realm into medieval cosmology, it will not be included here.
3. In the *Ymago*, d'Ailly relied heavily on Nicole Oresme's *Traitié de l'espere* [McCarthy], 1943.

> an eclipse of the moon, which is caused by the shadow of the earth, appears round. They say the earth is round, therefore, because it approximates to roundness.[4]

As we shall see, d'Ailly's remarks about the earth were commonplace.

I. The size of the earth

The earth's size was usually considered from two different vantage points, one absolute, the other relative. The former concerned the absolute size of the earth in terms of its circumference, measured in whatever units were in vogue; the latter was a measure of the earth's size in relation to the whole of the spherical universe. Despite the seeming importance of these two types of measure, scholastics seem not to have devoted any specific question to the earth's size.

Drawing on Alfraganus, John of Sacrobosco characterized the size of the earth as a mere point in comparison to the firmament. For if the earth were of any significant size as compared to the sphere of the fixed stars, we would be unable to see half of the heavens.[5] So small is the earth's magnitude in comparison to the size of the firmament that "an eye at the earth's center would see half the sky, and one on the earth's surface would see the same half" (Sacrobosco, *Sphere*, ch. 1, 1949, 122). According to Alfraganus, even a star is a mere point in comparison to the firmament. The earth, which is smaller than any star, must bear an even smaller ratio to that same firmament. Scholastics had no reason to disagree with any of these claims.

From such beliefs, it appears that medieval natural philosophers and astronomers conceived a universe that was quite large.[6] We should lay to rest the oft-repeated, but misleading, judgment that the medieval mind took comfort in a small, intimate universe, the coziness of which was shattered only in the seventeenth century with the gradual acceptance of its infinite extent. If any sense of coziness existed about the cosmos, it derived not from its size but from its assumed intelligibility.

Despite the conception of the earth as a point in relation to the universe, the earth had a measurable size, and at least three estimates were commonly known. In *De caelo* (2.14.298a.15–16), Aristotle reports that mathematicians estimate the earth's circumference at 400,000 stades. Although the exact value of Aristotle's stade is unknown, scholars estimate that his value of the earth's circumference was approximately twice that of the modern value

4. I quote my translation from Grant, 1974, 633. Aristotle's statement is from *Meteorology* 1.7.344a.9–23. For a longer and more systematic description of the three regions, see Johannes Velcurio [*Physics*], 1554, 79r–89r; on 82v–88r, Velcurio treats fire, air, water, and earth, in that order.
5. Sacrobosco [*Sphere*, ch. 1], 1949, 122. For the source of some of these ideas, see Alfraganus [Campani], 1910, 69, and Alfraganus [Carmody], 1943, 7.
6. For the rest of this section on the earth's size, I rely on Grant, 1971, 62.

of 24,902 English miles.[7] Another estimate, derived ultimately from Eratosthenes, a Greek of the third century A.D., was put at 252,000 stades, a good approximation to the true value.[8] Eratosthenes' value was popularized in the Middle Ages in Sacrobosco's *Sphere* and in d'Ailly's *Ymago mundi*. Because d'Ailly derived most of his report from Sacrobosco, their accounts are quite similar.

Anyone traveling south or north along a meridian until the polestar shifts its elevation by 1 degree would have traversed a distance of 700 stades, which, when multiplied by 360 degrees, yielded a figure of 252,000 stades. D'Ailly, but not Sacrobosco, equates 252,000 stades with 15,750 leagues, where a league equals 2 miles.[9] Using d'Ailly's figure, the earth's circumference would be 31,500 miles, a considerable improvement over Aristotle's value but still wide of the mark. The method described by Sacrobosco and d'Ailly is not the one used by Eratosthenes, who derived his figure geometrically, but is one devised by Arabs in the ninth century. According to d'Ailly,[10] the Arabs reckoned a degree along the meridian not in stades but in miles, 56 2/3 miles, to be precise. Based on this value, d'Ailly explains that Alfraganus and other Arabs determined a value of 20,400 miles for the earth's circumference (that is 360 × 56 2/3), a third estimate that was considerably less than the true figure.

In the same chapter in which he conveyed the estimate of 20,400 miles, d'Ailly explained that "According to Aristotle and Averroës at the end of the second book of *De caelo et mundo*, the end of the habitable earth toward the east and the end of the habitable earth toward the west are very close and there is a small sea between" (Grant, 1974, 639). D'Ailly's report of 20,400 miles as the Arabian value for the earth's circumference and his statement shortly afterward in the same chapter about a small intervening sea played a role in making Columbus conceive of a voyage of discovery to America as feasible. Convinced that he could reach India by sailing westward from Spain, Columbus required evidence to convince Spanish authorities of the viability of so bold and costly a venture. He found this in d'Ailly's *Ymago mundi*, which provided him with the support of Aristotle and Averroës, who were favorably disposed toward the idea that Spain and India were connected by the same ocean. In his personal, annotated copy of d'Ailly's book, which is preserved in Seville, Columbus seized every opportunity to emphasize that a degree is only 56 2/3 Roman miles and that, consequently, the earth's circumference is 20,400 miles, the smallest

7. Cohen and Drabkin, 1958, 150, n. 2, give 600 feet as the value of a stade. Because the length of a foot varied, so did that of the stade.
8. The method used by Eratosthenes and Posidonius was reported by Cleomedes in the latter's *On the Orbits of the Heavenly Bodies*. For Cleomedes' description, see Cohen and Drabkin, 1958, 149–153.
9. For Sacrobosco's description, see Sacrobosco, *Sphere*, ch.1, 1949, 122–123, and for d'Ailly's see Grant, 1974, 633–634.
10. D'Ailly conveys this in chapter 10 of the *Ymago mundi* (for the English translation, see Grant, 1974, 638).

of the available estimates. Since 56 2/3 yielded the smallest value for the earth's circumference and lent credibility to his claim, Columbus expressly declared that this number was an exact measurement of a degree and, in a few instances, drew a box around this all-important number in the text. A smaller earth made it likely that less ocean intervened between Spain and India. To buttress his case, however, Columbus exaggerated Aristotle's support by attributing to him not only the opinion that a single ocean joined Spain and India but also the belief that it was navigable in a few days because of its smallness. Far from marking a departure from ancient and medieval opinion, Columbus relied heavily on traditional views to support his daring proposal.

II. The earth's centrality

1. The three centers

When discussing the overall organization of the world, scholastic natural philosophers customarily located the earth at the center of the cosmos. With greater sophistication, however, they soon realized that what lay at the center of the world and what was meant by "center" were unclear and, at best, ambiguous. To clarify matters, three different kinds of centers were distinguished. Nicole Oresme presented them succinctly in a question on "Whether the earth is naturally at rest in the center of the universe" (Oresme [*De spera*, qu. 3], 1966a, 65): the first is the center of the whole world, "namely a point equidistant from all parts of the convex surface of the whole world," or the geometric center; the second kind is the earth's "center of magnitude," where the quantity of earth on one side of the center (presumably of a line drawn through the center) is equal to that on the other side (of the line);[11] finally, a third way is "center of gravity," where the heaviness on each side of the center (again, presumably of a line drawn through the center) is equal. These three definitions, sometimes in varying forms, were widely accepted even into the seventeenth century.[12]

The question of the earth's centrality became one of the doubts that Jean

11. Oresme, *De spera*, qu. 3, 1966a, 65–67, takes "center of magnitude" in two ways. Taken properly, it concerns only spherical or circular magnitudes where all the lines drawn from the center to the circumference are equal. In its improper signification, center of magnitude is the middle of anything, so that any magnitude will have a center. When John Major defined "center of magnitude," he did so in the "proper" way, confining himself to the center of magnitude of a sphere: "Centrum magnitudinis corporis est punctum a quo omnes linee recte ducte ad circumferentiam sunt equales." Major [*Sentences*, bk. 2, dist. 14, qu. 12], 1519b, 84v, col. 1.

12. Definitions similar to Oresme's were furnished by the Conimbricenses [*De coelo*, bk. 2, ch. 14, qu. 3, art. 1], 1598, 381; Mastrius and Bellutus [*De coelo*, disp. 4, qu. 3, art. 2], 1727, 3:560, col. 2, par. 105; and Aversa [*De caelo*, qu. 36, sec. 5], 1627, 225, col. 1. Christopher Clavius and Melchior Cornaeus gave different definitions for center of gravity. For references and further discussion, see Grant, 1984b, 21 and n. 65. In what follows, I shall draw heavily on the latter monograph, and shall frequently refer the reader to texts and notes in it.

Buridan felt he had to resolve in coping with the question "Whether the earth always is at rest in the center of the universe."[13] That the earth should be situated in the center of the world was evident on the basis of the Aristotelian principle of contrariety, in this case the pair of contraries up and down. If absolute up is taken as the concave surface of the lunar orb, then absolute down, which is the contrary of absolute up, must be the maximum distance from up. Absolute down must then be equivalent to the center of the world. But what is absolutely heavy, namely the earth, ought to be located in the place that is absolutely down, namely the center of the world. Hence the earth should be located at the geometric center of the world.[14]

Now if the earth were a perfectly homogeneous body, its center would coincide with the center of the world. Indeed, if the earth were homogeneous, its center of magnitude (*medium magnitudinis*) would coincide with its center of gravity (*medium gravitatis*), and both would coincide with the center of the world. But the earth is hardly a homogeneous body. One region of it is covered with water and uninhabited, whereas the other part is where plants and animals live and is relatively free of water. Thus the Sun and the air make the uncovered part hot and more rare and tenuous, while the other part, not so affected, remains denser and more compact. "Now if one body in one part is lighter and in an opposite part heavier, the center of gravity will not be the center of magnitude."[15] Herein lies a problem: Which center is the center of the world?[16]

Without hesitation, Buridan chooses the center of gravity. Heavy bodies are always moving toward the center of the world, with heavier parts displacing less heavy parts. But if the center of gravity is the real center of the world, it follows that the earth is not in the center with respect to its magnitude. Because of the uneven heating effects of the Sun and the uneven distribution of the earth's matter and surface waters, the center of the earth's magnitude could never coincide with the true geometric center of the universe. Indeed, "it follows that the earth is nearer to the heaven in the part

13. "Quaeritur consequenter: Utrum terra semper quiescat in medio mundi." Buridan [*De caelo*, bk. 2, qu. 22], 1942, 226. The translation is Clagett's in Clagett, 1959, 594; reprinted in Grant, 1974, 500–503, with additional commentary. As a glance at question 389 in the "Catalog of Questions" in Appendix I of this volume will reveal, this was one of the most popular questions in medieval cosmology.
14. Buridan, ibid., 230–231; for the translation, see Clagett, 1959, 596–597; Grant, 1974, 502.
15. Buridan, ibid., 231; Clagett, ibid., 597, par. 11; Grant, ibid., par. 11.
16. Aristotle distinguished a center of the world and a center of magnitude but made no mention of a center of gravity. In *De caelo* 2.14.296b15–18, Aristotle explained ([Guthrie], 1960) that "the earth and the Universe have the same centre, for the heavy bodies do move also towards the centre of the earth, yet only incidentally, because it has its centre at the centre of the Universe." Although Aristotle was aware of the irregular shape of the earth's surface, he treated the earth as if it were a homogeneous sphere. However, he imagined how it might be otherwise when, later in the same chapter (297a.31–33), he declared that "if, the earth being at the centre and spherical in shape, a weight many times its own were added to one hemisphere, the centre of the Universe would no longer coincide with that of the earth."

not covered with waters than in the covered part."[17] From the standpoint of its magnitude, then, the earth is not directly in the middle of the world. Nevertheless, "we commonly say, however, that it is in the center of the universe, because its center of gravity is the center of the universe."[18] In order to save the earth's cosmic centrality – and all were agreed that it did lie at the center of the world – it was essential that the earth's center of gravity be taken as equivalent to the center of the world.[19]

2. *Does the earth move with small rectilinear motions?*

But what a peculiar center was the earth's center of gravity. For, as Buridan reveals in the very next paragraph, the earth's center of gravity is incessantly changing. Geologic processes guarantee this. Earthly debris brought down from the mountains flows into the sea and settles to the bottom. From this incessant and inexorable process, the uncovered part of the earth diminishes, while the part covered by waters is increased. Consequently, the center of gravity is continually shifting. With each shift, however, the latest center of gravity seeks to coincide with the geometric center of the world. The center of the earth is thus variable. As a consequence, the earth is mobile, because with each shift of its center of gravity, the whole earth undergoes a small rectilinear motion as the new center of gravity seeks to coincide with the geometric center of the universe. In order that this process continue indefinitely and that earthy debris always be available for transmission via the rivers to the ocean, Buridan assumed that "mountains were consumed and destroyed sometimes, nay infinite times, if time were eternal."[20]

Buridan's conception of an earth undergoing incessant small rectilinear motions found some significant support from Albert of Saxony and John Major, who, as a result of the publication of their questions on *De caelo*, were instrumental in transmitting medieval ideas to early modern scholastics in the sixteenth and seventeenth centuries.[21] With somewhat less conviction, Nicole Oresme and Themon Judaeus also conceded the possibility that the earth could be moved locally,[22] while Pierre d'Ailly rejected the idea of

17. Clagett, 1959, 597, par. 12; Grant, 1974, 502, par. 12.
18. Clagett, ibid.; Grant, ibid. ("Sic igitur terra secundum suam magnitudinem non est directe in medio mundi; tamen dicimus communiter quod est in medio mundi, quia medium gravitatis eius est medium mundi.") Buridan, *De caelo*, bk. 2, qu. 22, 1942, 231.
19. For the manner in which the earth's centrality was reconciled with the assumed eccentricity of the earth in Ptolemaic astronomy, see Chapter 13, Section III.4.
20. Clagett, 1959, 598, par. 13; Grant, 1974, 502–503, par. 13. For the means by which this process was repeated indefinitely, see this chapter, Section IV, on the terraqueous globe.
21. Albert of Saxony, *De celo*, bk. 2, qu. 24, 1518, 117v, col. 1. In a fourth conclusion, Albert says "bene verisimile est quod semper quelibet pars terre totalis moveatur motu recto." He then describes how the earth's center of gravity continually shifts and seeks to coincide with the center of the world. For Major, see Major [*De celo*, bk. 2, qu. 9], 1526, 23, col. 2.
22. In the ninth of twelve conclusions included in his questions on whether the earth is naturally at rest in the center of the universe, Oresme says (*De spera*, qu. 3, 1966a, 73) that "it is possible for the whole earth to be moved locally, because, if a sufficient amount

small rectilinear motions.[23] D'Ailly repudiated Buridan's underlying thesis that because every part of the earth moves with a rectilinear motion, therefore the whole earth must do so.[24] By the sixteenth and seventeenth centuries, the suggestion that the earth might suffer small rectilinear motions was rarely mentioned.

Nevertheless, Buridan's idea of small rectilinear motions exercised some influence, even in the seventeenth century. Thus Raphael Aversa argued ([*De caelo*, qu. 36, sec. 6], 1627, 232, col. 2–234, col. 2) that the earth suffered continuous alterations that caused it to increase or decrease in weight. As a result, the earth's center of gravity, which is the earth's center, continually shifts, thereby causing an incessant sequence of rectilinear motions. Although such movements occur continually, Aversa considered them so minimal as to be imperceptible. Indeed, this "tenuous motion of the earth, which escapes our senses . . . , must be taken as if it did not exist."[25] Thus did Aversa have his motion and deny it at the same time, leaving his anti-Copernican credentials intact. Most scholastic authors, especially those who wrote after the condemnation of Copernicus's *De revolutionibus* in 1616, were reluctant to attribute any genuine motion to the earth.[26]

In 1622, Paul Guldin (1577–1643) published a dissertation in which he also assumed that the earth's center of gravity continually sought to coincide with the center of the universe, resulting in small but incessant rectilinear

of gravity were added to one part so that it would be more powerful than the resistance of the air opposing [the motion], then the earth would be moved." Among a number of possible ways for the earth to move, Themon Judaeus suggests that it could be moved by rising or falling – presumably rectilinearly – because of a greater heaviness on one side than on the other (quod tota terra moveatur ascendendo vel descendendo propter gravitatem maiorem ex una parte quam alia). Themon Judaeus [*Meteorology*, bk. 2, qu. 7], 1518, 174v, col. 1. Themon, however, does not think this possibility is really germane to the question (De isto nihil ad propositum quamvis sit possibile). Ibid.

23. D'Ailly, *14 Questions*, qu. 3, 1531, 150r.
24. In his counterargument, d'Ailly uses an example of a pillar, or column, composed of ten stones to show that although each of the ten stones might move rectilinearly and change positions, the pillar as a whole would remain in the same position and have the same center of gravity. This is achieved by taking the first stone at the top of the column and placing it below the lowest stone; then the second stone is taken and also placed below the lowest, or bottommost, stone; and then the third stone is taken and placed below the lowest stone of the column, and so on indefinitely. Although each stone will be in motion, the column as a whole will remain at rest in the same place, so that its center of gravity should also remain the same (Verbi gratia: In compositionibus fiat unum pilare compositum ex 10 lapidibus et capiatur lapis superior et ponatur sub inferiori pellendo inferiorem. Iterum capiatur secundus et ponatur sub, et sic continuando semper tunc in illo casu certum est quod quaelibet pars pilaris movetur et ascendit continue et tamen totum pilare in se quiescit). Ibid. Along with Albert of Saxony and Major, d'Ailly was another transmitter of significant medieval cosmological ideas to early modern scholastics.
25. Aversa, *De caelo*, qu. 36, sec. 6, 1627, 234, col. 2. The ellipsis replaces approximately fifteen lines of text.
26. Cornaeus seems to have allowed that minute motions – to the extent of a flea's leap – of the earth might occur to bring the earth's center of gravity into coincidence with the center of the world. But these motions were imperceptible and so minute that they caused no real physical changes. See Cornaeus [*De coelo*, tract. 4, disp. 3, qu. 2, dub. 12], 1657, 529–530. Cornaeus's position was essentially the same as Aversa's, and both were probably affected by the climate of opinion following the aftermath of the condemnation of 1616.

motions of the earth.[27] Even so slight an impact as the descent of a bird onto the earth's surface was sufficient to cause a change in the earth's center of gravity. With the republication of his treatise in 1635, Guldin's idea that even the flight of a bird could move the earth ran afoul of the climate of opinion created by the condemnation of Galileo in 1633. Guldin was compelled to say that the weight of a bird does not cause the earth to move physically but only in a mathematical sense.[28]

But if few accepted small rectilinear motions for the earth, most accepted, on almost empirical grounds, the idea that the earth's surface was in a continual process of change. How did these changes on the earth's surface affect scholastic thoughts about the shape of the earth? Was it round, as Aristotle had argued? Or was it some other shape, or perhaps without any definite shape at all? If number of discussants is a measure of popularity, then the question "Whether the whole earth is spherical" was clearly popular and very likely important (see qu. 383 in the "Catalog of Questions," Appendix I).

III. The shape of the earth

As we saw in Chapter 6, acceptance of a spherically shaped world was virtually unanimous. If pressed, Aristotelian natural philosophers would have described the world as a perfect sphere, by which they would have meant geometric sphericity. Just as scholastics asked about the shape of the world, so also did they ask about the shape of the earth. And again following Aristotle and others, they concluded that the earth was like a sphere but, unlike the heaven, was not a perfect sphere. In truth, it was not a sphere at all but was only "like a sphere," which made it "round."[29] Pierre d'Ailly defined both a bit more formally. "Properly speaking," he explains, "that is called spherical which is without any uneveness on its exterior surface and from the middle point of which all straight lines projected to the surface are equal."[30] Roundness, however, "is what tends toward sphericity, although it is not perfectly spherical; [as] for example, an apple is called round, but not spherical."[31] But d'Ailly adds that the two terms – roundness and

27. See H. L. L. Busard, "Guldin, Paul," in *Dictionary of Scientific Biography*, 5:588. The dissertation of 1622 was republished in 1635 in the first of four volumes with the title: *Centrobaryca seu de centro gravitatis trium specierum quantitatis continuae* (Vienna, 1635–1641).
28. See Schofield, 1981, 282–283.
29. Albert of Saxony, *De celo*, bk. 2, qu. 25, 1518, 118r, col. 1, offers the external surface of the world as an example of a perfect exterior sphericity without any unevennesses; as an example of an imperfectly round body, Albert cites an apple, which was commonly mentioned in this context.
30. "Pro primo puncto est advertendum quod sphaericum loquendo proprie dicitur quod in eius superficie exteriore est sine quacunque asperitate et a cuius puncto medio omnes lineae rectae ad illam superficiem ductae sunt sibi invicem aequales." D'Ailly, *14 Questions*, qu. 5, 1531, 153v.
31. Themon Judaeus also uses the apple as illustrative of roundness. See Themon Judaeus [*Tractatus de spera*, qu. 12] in Hugonnard-Roche, 1973, 121.

spherical – are often used interchangeably, and that he himself will use them in the same way.[32] This dichotomy was reflected in the formulation of questions about the earth's shape. Although most natural philosophers used the term spherical, as in the question "Whether the earth is spherical," some used the term round, perhaps because it more accurately depicted the physical earth.[33] Nevertheless, the two terms were used interchangeably.

That the earth was not perfectly spherical was patent to all. Its surface was covered with mountains and valleys. But those mountains and valleys were relatively small quantities in relation to the whole earth and did not alter the earth's overall spherical appearance.[34] The same high mountains could be used as evidence of the earth's rotundity. For if the earth was flat, high mountains ought to be seen from distances that are much farther away than the distances from which we actually see them. That they are not seen in this manner derives from the bulge of the earth that occurs between those mountains and the place from whence we view them.[35] At least four arguments were derived from Aristotle directly and were frequently repeated in the question on the earth's sphericity or roundness.[36] As a heavy body, every piece of earth falls naturally toward the center of the world. But although the parts of earth oppose each other as they approach the center, they sink toward the center little by little and, in the process, form themselves naturally into a spherical shape.[37] Again, wherever the earth is devoid of mountains or valleys, that is, where it is reasonably level, elevated heavy bodies fall toward the center of the world in nonparallel paths that form equal angles at the earth's surface.[38] This would not happen if the earth were plane rather than round. In another argument, Buridan, Themon

32. "Verumtamen rotundum et sphericum quandoque capiuntur large unum pro alio et ita capiuntur in proposito." D'Ailly, *14 Questions*, qu. 5, 1531, 153v.
33. Although most scholastics used the first mode of expression, those who used some version of the question "Utrum [or "an"] terra sit rotunda," are Michael Scot [*Sphere*, lec. 6], 1949, 294–295; Cecco d'Ascoli [*Sphere*], 1949, 366; and Cornaeus, *De coelo*, tract. 4, disp. 3, qu. 2, dub. 2, 1657, 516.
34. Johannes Versor declares that "even mountains and cavities of the earth are not notable quantities in relation to the whole earth. Thus they do not prevent us from calling the earth spherical in relation to the heaven" (Etiam montes et concavitates terre non sunt notabilis quantitatis in ordine ad totam terram. Ideo non impediunt quin terra posset dici sperica in ordine ad celum). Versor [*De celo*, bk. 2, qu. ult. (18)], 1493, 32r, col. 2. The same argument is cited by Cecco d'Ascoli, ibid., 367; Michael Scot, ibid., 296; Buridan, *De caelo*, bk. 2, qu. 23, 1942, 234; Albert of Saxony, *De celo*, bk. 2, qu. 25, 1518, 118r, col. 2; and Pierre d'Ailly, *Ymago mundi* (translation in Grant, 1974, 633).
35. See Albert of Saxony, ibid.
36. For Aristotle, see *De caelo* 2.14.297b.14–298b.20. Buridan was among those who repeated them (Buridan, ibid., 234–235). Themon Judaeus, *Tractatus de spera*, qu. 12, in Hugonnard-Roche, 1973, 121–126, and Albert of Saxony, ibid., also cite most of the arguments.
37. Themon Judaeus, ibid., 124, and Albert of Saxony, ibid., added to this example when they assumed that if all the parts of the earth tended equally toward the center of the world *and* if the earth were fluid like water so that no part supports another, the earth would shape itself into a perfectly spherical form.
38. That is, every downward rectilinear motion falls perpendicularly to a tangent at the point where the body impacts. Aristotle's discussion in *De caelo* 2.14.297b.17–20 forms the basis of this argument.

Judaeus, and Albert of Saxony, drawing ultimately from Aristotle, invoke lunar eclipses, where the earth's shadow forms an arc across the face of the Moon, separating the visible from the invisible part, a sure indication that the earth, interposed between Sun and Moon, is round.[39] Finally, they also mention the change of position of stars from north to south, and vice versa, as seen by travelers. If one travels from north to south, some stars at the south pole that were previously invisible become visible, and the polestar becomes less elevated. Similarly, as one moves from south to north, stars in the north that were previously invisible become visible, and stars that were previously visible to the south become invisible. If the earth were flat, the stars would always remain visible, whether one was moving to the north or to the south. Such changes could not occur unless the earth were spherical.[40] Although Aristotle did not include anything about roundness involving east and west, scholastics added certain experiences that lent further plausibility to a round earth. Their source was probably Sacrobosco's *Sphere*. According to Sacrobosco, the stars and the zodiacal signs do not set at the same time everywhere "but rise and set sooner for those in the east than for those in the west," a good indication that the earth is round.[41] Moreover, certain celestial events occur earlier for Orientals than for Westerners: for example, "one and the same eclipse of the moon which appears to us in the first hour of the night appears to orientals about the third hour of the night, which proves that they had night and sunset before we did, of which setting the bulge of the earth is the cause."[42]

Scholastic authors were unanimous in their acceptance of a round, though not perfectly spherical, earth. There was, however, a medieval tradition of a flat earth derived from Lactantius (ca. 250–ca. 325), who rejected a spherical earth since it necessitated the existence of antipodes, which he judged absurd, because things on the other side of a spherical earth would fall off the earth toward the heaven.[43] When pious Christians sometimes placed Jerusalem at the center of the earth, they seemed to imply, if they did not explicitly proclaim, a flat earth.[44] In at least one instance – in the *Travels* of John

39. For references to Aristotle, Buridan, Themon, and Albert, see note 36. This common argument was repeated at least twice by Oresme in *Le Livre du ciel*, bk. 2, ch. 31, 1968, 563, and in his *De spera*, qu. 5, 1966a, 101. It was also cited by d'Ailly (in his *Ymago mundi*; see Grant, 1974, 633); Versor, *De celo*, bk. 2, qu. ult. (18), 1493, 32v, col. 1; and Cornaeus, *De coelo*, tract. 4, disp. 3, qu. 2, dub. 2, 1657, 517.
40. The form of this argument was probably derived from Sacrobosco, *Sphere*, ch. 1, 1949, 121. It was also repeated by Cecco d'Ascoli, *Sphere*, 1949, 368, and many others.
41. Sacrobosco, ibid.
42. Ibid. Similar arguments appear in Cecco d'Ascoli, *Sphere*, 1949, 366; Albert of Saxony, *De celo*, bk. 2, qu. 25, 1518, 118r, col. 2; Velcurio, *Physics*, 1554, 87r; and, in the seventeenth century, in Cornaeus, *De coelo*, tract. 4, disp. 3, qu. 2, dub. 2, 1657, 517. As further evidence for a round earth, Versor, *De celo*, bk. 2, qu. ult. (18), 1493, 32v, col. 1, mentions a variant on the same theme, namely, that the Sun rises earlier in the east than in the west.
43. See Randles, 1980, 13. Lactantius proposes a flat earth in his *De divinis institutionibus*, 3:24.
44. Medieval "T and O" maps, which represented a Christian world, often located Jerusalem at the center of a circular, flat, earth. See Randles, 1980, 15.

Mandeville – the centrality of Jerusalem, and therefore the assumption of a flat earth, were accepted simultaneously with the concept of a spherical earth.[45] Although Mandeville's work was very popular, the flat-earth theory was dealt a heavy blow by the introduction of Ptolemy's *Geography* into the West in 1410 (the first edition was printed in 1475). Indeed, a few years before its publication in 1475, the Portuguese had crossed the equator to make many new discoveries along the coast of Africa and reinforce the concept of a spherical earth.[46]

Whatever may be said about the medieval flat-earth theory with Jerusalem at its center, it played no role in scholastic questions on the sphericity of the earth. In the relevant questions or sections by Cecco d'Ascoli, Buridan, Oresme, Albert of Saxony, Themon Judaeus, and Johannes de Magistris, there is no mention of a flat earth, even as an alternative to a spherical earth. Occasionally scholastic authors alluded to a flat earth, but they did not associate it with – or even mention, for that matter – Jerusalem or the Bible. The tenor of these citations of a flat earth is illustrated by John Major, who sought to refute the argument that "the earth appears plane to the sense; therefore it is not spherical." In response, Major explains that the greater the circle, the more a segment of it will appear plane; and the smaller the circle, the more that segment will tend to curvedness. "Thus if you take a portion of a great sphere, it will appear plane; but this will not happen if a portion is taken of a small circle.[47] But a man does not perceive the curvature of the earth by the sense [alone], but by judgment and sense."[48] Others introduce a flat earth because Aristotle mentioned and refuted it.[49]

We saw earlier that scholastic natural philosophers equated the earth's center of gravity with the geometric center of the world. Given the role that rivers and oceans played in altering the earth's center of gravity, how-

45. John Mandeville was an Englishman who wrote in French. He assumed these two irreconcilable concepts in 1366. See Mandeville, 1953, 1:129–130 (English), and 2:332–333 (French); also Randles, 1980, 17–18.
46. On Ptolemy's *Geography*, see Randles, 1980, 21–26.
47. This is a different argument from the one Aristotle presents (*De caelo* 2.13.293b.32–294a.9 [Guthrie], 1960) when he explains that while some think the earth is spherical, others consider it "flat and shaped like a drum. These latter adduce as evidence the fact that the sun at its setting and rising shows a straight instead of a curved line where it is cut off from view by the horizon, whereas were the earth spherical, the line of section would necessarily be curved. They fail to take into consideration either the distance of the sun from the earth, or the size of the earth's circumference, and the appearance of straightness which it naturally presents when seen on the surface of an apparently small circle a great distance away. This phenomenon therefore gives them no cogent ground for disbelieving in the spherical shape of the earth's mass."
48. Major initially concedes the antecedent of the argument, namely "the earth appears plane to the sense," and then proceeds: "Nam semper sphera quanto maior tanto apparet planior quanto circulus est maior tanto magis ad rectitudinem et quanto minor tanto ad curvitatem tendit, ut hec tibi manifestentur cape portionem maxime sphere et eam abscindet et apparebit plana non autem parve. Homo tamen non percipiet sensu terre curvitatem, sed iudicio et sensu." Major, *De celo*, bk. 2, qu. 8, 1526, 22, col. 1.
49. Albertus Magnus is one of them ([*De caelo*, bk. 2, tract. 4, ch. 3], 1971, 182, col. 2). For Aristotle's arguments against a flat earth, see *De caelo* 2.13.293b.32–294a.9 (for the passage, see this chapter, n. 47) and 294b.14–30.

ever, was their usage of the term "earth" here also intended to include water or just the earth alone? The debate that ensued produced the concept of the "terraqueous sphere."

IV. The terraqueous globe

In his famous *Commentary on the "Sphere" of Sacrobosco*, Christopher Clavius explained that although Sacrobosco located the earth in the center of the firmament, we should understand "the earth simultaneously with water. For although the author speaks expressly of the earth alone, the same arguments have the same force with respect to the whole aggregate of earth and water."[50] Although in fact Sacrobosco did not conceive of the earth as Clavius says he does, the treatment of earth and water as a single aggregate – though not a single sphere – may have begun in the fourteenth century.

Without emphasis or elaboration, Aristotle, in the manner presented by d'Ailly earlier in this chapter, had arranged the four elements in a series of concentric spheres, with the earth at the center of the universe, surrounded by the spheres of water, air, and fire, in that order (*Meteorology* 2.2.354b.23–27). Aristotle surely realized that the spheres of earth and water did not conform fully to his schema, since dry land extended above the waters and fire was sometimes visible on the earth's surface.[51] But the relations between the spheres of earth and water, on which Aristotle had provided little guidance, posed serious problems in the Middle Ages. In the *Sphere*, Sacrobosco described Aristotle's representation of the sublunar region as containing four concentric spheres, one for each element, but noted that in order to preserve some dry land for animate creatures, the sphere of water did not completely surround the earth. The relationship between the earth and surrounding waters also had a biblical explanation. After surrounding the earth with waters on the second day of creation, God, on the third day, commanded that "the waters under the heaven be gathered together into one place" and that "dry land appear."[52] As we saw in the preceding section, this was the standard assumption, at least in the thirteenth century.

1. The curious and minimal role of water in the determination of the earth's center of gravity

Further explanations were forthcoming in the fourteenth century, when a number of suggestions were made to explain more precisely the relationship

50. Clavius [*Sphere*, ch. 1], 1593, 151; Sacrobosco, *Sphere*, ch. 1, 1949, 84 (Latin), 122 (English). In this section on the terraqueous sphere, I draw heavily on Grant, 1984b, 22–32.
51. In his schematic representation of Aristotle's arrangement of the elements, H. D. P. Lee makes this assumption when he declares that Aristotle's "stratification is not rigid. Dry land rises above water, and fire burns on the earth" (Aristotle, *Meteorologica* [Lee], 24–26).
52. Genesis 1.9 and Psalms 103 (Vulgate).

between the spheres of earth and water. Buridan's explanation is revealing. Although he was convinced that the spheres of earth and water were concentric with respect to the center of the world, he also realized that the sphere of water did not completely surround the earth, since land existed above the water. Water was unable to surround the earth completely, because the latter had deep caverns which the water filled as it flowed naturally downward. Moreover, parts of the water evaporated and mixed with the air. The quantity of water was thus insufficient to cover the entire earth, and so part of the earth inevitably was left exposed above the water. The exposed part of the earth would be rarefied and lightened by the Sun's heat and the action of the air, whereas the submerged part would remain heavy and dense. It was this unhomogeneous arrangement of earth and water, as we saw earlier, that caused the earth's center of gravity to differ from its center of magnitude. In this physical arrangement, only the earth's center of gravity was coincident with the center of the world.

Buridan anticipated a serious objection. Over long periods of time, geologic processes – especially the effect of water flowing to the seas with earthy matter brought down from the mountains – would wear down the mountains and elevations until the earth was everywhere submerged below the water.[53] But this failed to occur. Why? According to Buridan's ingenious explanation, the earthy matter carried down from the mountains and elevations and continually deposited in the seas and waters makes the submerged portions of the earth heavier, which, in turn, causes the earth's center of gravity to shift continually. These minute but incessant rectilinear shifts of the earth's center of gravity cause previously submerged parts of the earth to rise above the surface of the seas and oceans to form new mountains.[54] Because this geologic process is cyclic and continuous, part of the earth will always remain elevated above the waters. Since the exposed part is more rarefied than the submerged part, more than half of the earth's magnitude will always lie above water.

But if more than half of the earth is exposed above the waters, Buridan, along with many others, assumed that only one-fourth of the whole earth was inhabited.[55] In yielding only one-quarter of the earth for habitation,

53. Buridan's analysis is contained in the question "Whether the whole earth is habitable." Buridan, *De caelo*, bk. 2, qu. 7, 1942, 154–155. For a translation of most of the question, see Grant, 1974, 621–624. For an earlier version of the text of this question and a brief commentary on its contents, see Moody, 1975.

54. Buridan, ibid., 159–160; Grant, ibid., 623. Part of this section is translated into French by Randles, 1980, 43. Albert of Saxony, *De celo*, bk. 2, qu. 24, 1518, 117v, col. 1, adopts a similar interpretation.

55. Buridan, ibid., 157, reports that some believe all of the quarters of the earth are habitable, while others assume that only one-fourth is habitable. In refuting the first opinion (158), Buridan declares that "all the seas which can be crossed and all habitable lands which can be found, are contained in this fourth of the earth which we inhabit" (Prima est, quia omnia maria quae ab aliquibus poterunt transiri, et omnes terrae habitabiles quae poterunt inveniri, continentur in ista quarta terrae quam habitamus). Buridan mentions no one who held the first opinion.

other factors besides water played a role, mainly the primary qualities of hotness and coldness as they affected the five zones into which the earth was usually divided, namely the two frigid zones, which were uninhabitable, the two temperate zones, only the northern one of which was thought habitable, and the torrid zone at the equator, which was thought uninhabitable (see Fig. 11).[56]

Where Buridan used natural explanations to account for the relationships between earth and water, Paul of Burgos (ca. 1350–1435) resorted to supernatural action. On the third day of creation, when the waters were gathered together upon divine command, Paul of Burgos imagined that God had lowered the sphere of water and thereby separated the latter's center of gravity from the earth's, an act that left all of the earth's dry land in its Northern Hemisphere while its Southern Hemisphere was perpetually submerged (see Fig. 12). Although they used different arguments, Buridan, Paul of Burgos, and others arrived at essentially the same configuration between water and earth and were also in general agreement that only one-fourth of the earth was habitable.

When Buridan spoke of the earth's center of gravity, did he also include the ubiquitous waters? No clear answer emerges. Water undoubtedly played a role in determining the earth's center of gravity – at the very least because it moved earth downward from the heights. Nevertheless, because Buridan considered water as having a separate sphere, he probably assumed that water had its own center of gravity and that the absence or presence of water probably would not affect the earth's center of gravity.[57]

Where Buridan was somewhat ambivalent or unclear, Albert of Saxony explicitly rejected the role of water in the determination of the earth's center of gravity. He argued that if all the water were removed from the earth's surface and interior, the earth's center of gravity would be unaffected. It would continue to coincide with the center of the world. The weight of the waters that rested on the earth and covered its uninhabited parts did not push the earth's center of gravity away from the center of the world.

To justify this strange conclusion, Albert insisted that earth is *essentially*

56. For the arguments, see Buridan, ibid., 155–157. Buridan reports (156) that Avicenna thought the torrid zone under the equinoctial circle was habitable; indeed, Avicenna believed a terrestrial paradise existed there. Opinions on the habitabilty of the world varied during the Middle Ages. Some – for example, Robert Grosseteste and Roger Bacon – thought that the region along the equator was habitable but that the Southern Hemisphere was not (see Wright, 1925, 163–165). Figure 11, a diagram of the globe, from Reisch's *Margarita philosophica*, shows both the southern and the northern temperate zones as habitable. By assuming only one-quarter of the earth as habitable, Buridan and those who shared the same opinion adopted the smallest estimate of the earth's size enunciated during the Middle Ages.

57. Hugonnard-Roche, 1973, 82–83, associates Oresme with Buridan in the assumption that the presence or absence of water has no effect on the earth's center of gravity. In *Meteorology*, bk. 1, qu. 5, 83, Themon Judaeus assumed that ocean waters would add their weight to that of the earth.

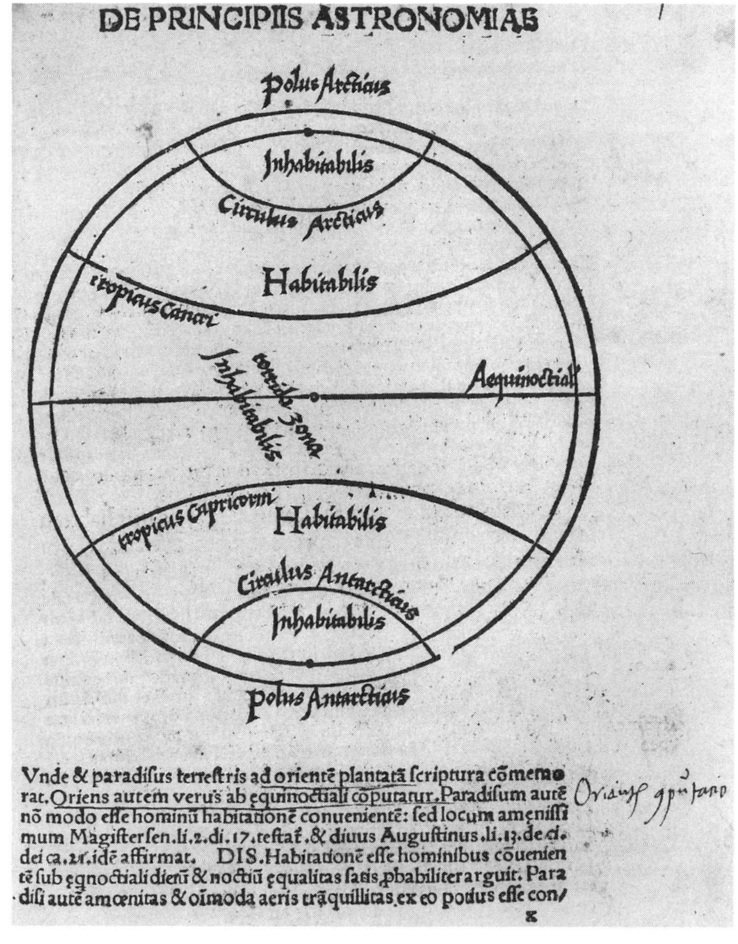

Figure 11. A late medieval representation of the habitable and uninhabitable parts of the earth, where it is assumed that not only the northern but also the southern temperate zone is habitable. (Gregor Reisch, *Margarita philosophica* (1504), fol. 162r. Courtesy Lilly Library, Indiana University, Bloomington.)

heavier than water, a fact made obvious by the descent of a small piece of earth through a large body of water; or, to put it another way, "water is essentially less heavy than earth." Since water cannot affect the behavior of the essentially heavier earth, only the earth's center of gravity can occupy the center of the world, not that of the earth and water combined.[58]

58. Albert of Saxony, *De celo*, bk. 2, qu. 23, 116v, col. 2–117r, col. 1. Themon Judaeus, in *Meteorology*, bk. 1, qu. 5, held a similar opinion (see Hugonnard-Roche, 1973, 84–85).

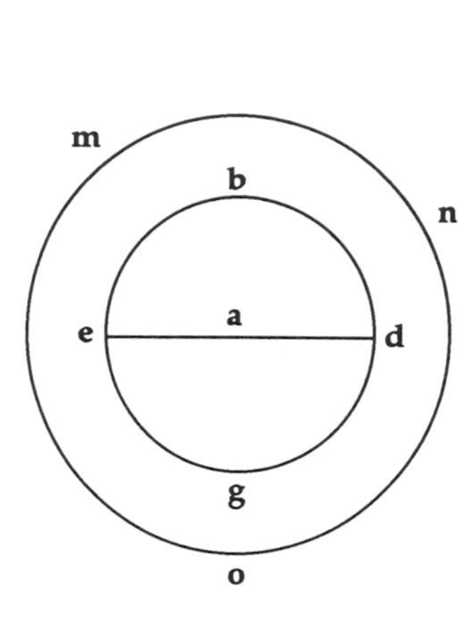

BEFORE

Circle *ebdg* represents the earth and circle *mno* the surrounding sphere of water. Point *a* is the common center that coincides with the center of the world.

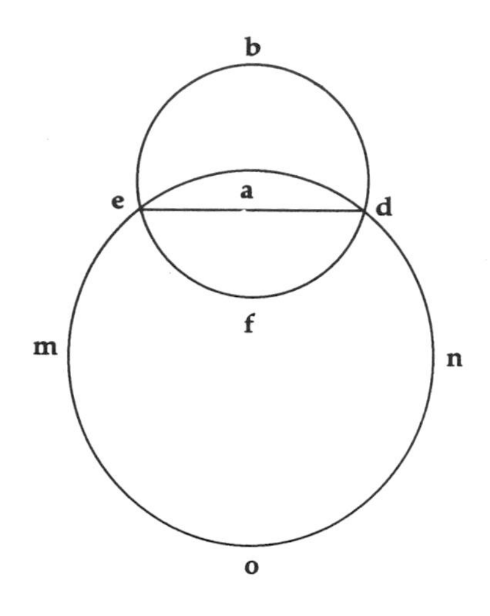

AFTER

Arc *ebd* represents the portion of dry land elevated above the sphere of water *mno*; the center of the sphere of water, *f*, is now separated from the earth's center (and the center of the universe), *a*.

Figure 12. Paul of Burgos's interpretation of the gathering of the waters appears in his *Postillae Nicolai de Lyra super totam bibliam cum additionibus Pauli burgensis et replicis Matthiae Doringk* (Nuremberg, 1481) and is reported by Randles, 1980, 29–30. (My alteration of Randles' figure is drawn from Grant, 1984b, 24, n. 74.)

2. *One center of gravity for the two separate spheres of water and earth*

But some time before, when he wrote his questions on Aristotle's *Physics*, Albert of Saxony attributed an entirely different role to water, one that gave it a new relationship with the earth.[59] What lies at the geometric center of the world is not the earth's center of gravity but rather the center of "the whole aggregate of earth and water, which makes one whole heaviness, the center of gravity of which is the center of the world."[60] Albert was here advocating a center of gravity based upon the idea of the two spheres of earth and water – that is, based on both of the heavy elemental bodies – and not of earth alone. Indeed, his contemporary, Themon Judaeus, did

59. For evidence that Albert's *Physics* preceded his *De celo*, see Sarnowsky, 1989, 51–52.
60. "Sexto dico quod conformiter intelligendum est de toto aggregato ex terra et aqua que forte faciunt unam totalem gravitatem cuius medium gravitatis est medium mundi." Albert of Saxony [*Physics*, bk. 4, qu. 5], 1518, 46r, col. 1. The Latin text with French translation is also given by Duhem, *Le Système*, 1913–1959, 9:213. On pages 205–219, Duhem discusses Albert's concepts of equilibrium of land and sea.

likewise.[61] But Albert abandoned this opinion in his questions on *De caelo*, and Themon, who upheld it in his questions on the *Sphere*, rejected it in his questions on the *Meteorology* in favor of the traditional opinion.[62]

If Albert and Themon were ambivalent, d'Ailly was not. Familiar with the opinions and theories of his Parisian predecessors, d'Ailly chose to support the interpretation that assigned a single center of gravity to the totality of earth and water. Unlike Buridan, he denied that either the earth's center of gravity or its center of magnitude occupied the center of the universe. That place was reserved for the center of gravity of the aggregate of earth and water (*centrum gravitatis aggregati ex aqua et terra*).[63] It is the natural tendency of every heavy body – and the composite of earth and water is a heavy body – to seek the center of the world and remain there. Thus the earth does not lie at the center of the world *per se*, but only as part of a composite that also includes the waters within it and on its surface. There is no reason to believe that d'Ailly thought of the spheres of earth and water as a single sphere. To the contrary, it is overwhelmingly likely that he conceived of them as two separate spheres that, when taken as a heterogeneous aggregate, yielded a single center of gravity. Thus d'Ailly retained the traditional opinion that the Southern Hemisphere was covered by water and therefore uninhabited.

3. The terraqueous globe: Earth and water form one sphere, with one center of gravity

The final step in the development of the concept of a terraqueous globe was to abandon the two spheres and to assume instead that water and earth formed a single sphere with a single center of gravity. This occurred only after the Portuguese explorations of the Southern Hemisphere, especially the voyage along the coast of Brazil in 1501, when Europeans learned of the wide distribution of land in that region. In a letter to Rudolf Agricola that was published in 1515 and a few times thereafter, Joachim Vadianus (1481–1551) of Switzerland took cognizance of the new knowledge to proclaim that not only did earth and water form a single globe, but their relationship was such that over the entire surface of that globe earth is partly submerged and partly elevated (Randles, 1980, 44–45). Here was perhaps the proper description of what would be called "the terraqueous globe," or "the terraqueous sphere," in the seventeenth century.[64]

61. Hugonnard-Roche, 1973, 81–82.
62. See ibid., 81, where Themon adopts the new theory, and 83, where he abandons it. Hugonnard-Roche (84) observes that both Themon and Albert hesitated between the two opinions.
63. D'Ailly's discussion occurs in his *14 Questions*, qu. 5, 1531, 155r. For the relevant Latin passages, drawn from a different edition, see Grant, 1984b, 26, nn. 83–84. Major seems to have adopted the same position in his *De celo*, bk. 2, qu. 9, 1526, 22, col. 2.
64. According to Randles, 1980, 63, the French Jesuit Georges Fournier first used the term "terraqueous globe" (*globe terraqué*) in 1643 in a work called *Hydrographie*. In 1651, Riccioli used the term *terraqueum*, which he said referred to a composite of earth and water.

The concept was thus formulated outside of the scholastic tradition.[65] It was adopted by Copernicus, who, in book 1, chapter 3, of *On the Revolutions*, explains "How Earth Forms a Single Sphere with Water" and shares with it the same center of gravity.[66] With Clavius's *Commentary on the Sphere of Sacrobosco*, the terraqueous globe entered scholastic cosmology in the late sixteenth century and was widely adopted in the seventeenth. Among its supporters were the Conimbricenses, Aversa, Amicus, Cornaeus, and Mastrius and Bellutus.[67] They all assumed that earth and water formed a single sphere, which Aversa and Mastrius and Bellutus called "the terrestrial globe" (*terrestris globus*).[68]

Seventeenth-century scholastics repeated most of the earlier arguments for a single globe of water and earth.[69] But just like those who earlier had assumed that the earth alone was the sphere at the center of the world, supporters of the terraqueous globe had to solve the problem of the three centers, that is, they had to determine whether or not the center of the universe and the centers of gravity and magnitude of the terraqueous sphere were identical. Fundamental to this determination was the conclusion that the terraqueous globe had a single, convex surface composed of water *and* earth commingled over the whole of the globe.[70] Because of the earth's mountains and prominences, the unified terraqueous globe was not considered a perfect geometric sphere. But, as his medieval predecessors did

65. Buridan, indeed, described an opinion that is very close to the terraqueous globe, only to reject it. See Grant, 1984b, 27, n. 88.
66. "Pouring forth its seas everywhere, then, the ocean envelops the earth and fills its deeper chasms. Both tend toward the same center because of their heaviness." Copernicus, *Revolutions*, bk. 1, ch. 3 [Rosen], 1978, 9. Surprisingly, Copernicus went a step farther and insisted that within this single globe the centers of gravity and magnitude were identical, a conclusion that committed him to a terraqueous sphere that is not only homogeneous in composition but "is perfectly round, as the philosophers held." For only if our sphere were "perfectly round" could the two centers be identical, although, in the heliocentric system, neither could coincide with the center of the world. For further discussion on the significance of Copernicus's departure from Aristotle's two-sphere system of earth and water, see Thomas Goldstein, 1972.
67. See Conimbricenses, *De coelo*, bk. 2, ch. 14, qu. 3, art. 2, 1598, 382–384; Aversa, *De caelo*, qu. 36, sec. 5, 1627, 227, col. 2–228, col. 1; Amicus [*De caelo*, tract. 8, qu. 4, dubit. 2], 1626, 582, col. 2–585, col. 1, dubit. 6, 598, col. 1; Cornaeus, *De coelo*, tract. 4, disp. 3, qu. 2, dub. 5–6, 1657, 518–520; and Mastrius and Bellutus, *De coelo*, disp. 4, qu. 3, art. 2, 1727, 3:560, col. 1–561, col. 2, pars. 105–111.
68. Aversa used the expression frequently, as, for example, in *De caelo*, qu. 36, sec. 5, 1627, 224, cols. 1–2; 226, col. 1; 227, col. 1; sec. 6, 231, col. 1; and 232, cols. 1–2. For instances of its use by Mastrius and Bellutus, see their *De coelo*, disp. 4, qu. 3, art. 3, 1727, 3:562, col. 1, par. 113, and 564, col. 1, par. 122. We should revise the claim by Randles, 1980, 63, that the term "terraqueous globe" was not replaced by the term "terrestrial globe" until the eighteenth century. The Conimbricenses, *De coelo*, bk. 2, ch. 14, qu. 3, art. 2, 1598, 384, did not name it but used the expression "unum globum ex terra et aqua," which signified a globe that had the same center of gravity and center of the universe (see also 382–383).
69. See Grant, 1984b, 28–29.
70. Clavius, *Sphere*, ch. 1, 1593, 142–143; Aversa, *De caelo*, qu. 36, sec. 5, 1627, 228, col. 2–229, col. 1; Amicus, *De caelo*, tract. 8, qu. 4, dubit. 3, art. 3, 1626, 588, col. 1; and Cornaeus, *De coelo*, tract. 4, disp. 3, qu. 2, dub. 6, 1657, 519.

with the earth, Clavius insisted that the inequality or difformity of the terrestrial globe was negligible when compared to the whole of it. It was therefore justifiably describable as round and spherical. Unlike his medieval predecessors, however, who argued that the earth's center of gravity, or the center of gravity of the two spheres of earth and water, was the real center, the one that coincided with the geometric center of the universe and which differed from the center of magnitude, Clavius concluded that the three centers were one and the same, a position that the Coimbra Jesuits, Aversa, and Cornaeus also adopted.[71]

Just as, during the Middle Ages, the earth was assumed to lie at the center of the world by means of its center of gravity, so also during the late sixteenth and the seventeenth century was the terraqueous globe assumed to lie naturally at the center of the universe. Physically, the compound terraqueous globe lies at the center of the universe, because both earth and water, as heavy bodies, move naturally downward to seek the center of the universe.[72] This and other arguments were largely the same as those presented earlier for the centrality of the earth.[73]

V. The earth's immobility: rejection of a daily axial rotation

Despite disagreement about relationships among the three centers relevant to the earth, scholastics placed the earth in the middle of the world. They were equally unanimous about its immobility. Although Buridan, and perhaps a few others, were prepared to allow slight rectilinear motions of the earth, scholastic authors of the Middle Ages upheld the traditional opinions of Aristotle and Ptolemy, and all of their lesser supporters, that the earth lay immobile at the center of the world.[74] Immobility seemed to preclude

71. Conimbricenses, *De coelo*, bk. 2, ch. 14, qu. 3, art. 2, 1598, 382–384; Aversa, ibid., 230, cols. 1–2; Cornaeus, ibid., 519–520. For further details, see Grant, 1984b, 29–30.
72. Mastrius and Bellutus, *De coelo*, disp. 4, qu. 4, art. 1, 1727, 3:559, col. 1, par. 97; Amicus, *De caelo*, tract. 8, qu. 4, dubit. 1, art. 4, 1626, 581, cols. 1–2; Cornaeus, ibid., dub. 15, 534.
73. For some of them, see Grant, 1984b, 31–32.
74. In a question "Whether the motion of the earth is possible," Themon Judaeus, *Meteorology*, bk. 2, qu. 7, 1518, 174v, cols. 1–2, distinguishes a number of ways in which the earth can be imagined to move. One is by a rectilinear motion caused by the rising and falling of the earth's center of gravity, but he says that this has nothing to do with what he wishes to discuss. A second way involves the earth's axial rotation around a resting center, an idea that he regards as false and contrary to what Aristotle says in book 2 of *De caelo*. The remaining ways, which are more appropriate to a work on meteorology, involve minor disturbances of parts of the earth, such as the motion of something on the earth's surface. For example, when water lies above compacted earth and men walk over it, water squirts out from the sides and, presumably, the earth yields. Or after coal has been extracted from the earth, the ground sometimes caves in; or when water makes a huge cavity underground, the earth collapses into it. And then there are hot and dry exhalations, which vigorously strive to issue forth from the earth's interior. (It is this motion of the earth, or *terremotus*, in which Themon is interested in *Meteorology*, bk. 2, qu. 7.)

a rotation around its own axis. But, according to Ptolemy, there were some who proposed an alternative opinion in which the heaven was assumed immobile and the earth was assigned a daily axial rotation from west to east.[75] Ptolemy argues against this alternative because it seemed absurd. With the translation of his *Almagest* into Latin in the twelfth century, he became one of the major sources for medieval discussions of possible axial rotation. In 1271, a second significant source was added when William of Moerbeke completed his translation from Greek to Latin of Simplicius's sixth-century commentary on Aristotle's *De caelo*. It is here that Simplicius said:

> There have been some, among them Heraclides of Pontus and Aristarchus, who thought that the phenomena could be accounted for by supposing the heaven and stars to be at rest, and the earth to be in motion about the poles of the equator from west [to east] making approximately one complete rotation each day.[76]

The impact of Simplicius's discussion is already evident some two or three years after Moerbeke's translation when, in his commentary on *De caelo*, Thomas Aquinas declared that

> the appearance of motion is caused either by the motion of the thing seen or by the motion of the one who sees it. For this reason some people, assuming that the stars and the whole sky rest, have posited that the earth on which we dwell is moved once daily from west to east around the equinoctial poles. Thus by our motion, it seems to us that the stars are moved in a contrary direction, which is what Heraclides of Pontus and Aristarchus are said to have posited.[77]

With these words, Thomas Aquinas could have become a potential stimulus for discussion. Unfortunately, his own consideration of the issue was muddled and unclear and seems to have played little, if any, role in the history of discussions about axial rotation.[78]

75. For Ptolemy's statement and rebuttal, see Ptolemy, *Almagest*, bk. 1, ch. 7 [Toomer], 1984, 44–45.
76. Translated by Cohen and Drabkin, 1958, 106–107. Although in *De caelo* 2.13.293b.30–32, Aristotle seems to attribute to Plato a belief in axial rotation and seems to mention it again vaguely in *De caelo* 2.14.296b.1–4, Aristotle's descriptions were too nondescript to count as genuine discussions. Thomas Aquinas, in his *De caelo*, was convinced that Plato was a firm believer in an immobile earth. Rather than Plato, Thomas believed that Aristotle had Heraclides of Pontus in mind, who, according to Thomas ([*De caelo*, bk. 2, lec. 21, par. 490], 1952, 245, col. 2), "assumed that the earth was moved in the center [of the world] and that the sky rested" (for translation and further discussion, see Grant, 1974, 496–497, n. 14).
77. Thomas Aquinas, ibid., lec. 11, par. 396, 197. The translation is mine, taken from Grant, 1974, 500. Thomas adds little to this statement that is useful for the problem of axial rotation.
78. The closest Thomas comes to a discussion of axial rotation appears in ibid., lec. 26, 262–265; translation of much of this *lectio* in Grant, 1974, 496–499.

1. The medieval scholastic case for and against the earth's axial rotation

By the end of the fourteenth century, however, Jean Buridan and Nicole Oresme had formulated two lengthy and profound analyses in favor of the earth's daily rotation. Although in the end both rejected axial rotation, for different reasons, they presented some compelling arguments in its behalf.[79] In his defense of axial rotation, Copernicus did not significantly add to the store of arguments proposed by his medieval predecessors.

a. Jean Buridan

Buridan, who probably wrote first, believed that the daily motion of the stellar sphere and planets could be saved either by an assumption of a stationary heavens and axially rotating earth, or the reverse.[80] On the first alternative, however, an additional assumption was essential, namely that despite the immobility of the heavens with respect to the east-to-west daily motion, the planetary spheres would continue their respective periodic motions from west to east against the background of the immobile sphere of the fixed stars. Only in this way could the planets change their positions relative to each other and the fixed stars. Thus the earth and planets would only move with west-to-east motions, and the earth would complete its rotation in a natural day, the Moon in a month, the Sun in a year, Mars in two years, and so on. In this way, all daily and periodic astronomical phenomena can be saved just as readily as on the alternative hypothesis. The celestial spheres formerly involved in the production of the daily motion would now be assumed at rest, while the remainder, whose motions generate the progressive and retrogressive movements of the planets along the zodiac, would function as before.

Buridan recognized that the problem was essentially one of relative motion. Although it appears to us that the earth on which we stand is at rest while the Sun is carried around us on its sphere, the reverse might be physically true, since the observed celestial phenomena would remain the same. Thus if the earth is moved circularly and the starry heaven rests, the earth will undergo a daily west-to-east motion that causes night and day. We would be as unaware of such a terrestrial rotation as a person on a

79. Albert of Saxony also considered the possibility in two different places in his *De celo*, first in book 2, question 6 ("Whether the motion of the heaven from east to west is regular"), 110r, col. 2 and 110v, col. 1, and again in book 2, question 24 ("Whether the earth always rests or is always moved in the middle [or center] of the heaven or in the middle [or center] of the world"), 117r, col. 2–117v, col. 2. On the whole, Albert's treatment seems derivative and omits the most significant arguments.

80. In this section, I follow my account in Grant, 1971, 64–66. Buridan's arguments are included in his question "Whether the earth always is at rest in the center of the universe." For the Latin text, see Buridan, *De caelo*, bk. 2, qu. 22, 1942, 226–232; for the translation of much of it, see Clagett, 1959, 594–599, and for a reprint of Clagett's translation, see Grant, 1974, 500–503.

moving ship that passes another ship at rest. If the observer on the moving ship imagines himself at rest, the ship at rest would appear to be in motion. Similarly, if the Sun were truly at rest and the earth rotated, we would perceive the opposite, that is, we would "see" that the Sun is in motion and believe that the earth, from which we view the Sun, is at rest.

On strictly astronomical grounds, Buridan was convinced, as was Ptolemy long before him, that either hypothesis could save the celestial phenomena.[81] Not even astronomers could resolve the issue and determine the physical truth. They were concerned with saving, or explaining, the celestial appearances and could employ whichever alternative seemed most convenient.

As further support for the earth's rotation, Buridan resorted to five "persuasive arguments" (*persuasiones*), that is, arguments that were not demonstrative but were intended to win over opponents by their reasonableness and cumulative impact.[82] Of these nonastronomical arguments, easily the most significant was the last, in which Buridan emphasized the desirability of saving the phenomena by the easiest means possible. It seemed better to assume that the relatively small earth turns with the swiftest speed while the uppermost and largest spheres remain at rest. To complete a daily rotation, the earth would require a much slower speed than the vastly larger celestial spheres. In this argument, simplicity and credulity were more adequately satisfied by an earth that rotated daily on its axis.[83] Oresme, Copernicus, and Galileo all found this argument worthy of inclusion in their defense of the earth's axial rotation.[84]

The other persuasions emphasized concepts and ideas about the world that were meaningful to those immersed in Aristotelian modes of thought. The heaven does not need the earth for anything, but the earth needs the heaven for the influences that enable it to operate. It is more reasonable, therefore, to assume that that which requires another thing would be moved to acquire it and that that which does not need another thing would not be so moved.[85] Therefore the earth ought to move, rather than the heaven.

81. Ptolemy says that there is "nothing in the celestial phenomena which would count against that hypothesis," namely the hypothesis that the earth revolves daily on its axis from west to east. See Ptolemy, *Almagest*, bk. 1, ch. 7 [Toomer], 1984, 45.

82. For the five, see Buridan, *De caelo*, bk. 2, qu. 22, 1942, 228–229. Of these, Albert of Saxony repeated three (he also called them *persuasiones*; see *De celo*, bk. 2, qu. 24, 1518, 117r, col. 2). For Albert's rebuttal of them, see 117v, col. 2.

83. The Latin text of this important idea reads: "Ultima persuasio est quia sicut melius est salvare apparentia per pauciora quam per plura, si hoc sit possibile, ita melius est salvare per viam faciliorem quam per viam difficiliorem. Modo facilius est movere parvum quam magnum; ideo melius est dicere quod terra, quae est valde parva, velocissime moveatur, et sphaera suprema quiescat, quam dicere e contrario." Buridan, *De caelo*, bk. 2, qu. 22, 1942, 228–229. This is a version of Ockham's razor, or a "simplicity" argument. (For more on simplicity arguments, see Section V.2.b of this chapter.)

84. Oresme, *Le Livre du ciel*, bk. 2, ch. 25, 1968, 535–537; see also this chapter, Sec. V.1.b, and Grant, 1974, 509; Copernicus, *Revolutions*, bk. 1, ch. 8 [Rosen], 1978, 15; and Galileo, *Dialogue*, Second Day [Drake], 1962, 115.

85. Albert of Saxony repeated this *persuasio* in *De celo*, bk. 2, qu. 24, 1518, 117r, col. 2. His rejection of it (117v, col. 2) was the same as Buridan's. Riccioli presents substantially the

Another *persuasio* holds that things require action in inverse proportion to their excellence. The best thing, the outermost sphere, requires no action and should remain at rest; the next-best thing, the sphere of Saturn, requires only a very small degree of action or motion; and the next sphere, Jupiter, requires more action than Saturn, and so on down to the Moon, which requires the fastest of the planetary motions; and finally there is the earth, which, being the least excellent, requires the fastest motion of all, the daily motion. A third persuasion is based on the assumption that rest is nobler than motion. Therefore the highest, or outermost sphere, ought to be at rest, because it is the noblest body. By inference, the least noble thing, the earth, ought to be in motion.[86]

Despite these strong arguments in favor of a daily terrestrial motion, Buridan opted for the traditional opinion. As part of his rebuttal of a rotating earth, he rejected each persuasion.[87] To cite only his rebuttal of the first *persuasio* that I discussed (the last or fifth in his order of presentation), Buridan says that if all things were equal, it would indeed be true that "it is easier to move a small body than a large [one]. But all things are not equal, because heavy, terrestrial bodies are unsuited for motion. It would be easier to move water than earth; and even easier to move air; and by ascending in this way, celestial bodies are, by their natures, most easily moved."[88] In similar fashion did Buridan dispatch the other persuasions.[89]

same argument in *Almagestum novum*, pars post., bk. 9, sec. 4, ch. 33, 1651, 467, cols. 1–2 (10th argument).

86. On the greater or lesser nobility of rest and motion, see Chapter 15, Section II.1 (for Buridan's arguments) and Section IV (for Amicus's). The fourth persuasion is quite obscure and depends on Aristotle's discussion in *De caelo*, book 2, chapter 2, of three pairs of opposite directions: above and below (or up and down), right and left, and front and back, directions which Aristotle believed were in the spherical heaven because the latter was a living thing. With respect to the primary motion of the heaven, namely the daily motion from east to west, Aristotle argued (2.2.285b.15–286a.1) that the north celestial pole was down, the south celestial pole up; the right side was in the east, from whence the stars rise; and the left side was in the west, where they set (in Buridan's discussion, front and back are ignored). Aristotle considered that those who live in the upper, or northern, hemisphere of the earth, as he and the Greeks did, lived in the lower hemisphere (because their pole, the north pole, was down) and to the left (because the daily motion begins to the right, in the east, and passes over us on its way to the west, or left). But these relationships are reversed with respect to the secondary revolution, namely the west-to-east periodic motions of the planets. In relation to these motions, however, we (presumably the Greeks) dwell in the upper and righthand part of the earth.

In his fourth *persuasio*, Buridan focuses on the west-to-east motions of the planets and also assumes a west-to-east daily motion of the earth. Therefore, in keeping with Aristotle's assertions, Buridan declares that, with respect to the heavens, we live upward and on the right side, just as Aristotle declared. And, he continues, "this seems very reasonable, because right ought to be more noble than left, and up than down. Now this section of the earth which is habitable is nobler than the other inhabitable parts; therefore it is reasonable that it should be toward the right. Even our pole seems nobler than the opposite pole, because it is surrounded by more and greater stars; therefore it is reasonable that it should be up." *De caelo*, bk. 2, qu. 22, 1942, 228.

87. Buridan not only rebutted each of the five persuasions but also formally repudiated four other arguments (or "principal reasons") in favor of the earth's axial rotation that he presented at the beginning of the question. (For these four arguments, ibid., 226; for the rebuttals, 232–233).

88. Ibid., 230. Amicus gives the same argument in *De caelo*, tract. 5, qu. 6, dubit. 1, art. 3,

To develop the fundamental argument against a rotating earth, Buridan used his own impetus theory and certain observational consequences derived from it. In his judgment, the earth's rotation could not explain why an arrow shot vertically upward always falls to the same spot from which it was projected. For if the earth rotates from west to east, it ought to rotate about a league to the east while the arrow is in the air. Consequently, the arrow should fall to the ground about a league to the west. But we fail to observe such a consequence.

Buridan realized that a supporter of the earth's rotation might have a ready reply. The air moves along with the rotating earth and carries the arrow with it, thus explaining why the arrow falls to the same place from whence it was shot. By virtue of the common rotatory motion shared by earth, air, arrow, and observer, the arrow's circular motion would go undetected.

Because of certain consequences that followed from his impetus theory, Buridan found this explanation unacceptable. When the arrow is projected, a sufficient quantity of impetus is impressed into it to enable the arrow to resist the lateral motion of the air as the latter accompanies the earth in rotatory motion. In resisting the motion and push of the air, the arrow should lag behind earth and air and drop noticeably to the west of the place from which it was launched. Since this is contrary to experience, Buridan concludes that the earth is at rest.

b. Nicole Oresme

In what seems an even more brilliant discussion, Nicole Oresme arrived at the same conclusion.[90] Near the beginning of his lengthy discussion of the earth's possible axial rotation, Oresme declares (*Le Livre du ciel*, bk. 2, ch. 25, 1968, 521): "it seems to me that it is possible to embrace the argument and consider with favor . . . that the earth rather than the heavens has a diurnal or daily rotation. At the outset, I wish to state that it is impossible to demonstrate from experience at all that the contrary is true; second, that no argument is conclusive; and third, I shall demonstrate why this is so."

1626, 291, col. 2. Oresme [*De celo*, bk. 2, qu. 13], 1965, 694, explains somewhat differently why the heavens move and the earth rests. For incorporeal things, rest is nobler than motion. In bodies, an important distinction must be made. For those bodies that are capable of violent motion but under natural conditions are disposed to rest, it is more noble to rest (as would be the case with terrestrial bodies). But for celestial bodies, where the motion is eternal and without violence, then it is more noble to be moved. Thus it is nobler for the heaven to be in motion and for the earth to rest. For Buridan's different rationale for attributing nobility to rest and motion, see this volume, Chapter 15, Section II.1.

89. For the persuasions he apparently derived from Buridan, Albert of Saxony also seems to have borrowed Buridan's corresponding rebuttals. *De celo*, bk. 2, qu. 24, 1518, 117v, col. 2.

90. For my summary of Oresme's arguments, I rely heavily on Grant, 1971, 66–69. For the French text and translation, see Oresme, *Le Livre du ciel*, bk. 2, ch. 25 [Menut], 1968, 518–539. Most of Menut's English translation is reprinted in Grant, 1974, 503–510.

Under the first subdivision, where he argues that experience cannot demonstrate a daily rotation of the heavens, Oresme considers the problem from the standpoint of three experiences (ibid., 521–527): (1) we see the rising and setting of the stars and the planets, from which it should follow that it is the heavenly motion that moves these stars and planets; (2) if the earth is moved with a daily circular motion from west to east, we should feel the wind coming on strongly from the east and also hear the noise it makes; and (3) if the earth moves rapidly from west to east, then, according to Ptolemy, "If someone were in a boat moving rapidly toward the east and shot an arrow straight upward, it would not fall in the boat but far behind it toward the west," and "if someone threw a stone straight upward it would not fall back to the place from which it was thrown, but far to the west."[91]

The response to the first experience involves relativity of motion. We perceive local motion, says Oresme (523), "only if we can see that one body assumes a different position relative to another body."[92] But which body moves? Like Buridan, Oresme invokes the relative motion of ships to counter our impression that the heavens move because in ordinary experience we "see" the planets and stars rise and set. The relativity argument is further reinforced when Oresme observes (523) that if a man were carried around by a daily motion of the heavens and could see the earth in some detail, it would appear to him that the earth moved with a daily motion just as it seems to us that the heavens move with such a motion.

To the claim that if the earth turned from west to east a great and easily detectable wind should blow constantly from the east, Oresme counters that the air rotates with the earth and therefore does not blow toward us from the east.[93]

The third experience, which Oresme attributes to Ptolemy, is similar to Buridan's crucial arrow experience. The issue was this: If an arrow were shot upward, or a stone thrown upward, as the earth rotated from west to east, would the arrow or stone fall to the west of the place from whence it was launched? Since we do not observe such effects, Ptolemy concluded that the earth rests. Arguing from his impetus theory, Buridan concurred with Ptolemy.

91. Oresme, ibid., 521. Ptolemy says something like this in *Almagest*, bk. 1, ch. 7 [Toomer], 1984, 45, but makes no mention of boats, arrows, or stones.

92. Oresme, ibid., 523, attributes this concept of local motion to Witelo, who enunciated it in his *Perspectiva* (in bk. 4, prop. 110, as cited by Menut). Witelo derived it from the *Optics* of Alhazen, the great Arabian scientist of the eleventh century (see Grant, 1974, 505, n. 39).

93. Albert of Saxony countered *(De celo*, bk. 2, qu. 6, 1518, 110r, col. 2) that even if the air moved along with the rotating earth from west to east, anyone walking toward the west ought to feel the wind sweeping toward the east. Indeed, it also ought to be more difficult to throw a stone against the earth's rotation – that is, toward the west – than with it. But we experience none of these things and should conclude that the earth does not rotate. Oresme's reply would have been that the person walking is also moving from west to east along with the earth and the air.

Oresme, however, saw nothing incompatible about the return of the arrow or the stone and the rotation of the earth. Once again, a ship's motion is used to illustrate the key points. The movements within a ship sailing eastward occur precisely as when the ship is at rest. Thus, if a man drew his hand vertically downward in line with the ship's mast, his hand would appear to move with only a vertical motion, despite the fact that his arm actually undergoes two simultaneous rectilinear motions, vertical and horizontal. If we now assume that the earth, the ambient air, and all sublunar matter rotate daily from west to east, the arrow's return to the place from which it was shot can be explained by reference to its two simultaneous component motions, namely vertical and horizontally circular motion (rather than one vertical motion and one horizontally rectilinear motion, as with the arm's motion along the mast of the ship). Since the arrow shares the earth's circular motion and turns with it at the same speed, the arrow when shot upward will rise directly above the place from which it was shot and then fall back to it. To an observer, who also participates in the earth's circular motion, the arrow will appear to possess only a vertical motion. On the basis of this and the preceding arguments, Oresme concluded that it is impossible to determine by experience that the heavens have a daily motion and the earth does not.

In the second subdivision, Oresme moves from experiential evidence to rational arguments (527–531), which can no more demonstrate the superiority of either of the alternatives than could arguments from experience. Among seven arguments in this category, only a few can be mentioned. Oresme counters the traditional Aristotelian argument that circular motion is not natural to the earth by arguing that it could be natural to the earth as a whole, even though its parts possess natural rectilinear upward and downward motions when they are out of their natural place. Oresme also sought to neutralize Averroës' argument that "all local motion is relative to some body at rest," so that "the earth must be at rest in the center of the heavens." Oresme insists that the circular celestial motions do not require an immobile earth at the center of the world. For if the earth were rotating in the opposite direction to the heavens, or, indeed, if the earth were annihilated, this would have no effect on the celestial motions. The earth's mobility or immobility has no effect on the celestial region. To those who believed that a daily rotation of the earth would destroy astronomy, Oresme denies the claim by insisting that "the astronomical tables of the heavenly motions and all other books would remain as true as they are at present, save that, with respect to diurnal motion, one would say that it is *apparently* in the heavens, but *actually* in the earth; no other effect would follow or result from one theory more than from another" (531).

In the third subdivision, Oresme explains (531–537): "I want to present several opinions or reasons favorable to the theory that the earth moves as we have stated." To demonstrate that the earth's daily rotation might be more than an equally plausible alternative, Oresme also presents nonastronomical reasons in favor of the earth's axial rotation, just as Buridan had done. For example, a terrestrial rotation from west to east would contribute toward a more harmonious universe, since all bodies would move in the same direction in periods that increased from the earth outward. This alternative is more desirable than the other in which contrary simultaneous motions are ascribed to the heavens, east to west for the daily motion and west to east for the periodic motions.[94] In favor of the earth's rotation, Oresme also includes Buridan's argument about the greater simplicity of the earth's rotation than that of the heavens. The former would require a much slower daily speed than the enormous celestial orbs, whose speeds would have to be "far beyond belief and estimation." Thus would be avoided "the multiplication of operations so diverse and so outrageously great" that "it follows that God and nature must have created and arranged them for naught" (537). Oresme also sought to enlist God on the side of the earth's axial rotation by recalling the biblical miracle when God intervened on behalf of the army of Joshua (Joshua 10.12–14) by lengthening the day and commanding the Sun to stand still over Gibeon. Since the earth is like a mere point in comparison to the heavens, the same effect could have been achieved with a minimum of disruption by a temporary cessation of the earth's rotation, rather than by bringing to a halt the Sun and all the other planets. In view of the greater economy of effort, perhaps God performed the miracle in this way.

At the termination of an impressive array of arguments in favor of the possibility of a rotating earth, Oresme's conclusion comes almost as a non sequitur when he declares (537–539):

> everyone maintains, and I think myself, that the heavens do move and not the earth: For God hath established the world which shall not be moved, in spite of contrary reasons because they are clearly not conclusive persuasions. However, after considering all that has been said, one could then believe that the earth moves and not the heavens, for the opposite is not clearly evident. Nevertheless, at first sight, this seems as much against natural reason as, or more against natural reason than, all or many of the articles of our faith. What I have said by way of diversion or intellectual exercise can in this manner serve as a valuable means of refuting and checking those who would like to impugn our faith by argument.

In the absence of demonstrative arguments for the earth's rotation, Oresme falls back on the traditional interpretation, which was not only consonant with natural reason but also had biblical support. Thus did Oresme choose to reject the cumulative impact of his own good reasons for the earth's possible rotation and opt instead for the traditional opinion.

In the final analysis, Oresme rejected the earth's rotation on the basis of a biblical appeal, and therefore ultimately on theological grounds. Indeed,

94. For additional arguments, see Grant, 1971, 68.

he goes on to explain that his discussion was "by way of diversion" (*par esbatement*), or by way of a game. But, in Oresme's judgment (539), that diversion, or game, could serve as an aid to counter those "who would like to impugn our faith by argument" (qui voudroient nostre foy par raysons impugner).

How could a discussion on the possible axial rotation of the earth be used in defense of faith? For a good clue to his intentions, we must turn to Oresme's *Quodlibeta*, probably composed in 1370. Here Oresme declares that although faith assumes miracles, such as the Trinity, penetration of bodies, and the Resurrection, "it seems to me," he continues,

> that many things equally miraculous are assumed in philosophy and [are even] less demonstrated because they are customary [or familiar], as, for example, what is prime matter, and [how] the production of a new form [occurs] of which at first nothing exists, . . . ; how fire heats [or burns] and what is the nature and quiddity of things. . . . Surely, if you consider the matter properly, [you will realize that] many such natural things are unknown, more [unknown] than many articles of the faith. And therefore, indeed, I know nothing except that I know that I know nothing.[95]

In this extraordinary passage, Oresme insists that many things about natural knowledge, or natural philosophy, are more unknown than are "many articles of the faith." Seven years later, in his *Le Livre du ciel et du monde*, Oresme argues analogously that knowledge from natural philosophy, namely the possible axial rotation of the earth, "seems as much against natural reason as, or more against natural reason than, all or many of the articles of our faith" (bk. 2, ch. 25, 539). Although faith was based on revelation and natural philosophy on reason, Oresme strongly suggests that faith is as intelligible and reliable as natural philosophy, and in many instances more so.

However plausible and convincing are one's reasoned and empirical arguments, Oresme suggests that the world may be otherwise than those arguments indicate. Although the Joshua miracle could be interpreted as favorable to a rotating earth, the literal interpretation of that scriptural passage and many others point overwhelmingly to an immobile earth. Faith triumphs over reason. Pronouncements of Scripture and faith have to be assumed true even if arguments for the earth's rotation seem more plausible. Indeed, in connection with the earth's possible axial rotation, Oresme may have implicitly assumed the dictum of Pierre Ceffons that "nothing prevents some false propositions from being more probable than some true ones."[96] The earth's rotation may have seemed more probable and plausible than the traditional alternative, but it was nevertheless false, presumably because

95. Translated from the Latin quoted by Thorndike, 1923–1958, 3:469, n. 128.
96. Cited by Sylla, 1991, 214 from Weinberg, 1948, 116–117.

it was plainly at variance with Scripture. There could be no "double truth," where a principle of nature is deemed true in natural philosophy and science but false in faith, or vice versa.[97]

In this regard, Augustine's famous criteria for the relations between scientific conclusions and scriptural truths (see Ch. 5, Sec. III) are relevant. Where scientific arguments and claims clashed with traditional theological or scriptural truths, the former had to be demonstrated beyond doubt or declared false and abandoned. Oresme adopted the latter approach because he could not have demonstrated the earth's rotation, any more than Galileo did. At best, he could only have shown it to be more plausible or probable, alternatives which he also rejected. Had Oresme seriously sought to proselytize for the plausibility or probability of axial rotation, and even the physical truth of it, as Galileo had, would something akin to the Galileo affair have arisen in the late fourteenth or the early fifteenth century? Would we today speak of the Trial of Oresme rather than, or in addition to, the Trial of Galileo?

Both Buridan and Oresme deserve high praise for their extraordinary arguments in behalf of a rotating earth. Each, however, rejected it for different reasons, which illustrate how dissimilar their approaches and attitudes were. Buridan arrived at his conclusion on the basis of rational argument and the senses, completely ignoring Scripture and faith; Oresme ultimately decided the issue on the basis of Scripture and faith.[98]

2. *The debate over the earth's immobility after Copernicus*

Despite Buridan and Oresme's rejection of a daily rotation for the earth, some of their arguments in favor of axial rotation turn up in Copernicus's defense of the heliocentric system, where the earth is assigned both a daily rotation and an annual motion around the Sun.[99] Among these we find

97. The doctrine of the double truth was denounced by the bishop of Paris both in the prologue to his Condemnation of 1277 and specifically in article 90, concerning the eternity of the world. See Grant, 1974, 47 (and n. 6) and 48 (for article 90) and Maurer, 1955.

98. I have argued elsewhere (Grant, 1974, 510, n. 61) that in the discussion of the earth's possible axial rotation, Oresme "sought to humble reason and show that physical arguments could not establish a relatively simple physical problem." Sylla disagrees (1991, 217) but offers no explanation as to why Oresme concluded his lengthy discussion of the earth's axial rotation by invoking Scripture in favor of the earth's traditional immobility and why he thought his discussion was useful against those who would "impugn our faith." In light of my arguments in Grant, 1974, 510, n. 61; Grant, 1978c, and Grant, 1988, I continue to believe that my own interpretation agrees tolerably well with Oresme's complex approach as a theologian–natural philosopher who possessed an outlook quite different from that of Buridan. We might say that Buridan emphasized the positive, always seeking to determine what we can know with reasonable or sufficient certainty. By contrast, although Oresme was also interested in what we could know with reasonable or sufficient certainty, he also tended, as a theologian, to emphasize the limits of human knowledge and the unreliability of our senses. Twice, at least, he professed Socratic ignorance about natural knowledge (see Grant, 1978c, 111, and 121, n. 32).

99. In this section, I draw heavily on Grant, 1984b, often incorporating lengthy sections verbatim.

relativity of motion, as illustrated by the movement of ships; that it is better for the earth to complete a daily rotation with a very much slower velocity than would be required for the vast heavens; that the air shares the daily rotation of the earth; that the motion of bodies rising and falling with respect to a rotating earth results from a motion compounded of rectilinear and circular elements; and, finally, that since a state of rest is more noble than motion, it is more appropriate for the ignoble earth to rotate than it is for the nobler heavens.

Did Copernicus derive some, or all, of these arguments from Oresme and/or Buridan, especially the latter, whose works were known in eastern Europe and were perhaps even studied at the University of Krakow in the late fifteenth century, when Copernicus was a student there? Other than a striking similarity of arguments, there is as yet no evidence that Copernicus knew these treatises or derived his arguments from medieval sources.

Prior to the impact of the Copernican theory, the problem of the earth's possible axial rotation – usually considered within the framework of a question as to whether the earth lies at rest in the center of the world – did little more than offer an opportunity for some critical and sophisticated argumentation before one acquiesced in the traditional opinion. The advent of the Copernican theory changed all that. By the end of the sixteenth century, scholastic natural philosophers were confronted with contemporary advocates of the new heliocentric theory, within which the daily axial rotation of the earth was a vital element. The kinds of arguments that Buridan and Oresme had given for the axial rotation of the earth in the fourteenth century were now repeated, along with others, within a broader framework that removed the earth from the center of the universe and made of it just another planet revolving around the Sun, which now became the new center of the world.

It was not just Copernicus who had brought the earth's daily rotation to the forefront. Indeed, if Copernicus had written nothing, his contemporaries and successors would nevertheless have considered the possibility of the earth's axial rotation, and quite possibly its orbital motion as well. For not only were all the sources available that had been known in the Middle Ages – for example, passages in Aristotle's *De caelo*, where both possible terrestrial motions are mentioned; Ptolemy's *Almagest*; Seneca's *Natural Questions*; and Simplicius's *Commentary on the De caelo* – but so were some of the arguments devised by Buridan and Oresme (though admittedly not the best ones), since Albert of Saxony had incorporated them into his questions on *De caelo*, which was printed in at least six editions between 1481 and 1520.[100]

Of the sources that became available in the sixteenth century, Copernicus himself cited two (in the preface to the *De revolutionibus*, or *On the Revolutions*), namely Cicero's *Academica* (2.39.123), where, according to Theophrastus, the Syracusan Hicetas is said to have assumed the earth's axial

100. For references, see ibid., 1–2.

rotation and the immobility of the planets and stars,[101] and the *De placitis philosophorum* (bk. 3, ch. 13), falsely ascribed to Plutarch, where Philolaus the Pythagorean is said to have ascribed an orbital motion to the earth and Heraclides of Pontus and Ecphantus the Pythagorean assigned to it a west-to-east axial rotation.[102] In his treatise *Concerning the Face which Appears in the Orb of the Moon* (*De facie quae in orbe lunae apparet*), Plutarch also found occasion to mention the earth's axial and orbital motions when he reported a charge of impiety brought by Cleanthes against Aristarchus of Samos, who was "disturbing the hearth of the universe because he sought to save [the] phenomena by assuming that the heaven is at rest while the earth is revolving along the ecliptic and at the same time . . . rotating about its own axis."[103]

Supporters of the traditional Aristotelian cosmology were as knowledgeable about these new references to possible terrestrial motion as were their opponents, and they were also reasonably knowledgeable about the arguments of such opponents as Copernicus, Galileo, Kepler, and others, who espoused the cause of axial rotation. The new sources were frequently mentioned in Aristotelian commentaries of the sixteenth and seventeenth centuries, especially those on *De caelo*. The scope of traditional medieval commentaries was thus considerably expanded. Even had they never heard of the Copernican theory, Aristotelians would have been obliged to take cognizance of these claims. All available evidence indicates, however, that it was Copernicus's arguments, rather than the brief, unsupported, fragmentary statements from the sources just mentioned, that eventually posed the real challenge to Aristotelian cosmology. With a few exceptions, of whom Christopher Clavius was one, Aristotelians and the Catholic Church were slow to arouse themselves against the arguments in favor of the earth's motions in the first book of *On the Revolutions*.[104] As Koyré explains, "Only at a much later date, when it became evident that this work of Copernicus was not intended for mathematicians alone; when it became clear that the blow to the geocentric and anthropocentric Universe was deadly; when certain of its metaphysical and religious implications were developed in the writings of Giordano Bruno, only then did the old world react." It did so by attempting to suppress the new ideas of

101. In Rosen's translation, Hicetas is said to have believed "that the sky, sun, moon, stars, in short, all the heavenly bodies stand still, and that nothing in the universe moves except the earth." See Copernicus, *Revolutions* [Rosen], 1978, 341.

102. The Coimbra Jesuits mention Philolaus, Heraclides, and Ecphantus and cite *De placitis* as their source. Conimbricenses, *De coelo*, bk. 2, ch. 14, qu. 5, art. 1, 1598, 389.

103. For references to all of these works and additional information, see Grant, 1984b, 2, nn. 7–9.

104. Sometime between 1545 and 1547, very shortly after the publication of *De revolutionibus* in 1543 and long before Clavius, Giovanni Maria Tolosani (1470/1–1549), the Dominican theologian-astronomer, apparently obtained a copy of Copernicus's work. In a treatise which he perhaps hoped to publish, Tolosani severely criticized the work and even spoke of plans to condemn it. His treatise, however, remained unpublished and caused no public reaction. See Westman, 1986, 87–89.

the universe by "the condemnation of Copernicus in 1616 and of Galileo in 1632."[105] Amicus reveals the impact of Copernicus when he declares that "In our age Copernicus has raised this opinion [of the earth's daily axial rotation], which had been buried with the dead, in a work *De revolutionibus orbium caelestium*, where he says that the heavens are immobile and the earth is moved." Copernicus raised this old, but not quite dead, issue by the assumption of an "immobile firmament, with the Sun fixed in the center of the universe, and finally the earth, in the third heaven, which is moved by a triple motion, by [means of] which he attempts to save all the appearances."[106] The three terrestrial motions to which Amicus alludes are (1) a daily axial rotation; (2) an annual west-to-east motion around the Sun; and (3) what may best be described as "other motions," which could be a rectilinear movement; a trepidational, or axial, motion; and even earthquakes, depending on the author.[107]

Of the three motions, the daily rotation attracted the most attention. With an occasional exception such as Riccioli, most scholastics considered the daily and annual rotation together, although they concentrated on the daily rotation, as we shall do.

a. *Physical arguments based on the common motion*

None of the physical consequences derived from acceptance of the earth's axial rotation was more pervasive and perplexing than that of the common motion, which assumed that all bodies on and above the earth's surface shared in the earth's rotational motion. Ptolemy had already used the concept of a common motion to subvert belief in the earth's daily rotation, while Nicole Oresme, as we saw, defended it as plausible.[108] During the seventeenth century, however, it was Copernicus's version of the common motion argument that became the focal point of scholastic reaction. In *On the Revolutions*, Copernicus declares (bk. 1, ch. 8 [Rosen], 1978, 16) that "the motion of falling and rising bodies in the framework of the universe is twofold, being in every case a compound of straight and circular." Oresme had earlier analyzed rectilinear motion on a rotating earth in precisely the same way. Copernicus, however, went much farther: he abandoned the Aristotelian idea that rectilinear motion was natural for the elements. As long as earth and fire are on the earth, their natural motion is circular, for

105. Koyré, 1973, 17. The condemnation of Galileo, however, was in 1633, not 1632.
106. "Hanc opinionem sepultam a mortuis excitavit nostra aetate Copernicus opus *de revol. orb. cael.*, ubi ait caelos esse immobiles et terram moveri sed ponit firmamentum immobile, tum Solem fixum in centro universi; denique est terra in tertio caelo, quae triplici motu movetur, per quos conatur salvare omnes apparentias." Amicus, *De caelo*, tract. 5, qu. 6, dubit. 1, art. 1, 1626, 289, col. 1.
107. For examples, see Grant, 1984b, 34, n. 122.
108. Ptolemy had argued that even if the air was carried around with the earth, objects in the air would always seem to be left behind by the motion of earth and air. See Ptolemy, *Almagest*, bk. 1, ch. 7 [Toomer], 1984, 45.

they are carried around by the earth. Only "when they are separated from their whole and forsake its unity" do they move rectilinearly (ibid., 17). Even then, however, despite their detachment from the earth, watery and earthy things in the air, and the air itself, also share in the earth's rotational motion.

No one defended the earth's axial rotation better than Galileo, whose arguments were designed to refute the so-called absurdities that Aristotelians and anti-Copernicans had alleged to follow from the daily rotation. The thrust of those arguments was nicely described by Mastrius and Bellutus, who vigorously opposed Galileo. Galileo denied these absurdities, they explain,

> because not only is the earth moved innately [*ab instrinseco*] around the center with a circular motion, but also all bodies, whether animate or inanimate, whether united to the earth or separate, that exist in this elemental universe have this motion perpetually [and] innately so that they move simultaneously with the earth around the center of this elemental world. And because this motion is common to all, it is not perceptible to us except in relation to the fixed stars to which it does not apply. ([*De coelo*, disp. 4, qu. 4, art. 3], 1727, 3:562, col. 1, par. 114)

Scholastic arguments against the Copernicans have a familiar Ptolemaic ring. There were those, like Raphael Aversa, who argued against the earth's rotation as if no one had ever proposed the common motion argument. The earth's immobility, Aversa insisted, could be demonstrated from a variety of experiences. If the earth really turned from west to east daily, the clouds would appear "to be carried constantly from east to west and in no way to remain over the same place of the earth." When anyone projects a stone upward with great force, it ought to fall to the earth considerably to the west, "because the motion of the earth has, in the interim, continued from west to east." This fails to happen. Aversa presents another familiar argument that emerged from the medieval tradition when he asserts that if the earth rotates swiftly from west to east, we should be aware of a strong wind from east to west. But no such effect is perceived. For all these reasons, Aversa concludes that "it is surely not the earth that is revolved constantly with a daily motion."[109]

Although such arguments were frequently repeated, some scholastics were aware that Copernicans had attempted to meet them by assigning the earth's rotational motion to all things in the air that were above the earth's surface. Indeed, as we saw, Nicole Oresme had done this in the fourteenth century. He appears to have believed that arguments in favor of the earth's daily rotation based on the common motion concept were sound. Not many

109. The arguments in this paragraph appear in Aversa, *De caelo*, qu. 34, sec. 5, 1627, 142, col. 2. Amicus, *De caelo*, tract. 5, qu. 6, dubit. 1, art. 1, 1626, 289, col. 2, mentions the same argument but omits the Copernican rebuttal. Galileo mentions it in the Second Day of his *Dialogue* ([Drake], 1962, 131–132).

scholastics would argue that boldly in the seventeenth century, especially after the condemnation of the Copernican theory in 1616 and the condemnation of Galileo in 1633.

Riccioli was an interesting exception. His treatment of the question of the earth's immobility or mobility in the *New Almagest* of 1651 was probably the lengthiest, most penetrating, and authoritative analysis made by any author of the sixteenth and seventeenth centuries. Most, if not all, of the known arguments for and against the earth's immobility are probably included within the approximately 126 arguments that he inserted into his famous treatise.[110] And, as had become the custom in scholastic questions on Aristotelian natural philosophy by the late sixteenth century, he meticulously and scrupulously cites the arguments of earlier and contemporary authors, which marked a dramatic departure from medieval practice, where citation of names and treatises in Aristotelian *questiones* was far from customary. But what is most remarkable about Riccioli's *New Almagest* is the spirit in which the arguments are presented. Although, as a Catholic and a Jesuit, he was committed to a rejection of the earth's daily axial rotation and annual revolution as described in the Copernican system, he presented the physical and metaphysical arguments for and against the earth's motions in an unusually evenhanded manner. In Riccioli's ultimate acceptance of the immobility of the earth, biblical and theological arguments proved decisive.

He repeated the arguments that I have cited earlier and similar ones. In each case, however, Riccioli concludes with the Copernican response based on the common motion. For example, he describes the claim that if the earth rotates from west to east we should have greater difficulty moving toward the west because of the air's resistance as the earth sweeps past. Riccioli then adds that the Copernicans deny that such a resistance would develop, because "there is a common motion toward the east for bodies similar to the earth, just as with the air near the earth."[111] In these particular instances, though not in many others, Riccioli apparently chose not to resolve the argument in favor of the earth's immobility. Indeed, he appears to have deliberately avoided a resolution of the issue. Neverthless, the tendency in scholastic responses to the common motion argument, which represented the core of the Copernican defense of axial rotation, was to subvert it by whatever means possible.

110. The arguments appear in section 4 of book 9, which is titled: "On the System of a Moved Earth" (De systemate terrae motae).

111. See Riccioli, *Almagestum novum*, pars post., bk. 9, sec. 4, ch. 34, 1651, 474, col. 1 (13th argument). In his twenty-eighth argument, he also concludes with the Copernican argument. There he not only mentions a perpetual wind toward the west, which was commonly mentioned, but adds that we should also perceive "sounds and hissing from the air striking against trees, mountains, towers, etc." Since we do not perceive such things, the earth must rest. But in the conclusion of his argument, Riccioli seems to defer to the Copernicans by observing that they deny such effects by their insistence that the air near the earth, which is filled with exhalations and vapors, would move with the common motion of the earth. Ibid., 475, col. 1.

Riccioli himself furnished a number of arguments in favor of the earth's immobility and against its rotation in which he ignored Copernicus and appealed strongly to sense experience. Riccioli emphasized that heavy bodies "descend naturally by a straight line perpendicular to the earth" and that if projected upward "they would return over the same path to the same place." So obvious was this experience that it "could not be shown to be false by any more certain sensational, nor by any necessary *a priori* arguments, nor from things revealed by God." Only two alternatives are possible: either heavy bodies descend in a path that is a straight line, or they descend by means of a nonrectilinear line that only appears rectilinear. Those who argue against the senses and experience insist that the senses are false and misleading. Indeed, they hold that such a judgment must not be made on the basis of the senses. For Riccioli, however, who speaks here for all Aristotelian geocentrists, the physical evidence is not merely that of a few sensations and experiences "but [arises] from the sensation of all [and has been] repeated nearly an infinite number of times and which maintains its force as long as the contrary does not prevail." Surely, Riccioli concludes, "if it is not evident to the sense that heavy bodies descend through a straight line, nothing will be evident to it and the whole of physical science will be destroyed."[112]

To reinforce the case for terrestrial immobility, Riccioli also appeals to intuition, when he argues, as would most Aristotelians, that "the nature of heavy and light bodies demands that they be returned to their places and united to their whole by means of the shortest path."[113] On the assumption of the earth's rotation, however, the paths of heavy and light bodies would be curvilinear and longer, rather than perpendicular and shorter. With a seeming sense of contempt, Riccioli accuses the Copernicans of saving their hypothesis at any cost, even ignoring the nature of heavy bodies.

Some scholastics fastened onto Copernicus's claim that sublunar bodies possess an innate tendency to move around the center of the earth. By this inherent property, the earth rotates around its own center just as do watery, airy, and fiery bodies. Mastrius and Bellutus challenged this claim on traditional grounds ([*De coelo*, disp. 4, qu. 4, art. 3], 1727, 3:562, col. 2, par. 115). Each of these four types of sublunar body differed in species and genus from the others and would not all possess the same tendency to rotate with the earth.[114] Were Mastrius and Bellutus inconsistent here? The same argument would seem applicable to rectilinear motion. If the four elemental

112. Ibid., 473, col. 1 (6th argument).
113. Ibid., col. 2 (7th argument).
114. Further absurdities would follow from the fact that each of the four elements should have a determinate velocity and yet their speeds would have to vary with their latitude. Thus one and the same stone would rotate fastest at the equator and slowest at the pole (see Mastrius and Bellutus, *De coelo*, disp. 4, qu. 4, art. 3, 1727, 3:562, col. 2, par. 115; and Grant, 1984b, 38). They also present here arguments denying the alleged capability of the earth to communicate its rotatory property to every one of its parts.

bodies differed in species and genus, why should they all possess a common tendency for rectilinear motion?[115]

i. Ships and the common motion. Despite Oresme's astute analysis of motion aboard a moving ship (see Sec. V.1.b of this chapter), arguments about the relationship between a moving ship and objects dropped from its mast or hurled upward from its deck or carried within its cabins did not form part of the traditional medieval core of arguments about a rotating earth and the objects that moved upward and downward with regard to its surface. Such arguments did not become a regular feature of the controversy over terrestrial motion until after Copernicus utilized them in defense of his own position.[116] Galileo mentioned the various motions of animate and inanimate things located within the confines of a cabin below the decks of a ship. He insisted that in such a cabin the observed motions of flies, men, fish, and water dripping from one vessel to another would be the same, whether the ship was at rest or in motion, provided only that the ship's motion was uniform and without fluctuation.[117] With some embellishments and additions, Mastrius and Bellutus described Galileo's argument but did not offer a direct refutation.[118]

Riccioli also cited it but used it in favor of the earth's immobility. He introduced it following a discussion of the following proposition (*Almagestum novum*, pars post., bk. 9, sec. 4, ch. 21, 1651, 423, col. 2):

> If the earth were moved with a daily rotation, or even an annual translation, the clouds hanging in the air, the smoke that rises, and the birds that are suspended [in the air] or flying toward the east would always seem to be carried toward the west. But this is contrary to manifest experience. Therefore the earth is not moved with a daily rotation, and much less with an annual translation.

Following a brief consideration of this claim, Riccioli invokes Galileo's argument about objects in an enclosed cabin and does so with apparent approval (424, col. 1). For if the motions of the animate and inanimate objects in the enclosed cabin are precisely the same whether the ship moves or rests, a consequence that follows from the fact that the rest and motion of the ship are common to all, one may not infer the rest or motion of the ship from the motion of the objects in the cabin. But Riccioli's argument

115. Perhaps they might have argued that these rectilinear motions differ, one moving up, the other down.
116. See Copernicus, *Revolutions*, bk. 1, ch. 8 [Rosen], 1978, 16.
117. See Galileo, *Dialogue*, Second Day [Drake], 1962, 186–187. Some 250 years earlier, Oresme, *Le Livre du ciel*, bk. 2, ch. 25, 1968, 525, had made the same argument when he declared that "Inside the boat moved rapidly eastward, there can be all kinds of movements – horizontal, criss-cross, upward, downward, in all directions – and they seem to be exactly the same as those when the ship is at rest." The passage is included in Grant, 1974, 505.
118. For the embellishments and additions, see Grant, 1984b, 37–39.

subverts his own proposition about the birds in the air. By his seeming approval of Galileo's argument, Riccioli had, in effect, conceded that even if birds do not appear to move toward the west when flying eastward, we cannot properly infer from this that the earth is at rest. The birds may appear to move eastward rather than westward either because the earth is really at rest or because the birds share the eastward motion of a rotating earth, as Galileo and the Copernicans believed. And yet, in the proposition cited earlier, Riccioli did indeed infer the immobility of the earth from the eastward flight of the birds. Rather than uphold Riccioli's argument for the immobility of the earth, the Galilean argument demonstrated its inconclusiveness, a situation of which Riccioli seems to have been unaware.

As part of his use of shipboard experiences to defend the earth's rotational motion, Galileo, according to Mastrius and Bellutus, had argued that a stone projected upward in a cabin would fall at the projector's feet, because both projector and stone are moved with the ship. However, the stone does not fall with a perpendicular motion but follows the path of a slanting line (*linea transversalis*), which results from the perpendicular motion of the stone and the horizontal motion of the ship.[119] Mastrius and Bellutus declare further that Galileo used the common motion of the rotating earth to explain why an arrow projected upward would fall at the foot of the projector. Indeed, they inform us that Galileo declared that "he himself had experienced many times that a stone projected from the top of a mast always fell to the foot of the mast, never into the sea, whether the ship was at rest or was moved very quickly."[120] To refute Galileo, Mastrius and Bellutus appealed to Johannes Cottunius (1577–1658), a professor of philosophy and theology, who, in a commentary on Aristotle's *Meteorology* (bk. 1, lec. 16) claimed that he had witnessed the fall of stones from the mast of a ship and not once did any of them fall to the foot of the mast; rather they dropped into the water off the stern of the ship.[121] Thus did they counteract, or

119. Mastrius and Bellutus, *De coelo*, disp. 4, qu. 4, art. 3, 1727, 3:562, cols. 1–2, par. 114. Although Galileo did not describe the path of a descending body in the cabin of a moving ship as "slanting," he earlier declared that the path of a stone falling toward a rotating earth would be slanting (Galileo, *Dialogue*, Second Day [Drake], 1962, 173). Tycho Brahe rejected the common motion argument and insisted that a body projected upward would fall quite differently when the ship is at rest than when it is in motion. The greater the ship's velocity, the greater the distance that would separate a falling body from the place to which it fell when the ship was at rest from that to which it fell when the ship was in motion. See Overmann, 1974, 14, where the passage from Tycho's *Epistolarum astronomicarum liber primus* is translated from *Tychonis Brahe Dani Opera omnia*, ed. J. L. E. Dreyer (Copenhagen, 1919), 6:220, lines 16–21.

120. In the Second Day of his *Dialogue* ([Drake], 1962, 126), Galileo describes the typical Aristotelian interpretation of the ship experiment; on 180, Galileo has Salviati, his spokesman, declare that the anti-Copernicans have never dropped a body from the mast of a moving ship. However, elsewhere in the same Second Day of the *Dialogue*, Salviati first implies that he had performed the experiment ("For anyone who does [perform this experiment] will find that the experiment shows exactly the opposite of what is written," 144) but then, a few paragraphs later, appears to admit that he had not performed it (145).

121. Mastrius and Bellutus, *De coelo*, disp. 4, qu. 4, art. 3, 1727, 3:563, col. 1, par. 116.

neutralize, Galileo's claim. While admitting that they themselves had never observed such a demonstration, Mastrius and Bellutus were nevertheless convinced that reason (*ratio*) would yield the same result as Cottunius reported. If the earth rotated from west to east, they assumed that two eastward motions should be distinguishable in a ship moving eastward: (1) the common west-to-east motion of the earth, and (2) an eastward motion caused by the force of the wind.[122] In their analysis, however, they assume that as the stone falls, it is influenced only by the earth's rotation and not by the ship's own eastward, or proper, motion. They were therefore led to conclude that the stone will not fall at the foot of the mast. Nor would the stone fall to the foot of the mast if the ship's direction were westward, instead of eastward: it would fall in the water off the ship's stern.[123]

Like Mastrius and Bellutus, Amicus, who substitutes an arrow (*sagitta*) for a stone, assumed that an arrow's path was independent of the ship's motion, from which he concluded that the greater the velocity of the ship, the farther behind it would the arrow fall.[124] From such arguments many scholastics convinced themselves that the earth did not rotate. Unfortunately, we cannot properly evaluate the role played by the condemnation of the Copernican theory in 1616. Did it compel scholastics to argue against a rotating earth by whatever means possible, so that the basic common motion argument had to be ignored? Whatever the reason, some two hundred fifty years earlier medieval scholastics such as Buridan and Oresme had embraced the common motion argument as essentially correct and thus were able to argue more cogently than their scholastic successors. Their rejection of the earth's rotation was on other grounds.

ii. Cannonballs to east and west. During the Middle Ages, the manner in which a rotating earth might affect cannonballs fired either to east or west was not a problem that surfaced in discussions about the possible rotation of the earth. Cannons were already in use during the fourteenth century,[125] but no one linked them with a possible rotating earth. One of the first to do so, if not the first, was Tycho Brahe. In letters written to Christoph Rothmann (d. ca. 1608) between 1586 and 1590, Tycho denied the Copernican claim that a heavy body falls simultaneously with rectilinear and circular motions. Because the two motions would be natural, Tycho concluded that they would interfere with each other. Moreover, how could

Cottunius was a Greek who studied at the Greek College in Rome and even founded a college for indigent Greeks at Padua in 1653. In addition to philosophy and theology, he also earned a doctorate in medicine at Padua. His commentary on the *Meteorology* was apparently unpublished (see Lohr, 1988, 105).

122. Presumably, this is the ship's own eastward motion.
123. For more details, see Grant, 1984b, 41–42.
124. Amicus believed (*De caelo*, tract. 5, qu. 6, dubit. 1, art. 1, 1626, 289, col. 2) that an arrow shot upward from the deck of a moving ship would not return to the place from whence it was launched.
125. See Hall, 1954–1984, 2:726–727.

bodies that fell with a variety of rectilinear speeds move with the same rotational speed as the earth? Even if one conceded that a body detached from the earth's surface could somehow move with two such simultaneous motions and thus follow the earth's rotation, a third and violent motion, which would render the rotational hypothesis untenable, also had to be considered. Tycho imagined that first a lead, iron, or stone ball was fired toward the east, after which, from the same location, another ball of equal size and weight was fired toward the west. Each cannonball would be moved by three simultaneous motions: (1) a natural motion toward the earth's center; (2) a natural rotational motion following the earth; and (3) a violent motion, caused by the powder exploding in the cannon. Convinced, as most Aristotelians would have been, that the natural, downward motion of a projectile hurled upward cannot commence until the violent upward motion is destroyed, Tycho applied this reasoning to the cannonballs. If they possessed a natural rotational motion transmitted to them by a really rotating earth, that natural, circular motion would be impeded by the violent motion caused by the powder exploding in the cannon. Consequently, the ball fired eastward should advance hardly any distance from the cannon, because the latter would be carried swiftly eastward with the rotating earth whereas the cannonball would move only with its violent eastward motion. The two eastward motions would prevent much of a separation.[126]

By contrast, the cannonball shot westward should be far removed from the cannon, because the latter would be carried eastward by the rotating earth whereas the cannonball would move westward by virtue of its violent motion, which would also negate its circular motion. Experience reveals no such discrepancies but shows rather that the cannonballs would travel equal distances.

Because of Tycho's great prestige, his argument might have served the cause of traditionalists.[127] But scholastic natural philosophers did not cite it often, perhaps because it was more complicated than many others that could also be invoked. Mastrius and Bellutus, however, furnished a variant of the argument when they declared that a cannonball shot toward the west should have a greater impact than one shot toward the east. The earth's easterly rotation would cause the greater and lesser impacts. Thus if the cannonball fired westward struck a house, the latter, carried eastward by the earth's swift rotation, would meet the cannonball head-on and, as Mastrius and Bellutus put it, "the impetus toward the west would be as if doubled."[128] The circumstances are radically different toward the east, where the force

126. We observe that in this argument, the first of the three motions – a natural motion toward the earth's center – plays no role because the violent motion is operative. The same is true when the cannonball is shot westward.

127. Galileo attempted to meet Tycho's argument in a number of places. See Galileo, *Dialogue*, Second Day [Drake], 1962, 126–127, where north–south shots are also considered, and 168, 171, 174, and 180.

128. "Quia versus occasum veluti duplicaretur impetus." Mastrius and Bellutus, *De coelo*, disp. 4, qu. 4, art. 3, 1727, 3:562, col. 1, par. 113.

of impact is diminished because the house is moving away from the oncoming cannonball.

Riccioli found arguments about cannonballs fired toward the cardinal points of considerable interest. Indeed, he also used the concept of impetus to express his results. Thus Riccioli declares (*Almagestum novum*, pars post., bk. 9, sec. 4, ch. 21, 1651, 427, col. 2) that "if the earth were moved with a daily motion, or even an annual motion, the same ball that is thrust forward by the same force for the same distance once to an eastern target and then to a western target would strike the eastern target with a stronger impact than the western target." These results were the opposite of those arrived at by Mastrius and Bellutus.

Much of Riccioli's reasoning about cannon shots in the different cardinal directions was based on his conviction that if the earth rotated on its axis from west to east, that rotation must confer an impetus on all bodies rotating with the earth. A body moved in a direction opposed to the daily rotation had an impetus that conflicted with the impetus it received from the daily rotation. Thus, in Riccioli's view, and that of Tycho as well, the final location of a cannonball on a rotating earth would be determined by the mutual interaction of the motive forces within the cannonball. For example, a ball projected eastward would be aided by the earth's eastward motion, which would add to the impetus imparted by the cannon or projector. By contrast, a ball hurled or projected westward would be affected by two oppositely directed impetuses: the impetus driving the ball westward would be retarded by the impulse of the ball to follow the earth's rotation eastward; similarly, the impetus that would normally carry the ball eastward with the earth's rotation would be diminished, or interfered with, by the contrary impetus impelling the ball westward. In brief, for westward cannon shots, the two impetuses resist and interfere with each other; for eastward shots, they reinforce each other. Because such disparate effects were not observed in the behavior of cannonballs, Riccioli concluded that the earth does not rotate.

Despite its oblique path, Copernicans insisted that a cannonball had its own proper motion and struck a target directly because target, cannon, and cannonball equally possess the common circular motion of the earth's daily rotation. Copernicans were committed to an interpretation that demanded they analyze every terrestrial motion as if it were compounded of two motions, its own proper motion and the common motion that it shared with the earth and all other objects. The two component motions did not, however, interfere with each other, an interpretation that Riccioli could not accept because of his conviction that each motion of a body supplied a quantity of impetus to it. If two or more distinct motions were involved in the production of an observable motion, the impressed forces associated with those motions must necessarily interfere with each other. Such mutual interference was not confined to contrary forces but also occurred with forces that were impressed obliquely.[129] With his assumption that impetus was imparted to a cannonball by both the powder that was exploded in the cannon *and* the earth's eastward rotation, Riccioli was committed to a wholly different analysis from that of the Copernicans. The consequences he derived from the application of these ideas were contrary to experience. He thus felt justified in rejecting the earth's daily rotation.[130]

Because he was an astronomer with mathematical and technical competence, Riccioli was rare among scholastics. Indeed, he not only accepted and understood Galileo's distance-and-time formulation for falling bodies (that is $s \propto t^2$, where s is distance and t is time), but had himself experimentally verified it.[131] He was one of the few who could cope with arguments that utilized the assumption that bodies fell with uniformly accelerated motion. Thus did he offer a series of unusual arguments to show that uniformly accelerated fall was incompatible with a rotating earth. Bodies could not really accelerate in descending toward the earth if the earth rotated on its axis.[132] Although his arguments were based on misunderstandings, Riccioli's attempt to use the latest scientific concepts in defense of terrestrial immobility is noteworthy.

iii. Other physical arguments. Many other physical arguments with seemingly endless variations could be cited. Only a few will be mentioned. One that was widely discussed in the sixteenth and seventeenth centuries is traceable to Copernicus, who mistakenly ascribed to Ptolemy the opinion that if the earth rotated, "living creatures and any other loose weights would by no means remain unshaken."[133] As in many instances, it was Clavius who had installed this argument in the scholastic repertoire against a rotating earth. To those who countered – as did Galileo some years later (*Dialogue*, Second Day [Drake], 1962, 189–190) – that the earth's rotation would no more cause buildings to collapse than would the swift revolution of a vessel filled with water cause the water to be expelled, Clavius devised a response. "The whole impetus of the water," he explained ([*Sphere*, ch. 1], 1593, 213), "is impressed toward the lower parts of the vessel, not toward its orifice. But the impetus impressed on the buildings is toward the farthest parts of the earth." In these somewhat cryptic words, Clavius seems to say that the water remains in the vessel because the impetus, or force, impressed

129. "Certi enim sumus," Riccioli insists, "ex plurimis experimentis motum semel impressum ac motivum versus unam partem debilitari ac minui ab impetu non tantum in contrariam sed etiam in alienam partem, seu in transversum movente." Riccioli, *Almagestum novum*, pars post., bk. 9, sec. 4, ch. 21, 1651, 427, col. 2.
130. For some of Riccioli's other arguments on cannonballs, see Grant, 1984b, 45–50.
131. See Koyré, 1953, 229–232 and 1955, 349.
132. For details and references, see Grant, 1984b, 51–54.
133. Copernicus, *Revolutions*, bk. 1, ch. 7 [Rosen], 1978, 15. On the falsity of the attribution to Ptolemy, see Rosen's note to page 15, line 17, on p. 351 and also Galileo, *Dialogue* [Drake], 1962, 481–482 (note to p. 188). Galileo who also attributed the same argument to Ptolemy, embellished it by adding (188) that "if the earth turned upon itself with great speed, rocks and animals would necessarily be thrown toward the stars."

on the water is totally concentrated at the bottom of the vessel so that the water tends toward the bottom of the vessel and cannot depart. The earth's rotation, by contrast, causes the impetus to concentrate at its surface and, perhaps like an earthquake, to crumble the foundations of the buildings on it.[134]

Amicus took this argument a step farther, arguing that even if the buildings could stand for a time on a rotating earth, they must eventually collapse as a consequence of that rotation. A few years later, Galileo countered that those who believed that buildings would collapse on a rotating earth could not also believe that the earth had always rotated, for otherwise how could the buildings ever have been constructed? Those who upheld such an argument had to assume that the earth was initially at rest when the buildings were constructed. With the commencement of rotation, however, the buildings would quickly collapse.[135]

Mastrius and Bellutus ([*De coelo*, disp. 4, qu. 4, art. 3], 1727, 3:563, col. 2, par. 119) report yet another argument in which impetus was involved. Because fire and air are moved circularly, so also ought the globe of earth be moved circularly, presumably by an impetus transmitted from heaven to earth via the spheres of fire and air. They denied the physical feasibility of this claim, because a fluid body like air could not push a solid body like earth. The latter is not only too heavy to be rotated by the air but too distant from the heaven to be affected by any impetus transmitted by the heaven to fire or air.

Scholastics did not rely only on arguments that rejected the earth's rotation but offered positive reasons in defense of the earth's immobility. Amicus ([*De caelo*, tract. 8, qu. 4, dubit. 6, art. 3], 1626, 601, col. 1) agreed with Aristotle that the earth's heaviness caused it to rest in the center, which is the lowest place in the universe. Or, as Mastrius and Bellutus ([*De coelo*, disp. 4, qu. 4, art. 3], 1727, 3:564, col. 2, par. 121) put it, the earth rests in the middle of the universe "because it is in the lowest place." The earth remains motionless in the center of the world because any movement away from the center would be an ascent, which is repugnant to the earth's heaviness.[136] That heavy things always rest at the world's center was obvious to Amicus, who was convinced that if a stone were dropped through a hole imagined to extend from one side of the earth's diameter to the other, "it would not be moved except to the middle, and there it would naturally rest and not proceed beyond except by force."[137]

134. By a parity of reasoning, the impetus impressed on the water at the bottom of the vessel ought to cause the bottom to collapse. Amicus repeated the argument and added qualifications. Amicus, *De caelo*, tract. 5, qu. 6, dubit. 1, art. 1, 1626, 289, col. 2.

135. Galileo, *Dialogue*, Second Day [Drake], 1962, 189, posed this argument against Ptolemy's alleged claim that buildings on a rotating earth would collapse. We have already seen that this was not Ptolemy's argument.

136. Amicus, *De caelo*, tract. 8, qu. 4, dubit. 6, art. 3, 1626, 601, col. 2. Aversa expressed the same opinion (*De caelo*, qu. 36, sec. 6, 1627, 231, col. 2).

137. Amicus, ibid. Amicus's conclusion that a stone dropped into a hole through the earth's

b. *Metaphysical arguments: simplicity, order, and nobility*

The strongest links between medieval and seventeenth-century scholastic arguments about the earth's rotation lie in the area of metaphysical appeals based on simplicity, order, and nobility. The Copernican system was itself a powerful statement for simplicity and order. Scholastic natural philosophers frequently recognized this when they presented arguments in favor of a rotating earth, arguments which they subsequently had to refute. The principle of Ockham's razor was the essential statement in behalf of the simplicity argument. Scholastics often invoked some version of it, as when Amicus declared that "it is vain to do with many [things] what can be done equally well with fewer."[138] In light of the principle of simplicity, would it not be "easier and of less cost [or effort]," queried Riccioli, "to move the small [*pusillum*] globe of the earth than the immense machine of the heavens? Therefore God and Nature, which do what is easier, move the earth, rather than the heavens, with a daily motion."[139] Such arguments were also applicable to speed as well as weight. If the stars and planets turned daily, instead of the earth, they would move with incredible speeds despite their enormously greater heaviness than the earth.[140]

Like Buridan before him, Riccioli was not impressed with simplicity arguments.[141] The great velocities of the celestial spheres are of no conse-

center would come immediately to rest had already been rejected in the fourteenth century by Albert of Saxony and Oresme, who argued that the residual, uncorrupted impetus in the falling stone would cause it to proceed past the center and ascend toward the heavens. And, as Albert put it, "in so ascending, when the impetus would be spent, it would conversely descend. And in such a descent it would again acquire unto itself a certain small impetus by which it would be moved again beyond the center. When this impetus was spent, it would descend again. And so it would be moved, oscillating [*titubando*] around the center until there no longer would be any such impetus in it, and then it would come to rest." For the translations, see Clagett, 1959, 566; for Oresme's version, ibid., 570 and 533. Clagett's translation of Albert's question is from bk. 2, qu. 4 ("Whether every natural movement is swifter in the end than in the beginning"), of the Venice, 1492 edition of Albert's *De celo*. That question is not included in the 1518 edition of Albert's *De celo* used in this study.

138. "Quia frustra fit per plura, quod potest aeque bene fieri per pauciora." Amicus, *De caelo*, tract. 5, qu. 6, dubit. 1, art. 1, 1626, 288, col. 2. For equivalent statements, see Cornaeus, *De coelo*, tract. 4, disp. 3, qu. 2, dub. 14, 1657, 532, and Mastrius and Bellutus, *De coelo*, disp. 4, qu. 4, art. 3, 3:563, col. 2, par. 119. Galileo also expressed and used the principle of simplicity at least twice in his *Dialogue*, Second Day [Drake], 1962, 117, 123, giving the Latin text in the second reference ("frustra fit per plura quod potest fieri per pauciora").
139. Riccioli, *Almagestum novum*, pars post., bk. 9, sec. 4, ch. 33, 1651, 466, col. 2 (5th argument). See also Amicus, ibid.; Cornaeus, *De coelo*, tract. 4, disp. 3, qu. 2, dub. 14, 1657, 532, and dub. 16, 537; and Mastrius and Bellutus, *De coelo*, disp. 4, qu. 4, art. 3, 563, col. 2, par. 119. See Section V.1.a of this chapter for Buridan's almost identical argument (and also Grant, 1974, 501), and Grant, 1974, 509, for Oresme's version.
140. Riccioli, *Almagestum novum*, pars post., bk. 9, sec. 4, ch. 33, 1651, 467, col. 2 (11th argument). As will be obvious from the next paragraph, few, if any, scholastics would have taken this argument seriously.
141. For Buridan's presentation of a simplicity argument and his rebuttal, see Section V.1.a and note 83 (for the text of the argument) of this chapter.

quence as long as the spheres themselves can endure them (*Almagestum novum*, pars post., bk. 9, sec. 4, ch. 33, 1651, 467, col. 2). Because the great planetary speeds are regulated by intelligences, they do not adversely affect our senses or cause ill effects. The greater mass of the celestial spheres would pose serious problems only if the motive forces that continually moved them met more resistance than they could cope with. Even if such resistances existed, they could cause no difficulty for God or the intelligences. Although it might be easier for God to move the smaller earth than the larger heavenly spheres, Riccioli alludes to valid arguments (though he cites none) as to "why God and Nature do not wish to do that which seems, at first glance, easier, just as in many other matters what seems easier, or of less cost, is not followed" (ibid., 466, col. 2). Despite the earth's considerably smaller size than the heavens, which might suggest a greater inclination for motion, Amicus ([*De caelo*, tract. 5, qu. 6, dubit. 1, art. 3], 1626, 291, col. 2) insisted, in what is essentially a repeat of Buridan's much earlier argument (see Sec. V.1.a of this chapter) that the earth's heaviness made it more unsuited for motion than water, which was less suited for motion than air, which in turn was less suited for motion than fire, from which he inferred that superior celestial bodies are far better adapted for motion in their places than is the earth in its place.

The traditional scholastic conviction that rest is nobler than motion was used by Copernicans in defense of a rotating earth. If rest is nobler than motion, why should the imperfect earth rest while the more perfect and noble heavens rotate? Scholastics responded in a variety of ways and reveal once again their indebtedness to medieval arguments and interpretations. Amicus ([*De caelo*, tract. 8, qu. 4, dubit. 7], 1626, 604, cols. 1–2), for example, concedes that rest is generally more perfect because it is the goal or end of motion. Under certain circumstances, however, the reverse is true, namely, when motion produces more noble effects than rest.[142] Nature assigned motion to the heavens because the latter act as an agent to produce such terrestrial effects as the seasons, the variation of day and night, and the distribution of influences. Since the motion of the earth alone could not produce the various astronomical aspects and conjunctions necessary to generate these causes, nature assigned rest to the earth.[143] Mastrius and Bellutus ([*De coelo*, disp. 4, qu. 4, art. 3], 1727, 3:563, col. 2, par. 119) adopted a similar approach. Natural rest – that is, rest that terminates motion to a natural place – is more perfect than motion toward that natural place. But motion that does not move toward a natural place in order to come to rest there but seeks rather to communicate its power to inferior things is

142. We saw earlier (Ch. 15, Sec. II.1.) that for Buridan circumstances dictated the greater nobility of rest or motion. Under certain circumstances, rest was nobler; under other conditions, motion was nobler. Amicus's conditions are similar to Buridan's.

143. In an earlier part of his *De caelo*, Amicus had presented much the same argument. Amicus, *De caelo*, tract. 5, qu. 6, dubit. 1, art. 3, 1626, 292, col. 1.

more perfect than rest. The circular celestial motions, which operate for the good of the universe and do not come to rest, belong in this category.

The argument from nobility had been employed by Copernicans to argue for the Sun's, rather than the earth's, centrality. Riccioli, who reported numerous Copernican arguments favoring the Sun's centrality,[144] attacked that Copernican argument which assumed that the center of the world is the most noble place and from that inferred that the Sun, which was usually deemed nobler than the earth, must occupy it. Riccioli conceded that *in the natural order* the center is the most noble place, but not *in the supernatural order*, where the most noble place is the empyrean sphere, the highest place, whereas the lowest place (that is, the center), is the place of the damned. But even in the natural order, the Sun is not in the center, because "the earth with its living things, especially rational animals, is nobler than the Sun."[145] To save the earth's centrality, Riccioli was prepared to abandon the traditional opinion that the Sun is nobler than the earth. Moreover, he also denied that the Sun was the efficient cause of celestial motions, as Kepler had argued, or that it could be the cause of the elements and of new phenomena. Rather it is the earth that is the ultimate cause of all these motions and changes, because "the earth, with its human beings, is the final cause and objective of the motion of the stars [or planets]."[146]

The sphericity of the earth had also served the Copernicans, many of whom argued that the earth's spherical shape was more suited for circular motion than for rest.[147] For just as the spherically shaped celestial bodies move with circular motion, so also should the spherically shaped earth.[148] Under the Copernican threat, some scholastics denied any necessary connection between sphericity and circular motion. Amicus, for example ([*De caelo*, tract. 5, qu. 6, dubit. 1, art. 3], 1626, 291, col. 2–292, col. 1) insisted that the earth's sphericity was more appropriate for rest "because, by reason of heaviness, parts of the earth tend to the center equally; [therefore] it [the earth] ought necessarily to have, as much as it can, a spherical figure, so that all parts of its circumference are equally distant from the center." Although Mastrius and Bellutus conceded that circular motion was indeed

144. See Riccioli's "29 arguments in favor of the Sun's position in the center of the universe and [in favor] of the annual motion around the center of the universe simultaneously with the daily motion, and their solutions." *Almagestum novum*, pars post., bk. 9, sec. 4, ch. 33, 1651, 469, col. 1. The twenty-nine arguments extend over pages 469–472 (arguments 21–49). Except for the first argument, which is cited here, the arguments are overwhelmingly astronomical rather than physical or metaphysical.
145. "Tellus enim cum viventibus et animalibus praesertim rationalibus est nobilior Sole." Ibid., 1st argument.
146. Ibid., 4th argument.
147. See Amicus, *De caelo*, tract. 5, qu. 6, dubit. 1, art. 1, 1626, 288, col. 2; Mastrius and Bellutus, *De coelo*, disp. 4, qu. 4, art. 3, 1727, 3:563, col. 2, par. 119.
148. Riccioli, *Almagestum novum*, pars post., bk. 9, sec. 4, ch. 33, 1651, 466, col. 1 (1st argument) had a different version of this Copernican argument. The daily motion should be assigned to the spherical earth rather than to the heaven of the fixed stars because the sphericity of the latter was uncertain.

appropriate to the spherical earth, they denied that the earth had such a motion and offered supporting reasons ([*De coelo*, disp. 4, qu. 4, art. 3], 1727, 3:563, col. 2, par. 119). Rather than the earth, it was the *primum mobile*, or first movable sphere, that rotated with a daily motion. True, the *primum mobile* required an enormously greater velocity to complete its daily rotation than did the far smaller earth. But that greater velocity was a direct reflection of God's omnipotence and therefore produced no disastrous consequences. Indeed, so admirably does the *primum mobile* illustrate God's power that we ought not to reject its tremendous speed in favor of the earth's more imaginable daily rotational velocity. Moreover, if the earth rotated, as Copernicans argued, every stone projected upward would require two external forces: one to move it up, the other to move it along with the earth's rotation. The daily motion is, therefore, better placed in the heavens.

Scholastic natural philosophers also rejected the popular Copernican argument that circular motion is more natural to the elements, and therefore to the earth, than is rectilinear motion.[149] In Riccioli's account of this argument ([*Almagestum novum*, pars post., bk. 9, sec. 4, ch. 33], 1651, 466, cols. 1–2), circular motion is said to be more appropriate to things that are in their natural places, as when earth, water, air, and fire are in their natural places. Only when a part of an element leaves its natural place does it follow a rectilinear path. But such rectilinear motions represent disorder and disorganization, because those elemental bodies have departed from their natural places by violent motion. In his reply, Riccioli denies that circular motion is an inherent (*ab intrinseco*) property of the elements. Circular motion is imposed on the elements externally (*ab extrinseco*) and cannot, therefore, be characterized as "natural." Were the elements arranged absolutely in their natural places, they would be immobile rather than tend toward circular motion, as the Copernicans assume. But when the elements are not in that arrangement, heavy and light bodies are moved with a natural, finite rectilinear motion along a "perpendicular line [that is] always accelerated uniformly difformly toward a goal." These rectilinear motions of bodies out of their natural places agree with observed phenomena. "For these and other reasons [or causes]," declared Riccioli, "we have taught that the Peripatetic doctrine is far more solid in this than the Copernican or Galilean [doctrine]."[150]

c. Theological arguments

Although a few theological arguments were employed during the Middle Ages in scholastic discussions concerning the earth's immobility, they

149. Copernicus advances this argument in *Revolutions*, bk. 1, ch. 8 [Rosen], 1978, 16–17.

150. "Ob quas et alias causas docuimus multo solidiorem esse in hoc doctrinam Peripateticam quam Copernicanam aut Galilaeisticam." Riccioli, *Almagestum novum*, pars post., bk. 9, sec. 4, ch. 33, 1651, 466, col. 2.

played a subsidiary role. By the late sixteenth and the seventeenth century, however, numerous biblical passages were invoked, especially after the condemnation of the Copernican theory in 1616. The theological arguments were largely based on biblical texts. By the seventeenth century, biblical passages were regularly invoked in support of traditional geocentric cosmology. Clavius, whose influence on seventeenth-century scholastic cosmology was enormous, was an immediate source for at least three of these: Psalms 103.5,[151] Ecclesiastes 1.4–5,[152] and Psalms 18.6–7.[153]

Other biblical passages were also cited in behalf of a geocentric world.[154] To confirm the claim that the earth continually supports itself in the center of the world, Aversa appealed not only to Psalm 103 but also to Job 26.7, where God is said to have suspended the world over nothing, and Isaiah 40.12, where God is said to have "poised with three fingers the bulk of the earth, and weighed the mountains in scales, and the hills in a balance."[155] As evidence of the earth's immobility, Aversa cited 1 Chronicles 16.30, where God is said to have made the orb immobile,[156] and Psalm 92, which declares that God fixed the orb of the earth so that it does not move.[157] Amicus ([*De caelo*, tract. 5, qu. 6, art. 2], 1626, 290, col. 2–291, col. 1) and others found support for the earth's immobility in 2 Kings 20.9–11, where, as a sign to Hezekiah, the Lord made the shadow retreat 10 degrees. Had this been done by turning the earth 10 degrees, the suddenness would have been apparent to the senses. Thus did Amicus tacitly assume that God achieved his purpose by causing the Sun to retreat.[158]

151. "Who hast founded the earth on its own bases; it shall not be moved for ever and ever." This and all subsequent biblical quotations in English are drawn from the Douai Version. The passage was also cited by Amicus, *De caelo*, tract. 5, qu. 6, dubit. 1, art. 2, 1626, 290, col. 2; Aversa, *De caelo*, qu. 36, sec. 6, 1627, 232, col. 2; Cornaeus, *De coelo*, tract. 4, disp. 3, qu. 2, dub. 15, 1657, 535; and Riccioli, ibid., ch. 36, 480, col. 2.

152. 4. "*One* generation passeth away and *another* generation cometh; but the earth standeth for ever." 5. "The sun riseth, and goeth down, and returneth to his place; and there rising again. . . ." See also Amicus, ibid.; Aversa, *De caelo*, qu. 31, sec. 2, 1627, 5, col. 2, qu. 34, sec. 5, 142, col. 1, qu. 36, sec. 6, 232, col. 2; Cornaeus, ibid., 536; and Riccioli, ibid., cols. 1–2. In his discussion of the earth's axial rotation, Oresme cites the same passage (*Le Livre du ciel*, bk. 2, ch. 25, 1968, 527); also Grant, 1974, 506.

153. 6. "He hath set his tabernacle in the sun: and he, as a bridegroom coming out of his bride chamber, Hath rejoiced as a giant to run the way." 7. "His going out is from the end of heaven, And his circuit even to the end thereof: and there is no one that can hide himself from his heat." See also Amicus, ibid., and Aversa, ibid., 142, col. 1.

154. See also Schofield, 1981, 271–272.

155. Aversa, *De caelo*, qu. 36, sec. 6, 1627, 231, col. 2. Riccioli, *Almagestum novum*, pars post., bk. 9, sec. 4, ch. 36, 1651, 480, col. 2, also cited Job 26.7.

156. Aversa, ibid., 232, col. 2. For this line from the Vulgate, Aversa, who uses the title 1 Paralipomenon instead of 1 Chronicles, has: "Deus fundavit orbem immobilem" where the Latin Vulgate has "ipse enim fundavit orbem immobilem." Riccioli's citation of this text agrees with the Vulgate (Riccioli, ibid.).

157. Aversa, ibid., has "Firmavit orbem terrae qui non commovebitur," which agrees with the Vulgate. Although Riccioli starts with "Etenim," his citation of Psalm 92 also agrees with the Vulgate (Riccioli, ibid.). In his discussion of the possibility of the earth's axial rotation, Oresme, *Le Livre du ciel*, bk. 2, ch. 25, 1968, 527, also found occasion to quote this line; see also Grant, 1974, 506.

158. In Isaiah 38.8, where the same event is described, the Sun's motion is made explicit in

The most significant biblical passage introduced in the debate over axial rotation was Joshua 10.12–14, where Joshua commanded the Sun to stop in midheaven for nearly a day. Because it was the Sun, not the earth, that was halted by Joshua's command, Aversa, Amicus, and Cornaeus saw this as powerful evidence in favor of the earth's immobility.[159] Nicole Oresme and Galileo had earlier exercised their exegetical talents on this famous passage. Aware that Joshua had commanded the Sun, not the earth, to stand still, Oresme nevertheless argued that the same effect could have been achieved by causing the earth's rotation to cease, and he even suggested that the latter hypothesis was the more attractive.[160] But how could one ignore the plain intent of the text, which speaks of the Sun (and Moon) being stopped, but not the earth? An obvious solution was to avoid a literal interpretation, which is the advice Oresme proposes. The Joshua passage, he declared, "conforms to the customary usage of popular speech just as it [i.e., the Bible] does in many other places, for instance, in those where it is written that God repented, and He became angry and became pacified, and other such expressions which are not to be taken literally."[161] In a similar manner, Galileo insisted that "to attribute motion to the sun and rest to the earth was therefore necessary lest the common people should become confused, obstinate, and contumacious in yielding assent to the principal articles that are absolutely matters of faith."[162]

With the condemnation of the earth's motion in 1616, the argument that Scripture deliberately concealed physical and other truths in order to facilitate the understanding of the common people became untenable. Scriptural passages that spoke of the earth at rest in the center of the world, or the Sun moving around it, were thereafter to be taken literally. To say that "Scripture speaks according to the sense of the common man and not according to the truth" was, in Aversa's judgment, nothing less than "abominable," as indeed it was to most of his scholastic contemporaries who offered public opinions. Without hesitation, Aversa concluded that "for the

the following lines: "I will bring again the shadow of the lines, by which it is now gone down in the sun dial of Achaz with the sun, ten lines backward. And the sun returned ten lines by the degrees by which it was gone down." It was this passage that Aversa cited (*De caelo*, qu. 34, sec. 5, 1627, 142, col. 1). Riccioli cited both passages (*Almagestum novum*, pars post., bk. 9, sec. 4, ch. 36, 1651, 480, col. 1).

159. Aversa, ibid.; Amicus, *De caelo*, tract. 5, qu. 6, dubit. 1, art. 2, 1626, 290, col. 2; Cornaeus, *De coelo*, tract. 4, disp. 3, qu. 2, dub. 15, 1657, 536. Riccioli also cited it (Riccioli, ibid.).

160. Oresme, *Le Livre du ciel*, bk. 2, ch. 25, 1968, 531, 537; Grant, 1974, 507–508, 509. Oresme applied the same reasoning to Isaiah 38:8, explaining that although it appeared that Joshua stopped the Sun and that the Sun returned in the time of Hezekiah, "in fact, it was the earth which stopped moving in Joshua's time and which later in Hezekiah's time advanced or speeded up its movement; whichever occurrence we prefer to believe, the effect would be the same. The latter opinion seems more reasonable than the former, as we shall make clear later." Oresme, ibid., 531; Grant, 1974, 508.

161. Oresme, ibid., 531.

162. "Letter to Madame Christina of Lorraine Grand Duchess of Tuscany," in Drake, 1957, 200.

safety of the faith, the opposite opinion" – that the earth does not rest at the center of the universe – "cannot be tolerated."[163] "Justly and rightly," he continues, "did the Roman Congregation of the Index order the doctrine of Copernicus to be corrected and prohibited the book of Johannes Kepler, titled *Epitome Astronomiae Copernicanae*, and all books containing the same doctrine, which, in the edict issued in 1616, it rightly calls false and absolutely repugnant to divine Scripture."[164]

The numerous passages in favor of the traditional cosmology now took on an even more formidable aspect. No such passage could be defended by any explanation that required abandonment of the literal meaning of the text. The more relaxed liberal and allegorical interpretations of the Middle Ages were no longer tolerated. Scripture, with its many passages favorable to an immobile and central earth, became the most potent weapon in defense of the traditional geocentric cosmology. For, as Koyré explained in his analysis of Riccioli's arguments (1955, 395), no one "has been able to demonstrate that the earth is at rest. Indeed it is impossible to do so as in both cases – whether the earth moved or not – all the phenomena available to us, all the phenomena observable by us would be exactly the same."

d. Scholastic acceptance of nontraditional systems

During the seventeenth century, scholastic natural philosophers had at least three primary world systems from which to choose: the traditional Aristotelian-Ptolemaic system; the Copernican; and the Tychonic, which was devised by Tycho Brahe toward the end of the sixteenth century. Thus far we have only concerned ourselves with scholastics who vigorously defended the traditional system. They were clearly the overwhelming majority. But there were scholastic Aristotelians who were persuaded that contemporary astronomy demanded certain compromises concerning the traditional arrangement and order of the planets, as well as the behavior of the earth itself. If the condemnation of the Copernican system had never occurred, it seems reasonable to assume that some, and perhaps many or

163. Aversa, *De caelo*, qu. 31, sec. 2, 1627, 5, col. 2. Cornaeus, *De coelo*, tract. 4, disp. 3, qu. 2, dub. 15, 1657, 536, makes a similar declaration. Copernicans, he explains, say that Scripture should be accommodated to our manner of speaking and feeling, so the earth is only apparently at rest. Cornaeus, however, insists that we follow Saint Augustine and always interpret the Bible literally, unless "manifest reason and necessity" dictate otherwise. Authoritative appeals were also made to the Church Fathers, though to a lesser extent. For a few such references, see Amicus, *De caelo*, tract. 5, qu. 6, dubit. 1, art. 2, 1626, 291, col. 1.

164. Here is Aversa's complete statement: "Et alibi etiam passim Scriptura supponit terram quiescere in medio et solem circa terram moveri. Nefas autem erit dicere Scripturam loqui iuxta sensum vulgarium hominum et non iuxta veritatem. Opposita ergo sententia non potest tuta fide tolerari. Unde iuste et prudenter Romana Congregatio Indicis corrigi iussit doctrinam Copernici et prohibuit librum Ioannis Kepleri, inscriptum *Epitome Astronomiae Copernicanae*, et omnes libros eandem doctrinam continentes, quam merito vocat falsam et divinae Scriptura penitus repugnantem, Edicto edito anno 1616." Aversa, ibid.

most, scholastic natural philosophers and natural philosopher-theologians might have adopted it. Thus any scholar who seeks to assess seventeenth-century scholastic arguments in favor of the earth's centrality and immobility confronts a dilemma. Did the condemnation of Copernicus, Diego de Zuniga, and Paolo Foscarini in 1616 by the Congregation of Cardinals, and of Galileo in 1633, compel the falsification of arguments by those scholastic theologians who may have been dubious about the traditional Aristotelian position and open-minded about, and even receptive to, Copernicus's claims about the earth's mobility? This is a definite possibility, because, in the aftermath of the condemnation of Galileo in 1633, the Church pressured some Jesuit scientists "to reinforce the Decree of 1633 by publishing books on the controversy themselves emphasizing the religious aspect." "There was," consequently, "a spate of such books by Jesuit writers in which ostentatious reference to the decision of the Church was made" (Schofield, 1981, 281). Riccioli and Cornaeus, both of whom wrote after 1633, cited the condemnation of 1633, and Riccioli even included the text of it in his *Almagestum novum*.[165]

Was the sincerity of those who had to defend the immobility and centrality of the earth, but who may have been sympathetic to the Copernican interpretation, affected? Riccioli, of whom Delambre would say "without his robe he would be Copernican," may have labored under this double pressure. Although he argued vigorously against the Copernican system, and even characterized as unanswerable some of his own arguments for terrestrial immobility, Riccioli also rebutted some arguments favoring terrestrial immobility by invoking counterarguments from "the Copernicans," which seemingly left the earth's immobility in doubt.[166]

Even if Riccoli was a secret Copernican – and there is no substantial evidence that he was – and subtly attempted to undermine the anti-Copernican position whenever feasible, the arguments he presented, many of which were traditional and well known, must nevertheless be evaluated at face value. While the motives and innermost convictions of authors like Riccioli are important where they can be discerned, we are rarely able to determine them and must therefore accept the arguments as we find them.

i. The Tychonic system. Whether they were frustrated Copernicans or not, numerous Catholic natural philosophers, theologians, and astronomers found Tycho Brahe's attempt to reconcile traditional cosmology with Co-

165. For Riccioli, see *Almagestum novum*, pars post., bk. 9, sec. 4, ch. 40, 1651, 497–499. Riccioli also included the text of Galileo's abjuration (499–500). A translation of Galileo's sentence and abjuration appears in Santillana, 1955, 306–310, 312–313. See also Cornaeus, *De coelo*, tract. 4, disp. 3, qu. 2, dub. 15, 1657, 536–537. Mastrius and Bellutus, *De coelo*, disp. 4, qu. 4, art. 3, 1727, 3:562, col. 1, par. 112, mention only that the opinion opposed to the earth's centrality and immobility "was damned by the Sacred Congregation of Cardinals and assigned to the index of books" (hinc opposita opinio damnatur a Sacra card. Congreg. ad indicem librorum deputata).

166. For two instances of this, see Grant, 1984b, 14–15, n. 45.

pernicanism attractive. Tycho, who published a description of his system in 1588, assumed an immobile earth at the center of the world, with the planets circling the Sun while the Sun, in turn, moved around the earth. Tycho also rejected the earth's daily rotation and thus retained a 24-hour revolution of the fixed stars. Jesuits who found the old system flawed – indeed, many of them had initially been favorably disposed to Galileo's new astronomical discoveries – became supporters of the Tychonic system,[167] including Christopher Scheiner,[168] Libert Froidmont,[169] Giorgio Polacco, Riccioli, and Melchior Cornaeus.[170] Indeed, even the Jesuit Clavius, who is sometimes called "the last of their number to advocate the old geocentric system of Ptolemy," seems to have abandoned the old cosmology just before his death in 1612 and recommended that his fellow astronomers devise a new system.

By adopting Tycho's system, with its elimination of celestial orbs, Jesuits and other supporters could also accept the celestial origin of new stars and comets, and therefore accept the corruptibility of the celestial region. It also enabled them to retain Aristotelian terrestrial physics until Newton made it untenable.

ii. The Copernican system. In view of the status of Copernicanism in seventeenth-century astronomy and cosmology, support for the Tychonic system raised few eyebrows. Although Aristotelians who sought refuge in the Tychonic system had to accept drastic changes from the traditional cosmology, they did not have to abandon the inviolate principle of the earth's centrality and immobility. Tycho's system had the additional virtue of immunity from theological censure. Under these circumstances, we would be justifiably surprised if any Aristotelians publicly supported Copernican ideas, especially any rotational motion of the earth. And yet at least two instances of such support occurred, one long before the condemnation of 1616, the other somewhat after the condemnation of Galileo in 1633.

In his *Five Books of Peripatetic Questions*, where no mention of Copernicus can be found, Andrea Cesalpino (1519–1603), a physician and natural philosopher, declared that the earth itself had only one natural motion, which, as Aristotle had argued, was rectilinear and directed toward the center of the world. But it was possible for the earth to move with a circular motion

167. Schofield, 1981, 277. Schofield's book is important; see especially "Seventeenth-century debate over the world-system: III. Religious aspects of the controversy," 264–308.
168. In a letter to Mersenne, Descartes was of the opinion that Scheiner was a Copernican at heart (Schofield, 1981, 283–284).
169. Froidmont had been quite sympathetic to Copernicanism before the condemnation of Galileo. According to Schofield, 1981, 270, "The Italian Jesuit, Bonaventura Cavalieri (1598–1647), wrote to Gaileo in May 1631 that Froidmont has expounded the Copernican arguments so skilfully, and refuted them with so little force, that he seems to believe in them himself."
170. On Jesuit support for the Tychonic system, see Schofield, 1981, 277–289. Only Cornaeus is not mentioned by Schofield.

if that motion was imposed on it by an external force.[171] What could be the source of such a force? The surrounding air and ultimately the heaven. As the heaven, which Cesalpino envisioned as a single continuum, moved from east to west, its force also caused the elemental spheres of fire and air to move in the same direction, but at a much slower rate. The rotating sphere of air, in turn, would incessantly impact on the uneven and continually changing earth's surface and push or carry it in the same east-to-west direction.

Cesalpino was impressed by the motive force of air, which could move massive ships by merely pressing on their sails. If small parts of air could move large ships, the whole mass of air moving from east to west ought to be capable of pushing the entire earth with a circular motion in the same direction,[172] an east-to-west motion that would be the slowest in the universe, since the earth was farthest removed from the eighth sphere of the fixed stars. But if air possessed sufficient force to cause the earth to move with the slowest circular motion in the universe, it lacked the power to cause that same earth to move rectilinearly away from the center of the universe.[173] The earth's circular motion, or rotation, occurred while it was forever stationary at the center of the world. Thus did Cesalpino reconcile the earth's circular motion with the basic Aristotelian requirement that it lie immobile at the center of the world.

By this means, moreover, Cesalpino sought to account for the precession and trepidation of the equinoxes, conceiving these celestial motions as mere appearances derived from the earth's circular motion. The earth's slow east-to-west motion gives to the sphere of the fixed stars the appearance of a slow west-to-east motion, which would correspond to the motion of precession falsely ascribed to the sphere of the fixed stars. Because of the earth's uneven and continually changing surface, however, its east-to-west rotatory motion is irregular and unpredictable, thus producing an apparent trepidation in the stars.[174]

In this extraordinary manner Andrea Cesalpino assigned a circular motion to the earth, explained certain astronomical phenomena by that motion, and yet remained faithful to the traditional Aristotelian conception of an earth located in the center of the universe but unable to move itself circularly by naturally inherent properties.[175]

171. Cesalpino, bk. 3, qu. 4, 1571, 53r–59v, for the complete discussion.

172. Indeed, if the earth were not moved with, and by, the air, "the peaks of the highest mountains would be worn away by the continuous rotation of the air," a consequence that is not observed (Signum praeterea est moveri terram cum aere: nam si aeris cursum non consequeretur, altissimorum montium cacumina continua aeris rotatione attererentur). Ibid., 59r.

173. "Ex centro enim dimoveri impossibile est ne minimum quidem, non enim aer huiusmodi impulsum praebet." Ibid., 58v.

174. For more on Cesalpino's ideas of trepidation and precession, see Grant, 1984b, 7, n. 22.

175. For his unusual opinion, Cesalpino was severely criticized by Aversa, who insisted that these motions were only appropriate to the heavens and not the earth. In Aversa's

Not many Aristotelians went beyond Cesalpino. But if Thomas White (1593–1676), an English Catholic and sometime professor of philosophy and theology, is included among them – and he seems to have proclaimed membership in the company of Aristotelians – then it is possible to think the unthinkable: the most basic elements of the Copernican theory, namely the earth's daily rotation *and* its annual motion around a stationary Sun, are somehow reconcilable with Aristotelianism. The reconciliation appears in White's *Peripateticall Institutions*, where, in a section called "The Authours Design," he offers the following explanation of the work's title: "Why I have stiled them Institutions, the shortnesse and concise connection of the works sufficiently discover. I call them Peripateticall because, throughout they [i.e., the "Institutions," or foundations] subsist upon Aristotle's principles; though the conclusions sometimes dissent."[176] In the concluding sentence, British understatement may have had its finest hour, since White's "dissent" led him to the assumption of the truth of the Copernican theory.

In a manner reminiscent of Cesalpino, White, in lessons 14 and 15, explains the earth's daily axial rotation by an east-to-west sweep of the wind, which causes the upper part of the sea to begin a process that enables the lowest level of the sea, in direct contact with the seabed or earth, to produce a west-to-east motion of the earth. The earth's daily circular motion is possible in this manner because it is not contrary to the earth's natural gravity and therefore offers no resistance to the west-to-east force of the sea at the points of contact.[177]

But, as White elaborates further, "because 'tis almost impossible this impulse should be equall on all sides, and cause a pure rotation about the Centre; there will, of necessity, a Progressive motion be mixt with it." This motion, which represents the earth's annual orbit, must, however, be "in one line," because "all the motions which Astronomers assign the Earth must, of necessity, compose one line; and, if the lashing or impulse of the underwater advance the Earth in that line, 'twill be an adequate cause of the motion of the Earth."[178] Like Cesalpino, then, White refused, as an Aristotelian, to confer *natural* circular or orbital motions on the earth. How-

judgment, Cesalpino "shamefully erred" (Cesalpinus in hoc turpiter lapsus fuit) when he proclaimed this opinion. Aversa, *De caelo*, qu. 34. sec. 5, 1627, 141, cols. 1–2, for the description of Cesalpino's ideas, and 143, col. 1, for the criticism. Mastrius and Bellutus, *De coelo*, disp. 4, qu. 4, art. 3, 1727, 3:563, col. 1, par. 116, also denied Cesalpino's claim. Not only did Scripture assign all circular motions to the heavens and rest to the earth, but the earth has no internal capacity to move circularly, nor is there any external force that could cause its continual rotation. None of the other elements, including air, has the power to move it.

176. White, 1656, sig. a4v–a5r. The work was originally published in Latin at Lyons in 1646. According to Phillip Drennon's article on him (*Dictionary of Scientific Biography*, 14:301–302), White, an English Catholic, was "a devoted follower of Aristotle," although his "scientific treatises contain modifications and revisions of Aristotle's thought."

177. White, ibid., lesson 14, pp. 174–175, pars. 1–3.

178. Ibid., 175, pars. 4–5. It is noteworthy that nowhere is the name of Copernicus mentioned.

ever, because he was convinced of the truth of heliocentric astronomy and that the earth really moved as the Copernican theory required, he derived its motions by appeal to an external force.[179]

The explanations of Cesalpino and White represented a basic model for those few Aristotelians who sought an accommodation with the new geokinetic astronomy. While retaining the Aristotelian principle that the element earth could possess only one simple, natural motion, which was downward and rectilinear, they were yet prepared to allow that *external forces* could cause the earth to move with one or more circular motions. Even those Aristotelians who disagreed with this approach could see its attractions. Aversa ([*De caelo*, qu. 34, sec. 5], 1627, 141, col. 2–142, col. 1) conceded that changes observed in the celestial region might well be saved by the assumption of a terrestrial motion, especially an axial rotation of the earth every twenty-four hours which could properly account for the same motions that many attribute to the heaven itself. But Aversa rejects this approach because "every apparent local change around the celestial bodies really and truly happens to those bodies by a real and true motion; but no such change and motion occurs to the earth. This is the common sense of both wise and ordinary men."[180] Any observed motions that alter the relations between celestial bodies must be assumed to occur in the heaven itself. Mobility is proper to the heaven and immobility to the earth, which lies in the center of the world. On the principle that "to one simple body, [only] one motion is appropriate," the "appropriate" motion for the earth is downward and rectilinear. But Aversa concedes that this constraint would prove no obstacle "if circular motion were attributed to the earth from another extrinsic cause, or even from another motive power."[181]

3. Scholastic attitudes toward the heliocentric system

Until its triumph near the end of the seventeenth century, the heliocentric system was contested more on physical and cosmological grounds than on its astronomical merits. Before Newton's theory of gravitation made physical sense of heliocentrism, no arguments presented in its favor were sufficiently overriding to render traditional geocentrism completely untenable. Because the case for a rotating and orbiting earth had not yet developed to the point where it completely subverted the alternative position, scholastic

179. White's opinions and explanations of the earth's motions were presented earlier in his better-known *De mundo dialogi tres* (Paris, 1642). Thomas Hobbes severely criticized White's book (for more detail, see Grant, 1984b, 9, n. 26).

180. "Denique de facto omnis apparens mutatio localis circa corpora caelestia vere et realiter convenit ipsis corporibus caelestibus per verum et realem motum nullaque huiusmodi mutatio aut motus terrae convenit. Hic est communis tam sapientum quam vulgarium hominum sensus." Aversa, *De caelo*, qu. 34, sec. 5, 1627, 141, col. 2–142, col. 1.

181. "Uni autem corpori simplici unus motus competere debet. Sed hoc non obstaret, si motus circularis tribueretur terrae ab alia causa extrinseca sive etiam ab alia distincta virtute motiva." Ibid., 142, col. 1.

arguments in favor of an immobile earth at the center of the universe continued to command widespread support through much of the seventeenth century.

Scholastic Aristotelians of the late sixteenth and the seventeenth century were a diverse group about whom no easy generalizations are warranted. They range from staunch supporters of the status quo to the likes of Cesalpino and White. The condemnation of both the Copernican theory and its most dramatic supporter, Galileo, undoubtedly deterred some scholastics from abandoning geocentrism, but it cannot alone explain continued support for the old cosmology. Aristotelian geocentrism was the system they knew best and with which they were most comfortable. Until the Newtonian theory of gravitation took hold in the late seventeenth century and provided at long last a sound physical basis for the heliocentric system, most scholastics found little reason to abandon a whole complex of traditional interpretations that had served reasonably well for almost five centuries in order to embrace what had yet to be conclusively demonstrated. With the publication of Newton's *Mathematical Principles of Natural Philosophy* in 1687, the situation changed radically. After that famous date, scholastic authors who continued to uphold the old cosmology did so not on scientific merit but to comply with theological decrees. By then, geocentrism and terrestrial immobility had lost all credibility and gradually faded away.

CONCLUSION

Five centuries of scholastic cosmology

I. Tradition

During the late Middle Ages, Aristotelian cosmology was unrivaled and essentially unchallenged. By the time rival interpretations of any consequence appeared in the sixteenth century – for example, Platonism, atomism, Stoicism, Neoplatonism, Hermeticism,[1] and Copernicanism – Aristotelian cosmology, despite its multiplicity of opinions on most major issues, was entrenched in the educational system and culture of Europe, where it enjoyed a status comparable to Euclidean geometry in mathematics. By the seventeenth century, however, the cumulative impact of its new rivals, the most significant of which was, of course, the Copernican system, had placed Aristotelian cosmology under the most severe stress of its long history in western Europe.

Tradition and intellectual inertia at the medieval universities, which were the strongholds of Aristotelian natural philosophy, proved powerful ingredients in the prolongation of medieval cosmology. That tradition continued to manifest itself even when Copernicanism posed the first serious challenge in the sixteenth century.

A glance at the "Catalog of Questions" in Appendix I provides an indication of how tradition shaped and perpetuated medieval cosmology. Examination of the 400 questions reveals that approximately 75 were discussed from the thirteenth or fourteenth century to the seventeenth.[2] The

1. The humanism that had generated a new interest in Greek antiquity, and the influx (beginning in the fifteenth century) of Byzantine Greeks into the Latin West, touched off a new wave of translations, now directly from Greek manuscripts. Old works were retranslated, and new ones not previously known to the Latins were translated into Latin. It was from this new wave of translation, the likes of which had not been seen in Europe since the twelfth and thirteenth centuries, that the rival philosophies and cosmologies emerged.
2. Here are the questions included in my count: 1, 12, 62, 68, 69, 74, 79, 85, 91, 94, 97, 111, 112, 122, 126, 127, 128, 130, 132, 141, 154, 155, 159, 163, 169, 173, 177, 180, 183, 184, 190, 192, 195, 196, 209, 211, 216, 220, 222, 228, 230, 231, 237, 242, 255, 256, 261, 262, 270, 275, 276, 277, 283, 289, 300, 301, 308, 321, 325, 331, 334, 335, 339, 347, 348, 374, 378, 383, 384, 387, 388, 389, 393, 396. In a few questions, the Conimbricenses have been taken as representing the seventeenth century.

   Whether an investigation of the ratio of the number of medieval discussants to early modern or Renaissance discussants would yield anything of significance is not clear. If we take 1500 as the arbitrary date that divides the Middle Ages from early modern scholas-

substance of many other questions, perhaps the majority of them, was discussed throughout the centuries, although they may appear in the "Catalog of Questions" in only one period. The number of questions that linked medieval scholastic cosmology with the seventeenth century is thus quite large. The responses to most of those questions in the seventeenth century were to a considerable extent similar to the responses formulated during the late Middle Ages. Had the fourteenth-century scholastic author Jean Buridan reappeared in the seventeenth century, he would have had little difficulty understanding his scholastic successors. Conversely, had seventeenth-century scholastics such as Bartholomew Amicus and Raphael Aversa stepped into the fourteenth century, they would have found the cosmological arguments and theories familiar and intelligible.

II. Innovation

Although the same substantive questions and themes were frequently discussed from the thirteenth to the late seventeenth century, and although the responses were frequently similar, scholastic natural philosophers, contrary to popular misconceptions, did not rigidly adhere to the same cosmological opinions.[3] Innovative tendencies were also operative. If natural philosophers were agreed on the centrality and immobility of the earth (this is overwhelmingly true, despite the unusual opinions expressed by Andrea Cesalpino and Thomas White, described in Chapter 20), they responded to a host of common problems with a variety of arguments for each problem or question. Although a comparison of any two sets of responses usually reveals some overlap, we often find divergent presentations, as illustrated by the numerous adoptions and adaptations of Tycho Brahe's geoheliocentric astronomy and by Giovanni Baptista Riccioli's generous treatment of certain arguments favorable to heliocentrism. Moreover, where scientific

ticism, what significance attaches, for example, to question 12, where 4 medieval scholastics discussed the question and 11 early modern scholastics; or to question 122, where 10 medieval scholastics discussed the question as against only 3 in the early modern period? And what of those questions where we have only medieval authors as discussants, as in question 123 with 5 discussants; or those questions where we only have discussants for the early modern period, as in question 15 with 4 discussants? There may be little significance in such divisions, since we may not infer from the absence of, say, medieval authors, from a particular question that the substance of that question was not discussed by some of them in another context. At best, a division into medieval and early modern may reveal that a question ceased to be enunciated as a question after 1500, or that it only began to be enunciated as a question after 1500. For example, question 55 ("What are imaginary spaces?") has 9 early modern discussants and none from the Middle Ages. We know, of course, that Thomas Bradwardine, Jean de Ripa, and Nicole Oresme discussed the issue, but not in the form of a question. Hence we may infer that the issue became sufficiently important in the late sixteenth and the seventeenth century to lead at least 9 scholastics to devote special questions to it.

3. Much of what is said here is drawn from Grant, 1978b; Grant, 1985a; Grant, 1987d; and Grant, 1989.

issues arose in areas relevant to cosmology but which did not directly threaten the foundations of traditional Aristotelian cosmology, scholastic authors were not reluctant to adopt and absorb new ideas and theories, as we find with the concept of the terraqueous sphere. Indeed, some were even prepared to accept new ideas that conflicted with basic Aristotelian views, as happened with the acceptance of a corruptible celestial region and the abandonment of hard orbs in favor of a fluid celestial medium.

In the history of medieval Aristotelian cosmology from the thirteenth and into the seventeenth century, at least two kinds of change are distinguishable: departures from the cosmological ideas of Aristotle himself *and* changes that occurred within scholastic cosmology itself. In reality, the first kind may be perceived as a subset of the second. The thirteenth and fourteenth centuries witnessed significant departures from Aristotle in the shift from Aristotle's system of purely concentric orbs to one in which Ptolemy's eccentric orbs were incorporated within Aristotle's concentrics. Numerous new problems emerged from this transformation: Did eccentrics imply rectilinear motions for planets? Would heavy bodies fall to the center of the earth or to the geometric center of an eccentric orb? Would vacua occur between successive eccentric orbs that were of unequal thickness? And so on. A significant consequence of the doctrine of eccentric orbs, contrary to the position of Aristotle, was that planets could rotate around a geometric point as well as around a physical body (the earth). During the fourteenth century, other dramatic departures from Aristotle occurred when scholastic natural philosophers demonstrated that an infinite extracosmic void space might lie beyond the world itself; that motion in a hypothetical vacuum was feasible; that the existence of other worlds was possible; and that the daily axial rotation of the earth was an intelligible, astronomical concept, even though it was ultimately rejected.

The changes that occurred in scholastic Aristotelian cosmology during the seventeenth century as a consequence of Copernicus's heliocentric system and the new discoveries made by Tycho (the new star of 1572 and the comet of 1577) and Galileo (the satellites of Jupiter; sunspots; and numerous stars never seen before) were overwhelming, as we have seen in the course of this study. The doctrine of celestial incorruptibility gradually yielded to the doctrine of corruptibility and change, thus causing the abandonment of a fundamental Aristotelian principle. Aristotle's celestial ether and the belief in hard celestial orbs were also abandoned by numerous scholastics in favor of a fluid medium composed of one or more of the four elements. Scholastics who substituted a fluid medium for hard orbs had now to explain the manner in which planets that were no longer carried by physical orbs could move at all. Were they self-moved or moved by something else?

Many scholastics, especially among the Jesuits, chose to adopt the system of Tycho and thus, while locating the stationary earth in the center of a spherical universe, also assumed that at least two, and perhaps all, of the other planets moved around the Sun as their center of motion, with the

Sun, in turn, orbiting the earth. And one scholastic – Riccioli – even assumed that the earth was more perfect than the Sun, an idea that signified rejection of the fundamental Aristotelian principle that the celestial region is more perfect than the terrestrial. Nor should we omit mention of scholastic acceptance of the terraqueous sphere, which assumed that earth and water formed a single, unified sphere and thus signaled the abandonment of Aristotle's concept of two separate spheres of earth and water, which had been a standard feature of medieval cosmology.

From any standpoint, these are dramatic departures from Aristotelian cosmology and yet they were made by Aristotelian natural philosophers. How could this have happened? How could Aristotelian cosmology have remained Aristotelian in the face of such radical alterations? This was in no small measure a consequence of the nature of Aristotelianism. Opinions and theories in Aristotelian natural philosophy were easily multiplied, because "Aristotle's was the most capacious of philosophies," which "in principle . . . explained everything."[4] Because Aristotelian cosmology was, at the macro level, a tightly integrated system, departures from it, whether by design or inadvertence, produced glaring inconsistencies. Scholastic cosmologists developed ideas whose consequences were subversive of part or all of Aristotle's concepts of the physical cosmos. Such consequences, however, either went undetected or were ignored, and in the rare instances where cognizance was taken of them they were made compatible with the capacious system. No systematic efforts were made to purge these inconsistencies, which simply remained as part of the total world view. From the thirteenth and into the seventeenth century, Aristotelians seemed capable of living with serious anomalies that should have focused attention on inconsistencies within the system as a whole.

Because traditional medieval Aristotelian cosmology had been inherently capacious, early modern scholastic natural philosophers sought to adjust to the new cosmology of the seventeenth century. They did so by rejection and absorption. The absorption process indicates quite clearly that Aristotelian cosmology was far from a static and congealed body of medieval doctrine. In truth, it was a body of varied opinion in which some genuine effort was made to incorporate aspects of the new cosmology into the old. Far from taking a monolithic view, scholastic Aristotelians ranged from steadfast defenders of the status quo to those who came to reject important elements of the system itself, replacing them with new ideas and observations derived from their opponents.

Aristotelian scholastic cosmology was never reformed from within. Although numerous Aristotelians adopted some of the new ideas that came into vogue in the seventeenth century, they did not, any more than their medieval predecessors, use those new variations to strengthen Aristotelian cosmology as a whole. In the end, the variations and new ideas proved

4. Gillispie, 1969, 11.

neither helpful nor harmful: they were simply variations. Whether an Aristotelian developed a variation himself or drew it from an outside source, the variant idea simply formed part of that scholar's Aristotelianism. If others found that particular idea useful or better than what was available, they could make it part of their own respective versions of Aristotelian cosmology or part of a larger conception of Aristotelianism. But this process did nothing more than increase the number of ideas in Aristotelianism. Old and new ideas simply subsisted together. Whether the old or new idea improved Aristotelian cosmology was a question that rarely, if ever, arose.

During the seventeenth century, many compromises were made. Bits and pieces of Aristotelian cosmology were replaced by bits and pieces of the new cosmology. Strange cosmological mosaics developed, none of which could win much support. The seventeenth century became a period of transition: one system was passing away, another coming into being. To comprehend the momentous changes that occurred in the seventeenth century, it is essential to study the fate of the old cosmology as well as the new. Only then can we have a comprehensive picture of the cosmos as it was understood in the seventeenth century. Perhaps this study will contribute toward that end, so that we may learn, among many other things, why Aristotelian cosmology coexisted with its Copernican rival for at least 144 years (1543 to 1687) before it finally succumbed.

APPENDIX I

Catalog of Questions on Medieval Cosmology, 1200–1687

Although bits and pieces of cosmological information appear in other kinds of treatises, our knowledge of the structure and operation of the medieval cosmos is overwhelmingly dependent on the types of literature described in Chapter 2 and on the great variety of questions (*questiones*) that formed their basic content. As yet, no one, to my knowledge, has collected and analyzed the sorts of questions that scholastic authors posed about the creation, structure, and operation of the world. As a contribution toward this goal, I have compiled a list of 400 questions. The purpose of this "Catalog of Questions" is to convey a sense of the number and range of questions that were posed over the active lifetime of medieval Aristotelian cosmology, which stretched from the late Middle Ages through the Renaissance, that is, from the thirteenth and into the seventeenth century. The rationale for the selection of questions, a list of the 52 authors from whose works they were selected, and an explication of the catalog are provided in Appendix II.

Under each question, the names of authors are arranged chronologically, following the chronology in the list of authors given at the beginning of Appendix II. Citations in the catalog follow the same pattern as in the textual citations and the notes. As in the footnotes, bracketed citation forms are used to indicate the genre type of works: *De caelo* (or *De coelo*, or *De celo*, depending on the language used); *Sphere*; *Physics*; *Sentences*, etc. For the actual title of each work consult the Bibliography. The numbers associated with various subdivisions and parts that appear within many treatises – for example, book, distinction, question, section, article, *dubium*, and others that appear among the abbreviations listed at the beginning of the book – are repeated here. A star after an author's name signifies that his form of the Latin text for that question was chosen to represent all the other versions (e.g., in qu. 1 Galileo's Latin text functions as surrogate for all 15 authors, as does Buridan's in qu. 54). Divergences from the chosen version are sometimes given in the notes. For more on the variants of questions, see Appendix II, Section IV.2. Where only one author contributes a question, no star is used. For complete titles, see the Bibliography.[1]

1. See Chapter 1, note 7, for the relationship between the fourfold division of the "Catalog of Questions" and the twofold division of Chapters 4 through 20 of this study.

PART I. THE WORLD AS A WHOLE

I. On the eternity of the world and the corruptibility or incorruptibility of the celestial region

1. Whether the universe could have existed from eternity.

1. Michael Scot [*Sphere*, lec. 1], 1949, 250–252.[2]
2. Bacon [*Physics*, bk. 8], *Opera*, fasc. 13, 1935, 390–392.
3. Bonaventure [*Sentences*, bk. 2, dist. 1, pt. 1, art. 1, qu. 2], *Opera*, 1885, 2:19–25.
4. Godfrey of Fontaines, *Quodlibet II*, qu. 3, 1904, 68–80.
5. Marsilius of Inghen [*Physics*, bk. 8, qu. 1], 1518, 79r, col. 2–80r, col. 2.[3]
6. Dullaert [*Physics*, bk. 8, qu. 1], 1506, sig. oiii,v–ovii,r.
7. Major [*Physics*, bk. 8], 1526, 54v, col. 1–55r, col. 1.
8. Toletus [*Physics*, bk. 8, qu. 2], 1580, 215r, col. 2–219v, col. 2.
9. Galileo* [*De caelo*, qu. 4 (F)], 1977, 49–57.[4]
10. Conimbricenses [*Physics*, pt. 2, bk. 8, ch. 2, qu. 3], 1602, cols. 427–434.
11. Amicus [*De caelo*, tract. 2, qu. 2, dubit. 3], 1626, 30, col. 2–40, col. 2.[5]
12. Poncius [*De coelo*, disp. 21, qu. 3], 1672, 602, col. 1–603, col. 1.[6]
13. Oviedo [*Physics*, bk. 8, contro. 19, punc. 1], 1640, 445, col. 1–447, col. 1.
14. Compton-Carleton [*Physics*, disp. 39, secs. 1–2], 1649, 362, col. 1–363, col. 1.
15. Serbellonus [*De caelo*, disp. 1, qu.1, arts. 4–6], 1663, 2:8, col. 2–18, col. 1.[7]
16. Oddus [*De coelo*, disp. 1, dub. 6], 1672, 10, col. 2–13, col. 1,[8] and [*Physics*, disp. 17], 1667, 763, col. 1–769, col. 1.[9]

2. Whether things, or the world, could exist from eternity with respect to successive beings.

1. Johannes Canonicus [*Physics*, bk. 8, qu. 1], 1520, 64r, col. 1–65v, col. 2.[10]

2. Although Michael Scot's treatise is a commentary on the *Sphere* of Sacrobosco, many specific questions are posed throughout. Those that appear relevant are included in this catalog, as are questions from the commentaries of Robertus Anglicus and Cecco d'Ascoli.
3. Marsilius asks "Whether motion and the world have existed from eternity."
4. Galileo's questions on *De caelo* consists of two introductory questions and two separate works, each bearing the title *Treatise on the Heavens*. All told, they contain twelve questions, although the questions in each of the three brief works are numbered separately. Because they are obviously questions on *De caelo* and belong together, Wallace has assigned the questions letters from A to L, which I shall also employ. The twelve questions occupy pages 25–158. For convenience, I shall subsume the three treatises under the title *De caelo*.
5. Amicus also discussed other aspects of the eternity of world. See tract. 2, qu. 3, dubits. 1–7, pp. 41–79.
6. Poncius asks "Whether the world was in fact produced from eternity." In qu. 4 (603, col. 1–609, col. 1), he asks the related question "Whether the world could exist from eternity." Poncius's *De coelo* forms disputations 21 and 22 of his *Philosophiae ad mentem Scoti cursus integer* (1672).
7. In these articles, Serbellonus specifies the following questions: "Whether the creation of the heavens could have been from eternity"; "Whether the motion of the heavens could have been from eternity"; "Whether generations could have been from eternity."
8. Although Oddus considers whether the world could have existed from eternity, he titles his question "On the beginning of the world."
9. Here Oddus asks "Whether, in duration, any creature could exist without beginning and end or whether the world could exist from eternity."
10. Canonicus asks "Whether it is formally repugnant that any creatable or producible permanent or successive thing should have existed from eternity."

2. Hurtado de Mendoza [*Physics*, disp. 18, sec. 2], 1615, 359, col. 1–361, col. 2.[11]
3. Aversa [*De caelo*, qu. 31, secs. 7–8], 1627, 26, col. 1–30, col. 2, and 30, col. 2–35, col. 1.[12]
4. Oviedo [*Physics*, bk. 8, contro. 19, punc. 3], 1640, 448, col. 1–451, col. 2.
5. Cornaeus [*De coelo*, tract. 4, disp. 1, qu. 1, sec. 1, dub. 1–2], 1657, 472–475.[13]
6. Oddus* [*Physics*, bks. 5–8, disp. 1, art. 13 (ult.)], 1667, 769, col. 2–774, col. 2.

3. *Whether or not all things without exception could have existed from eternity.*
   1. Conimbricenses [*Physics*, pt. 2, bk. 8, ch. 2, qu. 7], 1602, cols. 449–458.

4. *Whether the universe is eternal.*
   1. Pseudo–Siger of Brabant [*Physics*, bk. 8, qu. 5], 1941, 197–202.[14]
   2. Bacon*[*Physics*, bk. 8], *Opera*, fasc. 13, 1935, 370–377.
   3. Thomas Aquinas [*Sentences*, bk. 2, dist. 1, qu. 1, art. 5], 1929–1947, 2:27–41.
   4. Marsilius of Inghen [*Physics*, bk. 8, qu. 2], 1518, 80r, col. 1–80v, col. 2.[15]
   5. Hurtado de Mendoza [*Physics*, disp. 16, sec. 1], 1615, 354, col. 2–355, col. 2.[16]

5. *Whether there is eternal motion.*
   1. John of Jandun [*Physics*, bk. 8, qu. 3], 1519, 122v, col. 1–125r, col. 1.
   2. Buridan [*Physics*, bk. 8, qu. 3], 1509, 110v, col. 2–112v, col. 1.
   3. Marsilius of Inghen [*Physics*, bk. 8, qu. 2], 1518, 80r, col. 1–80v, col. 2.
   4. Benedictus Hesse* [*Physics*, bk. 8, qu. 4], 1984, 697–701.

6. *Whether eternal motion depends on a mover just as [if that mover were] an agent or efficient [cause].*
   1. John of Jandun* [*Physics*, bk. 8, qu. 5], 1519, 126r, col. 2–127v, col. 1.
   2. Javelli [*Physics*, bk. 8, qu. 5], 1568, 1:600, col. 1–601, col. 1.

7. *Whether generations could have proceeded from eternity without a first generation.*
   1. Hurtado de Mendoza [*Physics*, disp. 18, sec. 1], 1615, 356, col. 1–359, col. 1.
   2. Oviedo [*Physics*, bk. 8, contro. 19, punc. 2], 1640, 447, col. 1–448, col. 1.
   3. Cornaeus* [*De coelo*, tract. 4, disp. 1, qu. 1, sec. 1, dub. 7], 1657, 480–483.

11. Hurtado considers "Whether a successive being could exist from eternity" (disp. 18) and argues that "No successive motion could exist from eternity" (sec. 2).
12. Aversa considered the possible existence from eternity of both successive and permanent entities.
13. Like Aversa, Cornaeus also considers the possibility that the world could have existed from eternity with respect to both permanent and successive entities.
14. The question about the eternity of the world could take a number of forms. Pseudo-Siger chose to ask "Whether motion is eternal." Delhaye's attribution of these questions to Siger of Brabant has been rejected. Albert Zimmermann, 1968, xxxviii and n. 60, has suggested Peter of Auvergne as their author. Lohr (1972, 345) lists this treatise in the doubtful category, under "Petrus de Alvernia (de Crocq)," or Peter of Auvergne. For convenience, I cite the author as "Pseudo–Siger of Brabant."
15. The question reads: "Whether the world is eternal and some motion, like the motion of the heaven, is eternal."
16. Hurtado asks "Whether, if the world were eternal, it would be [eternal] with respect to permanent parts" (disp. 16) and subsequently argues (sec. 1) that "the world could not be eternal."

8. Whether the world had a beginning with motion and time.
1. Toletus [*Physics*, bk. 8, qu. 1], 1580, 210v, col. 2–215r, col. 2.

9. Whether the world will end at some time.
1. Michael Scot* [*Sphere*, lec. 1], 1949, 254–255.
2. Bacon [*Physics*, bk. 8], *Opera*, fasc. 13, 1935, 382–383.
3. Aversa [*De caelo*, qu. 31, sec. 10], 1627, 35, col. 1–40, col. 2.[17]
4. Poncius [*De coelo*, disp. 21, qu. ult. (qu. 5)], 1672, 609, col. 1–610, col. 2.[18]
5. Cornaeus [*De coelo*; tract. 4, disp. 1, qu. 2, dub. 4], 1657, 488.[19]

10. Whether the world could endure perpetually under the general concourse [or control] of God alone.
1. Conimbricenses* [*De coelo*, bk. 1, ch. 12, qu. 1], 1598, 154–158.[20]
2. Cornaeus [*De coelo*, tract. 4, disp. 1, qu. 2, dub. 3], 1657, 487.
3. Bonae Spei [comment. 3, *De coelo*, disp. 2, dub. 4], 1652, 7, cols. 1–2.

11. Whether the world is generable and corruptible or ungenerable and incorruptible.[21]
1. Albert of Saxony* [*De celo*, bk. 1, qu. 12], 1518, 95v, col. 1–98r, col. 2.
2. Paul of Venice, *Liber celi*, 1476, 20, col. 2–22, col. 1.[22]
3. Versor [*De celo*, bk. 1, qu. 14], 1493, 12r, col. 2–12v, col. 2.[23]
4. Amicus [*De caelo*, tract. 3, qu. 4, dubit. 2], 1626, 123, col. 1–125, col. 2.

12. Whether the sky [or heaven] is generable and corruptible, augmentable and diminishable, and alterable.
1. John of Jandun [*De coelo*, bk. 1, qus. 14, 16–17], 1552, 10r, col. 2–11r, col. 1; 12r, cols. 1–2; and 12r, col. 2–12v, col. 2.[24]
2. Buridan* [*De caelo*, bk. 1, qu. 10], 1942, 44–49.
3. Oresme [*De celo*, bk. 1, qu. 10], 1965, 143–157.
4. Versor [*De celo*, bk. 1, qu. 6], 1493, 4v, col. 1–5r, col. 2.[25]
5. Galileo [*De caelo*, qu. 4 (J)], 1977, 93–102.[26]
6. Conimbricenses [*De coelo*, bk. 1, ch. 3, qu. 1], 1598, 62–71.[27]
7. Hurtado de Mendoza [*De coelo*, disp. 1, sec. 5], 1615, 366, col. 2–367, col. 2.
8. Aversa [*De caelo*, qu. 33, sec. 2], 1627, 80, col. 2–83, col. 2.[28]

17. Aversa asks "Whether the world will last into eternity or will end at some time."
18. Poncius asks "Whether, from its very nature, the world would last into eternity."
19. Cornaeus puts the question in the contrary way, namely "Whether the world could endure through eternity."
20. For the Coimbra Jesuit editions of both the *De caelo* and *Physics*, see Lohr, 1988, 98–99.
21. For a related but nonetheless different question, see question 12 of this catalog.
22. Neither folio numbers nor question numbers are included. Not even the signatures are of much use. I have arbitrarily paginated from the first page of the text of the *Liber celi et mundi*, which follows Paul's *Liber physicorum*. Where it is helpful, I also include the signature.
23. Versor asks only "Whether the whole world is generated and corruptible." In Versor's volume, the *De celo et mundo* is separately foliated and follows his questions on *De anima*. In this edition of Versor's *De celo*, questions on the third book are lacking.
24. Jandun used three questions to encompass the whole of question 12.
25. Versor asks only "Whether the heaven is ungenerable and incorruptible."
26. Galileo simply asks, "Are the heavens incorruptible?"
27. Since this question is relevant to question 15 in this catalog, it is also cited there.
28. Aversa's version inquires "Whether celestial bodies are hostile to generation and corruption, or rather [whether they are] naturally ungenerable and incorruptible."

9. Poncius [*De coelo*, disp. 22, qu. 4], 1672, 616, col. 2–620, col. 1.[29]
10. Oviedo [*De caelo*, contro. 1, punc. 1], 1640, 463, col. 2, par. 14–464, col. 2, par. 21.
11. Compton-Carleton [*De coelo*, disp. 1, sec. 2], 1649, 397, col. 2–398, col. 2.
12. Riccioli, *Almagestum novum*,[30] pars post., bk. 9, sec. 1, ch. 6, 1651, 237, col. 1–238, col. 2.
13. Bonae Spei [comment. 3, *De coelo*, disp. 3, dub. 4], 1652, 10, col. 1–11, col. 1.[31]
14. Cornaeus [*De coelo*, tract. 4, disp. 2, qu. 1, dub. 3], 1657, 489.[32]
15. Serbellonus [*De caelo*, disp. 1, qu. 2, art. 3], 1663, 2:22, col. 1–25, col. 1.[33]

13. *Whether the heaven is corruptible with respect to substance.*
   1. Amicus [*De caelo*, tract. 5, qu. 1, dubit. 1], 1626, 230, col. 1–247, col. 1.

14. *Whether it [that is, the heaven] is corruptible with respect to accidents.*
   1. Amicus [*De caelo*, tract. 5, qu. 1, dubit. 2], 1626, 247, col. 2–249, col. 1.

15. *Whether the world could actually be corrupted.*
   1. Conimbricenses [*De coelo*, bk. 1, ch. 3, qu. 1], 1598, 62–71.
   2. Hurtado de Mendoza [*De coelo*, disp. 3, sec. 7], 1615, 388, cols. 1–2.
   3. Amicus* [*De caelo*, tract. 3, qu. 4 (actually qu. 3), dubit. 5], 1626, 128, col. 1–130, col. 1.
   4. Oddus [*De coelo*, disp. 1, dub. 7], 1672, 13, col. 1–15, col. 1.[34]

16. *Whether God could create a motion anew before which there was neither a motion nor a mutation.*
   1. John of Jandun [*Physics*, bk. 8, qu. 4], 1519, 125r, col. 1–126r, col. 2.[35]
   2. Buridan* [*Physics*, bk. 8, qu. 2], 1509, 109v, col. 2–110v, col. 2.

17. *Whether something created anew could be perpetuated; and whether something eternal could be corrupted.*
   1. Albert of Saxony* [*De celo*, bk. 1, qu. 15], 1518, 100r, col. 1–100v, col. 2.
   2. Oresme [*De celo*, bk. 1, qu. 24], 1965, 409–428.[36]
   3. Bricot [*De celo*, bk. 1], 1486, 13v, cols. 1–2.[37]
   4. Marsilius of Inghen [*Physics*, bk. 8, qu. 3], 1518, 80v, col. 2–81v, col. 1.[38]

29. Poncius asks only "Whether the heaven is corruptible."
30. Although the title page indicates a three-volume work, only one volume appeared, which was divided into two parts (*pars prior* and *pars posterior*) that were separately paginated.
31. Where Galileo asked whether the heavens are incorruptible, Bonae Spei asks "Whether the heavens are corruptible."
32. Cornaeus asks "Whether the heavens are essentially incorruptible."
33. Serbellonus asks only "Whether the heaven is incorruptible."
34. Oddus discusses the problem under the heading of "On the end of the world."
35. Jandun asks "Whether an eternal, unchangeable mover, which moves by the intellect and will, could immediately create a new motion with no preceding transmutation."
36. Oresme's question is cited again in question 190 of this catalog.
37. Bricot asks only whether something newly created could be perpetuated and ignores the second part of the question. Although George of Brussels (Bruxellensis) wrote most of the questions, Thomas Bricot revised them and added some of his own. For convenience, I cite Bricot as the author of the treatise. The *De caelo* is separately paginated. What should have been folio 10 recto was mistakenly numbered 9 recto. Although this mistake effected the remainder of the *De caelo* part of the volume, I shall, for convenience, use

18. *Whether every generable thing is corruptible and conversely.*
    1. Johannes de Magistris [*De celo*, bk. 1, qu. 6, dub. 2], 1490, 15, cols. 1–2.[39]
    2. Versor [*De celo*, bk. 1, qu. 17], 1493, 14v, col. 1–15r, col. 1.[40]
    3. Major* [*De celo*, bk. 1, qu. 5], 1526, 6, col. 1–7, col. 2.[41]
    4. Amicus [*De caelo*, tract. 3, qu. 4, dubit. 3], 1626, 126, col. 1.

19. *Whether, on the assumption of eternity, it could be demonstrated that every uncreated thing is incorruptible and that every incorruptible thing is ungenerated.*
    1. Versor [*De celo*, bk. 1, qu. 16], 1493, 14r, cols. 1–2.

20. *Whether every corruptible thing is necessarily corrupted.*
    1. Versor [*De celo*, bk. 1, qu. 18], 1493, 15v, cols. 1–2.

II. On the creation of the world and the meaning of the term "world" (*mundus*)

21. *What is the world, insofar as it is a totality of organized bodies?*
    1. Aversa [*De caelo*, qu. 31, sec. 1], 1627, 2, col. 1–4, col. 1.

22. *What is creation?*
    1. Conimbricenses [*Physics*, pt. 2, bk. 8, ch. 2, qu. 1], 1602, cols. 417–424.[42]
    2. Serbellonus* [*De caelo*, disp. 1, qu. 1, art. 1], 1663, 2:1, col. 1–3, col. 2.

23. *Whether creation is possible.*
    1. Javelli [*Physics*, bk. 8, qu. 3], 1568, 1:598, col. 2.[43]
    2. Compton-Carleton* [*Physics*, disp. 41, sec. 1], 1649, 371, col. 1–372, col. 1.

the foliation as given. Since the questions are unnumbered, only book, folio, and column numbers will be given.

38. Marsilius frames the question as "Whether some new action could arise from an eternal and immutable mover."
39. The 1490 edition is unfoliated. The pagination is my own with the count beginning from the first page of *De celo*. Since the *De celo* is separately distinguished in de Magistris's volume, I shall hereafter cite it as *De celo*.
40. Versor expands the question to read: "Whether generated and corruptible things are convertible and similarly whether ungenerated and incorruptible [things are convertible]."
41. Although Major's questions on the *Physics* and *Metaphysics* are listed on the title page, his *De celo* is not. Indeed, the latter is also omitted from the list of Major's works in Lohr, 1988, 237–239 (Lohr spells his name as "Maior"). The *De celo* begins immediately after the conclusion of the *Physics*. Unfortunately, not only is the volume unfoliated and with few signatures, but the questions are not consistently numbered. With regard to the *Physics*, I have used folio numbers that appear to have been written in by a former owner. No similar numbers were inserted in the section devoted to *De celo*. I have therefore counted pages from the first page of *De celo*. Also included in this volume are questions on Aristotle's *De generatione et corruptione, Meteorology, De anima,* and *Parva naturalia*, none of which are mentioned in Lohr's catalog. Because these works occur in a volume attributed solely to Major, I have thought it reasonable to assign all of them, including *De celo*, to Major. But even if *De celo* is not by Major, it is an independent *questiones* on *De caelo*, and its questions are therefore legitimately a part of this appendix and of this study.
42. The Conimbricenses ask more broadly, "What is creation, and in what manner is it distinguished from preservation?"
43. Javelli links the question directly to Aristotle when he asks "Whether, according to Aristotle, creation is possible."

24. *Whether creation could be known by the light of nature and whether Aristotle knew it.*
   1. Conimbricenses [*Physics*, pt. 2, bk. 8, ch. 2, qu. 2], 1602, cols. 424–427.

25. *Whether the world had existence from itself or from God.*[44]
   1. John of Jandun [*De coelo*, bk. 1, qu. 29], 1552, 19r, col. 2–20r, col. 1.[45]
   2. Toletus [*Physics*, bk. 8, qu. 6], 1580, 234r, col. 2–235r, col. 1.[46]
   3. Bonae Spei [comment. 3, *De coelo*, disp. 2, dub. 2], 1652, 5, col. 2–6, col. 2.
   4. Oddus* [*De coelo*, disp. 1, dub. 5], 1672, 8, col. 2–10, col. 2.

26. *Whether the world was truly made and created from nothing by God.*
   1. Aversa [*De caelo*, qu. 31, sec. 6], 1627, 18, col. 1–21, col. 2.

27. *Whether it is necessary that those who say that the world was made in time assume that a vacuum preceded the generation of the world.*
   1. Javelli [*Physics*, bk. 4, qu. 4], 1568, 1:551, cols. 1–2.

28. *Whether God created the world from a necessity of nature.*
   1. Toletus [*Physics*, bk. 8, qu. 3], 1580, 219v, col. 2–222r, col. 2.
   2. Conimbricenses* [*Physics*, pt. 2, bk. 8, ch. 2, qu. 5], 1602, cols. 440–445.
   3. Oña [*Physics*, bk. 8, qu. 1, art. 3], 1598, 361r, col. 1–363v, col. 1.

29. *Whether something could have been created from eternity by God.*
   1. Conimbricenses* [*Physics*, pt. 2, bk. 8, ch. 2, qu. 6], 1602, cols. 445–449.
   2. Oña [*Physics* bk. 8, qu. 1, art. 6], 1598, 370v, col. 1–375v, col. 2.[47]

30. *Whether the world was produced [or created] whole all at once, or whether [it was created] in six days.*
   1. Hurtado de Mendoza [*De coelo*, disp. 3, sec. 1], 1615, 376, col. 2–378, col. 2.

31. *Toward what end was the world created?*
   1. Amicus [*De caelo*, tract. 2, qu. 4, dubit. 1], 1626, 79, col. 2–81, col. 1.

32. *On the time of the creation of the world.*
   1. Amicus* [*De caelo*, tract. 2, qu. 4, dubit. 2], 1626, 81, col. 1–83, col. 2.
   2. Cornaeus [*De coelo*, tract. 4, disp. 1, qu. 1, sec. 2, dub. 1], 1657, 484.

33. *On the method by which the world was created.*
   1. Amicus [*De caelo*, tract. 2, qu. 4, dubit. 3], 1626, 83, col. 2–91, col. 1.

34. *How the universe comes forth from the first [cause] and the kinds of things that were made by him or effected in another way.*
   1. William of Auvergne* *De universo*, pt. 1 of pt. 1, chs. 17–21, *Opera*, 1674, 1:611, col. 2–617, col. 2.
   2. Cornaeus [*De coelo*, tract. 4, disp. 1, qu. 1, sec. 2, dub. 4], 1657, 485.

44. Although this question is clearly related to questions about the eternity of the world (see qus. 1 and 4 in this catalog), I place it separately here because it has a somewhat different emphasis and approaches the problem from the standpoint of a specific act of creation.
45. Jandun's version reads: "Whether the world was generated [or created]."
46. Toletus asks "Whether God is the efficient cause of the world and [whether this was] also [the opinion] of Aristotle."
47. Oña's version reads: "Whether God could have produced creatures from eternity." In book 8 (qu. 1, art. 4, 363v, col. 2–368r, col. 2), Oña asks "Whether God produced all creatures, or the universe, from eternity."

III. On the order in which God created things

35. *Whether angels were created at the beginning of the world or before the creation of the world.*[48]
    1. Amicus [*De caelo*, tract. 2, qu. 5, dubit. 1], 1626, 91, cols. 1–2.

36. *Whether the creation of any definite body on the first day was signified by these words: "In the beginning God created heaven and earth and the earth was empty and void."*
    1. Riccioli, *Almagestum novum*, pars post., bk. 9, sec. 1, ch. 1, qu. 1, 1651, 193, col. 2–194, col. 1.

37. *What is Chaos amongst the poets, philosophers, and theologians?*
    1. Riccioli, *Almagestum novum*, pars post., bk. 9, sec. 1, ch. 1, qu. 5, 1651, 201, col. 2–204, col. 2.

38. *On the conception of the names "heaven"* [caeli] *and "firmament"* [firmamenti] *among the Latins and Hebrews.*
    1. Riccioli, *Almagestum novum*, pars post., bk. 9, sec. 1, ch. 1, qu. 2, 1651, 194, cols. 1–2.

39. *By the names of "heaven" and "earth," what did Moses signify was created by God in the first instant of the world?*
    1. Riccioli, *Almagestum novum*, pars post., bk. 9, sec. 1, ch. 1, qu. 3, 1651, 194, col. 1–200, col. 2.

40. *Whether the heavens and the elements were simultaneously and instantaneously created or whether they came into being by succession and generation.*
    1. William of Auvergne, *De universo*, pt. 1 of pt. 1, ch. 29, *Opera*, 1674, 1:624, col. 2–625, col. 1.

41. *Whether prime matter was produced at the inchoate beginning before the creation of all the simple and mixed bodies.*
    1. Amicus [*De caelo*, tract. 2, qu. 5, dubit. 2], 1626, 91, col. 2–94, col. 2.

42. *Whether an empyrean heaven should be assigned, and when was it made and what kind of a thing is it?*
    1. Amicus [*De caelo*, tract. 2, qu. 5, dubit. 3], 1626, 94, col. 2–95, col. 1.[49]
    2. Riccioli,* *Almagestum novum*, pars post., bk. 9, sec. 1, ch. 1, qu. 6, 1651, 204, col. 2–209, col. 2.

43. *Whether the ethereal heavens were created on the first day.*
    1. William of Auvergne, *De universo*, pt. 1 of pt. 1, ch. 31, *Opera*, 1674, 1:625, col. 2–627, col. 1.[50]
    2. Amicus* [*De caelo*, tract. 2, qu. 5, dubit. 4], 1626, 95, col. 1–96, col. 1.

44. *How many and which elements did the ancients – and especially the Fathers – become acquainted with from Moses?*

48. In commentaries on the *Sentences*, many authors took up the works of the six days in succession, treating a variety of topics under each day. Because such a subdivision would be uninformative, I have focused on specific questions or themes that fall within the first four days, the days primarily concerned with cosmogony and cosmology.
49. Amicus takes up one aspect of the question when he asks "Whether the empyrean heaven was created at the beginning."
50. William asks "Whether the first created heaven was the empyrean, or another."

1. Riccioli, *Almagestum novum*, pars post., bk. 9, sec. 1, ch. 1, qu. 4, 1651, 200, col. 2–201, col. 2.

45. *Whether all the elements were created at the beginning with the empyrean heaven.*
   1. Amicus [*De caelo*, tract. 2, qu. 5, dubit. 5], 1626, 96, col. 1–98, col. 1.

46. *How many and which heavens were created by God on the first day?*
   1. Riccioli, *Almagestum novum*, pars post., bk. 9, sec. 1, ch. 3, 1651, 224–225.

47. *Whether light was created on the first day.*
   1. Aureoli [*Sentences*, bk. 2, dist. 13, qu. 1, art. 1], 1596–1605, 2:180, col. 1–185, col. 2.

48. *What was the light made by God on the first day of the world?*
   1. Bonaventure [*Sentences*, bk. 2, dist. 13, art. 1, qu. 1], *Opera*, 1885, 2:311–313.[51]
   2. Riccioli,* *Almagestum novum*, pars post., bk. 9, sec. 1, ch. 1, qu. 7, 1651, 194, col. 1–200, col. 2.

49. *Whether the light that was created on the first day causes day and night by local circulation.*
   1. Richard of Middleton [*Sentences*, bk. 2, dist. 13, art. 1, qu. 1], 1591, 2:156, cols. 1–2.

50. *How that light [made on the first day] made day and night.*
   1. Bonaventure* [*Sentences*, bk. 2, dist. 13, art. 1, qu. 2], *Opera*, 1885, 2:314–316.
   2. Major [*Sentences*, bk. 2, dist. 13, qu. 1], 1519b, 66v, col. 2–67v, col. 1.

51. *Whether some heaven was made on the second day of the world from water condensed in the manner of ice or a crystal.*
   1. Riccioli, *Almagestum novum*, pars post., bk. 9, sec. 1, ch. 2, qu. 1, 1651, 216, col. 2–218, col. 2.

52. *Whether the elements made on the third day were made by the power of the heavens from the matter created on the first day.*
   1. Richard of Middleton [*Sentences*, bk. 2, dist. 14, art. 3, qu. 5], 1591, 2:181, cols. 1–2.

53. *Why the production of the stars was left to the fourth day.*
   1. Riccioli, *Almagestum novum*, pars post., bk. 9, sec. 1, ch. 4, qu. 1, 1651, 225–226.

IV. Does anything exist beyond our world?

A. God, void spaces, time, place, or animated beings?

54. *Whether there is something [that is, body, void, imaginary space, or animate creatures] beyond the sky [or heavens].*

51. Bonaventure made the question more specific when he asked "Whether the light made on the first day was corporeal or spiritual."

1. Bacon, *De celestibus*, pt. 3, ch. 3, *Opera*, fasc. 4, 1913, 377–84;[52] and [*Physics*, bk. 4], *Opera*, fasc. 13, 1935, 223–224.[53]
2. John of Jandun [*De coelo*, bk. 1, qu. 26], 1552, 18r, col. 1–col. 2.[54]
3. Buridan* [*De caelo*, bk. 1, qu. 20], 1942, 91–95.
4. Oresme [*De celo*, bk. 1, qu. 19], 1965, 279–295.
5. Themon Judaeus [*De spera*, qu. 8], in Hugonnard-Roche, 1973, 99–105.[55]
6. Johannes de Magistris [*De celo*, bk. 1, qu. 4, dub. 1], 1490, 11, col. 1.[56]
7. Case [*Physics*, bk. 8, ch. 10], 1599, 858–863.[57]

55. *What are imaginary spaces?*
1. Conimbricenses [*Physics*, pt. 2, bk. 8, ch. 10, qu. 2], 1602, cols. 580–586.[58]
2. Hurtado de Mendoza [*Physics*, disp. 14, sec. 2], 1615, 321, col. 2–323, col. 2.[59]
3. Amicus [*Physics*, tract. 21, qu. 5, dubit. 1], 1626–1629, 2:759, col. 1–764, col. 1.[60]
4. Aversa [*De caelo*, qu. 31, sec. 4], 1627, 10, col. 2.
5. Arriaga [*Physics*, disp. 14, sec. 3], 1632, 431, cols. 1–2.
6. Oviedo [*Physics*, bk. 4, contro. 15, punc. 2], 1640, 370, col. 2–371, col. 2.
7. Compton-Carleton [*Physics*, disp. 32, secs. 1–3], 1649, 335–337.
8. Bonae Spei* [comment. 2, *Physics*, disp. 5 ("De spatiis imaginariis"), dub. 1], 1652, 177, col. 1.
9. Serbellonus [*Physics*, disp. 5, art. 1], 1657, 1:798, col. 1–803, col. 2.[61]

56. *Whether imaginary space exists by means of the operation of the intellect, or whether indeed it really exists.*
1. Amicus [*Physics*, tract. 21, qu. 5, dubit. 2], 1626–1629, 2:764, col. 2–766, col. 2.

57. *Whether place [*ubi*] refers essentially to imaginary space.*
1. Cornaeus [*Physics*, bk. 4, disp. 3 ("De loco et vacuo"), qu. 1, sec. 2, dub. 7], 1657, 365–368.

58. *Whether God is in imaginary spaces beyond the heavens.*
1. Major [*Sentences*, bk. 1, dist. 37, qu. 1], 1519a, 93r, col. 1–93v, col. 1.[62]

52. Bacon's discussion centers on the claim "that a body cannot exist outside the world."
53. Bacon asks "Whether there is a vacuum beyond the heaven."
54. Jandun's version of this question took the form "Whether a vacuum exists outside the sky [or heaven]."
55. Although the question reads "Whether something should be assumed beyond the ninth sphere," Themon also considers whether there is anything beyond the world. This latter question is also relevant to the existence of an immobile sphere or, more particularly, an empyrean sphere (see qu. 107 in this catalog).
56. De Magistris's version asks "Whether a place could exist outside the heaven [or sky]."
57. Case asks "Whether there is an imaginary space beyond the heaven and whether, according to Aristotle, God and the blessed spirits are in it."
58. The Coimbra Jesuits actually ask "Whether or not God is beyond [or outside] the heaven." The third article of this question considers "What is imaginary space and in what manner does God exist in it?"
59. Here Hurtado asks, "What is it to be somewhere, and what is imaginary space?"
60. Amicus asks "Whether imaginary space exists and what it is."
61. Serbellonus discusses imaginary spaces under the question "What is place?"
62. Major expresses the question as "Whether God is everywhere and in an infinite, imaginary place beyond the heaven."

2. Hurtado de Mendoza [*Physics*, disp. 14, sec. 3], 1615, 323, col. 2–324, col. 2.
3. Arriaga [*Physics*, disp. 14, sec. 4], 1632, 431, col. 2–434, col. 2.
4. Oviedo [*Physics*, bk. 4, contro. 15, punc. 3], 1640, 372, col. 1–373, col. 2.[63]
5. Compton-Carleton [*Physics*, disp. 33, sec. 4], 1649, 337.
6. Bonae Spei [comment. 2, *Physics*, disp. 5 ("De spatiis imaginariis"), dub. 2], 1652, 177, col. 1–178, col. 2.[64]
7. Cornaeus [*Physics*, bk. 4, disp. 3 ("De loco et vacuo"), qu. 1, sec. 2, dub. 9], 1657, 369–370.
8. Oddus* [*Physics*, bk. 4, disp. 14, art. 9], 1667, 640, col. 1–645, col. 1.

59. *Whether, according to Aristotle, God is everywhere.*
 1. Toletus [*Physics*, bk. 8, qu. 7], 1580, 242v, col. 1–243v, col. 1.

60. *Whether beings beyond the sky [or heavens] are unchangeable.*
 1. John of Jandun [*De coelo*, bk. 1, qu. 28], 1552, 18v, col. 2–19r, col. 2.

61. *Whether beings that might exist outside the heaven [or sky] are corporeal.*
 1. Johannes de Magistris [*De celo*, bk. 1, qu. 4, dub. 2], 1490, 11, cols. 1–2.

B. Other worlds?

62. *Whether there are, or could be, more worlds.*
 1. Michael Scot [*Sphere*, lec. 1], 1949, 252–254.
 2. William of Auvergne, *De universo*, pt. 1 of pt. 1, chs. 11–16, *Opera*, 1674, 1:604–611.
 3. Bacon, *De celestibus*, pt. 3, ch. 2, *Opera*, fasc. 4, 1913, 373–377.
 4. John of Jandun [*De coelo*, bk. 1, qu. 24], 1552, 16v, col. 1–17r, col. 1.
 5. Buridan [*De caelo*, bk. 1, qu. 19], 1942, 87–90.
 6. Albert of Saxony* [*De celo*, bk. 1, qu. 11], 1518, 95r, col. 1–95v, col. 1.
 7. Oresme [*De celo*, bk. 1, qu. 18], 1965, 265–279.
 8. Paul of Venice, *Liber celi*, 1476, 7, col. 1–9, col. 1 [sigs. fv–f3r].
 9. Johannes de Magistris [*De celo*, bk. 1, qu. 4], 1490, 9, col. 2–11, col. 2.
 10. Versor [*De celo*, bk. 1, qu. 13], 1493, 11r, col. 2–11v, col. 2.[65]
 11. Bricot [*De celo*, bk. 1], 1486, 9v, col. 2–10r, col. 1.
 12. Galileo [*De caelo*, qu. 3 (E)], 1977, 43–45.
 13. Conimbricenses [*De coelo*, bk. 1, ch. 8, qu. 1], 1598, 102–107.
 14. Amicus [*De caelo*, tract. 3, qu. 1], 1626, 101, col. 1–109, col. 2.
 15. Aversa [*De caelo*, qu. 31, sec. 3], 1627, 7, col. 1–9, col. 2.
 16. Bonae Spei [comment. 3, *De coelo*, disp. 1, dub. 1], 1652, 1, col. 1–2, col. 1.
 17. Cornaeus [*De coelo*, tract. 4, disp. 1, qu. 1, sec. 2, dub. 3], 1657, 484–485.[66]
 18. Oddus [*De coelo*, disp. 1, dub. 3], 1672, 5, col. 2–7, col. 2.[67]

63. Oviedo's question is somewhat broader, since he asks "Whether God is in imaginary space, and also beyond the heaven; and what about other entities?"
64. Bonae Spei simply asks "Whether God is actually in imaginary spaces."
65. Versor asks "Whether it is possible [*possibile*] that more worlds exist."
66. Cornaeus asks "Whether only one world was made."
67. Oddus discusses the plurality of worlds in a section titled "On the Unity of the World" (*De unitate mundi*).

63. *Whether or not other worlds could exist by divine power.*
   1. Conimbricenses* [*De coelo*, bk. 1, ch. 9, qu. 1], 1598, 112–115.
   2. Amicus [*De caelo*, tract. 3, qu. 2], 1626, 110, col. 1–111, col. 2.
   3. Cornaeus [*De coelo*, tract. 4, disp. 1, qu. 1, sec. 1, dub. 8], 1657, 483–484.

64. *Whether God could make other worlds more and more perfect into infinity.*
   1. Conimbricenses [*De coelo*, bk. 1, ch. 9, qu. 3], 1598, 120–123.

65. *Whether beyond this world, God could make another earth of the same species as this world.*
   1. Godfrey of Fontaines, *Quodlibet IV*, qu. 6, 1904, 331–332.

66. *If there were several worlds, whether the earth of one would be moved naturally to the middle [or center] of another.*
   1. Buridan* [*De caelo*, bk. 1, qu. 18], 1942, 83–87.[68]
   2. Albert of Saxony [*De celo*, bk. 1, qu. 10], 1518, 94r, col. 1–95r, col. 1.
   3. Oresme [*De celo*, bk. 1, qu. 17], 1965, 243–263.
   4. Versor [*De celo*, bk. 1, qu. 12], 1493, 10r, col. 1–10v, col. 2.
   5. Bricot [*De celo*, bk. 1], 1486, 9r,[69] cols. 1–2.
   6. Galileo [*De caelo*, qu. 3 (E)], 1977, 44–45.

67. *Whether the world is one.*
   1. Poncius* [*De coelo*, disp. 21, qu. 1], 1672, 596, col. 2–597, col. 1.
   2. Oddus [*De coelo*, disp. 1, dub. 4], 1672, 7, col. 2–8, col. 1.[70]

V. On the perfection and finitude of the world

68. *Whether the world is perfect.*
   1. John of Jandun [*De coelo*, bk. 1, qu. 6], 1552, 5r, col. 1–5v, col. 2.
   2. Buridan* [*De caelo*, bk. 1, qu. 12], 1942, 54–57.
   3. Albert of Saxony [*De celo*, bk. 1, qu. 5], 1518, 90r, col. 1–90v, col. 1.
   4. Oresme [*De celo*, bk. 1, qu. 3], 1965, 37–51.
   5. Versor [*De celo*, bk. 1, qu. 2], 1493, 1v, col. 2–2r, col. 2.
   6. Bricot [*De celo*, bk. 1], 1486, 5r, cols. 1–2.
   7. Major [*De celo*, bk. 1, qu. 4], 1526, 5, col. 1–6, col. 2.
   8. Galileo [*De caelo*, qu. 3 (E)], 1977, 46–47.
   9. Conimbricenses [*De coelo*, bk. 1, ch. 1, qu. 1], 1598, 8–19.
   10. Amicus [*De caelo*, tract. 3, qu. 2, dubit. 1], 1626, 111, col. 2–115, col. 1.
   11. Aversa [*De caelo*, qu. 31, sec. 5], 1627, 14, col. 2–18, col. 1.
   12. Poncius [*De coelo*, disp. 21, qu. 2], 1672, 597, col. 2–602, col. 1.
   13. Bonae Spei [comment. 3, *De coelo*, disp. 1, dub. 2], 1652, 2, col. 1–3, col. 1.
   14. Cornaeus [*De coelo*, tract. 4, disp. 1, sec. 2, qu. 2, dub. 1], 1657, 485–486.

68. The translation is from Grant, 1974, 204. I have added the words in brackets.
69. In the edition, folio "9" (actually "ix") is repeated. The second "9" should be 10. The reference here is to the second folio 9, which also has signature "B1" at the bottom of the page.
70. Oddus is more specific when he asks "Whether the world is one *per se* or by accident."

69. *Whether the world could be made more perfect.*
1. William of Auvergne, *De universo*, pt. 1 of pt. 1, ch. 23, *Opera*, 1674, 1:617, col. 2–618, col. 2.[71]
2. Bonaventure [*Sentences*, bk. 1, dist. 44, art. 1, qus. 1–3], *Opera*, 1882, 1:781, col. 1–787, col. 2.
3. Thomas Aquinas [*Sentences*, bk. 1, dist. 44, qu. 1, art. 2], 1929–1947, 1:1018–1021.
4. Richard of Middleton [*Sentences*, bk. 1, dist. 44, art. 1, qu. 1], 1591, 1:389, col. 2–391, col. 1.
5. Conimbricenses [*De coelo*, bk. 1, ch. 9, qu. 2], 1598, 115–120.
6. Amicus* [*De caelo*, tract. 3, qu. 2, dubit. 2], 1626, 115, col. 1–120, col. 2.
7. Aversa [*De caelo*, qu. 31, sec. 5], 1627, 15, col. 1–18, col. 1.
8. Cornaeus [*De coelo*, tract. 4, disp. 1, qu. 2, dub. 2], 1657, 486–487.

70. *Whether the world is a finite or infinite magnitude.*
1. Bacon, *De celestibus*, pt. 3, ch. 1, *Opera*, fasc. 4, 1913, 369–373.[72]
2. Albert of Saxony [*De celo*, bk. 1, qu. 9], 1518, 93r, col. 2–94r, col. 1.
3. Paul of Venice,* *Liber celi*, 1476, 10, col. 2–12, col. 1.

71. *Whether a body that is moved circularly could be actually infinite.*
1. Buridan [*De caelo*, bk. 1, qu. 15], 1942, 67–73.
2. Albert of Saxony [*De celo*, bk. 1, qu. 7], 1518, 91r, col. 1–92r, col. 1.
3. Versor* [*De celo*, bk. 1, qu. 8], 1493, 6r, col. 2–6v, col. 2.

72. *Whether it is possible that a body moved rectilinearly could be infinite.*
1. Buridan* [*De caelo*, bk. 1, qu. 16], 1942, 73–77.
2. Versor [*De celo*, bk. 1, qu. 9], 1493, 7r, col. 2–7v, col. 2.

73. *Whether it is possible that some sensible body be actually infinite.*
1. John of Jandun [*De coelo*, bk. 1, qu. 22], 1552, 14r, col. 2–14v, col. 2.
2. Buridan [*De caelo*, bk. 1, qu. 17], 1942, 77–82 and [*Physics*, bk. 3, qu. 14], 1509, 55v, col. 2–57r, col. 2.
3. Albert of Saxony [*De celo*, bk. 1, qu. 6], 1518, 90v, col. 1–91r, col. 1.[73]
4. Oresme [*De celo*, bk. 1, qu. 16], 1965, 225–243.
5. Marsilius of Inghen [*Physics*, bk. 3, qu. 10], 1518, 44v, col. 2–45v, col. 2.
6. Johannes de Magistris [*De celo*, bk. 1, qu. 3], 1490, 6, col. 1–9, col. 2.
7. Versor* [*De celo*, bk. 1, qu. 10], 1493, 8v, col. 1–9v, col. 1.

VI. On the simple bodies of the world

74. *Whether there are five distinct simple bodies in the world that differ in species, namely the four elements and the heavens or fifth essence.*
1. Albert of Saxony* [*De celo*, bk. 1, qu. 2], 1518, 86v, col. 2–87v, col. 1.
2. Johannes de Magistris [*De celo*, bk. 1, qu. 1, dub. 2], 1490, 3, col. 2–4, col. 2, and qu. 2, dub. 1, 5, col. 2–6, col. 1.

71. William expressed essentially the same question somewhat differently when he asked "Whether the Creator could have created things better than he did."
72. Bacon expressed the problem as "On the finitude of the world."
73. Albert's version reads: "Whether there can be an infinite, immobile body."

3. Versor [*De celo*, bk. 1, qu. 4], 1493, 3r, col. 2–4r, col. 2.
4. Cornaeus [*De coelo*, tract. 4, disp. 2, qu. 1, dub. 4], 1657, 490–491.[74]
5. Oddus [*De coelo*, disp. 1, dub. 2], 1672, 3, col. 2–5, col. 2.

75. *Are the heavens one of the simple bodies or composed of them?*
 1. Galileo★ [*De caelo*, qu. 3 (I)], 1977, 81–92.
 2. Poncius [*De coelo*, disp. 22, qu. 2], 1672, 613, col. 1–615, col. 2.[75]
 3. Bonae Spei [comment. 3, *De coelo*, disp. 3, dub. 2], 1652, 8, col. 2–9, col. 2.

76. *Whether a sixth element should be assumed besides the four elements and the heaven.*
 1. Bacon, *De celestibus*, pt. 1, ch. 4, *Opera*, fasc. 4, 1913, 331–332.
 2. Albert of Saxony★ [*De celo*, bk. 1, qu. 3], 1518, 87r, col. 1–88v, col. 1.

77. *Whether the whole world that is aggregated from the aforementioned five simple bodies is one continuous being.*
 1. Albert of Saxony [*De celo*, bk. 1, qu. 4], 1518, 88v, col. 1–89v, col. 2.

PART II. THE CELESTIAL REGION

VII. On celestial matter and form

78. *Whether the sky [or heaven] has matter.*
 1. Buridan★ [*De caelo*, bk. 1, qu. 11], 1942, 49–54.
 2. Oresme [*De celo*, bk. 1, qu. 11], 1965, 159–173.
 3. Versor [*De celo*, bk. 1, qu. 5, dub. 2], 1493, 5r, col. 1.
 4. Bricot [*De celo*, bk. 1], 1486, 4v, cols. 1–2.
 5. Major [*De celo*, bk. 1, qu. 3], 1526, 4, col. 1–5, col. 1.

79. *Whether the heaven is composed of matter and form.*
 1. Aureoli [*Sentences*, bk. 2, dist. 14, qu. 1, arts. 1–2], 1596–1605, 2:186, col. 1–191, col. 2.
 2. John of Jandun★ [*De coelo*, bk. 1, qu. 23], 1552, 14v, col. 2–16v, col. 1.
 3. Johannes de Magistris [*De celo*, bk. 1, qu. 2], 1490, 4, col. 2–6, col. 1.
 4. Major [*Sentences*, bk. 2, dist. 12, qu. 3], 1519b, 65r, col. 2–66v, col. 1.
 5. Galileo [*De caelo*, qu. 5 (K)], 1977, 103–147.
 6. Conimbricenses [*De coelo*, bk. 1, ch. 2, qu. 4], 1598, 37–41.
 7. Hurtado de Mendoza [*De coelo*, disp. 1, sec. 1], 1615, 362, col. 2–364, col. 1.
 8. Amicus [*De caelo*, tract. 4, qu. 2, dubit. 2], 1626, 137, col. 1–143, col. 2.

74. The question Cornaeus asks is "Whether the starry heaven is a fifth simple essence distinct from the elements." Later Cornaeus asks, "What is an element?" (tract. 4, disp. 3, qu. 1, dub. 1) and "How many [terrestrial] elements are there?" (tract. 4, disp. 3, qu. 1, dub. 2).

75. Poncius asks "Whether the heaven is a simple body in the same way as the elements are simple bodies, or indeed is it something mixed?" In the preceding question (disp. 22, qu. 1), Poncius inquired "Whether the heaven is a simple body" (611, col. 1–613, col. 1). For a closely related question, see question 82 in this catalog.

9. Aversa [*De caelo*, qu. 33, sec. 5], 1627, 100, col. 1–105, col. 2.[76]
10. Mastrius and Bellutus [*De coelo*, disp. 2, qu. 2, art. 1], 1727, 3:490, col. 2, par. 40–492, col. 2, par. 55.[77]
11. Compton-Carleton [*De coelo*, disp. 3], 1649, 404, col. 1–407, col. 2.
12. Riccioli, *Almagestum novum*, pars post., bk. 9, sec. 1, ch. 5, qu. 1, 1651, 232, col. 2–233, col. 2 and qu. 2, 233, col. 2–235, col. 1.
13. Cornaeus [*De coelo*, tract. 4, disp. 2, qu. 1, dub. 1], 1657, 488–489.
14. Serbellonus [*De caelo*, disp. 1, qu. 2, art. 2], 1663, 2:20, col. 1–22, col. 1.
15. De Rhodes [*De coelo*, bk. 2, disp. 8, qu. 1, sec. 1, art. 1], 1671, 276, cols. 1–2.
16. Oddus [*De coelo*, disp. 1, dub. 8], 1672, 15, col. 1–22, col. 1.

80. What is the form of celestial bodies?: whether it is a soul.

1. Aversa [*De caelo*, qu. 33, sec. 7], 1627, 109, col. 1–114, col. 2.

81. Whether celestial bodies differ among themselves not only in form but also in matter.

1. Aversa [*De caelo*, qu. 33, sec. 9], 1627, 119, col. 1–123, col. 2.

82. Whether the heaven is a mixed body.

1. Serbellonus [*De caelo*, disp. 1, qu. 2, art. 1], 1663, 2:18, col. 1–19, col. 2.

83. From what matter are the stars [or planets] produced?

1. Amicus* [*De caelo*, tract. 5, qu. 7, dubit. 7], 1626, 347, col. 2–348, col. 1.
2. Riccioli, *Almagestum novum*, pars post., bk. 9, sec. 1, ch. 4, qu. 4, 1651, 230–231.[78]

84. Whether the nature of the stars [or planets] differs from that of the sublunary [world] and is the same as their [respective] orbs.

1. Conimbricenses [*De coelo*, bk. 2, ch. 7, qu. 1], 1598, 292–296.[79]

85. Whether celestial and elementary matter are of the same species.

1. Bonaventure [*Sentences*, bk. 2, dist. 12, art. 2, qu. 1], *Opera*, 1885, 2:302–303.
2. Thomas Aquinas [*Sentences*, bk. 2, dist. 14, qu. 1, art. 2], 1929–1947, 2:349–351.[80]
3. Major [*Sentences*, bk. 2, dist. 12, qu. 3], 1519b, 65r, col. 2–66v, col. 1.
4. Conimbricenses [*De coelo*, bk. 1, ch. 2, qus. 5–6], 1598, 41–47, 47–57.

76. Replacing "heaven" with "celestial bodies," Aversa asks "Whether celestial bodies are composed of matter and form" (Utrum corpora caelestia sint ex materia et forma composita).
77. The *De coelo*, along with the *De generatione* and *Meteors*, was first published in 1640. These disputations, along with many others on philosophy and natural philosophy, were subsequently published as a *cursus philosophicus* in 1678, with other editions in 1688, 1708, and 1727. See Crowley, 1948, 144, 146–147.
78. Riccioli asks, "From what matter are the luminaries and the other planets [made]?"
79. This question is also relevant to question 94 in this catalog.
80. By asking "Whether the firmament has the nature of inferior bodies" (Utrum firmamentum sit de natura inferiorum corporum), Thomas couches his question in narrow and specific terms. Weisheipl explains (1974, 358) that "strictly speaking, this is not a 'commentary' on Peter Lombard's *Sentences*, but rather 'writings' (*scripta*), or elaborations of the text in the form of questions and discussions of relevant themes arising from the text." Most commentaries on the *Sentences* were similar, being comprised largely of questions.

5. Hurtado de Mendoza [*De coelo*, disp. 1, sec. 2], 1615, 364, col. 1–365, col. 2.
6. Amicus★ [*De caelo*, tract. 4, qu. 3, dubit. 1], 1626, 144, col. 1–155, col. 2.
7. Aversa [*De caelo*, qu. 33, sec. 1], 1627, 76, col. 1–80, col. 2, and [sec. 6], 105, col. 2–109, col. 1.
8. Mastrius and Bellutus [*De coelo*, disp. 2, qu. 2, art. 2], 1727, 3:492, col. 2, par. 56–495, col. 1, par. 73.
9. Oviedo [*De caelo*, contro. 1, punc. 1], 1640, 462, col. 1–464, col. 2.
10. Compton-Carleton [*De coelo*, disp. 2, sec. 1], 1649, 401, col. 1–402, col. 1.
11. Riccioli, *Almagestum novum*, pars post., bk. 9, sec. 1, ch. 5, qu. 2, 1651, 233, col. 2–235, col. 1.
12. Bonae Spei [comment. 3, *De coelo*, disp. 3, dub. 3], 1652, 9, col. 2–10, col. 1.
13. Cornaeus [*De coelo*, tract. 4, disp. 2, qu. 1, dub. 2], 1657, 489.[81]
14. Oddus [*De coelo*, disp. 1, dub. 9], 1672, 22, col. 2–25, col. 2.

86. *Whether celestial quantity is distinct in species from sublunar [quantity].*
 1. Compton-Carleton [*De coelo*, disp. 2, sec. 2], 1649, 402, cols. 1–2.

87. *Whether celestial matter is more perfect than elementary matter.*
 1. Amicus [*De caelo*, tract. 4, qu. 3, dubit. 3], 1626, 158, col. 1–160, col. 2.

88. *Of which element, or elements, do the heavens consist?*
 1. Riccioli,★ *Almagestum novum*, pars post., bk. 9, sec. 1, ch. 5, qu. 3, 1651, 235, col. 1–236, col. 2.
 2. De Rhodes [*De coelo*, bk. 2, disp. 8, qu. 1, sec.1, art. 2], 1671, 276, col. 2–277, col. 1.

89. *Whether, with respect to its substance, the heavens depend on God as an efficient cause [*causa agente*].*
 1. Buridan [*Metaphysics*, bk. 12, qu. 7], 1518, 69v, col. 2–70v, col. 1.

VIII. On orbs and planets and their relations

90. *Whether there are several and distinct heavens and orbs in which planets [*astra*] exist: How many are there, and how ought they to be distinguished?*
 1. Aversa [*De caelo*, qu. 32, sec. 5], 1627, 58, col. 1–64, col. 2.

91. *Whether all celestial spheres and all stars, both wandering [that is, the planets] and fixed, are of the same ultimate species [*speciei specialissime*].*
 1. John of Jandun [*De coelo*, bk. 2, qu. 14], 1552, 30v, col. 2–31v, col. 1.
 2. Buridan [*De caelo*, bk. 2, qu. 14], 1942, 184, 192.[82]
 3. Oresme [*De celo*, bk. 2, qu. 10], 1965, 603–638.
 4. Albert of Saxony [*De celo*, bk. 2, qu. 19], 1518, 113r, col. 2–114v, col. 1.

81. Cornaeus's version asks "Whether they [that is, the heavens] consist of sublunary matter."
82. Buridan asked whether "all celestial spheres and all stars [or planets] are of the same ultimate species [*speciei specialissimae*]." The translation is from Grant, 1974, 204; I have added the words in brackets. Albert of Saxony phrased the question in similar terms (see discussant 4 of this question), even using the same Latin phrase.

5. Johannes de Magistris* [*De celo*, bk. 2, qu. 4], 1490, 23, col. 1–25, col. 1.
6. Major [*De celo*, bk. 2, qu. 5], 19, col. 2–20, col. 1.
7. Conimbricenses [*De coelo*, bk. 2, ch. 5, qu. 3 (mistakenly numbered 1)], 1598, 256–260.
8. Hurtado de Mendoza [*De coelo*, disp. 1, sec. 3], 1615, 365, col. 2–366, col. 1.
9. Amicus [*De caelo*, tract. 4, qu. 3, dubit. 2], 1626, 155, col. 1–158, col. 1.
10. Poncius [*De coelo*, disp. 22, qu. 7], 1672, 624, col. 2–625, col. 1.

92. *Whether all the [celestial] orbs are distinguished from each other in species.*
1. Michael Scot [*Sphere*, lec. 4], 1949, 284.
2. Bacon, *De celestibus*, pt. 4, ch. 4, *Opera*, fasc. 4, 1913, 393–395.
3. Bricot [*De celo*, bk. 2], 1486, 21v, col. 2–22r, col. 1.
4. Amicus* [*De caelo*, tract. 4, qu. 7, dubit. 1], 1626, 202, col. 1–205, col. 2.[83]

93. *On the distinction among the parts of one orb that surrounds [or circumscribes] the whole heaven.*
1. Amicus [*De caelo*, tract. 4, qu. 7, dubit. 2], 1626, 205, col. 1–206, col. 1.

94. *Whether the stars [i.e., planets] differ in species from their orbs.*
1. Bacon, *De celestibus*, pt. 4, ch. 7, *Opera*, fasc. 4, 1913, 401–403.
2. Conimbricenses [*De coelo*, bk. 2, ch. 7, qu. 1], 1598, 292–296.[84]
3. Amicus* [*De caelo*, tract. 4, qu. 7, dubit. 3], 1626, 206, col. 2–208, col. 2.
4. Oddus [*De coelo*, disp. 1, dub. 10], 1672, 25, col. 2–29, col. 1.[85]

95. *Whether all the planets [or stars;* astra*] are distinguished from each other in species.*
1. Bacon, *De celestibus*, pt. 4, ch. 6, *Opera*, fasc. 4, 1913, 397–401.
2. Amicus* [*De caelo*, tract. 4, qu. 7, dubit. 4], 1626, 208, col. 2–213, col. 1.

IX. On the creation, number, order, nobility, and hierarchical structure of the heavens, planets, and stars

96. *In what order and position are the principal parts of the world arranged?*
1. Aversa [*De caelo*, qu. 31, sec. 2], 1627, 4, col. 1–6, col. 2.

97. *On the number of spheres, whether there are eight or nine, or more or less.*
1. Thomas Aquinas [*Sentences*, bk. 2, dist. 14, qu. 1, art. 4], 1929–1947, 2:354–356.
2. Bacon, *De celestibus*, pt. 4, ch. 2, *Opera*, fasc. 4, 1913, 387–393.
3. Albert of Saxony* [*De celo*, bk. 2, qu. 6], 1518, 105r, col. 1–106r [mistakenly foliated 107r], col. 2.
4. Themon Judaeus [*De spera*, qu. 3], in Hugonnard-Roche, 1973, 69–75.[86]

83. This is the first of four consecutive questions, or doubts (for the other three, see the next three questions), that Amicus raises concerning distinctions among the celestial orbs themselves, among the parts of a single orb, between the orbs and stars, and among the planets themselves.
84. This question is also relevant to question 84 in this catalog.
85. Oddus's discussion is presented under the title "On the differences of the heavens among themselves."
86. Themon's treatise consists of a series of questions on the *Sphere* of Sacrobosco. Parts of the Latin text of the questions are published in Hugonnard-Roche, 1973, with accompanying summaries and analyses.

5. D'Ailly, *14 Questions*, qu. 2, 1531, 148v–149v.[87]
6. Bricot [*De celo*, bk. 2], 1486, 21r, col. 1–21v, col. 1.
7. Major [*Sentences*, bk. 2, dist. 14, qu. 1], 1519b, 71v, col. 1–72v, col. 1.
8. Galileo [*De caelo*, qus. 1–2 (G–H)], 1977, 59–70, 71–80.
9. Conimbricenses [*De coelo*, bk. 2, ch. 5, qu. 1], 1598, 246–252.
10. Hurtado de Mendoza [*De coelo*, disp. 2, sec. 1], 1615, 367, col. 2–370, col. 1.
11. Amicus [*De caelo*, tract. 4, qu. 6, dubit. 4], 1626, 192, col. 1–193, col. 2.
12. Aversa [*De caelo*, qu. 32, sec. 2], 1627, 44, col. 2–45, col. 2.
13. Arriaga [*De caelo*, disp. 1, sec. 4], 1632, 504, col. 1–505, col. 2.
14. Poncius [*De coelo*, disp. 22, qu. 6], 1672, 621, col. 2–624, col. 2, and [*Sentences*, bk. 2, dist. 14, qu. 2], 1639, 6, pt. 2:725–731.[88]
15. Oviedo [*De caelo*, contro. 1, punc. 4], 1640, 469, col. 1–472, col. 2.
16. Mastrius and Bellutus [*De coelo*, disp. 2, art. 2], 1727, 488, col. 1, par. 16–490, col. 1, par. 35.
17. Compton-Carleton [*De coelo*, disp. 4, sec. 1], 1649, 408, col. 1–409, col. 1.
18. Riccioli, *Almagestum novum*, pars post., bk. 9, sec. 3, ch. 1, 1651, 271, col. 1–276, col. 1.[89]
19. Bonae Spei [comment. 3, *De coelo*, disp. 3, dub. 8], 1652, 14, col. 1–15, col. 2.
20. Cornaeus [*De coelo*, tract. 4, disp. 2, qu. 2, dub. 1], 1657, 492–493.[90]
21. Serbellonus [*De caelo*, disp. 1, qu. 2, art. 5], 1663, 2:28, col. 1–32, col. 2.
22. Oddus [*De coelo*, disp. 1, dub. 14], 1672, 38, col. 1–42, col. 2.

98. *Whether there are other superior heavens besides the sidereal heaven in which the planets [*astra*] are.*
   1. Aversa [*De caelo*, qu. 32, sec. 4], 1627, 52, col. 1–57, col. 2.

99. *On the order of the heavens.*
   1. Major [*Sentences*, bk. 2, dist. 14, qu. 1], 1519b, 71v, col. 1–72v, col. 1.
   2. Amicus* [*De caelo*, tract. 5, qu. 4, dubit. 2], 1626, 264, col. 1–265, col. 1.
   3. Aversa [*De caelo*, qu. 32, sec. 3], 1627, 47, col. 1–52, col. 1.
   4. Cornaeus [*De coelo*, tract. 4, disp. 3, qu. 2, dub. 9], 1657, 525–529.[91]

87. Part of d'Ailly's question is also cited under question 106 in this catalog.
88. Although, in his *Sentences*, Poncius discusses the number of heavens, the question by Scotus to which he is responding simply inquires (p. 725) "Whether there is any mobile heaven besides the starry heaven" (Utrum aliquod caelum mobile aliud a caelo stellato?). The same question from Poncius's *De coelo* is also relevant to question 105 in this catalog.
89. As he often did, Riccioli presents a thorough history of the problem, titling his chapter "On the total number of heavens" (De numero caelorum totalium).
90. Later (disp. 2, qu. 4, dub. 6), Cornaeus asks, "How many stars are planetary?" For Cornaeus these two questions are quite distinct. During the Middle Ages and into the seventeenth century, a celestial sphere and a heaven were often treated as synonymous. But there was a tradition which spoke of a sidereal heaven (as did Thomas Aquinas) which included the sphere of the fixed stars and the seven planetary spheres. Cornaeus, who did not believe in the existence of material celestial spheres, followed this tradition and distinguished between heavens and planets. He assumed the existence of only three heavens (see p. 493), one of which, the sidereal, embraced the seven planets and all the fixed stars.
91. Cornaeus asks "How the system of this world is organized."

100. *On the generation of the heavens.*
   1. William of Auvergne, *De universo*, pt. 1 of pt. 1, ch. 42, *Opera*, 1674, 1:639, col. 1–644, col. 2.

101. *On the perfection of the heavens among themselves and with respect to sublunar things.*
   1. Amicus [*De caelo*, tract. 5, qu. 4, dubit. 2], 1626, 264, col. 1–265, col. 1.[92]
   2. Oddus* [*De coelo*, disp. 1, dub. 13], 1672, 35, col. 2–38, col. 1.[93]

102. *Whether the arrangement of the celestial orbs* in situ *is [also] their order with respect to perfection.*[94]
   1. Buridan* [*Metaphysics*, bk. 12, qu. 12], 1518, 74r, col. 2–75r, col. 1.
   2. Conimbricenses [*De coelo*, bk. 2, ch. 5, qu. 2], 1598, 252–256.[95]

103. *Whether the heavens surpass any sublunar body in the dignity of their nature.*
   1. Conimbricenses* [*De coelo*, bk. 2, ch. 1, qu. 2], 1598, 171–177.
   2. Oddus [*De coelo*, disp. 1, dub. 13], 1672, 35, col. 2–38, col. 1.

104. *Whether it is necessary to assume a ninth sphere.*
   1. Michael Scot [*Sphere*, lec. 2], 1949, 259–260.
   2. Robertus Anglicus, [*Sphere*, lec. 1], 1949, 146, 148 (Latin); 202–203 (English).
   3. Oresme* [*De spera*, qu. 12], 1966a, 264–281.

105. *How many fixed stars are there?*
   1. Aversa [*De caelo*, qu. 32, sec. 2], 1627, 42, col. 2–44, col. 2.
   2. Poncius [*De coelo*, disp. 22, qu. 6], 1672, 621, col. 2–624, col. 2.[96]
   3. Cornaeus* [*De coelo*, tract. 4, disp. 2, qu. 4, dub. 7], 1657, 509–510.

X. The theological heavens

106. *Whether there is an empyrean [or immobile] sphere beyond the mobile heavens.*
   1. Alexander of Hales, *Summa theologica*, inq. 3, tract. 2, qu. 2, tit. 1, memb. 1, ch. 1, art. 1, 1928, 327–328.
   2. Buridan [*De caelo*, bk. 2, qu. 6], 1942, 149–153.
   3. Albert of Saxony [*De celo*, bk. 2, qu. 8], 1518, 107r, col. 1–107v, col. 1.[97]

92. Amicus treats part of this question in a section titled "On the order of the heavens [by nobility and perfection]."
93. Oddus's question is also relevant to question 103 in this catalog.
94. See also question 197 in this catalog.
95. The Coimbra Jesuits expressed the same sentiment by asking "Whether the celestial globes that are higher are of a nobler nature."
96. The number of fixed stars is discussed specifically on page 624, column 1. The same question by Poncius is included in question 97 of this catalog.
97. Albert, Buridan, and Themon Judaeus do not mention the empyrean sphere, although all intend it. Buridan asks "Whether beyond the heavens that are moved, there should be assumed a heaven that is resting or unmoved"; Themon asks "Whether something should be assumed [to exist] beyond the ninth sphere"; while Albert inquires "Whether every heaven is mobile, or whether we must assume some heaven that is at rest." Only the empyrean heaven was regularly assumed to be at rest. As masters of arts without theological degrees or credentials, Buridan, Themon, and Albert may have deliberately omitted the term "empyrean" to avoid possible theological entanglements.

4. Themon Judaeus [*De spera*, qu. 8], in Hugonnard- Roche, 1973, 99–105.[98]
5. D'Ailly, *14 Questions*, qu. 2, 1531, 149v.[99]
6. Amicus* [*De caelo*, tract. 4, qu. 6, dubit. 5], 1626, 193, col. 2–197, col. 2.

107. Whether the empyrean heaven is a body.

1. Alexander of Hales, *Summa theologica*, inq. 3, tract. 2, qu. 2, tit. 1, memb. 1, ch. 1, art. 2, 1928, 328–329.

108. Whether the matter of the empyrean heaven is similar to the matters of other bodies.

1. William of Auvergne, *De universo*, pt. 1 of pt. 1, ch. 36, *Opera*, 1674, 1: 631, col. 2–632, col. 1.

109. Whether the empyrean heaven is moved.

1. Alexander of Hales,* *Summa theologica*, inq. 3, tract. 2, qu. 2, tit. 1, memb. 1, ch. 1, art. 3, 1928, 329–330.
2. William of Auvergne, *De universo*, pt. 1 of pt. 1, chs. 32–33, *Opera*, 1674, 1:627, col. 1–628, col. 2.[100]

110. Whether the empyrean should be called a heaven.

1. Alexander of Hales, *Summa theologica*, inq. 3, tract. 2, qu. 2, tit. 1, memb. 1, ch. 1, art. 4, 1928, 330–331.

111. On a comparison of the empyrean heaven to other heavens.

1. Alexander of Hales,* *Summa theologica*, inq. 3, tract. 2, qu. 2, tit. 1, memb. 1, ch. 1, art. 5, 1928, 331.
2. William of Auvergne, *De universo*, pt. 1 of pt. 1, ch. 34, *Opera*, 1674, 1:628, col. 2–630, col. 1.[101]
3. Cornaeus [*De coelo*, tract. 4, disp. 2, qu. 1, dub. 5], 1657, 491.[102]

112. Whether waters are above the heavens.

1. Alexander of Hales, *Summa theologica*, inq. 3, tract. 2, qu. 2, tit. 2, ch. 1, 1928, 336–339.
2. Thomas Aquinas* [*Sentences*, bk. 2, dist. 14, qu. 1, art. 1], 1929–1947, 2:346–349.
3. Hurtado de Mendoza [*De coelo*, disp. 3, sec. 3], 1615, 381, col. 2–382, col. 1.[103]
4. Cornaeus [*De coelo*, tract. 4, disp. 3, qu. 2, dub. 8], 1657, 524–525.[104]

98. The same question by Themon is cited above, in question 54 of this catalog.
99. D'Ailly did not make this a separate question but included it as part of his discussion on the possible number of spheres (see qu. 97 of this catalog).
100. William approaches the problem from the standpoint of rest, titling his chapter 32 "The stability of the tenth sphere, which is called the empyrean, and its stability and rest." Similarly, he titles his chapter 33: "That the extreme, or supreme, heaven is necessarily at rest." Indeed, he returns to this theme in chapter 43 (644, col. 2–648, col. 1) when he discusses "The stability of the empyrean heaven and other parts of the corporeal universe."
101. William's chapter title speaks of "The order, or position of the empyrean heaven [in relation] to other bodies."
102. Cornaeus effectively poses the same question when he asks "Whether the empyrean heaven has the same nature as the sidereal heaven."
103. Hurtado specifies "elemental" water, when he asks "Whether there is elemental water above the sidereal heaven."
104. Cornaeus asks "Whether above the sidereal heavens there are true elemental waters."

113. *What is the utility of the waters above the firmament?*
  1. William of Auvergne, *De universo*, pt. 1 of pt. 1, ch. 38, *Opera*, 1674, 1:632, col. 2–633, col. 2.

114. *Whether the superior waters are collected in the manner [or shape] of an orb.*
  1. Alexander of Hales, *Summa theologica*, inq. 3, tract. 2, qu. 2, tit. 2, ch. 2, 1928, 339.

115. *Whether these waters are moved with an orbicular motion.*
  1. Alexander of Hales, *Summa theologica*, inq. 3, tract. 2, qu. 2, tit. 2, ch. 3, 1928, 340.

116. *By what necessity are these waters fixed above the firmament?*
  1. Alexander of Hales, *Summa theologica*, inq. 3, tract. 2, qu. 2, tit. 2, ch. 4, 1928, 340–341.

117. *Whether the crystalline heaven has the nature of water.*
  1. Alexander of Hales, *Summa theologica*, inq. 3, tract. 2, qu. 2, tit. 2, ch. 5, 1928, 341.[105]
  2. Bonaventure* [*Sentences*, bk. 2, dist. 14, pt. 1, art. 1, qu. 1], *Opera*, 1885, 2:335–338.
  3. Richard of Middleton [*Sentences*, bk. 2, dist. 14, art. 1, qu. 1], 1591, 2:167, col. 1–168, col. 2.
  4. Durandus [*Sentences*, bk. 2, dist. 14, qu. 1], 1571, 155v, col. 2–156r, col. 2.
  5. Major [*Sentences*, bk. 2, dist. 14, qu. 5], 76v, col. 1–77r, col. 2.[106]

118. *Whether the crystalline heaven is moved.*
  1. Richard of Middleton [*Sentences*, bk. 2, dist. 14, art. 1, qu. 2], 1591, 2:168, col. 2–169, col. 1.

119. *What is the firmament?*
  1. Hurtado de Mendoza [*De coelo*, disp. 3, sec. 3], 1615, 380, col. 2–381, col. 2.

120. *Whether the firmament has the nature of fire.*
  1. Richard of Middleton [*Sentences*, bk. 2, dist. 14, art. 1, qu. 3], 1591, 2:169, col. 1–170, col. 2.

121. *Whether the firmament has a spherical shape.*
  1. Richard of Middleton [*Sentences*, bk. 2, dist. 14, art. 1, qu. 4], 1591, 2:170, col. 2–171, col. 1.

XI. On directionality in the heavens

122. *Whether in the heavens there are up and down; in front of and behind; [and] right and left.*
  1. Bacon [*Physics*, bk. 4], *Opera*, fasc. 13, 1935, 175–176.[107]

105. Alexander's version asks "Why these waters are called the 'crystalline' heaven."
106. Major asks simply "Whether there is a watery heaven."
107. The questions are unnumbered. In fascicule 8 (1928), Delorme published a version of Bacon's questions on the *Physics* that contained only the first four books.

2. Bonaventure [*Sentences*, bk. 2, dist. 14, pt. 1, art. 2, qu. 2], *Opera*, 1885, 2:342–345.[108]
3. Richard of Middleton [*Sentences*, bk. 2, dist. 14, art. 1, qu. 5], 1591, 2:171, col. 1–172, col. 1.[109]
4. John of Jandun [*De coelo*, bk. 2, qu. 3], 1552, 24v, col. 1–25r, col. 1.[110]
5. Buridan [*De caelo*, bk. 2, qu. 2], 1942, 133–137.
6. Albert of Saxony* [*De celo*, bk. 2, qu. 1], 1518, 102r, col. 1–103r, col. 1.
7. Oresme [*De celo*, bk. 2, qu. 3], 1965, 461–477.
8. Paul of Venice, *Liber celi*, 1476, 24, col. 1–26, col. 2 [sigs. g2–g3].
9. Johannes de Magistris [*De celo*, bk. 2, qu. 1, dub. 1], 1490, 17, col. 1.[111]
10. Versor [*De celo*, bk. 2, qu. 3], 1493, 18r, col. 1–18v, col. 1.[112]
11. Major [*De celo*, bk. 1, qu. 6], 1526, 12, col. 1–13, col. 1.
12. Amicus [*De caelo*, tract. 4, qu. 9], 1626, 221, col. 1–230, col. 1.
13. Serbellonus [*De caelo*, disp. 1, qu. 3, art. 4], 1663, 2:37, col. 2–38, col. 2.[113]

123. *Whether in the heavens the antarctic pole may be up and the arctic pole down and whether east is to the right and west to the left and the meridian in front and the opposite point, that is the observer's nadir, may be in back.*
   1. Buridan [*De caelo*, bk. 2, qu. 4], 1942, 142–144.[114]
   2. Albert of Saxony [*De celo*, bk. 2, qu. 3], 1518, 103v, col. 1–104r, col. 1.[115]
   3. Oresme* [*De celo*, bk. 2, qu. 4], 1965, 477–493.
   4. Johannes de Magistris [*De celo*, bk. 2, qu. 1, dub. 2], 1490, 17, col. 1.[116]
   5. Versor [*De celo*, bk. 2, qu. 4], 1493, 18v, col. 2–19r, col. 2.

124. *Whether the said differences of position are found in the heavens because of their very nature, or only in relation to us.*
   1. Buridan [*De caelo*, bk. 2, qu. 5], 1942, 144–148.
   2. Albert of Saxony* [*De celo*, bk. 2, qu. 4], 1518, 104r, col. 1–104v, col. 1.
   3. Bricot [*De celo*, bk. 2], 1486, 15v, col. 2–16v, col. 2.
   4. Major [*De celo*, bk. 1, qu. 6], 1526, 12, col. 1–13, col. 1.
   5. Conimbricenses [*De coelo*, bk. 2, ch. 2, qu. 1], 1598, 182–185.

108. Bonaventure asks only about two directions, right and left.
109. As Bonaventure did, Richard asks only about right and left.
110. Jandun expresses the question as "Whether there are only six differences of position."
111. De Magistris's version of the question is virtually the same as John of Jandun's.
112. Versor asks "Whether in every celestial body six differences of position are found distinct from the thing itself." In the preceding question (bk. 2, qu. 2), Versor posed the same question but omitted the word "celestial."
113. Serbellonus asks "Whether in the heavens there are differences of positions."
114. Both Buridan and Albert of Saxony (discussant 2 in this question) confine their questions to the arctic and antarctic poles.
115. Albert asks specifically "Whether the arctic pole, namely the pole apparent to us, is down; and the other, namely the antarctic [pole], is up." In the preceding question (bk. 2, qu. 2, 103r, col. 1–103v, col. 1), Albert asked "Whether in the heavens, up and down ought to be understood with respect to the poles, so that one of the poles is up, the other down."
116. De Magistris asks only "Whether the pole that appears above us is up or down."

XII. On the properties of celestial spheres and bodies

125. *Whether there is only one sidereal heaven in which all the planets [*astra*] exist and it is truly solid.*
   1. Aversa [*De caelo*, qu. 32, sec. 6], 1627, 64, col. 2–69, col. 1.

126. *Whether the heaven is spherical in shape.*
   1. Michael Scot [*Sphere*, lec. 2, 5], 1949, 257–259 and 291.[117]
   2. Bacon, *De celestibus*, pt. 2, ch. 1, *Opera*, fasc. 4, 1913, 342–349.
   3. Bonaventure [*Sentences*, bk. 2, dist. 14, pt. 1, art. 2, qu. 1], *Opera*, 1885, 2:341–342.
   4. Cecco d'Ascoli [*Sphere*, ch. 1], 1949, 364–366.
   5. Durandus [*Sentences*, bk. 2, dist. 14, qu. 4], 1571, 157r (mistakenly foliated as 175), col. 1–157v, col. 1.
   6. Albert of Saxony* [*De celo*, bk. 2, qu. 5], 1518, 104v, col. 1–105r, col. 1.
   7. Themon Judaeus [*De spera*, qu. 11], in Hugonnard-Roche, 1973, 119–121.
   8. D'Ailly, *14 Questions*, qu. 5, 1531, 153r–156r.[118]
   9. Johannes de Magistris [*De celo*, bk. 2, qu. 2, dub. 2], 1490, 19, col. 2–20, col. 1 (sigs. k4v–k5r).
   10. Versor [*De celo*, bk. 2, qu. 6], 1493, 20v, col. 1–21r, col. 1.
   11. Major [*De celo*, bk. 2, qu. 5], 1526, 19, col. 2.
   12. Conimbricenses [*De coelo*, bk. 2, ch. 4, qu. 1], 1598, 238–241.
   13. Hurtado de Mendoza [*De coelo*, disp. 2, sec. 2], 1615, 371, col. 1.
   14. Aversa [*De caelo*, qu. 31, sec. 4], 1627, 9, col. 2–14, col. 2, and [qu. 34, sec. 2], 128, col. 1–129, col. 1.
   15. Compton-Carleton [*De coelo*, disp. 4, sec. 3], 1649, 410, col. 2.
   16. Cornaeus [*De coelo*, tract. 4, disp. 2, qu. 2, dub. 2], 1657, 493–494.

127. *Whether the planets [*astra*] are spherical in shape.*
   1. Michael Scot [*Sphere*, lec. 15], 1949, 330–331.
   2. Bacon, *De celestibus*, pt. 4, ch. 10, *Opera*, fasc. 4, 1913, 412–413.
   3. Johannes de Magistris [*De celo*, bk. 2, qu. 4, dub. 2], 1490, 24, cols. 1–2 (sig. k7r).
   4. Versor* [*De celo*, bk. 2, qu. 15], 1493, 27v, col. 2–28r, col. 2.
   5. Cornaeus [*De coelo*, tract. 4, disp. 2, qu. 4, dub. 10], 1657, 512–513.

128. *Whether the heavens are animated.*
   1. Michael Scot [*Sphere*, lec. 5], 1949, 289.
   2. Richard of Middleton [*Sentences*, bk. 2, dist. 14, art. 3, qu. 4], 1591, 2:186, col. 2–188, col. 1.
   3. Aureoli [*Sentences*, bk. 2, dist. 14, qu. 3], 1596–1605, 2:200, col. 1–201, col. 2.
   4. John of Jandun [*De coelo*, bk. 2, qu. 4], 1552, 25r, col. 1–25v, col. 2.
   5. Benedictus Hesse [*Physics*, bk. 8, qu. 17], 1984, 735–737.

117. On pages 257–259, Michael asks what the shape of the world (*mundus*) is; on page 259, he asks "Whether the heaven (*celum*) is round."
118. In the same question, d'Ailly also asks whether the four elements are spherical (see below, qu. 335, n. 265 in this catalog).

6. Johannes de Magistris★ [*De celo*, bk. 2, qu. 1, dub. 3], 1490, 17, col. 2 (k3v).
7. Javelli [*Metaphysics*, bk. 12, qu. 16], 1568, 1:888, col. 1–889, col. 2.
8. Galileo [*De caelo*, qu. 6 (L)], 1977, 148–158.
9. Conimbricenses [*De coelo*, bk. 2, qu. 1], 1598, 162–171.
10. Hurtado de Mendoza [*De coelo*, disp. 1, sec. 4], 1615, 366, col. 2.
11. Amicus [*De caelo*, tract. 4, qu. 4, dubit. 2], 1626, 168, col. 1–173, col. 2.
12. Aversa [*De caelo*, qu. 33, sec. 7], 1627, 109, col. 1–112, col. 2.
13. Arriaga [*De caelo*, disp. 1, sec. 2], 1632, 498, cols. 1–2.
14. Poncius [*De coelo*, disp. 22, qu. 3], 1672, 615, col. 2–616, col. 2.
15. Oviedo [*De caelo*, contro. 1, punc. 1], 1640, 460, col. 1–462, col. 1.
16. Compton-Carleton [*De coelo*, disp. 1, sec. 1], 1649, 397, cols. 1–2.
17. Riccioli, *Almagestum novum*, pars post., bk. 9, sec. 1, ch. 8, 1651, 244, col. 1–246, col. 2.
18. Bonae Spei [comment. 3, *De coelo*, disp. 3, dub. 5], 1652, 11, col. 1–12, col. 2.
19. Oddus [*De coelo*, disp. 1, dub. 11], 1672, 29, col. 1–34, col. 2.

129. Whether the celestial substance is mobile per se.
1. Godfrey of Fontaines, *Quodlibet V*, qu. 7, 1914, 25–28.

130. Whether the sky [or heaven] is heavy or light.
1. John of Jandun [*De coelo*, bk. 1, qu. 13], 1552, 10r, cols. 1–2.
2. Buridan★ [*De caelo*, bk. 1, qu. 9], 1942, 40–44.
3. Versor★ [*De celo*, bk. 1, qu. 5], 1493, 4r, col. 2–4v, col. 1.[119]
4. Amicus [*De caelo*, tract. 5, qu. 3], 1626, 260, col. 1–262, col. 2.
5. Aversa [*De caelo*, qu. 34, sec. 4], 138, col. 1–139, col. 1.[120]
6. Serbellonus [*De caelo*, disp. 1, qu. 3, art. 3], 1663, 2:36, col. 1–37, col. 2.

131. Whether rarity and density exist in the heavens.
1. Aversa [*De caelo*, qu. 34, sec. 2], 1627, 128, col. 1–133, col. 1.[121]
2. Serbellonus★ [*De caelo*, disp. 1, qu. 3, art. 2], 1663, 2:34, col. 1–36, col. 1.

132. Whether the whole heaven from the convexity of the supreme [or outermost] sphere to the concavity of the lunar orb is continuous or whether the orbs are distinct from each other.[122]
1. Michael Scot [*Sphere*, lec. 4], 1949, 281–284.
2. Bacon, *De celestibus*, pt. 4, ch. 1, *Opera*, fasc. 4, 1913, 385–387.
3. Bonaventure [*Sentences*, bk. 2, dist. 14, pt. 2, art. 1, qu. 1], *Opera*, 1885, 2:351–352.[123]
4. Robertus Anglicus [*Sphere*, lec. 1], 1949, 146–148 (Latin); 201–203 (English).

119. Since Versor's Latin enunciation of the question is identical to Buridan's ("Utrum celum [or "caelum"] sit grave aut leve"), I have also placed an asterisk beside his name.
120. This discussion forms part of a larger treatment titled "On the place and local positions of celestial bodies," 138, col. 1–140, col. 2.
121. The full title of the section in which Aversa considers rarity and density is "On the shape of celestial bodies, their density and rarity, opacity and transparency."
122. In effect, "Are the celestial orbs continuous or contiguous?".
123. Bonaventure's version asks "Whether all the [celestial] luminaries [that is, all the planets and stars] are located in one continuous body."

5. Richard of Middleton [*Sentences*, bk. 2, dist. 14, art. 3, qu. 1], 1591, 2:183, col. 2–184, col. 2.
6. Cecco d'Ascoli* [*Sphere*, ch. 1], 1949, 352–353.
7. Simon de Tunstede (Pseudo-Scotus) [*Meteorology*, bk. 1, qu. 3, art. 1], 1639, 3:6, col. 2–7, col. 1.[124]
8. Albert of Saxony [*De celo*, bk. 1, qu. 4], 1518, 88v, col. 1–89v, col. 2.[125]
9. Paul of Venice, *Liber celi*, 1476, 5, col. 1–7, col. 1.[126]
10. Amicus [*De caelo*, tract. 4, qu. 5, dubit. 1], 1626, 182, col. 1–184, col. 2.
11. Aversa [*De caelo*, qu. 34, sec. 3], 1627, 138, col. 2.[127]
12. Oddus [*De coelo*, disp. 1, dub. 14], 1672, 40, col. 2–41, col. 2.[128]

133. Whether all celestial spheres are contiguous or water is interposed between them.
1. Compton-Carleton [*De coelo*, disp. 4, sec. 2], 1649, 409, cols. 1–2.

134. On the depth, or thickness, of the heavens.
1. William of Auvergne, *De universo*, pt. 1 of pt. 1, ch. 45, *Opera*, 1674, 1:654, cols. 1–2.

135. Whether the heavens are fluid or solid.[129]
1. Amicus [*De caelo*, tract. 5, qu. 5], 1626, 270, col. 1–288, col. 1.
2. Arriaga [*De caelo*, disp. 1, sec. 3], 1632, 499, col. 1–504, col. 1.
3. Poncius [*De coelo*, disp. 22, qu. 5], 1672, 620, col. 1–621, col. 2.
4. Oviedo [*De caelo*, contro. 1, punc. 3], 1640, 464, col. 2–469, col. 1.
5. Compton-Carleton [*De coelo*, disp. 1, sec. 3], 1649, 398, col. 2–401, col. 2.
6. Riccioli, *Almagestum novum*, pars post., bk. 9, sec. 1, ch. 7, 1651, 238, col. 2–244, col. 1.
7. Bonae Spei* [comment. 3, *De coelo*, disp. 3, dub. 7], 1652, 13, col. 2–14, col. 1.
8. Cornaeus [*De coelo*, tract. 4, disp. 2, qu. 2, dub. 3], 1657, 494–500.
9. Serbellonus [*De caelo*, disp. 1, qu. 2, art. 4], 1663, 2:25, col. 1–28, col. 1.
10. De Rhodes [*De coelo*, bk. 2, disp. 8, sec. 2, art. 2], 1671, 280, col. 1–281, col. 1.
11. Oddus [*De coelo*, disp. 1, dub. 12], 1672, 34, col. 2–35, col. 2.[130]

124. Although published by Luke Wadding in the *Opera* of Duns Scotus, the work is definitely not by Scotus. In his introduction to the treatise, Wadding suggests Simon de Tunstede (d. 1369) as the author. For convenience, I have adopted Wadding's suggestion.
125. Albert's version extends the question to the whole world ("Whether the whole world that is aggregated from the aforementioned five simple bodies is one continuous being").
126. In this section, Paul considers, among other things, whether the orbs are continuous or contiguous.
127. The continuity–contiguity problem forms a small part of a larger consideration "On the place and local positions of celestial bodies" (138, col. 1–140, col. 2).
128. Although *dubium* 14 is titled "On the number of the heavens," part of it is devoted to the problem of the continuity and contiguity of the spheres.
129. No such question was distinguished in the Middle Ages. Following Aristotle, however, scholastics did consider whether the stars and planets in the heavens were fixed to their spheres or moved independently, like fish in water (see, for example, William of Auvergne, *De universo*, pt. 1 of pt. 1, ch. 44, *Opera*, 1674, 1:650, col. 1). Johannes Versor's question as to whether the planets have the nature of fire (see qu. 140 in this catalog) is confined to the planets themselves and does not encompass the entire medium of the heavens in which the planets are embedded or immersed. Hence it is not really about the solidity or fluidity of the celestial region.
130. Oddus has "Whether the heavens are solid or fluid."

136. Whether the non-starry part of the heaven is visible to us.
1. Richard of Middleton [*Sentences*, bk. 2, dist. 14, art. 3. qu. 6], 1591, 2:189, col. 2–192, col. 2.

137. [What are the planets in the opinions of the Church Fathers and Peripatetics?]
1. Cornaeus [*De coelo*, tract. 4, disp. 2, qu. 3, dub. 1–3], 1657, 501–504.[131]

138. Whether, and in what way, celestial bodies differ among themselves in substance.
1. Aversa [*De caelo*, qu. 33, sec. 8], 1627, 114, col. 2–119, col. 1.

139. Whether celestial bodies affect each other and what [they do to each other].
1. Aversa [*De caelo*, qu. 35, sec. 1], 1627, 165, col. 1–167, col. 1.

*140. Whether the planets [or stars] [*astra*] have the nature of fire.*
1. Simon de Tunstede (Pseudo-Scotus) [*Meteorology*, bk. 1, qu. 4], 1639, 3:8, col. 2–10, col. 2.
2. Themon Judaeus [*Meteorology*, bk. 1, qu. 3], 1518, 157r, col. 2–158r, col. 1.
3. Versor* [*De celo*, bk. 2, qu. 10], 1493, 23r, col. 2–23v, col. 1.

141. Whether the spots appearing in the Moon arise from differences in parts of the Moon or from something external.
1. Buridan* [*De caelo*, bk. 2, qu. 19], 1942, 212–217.
2. Albert of Saxony [*De celo*, bk. 2, qu. 22], 1518, 116r, col. 2–116v, col. 1.
3. Oresme [*De celo*, bk. 2, qu. 12], 1965, 653–665.
4. Amicus [*De caelo*, tract. 6, qu. 5, dubit. 3], 1626, 363, col. 1–368, col. 1.
5. Aversa [*De caelo*, qu. 34, sec. 8], 1627, 151, col. 2–154, col. 2.
6. Oddus [*De coelo*, disp. 1, dub. 15], 1672, 42, col. 2–45, col. 2.[132]

142. On the Sun's spots.
1. Aversa [*De caelo*, qu. 34, sec. 9], 1627, 154, col. 2–160, col. 2.

143. Whether the Moon is the lesser luminary.
1. Major [*Sentences*, bk. 2, dist. 14, qu. 10], 1519b, 82r, col. 2–83v, col. 1.

*144. Why some planets [*astra*] twinkle, and some very little; and why one [twinkles] more than another.*
1. Amicus [*De caelo*, tract. 5, qu. 7, dubit. 3], 1626, 343, col. 2–345, col. 2.[133]
2. Aversa [*De caelo*, qu. 34, sec. 3], 1627, 133, col. 1–135, col. 2.
3. Oddus* [*De coelo*, disp. 1, dub. 21], 1672, 63, col. 1–64, col. 1.

145. Whether [celestial] luminaries [that is, the planets and stars] have differences of perfection.
1. Bonaventure [*Sentences*, bk. 2, dist. 14, pt. 2, art. 2, qu. 1], *Opera*, 1885, 2:357, 359.

131. I have formulated this composite question which is based on three doubts (*dubia*) raised by Cornaeus: (1) "What are the planets [*stellae*] as conceived in the minds of the ancient Fathers? (Quid sint stellae ex mente antiquorum Patrum); (2) "What are the planets [*stellae*] in the opinion of the Peripatetics?" (Quid sint stellae ex opinione Peripateticorum); and (3) "What should be understood about this?" (Quid in hac re sentiendum).
132. Oddus's *dubium* also embraces the Milky Way and the spots on the Sun.
133. Amicus asks "Whether a twinkling [or sparkling] motion is proper to the planets."

146. *On the quantity and magnitude of celestial bodies.*
1. Hurtado de Mendoza [*De coelo*, disp. 2, sec. 2], 1615, 370, col. 1–371, col. 1.
2. Aversa* [*De caelo*, qu. 34, sec. 1], 1627, 123, col. 1–128, col. 1.

147. *Whether celestial bodies possess qualities.*
1. Michael Scot* [*Sphere*, lec. 11], 1949, 319–320.[134]
2. Aversa [*De caelo*, qu. 34, sec. 3], 1627, 135, col. 2–137, col. 2.[135]
3. Oddus [*De coelo*, disp. 1, dub. 22], 1672, 65, col. 2–67, col. 2.

148. *Whether celestial accidents are of the same kind as sublunary [accidents].*
1. Serbellonus* [*De caelo*, disp. 1, qu. 3, art. 1], 1663, 2:32, col. 2–34, col. 1.
2. Oddus [*De coelo*, disp. 1, dub. 22], 1672, 65, col. 2–69, col. 1.[136]

149. *Whether in celestial bodies any chance or fortuitous things happen.*
1. John of Jandun [*Physics*, bk. 2, qu. 10], 1519, 41v, col. 2–42v, col. 1.

150. *Whether on occasion some new stars are observed and generated in the heavens and in what manner this could occur without a substantial mutation.*
1. Aversa [*De caelo*, qu. 33, sec. 3], 1627, 83, col. 2–91, col. 1.

151. *Against those who say that the planets are evil.*
1. William of Auvergne, *De universo*, pt. 1 of pt. 1, ch. 46, *Opera*, 1674, 1:654, col. 2–667, col. 2.

XIII. On celestial motions and their causes

152. *Whether, [based] on Aristotle's opinion, the prime mover is one.*
1. Oña [*Physics*, bk. 8, qu. 1, art. 8], 1598, 376v, col. 1–377, col. 2.
2. Amicus* [*Physics*, bk. 8, qu. 6, dubit. 1], 1626–1629, 1:1049, col. 2–1050, col. 1.

153. *To demonstrate that the first mover is a separated substance.*
1. Oña [*Physics*, bk. 8, qu. 1, art. 1], 1598, 355r, col. 1–356v, col. 2.

154. *Whether the perpetuity of motion is necessary to understand the existence of the first principle.*
1. John of Jandun* [*Physics*, bk. 8, qu. 2], 1519, 121v, col. 1–122v, col. 1.
2. Javelli [*Physics*, bk. 8, qu. 2], 1568, 1:598, cols. 1–2.
3. Oddus [*Physics*, bks. 5–8, disp. 17, art. 9], 1667, 752, col. 1–754, col. 1.[137]

134. The Latin text declares: "Utrum corpora celestia sint complexionata." Michael's question is clearly related to Serbellonus's version of question 148 in this catalog.
135. Aversa's discussion occurs in a general section titled "On the qualities of celestial bodies" (133, col. 1–138, col. 1).
136. Oddus asks (65, col. 2) in general "about certain other celestial accidents and whether they differ from sublunar things in species." Beginning with page 68, column 1, however, he turns particularly "to the other point on the specific distinction or identity of common celestial accidents with sublunary accidents, such as quantity, light, density, opacity, and diaphaneity."
137. Oddus's question is related but somewhat different: "Whether from force of motion, the existence of the prime mover could be demonstrated."

155. *Whether the prime mover, which is glorious God, is of infinite strength.*
1. John of Jandun [*Physics*, bk. 8, qu. 22], 1519, 145r, col. 1–147r, col. 1.
2. Buridan [*Physics*, bk. 8, qu. 11], 1509, 119r, col. 1–120r, col. 1.
3. Benedictus Hesse* [*Physics*, bk. 8, qu. 25], 1984, 757–762.
4. Javelli [*Physics*, bk. 8, qu. 16], and [*Metaphysics*, bk. 12, qu. 20], 1568, 1:612, col. 2, and 894, col. 1–897, col. 1.
5. Conimbricenses [*Physics*, pt. 2, bk. 8, ch. 10, qu. 3], 1602, cols. 586–590.[138]

156. *Whether the prime mover is actual or potential.*
1. Buridan [*Metaphysics*, bk. 12, qu. 3], 1518, 66r, col. 1–66v, col. 2.

157. *Whether the prime mover is absolutely simple.*
1. Buridan [*Metaphysics*, bk. 12, qu. 4], 1518, 66v, col. 1–67r, col. 1.

158. *Whether the prime mover is without fatigue.*
1. Benedictus Hesse [*Physics*, bk. 8, qu. 28], 1984, 769–772.

159. *Whether the prime mover is absolutely immobile.*
1. Pseudo–Siger of Brabant [*Physics*, bk. 8, qu. 21], 1941, 224–225.[139]
2. John of Jandun [*Physics*, bk. 8, qu. 13], 1519, 136r, col. 2–137r, col. 1.[140]
3. Buridan [*Physics*, bk. 8, qu. 6], 1509, 114v, col. 2–115r, col. 2.
4. Albert of Saxony [*Physics*, bk. 8, qu. 8], 1518, 81r, cols. 1–2.
5. Marsilius of Inghen [*Physics*, bk. 8, qu. 8], 1518, 84v, col. 1–85r, col. 2.
6. Benedictus Hesse* [*Physics*, bk. 8, qu. 16], 1984, 730–735.
7. Major [*Physics*, bk. 8], 1526, 55r, cols. 1–2.
8. Toletus [*Physics*, bk. 8, qu. 5], 1580, 233r, col. 1–234r, col. 2.
9. Oña [*Physics*, bk. 8, qu. 1, art. 9], 1598, 377v, col. 2–379v, col. 1.

160. *Whether the prime mover is absolutely uniform.*
1. Benedictus Hesse [*Physics*, bk. 8, qu. 29], 1984, 771–772.

161. *Whether the prime mover that moves the heaven is a form of the heaven or a soul.*
1. Bacon [*Physics*, bk. 8], *Opera*, fasc. 13, 1935, 427–428.

162. *Whether the prime mover moves the heaven as an end [or goal] that is loved.*
1. Javelli [*Metaphysics*, bk. 12, qu. 15], 1568, 1:884, col. 2–888, col. 1.

163. *Whether the prime mover is indivisible and has no magnitude.*
1. Buridan* [*Physics*, bk. 8, qu. 13], 1509, 121r, col. 2–121v, col. 2.
2. Benedictus Hesse [*Physics*, bk. 8, qus. 31–32], 1984, 774–778 and 778–780.[141]
3. Conimbricenses [*Physics*, pt. 2, bk. 8, ch. 6, qu. 1], 1602, cols. 514–520.[142]

164. *Whether the prime mover is in the circle and circumference of the last sphere, or in the east.*
1. Benedictus Hesse* [*Physics*, bk. 8, qu. 30], 1984, 772–774.

138. The Conimbricenses ask "Whether, according to the doctrine of Aristotle, the prime mover operates with infinite power."
139. Pseudo-Siger asks "Whether in movers and in things that are moved it is necessary to arrive at an immobile prime mover."
140. In his version, John of Jandun asks "Whether there is an absolutely immobile mover."
141. Benedictus puts into two questions what Buridan encompassed with one. Thus question 31 asks "Whether the prime mover is indivisible" and question 32 "Whether the prime mover has magnitude."
142. Specifically, the Conimbricenses ask "Whether, according to Aristotle, the prime mover is free of all mutation and magnitude."

165. *Whether there are several prime movers.*
1. John of Jandun [*Physics*, bk. 8, qu. 23], 1519, 147r, col. 1–147v, col. 1.[143]
2. Marsilius of Inghen* [*Physics*, bk. 8, qu. 9], 1518, 85r, col. 2–86r, col. 2.
3. Benedictus Hesse [*Physics*, bk. 8, qu. 27], 1984, 767–769.[144]
4. Javelli [*Physics*, bk. 8, qu. 17], 1568, 1:612, col. 2.

166. *Whether a finite mover could move [something] through an infinite time.*
1. John of Jandun* [*Physics*, bk. 8, qu. 19], 1519, 140v, col. 1–141v, col. 2.
2. Javelli [*Physics*, bk. 8, qu. 14], 1568, 1:610, col. 1–611, col. 1.

167. *Whether some body [must] always be moved [while] some [other] body always rests.*
1. Benedictus Hesse [*Physics*, bk. 8, qu. 11], 1984, 713–715.

168. *Whether there are three simple motions and no more, namely upward motion, downward motion, and circular motion.*
1. Buridan* [*De caelo*, bk. 1, qu. 6], 1942, 27–31.
2. Oresme [*De celo*, bk. 1, qu. 6], 1965, 81–97.
3. Johannes de Magistris [*De celo*, bk. 1, qu. 1, dub. 1], 1490, 3, col. 2.
4. Versor [*De celo*, bk. 1, qu. 3], 1493, 2v, col. 1–3r, col. 2.

169. *Whether only one simple motion inheres naturally in any simple body.*
1. John of Jandun [*De coelo*, bk. 1, qu. 10], 1552, 8v, col. 1–9r, col. 1.
2. Buridan [*De caelo*, bk. 1, qu. 6], 1942, 27–31.
3. Albert of Saxony* [*De celo*, bk. 1, qu. 1], 1518, 84r, col. 1–86v, col. 2.
4. Oresme [*De celo*, bk. 1, qu. 7], 1965, 97–111.
5. Paul of Venice, *Liber celi*, 1476, 3, col. 1–5, col. 1.
6. Bricot [*De celo*, bk. 1], 1486, 3v, col. 1–4r, col. 1.[145]
7. Major [*De celo*, bk. 1, qu. 1], 1526, 1, col. 1–2, col. 2.
8. Conimbricenses [*De coelo*, bk. 1, ch. 2, qu. 2], 1598, 29–31.
9. Amicus [*De caelo*, tract. 4, qu. 1, dubit. 1], 1626, 130, col. 1–135, col. 2.

170. *Whether the heavens are moved.*
1. Cecco d'Ascoli* [*Sphere*, ch. 1], 1949, 363.
2. Aversa [*De caelo*, qu. 34, sec. 5], 1627, 140, col. 2–143, col. 1.
3. Cornaeus [*De coelo*, tract. 4, disp. 2, qu. 2, dub. 4], 1657, 500.

171. *Whether this is true according to the Philosopher's [that is, Aristotle's] intention: "time is the motion of the heaven."*
1. Ockham [*Physics*, qu. 40], *Opera philosophica*, 1984, 6:502–504.

172. *Whether according to the intention of the Philosopher [that is, Aristotle], anyone who perceives time perceives the motion of the heavens.*
1. Ockham [*Physics*, qu. 45], *Opera philosophica*, 1984, 6:517–520.

173. *Whether celestial motion is natural.*
1. Michael Scot [*Sphere*, lec. 5], 1949, 288–289.[146]

143. In his version, John of Jandun asks "Whether there is only one first principle of motion."
144. Benedictus treats essentially the same question by asking "Whether there is only one prime mover."
145. The questions are unnumbered, but this is the first. Bricot was both a master of arts and a doctor of theology. For information on his works and life, see Lohr, 1973, 173–174; for the edition cited here, 177–178.
146. Michael Scot adds "or violent."

2. Durandus* [*Sentences*, bk. 2, dist. 14, qu. 2], 1571, 156v, cols. 1–2.
3. Johannes de Magistris [*De celo*, bk. 1, qu. 1, dub. 3], 1490, 4, col. 2.
4. Case [*Physics*, bk. 8, ch. 7], 1599, 825–834.[147]
5. Hurtado de Mendoza [*De coelo*, disp. 2, sec. 4], 1615, 373, cols. 1–2.
6. Oddus [*De coelo*, disp. 1, dub. 18], 1672, 49, col. 2–54, col. 2.

174. *Whether circular motion is natural to the heavens.*
1. Conimbricenses [*De coelo*, bk. 1, ch. 2, qu. 3], 1598, 31–37.[148]
2. Amicus* [*De caelo*, tract. 5, qu. 6, dubit. 7], 1626, 317, col. 1–320, col. 1.
3. Oddus [*De coelo*, disp. 1, dub. 16], 1672, 45, col. 2.[149]

175. *Whether circular motion is more perfect than rectilinear motion.*
1. John of Jandun [*De coelo*, bk. 1, qu. 11], 1552, 9r, cols. 1–2.

176. *Whether circular motion is perpetual.*
1. John of Jandun [*Physics*, bk. 8, qu. 18], 1519, 139v, col. 2–140r, col. 2.[150]
2. Benedictus Hesse* [*Physics*, bk. 8, qu. 19], 1984, 742–743.
3. Javelli [*Physics*, bk. 8, qu. 13], 1568, 1:609, col. 2–610, col. 1.

177. *Whether the circular motion of the heavens has a contrary.*
1. John of Jandun [*De coelo*, bk. 1, qu. 18], 1552, 12v, col. 2–13v, col. 1.
2. Buridan [*De caelo*, bk. 1, qu. 8], 1942, 36–40.
3. Oresme [*De celo*, bk. 1, qu. 12], 1965, 173–189.
4. Johannes de Magistris [*De celo*, bk. 1, qu. 2, dub. 2], 1490, 6, col. 1.
5. Versor [*De celo*, bk. 1, qu. 7], 1493, 5r, col. 2–5v, col. 2.
6. Bricot* [*De celo*, bk. 1], 1486, 4v, col. 2–5r, col. 1.
7. Conimbricenses [*De coelo*, bk. 1, ch. 4, qu. 1], 1598, 72–75.

178. *If the heaven exists, it is necessary that it be moved.*
1. Paul of Venice, *Liber celi*, 1476, 27, col. 1 (conclus. 1).

179. *Whether there is any cessation or turning back in celestial motion.*
1. Case [*Physics*, bk. 8, ch. 8], 1599, 835–842.

180. *Whether the sky [*caelum*] is always moved regularly [i.e., uniformly].*
1. Michael Scot [*Sphere*, lec. 5], 1949, 290–291.
2. John of Jandun [*De coelo*, bk. 2, qu. 7], 1552, 27r, col. 1–28r, col. 1.
3. Buridan* [*De caelo*, bk. 2, qu. 11], 1942, 172–175.
4. Albert of Saxony [*De celo*, bk. 2, qu. 13], 1518, 110r, col. 1–110v, col. 2.
5. Oresme [*De celo*, bk. 2, qu. 8], 1965, 575–591.[151]
6. D'Ailly, *14 Questions*, qu. 3, 1531, 149v–151v.
7. Johannes de Magistris [*De celo*, bk. 2, qu. 3], 1490, 20, col. 2–22, col. 2 (sigs. k5r–k6r).

147. Case offers the following version of this question: "Whether the motion by which God moves the first celestial sphere is natural."
148. For "circular motion," Conimbricenses use the term *conversio* rather than *motus circularis*, as Amicus has it.
149. At first glance, it appears that questions 173 and 174 should be conflated into a single question. But this has been avoided, because Oddus has treated them as distinct questions.
150. John of Jandun asks "Whether circular motion could be infinite and perpetual."
151. By asking "Whether the stars [that is, stars and planets] are moved regularly and with the motion of their own spheres," Oresme incorporated two questions that have been distinguished here (see also qu. 211 in this catalog, where Oresme's question is cited again). Although I have added the bracketed qualification, Droppers is the translator.

8. Versor [*De celo*, bk. 2, qu. 8], 1493, 22r, col. 1–22v, col. 1.[152]
9. Bricot [*De celo*, bk. 2], 1486, 18r, cols. 1–2.
10. Conimbricenses [*De coelo*, bk. 2, ch. 6, qu. 2], 1598, 287–291.[153]
11. Amicus [*De caelo*, tract. 5, qu. 6, dubit. 6], 1626, 314, col. 1–317, col. 1.
12. Cornaeus [*De coelo*, tract. 4, disp. 2, qu. 4, dub. 3], 1657, 504.[154]

181. Whether this consequence is sound: The heaven is always moved, therefore it is necessary that it be moved with several motions.

1. Albert of Saxony* [*De celo*, bk. 2, qu. 11], 1518, 108v, col. 2–109r, col. 2.
2. Oresme [*De celo*, bk. 2, qu. 6], 1965, 507–523.
3. Paul of Venice, *Liber celi*, 1476, 27, col. 2.

182. How many local motions of celestial bodies are there, and which ones are they?

1. Major [*Sentences*, bk. 2, dist. 14, qu. 4], 1519b, 75v, col. 1–76v, col. 1.[155]
2. Aversa* [*De caelo*, qu. 34, sec. 6], 1627, 143, col. 1–147, col. 1.

183. Whether all the celestial spheres below the primum mobile *are revolved simultaneously from east to west and from west to east.*

1. Bacon [*Physics*, bk. 8], *Opera*, fasc. 13, 1935, 420–421.
2. Major [*Sentences*, bk. 2, dist. 14, qus. 2–3], 1519b, 72v, col. 1–74v, col. 1, and 74v, col. 2–75v, col. 1.
3. Conimbricenses* [*De coelo*, bk. 2, ch. 5, qu. 4], 1598, 262–264.
4. Hurtado de Mendoza [*De coelo*, disp. 2, sec. 3], 1615, 371, col. 2–375, col. 1.
5. Oddus [*De coelo*, disp. 1, dub. 17], 1672, 46, col. 1–49, col. 2.

184. Whether any celestial sphere could be moved with several motions, just as the sphere of the Sun is assumed to be moved with a daily and a proper motion.

1. Bacon, *De celestibus*, pt. 5, ch. 11, *Opera*, fasc. 4, 1913, 429–433.
2. Albert of Saxony [*De celo*, bk. 2, qu. 14], 1518, 110v, col. 2–111r, col. 2.
3. Oresme* [*De spera*, qu. 11], 1966a, 242–263.[156]
4. Themon Judaeus [*De spera*, qu. 10], in Hugonnard-Roche, 1973, 113–119.[157]
5. D'Ailly, *14 Questions*, qu. 4, 1531, 151v–153r.
6. Bricot [*De celo*, bk. 2], 1486, 21v, col. 2.
7. Major [*De celo*, bk. 2, qu. 1], 1526, 14, col. 2–16, col. 1.[158]
8. Cornaeus [*De coelo*, tract. 4, disp. 2, qu. 4, dub. 4–5], 1657, 504–506, 506–509.[159]

152. Versor asks specifically whether the eighth sphere moves uniformly and regularly.
153. The Coimbra Jesuits inquired "Whether or not the motion of the heavens and stars is equable [i.e., uniform] and orderly."
154. Cornaeus asks "Whether they [the planets] are moved with one simple and equal [or uniform] motion."
155. Major asks "Whether one [motion] to one planet is sufficient."
156. I have slightly altered Droppers' translation on page 243.
157. Themon asks simply "Whether the inferior spheres are moved simultaneously and at once with several motions."
158. Major asks "Whether all the inferior heavens [or celestial spheres] within the *primum mobile* are moved with several motions."
159. In *dubium* 4, Cornaeus asks "Whether it is repugnant to the planets [*stellas*] that they be moved simultaneously with a contrary motion, for example, from east to west and conversely" and in *dubium* 5 he inquires "Whether the planets are in fact moved with a contrary motion from east to west and conversely." Because Cornaeus did not believe in the existence of material spheres, he speaks only of planets, rather than spheres.

185. *Whether the heaven must be moved toward one definite side, namely toward the anterior [or forward side] and [must be moved] from one definite side, namely from the right.*
   1. Versor [*De celo*, bk. 2, qu. 7], 1493, 21r, col. 2–21v, col. 1.

186. *Whether the superior orbs, and those nearer the supreme [or outermost] heaven, are moved with a quicker daily motion the more distant they are but with a slower proper motion.*
   1. Versor [*De celo*, bk. 2, qu. 14], 1493, 26r, col. 2–27r, col. 1.

187. *Whether the motions of the celestial orbs ought to increase in proportion to the distance of any orb to the outermost heaven, so that just as the first heaven is moved with one motion, and the second with two, and the third with three, and the fourth with four [motions], the increase of motions ought to occur proportionally in relation to the distance [of an orb] with respect to the first [or outermost] heaven [or sphere], so that the fifth heaven would be moved with five motions, the sixth with six, and so on.*
   1. Versor [*De celo*, bk. 2, qu. 16], 1493, 28r, col. 2–29r, col. 2.

188. *Whether one heaven [or sphere] could be moved with a motion contrary to that of another [sphere].*
   1. Michael Scot [*Sphere*, lec. 8], 1949, 306–307.

189. *Whether the lower spheres of the planets should be moved more quickly in their proper motions than the superior spheres.*
   1. Buridan* [*De caelo*, bk. 2, qu. 20], 1942, 218–223.
   2. Albert of Saxony [*De celo*, bk. 2, qu. 16], 1518, 111v, col. 2–112r, col. 2.

190. *Whether the precise measures of celestial motions are based on rational ratios [or, in other words, whether the celestial motions are commensurable or incommensurable].*
   1. Oresme [*De celo*, bk. 1, qu. 24], 1965, 409–428.[160]
   2. Amicus* [*De caelo*, tract. 5, qu. 6, dubit. 10], 1626, 332, col. 1–335, col. 1.[161]
   3. Riccioli, *Almagestum novum*, pars post., bk. 9, sec. 2, ch. 6, 1651, 269, col. 1–270, col. 2.[162]

191. *Whether the heaven is moved directly by God.*
   1. Bonaventure [*Sentences*, bk. 2, dist. 14, pt. 1, art. 3, qu. 1], *Opera*, 1885, 2:345–346.

160. The question from *De celo* is also cited above in question 17 of this catalog. Oresme devotes only a small part of the question ("Whether something created *de novo* could be perpetuated and also whether something that existed from eternity could be corrupted") to the problem of the commensurability or incommensurability of the celestial motions (see Oresme [*De celo*, bk. 1, qu. 24], 1965, 422–424). Despite an almost total absence of discussion of celestial incommensurability in his *questiones*, Oresme devoted entire treatises (at least two: the *Ad pauca respicientes* and *De commensurabilitate*) and parts of treatises (especially his *De proportionibus proportionum*) to this unusual topic and appears to have been the only scholastic author to do so. For references and discussion, see this volume, Chapter 18, Section I.3.a. For additional references, see Oresme, *Le Livre du ciel*, bk. 1, ch. 29, 1968, 197–203 and bk. 1, ch. 30, 209–215; also Oresme, *De commensurabilitate*, 1971, 412–413 (under *Le Livre du ciel*).
161. In this question, Amicus considers whether the celestial motions are mutually commensurable or incommensurable.
162. Riccioli asks "Whether the periods of celestial motions are mutually commensurable and consist of rational ratios, or whether indeed they consist of irrational ratios."

192. *Whether the* primum mobile *is moved directly by God.*
1. John of Jandun [*Physics*, bk. 8, qu. 21], 1519, 143v, col. 1–145r, col. 1.[163]
2. Buridan [*Metaphysics*, bk. 12, qu. 6], 1518, 67v, col. 1–69v, col. 2.[164]
3. Conimbricenses* [*De coelo*, bk. 2, ch. 5, qu. 6], 1598, 271–273.

193. *Whether God could move the last [or outermost] sphere with a rectilinear motion.*
1. Richard of Middleton [*Sentences*, bk. 2, dist. 14, art. 3, qu. 3], 1591, 2:186, cols. 1–2.

194. *By what are the celestial bodies moved?*
1. Poncius [*De coelo*, disp. 22, qu. 9], 625, col. 2–630, col. 1.[165]
2. Oviedo [*De caelo*, contro. 1, punc. 1], 1640, 461, col. 2, par. 9–462, col. 1, par. 11.
3. Serbellonus* [*De caelo*, disp. 1, qu. 4, art. 2], 1663, 2:40, col. 2–43, col. 1.

195. *Whether the heavens or planets [*sidera*] are moved by intelligences or intrinsically by a proper form or nature.*
1. Bonaventure [*Sentences*, bk. 2, dist. 14, pt. 1, art. 3, qu. 2], *Opera*, 1885, 2:347–350.[166]
2. Thomas Aquinas [*Sentences*, bk. 2, dist. 14, qu. 1, art. 3], 1929–1947, 2:351–354.
3. Richard of Middleton [*Sentences*, bk. 2, dist. 14, art. 1, qu. 6], 1591, 2:172, col. 1–174, col. 2.[167]
4. Aureoli [*Sentences*, bk. 2, dist. 14, qu. 2, arts. 1 and 2], 1596–1605, 2:191, col. 1–195, col. 2, and 195, col. 2–199, col. 2.[168]
5. Conimbricenses [*De coelo*, bk. 2, ch. 5, qu. 5], 1598, 264–270.[169]
6. Aversa [*De caelo*, qu. 34, sec. 7], 1627, 147, col. 1–151, col. 1.
7. Compton-Carleton* [*De coelo*, disp. 4, sec. 2], 1649, 409, col. 1.
8. Riccioli, *Almagestum novum*, pars post., bk. 9, sec. 2, ch. 1, 1651, 247, col. 1–251, col. 2.
9. Bonae Spei [comment. 3, *De coelo*, disp. 3, dub. 6], 1652, 12, col. 2–13, col. 2.
10. Cornaeus [*De coelo*, tract. 4, disp. 2, qu. 4, dub. 2], 1657, 504.[170]

163. John of Jandun's version reads: "Whether the prime mover moves the *primum mobile* directly."
164. Buridan asks "Whether it was the intention of Aristotle and the Commentator that God moves the *primum mobile* actively or only in the manner of a final cause."
165. For the enunciation of question 9, Poncius ignores the question format and simply has "De motu coelorum."
166. Bonaventure expresses the question as "Whether the motion of the heaven is [caused] by a proper form or by an intelligence."
167. Like Bonaventure, Richard asks whether the heaven is moved by an intelligence or a [natural] form.
168. In the two relevant questions Aureoli asks first "Whether a heaven is moved by an intelligence" and then "Whether an intelligence that moves a heaven is united to it as a mover only, or in some [other] way, [say] as a form."
169. The Coimbra Jesuits asked only "Whether or not the celestial orbs are moved by intelligences."
170. Cornaeus asks "Whether the planets are moved by their intrinsic forms." Because he rejected the existence of celestial orbs, Cornaeus once again speaks only of the motion of planets.

11. Serbellonus [*De caelo*, disp. 1, qu. 4, art. 3], 1663, 2:43, col. 1–44, col. 2.[171]
12. Oddus [*De coelo*, disp. 1, dub. 19], 1672, 54, col. 2–57, col. 2.[172]

196. Whether the celestial spheres are moved by one or by several intelligences.
1. Michael Scot [*Sphere*, lec. 4], 1949, 284–285.[173]
2. Cecco d'Ascoli [*Sphere*, ch. 1], 1949, 349–350.[174]
3. Conimbricenses* [*De coelo*, bk. 2, ch. 5, qu. 8], 1598, 276–279.

197. If there is an order of nobility in intelligences according to the arrangement of their mobiles with respect to position.[175]
1. Javelli [*Metaphysics*, bk. 12, qu. 22], 1568, 1:898, col. 1–900, col. 1.

198. If the Philosopher thought that all intelligences are of infinite strength.
1. Javelli [*Metaphysics*, bk. 12, qu. 21], 1568, 1:897, col. 1–898, col. 1.

199. If an intelligence that moves a heaven can be a true form that gives being to that heaven.
1. Javelli [*Metaphysics*, bk. 12, qu. 17], 1568, 1:889, col. 2–890, col. 2.

*200. Whether the same intelligence moves the same mobile as an efficient cause [*causa agens*] and as a final cause [*causa finalis*].*
1. Buridan [*Metaphysics*, bk. 12, qu. 5], 1518, 67r, col. 1–67v, col. 1.

201. On the number of movers of the orbs.
1. Cecco d'Ascoli [*Sphere*, ch. 1], 1949, 350–351.[176]
2. Oddus* [*De coelo*, disp. 1, dub. 20], 1672, 57, col. 1–60, col. 2.

202. Whether there are as many celestial motions as there are intelligences; and conversely, namely, that there are no more celestial motions than intelligences, nor more intelligences than there are celestial motions.
1. Buridan [*Metaphysics*, bk. 12, qu. 9], 1518, 71v, col. 1–73r, col. 1.

203. Whether the mover of a heaven [or sphere] is moved.
1. Michael Scot* [*Sphere*, lec. 12], 1949, 324–325.
2. Pseudo–Siger of Brabant [*Physics*, bk. 8, qu. 27], 1941, 233–234.

204. Whether the movers of celestial bodies move [that is, cause motion] per accidens.
1. John of Jandun [*Physics*, bk. 8, qu. 14], 1519, 137r, col. 1–137v, col. 2.

205. Whether the movers of the planetary orbs are [themselves] moved per accidens.
1. Javelli [*Physics*, bk. 8, qu. 10], 1568, 1:607, col. 1–608, col. 1.

206. Whether the faculty by which the intelligences move the celestial sphere differs from their understanding and will.
1. Conimbricenses [*De coelo*, bk. 2, ch. 5, qu. 7], 1598, 273–276.

171. Serbellonus asks only, "How can an intelligence move celestial bodies?"
172. Oddus: "Whether the heaven is moved by itself or by an intelligence."
173. Michael Scot speaks of movers generally rather than intelligences specifically.
174. Cecco's question, like Michael Scot's, is broader, since he asks "Whether the spheres are moved by one mover or by several." Although Cecco's work is a commentary, not a *questiones*, he often framed responses in question form.
175. See also question 102 in this catalog.
176. Cecco asks, "How many movers are there, and of what kind are they?"

207. *Whether this consequence or conditional [statement] is true: If the motor [or mover] of the heaven were in a magnitude, it would move [that is, it would cause the magnitude to move] in an instant.*[177]
   1. Oresme [*De spera*, qu. 9], 1966a, 174–217.

208. *Whether the eighth sphere is moved by itself.*
   1. Ockham [*Physics*, qu. 80], *Opera philosophica*, 1984, 6:614–616.

209. *Whether the heaven is moved by itself or by another.*
   1. Bacon [*Physics*, bk. 8], *Opera*, fasc. 13, 1935, 410–411.
   2. Pseudo-Siger of Brabant [*Physics*, bk. 8, qu. 23], 1941, 226–228.[178]
   3. Hurtado de Mendoza [*De coelo*, disp. 2, sec. 4], 1615, 373, col. 2–374, col. 1.[179]
   4. Serbellonus* [*De caelo*, disp. 1, qu. 4, art. 1], 1663, 2:39, col. 1–40, col. 2.

210. *Whether nature is the principle of celestial motion.*
   1. Ockham [*Physics*, qu. 123], *Opera philosophica*, 1984, 6:729–731.

211. *Whether the stars are self-moved or are moved only by the motions of their orbs.*
   1. Bacon, *De celestibus*, pt. 4, ch. 9, *Opera*, fasc. 4, 1913, 403–408.
   2. Bonaventure [*Sentences*, bk. 2, dist. 14, pt. 2, art. 1, qu. 2], *Opera*, 1885, 2:353–354.[180]
   3. Richard of Middleton [*Sentences*, bk. 2, dist. 14, art. 3, qu. 2], 1591, 2:184, col. 2–186, col. 1.[181]
   4. Durandus [*Sentences*, bk. 2, dist. 14, qu. 3], 1571, 156v, col. 2–157r (mistakenly foliated as 175), col. 1.
   5. John of Jandun [*De coelo*, bk. 2, qu. 13], 1552, 30v, cols. 1–2.
   6. Buridan [*De caelo*, bk. 2, qu. 18], 1942, 209–212.
   7. Albert of Saxony* [*De celo*, bk. 2, qu. 18], 1518, 112v, col. 2–113r, col. 2.
   8. Oresme [*De celo*, bk. 2, qu. 8], 1965, 575–591.[182]
   9. Versor [*De celo*, bk. 2, qu. 12], 1493, 24v, col. 1–25v, col. 1.[183]
   10. Bricot [*De celo*, bk. 2], 1486, 21v, col. 1.
   11. Amicus [*De caelo*, tract. 5, qu. 7, dubit. 2], 1626, 338, col. 1–343, col. 2.

212. *Whether a heaven is moved more quickly than [the body of] a star [i.e., planet].*
   1. Michael Scot [*Sphere*, lec. 5], 1949, 290.

213. *Whether or not the stars [that is, planets] are moved, with the heaven [remaining] unmoved.*
   1. Conimbricenses [*De coelo*, bk. 2, ch. 8, qu. 1], 1598, 324–327.

177. The bracketed words are my additions to Droppers' translation.
178. Pseudo-Siger asks "Whether the mover of a heaven is an immobile mover."
179. Hurtado asks "Whether the heavens are moved by some intrinsic principle or by an external mover."
180. Bonaventure asks "Whether the luminaries [that is, planets] are moved in their orbs with a proper motion."
181. Richard's question took the form "Whether, except for the motion of the spheres, any celestial bodies [*luminaria*] are moved with a proper motion."
182. This same question is also cited in question 180 of this catalog.
183. Versor: "Whether the planets are moved with a proper motion [that is] distinct from the motion of their orbs."

214. *Whether any orb that lacks a star [or planet] can be moved.*
   1. Bonaventure [*Sentences*, bk. 2, dist. 14, pt. 2, art. 1, qu. 3], *Opera*, 1885, 2:354–357.

215. *Whether any heavenly body is moved circularly [with a proper motion].*
   1. Oresme* [*De spera*, qu. 13], 1966a, 282–303.
   2. Major [*De celo*, bk. 2, qu. 3], 1526, 18, col. 2–19, col. 1.

216. *Whether the heaven is moved with exertion and fatigue.*
   1. John of Jandun* [*De coelo*, bk. 2, qu. 2], 1552, 23v, col. 2–24r, col. 2.
   2. Buridan [*De caelo*, bk. 2, qu. 1], 1942, 129–133.
   3. Albert of Saxony [*De celo*, bk. 2, qu. 9], 1518, 107v, col. 1–108r, col. 2.
   4. Oresme [*De celo*, bk. 2, qu. 2], 1965, 445–461.
   5. Paul of Venice, *Liber celi*, 1476, 32, col. 1–33, col. 2.
   6. Johannes de Magistris [*De celo*, bk. 2, qu. 1], 1490, 15, col. 2–17, col. 2 (sigs. k2v–k3v).
   7. Versor [*De celo*, bk. 2, qu. 1], 1493, 16r, col. 2–16v, col. 2.[184]
   8. Bricot [*De celo*, bk. 2], 1486, 16v, col. 2–17r, col. 1.
   9. Amicus [*De caelo*, tract. 5, qu. 6, dubit. 5 (mistakenly numbered 7)], 1626, 311, col. 1–314, col. 2.

217. *Whether violence or contrariety is the cause of fatigue.*
   1. Oresme [*De celo*, bk. 2, qu. 1], 1965, 427–445.

218. *Whether by its motion the sphere of the Sun makes inferior [that is, sublunar] things hotter than [the motions] of other spheres.*
   1. Buridan [*De caelo*, bk. 2, qu. 17], 1942, 205–208.[185]

219. *Whether the Sun and Moon ought to be moved with fewer motions than other planets.*
   1. Buridan* [*De caelo*, bk. 2, qu. 21], 1942, 223–225.
   2. Albert of Saxony [*De celo*, bk. 2, qu. 17], 1518, 112r, col. 2–112v, col. 2.

220. *Whether for saving the celestial motions of the planets it is necessary to assume eccentrics and epicycles.*[186]
   1. Bacon, *De celestibus*, pt. 5, ch. 13, *Opera*, fasc. 4, 1913, 433–445.
   2. Buridan [*Metaphysics*, bk. 12, qus. 10–11], 1518, 73r, col. 1–73v, col. 1, and 73v, col. 1–74r, col. 1.[187]
   3. Albert of Saxony* [*De celo*, bk. 2, qu. 7], 1518, 106r, col. 2–107r, col. 1.
   4. Themon Judaeus [*De spera*, qu. 23], in Hugonnard-Roche, 1973, 144–145.
   5. D'Ailly, *14 Questions*, qu. 13, 1531, 163v–164v.
   6. Paul of Venice, *Liber celi*, 1476, 30, col. 1–32, col. 1.
   7. Johannes de Magistris [*De celo*, bk. 2, qu. 3, dub. 1–3], 1490, 21, col. 2–22, col. 2 (sigs. k5v–k6r).

184. Versor's question differs from most because he asks "Whether the heaven is eternal, immortal, moved eternally, and without fatigue."
185. I have slightly revised and expanded my earlier translation of this question in Grant, 1974, 205.
186. Cecco d'Ascoli considers this question in his *De eccentricis et epicyclis*, which is not in question form (see Cecco d'Ascoli, *De eccentricis* [Boffito], 1906, 161–167).
187. Buridan devotes one question (10) to epicycles ("Whether epicycles must be assumed in celestial bodies") and the other (11) to eccentrics ("Whether eccentric orbs must be assumed in the heavens").

8. Versor [*De celo*, bk. 2, qu. 9], 1493, 22v, col. 1–23r, col. 1.
9. Bricot [*De celo*, bk. 2], 1486, 21v, col. 2, and 29r, cols. 1–2.
10. Javelli [*Metaphysics*, bk. 12, qu. 23], 1568, 1:900, col. 1–901, col. 1.
11. Major [*De celo*, bk. 2, qu. 2], 1526, 16, col. 1–18, col. 1.
12. Amicus [*De caelo*, tract. 5, qu. 5, dubit. 3], 1626, 265, col. 1–270, col. 2.
13. Aversa [*De caelo*, qu. 32, sec. 7], 1627, 69, col. 1–76, col. 2.
14. Cornaeus [*De coelo*, tract. 4, disp. 2, qu. 2, dub. 3], 1657, 494–498.[188]

221. *Many think that, just as in inferior bodies, heat and sound exist in celestial bodies by reason of motion and light.*
1. Paul of Venice, *Liber celi*, 1476, 33, col. 2–35, col. 1.

222. *Whether celestial bodies cause sound by their motions.*
1. Bacon, *De celestibus*, pt. 4, ch. 9, *Opera*, fasc. 4, 1913, 408–410.
2. Albert of Saxony* [*De celo*, bk. 2, qu. 15], 1518, 111r, col. 2–111v, col. 2.
3. Oresme [*De celo*, bk. 2, qu. 9], 1965, 591–603.
4. Versor [*De celo*, bk. 2, qu. 13], 1493, 25v, col. 2–26r, col. 1.[189]
5. Bricot [*De celo*, bk. 2], 1486, 20r, col. 1–20v, col. 1.
6. Major [*De celo*, bk. 2, qu. 4], 1526, 19, cols. 1–2.
7. Conimbricenses [*De coelo*, bk. 2, ch. 9, qu. 1], 1598, 330–335.
8. Amicus [*De caelo*, tract. 5, qu. 6, dubit. 11], 1626, 335, col. 1–337, col. 1.
9. Aversa [*De caelo*, qu. 34, sec. 3], 1627, 137, col. 1–138, col. 1.[190]

223. *Whether the ethereal region is moved around the elemental.*
1. Michael Scot [*Sphere*, lec. 4], 1949, 279–280.

224. *Whether the earth is necessary for the motion of the heavens.*
1. John of Jandun [*De coelo*, bk. 2, qu. 6], 1552, 26v, cols. 1–2.

225. *Whether the following consequence is valid: If the heaven is moved, therefore the earth is necessarily at rest.*
1. Albert of Saxony [*De celo*, bk. 2, qu. 10], 1518, 108r, col. 2–108v, col. 2.
2. Oresme* [*De celo*, bk. 2, qu. 5], 1965, 493–507.[191]
3. Paul of Venice, *Liber celi*, 1476, 27, col. 1 (conclus. 2).

XIV. Celestial light

226. *What light was meant when it was said, "Let there be light"* [Fiat lux].
1. William of Auvergne, *De universo*, pt. 1 of pt. 1, ch. 41, *Opera*, 1674, 1:635, col. 2–639, col. 1.

227. *How light exists in the heaven.*
1. Serbellonus [*De caelo*, disp. 1, qu. 5, art. 1], 1663, 2:46, col. 1–48, col. 1.

188. Cornaeus includes a discussion of eccentrics and epicycles within the framework of a question on whether the heavens are hard and solid.
189. More specifically, Versor inquires whether the motions of celestial bodies and their orbs produce "harmonic sounds" (*sonos armonicos*).
190. Aversa's discussion of sound forms part of a larger consideration of qualities in celestial bodies.
191. Oresme's enunciation of the question is almost identical to Albert of Saxony's.

228. On corporeal light diffused in the empyrean heaven.

1. Alexander of Hales,★ *Summa theologica*, inq. 3, tract. 2, qu. 2, tit. 1, memb. 1, ch. 1, arts. 1–5, 1928, 327–331.
2. Hurtado de Mendoza [*De coelo*, disp. 2, sec. 5], 1615, 374, col. 2–375, col. 1.[192]

*229. Whether or not the light [*lux*] of the stars is a substantial form or even a body.*

1. Conimbricenses [*De coelo*, bk. 2, ch. 7, qu. 2], 1598, 296–299.

230. Whether or not the light of all the stars, and all light taken as a whole, are of the same species.

1. Richard of Middleton [*Sentences*, bk. 2, dist. 13, art. 1, qu. 4], 1591, 2:158, col. 2–159, col. 2.
2. Conimbricenses★ [*De coelo*, bk. 2, ch. 7, qu. 3], 1598, 299–303.

*231. Whether all the planets [*astra*], except the Sun, receive their light from the Sun or from themselves.*

1. Albert of Saxony★ [*De celo*, bk. 2, qu. 20], 1518, 114v, col. 1–115v, col. 1.
2. Oresme [*De celo*, bk. 2, qu. 11], 1965, 637–653
3. Paul of Venice, *Liber celi*, 1476, 35, col. 2.
4. Versor [*De celo*, bk. 2, qu. 11], 1493, 23v, col. 2–24r, col. 1.
5. Bricot [*De celo*, bk. 2], 1486, 22r, col. 1.
6. Major [*De celo*, bk. 2, qu. 7], 1526, 20, col. 1–21, col. 1.
7. Conimbricenses [*De coelo*, bk. 2, ch. 7, qu. 4], 1598, 303–307.
8. Aversa [*De caelo*, qu. 35, sec. 2], 1627, 167, col. 1–171, col. 2 and [sec. 3], 171, col. 2–174, col. 1.[193]
9. Poncius [*De coelo*, disp. 22, qu. 8], 1672, 625, cols. 1–2.[194]
10. Compton-Carleton [*De coelo*, disp. 2, sec. 3], 1649, 402, col. 2–404, col. 2.
11. Cornaeus [*De coelo*, tract. 4, disp. 2, qu. 4, dub. 9], 1657, 512.

*232. Whether all the heavens [that is, spheres] and planets [*astra*] have light, or [whether] only the planets [do]; and of the planets, whether all of them have light.*

1. Oddus [*De coelo*, disp. 1, dub. 21], 1672, 61, col. 2–63, col. 1.

*233. Whether the light [*lumen*] received in the heavens produces colors in the heavens.*

1. Amicus★ [*De caelo*, tract. 6, qu. 5, dubit. 4], 1626, 368, col. 1–369, col. 2.
2. Aversa [*De caelo*, qu. 34, sec. 3], 1627, 136, col. 2–137, col. 2.[195]

*234. Whether, by means of light, all planets [*astra*] generate heat in inferior things.*[196]

1. Buridan [*De caelo*, bk. 2, qu. 15], 1942, 192–199.
2. Albert of Saxony★ [*De celo*, bk. 2, qu. 21], 1518, 115v, col. 1–116r, col. 1.

192. Although the topic under consideration in section 5 is "On the light of the heavens," Hurtado de Mendoza devotes most of his discussion to the light of the empyrean sphere.
193. In the first of these two questions, Aversa asks "Whether the Sun truly illuminates the Moon, so that the whole Moon receives light from the Sun" and in the second "Whether the Sun contributes light equally to the other planets."
194. Poncius asks simply "Whether the stars [i.e., planets and perhaps stars] have light from themselves."
195. Aversa considers the colors of celestial bodies as part of a general section "On the qualities of celestial bodies."
196. Some scholastics considered this topic tangentially under the question "Whether light

PART III. QUESTIONS RELEVANT TO THE CELESTIAL AND TERRESTRIAL REGIONS

XV. On the influence of the celestial region on the terrestrial

235. *Whether it is necessary that this inferior world be continuous with the superior motions so that the total power [of the inferior world] could be governed from there [that is, from the celestial region].*
   1. Simon de Tunstede (Pseudo-Scotus) [*Meteorology*, bk. 1, qu. 3], 1639, 3:6, col. 1–8, col. 1.
   2. Themon Judaeus [*Meteorology*, bk. 1, qu. 1], 1518, 155v, col. 1–156r, col. 2.
   3. Johannes de Magistris* [*Meteorology*, bk. 1, qu. 2], 1490, 5, col. 1–9, col. 1.[197]

236. *Whether celestial bodies are the causes of these inferior generable and corruptible things.*
   1. John of Jandun [*De coelo*, bk. 1, qu. 1], 1552, 2r, col. 1–2v, col. 2.

237. *Whether or not celestial bodies act on the sublunar world.*
   1. Thomas Aquinas [*Sentences*, bk. 2, dist. 15, qu. 1, art. 2], 1929–1947, 2:370–373.[198]
   2. Durandus [*Sentences*, bk. 2, dist. 15, qu. 1], 1571, 157v, col. 1.
   3. Javelli [*Metaphysics*, bk. 12, qu. 13], 1568, 1:878, col. 1–881, col. 1.
   4. Conimbricenses* [*De coelo*, bk. 2, ch. 3, qu. 1], 1598, 188–193.
   5. Hurtado de Mendoza [*De coelo*, disp. 2, sec. 6], 1615, 375, col. 2–376, col. 1.
   6. Amicus [*De caelo*, tract. 6, qu. 1], 1626, 348, col. 1–351, col. 2.
   7. Aversa [*De caelo*, qu. 35, sec. 4], 1627, 174, col. 1–176, col. 2.
   8. Poncius [*Sentences*, bk. 2, dist. 14, qu. 3], 1639, 6, pt. 2:734–738, 742–744, and [*De coelo*, disp. 22, qu. ult. (qu.10)], 1672, 630, col. 1–634, col. 2.[199]
   9. Compton-Carleton [*De coelo*, disp. 2, sec. 4], 1649, 404, cols. 1–2.
   10. Cornaeus [*De coelo*, tract. 4, disp. 2, qu. 2, dub. 5], 1657, 500.
   11. Serbellonus [*De caelo*, disp. 1, qu. 5, art. 2], 1663, 2:48, col. 1–49, col. 2.
   12. Oddus [*De coelo*, disp. 1, dub. 23], 1672, 69, col. 1–71, col. 2.

238. *Whether the influence of the heavens and the stars [or planets] rather than the power of the mover effects the manifold changes of inferior things.*
   1. Case [*Physics*, bk. 8, ch. 6], 1599, 824–825.

239. *Which celestial bodies affect inferior things?*
   1. Aversa [*De caelo*, qu. 35, sec. 5], 1627, 176, col. 2–181, col. 2.

240. *Whether the governance of inferior things depends on the* primum mobile, *or on the eighth sphere, or on the other seven spheres.*
   1. Cecco d'Ascoli [*Sphere*, ch. 1], 1949, 353–354.

causes heat." (See John of Jandun [*De coelo*, bk. 2, qu. 10], 1552, 29r, col. 1–29v, col. 1, and Conimbricenses [*De coelo*, bk. 2, ch. 7, qu. 5], 1598, 307–312.)

197. Since the volume is unpaginated, I have numerated each page sequentially from the beginning of the *Meteorology*. De Magistris's *De celo* is also in this volume.
198. Thomas's question takes the form "Whether celestial bodies have any effect on inferior bodies." In the Latin text, *Unum* is inadvertently substituted for *Utrum*.
199. The same question from Poncius's *De coelo* is cited in question 250 of this catalog.

241. *Whether or not the empyrean heaven influences bodies subjected to it.*
1. Conimbricenses* [*De coelo*, bk. 2, ch. 3, qu. 2], 1598, 193–196.
2. Oddus [*De coelo*, disp. 1, dub. 24], 1672, 71, col. 2–74, col. 1.

242. *Whether the stars influence sublunar things by motion, light, and occult qualities.*
1. Michael Scot [*Sphere*, lec. 9], 1949, 314–315.
2. Durandus [*Sentences*, bk. 2, dist. 15, qu. 3], 1571, 158r, col. 2–158v, col. 2.[200]
3. Major [*Sentences*, bk. 2, dist. 15, qus. 6–7], 1519b, 77v, col. 1–78r, col. 2, and 78r, col. 2–79v, col. 1.
4. Amicus [*De caelo*, tract. 6, qu. 6, dubit. 1 and qu. 7], 1626, 376, col. 1–378, col. 2, and 391 (given erroneously as 381), col. 1–398, col. 2.[201]
5. Aversa [*De caelo*, qu. 35, secs. 6–7], 1627, 181, col. 2–187, col. 1, and 187, col. 2–192, col. 2.[202]
6. Bonae Spei [comment. 3, *De coelo*, disp. ult. (4), dub. 1], 1652, 15, col. 1–16, col. 1.[203]
7. Oddus* [*De coelo*, disp. 1, dub. 29], 1672, 100, col. 2–109, col. 1.

243. *In how many ways is motion necessary for producing effects on inferior things?*
1. Amicus [*De caelo*, tract. 6, qu. 6, dubit. 1], 1626, 376, col. 1–378, col. 2.

244. *Whether this consequence is sound: If it is necessary that generations and corruptions be here below, it is [therefore] necessary that there be several celestial motions.*
1. Buridan* [*De caelo*, bk. 2, qu. 10], 1942, 168–172.
2. Albert of Saxony [*De celo*, bk. 2, qu. 12], 1518, 109v, cols. 1–2.
3. Paul of Venice, *Liber celi*, 1476, 27, cols. 1–2.
4. Johannes de Magistris [*De celo*, bk. 2, qu. 2], 1490, 17, col. 2–20, col. 2 (sigs. k3v–k5r).
5. Versor [*De celo*, bk. 2, qu. 5], 1493, 19v, col. 1–20r, col. 1.
6. Major [*De celo*, bk. 1, qu. 7], 1526, 13, col. 1–14, col. 2.

245. *Whether the first orb is more the cause of generation and corruption than the motion of the planets in the oblique circle [or ecliptic].*
1. Michael Scot [*Sphere*, lec. 9], 1949, 311–313.

246. *On the effects which are produced by the planets [*astris*].*
1. Oddus [*De coelo*, disp. 1, dub. 25], 1672, 74, col. 1–78, col. 1.

247. *Whether different celestial bodies [*luminaria*] could have different effects on corporeal bodies.*
1. Bonaventure [*Sentences*, bk. 2, dist. 14, pt. 2, art. 2, qu. 2], *Opera*, 1885, 2:359–361.

200. Michael Scot and Durandus speak only of the influence of motion.
201. In question 6, Amicus considers only the effect of celestial motion ("Whether local motion is the instrument of the heavens for producing inferior effects"), while in question 7 he concentrates on occult qualities or influences ("Whether, besides light and motion in the heavens, virtues and occult qualities are to be admitted as influences").
202. In these two questions, Aversa asks respectively "Whether, and in what manner, celestial bodies affect inferior things by means of motion" and "Whether, and in what manner, celestial bodies affect [inferior things] not only by light but by other occult powers and influences."
203. Bonae Spei focuses on occult qualities – that is, "influences" ("Whether celestial bodies act on inferior things by means of influences [*influentias*]").

248. *Whether the orb of the sun could be called the cause of generation and corruption.*
1. Michael Scot [*Sphere*, lec. 9], 1949, 313.

249. *Whether celestial bodies have [that is, produce] natural effects that differ in species with respect to corruptible bodies.*
1. Richard of Middleton [*Sentences*, bk. 2, dist. 14, art. 3, qu. 5], 1591, 2:188, col. 1–189, col. 2.

250. *Whether living things could be generated by the power of celestial bodies.*
1. Conimbricenses* [*De coelo*, bk. 2, ch. 3, qu. 6], 1598, 209–216.
2. Amicus [*De caelo*, tract. 6, qu. 8, dubit. 6], 1626, 425, col. 1–439, col. 1.[204]
3. Aversa [*De caelo*, qu. 35, sec. 8], 1627, 194, col. 1–197, col. 2.[205]
4. Poncius [*De coelo*, disp. 22, qu. ult. (qu. 10)], 1672, 631, col. 2–633, col. 1.[206]

251. *Whether animals that are generated by the force of putrid matter are of the same species as those that bear the same name but are generated by [means of] seed.*
1. Conimbricenses [*De coelo*, bk. 2, ch. 3, qu. 7], 1598, 216–219.

252. *Whether, besides light and motion, occult powers and qualities in the heavens influence things.*
1. Conimbricenses [*De coelo*, bk. 2, ch. 3, qu. 3], 1598, 196–200.
2. Amicus* [*De caelo*, tract. 6, qu. 7], 1626, 391, col. 2–398, col. 2.

253. *Whether inferior things depend on the heavens for their preservation.*
1. Amicus [*De caelo*, tract. 6, qu. 8, dubit. 1], 1626, 398, col. 2–404, col. 1.

254. *Whether the actions of inferior agents depend on the motion and influence of the heavens.*
1. Case [*Physics*, bk. 8, ch. 7], 1599, 828–831.
2. Amicus* [*De caelo*, tract. 6, qu. 8, dubit. 2], 1626, 404, col. 1–412, col. 1.

255. *Whether celestial bodies conserve inferior things in their being and directly join in all of their operations.*
1. Hervaeus Natalis, *De materia celi*, qu. 6, in *Quolibeta Hervei*, 1513, 45r, col. 2–47v, col. 1.
2. Aversa* [*De caelo*, qu. 35, sec. 9], 1627, 197, col. 2–201, col. 2.

256. *Whether all motions and actions of inferior bodies would cease if the celestial motions ceased.*
1. Richard of Middleton [*Sentences*, bk. 2, dist. 14, art. 3, qu. 6], 1591, 2:182, col. 1–183, col. 2.[207]
2. Hervaeus Natalis, *De materia celi*, qu. 7, in *Quolibeta Hervei*, 1513, 47v, col. 1–49r, col. 1.
3. Conimbricenses* [*De coelo*, bk. 2, ch. 3, qu. 4], 1598, 201–205.

204. Although the pages are incorrectly numbered at this point, *dubitatio* 6 begins on a page numbered 425 and ends on one numbered 439. Amicus's version asks "Whether the heavens can produce perfect or imperfect living things."
205. The possible production of living things by the power of celestial bodies constitutes the most extensive part of Aversa's general question "Which effects do celestial bodies cause on inferior things" (192, col. 2–197, col. 2).
206. The same question is cited above in question 237.
207. Richard's version of this question asks "Whether the elements could operate if the heavens exerted no influence on them."

4. Hurtado de Mendoza [*De coelo*, disp. 2, sec. 4], 1615, 374, cols. 1–2.
5. Aversa [*De caelo*, qu. 35, sec. 9], 1627, 198, col. 2.[208]
6. Poncius [*De coelo*, disp. 22, qu. 9], 1672, 630, col. 1.[209]
7. Compton-Carleton [*De coelo*, disp. 4, sec. 3], 1649, 410 (mistakenly given as 310), col. 1.

257. *Whether all mixed bodies would naturally cease to exist if the celestial motions ceased.*
   1. Hervaeus Natalis, *De materia celi*, qu. 8, in *Quolibeta Hervei*, 1513, 49r, col. 1–51v, col. 1.

258. *Every natural transmutation of inferior things can be traced to a celestial cause.*
   1. Hervaeus Natalis, *De materia celi*, qu. 5, in *Quolibeta Hervei*, 1513, 43r, col. 2–45r, col. 2.
   2. Amicus* [*De caelo*, tract. 6, qu. 8, dubit. 3], 1626, 412, col. 1–418, col. 1.

259. *With respect to the heavens, no natural effect occurs by chance.*
   1. Conimbricenses [*De coelo*, bk. 2, ch. 3, qu. 5], 1598, 206–208.

260. *Whether some celestial bodies cause coldness in inferior things.*
   1. John of Jandun [*De coelo*, bk. 2, qu. 12], 1552, 29v, col. 2–30r, col. 2.

261. *Whether motion* per se *produces heat rather than cold.*
   1. John of Jandun [*De coelo*, bk. 2, qu. 9], 1552, 28r, col. 1–28v, col. 2.[210]
   2. Buridan [*De caelo*, bk. 2, qu. 16], 1942, 199–205.
   3. Amicus* [*De caelo*, tract. 6, qu. 6, dubit. 2], 1626, 378, col. 2–381, col. 2.

262. *Whether celestial motion causes hotness in inferior things.*
   1. Themon Judaeus* [*Meteorology*, bk. 1, qu. 6], 1518, 160v, col. 1–161r, col. 1.
   2. Amicus [*De caelo*, tract. 6, qu. 6, dubit. 4], 1626, 387, col. 2–391, col. 2.[211]
   3. Serbellonus [*De caelo*, disp. 1, qu. 4, art. 4], 1663, 2:44, col. 2–46, col. 1.

263. *Whether every luminous thing causes heat by its light.*
   1. Themon Judaeus [*Meteorology*, bk. 1, qu. 8], 1518, 162r, col. 1–162v, col. 2.

264. *Whether the Sun heats inferior [that is, lower] things [or bodies].*
   1. John of Jandun [*De coelo*, bk. 2, qu. 11], 1552, 29v, cols. 1–2.

265. *Whether the sphere of the Sun heats inferior things more by its motion [than do] other celestial spheres.*
   1. Simon de Tunstede (Pseudo-Scotus) [*Meteorology* bk. 1, qu. 12], 1639, 3:28, col. 1–29, col. 2.

266. *Whether we receive heat from the Sun, or from the heaven, or by the motion of the orb of a planet [*stelle*].*
   1. Michael Scot [*Sphere*, lec. 15], 1949, 330.

208. The theme of this question forms a small part of Aversa's general question "Whether celestial bodies conserve inferior things in their being and directly join in all of their operations" (197, col. 2–201, col. 2).
209. The same question is cited in question 194 of this catalog.
210. John of Jandun: "Whether local motion causes hotness" (An motus localis habeat calefacere).
211. Amicus asks "How heat is caused in inferior things by the motion of the heaven." The text mistakenly numbers page 391 as 381.

267. *Whether the action of the heavens extends to all four qualities of the elements.*
   1. Amicus [*De caelo*, tract. 6, qu. 8, dubit. 4], 1626, 418, col. 1–426, col. 2.

268. *Whether the planets [*astra*] influence the vegetative and sensitive soul.*
   1. Oddus [*De coelo*, disp. 1, dub. 26], 1672, 78, col. 1–87, col. 1.

269. *Whether celestial bodies impose necessity on the effects they cause [even] with the action of the agent limited by free will.*
   1. Hervaeus Natalis, *De materia celi*, qu. 9, in *Quolibeta Hervei*, 1513, 51v, col. 1–53v, col. 1.

270. *What do celestial bodies especially effect with regard to humans, and what can be predicted from them [that is, celestial bodies] concerning human affairs?*
   1. Thomas Aquinas [*Sentences*, bk. 2, dist. 15, qu. 1, art. 3], 1929–1947, 2:374–377.[212]
   2. Conimbricenses [*De coelo*, bk. 2, ch. 3, qu. 8], 1598, 219–221.[213]
   3. Aversa* [*De caelo*, qu. 35, sec. 10], 1627, 201, col. 2–207, col. 2.
   4. Oddus [*De coelo*, disp. 1, dub. 27], 1672, 87 col. 1–92, col. 1.[214]

271. *Whether the heaven acts on the human intellect and will.*
   1. Pseudo-Siger of Brabant [*Physics*, bk. 8, qus. 13–14], 1941, 211–214.[215]
   2. John of Jandun [*Physics*, bk. 8, qu. 6], 1519, 127v, col. 1–128r, col. 2.
   3. Benedictus Hesse* [*Physics*, bk. 8, qu. 10], 1984, 710–713.

272. *Whether by means of the influences of the stars an astronomer can predict the future by the influence which the heaven exerts on sensitive and intellective powers.*
   1. Major [*Sentences*, bk. 2, dist. 14, qu. 8], 1519b, 79v, col. 1–81r, col. 1.

273. *Whether astrologers can predict effects that are dependent on observation of the stars.*
   1. Conimbricenses [*De coelo*, bk. 2, ch. 3, qu. 9], 1598, 222–233.[216]
   2. Amicus* [*De caelo*, tract. 6, qu. 8, dubit. 9], 1626, 450, col. 2–474, col. 2.
   3. Cornaeus [*De coelo*, tract. 4, disp. 2, qu. 4, dub. 11], 1657, 513–514.[217]
   4. Oddus [*De coelo*, disp. 1, dub. 28], 1672, 92, col. 1–100, col. 2.[218]

XVI. Place in the celestial and/or terrestrial regions

274. *Whether place exists.*
   1. Toletus* [*Physics*, bk. 4, qu. 1], 1580, 104v, col. 2–112r, col. 2.
   2. Amicus [*Physics*, tract. 20, qu. 2], 1626–1629, 2:665, cols. 1–2.

212. Thomas asks specifically "Whether superior bodies exercise causality over the movement of the free will."
213. The Coimbra Jesuits expressed the question more narrowly as "Whether or not celestial bodies influence the human will."
214. Oddus's version asks "Whether the celestial bodies influence the rational soul and its powers."
215. In these two questions, Pseudo-Siger asks "Whether by their motion superior [that is, celestial] bodies cause the operation of the intellect" and "Whether the motion of the superior [that is, celestial] bodies are *per se* the cause of the will."
216. The Coimbra Jesuits ask "Whether astrologers can predict future contingents on the basis of observations of the stars."
217. Cornaeus asks "Whether from a conjunction of stars, future things could be predicted."
218. Oddus's version: "What could astrologers know from observation of the stars?"

275. *We inquire what place is: Whether it is matter or form or the space between the sides of the container.*
1. Bacon★ [*Physics*, bk. 4], *Opera*, fasc. 13, 1935, 178–182.[219]
2. Ockham [*Physics*, qu. 74], *Opera philosophica*, 1984, 6:602–604.
3. Marsilius of Inghen [*Physics*, bk. 4, qu. 2], 1518, 46v, col. 1–47r, col. 1.[220]
4. Benedictus Hesse [*Physics*, bk. 4, qu. 6], 1984, 406–407.
5. Oña [*Physics*, bk. 4, qu. 1, art. 2], 1598, 207v, col. 1–210r, col. 1.
6. Hurtado de Mendoza [*Physics*, disp. 14, sec. 1], 1615, 321, cols. 1–2.[221]
7. Amicus [*Physics*, tract. 20, qu. 3, dubit. 1], 1626–1629, 2:666, col. 1–672, col. 2.[222]
8. Cornaeus [*Physics*, bk. 4, disp. 3 ("De loco et vacuo"), qu. 1, sec. 1, dub. 1], 1657, 358–359.
9. Serbellonus [*Physics*, disp. 5, qu. 1, art. 1], 1657, 1:798, col. 2–803, col. 2.
10. Oddus [*Physics*, bk. 4, disp. 14, art. 1], 1667, 601, col. 1–612, col. 1.

276. *Whether place is the ultimate [surface or terminus] of the container.*
1. Pseudo-Siger of Brabant★ [*Physics*, bk. 4, qu. 8], 1941, 154–155.
2. John of Jandun [*Physics*, bk. 4, qu. 4], 1519, 58v, col. 1–59v, col. 2.
3. Johannes Canonicus [*Physics*, bk. 4, qu. 1], 1520, 39v, col. 1–40v, col. 2.[223]
4. Buridan [*Physics*, bk. 4, qu. 2], 1509, 67v, col. 1–68v, col. 2.[224]
5. Albert of Saxony [*Physics*, bk. 4, qu. 1], 1518, 42v, col. 2–43v, col. 2.[225]
6. Marsilius of Inghen [*Physics*, bk. 4, qu. 3], 1518, 47r, col. 1–48r, col. 1.
7. Benedictus Hesse [*Physics*, bk. 4, qu. 12], 1984, 416–420.
8. Toletus [*Physics*, bk. 4, qu. 3], 1580, 116v, col. 1–119r, col. 1.
9. Amicus [*Physics* tract. 20, qu. 4, dubits. 2–3], 1626–1629, 2:680, col. 2–686, col. 2.[226]

277. *Whether place is immobile.*
1. John of Jandun [*Physics*, bk. 4, qu. 6], 1519, 61v [mistakenly foliated 59v], col. 1–62v, col. 1.
2. Ockham [*Physics*, qu. 78], *Opera philosophica*, 1984, 6:610–611.
3. Buridan★ [*Physics*, bk. 4, qu. 3], 1509, 68v, col. 2–70r, col. 1.
4. Albert of Saxony [*Physics*, bk. 4, qu. 3], 1518, 44r, col. 2–45r, col. 1.
5. Benedictus Hesse [*Physics*, bk. 4, qu. 13], 1984, 420–422.
6. Toletus [*Physics*, bk. 4, qu. 5], 1580, 120r, col. 2–121, col. 2.

219. Although Bacon considers some forty-five questions on the concept of place, only a few are germane to this study.
220. The versions of Ockham and Marsilius are identical and ask only "Whether place is the space between the sides of a container."
221. Hurtado asks, "What is an external place?"
222. Amicus asks "Whether place is an interval, or space."
223. Canonicus asks "Whether place is some absolute entity, essentially the same as a surface."
224. Later Buridan asks "Whether the definition of place which Aristotle assigned is sound in which place is called the terminus of an immobile container" (*Physics*, bk. 4, qu. 4, 70r, col. 1–70v, col. 1).
225. Like Buridan, by whom he may have been influenced, Albert of Saxony also asked (*Physics*, bk. 4, qu. 4, 45r, col. 1–45v, col. 1) "Whether the definition of place is sound, namely that place is the terminus of the first immobile containing body."
226. In *dubitatio* 2 (680, col. 2–683, col. 1), Amicus asks "How surface is included in the concept of place," and in the third (683, col. 1–686, col. 2) inquires "How place includes surface in its concept, whether primarily or secondarily."

7. Oña [*Physics*, bk. 4, qu. 1, art. 4], 1598, 214r, col. 2–217r [mistakenly printed as 117], col. 2.
8. Amicus [*Physics*, tract. 20, qu. 1, dubit. 9], 1626–1629, 2:658, col. 2–664, col. 1.[227]
9. Cornaeus [*Physics*, bk. 4, disp. 3 ("De loco et vacuo"), qu. 1, sec. 1, dub. 2], 1657, 359–360.
10. Serbellonus [*Physics*, disp. 5, qu. 1, art. 2], 1657, 1:804, col. 1–805, col. 1.

278. *What is* ubi?
1. Cornaeus [*Physics*, bk. 4, disp. 3 ("De loco et vacuo"), qu. 1, sec. 2, dub. 1], 1657, 360.

279. *Whether* ubi *is something absolute.*
1. Cornaeus [Physics, bk. 4, disp. 3 ("De loco et vacuo"), qu. 1, sec. 2, dub. 4], 1657, 364–365.

280. *Whether besides external place, an internal* ubi *ought to be admitted in the located thing.*
1. Oña [*Physics*, bk. 4, qu. 1, art. 3], 1598, 210r, col. 1–214r, col. 2.[228]
2. Amicus [*Physics*, tract. 20, qu. 6], 1626–1629, 2:708, col. 1–710, col. 2.[229]
3. Oviedo [*Physics*, bk. 4, contro. 15, punc. 4], 1640, 373, col. 2–375, col. 2.
4. Compton-Carleton [*Physics*, disp. 34, secs. 1–5], 1649, 341–345.[230]
5. Cornaeus [*Physics*, bk. 4, disp. 3 ("De loco et vacuo"), qu. 1, sec. 2, dub. 3], 1657, 361–364.[231]
6. Serbellonus [*Physics*, disp. 5, qu. 1, art. 3], 1657, 1:805, col. 2–809, col. 1.[232]
7. Oddus* [*Physics*, bk. 4, disp. 14, art. 4], 1667, 621, col. 1–624, col. 1.

281. *Whether* ubi *essentially concerns place and an external body.*
1. Cornaeus [*Physics*, bk. 4, disp. 3 ("De loco et vacuo"), qu. 1, sec. 2, dub. 5], 1657, 365.[233]

282. *Whether, by the absolute power of God, any creature could lack an* ubi *and be nowhere.*
1. Cornaeus [*Physics*, bk. 4, disp. 3 ("De loco et vacuo"), qu. 1, sec. 2, dub. 10], 1657, 370–371.

283. *Whether every being is in a place.*
1. John of Jandun [*Physics*, bk. 4, qu. 1], 1519, 57r, col. 2–57v, col. 1.
2. Albert of Saxony* [*Physics*, bk. 4, qu. 7], 1518, 47v, col. 1–48r, col. 1.
3. Marsilius of Inghen [*Physics*, bk. 4, qu. 7], 1518, 50r, col. 1–50v, col. 2.
4. Benedictus Hesse [*Physics*, bk. 4, qu. 2], 1984, 400–402.

227. In *tractatio* 20, *questio* 4, *dubitatio* 4, Amicus also asks, "How is immobility included in the concept of place?"
228. Oña asks "Whether place is *ubi*, or quantity."
229. Amicus asks, "What is *ubi*, or local presence, and how many [aspects of it are there]?"
230. Compton-Carleton poses five questions on the various aspects of *ubi*.
231. Cornaeus asks "Whether *ubi* is distinguished more by reason than by the things located [*ubicata*]."
232. Serbellonus asks "Whether a distinct *ubi* could be assigned."
233. In the next *dubium*, no. 6 (p. 365), Cornaeus asks again about *ubi*: "Whether *ubi* essentially concerns the thing of which it is the *ubi*."

5. Amicus [*Physics*, tract. 20, qu. 5, dubit. 1], 1626–1629, 2:692, col. 2–698, col. 1.[234]
6. Compton-Carleton [*Physics*, disp. 33, sec. 5], 1649, 338.[235]
7. Cornaeus [*Physics*, bk. 4, disp. 3 ("De loco et vacuo"), qu. 2, sec. 1, dub. 1], 1657, 371.[236]

284. *Whether all spiritual substances are in a place.*
1. Amicus [*Physics*, tract. 20, qu. 5, dubit. 3], 1626–1629, 2:702, col. 1–703, col. 2.
2. Oddus* [*Physics*, bk. 4, disp. 14, art. 7], 1667, 631, col. 1–633, col. 2.

285. *Whether the whole universe has a place.*[237]
1. Bacon [*Physics*, bk. 4], *Opera*, fasc. 13, 1935, 221.

286. *Whether the heaven has a place.*
1. Bacon [*Physics*, bk. 4], *Opera*, fasc. 13, 1935, 216.

287. *What is the place of the heaven?*
1. Bacon [*Physics*, bk. 4], *Opera*, fasc. 13, 1935, 217–220.

288. *Whether, in some way, the heaven has a place "in which."*
1. Bacon [*Physics*, bk. 4], *Opera*, fasc. 13, 1935, 220–221.

289. *Whether the last sphere, namely the supreme [sphere], is in a place.*[238]
1. Bacon [*Physics*, bk. 4], *Opera*, fasc. 13, 1935, 216–217.[239]
2. Pseudo–Siger of Brabant [*Physics*, bk. 4, qu. 18], 1941, 168–170.[240]
3. John of Jandun [*Physics*, bk. 4, qu. 9], 1519, 64v, col. 1–67v, col. 2.
4. Ockham [*Physics*, qu. 79], *Opera philosophica*, 1984, 6:612–614.[241]
5. Buridan* [*Physics*, bk. 4, qu. 6], 1509, 72r, col. 2–72v, col. 2.
6. Albert of Saxony [*Physics*, bk. 4, qu. 7], 1518, 47v, col. 1–48r, col. 1.[242]
7. Marsilius of Inghen [*Physics*, bk. 4, qu. 7], 1518, 50r, col. 1–50v, col. 1.
8. Benedictus Hesse [*Physics*, bk. 4, qus. 2 and 16], 1984, 400–402 and 433–437.[243]
9. Javelli [*Physics*, bk. 4, qu. 14], 1568, 1:557, col. 2–559, col. 2.

234. Amicus expressed substantially the same question in a radically different way when he asked "Whether some being could exist that exists [i.e., is located] nowhere."
235. Like Amicus, Compton-Carleton asked "Whether a thing could exist and be nowhere."
236. Cornaeus asks "Whether every body is in some external place."
237. To ask whether the world is in a place is much the same as asking whether the last sphere is in a place (see qu. 289 in this catalog). Although questions 285 through 289 might have been conflated into one question, I keep them separate because Roger Bacon chose to list them as distinct questions.
238. A question tantamount to asking whether the world, or universe, is in a place (see qu. 285 in this catalog).
239. Bacon asks "Whether the heaven has a place *per se* or *per accidens*."
240. Pseudo-Siger distinguishes the place of the last sphere as either *per se* or *per accidens*. Thus in question 18, he asks "Whether the last [or outermost] sphere is in a place *per se*" and in question 19 "Whether the last [or outermost] sphere is in a place *per accidens*."
241. Ockham poses much the same question as did Bacon: "Whether the eighth sphere is in place *per se* or *per accidens*."
242. Albert considers the problem under the question "Whether every being is in a place" (Utrum omne ens sit in loco), as did Marsilius of Inghen and Benedictus Hesse (discussants 7 and 8). See also note 243.
243. In book 4, question 2, Benedictus Hesse considers "Whether every being is in a place" and in question 16 asks directly "Whether the last sphere is in a place."

10. Toletus [*Physics*, bk. 4, qu. 7], 1580, 121v, col. 1–122v, col. 2.
11. Conimbricenses [*Physics*, pt. 2, bk. 4, ch. 5, qu. 1], 1602, cols. 40–44.
12. Oña [*Physics*, bk. 4, qu. 1, art. 6], 1598, 221r, col. 2–223v, col. 2.
13. Case [*Physics*, bk. 4, ch. 5], 1599, 463–473.
14. Amicus [*Physics*, tract. 20, qu. 5, dubit. 2], 1626–1629, 2:698, col. 1–702, col. 1.
15. Oddus [*Physics*, bk. 4, disp. 14, art. 6], 1667, 628, col. 1–631, col. 1.

290. Whether the orbs of the planets have a place per se *or* per accidens.
1. Bacon [*Physics*, bk. 4], *Opera*, fasc. 13, 1935, 221–222.

291. Whether the lower spheres are in a place.
1. Pseudo-Siger of Brabant [*Physics*, bk. 4, qu. 17], 1941, 167–168.

292. Whether or not each particular element in the whole world assumes a proper place for itself.
1. Conimbricenses [*Physics*, pt. 2, bk. 4, ch. 5, qu. 3], 1602, cols. 44–51.

293. Whether the concavity of the lunar orb is the place of fire.
1. Bacon [*Physics*, bk. 4], *Opera*, fasc. 13, 1935, 212–213.[244]
2. Pseudo-Siger of Brabant* [*Physics*, bk. 4, qu. 14], 1941, 162–163.
3. John of Jandun [*Physics*, bk. 4, qu. 8], 1519, 63v, col. 1–64v, col. 1.
4. Albert of Saxony [*Physics*, bk. 4, qu. 6], 1518, 46v, col. 1–47r, col. 2.
5. Marsilius of Inghen [*Physics*, bk. 4, qu. 6], 1518, 49v, col. 1–50r, col. 1.
6. Major [*Physics*, bk. 4], 1526, 36v, cols. 1–2.[245]

294. We inquire what the place of air is, whether fire or the orb (of fire) is the place of air.
1. Bacon [*Physics*, bk. 4], *Opera*, fasc. 13, 1935, 213–214.

295. Whether the ultimate [surface, that is, the concave surface] of air is the place of water.
1. Bacon [*Physics*, bk. 4], *Opera*, fasc. 13, 1935, 214.
2. Pseudo-Siger of Brabant* [*Physics*, bk. 4, qu. 13], 1941, 162.

296. Whether the proper place of the earth is the concave surface of water.
1. Bacon [*Physics*, bk. 4], *Opera*, fasc. 13, 1935, 215–216.
2. Pseudo-Siger of Brabant [*Physics*, bk. 4, qu. 11], 1941, 159–160.
3. John of Jandun [*Physics*, bk. 4, qu. 7], 1519, 62v, col. 1–63v, col. 1.
4. Buridan [*Physics*, bk. 4, qu. 5], 1509, 70v, col. 1–72r, col. 2.
5. Albert of Saxony [*Physics*, bk. 4, qu. 5], 1518, 45v, col. 1–46r, col. 2.
6. Marsilius of Inghen [*Physics*, bk. 4, qu. 5], 1518, 48v, col. 2–49v, col. 1.
7. Benedictus Hesse* [*Physics*, bk. 4, qu. 15], 1984, 430–433.

297. Whether water and air are the natural places of earth.
1. Major [*Physics*, bk. 4], 1526, 36v, cols. 1–2.

298. Whether the place of water and earth is the same.
1. Pseudo-Siger of Brabant [*Physics*, bk. 4, qu. 12], 1941, 160–161.

299. Whether the place of the earth is the center [of the world] itself.
1. Pseudo-Siger of Brabant [*Physics*, bk. 4, qu. 10], 1941, 157–159.

244. Bacon's version is somewhat broader, since he asks "Whether the ultimate [containing surface] of the heaven or the ultimate [surface] of the Moon [that is, the concave surface of the lunar sphere] is the place of fire."
245. The questions in Major's *Physics* are unnumbered.

300. By what rationale is place said to attract the located thing?
1. John of Jandun [*De coelo*, bk. 4, qu. 2 (numbered as 19)], 1552, 2r, col. 1–2v, col. 2.[246]
2. Javelli [*Physics*, bk. 4, qu. 2], 1568, 1:549, col. 2–550, col. 2.[247]
3. Amicus* [*Physics*, tract. 20, qu. 1, dubit. 7], 1626–1629, 2:657, col. 2–658, col. 1.

301. Whether a place has the power to preserve the thing located [in it].
1. Pseudo-Siger of Brabant* [*Physics*, bk. 4, qu. 4], 1941, 150–151.
2. John of Jandun [*Physics*, bk. 4, qu. 2], 1519, 57v, col. 2–58r, col. 1.
3. Benedictus Hesse [*Physics*, bk. 4, qu. 4], 1984, 403–405.
4. Javelli [*Physics*, bk. 4, qu. 2], 1568, 1:549, col. 2–550, col. 2.[248]
5. Oña [*Physics*, bk. 4, qu. 1, art. 5], 1598, 217r [mistakenly printed 117], col. 2–221r, col. 2.
6. Amicus [*Physics*, tract. 20, qu. 1, dubit. 8], 1626–1629, 2:658, cols. 1–2.

302. Whether the difference between proper place and common place is rightly assumed.
1. Benedictus Hesse [*Physics*, bk. 4, qu. 5], 1984, 405–406.

303. Whether up and down are species, that is, whether they are the extremities of a longest line.
1. Javelli [*Physics*, bk. 4, qu. 3], 1568, 1:550, col. 2–551, col. 1.

XVII. Vacuum in the celestial and/or terrestrial regions

304. What is vacuum and whether it can be assigned [or exist].
1. Cornaeus [*Physics*, bk. 4, disp. 3 ("De loco et vacuo"), qu. 3, dub. 1], 1657, 378.
2. Serbellonus* [*Physics*, disp. 5, qu. 2, art. 1], 1657, 1:822, col. 1–825, col. 1.

305. Whether from its very nature vacuum signifies dimensions.
1. Pseudo-Siger of Brabant [*Physics*, bk. 4, qu. 22], 1941, 175–176.

306. Whether there can be knowledge about a vacuum.
1. Pseudo-Siger of Brabant [*Physics*, bk. 4, qu. 23], 1941, 176–177.

307. Whether it is necessary that a vacuum exist.
1. John of Jandun [*Physics*, bk. 4, qu. 10], 1519, 67v, col. 2–70r, col. 2.

308. Whether it is possible that a vacuum can exist naturally.
1. Bacon [*Physics*, bk. 4], *Opera*, fasc. 13, 1935, 230.
2. Pseudo-Siger of Brabant [*Physics*, bk. 4, qu. 24], 1941, 177–180.
3. Johannes Canonicus [*Physics*, bk. 4, qu. 4], 1520, 42v, col. 1–43r, col. 2.
4. Buridan [*Physics*, bk. 4, qu. 7], 1509, 72v, col. 2–73v, col. 2.

246. Jandun's version asks "Whether an inanimate body is also moved by the natural power existing in a place."
247. Javelli encapsulates two questions in one (this and the next) when he asks "Whether natural place has the power to draw to itself the located thing and to preserve and perfect it."
248. See note 247, on the preceding question.

5. Albert of Saxony [*Physics*, bk. 4, qu. 8], 1518, 48r, col. 2–48v, col. 2.
6. Marsilius of Inghen [*Physics*, bk. 4, qu. 13], 1518, 55r, col. 1–55v, col. 2.
7. Benedictus Hesse* [*Physics*, bk. 4, qu. 17], 1984, 437–442.
8. Toletus [*Physics*, bk. 4, qu. 10], 1580, 130v, col. 2–133r, col. 1.
9. Conimbricenses [*Physics*, pt. 2, bk. 4, ch. 9, qu. 1], 1602, cols. 89–96.
10. Oña [*Physics*, bk. 4, qu. 2, art. 1], 1598, 227r, col. 2–228r, col. 1.
11. Amicus [*Physics*, tract. 21, qu. 2, dubit. 1], 1626–1629, 2:740, col. 1–743, col. 1.
12. Cornaeus [*Physics*, bk. 4, disp. 3 ("De loco et vacuo"), qu. 3, dub. 2], 1657, 378.
13. Oddus [*Physics*, bk. 4, disp. 15, art. 1], 1667, 674, col. 2–679, col. 2.

309. Whether, if a vacuum existed, it would be a privation.
1. John of Jandun [*Physics*, bk. 4, qu. 14], 1519, 73r, col. 1–73v, col. 1.

310. By what are bodies moved to fill a vacuum and whether such motion is natural.
1. Amicus [*Physics*, tract. 21, qu. 3, dubit. 2], 1626–1629, 2:749, col. 1–752, col. 2.[249]
2. Oddus* [*Physics*, bk. 4, disp. 15, art. 2], 1667, 679, col. 2–683, col. 2.

311. Whether it is possible that a vacuum be made by a natural power.
1. Amicus [*Physics*, tract. 21, qu. 2, dubit. 2], 1626–1629, 2:743, col. 1–746, col. 1.

312. On two new experiments of the Vacuists.
1. Cornaeus [*Physics*, bk. 4, disp. 3 ("De loco et vacuo"), qu. 4], 1657, 383–384.

313. For what reason is the first experiment [the fall of mercury in a tube closed at the top] made?
1. Cornaeus [*Physics*, bk. 4, disp. 3 ("De loco et vacuo"), qu. 4, sec. 1, dub. 1], 1657, 384.

314. How the authors of this experiment prove that a vacuum exists in the upper part of the glass tube.
1. Cornaeus [*Physics*, bk. 4, disp. 3 ("De loco et vacuo"), qu. 4, sec. 1, dub. 3], 1657, 387–388.

315. Whether in this tube experiment there is a true vacuum.
1. Cornaeus [*Physics*, bk. 4, disp. 3 ("De loco et vacuo"), qu. 4, sec. 1, dub. 4], 1657, 388–389.

316. How to respond to the arguments of the adversaries [of the tube experiment].
1. Cornaeus [*Physics*, bk. 4, disp. 3 ("De loco et vacuo"), qu. 4, sec. 1, dub. 6], 1657, 392–394.

317. How is the [second] experiment [concerning the violent exhaustion of air] to be taken?
1. Cornaeus [*Physics*, bk. 4, disp. 3 ("De loco et vacuo"), qu. 4, sec. 2, dub. 1], 1657, 395–396.

318. How the Vacuists attempt to prove the vacuum from this experiment.
1. Cornaeus [*Physics*, bk. 4, disp. 3 ("De loco et vacuo"), qu. 4, sec. 2, dub. 3], 1657, 398–399.

249. Amicus asks only, "By what power are bodies moved to fill a vacuum?"

319. *Whether a vacuum is found in this [second] experiment [concerning the violent exhausion of air].*
   1. Cornaeus [*Physics*, bk. 4, disp. 3 ("De loco et vacuo"), qu. 4, sec. 2, dub. 4], 1657, 399–404.

320. *How the arguments of the Vacuists are to be resolved.*
   1. Cornaeus [*Physics*, bk. 4, disp. 3 ("De loco et vacuo"), qu. 4, sec. 2, dub. 7], 1657, 408–410.

321. *Whether it is possible that a vacuum [come to] exist by means of supernatural [that is, divine] power.*
   1. Buridan [*Physics*, bk. 4, qu. 8], 1509, 73v, col. 2–74r, col. 2.[250]
   2. Benedictus Hesse* [*Physics*, bk. 4, qu. 18], 1984, 442–446.
   3. Major [*Physics*, bk. 4], 1526, 37r, col. 1–38r, col. 1.[251]
   4. Conimbricenses [*Physics*, pt. 2, bk. 4, ch. 9, qu. 2], 1602, cols. 97–98 [mistakenly numbered 87–88].
   5. Amicus [*Physics*, tract. 21, qu. 2, dubit. 3], 1626–1629, 2:746, col. 1–749, col. 1.
   6. Cornaeus [*Physics*, bk. 4, disp. 3 ("De loco et vacuo"), qu. 3, dub. 3], 1657, 378.

322. *Whether angelic power could produce a vacuum in nature.*
   1. Conimbricenses [*Physics*, pt. 2, bk. 4, ch. 9, qu. 3], 1602, cols. 98–102 [mistakenly numbered 88–92].
   2. Cornaeus* [*Physics*, bk. 4, disp. 3 ("De loco et vacuo"), qu. 3, dub. 4], 1657, 378–379.

323. *Whether a vacuum can be assumed below the heavens.*
   1. Bacon [*Physics*, bk. 4], *Opera*, fasc. 13, 1935, 224–230.

324. *Whether, if a vacuum existed, up or down could be in it.*
   1. Benedictus Hesse [*Physics*, bk. 4, qu. 21], 1984, 447.

325. *Whether, if a vacuum existed, a heavy body could be moved in it.*
   1. Pseudo-Siger of Brabant [*Physics*, bk. 4, qu. 25], 1941, 180–183.[252]
   2. John of Jandun [*Physics*, bk. 4, qu. 13], 1519, 72v, col. 1–73r, col. 1.[253]
   3. Buridan* [*Physics*, bk. 4, qu. 10], 1509, 76v, col. 2–77v, col. 1.
   4. Albert of Saxony [*Physics*, bk. 4, qu. 11], 1518, 50r, col. 2–50v, col. 2.
   5. Marsilius of Inghen [*Physics*, bk. 4, qu. 12], 1518, 54r, col. 2–55r, col. 1.
   6. Benedictus Hesse [*Physics*, bk. 4, qu. 26], 1984, 457–461.
   7. Conimbricenses [*Physics*, pt. 2, bk. 4, ch. 9, qu. 4], 1602, cols. 102–104 [mistakenly numbered 92–94].
   8. Oña [*Physics*, bk. 4, qu. 2, art. 2], 1598, 228r, col. 1–229r, col. 2.

250. Although Buridan asks generally "Whether it is possible for a vacuum to exist by means of some power," his concern is specifically with God's supernatural power to create a vacuum.
251. By asking "Whether it is possible to assume a vacuum with regard to some power," Major phrases the question in much the same manner as did Buridan.
252. Pseudo-Siger omits mention of bodies and asks only "Whether a motion could occur in a vacuum."
253. Jandun replaces "body" with "animal" when he asks "Whether, if a vacuum existed, the local motion of an animal could occur."

9. Amicus [*Physics*, tract. 21, qu. 4, dubit. 1], 1626–1629, 2:752, col. 2–754, col. 2.[254]
10. Cornaeus [*Physics*, bk. 4, disp. 3 ("De loco et vacuo"), qu. 3, dub. 5], 1657, 379–380.[255]
11. Serbellonus [*Physics*, disp. 5, qu. 2, art. 2], 1657, 1:825, col. 1–829, col. 1.[256]
12. Oddus [*Physics*, bk. 4, disp. 15, art. 3], 1667, 684, col. 1–688, col. 2.[257]

326. Whether, on the assumption of a vacuum, a body would be moved in an instant or in time.

1. Dullaert [*Physics*, bk. 4, qu. 2,], 1506, sig. oiii,v–ovii,r.
2. Javelli [*Physics*, bk. 4, qu. 17], 1568, 1:561, col. 2–564, col. 1.
3. Major [*Physics*, bk. 4], 1526, 38r, col. 1–39r, col. 1.[258]
4. Toletus [*Physics*, bk. 4, qu. 9], 1580, 129r, col. 2–130r, col. 1.
5. Conimbricenses* [*Physics*, pt. 2, bk. 4, ch. 9, qu. 5], 1602, cols. 104 [mistakenly numbered 94]–108 [mistakenly numbered 98].
6. Oña [*Physics*, bk. 4, qu. 2, art. 3], 1598, 228r, col. 1–229r, col. 2.
7. Amicus [*Physics*, tract. 21, qu. 4, dubit. 2], 1626–1629, 2:754, col. 2–758, col. 2.[259]
8. Oddus [*Physics*, bk. 4, disp. 15, art. 3], 1667, 684, col. 1–688, col. 2.

327. Whether a violent motion could occur in a vacuum.

1. Bacon [*Physics*, bk. 4], *Opera*, fasc. 13, 1935, 234.

328. Whether, on the assumption of a vacuum in the concavity of the heaven [or sky], the sides of the heaven would come together.

1. Pseudo–Siger of Brabant [*Physics*, bk. 4, qu. 26], 1941, 183–184.

329. Whether, on the assumption of a vacuum in the concavity of the heaven [or sky], the heaven could be moved.

1. Pseudo–Siger of Brabant [*Physics*, bk. 4, qu. 27], 1941, 184–185.

330. Whether one should assume a vacuum diffused in bodies [that is, whether one should assume interstitial vacua].

1. Bacon [*Physics*, bk. 4], *Opera*, fasc. 13, 1935, 236–237.

PART IV. THE TERRESTRIAL [OR SUBLUNAR] REGION

XVIII. Between earth and moon: the elements fire, air, and water

331. On the number of [sublunar] elements.

1. Michael Scot* [*Sphere*, lec. 3], 1949, 265–270.

254. Like Pseudo-Siger, Amicus asks only "Whether, if a vacuum were assumed, motion could occur in it."
255. Cornaeus asks only "Whether a motion could occur in a vacuum, if one existed."
256. Like many others, Serbellonus asked only "Whether a motion could occur in a vacuum."
257. Oddus incorporates two separate questions in one: "Whether a body in a vacuum could be moved and whether such a motion would be instantaneous or successive." The first question is relevant here; the second corresponds to the next question (326) in this catalog.
258. Specifically, Major asks "Whether a simple mobile – that is, a heavy or light simple [body] – assumed in a vacuum will be moved in an instant or successively."
259. Like Major, Amicus asks "Whether a motion that might occur in a vacuum would be successive or instantaneous."

2. Conimbricenses [*De coelo*, bk. 3, ch. 5, qu. 1], 1598, 424–429.
3. Hurtado de Mendoza [*On Generation and Corruption*, disp. 6, sec. 1], 1615, 464, cols. 1–2.
4. Aversa [*De caelo*, qu. 36, sec. 1], 1627, 208, col. 1–210, col. 2.
5. Cornaeus [*De coelo*, tract. 4, disp. 3, qu. 1, dub. 2], 1657, 515.

332. *What causes the location of these elements?*
1. Amicus [*De caelo*, tract. 7, qu. 5, dubit. 2], 1626, 514, col. 1–515, col. 1.

333. *On the dignity of the elements in relation to one another.*
1. Conimbricenses* [*De coelo*, bk. 3, ch. 5, qu. 2], 1598, 429–434.
2. Amicus [*De caelo*, tract. 7, qu. 5, dubit. 3], 1626, 515, col. 1–517, col. 1.

334. *On the magnitude of each of the elements.*
1. Simon de Tunstede (Pseudo-Scotus) [*Meteorology*, bk. 1, qu. 13], 1639, 3:29, col. 2–34, col. 1.[260]
2. Themon Judaeus [*Meteorology*, bk. 1, qu. 5], 1518, 158v, col. 2–160r, col. 2.
3. Conimbricenses [*De coelo*, bk. 3, ch. 5, qu. 3], 1598, 435–439.[261]
4. Amicus [*De caelo*, tract. 7, qu. 6, dubit. 3], 1626, 522, col. 2–525, col. 1.[262]
5. Aversa* [*De caelo*, qu. 36, sec. 7], 1627, 235, col. 1–239, col. 2.

335. *On the shape of the elements.*
1. Michael Scot [*Sphere*, lec. 6], 1949, 293–294.[263]
2. Themon Judaeus [*De spera*, qu. 13], in Hugonnard-Roche, 1973, 130–132.[264]
3. D'Ailly, *14 Questions*, qu. 5, 1531, 153r–156r.[265]
4. Conimbricenses [*De coelo*, bk. 3, ch. 8, qu. 1], 1598, 450–456.
5. Hurtado de Mendoza [*De coelo*, disp. 3, sec. 5, subsec. 1], 1615, 383, col. 1–384, col. 1.[266]
6. Amicus [*De caelo*, tract. 7, qu. 6, dubit. 1], 1626, 517, col. 1–521, col. 2.
7. Aversa* [*De caelo*, qu. 36, sec. 8], 239, col. 2–243, col. 2.

336. *Whether the four elements determine their shapes naturally.*
1. Bacon, *De celestibus*, pt. 2, ch. 3, *Opera*, fasc. 4, 1913, 349–355.
2. Albert of Saxony* [*De celo*, bk. 3, qu. 13], 1518, 126v, col. 2–128r, col. 2.[267]
3. Oresme [*De celo*, bk. 3, qu. 2], 1965, 713–743.

260. Tunstede (and Themon Judaeus, discussant 2), asked "Whether the four elements are continuously proportional."
261. The Coimbra Jesuits asked "Whether any element exceeds in magnitude the next element immediately below it."
262. Amicus asks essentially the same question as did the Coimbra Jesuits: "Whether the element that contains another is greater than it."
263. In this question, Michael ignores the heavens and considers only the shape of the four elements. Compare d'Ailly's version in note 265.
264. Themon asks "Whether fire, water, and air are of a spherical shape."
265. D'Ailly asks "Whether the heaven and the four elements are spherical" (see n. 118).
266. The entire subsection is devoted to "The location, shape, heaviness, and lightness of the elements."
267. Although Albert divided his treatise into three parts or books (see this volume, beginning of Ch. 20; Ch. 1, n. 7) and the question cited here was numbered the thirteenth of the third book, he began a new ordinal numeration with question 7, which he begins with the term "firstly." He then begins the succeeding questions with "secondly," thirdly,"

337. *Whether one element is the natural place of another element.*
1. Johannes de Magistris [*Meteorology*, bk. 1, qu. 3], 1490, 10, col. 1–13, col. 1.[268]
2. Javelli* [*Physics*, bk. 4, qu. 15], 1568, 1:559, col. 2–560, col. 2.

338. *Whether one element that is located in another, say as air in fire, suffers from it and conversely.*
1. Javelli [*Physics*, bk. 4, qu. 16], 1568, 1:560, col. 2–561, col. 2.

339. *Whether the four prime qualities are substantial forms of the elements.*
1. Aureoli [*Sentences*, bk. 2, dist. 15, art. 1], 1596–1605, 2:201, col. 1–204, col. 2.
2. Hurtado de Mendoza [*On Generation and Corruption*, disp. 6, sec. 3], 1615, 466, col. 2–467, col. 2.
3. Cornaeus* [*De coelo*, tract. 4, disp. 3, qu. 1, dub. 3], 1657, 515.

340. *Heaviness and lightness cannot be substantial forms of elements.*
1. Paul of Venice, *Liber celi*, 1476, 41, col. 1 (conclus. 2).

341. *Whether the particular elements have a fixed and proper place in the universe.*
1. Hurtado de Mendoza [*De coelo*, disp. 3, sec. 5, subsec. 1], 1615, 383, col. 1–384, col. 2.
2. Amicus [*De caelo*, tract. 7, qu. 5, dubit. 1], 1626, 509, col. 2–514, col. 1.[269]
3. Cornaeus* [*De coelo*, tract. 4, disp. 3, qu. 2, dub. 1], 1657, 516.

342. *What is the place of meteors, and where do they occur?*
1. Cornaeus [*Tractatus Physicus* VI: *Meteorology*, disp. 1, qu. 1, dub. 6], 1657, 5–6.[270]

343. *What is the final cause of meteors?*
1. Cornaeus [*Tractatus Physicus* VI: *Meteorology*, disp. 1, qu. 1, dub. 7], 1657, 6–7.

344. *Whether the matter of all meteorological impressions is a vapor or an exhalation.*
1. Themon Judaeus [*Meteorology*, bk. 1, qu. 9], 1518, 162v, col. 2–163v, col. 1.

345. *What is a vapor and exhalation?*
1. Cornaeus [*Tractatus Physicus* VI: *Meteorology*, disp. 1, qu. 1, dub. 3], 1657, 4.

346. *Whether a vapor and exhalation are distinguished in species from the water and earth from which they come.*

and so on, until arriving at the final question of the treatise, which he describes as "lastly" (*ultimo*) but which is really "seventhly." Questions 7 through 13 of book 3 therefore may be appropriately considered as questions 1 through 7 of book 4, because there Albert treats of the motions of heavy and light bodies, whereas in the first six questions he confines himself to the heaviness and lightness of bodies. This division corresponds to books 3 and 4, respectively, of Aristotle's *De caelo*. For a full discussion, see Grant, 1991b, 211–214.

268. De Magistris's version of this question reads: "Whether any inferior element is naturally located in the concave surface of the superior element."
269. Amicus's version takes the form of a topical theme: "On the natural place of elements."
270. With the *Meteorology*, in *Tractatus Physicus VI*, the pagination of Cornaeus's book begins anew.

1. Cornaeus [*Tractatus Physicus* VI: *Meteorology*, disp. 1, qu. 1, dub. 4], 1657, 4–5.

347. Whether a comet is a terrestrial vapor.

1. Simon de Tunstede* (Pseudo-Scotus) [*Meteorology*, bk. 1, qu. 18], 1639, 3:42, col. 1–44, col. 1.
2. Cornaeus [*Tractatus Physicus* VI: *Meteorology*, disp. 2, qu. 1, dub. 13], 1657, 12.[271]

348. Whether a comet [stella comata] has a celestial nature.

1. Simon de Tunstede* (Pseudo-Scotus) [*Meteorology*, bk. 1, qu. 17], 1639, 3:40, col. 1–42, col. 1.
2. Themon Judaeus [*Meteorology*, bk. 1, qu. 12], 1518, 164v, col. 2–165v, col. 1.
3. Johannes de Magistris [*Meteorology*, bk. 1, qu. 5], 1490, 16, col. 2–20, col. 1.
4. Aversa [*De caelo*, qu. 33, sec. 4], 1627, 91, col. 1–100, col. 1.[272]
5. Cornaeus [*Tractatus Physicus* VI: *Meteorology*, disp. 2, qu. 1, dub. 14], 1657, 12–13.

349. Whether some comets are not sometimes of a sublunary matter.

1. Cornaeus [*Tractatus Physicus* VI: *Meteorology*, disp. 2, qu. 1, dub. 15], 1657, 13.

350. Whether a comet signifies the death of rulers.

1. Simon de Tunstede (Pseudo-Scotus) [*Meteorology*, bk. 1, qu. 19], 1639, 3:44, col. 1–46, col. 2.[273]
2. Themon Judaeus [*Meteorology*, bk. 1, qu. 13], 1518, 165v, col. 1–166r, col. 1.[274]
3. Major* [*Sentences*, bk. 2, dist. 14, qu. 11], 1519b, 83v, col. 1–84v, col. 1.

351. Whether war and plague can be foreknown from comets and meteors.

1. Cornaeus [*Tractatus Physicus* VI: *Meteorology*, disp. 1, qu. 1, dub. 8], 1657, 7.

352. Whether the Milky Way has a celestial nature.

1. Simon de Tunstede* (Pseudo-Scotus) [*Meteorology*, bk. 1, qu. 18], 1639, 3:46, col. 2–48, col. 1.
2. Themon Judaeus [*Meteorology*, bk. 1, qu. 14], 1518, 166r, col. 1–166v, col. 1.

353. Whether fire exists in the vicinity of the Moon.

1. Hurtado de Mendoza* [*De coelo*, disp. 3, sec. 4], 1615, 382, col. 1–383, col. 1.

271. Cornaeus phrases the question in its most general form: "What is a comet?"
272. Aversa asks "Whether comets are also generated and exist in the heaven."
273. In asking "Whether a comet signifies the death of princes," Simon uses *stella comata*, instead of *cometa*, for "comet" and "princes" (*principum*) for "rulers" (*regum*).
274. Themon uses both *cometa* and *stella comata* for "comet" when he asks "Whether a comet, or 'bearded star,' signifies the death of princes (*principum*), dryness, and winds, and other bad things."

2. Amicus [*De caelo*, tract. 8, qu. 1, dubit. 1], 1626, 552, col. 2–559, col. 1.
3. Aversa [*De caelo*, qu. 36, sec. 2], 1627, 210, col. 2–217, col. 1.[275]

354. *Whether fire is moved circularly in the concave orb of the moon.*
1. Simon de Tunstede (Pseudo-Scotus) [*Meteorology*, bk. 1, qu. 7], 1639, 3:14, col. 1–16, col. 1.

355. *Whether fire is hot and dry.*
1. Amicus [*De caelo*, tract. 8, qu. 1, dubit. 4], 1626, 561, col. 2.

356. *Whether the form constituting fire is light [*lux*].*
1. Amicus [*De caelo*, tract. 8, qu. 1, dubit. 5], 1626, 561, col. 2–562, col. 1.

357. *Whether the fire above [that is, in its natural place just below the Moon] is the same in species as the fire below [that is, in the regions below its natural place].*
1. Amicus [*De caelo*, tract. 8, qu. 1, dubit. 2], 1626, 559, col. 2–561, col. 1.

358. *Whether the element of air truly exists and in what manner it is distributed in several regions.*
1. Aversa [*De caelo*, qu. 36, sec. 3], 1627, 217, col. 1–218, col. 2.

359. *Whether air has a nature distinct from other bodies.*
1. Amicus [*De caelo*, tract. 8, qu. 2, dubit. 1], 1626, 565, col. 2–566, col. 2.

360. *On the circular motion of air.*
1. Amicus [*De caelo*, tract. 8, qu. 2, dubit. 2], 1626, 566, col. 2–563, col. 1.

361. *Whether air is naturally hot and humid.*
1. Simon de Tunstede (Pseudo-Scotus) [*Meteorology*, bk. 1, qu. 8], 1639, 3:16, col. 1–17, col. 2.

362. *Whether the middle region of air is always cold.*
1. Simon de Tunstede* (Pseudo-Scotus) [*Meteorology*, bk. 1, qu. 9], 1639, 3:18, col. 1–23, col. 1.
2. Themon Judaeus [*Meteorology*, bk. 1, qu. 7], 1518, 161r, col. 1–162r, col. 1.
3. Johannes de Magistris [*Meteorology*, bk. 1, qu. 4], 1490, 13, col. 1–16, col. 2.

363. *Whether the relatively heavy or light could rest anywhere naturally.*
1. Oresme [*De celo*, bk. 4, qu. 3], 1965, 823–843.

364. *Whether water is moved toward the center.*
1. Amicus [*De caelo*, tract. 8, qu. 3, dubit. 4], 1626, 573, col. 2.

365. *Whether anything heavy or light can rest naturally in the middle of another element of another species.*
1. Oresme [*De celo*, bk. 4, qu. 4], 1965, 843–865.

366. *Whether something absolutely light, such as fire, can be naturally at rest down.*
1. Oresme [*De spera*, qu. 6], 1966a, 114–131.[276]

275. Saying substantially the same thing, Aversa inquired "Whether directly under the heaven and [directly] above the air, the great element of fire exists."
276. By "down" Oresme means at the center of the world, which was usually taken to coincide with the center of the earth (or the earth's center of gravity).

367. By what principle are the elements moved?
1. Amicus [*De caelo*, tract. 7, qu. 7, dubit. 1], 1626, 525, col. 2–536, col. 1.

368. Whether the elements are moved with a rectilinear or circular motion.
1. Michael Scot [*Sphere*, lec. 6], 1949, 293–294.

369. Whether the number of four elements could be derived in terms of [or by arguing from] heaviness and lightness.
1. Buridan★ [*De caelo*, bk. 4, qu. 8], 1942, 269–272.
2. Albert of Saxony [*De celo*, bk. 3, qu. 6], 1518, 123r, col. 1–123v, col. 1.
3. Versor [*De celo*, bk. 4, qu. 5], 1493, 40v, col. 1–41r, col. 1.[277]

370. What causes the disparity of speed in the motions of sublunary bodies?
1. Conimbricenses [*De coelo*, bk. 2, ch. 6, qu. 1], 1598, 282–286.

*371. Whether every compound [*elementatum*] is moved in accordance with the nature of its predominant element.*
1. John of Jandun [*De coelo*, bk. 1, qu. 9], 1552, 7v, col. 2–8v, col. 1.
2. Major★ [*De celo*, bk. 1, qu. 2], 1526, 2, col. 2–4, col. 1.

372. Whether plenum and void are the causes of heaviness and lightness.
1. Versor [*De celo*, bk. 4, qu. 2], 1493, 38v, col. 1–39r, col. 1.

373. Whether the natural places of heavy and light bodies are the causes of their motions.
1. Buridan★ [*De caelo*, bk. 4, qu. 2], 1942, 248–250.
2. Versor [*De celo*, bk. 4, qu. 4], 1493, 40r, cols. 1–2.

374. Whether heavy and light bodies are actively moved by heaviness and lightness [respectively].
1. Buridan★ [*De caelo*, bk. 4, qu. 4], 1942, 254–256.
2. Albert of Saxony [*De celo*, bk. 4, qu. 3 (erroneously given as bk. 3, qu. 9)], 1518, 124v, col. 1–125r, col. 1.
3. Major [*De celo*, bk. 4, qu. 3], 1526, 27, cols. 1–2.
4. Amicus [*De caelo*, tract. 7, qu. 2, dubit. 2], 1626, 492, col. 2–498, col. 2.[278]

375. Whether elements gravitate and levitate into [their] proper places.
1. Amicus [*De caelo*, tract. 7, qu. 7, dubit. 3], 1626, 537, col. 2–543, col. 1.

376. Whether the shapes of absolutely heavy and light bodies are the causes of their upward and downward motions.
1. Versor [*De celo*, bk. 4, qu. 3], 1493, 39v, col. 1–40r, col. 1.

377. Whether part of any element that is assumed outside of its [natural] place could be moved to its [natural] place.
1. Richard of Middleton [*Sentences*, bk. 2, dist. 14, art. 3, qu. 4], 1591, 2:179, col. 2–181, col. 1.

277. Instead of heaviness and lightness, Versor speaks of "motive qualities." This appears to be the fifth question of the fourth book, but is listed as the third.
278. Amicus asks "Whether and how motive qualities arise from active [powers]." Since the motive qualities are heaviness and lightness, Amicus does discuss our question 374. Indeed, he includes seven questions, or *dubitationes*, on motive qualities, extending over pages 491 to 509.

378. *Whether or not elements have heaviness or lightness in their [natural] places.*
1. Buridan [*De caelo*, bk. 4, qu. 7], 1942, 264–269.[279]
2. Albert of Saxony [*De celo*, bk. 3, qu. 3], 1518, 121r, col. 2–122r, col. 1.
3. Oresme [*De celo*, bk. 4, qu. 1], 1965, 773–789.[280]
4. Major [*De celo*, bk. 4, qu. 1], 1526, 25, col. 2–26, col. 2.
5. Conimbricenses* [*De coelo*, bk. 4, ch. 6, qu. 2], 1598, 485–486.

379. *Wherever any heavy or light body is placed, it is equally heavy or light.*
1. Paul of Venice, *Liber celi*, 1476, 41, col. 1 (conclus. 3).

380. *Whether any body that is heavier than another in air is [also] heavier than the same [body] in water.*
1. Albert of Saxony* [*De celo*, bk. 3, qu. 2], 1518, 120v, col. 2–121r, col. 2.
2. Paul of Venice, *Liber celi*, 1476, 41, col. 1 (conclus. 4).[281]

381. *Whether there is any simple element not absolutely heavy or light.*
1. Buridan [*De caelo*, bk. 4, qu. 1], 1942, 244–247.
2. Albert of Saxony [*De celo*, bk. 3, qu. 1], 1518, 119v [mistakenly given as 117], col. 2–120v, col. 2.
3. Oresme* [*De celo*, bk. 4, qu. 2], 1965, 789–823.

382. *Whether there is any absolutely heavy body and absolutely light body; and [whether there are] other bodies that are relatively heavy and light.*
1. Versor* [*De celo*, bk. 4, qu. 1], 1493, 38r, col. 2–38v, col. 1.
2. Conimbricenses [*De coelo*, bk. 4, ch. 6, qu. 1], 1598, 482–485.[282]

XIX. The earth as a whole and as the center of the cosmos

383. *Whether the whole earth is spherical.*
1. Michael Scot [*Sphere*, lec. 6], 1949, 294–296.
2. Cecco d'Ascoli [*Sphere*, ch. 1], 1949, 367–368.
3. Buridan [*De caelo*, bk. 2, qu. 23], 1942, 233–235.
4. Albert of Saxony* [*De celo*, bk. 2, qu. 25], 1518, 117v, col. 2–118v, col. 1.
5. Oresme [*De spera*, qu. 5], 1966a, 100–113.
6. Themon Judaeus [*De spera*, qu. 12], in Hugonnard-Roche, 1973, 121–130.
7. Johannes de Magistris [*De celo*, bk. 2, qu. 5, dub. 2], 1490, 26, col. 2 (sig. k8r).
8. Versor [*De celo*, bk. 2, qu. 18], 1493, 32r, col. 2–32v, col. 2.

279. Buridan's version is restricted to air when he asks "Whether in its proper region [or place], air is heavy or light; or neither heavy nor light." Translation from Grant, 1974, 205.
280. Oresme asks "Whether any element may be heavy in its own proper place." I have added "proper" to Kren's translation (the text has "in proprio loco").
281. Paul concludes that "If some body is heavier or lighter than another in some element, the same body will be heavier or lighter in any other element."
282. In their version, the Coimbra Jesuits ask "Whether [of] two light elements [and of] two heavy elements, one is the lightest [body], the other the heaviest [body]."

9. Major [*De celo*, bk. 2, qu. 8], 1526, 21, col. 1–22, col. 2.
10. Cornaeus [*De coelo*, tract. 4, disp. 3, qu. 2, dub. 2], 1657, 516–518.[283]

384. Whether the earth is like a point in comparison to the heavens.
1. Michael Scot [*Sphere*, lec. 7], 1949, 298.[284]
2. Themon Judaeus [*De spera*, qu. 6], in Hugonnard-Roche, 1973, 93–94.
3. Conimbricenses* [*De coelo*, bk. 2, ch. 14, qu. 2], 1598, 378–381.
4. Amicus [*De caelo*, tract. 8, qu. 4, dubit. 12], 1626, 607, cols. 1–2.
5. Cornaeus [*De coelo*, tract. 4, disp. 3, qu. 2, dub. 7], 1657, 522–524.

385. Whether the mass of the whole earth is a quantity or magnitude that is much smaller than certain stars [or planets].
1. Themon Judaeus [*Meteorology*, bk. 1, qu. 2], 1518, 156r, col. 2–157r, col. 2.

386. How can the generation of mountains be reconciled with the spherical figure of the earth?
1. Johannes de Magistris* [*De celo*, bk. 2, qu. 5, dub. 3], 1490, 27, col. 1 (sig. k8v).
2. Amicus [*De caelo*, tract. 8, qu. 4, dubit. 8], 1626, 604, col. 2–605, col. 1.

387. Whether the earth is fixed in the middle of the world and has the same center of gravity and magnitude.
1. Albert of Saxony [*De celo*, bk. 2, qu. 23], 1518, 116v, col. 1–117r, col. 2.[285]
2. Paul of Venice, *Liber celi*, 1476, 37, col. 1–38, col. 2.[286]
3. Bricot [*Decelo*, bk. 2], 1486, 24r, col. 2.[287]
4. Major [*Sentences*, bk. 2, dist. 14, qu. 12], 1519b, 84v, col. 1–85v, col. 2.[288]
5. Conimbricenses* [*De coelo*, bk. 2, ch. 14, qu. 3], 1598, 381–384.
6. Amicus [*De caelo*, tract. 8, qu. 4, dubits. 1–2], 1626, 578, col. 1–585, col. 1.[289]

388. [The terraqueous sphere:] Whether water and earth make one globe.
1. Albert of Saxony [*Physics*, bk. 4, qu. 5], 1518, 46r, col. 2.[290]
2. D'Ailly, *14 Questions*, qu. 5, 1531, 153r–156r.
3. Clavius [*Sphere*, ch. 1], 1593, 133–151 and *Opera*, 1611, 3:57–66.
4. Amicus* [*De caelo*, tract. 7, qu. 6, dubit. 2], 1626, 521, col. 2–522, col. 1.

283. Cornaeus asks "Whether the earth is round [*rotunda*]."
284. Michael Scot says substantially the same thing in somewhat different words than the Conimbricenses, whose version I have used.
285. Albert asks "Whether the earth is in the middle [or center] of the world, located naturally like a point with respect to the heavens."
286. Although Paul does not take up the question in this explicit form, he does discuss the relations between the earth's center of gravity and its magnitude.
287. Bricot's version reads: "Whether it was Aristotle's intention that the center of the earth's gravity be the center of the world."
288. Major asks only "Whether the earth's center of gravity coincides with its center of magnitude."
289. Amicus takes two questions, or *dubitationes*, to ask the equivalent of question 387. In *dubitatio* 1, he is concerned about "the place and location of the earth," where he considers whether the earth is in the center of the world. The second *dubitatio* asks "whether the center of magnitude of the earth is the same as its center of gravity."
290. The question in which Albert discusses the terraqueous sphere is titled "Whether the earth is in water or in the concave surface of this water."

5. Aversa [*De caelo*, qu. 36, secs. 4–5], 1627, 218, col. 2–224, col. 1, and 224, col. 1–231, col. 1.[291]
6. Riccioli, *Almagestum novum*, pars prior, bk. 2, ch. 1, 1651, 47–49.
7. Cornaeus [*De coelo*, tract. 4, disp. 3, qu. 1, dub. 6], 1657, 519–522.[292]

389. Whether the earth always rests or is always moved in the middle [or center] of the heavens or world.

1. Michael Scot [*Sphere*, lec. 7], 1949, 298–299.[293]
2. Cecco d'Ascoli [*Sphere*, ch. 1], 1949, 358, 371–372.[294]
3. John of Jandun [*De coelo*, bk. 2, qu. 16], 1552, 31v, col. 2–32r, col. 2.
4. Buridan [*De caelo*, bk. 2, qu. 22], 1942, 226–233.
5. Albert of Saxony* [*De celo*, bk. 1, qu. 24], 1518, 117r, col. 2–117v, col. 2.
6. Oresme [*De celo*, bk. 2, qu. 13], 1965, 667–695, and [*De spera*, qu. 3], 1966a, 44–81.
7. Themon Judaeus [*Meteorology*, bk. 2, qu. 7], 1518, 174v, col. 1–175r, col. 1, and [*De spera*, qus. 4 and 9], in Hugonnard-Roche, 1973, 76–86 and 105–113.[295]
8. Johannes de Magistris [*De celo*, bk. 2, qu. 5], 1490, 25, col. 1–27, col. 1 (sigs. k7v–k8v), and [*Meteorology*, bk. 2, qu. 4], 37, col. 1–40, col. 1.
9. Versor [*De celo*, bk. 2, qu. 17], 1493, 31r, col. 2–32r, col. 2.
10. Bricot [*De celo*, bk. 2], 1486, 23v, col. 2–24r, col. 2.
11. Major [*De celo*, bk. 2, qu. 9], 1526, 22, col. 2–24, col. 1.
12. Conimbricenses [*De coelo*, bk. 2, ch. 14, qu. 5], 1598, 389–391.
13. Aversa [*De caelo*, qu. 36, sec. 6], 1627, 231, col. 1–234, col. 2.
14. Riccioli, *Almagestum novum*, pars prior, bk. 2, ch. 2, 1651, 49, col. 1–50, col. 1 (for the earth's centrality); ibid., ch. 3, 51, col. 1–52, col. 1; ibid., pars post., bk. 9, sec. 4, 408–479 (for numerous arguments for and against the earth's daily motion).
15. Cornaeus [*De coelo*, tract. 4, disp. 3, qu. 2, dub. 10], 1657, 529.[296]

390. On the cause of the earth's immobility.

1. Amicus [*De caelo*, tract. 8, qu. 4, dubit. 6], 1626, 597, col. 2–604, col. 1.

391. Since it is more perfect to cease from local motion than to be moved, why should the more imperfect earth rest?

1. Amicus [*De caelo*, tract. 8, qu. 4, dubit. 7], 1626, 604, cols. 1–2.

392. Whether the earth shakes from the motion of other things.

1. Hurtado de Mendoza* [*De coelo*, disp. 3, sec. 5, subsec. 2], 1615, 384, col. 2–386, col. 2.

291. Aversa's two questions ask respectively "Whether, and in what manner, earth and water simultaneously exist and form one globe" and "Whether, and in what manner, the compacted globe of earth and water is fixed in the middle of the world."
292. Cornaeus discusses the terraqueous sphere in a question titled "Whether earth and water have the same center of gravity and magnitude."
293. Michael only asks "Whether the earth rests naturally around the center."
294. Cecco restricts the question to "Whether the earth is in the middle of the heaven."
295. In the *Meteorology* Themon asks "Whether the earth's motion [*terremotus*] is possible"; in the *De spera*, he asks first (qu. 4) "Whether the earth rests naturally in the middle of the world" and later (qu. 9) "Whether it is more reasonable to assume that the earth is moved and the heaven rests than conversely."
296. Cornaeus's version asks only "Whether the earth rests in the center of the world."

2. Cornaeus [*Tractatus Physicus* VI: *Meteorology*, disp. 2, qu. 4, dub. 1], 1657, 29–30.[297]

393. Whether the earth is moved circularly [that is, around its axis].

1. Oresme★ [*De spera*, qu. 8], 1966a, 154–173.
2. Johannes de Magistris [*De celo*, bk. 2, qu. 5, dub. 1], 1490, 26, cols. 1–2 (sig. k8r).
3. Riccioli, *Almagestum novum*, pars prior, bk. 2, ch. 3, 1651, 51, col. 1–52, col. 1; ibid., pars post., bk. 9, sec. 4, 408–479 (for numerous arguments for and against the earth's daily motion).
4. Cornaeus [*De coelo*, tract. 4, disp. 3, qu. 2, dub. 11], 1657, 529.

394. Whether the earth is moved with an oscillating motion even to the extent of a flea's leap.

1. Cornaeus [*De coelo*, tract. 4, disp. 3, qu. 2, dub. 12], 1657, 529–530.

395. Whether the earth could be moved with a Copernican motion and what is a Copernican motion for the earth.

1. Cornaeus [*De coelo*, tract. 4, disp. 3, qu. 2, dub. 13], 1657, 530–532.

396. Whether the whole earth is habitable.

1. Michael Scot [*Sphere*, lec. 11], 1949, 262, 317–318.[298]
2. Cecco d'Ascoli [*Sphere*, ch. 1], 1949, 356–357.
3. Buridan★ [*De caelo*, bk. 2, qu. 7], 1942, 154–160.
4. Albert of Saxony [*De celo*, bk. 2, qu. 26], 1518, 118v, col. 1–119v [for 117], col. 2.
5. Themon Judaeus [*De spera*, qu. 22], in Hugonnard-Roche, 1973, 142–144.[299]
6. D'Ailly, *14 Questions*, qu. 11, 1531, 162r–163r.[300]
7. Paul of Venice, *Liber celi*, 1476, 39, col. 1–40, col. 1.
8. Major [*De celo*, bk. 2, qu. 10], 1526, 24, col. 1–25, col. 2.
9. Conimbricenses [*De coelo*, bk. 2, ch. 14, qu. 1], 1598, 370–378.[301]
10. Amicus [*De caelo*, tract. 8, qu. 4, dubit. 4], 1626, 588, col. 2–594, col. 2.

297. Since he is offering questions on Aristotle's *Meteorology*, Cornaeus's discussion of the earth's motions concerns earthquakes and volcanic activity rather than motions relevant to astronomy, such as, for example, axial rotation or rectilinear motions of the whole earth.
298. On page 262, Michael first asks only whether the region under the equator is habitable; later he asks about the entire earth.
299. In his version, Themon asks "Whether of the four quarters [of the earth] only the northern is habitable."
300. Folio 162r is mistakenly foliated as 170r. D'Ailly asks "Whether only one of the northern quarters [of the earth] is habitable" (utrum solum una quartarum septentrionalium sit habitabilis), by which he seems to mean whether only one of the four quarters of the earth is habitable. The question is based on Sacrobosco's division of the earth into four equal quarters by a circle following the equatorial circle of the earth which is intersected at right angles by another circle on the earth's surface running from east to west and also passing through the poles of the world. Sacrobosco declares that only one of these quarters is habitable (see Sacrobosco, *Sphere*, ch. 3, 1949, 110 [Latin] and 138–139 [English]). D'Ailly seems to inquire whether Sacrobosco is correct.
301. The Coimbra Jesuits enunciated a broader question than most when they inquired, "What is the earth's magnitude, what are the earth's divisions, and what are its inhabited parts?"

397. *Why does every part of the earth not produce all things?*
   1. Amicus [*De caelo*, tract. 8, qu. 4, dubit. 11], 1626, 607, col. 1.

398. *Whether or not the earth is more depressed than the sea.*
   1. Conimbricenses* [*De coelo*, bk. 2, ch. 14, qu. 4], 1598, 384–388.
   2. Amicus [*De caelo*, tract. 8, qu. 4, dubit. 3], 1626, 585, col. 1–588, col. 2.

399. *What produces the ebb and flow of the sea?*
   1. Bonae Spei [comment. 3, *De coelo*, disp. ult. (4), dub. 2], 1652, 16, cols. 1–2.

400. *Whether man could [properly] be called a microcosm [that is, a small world,* minor mundus*].*
   1. Michael Scot [*Sphere*, lec. 8], 1949, 308–309.

APPENDIX II

The anatomy of medieval cosmology – the significance of the "Catalog of Questions" in Appendix I

Before discussing the scope and significance of the list of questions in Appendix I, I shall describe the basis of its construction. This is best achieved by first listing the authors and works from which the questions were drawn. To the extent feasible, their names are presented here chronologically, in the order in which their relevant work(s) were composed or published. Where dates of composition or publication of medieval works are unknown, as is often the case, the birth and death dates of the authors determine the order. For printed works published between the late fifteenth and the seventeenth century, dates of publication are provided, in parentheses, immediately following each title. Although little biographical data are available for most authors, a source for such data is furnished, following the date for each author and within the same parentheses. (Among the sources, *DSB* represents the *Dictionary of Scientific Biography*). References to numbered questions in the "Catalog of Questions" in Appendix I are in the form "qu. *n*," where "qu." represents "question" and *n* is any number between 1 and 400.

I. Chronological list of authors and works on which the "Catalog of Questions" is based

Thirteenth century

1. Michael Scot (d. ca. 1235; *DSB*, 9:361–365)
   Commentary on the Sphere of Sacrobosco
2. Alexander of Hales (ca. 1170–1245; Gilson, 1955, 327–329)
   Summa theologica

3. William of Auvergne (ca. 1180–1249; *DSB*, 14:388–389)
   De universo
4. Roger Bacon (ca. 1219–ca. 1292; *DSB*, 1:377–385)
   De celestibus
   Questions on the Eight Books of the Physics
5. Saint Bonaventure (1221–1274; Gilson, 1955, 685, n. 7)
   Commentary on the Sentences
6. Thomas Aquinas (ca. 1224/25–1274; *DSB*, 1:196–200)
   Commentary on the Sentences
7. Pseudo–Siger of Brabant (fl. 2nd half 13th c.; see Ch. 4, n. 14)
   Questions on the Physics
8. Robertus Anglicus (fl. ca. 1271; Sarton, 1927–1948, vol. 2, pt. 2, pp. 993–994)
   Commentary on the Sphere of Sacrobosco
9. Richard of Middleton (fl. 2nd half 13th c.; Gilson, 1955, 695–696, n. 46)
   Commentary on the Sentences
10. Godfrey of Fontaines (d. 1306; Gilson, 1955, 739, n. 95)
   Quodlibets II, IV, and V

Fourteenth century

11. Peter Aureoli (d. 1322; Gilson, 1955, 476–480, 777, n. 91)
   Commentary on the Sentences
12. Hervaeus Natalis [Harvey of Nedellec] (ca. 1260–1323; Gilson, 1955, 747–748, n. 124)
   De materia celi
13. Cecco d'Ascoli (1269–1327; Sarton, 1927–1948, vol. 3, pt. 1, pp. 643–645)
   Commentary on the Sphere of Sacrobosco
14. Durandus de Sancto Porciano (1270/75–1334; Gilson, 1955, 473–476, 774, n. 80)
   Commentary on the Sentences
15. John of Jandun (ca. 1285–1328; Gilson, 1955, 522–524, 797, n. 62)
   Questions on De coelo
   Questions on the Physics
16. Johannes Canonicus (1st half 14th c.; Lohr, 1970, 183–184)
   Questions on the Physics
17. William of Ockham (ca. 1290–1349; *DSB*, 10:171–175)
   Questions on the Physics
18. Jean Buridan (ca. 1300-ca. 1358; *DSB*, 2:603–608)
   Questions on De caelo
   Questions on the Physics
   Questions on the Metaphysics
19. Simon de Tunstede (Pseudo-Scotus) (d. 1361; Sarton, 1927–1948, vol. 3, pt. 2, pp. 1568–1569)
   Questions on the Meteorology
20. Albert of Saxony (ca. 1316–1390; *DSB*, 1:93–95)
   Questions on De caelo
   Questions on the Physics

21. Nicole Oresme (ca. 1320–1382; *DSB*, 10:223–230)
    Questions on De celo
    Questions on the Sphere
22. Themon Judaeus (ca. 1330-d. after 1371; Hugonnard-Roche, 1973, 11–23; Sarton, 1927–1948, vol. 3, pt. 2, 1539–1540)
    Questions on the Meteorology
    Questions on the Sphere
23. Marsilius of Inghen (d. 1396; *DSB*, 9:136–138)
    Questions on the Physics[1]
24. Pierre d'Ailly (1350–1420; *DSB*, 1:84)
    Questions on the Sphere of Sacrobosco

Fifteenth century

25. Paul of Venice (1369/72–1429; *DSB*, 10:419–421)
    Questions on De celo (*Liber celi et mundi*)
26. Benedictus Hesse of Krakow (fl. 1st half 15th c.; Lohr, 1974, 126–128)
    Questions on the Physics
27. Johannes de Magistris (fl. 2nd half 15th c.; Lohr, 1971, 257–261)
    Questions on De celo
    Questions on the Meteorology
28. Johannes Versor (d. after 1482; Lohr, 1971, 290–299)
    Questions on De celo
29. Thomas Bricot (d. 1516; Lohr, 1973, 173–178)
    Questions on De celo (1486)

Sixteenth century

30. Johannes Dullaert (ca. 1470–1513; *DSB*, 4:237–238)
    Questions on the Physics (ed. 1506)
31. Chrysostomus Javelli (1470/72-ca. 1538; Lohr, 1988, 202–204)
    Questions on the Physics (completed 1533; *Opera*, 1568)
    Questions on the Metaphysics (*Opera*, 1568)
32. John Major (1467/68–1550; *DSB*, 9:32–33, where the birthdate is given as 1469)
    Commentary on the Sentences (1519)
    Questions on the Physics (1526)
    Questions on De celo (1526; *Physics* and *De celo* are in same publication)
33. Christopher Clavius (1537–1612; *DSB*, 3:311–312)

1. Despite the attribution of the printed edition of Lyon, 1518 (see Bibliography), to Marsilius, doubts have been raised about his authorship. (Clagett, 1959, 615, n., says cautiously that the text is "believed by some to be by Marsilius of Inghen"; Lohr, 1971, 333, includes it in the "doubtful" category of Marsilius's works.) Indeed, the same text was also falsely attributed to Duns Scotus in the Luke Wadding edition of Scotus's *Opera omnia* (Lyon, 1639), vol. 2. I am grateful to Dr. Hans Thijssen (Katholieke Universiteit, Nijmegen) for first alerting me to the improbability of Marsilius's authorship. For convenience, however, it will be assumed here that the text is by Marsilius and was composed in the fourteenth century.

Commentary on the Sphere of Sacrobosco (4th ed., 1593; *Opera* 1611, vol. 3; 1st ed., 1570)

34. Franciscus Toletus (1532–1596; Lohr, 1988, 458–461)
    Questions on the Physics (1580; 1st ed. 1574)
35. Galileo Galilei (1564–1642; *DSB*, 5:237–249)
    Questions on De caelo (ca. 1590)
36. Conimbricenses, or Coimbra Jesuits (Lohr, 1988, 98–99)
    Questions on De coelo (1598; 1st ed. 1592)[2]
    Questions on the Physics (1602; 1st ed. 1592)
37. Peter de Oña (1560–1626; Lohr, 1988, 294)
    Questions on the Physics (1598)
38. John Case (Johannes Casus) (ca. 1546–1600; Lohr, 1988, 85–86)
    Questions on the Physics (1599)

Seventeenth century

39. Pedro Hurtado de Mendoza (1578–1651; Lohr, 1988, 194–195)
    Questions on De coelo (1615)
    Questions on the Physics (1615)
    Questions on On Generation and Corruption (1615; all are in the same publication)
40. Bartholomew Amicus (1562–1649; Lohr, 1988, 13–14)
    Questions on De caelo (1626)
    Questions on the Physics (1626–1629)
41. Raphael Aversa (ca. 1589–1657; Lohr, 1988, 24–25)
    Questions on De caelo (1627 [vol.2])
42. Roderigo de Arriaga (1592–1667; Lohr, 1988, 21–22)
    Questions on De caelo (1632)
    Questions on the Physics (1632; both works in same publication)
43. Johannes Poncius (1599–1661; Lohr, 1988, 362)
    Commentary on the Sentences (1639)
    Questions on De coelo (1672)
44. Bartholomaeus Mastrius (1602–1673; Lohr, 1988, 249–250) and Bonaventura Bellutus (1600–1676; Lohr, 1988, 37–38)
    Questions on De coelo (1640)
45. Franciscus de Oviedo (1602–1651; Lohr, 1988, 295)
    Questions on De caelo (1640)
    Questions on the Physics (1640; both works in same publication)
46. Thomas Compton-Carleton (ca. 1591–1666; Lohr, 1988, 97–98)
    Questions on De coelo (1649)
    Questions on the Physics (1649; both works in same publication)
47. Giovanni Baptista Riccioli (1598–1671; *DSB*, 11:411–412)
    Almagestum novum (1651)

2. Although the Coimbra Jesuits published their Aristotelian commentaries under the collective authorship of the name "Conimbricenses" between 1592 and 1598, the *De caelo* commentary was actually the work of Emmanuel de Goes (1542–1597) and first appeared in 1592.

48. Franciscus Bonae Spei (1617–1677)[3]
    Questions on De coelo (1652)
    Questions on the Physics (both works in same publication)
49. Melchior Cornaeus (1598–1665; Sommervogel, 1890–1911, 2:cols. 1467–1472)
    Questions on De coelo (1657)
    Questions on the Physics (1657)
    Questions on the Meteorology (1657; all three treatises or sections in same publication)
50. Sigismundus Serbellonus (d. ca. 1660)[4]
    Questions on the Physics (vol. 1, 1657)
    Questions on De caelo (vol. 2, 1663)
51. George de Rhodes (1597–1661; Sommervogel, 1890–1911, 6:cols. 1721–1722)
    Questions on De coelo (1671)
52. Illuminatus Oddus (d. 1683)[5]
    Questions on the Physics (1667)
    Questions on De caelo (1672)

The 52 authors just listed furnished 400 questions for the "Catalog of Questions" from a total of 67 treatises, a number that expands to 76 if we count as independent works the separate sections of the various single publications devoted to questions on two or more of Aristotle's relevant works.[6] To determine whether we have an appropriately representative sampling of authors, works, and questions

3. According to the *Biographie nationale de Belgique*, 30 vols. (Brussels, 1866–1959), 3:500, Franciscus Bonae Spei, or Françoise de Bonne-Espérance, was actually named Crespin. He was born in Lille and died in Brussels. In 1635, he entered the Carmelite Order, and taught theology and philosophy at Louvain. He is the author of numerous works on theology and natural philosophy. I am grateful to Danny Burton for discovering this entry.
4. Of the man Sigismundus Serbellonus, virtually nothing is known other than what can be gleaned from the title pages and prefatory material of his two-volume treatise. There is an uninformative mention of him in Zedler, 1732–1750, 37:col. 340, where we read "Serbellonus (Siegmund) von ihm ist in Druck vorhanden Philosophia, II Tomi Mayland [i.e., Milan] 1657 in Fol." From the title page of Serbellonus's two-volume work *Philosophia Ticinensis*, we learn that he was in the order of the Regular Barnabite Clerics of the Congregation of Saint Paul ("ex clericis regularibus Barnabitis Congreg. S. Pauli") of Ticino (in Switzerland, near the Italian border); indeed, he was a professor of sacred theology in the college of that order in Ticino. In the preface to the second volume in 1663, written by the bishop of Ticino, we are informed of Serbellonus's death. Because the prefatory sections of the first volume were written by Serbellonus himself, we may conclude that he was probably alive when the first volume was printed in 1657 and that his death occurred sometime between the publication of the first and second volumes, perhaps around 1660. The work is unmistakably a *cursus philosophicus* (see n. 6 of this appendix) in the seventeenth-century mode.
5. Iluminatus Oddus was a Sicilian from Collesano who became a member of the Capucine order and, for a while, taught philosophy and scholastic theology at Messina. He also preached for his order. It is said that he was nearly deaf and could hear only if a tube was placed in his ear. See Zedler, 1732–1750, 25:col. 445. I am grateful to Professor Charles Lohr for sending me the reference to Oddus.
6. For example, Thomas Compton-Carleton, who, in his *Philosophia universa*, included questions in separate, abbreviated sections on Aristotle's *De caelo* and *Physics*; or Melchior Cornaeus, who, in his *Curriculum philosophiae peripateticae*, included separate sections on the *De caelo, Physics*, and *Meteorology*. A book that included questions or commentaries on a range of Aristotle's works was known as a *cursus philosophicus*. Such works were fairly common in the seventeenth century.

Figure 13. Saint Bonaventure. From a fresco painted by Fra Angelico around 1450 and now in the Vatican.

over the span of five centuries, it will be helpful to examine each of these categories.

II. Authors

From the chronological list just given, we may conclude that the spread of authors over the five centuries is reasonably balanced: 10 in the thirteenth; 14 in the fourteenth; 5 in the fifteenth; 9 in the sixteenth; and 14 in the seventeenth century. Although only 5 authors have been included from the fifteenth century, 4 are well-known scholastic names and typical representatives of scholastic thought. If we use the year 1500 as an arbitrary, but not inappropriate, point of division between the medieval and Renaissance periods, 29 of our authors fall into the former category and 23 into the latter.

As with other physical subjects, medieval opinions on cosmological matters were largely shaped by famous masters, most of whom were theologians trained in both natural philosophy and theology.[7] Their opinions were appropriated, disseminated, and perpetuated by hordes of lesser, often obscure, teachers. For this reason, the list includes many of the great names in scholastic natural philosophy. Thus for the

7. Three notable exceptions are John of Jandun, Jean Buridan, and Albert of Saxony, all of whom were masters of arts.

Figure 14. Vincent of Beauvais. "Miniature from Jean de Vignay's translation of Vincent de Beauvais' *Miroir Historial*, Vincent in his study (enlarged); Flemish, late 15th c. MS. Roy.14.E.I. (vol. I), fol. 3r. *Trustees of the British Museum*." (As described in Joan Evans, ed., *The Flowering of the Middle Ages* [London: Thames & Hudson, 1966], 192.)

thirteenth and fourteenth centuries, Roger Bacon, Thomas Aquinas, John of Jandun, William of Ockham, Jean Buridan, Nicole Oresme, Albert of Saxony, and Pierre d'Ailly are included. Indeed, this is also true for the sixteenth century, which is represented by John Major, Christopher Clavius, Galileo,[8] and the widely known

8. Although Galileo is not usually thought of as a scholastic author, he did write scholastic

Figure 15. Saint Thomas Aquinas. From a fifteenth-century painting by Justus of Ghent, which was based on an earlier copy.

Coimbra Jesuits (Conimbricenses), whose commentaries and questions on the works of Aristotle were popular in the seventeenth century. Not only were these authors well known in the Middle Ages and Renaissance, but modern scholars have also shown a continuing interest in their ideas.

By contrast, of the 14 authors representing the seventeenth century, only Giovanni Baptista Riccioli, who was an eminent astronomer and physicist, is well known to modern historians of science.[9] This is partly attributable to the fact that the treatise on which his fame rests, the *Almagestum novum*, is an astronomical work, not an Aristotelian commentary, although it includes numerous questions in the scholastic mode. The other 13 authors are virtually unknown today and were even in their

treatises in his earlier career. For that reason, and because of his fame, he is included in our list.

9. A substantial discussion of Riccioli's treatment of the problem of fall in the latter's *Almagestum novum* appears in Koyré, 1955, 349–354. For Koyré's earlier paper on Riccioli, see Koyré, 1953, 222–237.

Figure 16. Nicole Oresme in his study, with an armillary sphere in the foreground. Miniature from Oresme's French translation of Aristotle's *De caelo*. (Bibliothèque Nationale, MS. fr. 565, fol. 1r.)

own time known largely, if not solely, in scholastic circles. The disappearance of these authors from works of modern scholarship may be attributed to a lack of interest in Renaissance Aristotelianism, which scholars of the past two centuries have usually portrayed – when they have considered it at all – as a rigid, monolithic body of traditional medieval ideas that in the seventeenth century gained notoriety solely for its obstinate opposition to the emerging new science. Few did more to promote and foster these attitudes than Galileo, who, in the *Dialogue on the Two Chief World Systems*, his major attack on Aristotelian cosmology, made his character Simplicio the stereotype of the dull-witted, unimaginative, unyielding scholastic defender of the indefensible.[10]

10. For details, see Grant, 1985a, 418–419.

Following Galileo's assault and the triumph of Copernicanism over Aristotelian cosmology in the second half of the seventeenth century, scholastic cosmological treatises were little read; by the twentieth century, they were neither read nor studied. Historians of science and natural philosophy know little more about them than the few arguments that had been refuted by the victors and accidentally preserved.[11] In the course of this study, I have sought to show that scholastic Aristotelians ranged from steadfast defenders of the status quo to those who actually came to oppose important elements of the Aristotelian system, replacing them with new ideas and observations derived from their opponents. Aristotelian cosmology in the seventeenth century was far from monolithic.

Overall, the authors on whose works I have drawn to represent scholastic cosmology constitute a good sampling of opinions on relevant cosmological problems for the nearly five centuries during which the medieval cosmos functioned as the dominant world view in western Europe.

III. Works

In Chapter 2, I briefly described the kinds of literature that contained relevant cosmological questions. Contributions from most of them appear among the 400 questions in Appendix I. From the Aristotelian corpus, questions have been drawn from *questiones* on the *De caelo, Physics, Meteorology, Metaphysics*, and even two questions concerning *On Generation and Corruption*. Questions have also been extracted from commentaries on the *Sentences* of Peter Lombard and from two types of treatises with the basic title *Questions on the Sphere*. The first type follows the order and content of Sacrobosco's *On the Sphere* and is therefore called *Questions on the Sphere of Sacrobosco*. The second type has little in common with Sacrobosco's treatise other than a similarity of titles. Although works in this second type may include some of Sacrobosco's themes, they do not follow the order of his treatment and contain numerous questions that have no counterpart in Sacrobosco's treatise. Works in this group are referred to simply as *Questions on the Sphere*.[12] Other types of *questiones*, bearing titles such as *De materia celi* (*On Celestial Matter*),[13] *Quodlibet*,[14]

11. In a relevant and perceptive passage, James R. Moore declares (1979, 114): "In science as in war, history is written by the victors. Those who first embraced a new science are styled as precursors of the latest orthodoxy. Those who stubbornly clung to the old are featured as historical curiosities. One group is absorbed, the other is absurd." Seventeenth century scholastics did not even rise to the status of "historical curiosities." They paid the ultimate price: banishment from the pages of history and virtual oblivion. See Grant, 1984b, 65 and 1985a, 427.
12. For example, by Nicole Oresme and Themon Judaeus. Sacrobosco's treatise was not in the form of questions; only some of the commentaries on it were. For the difference between Oresme's *Sphere* and Sacrobosco's, see Droppers' remarks in Oresme [*De spera*], 1966a, 304–308.
13. The treatise is by Hervaeus Natalis.
14. From Godfrey of Fontaines. In truth, the *De materia celi* of Hervaeus Natalis is a quodlibetal question, one of a number, as is evident from the title of his work *Quolibeta Hervei* (see Hervaeus Natalis, 1513).

and *Summa theologica*,[15] are represented by one treatise each. Relevant questions have also been extracted from separate treatises that are not formally in the genre of *questiones* but which do contain questions. In this group are the treatises *De universo* by William of Auvergne and Riccioli's *Almagestum novum*.

Despite their importance, *questiones* are by no means the exclusive source for medieval cosmology. Straightforward commentaries on the works of Aristotle also play a significant role. Although commentators frequently remained content merely to explain the meaning of Aristotle's text, some integrated ideas from other sources and generally enriched the commentary tradition. The most influential was the corpus of Aristotelian commentaries by the Arab commentator Averroës (ibn Rushd), whose commentaries on all of the works relevant to cosmology (*De caelo, Physics, Metaphysics*, and *Meteorology*) were translated from Arabic to Latin.[16] Among commentaries by Latin authors, two of the most important were by Albertus Magnus and Thomas Aquinas.[17] On occasion, a commentary contained much that was original, as in the case of Nicole Oresme's French commentary on his own French translation of Aristotle's *De caelo* (see Oresme, *Le Livre du ciel*, 1968). Many issues that were taken up as particular questions in the *questiones* literature were also discussed in commentaries. Problems treated in commentaries are therefore well represented in the *questiones*.[18]

Of the 76 treatises examined,[19] 11 were written in the thirteenth century, 20 in the fourteenth, 6 in the fifteenth, 13 in the sixteenth, and 26 in the seventeenth. More important, however, is the range of representation and temporal distribution of the works. For the most important cosmological treatise, *De caelo*, 24 different authors contributed 24 different versions and a total of 221 questions. These treatises are distributed over the centuries as follows: 1 in the thirteenth;[20] 4 in the fourteenth; 4 in the fifteenth; 3 in the sixteenth; and 12 in the seventeenth century (for titles and authors, see the list in Section I of this appendix). Judging from Lohr's catalog, no treatise classifiable as a *questiones* on *De caelo* has survived from the thirteenth century.[21] Thus Roger Bacon's *De celestibus*, which is akin to a questions treatise on

15. By Alexander of Hales.
16. See Averroës, *Aristotelis opera*, 1562–1574, vols. 4, 5, and 8.
17. Both wrote important commentaries on *De caelo*, with Albertus providing an unusually rich mass of detail, drawn from a variety of sources. See Albertus Magnus [*De caelo*], *Opera*, 1971, vol. 5, pt. 1, and Thomas Aquinas [*De caelo*], 1952. In the fourteenth century, Walter Burley wrote an important commentary on the *Physics* (1501). Although Aquinas's commentaries bear the title *Expositio* rather than *Commentaria*, the difference between an exposition and a commentary was often difficult to determine. Where the commentator was content merely to explain the text, the work might be called an "exposition" (*expositio*); where he offered his own opinions, the text might be more appropriately described as a commentary. The distinction, however, is only marginally useful, because the terms *expositio* and *commentaria* were often used interchangeably.
18. The encyclopedias mentioned in Chapter 2, Section III.6, are also important sources of cosmological discussion. So too are occasional summary accounts, as for example, that by Johannes Bernardi Velcurio (Veltkirchius) (d. 1534), who provided a clear and concise description of the universe (see Velcurio [*Physics*], 1554).
19. Where questions on more than one work of Aristotle's are included in the same volume, each is counted as a separate treatise.
20. Roger Bacon's *De celestibus*.
21. A *Questiones super libros de celo et mundo* exists in a single manuscript (MS. 509 in Gonville

De caelo but is not so titled, stands as the sole representative for the thirteenth century. A number of the treatises listed under *De caelo* for the seventeenth century are not full-blown, independent *questiones* but briefer, selective treatments in which a smaller number of questions is discussed.[22]

Although 25 *Physics* treatises are included in our sample – one more than the total of *De caelo* treatises – the latter group contributed 221 questions to only 116 for the former, from which we may conclude that during the late Middle Ages and Renaissance questions on the *Physics* were second in importance to questions on *De caelo*. Of the 25 *Physics* treatises, 2 were in the thirteenth century, 6 in the fourteenth, 1 in the fifteenth, 7 in the sixteenth, and 9 in the seventeenth.

The temporal distribution of both of these crucial treatise types, *De caelo* and *Physics*, and the number and quality of the authors who composed them are more than adequate to guarantee inclusion of the kinds of cosmological questions that scholastic natural philosophers considered important enough to insert in their treatises over a period of some five centuries. Because of the strongly traditional nature of *questiones*, inspection of additional treatises is not likely to reveal many more significant questions beyond those already incorporated in the "Catalog of Questions." The number of works, their temporal distribution, and the quality of the authors provide assurance that the questions selected from *De caelo* and *Physics* treatises appropriately represent cosmology as it was typically depicted in those treatises.

Of the remaining types and genres, the most significant are the questions from commentaries on the *Sentences* of Peter Lombard and the questions on the *Sphere*, especially those on Sacrobosco's *Sphere*. The number of treatises representing each

and Caius College, Cambridge) attributed to Thomas de Bungeye (fl. 1275), a Franciscan friar (see Lohr, 1973, 178–179). Because very few questions are specified or identifiable in Bernard Parker's edition of the first book of this treatise (see Thomas de Bungeye [Parker], 1969), it does not seem to qualify as a genuine *questiones* treatise. In his description of the contents of the first book, Parker regularly characterizes the treatise as a commentary and so identifies it in the title of his edition. Roger Bacon's *De celestibus* is far more a *questiones* treatise than is Bungeye's work. Parker also mentions (15) another apparent thirteenth-century manuscript treatise (MS. 344, also in Gonville and Caius College, Cambridge) that bears the title *Questiones super librum De celo et mundo*. I have not examined this unedited work.

22. For example, Sigismundus Serbellonus subdivided his section, or single disputation ("Disputatio unica"), on *De caelo* into five major questions, which were then subdivided into twenty-two articles phrased in the form of questions; Thomas Compton-Carleton presented fifteen questions (each called a *sectio*) spread over four *disputationes* on *De caelo*. In his section on *De caelo*, Franciscus de Oviedo offered a *controversia unica*, which he subdivided into four basic questions, or *puncta*. Under a section titled "Disputations on Incorruptible, Corporeal Substance, or On the Heavens" ("Disputationes de substantia corporea incorruptibili, sive De caelo"), Pedro Hurtado de Mendoza included three *disputationes*. The first two consider the nature, essence, and properties of the celestial region and contain eleven questions, or *sectiones*. The third disputation, which contains seven questions, or *sectiones*, is on the Creation in six days and bears the heading "De coelorum et mundi productione sive de opere sex dierum." Because Hurtado chose to place his discussion on the Creation within the category of questions on *De caelo* (*On the Heavens*), I shall also subsume those seven questions under *De caelo*. What has been said here about *De caelo* in the seventeenth century applies also to the rest of Aristotle's works on natural philosophy.

type, however, is relatively small, 7 for each. The temporal distribution is also irregular. For the commentaries on the *Sentences*, 3 were composed in the thirteenth century, 2 in the fourteenth, 1 in the sixteenth, and 1 in the seventeenth. The fifteenth century is unrepresented. A similar situation obtains for questions on the *Sphere*. Two were composed in the thirteenth century, 4 in the fourteenth (of these, 2 are general commentaries, and 2 are commentaries on Sacrobosco's *Sphere*), and 1 in the sixteenth, leaving gaps in the fifteenth and seventeenth centuries. By the seventeenth century few, if any, questions on the *Sphere* were written.[23] But even with these temporal gaps, a sufficient number of both kinds of treatises are included and adequately spread over the five centuries of this study, so that we can be reasonably confident of a proper sampling of questions in the "Catalog of Questions." Although some of the questions in the remaining treatises are important, the works themselves played a much smaller role in medieval cosmology, as evidenced by the relatively modest number of questions which they contribute. Within this group are the *questiones* on the Aristotelian treatises *Meteorology*, the *On Generation and Corruption*, and the *Metaphysics*; the *Quodlibeta*, which also includes Hervaeus's *De materia celi*; and the independent works by William of Auvergne, Alexander of Hales, and Giovanni Baptista Riccioli that have been mentioned. Of these, Riccioli's *Almagestum novum* provides the largest number of questions, most of which are akin to questions on *De caelo*. Indeed, numerous questions from these authors are similar to questions drawn from some of the major categories of treatises. For example, many questions in William of Auvergne's *De universo* are appropriate to *De caelo*, while those in Alexander of Hales's *Summa theologica* differ little, if at all, from those in the second book of a typical commentary on the *Sentences*.

Enough has now been said about the works that were consulted. We must now see what themes the questions contained, for only then can we determine the foundations on which this study has been based.

IV. The questions

For convenience, the 400 questions listed in Appendix I have been organized in terms of nineteen topics that encompass most, if not all, of the chief themes of medieval cosmology. These topics fall into four major subdivisions which correspond to Parts I and II of this study. (For an explanation, see Ch. 1, Sec. II.1 and n. 7.) With the exception of a few questions on celestial corruptibility or incorruptibility (qus. 12–14), the creation of specific heavens and stars (qus. 51, 53), and the composition of the heavens (qu. 75), themes I through VI, comprising Part I, include 77 questions of which 71 (i.e., 77 − 6) are on the world, or universe, as a totality and address questions about its eternity, creation, the possible existence of things beyond it, including other worlds and void spaces, its perfection and finitude, as well as the kinds of bodies that fill its different parts.

The remaining topics (VII–XIX) concern the celestial region and/or the terrestrial

23. The same may be said of commentaries on the *Sentences* (see Ch. 2, n. 54).

region. The division of the world into celestial and terrestrial parts, separated by the concave surface of the lunar sphere, which represented the innermost, or lowest, part of the celestial region, and the convex surface of the sphere of fire, which represented the uppermost surface of the terrestrial zone, was undoubtedly the most significant physical distinction that Aristotle made. Indeed, he divided the *De caelo* itself along these lines, with its first two chapters devoted to the celestial region and the last two to the terrestrial. That Aristotle emphasized the celestial over the terrestrial region is evidenced by the fact that the former part is approximately two and a half times longer than the latter. This disparity was preserved, and even extended, by most scholastic commentators. For example, in their questions on *De caelo*, John of Jandun considered a total of 57 questions, 54 of which were confined to the first two books, thus generating a ratio of 18 to 1; Jean Buridan devoted 49 questions to the first two books and only 10 to the third and fourth, maintaining nearly a 5 to 1 ratio; and, finally, with a 43 to 13 split, Albert of Saxony represents a slightly better than 3 to 1 ratio. This disparity was preserved in the Renaissance when the Coimbra Jesuits, at the end of the sixteenth century, included 56 questions on the first two books and only 6 on the last two.

In the catalog, the large imbalance between celestial to terrestrial questions is not as great. Themes VII to XIV are exclusively devoted to the celestial region and include 157 questions (qus. 78–234, inclusive). The final themes, XVIII and XIX, are concerned with the terrestrial region and contain 70 questions (qus. 331–400, inclusive). The ratio of celestial to terrestrial questions is thus 2.3 to 1 and while it is less than that found in the *De caelo* questions of Buridan and Albert of Saxony,[24] it closely reflects Aristotle's emphasis of the celestial over the terrestrial. Although the 2.3 to 1 ratio is considerably less than that found in quite a number of questions on *De caelo*, it nonetheless preserves the medieval cosmological tradition, with its greater emphasis on the celestial region than on the terrestrial.

Although the 400 questions assembled here represent the scope and depth of this study, there is obviously no canonical number of questions that comprises the raw data for a proper understanding of medieval cosmology. The 400 questions represent the end result of a selection process that continually required decisions for inclusion or exclusion. It is now time to explain that process, as well as the overall organization of the questions.

1. Organization of questions

As already noted, the questions included in the catalog are drawn from many authors, whose works span eight distinct categories of scholastic literature, and from three

24. The smaller ratio of celestial to terrestrial questions in my list is perhaps attributable to the fact that most seventeenth-century scholastics who wrote questions on *De caelo* (with the notable exception of Amicus and perhaps also Aversa) included far fewer questions than their medieval predecessors. In numerous instances, their questions on *De caelo* formed but one part of a work in which questions were also posed about three or four other Aristotelian treatises. If an author included a total of 12 or 13 questions on *De caelo*, 3 or 4 might be allocated to the terrestrial region, leaving only about 8 or 9 for the celestial region. The ratio of celestial to terrestrial questions thus decreased dramatically.

treatises that lie outside those categories but play a role in this study. Similar and related questions are grouped together in the catalog under nineteen convenient headings. Of 400 questions, 186 were discussed by more than 1 author. For each such question, I cite the authors who discussed it, using the roughly chronological order in which those names appear in the list of authors presented in Section I of this appendix. For example, 22 authors treated question 97 ("On the number of spheres, whether there are eight or nine, or more or less") and gave twenty-three responses. (Poncius took up the question in two different types of treatises). The authors range from Thomas Aquinas in the 1250s to Illuminatus Oddus in 1672, a period of over 400 years. Furthermore, the same question was discussed in four kinds of treatises, namely commentaries on the *Sentences*, tractates on the *Sphere*, questions on *De caelo*, and Riccioli's *Almagestum novum*.

Works are cited in essentially the same short form used in the notes. The full bibliographical information for an author's work appears only in the Bibliography. For convenience of reference, the questions have been numbered consecutively from the first to the last topic, rather than sequentially within each topic.

2. *On similarities and differences among questions*

Among the discussants of a given question, the name of the author whose question has been selected to represent that question in the catalog is indicated by an asterisk, since the forms in which authors framed their versions of the question often differ.[25] Thus in question 97 the translation of the model question was made from Albert of Saxony's version, and in question 78 from Jean Buridan's. These differences are sometimes trivial, as in question 173 where Durandus de Sancto Porciano asks "Whether celestial motion is natural," while Michael Scot asks "Whether celestial motion is natural or violent" and John Case inquires "Whether the motion by which God moves the first celestial sphere is natural"; or in question 159, where Benedictus Hesse asks "Whether the prime mover is absolutely immobile," while Pseudo-Siger expresses the same question more fully in the form "Whether in movers and in things that are moved it is necessary to arrive at an immobile prime mover," and John of Jandun asks "Whether there is an absolutely immobile mover," omitting the word "prime."[26]

On numerous occasions, however, the differences are more substantial, as in question 106: "Whether there is an empyrean [or immobile] sphere beyond the mobile heavens." In their versions of this question, Jean Buridan, Themon Judaeus, and Albert of Saxony make no mention of the empyrean sphere, although all implied it. Buridan asks "Whether beyond the heavens that are moved, there should be assumed a heaven that is resting or unmoved"; Themon asks "Whether something

25. Because it would be superfluous, no asterisk appears in any of the 214 questions which have only one discussant.

26. Another good illustration appears in question 276. For the variety of forms, see notes 223–226 to that question.

should be assumed [to exist] beyond the ninth sphere";[27] while Albert inquires "Whether every heaven is mobile, or whether we must assume some heaven that is at rest." Because only the empyrean heaven was regularly thought to be at rest, and all 6 authors who treated the question were concerned with the possible existence of an immobile heaven, it seems reasonable to assume that despite the absence of the term "empyrean," they were all treating the same substantive question.[28] The reader may therefore assume that all authors listed under a particular question have devoted all or part of that question to the same problem, although they may have expressed the question itself in a variety of ways.[29]

By contrast, some questions seem at first glance similar in content but differ sufficiently to warrant separate entries. This is especially true of questions dealing with the first theme, concerning the eternity of the world and the corruptibility or incorruptibility of the celestial region. Although the first four questions appear substantially the same, they have been accorded separate status because of somewhat different emphases.

It is apparent that in compiling this lengthy list of questions, two kinds of judgments had constantly to be made: whether two differently worded questions were substantially the same, and therefore to be recorded under the same question; or whether they were sufficiently different to require separate entries. This was a difficult task, and some errors of judgment may have been made.

3. Criteria for inclusion and exclusion of questions

Before deciding similarities and differences among questions, it was essential to determine whether they were appropriate at all for a list of questions on medieval cosmology. Here, of course, my own concept of medieval cosmology is crucial. In the broadest sense, one might include everything that happens in the cosmos. The absurdity of such an approach hardly needs elaboration. To present a comprehensive and manageable account in a single volume, restrictions on the the scope of the study are essential. Although it is extremely difficult, if not virtually impossible, to define the precise limits of medieval cosmology so that any given question could be readily included or excluded on the basis of that definition, I have been guided by Albert of Saxony's threefold division (see the beginning of Chapter 20), taking first the world as a whole and then the structure and operation of its two radically different but fundamental parts, the nobler celestial region and the less noble sublunar, or elemental, region. In all of the kinds of treatises from which cosmological

27. In the course of his discussion, Themon concludes that "an immobile tenth sphere should be assumed beyond the ninth sphere" (pono istam conclusionem probabilem, scilicet quod ultra nonam speram est ponenda decima spera immobilis). See Hugonnard-Roche, 1973, 99–103.
28. As masters of arts without theological degrees or credentials, Buridan, Themon, and Albert may have deliberately omitted the term "empyrean" to avoid possible theological entanglements.
29. In notes, I have often provided translations of alternative versions of the representative question.

questions have been drawn, some, and often many, questions from each were deliberately excluded as irrelevant.

For example, from a total of 59 questions in Buridan's *Questions on De caelo*, 17, or approximately 28 percent, were omitted from the catalog. Among those excluded were such questions as "Whether in the same body the dimensions length, width, and depth are mutually distinct" (bk. 1, qu. 2);[30] "Whether a power [or force] ought to be defined by its maximum capability" (bk. 1, qu. 21); "We inquire whether natural motion ought to be quicker at the end than at the beginning" (bk. 2, qu. 12);[31] "Whether after departure from a projector, a stone that is projected, or an arrow that is shot from a bow, and so on in similar cases, is moved by an internal principle or an external principle" (bk. 3, qu. 2);[32] and, finally, "Whether heaviness and lightness are substantial forms of heavy and light bodies" (bk. 4, qu. 5).[33]

The largest number of questions deliberately omitted from Buridan's *Questions on De caelo* concerns problems of natural and violent motion, as in book 2, questions 12 and 13, and book 3, question 2, three questions in which Buridan discusses his famous impetus theory of projectile motion (even though impetus plays a significant role in Chapter 18, Section II of this book). Also eliminated were questions 3 through 6 in book 4, which are concerned with the motion of heavy and light bodies. While all of these questions are appropriate for physics, they are not of direct relevance for understanding cosmology.[34] Other questions were largely of a logico-mathematical character, with no apparent connection to cosmology (bk. 1, qus. 21–22).[35] Questions about indivisibles (such as bk. 3, qu. 1) seem equally remote from cosmology, as do questions on hypothetically infinite bodies (bk. 1, qus. 13–14). Because no one believed in the existence of an infinite body, the questions and problems are mainly of a paradoxical kind that often intrigued medieval scholastics, but they shed no light whatever on the structure or operation of the medieval cosmos.

Virtually every author of a *questiones* on a relevant treatise by Aristotle included different but similarly inappropriate questions that I have ignored. John of Jandun asked "Whether a potency or power is determined by excellence" and "Whether the power of a finite body ought to be finite."[36] Nicole Oresme asked "Whether the substantial forms of the elements are intended and remitted, or whether they

30. Translations of the titles of Buridan's questions are drawn from Grant, 1974, 203–205.
31. A similar question is posed by Nicole Oresme [*De celo*, bk. 2, qu. 7], 1965, 525–575.
32. I have slightly altered the translation in Grant, 1974, 205. Albert of Saxony asked substantially the same question ([*De celo*, bk. 3, qu. 12], 1518, 126r, col. 2–126v, col. 2) in the form "Whether the air is moved naturally in the projection of a stone or whether [the stone is moved] by another thing that is projected into it."
33. Albert of Saxony asks the same question ([*De celo*, bk. 3, qu. 4], 1518, 122r, col. 1–122v, col. 1).
34. Not all questions on sublunar local motion were ignored. In theme XVIII ("Between Earth and Moon: The Elements Fire, Air, and Water") of the catalog, questions 366–368, 370–371, and 373–377 are concerned with motion but, in my judgment, seem more relevant to cosmology because they concern the nature of the four elements and their role in motion and are thus conducive to a better understanding of the overall workings of the sublunar region.
35. For a sense of the content of these questions, see Grant, 1974, 360–361, n. 2.
36. [*De caelo*, bk. 1, qu. 31], 1552, 20r, col. 2–20v, col. 1, and bk. 2, qu. 15, 31v, cols. 1–2.

admit of more and less"; Albert of Saxony inquired "Whether, if there were several infinites, one infinite could be greater than another, or whether one would be comparable to another"; "Whether every corruptible thing is necessarily corrupted"; and "Whether heavy things seek to descend according to the shortest lines [or paths]."[37] Finally, Johannes de Magistris asked "Whether an infinite body could be affected by a finite body"; "Whether an infinite circular body could have a center"; and "Whether the place of a whole and [its] part are the same."[38]

Although some broad questions on the four elements have been included, many were deemed too narrow for inclusion. In this category are numerous questions posed by Bartholomew Amicus in his seventeenth-century questions and commentary on *De caelo*. I have ignored such questions as "Why air that is blown when the mouth is compressed appears cold but appears warm when the mouth is open"; "Whether air is seen when illuminated"; "Why air, water, glass, and similar things are transparent, unlike metals, stones, and wood"; "Whether air has the power to nourish"; "Since the place of water is to be above the earth, why is it found within caverns of the earth?"; "Since water naturally rests on [or above] the earth as in its proper place, why do rivers flow?"; "Why water spread over the earth appears dark, but spread over oil appears white"; and "Why rain water is the most excellent of all."[39]

Certain categories of cosmological questions have also been omitted. Although a number of questions that treat of the eternity of the world in its different manifestations has been included, those that compare eternity with time or the ages have been ignored.[40] Indeed, time has generally been ignored, especially as discussed in the fourth book of Aristotle's *Physics*. Purely astronomical questions, some of which appear in questions on the *Sphere*, have also been omitted.[41] Except for questions about the celestial light of stars and planets or about the creation of light as described in Genesis, most questions about the general nature of light in the perspectivist or

37. For the enunciations of the questions (cited here in the order just given), see Oresme [*De celo*, bk. 3, qu. 3], 1965, 743–771, and Albert of Saxony, *De celo*, bk. 1, qu. 8, 1518, 92r, col. 1–93r, col. 2; qu. 14, 99r, col. 2–100r, col. 1; and bk. 3, qu. 11, 125v, col. 2–126r, col. 2. In the seventeenth century, Amicus posed the same question ([*De caelo*, tract. 7, qu. 7, dubit. 2], 1626, 536, col. 1–537, col. 2).
38. For the enunciations of these three questions (cited here in the order just given), see Johannes de Magistris [*De celo*, bk. 1, qu. 3, dub. 1], 1490, 8, col. 2–9, col. 1; ibid., dub. 2, 9, col. 1; and ibid., dub. 3, 9, col. 1.
39. For the references to these eight questions in the order in which they are given, see Amicus [*De caelo*, tract. 8, qu. 2, dubit. 9], 1626, 568, cols. 1–2; ibid., dubit. 11, 568, col. 2; dubit. 16, 570, col. 1; dubit. 20, 571, cols. 1–2; ibid., qu. 3, dubit. 1, 571, col. 1–573, col. 1; dubit. 2, 573, col. 1; dubit. 7, 574, cols. 1–2; and dubit. 25, 577, cols. 1–2.
40. For example, "Whether the eternity of the world is the same or is distinguished from time," a question that appears in John of Jandun, *De coelo*, bk. 1, qu. 27, 1552, 18v, cols. 1–2.
41. Of the questions in Pierre d'Ailly's *14 Questions*, only 5 were judged relevant to this study. Many of the rest are astronomical in character; for example, the following: "Whether the latitude of a region could be investigated by means of the elevation of the pole above the horizon" (qu. 8, 159r–159v); "Whether the distances of the poles of the zodiac from the poles of the world are equal to the greatest declination of the Sun, namely, in the sense that the distance of the zodiacal pole from the arctic pole is equal to the greatest northern declination of the Sun" (qu. 9, 159v–160r); and "Whether natural days are mutually unequal" (qu. 11, 161v–162r).

optical tradition, or about spiritual light per se, have been disregarded:[42] for example, questions like "Whether light [*lumen*] is a substantial or accidental form";[43] "Whether light is properly found in spiritual things";[44] "Whether light is an accident";[45] and "Whether light is a body, or the form of a body."[46]

It seems apparent that any attempt to achieve total coverage would inevitably result in a considerable increase in the number of questions, perhaps by as many as one or two hundred, or more. Such an uncritical increase would have overwhelmed this study with a host of minor details without substantially adding to our understanding of the larger cosmological picture.

In addition to questions that were discussed by scholastic authors but which I have deliberately ignored, I must mention at least one potential cosmological subject on which the scholastics might have asked questions but did not: the size and dimensions of the world. Cosmic dimensions – that is, the diameters of the planets, as well as their volumes and distances from the earth as measured in earth radii – reached the Latin West from Arabic sources, especially in two different translations of al-Farghani's summary account of Ptolemy's *Almagest.*[47] Knowledge of the cosmic dimensions soon became commonplace, appearing in astronomical treatises, general philosophical literature (for example, in Roger Bacon's *Opus maius* and in the Pseudo-Grosseteste *Summa philosophiae*), and in popular literature.[48] Albert Van Helden concludes that

> the cosmic dimensions that originated with Ptolemy can . . . be found on all levels of sophistication in medieval literature. This should not surprise us, for as an accepted part of the quadrivium, the regular mathematical subjects in the undergraduate curriculum, this scheme of sizes and distances was learned by virtually all university students. We may assume that it also frequently found its way, in full or in part, into the experience of the lay public.[49]

And yet in all the 400 questions in the catalog, not one asks about the size and dimensions of the universe. How did it happen that such a popular cosmological

42. For the kind of questions included, see the "Catalog of Questions," theme XIV ("Celestial Light").
43. Bonaventure [*Sentences*, bk. 2, dist. 13, art. 3, qu. 2], *Opera*, 1885, 2:327–329. Thomas of Strasbourg [*Sentences*, bk. 2, dist. 13, qu. 1], 1564, 155r, col. 2–156v, col. 1, asked "Whether light [*lumen*] is a real form." Although this general question about light has been excluded, a more specific version relevant to stars has been included (qu. 229: "Whether or not the light [*lux*] of the stars is a substantial form or even a body"). During the Middle Ages, a distinction was sometimes made between *lux* and *lumen*, which are both translated as light. *Lux* was "the brightness that one observes in fire or the sun," whereas "*lumen* might be thought of as the effect of *lux* on the adjacent medium and surrounding objects" (see Lindberg, 1978b, 356). Thus Bonaventure not only asked about *lumen* in the questions cited, but he earlier asked precisely the same question about *lux* (Bonaventure, ibid., art. 2, qu. 2, 2:319–322).
44. Thomas Aquinas [*Sentences*, bk. 2, dist. 13, qu. 1, art. 2], 1929–1947, 2:328–330.
45. Ibid., art. 3, 2:331–337. In this question, Thomas discusses the difference between *lux* and *lumen*.
46. Bonaventure, *Sentences*, bk. 2, dist. 13, art. 2, qu. 1, *Opera*, 1885, 2:317–318.
47. I rely here on Van Helden, 1985, 29, 33.
48. Ibid., 29, 33–40.
49. Ibid., 40.

theme was ignored in the literature which forms the basis of this study?[50]

At least two explanations seem plausible. Apart from mere recitation of the distances and sizes of planets and spheres, any question about cosmic dimensions would have required technical knowledge of astronomy, which few scholastic natural philosophers possessed. Such a question would more than likely have been perceived as astronomical rather than cosmological. Moreover, Aristotle did not discuss the size of the universe (although he discussed the size of the earth) but only whether it was finite or infinite. Once he decided on a finite cosmos, he did not inquire further about its precise size. Thus natural philosophers would have had little occasion to pose any particular question about the size and dimension of the universe and the distances and sizes of its planets and spheres.

Despite the absence of questions about cosmic dimensions in the *questiones* literature, a few scholastics found occasion to introduce comparative dimensions into their cosmological discussions. For this reason, and because of the importance of dimensions in a study of cosmology, I have devoted Section III in Chapter 17 to this subject.

4. On the adequacy of the sample of authors, works, and questions

I have by now discoursed enough about questions and themes that have deliberately been ignored. Let us return to the question raised earlier: Do the authors, works, and questions included in the catalog represent a proper sampling of nearly five centuries of medieval cosmology? Or would it be more desirable, and perhaps even necessary, to include more, perhaps many more, authors and questions? Indeed, should some or many of the omitted questions have been included?

The 52 authors and 400 questions represent a rather large sample of authors and questions. To have significantly increased the number of authors, and therefore the number of questions, on the basis of the principles of selection and exclusion enunciated earlier, would have added only to the quantity of the sample, not to its quality. By including at least 5 authors from each century covered in this study, I have almost certainly arrived at a representative sampling of the questions that were raised during this period.

Although I have sought to convey a reasonable sense of the methodology that guided my selection of authors, works, and questions, it would be rash to claim that my choices were definitive or beyond reproach. No two scholars are likely to follow the same path. The results of my research, however, led me to the realization that *questiones* on *De caelo* formed the principal type of cosmological treatise produced by scholastic natural philosophers. It therefore served as a useful model and led me to lay much greater stress on the celestial region than the terrestrial. In turning to the other relevant treatises, I sought, as much as was feasible, questions that emphasized the celestial region and shed light on its structure and operation.

50. Robertus Anglicus, who is included in our study, devotes a single paragraph of his commentary on the *Sphere* of Sacrobosco to a mere recital of the magnitudes of the planets and the distances of their spheres from the earth's center. He offers neither a question nor any discussion of the subject. See Robertus Anglicus [*Sphere*, lec. 14], 1949, 195 (Latin), 243 (English), and Van Helden, 1985, 34.

While one may challenge the choice of authors and questions, no one, I believe, will doubt that the themes and details embodied in the 400 questions are far more than can be accommodated in a study of the present length. Even if space were not a factor, the questions assembled here are far more than required. Many, if not the majority, of those in the catalog have played little if any role in my discussion. Of the relevant questions, those that were most frequently treated by the authors played a larger role in my discussion than those that were treated by only 1 or 2 authors.[51] Indeed, questions that were considered by 5 or more of the 52 authors formed the solid core of this study. The catalog made selection of the most frequently considered questions easy to determine and provided a rough measure – perhaps the only measure – for identifying questions that the authors themselves thought important.

5. *What the "Catalog of Questions" reveals*

Before turning to the most frequently discussed questions, however, let us first see what other information can be gleaned from the catalog and the two tables that follow at the end of Appendix II.

The 400 questions elicited a total of 1,176 responses,[52] thus averaging slightly under 3 responses per question, but ranging from 1 discussant per question (in 214 questions) to a high of 22 discussants for one question (qu. 97). At the end of this appendix, Table A.1 provides useful information about the contributions of individual authors, who are organized in the same chronological order as in the list presented in Section I of this appendix. The numbers of the questions discussed by each author are cited for each work, where there is more than one work. The total number of questions discussed in each work is then supplied. By arranging these totals in descending order, we can identify the major and minor contributors among all authors. In the list that follows, the total number of questions from each author is given first. If that total was drawn from more than one treatise, the number of questions from each treatise is given within parentheses. For example, Amicus, the first author, contributed 90 questions from his *De caelo* and 11 from his *Physics*, for a total of 101. Here is what Table A.1 reveals:

1. Amicus	101 (90 + 11)
2. Cornaeus	72 (39 + 24 + 9)
3. Conimbricenses	69 (55 + 14)
4. Buridan	55 (36 + 11 + 8)
5. Albert of Saxony	54 (44 + 10)
6. Aversa	54

51. With at least one exception, namely theme XI, "On Directionality in the Heavens." Although the 3 questions that fall within this theme (qus. 122–124) had a total of 23 discussants, with qu. 122 receiving the most attention (13 discussants), these questions, and the theme as a whole, received relatively little emphasis in this study.

52. This total includes three responses in which authors treated the same question in two different treatises. Thus, Buridan responded to question 73 in both his *De caelo* and *Physics*; Javelli considered question 155 in his *Physics* and *Metaphysics*; and Oresme treated question 389 in his *De celo* and *De spera*.

7. John of Jandun	48 (26 + 22)
8. Oddus	43 (11 + 32)
9. Major	40 (14 + 6 + 20)
10. Oresme	38 (30 + 8)
11. Versor	38
12. Bacon	37 (16 + 21)
13. Hurtado de Mendoza	30 (23 + 5 + 2)
14. Johannes de Magistris	29 (24 + 5)
15. Michael Scot	28
16. Themon Judaeus	23 (12 + 11)
17. Javelli	24 (16 + 8)
18. Benedictus Hesse	23
19. Riccioli	23
20. Serbellonus	23 (17 + 6)
21. Pseudo–Siger of Brabant	20
22. Paul of Venice	19
23. Bricot	19
24. Compton-Carleton	19 (13 + 6)
25. Poncius	18 (2 + 16)
26. Richard of Middleton	18
27. Bonaventure	15
28. Bonae Spei	15 (13 + 2)
29. Marsilius of Inghen	14
30. William of Auvergne	13
31. Petrus de Oña	13
32. Simon de Tunstede	12
33. Toletus	12
34. Oviedo	12 (6 + 6)
35. Alexander of Hales	11
36. Cecco d'Ascoli	9
37. Pierre d'Ailly	9
38. Galileo	9
39. Thomas Aquinas	8
40. William of Ockham	7
41. Durandus de Sancto Porciano	6
42. Case	6
43. Aureoli	5
44. Hervaeus Natalis	5
45. Arriaga	5 (3 + 2)
46. Godfrey of Fontaines	3
47. Johannes Canonicus	3
48. Mastrius and Bellutus	3
49. George de Rhodes	3
50. Robertus Anglicus	2
51. Dullaert	2
52. Clavius	1

Of the 52 authors, it is apparent that 3 Jesuit authors from the late sixteenth and the seventeenth century made by far the largest contributions. Amicus leads with 101 questions drawn from two treatises, followed by Cornaeus with 72 drawn from three treatises and the Conimbricenses, or Coimbra Jesuits, with 69 drawn from two treatises. The most significant medieval authors lagged considerably behind their Jesuit successors: Jean Buridan contributed 55 questions from three treatises; Albert of Saxony responded to 54 questions from two treatises; and Nicole Oresme to 37 questions from two treatises. The remaining authors are ranged in a gradual descent all the way to 1 response by Clavius.

Of the eleven genres of works from which questions were extracted, two – *De caelo* and *Physics* – play a predominant role. Questions on *De caelo* included 221 of the 400 questions, or approximately 55 percent, while questions on the *Physics* contained 116 questions, or about 29 percent of the total (see Table A.2, at the end of this appendix). The remaining nine types lag far behind: 45 questions, or approximately 11 percent, appear in *Sentences* commentaries; 41 questions, or 10 percent of the total, are found in treatises on the *Sphere*; 25 questions, or 6 percent, turn up in questions on the *Meteorology*; 23, or approximately 6 percent, in Riccioli's *Almagestum novum;* 15, or nearly 4 percent, appear in questions on the *Metaphysics*; 13, or around 3 percent, in William of Auvergne's *De universo*; 11, or nearly 3 percent, in Alexander of Hales's *Summa theologica*; 8, or 2 percent, in the treatises with quodlibetal questions; and, finally, 2 questions, or one-half of 1 percent, in the sole treatise representing questions on *On Generation and Corruption*. The numbers of questions in the different categories sum to 547, thus signifying that numerous questions – 84, to be precise – were discussed in two or more genres. Indeed, one question – number 128 – was treated in 6 of the 11 genres distinguished here.

Of the 1,176 responses to the 400 questions, 638, or approximately 54 percent, were made in treatises on *De caelo* and 292 in treatises on the *Physics*, or almost 25 percent, with the other nine treatises contributing the remaining 246 responses.[53] The dominance of *De caelo* treatises in medieval cosmology is obvious.

Perhaps the most useful information derivable from the catalog concerns the frequency with which questions were discussed. At the very least, such data provide a measure for the popularity, and perhaps importance, of a question as compared to other questions. If we arbitrarily assume that a question is popular if it had at least 5 discussants, then 327 of the 400 questions, which were discussed by only 1 to 4 discussants, found moderate to minimal appeal among natural philosophers. We are therefore left with 73 questions that attracted anywhere from 5 to 22 discussants. These 73 questions form the significant core of the various themes of medieval cosmology. By attracting the largest number of discussants, 22, question 97 ("On the number of spheres, whether there are eight or nine, or more or less") was perhaps the most popular question in medieval cosmology, if popularity may

53. In descending order, the responses are distributed as follows: *Sentences*, 69; *Sphere*, 66; *Meteorology*, 38; *Almagestum novum* (Riccioli), 23; *Metaphysics*, 16; *De universo* (William of Auvergne), 13; Alexander of Hales (*Summa theologica*), 11; quodlibetal questions, 8; and *On Generation and Corruption*, 2.

be gauged by the number of discussants. In the following list, all questions that were discussed by 5 or more discussants are cited, along with their respective themes as set forth in the "Catalog of Questions."

Theme I (qus. 1–20)	Qus. 1, 2, 4, 9, 12
Theme II (qus. 21–34)	None
Theme III (qus. 35–53)	None[54]
Theme IV A (qus. 54–61)	Qus. 54, 55, 58
Theme IV B (qus. 62–67)	Qus. 62, 66
Theme V (qus. 68–73)	Qus. 68, 69, 73
Theme VI (qus. 74–77)	Qu. 74
Theme VII (qus. 78–89)	Qus. 78, 79, 85
Theme VIII (qus. 90–95)	Qu. 91
Theme IX (qus. 96–105)	Qu. 97
Theme X (qus. 106–121)	Qus. 106, 117
Theme XI (qus. 122–124)	Qus. 122, 123, 124
Theme XII (qus. 125–151)	Qus. 126, 127, 128, 130, 132, 135, 141
Theme XIII (qus. 152–225)	Qus. 155, 159, 169, 173, 177, 180, 183, 184, 195, 211, 216, 220, 222
Theme XIV (qus. 226–234)	Qu. 231
Theme XV (qus. 235–273)	Qus. 237, 242, 244, 256
Theme XVI (qus. 274–303)	Qus. 275, 276, 277, 280, 283, 289, 293, 296, 301
Theme XVII (qus. 304–330)	Qus. 308, 321, 325, 326
Theme XVIII (qus. 331–382)	Qus. 331, 334, 335, 348, 378
Theme XIX (qus. 383–400)	Qus. 383, 384, 387, 388, 389, 396

Because they clearly indicate the specific questions thought most worthy of treatment by scholastic authors, the 73 questions furnish an excellent overview of medieval cosmology. To illustrate this, we need only refer to the most discussed questions in themes XII ("On the Properties of Celestial Spheres and Bodies") and XIII ("On Celestial Motions and Their Causes"), two of the most important cosmological topics distinguished in this study. A glance at the 7 questions cited in the list just given for theme XII and the 13 for theme XIII will reveal how fundamental were most of those questions.

The catalog has thus provided a means of identifying the most significant questions to which a study of medieval cosmology ought to be addressed. Unfortunately, space considerations prevented the inclusion of all of them. For example, no occasion was found to treat the question concerning the production of sounds by the motions of celestial bodies (qu. 222), and only a cursory mention has been accorded to the questions about possible absolute directions in the world (qus. 122–124).

Emphasis on the 73 most frequently considered cosmological questions should not be taken to imply that other questions have been ignored. Within a particular

54. Since themes II and III lack questions that have 5 or more discussants, I have relied on questions that were less widely discussed.

theme, not only are many questions with only 1 or 2 discussants interrelated and mutally reinforcing with respect to the more popular questions, but numerous questions within a single theme, however popular, contain varying degrees of overlapping discussion. Many single-discussant questions (and there are 214 of them) provide useful additional information relevant to the topics treated in the more popular questions. Nor should we ignore the fact that some, and perhaps many, questions with fewer than 5 discussants include points of view and insights that are lacking in the more popular questions. The quality of an author's thought also plays a role. Single questions authored by scholastics of the stature of Jean Buridan and Nicole Oresme must always be carefully studied for special or additional insights into more popular but related questions. For these and other reasons, my attention has not been confined to the 73 most popular questions.

Although the catalog, with its authors and works, is obviously of great importance to this study, it is only a sample, though a reasonably large one. Nevertheless, just as many questions from the sample had to be ignored, so also is it necessary to stress that information and arguments from questions and commentaries excluded from the catalog were included where it seemed appropriate.

6. *Within the compass of a single treatise, were some questions judged more important than others?*

If, by a statistical count, I have been able to determine which questions appear to have been most popular over the centuries, examination of single treatises provides little insight into any particular author's attitude toward any given question. Scholastic authors give little or no indication of the significance of questions. They simply take up one question after another in the same dispassionate manner, without providing any clue that this or that question might have greater relative or absolute importance than some other question or group of questions. Although Albert of Saxony was one of 22 authors who considered question 97, on the number of spheres, we have no reason to believe that he thought it more important than question 141, which was concerned with the cause of lunar spots and of which he was one of 6 discussants; or more important than question 225 of which he was one of 3 discussants. In the same manner, Jean Buridan does not specifically identify any of the 59 questions which he included in his *Questions on De caelo* as being more important than any of the other 58.

It is thus only with hindsight and by a statistical count that we can derive the popularity of a question from the frequency with which it was discussed. Superficially, we may infer the popularity of a question in proportion to the number of its discussants. But we may not also infer that a statistically popular question was considered important by a particular author who included it in his treatise. Nor is the length of a question a measure of its significance. Despite the temptation, we should resist the urge to infer that an author who devotes two pages to a question must have thought it more important in some sense than one to which he devoted only a single page.

Table A.1 *Chronological list of authors and the questions they contributed to the "Catalog of Questions"*

Thirteenth century

1. Michael Scot (d. ca. 1235)
   Commentary on the Sphere of Sacrobosco
   1, 62, 92, 104, 126, 127, 128, 132, 147, 173, 180, 188, 196, 203, 212, 223, 242, 245, 248, 266, 331, 335, 368, 383, 384, 389, 396, 400.
   Total = 28

2. Alexander of Hales (ca. 1170–1245)
   Summa theologica
   106, 107, 109, 110, 111, 112, 114, 115, 116, 117, 228.
   Total = 11

3. William of Auvergne (ca. 1180–1249)
   De universo
   34, 40, 43, 62, 69, 100, 108, 109, 111, 113, 134, 151, 226.
   Total = 13

4. Roger Bacon (ca. 1219–ca. 1292)
   De celestibus
   54, 62, 70, 76, 92, 94, 95, 97, 126, 127, 132, 184, 211, 220, 222, 336.
   Total = 16

   Questions on the Eight Books of the Physics
   1, 4, 122, 161, 183, 209, 275, 285, 286, 287, 288, 289, 290, 293, 294, 295, 296, 308, 323, 327, 330.
   Total = 21

5. Saint Bonaventure (1221–1274)
   Commentary on the Sentences
   1, 48, 50, 69, 85, 117, 122, 126, 132, 145, 191, 195, 211, 214, 247.
   Total = 15

6. Thomas Aquinas (ca. 1224/25–1274)
   Commentary on the Sentences
   4, 69, 85, 97, 112, 195, 237, 270.
   Total = 8

7. Pseudo-Siger of Brabant (fl. 2nd half 13th c.)
   Questions on the Physics
   4, 159, 203, 209, 271, 276, 289, 291, 293, 295, 296, 298, 299, 301, 305, 306, 308, 325, 328, 329.
   Total = 20

8. Robertus Anglicus (fl. ca. 1271)
   Commentary on the Sphere of Sacrobosco
   104, 132.
   Total = 2

9. Richard of Middleton (fl. 2nd half 13th c.)
   Commentary on the Sentences

49, 52, 69, 117, 118, 120, 121, 122, 128, 132, 136, 193, 195, 211, 230, 249, 256, 377.
Total = 18

10. Godfrey of Fontaines (d. 1306)
Quodlibets II, IV, and V
1, 65, 129.
Total = 3

Fourteenth century

11. Peter Aureoli (d. 1322)
Commentary on the Sentences
47, 79, 128, 195, 339.
Total = 5

12. Hervaeus Natalis [Harvey of Nedellec] (ca. 1260–1323)
De materia celi
255, 256, 257, 258, 269.
Total = 5

13. Cecco d'Ascoli (1269–1327)
Commentary on the Sphere of Sacrobosco
126, 132, 170, 196, 201, 240, 383, 389, 396.
Total = 9

14. Durandus de Sancto Porciano (1270/75–1334)
Commentary on the Sentences
117, 126, 173, 211, 237, 242.
Total = 6

15. John of Jandun (ca. 1285–1328)
Questions on De coelo
12, 25, 54, 60, 62, 68, 73, 79, 91, 122, 128, 130, 169, 175, 177, 180, 211, 216, 224; 236, 260, 261, 264, 300, 371, 389.
Total = 26

Questions on the Physics
6, 16, 149, 154, 155, 159, 165, 166, 176, 192, 204, 271, 276, 277, 283, 289, 293, 296, 301, 307, 309, 325.
Total = 22

16. Johannes Canonicus (1st half 14th c.)
Questions on the Physics
2, 276, 308.
Total = 3

17. William of Ockham (ca. 1290–1349)
Questions on the Physics
171, 172, 208, 210, 275, 277, 289.
Total = 7

18. Jean Buridan (ca. 1300–ca. 1358)
 Questions on De caelo
 12, 54, 62, 66, 68, 71, 72, 73, 78, 91, 106, 122, 123, 124, 130, 141, 168, 169, 177, 180, 189, 211, 216, 218, 219, 234, 244, 261, 369, 373, 374, 378, 381, 383, 389, 396.
 Total = 36

 Questions on the Physics
 16, 73, 155, 159, 163, 276, 277, 289, 296, 308, 321, 325.
 Total = 12

 Questions on the Metaphysics
 89, 102, 156, 157, 192, 200, 202, 220.
 Total = 8

19. Simon de Tunstede (Pseudo-Scotus) (d. 1361)
 Questions on the Meteorology
 132, 140, 235, 265, 334, 347, 348, 350, 352, 354, 361, 362.
 Total = 12

20. Albert of Saxony (ca. 1316–1390)
 Questions on De celo
 11, 17, 62, 66, 68, 70, 71, 73, 74, 76, 77, 91, 97, 106, 122, 123, 124, 126, 132, 141, 169, 180, 181, 184, 189, 211, 216, 219, 220, 222, 225, 231, 234, 244, 336, 369, 374, 378, 380, 381, 383, 387, 389, 396.
 Total = 44

 Questions on the Physics
 159, 276, 277, 283, 289, 293, 296, 308, 325, 388.
 Total = 10

21. Nicole Oresme (ca. 1320–1382)
 Questions on De celo
 12, 17, 54, 62, 66, 68, 73, 78, 91, 122, 123, 141, 168, 169, 177, 180, 181, 190, 211, 216, 217, 222, 225, 231, 336, 363, 365, 378, 381, 389.
 Total = 30

 Questions on the Sphere
 104, 184, 207, 215, 366, 383, 389, 393.
 Total = 8

22. Themon Judaeus (ca. 1330–d. after 1371)
 Questions on the Meteorology
 140, 235, 262, 263, 334, 344, 348, 350, 352, 362, 385, 389.
 Total = 12

 Questions on the Sphere
 54, 97, 106, 126, 184, 220, 335, 383, 384, 389, 396.
 Total = 11

23. Marsilius of Inghen (d. 1396)
 Questions on the Physics
 1, 4, 17, 73, 159, 165, 275, 276, 283, 289, 293, 296, 308, 325.
 Total = 14

24. Pierre d'Ailly (1350–1420)
    Questions on the Sphere of Sacrobosco
    97, 106, 126, 180, 184, 220, 335, 388, 396.
    Total = 9

Fifteenth century

25. Paul of Venice (1369/72–1429)
    Questions on De celo (Liber celi et mundi)
    11, 62, 70, 122, 132, 169, 178, 181, 216, 220, 221, 225, 231, 244, 340, 379, 380, 387, 396.
    Total = 19

26. Benedictus Hesse of Krakow (fl. 1st half 15th c.)
    Questions on the Physics
    128, 155, 158, 159, 160, 163, 164, 165, 167, 176, 271, 275, 276, 277, 283, 289, 296, 301, 302, 308, 321, 324, 325.
    Total = 23

27. Johannes de Magistris (fl. 2nd half 15th c.)
    Questions on De celo
    18, 54, 61, 62, 73, 74, 79, 91, 122, 123, 126, 127, 128, 168, 173, 177, 180, 216, 220, 244, 383, 386, 389, 393.
    Total = 24

    Questions on the Meteorology
    235, 337, 348, 362, 389.
    Total = 5

28. Johannes Versor (d. after 1482)
    Questions on De celo
    11, 12, 18, 19, 20, 62, 66, 68, 71, 72, 73, 74, 78, 122, 123, 126, 127, 130, 140, 168, 177, 180, 185, 186, 187, 211, 216, 220, 222, 231, 244, 369, 372, 373, 376, 382, 383, 389.
    Total = 38

29. Thomas Bricot (d. 1516)
    Questions on De celo
    17, 62, 66, 68, 78, 92, 97, 124, 169, 177, 180, 184, 211, 216, 220, 222, 231, 387, 389.
    Total = 19

Sixteenth century

30. Johannes Dullaert (ca. 1470–1513)
    Questions on the Physics
    1, 326.
    Total = 2

31. Chrysostomus Javelli (1470/72–ca. 1538)
    Questions on the Physics
    6, 23, 27, 154, 155, 165, 166, 176, 205, 289, 300, 301, 303, 326, 337, 338.
    Total = 16

Questions on the Metaphysics
128, 155, 162, 197, 198, 199, 220, 237.
Total = 8

32. John Major (1467/68–1550)
Commentary on the Sentences
50, 58, 79, 85, 97, 99, 117, 143, 182, 183, 242, 272, 350, 387.
Total = 14

Questions on the Physics
1, 159, 293, 297, 321, 326.
Total = 6

Questions on De celo (in same publication as *Physics*)
18, 68, 78, 91, 122, 124, 126, 169, 184, 215, 220, 222, 231, 244, 371, 374, 378, 383, 389, 396.
Total = 20

33. Christopher Clavius (1537–1612)
Commentary on the Sphere of Sacrobosco
388.
Total = 1

34. Franciscus Toletus (1532–1596)
Questions on the Physics
1, 8, 25, 28, 59, 159, 274, 276, 277, 289, 308, 326.
Total = 12

35. Galileo Galilei (1564–1642)
Questions on De caelo
1, 12, 62, 66, 68, 75, 79, 97, 128.
Total = 9

36. Conimbricenses (Coimbra Jesuits)
Questions on De coelo
10, 12, 15, 62, 63, 64, 68, 69, 79, 84, 85, 91, 94, 97, 102, 103, 124, 126, 128, 155, 163, 169, 174, 177, 180, 183, 192, 195, 196, 206, 213, 222, 229, 230, 231, 237, 241, 250, 252, 256, 259, 270, 273, 331, 333, 334, 335, 370, 378, 382, 384, 387, 389, 396, 398.
Total = 55

Questions on the Physics
1, 3, 22, 24, 28, 29, 55, 289, 292, 308, 321, 322, 325, 326.
Total = 14

37. Peter de Oña (1560–1626)
Questions on the Physics
28, 29, 152, 153, 159, 275, 277, 280, 289, 301, 308, 325, 326.
Total = 13

38. John Case (Johannes Casus) (ca. 1546–1600)
Questions on the Physics
54, 173, 179, 238, 254, 289.
Total = 6

Seventeenth century

39. Pedro Hurtado de Mendoza (1578–1651)

Questions on De coelo

12, 15, 30, 79, 85, 91, 97, 112, 119, 126, 128, 146, 173, 183, 209, 228, 237, 256, 275, 335, 341, 353, 392.

Total = 23

Questions on the Physics

2, 4, 7, 55, 58.

Total = 5

Questions on On Generation and Corruption (all are in the same publication)

331, 339.

Total = 2

40. Bartholomew Amicus (1562–1649)

Questions on De caelo

1, 11, 13, 14, 15, 18, 31, 32, 33, 35, 41, 42, 43, 45, 62, 63, 68, 69, 79, 83, 85, 87, 91, 92, 93, 94, 95, 97, 99, 101, 106, 122, 128, 130, 132, 135, 141, 144, 169, 174, 180, 190, 211, 216, 220, 222, 233, 237, 242, 243, 250, 252, 253, 254, 258, 261, 262, 267, 273, 274, 275, 276, 277, 280, 284, 301, 332, 333, 334, 335, 341, 353, 355, 356, 357, 359, 360, 364, 367, 374, 375, 384, 386, 387, 388, 390, 391, 396, 397, 398.

Total = 90

Questions on the Physics

55, 56, 152, 283, 289, 308, 310, 311, 321, 325, 326.

Total = 11

41. Raphael Aversa (ca. 1589–1657)

Questions on De caelo

2, 12, 21, 26, 55, 62, 68, 69, 79, 80, 81, 85, 90, 96, 97, 98, 99, 105, 125, 126, 128, 130, 131, 132, 138, 139, 141, 142, 144, 146, 147, 150, 170, 182, 195, 220, 222, 231, 233, 237, 239, 242, 250, 255, 256, 270, 331, 334, 335, 348, 353, 358, 388, 389.

Total = 54

42. Roderigo de Arriaga (1592–1667)

Questions on De caelo

97, 128, 135.

Total = 3

Questions on the Physics (in same publication as *De caelo*)

55, 58.

Total = 2

43. Johannes Poncius (1599–1661)

Commentary on the Sentences

97, 237.

Total = 2

Questions on De coelo

1, 9, 12, 67, 68, 75, 91, 97, 105, 128, 135, 194, 231, 237, 250, 256.

Total = 16

44. Bartholomaeus Mastrius (1602–1673) and Bonaventura Bellutus (1600–1676)
Questions on De coelo
79, 85, 97.
Total = 3

45. Franciscus de Oviedo (1602–1651)
Questions on De caelo
12, 85, 97, 128, 135, 194.
Total = 6

Questions on the Physics (both works in same publication)
1, 2, 7, 55, 58, 280.
Total = 6

46. Thomas Compton-Carleton (ca. 1591–1666)
Questions on De coelo
12, 79, 85, 86, 97, 126, 128, 133, 135, 195, 231, 237, 256.
Total = 13

Questions on the Physics (both works in same publication)
1, 23, 55, 58, 280, 283.
Total = 6

47. Giovanni Baptista Riccioli (1598–1671)
Almagestum novum
12, 36, 37, 38, 39, 42, 44, 46, 48, 51, 53, 79, 83, 85, 88, 97, 128, 135, 190, 195, 388, 389, 393.
Total = 23

48. Franciscus Bonae Spei (1617–1677)
Questions on De coelo (1652)
10, 12, 25, 62, 68, 75, 85, 97, 128, 135, 195, 242, 399.
Total = 13

Questions on the Physics (in same publication as *De coelo*)
55, 58.
Total = 2

49. Melchior Cornaeus (1598–1665)
Questions on De coelo
2, 7, 10, 12, 32, 62, 63, 68, 69, 74, 79, 85, 97, 99, 105, 111, 112, 126, 127, 135, 137, 170, 180, 183, 195, 220, 231, 237, 273, 331, 339, 341, 383, 384, 388, 389, 393, 394, 395.
Total = 39

Questions on the Physics
57, 58, 275, 277, 278, 279, 280, 281, 282, 283, 304, 308, 312, 313, 314, 315, 316, 317, 318, 319, 320, 321, 322, 325.
Total = 24

Questions on the Meteorology (all three treatises or sections are in the same publication)
342, 343, 345, 346, 347, 348, 349, 351, 392.
Total = 9

50. Sigismundus Serbellonus (d. ca. 1660)

Questions on De caelo (vol. 1, 1657)

1, 12, 22, 79, 82, 97, 122, 130, 131, 135, 148, 194, 195, 209, 227, 237, 262.
Total = 17

Questions on the Physics (vol. 2, 1663)

55, 275, 277, 280, 304, 325.
Total = 6

51. George de Rhodes (1597–1661)

Questions on De coelo

79, 88, 135.
Total = 3

52. Illuminatus Oddus (d. 1683)

Questions on the Physics

2, 58, 154, 275, 280, 284, 289, 308, 310, 325, 326.
Total = 11

Questions on De coelo

1, 15, 25, 62, 67, 74, 79, 85, 94, 97, 101, 103, 128, 132, 135, 141, 144, 147, 148, 173, 174, 183, 195, 201, 232, 237, 241, 242, 246, 268, 270, 273.
Total = 32

Table A.2 *Treatises and the questions drawn from them*

I. De caelo

1, 2, 7, 9, 10, 11, 12, 13, 14, 15, 17, 18, 19, 20, 21, 22, 25, 26, 30, 31, 32, 33, 34, 35, 41, 42, 43, 45, 54, 55, 60, 61, 62, 63, 64, 66, 67, 68, 69, 70, 71, 72, 73, 74, 75, 76, 77, 78, 79, 80, 81, 82, 83, 84, 85, 86, 87, 88, 90, 91, 92, 93, 94, 95, 96, 97, 98, 99, 101, 102, 103, 105, 106, 111, 112, 119, 122, 123, 124, 125, 126, 127, 128, 130, 131, 132, 133, 135, 137, 138, 139, 140, 141, 142, 144, 146, 147, 148, 150, 168, 169, 170, 173, 174, 175, 177, 178, 180, 181, 182, 183, 184, 185, 186, 187, 188, 189, 190, 192, 194, 195, 196, 201, 206, 209, 211, 213, 215, 216, 217, 218, 219, 220, 221, 222, 224, 225, 227, 228, 229, 230, 231, 232, 233, 234, 236, 237, 239, 241, 242, 243, 244, 246, 250, 251, 252, 253, 254, 255, 256, 258, 259, 260, 261, 262, 264, 267, 268, 270, 273, 300, 331, 332, 333, 334, 335, 336, 339, 340, 341, 348, 353, 355, 356, 357, 358, 359, 360, 363, 364, 365, 367, 369, 370, 371, 372, 373, 374, 375, 376, 378, 379, 380, 381, 382, 383, 384, 386, 387, 388, 389, 390, 391, 392, 393, 394, 395, 396, 397, 398, 399.
Total = 221

II. Physics

1, 2, 3, 4, 5, 6, 7, 8, 9, 16, 17, 22, 23, 24, 25, 27, 28, 29, 54, 55, 56, 57, 58, 59, 73, 122, 128, 149, 152, 153, 154, 155, 158, 159, 160, 161, 163, 164, 165, 166, 167, 171, 172, 173, 176, 179, 183, 192, 203, 204, 205, 208, 209, 210, 238, 254, 271, 274, 275, 276, 277, 278, 279, 280, 281, 282, 283, 284, 285, 286, 287, 289, 290, 291, 292, 293, 294, 295, 296, 297, 298, 299, 300, 301, 302, 303, 304, 305, 306, 307, 308, 309, 310, 311, 312, 313, 314, 315, 316, 317, 318, 319, 320, 321, 322, 323, 324, 325, 326, 327, 328, 329, 330, 337, 338, 388.
Total = 116

III. Questions on the Sphere

1, 9, 54, 62, 97, 104, 106, 126, 127, 128, 132, 147, 170, 173, 180, 184, 188, 196, 201, 203, 207, 212, 215, 220, 223, 240, 242, 245, 248, 266, 331, 335, 366, 368, 383, 384, 388, 389, 393, 396, 400.

Total = 41

IV. Commentaries on the Sentences

1, 4, 47, 48, 49, 50, 52, 58, 69, 79, 85, 97, 99, 112, 117, 118, 120, 121, 122, 126, 128, 132, 136, 143, 145, 173, 182, 183, 191, 193, 195, 211, 214, 230, 237, 242, 247, 249, 256, 270, 272, 339, 350, 377, 387.

Total = 45

V. Questions on the Meterology

132, 140, 235, 262, 263, 265, 334, 337, 342, 343, 344, 345, 346, 347, 348, 349, 350, 351, 352, 354, 361, 362, 385, 389, 392.

Total = 25

VI. Questions on the Metaphysics

89, 102, 128, 155, 156, 157, 162, 192, 197, 198, 199, 200, 202, 220, 237.

Total = 15

VII. Questions on On Generation and Corruption

331, 339.

Total = 2

VIII. Quodlibetal Questions

1, 65, 129, 255, 256, 257, 258, 269.

Total = 8

IX. Almagestum novum *(Riccioli)*

12, 36, 37, 38, 39, 42, 44, 46, 48, 51, 53, 79, 83, 85, 88, 97, 128, 135, 190, 195, 388, 389, 393.

Total = 23

X. De universo *(William of Auvergne)*

34, 40, 43, 62, 69, 100, 108, 109, 111, 113, 134, 151, 226.

Total = 13

XI. Summa theologica *(Alexander of Hales)*

106, 107, 109, 110, 111, 112, 114, 115, 116, 117, 228.

Total = 11

Bibliography

Publications in the same year are cited in order of appearance. If, for example, the year of publication is 1978, the first publication in that year is cited as 1978a, the second as 1978b, and so on. In the bibliography, the letters a, b, c, etc., are placed after the relevant works. In multivolume works by scholastic authors, the volume concerning cosmology is specified when necessary. In early printed books, where sections devoted to questions on one or more of Aristotle's relevant cosmological works (*De caelo, Physics, Metaphysics*, and so on) occur but are not reflected in the title, they are mentioned following the entry (for example, see Hurtado de Mendoza).

Achillini, Alessandro. *Liber de orbibus*. In *Alexandri Achillini Bononiensis Philosophi celeberrimi Opera omnia unum collecta. Cum annotationibus excellentissimi doctoris Pamphili Montij, Bononiensis scholae Patavinae publici professoris. Omnia post primas editiones nunc primum emendatiora in lucem prodeunt*. Venice: apud Hieronymum Scotum, 1545.

Aegidius Romanus (Giles of Rome). *Egidii Romani in libros De physico auditu Aristotelis commentaria accuratissime emendata et in marginibus ornata quotationibus textuum et commentorum ac alijs quamplurimis annotationibus. Cum tabula questionum in fine. Eiusdem questio de gradibus formarum*. Venice, 1502. (a)

Gaetani expositio in libro De celo et mundo.Cum questione Domini Egidii De materia celi nuperrime impressa et quam diligentissime emendata. Venice, 1502. (b)

In primum librum Sententiarum. Correctus a reverendo magistro Augustino Montifalconio. Venice: Octavianus Scotus, 1521. Facsimile, Frankfurt: Minerva, 1968.

Opus Hexaemeron, sive De mundo sex diebus condito, in *Primus tomus operum D. Aegidii Romani, Bituricensis Archiepiscopi, Ordinis Fratrum Eremitarum Sancti Augustini*. Rome: apud Antonium Bladum, 1555.

Giles of Rome, Errores philosophorum: Critical Text with Notes and Introduction. Ed. Josef Koch; tr. John O. Riedl. Milwaukee: Marquette University Press, 1944.

Ailly, Pierre d'. *14 Quaestiones*. In *Spherae tractatus Ioannis de Sacro Busto Anglici viri clariss.; Gerardi Cremonensis Theoricae planetarum veteres; Georgii Purbachii Theoricae planetarum novae; Prosdocimo de Beldomando Patavini super tractatus sphaerico commentaria, nuper in lucem diducta per L.GA. nunquam amplius impressae . . . Campani compendium super Tractatu de sphera; Eiusdem Tractatulus de modo fabricandi spheram solidam; Petri Cardin. de Aliaco episcopi Camaracensis 14 Quaestiones . . . Alpetragii Arabi Theorica planetarum nuperrime Latinis mandata literis a Calo Calonymos Hebreo Neapolitano, ubi nititur salvare apparentias in motibus planetarum absque eccentricis et epicyclis*. Venice, 1531.

Ymago Mundi de Pierre d'Ailly. French translation. Ed. and tr. Edmond Buron. Paris: Maisonneuve Frères, 1930.

Aiton, E. J. "Celestial Spheres and Circles." *History of Science* 19 (June 1981), 75–114.

Albert of Saxony. *Questiones et decisiones physicales insignium virorum: Alberti de Saxonia in octo libros Physicorum; tres libros De celo et mundo; duos libros De generatione et corruptione; Thimonis in quatuor libros Meteororum; Buridani in tres libros De anima;*

librum De sensu et sensato; librum De memoria et reminiscentia; librum De somno et vigilia; librum De longitudine et brevitate vite; librum De iuventute et senectute Aristotelis. Recognitae rursus et emendatae summa accuratione et iudicio Magistri Georgii Lokert Scotia quo sunt tractatus proportionum additi. Paris: vaenundantur in aedibus Iodici Badii Ascensii et Conradi Resch, 1518.

Albertus Magnus. *Alberti Magni Opera omnia.* Ed. Bernhard Geyer.

Vol. 16, pt. 2: *Metaphysica,* bks. 6–13. Ed. Bernhard Geyer. Aschendorff: Monasterii Westfalorum, 1964.

Vol. 5, pt. 1: *De caelo et mundo.* Ed. Paul Hossfeld. Aschendorff: Monasterii Westfalorum, 1971.

Albumasar. *Introductorium in astronomiam Albumasaris abalachi octo continens libros partiales.* Venice: per Jacobum pentium Leucensem, 1506. Unfoliated.

Alcuin. *Interrogationes et responsiones in Genesim.* Vol. 100 in J. P. Migne, ed., *Patrologiae cursus completus, series Latina.* Paris, 1863.

Alexander of Hales. *Summa Theologica.* Tomus II: Prima pars secundi libri. Florence: Collegium S. Bonaventurae, 1928.

Alfraganus. *Alfragano (al-Fargani) Il "Libro dell'aggregazione delle stelle" (Dante, Conv., II, VI–134). Secondo il codice Mediceo-Laurenziano Pl. 29-Cod. 9 contemporaneo a Dante.* Ed. Romeo Campani. Castello: Lapi. Vols. 87–90 in G. L. Passerini, ed., *Collezione di opuscoli danteschi indediti o rari.* Florence, 1910.

Al Farghani Differentie scientie astrorum. Ed. Francis J. Carmody. Berkeley: 1943.

Alhazen. *See* Haytham, ibn al-.

Ambrose, Saint. *Hexameron, Paradise, and Cain and Abel.* Tr. John J. Savage. Vol. 42 in *The Fathers of the Church, A New Translation.* New York: Fathers of the Church, 1961.

Amicus, Bartholomew. *In Aristotelis libros De caelo et mundo dilucida textus explicatio et disputationes in quibus illustrium scholarum Averrois, D. Thomae, Scoti, et Nominalium sententiae expenduntur earumque tuendarum probabiliores modi afferuntur.* Naples: apud Secundinum Roncaliolum, 1626.

In Aristotelis libros De physico auditu dilucida textus explicatio et disputationes in quibus illustrium scholarum Averrois, D. Thomae, Scoti, et Nominalium sententiae expenduntur earumque tuendarum probabiliores modi afferuntur. 2 vols. Naples: apud Secundinum Roncaliolum, 1626–1629.

It appears that Amicus's commentaries and *questiones* on Aristotle's *De caelo* and *Physics* formed part of a seven-volume set of commentaries with the following general title: *In universam Aristotelis philosophiam notae et disputationes, quibus illustrium scholarum Averrois, D. Thomae, Scoti et Nominalium sententiae expenduntur earumque tuendarum probabiles modi afferuntur.* Naples, 1623–1648.

Anonymous. *See* "Compendium of Six Books."

Aquinas. *See* Thomas Aquinas.

Aristotle. *Aristoteles Latinus XI 1.2, De mundo.* Ed. W. L. Lorimer. Rome: Libreria dello stato, 1951.

On the Heavens. Tr. W. K. C. Guthrie. Loeb Classical Library. 1960.

Meteorologica. Tr. H. D. P. Lee. Loeb Classical Library. 1962.

The Complete Works of Aristotle. 2 vols. Rev. ed. Ed. Jonathan Barnes. Princeton: Princeton University Press, 1984.

Categories, tr. J. L. Ackrill.

On Generation and Corruption, tr. H. H. Joachim.

Generation of Animals, tr. A. Platt.

On the Heavens, tr. J. L. Stocks.

Physics, tr. R. P. Hardie and R. K. Gaye.

Metaphysics, tr. W. D. Ross.

Meteorology, tr. E. W. Webster.

De mundo, tr. E. S. Forster.

Nicomachean Ethics tr. W. D. Ross, rev. J. O. Urmson.

Arriaga, Roderigo de. *Cursus philosophicus*. Antwerp: Plantiniana Balthasaris Moreti, 1632.

Augustine, Saint. *The City of God*. Ed. and tr. Marcus Dods. 2 vols. New York: Hafner, 1948.

St. Augustine The Literal Meaning of Genesis: De Genesi ad litteram. 2 vols. Ed. and tr. John Hammond Taylor. Vols. 41–42 in Johannes Quasten, Walter J. Burghardt, and Thomas Comerford Lawler, eds., *Ancient Christian Writers: The Works of the Fathers in Translation*. New York: Newman, 1982.

Aureoli, Peter. *Petri Aureoli Verberii Ordinis Minorum Archiepiscopi Aquensis S.R.E. Cardinalis Commentariorum in primum [-quartum] librum Sententiarum pars prima [-quarta]*. 2 vols. Rome: Aloysius Zannetti, 1596–1605.

Averroës. *Averroes' Tahafut al-Tahafut (The Incoherence of the Incoherence)*. 2 vols. Tr. Simon Van den Bergh. London: Luzac, 1954.

Aristotelis opera cum Averrois commentariis. 9 vols. in 11 parts plus 3 supplementary vols. Venice: Junctas, 1562–1574. Facsimile, Frankfurt: Minerva, 1962.

Vol. 4: *Aristotelis De physico auditu libri octo cum Averrois Cordubensis variis in eosdem commentariis*.

Vol. 5: *Aristotelis De coelo, De generatione et corruptione, Meteorologicorum, De plantis cum Averrois Cordubensis in eosdem commentariis*.

Vol. 7: *De causis libellus proprietatum elementorum Aristoteli ascriptus*.

Vol. 8: *Aristotelis Metaphysicorum Libri XIII cum Averrois Cordubensis in eosdem commentariis et Epitome*.

Vol. 9: *De substantia orbis*.

"A Treatise Concerning the Substance of the Celestial Sphere." [A translation of *De substantia orbis*.] Tr. Arthur Hyman. In Arthur Hyman and James J. Walsh, eds., *Philosophy in the Middle Ages: The Christian, Islamic, and Jewish Traditions*. Indianapolis: Hackett, 1973, 307–314.

Aversa, Raphael. *Philosophia metaphysicam physicamque complectens quaestionibus contexta*. 2 vols. Vol. 2. Rome: apud Iacobum Mascardum, 1627. Vol. 1 was published in 1625, but has not been used here. Vol. 2 includes questions on what Aversa calls *De mundo*, but which is cited throughout this study as *De caelo*.

Avicenna. *Avicenne perhypatetici philosophi medicorum ac facile primi opera in lucem redacta ac nuper quantum ars niti potuit per canonicos emendata: Logyca; Sufficientia; De celo et mundo; De anima; De animalibus; De intelligentiis; Alpharabius de intelligentiis; Philosophia prima*. Venice: Octavianus Scotus, 1508. Facsimile, Frankfurt: Minerva, 1961.

Bacon, Roger. *The "Opus Majus" of Roger Bacon*. 2 vols. Ed. John Henry Bridges. London: Williams & Norgate, 1900.

Opera hactenus inedita. 16 fascicules. Ed. Robert Steele and Ferdinand M. Delorme. Oxford: Clarendon Press, 1905–1940.

Fasc. 3: *Liber primus communium naturalium Fratris Rogeri*. Pts. 3–4. Ed. Robert Steele. 1911.

Fasc. 4: *Liber secundus communium naturalium: De celestibus*. Pt. 5. Ed. Robert Steele. 1913.

Fasc. 13: *Questiones supra libros octo Physicorum Aristotelis*. Ed. Ferdinand Delorme and Robert Steele. 1935.

Un fragment inédit de l'Opus Tertium de Roger Bacon precedé d'une étude sur ce fragment. Ed. Pierre Duhem. Florence: Collegium S. Bonaventurae, 1909.

The Opus majus of Roger Bacon. 2 vols. Tr. Robert B. Burke. Philadelphia: University of Pennsylvania Press, 1928. Reprinted New York: Russell & Russell, 1962.

Roger Bacon's Philosophy of Nature, A Critical Edition, with English Translation,

Introduction, and Notes, of "De multiplicatione specierum" and "De speculis comburentibus." Ed. and tr. David C. Lindberg. Oxford: Clarendon Press, 1983.

Baldini, Ugo. "L'Astronomia del Cardinale Bellarmino." In P. Galluzzi, ed., *Novità celesti e crisi del sapere.* Atti del Convegno internazionale di studi Galileiani. Florence: Giunti Barbèra, 1984, 293–305.

Bartholomew the Englishman. *Bartholomaei Anglici De genuinis rerum coelestium, terrestrium et inferarum proprietatibus.* Frankfurt: Apud Wolfgang Richterum, impensis Nicolai Steinii, 1601. Facsimile, with title *De rerum proprietatibus.* Frankfurt: Minerva, 1964.

Basil, Saint. *Exegetic Homilies.* Tr. Agnes Clare Way. Vol. 46 in *The Fathers of the Church, A New Translation.* Washington, D.C.: Catholic University of America Press, 1963. This is Basil's *Hexaemeron.*

Bechler, Zev. "The Essence and Soul of the Seventeenth-Century Scientific Revolution." *Science in Context* 1 (1987), 87–101.

Bede, Venerable. *Bedae Venerabilis Opera. Pt. II: Opera exegetica, 1: Libri quatuor in principium Genesis usque ad nativitatem Isaac et eiectionem Ishmahelis adnotationem.* Ed. Charles W. Jones. In *Corpus Christianorum, series Latina*, vol. 118a. Turnholt: Brepols, 1967.

Bellarmine, Robert. "The Louvain Lectures (Lectiones Lovanienses) of Bellarmine and the Autograph Copy of his 1616 Declaration to Galileo." Ed. and tr. Ugo Baldini and George V. Coyne. Latin or Italian texts with English translation. *Vatican Observatory Publications, Special Series: Studi Galileiani*, vol. 1, no. 2. Vatican City, 1984.

Bellutus, Bonaventura. *See* Mastrius, Bartholomaeus.

Benedictus Hesse. *Quaestiones super octo libros "Physicorum" Aristotelis.* Ed. Stanislaw Wielgus. Wroclaw: Polish Academy of Sciences, 1984.

Bernard of Verdun. *Birnardus de Virduno, Ordinis Fratrum Minorum, Tractatus super totam astrologiam.* Ed. P. Polykarp Hartmann. Werl: Dietrich-Coelde-Verlag, 1961.

Bible. *Biblia Sacra Vulgatae editionis Sixti V Pont. Max. Iussu recognita et Clementis VIII auctoritate edita.* Rome: Marietti, 1965.

The New English Bible with the Apocrypha, Oxford Study Edition. Ed. M. Jack Suggs (New Testament) and Arnold J. Tkacik. General editor, Samuel Sandmel. New York: Oxford University Press, 1976.

Biel, Gabriel. *Collectorium circa quattuor libros Sententiarum, Prologus et Liber primus.* Ed. Wilfrid Werbeck and Udo Hofmann. Tübingen: Mohr [Siebeck], 1973.

Bitrūjī, al-. *De motibus celorum*, Critical edition of the Latin translation of Michael Scot. Ed. Francis J. Carmody. Berkeley/Los Angeles: University of California Press, 1952.

Al-Bitrūjī: On the Principles of Astronomy. Arabic and Hebrew texts with translation, analysis, and Arabic-Hebrew-English glossary by Bernard R. Goldstein. New Haven: Yale University Press, 1971.

Blund, John. *Iohannes Blund, Tractatus de anima.* Ed. D. A. Callus and R. W. Hunt. Auctores Britannici Medii Aevi, no. 2. London: For the British Academy by Oxford University Press, 1970.

Boethius of Dacia. *Quaestiones super libros Physicorum.* In *Boethii Daci Opera.* Vol. 5, pt. 2. Ed. Géza Sajó. Copenhagen: 1974.

De aeternitate mundi. In *Boethii Daci Opera, Topica-Opuscula.* Vol. 6, pt. 2. Ed. Nicolaus Georgius Green-Pedersen. Copenhagen: 1976.

Bonae Spei, Franciscus. *R.P. Francisci Bonae Spei, Carmelitae Reformati, Commentarii tres in universam Aristotelis philosophiam. Commentarius tertius in libros De coelo; De generatione et corruptione; De anima et Metaphysicam.* Brussels: apud Franciscum Vivienum, 1652.

Bonaventure, Saint. *Opera omnia*. Ad Claras Aquas (Quaracchi): Collegium S. Bonaventurae, 1882–1901.

Vol. 1: *Commentaria in quatuor libros Sententiarum Magistri Petri Lombardi: In Primum librum Sententiarum* (1882).

Vol. 2: *Commentaria in quatuor libros Sententiarum Magistri Petri Lombardi: In Secundum librum Sententiarum* (1885).

On the Eternity of the World (*De aeternitate mundi*). Tr. Paul M. Byrne. Milwaukee: Marquette University Press, 1964, 99–117.

See also Siger of Brabant, 1964; Thomas Aquinas, 1964.

Bradwardine, Thomas. *De causa Dei contra Pelagium et de virtute causarum ad suos Mertonenses libri tres . . . opera et studio Dr. Henrici Savili Collegii Mertonensis in Academia Oxoniensis custodis, ex scriptis codicibus nunc, primum editi*. London: apud Ioannem Billium, 1618. Facsimile, Frankfurt: Minerva, 1964.

Brahe, Tycho. *Tychonis Brahe Dani Opera omnia*. 15 vols. Ed. J. L. E. Dreyer. Copenhagen, 1913–1929.

De Mundi aetherei recentioribus Phaenomenis. Vol. 4 in J. L. E. Dreyer, ed., *Tychonis Brahe Dani Opera omnia*. Copenhagen, 1922. First published Uraniborg, 1588; then Prague, 1603; Frankfurt, 1610.

Bricot, Thomas. *De celo et mundo*. In *Textus abbreviatus Aristotelis super octo libris Phisicorum et tota naturali philosophia nuper a Magistro Thoma Bricot vernantissimarum artium atque divine legis claro professore compilatus una cum continuatione textus magistri Georgii et questionibus eiusdem de recenti ab eodem Thoma Bricot revisus atque dililgentissime emendatus*. Lyon, 1486.

Buridan, Jean. *Acutissimi philosophi reverendi Magistri Johannis Buridani subtilissime questiones super octo Phisicorum libros Aristotelis diligenter recognite et revise a Magistro Johanne Dullaert de Gandavo antea nusquam impresse*. Paris, 1509. Facsimile, entitled *Johannes Buridanus, Kommentar zur Aristotelischen Physik*. Frankfurt: Minerva, 1964.

In Metaphysicen Aristotelis. Questiones argutissimae Magistri Ioannis Buridani in ultima praelectione ab ipso recognitae et emissae ac ad archetypon diligenter repositae cum duplice indicio materiarum videlicet in fronte et quaestionum in operis calce. Paris: Vaenundantur Badio, 1518. Facsimile, Frankfurt: Minerva, 1964. The added title page of the facsimile gives 1588 as the date of publication.

Iohannis Buridani Quaestiones super libris quattuor De caelo et mundo. Ed. Ernest A. Moody. Cambridge, Mass.: Mediaeval Academy of America, 1942.

Burley, Walter. "Questiones circa libros Physicorum." Basel, Universitätsbibliothek, MS. F. V. 12, fols. 108r–171v.

Burleus super octo libros Phisicorum. Venice, 1501. Facsimile, entitled *Walter Burley, In Physicam Aristotelis expositio et quaestiones*. Hildesheim: Olms, 1972.

Campanus of Novara. *Tractatus de sphera*. *See* d'Ailly, 1531.

Campanus of Novara and Medieval Planetary Theory: Theorica planetarum. Ed.and tr. Francis S. Benjamin, Jr., and G. J. Toomer. Madison: University of Wisconsin Press, 1971.

Čapek, Milič, ed. *The Concepts of Space and Time, Their Structure and Development*. Dordrecht: Reidel, 1976.

Carmody, Francis J. "The Planetary Theory of Ibn Rushd." *Osiris* 10 (1952), 556–586.

Caroti, Stefano. "Nicole Oresme's Polemic against Astrology, in his *Quodlibeta*." In Patrick Curry, ed., *Astrology, Science and Society: Historical Essays*. Woodbridge, U.K.: Boydell, 1987, 75–93.

Case, John (Johannes Casus). *Lapis philosophicus seu commentarius in 8 libros Physicorum Aristotelis in quo arcana physiologiae examinantur*. Oxford: Josephus Barnesius, 1599.

Cavellus, Hugo. *R.P.F. Ioannis Duns Scoti, Doctoris Subtilis, Ordinis Minorum, In*

XII libros Metaphysicorum Aristotelis Expositio. Cum summariis, notis, et scholiis R.P.F. Hugonis Cavelli, Hiberni. Item eiusdem Doctoris In Metaphysicam quaestiones subtilissima, cum annotationibus R.P.F. Mauritii de Portu Hiberni. Vol. 4. Lyon, 1639.

This is the Luke Wadding edition of the *Opera omnia* of John Duns Scotus. Facsimile, Hildesheim: Olms, 1968. Although the *Expositio* and questions on the twelfth book of the *Metaphysics* are attributed to Duns Scotus, they are not his. Cavellus's comments on those books are nevertheless relevant.

Cecco d'Ascoli. *De eccentricis et epicyclis.* In G. Boffito, "Il De eccentricis et epicyclis di Cecco d'Ascoli novamente scoperto e illustrato." In Leo S. Olschki, ed., *La Bibliofilia rivista dell'arte antica in libri, stampe, manoscritto, autografi e legature.* Vol. 7. Florence: Leo S. Olschki, 1906, 150–167.

"Commentary on the *Sphere* of Sacrobosco." In Lynn Thorndike, ed. and tr., *The "Sphere" of Sacrobosco and Its Commentators.* Chicago: University of Chicago Press, 1949, 343–411.

Cesalpino, Andrea. *Andreae Cesalpini Aretini medici atque philosophi subtilissimi peritissimique peripateticarum quaestionum libri quinque.* Venice: apud Iuntas, 1571.

Chalcidius. *Timaeus a Calcidio translatus commentarioque instructus.* Ed. J. H. Waszink, and P. J. Jensen. Vol. 4 in Raymond Klibansky, ed., *Plato Latinus.* London: Warburg Institute, 1962.

Chenu, M. D. *Nature, Man, and Society in the Twelfth Century: Essays on New Theological Perspectives in the Latin West.* Ed. and tr. Jerome Taylor and Lester K. Little. Chicago: University of Chicago Press, 1968; originally published in French, 1957.

Chrysostom, John. *Homilies on Genesis 1–17.* Tr. Robert C. Hill. Vol. 74 in *The Fathers of the Church, A New Translation.* Washington, D.C.: Catholic University of America Press, 1986.

Cicero. *De natura deorum; Academica.* Tr. Harris Rackham. Loeb Classical Library. 1933.

Clagett, Marshall. *The Science of Mechanics in the Middle Ages.* Madison: University of Wisconsin Press, 1959.

Vol 1: *Archimedes in the Middle Ages. The Arabo-Latin Tradition.* Madison: University of Wisconsin Press, 1964.

Clavius, Christopher. *Christophori Clavii Bambergensis ex Societate Iesu in Sphaeram Iohannis de Sacro Bosco Commentarius.* 4th ed. Lyon: sumptibus Fratrum de Gabiano, 1593. First ed. Rome: apud Victorium Helianum, 1570.

Christophori Clavii Bambergensis e Societate Iesu Opera Mathematica V tomis distributa ab auctore nunc denuo correcta et plurimis locis aucta. Ad Reverendiss. et Illustriss. Principem ac Dominum. D. Joannem Godefridvm Episcopum Bambergensem etc. Mogvntiae: sumptibus Antonij Hierat Excudebat Reinhardus Eltz, 1612. Tomus tertius complectens commentarium In sphaeram Ioannis de Sacro Bosco, et Astrolabium. Moguntiae [Mainz]: sumptibus Antonii Hierat excudebat Reinhardus Eltz, 1611.

Cobban, Alan B. *The Medieval English Universities: Oxford and Cambridge to c.1500.* Berkeley/Los Angeles: University of California Press, 1988.

Cohen, Morris R., and I. E. Drabkin. *A Source Book in Greek Science.* 1st ed. New York: McGraw-Hill, 1948. Reprinted Cambridge, Mass.: Harvard University Press, 1958.

"Compendium of Six Books on Metaphysics and Natural Philosophy." Paris, Bibliothèque Nationale, fonds latin, MS. 6752. Anonymous. Composed after 1323. 239 leaves. Book 6, the last book, concerns the spheres, intelligences, and first cause and begins on fol. 213v.

Compton-Carleton, Thomas. *Philosophia universa.* Antwerp: apud Iacobum Meursium, 1649.

Conimbricenses. *Commentarii Collegii Conimbricensis Societatis Iesu In quatuor libros De coelo Aristotelis Stagiritae*. 2nd ed. Lyon: ex officina iuntarum, 1598. 1st ed. Coimbra, 1592.

Commentariorum Collegii Conimbricensis Societatis Iesu In octo libros Physicorum Aristotelis Stagiritae. 2 parts with separate title pages. Lyon: sumptibus Horatii Cardon, 1602. 1st ed. Coimbra, 1592.

Each part is separately numbered by columns. Part 1 contains the commentary on books 1 through 3 of the *Physics*; part 2, the commentary on books 4 through 8.

Collegii Conimbricensis Societatis Iesu In duos libros De generatione et corruptione Aristotelis Stagiritae. Hac secunda editione, Graeci contextus Latino e regione respondentis accessione auctiores et emendatiores. Lyon: sumptibus Horatij Cardon, 1606.

Copernicus, Nicholas. *Nicholas Copernicus On the Revolutions*. Ed. Jerzy Dobrzycki; tr. and commentary by Edward Rosen. London: Macmillan, 1978. First published 1543.

Copleston, Frederick. *Mediaeval Philosophy, Augustine to Scotus*. Vol. 2 in *A History of Philosophy*. Westminster, Md.: Newman, 1957.

Cornaeus, Melchior. *Curriculum philosophiae peripateticae, uti hoc tempore in scholis decurri solet . . . autore R. P. Melchiore Cornaeo, Soc. Iesu, SS. Theologiae Doctore eiusdemque in Alma Universitate Herbipolensi Professore ordinario*. Herbipoli [Würzburg]: sumptibus et typis Eliae Michaelis Zinck, 1657.

Courtenay, William J. *Capacity and Volition: A History of the Distinction of Absolute and Ordained Power*. Bergamo: Pierluigi Lubrina Editore, 1990.

Crombie, A. C. *Medieval and Early Modern Science*. 2 vols. Garden City, N.Y.: Doubleday, 1959.

Crowley, Bonaventure. "The Life and Works of Bartholomew Mastrius, O.F.M. Conv. 1602–1673." *Franciscan Studies* 8 (1948), 97–152.

Dales, Richard C. "The De-animation of the Heavens in the Middle Ages." *Journal of the History of Ideas* 41 (1980), 531–550.

"The Origin of the Doctrine of the Double Truth." *Viator* 15 (1984), 169–179.

Denifle, Heinrich, and Emil Chatelain. *Chartularium Universitatis Parisiensis*. 4 vols. Paris: Fratrum Delalain, 1889–1897.

Dick, Steven J. *Plurality of Worlds: The Origins of the Extraterrestrial Life Debate from Democritus to Kant*. Cambridge: Cambridge University Press, 1982.

Dicks, D. R. *Early Greek Astronomy to Aristotle*. Ithaca: Cornell University Press, 1970.

Dictionary of Scientific Biography. 16 vols. Ed. Charles C. Gillispie. New York: Scribner's, 1970–1980.

Dictionnaire de théologie catholique. 15 vols. Ed. A. Vacant, E. Mangenot, and E. Amann. Paris: Librairie Letouzey et Aré, 1926–1950.

Dominicus de Flandria. *Questionum super xii libros Methaphisice*. N.p., 1523. Facsimile, Frankfurt: Minerva, 1967.

Donahue, William H. *The Dissolution of the Celestial Spheres, 1595–1650*. New York: Arno Press, 1981. Ph.D. diss., Cambridge University, 1972.

"The Solid Planetary Spheres in Post-Copernican Natural Philosophy." In Robert S. Westman, ed., *The Copernican Achievement*. Berkeley/Los Angeles: University of California Press, 1975, 244–275.

Drake, Stillman, and C. D. O'Malley. *The Controversy on the Comets of 1618*. Tr. Stillman Drake and C. D. O'Malley. Philadelphia: University of Pennsylvania Press, 1960.

Dreyer, J. L. E. *A History of Astronomy from Thales to Kepler*. 2nd ed., rev. W. H. Stahl. New York: Dover, 1953. 1st ed. titled *History of the Planetary Systems from Thales to Kepler*, 1906.

DSB. See *Dictionary of Scientific Biography*.

Duhem, Pierre. *Etudes sur Léonard de Vinci, ceux qu'il a lus et ceux qui l'ont lu.* 3 vols. Paris: Hermann, 1906–1913. Vol. 2, 1909.

Le système du monde. Histoire des doctrines cosmologiques de Platon à Copernic. 10 vols. Paris: Hermann, 1913–1959.

To Save the Phenomena: An Essay on the Idea of Physical Theory from Plato to Galileo. Tr. Edmund Doland and Chaninah Maschler. Chicago: University of Chicago Press, 1969. First published in French, 1908.

Medieval Cosmology: Theories of Infinity, Place, Time, Void, and the Plurality of Worlds. Ed. and tr. Roger Ariew. Chicago: University of Chicago Press, 1985.

Dullaert, Johannes. *Questiones super octo libros Physicorum Aristotelis necnon super libros De coelo et mundo.* Paris: Oliverius Senant, 1506.

Durandus de Sancto Porciano. *D. Durandi a Sancto Porciano, Ord. Praed. et Meldensis Episcopi, in Petri Lombardi Sententias theologicas commentariorum libri IIII.* Venice: Guerraea, 1571.

Eastwood, Bruce S. "The 'Chaster Path of Venus' (*Orbis Veneris Castior*) in the Astronomy of Martianus Capella." *Archives Internationales d'Histoire des Sciences* 32 (1982), 145–158.

"Heraclides and Heliocentrism: Texts, Diagrams, and Interpretations." *Journal for the History of Astronomy* 23 (1992), 233–260.

Eisenstein, Elizabeth L. *The Printing Press as an Agent of Change.* Cambridge: Cambridge University Press, 1979.

Finocchiaro, Maurice A. *The Galileo Affair: A Documentary History.* Berkeley/Los Angeles: University of California Press, 1989.

Fortin, Ernest L., and Peter D. O'Neill. Translation of Condemnation of 1277, in Ralph Lerner and Muhsin Mahdi, eds., *Medieval Political Philosophy: A Sourcebook.* New York: Free Press of Glencoe, 1963, 337–354.

Gaietanus de Thienis. *Recollecte . . . super octo libros Physicorum cum annotationibus textuum.* Venice, 1496.

Galileo Galilei. *Le Opere di Galileo Galilei.* Edizione Nazionale. 23 vols. Ed. Antonio Favaro. Florence, 1891–1909.

Discoveries and Opinions of Galileo. Including The Starry Messenger (1610), Letters on Sunspots (1613), Letter to the Grand Duchess Christina (1615), and Excerpts from The Assayer (1623). Ed. and tr. Stillman Drake. Garden City, N.Y.: Doubleday, 1957.

Galileo: Dialogue Concerning the Two Chief World Systems. Tr. Stillman Drake. Berkeley/Los Angeles: University of California Press, 1962.

Galileo's Early Notebooks: The Physical Questions. A Translation from the Latin, with Historical and Paleographical Commentary. Ed. and tr. William A. Wallace. Notre Dame, Ind.: University of Notre Dame Press, 1977. Includes questions on *De caelo.*

Sidereus Nuncius or The Sidereal Messenger: Galileo Galilei. Tr. Albert Van Helden. Chicago: University of Chicago Press, 1989.

Gesner, Conrad. *Physicarum meditationum, annotationum et scholiorum libri V.* Zurich, 1586.

Gillispie, Charles C. *The Edge of Objectivity, An Essay in the History of Scientific Ideas.* Princeton: Princeton University Press, 1969.

See also Dictionary of Scientific Biography.

Gilson, Etienne. *A History of Christian Philosophy in the Middle Ages.* London: Sheed & Ward, 1955.

Ginzburg, Carlo. *The Cheese and the Worms: The Cosmos of a Sixteenth-Century Miller.* Tr. John and Anne Tedeschi. Harmondsworth, U.K.: Penguin, 1982. First published in Italian, 1976.

Glorieux, Palémon. *La Littérature quodlibétique.* 2 vols. Vol. 1: Le Saulchoir, Kain (Belgium), 1925; Vol. 2: Paris, Vrin, 1935.

"Sentences (commentaires sur les)." In A. Vacant, E. Mangenot, and E. Amann, eds., *Dictionnaire de théologie catholique.* Paris: Librairie Letouzey et Ané, 1941, vol. 14, cols. 1860–1884.

Godfrey of Fontaines. *Les quatre premiers Quodlibets de Godefroid de Fontaines.* Ed. M. De Wulf and A. Pelzer. Vol. 2 in M. De Wulf and Auguste Pelzer, eds., *Les Philosophes belges, Texte et études.* Louvain: Éditions de l'Institut Supérieur de Philosophie de l'Université, 1904.

Les Quodlibets cinq, six et sept de Godefroid de Fontaines. Ed. M. DeWulf and J. Hoffmans. Vol. 3 in M. De Wulf and Auguste Pelzer, eds., *Les Philosophes belges, Texte et études.* Louvain: Éditions de l'Institut Supérieur de Philosophie de l'Université, 1914.

Goldstein, Bernard R. "Levi ben Gerson's Theory of Planetary Distances." *Centaurus* 29 (1986), 272–313.

Goldstein, Thomas. "The Renaissance Concept of the Earth in its Influence upon Copernicus." *Terrae Incognitae* 4 (1972), 19–51.

Grant, Edward. "Jean Buridan: A Fourteenth-Century Cartesian." *Archives Internationales d'Histoires des Sciences* 64 (1963), 251–255.

Physical Science in the Middle Ages. New York: Wiley, 1971. Reprinted Cambridge: Cambridge University Press, 1977.

"Medieval Explanations and Interpretations of the Dictum that 'Nature Abhors a Vacuum.'" *Traditio* 29 (1973), 327–355.

A Source Book in Medieval Science. Cambridge, Mass.: Harvard University Press, 1974.

"Place and Space in Medieval Physical Thought." In Peter K. Machamer and Robert G. Turnbull, eds., *Motion and Time, Space and Matter.* Columbus: Ohio State University Press, 1976, 137–167. (a)

"The Concept of *Ubi* in Medieval and Renaissance Discussions of Place." In Nancy G. Siraisi and Luke Demaitre, eds., *Science, Medicine, and the University: 1200–1550, Essays in Honor of Pearl Kibre.* Pt. 1. *Manuscripta* 20 (1976), 71–80. (b)

"Cosmology." In David C. Lindberg, ed., *Science in the Middle Ages.* Chicago: University of Chicago Press, 1978, 265–302. (a)

"Aristotelianism and the Longevity of the Medieval World View." *History of Science* 16 (1978), 93–106. (b)

"Scientific thought in Fourteenth-Century Paris: Jean Buridan and Nicole Oresme." In Madeleine Pelner Cosman and Bruce Chandler, eds., *Machaut's World: Science and Art in the Fourteenth Century.* Annals of the New York Academy of Sciences, no. 314. New York: New York Academy of Sciences, 1978, 105–124. (c)

"The Principle of the Impenetrability of Bodies in the History of Concepts of Separate Space from the Middle Ages to the Seventeenth Century." *Isis* 69 (1978), 551–571. (d)

"The Condemnation of 1277, God's Absolute Power, and Physical Thought in the Late Middle Ages." *Viator* 10 (1979), 211–244.

Much Ado about Nothing: Theories of Space and Vacuum from the Middle Ages to the Seventeenth Century. Cambridge: Cambridge University Press, 1981. (a)

"The Medieval Doctrine of Place: Some Fundamental Problems and Solutions." In A. Maierù and A. Paravicini Bagliani, eds., *Studi sul xiv secolo in memoria di Anneliese Maier.* Rome: Edizioni di Storia e Letteratura, 1981, 57–79. (b)

Studies in Medieval Science and Natural Philosophy. London: Variorum Reprints, 1981. (c)

"Celestial Matter: A Medieval and Galilean Cosmological Problem." *Journal of Medieval and Renaissance Studies* 13 (1983), 157–186.

"Science and the Medieval University." In James M. Kittelson and Pamela J.

Transue, eds., *Rebirth, Reform and Resilience: Universities in Transition 1300–1700*. Columbus: Ohio State University Press, 1984, 68–102. (a)

"In Defense of the Earth's Centrality and Immobility: Scholastic Reaction to Copernicanism in the Seventeenth Century." *Transactions of the American Philosophical Society* 74, pt. 4. Philadelphia: American Philosophical Society, 1984, 1–69. (b)

"A New Look at Medieval Cosmology, 1200–1687." *Proceedings of the American Philosophical Society* 129 (1985), 417–432. (a)

"Issues in Natural Philosophy at Paris in the Late Thirteenth Century." *Medievalia et Humanistica*, n.s., 13 (1985), 75–94. (b)

"Celestial Perfection from the Middle Ages to the Late Seventeenth Century." In Margaret J. Osler and Paul L. Farber, eds., *Religion, Science, and Worldview: Essays in Honor of Richard S. Westfall*. Cambridge: Cambridge University Press, 1985, 137–162. (c)

"Celestial Orbs in the Latin Middle Ages." *Isis* 78 (1987), 153–173. (a)

"Eccentrics and Epicycles in Medieval Cosmology." In Edward Grant and John E. Murdoch, eds., *Mathematics and Its Applications to Science and Natural Philosophy in the Middle Ages*. Cambridge: Cambridge University Press, 1987, 189–214. (b)

"Medieval and Renaissance Scholastic Conceptions of the Influence of the Celestial Region on the Terrestrial." *Journal of Medieval and Renaissance Studies* 17 (1987), 1–23. (c)

"Ways to Interpret the Terms 'Aristotelian' and 'Aristotelianism' in Medieval and Renaissance Natural Philosophy." *History of Science* 25 (1987), 335–358. (d)

"Nicolas Oresme on Certitude in Science and Pseudo-Science." In P. Souffrin and A. Ph. Segonds, eds., *Nicolas Oresme: Tradition et innovation chez un intellectuel du XIVe siècle*. Paris: Belles Lettres, 1988, 31–43.

"Medieval Departures from Aristotelian Natural Philosophy." In Stefano Caroti, ed., *Studies in Medieval Natural Philosophy*. Biblioteca di Nuncius. Florence: Olschki, 1989, 237–256.

Review of Alan B. Cobban, *The Medieval English Universities. Libraries and Culture* 25 (1990), 276–277.

"Celestial Incorruptibility in Medieval Cosmology, 1200–1687." In Sabetai Unguru, ed., *Physics, Cosmology and Astronomy, 1300–1700: Tension and Accommodation*. Dordrecht: Kluwer, 1991, 101–127. (a)

"The Unusual Structure and Organization of Albert of Saxony's *Questions on De coelo*." In Joël Biard, ed., *Itinéraires d'Albert de Saxe Paris-Vienne au XIVe Siècle*. Paris: Vrin, 1991, 205–217. (b)

See also Oresme, Nicole.

Greenberg, Sidney. *The Infinite in Giordano Bruno with a translation of his Dialogue "Concerning the Cause Principle, and One."* New York: King's Crown Press, 1950.

Grosseteste, Robert. *Die philosophischen Werke des Robert Grosseteste, Bischofs von Lincoln*. Ed. Ludwig Baur. *Beiträge zur Geschichte der Philosophie des Mittelalters, Texte und Untersuchungen*, vol. 9. Münster: Aschendorfsche Verlagsbuchhandlung, 1912.

De sphaera, 10–32.

De generatione stellarum, 32–36.

De intelligentiis, 112–119.

Summa philosophiae, 275–643.

On Light [De luce]. Tr. Clare C. Riedl. Milwaukee: Marquette University Press, 1942.

Roberti Grosseteste Episcopi Lincolniensis Commentarius in VIII libros Physicorum Aristotelis. Ed. Richard C. Dales. Boulder: University of Colorado Press, 1963.

Hexaëmeron. Eds. Richard C. Dales and Servus Gieben. London: Oxford University Press for British Academy, 1982.

Guericke, Otto von. *Experimenta nova [ut vocantur] Magdeburgica de vacuo spatio.* Amsterdam, 1672. Facsimile, Aalen: Zeller, 1962.

Hall, A. R. "Military Technology." In *The Mediterranean Civilizations and the Middle Ages c. 700* B.C. to c. A.D. 1500. Vol. 2 in Charles Singer et al., *A History of Technology*. 8 vols. Oxford: Clarendon Press, 1954–1984, 695–730. 1956.

Hanson, Norwood Russell. *Constellations and Conjectures*. Ed. Willard C. Humphreys, Jr. Dordrecht: Reidel, 1973.

Haytham, ibn al-. *Las traducciones orientales en los manuscritos de la Biblioteca Catedral de Toledo*. Ed. José Maria Millás Vallicrosa. Madrid: 1942. Appendix 2: *Tratado de astronomía de al-Hasan ben al-Haytam, en traducción latina desconocida*, 285–312.

Heath, Thomas. *Aristarchus of Samos, The Ancient Copernicus. A History of Greek Astronomy to Aristarchus together with Aristarchus's Treatise on the Sizes and Distances of the Sun and Moon. A New Greek Text with Translation and Notes*. Oxford, Clarendon Press, 1913.

Hefele, Charles-Joseph. *Histoire des Conciles d'après les documents originaux*. Tr. Dom H. LeClercq from 2nd German ed. Vol. 5, pt. 2. Paris: Letouzey et Ané, 1913.

Hellman, C. Doris. *The Comet of 1577: Its Place in the History of Astronomy*. New York: Columbia University Press, 1944.

Heninger, S. K., Jr. *Touches of Sweet Harmony: Pythagorean Cosmology and Renaissance Poetics*. San Marino, Calif.: Huntington Library, 1974.

Hervaeus Natalis. *Quolibeta Hervei. Subtilissima Hervei Natalis Britonis theologi acutissimi quolibeta undecim cum octo ipsius profundissimis tractatibus infra per ordinem descriptis . . . Tractatus VIII videlicet: De beatitudine; De verbo; De eternitate mundi; De materia celi; De relationibus; De pluralitate formarum; De virtutibus; De motu angeli*. Venice, 1513. Facsimile, Ridgewood, N.J.: Gregg, 1966.

Hervei Natalis Britonis, Doctoris theologi Parisiensis, praestantissimi et subtilissimi ac quondam Ordinis Fratrum Praedicatorum, Generalis Magistri, In Quatuor Libros Sententiarum Commentaria. Quibus adiectus est eiusdem auctoris Tractatus de potestate Papae. Paris: apud viduam Dyonisii Moreau et Dyonisium Moreau filium, via Iacobaea sub signo Salamandrae Argenteae, 1647.

Hidalgo, Juan. *Cursus philosophicus ad mentem B. Aegidii Col. Rom., doctoris fundatissimi, Ordinis Eremitarum*. 2 vols. Cordova: in vico Cisterciensi, per Petrum Arias. Vol. 1, 1736; vol. 2, 1737.

Hissette, Roland. *Enquête sur les 219 articles condamnés à Paris le 7 Mars 1277*. Louvain: Publications Universitaires, 1977.

Holkot, Robert. *In quattuor libros Sententiarum quaestiones*. Lyon, 1518. Facsimile, Frankfurt: Minerva, 1967. Unfoliated.

Hugonnard-Roche, Henri. *L'oeuvre astronomique de Themon Juif, Maître parisien du xive siècle*. Geneva: Droz, 1973.

Hurtado de Mendoza, Pedro. *Universa philosophia*. Valladolid, 1615. Other editions: Lyon, 1617, and later. Includes questions on *De coelo* and *Physics*.

Hyatte, Reginald, and Maryse Ponchard-Hyatte, eds. *L'Harmonie des sphères*. Encyclopédie d'astronomie et de musique extraite du commentaire sur *Les Echecs amoreux* (XVe s.). Edition critique d'après les mss de la Bibliothèque Nationale de Paris. New York: Lang, 1985.

Hyman, Arthur, and James J. Walsh, eds. *Philosophy in the Middle Ages: The Christian, Islamic, and Jewish Traditions*. Indianapolis: Hackett, 1973.

See also Averroës.

Jaki, Stanley L. *Uneasy Genius: The Life and Work of Pierre Duhem*. The Hague: Nijhoff, 1984.

Jardine, Nicholas. "The Significance of the Copernican Orbs." *Journal for the History of Astronomy* 13 (1982), 168–194.

The Birth of History and Philosophy of Science: Kepler's "A Defence of Tycho against Ursus," with Essays on Its Provenance and Significance. Cambridge: Cambridge University Press, 1984.

Javelli, Chrysostom. *Chrysostomi Iavelli Canapicii, Ordinis Praedicatorum, Philosophi et theologi inprimis nostra aetatis eruditissimi, Totius rationalis, naturalis divinae ac moralis philosophiae compendium innumeris fere locis castigatum et in duos tomos digestum quorum qua tomus primus contineat sequens pagina indicabit. His adiecimus in libros Physicorum, De anima, Metaphysicorum eiusdem quaestiones, quarum omnium conclusiones mira brevitate nuper collectas in margine subiecimus.* 2 vols. Lyon: apud heredes Iacobi Iunctae, 1568.

Jervis, Jane L. *Cometary Theory in Fifteenth-Century Europe.* Dordrecht: Reidel, 1985.

Johannes Canonicus. *Quaestiones super VIII libros Physicorum Aristotelis.* Venice: mandato heredum quondam domini Octaviani Scoti, 1520.

Johannes de Magistris. *Questiones perutiles supra tota philosophia [naturali] magistri Johannis Magistri[s] cum explanatione textus Aristotelis secundum mentem doctoris subtilis Scoti.* Venice, 1490. Unfoliated. Other editions: Parma, 1481, and Venice, 1487. Includes questions on *De celo* and the *Meteorology.*

John Damascene (Saint John of Damascus). *De fide orthodoxa: Versions of Burgundio and Cerbanus.* Ed. Eligius M. Buytaert. Franciscan Institute Publications, Text Series, no. 8. St. Bonaventure, N.Y.: Franciscan Institute, 1955.

The Orthodox Faith. In *Saint John of Damascus: Writings.* Tr. Frederic H. Chase, Jr. Vol. 37 in *The Fathers of the Church, A New Translation.* New York: Fathers of the Church, 1958.

John of Jandun. *Quaestiones Joannes de Janduno De physico auditu noviter emendatae. Helie Hebrei Cretensis questiones: De primo motore; De efficientia mundi; De esse essentia et uno; annotationes in plurima dicta commentatoris.* Venice, 1519.

Ioannis de Ianduno In libros Aristotelis De coelo et mundo quae extant quaestiones subtilissimae . . . De substantia orbis cum eiusdem Ioannis Commentario ac Quaestionibus . . . Venice: apud Iuntas, 1552.

Quaestiones in duodecim libros Metaphysicae. Venice, 1553. Facsimile, Frankfurt: Minerva, 1966.

Kelly, Mary Suzanne. "Celestial Motors: 1543–1632." Ph.D. diss., University of Oklahoma, 1964.

Kepler, Johannes. *Epitome of Copernican Astronomy*, bks. 4 and 5. Tr. Charles Glenn Wallis. In *Encyclopaedia Britannica*, 1952, vol. 16, 839–1004.

Kepler's "Somnium," "The Dream," or Posthumous Work on Lunar Astronomy. Tr. Edward Rosen. Madison: University of Wisconsin Press, 1967.

Johannes Kepler: "Mysterium cosmographicum," The Secret of the Universe. Ed. E. J. Aiton, tr. A. M. Duncan. New York: Abaris, 1981.

Kibre, Pearl. *See* Thorndike and Kibre.

Kilwardby, Robert. *De ortu scientiarum.* Ed. Albert G. Judy. Oxford: British Academy, 1976.

Koyré, Alexandre. "Le vide et l'espace infini au xive siècle." *Archives d'Histoire Doctrinale et Littéraire du Moyen Âge* 24 (1949), 45–91.

"An Experiment in Measurement." *Proceedings of the American Philosophical Society* 97, no. 2 (1953), 222–237.

"A Documentary History of the Problem of Fall from Kepler to Newton, *De motu gravium naturaliter cadentium in hypothesi terrae motae.*" *Transactions of the American Philosophical Society*, n.s., vol. 45, pt. 4 (1955), 329–395.

The Astronomical Revolution: Copernicus, Kepler, Borelli. Tr. R. E. W. Maddison. Ithaca: Cornell University Press, 1973. Originally published in French, 1961.

Leff, Gordon. *Paris and Oxford Universities in the Thirteenth and Fourteenth Centuries: An Institutional and Intellectual History.* New York: Wiley, 1968.

Lemay, Helen R. "Science and Theology at Chartres: The Case of the Supracelestial Waters." *British Journal for the History of Science* 10 (1977), 226–236.

Lemay, Richard. *Abu Ma 'shar and Latin Aristotelianism in the Twelfth Century: The Recovery of Aristotle's Natural Philosophy through Arabic Astrology*. Beirut: American University of Beirut, 1962.

Leonardo da Vinci. *The Literary Works of Leonardo da Vinci*. 2 vols. 3rd ed. Ed. Jean Paul Richter. New York: Phaidon, 1970. First published London, 1883.

Lerner, Michel-Pierre. "Le problème de la matière céleste après 1550: Aspects de la bataille des cieux fluides." *Revue d'Histoire des Sciences* 42 (1989), 255–280.

Lindberg, David C. *Theories of Vision from Al-Kindi to Kepler*. Chicago: University of Chicago Press, 1976.

"The Transmission of Arabic Learning to the West." In David C. Lindberg, ed., *Science in the Middle Ages*. Chicago: University of Chicago Press, 1978, 52–90. (a)

"The Science of Optics." In David C. Lindberg, ed., *Science in the Middle Ages*. Chicago: University of Chicago Press, 1978, 338–368. (b)

Litt, Thomas. *Les corps célestes dans l'univers de Saint Thomas d'Aquin*. Louvain: Publications Universitaires, 1963.

Lohr, Charles H. "Medieval Latin Aristotle Commentaries," *Traditio:*
"Authors: A–F." 23 (1967), 313–413;
"Authors: G–I." 24 (1968), 149–245;
"Authors: Jacobus-Johannes Juff." 26 (1970), 135–216;
"Authors: Johannes de Kanthi-Myngodus." 27 (1971), 251–351;
"Authors: Narcissus-Richardus." 28 (1972), 281–396;
"Authors: Robertus-Wilgelmus." 29 (1973), 93–197;
"Supplementary Authors." 30 (1974), 119–144.

Latin Aristotle Commentaries, II: Renaissance Authors. Florence: Olschki, 1988.

Lovejoy, Arthur O. *The Great Chain of Being: A Study in the History of an Idea*. New York: Harper & Row [Harper Torchbooks], 1960. 1st ed., 1936.

McColley, Grant, and H. W. Miller. "Saint Bonaventure, Francis Mayron, William Vorilong, and the Doctrine of a Plurality of Worlds." *Speculum* 12 (1937), 386–389.

McEvoy, James. *The Philosophy of Robert Grosseteste*. Oxford: Clarendon Press, 1982.

McLaughlin, Mary Martin. *Intellectual Freedom and Its Limitations in the University of Paris in the Thirteenth and Fourteenth Centuries*. New York: Arno Press, 1977. Ph.D. diss., Columbia University, 1952.

McMullin, Ernan. "Augustine of Hippo, Saint." In *Dictionary of Scientific Biography*, 1, 1970, 333–338.

Macrobius. *Commentary on the Dream of Scipio*. Tr. William Stahl. New York: Columbia University Press, 1952.

Maier, Anneliese. *Die Vorläufer Galileis im 14. Jahrhundert. Studien zur Naturphilosophie der Spätscholastik*. Rome: Edizioni di Storia e Letteratura, 1949.

Zwei Grundprobleme der scholastischen Naturphilosophie: Das Problem der Intensiven Grösse; Die Impetustheorie. 2nd ed. Rome: Edizioni di Storia e Letteratura, 1951.

Zwischen Philosophie und Mechanik. Studien zur Naturphilosophie der Spätscholastik. Rome: Edizioni di Storia e Letteratura, 1958.

Ausgehendes Mittelalter: Gesammelte Aufsätze zur Geistesgeschichte des 14. Jahrhunderts. Rome: Edizioni di Storia e Letteratura, 1964.

On the Threshold of Exact Science: Selected Writings of Anneliese Maier on Late Medieval Natural Philosophy. Ed. and tr. Steven D. Sargent. Philadelphia: University of Pennsylvania Press, 1982.

Maignan, Emanuel. *Cursus philosophicus*. Lyon, 1673.

Maimonides, Moses. *The Guide of the Perplexed*. 2 vols. Tr. Shlomo Pines. Chicago: University of Chicago Press, 1963.

Major, John. *In Primum Sententiarum ex recognitione Io. Badii.* Paris: apud Badium, 1519. (a)

Editio secunda Johannis Maioris Doctoris Parisiensis In secundum librum Sententiarum nunquam antea impressa. Paris: apud Iohannem Granion, 1519. (b)

Octo libri Physicorum cum naturali philosophia atque Metaphysica Johannis Maioiris Hadingtonani Theologi Parisiensis. Paris: Venundantur Parrhisiis in vico sancti Jacobia Johanne Parvo sub intersignio Lilii Aurei et ab Egidio Gormontio scuto trium coronatum Colonie indice, 1526. Includes *Questions on De celo,* which is not mentioned on title page.

Le Traité "De l'infini" de Jean Mair. New ed. Ed. and tr. Hubert Elie. Paris: Vrin, 1938.

Malebranche, Nicholas. *Dialogues on Metaphysics and on Religion.* Tr. Morris Ginsberg. London: Allen & Unwin, 1923.

Oeuvres complètes de Malebranche. 21 vols. Ed. André Robinet. Paris: Vrin, 1958–1970.

Vol. 12: *Entretiens sur la Métaphysique et sur la Religion.* Paris: Vrin, 1965.

Mandeville, John. *Mandeville's Travels, Texts and Translations.* 2 vols. Ed. and tr. Malcolm Letts. Hakluyt Society, 2nd ser., nos. 100–101. London: Hakluyt Society, 1953.

Marsilius of Inghen. *Questiones Marsilii super quatuor libros Sententiarum.* Strasbourg, 1501.

Questiones subtilissime Johannis Marcilii Inguen super octo libros Physicorum secundum nominalium viam. Cum tabula in fine libri posita suum in lucem primum sortiuntur effectum. Lyon, 1518. Facsimile, Frankfurt: Minerva, 1964.

Martin of Dacia. *Martini de Dacia Opera.* Ed. Henry Roos. Corpus Philosophorum Danicorum Medii Aevi, no. 2. Copenhagen: G. E. C. Gad, 1961.

Mastrius, Bartholomaeus, and Bonaventura Bellutus. *In Arist. Stag. libros Physicorum quibus ab adversantibus veterum tum recentiorum laculis Scoti philosophia vindicatum. A P.P. Magistris Bartholomeo Mastrio de Meldula . . . et Bonaventura Bellutos de Catana . . . In hac secunda editione ab eisdem ubique sedulo recisae, novisque additionibus, ac copiosiori indice locupletatae.* Venice: Typis Marci Ginammi, 1644.

RR. PP. Bartholomaei Mastrii de Meldula et Bonaventurae Belluti . . . Philosophiae ad mentem Scoti cursus integer. Editio novissima a mendis expurgata. 4th ed. 5 vols. Venice: apud Nicolaum Pezzana, 1727. Earlier editions in 1678, 1688, and 1708.

Vol. 2: *Continens disputationes ad mentem Scoti in Aristotelis Stagiritae libros: Physicorum.*

Vol. 3: *Continens disputationes ad mentem Scoti in Aristotelis Stagiritae libros: De anima, De generatione et corruptione, De coelo, et Metheoris.* Disputations in vol. 3 first published 1640.

Maurach, Gregor. *Coelum Empyreum, Versuch einer Begriffsgeschichte. Boethius: Texte und Abhandlungen zur Geschichte der exakten Wissenschaften.* Vol. 8. Wiesbaden: Steiner, 1968.

Maurer, Armand. "Boetius of Dacia and the Double Truth." *Mediaeval Studies* 17 (1955), 233–239.

Melanchthon, Philip. *Initia doctrinae physicae dictata in Academiia Vuitebergensi.* Wittenberg: per Iohannem Lufft, 1550.

Michael Scot. "Commentary on the *Sphere* of Sacrobosco." In Lynn Thorndike, ed. and tr., *The "Sphere" of Sacrobosco and Its Commentators.* Chicago: University of Chicago Press, 1949, 247–342.

Minio-Paluello, Laurence, and Bernard G. Dod, eds. *Aristoteles Latinus I 6–7, Categoriarum Supplementa: Porphyrii Isagoge translatio Boethii et Anonymis Fragmentum vulgo vocatum 'Liber sex principiorum.'* Bruges: Desclée de Brouwer, 1966.

Moody, Ernest A. "John Buridan on the Habitability of the Earth." In Ernest A. Moody, *Studies in Medieval Philosophy, Science, and Logic, Collected Papers 1933–*

1969. Berkeley/Los Angeles: University of California Press, 1975, 111–125. First published in *Speculum* 16 (1941).

Moore, James R. *The Post-Darwinian Controversies: A Study of the Protestant Struggle to Come to Terms with Darwin in Great Britain and America, 1870–1900*. Cambridge: Cambridge University Press, 1979.

Murdoch, John E. "Mathesis in philosophiam scholasticam introducta: The Rise and Development of the Application of Mathematics in Fourteenth-Century Philosophy and Theology." In *Arts Libéraux et Philosophie au Moyen Âge*. Montréal: Institut d'Études Médiévales, 1969, 215–254.

"Infinity and Continuity." Ch. 28 in Norman Kretzmann, Anthony Kenny, Jan Pinborg, Eleonore Stump, eds., *The Cambridge History of Later Medieval Philosophy from the Rediscovery of Aristotle to the Disintegration of Scholasticism, 1100–1600*. Cambridge: Cambridge University Press, 1982, 564–591.

Album of Science: Antiquity and the Middle Ages. I. B. Cohen, general ed. *Albums of Science*. New York: Scribner's, 1984.

North, John. "Celestial Influence: the Major Premiss of Astrology." In Paola Zambelli, ed., *"Astrologi hallucinati": Stars and the End of the World in Luther's Time*. Berlin: de Gruyter, 1986, 45–100.

"Medieval Concepts of Celestial Influence: A Survey." In Patrick Curry, ed., *Astrology, Science and Society, Historical Essays*. Woodbridge, U.K.: Boydell, 1987, 5–17.

Ockham. *See* William of Ockham.

Oddus, Illuminatus. *Physica peripatetica ad mentem Scoti. Qua subtilissimi doctrina adhuc magis dilucidatur, defenditur, roboratur; opera et industria R. P. Illuminati de Oddo a Collisano Capuccini, Sacrae Theologiae Praelectoris, etc.* Messanae: apud Haeredes Petri Bree(?), 1667.

Disputationes De generatione et corruptione ad mentem Scoti cuius doctrina adhuc magis elucidatur, defenditur, roboratur. Opera et industria R. P. Illuminati a Collisano, Capuccini. Cum resolutione aliquorum dubiorum ad libros De coelo et Meteoris spectantium . . . Naples: apud Andream Colicchia, 1672.

Oña, Peter de. *Fratris Petri de Oña Burgensis,in sacra theologia Magistri, Ordinis Regalis S. Mariae de Mercede, Redemptionis captivorum de Observantia, in provincia Castellae Provincialis. Super octo libros Aristotelis de Physico auditu commentaria una cum quaestionibus . . . Secunda editio, in qua pene infinita loca huius operis ab Auctore sunt recognita, emendata. . . .* Complutense [Alcalá]: ex officina Ioannis Gratiani, 1598.

Oresme, Nicole. "Maistre Nicole Oresme, Traitié de l'espere." Ed. Lillian M. McCarthy. Ph.D diss., University of Toronto, 1943.

"The Questiones super De celo of Nicole Oresme." Ed. and tr. Claudia Kren. Ph.D diss., University of Wisconsin, 1965.

"The Questiones de spera of Nicole Oresme." Latin text with English translation. Ed. and tr. Garrett Droppers. Ph.D diss., University of Wisconsin, 1966. (a)

Nicole Oresme: "De proportionibus proportionum" and "Ad pauca respicientes." Ed. and tr. Edward Grant. Madison: University of Wisconsin Press, 1966. (b)

Le Livre du ciel et du monde. Ed. Albert D. Menut and Alexander J. Denomy. Tr. Albert D. Menut. Madison: University of Wisconsin Press, 1968.

Nicole Oresme and the Kinematics of Circular Motion. "Tractatus de commensurabilitate vel incommensurabilitate motuum celi." Ed. and tr. Edward Grant. Madison: University of Wisconsin Press, 1971.

"Nicole Oresme, *Quaestio contra divinatores horoscopios*." Ed. S. Caroti. *Archives d'Histoire Doctrinale et Littéraire du Moyen Age* 43, Paris, 1977, 201–310.

Nicole Oresme and the Marvels of Nature. Critical edition and translation of *"De causis mirabilium."* Ed. and tr. Bert Hansen. Pontifical Institute of Mediaeval Studies, Studies and Texts, no. 68. Toronto: Pontifical Institute of Mediaeval Studies, 1985.

See also Caroti.

Overbye, Dennis. *Lonely Hearts of the Cosmos: The Story of the Scientific Quest for the Secret of the Universe.* New York: Harper Collins, 1991.

Overmann, Ronald J. "Theories of Gravity in the Seventeenth Century." Ph.D diss., Indiana University, 1974.

Oviedo, Franciscus de. *Integer cursus philosophicus ad unum corpus redactus in summulas, logicam, physicam, De caelo, De generatione, De anima, & Metaphysicam distributus.* 2 vols. Vol. 1: *Complectens Summulas, Logicam, Physicam, libros De caelo, et De generatione.* Lyon: sumptibus Petri Prost, 1640.

Palacio, Michael de. *Magistri Michaelis de Palacio, Granatensis philosophi atque theologi, civitatensis ecclesiae a sacris concionibus. In primum librum Magistri Sententiarum, disputationes gravissimae abstrusos quaestionum theologiarum sensus enodantes.* Salamanca: In aedibus Gasparis à Portonariis, 1574.

Patrizi, Francesco. *Francisci Patricii Nova de universis philosophia. In qua Aristotelica methodo non per motum, sed per lucem, et lumina, ad primam causam ascenditur; deinde propria Patricii methodo, tota in contemplationem venit Divinitas.* Ferrara: apud Benedictum Mammarellum, 1591.

Part 4: *Francisci Patricii Pancosmiae mundi corporei principia et constitutio.*

Paul of Venice. *Liber celi et mundi.* In *Summa naturalium.* Venice, 1476.

Pedersen, Olaf. *A Survey of the Almagest.* Acta Historica Scientiarum Naturalium et Medicinalium, no. 30. Odense: Odense University Press, 1974.

"Astronomy." In David C. Lindberg, ed., *Science in the Middle Ages.* Chicago: University of Chicago Press, 1978, 303–337.

Peter of Abano. *Il Lucidator Dubitabilium Astronomiae di Pietro d'Abano, Opere scientifiche inedite.* Ed. Graziella Federici Vescovini. Padua: Programma e 1 + 1 Editori, 1988.

Peter Lombard. *Magistri Petri Lombardi, Parisiensis Episcopi, Sententiae in IV libris distinctae.* 3rd ed. Grottaferrata [Rome]: Editiones Collegii S. Bonaventurae ad Claras Aquas, 1971. Tom I, pars II, liber I et II.

Peurbach, Georg. "Peurbach's *Theoricae novae planetarum:* A Translation with Commentary by E. J. Aiton." *Osiris,* 2nd ser., 3 (1987), 5–43.

See also d'Ailly 1531 for a Latin edition of the *Theorica.*

Philo Judaeus. *Philo with an English translation.* 12 vols. Ed. and tr. F. H. Colson, G. H. Whitaker, and R. Marcus. Loeb Classical Library. 1929–1962.

Vol. 1: *De opificio mundi,* tr. F. H. Colson and G. H. Whitaker, 1929.

Philoponus, John. *Johannes Philoponos Grammatikos von Alexandrien [6. Jh. n. Chr.], Christliche Naturwissenschaft im Ausklang der Antike, Vorläufer der modernen Physik, Wissenschaft und Bibel, Ausgewählte Schriften.* Ed. and tr. Walter Böhm. Munich: Schöningh, 1967.

Plato. *Plato's Cosmology. The Timaeus,* ed. and tr. Francis MacDonald Cornford, with running commentary. New York: Liberal Arts, 1957.

See Chalcidius.

Pliny. *Natural History.* 10 vols. Tr. Harris Rackham (vols. 1–5, 9); W. H. S. Jones (vols. 6–8), and D. E. Eichholz (vol. 10). Loeb Classical Library. 1938–1963.

Plutarch. *Plutarch's Moralia.* Vol. 12, 920A-999B. Tr. Harold Cherniss and William C. Helmbold. Loeb Classical Library. 1957.

Polacco, Giorgio. *Anticopernicus Catholicus seu De terrae statione et de solis motu contra systema Copernicanum Catholicae assertiones.* Venice: apud Guerilios, 1644.

Poncius, Johannes (John Punch). *R.P.F. Ioannis Duns Scoti, Doctoris Subtilis, Ordinis Minorum, Quaestiones in Lib. II Sententiarum . . . cum commentariis Rmi P.F. Francisci Lycheti, Brixiensis, Ordinis Minorum Regularis Observantiae olim Ministri Generalis, et supplemmento R.P.F. Ioannis Poncii* [John Punch] *Hiberni, eiusdem Ordinis, in Collegio Romano Hibernorum Theologie primarij Professoris.* Lyon, 1639. Vol. 6, pt. 2. Facsimile, Hildesheim: Olms, 1968.

Philosophiae ad mentem Scoti cursus integer primum quidem editus in collegio Romano Fratrum Minorum Hibernorum. Nunc vero demum ab ipso authore in conventu magno Parisiensi recognitus... Lyon, 1672.

See also Scotus.

Pseudo-Grosseteste. *Summa philosophiae. See* Grosseteste, Robert (to whom it is falsely ascribed).

Pseudo-Siger of Brabant. *Siger de Brabant Questions sur la Physique d'Aristote*. Ed. Philippe Delhaye. Louvain: Editions de l'Institut Supérieur de Philosophie, 1941. Includes only books 1–4 and 8.

Ptolemy, Claudius. *Hypotheses of the Planets*. In *Opera quae exstant omnia*, vol. 2: *Opera astronomica minora*. Ed. J. L. Heiberg. Leipzig: Teubner, 1907, 69–145.

Tetrabiblos. Ed. and tr. F. E. Robbins. Loeb Classical Library. 1948.

The Almagest. Tr. R. Catesby Taliaferro. Vol. 16 in Robert Maynard Hutchins, ed., *Ptolemy, Copernicus, Kepler: Great Books of the Western World*. Chicago: Encyclopaedia Britannica, 1952, 5–495.

"The Arabic Version of Ptolemy's *Planetary Hypotheses*." Tr. Bernard R. Goldstein. *Transactions of the American Philosophical Society*, n.s., vol. 57, pt. 4 (1967). Philadelphia: American Philosophical Society, 1967.

Ptolemy's Almagest. Tr. G. J. Toomer. New York: Springer, 1984.

Randles, W. G. L. *De la Terre plate au globe terrestre, une mutation épistémologique rapide, 1480–1520*. Paris: Colin, 1980.

Reisch, Gregor. *Margarita philosophica*. Basel, 1517. 1st ed., 1503. Facsimile, Düsseldorf: Stern-Verlag Janssen, 1973.

Rhodes, George de. *R.P. Georgii de Rhodes Avenionensis, e Societate Iesu Philosophia peripatetica ad veram Aristotelis mentem libris quatuor digesta et disputata Pharus ad theologiam scholasticam nunc primum in lucem prodit*. Lyons: sumptibus Antonii Huguetan et Guilelmi Barbier, 1671. Includes questions on *De caelo*.

Riccioli, Giovanni Baptista. *Almagestum novum astronomiam veterem novamque complectens observationibus aliorum et propriis novisque theorematibus, problematibus ac talibus promotam; in tres tomos distributam quorum argumentum sequens pagina explicabit*. Bologna: Ex typograpia haeredis Victorii Benatii, 1651.

14 → Richard of Middleton. *Clarissimi theologie magistri Ricardi de Media Villa... Super quatuor libros Sententiarum Petri Lombardi questiones subtilissimae*. 4 vols. Brixiae [Brescia], 1591. Facsimile, Frankfurt: Minerva, 1963.

Ripa, Jean de. "Jean de Ripa I Sent. Dist. XXXVII: De modo inexistendi divine essentie in omnibus creaturis." Ed. André Combes and Francis Ruello. *Traditio* 23 (1967), 191–267.

Robertson, Howard Percy. "Cosmology." *Encyclopaedia Britannica*, 1968. 2:582–587.

Robertus Anglicus. "Commentary on the *Sphere* of Sacrobosco." In Lynn Thorndike, ed. and tr., *The "Sphere" of Sacrobosco and Its Commentators*. Chicago: University of Chicago Press, 1949, 143–198 [Latin], 199–246 [English].

Rosen, Edward. "The Dissolution of the Celestial Spheres." *Journal of the History of Ideas* 46 (1985), 13–31.

Ross, W. D. *Aristotle*. 5th ed., rev. London: Methuen, 1949.

Sacrobosco, John of. *The "Sphere" of Sacrobosco and Its Commentators*. Ed. and tr. Lynn Thorndike. Chicago: University of Chicago Press, 1949.

Sambursky, Samuel. *Physics of the Stoics*. New York: Macmillan, 1959.

The Physical World of Late Antiquity. New York: Basic Books, 1962.

Santillana, Giorgio de. *The Crime of Galileo*. Chicago: University of Chicago Press, 1955.

Sarnowsky, Jürgen. *Die aristotelisch-scholastische Theorie der Bewegung. Studien zum Kommentar Alberts von Sachsen zur Physik des Aristoteles*. Beiträge zur Geschichte der Philosophie und Theologie des Mittelalters, n.s., no. 32. Münster: Aschendorff, 1989.

Sarton, George. *Introduction to the History of Science.* 3 vols. Baltimore: Williams & Wilkins, 1927–1948.

Scaliger, Julius Caesar. *Exotericarum exercitationum liber XV de subtilitate ad Hieronymum Cardanum.* Lyon: sumptibus Viduae Antonii de Hersy, 1615.

Schedel, Hartmann. *Liber chronicarum.* Pt. 1: Nuremberg, 1493. Latin text and translation in Edward Rosen, ed. and tr., *The Story of the Creation of the World.* New York: Burndy Library, 1948.

Scheiner, Christopher. *Rosa Ursina sive Sol ex admirando facularum et macularum suarum phoenomeno varius, necnon circa centrum suum et axem fixum ab occasu in ortum annua, circaque alium axem mobilem ab ortu in occasum conversione quasi menstrua super polos proprios libris quatuor mobilis ostensus.* Bracciani: apud Andream Phaeum typographum Ducalem, 1630.

Schmitt, Charles B. *Aristotle and the Renaissance.* Cambridge, Mass.: Harvard University Press for Oberlin College, 1983.

Schmitt, Charles B., and Dilwyn Knox. *Pseudo-Aristoteles Latinus: A Guide to Latin Works Falsely Attributed to Aristotle before 1500.* London: Warburg Institute, 1985.

Schneider, Notker. *Die Kosmologie des Franciscus de Marchia. Texte, Quellen und Untersuchungen zur Naturphilosophie des 14. Jahrhunderts.* Leiden: Brill, 1991.

Schofield, Christine Jones. *Tychonic and Semi-Tychonic World Systems.* New York: Arno, 1981. Ph.D diss., Cambridge University, 1964.

Scotus, John Duns. *Opera omnia.* Ed. Luke Wadding. 12 vols. Lyon, 1639. Facsimile, Hildesheim: Olms, 1968.

Vol. 3: *Meteorologicorum libri quatuor* (the first of a number of works in vol. 3). Falsely ascribed to Duns Scotus; perhaps by Simon de Tunstede (Pseudo-Scotus) (d. 1369).

Vol. 6, pt. 2: *R.P.F. Ioannis Duns Scoti, Doctoris Subtilis, Ordinis Minorum, Quaestiones in Lib. II Sententiarum . . . cum commentariis Rmi P. F. Francisci Lycheti, Brixiensis, Ordinis Minorum Regularis Observantiae olim Ministri Generalis, et supplemmento R.P.F. Ioannis Poncii* [John Punch] *Hiberni, eiusdem Ordinis, in Collegio Romano Hibernorum Theologie primarij Professoris.*

The questions on Peter Lombard's *Sentences*, bk. 2, are those of Scotus, but the commentaries are by Franciscus Lychetus and Johannes Poncius (John Punch).

Vol. 8: *R.P.F. Ioannis Duns Scoti, Doctoris Subtilis, Ordinis Minorum, In Lib. IV Sententiarum . . . cum commentario R.P.F. Antonii Hiquaei Hiberni, eiusdem Ordinis S. Theologiae Lectoris emeriti.*

Opera omnia. General ed. Carl Balić. Vol. 4: *Ordinatio*, bk. 1, 4th to 10th dist. Vatican City: Typis Polyglottis Vaticanis, 1956.

Serbellonus, Sigismundus. *Philosophia Ticinensis.* 2 vols. Milan: ex typographia Ludovici Montiae in Collegio S. Alexandrii PP. Barnabitarum, 1657–1663.

Vol. 1 (1657) contains the *Physics.*

Vol. 2 (1663) contains *De caelo.*

Shea, William R. "Galileo, Scheiner, and the Interpretation of Sunspots." *Isis* 61, (1970), 498–519.

Siger of Brabant. *On the Eternity of the World* (*De aeternitate mundi*). Tr. Lottie H. Kendzierski. Milwaukee: Marquette University Press, 1964, 75–98.

Les Quaestiones super Librum de causis. Ed. Antonio Marlasca. Paris, 1972.

See also Bonaventure, 1964; Pseudo-Siger; Thomas Aquinas, 1964.

Simon de Tunstede (Pseudo-Scotus). *See* Scotus, *Opera*, vol. 3.

Simplicius. *Simplicii philosophi acutissimi commentaria in quatuor libros De celo Aristotelis, Guillermo Morbeto interprete.* Venice, 1540.

Singer, Charles. "The Scientific Views and Visions of Saint Hildegarde (1098–

1180)." In Charles Singer, ed., *Studies in the History and Method of Science.* Oxford: Clarendon Press, 1917, 1–55.

Siraisi, Nancy. *Avicenna in Renaissance Italy: The "Canon" and Medical Teaching in Italian Universities after 1500.* Princeton: Princeton University Press, 1987.

Solmsen, Friedrich. *Aristotle's System of the Physical World: A Comparison with His Predecessors.* Cornell Studies in Classical Philology, no. 33. Ithaca: Cornell University Press, 1960.

Sommervogel, Carlos. *Bibliothèque de la Compagnie de Jésus.* 12 vols. New ed. Ed. Carlos Sommervogel. Brussels: Oscar Schepens, 1890–1911.

Sorabji, Richard. *Time, Creation and the Continuum: Theories in Antiquity and the Early Middle Ages.* Ithaca: Cornell University Press, 1983.

"Infinity and Creation." In Richard Sorabji, ed., *Philoponus and the Rejection of Aristotelian Science.* Ithaca: Cornell University Press, 1987, 164–178.

Stahl, William H. *Roman Science: Origin, Development and Influence to the Later Middle Ages.* Madison: University of Wisconsin Press, 1962.

Stegmüller, Friedrich. *Reportorium Commentatorium in Sententias Petri Lombardi.* 2 vols. Würzburg: Schöningh, 1947.

Steneck, Nicholas. *Science and Creation in the Middle Ages: Henry of Langenstein (d. 1397) on Genesis.* Notre Dame, Ind.: University of Notre Dame Press, 1976.

Stewart, David. "Aristotle's Doctrine of the Unmoved Mover." *Thomist* 37 (1973), 522–547.

Stiefel, Tina. *The Intellectual Revolution in Twelfth-Century Europe.* New York: St. Martin's Press, 1985.

Stock, Brian. *Myth and Science in the Twelfth Century: A Study of Bernard Silvester.* Princeton: Princeton University Press, 1972.

Suarez, Francisco. *Disputationes metaphysicae.* 2 vols. Paris, 1866. Facsimile, Hildesheim: Olms, 1965. First published 1597.

Swerdlow, Noel. "Pseudoxia Copernicana: or, Enquiries into very many received tenents and commonly presumed truths, mostly concerning spheres." *Archives Internationales d'Histoire des Sciences*26 (1976), 108–158.

Sylla, Edith. "Galileo and Probable Arguments." In Daniel O. Dahlstrom, ed., *Nature and Scientific Method.* Studies in Philosophy and the History of Philosophy, no. 22. Washington, D.C.: Catholic University of America Press, 1991, 211–234.

Thabit ibn Qurra. *The Astronomical Works of Thabit b.Qurra.* Ed. Francis J. Carmody. Berkeley/Los Angeles: University of California Press, 1960.

Themistius. *Libri Paraphraseos Themistii Peripatetici acutissimi: In Posteriora; In libros De anima; . . . interprete Hermolao Barbaro.* Venice, 1499.

Themon Judaeus. *In quatuor libros Meteororum.*
See Albert of Saxony.

Thomas Aquinas. *S. Thomae Aquinatis Quaestiones disputatae.* Ed. P. Mandonnet. 3 vols. Paris: P. Lethielleux, 1925.
Vol. 2: *De potentia*; *De malo.*

Scriptum super libros Sententiarum Magistri Petri Lombardi Episcopi Parisiensis. 4 vols. New ed. Paris: Lethielleux, 1929–1947.
Bks. 1–2: ed. R. P. Mandonnet, in vols. 1–2 (1929)
Bks. 3–4, dist. 22: ed. R. P. Maria Fabianus Moos, in vols. 3–4 (1933, 1947).

The Letter of Saint Thomas Aquinas "De occultis operibus naturae ad quemdam militem ultramontanum." Tr. Joseph B. McAllister. Catholic University of America Philosophical Studies, no. 42. Washington, D.C.: Catholic University of America, 1939. Ph.D. diss.

Quaestiones Quodlibetales. Ed. Raymundus Spiazzi. Turin: Marietti, 1949.

S. Thomae Aquinatis In Aristotelis libros De caelo et mundo; De generatione et corrup-

tione; Meteorologicorum Expositio cum textu ex recensione leonina. Ed. Raymundus M. Spiazzi. Turin: Marietti, 1952.

S. Thomae Aquinatis In octo libros De physico auditu sive Physicorum Aristotelis commentaria. New ed. Ed. Angeli-M. Pirotta. Naples: D'Auria Pontificus, 1953.

Commentary on the Metaphysics of Aristotle. 2 vols. Tr. John P. Rowan. Chicago: Regnery, 1961.

Treatise on Separate Substances. Latin text with English translation. Ed. and tr. Francis J. Lescoe. West Hartford: Saint Joseph College, 1963.

On the Eternity of the World (*De aeternitate mundi*). Tr. Cyril Vollert. Milwaukee: Marquette University Press, 1964, 1–73.

See also Bonaventure, 1964; Siger of Brabant, 1964.

Summa theologiae. 60 vols. Latin text with English translation. Ed. and tr. T. Gilby et al. New York: Blackfriars/McGraw-Hill, 1964–1976.

Vol. 8 (1967): *Creation, Variety and Evil* (44–49). Tr. Thomas Gilby.

Vol. 10 (1967): *Cosmogony* (65–74). Tr. William A. Wallace.

Thomas de Bungeye. "Thomas de Bungeye's Commentary on the First Book of Aristotle's De caelo." Ed. Bernard Street Parker. Ph.D. diss., Tulane University, 1968.

Thomas of Strasbourg. *Thomae ab Argentina, Eremitarum divi Augustini prioris generalis qui floruit anno Christi 1345, Commentaria in IIII libros Sententiarum*. Venice: ex officina Stellae, Iordani Ziletti, 1564. Facsimile, Ridgewood, N.J.: Gregg, 1965.

Thoren, Victor E. "The Comet of 1577 and Tycho Brahe's System of the World." *Archives Internationales d'Histoire des Sciences* 29 (1979), 53–67.

The Lord of Uraniborg: A Biography of Tycho Brahe. Cambridge: Cambridge University Press, 1990.

Thorndike, Lynn. *A History of Magic and Experimental Science*. 8 vols. New York: Columbia University Press, 1923–1958.

University Records and Life in the Middle Ages. New York: Columbia University Press, 1944.

Thorndike, 1949. *See* Sacrobosco, John of.

Thorndike, Lynn, and Pearl Kibre. *A Catalogue of Incipits of Mediaeval Scientific Writings in Latin*. Rev. ed. Cambridge, Mass.: Mediaeval Academy of America, 1963.

Toletus, Franciscus. *D. Francisci Toleti Societatis Iesu Commentaria una cum Questionibus in octo libros Aristotelis De physica auscultatione, nunc secundo in lucem edita*. Venice: apud Iuntas, 1580.

Toynbee, Paget. *A Dictionary of Proper Names and Notable Matters in the Works of Dante*. Rev. ed. Ed. Charles S. Singleton. Oxford: Clarendon Press, 1968.

Trinkaus, Charles. "The Astrological Cosmos and Rhetorical Culture of Giovanni Gioviano Pontano." *Renaissance Quarterly* 33 (1985), 446–472.

Van Helden, Albert. *Measuring the Universe: Cosmic Dimensions from Aristarchus to Halley*. Chicago: University of Chicago Press, 1985.

Velcurio (Veltkirchius), Johannes Bernardi. *Commentarii in universam Physicen Aristotelis distincti libris IIII, iam recens accuratissime recogniti*... Lyon: apud Io. Franciscum de Gabiano, 1554.

Versor, Johannes. *Questiones subtilissime in via sancti Thome magistri Johannes Versoris super libros De celo et mundo Arestotelis* [sic]... *Questiones Versoris super Parva Naturalia cum textu Arestotelis*... *Tractatus compendiosus sancti Thome De ente et essentia*... *Tractatus*... *Gerhardi de Monte ostendens sanctum Thomam et venerabilem Albertum*... *non esse contrarios*... *Questiones magistri Johannis Versoris super libros De generatione et corruptione*... *super libros Metheororum*... Cologne: H. Quentell, ca. 1493.

Vincent of Beauvais. *Bibliotheca mundi seu venerabilis viri Vincentii Burgundi ex Ordine Praedicatorum, episcopi Bellovacensis, Speculum Quadruplex: naturale, doctrinale, mo-*

rale, historiale. 2 vols. Douai: Ex officina typographica et sumptibus Balthazaris Belleri in Circino aureo, 1624.
Vol. 1: *Speculum naturale.*

Von Plato, Jan. "Nicole Oresme and the Ergodicity of Rotations." In Ingmar Pörn, ed., *Essays in Philosophical Analysis Dedicated to Erik Stenius on the Occasion of his 70th Birthday.* Acta Philosophica Fennica, no. 32 (1981), 190–197.

Wallace, William A. "Vincent of Beauvais." *Dictionary of Scientific Biography,* 1976, 14: 34–36.

Prelude to Galileo: Essays on Medieval and Sixteenth-Century Sources of Galileo's Thought. Boston Studies in the Philosophy of Science, no. 62. Dordrecht: Reidel, 1981.

Wallace-Hadrill, D. S. *The Greek Patristic View of Nature.* New York: Barnes & Noble, 1968.

Weinberg, Julius. *Nicolaus of Autrecourt: A Study in Fourteenth-Century Thought.* Princeton: Princeton University Press, 1948.

Weisheipl, James A. "The Problemata Determinata XLIII Ascribed to Albertus Magnus (1271)." *Mediaeval Studies* 22 (1960), 303–354.

"The Celestial Movers in Medieval Physics." *Thomist* 24 (1961), 286–326.

"The Concept of Matter in Fourteenth-Century Science." In Ernan McMullin, *The Concept of Matter in Greek and Medieval Philosophy.* Notre Dame, Ind.: University of Notre Dame Press, 1963.

"Curriculum of the Faculty of Arts at Oxford in the Early Fourteenth-Century." *Mediaeval Studies* 26 (1964), 143–185.

Friar Thomas D'Aquino, His Life, Thought, and Work. Garden City, N.Y.: Doubleday, 1974.

Albertus Magnus and the Sciences: Commemorative Essays, 1980. Ed. James A. Weisheipl. Toronto: Pontifical Institute of Mediaeval Studies, 1980.

Westman, Robert S. "The Copernicans and the Churches." In David C. Lindberg and Ronald L. Numbers, eds., *God and Nature: Historical Essays on the Encounter between Christianity and Science.* Berkeley/Los Angeles: University of California Press, 1986, 76–113.

White, Lynn, Jr. "Cultural Climates and Technological Advance in the Middle Ages." In Lynn White, Jr., *Medieval Religion and Technology, Collected Essays.* Berkeley/Los Angeles: University of California Press, 1978, 217–253.

White, Thomas. *Peripateticall Institutions in the way of that eminent Person and excellent Philosopher Kenelm Digby. The Theoreticall part. Also a Theological Appendix of the Beginning of the World.* London: Printed by R. D. and are to be sold by John Williams at the sign of the Crown in S. Paul's Churchyard, 1656.

Whitrow, G. J. "On the Impossibility of an Infinite Past." *British Journal for the Philosophy of Science* 29 (1978), 39–45.

William of Auvergne. *Guilelmi Alverni Opera omnia.* 2 vols. Paris, 1674. Facsimile, Frankfurt: Minerva, 1963.
Vol. 1: *De universo*

William of Ockham. *Guilelmi de Ockham Opera philosophica et theologica ad fidem codicum manuscriptorum edita.* Ed. Franciscan Institute. St. Bonaventure, N.Y.: St. Bonaventure University:

Opera theologica, vol. 4. *Venerabilis Inceptoris Guillelmi de Ockham Scriptum in librum primum Sententiarum Ordinatio.* Dists. 19–48. Ed. Girard I. Etzkorn and Francis E. Kelly. 1979.

Opera theologica, vol. 5. *Venerabilis Inceptoris Guilelmi de Ockham Quaestiones in librum secundum Sententiarum (Reportatio).* Ed. Gedeon Gál and Rega Wood. 1981.

Opera philosophica, vol. 6. *Venerabilis Inceptoris Guillelmi de Ockham Brevis Summa libri Physicorum, Summula philosophiae naturalis, et Quaestiones in libros Physicorum Aristotelis.* Ed. Stephen Brown. 1984.

Wilson, Curtis. *William Heytesbury: Medieval Logic and the Rise of Mathematical Physics*. Madison: University of Wisconsin Press, 1956.

Wippel, John F. *The Metaphysical Thought of Godfrey of Fontaines: A Study in Late Thirteenth-Century Philosophy*. Washington, D.C.: Catholic University of America Press, 1981. (a)

"Did Thomas Aquinas Defend the Possibility of an Eternally Created World?" *Journal of the History of Philosophy* 19 (1981), 21–37. (b)

Wolfson, Harry A. "The Plurality of Immovable Movers in Aristotle and Averroes." In Isadore Twersky and George H. Williams, eds. *Studies in the History of Philosophy and Religion*. Cambridge, Mass.: Harvard University Press, 1973, 1–21. (a)

"The Problem of the Souls of the Spheres from the Byzantine Commentaries on Aristotle through the Arabs and St. Thomas to Kepler." In Isadore Twersky and George H. Williams, eds., *Studies in the History of Philosophy and Religion*. Cambridge, Mass.: Harvard University Press, 1973, 22–59. (b)

The Philosophy of the Kalam. Cambridge, Mass.: Harvard University Press, 1976.

Wright, John Kirtland. *The Geographical Lore of the Time of the Crusades: A Study in the History of Medieval Science and Tradition in Western Europe*. American Geographical Society Research Series, no. 15. New York: American Geographical Society, 1925.

Yates, Frances A. *Giordano Bruno and the Hermetic Tradition*. New York: Random House, 1969. First published 1964.

Zedler, Johann Heinrich. *Grosses vollständiges Universal-Lexikon*. 64 vols. Halle, 1732–1750. Facsimile, Graz: Akademische Druck, 1961–1964.

Zimmermann, Albert, ed. *Ein Kommentar zur Physik des Aristoteles aus der Pariser Artistenfakultät um 1273*. Berlin: de Gruyter, 1968.

Index

Page numbers cited directly after a semicolon following a final textual subentry refer to relatively minor mentions in the text of the main entry.